Quantum Theory

The ideal text for a two-semester graduate course on quantum mechanics. Fresh, comprehensive, and clear, it strikes the optimal balance between covering traditional material and exploring contemporary topics. Focusing on the probabilistic structure of quantum mechanics and the central role of symmetries to unify principles, this textbook guides readers through the logical development of the theory. Students will also learn about the more exciting and controversial aspects of quantum theory, with discussions on past interpretations and the current debates, including on cutting-edge concepts such as quantum information and entanglement, open quantum systems, and quantum measurement theory.

The book contains two types of content: Type A material is more elementary and is fully self-contained, functioning like a separate book within the book, while Type B content is at the level of a graduate course. Requiring minimal physics background, this textbook is appropriate for mathematics and engineering students, in addition to physicists. Introducing cutting-edge topics in the field, the book features about 150 concept-checking questions, 300 homework problems and a solutions manual.

Charis Anastopoulos is an associate professor in the department of physics at the University of Patras, Greece. He holds a Ph.D. from Imperial College, London and has worked in various academic positions at the universities of Barcelona, Maryland, and Utrecht. His research focuses on quantum foundations, quantum information and general relativity. He is also the author of *Particle or Wave: The Evolution of the Concept of Matter in Modern Physics* (Princeton University Press, 2008).

Quantum Theory
A Foundational Approach

Charis Anastopoulos
University of Patras, Greece

CAMBRIDGE
UNIVERSITY PRESS

Shaftesbury Road, Cambridge CB2 8EA, United Kingdom

One Liberty Plaza, 20th Floor, New York, NY 10006, USA

477 Williamstown Road, Port Melbourne, VIC 3207, Australia

314–321, 3rd Floor, Plot 3, Splendor Forum, Jasola District Centre, New Delhi – 110025, India

103 Penang Road, #05–06/07, Visioncrest Commercial, Singapore 238467

Cambridge University Press is part of Cambridge University Press & Assessment,
a department of the University of Cambridge.

We share the University's mission to contribute to society through the pursuit of
education, learning and research at the highest international levels of excellence.

www.cambridge.org
Information on this title: www.cambridge.org/highereducation/isbn/9781316518595

DOI: 10.1017/9781009000871

First published 2024

A catalogue record for this publication is available from the British Library.

A Cataloging-in-Publication data record for this book is available from the Library of Congress.

ISBN 978-1-316-51859-5 Hardback

Additional resources for this publication at www.cambridge.org/anastopoulos.

To my mother, Maria

Contents

Preface

This book was born out of my frustrations with quantum theory as a student. I wanted to understand the mathematical and conceptual foundations of quantum theory, but available textbooks paid little or no attention to foundational issues. No matter how technically proficient one became, speaking intelligibly, let alone intelligently, about quantum foundations, required a lot of additional study, other books, or even research papers.

In later years, as quantum foundations occupied an increasingly larger part of my research, I came to the conclusion that a mathematically and conceptually rigorous exposition of the fundamental principles should allow the student to deal with both foundational issues and applications. I followed this logic when I started teaching quantum mechanics. I found that the emphasis on the abstract mathematical structure of quantum theory is initially surprising for the students, but once they get used to it, they find it easier to learn more advanced techniques, and at the end they gain better understanding of what quantum theory is about.

This book grew out of these experiences, but also out of many intellectual influences on the subject, two of which have been crucial. The first major influence is my friend and collaborator Ntina Savvidou. Her work on the two aspects of time in physics, and their implications, greatly affected my thinking on quantum foundations, my perspective on what is important and what is not, and of course, my own research. Time, after all, is the central concept of physics, and in this book I always try to describe quantum phenomena as they unfold in time. The second major influence has been Chris Isham. His course on quantum mechanics that I had attended as a graduate student showed me that the combination of foundational issues with the development of technical skills in a single course is feasible. Besides that, his ideas about the role of symmetry in quantum theory – which culminated in the canonical group program – form the basis of my understanding of symmetry and, hence, of the way that symmetries are presented here.

Certainly, this book is not only about quantum foundations; its primary aim is to instruct students (graduates and advanced undergraduates) to *use* quantum theory as a practical tool. My strategy is to make students aware of foundational problems from the very beginning, but at the same time, I urge them to adopt a pragmatic attitude towards the quantum formalism. I explain why they must abandon all mental habits inherited from classical physics when they think about microscopic systems. When the students learn to do this consistently, they can build their physical intuition and their technical skills without worry. By the end of the book, they will have developed a solid grasp of quantum theory, both in relation to modern-day research and as general background knowledge.

In recent years, our perspective about what constitutes basic knowledge on quantum theory has shifted. Topics such as quantum information, open quantum systems, and quantum measurements have become increasingly important for research in different branches of physics. We need accounts of quantum theory that present both new topics and traditional material, and it is a challenge to fit them together. Here, the unifying perspective is the presentation of quantum theory in terms of a set of fundamental principles that function like axioms. In every single application of quantum theory, the link to the axioms is made explicit. Thus, the totality of quantum theory, from foundations to applications, is presented as a logically coherent system.

The book consists of five parts. Part I is a review of classical physics, including probability theory, and a historical introduction to quantum theory. Part II presents quantum theory in terms of six fundamental principles. The principles are presented together with the necessary mathematical background and with applications to simple systems. Part III analyzes the properties of quantum particles. It starts with the description of symmetries, continues to nonrelativistic particles, the introduction of spin, and particle statistics. It concludes with a symmetries-based introduction to relativistic quantum systems and with the first steps towards relativistic quantum field theory. Part IV presents techniques adapted to specific problems: the structure and spectra of composite systems, decays and transitions, particle scattering, and open quantum systems. Finally, Part V revisits quantum foundations: I present quantum measurement theory, the major interpretations of quantum theory, and the frontiers of quantum theory in contemporary research.

Since the field of quantum foundations is replete with controversies, the reader needs to be aware of my biases. While I like (and sometimes I strongly support) specific ideas from different interpretations, I do not commit to any interpretation. In fact, I find all existing interpretations to have serious flaws. Certainly, their supporters strive to cure them, and it is possible that one school of thought will eventually succeed, and that one out of the many current interpretations will turn out to be the correct one. But, in my opinion, the foundational problems of quantum theory will only be resolved by a new theory that will be based on physical principles of which we now have just faint hints. And I believe that theoretical and experimental research in the frontiers of physics is the best bet for discovering those principles.

Acknowledgments

My strongest thanks go to Ntina Savvidou for her encouragement and feedback throughout this project, from inception to conclusion. I am grateful to Andreas Zoupas, Eirini Sourtzinou, Maria Papageorgiou, Pelagia Zogogianni, and Vassilis Letsios for comments and/or suggestions on the manuscript. Also, many thanks to Nikos Kollas and Agapi Chrysanthakopoulou for the concept of the catty Q.

Special thanks to my CUP editors: Arya Thampi, for supporting this book from its early stages, and Maggie Jeffers for many suggestions that improved the book and for organizing the feedback process. I am also grateful to a large number of anonymous reviewers, whose feedback has been invaluable.

Finally, I want to thank my students in quantum theory courses during the past eight years, who had to work through the half-baked lecture notes from which this book emerged. It would have been a poorer book without their collective input.

Important Information

How to Use this Book

This book is primarily intended for physics students taking up a graduate course in quantum mechanics. It covers more material than the core quantum mechanics course in North American curricula (Quantum Mechanics I and II). The additional material can be used for a more advanced course (e.g., Quantum Mechanics III). Parts of the book can also be used in advanced undergraduate courses or in a specialist course on quantum foundations.

The book contains material of type A and of type B, as designated at the top corner of each page. Type A material is intended for a less demanding graduate course that skims in-depth analysis of some topics, and it also avoids the most complex interpretational issues. Material of type A can also be used in an advanced undergraduate course. Readers that follow only type A material will encounter no continuity problems. Material of type B is more advanced, and it is intended solely for graduate courses. Some chapters contain only material of type A, and some chapters contain only material of type B. A few chapters contain both A-type and B-type material; the latter almost always following the former. (Chapter 13 is the only exception.) In these chapters, questions and problems that employ B-type material are distinguished.

The book also contains some material in boxes that can be skipped at first reading. This material includes technical proofs, historical facts, and mathematical digressions, and it is important for the reader's overall learning experience.

There are several possible ways to use this book for a standard two-semester graduate course on quantum mechanics (100 hours or more). The material typically covered in such courses is fully contained in Chapters 1–15 and 17–19; the instructor may choose to focus solely on those chapters. Alternatively, the instructor may choose to cover only the type A material for 70 percent to 80 percent of the course and to supplement it with sections of type B material of his/her choice. Note that each of the Chapters 16 and 19–21 functions as an independent module, in the sense that it is not needed by subsequent chapters. Chapter 12 requires a basic knowledge of group theory, which is presented in Appendix C. This chapter can be skipped without loss of continuity, but the reader should also skip the B part of Chapter 14 and the whole of Chapter 16.

Other ways to use this book for coursework include the following.

- An advanced undergraduate course on quantum theory from type-A material in Chapters 3–11, 13–15, and 17–18 (up to 60 hours).
- An advanced introduction to quantum theory for mathematics/engineering students with an interest in quantum technologies from Chapters 1–9 (40 hours).
- A quantum foundations course from Chapters 1–10, 21, and 22 (up to 50 hours).

One can also use this book for self-study. Necessary mathematical background for type-A material is university-level calculus, linear algebra, complex analysis, and differential equations. Some knowledge of classical mechanics and electromagnetism at the same level is also needed. For type-B material, a familiarity with basic notions of group theory is useful for Chapters 12 and 16, and a familiarity with special relativity for Chapter 16.

The book contains about 150 questions and 300 problems. The questions test the students' understanding of the main issues, and urge them to develop a critical perspective towards the meaning of quantum theory. Problems vary in difficulty from simple algorithmic calculations to research-level analyses of physical systems.

Conventions

In most of this book, I use the natural system (NS) of units, in which Planck's constant $\hbar = 1$ and the speed of light $c = 1$. In the International System (SI), any magnitude A can be expressed in terms of units of length L, mass M, and time T, as $[L]^{\alpha}[M]^{\beta}[T]^{\gamma}$. To convert the magnitude A from SI to NS, use the following formula

$$A_{SI} = A_{NS}\hbar^{\beta}c^{-\beta-\gamma},$$

and A_{NS} has dimension of $[L]^{\alpha-\beta+\gamma}$.

For electromagnetism, I use the Lorentz–Heaviside system of units, in which the Coulomb force between two point charges q_1 and q_2 at distance r is $\frac{q_1 q_2}{4\pi r^2}$. To transform to the SI, simply divide any electric charge by the square root of the vacuum permittivity ϵ_0.

A dot over any quantity denotes the time-derivative of that quantity. Primes are often used to denote derivatives with respect to quantities other than time. Spatial three-vectors are indicated by symbols in boldface.

Some specific notations may be unfamiliar to readers. I write := rather than the equality symbol, in order to denote the defining equality for a quantity. For any function $F(\mathbf{q})$ of a vector $\mathbf{q}$, I write $\frac{\partial F}{\partial \mathbf{q}}$ to stand for the vector with components $\frac{\partial F}{\partial q_i}$. Occasionally, I will use the symbol ∂_x for ∂/∂_x. I also write $O(x^n)$ to denote small terms of the order of x^n, for $|x| << 1$.

Part I

Introduction

1 The Classical World

We ought to regard the present state of the universe as the effect of its antecedent state and as the cause of the state that is to follow. An intelligence knowing all the forces acting in nature at a given instant, as well as the momentary positions of all things in the universe, would be able to comprehend in one single formula the motions of the largest bodies as well as the lightest atoms in the world, provided that its intellect were sufficiently powerful to subject all data to analysis; to it nothing would be uncertain, the future as well as the past would be present to its eyes.

Laplace (1814)

1.1 Newton's Theory

Newtonian mechanics was the first great synthesis of modern physics. It provided the main theory about the workings of the physical world from the date of its first presentation (1687) until the beginnings of the twentieth century. The main principles of Newton's theory were the following.

1. *Absolute time* and *absolute space* exist at the background of all physical events. This means that any physical event is uniquely identified by (i) a real number $t \in \mathbb{R}$ that specifies when the event happens, and (ii) a position vector $\mathbf{r} = (x_1, x_2, x_3) \in \mathbb{R}^3$ that specifies the point of space where the event happens.
2. The world consists of particles that can be viewed as pointlike objects. A system of N particles at a single moment of time t is determined by N position vectors $\mathbf{r}_i$, where $i = 1, 2, \ldots, N$. In order to describe the time evolution of the particles, we must express the position vectors as a function of time, that is, we must determine the N paths $\mathbf{r}_i(t)$.
3. Particles act upon each other through their accelerations. This means that the paths $\mathbf{r}_i(t)$ obey *Newton's second law*

$$m_i \frac{d^2 \mathbf{r}_i}{dt^2} = \mathbf{F}_i, \tag{1.1}$$

where m_i is the *inertial mass* of the ith particle, and the vector $\mathbf{F}_i$ is the total *force* acting on the ith particle.
4. If all forces originate from the action of one particle upon the other, then $\mathbf{F}_i = \sum_{j \neq i} \mathbf{F}_i(\mathbf{r}_j)$, where $\mathbf{F}_i(\mathbf{r}_j)$ is the force acted by the jth particle upon the ith particle. These forces satisfy *Newton's third law*.

3

$$\mathbf{F}_i(\mathbf{r}_j) = -\mathbf{F}_j(\mathbf{r}_i). \tag{1.2}$$

Newton's equations for N particles form a system of $3N$ ordinary second-order differential equations. According to the *fundamental theorem of ordinary differential equations*, the knowledge of the position vectors $\mathbf{r}_i(t_0)$ and of their first derivatives $\dot{\mathbf{r}}_i(t_0)$ at an initial moment of time t_0 allows us to find a unique solution $\mathbf{r}_i(t)$ for all $t \in \mathbb{R}$. Hence, Newton's theory describes a deterministic world. The full knowledge of the state of a system at a time t_0 provides full knowledge of the system at all times t in the past or in the future of t_0.

1.2   Hamiltonian Mechanics

The Newtonian notion of force turned out to be inadequate for describing gravitational and electromagnetic interactions, and for this reason it has been abandoned at the level of fundamental physics. Today, when one talks about fundamental forces (nuclear, gravitational and so on) the word "force" is not used as a technical term, that is, referring to the vectors that appear in the right-hand side of Newton's second law, but as a colloquial term.

The notion of the force was superseded by that of energy. In particular, the notion of energy, as expressed in a mathematical formalism developed by William Hamilton, offers an elegant reformulation of classical physics and it can also be transferred to quantum physics.

1.2.1   Hamilton's Equations

We proceed to the presentation of Hamilton's formalism for an N-particle system. We denote the $3N$ coordinates that specify the N position vectors of the system by x_a, where $a = 1, 2, \ldots, 3N$. The variables x_a define the *configuration space* Q of the system. Given that the x_a can take any real value, $Q = \mathbb{R}^{3N}$.

Each particle is also characterized by a momentum vector, $\mathbf{p}_i = m_i \dot{\mathbf{r}}_i$. In the Hamiltonian formalism, we take momenta as independent variables that are not fundamentally defined by velocities. The N momentum vectors correspond to $3N$ momentum coordinates, which we will denote as p_a. The label a is common to position and momentum coordinates, so that each momentum coordinate corresponds to a unique position coordinate and vice versa.

Suppose we specify the $3N$ position coordinates and the $3N$ momentum coordinates of a system at a single moment of time t_0. Then, we can use Newton's second law in order to specify the position and momentum coordinates at any other moment of time t. Hence, we describe the time evolution of the system as a succession of states, where each state corresponds to an element

$$\xi = (x_1, p_1, x_2, p_2, \ldots, x_{3N}, p_{3N}).$$

The elements ξ form a set $\Gamma = \mathbb{R}^{6N}$, the *state space* of the N-particle system. Another name for Γ is "*phase space.*"

We define the *Hamiltonian* of the system as a function $H : \Gamma \to \mathbb{R}$,

$$H(x_a, p_a) := \sum_{a=1}^{3N} \frac{p_a^2}{2m_a} + V(x_1, x_2, \ldots, x_{3N}), \tag{1.3}$$

where $V : Q \to \mathbb{R}$ is the potential energy.

The values of the Hamiltonian correspond to the energy E of the physical system. Note that the Hamiltonian H is a function, while the energy E is a number, the value of the function H. This distinction is crucial, and it persists in quantum theory.

The Hamiltonian incorporates all information about the time evolution of the system in a compact form. To see this, we note that the system of equations

$$\dot{x}_a = \frac{\partial H}{\partial p_a}, \qquad \dot{p}_a = -\frac{\partial H}{\partial x_a} \tag{1.4}$$

leads to the second-order equation

$$m_a \ddot{x}_a = -\frac{\partial V}{\partial x_a}, \tag{1.5}$$

that is, Newton's second law, Eq. (1.1), for forces of the form $F_a = -\partial V / \partial x_a$. Forces of this type are called *conservative*, because they conserve energy. Equations (1.4) are called *Hamilton's equations*, and they can be generalized to Hamiltonians that are not of the form (1.3).

1.2.2 Poisson Brackets

Any physical magnitude for an N-particle system can be expressed as a function of the particle's positions and momenta, namely, as a function $F : \Gamma \to \mathbb{R}$. Any functions F on Γ evolve in time according to the equation

$$\dot{F} = \sum_a \left(\frac{\partial F}{\partial x_a} \dot{x}_a + \frac{\partial F}{\partial p_a} \dot{p}_a \right) = \sum_a \left(\frac{\partial F}{\partial x_a} \frac{\partial H}{\partial x_a} - \frac{\partial F}{\partial p_a} \frac{\partial H}{\partial q_a} \right) = \{F, H\}. \tag{1.6}$$

In Eq. (1.6), the *Poisson bracket* between two functions F and G on Γ is defined as another function of Γ,

$$\{F, G\} := \sum_a \left(\frac{\partial F}{\partial x_a} \frac{\partial G}{\partial p_a} - \frac{\partial F}{\partial p_a} \frac{\partial G}{\partial x_a} \right). \tag{1.7}$$

From this definition, we obtain the fundamental Poisson brackets,

$$\{x_a, x_b\} = 0, \qquad \{p_a, p_b\} = 0, \qquad \{x_a, p_b\} = \delta_{ab}, \tag{1.8}$$

where δ_{ab} is the *Kronecker delta* (Kronecker), defined by

$$\delta_{ab} = \begin{cases} 1 & \text{if } a = b \\ 0 & \text{if } a \neq b \end{cases}. \tag{1.9}$$

Any pair of variables F and G that satisfies $\{F, G\} = 1$ is called a *conjugate pair* of variables. Clearly x_a and p_a form a conjugate pair for all variables.

The Poisson brackets satisfy the following identities.

1. *Antisymmetry*: $\{G, F\} = -\{F, G\}$.
2. *Leibniz's rule*: $\{F, GH\} = \{F, G\}H + \{F, H\}G$.
3. *Jacobi's identity*: $\{\{F, G\}, H\} + \{\{H, F\}, G\} + \{\{G, H\}, F\} = 0$.

Here, F, G, and H are functions on Γ. The proof of all three identities follows from the definition in a straightforward way.

1.2.3 Liouville's Theorem

Liouville's theorem highlights a crucial symmetry of Hamilton's equations. Let U be a subset of the state space Γ ($U \subset \Gamma$). For example, U may describe the initial conditions for a particle that have been determined with some uncertainty. The volume $[U]$ of U is defined as $[U] := \int_U d^{3N}x\,d^{3N}p$.

Positions and momenta evolve according to Hamilton's equations. A point of U with coordinates x_a and p_a at time t, will evolve to another point with coordinates $x'_a = x_a + \delta t\,\partial H/\partial p_a$ and $p'_a = p_a - \delta t\,\partial H/\partial x_a$ at time $t + \delta t$. It can be proven that the Jacobian of this transformation is unity (Goldstein et al., 2002),

$$\frac{\partial(x'_a, p'_a)}{\partial(x_a, p_a)} = 1. \tag{1.10}$$

Hence, the volume of U at time $t + \delta t$ is

$$[U']_{t+\delta t} = \int_U d^{3N}x'\,d^{3N}p' = \int_U \frac{\partial(x'_a, p'_a)}{\partial(x_a, p_a)} d^{3N}x\,d^{3N}p$$

$$= \int_U d^{3N}x\,d^{3N}p = [U]_t. \tag{1.11}$$

We conclude that the volume of any subset of the state space is constant in time under evolution through Hamilton's equations. This proposition is known as *Liouville's theorem*.

Liouville's theorem provides a key mathematical criterion for defining the notion of a closed physical system.

> In classical physics, a physical system is closed if its time evolution preserves energy and state-space volume.

1.2.4 Lagrangian Mechanics

Classical mechanics can also be described in the Lagrangian formalism. In this formalism, the basis variables are coordinates and velocities, rather than coordinates and momenta. We define the *Lagrangian state space* $\tilde{\Gamma} = \mathbb{R}^{6N}$ for a system of N particles, with points

$$y = (x_1, v_1, x_2, v_2, \ldots, x_{3N}, v_{3N}).$$

The variable v_a is the velocity associated to coordinate x_a. For any path $x_a(\cdot)$ on the configuration space Q, $v_a(t) := \dot{x}_a(t)$.

The dynamics are encoded in the Lagrangian $L : \Gamma \to \mathbb{R}$, defined by

$$L(x_a, v_a) := \sum_{a=1}^{3N} \frac{1}{2} m_a v_a^2 + V(x_1, x_2, \dots, x_{3N}). \tag{1.12}$$

The evolution equations are the *Lagrange equations*,

$$\frac{d}{dt}\frac{\partial L}{\partial v_a} = \frac{\partial L}{\partial x_a}, \tag{1.13}$$

which, together with the condition $v_a = \dot{x}_a$, lead to Eq. (1.5).

The momenta p_a conjugate to x_a are defined by

$$p_a := \frac{\partial L}{\partial v_a}. \tag{1.14}$$

Equation (1.14) defines a natural map $F_L : \tilde{\Gamma} \to \Gamma$, known as the Legendre transform. Under the Legendre transform, the Hamiltonian is expressed as

$$H(x_a, p_a) = \sum_{a=1}^{3N} p_a v_a - L(x_a, v_a), \tag{1.15}$$

where the velocities v_a are expressed in terms of the momenta by solving Eq. (1.14).

1.2.5 Hamilton's Action Principle

The Lagrange equations can be obtained by an extremization condition, known as *Hamilton's principle*. Consider the space Π of all configuration space paths $x_a(t)$, that is, of all differentiable maps from a time interval $[t_i, t_f]$ to Q. For a given Lagrangian function L on $\tilde{\Gamma}$, we define the *action* $S : \Pi \to \mathbb{R}$ by

$$S[x_a(\cdot)] := \int_{t_1}^{t_f} dt\, L(x_a, \dot{x}_a). \tag{1.16}$$

Then the maximum of S over all paths with fixed endpoints $x_a(t_i)$ and $x_a(t_f)$ is obtained for the paths that satisfy the Lagrange equations (1.13).

For the proof, we consider variations $\delta x(\cdot)$ of the paths. For such variations,

$$\begin{aligned}
\delta S &= \int_{t_1}^{t_f} dt \sum_a \left(\frac{\partial L}{\partial x_a}\delta x_a(t) + \frac{\partial L}{\partial v_a}\delta \dot{x}_a(t) \right) = \int_{t_1}^{t_f} dt \sum_a \left(\frac{\partial L}{\partial x_a}\delta x_a(t) + \frac{\partial L}{\partial v_a}\frac{d}{dt}\delta x_a(t) \right) \\
&= \sum_a \int_{t_1}^{t_f} dt \frac{d}{dt}\left[\frac{\partial L}{\partial v_a}\delta x_a(t) \right] + \int_{t_1}^{t_f} dt \sum_a \delta x_a(t)\left(\frac{\partial L}{\partial x_a} - \frac{d}{dt}\frac{\partial L}{\partial v_a} \right) \\
&= \sum_a \left[\frac{\partial L}{\partial v_a}(t_f)\delta x_a(t_f) - \frac{\partial L}{\partial v_a}(t_i)\delta x_a(t_i) \right] + \int_{t_1}^{t_f} dt \sum_a \delta x_a(t)\left(\frac{\partial L}{\partial x_a} - \frac{d}{dt}\frac{\partial L}{\partial v_a} \right).
\end{aligned}$$

For variations with fixed endpoints, $\delta x_a(t_f) = \delta x_a(t_i) = 0$, hence,

$$\delta S = \sum_a \int_{t_1}^{t_f} dt \delta x_a(t) \left(\frac{\partial L}{\partial x_a} - \frac{d}{dt} \frac{\partial L}{\partial v_a} \right). \tag{1.17}$$

The maximum of S occurs for $\delta S = 0$, and since the variations are arbitrary, the maximizing path satisfies Eq. (1.13).

1.3 Classical Electromagnetism

We mentioned earlier that Newtonian mechanics describes a world that consists of mutually interacting particles. However, during the nineteenth century a new point of view was developed, that forces are autonomous entities, not dependent upon particles for their definition and characterized by their own time evolution laws. This idea originates from Faraday and it came into fruition in the work of Maxwell, who achieved the unification of electric, magnetic, and optical phenomena in a theory that is nowadays referred to as *classical electromagnetism*.

The fundamental variables of classical electromagnetism are two vector fields on the physical space, the electric field $\mathbf{E}$ and the magnetic field $\mathbf{B}$. A particle with charge q in the presence of electric and magnetic field is acted upon by a force $\mathbf{F}$, the *Lorentz force*,

$$\mathbf{F} = q \left(\mathbf{E} + \frac{1}{c} \mathbf{v} \times \mathbf{B} \right), \tag{1.18}$$

where $\mathbf{v}$ is the particle's velocity.

The properties and dynamics of the electric and the magnetic field are described compactly by Maxwell's equations

$$\nabla \cdot \mathbf{E} = \rho \tag{1.19}$$

$$\nabla \cdot \mathbf{B} = 0 \tag{1.20}$$

$$\nabla \times \mathbf{E} = -\frac{1}{c} \frac{\partial \mathbf{B}}{\partial t} \tag{1.21}$$

$$\nabla \times \mathbf{B} = \frac{1}{c} \left(\frac{\partial \mathbf{E}}{\partial t} + \mathbf{j} \right), \tag{1.22}$$

where ρ is the charge density, $\mathbf{j}$ is the current density, and c is the speed of light.

The electric and magnetic fields are expressed in terms of the electric potential ϕ and the magnetic potential $\mathbf{A}$ as

$$\mathbf{E} = -\frac{1}{c} \frac{\partial \mathbf{A}}{\partial t} - \nabla \phi \qquad \mathbf{B} = \nabla \times \mathbf{A}. \tag{1.23}$$

Substituting Eqs. (1.23) into Maxwell's equations, we find that Eqs. (1.20) and (1.21) are satisfied identically. We also note that the fields $\mathbf{E}$ and $\mathbf{B}$ are invariant under the gauge transformation

$$\phi \to \phi' = \phi - \frac{1}{c} \frac{\partial f}{\partial t} \qquad \mathbf{A} \to \mathbf{A}' = \mathbf{A} + \nabla f, \tag{1.24}$$

for any scalar function f.

In a vacuum, $\rho = \mathbf{j} = 0$; Maxwell's equations lead to the wave equation for the propagation of electromagnetic (EM) waves

$$\nabla^2 \mathbf{E} = \frac{1}{c^2} \frac{\partial^2 \mathbf{E}}{\partial t^2}. \tag{1.25}$$

Any solution to the wave equation can be written as a superposition of plane waves

$$\mathbf{E}(\mathbf{x}, t) = \mathbf{E}_0 e^{i\mathbf{k}\cdot\mathbf{x} - i\omega t}, \tag{1.26}$$

where $\mathbf{E}_0$ is the wave's amplitude, $\mathbf{k}$ is the wavenumber vector, and ω is the angular frequency. Substituting in Eq. (1.25), we obtain the *dispersion relation* for EM waves in a vacuum

$$\omega = c|\mathbf{k}|. \tag{1.27}$$

Furthermore, substituting Eq. (1.26) into Eq. (1.19) for $\rho = 0$, we obtain $\mathbf{k} \cdot \mathbf{E}_0 = 0$. This means that the EM wave is *transverse*, that is, the field oscillates in a direction normal to the propagation direction. The possible directions of oscillation span a two-dimensional plane and correspond to the *polarizations* of the EM wave.

1.4   Classical Probability Theory

Probability theory has become an indispensable component of contemporary physics, both as a practical tool and at the level of fundamental theories. Probabilities appear in both classical and quantum physics. Probabilities in quantum theory have strong mathematical and interpretational differences from the probabilities that are used not only in classical physics, but in all other sciences. For this reason, it is useful to distinguish between classical and quantum probability theories. In this section, we present the former.

1.4.1   Sample Space

Probabilities always refer to some alternatives. For example, we may be interested in the probability to "get two sixes the next time we throw a pair of dice," or in the probability that "Brazil wins the next World Cup." Moreover, probabilities are meaningful only if we compare different alternatives; hence, we must have a *set of possible alternatives*.

We must also distinguish between elementary and composite alternatives. When rolling a single die, the elementary alternatives correspond to one of the integers 1, 2, 3, 4, 5, or 6. However, we can easily define the alternative $A =$ "get an odd number" or $B =$ "get 5 or 6." The elementary alternatives correspond to sets with a single element $\{1\}$, $\{2\}$, $\{3\}$, $\{4\}$, $\{5\}$, and $\{6\}$; the alternative A corresponds to the set $\{1, 3, 5\}$; and the alternative B corresponds to the set $\{5, 6\}$. Hence, any alternative corresponds to a subset of the set $\{1, 2, 3, 4, 5, 6\}$.

We call *random experiment* any process that leads to the realization of one out of many alternatives. The set of all elementary alternatives in an experiment is the *sample space* of the experiment. We denote sample spaces by capital Greek letters (for example Γ or Ω).

Any alternative, whether elementary or composite, corresponds to a subset of a sample space. We will denote such subsets by capital italic Latin letters.

We define the *complement* of any alternative A as the set $\bar{A} = \Omega - A$, that is, as the unique subset $\bar{A}$ of Ω that satisfies $A \cup \bar{A} = \Omega$ and $A \cap \bar{A} = \emptyset$.

Two alternatives A and B are called *mutually exclusive*, if $A \cap B = \emptyset$. Obviously, two elementary alternatives are always mutually exclusive. It is impossible to roll a single die and obtain both one and six.

Example 1.1 Consider the simultaneous roll of two dice as a random experiment. The sample space Ω consists of 36 pairs of integers,

$$\Omega = \{(1,1),(1,2),(1,3),(1,4),(1,5),(1,6),(2,1),(2,2),(2,3),(2,4),(2,5),(2,6),$$
$$(3,1),(3,2),(3,3),(3,4),(3,5),(3,6),(4,1),(4,2),(4,3),(4,4),(4,5),(4,6),$$
$$(5,1),(5,2),(5,3),(5,4),(5,5),(5,6),(6,1),(6,2),(6,3),(6,4),(6,5),(6,6)\}.$$

The following alternatives are possible.

1. $A = $ "sixes" $= \{(6,6)\}$ (one member).
2. $B = $ "ace-deuce" $= \{(1,2),(2,1)\}$ (two members).
3. $C = $ "six" $= \{(1,5),(5,1),(2,4),(4,2),(3,3)\}$ (five members).
4. $D = $ "seven" $= \{(1,6),(6,1),(2,5),(5,2),(3,4),(4,3)\}$ (six members).
5. The empty set is trivially an alternative (the alternative that never happens), and so is the set Ω itself ("something happened").

In physics, and particularly in quantum theory, the elementary alternatives are referred to as *fine-grained* and the composite ones as *coarse-grained* . We will employ this terminology from now on.

1.4.2   Probability Distributions

Let Ω be a sample space. A *probability distribution* on Ω is the assignment of probabilities to each of the alternatives associated to Ω, namely, a function Prob that assigns to each subset A of Ω a number $\mathrm{Prob}(A) \in [0,1]$. The value $\mathrm{Prob}(A)$ is called the *probability* of the alternative A.

A probability distribution must satisfy the following properties, known as the *Kolmogorov axioms*.

1. For every $A, B \subset \Omega$ such that $A \cap B = \emptyset$, $\mathrm{Prob}(A \cup B) = \mathrm{Prob}(A) + \mathrm{Prob}(B)$.
2. $\mathrm{Prob}(\emptyset) = 0$.
3. $\mathrm{Prob}(\Omega) = 1$.

The following are consequences of the Kolmogorov axioms.

1. $\mathrm{Prob}(\bar{A}) = 1 - \mathrm{Prob}(A)$, for every $A \subset \Omega$.
2. Let $B \subset A$. The sets B and $A - B$ are disjoint and they satisfy $B \cup (A - B) = A$. By the first Kolmogorov axiom,

$$\text{Prob}(A - B) = \text{Prob}(A) - \text{Prob}(B) \tag{1.28}$$

for every $B \subset A$.

3. $\text{Prob}(A \cup B) = \text{Prob}(A) + \text{Prob}(B) - \text{Prob}(A \cap B)$, for every $A, B \subset \Omega$.

Proof The three sets $A' = A - (A \cap B)$, $B' = B - (A \cap B)$ and $C = A \cap B$ are mutually exclusive and they satisfy $A' \cup B' \cup C = A \cup B$. By the first Kolmogorov axiom, $\text{Prob}(A \cup B) = \text{Prob}(A - (A \cap B)) + \text{Prob}(B - (A \cap B)) + \text{Prob}(A \cap B)$. Then, we use Eq. (1.28) to obtain the desired result.

1.4.3   Properties of Probability Distributions

Consider a sample space $\Omega = \{x_1, x_2, \ldots, x_n\}$, with n elements. We define a probability distribution on Ω by specifying n positive numbers $p_i = \text{Prob}(\{x_i\})$, $i = 1, \ldots n$, each defining the probability of the fine-grained event $\{x_i\}$. The numbers p_i satisfy the *normalization condition*

$$\sum_{i=1}^{n} p_i = 1. \tag{1.29}$$

We can represent the n numbers p_i as a *probability vector*

$$\mathbf{w} = (p_1, p_2, \ldots, p_n). \tag{1.30}$$

Then, we use the Kolmogorov axioms in order to construct the probability of any coarse-grained alternative. For example,

$$\text{Prob}(\{x_1, x_3\}) = \text{Prob}(\{x_1\}) + \text{Prob}(\{x_3\}) = p_1 + p_3.$$

Let $f : \Omega \to \mathbb{R}$ be a function on the sample space

$$\Omega = \{x_1, x_2, \ldots, x_n\}.$$

Given a probability distribution, defined in terms of the positive numbers p_i, $i = 1, 2, \ldots, n$, we define

- the *mean value* of f: $\langle f \rangle := \sum_{i=1}^{n} p_i f(x_i)$,
- the *nth moment* of f: $\langle f^n \rangle := \sum_{i=1}^{n} p_i f(x_i)^n$,
- the *variance* $(\Delta f)^2$ of f: $(\Delta f)^2 := \langle f^2 \rangle - \langle f \rangle^2$,
- the *mean deviation* Δf of f: $\Delta f := \sqrt{(\Delta f)^2}$.

A key property of probability distributions is that they satisfy Jensen's inequality,

$$\langle F(f) \rangle \geq F(\langle f \rangle), \tag{1.31}$$

for every *convex function* $F : \mathbb{R} \to \mathbb{R}$ and every function $f : \Omega \to \mathbb{R}$.

The proof is elementary. We first consider a sample space with two elements, $\Omega = \{x_1, x_2\}$. We write $y_1 = f(x_1)$ and $y_2 = f(x_2)$, $p = p_1, p_2 = 1 - p_1$. Then Eq. (1.31) becomes $pF(y_1) + (1 - p)F(y_2) \geq F(py_1 + (1 - p)y_2)$. Hence the straight line connecting the points $(y_1, F(y_1))$ and $(y_2, F(y_2))$ lies above the graph of the function F, which is the actual definition of a convex function. The generalization to $n > 2$ via induction is straightforward.

Example 1.2 We revisit the random experiment of rolling a pair of dice that was described in example 1.1. Consider a probability distribution that assigns equal probabilities $p = 1/36$ to all 36 members of Ω. Then, for the sets A, B, C, and D of Example 1.1,

$$\text{Prob}(A) = \frac{1}{36}, \quad \text{Prob}(B) = \frac{2}{36} = \frac{1}{18}, \quad \text{Prob}(C) = \frac{5}{36}, \quad \text{Prob}(D) = \frac{6}{36} = \frac{1}{6}.$$

1.4.4 Marginal Probabilities

Consider a sample space Ω of the form $\Omega_1 \times \Omega_2$, where $\Omega_1 = \{x_1, x_2, \ldots, x_n\}$ and $\Omega_2 = \{y_1, y_2, \ldots, y_m\}$. The sample space of Example 1.1 was of this form with $\Omega_1 = \Omega_2 = \{1, 2, 3, 4, 5, 6\}$. A fine-grained alternative corresponds to a pair (x_i, y_a), where $i = 1, \ldots, n$ and $a = 1, \ldots m$. The associate probabilities are expressed in terms of the positive numbers $p_{ia} := \text{Prob}[\{(x_i, y_a)\}]$.

We define the *marginal probabilities* p_i^1 and p_a^2 as

$$p_i^1 = \sum_{a=1}^{m} p_{ia} = \text{Prob}(\{x_i\} \times \Omega_2) \tag{1.32}$$

$$p_a^2 = \sum_{i=1}^{n} p_{ia} = \text{Prob}(\Omega_1 \times \{y_a\}). \tag{1.33}$$

The numbers p_i^1 define a probability distribution on Ω_1 and the numbers p_a^2 a probability distribution on Ω_2. These two distributions contain all the information about the alternatives associated separately to Ω_1 and Ω_2, but they contain no information about the relations between the alternatives of Ω_1 and Ω_2.

1.4.5 Conditional Probability

Suppose we employ a probability distribution in order to describe a random experiment with sample space Ω. At some moment, we obtain some new information about our system and we want to change the probability distribution accordingly. This is achieved through the following definition of conditional probability.

For all alternatives $A, B \subset \Omega$, the conditional probability of A *given* B, $\text{Prob}(A|B)$, is defined as

$$\text{Prob}(A|B) = \frac{\text{Prob}(A \cap B)}{\text{Prob}(B)}. \tag{1.34}$$

Example 1.3 Consider the random experiment described in Examples 1.1 and 1.2. Let $D =$ "total seven" and $E =$ "no one or two in any die" and we want to evaluate the conditional probability $\text{Prob}(D|E)$. Since E has been realized, it constitutes our new sample space

$$E = \{(3, 3), (3, 4), (3, 5), (3, 6), (4, 3), (4, 4), (4, 5), (4, 6),$$
$$(5, 3), (5, 4), (5, 5), (5, 6), (6, 3), (6, 4), (6, 5), (6, 6)\}.$$

E has 16 elements. Among those elements, only $(3,4)$ and $(4,3)$ have total seven. Hence, we expect the conditional probability to be $\text{Prob}(D|E) = 2/16 = 1/8$. On the other hand,

$$D \cap E = \{(3, 4), (4, 3)\},$$

therefore, $\text{Prob}(D \cap E) = 2/36 = 1/18$. Furthermore, $\text{Prob}(E) = 16/36 = 4/9$. Consequently, $\text{Prob}(D \cap E)/\text{Prob}(E) = (1/18)/(4/9) = 1/8$, thereby verifying the consistency of the definition of conditional probability.

If $\text{Prob}(A|B) = \text{Prob}(A)$ for two alternatives A and B, then A and B are *statistically independent*. A is not affected whether B happens or not. In this case,

$$\text{Prob}(A \cap B) = \text{Prob}(A)\text{Prob}(B). \tag{1.35}$$

Example 1.4 Consider the following game of chance. A player rolls a pair of dice three times. The player wins if he or she rolls a seven at least once. Assuming that the dice are fair, what is the probability of winning?

Answer Let A_i be the alternative "not seven in the ith roll," where $i = 1, 2, 3$. According to Example 1.2, the probability of rolling seven is $\frac{1}{6}$, hence, $p(A_i) = 1 - \frac{1}{6} = \frac{5}{6}$. The alternative L that the player loses corresponds to not rolling seven in any attempt, hence, $L = A_1 \cap A_2 \cap A_3$. Assuming that the three rolls are statistically independent, $\text{Prob}(L) = \text{Prob}(A_1 \cap A_2 \cap A_3) = \text{Prob}(A_1)\text{Prob}(A_2)\text{Prob}(A_3) = (5/6)^3 = \frac{125}{216}$. The alternative W that the player wins is the complement of L; therefore, $\text{Prob}(W) = 1 - \text{Prob}(L) = 1 - \frac{125}{216} = \frac{91}{216} \simeq 0.42$.

Example 1.5 Consider a random experiment with two elementary alternatives: Alternative 1 with probability p and alternative 0 with probability $q = 1 - p$. A random experiment of this type is called *dichotomic*. Suppose this experiment is repeated N times. The probability $\text{Prob}(n)$ that 1 happens exactly n times (and hence, 0 happens $N - n$ times) is given by the *binomial distribution*

$$\text{Prob}(n) = p^n q^{N-n} \frac{N!}{(N-n)!\,n!}. \tag{1.36}$$

The associated mean value is

$$\langle n \rangle = \sum_{n=0}^{N} n p^n q^{N-n} \frac{N!}{(N-n)!\,n!} = p\frac{\partial}{\partial p} \sum_{n=0}^{N} p^n q^{N-n} \frac{N!}{(N-n)!\,n!}$$

$$= p\frac{\partial}{\partial p}(p+q)^N = Np(p+q)^{N-1} = Np. \tag{1.37}$$

Similarly, we evaluate

$$\langle n^2 \rangle = \sum_{n=0}^{N} n^2 p^n q^{N-n} \frac{N!}{(N-n)!\,n!} = p\frac{\partial}{\partial p}p\frac{\partial}{\partial p} \sum_{n=0}^{N} p^n q^{N-n} \frac{N!}{(N-n)!\,n!}$$

$$= p\frac{\partial}{\partial p}p\frac{\partial}{\partial p}(p+q)^N = N(N-1)p^2 + Np. \tag{1.38}$$

Hence, the variance of the binomial distribution is $(\Delta n)^2 = Np(1 - p)$.

Example 1.6 Consider the binomial distribution of Example 1.4 in the regime where alternative 1 is highly improbable ($p \ll 1$), but the experiment is run many times. We take the limits $N \to \infty$ so that the mean value $\lambda := Np$ stays constant. Then,

$$\text{Prob}(n) = \frac{\lambda^n}{n!} \left(\frac{N!}{(N-n)!\, N^n} \right) \left(1 - \frac{\lambda}{N} \right)^{-n} \left(1 - \frac{\lambda}{N} \right)^N . \tag{1.39}$$

As $N \to \infty$, the three terms in the parentheses in Eq. (1.39) converge to 1, 1, and $e^{-\lambda}$ respectively. Thus, we derived the *Poisson distribution*,

$$\text{Prob}(n) = \frac{\lambda^n e^{-\lambda}}{n!} . \tag{1.40}$$

1.4.6   Continuous Sample Space

The definitions above apply to samples spaces with a finite number of elements. While it is possible to claim that the description of actual experiments does not require infinite sample spaces, such a restriction does not allow the formulation of fundamental physical theories that rely on continuous variables, such as time and space. For this reason, it is necessary to generalize Kolmogorov's axioms in order to describe sample spaces defined in terms of continuous variables.

Consider a sample space $\Omega = \mathbb{R}$, that is, that fine-grained alternatives correspond to points x of the real line (e.g., a particle's position). Coarse-grained alternatives correspond to subsets of $\mathbb{R}$. The probabilities associated to a subset A are expressed as

$$\text{Prob}(A) = \int_A dx\, p(x), \tag{1.41}$$

where $p(x)$ stands for a positive function $p : \Omega \to \mathbb{R}^+$ that is normalized to unity: $\int_{\mathbb{R}} dx\, p(x) = 1$. The function $p(x)$ is a *probability density* on Ω.

For any $A \subset \Omega$, we define an associated *characteristic function* $\chi_A : \Omega \to \{0, 1\}$,

$$\chi_A(x) := \begin{cases} 1 & x \in A \\ 0 & x \notin A. \end{cases} \tag{1.42}$$

Then, Eq. (1.41) can be expressed as

$$\text{Prob}(A) = \int dx\, p(x) \chi_A(x). \tag{1.43}$$

For any function $f : \Omega \to \mathbb{R}$, we define

- the *mean value* of f: $\langle f \rangle := \int_{\mathbb{R}} dx p(x) f(x)$,
- the *nth moment* of f: $\langle f^n \rangle := \int_{\mathbb{R}} dx p(x) f(x)^n$,
- the *mean deviation* Δf of f: $(\Delta f)^2 := \langle f^2 \rangle - \langle f \rangle^2$,
- the *moment generating functional* $Z_f(\mu) := \int_{\mathbb{R}} dx\, p(x) e^{i\mu f(x)}$, which allows us to compute the moments as $\langle f^n \rangle = (-i)^n dZ_f(0)/d\mu$.

Jensen's inequality (1.31) is also satisfied by continuous probability distributions.

Next, consider a probability distribution $p(x, y)$ on a two-dimensional sample space $\Omega = \mathbb{R}^2$, defined in terms of two continuous variables x and y. The *marginal probability distributions* are defined as

$$p_1(x) = \int dy p(x, y), \quad p_2(y) = \int dx p(x, y). \tag{1.44}$$

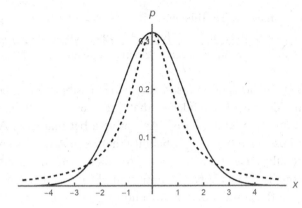

Example 1.7 The most commonly employed probability distributions on $\mathbb{R}$ are the Gaussian and the Lorentzian (see Fig. 1.1). The former is defined by the probability density

$$p_G(x) := \frac{1}{\sqrt{2\pi\sigma^2}} e^{-\frac{(x-x_0)^2}{2\sigma^2}}. \tag{1.45}$$

The Gaussian distribution depends on two parameters, the real number x_0 and the positive number σ. Using Eqs. (A.1) and (A.2), we find that for the Gaussian distribution

$$\langle x \rangle = x_0, \qquad \Delta x = \sigma. \tag{1.46}$$

Example 1.8 The Lorentzian distribution is defined by the probability density

$$p_L(x) := \frac{1}{\pi} \frac{\gamma}{\gamma^2 + (x - x_0)^2}, \tag{1.47}$$

which depends on the real number x_0 and the positive number γ. The mean value of x in the Lorentzian distribution is equal to x_0. The parameter γ is a measure of the distribution's spread, but it is not related to its variance. The latter is infinite, because $\langle x^2 \rangle = \infty$.

1.5 Probability in Physics

In this section, we describe how probability is used in physics. First, we explain the different meanings and interpretations of the probability concept. Then, we present the main ideas of statistical mechanics, which was the main field of physics to employ probabilities prior to quantum theory. We conclude with a probabilistic description of measurements in classical physics, which sets a basis for comparisons with the description of measurements in quantum physics that will be taken up in chapter 4 and 6.

1.5.1 Interpretations and Uses of the Notion of Probability

The basic principles of probability theory were given in Section 1.4.2. However, these principles are purely mathematical and they offer no guidelines about the application of probabilities in describing or predicting events in nature. To this end, it is necessary to distinguish between different notions of probability that refer to different things even if they share a common mathematical description. We distinguish between subjective, epistemic, and physical probabilities.

Subjective probabilities refer to the evaluations of a person about the possibility of a specific event, and they express the person's degree of confidence that these events will happen. Subjective probabilities are most commonly used in betting. Accepting a bet that team A will win at three-to-one odds, means that I estimate that the probability that team A wins is greater than $1/(3 + 1) = 1/4$. In spite of their subjective character, these probabilities are not arbitrary. If I make a practice of betting irrationally or incoherently, I am bound to lose all my money. Subjective probabilities appear rarely in the context of physics, and they will not be addressed in this book.

Epistemic probabilities refer to the use of data in order to make rational estimates, for example, the estimate of the true value of a physical magnitude given the outcomes of an experiment. When a student applies the least-squares method in the laboratory in order to construct a straight line that best fits the experimental data, the student employs probabilities in an epistemic sense. The use of probability theory in polling is similar: one estimates the voting intentions of a population given measurements in a small sample. Epistemic probabilities are very important in physics, but they are used at the level of data processing, not at the level of fundamental theories. For this reason, we shall not consider them further.

Physical probabilities refer to the behavior of a physical system, and not to the processing of information we have about this system. They may originate from the inability to control the initial conditions of a physical system, from random effects due to interaction with an environment, or from a physical system being intrinsically random. In this book, we will exclusively deal with physical probabilities.

Probabilities can also be interpreted in two different ways: as referring to *event frequencies* or as referring to *logical propositions* about a system. The first case corresponds to the *frequency* interpretation of probability; the second case corresponds to the *logical interpretation* of probability.

In the frequency interpretation of probabilities, one does not assign probabilities to individual random experiments. Instead, one defines a *statistical ensemble* that consists of N repetitions of the same random experiments. Let A be an alternative of the random experiment that is realized n_A times in the ensemble. We define the *relative frequency* of A as the ratio n_A/N. According to the frequency interpretation, the relative frequency n_A/N converges to the probability $\text{Prob}(A)$ as $N \to 0$. If the frequency does not converge, then the system is too "unstable" to be described by probabilities.

The frequency interpretation is the most common in physics and it applies to almost all experiments. In order to relate theoretical predictions to experiment, it is necessary to repeat the experiment many times. This requires a large number of copies of the same system prepared with identical procedures.

According to the *logical interpretation* of probability, it is meaningful to employ probabilities even for individual systems, in order to predict outcomes of individual random experiments. For example, one may want to predict how Earth's climate will evolve in the coming centuries.

Obviously, no exact copies of Earth are available to form a statistical ensemble. Predictions of global warming employ the logical interpretation of probabilities.

The key tool for describing individual systems is the notion of conditional probability $\text{Prob}(A|B)$; this allows for estimates about a random experiment given information B. Of particular interest is the case where one can find alternatives A and B such that $\text{Prob}(A|B) = 1$. In this case, probabilities lead to a *logical implication* $B \to A$. If we find out that B has been realized, we can be certain that A will be (or has been) also realized. We can also talk about approximate implications $B \to A$, of accuracy ϵ, if $\text{Prob}(A|B) = 1 - \epsilon$, for $0 < \epsilon << 1$.

The interpretations of probability given above are not mutually exclusive. One may employ one in one case and the other in another, and many people propose probability types different to those presented here. This discussion concerns the foundations of probability theory, whose details lie outside the scope of this book. Nonetheless, it is worth noting that, in quantum theory, we encounter a different facet of this discussion. The difference between the frequency and the logical interpretation of probabilities sets the fundamental question, whether the notion of a quantum state (see Chapter 3) refers to an individual quantum system (e.g., a single electron), or if it makes sense only in a statistical ensemble of such systems.

1.5.2 Classical Statistical Mechanics

Statistical mechanics is a synthesis of Hamiltonian mechanics and probability theory. It provides a microscopic explanation of the laws of thermodynamics that apply to macroscopic systems. Since statistical mechanics incorporates probabilistic notions and the distinction between the microscopic and macroscopic level of description, it is the most appropriate theory of classical physics to compare with quantum mechanics.

Next, we present the basic concepts of statistical mechanics that demonstrate how probabilities are employed in a fundamental physical theory.

Consider a system of N particles, described by the state space $\Gamma = \mathbb{R}^{6N}$, for $N >> 1$. In a deterministic theory, the knowledge of the system's initial conditions allows us fully and uniquely to specify the outcome of any measurement. However, either because the knowledge of the system's initial conditions at the microscopic level is impossible, or because any system is subject to small external influences that cumulatively destroy determinism, the measurement outcomes will not in general be the same in different runs of the experiment. This implies that *a physical measurement can be identified with a random experiment*, even in a fully deterministic world.

In classical physics there is no restriction on the accuracy of measurements. Ideally, one can determine the position and momentum of each particle with unlimited precision. This implies that the points of Γ define the fine-grained alternatives for this system. It is therefore reasonable to identify the state space of the Hamiltonian description with the sample space for measurements. With this identification, the points of Γ are referred to as *microstates*. Different microstates are mutually exclusive (at a single moment of time): It is impossible for a molecule to have momentum 10^{-25} kg m/s and at the same time to have momentum 10^{-24} kg m/s.

Hence, the following concepts can be identified:

- measurement = random experiment,
- state space = sample space,
- microstate = fine-grained alternative.

The identifications above suggest the definition of a probability density $\rho : \Gamma \to \mathbb{R}^+$, such that the quantity

$$\rho(q_1, p_1, q_2, p_2 \ldots, q_{3N}, p_{3N}) d^{3N}q \, d^{3N}p \tag{1.48}$$

defines the probability that the system is found within an elementary volume $d^{3N}q \, d^{3N}p$ of Γ. The probability density ρ is normalized,

$$\int d^{3N}q \, d^{3N}p \, \rho(q_1, p_1, q_2, p_2 \ldots, q_{3N}, p_{3N}) = 1. \tag{1.49}$$

We will employ the notation $\xi = (q_1, p_1, \ldots, q_{3N}, p_{3N})$ for the points of Γ, so that Eq. (1.49) reads $\int d\xi \rho(\xi) = 1$, where $d\xi = d^{3N}q \, d^{3N}p$.

We assume that the system is closed; hence, it is described by Hamilton's equations for some Hamiltonian H. Then, the probability density ρ satisfies Liouville's equation

$$\frac{\partial \rho}{\partial t} = \{H, \rho\}. \tag{1.50}$$

Equation (1.50) follows from the requirement that the normalization condition (1.49) holds at all times t, that is, $d\rho/dt = 0$. Equivalently,

$$\frac{\partial \rho}{\partial t} + \sum_a \left(\frac{\partial \rho}{\partial q_a} \dot{q}_a + \frac{\partial \rho}{\partial p_a} \dot{p}_a \right) = 0. \tag{1.51}$$

We substitute $\dot{q}_a, \dot{p}_a$ from Eqs. (1.4) and use the definition (1.7) to obtain Eq. (1.50).

The formulation of statistical mechanics by Gibbs is based on the following hypothesis. For any system in contact with a heat bath at constant temperature T, the probability density ρ approaches at long times the *canonical distribution*

$$\rho_{can}(\xi) = Z^{-1} e^{-\beta H(\xi)}, \tag{1.52}$$

where $\beta := 1/(k_B T)$ with k_B Boltzmann's constant, and $Z(T) := \int d\xi e^{-\beta H(\xi)}$ is the *partition function*. The mean energy of the system is given by

$$\langle E \rangle = \frac{\int d\xi \, H(\xi) e^{-\beta H(\xi)}}{\int d\xi \, e^{-\beta H(\xi)}} = -\frac{\partial \ln Z}{\partial \beta}. \tag{1.53}$$

1.5.3   Measurements

Measurements of macroscopic systems involve alternatives that are highly coarse-grained in comparison with the microstates that are defined at the molecular level. Consider, for example, the roll of a single die. The state space Γ of the die describes the motions of all molecules that constitute it. Of course, we are not interested in that detail, we only want to know which of its six faces will turn out to be on top when the die's center of mass stops moving. The outcome of a measurement is one value of $\lambda \in \{1, 2, 3, 4, 5, 6\}$.

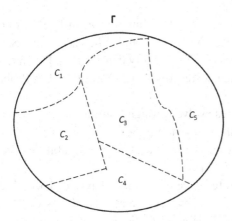

Fig. 1.2 A measurement in classical physics corresponds to a partition of the state space Γ in mutually exclusive and exhaustive subsets C_λ.

Evidently, there are many microstates that correspond to a single value of λ. This means that the measurement of λ partitions the state space into subsets C_λ, one for each value of λ. These subsets are the *value sets* of the measurement, and they satisfy the following properties.

1. $C_\lambda \cap C_{\lambda'} = \emptyset$,   if $\lambda \neq \lambda'$ (*mutually exclusive* alternatives).
2. $\cup_\lambda C_\lambda = \Gamma$ (alternatives that *exhaust* Γ).

The above observation generalizes straightforwardly. Any measurement splits the state space Γ into a family of mutually exclusive and exhaustive subsets C_λ. Furthermore, all information about the measurements encoded in a function $F : \Gamma \to \mathbb{R}$, defined by

$$F(\xi) := \sum_\lambda \lambda \chi_{C_\lambda}(\xi), \quad x \in \Gamma, \tag{1.54}$$

where χ_{C_λ} is the characteristic function of the subset C_λ and $\xi \in \Gamma$.

Conversely, any function $F : \Gamma \to \mathbb{R}$, defines a family of values sets, that is, of mutually exclusive and exhaustive alternatives C_λ, by

$$C_\lambda = \{\xi \in \Gamma | F(\xi) = \lambda\}. \tag{1.55}$$

Hence, we conclude with the following.

In classical physics, a measurement corresponds to a function F on the state space of the measured system.

1.6   The Mathematical Structure of Classical Physics

We summarize the main contents of this chapter by describing the key mathematical principles of classical physics, in a way that allow for a direct comparison with the principles of quantum physics that will be presented in following chapters.

1. *The fundamental space.* Any physical system is characterized by a *state space* Γ. Each point $\xi \in \Gamma$ is called a microstate and corresponds to the finest possible description of the system.

2. *Measurement*. A measurement procedure corresponds to a function F on Γ. By Eq. (1.55), any function F partitions Γ in a family of mutually exclusive and exhaustive subsets.

3. *Probabilities*. Probabilities are defined in terms of a probability density ρ on the state space. The probability of a measurement outcome that corresponds to a subset U of Γ is $\text{Prob}(U) = \int d\xi \rho(\xi) \chi_U(\xi)$.

4. *Time evolution*. In a closed system, microstates evolve through Hamilton's equations (1.4), and the probability density ρ through Liouville's equation (1.50).

5. *Incorporation of information*. New information is incorporated in probabilities proceeds through the conditional probability rule, Eq. (1.34).

6. *Combination of subsystems*. Let Γ_1 be the state space of system 1, and let Γ_2 be the state space of system 2. The composite system formed by subsystems 1 and 2 is described by the Cartesian product $\Gamma_1 \times \Gamma_2$. (This principle is so self-evident that it would not have to be mentioned except for the fact that quantum theory has a very different rule for the combination of subsystems.)

QUESTIONS

1.1 You are in a casino and you hear someone say "Since the wheel landed 14 times in a row on red, I am going to bet on black." Is this person right?

1.2 Alice wants to travel abroad but has a pathological fear that terrorists may put a bomb in her plane. Her friend Bob tells her to take her own bomb on the plane, as the probability of having two bombs on the same flight is negligible. Is this a joke or sound probabilistic advice?

1.3 Passedix was a popular game in Galileo's time that involved throwing three dice. Galileo was asked by his patron, the Grand Duke of Tuscany, to explain "why experience shows that the chance of throwing a total of 9 was less than that of throwing a total of 10, despite the fact that there is an equal number of partitions of 9 and 10 in three dice." How would you answer?

1.4 A numerical simulation of the Solar System analyzed many different planetary orbits (solutions of Hamilton's equations) that are compatible with current observations (Laskar and Gastineau, 2009). In 1 percent of those orbits a planetary collision involving Earth occurs within the next 5 billion years, but no such collision occurs in the remaining 99 percent of orbits. The researchers conclude that the probability of Earth colliding within the next 5 billion years is 0.01. Which notion of probability are they using?

1.5 An experimental high-energy physics group makes a press release about the discovery of a particle X. The release contains the following phrases.

 (i) The mass of X is between 350 and 400 GeV with a 95 percent certainty.

 (ii) Thirty percent of X's decays occur through reaction A and 70 percent occur through reaction B.

 (iii) We will likely have a sharper measurement of X's mass before July 2025.

 Identify the type of probability that is being employed in each phrase.

PROBLEMS

1.1 A system is described by a Lagrangian $L = \frac{1}{2}m(x)v^2 + b(x)v - V(x)$, for some functions m, b of x and a potential V. Identify the Hamiltonian and write Hamilton's equations.

1.2 Consider a particle of mass m in a central potential $V(r)$. (i) Using the standard spherical coordinates (r, θ, ϕ) for the particle's position vector, show that the kinetic energy is $T = \frac{1}{2}m(\dot{r}^2 + r^2\dot{\theta}^2 + r^2\sin^2\theta\dot{\phi}^2)$. (ii) Write down the Lagrange equations, and show that you can always choose coordinates so that the orbit lies in the $\theta = \frac{\pi}{2}$ plane and that $\ell := mr^2\dot{\phi}$ is a constant (the angular momentum). (iii) Show that the particle's orbit satisfies the differential equation

$$\frac{dr}{d\phi} = \frac{mr^2}{\ell}\sqrt{\frac{2}{m}\left[E - V(r) - \frac{\ell^2}{2mr^2}\right]}, \tag{1.56}$$

where E is the particle's energy. (iv) Consider the Kepler potential, $V = -\frac{\kappa}{r}$, where κ is a constant. Set $r = u^{-1}$, and use Eq. (A.12) to show that

$$\frac{1}{r} = \frac{m\kappa}{\ell^2}[1 + e\cos(\phi - \phi_0)], \tag{1.57}$$

where $e = \sqrt{1 + \frac{2E\ell^2}{m\kappa^2}}$ and ϕ_0 a constant of integration. For $e > 1$ the orbit is a hyperbola, for $e = 1$ it is a parabola, and for $e < 1$ it is an ellipse.

1.3 A random experiment consists of the simultaneous throw of two dice with the apparently paradoxical property that they are indistinguishable (bosonic dice). This means that the alternatives (1,2) and (2,1) correspond to a single elementary alternative rather than to two. (i) Identify the sample space, and evaluate the probability of rolling seven. Assume equal probabilities for all fine-grained alternatives. (ii) Then, consider a different pair of dice, with the additional property that they can never show the same number when thrown (fermionic dice). Identify the sample space and evaluate the probability of rolling seven.

1.4 (i) Show that for any probability distribution p_i, where $i = 1, \ldots, n$ the *purity* $\gamma := \sum_{i=1}^{n} p_i^2$, satisfies $\frac{1}{n} \leq \gamma \leq 1$. Then, argue that $1 - \gamma$ is a measure of the spread of the distribution. (ii) Evaluate γ for the binomial and the Poisson distributions. (ii) For a continuous probability distribution, the purity is defined by $\gamma := \int dx p^2(x)$. Evaluate γ for the Gaussian and for the Lorentzian probability distributions. Does γ have an upper bound?

1.5 A particle can decay through two channels, A and B. The B-channel decays are extremely rare. They have never been observed, but the theory predicts that they happen with probability $p = 0.00001$. We construct a detector that is sensitive to the B process with 99 percent reliability. Suppose that the detector finds one record for the process B. What is the probability that B truly occurred? How many records of B are needed in order to be 90 percent certain that B occurred at least once?

1.6 The probability that a nucleus decays within any time interval of width τ is constant and equal to $w << 1$. Show that the probability that a nucleus has not decayed at time $t >> \tau$ approximately equals $e^{-wt/\tau}$.

1.7 A random walker decides her steps on the flips of a coin. If the coin lands heads she goes left, if the coin lands tails, she goes right. Let us use the index $i \in \mathbb{Z}$ to label the walker's possible positions. Then, the probability distribution $p_i(n)$ that she is found at i after her nth step satisfies the difference equation $p_i(n+1) = \frac{1}{2}[p_{i+1}(n) + p_{i-1}(n)]$. Suppose that the walker takes her first step from $i = 0$. Evaluate $p_i(n)$ and the mean value $\langle i(n) \rangle = \sum_i p_i(n)i$.

1.8   Take the continuum limit for the random walker of Problem 1.7. To this end, assume that each step of the walker covers distance δ and that the time between two flips of the coin is τ. Then, describe the walker by the continuous probability distribution $p(x,t)$ (x is the random variable, t a parameter). (i) Taking $\delta \to 0$ and $\tau \to 0$ with constant $D := \delta^2/\tau$, show that $p(x,t)$ satisfies the *diffusion equation*,

$$\frac{\partial p(x,t)}{\partial t} = \frac{D}{2} \frac{\partial^2 p(x,t)}{\partial x^2}.$$

(ii) Assuming that the walker was at $x = 0$ at time t, show that

$$p(x,t) = \frac{1}{\sqrt{2\pi Dt}} e^{-\frac{x^2}{2Dt}}.$$

(iii) Find the mean distance of the walker from the origin $\langle |x| \rangle$ at time t.

1.9   In Problem 1.7, we constructed a probability distribution $p_i(n)$ in which the location i of the walker is the random variable, and the step n is a parameter. We can also construct the probability distribution $w_n(k)$ that the walker *first* arrives at the location $i = k$ with her nth step. Here, n is the random variable and k is a parameter. (i) Evaluate $w_n(1)$ for a walker that starts at $i = 0$. (ii) What is the probability that the walker never reaches $i = 1$?

1.10   The uncertainties in measuring the position and momentum of a classical particle of mass m are encoded in the probability distribution $\rho_0(x,p)$ in which x and p are statistically independent. (i) Suppose that the particle moves in a harmonic oscillator potential of frequency ω, $V(x) = \frac{1}{2}m\omega^2 x^2$. Evaluate the variances $(\Delta x)^2(t)$ and $(\Delta p)^2(t)$ as a function of the time t. Then, show that

$$\sigma_x \sigma_p \leq \Delta x(t) \Delta p(t) \leq \frac{1}{2}(\sigma_x^2 + \sigma_p^2),$$

where $\sigma_x = \sqrt{m\omega}\Delta x(0)$ and $\sigma_p = \Delta p(0)/\sqrt{m\omega}$. (ii) Suppose that the particle moves in an inverse harmonic oscillator potential $V(x) = -\frac{1}{2}m\eta^2 x^2$, where η is a constant. Evaluate the variances $(\Delta x)^2(t)$ and $(\Delta p)^2(t)$ as a function of the time t. Then, show that $\Delta x(t)\Delta p(t) > \frac{1}{2}(s_x^2 + s_p^2)\sinh(\eta t)$, where $s_x = \sqrt{m\eta}\Delta x(0)$ and $s_p = \Delta p(0)/\sqrt{m\eta}$. How does the exponential growth of the "uncertainty" $\Delta x(t)\Delta p(t)$ reconcile with Liouville's theorem?

1.11   The Maxwell–Boltzmann distribution of velocity vectors $\mathbf{v}$ of particles in a gas at temperature T is

$$f(\mathbf{v}) = \left(\frac{m\beta}{2\pi}\right)^{3/2} e^{-\frac{\beta m v^2}{2}},$$

where $\beta = (k_B T)^{-1}$. (i) Write the probability distribution $p(v)$ for the norm $v := |\mathbf{v}|$ of the velocity vector. (ii) Find the mean value $\langle v \rangle$ and the variance $(\Delta v)^2$ of v.

Bibliography

- For classical mechanics and Hamilton's formalism in particular, see Goldstein et al. (2002). For classical EM, see Griffiths (2017). For classical statistical mechanics, see chapters 6 and 7 of Huang (1987).
- For an introduction to probability theory, see Bertsekas and Tsitsiklis (2008), and Hacking (2001) for interpretational issues.
- For the notion of state space and the mathematical and conceptual structure of classical physics, see chapter 4 of Isham (1995).

2   The Birth of Quantum Theory

To those of us who were educated after light and reason had struck in the final formulation of quantum mechanics, the subtle problems and the adventurous atmosphere of these pre-quantum mechanics days, at once full of promise and despair, seem to take on an almost eerie quality. We could only wonder what it was like when to reach correct conclusions through reasonings that were manifestly inconsistent constituted the art of the profession.

Yang (1961)

2.1   Energy Quanta

The first step towards quantum theory was a response to a problem that could not be addressed by the concepts and methods of classical physics: the radiation from black bodies.

A body is black if it absorbs all light that falls upon it, while reflecting nothing. Objects commonly referred to as black are not really black bodies, because they reflect some light. A better example of a black body is a hole. Consider a wooden box with thick walls and an interior cavity with rough walls. Suppose we open a small hole on one side of the box. The light that enters the cavity from this hole is absorbed by the walls. If a light beam is not absorbed at once, it will be absorbed after one reflection, or after a second reflection, and so on. It will certainly be absorbed before it finds its way out of the hole through which it entered. Hence, the hole absorbs all the light that falls within it and is completely black.

However, it was discovered that black bodies emit EM waves, mainly in the infrared portion of the spectrum. This phenomenon can be explained by elementary thermodynamical arguments. The box is never completely isolated. It exchanges heat with its surroundings, until it reaches a constant temperature. This implies that electrons bound to the walls oscillate with a kinetic energy determined by temperature. These oscillations give rise to EM waves. These waves eventually exit the hole and are observed.

Let ω be the angular frequency of oscillation of an electron in the walls. This oscillation generates an EM field mode of the same frequency, and with energy proportional to the mean energy $\langle E \rangle$ of the oscillations. The number of oscillating modes with frequency in the interval $[\omega, \omega + \delta\omega]$ equals $g(V, \omega)\delta\omega$, where V is the volume of the cavity. The quantity $g(V, \omega)$ is known as the *density of states* for the EM field in the cavity. In Problem 2.1, it is shown that

$$g(V, \omega) = \frac{V\omega^2}{\pi^2 c^3}.$$

(2.1)

Then, the energy density of EM waves with frequency in the interval $[\omega, \omega + \delta\omega]$ in a black body of volume V is $\epsilon(\omega)\delta\omega$, where

$$\epsilon(\omega) = V^{-1}g(V,\omega)\langle E \rangle. \tag{2.2}$$

A harmonic oscillator of mass m and frequency ω is described by the Hamiltonian

$$H = \frac{p^2}{2m} + \frac{1}{2}m\omega^2 x^2.$$

The corresponding partition function at temperature T is straightforwardly evaluated to $Z = \frac{2\pi k_B T}{\omega}$. Using Eq. (1.53) we obtain $\langle E \rangle = k_b T$. Hence, Eq. (2.2) becomes

$$\epsilon(\omega) = \frac{\omega^2 k_B T}{\pi^2 c^3}. \tag{2.3}$$

Equation (2.3) is problematic. It predicts infinite total EM energy in the cavity,

$$\int_0^\infty d\omega \epsilon(\omega) \sim \int_0^\infty d\omega \omega^2.$$

Moreover, it contradicts experiment. Measurements of the intensity of black body radiation identify $\epsilon(\omega)$ as a function with a single maximum that vanishes for large frequencies.

This was a severe problem. The predictions about black body radiation were based upon elementary assumptions within well-accepted physical theories. The fact that they led to predictions that were *qualitatively* wrong implied that something fundamental was missing in the understanding of this system.

The problem was resolved by Max Planck in 1900. Planck found a semi-empirical formula – see Eq. (2.7) below – that accounts for the experimental curve of $\epsilon(\omega)$, by requiring that the fundamental principles of thermodynamics were respected (Planck, 1900a). But then, the problem was transferred to finding an explanation for Planck's formula. There was no such explanation in known physics.

Planck came up with a radically novel explanation: He assumed that the energy of the oscillations at the walls of the cavity takes discrete values, rather than continuous (Planck, 1900b). For each oscillator of frequency ω, there is a minimum value of energy equal to $\hbar\omega$, a *quantum* of energy as Planck called it. The oscillator energies can only be multiples of the quantum, that is, of the form $n\hbar\omega$, where $n = 0, 1, 2, \ldots$. The constant $\hbar$ is a new fundamental constant – now called *Planck's constant* – with value 1.0545×10^{-34} m^2 kg/s.

Accepting the quantum hypothesis, we generalize the canonical distribution for discrete values of energy. The probability distribution for the number n of quanta in a harmonic oscillator of frequency ω at temperature β^{-1} is

$$p_n = \frac{e^{-n\beta\hbar\omega}}{Z}, \tag{2.4}$$

where now the partition function is

$$Z = \sum_{n=0}^\infty e^{-n\beta\hbar\omega} = \frac{1}{1 - e^{-\beta\hbar\omega}}. \tag{2.5}$$

Equation (2.4) defines the *Planck distribution*.

Fig. 2.1 Indicative plot of the energy density of a black body as a function of the frequency, according to Eq. (2.7). Temperature increases from a to b to c.

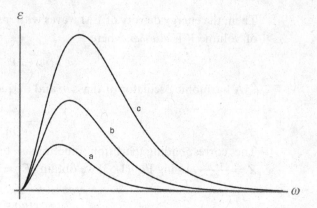

The mean energy of one oscillator is

$$\langle E \rangle = \frac{\sum_{n=0}^{\infty}(n\hbar\omega)e^{-n\beta\hbar\omega}}{Z} = -\frac{\partial}{\partial\beta}\ln Z = \frac{\hbar\omega}{e^{\beta\hbar\omega}-1}. \tag{2.6}$$

Hence, the energy distribution function $\epsilon(\omega)$ becomes

$$\epsilon(\omega) = \frac{\hbar\omega^3}{\pi^2 c^3 (e^{\beta\hbar\omega}-1)}, \tag{2.7}$$

in excellent agreement with experiment. The function $\epsilon(\omega)$ is plotted in Fig. 2.1. We note that for $e^{\beta\hbar\omega} \simeq 1 + \beta\hbar\omega$ for short frequencies ($\hbar\omega << T$), and Eq. (2.7) coincides with the classical expression (2.3).

Planck's theory was very successful in describing black body radiation, but it opened Pandora's box. Where did the quantum hypothesis originate from? It appeared incompatible with all known physics, including Newton's laws.

2.2   Light Quanta

The next step towards the development of quantum theory was made by Albert Einstein, in his "annus mirabilis" (1905), in which he also developed special relativity. Einstein proposed that

...[all] phenomena connected with the emission or transformation of light are more readily understood if one assumes that the energy of light is discontinuously distributed in space. In accordance with the assumption to be considered here, the energy of a light ray spreading out from a point source is not continuously distributed over an increasing space but consists of a finite number of energy quanta which are localized at points in space, which move without dividing, and which can only be produced and absorbed as complete units. (Einstein, 1905)

Einstein's most important motive for introducing quanta of light was that he perceived an incompatibility between the wave theory of light and the atomic description of matter. How can an atom interact with an EM wave, when the latter extends to a much larger region of space? The typical wavelength of light is 10^3–10^4 times larger than the size of an atom.

According to Einstein's hypothesis, a monochromatic EM wave of frequency ω consists of individual light quanta, each carrying energy

$$E = \hbar\omega. \tag{2.8}$$

Einstein later realized that these light quanta must also have momentum

$$\mathbf{p} = \hbar\mathbf{k}, \tag{2.9}$$

where $\mathbf{k}$ is the associated wavenumber. Since light quanta have both energy and momentum, they must be particles.

It took a long time for the physics community to accept this conclusion. It contradicted the phenomenon of interference (see Problem 2.7) that had been used as a demonstration of the wave nature of light ever since the nineteenth century. Indeed, the notion of photons is only consistent within the full quantum theory that emerged more than 20 years after Einstein's hypothesis. Particles described by classical physics are not compatible with the wave description of light according to Maxwell's equations. Eventually, the light quanta were named *photons* (Lewis, 1926).

The crucial result that led to the acceptance of photons was Arthur Compton's experimental study of the elastic scattering of X-rays by electrons (Compton, 1923). The experiment shows that the wavelength of scattered X-rays depends on the scattering angle, a conclusion that is inexplicable by classical EM theory. In contrast, the treatment of photons as particles, with energy given by Eq. (2.8) and momentum by Eq. (2.9), leads to the relation (see Problem 2.3)

$$\lambda' - \lambda = \frac{2\pi\hbar}{m_e c}(1 - \cos\theta), \tag{2.10}$$

where λ is the wavelength of incoming X-rays, λ' is the wavelength after scattering, m_e is the electron's mass, and θ is the scattering angle (in the reference frame where the electrons are stationary). Equation (2.10) turned out to be in excellent agreement with the experiment.

However, even Compton's experiment does not provide a *direct* experimental proof of the existence of photons. A direct experimental proof has been made possible with modern technology – see Box 2.1.

Box 2.1 **Experimental Proof of Photons**

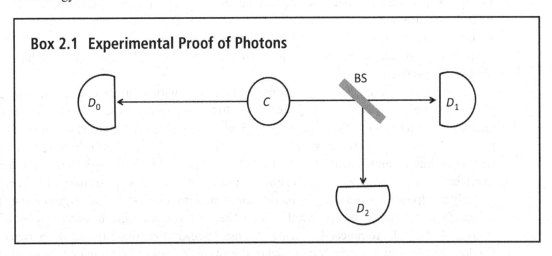

The figure in this box describes an experiment that proves the existence of photons as discrete entities.

- C is a down converter. This is a particular type of crystal: its input is a photon in violet or ultraviolet frequencies and its output is a pair of lower frequency photons. The converter's efficiency is typically very low: only one out of 10^{10}–10^{12} photons is split. Hence, for low intensities of incoming light, there is a delay between one pair of outcoming photons and the next.
- BS is a beam splitter, that is, a half-silvered mirror: Half of any incoming light beam is reflected at an angle of $90°$ by the mirror and half is transmitted through the mirror.
- D_0, D_1, and D_2 are photodetectors, that is, highly sensitive diodes that can detect even individual photons.

Consider one photon pair exiting C. One photon moves towards the detector D_0 and the other towards the beam splitter. If the second photon is transmitted without reflection it is detected at D_1; if it is reflected it is detected at D_2. This means that the detectors D_1 and D_2 cannot click simultaneously. In contrast, if the EM radiation emitted by C was a classical wave, its second component would be split at the beam splitter and one part would be recorded at D_1 and one part at D_2. Hence, we would expect simultaneous clicks at D_1 and D_2. In the experiment, we observe simultaneous detections of D_0 and D_1, of D_0 and D_2, but never of D_1 and D_2. This proves the existence of photons as discrete entities that are localized in space and in time, namely, particles.

2.3   Bohr's Atom

The development of quantum theory coincided with the first glimpses to the atomic world allowed by modern technology. In 1895, Wilhelm Röntgen discovered X-rays; in 1896, Joseph Thompson and Antoine Bequerel discovered the electron and radioactivity, respectively; in 1900, Ernest Rutherford and Frederick Soddy showed that radioactivity is accompanied by transmutation of elements. The radiation of radioactive substances was split into the types alpha, beta, and gamma, which we know identify as helium nuclei, electrons, and high-energy photons, respectively.

This succession of great discoveries reached its peak with a famous experiment that took place in Rutherford's laboratory in Manchester. In this experiment, Geiger and Marsden (1909) directed alpha particles to thin foils of gold. They observed backscattering, that is, some alpha particles scattered at angles greater than $90°$. This behavior was inexplicable by any atomic model with a continuous charge distribution. It led Rutherford (1911) to the hypothesis that there is a small region inside atoms with a high concentration of mass and positive charge, which repulses strongly any (positively charged) alpha particles that come close to it. This region is the atomic nucleus. The atom has to be electrically neutral, hence it also contains the exact number of electrons needed in order to cancel the nucleus' charge. The electrons rotate around the nucleus under Coulomb forces. Thus, the modern model of the atom, as a solar system in miniature, was born.

No sooner had Rutherford's model been proposed than it ran aground. If electrons move under the Coulomb forces, they are accelerated, hence they emit EM waves, and they lose energy. A simple calculation – see Problem 2.5 – demonstrates that the electron in the hydrogen atom crashes on the nucleus within a timescale of the order of 10^{-11} s. Atoms are not stable in Rutherford's model.

A solution to this problem was given by Niels Bohr's model of the atom (Bohr, 1913) that also provided an answer to a different long-standing problem of physics. The hydrogen atoms, like all known elements, emit EM radiation only at specific discrete frequencies. There was no known way to explain this discreteness of the atomic spectra or to predict the values of the emitted frequencies.

Bohr's proposal was to incorporate the ideas of quanta into atomic physics. Spectroscopy implies that photons emitted by atoms have specific discrete values of energy. It is reasonable to assume that the emission of a photon corresponds to an electron that transitions from one orbit to another of lower energy (closer to the nucleus). Since the photon energies can take only discrete values, only some orbits out of all conceivable ones can be realized and all others must be forbidden. Then, the electron can avoid crashing to the nucleus, if the corresponding orbit is a forbidden one. With this reasoning, the key challenge was to find a simple model for identifying the allowed orbits.

It was easier to find such a model in the hydrogen atom, where only one electron rotates around the nucleus. Bohr proposed the *quantization of angular momentum*, namely, that angular momentum ℓ is an integer multiple of $\hbar$,

$$\ell = n\hbar, \quad n = 1, 2, \dots. \tag{2.11}$$

The integer n in Eq. (2.11) is referred to as a *quantum number*.

For simplicity, we consider circular electron orbits around the nucleus. Newton's second law for the Coulomb force $m_e v^2 / r = e^2/(4\pi r^2)$ allows us to express the electron's rotational velocity as a function of the orbit radius r,

$$v^2 = \frac{e^2}{4\pi m_e r}. \tag{2.12}$$

Then, the electron's energy $E = \frac{1}{2} m v^2 - e^2/(4\pi r)$ becomes

$$E = -\frac{e^2}{8\pi r}. \tag{2.13}$$

Equation (2.11) for the angular momentum $L = m_e v r$ leads to a quantization condition for the radius,

$$r_n = \frac{4\pi \hbar^2}{m_e e^2} n^2. \tag{2.14}$$

Substituting Eq. (2.14) into Eq. (2.13) we obtain quantized values of energy

$$E_n = -\frac{e^4 m_e}{32\pi^2 \hbar^2 n^2} = -\frac{E_1}{n^2}, \tag{2.15}$$

where $E_1 = \frac{e^4 m_e}{32\pi^2 \hbar^2} \simeq 13.6\,\text{eV}$ is the absolute value of minimum energy. For the orbit of the minimum energy, Eq. (2.14) gives $r_1 = a_0$, where the constant

$$a_0 = \frac{4\pi \hbar^2}{m_e e^2} \simeq 0,5 \times 10^{-10}\,\text{m} \qquad (2.16)$$

is called *Bohr radius* and it specifies the characteristic length scale of atoms.

Equation (2.15) implies the following. First, given that there is an orbit of minimum energy, orbits of the electron crashing to the nucleus ($E \to -\infty$) are forbidden. Second, the energy of an emitted photon is the difference $E_m - E_n$ of orbit energies, where n and m are integers, and $n < m$. Therefore, the emitted frequencies are of the form

$$\hbar \omega_{m,m} = E_m - E_n = E_1 \left(\frac{1}{n^2} - \frac{1}{m^2} \right). \qquad (2.17)$$

Remarkably, Eq. (2.17) coincided with well-known empirical formulas for the emission spectrum of the hydrogen atom!

Hence, Bohr's theory vindicated Rutherford's model, explained the spectroscopic data, and most importantly, it made the new ideas about the quantum nature of the microcosm widely accepted.

However, there was a price to pay for the new theory. "There appears to me one great difficulty in your hypothesis", wrote Rutherford (1913) to Bohr, "...how does an electron decide what frequency it is going to vibrate at when it passes from one stationary state to the other? It seems to me that you would have to assume that the electron knows beforehand where it is going to stop." Rutherford had discerned the crucial feature of the nascent quantum theory, a feature that would also persist in its final form. The quantum laws seem to violate all notions of causality and determinism of classical physics.

Bohr believed that his theory was nothing but a first approximation. After all, his model worked only for the hydrogen atom. His hypothesis did not offer a general physical principle that could be applied to, say, more complex atoms or molecules. For this reason, he viewed his work as a first step towards the formation of a new radical theory. He did not conceive the electron's transition from one orbit into another as a continuous process, similar to those of classical physics. Instead, he thought of it as a *jump*, an unpredictable and noncausal process that cannot be described by the usual imagery of motion in space and in time.

In contrast to his predecessors, Bohr did not look for a union between quantum effects and classical physics. He sought a formulation of the quantum concepts that would be self-contained, in the same way that the concepts of classical physics are self-contained. The two physical descriptions should be related by a rule that he proposed, the *correspondence principle*: At the limit of large quantum numbers, the predictions of quantum theory coincide with the predictions of classical physics.

In all systems we have encountered so far (harmonic oscillator, hydrogen atom), energies depend on a single quantum number n. Let $\Delta E_n = E_{n+1} - E_n$ be the difference between two consecutive energy values. According to the correspondence principle,

$$\frac{\Delta E_n}{E_n} \to 0, \text{as } n \to \infty, \qquad (2.18)$$

so that the energy is almost continuous at the limit of large n. Indeed, Eq. (2.18) applies for both the harmonic oscillator ($\Delta E_n/E_n \sim n^{-1}$) and the hydrogen atom ($\Delta E_n/E_n \sim n^{-1}$).

2.4 Particle or Wave?

The next fundamental building block of quantum theory was the extension of the duality between the wave and particle aspects of light to all particles. This idea was proposed in 1924 by Louis de Broglie; de Broglie (1929) described his reasoning as follows.

When I started to ponder these difficulties two things struck me in the main. Firstly the light-quantum theory cannot be regarded as satisfactory since it defines the energy of a light corpuscle by the relation $E = \hbar\omega$, which contains a frequency ω. Now a purely corpuscular theory does not contain any element permitting the definition of a frequency. This reason alone renders it necessary in the case of light to introduce simultaneously the corpuscle concept and the concept of periodicity.

On the other hand the determination of the stable motions of the electrons in the atom involves whole numbers, and so far the only phenomena in which whole numbers were involved in physics were those of interference and of eigenvibrations. That suggested the idea to me that electrons themselves could not be represented as simple corpuscles either, but that a periodicity had also to be assigned to them too.

I thus arrived at the following overall concept which guided my studies: for both matter and radiations, light in particular, it is necessary to introduce the corpuscle concept and the wave concept at the same time. In other words the existence of corpuscles accompanied by waves has to be assumed in all cases. However, since corpuscles and waves cannot be independent [...], it must be possible to establish a certain parallelism between the motion of a corpuscle and the propagation of the associated wave.

Following de Broglie's reasoning, we consider the relation between the velocity of a particle and the velocity of propagation of a wave. The latter is the so-called *group velocity*, defined as (see Box 2.2)

$$\mathbf{v}_g = \frac{\partial \omega}{\partial \mathbf{k}}, \tag{2.19}$$

where ω is the angular frequency and $\mathbf{k}$ the wavenumber.

On the other hand, the velocity of a free particle of mass m is given by

$$\mathbf{v} = \frac{\mathbf{p}}{m} = \frac{\partial E}{\partial \mathbf{p}}, \tag{2.20}$$

where $E = \frac{\mathbf{p}^2}{2m}$ is the kinetic energy.

The wave–particle duality can only make sense if the particle and the corresponding wave must move together, that is, if $\mathbf{v}_g = \mathbf{v}$. Given Planck's relation $E = \hbar\omega$, the comparison of Eqs. (2.19) and (2.20) leads to the conclusion

$$\mathbf{p} = \hbar\mathbf{k}. \tag{2.21}$$

Equation (2.21) implies that the wavelength $\lambda = 2\pi/|\mathbf{k}|$ corresponding to a particle is

$$\lambda = 2\pi \frac{\hbar}{|\mathbf{p}|}. \tag{2.22}$$

Equation (2.22) defines a particle's *de Broglie wavelength*.

Box 2.2 Group Velocity

Consider a wave that propagates in one spatial dimension and that is described by the wave function $A(x,t)$. The latter is expressed in terms of plane waves through a Fourier transform

$$A(x,t) = \int dk \tilde{A}(k) e^{ikx - i\omega(k)t},$$

where the angular frequency ω is a function of the wavenumber k.

We assume that the wave is almost monochromatic, that is, the amplitude function $\tilde{A}(k)$ is nonzero only for a narrow band of values of k close to k_0. With this assumption, we approximate the angular frequency as $\omega_k = \omega_0 + v_g(k - k_0)$, where $\omega_0 := \omega(k_0)$ and $v_g := (\partial\omega/\partial k)_{k=k_0}$ is the group velocity. Then,

$$A(x,t) = e^{i(v_g k_0 - \omega_0)t} \int dk \tilde{A}(k) e^{ik(x - v_g)t} = A(x - v_g t, 0) e^{i(v_g k_0 - \omega_0)t}.$$

It follows that $|A(x,t)| = |A(x - v_g t, 0)|$. Since the energy density of a wave is proportional to $|A(x,t)|^2$, we conclude that the wave's energy propagates with the group velocity v_g.

One expects that characteristic wave effects such as diffraction are stronger when the wave passes through slits of size d of the order of the wavelength. Therefore, de Broglie made a concrete prediction of diffraction by "matter waves." This prediction was soon verified experimentally, first with elementary particles and later with composite ones (see Box 2.3).

Box 2.3 Matter Waves

The wave behavior of particles was first confirmed through the observation of diffraction in "slits" of width d that is of the same order of magnitude with the particles' de Broglie wavelength. The first experimental confirmation was made by Davisson and Germer (1928), who observed electron diffraction through a nickel crystal. As a matter of fact, they had observed this effect a couple of years before that, but they were not aware of de Broglie's ideas at that time. In the Davisson–Germer experiment, the width d of the "slits" coincides with the spacing of the crystalline planes of nickel ($d \simeq 10^{-10}$ m).

Estermann and Stern (1930) observed the diffraction of hydrogen and helium atoms on NaCl crystals, thus verifying de Broglie's relation also for composite particles. Later, Shull and Wollan (1947) achieved neutron diffraction in crystalline materials.

The de Broglie wavelength is inversely proportional to particle mass, hence, it decreases for heavy particles. For de Broglie wavelength significantly smaller than 10^{-11} m, it is impossible to get diffraction on crystalline planes. A new type of experiments was pioneered in the 1960s, where the wave behavior of particles was observed in set-ups very similar to the two-slit experiment of classical optics. The key idea was to make available two possible paths for the particles of a beam. The phase difference between those paths leads to an interference pattern.

Experiments of this type have been carried out with electrons (Jöhnsoon, 1961; Tonomura et al., 1989), with neutrons (Zeilinger et al., 1988), with atoms (Keith et al., 1988; Carnal and Mlynek, 1991) and with large molecules (Arndt et al., 1999). An example of a large particle that has been found to behave like a wave is the macromolecule of phthalocyanine ($C_{48}H_{26}F_{24}N_8O_8$), which contains 114 atoms and has mass equal to 1,298 amu (Juffmann et al., 2012). Remarkably, the de Broglie wavelength of those molecules ($\lambda \simeq 5\,pm$) is much smaller than their size (about 1 nm)!

As of 2022, the current record for the mass of interfering particles is 25.000 amu $\sim 4 \cdot 10^{-23}\,kg$ (Fein et al., 2019).

The quintessential experiment that demonstrates the wave behavior of particles is the two-slit-experiment – see Fig. 2.2. It is the analogue of the classic nineteenth-century experiment that demonstrated the wave nature of light (see Problem 2.7). In what follows, we describe a *contemporary* version of the two-slit experiment with electrons, that was not possible in the early years of quantum theory.

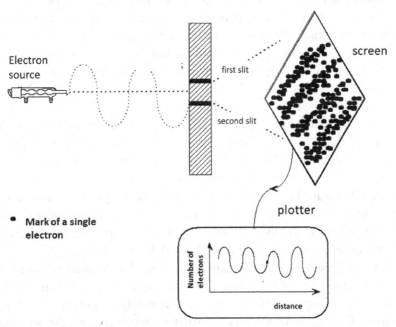

Fig. 2.2 The two-slit experiment. A slit emits electrons towards a screen with two slits. Each electron passes through one of the slits and is recorded at a membrane behind the slits. As the number of electrons increases, an interference pattern appears: The plot of the number of electrons as a function of their position on the membrane varies periodically.

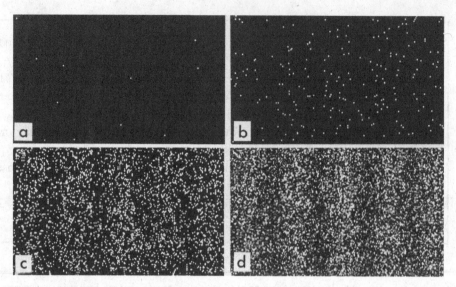

Fig. 2.3 The gradual build-up of the interference pattern in the two-slit experiment of electrons in Tonomura et al. (1989). The photographs show records of detected electrons during the experiment: (a) 11 electrons, (b) 200 electrons, (c) 6000 electrons, (d) 40,000 electrons. [Slightly modified form of a photo provided from Dr. Tonomura, under the GNU Free Documentation License.]

Consider a source that emits electrons towards a screen with two slits.[1] Behind the screen is placed a surface with a recording membrane so that the arrival of an electron is recorded by a dot on the membrane. We also assume that we can control the source so finely that we can send one electron at a time. Suppose we send the first electron. If it is not absorbed by the screen, it will pass through one of the slits and it will be recorded as a dot on the membrane. The second electron will leave another dot. The two electrons may have passed through the same slit, or they may have not. There is no way to tell in this experiment.

We continue sending electrons one after the other. The more we send, the more dots appear on the membrane. Initially, the distribution of the dots appears completely random. After about 10,000 recorded electrons, we start discerning a pattern. At 50,000 electrons the pattern is clear – see Fig. 2.3. The dots on the membrane create an interference pattern, exactly what one would have seen if a wave had been driven towards the two slits. Regions on the membrane with high dot concentration and regions with low dot concentration succeed each other periodically. Nonetheless, if we repeat the experiment with a single open slit, we observe no interference: only a higher concentration of dots directly behind the open slit.

What does one conclude from this experiment? Individual electrons behave like particles since they leave a pointlike record. However, collectively they behave like a wave, as seen in the interference pattern. This behavior is completely at odds with the way we expect particle beams to behave. A beam of classical particles directed towards the two slits would result to two concentrations of dots, one behind each slit.

[1] For electrons, the notion of a screen with slits is an idealization. In actual experiments, the role of the slits is played by a complex combination of electric and magnetic fields.

At this point, we must stress a fine point that was not evident in de Broglie's time. In the early days of quantum theory it was impossible to conduct experiments with individual electrons; the sources could only emit beams with a large number of electrons.[2] It was therefore impossible to distinguish the individual electron detection events that collectively construct the interference pattern. For this reason, de Broglie made much stronger statements about the relation between particles and waves than can be justified by contemporary experiments, or by our present understanding of quantum theory. For example, he proposed that every individual electron behaves like a wave, or even that each particle is accompanied by a wave that guides its motion.

In modern two-slit experiments, we observe that wave properties are manifested only when we consider the position records of many particles (i.e., in a statistical ensemble of position measurements). An individual particle is always recorded as a pointlike object; it cannot exhibit any type of wave behavior.

2.5 Heisenberg's Matrix Mechanics

The ideas of quantum theory mentioned in the past sections were foundational, but rather disconnected. The conception that would place them in a logical order was missing. Bragg (1922) was complaining that

...no known theory can be distorted so as to provide even an approximate explanation [of wave-particle duality]. There must be some fact of which we are entirely ignorant and whose discovery may revolutionize our views of the relations between waves and ether and matter. For the present we have to work on both theories. On Mondays, Wednesdays, and Fridays we use the wave theory; on Tuesdays, Thursdays, and Saturdays we think in streams of flying energy quanta or corpuscles.

The missing fact that was alluded to by Bragg was eventually discovered by Werner Heisenberg, a few years later. Heisenberg's starting point was the form of the angular frequencies of emitted photons in Bohr's model of the hydrogen atom (Heisenberg, 1925)

$$\omega_{nm} = \frac{E(n) - E(m)}{\hbar}, \tag{2.23}$$

where n and m are integers and $E(n)$ a function, whose explicit form is irrelevant to the argument.

Consider an electron moving along one of Bohr's orbits with quantum number n. Let $T(n)$ be the associated period. By Fourier analysis, we can express the electron's path $\mathbf{x}(n, t)$ as

$$\mathbf{x}(n, t) = \sum_{m=-\infty}^{\infty} \mathbf{x}_m(n) e^{im\omega(n)t}, \tag{2.24}$$

where $\omega(n) = 2\pi/T(n)$.

---

[2] In the first two-slit experiment with distinguishable individual electrons (Tonomura et al., 1989), the source emitted on average one electron per 10^{-3} s with a velocity of 1.5×10^8 m/s. Hence, the mean distance between two successive electrons was 150 km, while the target was at a distance of only 1.5 m. It was practically impossible for two successive electrons to approach each other, hence their detections can be viewed as statistically independent events.

Heisenberg noted that, according to EM theory, any radiation from the hydrogen atom ought to be emitted in the frequencies $\omega_{n,m} = m\omega(n)$ that appear in Eq. (2.24). However, such frequencies are very different from the observed ones that are given by Eq. (2.23). It follows that the physics of the hydrogen atom cannot be expressed in terms of particles traveling along well-defined paths.

Heisenberg insisted on the necessity to describe quantum theory in terms of quantities that reflect the observational data, and not on hypothetical objects such as electrons' paths. Hence, the fundamental objects of the theory must incorporate Eq. (2.23). Following this line of thought, he came to the following conclusions.

First, the fundamental objects of the quantum description are not particle paths as in classical physics, but quantities A_{nm} labeled by two integers m and n, like ω_{nm} of Eq. (2.23). These quantities evolve in time according to the rule

$$A_{nm} \rightarrow A_{nm}e^{i\omega_{nm}t}. \tag{2.25}$$

Second, there is a correspondence between the quantum variables A_{nm} and classical physical quantities. If the classical variable a corresponds to the quantum variable A_{nm}, then the classical quantity $b := a^2$ corresponds to the quantum variable

$$B_{nm} = \sum_l A_{nl}A_{lm}, \tag{2.26}$$

which also evolves according to Eq. (2.25). Equation (2.26) is the well-known rule for *multiplication of matrices*.[3] Hence, the variables A_{mn} are nothing but matrices, except for the fact that their indices may range over an infinite set. Hence, Heisenberg's theory was named "matrix mechanics."

Third, there is a crucial algebraic relation between the quantum variables for position and momentum. Let X_{nm} be the quantum variable that corresponds to a position coordinate x of a particle, and let P_{nm} be the quantum variable that corresponds to the conjugate momentum p. Then,

$$\sum_l (X_{nl}P_{lm} - P_{nl}X_{lm}) = i\hbar\delta_{nm}, \tag{2.27}$$

or, using a matrix notation,

$$XP - PX = i\hbar I, \tag{2.28}$$

where I is the unit matrix. Hence, physical quantities are described by mathematical objects that do not satisfy a *commutative* multiplication rule.

Heisenberg arrived at the above principles with a mixture of intuition and complex argumentation that cannot be easily reproduced outside the context of the 1920s physics research. He used these principles to derive Planck's quantization condition for the harmonic oscillator – we will present this proof in Section 5.2. A few months later, Heisenberg's friend and collaborator Wolfgang Pauli showed that Heisenberg's principles allow us to derive the spectrum of the hydrogen atom (Pauli, 1926) – we will present this derivation in Section 13.2.

---

[3] Heisenberg did not know that; in his day, matrix theory was still an esoteric branch of modern algebra.

Heisenberg's ideas required abandoning the edifice of classical physics. The fundamental physical magnitude would not be described by numbers or functions, but by some mysterious mathematical objects with no relation to our direct experience, or our geometric notions about particle motion. A new level of physical reality appeared that was inaccessible to classical physics.

Obviously, Heisenberg's ideas met with intense distrust. Few physicists were enthusiastic. Bohr was among them, because Heisenberg's ideas fit his way of thinking. Other supporters included Max Born and Pascuale Jordan, who placed Heisenberg's ideas in solid mathematical footing (Born and Jordan, 1925; Born et al., 1926), Wolfgang Pauli, who was mentioned earlier, and Paul Dirac who elaborated on the meaning of the fundamental equation (2.28) (Dirac, 1926).

2.6 Schrödinger's Wave Mechanics

A few months after the presentation of matrix mechanics, a different approach towards quantum theory was pioneered by Erwin Schrödinger (Schrödinger, 1926). Schrödinger followed a very different rationale, he looked for a mathematical representation of de Broglie's ideas on wave–particle duality. His theory appeared initially to contradict that of Heisenberg, as it provided a continuous description of quantum phenomena, while Heisenberg's emphasized their discrete character.

2.6.1 Schrödinger's Equation

Consider a free particle of mass m moving in a single spatial dimension. By wave–particle duality, the particle can be described by the wave function of a plane wave $\psi(t, x) = e^{ikx - i\omega t}$. The wavenumber k and the angular frequency ω are related to the particle's momentum p and energy E by the de Broglie relations $E = \hbar\omega$ and $p = \hbar k$. We observe that the action of the differential operator $i\hbar\frac{\partial}{\partial t}$ on $\psi(t, x)$ corresponds to the multiplication of $\psi(t, x)$ by E, and that the action of the operator $-i\hbar\frac{\partial}{\partial x}$ on $\psi(t, x)$ corresponds to multiplication by p.

Given that $E = \frac{p^2}{2m}$, for a nonrelativistic particle, the wave function $\psi(t, x)$ satisfies the differential equation

$$i\hbar\frac{\partial\psi(x, t)}{\partial t} = -\frac{\hbar^2}{2m}\frac{\partial^2\psi(x, t)}{\partial x^2}. \tag{2.29}$$

Equation (2.29) applies to free particles. What equation applies for particles in an external potential $V(x)$?

Schrödinger made the following guess. Given that the Hamiltonian of a particle in an external potential $V(x)$ is

$$H = \frac{p^2}{2m} + V(x),$$

the simplest thing to do is to substitute the differential operator $-\frac{\hbar^2}{2m}\frac{\partial^2}{\partial x^2}$ of Eq. (2.29) – that corresponds to the kinetic energy – by a differential operator $-\frac{\hbar^2}{2m}\frac{\partial^2}{\partial x^2} + V(x)$ that corresponds to the sum of kinetic and potential energy.

Hence, Schrödinger proposed the differential equation (nowadays named after him)

$$i\hbar\frac{\partial\psi(x,t)}{\partial t} = \left[-\frac{\hbar^2}{2m}\frac{\partial^2}{\partial x^2} + V(x)\right]\psi(x,t). \tag{2.30}$$

For a particle moving in three dimensions, the same arguments give

$$i\hbar\frac{\partial\psi(\mathbf{x},t)}{\partial t} = \left[-\frac{\hbar^2}{2m}\nabla^2 + V(\mathbf{x})\right]\psi(\mathbf{x},t). \tag{2.31}$$

Consider solutions of Schrödinger's equation that describe a wave of constant angular frequency ω or, equivalently, a particle of constant energy $E = \hbar\omega$. Substituting the ansatz

$$\psi(\mathbf{x},t) = \psi_E(\mathbf{x})e^{-iEt/\hbar} \tag{2.32}$$

in Eq. (2.31), we obtain the *time-independent Schrödinger equation*

$$\left[-\frac{\hbar^2}{2m}\nabla^2 + V(\mathbf{x})\right]\psi_E(\mathbf{x}) = E\psi_E(\mathbf{x}). \tag{2.33}$$

Similarly, setting $\psi(x,t) = \psi_E(x)e^{-iEt/\hbar}$ in Eq. (2.30), we obtain the time-independent Schrödinger equation for motion in one dimension

$$\left[-\frac{\hbar^2}{2m}\frac{\partial^2}{\partial x^2} + V(x)\right]\psi_E(x) \doteq E\psi_E(x). \tag{2.34}$$

2.6.2 The Harmonic Oscillator

Next, we show that the time-independent Schrödinger equation for a harmonic oscillator reproduces Planck's quantization condition for energy.

Consider a particle in one dimension under the harmonic oscillator potential $V(x) = \frac{1}{2}m\omega^2 x^2$, where ω is the classical oscillator's angular frequency. Equation (2.33) becomes

$$\left(-\frac{\hbar^2}{2m}\frac{\partial^2}{\partial x^2} + \frac{1}{2}m\omega^2 x^2\right)\psi_E(x) = E\psi_E(x). \tag{2.35}$$

In classical physics, a harmonic oscillator's orbit is bounded in space; the particle cannot escape to infinity. We expect that quantum oscillators have the same property. Hence, we restrict to solutions of Eq. (2.35) that satisfy the boundary condition

$$\lim_{x\to\pm\infty}\psi(x) = 0. \tag{2.36}$$

The parameters m, ω and $\hbar$ that appear in Eq. (2.35) define a length scale $x_0 = \sqrt{\hbar/(m\omega)}$.

Changing the independent variable from x to $\xi = x/x_0$, Eq. (2.35) becomes

$$\psi_E'' + (2\epsilon - \xi^2)\psi_E = 0, \tag{2.37}$$

where the prime denotes derivative with respect to ξ, and $\epsilon = E/(\hbar\omega)$.

In the limit where $\xi \gg \sqrt{\epsilon}$, Eq. (2.37) becomes

$$\psi_E'' - \xi^2\psi_E = 0. \tag{2.38}$$

We set $s = \frac{1}{2}\xi^2$, so that Eq. (2.38) becomes $d^2\psi_E/ds^2 + \psi_E = 0$. This equation admits the independent solutions e^s and e^{-s}. Hence, Eq. (2.38) admits a pair of independent solutions $e^{\pm\xi^2/2}$; only the solution $e^{-\xi^2/2}$ is compatible with the boundary conditions (2.36).

We substitute $\psi_E(\xi) = e^{-\xi^2/2}u_E(\xi)$ in Eq. (2.37), and obtain a differential equation for u_E,

$$u_E'' - 2\xi u_E' + (2\epsilon - 1)u_E = 0. \tag{2.39}$$

Equation (2.39) is known as Hermite's equation. In Box 2.4, it is proven that Hermite's equation is compatible with the boundary conditions (2.36), only if

$$\epsilon = n + \frac{1}{2}, \quad n = 0, 1, 2, \dots \tag{2.40}$$

Then, solutions to Eq. (2.39) are nth-degree polynomials $H_n(\xi)$, known as *Hermite polynomials*. Hermite polynomials are even for even n and odd for odd n. Their properties are described in the Appendix B.4.

We conclude that the solutions to the time-independent Schrödinger equation for a harmonic oscillator depend on a quantum number $n = 0, 1, 2, \dots$, and they are of the form

$$\psi_n(x) = C_n \exp\left[-\frac{m\omega}{2\hbar}x^2\right] H_n\left(\sqrt{\frac{m\omega}{\hbar}}x\right), \tag{2.41}$$

where C_n is a constant. The constant C_n is usually determined by the requirement that $\int dx |\psi_n(x)|^2 = 1$. Using the normalization properties of the Hermite polynomials (see the Appendix B.4), we obtain

$$C_n = \frac{1}{\sqrt{2^n n!}}\left(\frac{m\omega}{\pi\hbar}\right)^{1/4}. \tag{2.42}$$

The functions $\psi_n(x)$ are plotted in Fig. 2.4. Explicit expressions for the first few values of n are given in Table 2.1. The corresponding values of energy (the energy eigenvalues) are given by

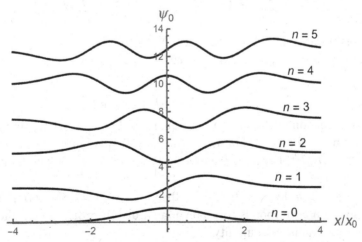

Fig. 2.4 The solutions (2.41) of the time-independent Schrödinger equation for a harmonic oscillator for $n = 0, 1, 2, \dots, 5$.

Table 2.1 Solutions to Schrödinger's equation for a harmonic oscillator potential. We have set $\xi = x/x_0$ and $x_0 = \sqrt{\hbar/(m\omega)}$

n	$\psi_n(x)(\pi x_0)^{\frac{1}{4}}$
0	$e^{-\frac{\xi^2}{2}}$
1	$\xi e^{-\frac{\xi^2}{2}}$
2	$\frac{1}{\sqrt{2}}(2\xi^2 - 1)e^{-\frac{\xi^2}{2}}$
3	$\frac{1}{\sqrt{3}}(2\xi^3 - 3\xi)e^{-\frac{\xi^2}{2}}$
4	$\frac{1}{2\sqrt{6}}(4\xi^4 - 12\xi^2 + 3)e^{-\frac{\xi^2}{2}}$
5	$\frac{1}{2\sqrt{15}}(4\xi^5 - 20\xi^3 + 15\xi)e^{-\frac{\xi^2}{2}}$

$$E_n = \left(n + \frac{1}{2}\right)\hbar\omega. \tag{2.43}$$

Hence, the solution of the time-independent Schrödinger equation leads to Planck's quantization condition for a harmonic oscillator. The only difference is the constant term $\frac{1}{2}\hbar\omega$, that can often be absorbed in redefinition of the zero of the potential.

Box 2.4 Solutions to Hermite's Equation

We express the function $u_E(\xi)$ of Eq. (2.39) as a series $u_E(\xi) = \sum_{k=0}^{\infty} a_k \xi^k$, for some coefficients a_k. The first two derivatives of u_E are

$$u'_E(\xi) = \sum_{k=0}^{\infty} a_{k+1}(k+1)\xi^k, \qquad u''_E(\xi) = \sum_{k=0}^{\infty} a_{k+2}(k+1)(k+2)\xi^k.$$

Substituting in Eq. (2.39), we obtain $\sum_{k=0}^{\infty}\left[(k+1)(k+2)a_{k+2} - 2(k - \epsilon + \frac{1}{2})a_k\right]\xi^k = 0$, which leads to a recursion relation between the coefficients a_k,

$$a_{k+2} = \frac{2(k - \epsilon + \frac{1}{2})}{(k+1)(k+2)}a_k.$$

This recursion relation has a step equal to 2. Hence, by specifying a_0 and a_1 we find all coefficients a_k for $k > 1$. For $a_1 = 0$ and $a_0 \neq 0$, we obtain a series with only even powers of x: $\sum_{l=0}^{\infty} a_{2l}\xi^{2l}$. For $a_0 = 0$ and $a_1 \neq 0$, we obtain a series with only odd powers of x: $\sum_{l=0}^{\infty} a_{2l+1}\xi^{2l+1}$.

If $\epsilon - \frac{1}{2} = n$, for $n = 0, 1, 2, \ldots$, then $a_{n+2} = 0$ and the series terminates at $k = n$. The solution corresponds to a polynomial of nth order. Hence, the function $\psi_E(\xi) = e^{-\xi^2/2}u_E(\xi)$ vanishes at infinity.

For other values of ϵ, the series has infinite nonzero terms. For large values of k, $a_{k+2}/a_k \sim 2/k$. We compare with the power series $e^{\xi^2} = \sum_{l=0}^{\infty} a_{2l}\xi^{2l}$ where $a_{2l} = 1/l!$.

Then, $a_{2l+2}/a_{2l} \sim \frac{1}{l}$ or, equivalently, for $k = 2l$, $a_{k+2}/a_k \sim 2/k$. The asymptotic behavior of the series for e^{ξ^2} is the same with the series for $u_E(\xi)$. Hence, at large ξ, $u_E(\xi) \sim e^{\xi^2}$ and $\psi_E(\xi) \sim e^{\xi^2/2}$.

We conclude that the boundary condition $\lim_{x\to\pm\infty} \psi(x) = 0$ enforces the relation $\epsilon - \frac{1}{2} = n$ and, consequently, the quantization of energy according to Eq. (2.43).

2.6.3 The Limitations of Wave Mechanics

Schrödinger's theory was highly successful. Like Heisenberg's theory, it reproduced Planck's quantization condition for harmonic oscillators and Bohr's spectrum of the hydrogen atom – see Section 13.4 for details. More importantly, it offered a general methodology for the study of more complex systems. Schrödinger's equation can be generalized for N particles of masses m_i, $i = 1, 2, \ldots, N$,

$$i\frac{\partial \psi}{\partial t}(\mathbf{x}_1, \ldots, \mathbf{x}_N) = \left[-\sum_{i=1}^{N} \frac{\hbar^2}{2m_i} \nabla_i^2 + V(\mathbf{x}_1, \ldots, \mathbf{x}_N) \right] \psi(\mathbf{x}_1, \ldots, \mathbf{x}_N), \qquad (2.44)$$

where $V(\mathbf{x}_1, \ldots, \mathbf{x}_N)$ is an interaction potential. In principle, the solution of Eq. (2.44) (or of its time-independent version) should provide a consistent mathematical description of the many-particle system. An added benefit was that Schrödinger's theory was based on differential equations, which were much more familiar to physicists in the 1920s than matrix theory.

Schrödinger initially preferred to interpret his equation as describing real waves that propagate in space. In this interpretation, particles correspond to spatially localized wave packets. He was deeply skeptical about the theories of Bohr and Heisenberg, and he did not like the discontinuities inherent in the notion of the quantum jumps.

However, the interpretation of the wave function ψ as a physical wave cannot survive even elementary scrutiny. First, the wave function ψ of an N-particle systems depends on the position vectors of all particles. Hence, ψ is a function on the configuration space $Q = \mathbb{R}^{3N}$ of the N-particle system. In contrast, physical waves can only be described by wave functions on the physical space $\mathbb{R}^3$.

Second, Schrödinger's equation is incompatible with the spontaneous emission of photons from atoms. Consider a system that consists of N_0 excited atoms at time $t = 0$. Experiment shows that the deexcitation of those atoms through the emission of photons follows an exponential decay law: The number $N(t)$ of excited atoms at time t is $N(t) = N_0 e^{-\Gamma t}$, where Γ is a constant. The function $N(t)$ is not time invertible: The function $N(-t)$ is not an acceptable description of the de-excitation process. In contrast, Schrödinger's equation is time invertible: If $\psi(\mathbf{x}, t)$ is a solution to Eq. (2.31), then $\psi^*(\mathbf{x}, -t)$ is also a solution. If Schrödinger's equation is interpreted as an equation of a physical wave, it cannot describe irreversible phenomena such as spontaneous emission.

2.7   Probabilities and Uncertainty

After the completion of Schrödinger's theory there were two candidates for a fundamental quantum theory, Heisenberg's matrix mechanics and Schrödinger's wave mechanics. At first glance, they appeared radically different: The former introduced a novel mathematical description, whereas the latter employed familiar notions from classical wave theories. However, this difference turned out to be superficial (see Box 2.5).

To see this, consider solutions $\psi_n(x)$ to the time-independent Schrödinger equation with corresponding energies E_n, where the quantum number $n = 0, 1, 2, \ldots$. Let $\hat{X}$ and $\hat{P}$ be operators defined by their action on wave functions $\psi(x)$, as

$$\hat{X}\psi(x) = x\psi(x), \qquad \hat{P}\psi(x) = -i\frac{\partial}{\partial x}\psi(x). \tag{2.45}$$

We observe that these operators satisfy the fundamental relation (2.28) in the sense that $\hat{X}\hat{P}\psi(x) - \hat{P}\hat{X}\psi(x) = i\psi(x)$.

Box 2.5   The Disagreements about Schrödinger's Theory

Heisenberg (1967) recalls:

As far as I remember these discussions took place in Copenhagen around September 1926 and in particular they left me with a very strong impression of Bohr's personality. For though Bohr was an unusually considerate and obliging person, he was able in such a discussion, which concerned epistemological problems which he considered to be of vital importance, to insist fanatically and with almost terrifying relentlessness on complete clarity in all arguments. He would not give up, even after hours of struggling, before Schrödinger had admitted that this interpretation was insufficient, and could not even explain Planck's law. Every attempt from Schrödinger's side to get round this bitter result was slowly refuted point by point in infinitely laborious discussions.

It was perhaps from over-exertion that after a few days Schrodinger became ill and had to lie abed as a guest in Bohr's home. Even here it was hard to get Bohr away from Schrodinger's bed and the phrase, "But, Schrodinger, you must at least admit that..." could be heard again and again. Once Schrodinger burst out almost desperately, "If one has to go on with these damned quantum jumps, then I'm sorry that I ever started to work on atomic theory." To which Bohr answered, "But the rest of us are so grateful that you did, for you have thus brought atomic physics a decisive step forward." Schrödinger finally left Copenhagen rather discouraged, while we at Bohr's Institute felt that at least Schrödinger's interpretation of quantum theory, an interpretation rather too hastily arrived at using the classical wave-theories as models, was now disposed of, but that we still lacked some important ideas before we could really reach a full understanding of quantum mechanics.

We define the infinite matrices

$$X_{mn} = \int dx\,\psi_m^*(x, t)\,x\,\psi_n(x, t) \tag{2.46}$$

$$P_{mn} = \int dx\,\psi_m^*(x, t)(-i\partial/\partial x)\psi_n(x, t). \tag{2.47}$$

Equation (2.32) implies that $\psi_n(x,t) = \psi_n(x)e^{-iE_n t/\hbar}$. Hence, both matrices X_{mn} and P_{mn} evolve in time as

$$X_{mn}(t) = X_{mn}(0)e^{i(E_m - E_n)t/\hbar}, \quad P_{mn}(t) = P_{mn}(0)e^{i(E_m - E_n)t/\hbar},$$

in full agreement with Heisenberg's fundamental postulate, Eq. (2.25).

Hence, the two theories were mathematically equivalent. They led to the same physical predictions to all problems that they could handle at that time, namely, the calculation of quantized energy values. However, in absence of a consistent physical interpretation of Schrödinger's wave functions, the theories could not be unified.

2.7.1   The Uncertainty Principle

Heisenberg made the next crucial discovery for the correct interpretation of the quantum formalism. He discovered the celebrated *uncertainty principle* (Heisenberg, 1927). His most important argument was the following.

Consider the accuracy of possible measurements on macroscopic particles. In order to determine the position of one particle, we must interact with it. For example, we can determine the particle by the scattering of EM waves upon it. However, no EM wave can discern structures at length scales significantly smaller than its wavelength, because of diffraction effects. In particular, a microscope that functions with light of wavelength λ can distinguish objects at a scale

$$\Delta x \geq w\lambda, \tag{2.48}$$

where w is of order unity.[4]

However, we can use the same EM field in order to specify the particle's momentum. If the field were continuous, it would be possible to specify momentum with infinite accuracy. However, the EM field consists of photons, each with momentum $p = 2\pi\hbar/\lambda$. The accuracy in measuring the momentum of a particle cannot be better than the momentum of a single photon, hence,

$$\Delta p \geq 2\pi\frac{\hbar}{\lambda}. \tag{2.49}$$

We multiply the inequalities (2.48) and (2.49), to obtain

$$\Delta x \Delta p \geq (2\pi w)\hbar \sim \hbar. \tag{2.50}$$

Equation (2.50) is the famous *Heisenberg's uncertainty principle*. In the form presented here, it refers to the impossibility of simultaneously measuring the position and momentum of a particle with arbitrary accuracy (see Fig. 2.5). The quantities Δx and Δp represent *random errors* associated to measurements on a single particle.

---

[4] According to the theory of the microscope – see section 8.6.3 of Born and Wolf (1999) – the optimum value is $w = 0.61/NA$ where NA is the numerical aperture, a manufacturing element of a microscope. NA can never be larger than 2 – its maximum value in real microscopes is 1.45. Hence, setting $w = 0.42$ gives an accurate estimate of the maximum resolution of a microscope. Then, the uncertainty relation (2.50) becomes $\Delta x \Delta p \gtrsim 2.6\hbar$.

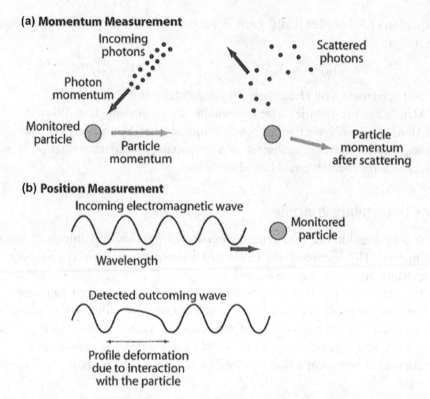

Fig. 2.5 (a) A particle's momentum is measured by specifying the total energy of photons that scatter upon the particle. Our accuracy is limited by the momentum carried by single photons. (b) In the same set-up, the wave properties of the EM field are employed in order to determine the particle's position. The accuracy cannot be greater than the wavelength of the EM radiation. De Broglie's relation between wavelength and momentum leads to the uncertainty principle. [Figure taken from Anastopoulos (2008), courtesy of Princeton University Press.]

Other versions of Eq. (2.50) have different interpretations (Busch et al., 2007). In Chapter 6, we will see that Δx and Δp may be interpreted as mean deviations for a statistical ensemble of measurements of position and momentum in *different experiments*. In yet other versions, Δx may correspond to the precision in the measurement of position and Δp to the change of momentum of a system by the measurements (Busch et al., 2013). Today, such distinctions are crucial because quantities like Δx and Δp are experimentally accessible, and different definitions refer to different measurements. In Heisenberg's days these distinctions were less important and the term "uncertainty relation" covered all such cases.[5]

Equation (2.50) leads to another uncertainty relation, between energy and time. Consider a particle detector placed behind a screen with a hole. The hole is covered by a movable shutter.

---

[5] For example, we may reinterpret the analysis leading to Eq. (2.50), by identifying Δp with the momentum change in the particle after a measurement of position through light of wavelength λ. Then, Eq. (2.50) refers to a *position measurement*, and it relates the error Δx of the measurement to the disturbance Δp of the particle's momentum.

We direct a particle towards the detector, and we leave the hole uncovered for a time interval Δt. If the particle is recorded, then we measured the time it passed through the hole with accuracy Δt. Equivalently, there is an uncertainty $\Delta x = v\Delta t$ in specifying the particle's position, where v is the particle's velocity. The uncertainty in determining the energy E is $\Delta E = \frac{\partial E}{\partial p}\Delta p = v\Delta p$. By Eq. (2.50),

$$\Delta E \Delta t \gtrsim \hbar. \tag{2.51}$$

Heisenberg believed that the uncertainty principle vindicated his overall point of view. Particle paths cannot be fundamental notions in quantum physics, because they involve an idealized description of physical variables – namely, the simultaneous definition of position and momentum – that contradicts the uncertainty principle. The latter also implied the end of the determinism of Newtonian physics. In Heisenberg's words (Heisenberg, 1999),

In the sharp formulation of the law of causality – "if we know the present exactly, we can calculate the future" – it is not the conclusion that is wrong but the premise.

The uncertainty principle became the banner of the radical approach to quantum theory that was advanced by Bohr and Heisenberg. It was simple, provocative, and convincing. For this reason it was the primary target of all sceptics. There were many attempts to find counterexamples, but they all turned out to be fruitless. The most important counterexample was proposed by Einstein. It is the so-called *Einstein's box*, described in Box 2.6. Its eventual demise left a strong impression to its contemporaries, and it sealed the broad acceptance of the nascent quantum theory.

Box 2.6 Einstein's Box

Einstein proposed an experiment in which the uncertainty relation for energy and time, Eq. (2.51), could be violated. Consider a box that contains EM radiation and a clock that controls the opening of a small door that covers a hole in the box. The door opens for a time interval T that can be made arbitrarily small. We assume that one photon exits the box when the door is open.

We have placed a scale beneath the box. The scale records the mass m_1 of the box before the door opens and the mass m_2 after the door has been opened and closed. Then, the energy $E = (m_1 - m_2)c^2$ of the emitted photon can be specified with arbitrary accuracy, because there is no limitation in the accuracy of measuring a mass. The time of the photon's emission is specified with accuracy $\Delta t = T$. Hence, the product $\Delta E \Delta t$ can be made arbitrarily small; it is not limited by the uncertainty principle.

At first, Einstein's argument appeared unassailable. Leon Rosenfeld, who was present in these discussions, describes (Rosenfeld, 1968):

It was quite a shock for Bohr ... he did not see the solution at once. During the whole evening he was extremely unhappy, going from one to the other and trying to persuade them that it couldn't be true, that it would be the end of physics if Einstein were right; but he couldn't produce any refutation. I shall never forget the vision of the two antagonists leaving the club: Einstein a tall

---

majestic figure, walking quietly, with a somewhat ironical smile, and Bohr trotting near him, very excited ... . The next morning came Bohr's triumph.

Bohr's reply was astounding. Not only did he find the error in Einstein's argument, but he did so by using one of the great ideas of Einstein himself, that gravity influences the readings of clocks. Bohr argued that we determine the mass m of an object through the gravitational force $F = mg$ that it exerts on a scale; g is the gravitational acceleration. Hence, the uncertainty Δm in measuring the mass equals $\Delta F/g$, where ΔF is uncertainty of the force. By Newton's law, $\Delta F = \Delta p_z/T$ where Δp_z is the momentum uncertainty in the vertical direction. In Einstein's experiment, T is the time that the door is open. Hence, the energy of the emitted photon is determined with an error $\Delta E := \Delta mc^2 = \Delta p_z c^2/(Tg)$.

According to general relativity, the time measured by a clock in a gravitational field depends on the gravitational potential at the point where the clock is located. Two clocks, one located at height z and the other at height $z + \Delta z$ measure durations T and $T + \Delta T$, respectively, where $\Delta T = g\Delta z/c^2 T$. Hence, the minimum uncertainty for the time of photon emission is ΔT. It follows that

$$\Delta E \Delta T \sim \frac{\Delta p_z c^2}{Tg} \frac{g\Delta z T}{c^2} = \Delta p_z \Delta z > \hbar,$$

where in the last step the position–momentum uncertainty relation was employed. Einstein's argument evidently fails.

The Einstein box debate is found in Bohr (1998); it has also been reprinted in Wheeler and Zurek (1983). For a modern critical account, see de la Torre et al. (2000).

2.7.2 Born's Probability Interpretation

The uncertainty principle offered an appropriate conceptual background for interpreting Schrödinger's wave functions. Since uncertainty is an inherent feature of the theory, it is necessary to employ probabilities. Therefore, it is necessary to associate quantum objects, either Heisenberg's matrices or Schrödinger's wave functions, with physical probabilities.

The correct association was identified by Max Born, in his *statistical interpretation of the wave function* (Born, 1926): If $\psi(\mathbf{x}, t)$ is a solution to Schrödinger's equation, then $|\psi(\mathbf{x}, t)|^2$ is a probability density with respect to the particle's position $\mathbf{x}$ at time t. Hence, if $U \subset \mathbb{R}^3$ is a region of space, the probability $\mathrm{Prob}(U, t)$ that the particle is found within U at time t is

$$\mathrm{Prob}(U, t) = \int_U d^3x |\psi(\mathbf{x}, t)|^2. \tag{2.52}$$

The quantity $|\psi(\mathbf{x}, t)|^2$ can function as a probability density only if it is normalized to unity,

$$\int d^3x |\psi(\mathbf{x}, t)|^2 = 1. \tag{2.53}$$

In Chapter 6, we will show that Born's probability interpretation can be generalized to any physical system and for any physical magnitude.

The normalization condition (2.53) is compatible with Schrödinger's equation. To see this, we differentiate $\rho(\mathbf{x}, t) := |\psi(\mathbf{x}, t)|^2$ with respect to time, and we use Eq. (2.31) to obtain,

$$\frac{\partial \rho}{\partial t} = \psi^* \frac{\partial \psi}{\partial t} + \psi \frac{\partial \psi^*}{\partial t} = -\frac{i}{\hbar} \psi^* \left(-\frac{\hbar^2}{2m} \nabla^2 \psi + V\psi \right)$$

$$+ \frac{i}{\hbar} \psi \left(-\frac{\hbar^2}{2m} \nabla^2 \psi^* + V\psi^* \right) = \frac{i\hbar}{2m} (\psi^* \nabla^2 \psi - \psi \nabla^2 \psi^*). \tag{2.54}$$

Equation (2.54) implies the continuity equation

$$\frac{\partial \rho}{\partial t} + \mathbf{\nabla} \cdot \mathbf{J}_S = 0, \tag{2.55}$$

where $\mathbf{J}_S$ is the *Schrödinger current*,

$$\mathbf{J}_S := \frac{i\hbar}{2m} \left(\psi \mathbf{\nabla} \psi^* - \psi^* \mathbf{\nabla} \psi \right). \tag{2.56}$$

We integrate both sides of Eq. (2.55) within a spatial region U, to obtain

$$\frac{\partial}{\partial t} \mathrm{Prob}(U, t) = -\oint_{\partial U} d^2\boldsymbol{\sigma} \cdot \mathbf{J}_S, \tag{2.57}$$

where ∂U stands for the boundary U. If $U = \mathbb{R}^3$, that is, if integration is over all space, ∂U is at infinity, where the wave functions vanish. The Schrödinger current also vanishes. Therefore,

$$\frac{\partial}{\partial t} \int d^3 x |\psi(\mathbf{x}, t)|^2 = 0. \tag{2.58}$$

We conclude that time evolution according to Schrödinger's equation preserves the normalization (2.53).

We emphasize that $|\psi(\mathbf{x}, t)|^2$ is a probability density only with respect to the position $\mathbf{x}$, and *not with respect to the time t*. The normalization condition involves an integral only with respect to space. The probabilities $\mathrm{Prob}(U, t)$ describe alternatives that are defined at a single fixed moment of time t. Hence, t is a *parameter* of the probabilities and not a random variable.

Quantum theory treats time and space asymmetrically. This asymmetry does not originate from Schrödinger's equation. Time and space do appear asymmetrically there, but we can generalize Schrödinger's equation to relativistic systems in a way that treats time and space at equal footing. Even then, the asymmetry in the probability assignment persists. This asymmetry is a brute fact about quantum theory, and it is incorporated as such into its fundamental principles.

2.8 The Great Synthesis

Heisenberg's uncertainty principle and Born's statistical interpretation of the wave function were the last pieces of the puzzle in the formation of a completed quantum theory. At least, this is what Bohr and his team – the so-called *Copenhagen school* – asserted in the fifth Solvay conference that took place in Brussels in October 1927. This conference is traditionally considered as the turning point for the acceptance of quantum theory by the physics community.

At the end of their common presentation of matrix mechanics, Heisenberg and Born made a bold assertion: "We claim that quantum mechanics is a complete theory. Its fundamental physical and mathematical assumptions do not accept further modification." As it turned out, they were right. Contemporary quantum theory has extended and clarified its mathematical background, it has passed through experimental tests with a precision that was unthinkable in 1927, but its foundational framework is essentially the same. At the end of the Solvay conference, the majority of the participants had adopted the new theory and its interpretation by the Copenhagen school. However, this majority did not include Einstein, de Broglie, and Schrödinger.

The end result was a complete vindication of Bohr and of his characteristic attitude towards quantum phenomena ever since he started thinking on the topic. His persistence that the rules of classical physics must be abandoned was the most crucial factor in the development of quantum mechanics.

The key point in Bohr's thought was the relation between language and reality. No description can fully represent reality; all descriptions are at most partially true. The particle–wave duality must be understood in this sense. The wave picture and the particle picture are two different attempts to interpret experimental results. The former emphasizes the continuous spacetime aspects of a physical system; the latter emphasizes the discrete and dynamical ones. Bohr believed those two descriptions to be *complementary*, in the sense that they cannot both be valid at the same time.

The complementarity principle is related to the fact that it is possible to measure a physical system only if we interact with it. The rules of quantum theory make impossible a sharp distinction between a microscopic system and the device that measures it. If the same system is measured in different experiments, it will manifest different properties. For this reason, we can say nothing about a microscopic system without referring explicitly to the experimental set-up that is employed for measuring the system's properties (see Fig. 2.6).

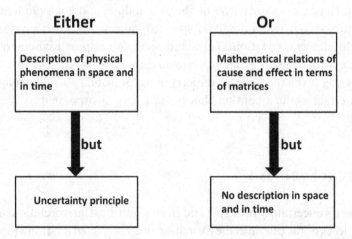

Fig. 2.6 A sketch of Bohr's complementarity principle. Adapted from Heisenberg (1949).

Going back to 1927, there were still many open issues in quantum theory, in particular, regarding its mathematical foundation and its logical structure. These issues were addressed primarily by Dirac (1930a) and von Neumann (1932a), who unified matrix mechanics and wave mechanics within a general mathematical framework. In this framework, a physical system is described by a complex vector space with an inner product, a *Hilbert space*, and the physical principles of quantum theory are expressed in terms of Hilbert space objects.

The Hilbert space formalism allows for the quantum description of systems much more complex than the ones considered by the founders of the theory. It has led to a theory of quantum fields that describes the electromagnetic and nuclear interactions, to a complete explanatory framework for chemical phenomena, to a theory of quantum statistical mechanics for bulk matter, to a deep understanding of the role of symmetries in physics, and to theories for open quantum systems, quantum information and computation, and quantum thermodynamics.

In Chapters 3–9, we will present the foundations of quantum mechanics in terms of six physical principles, analogous to the principles of classical physics that were summarized in Section 1.6.

In spite of the simplicity and universality of its foundations, quantum theory is not a completed physical theory. It gives infuriatingly cryptic answers about the fundamental nature of the objects that it purports to describe, providing only an abstract mathematical description. There is a minimal interpretation of the mathematical formalism that enables concrete experimental predictions and first-principles explanations for many physical phenomena. In fact, by predictive accuracy and explanatory power, quantum theory is *the most successful scientific theory ever*. However, even the smallest issue that goes beyond the minimal interpretation is subject to strong disagreements and conflicts.

On the other hand, one cannot say that quantum theory is a mere mathematical formalism that happens to be successful. The theory makes very strong and explicit statements about the deeper nature of the physical world, many of which can be verified experimentally. Quantum theory is *sui generis* as a scientific theory; there has never been anything comparable in the history of science. As such, it sets its own standard about what a scientific theory can be.

QUESTIONS

2.1 Interference in the two-slit experiment for light has seemed to disprove any corpuscular theory of light since the nineteenth century. Why did it not disprove photons?

2.2 Planck (1918) wrote about Bohr's theory:

> The fact that the quite sharply defined frequency of an emitted photon should be different from the frequency of the emitting electron must seem to a theoretical physicist, brought up in the classical school, at first sight to be a monstrous and, for the purpose of a mental picture, a practically intolerable demand.

Why does Planck say this?

2.3 The two-slit experiment can also be conducted with individual photons, in full analogy to the experiment with electrons described in Section 2.4. It is often stated that, in the two-slit experiment "a photon interferes with itself." Explain why this phrase makes no literal sense. What does it really mean?

2.4 Explain why the phrase "in the two-slit experiment, an electron passes through both slits simultaneously" is wrong.

2.5 Is the de Broglie wavelength a measure of the size of microscopic particles?

2.6 Do the wave properties of matter appear in composite particles, or only in elementary ones?

2.7 Explain why the wave function of Schrödinger's equation cannot be interpreted as a physical wave.

2.8 "Suppose we measure a free particle's position x_1 at time t_1, and then its position x_2 again at time t_2. There is no restriction in how accurately x_1 and x_2 can be determined, hence, Δx_1 and Δx_2 can be made arbitrarily small. But then, the uncertainty in the momentum $p = m\frac{x_2 - x_1}{t_2 - t_1}$ can also be made arbitrarily small. Hence, the uncertainty principle is violated." What is the problem with this argument?

2.9 What are the units of the wave function for a particle in three dimensions, and what are the units of the corresponding Schrödinger current $\mathbf{J}_S$?

2.10 Suppose that the energy levels E_n of a quantum systems are determined by a single quantum number $n = 1, 2, 3, \ldots$. How fast must E_n grow with n, so that the correspondence principle fails?

PROBLEMS

2.1 Consider stationary EM waves inside a cube of side L and volume $V = L^3$. Show that there are two different oscillation modes (the two polarizations) in every volume $(\pi/L)^3$ in the space of wavenumbers. Then, derive the density of states (2.1).

2.2 Show that the maximum of the distribution (2.7) corresponds to $\omega/T = b$, where b is a constant.

2.3 Prove Eq. (2.10) for photons scattering on electrons by requiring conservation of energy and momentum. Assume that the electron is initially at rest and that it satisfies the relativistic relation between energy E and momentum $\mathbf{p}$.

2.4 Consider scattering of a charged particle by a Coulomb potential, as in Rutherford's experiment. (i) Use the results of Problem 1.2 for $\kappa < 0$, to show that the closest distance b that a particle approaches the center is related to the scattering angle θ (the angle between incoming and outgoing momentum) by the relation

$$b = \frac{|\kappa|}{2E} \cot \frac{\theta}{2}.$$

(ii) Consider a beam of incoming particles with the same momentum. Assume that the beam is cylindrical with width a, and that the particles are homogeneously distributed in relation to b. Show that the probability distribution $p(\theta)$ for the scattering angle is given by Rutherford's formula

$$p(\theta) = (\pi a^2)^{-1} \frac{\kappa^2}{16E^2 \sin^4 \frac{\theta}{2}}.$$

2.5 A particle of charge q under acceleration a loses energy E at a rate given by Larmor's formula

$$\frac{dE}{dt} = -\frac{2q^2 a^2}{3c^3}. \tag{2.59}$$

Assume that the electron in the hydrogen atom moves at a circular orbit of radius r. Show that by Eq. (2.59) the radius decreases according to the equation

$$\frac{dr}{dt} = -\frac{e^4}{3\pi c^3 m_e^2 r^2}. \tag{2.60}$$

Solve Eq. 2.60 for $r(0) = r_0$, and show that the time τ it takes for the electron to crash on the nucleus ($r = 0$) is $\tau = \pi m_e^2 c^3 r_0^3 / e^4$. Evaluate τ for $r_0 \simeq 10^{-10}$ m.

2.6 Verify Eq. (2.20) for a relativistic particle of mass m. The energy E, the momentum $\mathbf{p}$ and the velocity $\mathbf{v}$ are related by the equations

$$E = c\sqrt{\mathbf{p}^2 + m^2 c^2} \quad \text{and} \quad \mathbf{p} = \frac{m\mathbf{v}}{\sqrt{1 - \mathbf{v}^2/c^2}}.$$

2.7 Consider the two-slit experiment for a classical wave, as in Fig. 2.7. Assume that the two slits are almost pointlike and that they act as distinct wave sources. For a monochromatic incoming wave, the waves emitted from the slits have wavenumber vectors $\mathbf{k}$ with the same length $k := |\mathbf{k}|$. Consider the detection at a screen at distance $L \gg d$ from the slits, where d is the distance between the slits. (i) Show that the phase difference between the two partial waves at the detection point $\mathbf{x}$ equals $kd\sin\theta$, where θ is defined in Fig. 2.7. (ii) Assuming that both slits emit waves with the same amplitude, show that total intensity of the wave $I(y) = |A(t, \mathbf{x})|^2$ is given by $I(y) = 4|A|^2 \cos^2\left(\frac{1}{2}kd\sin\theta\right)$, where y is shown in Fig. 2.1. (iii) Identify the local maxima and minima of the wave intensity. Evaluate the distance between two successive local maxima and minima for small angles θ.

2.8 Consider a harmonic oscillator of mass m and angular frequency ω in one spatial dimension. Show that the Gaussian wave functions

$$\psi(x, t) = C(t) \exp\left[-\frac{m\omega}{2\hbar}[x - q(t)]^2 + ip(t)x/\hbar\right] \tag{2.61}$$

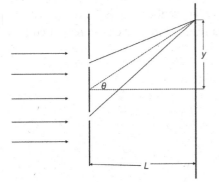

Fig. 2.7 The geometry of the classical two-slit experiment, as described in Problem 2.7.

satisfy the time-dependent Schrödinger's equation if the functions $q(t)$ and $p(t)$ satisfy Hamilton's equations for a harmonic oscillator. Hence, the center of the wave packet moves along the classical equations of motion.

2.9   The lowest energy for a quantum oscillator of angular frequency ω appears to be $\frac{1}{2}\hbar\omega$. Certainly, we have the option of shifting the energy levels by a constant, so that the lowest energy vanishes. To see whether this should be done for the modes of the EM field, consider the total ground-state energy $E_0 = \int_0^\infty d\omega g(V,\omega)\left(\frac{1}{2}\hbar\omega\right)$ for the EM field in a cavity. The integral diverges. It can be made finite by carrying out the integration up to a maximum frequency Ω_m, assuming that for higher frequencies/energies, our current understanding of the EM field breaks down. (i) Today, we know that the EM theory works at least up to energies of 10 TeV. Choose the appropriate Ω_m, and estimate the corresponding energy density E_0/V. Use mass–energy equivalence to compare with the mass density of water, $\rho = 1000 \text{ kg/m}^3$. (ii) If the ground-state energy exists, then it should gravitate, and perhaps manifest at the cosmological level. If this is the case, then, cosmological observations constrain the energy density to be about $2.2 \times 10^{-27} \text{ kg/m}^3$. What would Ω_m be in this case? Compare with the frequencies of the optical part of the spectrum. (iii) Overall, does it make sense to treat the ground-state energy of the EM field modes as physical?

Bibliography

- For a brief account of the historical development of quantum theory, see Chapter 5 of Anastopoulos (2008); for more details, see Kragh (2002) and Jammer (1989). The first papers on quantum theory have been reprinted in ter Haar (1967) and in van der Waerden (2007). Of particular interest are the Nobel lectures of Bohr (1922), de Broglie (1929), Heisenberg (1932), Schrödinger (1933), and Born (1954).

- For nontechnical descriptions of modern two-slit experiments, see Rodger (2002). See also the videos for Hitachi's two-slit experiment with electrons (www.youtube.com/watch?v=PanqoHa_B6c) and Vienna's two-slit experiment with molecules (www.youtube.com/watch?v=vCiOMQIRU7I).

- The Bohr–Einstein debates as described in Bohr (1998) make fascinating reading; see also Chapter 25 in Pais (2005).

- For the early history of the uncertainty principle, see Hilgevoord and Uffink (2016). For the Copenhagen interpretation of quantum theory, see Heisenberg (1999) and Faye (2019). Many important early papers on the foundations of quantum theory have been reprinted by Wheeler and Zurek (1983).

Part II

The Principles of Quantum Theory

3 Hilbert Spaces and the Superposition Principle

Quantum phenomena do not occur in a Hilbert space. They occur in a laboratory.

Peres (2002)

3.1 Hilbert Spaces

From now on, we will be using natural units, setting $\hbar = c = 1$.

The introduction of the Hilbert space as the essential mathematical structure for the formulation of quantum theory was motivated by the following facts.

Schrödinger's equation is a linear differential equation. Hence, if $\psi_1(\mathbf{r}, t)$ and $\psi_2(\mathbf{r}, t)$ are solutions, then $c_1\psi_1(\mathbf{r}, t) + c_2\psi_2(\mathbf{r}, t)$ is also a solution, for any complex numbers c_1 and c_2. This means that solutions to Schrödinger's equations define a vector space with complex coefficients.

Heisenberg's matrix mechanics also suggests the introduction of a complex vector space for the description of quantum system. After all, matrices are *defined* in terms of their action on vector spaces. Furthermore, Eq. (2.47), which relates wave mechanics to matrix mechanics, involves expressions of the form $\int dx\,\psi^*(x)\phi(x)$ for some wave functions $\psi(x)$ and $\phi(x)$. These expressions define an inner product in the vector space of the wave functions.

Hence, the mathematical foundation of quantum theory should involve a complex vector space with an inner product. This is exactly how one defines a Hilbert space.

3.1.1 Vector Spaces

Definition 3.1 *A complex vector space V* is a set V equipped with two operations: (i) addition: $\phi, \chi \in V \rightarrow \phi + \chi \in V$, and (ii) scalar multiplication: $\lambda \in \mathbb{C}, \phi \in V \rightarrow \lambda\phi \in V$, such that the following properties hold.

1. For every $\phi, \chi, \psi \in V$, $(\phi + \chi) + \psi = \phi + (\chi + \psi)$.
2. For every $\phi, \chi \in V$, $\phi + \chi = \chi + \phi$.
3. There is $0 \in V$, such that $\phi + 0 = \phi$, for every $\phi \in V$.
4. For every $\phi \in V$, there is $-\phi \in V$, such that $\phi + (-\phi) = 0$.
5. For every $\lambda_1, \lambda_2 \in \mathbb{C}$ and $\phi \in V$, $\lambda_1(\lambda_2\phi) = (\lambda_1\lambda_2)\phi$.
6. For every $\phi \in V$, $1\phi = \phi$.
7. For every $\lambda \in \mathbb{C}$ and $\phi, \chi \in V$, $\lambda(\phi + \chi) = \lambda\phi + \lambda\chi$.
8. For every $\lambda_1, \lambda_2 \in \mathbb{C}$ and $\phi \in V$, $(\lambda_1 + \lambda_2)\phi = \lambda_1\phi + \lambda_2\phi$.

Note that we employ a common abuse of notation in using the same symbol (0) for the number zero and for the zero vector of V.

A *basis* in a vector space V is a *minimal* subset $B = \{e_1, e_2, \ldots, e_n\}$ of V, such that every $\phi \in V$ can be expressed as a linear combination of the elements e_i of V, that is, $\phi = \sum_{i=1}^{n} \lambda_i e_i$, for some complex numbers λ_i. That B is a minimal subset means that if one of its elements is removed, then there are vectors $\phi \in V$ that cannot be written as linear combinations of the remaining elements of B.

There are many different bases in any vector space. However, all bases contain the same number of vectors. The number of vectors in a basis of a vector space V is called the *dimension* of the vector space V; it is denoted as dim V. If a basis contains an infinity of elements, the vector space is called infinite-dimensional.

The simplest example of a vector space is $\mathbb{C}^n$, that is, the set of all column vectors with n entries

$$\phi = \begin{pmatrix} \phi_1 \\ \phi_2 \\ \ldots \\ \phi_n \end{pmatrix}, \tag{3.1}$$

under the usual operation of vector addition and multiplication.

An obvious basis of $\mathbb{C}^n$ consists of the n vectors

$$e_1 = \begin{pmatrix} 1 \\ 0 \\ \ldots \\ 0 \end{pmatrix} \quad e_2 = \begin{pmatrix} 0 \\ 1 \\ \ldots \\ 0 \end{pmatrix} \quad \ldots \quad e_n = \begin{pmatrix} 0 \\ 0 \\ \ldots \\ 1 \end{pmatrix}. \tag{3.2}$$

3.1.2 Inner Product

Definition 3.2 An *inner product* in a complex vector space V is a function that assigns any two vectors $\phi, \psi \in V$ to a complex number (ϕ, ψ) so that,

(a) for every $\lambda \in \mathbb{C}$ and $\phi, \chi, \psi \in V$, $(\lambda\phi + \chi, \psi) = \lambda(\phi, \psi) + (\chi, \psi)$; (*linearity with respect to the first element*)
(b) for every $\phi, \psi \in V$, $(\psi, \phi) = (\phi, \psi)^*$; (*hermiticity*)
(c) for every $\phi \in V$, $(\phi, \phi) \geq 0$. Also, $(\phi, \phi) = 0$ if and only if $\phi = 0$. (*positivity*)

An immediate consequence of Definition 3.2 is that the inner product is *antilinear* with respect to the second element, namely, $(\phi, \lambda\chi + \psi) = \lambda^*(\phi, \chi) + (\phi, \psi)$.

The natural inner product between two vectors $\phi, \chi \in \mathbb{C}^n$,

$$\phi = \begin{pmatrix} \phi_1 \\ \phi_2 \\ \ldots \\ \phi_n \end{pmatrix} \quad \chi = \begin{pmatrix} \chi_1 \\ \chi_2 \\ \ldots \\ \chi_n \end{pmatrix}, \tag{3.3}$$

is defined as

$$(\phi, \chi) = \sum_{i=0}^{n} \phi_i \chi_i^*. \tag{3.4}$$

Having defined the inner product,

- we define the *norm* $\|\phi\|$ of a vector $\phi \in V$: $\|\phi\| = \sqrt{(\phi, \phi)}$;
- we define the *distance* $D(\phi, \psi)$ of two vectors $\phi, \psi \in V$: $D(\phi, \psi) = \|\phi - \psi\|$;
- we call a vector ϕ *unit vector* or *normalized vector*, if $\|\phi\| = 1$;
- we call two nonzero vectors ϕ and ψ *orthogonal*, if $(\phi, \psi) = 0$;
- we call two nonzero vectors ϕ and ψ *collinear*, if $\phi = \lambda \psi$, for some complex number $\lambda \neq 0$.

3.1.3 Definition of a Hilbert Space

A Hilbert space $\mathcal{H}$ is a vector space with an inner product. If $\mathcal{H}$ is of finite dimension, this definition is sufficient. However, in quantum theory we are interested in vector spaces with elements wave functions. These vector spaces are infinite-dimensional, and in this case, an additional condition is required in the definition of a Hilbert space.

Definition 3.3 A *Hilbert space* $\mathcal{H}$ is a *complete* complex vector space with an inner product.

A vector space is complete if the limit of any convergent sequence of vectors in $\mathcal{H}$ has a limit that is also a vector $\mathcal{H}$. For a precise definition, see Box. 3.1. The requirement of completeness is essential in order to prove important theorems about Hilbert space operators; it guarantees the rigorous description of continuous physical magnitudes like position and momentum.

Box 3.1 Convergence in a Hilbert Space

The study of infinite-dimensional Hilbert spaces requires a precise description of limits. In any vector space V with inner product we define the following.

Strong limit of a sequence $\{\phi_n | n = 1, 2, \ldots\}$ of vectors of V is a vector $\bar{\phi} \in V$ that satisfies the following. For every $\epsilon > 0$, there is an integer N such that $\|\phi_n - \bar{\phi}\| < \epsilon$, for every $n > N$.

In this definition, we can show that a sequence converges only if we know its limit beforehand. The Cauchy convergence criterion allows us to establish whether sequence converges without prior knowledge of the limit.

Cauchy convergence A sequence $\{\phi_n | n = 1, 2, \ldots\}$ of vectors of V is *Cauchy convergent* if for every $\epsilon > 0$ there is an integer N, such that $\|\phi_n - \phi_m\| < \epsilon$, for all $n, m > N$.

It is possible that a Cauchy-convergent function has no limit in V. For example, let V be the vector space of *continuous* square integrable functions $\phi : \mathbb{R} \to \mathbb{C}$. All elements of the sequence $\phi_n(x) = \tanh(nx)/(1 + x^2)$ are in V, and the sequence can be proved to be Cauchy convergent. However, its limit is the function $\bar{\phi}(x) = \mathrm{sgn}(x)/(1 + x^2)$ that is not continuous at $x = 0$. Hence, the limit of the sequence ϕ_n does not lie in V. In order to avoid such inconvenient behaviors, we incorporate the requirement of *completeness* into the definition of a Hilbert space.

A vector space with inner product V is *complete*, if every Cauchy-convergent sequence of vectors of V has a limit in V.

We can also define a different notion of convergence, *weak convergence*. This notion is essential for describing the spectral properties of operators that are analyzed in Chapter 4.

Weak limit of a sequence $\{\phi_n | n = 1, 2, \ldots\}$ of elements of a Hilbert space V is a vector $\bar{\phi} \in V$ that satisfies the following. For every $\chi \in V$ and every $\epsilon > 0$, there is an integer N such that $|(\phi_n, \chi) - (\bar{\phi}, \chi)| < \epsilon$ for every $n > N$.

A sequence that converges strongly also converges weakly. The converse does not hold. To see this, consider the sequence $\phi_n(x) = \sin(nx)$ on the Hilbert space $L^2([0, 2\pi], dx)$. The sequence ϕ_n has no strong limit in $L^2([0, 2\pi], dx)$. However, it can be proven that $\lim_{n \to \infty}(\phi_n, \chi) = 0$ for any $\chi \in L^2([0, 2\pi], dx)$. Hence, the sequence ϕ_n converges weakly to the zero vector.

3.1.4 Examples of Hilbert Spaces

A complex-valued sequence $\phi = \{\phi_i, i = 1, 2, \ldots\}$ is *square summable* if $\sum_{i=1}^{\infty} |\phi_i|^2 < \infty$. The vector space ℓ^2 of all square summable sequences under the usual addition and scalar multiplication is a Hilbert space with inner product $(\phi, \chi) = \sum_{i=1}^{\infty} \chi_i^* \phi_i$.

A vector of ℓ^2 can be represented as a column vector with infinite elements,

$$\phi = \begin{pmatrix} \phi_1 \\ \phi_2 \\ \phi_3 \\ \ldots \\ \ldots \end{pmatrix}. \tag{3.5}$$

A function $\psi : \mathbb{R} \to \mathbb{C}$ is *square integrable* if $\int_{-\infty}^{\infty} dx \, |\psi(x)|^2 < \infty$. The vector space $L^2(\mathbb{R}, dx)$ of all square integrable functions is a Hilbert space with inner product

$$(\psi, \chi) = \int_{-\infty}^{\infty} dx \psi(x) \chi^*(x). \tag{3.6}$$

As shown in Chapter 2, Schrödinger's equation in one dimension describes the time evolution of such a function.

The Hilbert spaces $L^2(\mathbb{R}^n, d^n x)$ of square integrable functions in n dimensions are similarly defined. In particular, the Hilbert space $L^2(\mathbb{R}^3, d^3 x)$ corresponds to the motion of one particle in physical space.

We can also define the Hilbert space $L^2[U, d\mu(x)]$ of square integrable functions on $U \subset \mathbb{R}^n$ with respect to some integration measure $d\mu(x)$.[1] The Hilbert space $L^2(U, d\mu(x))$ contains all functions $\phi : U \to \mathbb{C}$ that satisfy

$$\int_U d\mu(x) |\phi(x)|^2 < \infty$$

---

[1] One can think of the integration measure $d\mu(x)$ as a term $f(x) d^n x$ in the integral for some nonnegative function f.

and it is equipped with the inner product

$$(\phi, \chi) = \int_U d\mu(x)\phi(x)\chi^*(x). \tag{3.7}$$

In general, if the integration measure is obvious in a given context, we do not include it in the designation of the Hilbert space. Then, we write $L^2(\mathbb{R})$, $L^2(\mathbb{R}^n)$, and so on.

Hilbert spaces can also be defined for functions that satisfy specific boundary conditions. For example, the Hilbert space $L_D^2([0,a])$ consists of all square integrable functions $\psi : [0,a] \to \mathbb{C}$ on the interval $[0,a]$ that satisfy Dirichlet boundary conditions: $\psi(0) = \psi(a) = 0$. The inner product is defined by Eq. (3.7) for $U = [0,a]$ and $d\mu = dx$.

3.2   Quantum States

The notion of the Hilbert space is essential for the mathematical foundation of quantum theory. The association of quantum states to Hilbert space vectors is a Foundational Principle (FP) of quantum theory.

> **FP1** A physical system is described by a Hilbert space. A state of the system at a single moment of time is represented by a Hilbert space vector, called the *state vector*.

In Chapter 6, we will see that FP1 must be modified; there are states that are not described by Hilbert space vectors, but can nonetheless be described by mathematical objects defined on the Hilbert space of the system.

The term "state" appears in FP1 without prior definition. The state is a *primitive* notion in quantum theory. It is supposed to be understood in advance, or through informal descriptions, or to be demonstrated through examples, but it is not defined.

The notion of a physical system is also primitive. This is easier to comprehend intuitively, because it is the same with classical physics. In contrast, the notion of the quantum state is very different from the analogous notion in classical physics. In fact, it was fully developed only after the mathematical formulation of quantum theory had been completed.

The notion of a state was first introduced in statistical mechanics. A *microstate* of a mechanical system corresponds to a point in the system's state space, that is, to the precise specification of the positions and momenta of all particles in the system at a moment of time. Hence, a microstate incorporates the most precise information possible about a classical system. Once we know the microstate at a moment of time, we fully specify the time evolution of a system by solving Hamilton's equation.

Quantum states are very different from microstates. Microstates correspond to mutually exclusive alternatives. If a classical system is in the microstate ξ_1, no measurement can find the system in a different microstate $\xi_2 \neq \xi_1$. In contrast, in quantum theory, if we describe the state of the system by a vector ψ_1 of a Hilbert space, there is a nonzero probability that a measurement will "find" the system in a state described by a different vector ψ_2 (we will explain precisely what this means in Section 6.3.2). Hence, different vectors of the Hilbert space do not, in general, correspond to exclusive alternatives. This implies that the Hilbert space in quantum theory plays a very different role from that of the state space in classical physics. This point is crucial; the confusion between the notion of a quantum state and the classical notion of microstates underlies many of the so-called paradoxes of quantum theory.

> Hilbert space vectors do not describe microstates.

The quantum state is best understood in informational terms. It is the mathematical object that incorporates *all information about the preparation of a physical system* that will undergo measurements. The information is encoded in a form that provides predictions about the outcome of any measurement on the system. Hence, a quantum state encodes the specific ways that we can prepare a physical system in the lab, or specific and distinctive ways that this system appears in nature. Different preparations of the same system correspond to different states.

Quantum theory associates an input–output process to each experiment, as in Fig. 3.1. The input data is of two types. Type A input consists of all actions on the physical system under study that take place prior to the act of measurement. It is codified into the specification of a quantum state. Type B input is the specification of the measurement to be carried out on the physical system. All information about the choice of measurement is expressed in terms of Hilbert space operators, in a way that will be described in Chapter 4.

The output consists of all possible measurement outcomes and of the probability associated to each outcome. The output is obtained by processing the codified input, namely, by mathematical operations that involve the quantum state and the operators that describe the measurement. Quantum theory tells you how to properly codify the information for the two types of input, and how to obtain a unique output.[2]

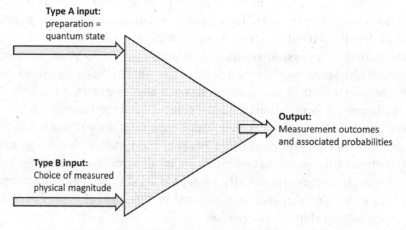

Fig. 3.1 Quantum theory as an input–output process.

---

[2] Classical measurements, described in Section 1.5.3, also define input–output processes for experiments.

The preceding characterization of a quantum state can be widely and successfully employed without leading to any paradoxes. For this reason, it is almost universally accepted, at least as a starting point. The fact that it places measurement as a central point may be counterintuitive to people trained in classical physics, where physical theories refer to properties of physical systems and not to measurement outcomes. Nonetheless, we recommend that the reader provisionally adopts this viewpoint. At this stage, other ways of thinking – especially those from classical physics – will almost certainly lead to physical or conceptual errors. We will revisit the interpretation of the quantum state in Chapter 6.

Given FP1, we will henceforward employ the terms "state" and "state vector" interchangeably. Hence, we will refer to elements of a system's Hilbert space as states.

3.3 The Quantum Superposition Principle

The choice of the Hilbert space as the fundamental space for describing a quantum system implies the acceptance of the so-called *quantum superposition principle*.

Quantum superposition principle Let a quantum system be described by a Hilbert space $\mathcal{H}$. If $\psi, \phi \in \mathcal{H}$ are different states of the system, then also $\lambda \psi + \phi$ (for $\lambda \in \mathbb{C}$) is a state of the system.

Hence, if we can prepare a physical system in states ϕ and ψ, then we can also prepare the system in a state $\psi + \phi$, at least in principle. However, quantum theory makes no statement about *how* $\psi + \phi$ will be prepared, or whether such a preparation is feasible with the available technologies.

It is important to emphasize that the properties of the superposition state $\psi + \phi$ *are not intermediate between the properties of ψ and ϕ*. For example, if all measurements of energy on the state ψ give the same value of energy E_1, and all measurements of energy on the state ϕ give the same value of energy E_2, measurements of energy on the state $\psi + \phi$ *will not* give values of energy between E_1 and E_2. The measurement outcomes will be *either E_1 or E_2*, and their relative frequency will be specified by a probability distribution.

3.3.1 Constructing Superposition States

Special techniques have been developed for the construction of superpositions of known states in various systems. They often employ *interferometers*, that is, devices that allow for the observation of interference phenomena, analogous to the two-slit experiment. One of the most commonly employed interferometers in optical systems is the Mach–Zehnder interferometer that is depicted in Fig. 3.2.

The Mach–Zehnder interferometer consists of two ordinary mirrors and two beam splitters. A beam splitter is an optical element that reflects a fraction of incoming photons (usually

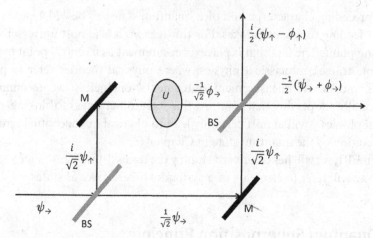

Fig. 3.2 The Mach–Zehnder interferometer consists of two ordinary mirrors (M) and two half-silvered mirrors that act as beam splitters (BS). Incoming photons with a state vector $\psi_\rightarrow$ exit with either state vector $\psi_\rightarrow + \phi_\rightarrow$ or state vector $\psi_\uparrow - \phi_\uparrow$.

50 percent) and lets the remaining photons pass with no change. The most commonly employed beam splitter is a half-silvered mirror, that is, a glass plate with a thin layer of either a metal or a dielectric. The properties of the beam splitter depend on the type and characteristics of the layer. Half-silvered mirrors and common mirrors transform the photon state vector in the following way.

Let us denote the propagation directions of a photon by $\rightarrow$ (horizontal) and $\uparrow$ (vertical). A mirror placed at an angle of $45°$ with respect to the incoming photon beam changes the propagation direction by $90°$, in accordance with the reflection law. It also leads to a multiplication of the state vector by i, namely, a phase $e^{i\pi/2}$. This means that a mirror corresponds to the following transformations

$$\psi_\rightarrow \to i\psi_\uparrow, \qquad \psi_\uparrow \to i\psi_\rightarrow. \tag{3.8}$$

A half-silvered mirror lets half of the incoming photons pass with no change, while the other half are reflected according to the rule (3.8). Overall, it transforms the state vectors as

$$\psi_\rightarrow \to \frac{1}{\sqrt{2}}(i\psi_\uparrow + \psi_\rightarrow), \qquad \psi_\uparrow \to \frac{1}{\sqrt{2}}(\psi_\uparrow + i\psi_\rightarrow), \tag{3.9}$$

where we assume that $(\psi_\rightarrow, \psi_\rightarrow) = (\psi_\uparrow, \psi_\uparrow) = 1$ and $(\psi_\rightarrow, \psi_\uparrow) = 0$. The factor $1/\sqrt{2}$ enters so that the transformation (3.9) preserves the norm of the state vector.

Let us also assume that we have a process U operating on the state vector $\psi_\rightarrow$ that transforms it into a different state vector $\phi_\rightarrow$. For example, U may be an optically active medium that rotates the photon's polarization.

Figure 3.2 shows how photons that enter a Mach–Zehnder interferometer with a state vector $\psi_\rightarrow$ exit with either a state vector $-\frac{1}{2}(\psi_\rightarrow + \phi_\rightarrow)$ or with a state vector $\frac{i}{2}(\psi_\uparrow - \phi_\uparrow)$. Hence, if we can physically implement a transformation $U : \psi_\rightarrow \to \phi_\rightarrow$ we can prepare a system in a superposition of the state vectors ψ and ϕ.

The key ingredient in preparing a superposition state is the beam splitter. In optical systems, beam splitters can be implemented with half-silvered mirrors. For neutrons, specially constructed crystals are commonly employed, while for electrons beam splitting is implemented by the action of an external field.

These experiments can be carried out with individual particles crossing the interferometer. In this sense, we can say that the quantum state of a single particle can be delocalized at a scale of the distance D between the interferometer's two arms. There is no fundamental restriction on how large D may be. With current technologies, D can be as large as tens of centimeters for neutrons and atoms, and it can reach hundreds of kilometers for photons.

3.3.2 Superselection Rules

The superposition principle does not hold universally. The existence of specific superposition states could lead to the violation of fundamental physical principles. In particular, we cannot construct superpositions of states with different values of the electric field. For example, there cannot exist a superposition of the state ψ_{5e} = "five electrons" with the state ψ_{7e} = "seven electrons."

The reason is that the existence of such states violates the law of charge conservation. Suppose that a source prepares electrons in a state $\psi_{5e} + \psi_{7e}$. Then, whenever we make a measurement of the total charge of the system, we will sometimes obtain $5e$ and sometimes $7e$, even if we can perform measurements that do not add or subtract electric charge from the system. This behavior contradicts our intuition about how the principle of charge conservation works in nature. More importantly, no such state has ever been produced.

Therefore, we accept the absence of superpositions between states with different total charge. This principle is known as *superselection rule* for the electric charge. In Section 7.2, we explain how superselection rules are related to conserved quantities.

3.3.3 Schrödinger Cat States

An important open issue in quantum theory is whether superpositions of states with different *macroscopic features* are possible. The debate on this issue starts from a critique of quantum theory by Schrödinger (1935).

One can even set up quite ridiculous cases. A cat is penned up in a steel chamber, along with the following device (which must be secured against direct interference by the cat): in a Geiger counter there is a tiny bit of radioactive substance, so small, that perhaps in the course of the hour one of the atoms decays, but also, with equal probability, perhaps none; if it happens, the counter tube discharges and through a relay releases a hammer which shatters a small flask of hydrocyanic acid. If one has left this entire system to itself for an hour, one would say that the cat still lives if meanwhile no atom has decayed. The psi-function of the entire system would express this by having in it the living and dead cat (pardon the expression) mixed or smeared out in equal parts.

It is typical of these cases that an indeterminacy originally restricted to the atomic domain becomes transformed into macroscopic indeterminacy, ...

Nothing in quantum theory forbids the formation of superpositions of distinct macroscopic states. Certainly, the preparation of such states is technically difficult, because the environment (e.g., air molecules) tends to destroy such superpositions rapidly. However, Schrödinger's superposition of a living and a dead cat is, in principle, possible.

Like all superpositions, Schrödinger's cat is not half alive and half dead. It is *either* alive or dead, and the state vector contains the probabilities for either alternative. In particular, the state vector does not describe a "state" between life and death; such an interpretation would be meaningful only if state vectors described microstates of a physical system. We have emphasized that they do not.

As long as we view the state vector as a carrier of information about measurement outcomes, the possibility of Schrödinger's cat does not lead to any logical paradox. This does not mean that we are comfortable with it. Quantum theory routinely predicts that microscopic superpositions can be amplified to a macroscopic scale. In fact most states available in the Hilbert space of a system that is large enough to accommodate a cat involve such superpositions. Why, then, do we not observe them? Why is quantum behavior so conspicuously absent from the macroscopic scale, when the theory involves no such restriction?

One possible answer is that, indeed, the validity of the superposition principle is restricted. Perhaps there is a (yet unknown) physical principle that forbids the superposition of states with different macroscopic properties.[3] This thesis conflicts with standard quantum theory, as it asserts that Schrödinger cats (possibly even Schrödinger microbes) are impossible, not just difficult to construct. This gives as a strong motivation to construct superpositions of states with different macroscopic properties, nowadays referred to as *cat states*. Some examples are given in Box 3.2. In spite of dramatic progress in the past couple of decades, we are still far from a definitive answer.

Box 3.2 Schrödinger Cat States in the Laboratory

The following is a short list of important experiments that constructed cat states.

1. Brune et al. (1996) prepared a superposition of different states of the electromagnetic field $\psi = c_1\alpha + c_2\beta$, where α and β correspond to the amplitude of an electromagnetic wave in a cavity. The numbers of photons "contained" in the states α and β differ by about 5 to 10.
2. Monroe et al. (1996) prepared a superposition state of a Be^+ ion of the form $\psi(\mathbf{x}) = [\psi_\sigma(\mathbf{x} - \mathbf{x_1}) + \psi_\sigma(\mathbf{x} - \mathbf{x_2})]/\sqrt{2}$, where ψ_σ is a wave packet of width σ, centered around $\mathbf{x} = 0$. In this experiment, the distance between the centers of the two wave packets $|\mathbf{x_1} - \mathbf{x_2}| \simeq 80\,\mathrm{nm}$ was much larger than both the width $\sigma \simeq 7\,\mathrm{nm}$ of the wave packet, and the size of the ion (about 0.1 nm). This was a Schrödinger's cat in the nanoscale.

---

[3] There is no clear-cut definition of macroscopic systems. Usually, lengths larger than 1 μm, or masses larger than that of a small cell (10^{-15} kg) are deemed macroscopic, but the borderline with the atomic scale is fuzzy.

3. Van der Wal et al. (2000) and Friedman et al. (2002) prepared superpositions of two states with different values of the electric current in a superconducting circuit. One state corresponds to a current of some microamps (μA) in one direction and the other to a current of the same magnitude in the opposite direction. The initial estimate was that about 10^{10} electrons contribute to each current state; however, there are arguments that this number is as low as 10^3.

4. Matter wave interferometry, as described in Box 2.3, has also led to the creation of macroscopically extended quantum states. Kovachy et al. (2015) delocalized a wave packet of rubidium atoms over a separation of 54 cm, while Arndt et al. (1999) did the same for fullerene molecules (C_{60}) over a separation of 100 μm.

To summarize, with the exception of superselection rules, quantum theory suggests that every Hilbert space vector corresponds to a physically allowed state. Whether this is true or not will eventually be settled by experiment. We know that the converse does not hold: There are preparations of a system that cannot be described by state vectors. They require the use of a different mathematical object, the *density matrix*, which will be introduced in Chapter 6.

3.4 Hilbert Space Structure

In this section, we present the most important mathematical properties of Hilbert space vectors, and we introduce the key notions of orthonormal bases, direct sums, linear subspaces and dual Hilbert spaces.

3.4.1 Identities and Inequalities

The following propositions hold for any vectors ϕ, ψ of a Hilbert space $\mathcal{H}$.

Proposition 3.4 (Pythagoras' theorem) *If $Re(\phi, \psi) = 0$, then $\|\phi + \psi\|^2 = \|\phi\|^2 + \|\psi\|^2$.*

Proof $\|\phi + \psi\|^2 = (\phi + \psi, \phi + \psi) = (\phi, \phi) + (\psi, \psi) + (\phi, \psi) + (\phi, \psi) = \|\phi\|^2 + \|\psi\|^2 + 2Re(\phi, \psi) = \|\phi\|^2 + \|\psi\|^2$. $\qquad\square$

Note that Pythagoras' theorem in a complex Hilbert space does not require that $(\phi, \psi) = 0$, but it follows from the weaker condition $Re(\phi, \psi) = 0$.

Proposition 3.5 (Cauchy–Schwarz inequality) $|(\phi, \psi)| \leq \|\phi\| \|\psi\|$. *Equality holds only if $\phi = \lambda\psi$, for some $\lambda \in \mathbb{C}$.*

Proof If $\psi = 0$, the statement holds identically. For $\psi \neq 0$, we define $\chi := \phi - \frac{(\phi, \psi)}{(\psi, \psi)}\psi$. We evaluate $(\chi, \psi) = (\phi, \psi) - (\phi, \psi)(\psi, \psi)/(\psi, \psi) = 0$. Hence by Pythagoras' theorem,

$$\|\phi\|^2 = \left|\frac{(\phi, \psi)}{(\psi, \psi)}\right|^2 (\psi, \psi) + \|\chi\|^2 = \frac{|(\phi, \psi)|^2}{\|\psi\|^2} + \|\chi\|^2 \geq \frac{|(\phi, \psi)|^2}{\|\psi\|^2}.$$

We obtain $\|\phi\|^2\|\psi\|^2 \geq |(\phi,\psi)|^2$; the Cauchy–Schwarz inequality follows immediately. Equality holds if $\chi = 0$, whence $\phi = \lambda\psi$ for $\lambda = (\phi,\psi)/(\psi,\psi) \in \mathbb{C}$. $\square$

Proposition 3.6 (Triangle inequality) $|\|\phi\| - \|\psi\|| \leq \|\phi + \psi\| \leq \|\phi\| + \|\psi\|$.

Proof $(\|\phi\| + \|\psi\|)^2 - \|\phi + \psi\|^2 = \|\phi\|^2 + \|\psi\|^2 + 2\|\phi\|\cdot\|\psi\| - \|\phi\|^2 - \|\psi\|^2 - (\phi,\psi) - (\psi,\phi) = 2\|\phi\|\cdot\|\psi\|\left(1 - \frac{\mathrm{Re}(\phi,\psi)}{\|\phi\|\cdot\|\psi\|}\right) \geq 2\|\phi\|\cdot\|\psi\|\left(1 - \frac{|(\phi,\psi)|}{\|\phi\|\cdot\|\psi\|}\right) \geq 0$, where in the last step we used the Cauchy–Schwarz inequality. We conclude that $\|\phi + \psi\| \leq \|\phi\| + \|\psi\|$. The second part of the triangle inequality is proved similarly. $\square$

Proposition 3.7 (Polarization identity)

$$(\phi,\psi) = \frac{1}{4}\left(\|\phi + \psi\|^2 - \|\phi - \psi\|^2 + i\|\phi + i\psi\|^2 - i\|\phi - i\psi\|^2\right).$$

Proof Straightforward: start working from the right-hand side, and use the definition of the norm. $\square$

The polarization identity implies that the norm on a Hilbert space contains all information about the inner product. In Hilbert space, the norm and the inner product are equivalent; if we know one, then we specify the other uniquely.

Example 3.1 Let f^a be a sequence with elements $f_n^a = \frac{1}{n^a}$, where $n = 1, 2, \ldots$ and $a \in \mathbb{R}$. For $a > \frac{1}{2}$, $\|f^a\|^2 = \sum_{n=1}^{\infty} n^{-2a} = \zeta(2a)$, where ζ is the Riemann zeta function – see Appendix B.2. If $a \leq \frac{1}{2}$, then $\|f^a\|^2 = \infty$. Hence, $f^a \in \ell^2$, if $a > \frac{1}{2}$.

Next, we calculate the inner product (f^a, f^b), for $a, b > \frac{1}{2}$. We obtain $(f^a, f^b) = \sum_{n=1}^{\infty} n^{-(a+b)} = \zeta(a+b)$. Then, the Cauchy–Schwarz inequality implies a nontrivial property of the zeta function,

$$\zeta(a+b) \leq \sqrt{\zeta(2a)\zeta(2b)}.$$

Example 3.2 For any $\phi = \begin{pmatrix} z_1 \\ z_2 \end{pmatrix} \in \mathbb{C}^2$, we define the norm $\|\phi\| = |z_1| + |z_2|$. This norm does *not* define an inner product. To show this, consider the vectors $\phi_1 = \begin{pmatrix} 1 \\ 1 \end{pmatrix}$ and $\phi_2 = \begin{pmatrix} 0 \\ 1 \end{pmatrix}$. Using the polarization identity, we find that the associated inner product should satisfy $(\phi_1, \phi_2) = \frac{1}{2}$, and also $(\phi_1, 2\phi_2) = \frac{1}{2}$. This contradicts the defining properties of an inner product.

3.4.2 Orthonormal Bases

Definition 3.8 An *orthonormal set* is any set of vectors $\{e_i\}$ that satisfies $(e_i, e_j) = \delta_{ij}$, for all i, j.

Kronecker's delta δ_{ij} was defined by Eq. (1.9).

Definition 3.9 An *orthonormal basis* $\{e_i\}$ in a Hilbert space $\mathcal{H}$ is an orthonormal set that defines a vector space basis on $\mathcal{H}$.

In any finite-dimensional Hilbert space ($\mathbb{C}^n$), the number of vectors in an orthonormal basis equals the dimension of the Hilbert space. In infinite-dimensional Hilbert spaces, each orthonormal basis has an infinite number of vectors.

Definition 3.10 A Hilbert space is *separable* if it has one orthonormal basis with a denumerable number of vectors.

Nonseparable Hilbert spaces are essentially "too large," and calculations in them are more complex than calculations in separable Hilbert spaces. Thankfully, all known physical systems are described by separable Hilbert spaces. In many textbooks, separability is part of the definition of a Hilbert space.

From now on we will assume that all Hilbert spaces that we encounter are separable. Hence, the indices i that label the elements of an orthonormal set $\{e_i\}$ are integers. They either take a finite number of values ($i \in \{1, 2, \ldots, n\}$), or an infinite one ($i \in \{1, 2, \ldots\}$). In what follows, we will not specify the number of values. We will write $\sum_i$ for both the finite sum $\sum_{i=1}^{n}$ and for the series $\sum_{i=1}^{\infty}$. Conflating these notions is a serious mathematical abuse: A series is not a sum with infinite terms, but the limit of a sequence that is defined in terms of successive additions. However, most series that arise in this book are rather "tame," and this abuse will not lead to problems.

We can use the inner product in order to evaluate the coefficients in the expansion of a vector with respect to any basis. Let $\{e_i\}$ be an orthonormal basis on a Hilbert space $\mathcal{H}$. Any vector $\phi \in \mathcal{H}$ can be expressed as $\phi = \sum_i c_i e_i$, for some complex constants c_i. We take the inner product of both terms in the equation with a vector e_j. Then, $(\phi, e_j) = \sum_i c_i (e_i, e_j) = \sum_i c_i \delta_{ij} = c_j$.

Hence, we derived the so-called *resolution of the unity*

$$\phi = \sum_i (\phi, e_i) e_i, \tag{3.10}$$

which is one of the most common identities that will be used in this book.

Proposition 3.11 (Parseval identity) *For any orthonormal basis $\{e_i\}$ in a Hilbert space $\mathcal{H}$ and for every $\phi \in \mathcal{H}$, $\|\phi\|^2 = \sum_i |(e_i, \phi)|^2$.*

Proof We write $\phi = \sum_i c_i e_i$, whence $\|\phi\|^2 = (\phi, \phi) = \sum_i \sum_j c_i^* c_j (e_i, e_j) = \sum_i \sum_j c_i^* c_j \delta_{ij} = \sum_i |c_i|^2 = \sum_i |(e_i, \phi)|^2$. $\qquad\square$

Example 3.3 We will construct an orthonormal basis $\phi_\ell(x)$ on $L^2([-1, 1])$, where $\ell = 0, 1, 2, \ldots$, such that $\phi_\ell(x)$ is a polynomial of order ℓ. In this problem, it is convenient to choose orthogonal but not normalized vectors, the so-called *Legendre polynomials* $P_\ell(x)$. Their scale is determined by the requirement that $P_\ell(1) = 1$.

To construct this basis, we start with the constant polynomial $P_0(x) = 1$. By definition, $P_1(x)$ is of the form $ax+b$, and satisfies two equations: $P_1(1) = 1$ and $\int_{-1}^{1} dx P_0 P_1 = 0$. These equations specify the constants a and b: They give $a+b = 1$ and $2b = 0$, whence $P_1(x) = x$. Similarly $P_2(x)$ is of the form ax^2+bx+c, with defining equations $P_2(1) = 1$, $\int_{-1}^{1} dx P_0 P_2 = 0$, and $\int_{-1}^{1} dx P_1 P_2 = 0$. These yield $a+b+c = 1$, $\frac{2}{3}a + 2c = 0$, and $b = 0$, respectively, with solution $a = \frac{3}{2}, b = 0, c = -\frac{1}{2}$. Hence, $P_2(x) = \frac{1}{2}(3x^2 - 1)$.

Any Legendre polynomial can be constructed by a repeated application of this procedure. By construction, Legendre polynomials satisfy

$$P_\ell(-x) = (-1)^\ell P_\ell(x). \tag{3.11}$$

A closed expression for the Legendre polynomials is provided by the Rodriguez formula

$$P_\ell(x) = \frac{1}{2^\ell \ell!} \frac{d^\ell}{dx^\ell}(x^2 - 1)^\ell, \tag{3.12}$$

which leads to the normalization condition

$$\int_{-1}^{1} dx P_\ell(x) P_{\ell'}(x) = \frac{2}{2\ell + 1} \delta_{\ell\ell'}. \tag{3.13}$$

3.4.3 Isometric Hilbert Spaces

Let us choose an orthonormal basis $\{e_i\}, i = 1, 2, \ldots n$ on a Hilbert space $\mathcal{H}$ of finite dimension n. We associate to every $\phi \in \mathcal{H}$ a column vector, defined by the components of ϕ in the basis

$\{e_i\}$: $\begin{pmatrix} c_1 \\ c_2 \\ \ldots \\ c_n \end{pmatrix} \in \mathbb{C}^n$, where $c_i = (\psi, e_i)$. The inner product on $\mathcal{H}$ fully reproduces to the inner

product of $\mathbb{C}^n$, since for $\phi = \sum_{i=1}^n c_i e_i$ and $\psi = \sum_{i=1}^n d_i e_i$,

$$(\phi, \psi) = \sum_{i=1}^n c_i d_i^*. \tag{3.14}$$

This means that all Hilbert spaces of dimension n are identical with $\mathbb{C}^n$; we say that they are *isometric* to $\mathbb{C}^n$.

Similarly, in a Hilbert space of infinite dimension, we can use an orthonormal basis $\{e_i\}, i = 1, 2, \ldots$ in order to associate a sequence $\{c_1, c_2, \ldots\} \in \ell^2$ to any vector $\phi \in \mathcal{H}$, by setting $c_i = (\psi, e_i)$. Again, the inner product on $\mathcal{H}$ is perfectly mapped to the inner product of ℓ^2, since for $\phi = \sum_{i=1}^\infty c_i e_i$ and $\psi = \sum_{i=1}^\infty d_i e_i$,

$$(\phi, \psi) = \sum_{i=1}^\infty c_i d_i^*. \tag{3.15}$$

Hence, all infinite-dimensional Hilbert spaces are isometric to ℓ^2.

The fact that all infinite-dimensional Hilbert spaces are identical does not mean that they describe the same physics. The physics is determined by the physical magnitudes that are represented on those Hilbert spaces, and these may be different even if the Hilbert spaces are isometric.

3.4.4 Direct Sum

Consider two Hilbert spaces $\mathcal{H}_1$ and $\mathcal{H}_2$, with vectors

$$\phi = \begin{pmatrix} c_1 \\ c_2 \\ \ldots \\ c_n \end{pmatrix} \in \mathcal{H}_1 \qquad \psi = \begin{pmatrix} d_1 \\ d_2 \\ \ldots \\ d_m \end{pmatrix} \in \mathcal{H}_2. \tag{3.16}$$

We define the *direct sum* $\mathcal{H}_1 \oplus \mathcal{H}_2$ as the Hilbert space that contains all vectors of the form

$$\phi \oplus \psi = \begin{pmatrix} c_1 \\ \cdots \\ c_n \\ d_1 \\ \cdots \\ d_m \end{pmatrix}. \tag{3.17}$$

This definition suffices for all finite-dimensional Hilbert spaces.

If $\{e_i\}$ is an orthonormal basis on $\mathcal{H}_1$ and $\{o_j\}$ is an orthonormal basis on $\mathcal{H}_2$, then the set of vectors $\{e_i \oplus 0, 0 \oplus o_j\}$ forms a *natural* basis on $\mathcal{H}_1 \oplus \mathcal{H}_2$. Let $\mathcal{H}_1 = \mathbb{C}^n$ and $\mathcal{H}_2 = \mathbb{C}^m$. Then, $\{e_i\}$ consists of n vectors, $\{o_j\}$ of m vectors, which implies that $\{e_i \oplus 0, 0 \oplus o_j\}$ consists of $n + m$ vectors. Hence,

$$\mathbb{C}^n \oplus \mathbb{C}^m = \mathbb{C}^{n+m}. \tag{3.18}$$

For the more general case where at least one of the two Hilbert spaces $\mathcal{H}_1$ and $\mathcal{H}_2$ may be infinite-dimensional, a more elaborate definition of the direct sum is required.

Definition 3.12 The *direct sum* $\mathcal{H}_1 \oplus \mathcal{H}_2$ of two Hilbert spaces $\mathcal{H}_1$ and $\mathcal{H}_2$ is the Hilbert space $\mathcal{H}_1 \times \mathcal{H}_2$, with elements in ordered pairs $\phi \oplus \psi$, for $\phi \in \mathcal{H}_1$ and $\psi \in \mathcal{H}_2$. Operations on $\mathcal{H}_1 \oplus \mathcal{H}_2$ are defined as: $\lambda(\phi_1 \oplus \psi_1) + (\phi_2 \oplus \psi_2) = (\lambda\phi_1 + \phi_2) \oplus (\lambda\psi_1 + \psi_2)$. The inner product is $(\phi_1 \oplus \psi_1, \phi_2 \oplus \psi_2) = (\phi_1, \phi_2) + (\psi_1, \psi_2)$.

It is straightforward to show that $\mathcal{H}_1 \oplus \mathcal{H}_2$, as defined above, satisfies all axioms of a Hilbert space.

3.4.5 Closed Linear Subspaces

Definition 3.13 A *linear subspace* V of a Hilbert space $\mathcal{H}$ is a subset of $\mathcal{H}$ that is closed with respect to the Hilbert space operations. This means that for every $\phi, \psi \in V$ and $\lambda \in \mathbb{C}$, $\lambda\phi + \psi \in V$.

We are mainly interested in *closed* linear subspaces. A subspace V is closed if, for any sequence of vector $\phi_n \in V$ that converges, its limit is also in V. This means that V is complete and, hence, a Hilbert space. If a linear subspace V has finite dimension it is always closed; only infinite-dimensional subspaces may fail to be closed. Here, we present proofs that apply to subspaces of finite dimension, but the results also apply to infinite-dimensional subspaces. From now on, when we will use the word "subspace" to refer to closed linear subspaces (see Fig. 3.3).

Consider n arbitrary vectors $\phi_1, \ldots, \phi_n$ of some Hilbert space $\mathcal{H}$. The set of all linear combinations $\sum_{i=1}^{n} c_i\phi_i$ defines a subspace V of $\mathcal{H}$, which is called the *linear span* of the vectors $\phi_1, \ldots, \phi_n$. If $\phi_1, \ldots, \phi_n$ form an orthonormal set, then they define an orthonormal basis on V. Hence, V has dimension n. If the vectors $\phi_1, \ldots, \phi_n$ are not orthonormal, then V has dimension smaller than or equal to n.

The *complement* $V^\perp$ of a subspace V consists of all vectors normal to V, that is, it consists of all vectors $\phi \in \mathcal{H}$ such that $(\phi, \psi) = 0$ for all $\psi \in V$. The only vector that is common to both V

Fig. 3.3 Projection of a vector into a subspace in $\mathbb{R}^3$. The vector ϕ splits as $\phi_V + \phi_{V\perp}$, where V is the horizontal plance. The vector ϕ equals $(\phi, e_1)e_1 + (\phi, e_2)e_2$, with respect to two orthonormal axes on V.

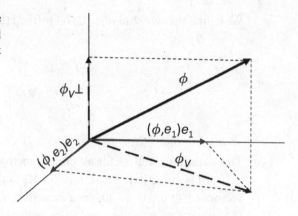

and $V^\perp$ is the zero vector. It is straightforward to show that the complement of a complement is the original subspace: $V^{\perp\perp} = V$.

Let $\{e_i\}$ be an orthonormal basis on a subspace V of dimension m. For any vector $\phi \in \mathcal{H}$, we define the *projection* of $\phi_V \in V$ as

$$\phi_V = \sum_{i=1}^{m} (\phi, e_i)e_i. \tag{3.19}$$

By construction, the vector $\phi - \phi_V$ is normal to all basis vectors e_i, and hence it belongs to the complement $V^\perp$. We express it as $\phi_{V\perp}$. We conclude that all vectors $\phi \in \mathcal{H}$ can be expressed as $\phi = \phi_V + \phi_{V\perp}$, where $\phi_V \in V$ and $\phi_{V\perp} \in V_\perp$. This implies that $\mathcal{H} = V \oplus V^\perp$.

The definition (3.19) depends on the choice of basis $\{e_i\}$, but ϕ_V is independent of this choice. Suppose there are other vectors $\phi'_V \in V$ and $\phi'_{V\perp} \in V_\perp$ that satisfy $\phi = \phi'_V + \phi'_{V\perp}$. Then, $\phi_V - \phi'_V = \phi_{V\perp} - \phi'_{V\perp}$. The left-hand side of this equality is an element of V and the right-hand side is an element of $V^\perp$. This is only possible if $\phi_V - \phi'_V = \phi_{V\perp} - \phi'_{V\perp} = 0$.

Example 3.4 Let V be the subspace of the Hilbert space $L^2([-1, 1])$ that consists of all polynomials up to second order. An orthonormal basis ϕ_i on V is defined by the first three Legendre polynomials, normalized according to Eq. (3.13): $\phi_0 := \frac{1}{\sqrt{2}}P_0$, $\phi_1 := \sqrt{\frac{3}{2}}P_1$, and $\phi_2 := \sqrt{\frac{5}{2}}P_2$.

Suppose we want to project the function $f(x) = e^{i\pi x}$ to V. We evaluate

$$(\phi_0, f) = \frac{1}{\sqrt{2}} \int_{-1}^{1} dx\, e^{i\pi x} = 0, \quad (\phi_1, f) = \sqrt{\frac{3}{2}} \int_{-1}^{1} dx\, x\, e^{i\pi x} = \frac{\sqrt{6}}{\pi}i,$$

$$(\phi_2, f) = \sqrt{\frac{5}{2}} \int_{-1}^{1} dx\, e^{i\pi x} \frac{1}{2}(3x^2 - 1) = -\frac{3\sqrt{10}}{\pi^2}.$$

Hence, the projected vector f_V is

$$f_V(x) = \sum_{i=0}^{2} (\phi_i, f)\phi_i(x) = \frac{3i}{\pi}P_1(x) - \frac{15}{\pi^2}P_2(x) = -\frac{45}{2\pi^2}x^2 + \frac{3i}{\pi}x + \frac{15}{2\pi^2}.$$

3.4.6 Dual Hilbert Space

Let $\mathcal{H}$ be a finite-dimensional Hilbert space, with elements represented by column vectors

$$\phi = \begin{pmatrix} \phi_1 \\ \phi_2 \\ \cdots \\ \phi_n \end{pmatrix} \in \mathcal{H}, \tag{3.20}$$

and an inner product given by Eq. (3.4). The *dual space* $\mathcal{H}^*$ of $\mathcal{H}$ consists of row vectors

$$\alpha = (\alpha_1, \alpha_2, \ldots, \alpha_n). \tag{3.21}$$

The inner product between two row vectors α and β is

$$(\alpha, \beta)_{\mathcal{H}^*} = \sum_i \alpha_i \beta_i^*. \tag{3.22}$$

We note that each $\alpha \in \mathcal{H}^*$ defines a unique linear function $\alpha : \mathcal{H} \to \mathbb{C}$, such that for every $\phi \in \mathcal{H}$, $\alpha(\phi) = \sum_i \alpha_i \phi_i \in \mathcal{H}$. This means that we can define the dual space $\mathcal{H}^*$ as the set of all linear functions on $\mathcal{H}$.

For any column vector $\phi \in \mathcal{H}$, as in Eq. (3.20), there is a unique row vector $\alpha_\phi = (\phi_1^*, \phi_2^*, \ldots, \phi_n^*) \in \mathcal{H}^*$. The correspondence $\phi \to \alpha_\phi$ is a bijection between $\mathcal{H}$ and $\mathcal{H}^*$. It also satisfies

$$(\alpha_\phi, \alpha_\psi)_{\mathcal{H}^*} = (\psi, \phi)_{\mathcal{H}}, \tag{3.23}$$

for all $\phi, \psi \in \mathcal{H}$. This statement is known as *Riesz's theorem*.

The preceding description of dual spaces applies to finite-dimensional Hilbert spaces. In infinite-dimensional Hilbert spaces, definitions and proofs are more technical – see Box 3.3 – but the main results are the same.

Box 3.3 Dual Hilbert Space

The dual $\mathcal{H}^*$ of a Hilbert space $\mathcal{H}$ consists of all continuous linear functions on $\mathcal{H}$, that is, of all functions $\alpha : \mathcal{H} \to \mathbb{C}$, such that $\alpha(\lambda\phi + \psi) = \lambda\alpha(\phi) + \alpha(\psi)$, for all $\phi, \psi \in \mathcal{H}$ and $\lambda \in \mathbb{C}$.

Operations on $\mathcal{H}^*$ are defined by $(\lambda\alpha_1 + \alpha_2)(\phi) := \lambda\alpha_1(\phi) + \alpha_2(\phi)$, for all $\alpha_1, \alpha_2 \in \mathcal{H}^*$, $\phi \in \mathcal{H}$ and $\lambda \in \mathbb{C}$. The norm of a linear function α is defined as $\|\alpha\| := \sup_{\phi \in \mathcal{H}} \frac{|\alpha(\phi)|}{\sqrt{(\phi,\phi)}}$.

Having defined the norm, we can use the polarization identity in order to define the inner product on $\mathcal{H}^*$. In finite-dimensional Hilbert spaces, the norm is always finite. In infinite-dimensional Hilbert spaces the requirement that the linear functions are continuous implies that the norm must be finite.

Riesz's theorem For any $\alpha \in \mathcal{H}^*$, there is a unique vector $\psi_\alpha \in \mathcal{H}$, such that $\alpha(\phi) = (\phi, \psi_\alpha)$ for all $\phi \in \mathcal{H}$.

Proof   Let $\{e_i\}$ be an orthonormal basis on $\mathcal{H}$. We define $\gamma_i := \alpha(e_i)$ for each i, and then the vector $\psi_\alpha := \sum_i \gamma_i^* e_i$. For every $\phi \in \mathcal{H}$,
$(\phi, \psi_\alpha) = \sum_i \gamma_i(\phi, e_i) = \sum_i \alpha(e_i)(\phi, e_i) = \alpha\left(\sum_i(\phi, e_i)e_i\right) = \alpha(\phi)$. The vector ψ_α does not depend on the choice of basis. If another vector ψ_α' satisfies $\alpha(\phi) = (\phi, \psi_\alpha')$ for all $\phi \in \mathcal{H}$, then $(\phi, \psi_\alpha - \psi_\alpha') = 0$, for all $\phi \in \mathcal{H}$. For $\phi = \psi_\alpha - \psi_\alpha'$, we obtain $\|\psi_\alpha - \psi_\alpha'\| = 0$. Hence, $\psi_\alpha = \psi_\alpha'$.

3.5   Dirac Notation

Riesz's theorem provides the mathematical justification of the very useful Dirac notation.

In Dirac notation, a vector $\psi \in \mathcal{H}$ is represented by a *ket* $|\psi\rangle$. The ket contains any reasonable characterization of the vector, such as a number, a symbol, or a phrase. For example, the expressions

$$|\text{four-electron state}\rangle, \qquad \frac{1}{\sqrt{2}}\left(|\text{alive cat}\rangle + |\text{dead cat}\rangle\right)$$

constitute acceptable kets. Equally acceptable is the ket $\frac{1}{\sqrt{2}}|\text{☺}\rangle + \frac{1}{\sqrt{2}}|\text{☹}\rangle$ for a person whose cat becomes Schrödinger's cat.

An element of $\mathcal{H}^*$ is represented by a *bra* $\langle\cdot|$. Thanks to Riesz's theorem, the possible characterizations of a bra coincide with the possible characterizations of a ket. Hence, we write $\langle\psi|$ meaning the vector of $\mathcal{H}^*$ that corresponds to the ket $|\psi\rangle$.

Again, due to Riesz's theorem, we can represent the inner product on a Hilbert space $\mathcal{H}$ as a *braket*: $\langle\phi|\psi\rangle$. A *ketbra* corresponds to an operator $(|\phi\rangle\langle\chi|)|\psi\rangle = \langle\chi|\psi\rangle|\phi\rangle$.

Overall, we employ the following rules of correspondence (in finite dimensions):

- ket = column vector = $n \times 1$ matrix;
- bra = row vector = $1 \times n$ matrix;
- braket = ($1 \times n$ matrix) times ($n \times 1$ matrix) = 1×1 matrix = complex number;
- ketbra = ($n \times 1$ matrix) times ($1 \times n$ matrix) = $n \times n$ matrix = operator.

In problems that require the consideration of several different Hilbert spaces, we specify the Hilbert space we refer to as an index in bras and kets. For example, we can write $|\psi\rangle_\mathcal{H}$ or $_\mathcal{H}\langle\phi|$.

An orthonormal set $\{e_n\}$ is represented by the set of kets $|n\rangle$, $n = 1, 2, \ldots$. If the orthonormal set is enumerated by several integers, then all such integers enter the ket.

The expansion of a vector $|\psi\rangle$ with respect to an orthonormal basis $|n\rangle$ is written as $|\psi\rangle = \sum_n |n\rangle\langle n|f\rangle$. Equivalently, we write the *resolution of the unity*

$$\hat{I} = \sum_n |n\rangle\langle n|, \tag{3.24}$$

where $\hat{I}$ is the identity operator – see Section 4.1.

Note!. The inner product (ϕ, ψ) is written as $\langle \psi | \phi \rangle$ in Dirac notation: The order of the vectors is the opposite.

We will continue to use the same notation as before until Section 4.6, where we will switch to Dirac notation. The reason is that some proofs in operator theory are awkward with Dirac notation, because the Dirac notation takes as given the properties of operators we will have to prove.

QUESTIONS

3.1 What is the difference between a classical microstate and a quantum state?

3.2 We can split a complex wave function $\psi = \psi_1 + i\psi_2$, where ψ_1 and ψ_2 are real functions. Then, Schrödinger's equation for a particle of mass m becomes

$$\frac{\partial}{\partial t} \begin{pmatrix} \psi_1 \\ \psi_2 \end{pmatrix} = \begin{pmatrix} -\frac{1}{2m}\nabla^2\psi_2 + V(\mathbf{x})\psi_2 \\ \frac{1}{2m}\nabla^2\psi_1 - V(\mathbf{x})\psi_1 \end{pmatrix},$$

that is, the complex unit i does not appear. Why then do we formulate quantum theory with complex Hilbert spaces and not real ones?

3.3 Change a logical connective and make the logically self-contradictory statement "Schrödinger's cat is both living and dead" meaningful.

3.4 The following is a description of quantum superposition from a popular science article (Leftzer, 2019).

Every particle or group of particles in the universe is also a wave – even large particles, even bacteria, even human beings, even planets and stars. And waves occupy multiple places in space at once. So any chunk of matter can also occupy two places at once. Physicists call this phenomenon "quantum superposition," and for decades, they have demonstrated it using small particles.

Is this description of quantum superposition accurate? How would you explain this concept to an audience of nonphysicists?

3.5 Explain in detail how the construction method of Legendre polynomials leads to Eq. (3.11).

PROBLEMS

3.1 Prove that the functions $\frac{\sin x}{x}, \frac{1-\cos x}{x}, \frac{x}{\sinh x}$, and $\frac{1}{|x|^{1/4}(x^2+1)}$ are square integrable in $\mathbb{R}$.

3.2 Prove that the functions x^{-2} and $\frac{1}{\sqrt{|x|}+1}$ are not square integrable in $\mathbb{R}$.

3.3 Consider the Hilbert space $\mathcal{H} = L^2([-\pi, \pi])$. Evaluate $\|\cos x\|$. Then, find nontrivial $\phi, \psi \in \mathcal{H}$, such that $(\phi, \cos x) = (\psi, \cos x) = (\phi, \psi) = 0$.

3.4 Let $\phi(x) = |\sin x|$ and $\psi(x) = |\cos x|$ be elements of $L^2([-\pi, \pi])$. Evaluate $\|\phi - \psi\|$.

3.5 Let ϕ and ψ be vectors on a Hilbert space V. Show that the following propositions are equivalent. (i) $(\phi, \psi) = 0$. (ii) $\|\phi\| \leq \|\phi + \lambda\psi\|$, for any $\lambda \in \mathbb{C}$. (iii) $\|\phi + \lambda\psi\| = \|\phi - \lambda\psi\|$, for any $\lambda \in \mathbb{C}$.

3.6 Show that in the Hilbert space $L_D^2([0, 1])$, $\|\phi\| \leq \|\phi'\|$, where ϕ' is the derivative of ϕ.

3.7 Show that the distance between two unit vectors ϕ and ψ takes values between 0 and 2. What is the minimum value of the distance $D(\phi, \psi)$ that we can achieve when transforming ϕ by a phase ($\phi \rightarrow e^{i\theta}\phi$)?

3.8 (i) Prove that any polynomial that satisfies the Rodriguez formula (3.12) also satisfies Eq. (3.13). (ii) Use this result in order to prove that the polynomials defined by the Rodriguez formula are actually the Legendre polynomials.

3.9 The method for constructing the Legendre-polynomial basis in the Hilbert space $L^2([-1, 1])$ can be adapted to other Hilbert spaces. (i) Find the first five vectors of the polynomial basis on the Hilbert space $L^2(\mathbb{R}, e^{-x^2}dx)$. These are the Hermite polynomials, encountered in the study of the quantum harmonic oscillator in Chapter 2. (ii) Find the first four vectors of the polynomial basis on the Hilbert space $L^2(\mathbb{R}^+, e^{-x}dx)$. These are the Laguerre polynomials; we will encounter a variant in the study of the hydrogen atom in Chapter 13. (iii) Choose an appropriate Hilbert space and define your own polynomial basis.

3.10 Let V be a subspace of $\mathbb{C}^4$ spanned by the vectors $\phi_1 = (1, 0, 0, 0)$ and $\phi_2 = (1, 2, 2, 1)$. Find the projection of the vector $\psi = (3, 4, 0, 0)$ to V.

3.11 Let V be the subspace of the Hilbert space $L^2([0, 1], dx)$ that consists of all polynomial functions up to second order. Show that the function $f(x) = \ln x$ is square integrable in $[0, 1]$ and evaluate its projection to V.

Bibliography

- For an introduction to vector spaces, see Axler (2004). For a mathematical introduction to Hilbert spaces, see chapters 1–6 from Young (1988) or chapters 1–4 from Cohen (1989). For the Hilbert space in quantum theory, see chapter 2 of Isham (1995) and chapter 3 of Peres (2002).

- For an elementary introduction to quantum interferometry, see Scarani (2006). See Holbrow et al. (2002) for foundational quantum experiments with the Mach–Zehnder interferometer.

- For the construction of macroscopic superpositions and their implications, see Leggett (2002a) and the Nobel lectures by Wineland (2012) and Haroche (2012).

4 Operators I

General Theory

Quantum mechanics requires the introduction into physical theory of a vast new domain of pure mathematics – the whole domain connected with non-commutative multiplication. This ... indicates a trend which we may expect to continue. We may expect that in the future further big domains of pure mathematics will have to be brought in to deal with the advances in fundamental physics.

Dirac (1939)

4.1 Fundamental Notions

Given the description of a quantum state in terms of Hilbert space vectors, physical magnitudes (Heisenberg's matrices) correspond to linear operators on the Hilbert space. A linear operator (or simply, operator) is a linear map of a Hilbert space to itself.

Definition 4.1 An *operator* $\hat{A}$ on a Hilbert space $\mathcal{H}$ is a function $\hat{A} : \mathcal{H} \rightarrow \mathcal{H}$, that assigns to each $\phi \in \mathcal{H}$ a vector $\hat{A}\phi \in \mathcal{H}$, so that $\hat{A}(\lambda\phi + \psi) = \lambda\hat{A}\phi + \hat{A}\psi$, for all $\phi, \psi \in \mathcal{H}$ and $\lambda \in \mathbb{C}$.

In physics, we often employ the convention of placing a hat on top of operators in order to distinguish them from classical quantities. We will conform to this convention.

4.1.1 Definitions

We define the following operations between Hilbert space operators.

1. *Addition* $(\hat{A} + \hat{B})\phi := \hat{A}\phi + \hat{B}\phi$, for all $\phi \in \mathcal{H}$.
2. *Scalar product* $(\lambda\hat{A})\phi := \lambda\hat{A}\phi$, for all $\lambda \in \mathbb{C}$ and $\phi \in \mathcal{H}$.
3. *Operator product* $(\hat{A}\hat{B})\phi := \hat{A}(\hat{B}\phi)$, or all $\phi \in \mathcal{H}$.

The operations above satisfy the following properties.

1. Addition and scalar product satisfy the conditions of Definition 4.1. Hence, the operators of Hilbert space $\mathcal{H}$ define a vector space.
2. *Associativity* For all operators $\hat{A}, \hat{B}$ and $\hat{C}$, $(\hat{A}\hat{B})\hat{C} = \hat{A}(\hat{B}\hat{C})$.
3. *Bilinearity* For all operators $\hat{A}, \hat{B}$ and $\hat{C}$ and for all $\lambda \in \mathbb{C}$,

$$\hat{A}(\lambda\hat{B} + \hat{C}) = \lambda\hat{A}\hat{B} + \hat{A}\hat{C}, \quad \text{and} \quad (\lambda\hat{B} + \hat{C})\hat{A} = \lambda\hat{B}\hat{A} + \hat{C}\hat{A}.$$

4. *Existence of unity* The operator $\hat{I}$, defined by $\hat{I}\phi = \phi$ for all $\phi \in \mathcal{H}$, satisfies $\hat{I}\hat{A} = \hat{A}\hat{I} = \hat{A}$ for all operators $\hat{A}$.

5. *Existence of zero* The operator $\hat{0}$, defined by $\hat{0}\phi = 0$ for all $\phi \in \mathcal{H}$, satisfies $\hat{0}\hat{A} = \hat{A}\hat{0} = \hat{0}$ and $\hat{A} + \hat{0} = \hat{A}$, for all operators $\hat{A}$.

The above properties imply that the Hilbert space operators define a mathematical structure known as *associative algebra*.

We will denote the zero operator simply by 0. We will often express multiples $\lambda\hat{I}$ of the unit operator simply as λ.

We define the nth power of an operator $\hat{A}$ as the operator

$$\hat{A}^n = \underbrace{\hat{A}\hat{A}\ldots\hat{A}}_{n \text{ times}}.$$

For any nth degree polynomial $h_n(x) = \sum_{k=0}^{n} c_k x^k$ with complex coefficients c_k, we define the associated polynomial function of an operator $\hat{A}$ as

$$h_n(\hat{A}) = \sum_{k=0}^{n} c_k \hat{A}^k.$$

Operator inverse If there exists an operator $\hat{B}$, such that $\hat{A}\hat{B} = \hat{B}\hat{A} = \hat{I}$ for some operator $\hat{A}$, then we call $\hat{B}$ *the inverse* of $\hat{A}$, and we write it as $\hat{A}^{-1}$. Sometimes, we will use the symbol $1/\hat{A}$ for $\hat{A}^{-1}$.

If two operators $\hat{A}$ and $\hat{B}$ have an inverse, then their product $\hat{A}\hat{B}$ also has an inverse, and

$$(\hat{A}\hat{B})^{-1} = \hat{B}^{-1}\hat{A}^{-1}. \tag{4.1}$$

The proof of Eq. (4.1) follows straightforwardly from the definition: $(\hat{A}\hat{B})\hat{B}^{-1}\hat{A}^{-1} = \hat{A}(\hat{B}\hat{B}^{-1})\hat{A}^{-1} = \hat{A}\hat{I}\hat{A}^{-1} = \hat{A}\hat{A}^{-1} = \hat{I}$.

Adjointness The *adjoint* of an operator $\hat{A}$ is the operator $\hat{A}^\dagger$ that satisfies

$$(\hat{A}\phi, \psi) = (\phi, \hat{A}^\dagger\psi), \tag{4.2}$$

for all $\phi, \psi \in \mathcal{H}$.

The following identities follow straightforwardly from the definition above.

1. $(c\hat{A} + \hat{B})^\dagger = c^*\hat{A}^\dagger + \hat{B}^\dagger$, for all operators $\hat{A}$ and $\hat{B}$ and every $c \in \mathcal{C}$.
2. $(\hat{A}^\dagger)^\dagger = \hat{A}$.
3. $(\hat{A}\hat{B})^\dagger = \hat{B}^\dagger\hat{A}^\dagger$.

Proof $(\phi, (\hat{A}\hat{B})^\dagger\psi) = (\hat{A}\hat{B}\phi, \psi) = (\hat{B}\phi, \hat{A}^\dagger\psi) = (\phi, \hat{B}^\dagger\hat{A}^\dagger\psi)$, for all $\phi, \psi \in \mathcal{H}$. □

Operators as matrices For every operator $\hat{A}$, we define the associated matrix $A_{mn} := (\hat{A}e_n, e_m)$, where $\{e_n\}$ is an orthonormal basis on $\mathcal{H}$. For any vector $\phi = \sum_n c_n e_n$,

$$\hat{A}\phi = \sum_n c_n \hat{A}e_n = \sum_{n,m} c_n(\hat{A}e_n, e_m)e_m = \sum_m \left(\sum_n A_{mn}c_n\right) e_m, \tag{4.3}$$

where we used the resolution of the unity. This means that the map $\phi \to \hat{A}\phi$ corresponds to the transformation $c_m \to c'_m = \sum_n A_{mn}c_n$ to the components of ϕ with respect to the basis $\{e_n\}$.

Given that the components c_n define a column vector, the operator mapping is mathematically identical to the action of the matrix A_{mn} to column vectors.

In Dirac notation, we write $A_{mn} = \langle m|\hat{A}|n\rangle$. We also assume that an operator always acts on kets, so we represent the action of any operator on a bra by the action of its adjoint on a ket. For example, the quantity $(\hat{A}\phi, \hat{B}\hat{C}\psi)$ equals $(\hat{C}^\dagger \hat{B}^\dagger \hat{A}\phi, \psi)$, which corresponds to the braket $\langle \psi|\hat{C}^\dagger \hat{B}^\dagger \hat{A}|\phi\rangle$.

Direct sums of operators Let the operator $\hat{A}$ be defined on the Hilbert space $\mathcal{H}$ and let the operator $\hat{B}$ be defined on the Hilbert space $\mathcal{V}$. Then, the direct sum of $\hat{A}$ and $\hat{B}$ is the operator $\hat{A} \oplus \hat{B}$ on the Hilbert space $\mathcal{H} \oplus \mathcal{V}$, defined by

$$(\hat{A} \oplus \hat{B})(\phi \oplus \psi) = \hat{A}\phi \oplus \hat{B}\psi, \tag{4.4}$$

for all $\phi \in \hat{H}$ and $\psi \in \hat{V}$. The matrix elements of $\hat{A} \oplus \hat{B}$ in a natural basis of $\mathcal{H} \oplus \mathcal{V}$ are of a block-diagonal form, that is, in matrix notation,

$$\hat{A} \oplus \hat{B} = \left(\begin{array}{c|c} \hat{A} & 0 \\ \hline 0 & \hat{B} \end{array}\right). \tag{4.5}$$

It is straightforward to show that $(\hat{A} \oplus \hat{B})(\hat{C} \oplus \hat{D}) = \hat{A}\hat{C} \oplus \hat{B}\hat{D}$.

Example 4.1 Consider the Hilbert space $L^2(\mathbb{R}, dx)$ describing the motion of a particle in one dimension. We define the position operator $\hat{x}$ and the momentum operator $\hat{p}$ by

$$\hat{x}\psi(x) = x\psi(x) \tag{4.6}$$

$$\hat{p}\psi(x) = -i\frac{\partial}{\partial x}\psi(x). \tag{4.7}$$

The time-dependent Schrödinger equation (2.30) is expressed as

$$i\frac{\partial}{\partial t}\psi = \hat{H}\psi, \tag{4.8}$$

where

$$\hat{H} = \frac{1}{2m}\hat{p}^2 + V(\hat{x}) \tag{4.9}$$

is a linear operator on $L^2(\mathbb{R}, dx)$, the *Hamiltonian* operator.

4.1.2 Eigenvalues and Eigenvectors

Let $\hat{A}$ be an operator on a Hilbert space $\mathcal{H}$. Any nonzero vector $\psi \in \mathcal{H}$ that satisfies $\hat{A}\psi = a\psi$, for some $a \in \mathbb{C}$, is called an *eigenvector* of $\hat{A}$. The number a is called an *eigenvalue* of $\hat{A}$. For example, the time-independent Schrödinger equation (2.34) is the eigenvalue equation of the operator $\hat{H}$ of Eq. (4.9).

If ϕ and ψ are two eigenvectors of $\hat{A}$ with the same eigenvalue, then any linear combination $c_1\phi + c_2\psi$, for $c_1, c_2 \in \mathbb{C}$, is an eigenvector of $\hat{A}$ with the same eigenvalue. Hence, all eigenvectors of $\hat{A}$ with the same eigenvalue a define a subspace V_a that is called the *eigenspace* of $\hat{A}$ for the eigenvalue a. If an operator $\hat{A}$ has the same eigenvalue a for two noncollinear eigenvectors, the eigenvalue is called *degenerate*; we refer to $\dim V_a$ as the *degeneracy* of the eigenvalue a.

Let $\hat{A}$ satisfy the eigenvalue equation $\hat{A}\psi = a\psi$. Then, the following hold.

1. $\hat{A}^2\psi = \hat{A}(a\psi) = a\hat{A}\psi = a^2\psi$. The proof straightforwardly generalizes to $\hat{A}^n\psi = a^n\psi$, for all $n = 1, 2, 3, \ldots$.
2. For any polynomial $f(x) = \sum_{k=0}^{n} c_k x^k$, $f(\hat{A})\psi = f(a)\psi$.
3. If $\hat{A}$ has an inverse, then $a \neq 0$. (If $a = 0$, then $\hat{A}^{-1}\hat{A}\psi = 0$, hence, $\psi = 0$.)
4. If $\hat{A}^{-1}$ exists, $\hat{A}^{-1}|\psi\rangle = a^{-1}|\psi\rangle$.

Suppose that the operator $\hat{A}_\lambda$ depends on some parameter $\lambda \in \mathbb{R}$. Then the eigenvalues a_λ and the eigenvectors ψ_λ also depend on λ. Normalizing the eigenvectors so that $(\psi_\lambda, \psi_\lambda) = 1$, we write $a_\lambda = (\hat{A}_\lambda\psi_\lambda, \psi_\lambda)$. Then, $da_\lambda = (d\hat{A}_\lambda\psi_\lambda, \psi_\lambda) + (\hat{A}_\lambda d\psi_\lambda, \psi_\lambda) + (\hat{A}_\lambda\psi_\lambda, d\psi_\lambda) = (d\hat{A}_\lambda\psi_\lambda, \psi_\lambda) + d(\psi_\lambda, \psi_\lambda)$. The last term vanishes, hence, we obtain

$$\frac{da_\lambda}{d\lambda} = \left(\frac{d\hat{A}_\lambda}{d\lambda}\psi_\lambda, \psi_\lambda \right). \tag{4.10}$$

Equation (4.10) is known as the *Feynman–Hellmann theorem*.

4.1.3 Operator Norm

Definition 4.2 The *norm* $\|\hat{A}\|$ of an operator $\hat{A}$ is defined as

$$\|\hat{A}\| := \sup_{\psi \in \mathcal{H}} \frac{\|\hat{A}\psi\|}{\|\psi\|}. \tag{4.11}$$

If no supremum exists in Eq. (4.11), then the operator has infinite norm.

We employ the same symbol ($\|\cdot\|$) for both the norm of a vector and the norm of an operator. The distinction is always clear by the context.

An operator with finite norm is called *bounded*. An operator with infinite norm is called *unbounded*. The set of all bounded operators on a Hilbert space $\mathcal{H}$ is denoted by $B(\mathcal{H})$.

For any bounded operators $\hat{A}$ and $\hat{B}$ in the Hilbert space $\mathcal{H}$, the following hold.

1. If $\|\hat{A}\| = 0$, then $\hat{A} = 0$.
2. $\|\lambda\hat{A}\| = |\lambda|\|\hat{A}\|$, for all $\lambda \in \mathbb{C}$.
3. $\|\hat{A}\psi\| \leq \|\hat{A}\| \cdot \|\psi\|$, for all $\psi \in \mathcal{H}$.
4. $\|\hat{A} + \hat{B}\| \leq \|\hat{A}\| + \|\hat{B}\|$.

   Proof For all $\psi \in \mathcal{H}$, $\|(\hat{A} + \hat{B})\psi\| = \|\hat{A}\psi + \hat{B}\psi\| \leq \|\hat{A}\psi\| + \|\hat{B}\psi\| \leq (\|\hat{A}\| + \|\hat{B}\|)\|\psi\|$. The desired inequality follows from Eq. (4.11). □

5. (Cauchy–Schwarz inequality for operators.) $\|\hat{A}\hat{B}\| \leq \|\hat{A}\| \cdot \|\hat{B}\|$.

   Proof $\|\hat{A}\hat{B}\psi\| \leq \|\hat{A}\|\|\hat{B}\psi\| \leq \|\hat{A}\|\|\hat{B}\|\|\psi\|$, for all $\psi \in \mathcal{H}$. By Eq. (4.11), the Cauchy–Schwarz inequality follows. □

In finite-dimensional Hilbert spaces ($\mathcal{H} = \mathbb{C}^n$), an operator is just a complex $n \times n$ matrix. The norm of an $n \times n$ matrix A_{ij}, where $i, j = 1, 2, \ldots, n$ is always finite: All operators on $\mathbb{C}^n$ are bounded.

Unbounded operators exist in infinite-dimensional Hilbert spaces. One example is the position operator $\hat{x}\psi(x) = x\psi(x)$ on $L^2(\mathbb{R})$. To show that $\hat{x}$ is unbounded, we consider a Gaussian function $\phi_0(x) = e^{-x^2/2} \in L^2(\mathbb{R})$, and then we define a family of functions $\phi_a(x) = \phi_0(x-a)$. The functions ϕ_a are square integrable for any a. We readily find that the ratio $\|\hat{x}\phi_a\|/\|\phi_a\| = \sqrt{a^2 + \frac{1}{2}}$ can become unlimitedly large for large a. Hence, the operator $\hat{x}$ has no finite norm.

Unbounded operators often lead to complications in analysis. The reason is that we cannot define their action on all Hilbert space vectors. For example, the action of the position operator $\hat{x}$ on the square integrable function $\psi_0(x) = 1/\sqrt{x^2 + 1}$ yields the function $x/\sqrt{x^2 + 1}$, which is not square integrable.

The *domain* $D_{\hat{A}}$ of an unbounded operator $\hat{A}$ is the set of all vectors of the Hilbert space $\mathcal{H}$ upon which $\hat{A}$ is well defined. An operator $\hat{A}$ is useful if its domain is "densely" distributed in the Hilbert space. By this we mean that, for any vector ϕ of $\mathcal{H}$ outside $D_{\hat{A}}$, we can find neighboring vectors within $D_{\hat{A}}$, using which we can approximate the matrix elements $(\hat{A}\phi, \psi)$ for all $\psi \in D_{\hat{A}}$. For example, the action of $\hat{x}$ may be ill defined on $\psi_0(x) = 1/\sqrt{x^2 + 1}$, but it is well defined on $\psi_\epsilon = e^{-\epsilon x^2}/\sqrt{x^2 + 1}$, for all $\epsilon > 0$. Hence, we can specify any matrix element $(\hat{x}\psi_0, \phi)$ as the limit $\lim_{\epsilon \to 0}(\hat{x}\psi_\epsilon, \phi)$.

In this book, we will usually ignore complications related to the domain of unbounded operators. This does not imply that such complications are not important, only that they can be successfully addressed in the problems we will consider in this book. We sacrifice mathematical precision, which is often time consuming, in order to focus on the physical meaning of our results and on the development of practical calculational tools. We will follow a "presumption of innocence" for operators: We will assume that any operator with an explicit defining formula is well defined, unless it is proven otherwise. Proper mathematical procedure requires a "presumption of guilt": No operator should be assumed to be well defined, unless it is proven to be so.

4.2   Operator Types

In this section, we present the most common types of operator that appear in quantum theory, and we analyze their properties.

4.2.1   Normal Operators

Definition 4.3 An operator $\hat{A}$ such that $\hat{A}\hat{A}^\dagger = \hat{A}^\dagger\hat{A}$ is called *normal*.

Proposition 4.4 *If $\hat{A}$ is normal, then $\hat{A}^\dagger$ has the same eigenvectors with $\hat{A}$ and the complex conjugate eigenvalues.*

Proof Let $\hat{A}\psi = a\psi$ for some vector ψ and $a \in \mathbb{C}$, then $\hat{A}^\dagger\psi = a^*\psi$. $\|(\hat{A}^\dagger - a^*\hat{I})\psi\|^2 = ((\hat{A}^\dagger - a^*\hat{I})\psi, (\hat{A}^\dagger - a^*\hat{I})\psi) = ((\hat{A} - a\hat{I})(\hat{A}^\dagger - a^*\hat{I})\psi, \psi) = ((\hat{A}^\dagger - a^*\hat{I})(\hat{A} - a\hat{I})\psi, \psi) = 0$, since $(\hat{A} - a\hat{I})\psi = 0$, by definition. We conclude that $\hat{A}^\dagger\psi = a^*\psi$.   □

Obviously, $\hat{A}^\dagger \hat{A}\psi = |a|^2\psi$. Furthermore, $\|\hat{A}\psi\|^2 = (\hat{A}^\dagger \hat{A}\psi, \psi)$. In a finite-dimensional Hilbert space, the maximum of $\|\hat{A}\psi\|^2$ is achieved for the maximum eigenvalue $|a|^2$ of $\hat{A}^\dagger \hat{A}$. It follows that, $\|\hat{A}\| = \max_a |a|$.

Normal operators are the most important in quantum theory. In what follows, we will consider special cases of normal operators of particular importance in quantum theory.

4.2.2 Self-Adjoint Operators

Definition 4.5 Any operator $\hat{A}$ that satisfies $\hat{A} = \hat{A}^\dagger$ is called *self-adjoint*.

Self-adjoint operators are obviously normal, and they have two properties that are crucial for quantum theory.

Proposition 4.6 *The eigenvalues of a self-adjoint operator $\hat{A}$ are real numbers.*

Proof Let $\hat{A}\psi = a\psi$. For a self-adjoint operator $\hat{A}^\dagger\psi = a\psi$. Since $\hat{A}$ is normal, $\hat{A}^\dagger\psi = a^*\psi$. Hence, $a = a^*$. $\qquad\square$

Proposition 4.7 *Vectors from different eigenspaces of $\hat{A}$ are normal to each other: If $\hat{A}\psi = a\psi$, $\hat{A}\phi = a'\phi$, and $a \neq a'$, then $(\phi, \psi) = 0$.*

Proof The hypothesis implies that $(\hat{A}\psi, \phi) = a(\psi, \phi)$ and $(\hat{A}\phi, \psi) = a'(\phi, \psi)$. The latter equation can be written as $(\phi, \hat{A}\psi) = a'(\phi, \psi)$. Since a' is real, $(a - a')(\phi, \psi) = 0$. By hypothesis, $a \neq a'$; hence, $(\phi, \psi) = 0$. $\qquad\square$

We can choose an orthonormal basis in each eigenspace V_a of the operator $\hat{A}$, consisting, say, of vectors $e_{a,1}, e_{a,2}, \ldots, e_{a,g_a}$, where g_a is the degeneracy of a. The set of all vectors $e_{a,i}$ for all a and all i defines an orthonormal set of eigenvectors of $\hat{A}$. In finite-dimensional Hilbert spaces, this orthonormal set defines a basis.[1] To see this, recall that the eigenvalues of a matrix A are obtained by solving the characteristic equation $\det(\hat{A} - a\hat{1}) = 0$ for a. In $\mathbb{C}^n$, this equation is a polynomial of nth degree. Therefore, it has n solutions. The corresponding orthonormal set consists of n vectors, hence, a basis on $\mathbb{C}^n$. See Box 4.1 for complications in defining self-adjoint operators.

Box 4.1 Complications in Defining Self-Adjoint Operators

In Section 4.1, we defined the adjoint $\hat{A}^\dagger$ of an operator $\hat{A}$ by requiring that $(\hat{A}\phi, \psi) = (\phi, \hat{A}^\dagger\psi)$ for all $\phi, \psi \in \mathcal{H}$. This definition is fine if $\hat{A}$ is bounded. But for a nonbounded operator, ϕ belongs in the domain $D_{\hat{A}}$ of $\hat{A}$, while ψ belongs in the domain $D_{\hat{A}^\dagger}$ of $\hat{A}^\dagger$. The two domains do not necessarily coincide. Even if $D_{\hat{A}}$ is a dense subset of $\mathcal{H}$, there is no guarantee that $D_{\hat{A}^\dagger}$ is also a dense subset, or vice versa.

---

[1] This is not always true in infinite-dimensional Hilbert spaces.

The definition of a self-adjoint operator, $\hat{A} = \hat{A}^\dagger$ includes the equality of the associated domains, namely, that $D_{\hat{A}} = D_{\hat{A}^\dagger}$. This is a highly nontrivial assumption. To see this, we consider the operator $\hat{p} = -i\frac{\partial}{\partial x}$, defined on the Hilbert space $\mathcal{H} = L^2([0,a])$ of square integrable functions on $[0,a]$. The domain of $\hat{p}$ is dense in $\mathcal{H}$. However, the domain of $\hat{p}^\dagger$, also written as $-i\frac{\partial}{\partial x}$, is not. To see this, let us evaluate $\Delta := (\hat{p}\phi, \psi) - (\phi, \hat{p}^\dagger \psi)$. By definition,

$$\Delta = -i \int_0^a dx(\psi^* \phi' + \psi^{*\prime}\phi) = -i \int_0^a (\psi^* \phi)' = -i[\psi^*(a)\phi(a) - \psi^*(0)\phi(0)].$$

For generic $\phi(0)$ and $\phi(a)$, $\Delta = 0$ only if $\psi(0) = \psi(a) = 0$. But then, the domain of $\hat{p}^\dagger$ (where ψ belongs) is a small subset of the domain of $\hat{p}$: $D_{\hat{p}^\dagger} \subset D_{\hat{p}}$. On the other hand, substituting $\hat{p}^\dagger$ for $\hat{p}$ in the definition of Δ, we find no restriction for $\psi(0)$ and $\psi(a)$. Hence, $D_{\hat{p}^\dagger} = D_{\hat{p}^{\dagger\dagger}}$.

An operator $\hat{A}$ is *symmetric* if $(\hat{A}\phi, \psi) = (\phi, \hat{A}^\dagger \psi)$, $D_{\hat{A}}$ is dense, and $D_{\hat{A}} \subseteq D_{\hat{A}^\dagger}$. Obviously, $\hat{p}^\dagger$ is symmetric, while $\hat{p}$ is not symmetric.

A self-adjoint operator is always symmetric, but the converse does not hold. We can extend the domain of a symmetric operator, in order to define a self-adjoint operator, but such extensions may not be unique.

For example, consider the operators $\hat{p}_\theta = -i\frac{\partial}{\partial x}$ with domain all square integrable functions ϕ on $[0,a]$ such that $\phi(a) = e^{i\theta}\phi(0)$ for some phase θ. For such functions, $\Delta = 0$ if $\psi(a) = e^{i\theta}\psi(0)$. Hence, $D_{\hat{p}_\theta} = D_{\hat{p}_\theta^\dagger}$. We conclude that $\hat{p}_\theta$ is self-adjoint on the Hilbert space $\mathcal{H}_\theta$ of square integrable functions that satisfy $\phi(a) = e^{i\theta}\phi(0)$. The key point is the two operators $\hat{p}_\theta$ and $\hat{p}_{\theta'}$ with $\theta \neq \theta'$ have domains with no common element. In quantum theory, different extensions of a symmetric operator correspond to different Hilbert spaces and, hence, to different physical systems.

Example 4.2 The operator $\hat{p} = -i\frac{\partial}{\partial x}$ on the Hilbert space $L^2(\mathbb{R})$ is self-adjoint:

$$(\hat{p}\phi, \psi) = \int_{-\infty}^{\infty} dx\,\psi^*(x)i\phi'(x) = \psi^*(x)\phi(x)|_{-\infty}^{\infty} - i\int_{-\infty}^{\infty} dx(\psi^*)'(x)\phi(x)$$

$$= \int_{-\infty}^{\infty} dx(i\psi^*)'(x)\phi(x) = (\phi, \hat{p}\psi).$$

In the derivation, we used the fact that the square integrable functions ϕ, ψ vanish at $\pm\infty$.

Example 4.3 The Legendre operator $\hat{\Lambda}$ on the Hilbert space $L^2([-1,1])$ is defined as

$$\hat{\Lambda}\psi(x) := -\frac{d}{dx}\left[(1-x^2)\frac{d\psi(x)}{dx}\right]. \tag{4.12}$$

We examine whether $\hat{\Lambda}$ accepts polynomial eigenfunctions $\psi_\ell(x) = \sum_{n=0}^\ell c_n x^n$ of degree ℓ. Substituting into the eigenvalue equation $\hat{\Lambda}\psi = \lambda\psi$, we obtain

$$\sum_{n=0}^\ell \left[[(n(n+1) - \lambda]c_n - (n+1)(n+2)c_{n+2} \right] x^n = 0. \tag{4.13}$$

Equation (4.13) provides a recursive relation for c_{n+2} given c_n. For a polynomial solution, the coefficient of c_ℓ must vanish, namely, $\lambda = \ell(\ell+1)$. Hence, $\hat{\Lambda}$ accepts polynomials of order ℓ as eigenvectors, for all $\ell = 0, 1, 2, \ldots$. Given that $\hat{\Lambda}$ is self-adjoint, its eigenvectors define an orthonormal set. But the only such set in $L^2([-1, 1])$ consists of Legendre polynomials – obviously, the name of the operator is not a coincidence. We conclude that

$$\hat{\Lambda}P_\ell(x) = \ell(\ell+1)P_\ell(x). \tag{4.14}$$

Given that the Legendre polynomials define an orthonormal basis, no other square integrable solutions to the eigenvalue equation exist.

4.2.3 Unitary Operators

Definition 4.8 Any operator $\hat{U}$ that satisfies $\hat{U}^\dagger = \hat{U}^{-1}$ is called *unitary*.

Any unitary operator $\hat{U}$ is normal, since $\hat{U}^\dagger \hat{U} = \hat{U}\hat{U}^\dagger = \hat{I}$. It is straightforward to show that $\|\hat{U}\| = 1$.

Proposition 4.9 *All eigenvalues λ of a unitary operator $\hat{U}$ lie on the unit circle on the imaginary plane:* $|\lambda| = 1$.

Proof The eigenvalue equation $\hat{U}\psi = \lambda\psi$ implies that $\lambda^{-1}\psi = \hat{U}^{-1}\psi$. Since $\hat{U}^\dagger = \hat{U}^{-1}$, λ^{-1} is an eigenvalue of $\hat{U}^\dagger$ with eigenvector ψ. Since $\hat{U}$ is normal, the eigenvalue of $\hat{U}^\dagger$ for ψ is λ^*. It follows that the eigenvalues of $\hat{U}$ satisfy $\lambda^* = \lambda^{-1}$, and hence, $|\lambda| = 1$. $\qquad\square$

The following theorem presents a crucial property of unitary operators: They implement changes of basis on the Hilbert space.

Theorem 4.10 *For any two orthonormal bases $\{e_i\}$ and $\{e_i'\}$, there is a unique unitary operator $\hat{U}$ such that $e_i' = \hat{U}e_i$, for all i.*

Proof We define $\hat{U}\psi := \sum(\psi, e_i)e_i'$, for any vector ψ. $\hat{U}$ is unitary, because for any pair of vectors ϕ and ψ,

$$(\hat{U}^\dagger \hat{U}\psi, \phi) = (\hat{U}\psi, \hat{U}\phi) = \sum_{i,j}(\psi, e_i)(\phi, e_j)^*(e_i', e_j') = \sum_i (\psi, e_i)(e_j, \phi) = (\psi, \phi).$$

Hence, $\hat{U}^\dagger \hat{U} = \hat{I}$. The proof that $\hat{U}\hat{U}^\dagger = \hat{I}$ proceeds similarly. $\qquad\square$

Unitary operators define transformations that preserve an operator's eigenvalues. Consider an operator $\hat{A}$ and its eigenvalue equation $\hat{A}\psi = a\psi$. We define $\psi' = \hat{U}^\dagger\psi$. Then, $\hat{U}^\dagger\hat{A}\hat{U}\psi' = \hat{U}^\dagger\hat{A}\hat{U}\hat{U}^\dagger\psi = \hat{U}^\dagger\hat{A}\psi = a\hat{U}^\dagger\psi = a\psi'$, that is, ψ' is an eigenvector of $\hat{U}^\dagger\hat{A}\hat{U}$ with eigenvalue a. Hence, the operators $\hat{U}^\dagger\hat{A}\hat{U}$ and $\hat{A}$ have the same eigenvalues.

The product of two unitary operators is a unitary operator. If $\hat{U}_1$ and $\hat{U}_2$ are unitary, then $\hat{U}_1\hat{U}_2(\hat{U}_1\hat{U}_2)^\dagger = \hat{U}_1\hat{U}_2\hat{U}_2^\dagger\hat{U}_1^\dagger = \hat{U}_1\hat{U}_1^\dagger = \hat{I}$. Hence, the set $U(\mathcal{H})$ of all unitary operators on a

Hilbert space $\mathcal{H}$ is closed with respect to multiplication. It defines a group, known as the *unitary group* of $\mathcal{H}$.

4.2.4 Positive Operators

Definition 4.11 Any operator $\hat{A}$ that satisfies $(\hat{A}\psi, \psi) \geq 0$ for all $\psi \in \mathcal{H}$ is called *positive*, and we write $\hat{A} \geq 0$.

Positive operators are self-adjoint – see Problem 4.3 – with non negative eigenvalues. If $\hat{A}\psi = a\psi$ for $a < 0$, then, $(\hat{A}\psi, \psi) = a(\psi, \psi) < 0$.

If $\lambda > 0$ and $\hat{A} \geq 0, \hat{B} \geq 0$, then $\lambda \hat{A} + \hat{B} \geq 0$. Any operator of the form $\hat{A}^\dagger \hat{A}$ for some operator $\hat{A}$ is positive, since $(\hat{A}^\dagger \hat{A}\phi, \phi) = (\hat{A}\phi, \hat{A}\phi) = \|\hat{A}\phi\|^2 \geq 0$ for all ϕ.

We also define the notion of *inequality* between two self-adjoint operators $\hat{A}$ and $\hat{B}$: $\hat{A} \geq \hat{B}$, if $\hat{A} - \hat{B} \geq 0$.

4.2.5 Projection Operators

Definition 4.12 Any self-adjoint operator $\hat{P}$ that satisfies $\hat{P}^2 = \hat{P}$ is called a *projection operator*, or simply a *projector*.

A projector's eigenvalues are either 0 or 1. If $\hat{P}\psi = a\psi$, then $\hat{P}^2\psi = a^2\psi$. For projectors $\hat{P}^2 = \hat{P}$, hence $a^2 = a$. It follows that either $a = 0$ or $a = 1$. All projectors are positive operators.

For any normalized vector $|\psi\rangle$ on $\mathcal{H}$, the operator $\hat{P}_\psi$ defined by $\hat{P}_\psi \phi = (\phi, \psi)\psi$ for all $\phi \in \mathcal{H}$ is a projector. In Dirac notation, we write $\hat{P}_\psi = |\psi\rangle\langle\psi|$. We call operators of the form $\hat{P}_\psi$ one-dimensional projectors or *rays*.

We extend this definition to any orthonormal set $S = \{e_1, \ldots, e_n\}$ of n vectors. We define $\hat{P}_S \phi := \sum_{i=1}^{n}(\phi, e_i)e_i$ for all $\phi \in \mathcal{H}$. It is easy to show that $\hat{P}_S$ is a projector. In Dirac notation, we write $\hat{P}_S = \sum_{i=1}^{n}|i\rangle\langle i|$.

More generally, we can define a projector $\hat{P}_V$ for every subspace V of $\mathcal{H}$. We define $\hat{P}_V \phi := \phi_V$, where ϕ_V is the projection of $\phi \in \mathcal{H}$ into V, given by Eq. (3.19). The complement $V^\perp$ corresponds to the projector $\hat{P}_{V\perp} := \hat{I} - \hat{P}_V$. Conversely, we can associate each projector $\hat{P}$ with the eigenspace V_1 of its unit eigenvalue. We conclude that there is a unique correspondence between projectors and subspaces in every Hilbert space (see Fig. 4.1).

For any operator $\hat{A}$, $\hat{P}_V \hat{A} \hat{P}_V$ is an operator on the subspace V, the projection of $\hat{A}$ onto V.

Example 4.4 Consider an infinite-dimensional Hilbert space $\mathcal{H}$ with an orthonormal basis e_n, for $n = 0, 1, 2, 3, \ldots$. We define the operator $\hat{V}$ by $\hat{V}e_n = e_{n+1}$ for all n. Its adjoint satisfies $(\hat{V}^\dagger e_n, e_m) = (e_n, \hat{V}e_m) = (e_n, e_{m+1}) = \delta_{n,m+1} = \delta_{n-1,m}$. This means that for $n = 0$, $\hat{V}^\dagger e_0 = 0$, since $\delta_{0,m+1}$ vanishes for all m. For $n > 0$, $\hat{V}^\dagger e_n = e_{n-1}$.

It follows that $\hat{V}^\dagger \hat{V} e_n = e_n$, for all n; hence, $\hat{V}^\dagger \hat{V} = \hat{I}$. Still, $\hat{V}$ is not unitary, because $\hat{V}\hat{V}^\dagger e_n = e_n$ only for $n > 0$. For $n = 0$, $\hat{V}\hat{V}^\dagger e_0 = 0$. Hence, $\hat{V}\hat{V}^\dagger = \hat{I} - \hat{P}_0$, where $\hat{P}_0$ is the projector into the vector e_0. Since $\hat{V}\hat{V}^\dagger - \hat{V}^\dagger\hat{V} = -\hat{P}_0$, $\hat{V}$ is not a normal operator.

Operators that satisfy the condition $\hat{V}^\dagger \hat{V} = \hat{I}$, but not necessarily $\hat{V}\hat{V}^\dagger = \hat{I}$, are known as isometric operators.

4.3   Functions of Operators

In Section 4.1, we showed how to define polynomial functions of an operator. Many functions are well approximated by polynomials, so we can extend the definition of functions of operators by taking limits.

All analytic functions on $\mathbb{R}$ can be expressed as limits of polynomials, using their Taylor expansion. For example, the exponential function $f(x) = e^x$ is the limit of polynomials $h_n(x) = \sum_{k=1}^{n} x^k/k!$, as $n \to \infty$. It can be proven that we can extend the definition of functions of operators $f(\hat{A})$ to all analytic functions $f : \mathbb{R} \to \mathbb{C}$ and self-adjoint operators $\hat{A}$.

Furthermore, we can also consider limits of analytic functions. For example, the sequence of functions $f_n(x) = \frac{1}{2}(1 + \tanh(nx))$ converges as $n \to \infty$ to the step function

$$\theta(x) = \begin{cases} 0 & x < 0 \\ 1 & x > 0 \end{cases}. \tag{4.15}$$

Similarly, the sequence of functions

$$g_n(x) = f_n(x - a)f_n(b - x) \tag{4.16}$$

converges to the characteristic function $\chi_{[a,b]}(x)$ of the interval $[a, b] \subset \mathbb{R}$.

Hence, by taking appropriate limits we can define $f(\hat{A})$ for functions f that are not analytic, and they have points of discontinuity. The most general functions for which it is possible to define $f(\hat{A})$ for normal operators $\hat{A}$ are called *measurable*. In this book, any function $f : \mathbb{R} \to \mathbb{C}$ that we encounter is assumed to be measurable. Then, the eigenvalue equation $\hat{A}\psi = a\psi$ implies that $f(\hat{A})\psi = f(a)\psi$. If f is real-valued and $\hat{A}$ is self-adjoint, then $f(\hat{A})$ is self-adjoint. In general, any algebraic property of a function $f(x)$, is also satisfied by the operators $f(\hat{A})$.

One of the most useful functions of operators is the *exponential*: $e^{\hat{A}} = \sum_{n=1}^{\infty} \frac{1}{n!}\hat{A}^n$. Since $e^{ix}e^{-ix} = 1$ for all real x, $e^{i\hat{A}}e^{-i\hat{A}} = \hat{I}$ for all self-adjoint operators $\hat{A}$. Because of self-adjointness, $(e^{i\hat{A}})^\dagger = e^{-i\hat{A}}$, hence, $e^{i\hat{A}}(e^{i\hat{A}})^\dagger = \hat{I}$. The operator $e^{i\hat{A}}$ is unitary if $\hat{A}$ is self-adjoint. Of particular relevance is the following theorem (Stone, 1930, 1932; von Neumann, 1932b).

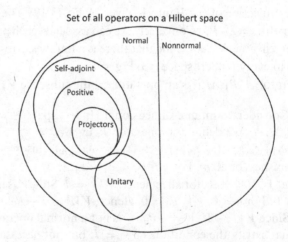

Fig. 4.1 The set of all operators on a Hilbert space and some of its subsets.

Theorem 4.13 (Stone's theorem) *If a family of unitary operators $\hat{U}_s$ exists, such that (i) $\hat{U}_s\hat{U}_{s'} = \hat{U}_{s+s'}$ for all $s, s' \in \mathbb{R}$, and (ii) all matrix elements of $\hat{U}_s$ are continuous with respect to s, then $\hat{U}_s = e^{is\hat{A}}$ for some self-adjoint operator $\hat{A}$. The operator $\hat{A}$ is called the* generator *of the family $\hat{U}_s$.*

Stone's theorem is of great importance when discussing symmetries in quantum theory (Chapter 12). Furthermore, it provides a widely used method for *rigorously* defining self-adjoint operators. The construction of the family $\hat{U}_s$ is often easier than the rigorous definition of an unbounded self-adjoint operator, as the latter may be plagued by the problems described in Box 4.1.

Of importance are also the characteristic functions χ_C of subsets C of $\mathbb{R}$, defined by Eq. (1.42). Since χ_C takes only values 0 and 1, $\chi_C^2 = \chi_C$. Hence, for any self-adjoint $\hat{A}$, $\chi_C(\hat{A})$ is self-adjoint, and $\chi_C(\hat{A}) = \chi_C(\hat{A})^2$. The operators $\chi_C(\hat{A})$ are projectors.

Furthermore, if $C_1, C_2 \subset \mathbb{R}$, such that $C_1 \cap C_2 = \emptyset$, then $\chi_{C_1}(x)\chi_{C_2}(x) = 0$ and $\chi_{C_1 \cup C_2}(x) = \chi_{C_1}(x) + \chi_{C_2}(x)$. It follows that $\chi_{C_1}(\hat{A})\chi_{C_2}(\hat{A}) = 0$ and $\chi_{C_1 \cup C_2}(\hat{A}) = \chi_{C_1}(\hat{A}) + \chi_{C_2}(\hat{A})$. Hence, for any partition of $\mathbb{R}$ into subsets C_λ that are mutually exclusive ($C_\lambda \cap C_{\lambda'} = \emptyset$, if $\lambda \neq \lambda'$) and exhaustive ($\cup_\lambda C_\lambda = \mathbb{R}$), the associated projectors $\hat{P}_\lambda := \chi_{C_\lambda}(\hat{A})$ satisfy

- the mutual exclusion property: $\hat{P}_\lambda \hat{P}_{\lambda'} = \delta_{\lambda\lambda'}\hat{P}_\lambda$, and
- the exhaustion property: $\sum_\lambda \hat{P}_\lambda = \hat{I}$.

Other useful operator functions are the nth root of a positive operator $\hat{A}$: $\sqrt[n]{\hat{A}}$, and the *absolute value* of an operator $|\hat{A}| := \sqrt{\hat{A}^\dagger \hat{A}}$.

Finally, we mention that the geometric series identity $(1 - x)^{-1} = 1 + x + x^2 + x^3 + \cdots$ for $|x| < 1$, has an operator analogue for operators $\hat{A}$ with $||\hat{A}|| < 1$,

$$(\hat{I} - \hat{A})^{-1} = 1 + \hat{A} + \hat{A}^2 + \hat{A}^3 + \cdots. \tag{4.17}$$

4.4 Commutators

Next, we present the crucial notion of an operator commutator, and we analyze its properties. We pay particular emphasis to the fundamental commutator between position and momentum that was postulated by Heisenberg.

4.4.1 Properties

Definition 4.14 The commutator of two operators $\hat{A}$ and $\hat{B}$ is the operator $[\hat{A}, \hat{B}] := \hat{A}\hat{B} - \hat{B}\hat{A}$.

If $\hat{A}$ and $\hat{B}$ are self-adjoint, then the operator $\hat{C} = i[\hat{A}, \hat{B}]$ is also self-adjoint. Indeed, $\hat{C}^\dagger = -i(\hat{A}\hat{B} - \hat{B}\hat{A})^\dagger = -i(\hat{B}\hat{A} - \hat{A}\hat{B}) = -i[\hat{B}, \hat{A}] = \hat{C}$.

The following properties of commutators are straightforward consequences of the definition.

1. $[\hat{A}, \hat{I}] = 0$.
2. $[\hat{A}, \hat{B}] = -[\hat{B}, \hat{A}]$.
3. $[\hat{A}, \hat{B} + \hat{C}] = [\hat{A}, \hat{B}] + [\hat{A}, \hat{C}]$.

4. $[\hat{A}, \hat{B}\hat{C}] = [\hat{A}, \hat{B}]\hat{C} + \hat{B}[\hat{A}, \hat{C}]$.

5. $[\hat{A}, [\hat{B}, \hat{C}]] + [\hat{C}, [\hat{A}, \hat{B}]] + [\hat{B}, [\hat{C}, \hat{A}]] = 0$. (Jacobi identity).

If $[\hat{A}, \hat{B}] = 0$, we say that the operators $\hat{A}$ and $\hat{B}$ *commute*.

Identity 4 implies that if $[\hat{A}, \hat{B}] = 0$, then $[\hat{A}^2, \hat{B}] = \hat{A}[\hat{A}, \hat{B}] + [\hat{A}, \hat{B}]\hat{A} = 0$. By induction we find that $[\hat{A}^n, \hat{B}] = 0$, for all integers n. Hence, for all functions f that can be obtained by successive limits of polynomials, $[f(\hat{A}), \hat{B}] = 0$.

Consider two self-adjoint operators $\hat{A}$ and $\hat{B}$, such that $[\hat{A}, \hat{B}] = 0$. If $\hat{A}\phi = a\phi$, then $\hat{A}\hat{B}\phi = \hat{B}\hat{A}\phi = a\hat{B}\phi$. This means that $\hat{B}\phi$ is an eigenvector of $\hat{A}$. If the eigenvalue a is nondegenerate, then $\hat{B}\phi = b\phi$ for some b. If a is degenerate, we can always find an eigenvector ϕ of $\hat{A}$ such that $\hat{B}\phi = b\phi$. The proof is the following.

Let V_a be the eigenspace associated to a, and let e_i be an orthonormal basis on V_a. We construct the matrix $B_{ij} = (\hat{B}e_i, e_j)$. Consider a eigenvector c of B_{ij} with components c_i. The latter satisfy $\sum_i B_{ij}c_j = bc_j$ for some b. We define $\phi := \sum_i c_i e_i$. Then,

$$\hat{B}\phi = \sum_j (\hat{B}\phi, e_j)e_j = \sum_{i,j} c_i(\hat{B}e_i, e_j)e_j = \sum_{i,j} c_i B_{ji}e_j = b\sum_j c_j e_j = b\phi. \qquad (4.18)$$

We conclude that there are vectors $\phi_{a,b}$ that satisfy $\hat{A}\phi_{a,b} = a\phi_{a,b}$ and $\hat{B}\phi_{a,b} = b\phi_{a,b}$ for all eigenvalues a of $\hat{A}$ and b of $\hat{B}$. Hence, for any pair (a,b), there is a subspace $\hat{V}_{a,b}$, the *common eigenspace* of $\hat{A}$ and $\hat{B}$.

4.4.2 Key Identities

The following two theorems turn out to be very helpful in the analysis of operators.

Theorem 4.15 (Baker–Campbell–Hausdorff (BCH) identity) *For any operators $\hat{A}$ and $\hat{B}$,*

$$e^{\hat{A}}e^{\hat{B}} = \exp\left[\hat{A} + \hat{B} + \frac{1}{2}[\hat{A}, \hat{B}] + \frac{1}{12}[\hat{A}, [\hat{A}, \hat{B}]] - \frac{1}{12}[\hat{B}, [\hat{A}, \hat{B}]] + \cdots\right],$$

where higher-order terms involve nested commutators with $\hat{A}$ and $\hat{B}$.

If $\|\hat{A}\|$ and $\|\hat{B}\|$ are small, we can derive the first few terms in the BCH expansion by using the Taylor expansion of the exponential and the logarithmic function. However, this procedure does not guarantee that *all* higher-order terms can be expressed in terms of commutators. For a general proof of the BCH identity, see chapter 3 of Hall (2004).

Note that if $[\hat{B}, [\hat{A}, \hat{B}]] = [\hat{A}, [\hat{A}, \hat{B}]] = 0$, the series for $\hat{C}$ terminates after the third term, hence,

$$e^{\hat{A}}e^{\hat{B}} = e^{\hat{A}+\hat{B}+\frac{1}{2}[\hat{A}, \hat{B}]}. \qquad (4.19)$$

Theorem 4.16 (Hadamard's identity)

$$e^{\hat{A}}\hat{B}e^{-\hat{A}} = \hat{B} + [\hat{A}, \hat{B}] + \frac{1}{2!}[\hat{A}, [\hat{A}, \hat{B}]] + \frac{1}{3!}[\hat{A}, [\hat{A}, [\hat{A}, \hat{B}]]] + \cdots.$$

Proof We consider the family of operators $\hat{C}(s) = e^{s\hat{A}}\hat{B}e^{-s\hat{A}}$ and Taylor expand around $s = 0$. At the end, we set $s = 1$. The first term in the Taylor expansion is $\hat{C}(0) = \hat{B}$. We calculate

$\frac{d}{ds}\hat{C}(s) = \hat{A}e^{s\hat{A}}\hat{B}e^{-s\hat{A}} - e^{s\hat{A}}\hat{B}e^{-s\hat{A}}\hat{A} = \hat{A}\hat{C}(s) - \hat{C}(s)\hat{A}$, hence, the second term in the Taylor expansion is $\hat{C}'(0) = [\hat{A}, \hat{C}(0)] = [\hat{A}, \hat{B}]$. Similarly, for the third term, we find $\hat{C}''(0) = [\hat{A}, \hat{C}'(0)]$, and, in general, $\hat{C}^{(n)}(0) = [\hat{A}, \hat{C}^{(n-1)}(0)]$. The identity follows. $\square$

4.4.3 Heisenberg's Commutator

Of particular importance is the case of the commutator between position $\hat{x}$ and momentum $\hat{p}$, given by Heisenberg's relation (2.28), $[\hat{x}, \hat{p}] = i\hat{I}$. For this commutator,

$$[\hat{x}^n, \hat{p}] = in\hat{x}^{n-1}. \tag{4.20}$$

The term on the right-hand side is the derivative of x^n. Hence, for any function $f(x)$, defined through limits of polynomials,

$$[f(\hat{x}), \hat{p}] = if'(\hat{x}). \tag{4.21}$$

Similarly, we obtain $[\hat{x}, f(\hat{p})] = if'(\hat{p})$.

The unitary operators

$$\hat{V}(a, b) = \exp\left(ia\hat{x} - ib\hat{p}\right), \tag{4.22}$$

for real a and b, are called *Weyl operators*.

The product of two Weyl operators $\hat{V}(a_1, b_1)$ and $\hat{V}(a_2, b_2)$ is computed with the BCH identity. We calculate $[ia_1\hat{x} - ib_1\hat{p}, ia_2\hat{x} - ib_2\hat{p}] = i(a_1b_2 - a_2b_1)\hat{I}$. Since the first commutator in the expansion is proportional to unity, terms with more than two commutators in the BCH identity vanish. Hence, Eq. (4.19) applies. We obtain

$$\hat{V}(a_1, b_1)\hat{V}(a_2, b_2) = \hat{V}(a_1 + a_2, b_1 + b_2)e^{\frac{i}{2}(a_1b_2 - a_2b_1)}. \tag{4.23}$$

For the physical interpretation of the Weyl operators, we use the Hadamard identity, to obtain

$$\hat{V}^\dagger(a, b)\hat{x}\hat{V}(a, b) = \hat{x} + b\hat{I}, \qquad \hat{V}^\dagger(a, b)\hat{p}\hat{V}(a, b) = \hat{p} + a\hat{I}. \tag{4.24}$$

We see that $\hat{V}(a, b)$ generates a transformation that translates position by a constant amount b and momentum by a constant amount a. We will revisit the Weyl operators in a later chapter.

This feature of the Weyl operator is confirmed by its action on wave functions. Eq. (4.23) for $a_1 = a$, $a_2 = 0$, $b_1 = 0$, and $b_2 = b$ gives $\hat{V}(a, b) = e^{-\frac{i}{2}ab}\hat{V}(a, 0)\hat{V}(0, b)$, or,

$$e^{ia\hat{x} - ib\hat{p}} = e^{-\frac{i}{2}ab}e^{ia\hat{x}}e^{-ib\hat{p}}. \tag{4.25}$$

The action of $e^{-ib\hat{p}}$ on a wave function ψ is

$$e^{-ib\hat{p}}\psi(x) = \sum_{n=0}^{\infty}\frac{(-ib)^n}{n!}\hat{p}^n\psi(x) = \sum_{n=0}^{\infty}\frac{(-b)^n}{n!}\frac{d^n\psi(x)}{dx^n} = \psi(x - b), \tag{4.26}$$

where in the last term we used the formula for the Taylor series. By Eq. (4.25),

$$\hat{V}(a, b)\psi(x) = e^{-\frac{i}{2}ab + iax}\psi(x - b). \tag{4.27}$$

4.5   The Spectral Theorem in Finite Dimensions

From now on, we will be using Dirac notation. With the spectral theorem, we switch to Dirac notation. This notation would have been unwieldy for the proofs in the previous sections, but it is very helpful for describing the spectrum of operators.

The physical significance of the spectral theorem originates from Heisenberg's thesis that physical magnitudes correspond to operators. However, any measurement of a physical quantity ends up in the determination of a single number, its value. The only natural way of obtaining such numbers from the structure of operators is through their eigenvalues or, more generally, their spectrum.

We first examine the spectral theorem in finite-dimensional Hilbert spaces. In this case, the main results are known from elementary matrix theorem, and we just have to phrase them in a way appropriate for quantum theory.

4.5.1   Proof of the Spectral Theorem

Let $\hat{A}$ be a self-adjoint operator on $\mathbb{C}^N$. We first examine the case of no degeneracy, that is, we assume that $\hat{A}$ has N distinct eigenvalues a_n, for $n = 1, 2, \ldots, N$. Each eigenvalue corresponds to a normalized eigenvector$|n\rangle$, unique up to a phase change $|n\rangle \to e^{i\theta}|n\rangle$.

The set of the N eigenvectors $|n\rangle$ defines an orthonormal basis on $\mathbb{C}^N$, hence, we have the resolution of the unity $\hat{I} = \sum_{n=1}^{N} |n\rangle\langle n|$. Furthermore, for any vector $|\psi\rangle$,

$$\hat{A}|\psi\rangle = \hat{A}\sum_{n=1}^{N}\langle n|\psi\rangle|n\rangle = \sum_{n=1}^{N}\langle n|\psi\rangle\hat{A}|n\rangle = \sum_{n=1}^{N}a_n|n\rangle\langle n|\psi\rangle. \tag{4.28}$$

We conclude that

$$\hat{A} = \sum_{n=1}^{N} a_n|n\rangle\langle n|. \tag{4.29}$$

Consider now the most general case, where the self-adjoint operator $\hat{A}$ on $\mathbb{C}^N$ has degenerate eigenvalues. Then, there are K distinct eigenvalues, with $K < N$. Each eigenvalue a_n, $n = 1, 2, \ldots, K$ corresponds to an eigenspace V_n, of dimension D_n, such that $\sum_{n=1}^{K} D_n = N$.

We define an orthonormal basis in each eigenspace V_n. We denote the basis vectors by $|n, i_n\rangle$, $i_n = 1, 2, \ldots, D_n$. By definition, $\langle n, i_n|n, j_n\rangle = \delta_{i_n j_n}$. The projector associated to V_n is

$$\hat{P}_n = \sum_{i_n} |n, i_n\rangle\langle n, i_n|. \tag{4.30}$$

We will call the projectors $\hat{P}_n$ *spectral projectors* of $\hat{A}$.

All vectors $|n, i_n\rangle$ are eigenvectors of $\hat{A}$. Two vectors $|n, i_n\rangle$ and $|m, i_m\rangle$ for $n \neq m$ belong in different eigenspaces, hence, they are orthogonal $\langle n, i_n|m, j_m\rangle = 0$. This means that $\hat{P}_n\hat{P}_n = 0$ for $n \neq m$. Since $\hat{P}_n^2 = \hat{P}_n$, we obtain the mutual exclusion property, $\hat{P}_n\hat{P}_m = \hat{P}_n\delta_{nm}$.

The set of all vectors $|n, i_n\rangle$ defines an orthonormal basis on $\mathbb{C}^N$, since it contains $\sum_{n=1}^{K} D_n = N$ normalized and mutually orthogonal vectors. The resolution of the unity

$$\sum_{n=1}^{K}\sum_{i_n=1}^{D_n} |n, i_n\rangle\langle n, i_n| = \hat{I},$$ (4.31)

implies the exhaustion property, $\sum_{n=1}^{K} \hat{P}_n = \hat{I}$.

Acting $\hat{A}$ on a vector $|\psi\rangle$, and using Eq. (4.31), we obtain

$$\hat{A} = \sum_{n=1}^{K} a_n \hat{P}_n,$$ (4.32)

which generalizes Eq. (4.29).

Since $f(\hat{A})$ has eigenvalues $f(a)$ and the same eigenspaces,

$$f(\hat{A}) = \sum_{n=1}^{K} f(a_n)\hat{P}_n.$$ (4.33)

Hence, we obtained the following result.

Theorem 4.17 Spectral theorem in a finite-dimensional Hilbert space. *Every self-adjoint operator $\hat{A}$ has real eigenvalues a_n with associated spectral projectors $\hat{P}_n$, where $n = 1, 2, \ldots, K$. The spectral projectors satisfy the properties of mutual exclusion ($\hat{P}_n\hat{P}_m = \hat{P}_n\delta_{nm}$) and exhaustion $\sum_{n=1}^{K}\hat{P}_n = \hat{I}$. For any function $f : \mathbb{R} \to \mathbb{C}, f(\hat{A}) = \sum_{n=1}^{K}f(a_n)\hat{P}_n.$*

In what follows, whenever we write a self-adjoint operator $\hat{A}$ as $\hat{A} = \sum_n a_n\hat{P}_n$, a_n stands for the operator's eigenvalues and $\hat{P}_n$ stands for the associated spectral projectors.

Example 4.5 Consider the following matrix on $\mathbb{C}^3$,

$$\hat{A} = \begin{pmatrix} -2 & 0 & 0 \\ 0 & 1 & 3i \\ 0 & -3i & 1 \end{pmatrix}.$$

The characteristic equation $\det(\hat{A} - \lambda\hat{I}) = 0$ becomes $(\lambda + 2)(\lambda^2 - 2\lambda - 8) = 0$. This equation has a doubly degenerate solution $\lambda = -2$ and a simple solution $\lambda = 4$.

The normalized vector association to $\lambda = 4$ is

$$|4\rangle = \frac{1}{\sqrt{2}}\begin{pmatrix} 0 \\ i \\ 1 \end{pmatrix}.$$

The eigenvalue $\lambda = -2$ has associated eigenvectors of the form

$$\begin{pmatrix} c_1 \\ c_2 \\ ic_2 \end{pmatrix},$$

where $c_1, c_2 \in \mathbb{C}$. These vectors define a two-dimensional eigenspace V_{-2}. We choose an orthonormal basis on V_{-2} with vectors

$$|-2,a\rangle = \begin{pmatrix} 1 \\ 0 \\ 0 \end{pmatrix} \qquad |-2,b\rangle = \frac{1}{\sqrt{2}} \begin{pmatrix} 0 \\ 1 \\ i \end{pmatrix}.$$

The spectral projector associated to the eigenvalue $\lambda = 4$ is

$$\hat{P}_4 = |4\rangle\langle 4| = \frac{1}{2} \begin{pmatrix} 0 \\ i \\ 1 \end{pmatrix} \begin{pmatrix} 0 & -i & 1 \end{pmatrix} = \frac{1}{2} \begin{pmatrix} 0 & 0 & 0 \\ 0 & 1 & i \\ 0 & -i & 1 \end{pmatrix}.$$

The spectral projector associated to the eigenvalue $\lambda = -2$ is

$$\hat{P}_{-2} = |-2,a\rangle\langle -2,a| + |-2,b\rangle\langle -2,b|$$

$$= \begin{pmatrix} 1 \\ 0 \\ 0 \end{pmatrix} \begin{pmatrix} 1 & 0 & 0 \end{pmatrix} + \frac{1}{2} \begin{pmatrix} 0 \\ 1 \\ i \end{pmatrix} \begin{pmatrix} 0 & 1 & -i \end{pmatrix} = \begin{pmatrix} 1 & 0 & 0 \\ 0 & \frac{1}{2} & -\frac{i}{2} \\ 0 & \frac{i}{2} & \frac{1}{2} \end{pmatrix}.$$

Hence, the spectral analysis of $\hat{A}$ gives

$$\hat{A} = (-2) \begin{pmatrix} 1 & 0 & 0 \\ 0 & \frac{1}{2} & -\frac{i}{2} \\ 0 & \frac{i}{2} & \frac{1}{2} \end{pmatrix} + (4)\frac{1}{2} \begin{pmatrix} 0 & 0 & 0 \\ 0 & 1 & i \\ 0 & -i & 1 \end{pmatrix}.$$

We straightforwardly evaluate the operator $e^{i\hat{A}x}$,

$$e^{i\hat{A}x} = e^{-2ix} \begin{pmatrix} 1 & 0 & 0 \\ 0 & \frac{1}{2} & -\frac{i}{2} \\ 0 & \frac{i}{2} & \frac{1}{2} \end{pmatrix} + e^{4ix}\frac{1}{2} \begin{pmatrix} 0 & 0 & 0 \\ 0 & 1 & i \\ 0 & -i & 1 \end{pmatrix}.$$

4.5.2 Eigenvalue Ordering

An important theorem, due to Weyl, asserts that inequality between operators induces inequality between their eigenvalues.

Let $\hat{A}$ be a self-adjoint operator on $\mathbb{C}^N$. We define the *increasing order* of its eigenvalues as the sequence of N real numbers a_n, for $n = 1, 2, \ldots, N$, such that every eigenvalue of $\hat{A}$ with degeneracy D appears D successive times, and

$$a_n \leq a_m \quad \text{for all} \quad n < m.$$

Theorem 4.18 (Weyl's ordering theorem) *If $\hat{A}$ and $\hat{B}$ are self-adjoint operators on $\mathbb{C}^N$, with increasing orders of eigenvalues a_n and b_n, respectively, and $\hat{A} \geq \hat{B}$, then $a_n \geq b_n$, for all $n = 1, 2, \ldots, N$.*

Hence, the lowest eigenvalue of $\hat{A}$ is larger than or equal to the lowest eigenvalue of $\hat{B}$, the second lowest eigenvalue of $\hat{A}$ is larger than or equal to the second lowest eigenvalue of $\hat{B}$, and so on. Weyl's ordering theorem allows us to use operator inequalities in order to establish properties of an operator's eigenvalues even if the latter are not explicitly determined.

4.6 The General Spectral Theorem

It is necessary to revisit the notion of an eigenvalue and of an eigenvector for infinite-dimensional Hilbert spaces, in order for them to be compatible with our physical intuition that eigenvalues must correspond to the measured values of a physical magnitude.

Consider, for example, the momentum operator $\hat{p} = -i\partial/\partial x$ on $L^2(\mathbb{R}, dx)$. The functions $\phi_p(x) = e^{ipx}$ satisfy the eigenvalue equation $\hat{p}\phi_p = p\phi_k$, but they are not square integrable. Hence, they do not belong in the Hilbert space. In this sense, the momentum operator has no eigenvalues. This is physically absurd; it is impossible that a fundamental physical quantity does not have measured values. Dirac proposed that we can somehow write a ket $|p\rangle$ satisfying $\hat{p}|p\rangle = p|p\rangle$, so that we can talk about the *generalized eigenvalues* of momentum, and the momentum operator having *continuous spectrum*.

We want to define kets such as $|p\rangle$ so that they make mathematical sense. There are two possibilities. The first one involves a generalization of the notion of the eigenvalue. We define the *spectrum* $\sigma(\hat{A})$ of an operator $\hat{A}$ as a set that contains the eigenvalues of $\hat{A}$ (the operator's discrete spectrum), but also other elements that correspond to continuous variables (the operator's continuous spectrum) – for details, see Box 4.2. Then, kets such as $|p\rangle$ are not actual vectors, but they are a shorthand notation for expressions that involve limits of sequences. The second possibility is to enlarge the Hilbert space towards a larger vector space that accepts Dirac's kets as elements. This method is described in Box 4.3.

Here, we will employ the first method, but we will forego rigorous proofs. Instead, we will focus an important special case, the momentum operator.

Box 4.2 Operator spectrum

The *spectrum* $\sigma(\hat{A})$ of an operator $\hat{A}$ on a Hilbert space $\mathcal{H}$ consists of all complex numbers λ such that the operator $\hat{A} - \lambda\hat{I}$ has no bounded inverse.

Examples

1. Every eigenvalue λ of $\hat{A}$ is an element of $\sigma(\hat{A})$. The operator $\hat{A} - \lambda\hat{I}$ has one zero eigenvalue; hence, it has no inverse.
2. Consider the position operator $\hat{x}$ on $L^2(\mathbb{R}, dx)$. The operator $\hat{x} - \lambda\hat{I}$ acts by multiplication on wave functions $\psi(x)$. Its inverse corresponds to the multiplication of $\psi(x)$ with $(x - \lambda)^{-1}$. Writing $\lambda = \lambda_R + i\lambda_I$, we find

$$\|(\hat{x} - \lambda)^{-1}\psi\|^2 = \int dx \frac{|\psi(x)|^2}{(x - \lambda_R)^2 + \lambda_I^2} \leq \frac{1}{|\lambda_I|^2} \int dx |\psi(x)|^2 = \frac{\|\psi\|^2}{|\lambda_I|^2},$$

hence, $(\hat{x} - \lambda)^{-1}$ is bounded for $\lambda_I \neq 0$. For $\lambda_I = 0$, $\|(\hat{x} - \lambda)^{-1}\psi\|^2$ diverges for all functions ψ that do not vanish at $x = \lambda_R$. It follows that $(\hat{x} - \lambda)^{-1}$ is not bounded if $\lambda_I = 0$. By definition, $\sigma(\hat{x}) = \mathbb{R}$.

3. Consider the operator $\hat{H} = -\partial_x^2$ in $L^2(\mathbb{R}, dx)$. Let $\tilde{\psi}(p)$ be the Fourier transform of the wave function $\psi(x)$. Then, $\hat{H}\psi(x) = \int dp e^{-ipx} p^2 \tilde{\psi}(p)$, hence, $(\hat{H} - \lambda 1)^{-1}\psi(x) = \int dp e^{-ipx}(p^2 - \lambda)^{-1}\tilde{\psi}(p)$. The denominator vanishes for some value of p, if $\lambda \geq 0$. Hence, the operator $(\hat{H} - \lambda\hat{I})^{-1}$ is unbounded for $\lambda \geq 0$. It follows that $\sigma(\hat{H}) = \mathbb{R}^+$.

Types of spectrum

- The eigenvalues of the operator $\hat{A}$ define the subset $\sigma_p(\hat{A})$ of $\sigma(\hat{A})$, called the *discrete spectrum* or *point spectrum* of $\hat{A}$. If $\lambda \in \sigma_p(\hat{A})$, the vector $(\hat{A} - \lambda\hat{I})^{-1}\psi$ is not defined for any $\psi \in V_\lambda$, where V_λ is the corresponding eigenspace.

- The *continuous spectrum* $\sigma_c(\hat{A})$ of $\hat{A}$ consists of all points of $\sigma(\hat{A})$ for which the range of the operator $(\hat{A} - \lambda\hat{I})$ is dense in the Hilbert space. Equivalently, $\lambda \in \sigma_c(\hat{A})$, if the domain of $(\hat{A} - \lambda\hat{I})^{-1}$ is dense in the Hilbert space.

This means that we cannot find any subspace V of $\mathcal{H}$, such that $(\hat{A} - \lambda\hat{I})^{-1}\psi$ is not defined for all $\psi \in V$. This is the crucial difference from the discrete spectrum.

In Example 2 in this box, the vector $(x - \lambda)^{-1}\psi(x)$ for $\lambda \in \mathbb{R}$ is not defined for any wave function ψ such that $\psi(\lambda) \neq 0$. Since this condition refers to a single point, we can find a function $\psi_\epsilon(x)$ that coincides with $\psi(x)$ everywhere except for a region of size ϵ around λ, where ψ_ϵ vanishes. Then, the vector $(x - \lambda)^{-1}\psi_\epsilon(x)$ is well defined. We can approximate $(x - \lambda)^{-1}\psi(x)$ with arbitrary accuracy taking sufficiently small ϵ. Hence, the operator $(x - \lambda)^{-1}$ is densely defined on the Hilbert space, and the spectrum of $\hat{x}$ is continuous.

- The *singular spectrum* $\sigma_s(\hat{A})$ corresponds to "weird" subsets of the real numbers, such as Cantor's set or the so-called fractals. Operators with singular spectrum appear very rarely in physical applications. We encounter none in this book.

The total spectrum of an operator is the union of its three components, $\sigma(\hat{A}) = \sigma_p(\hat{A}) \cup \sigma_c(\hat{A}) \cup \sigma_s(\hat{A})$. The Hilbert space $\mathcal{H}$ also splits into three subspaces as $\mathcal{H}_p \oplus \mathcal{H}_c \oplus \mathcal{H}_s$.

Box 4.3 Rigged Hilbert Space

In Section 3.4, we saw that the dual space $\mathcal{H}^*$ of a Hilbert space $\mathcal{H}$ is the space of linear functions from $\mathcal{H}$ to $\mathbb{C}$. We also saw that Riesz's theorem establishes a bijection between a Hilbert space and its dual.

The notion of a dual space applies to generic vector spaces; however, in general, there is no bijection between a vector space and its dual. The following property is important. If Φ is a vector space that is a subset of a Hilbert space $\mathcal{H}$, then the corresponding dual spaces satisfy $\mathcal{H}^* \subset \Phi^*$. Since by Riesz's theorem $\mathcal{H}^*$ can be identified with $\mathcal{H}$,

$$\Phi \subset \mathcal{H} \subset \Phi^*.$$

This relation allows us to define generalized eigenvectors $|p\rangle$ as elements of Φ^*. For example, let $\mathcal{H} = L^2(\mathbb{R})$ and Φ the set of smooth functions on $\mathbb{R}$ that vanish at infinity faster than $|x|^{-a}$, for some $a > 0$. Then the delta function $\delta(x)$ is an element of Φ^*, because $f \in \Phi \to \int dx \delta(x) f(x) = f(0)$ is a linear map from Φ to $\mathbb{C}$. This way, all generalized eigenvectors of physical interest can be expressed as elements of Φ^*.

The triplet $(\Phi, \mathcal{H}, \Phi^*)$ that makes possible the definition of generalized eigenvectors is called a *rigged Hilbert space*.

4.6.1 The Delta Function

We study the momentum operator $\hat{p} = -i\partial_x$ on the Hilbert space $\mathcal{H} = L^2(\mathbb{R}, dx)$. Consider the family of functions

$$f_{p,\epsilon}(x) = \frac{1}{\sqrt{2\pi}} e^{ipx - \frac{1}{2}\epsilon x^2}, \tag{4.34}$$

for a fixed value p of momentum and $\epsilon > 0$. These functions are square integrable, hence, they define a ket $|p, \epsilon\rangle$. We define the ket $|p\rangle$ as

$$\langle p|\psi\rangle := \lim_{\epsilon \to 0} \langle p, \epsilon|\psi\rangle = \lim_{\epsilon \to 0} \int \frac{dx}{\sqrt{2\pi}} e^{-ipx - \frac{1}{2}\epsilon x^2} \psi(x) \tag{4.35}$$

for all $|\psi\rangle \in \mathcal{H}$. Since the limit can enter the integral, the bracket $\langle p|\psi\rangle$ coincides with the Fourier transform of ψ: $\tilde{\psi}(p) = \frac{1}{\sqrt{2\pi}} \int dx e^{-ipx} \psi(x)$. Hence, the limit (4.35) is well defined.

We also find that $\lim_{\epsilon \to 0}(-i\partial_x f_{p,\epsilon} - p f_{p,\epsilon}) = 0$. This means that

$$\lim_{\epsilon \to 0} \langle \psi|(\hat{p} - p\hat{I})|p, \epsilon\rangle = 0, \tag{4.36}$$

for all vectors $|\psi\rangle$.

We write Eq (4.36) as

$$\hat{p}|p\rangle = p|p\rangle. \tag{4.37}$$

Equation (4.37) is the generalized eigenvalue equation for the momentum operator. We emphasize that it is a shorthand for Eq. (4.36). The kets $|p\rangle$ are not actual Hilbert space vectors, but mathematical objects that are defined as limits of Eq. (4.35).

The inner product $\langle p, \epsilon|p', \epsilon\rangle$ is found equal to $\delta_\epsilon(p - p')$, where

$$\delta_\epsilon(\xi) = \sqrt{\frac{1}{4\pi\epsilon}} e^{-\frac{\xi^2}{4\epsilon}}. \tag{4.38}$$

The function $\delta_\epsilon(\xi)$ is positive and even. As $\epsilon \to 0$, $\delta_\epsilon(\xi) \to 0$ for $\xi \neq 0$ and $\delta_\epsilon(0) = \infty$ for $\xi = 0$.

Equations (A.1) and (A.2) imply that $\int d\xi \delta_\epsilon(\xi) = 1$, and that $\lim_{\epsilon \to 0} \int d\xi \xi^n \delta_\epsilon(\xi) = 0$, for all integers $n \geq 1$. It follows that for any analytic function $f(\xi) = \sum_{k=0}^\infty c_k \xi^k$,

$$\lim_{\epsilon \to 0} \int d\xi f(\xi) \delta_\epsilon(\xi) = c_0 = f(0).$$

Fig. 4.2 The function δ_ϵ of Eq. (4.38) for decreasing values of ϵ.

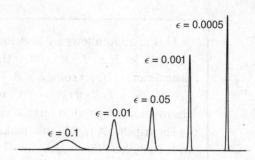

The limit of the functions $\delta_\epsilon(\xi)$ as $\epsilon \to 0$ is written as $\delta(\xi)$ and it is called *Dirac's delta function*. The limiting procedure is visualized in Fig. 4.2. The delta function is not literally a function, but a mathematical object called a *distribution* that satisfies many properties of ordinary functions.

Distributions are usually defined in terms of their action on ordinary functions. That is, we define a distribution ω as a rule that assigns a number $\omega(f)$ to each smooth square integrable function f, so that the linearity condition is satisfied, $\omega(af + g) = a\omega(f) + \omega(g)$. Since linearity implies addition or integration, typical distributions can be expressed as $\omega(f) = \int dx\rho(x)f(x)$, for some function $\rho(x)$.

The key point is that finite values for $\omega(f)$ can also be obtained through a limiting procedure. Consider a family of functions $\rho_\epsilon(x)$ indexed by a parameter $\epsilon > 0$. We assume that $\rho_\epsilon(x)$ does not converge to some function $\rho(x)$, as $\epsilon \to 0$, like, for example, the functions $\delta_\epsilon(x)$ of Eq. (4.38). However, if the limit $\lim_{\epsilon \to 0} \int dx\rho_\epsilon(x)f(x)$ is finite for all analytic functions f, then the family $\rho_\epsilon(x)$ defines a new distribution $\omega(f)$ at this limit. In an abuse of notation, we write $\omega(f) = \int dx\rho(x)f(x)$, that is, we make use of a quantity $\rho(x)$ that we treat as a function, even if it is not. In an abuse of terminology, we also refer to the nonfunction $\rho(x)$ also as a "distribution."

The preceding reasoning is reflected in the definition of the delta function that follows.

Definition 4.19 The *delta function* is a distribution $\delta(x)$, such that

$$\int dxf(x)\delta(x) = f(0) \tag{4.39}$$

for all square integrable analytic functions f.

Equation (4.39) implies that the delta function arises as a limit of functions like δ_ϵ of Eq. (4.38). Hence, we can express the inner product $\langle p,\epsilon|p',\epsilon\rangle$ at the limit $\epsilon \to 0$, as

$$\langle p|p'\rangle = \delta(p - p'). \tag{4.40}$$

Many families of functions, other than the Gaussians (4.38), can be used to define the delta function as a limit. Examples include the Lorentzian distribution at the limit of vanishing width,

$$\lim_{\gamma \to 0} \frac{\gamma}{\pi(\gamma^2 + x^2)} = \delta(x), \tag{4.41}$$

and functions with decaying oscillations at the limit of vanishing period,

$$\lim_{\epsilon \to 0} \frac{\epsilon \sin^2 \frac{x}{\epsilon}}{\pi x^2} = \delta(x). \tag{4.42}$$

Another useful representation of the delta function is

$$\delta(x) = \frac{1}{2\pi} \int_{-\infty}^{\infty} dp e^{ipx}. \tag{4.43}$$

This is simply obtained by taking the limit $\epsilon \to 0$ to the identity $\int_{-\infty}^{\infty} dp e^{ipx - \epsilon p^2} = \sqrt{\frac{\pi}{\epsilon}} e^{-\frac{x^2}{4\epsilon}} = 2\pi \delta_{\epsilon}(x)$.

The properties of the delta function do not depend on the choice of functions used in its definition, but are derived solely from the definition (4.39). For example, to prove that the delta function is even, we substitute $\delta(x)$ with $\delta(-x)$ in Eq. (4.39). We obtain $\int dx \delta(-x) f(x) = \int dy \delta(y) f(-y) = \delta(0) = \int dx \delta(x) f(x)$. Since the derived equality holds for all f, we conclude that

$$\delta(x) = \delta(-x). \tag{4.44}$$

We evaluate the delta function of a differentiable function $g : \mathbb{R} \to \mathbb{R}$ with the following theorem.

Theorem 4.20 $\delta[g(x)] = \sum_i \frac{\delta(x - x_i)}{|g'(x_1)|}$, where x_i are solutions to the equation $g(x) = 0$, such that $g'(x_i) \neq 0$.

Proof Since the delta function vanishes everywhere except at zero, we are interested only in the behavior of $g(x)$ in the neighborhood of its roots. In a small region U_i around the root x_i, g is one-to-one. By restricting its range to $V_i = g(U_i)$, we define its inverse $g^{-1} : V_i \to U_i$, which satisfies $g^{-1}(0) = x_i$. Hence,

$$\int_{U_i} dx \delta(g(x)) f(x) = \int_{V_i} dy \delta(y) \frac{f(g^{-1}(y))}{|g'(g^{-1}(y))|}$$

$$= \frac{f(g^{-1}(0))}{|g'(g^{-1}(0))|} = \frac{f(x_i)}{|g'(x_i)|} = \frac{1}{|g'(x_i)|} \int_{U_i} dx \delta(x - x_i) f(x).$$

where in the first step we switched the integration variable to $y = g(x)$.

It follows that

$$\int dx \delta[g(x)] f(x) = \sum_i \int_{U_i} dx \delta[g(x)] f(x) = \sum_i |g'(x_i)|^{-1} \int dx \delta(x - x_i) f(x)$$

for all f, which implies the desired result.   □

For $g(x) = ax$, Theorem 4.20 implies that $\delta(ax) = \frac{1}{|a|} \delta(x)$.

Example 4.6 Let $g(x) = \sin x$. The roots to the equation $\sin x = 0$ are $x_n = n\pi$, for $n = 0, \pm 1, \pm 2, \ldots$. Since $(\sin x)' = \cos x$, $g'(x_n) = \cos(n\pi) = (-1)^n$. Theorem 4.20 implies that $\delta(\sin x) = \sum_{n=-\infty}^{\infty} \delta(x - n\pi)$.

Example 4.7 The delta function also satisfies the identity

$$\frac{d}{dx} \theta(x) = \delta(x), \tag{4.45}$$

where $\theta(x)$ is the step function: $\theta(x) = 0$ for $x < 0$ and $\theta(x) = 1$ for $x \geq 0$. To see this, note that for any analytic square integrable function $f(x)$, $\int_{-\infty}^{\infty} dx\theta'(x)f(x) = -\int_{-\infty}^{\infty} dx\theta(x)f'(x) = -\int_{0}^{\infty} f'(x)dx = -[f(\infty) - f(0)] = f(0)$.

Example 4.8 Consider the integral $\int_{0}^{\infty} dse^{i\omega s}$. We evaluate it as the limit of $I_\epsilon = \int_{0}^{\infty} dse^{i\omega s - \epsilon s}$ for $\epsilon \to 0^+$. We readily evaluate

$$I_\epsilon = \frac{1}{\epsilon - i\omega} = \frac{\epsilon + i\omega}{\epsilon^2 + \omega^2}.$$

The real part of I_ϵ is a Lorentzian. By Eq. (4.41), it converges to $\pi\delta(\omega)$ as $\epsilon \to 0$. The imaginary part of I_ϵ converges to another distribution, the *Cauchy principal value* of x^{-1}, denoted as $\left(\frac{1}{x}\right)_{pv}$.

The integral of $\left(\frac{1}{x}\right)_{pv}$ with a smooth function f is defined by

$$\int_{-\infty}^{\infty} dx \left(\frac{1}{x}\right)_{pv} f(x) = \lim_{\epsilon^2 \to 0^+} \int_{-\infty}^{\infty} dx \frac{xf(x)}{x^2 + \epsilon^2}.$$

We conclude that

$$\int_{0}^{\infty} dse^{i\omega s} = \pi\delta(\omega) + i\left(\frac{1}{x}\right)_{pv}. \tag{4.46}$$

4.6.2 Spectral Theorem for Momentum Operator

The main result about the spectrum of the momentum operator is the following.

Theorem 4.21 *For all functions $f : \mathbb{R} \to \mathbb{C}$,*

$$f(\hat{p}) = \int dp f(p)|p\rangle\langle p|, \tag{4.47}$$

where the kets $|p\rangle$ are defined by Eq. (4.35).

Proof It suffices to prove Eq. (4.47) for $f(p) = p^n$, for $n = 0, 1, 2, \ldots$. Then, Eq. (4.47) will hold for all polynomials and all functions that can be obtained by limits of polynomials.

By Eq. (4.35) $\langle p|\psi\rangle = (2\pi)^{-1/2} \int dxe^{-ipx}\psi(x)$. Let $|\psi\rangle, |\phi\rangle \in L^2(\mathbb{R})$. Define

$$I = \int dp p^n \langle\phi|p\rangle\langle p|\psi\rangle = \int \frac{dp}{2\pi} dxdx' p^n e^{ip(x'-x)} \phi^*(x')\psi(x).$$

Since $\int \frac{dp}{2\pi} p^n e^{ipx} = (-i\partial_x)^n \int \frac{dp}{2\pi} e^{ipx} = (-i\partial_x)^n \delta(x)$, we obtain

$$I = \int dxdx' \phi^*(x')\psi(x)(-i\partial_{x'})^n \delta(x' - x).$$

Integrating by parts, we obtain

$$I = \int dxdx' \delta(x - x')(-i\partial_{x'})^n \phi^*(x')\psi(x) = \int dx(-i\partial_x)^n \phi^*(x)\psi(x)$$

$$= \int dx\phi^*(x)(-i\partial_x)^n \psi(x) = \langle\phi|\hat{p}^n|\psi\rangle,$$

where all terms evaluated at $\pm\infty$ vanish for square integrable functions. This above relation applies to all vectors $|\psi\rangle$ and $|\phi\rangle$; the desired result follows. $\square$

The analogy to the spectral theorem in finite-dimensional Hilbert spaces is obvious. Of course, there are differences. Here, we have integration rather than addition, while the ketbra $|p\rangle\langle p|$ is not an actual projector, but it is to be understood as a limit of projectors.

For $f(p) = 1$, Eq. (4.47) gives the resolution of the unity $\int dp|p\rangle\langle p| = \hat{I}$. The kets $|p\rangle$ define a *generalized basis*, on $L^2(\mathbb{R}, dx)$, since any vector $|\psi\rangle$ can by analyzed as $|\psi\rangle = \int dp\tilde{\psi}(p)|p\rangle$, where $\tilde{\psi}(p) = \langle p|\psi\rangle$.

For $f(p) = p$, we obtain $\int dpp|p\rangle\langle p| = \hat{p}$. For any subset U of $\mathbb{R}$, we define the projector

$$\hat{P}_U = \chi_U(\hat{p}) = \int_U dp|p\rangle\langle p|. \tag{4.48}$$

We will see that this projector represents the alternative that the result of a momentum measurement lies in U. The projectors $\hat{P}_U$ are called *spectral projectors* of momentum.

4.6.3   General Form of the Spectral Theorem

The case of momentum is indicative for operators with *continuous spectrum*. In the most general case, a self-adjoint operator has both continuous and discrete spectrum. The good thing is that the spectra separate: The Hilbert space $\mathcal{H}$ can be written as $\mathcal{H}_c \oplus \mathcal{H}_d$, where the subspaces $\mathcal{H}_c$ and $\mathcal{H}_d$ correspond to continuous and discrete spectrum, respectively.

The discrete spectrum of a self-adjoint operator $\hat{A}$ consists of the operator's eigenvalues. Usually its description differs little from that of operators in finite-dimensional Hilbert spaces, except for the fact that the number of eigenvalues may be infinite. Weyl's ordering theorem also applies to the discrete spectrum in infinite dimensions – see section 13.1 in Reed and Simon (1978) for a proof. It is worth pointing out that if $\hat{A}$ has only a discrete spectrum, we can construct an orthonormal basis from the eigenvectors of $\hat{A}$.

The continuous spectrum may be significantly more complicated than that of the momentum operator. For example, there may be degenerate generalized eigenvalues. However, the common point is the existence of spectral projectors. Let $\sigma_c(\hat{A}) \subset \mathbb{R}$ be the continuous spectrum of $\hat{A}$, that is, the set of all generalized eigenvalues of $\hat{A}$. For every $U \subset \sigma_c(\hat{A})$ there is a spectral projector $\hat{P}_U = \chi_U(\hat{A})$, exactly as in the case of momentum. In the general case, the form of $\hat{P}_U$ is more complex than Eq. (4.48).

Given the spectral projectors of $\hat{A}$, we can define an "operator" that projects to the point λ of the spectrum, as

$$\hat{P}_\lambda \delta\lambda = \hat{P}_{[\lambda, \lambda+\delta\lambda]} \tag{4.49}$$

at the limit where $\delta\lambda \to 0$. $\hat{P}_\lambda$ is not an actual operator, it is like the expression $|p\rangle\langle p|$ for the momentum operator; it is defined as a limit. Using $\hat{P}_\lambda$, we can write the spectral theorem for a general operator $\hat{A}$ with a continuous spectrum as

$$f(\hat{A}) = \int f(\lambda)\hat{P}_\lambda d\lambda, \tag{4.50}$$

Example 4.9 Consider the kinetic energy operator $\hat{H} = \frac{\hat{p}^2}{2m}$ for a particle of mass m. By the spectral theorem, $\hat{H} = \int_{-\infty}^{\infty} dp \frac{p^2}{2m} |p\rangle\langle p|$. We write the generalized eigenvalues of $\hat{H}$ as ϵ, then we can solve $p = \pm p_\epsilon$, where $p_\epsilon = \sqrt{2m\epsilon}$. Changing the integration variable to ϵ, we find

$$\hat{H} = \int_0^{\infty} d\epsilon \sqrt{\frac{m}{2\epsilon}} \epsilon (|p_\epsilon\rangle\langle p_\epsilon| + |-p_\epsilon\rangle\langle -p_\epsilon|).$$

Hence, $\hat{H} = \int_0^{\infty} d\epsilon\, \epsilon \hat{P}_\epsilon$, where $\hat{P}_\epsilon = \sqrt{\frac{m}{2\epsilon}}(|p_\epsilon\rangle\langle p_\epsilon| + |-p_\epsilon\rangle\langle -p_\epsilon|)$, in accordance with the spectral theorem.

4.6.4 Position Operator

The position operator $\hat{x}$ on $L^2(\mathbb{R}, dx)$ is written as

$$\hat{x} = \int dx x |x\rangle\langle x|, \tag{4.51}$$

where the kets $|x\rangle$ satisfy

$$\langle x|x'\rangle = \delta(x - x'). \tag{4.52}$$

We will define the kets $|x\rangle$ as limits of Hilbert space vectors. Consider the functions

$$f_\epsilon(x) = \frac{1}{\sqrt{2\pi\epsilon^2}} \exp\left[-\frac{x^2}{2\epsilon^2}\right], \tag{4.53}$$

that satisfy $\lim_{\epsilon \to 0} \hat{x} f_\epsilon(x - x_0) = x_0 f_\epsilon(x - x_0)$. Hence, we will identify the limit of $f_\epsilon(\cdot - x_0)$ as $\epsilon \to 0$ with the ket $|x_0\rangle$. Since at the limit $\epsilon \to 0$, $f_\epsilon(\cdot)$ becomes a delta function, we identify the ket $|x_0\rangle$ with the function $\delta(x - x_0)$.

With this definition

$$\langle x_0|\psi\rangle = \lim_{\epsilon \to 0} \int dx f_\epsilon(x - x_0)\psi(x) = \int dx \delta(x - x_0)\psi(x) = \psi(x_0), \tag{4.54}$$

that is, the values $\psi(x)$ of a function ψ can be written as $\langle x|\psi\rangle$, for all x.

Some caution is required here. When we say that the wave function ψ coincides with the ket $|\psi\rangle$, we refer to the function as a map $\psi : \mathbb{R} \to \mathbb{C}$. We do not mean the individual values $\psi(x)$. Here, the common abusive symbolism of writing functions ψ as $\psi(x)$, in order to show the argument, may cause confusion.

Given that a generalized eigenvector $|p\rangle$ corresponds to $\frac{1}{\sqrt{2\pi}} e^{ipx}$, we obtain the key formula relating position and momentum kets

$$\langle x|p\rangle = \frac{1}{\sqrt{2\pi}} e^{ipx}. \tag{4.55}$$

The identification of momentum with the operator $-i\frac{\partial}{\partial x}$ means that

$$\langle x|\hat{p}|\psi\rangle = -i\frac{\partial}{\partial x}\langle x|\psi\rangle. \tag{4.56}$$

The resolution of the unity of any orthonormal basis $\psi_n(x) = \langle x|n\rangle$ on $L^2(\mathbb{R}, dx)$ becomes

$$\sum_n \psi_n(x)\psi_n^*(x') = \sum_n \langle x|n\rangle\langle n|x'\rangle = \langle x|x'\rangle = \delta(x - x'). \tag{4.57}$$

We can straightforwardly generalize these results to more complex systems. For example, consider the Hilbert space $L^2(\mathbb{R}^3, dx_1 dx_2 dx_3)$ that describes a particle in three dimensions. There exist generalized eigenvectors $|x_1, x_2, x_3\rangle$, so that the wave functions are written as

$$\psi(x_1, x_2, x_3) = \langle x_1, x_2, x_3 | \psi \rangle, \tag{4.58}$$

and $\langle x_1, x_2, x_3 | x_1', x_2', x_3' \rangle = \delta(x_1 - x_1')\delta(x_2 - x_2')\delta(x_3 - x_3')$.

The operators that correspond to position coordinates are written as

$$\hat{x}_i = \int d^3x\, x_i |x_1, x_2, x_3\rangle\langle x_1, x_2, x_3|. \tag{4.59}$$

The "projectors" $\hat{P}_{x_1}$ for $\hat{x}_1$ are

$$\hat{P}_{x_1} = \int dx_2 dx_3 |x_1, x_2, x_3\rangle\langle x_1, x_2, x_3|. \tag{4.60}$$

Similar expressions apply for $\hat{P}_{x_2}$ and $\hat{P}_{x_3}$.

We use the following compact notation. We write the vector operator $\hat{\mathbf{x}}$ for the position coordinates, so that the eigenvalue equation becomes $\hat{\mathbf{x}}|\mathbf{x}\rangle = \mathbf{x}|\mathbf{x}\rangle$, where $\langle \mathbf{x}|\mathbf{x}'\rangle = \delta^{(3)}(\mathbf{x} - \mathbf{x}')$.

Example 4.10 We can also define a delta function $\delta(\xi)$ for functions f on the interval $[-1, 1]$, such that $\int_{-1}^{1} d\xi\, \delta(\xi) f(\xi) = f(0)$. Then, the resolution of the unity for the basis defined by the Legendre polynomials yields

$$\frac{1}{2}\sum_{\ell=0}^{\infty}(2\ell + 1)P_\ell(\xi)P_\ell(\xi') = \delta(\xi - \xi'). \tag{4.61}$$

4.6.5 Different Representations of an Operator

We observe that a ket $|\psi\rangle$ can be represented either as a square integrable function of position $\psi(x) = \langle x|\psi\rangle$, or as a square integrable function of momentum $\tilde{\psi}(p) = \langle p|\psi\rangle$. In the former case, the momentum operator $\hat{p}$ acts differentially on $\psi(x)$: $\hat{p}\psi(x) = -i\frac{\partial\psi}{\partial x}$. In the latter case,

$$\hat{p}|\psi\rangle = \int dp\, p\langle p|\psi\rangle |p\rangle = \int dp\, p\tilde{\psi}(p)|p\rangle, \tag{4.62}$$

that is, $\hat{p}$ acts multiplicatively on $\tilde{\psi}(p)$: $\hat{p}\tilde{\psi}(p) = p\tilde{\psi}(p)$.

Furthermore,

$$\tilde{\psi}(p) = \int \langle p|x\rangle\langle x|\psi\rangle = \frac{1}{\sqrt{2\pi}}\int dx\, e^{-ipx}\psi(x), \tag{4.63}$$

which means that $\tilde{\psi}$ is the Fourier transform of the position wavefunction ψ.

We emphasize that there is no physical or mathematical necessity to employ position wave functions for describing particles. Momentum wave functions or wave functions with respect to any other quantity are equally good.

> Position wave functions in quantum theory do not describe physical waves in space. They provide one way of describing state vectors that is convenient or some applications.

4.7 Trace

Definition 4.22 The *trace* of an operator $\hat{A}$ on a Hilbert space $\mathcal{H}$ is the complex number

$$\text{Tr}\hat{A} := \sum_n \langle n|\hat{A}|n\rangle, \tag{4.64}$$

where $|n\rangle$ is an orthonormal basis on $\mathcal{H}$.

Despite the arbitrary choice of an orthonormal basis in Definition 4.22, the trace is basis-independent. Consider a different basis $|n'\rangle$. The trace Tr' with respect to this basis is

$$\text{Tr}'\hat{A} = \sum_{n'}\langle n'|\hat{A}|n'\rangle = \sum_{n'}\sum_{m,n}\langle n'|n\rangle\langle n|\hat{A}|m\rangle\langle m|n'\rangle$$

$$= \sum_{m,n}\langle m|n\rangle\langle n|\hat{A}|m\rangle = \sum_n\sum_n\langle n|\hat{A}|n\rangle = \text{Tr}\hat{A}. \tag{4.65}$$

By definition, the trace is a linear function: $\text{Tr}(\lambda\hat{A} + \hat{B}) = \lambda\text{Tr}\hat{A} + \text{Tr}\hat{B}$, for all $\lambda \in \mathbb{C}$.

The trace is also symmetric:

$$\text{Tr}(\hat{A}\hat{B}) = \text{Tr}(\hat{B}\hat{A}). \tag{4.66}$$

Indeed, $\text{Tr}(\hat{A}\hat{B}) = \sum_n\langle n|\hat{A}\hat{B}|n\rangle = \sum_n\sum_m\langle n|\hat{A}|m\rangle\langle m|\hat{B}|n\rangle = \sum_m\langle m|\hat{B}\hat{A}|m\rangle = \text{Tr}(\hat{B}\hat{A})$.

Furthermore, $\text{Tr}(|\psi\rangle\langle\phi|) = \sum_n\langle n|\psi\rangle\langle\phi|n\rangle = \langle\phi|\hat{I}|\psi\rangle = \langle\phi|\psi\rangle$. The trace of the ketbra is the braket.

If $\hat{P}$ is a projector to an N-dimensional subspace V, then $\text{Tr}\hat{P} = N$. Indeed, $\hat{P}$ can be written $\sum_{i=1}^N |i\rangle\langle i|$, for some orthonormal basis $|i\rangle$ on V. Hence, $\text{Tr}\hat{P} = \sum_{i=1}^N \langle i|i\rangle = N$.

Theorem 4.23 *If* $\text{Tr}(\hat{A}^\dagger\hat{A}) < \infty$ *and* $\text{Tr}(\hat{B}^\dagger\hat{B}) < \infty$, *then*

$$|\text{Tr}(\hat{A}\hat{B}^\dagger)|^2 \leq \text{Tr}(\hat{A}^\dagger\hat{A})\text{Tr}(\hat{B}^\dagger\hat{B}). \tag{4.67}$$

Proof We define an inner product for operators $\hat{A}$ and $\hat{B}$, as $\langle\hat{A}, \hat{B}^\dagger\rangle = \text{Tr}(\hat{A}\hat{B}^\dagger)$. Since the set of operators is a vector space, all operators $\hat{A}$ with finite norm, $(\text{Tr}(\hat{A}^\dagger\hat{A}) < \infty)$ define a Hilbert space with respect to this inner product. Equation (4.67) is just the Cauchy–Schwarz inequality for this inner product. $\square$

In finite-dimensional Hilbert spaces, there is a useful relation between the trace and the determinant of operators. The determinant of an operator $\hat{A}$ is the product of its eigenvalues a_i, while the trace is its sum:

$$\det\hat{A} = \prod_i a_i = \prod_i e^{\ln a_i} = e^{\sum_i \ln a_i} = e^{\text{Tr}\ln\hat{A}}. \tag{4.68}$$

Finally, we note that operators with continuous spectrum have infinite trace. For example, the spectral projector of momentum $\hat{P}_U = \int_U dp|p\rangle\langle p|$ satisfies

$$\text{Tr}\hat{P}_U = \int_U dp\langle p|p\rangle = \int_U dp\,\delta(0) = \infty.$$

4.8   Physical Interpretation of Operators

In order to explain the role played by operators in quantum theory, we must recall the description of classical measurements, in Section 1.5.

A measurement in a classical system corresponds to a partition of its state space Γ in mutually exclusive and exhaustive subsets C_λ, each of which corresponds to a different measurement outcome λ. Using the characteristic functions $\chi_\lambda := \chi_{C_\lambda}$ of the sets C_λ, we write the classical relation for the partition of the unity

$$\sum_\lambda \chi_\lambda = 1. \tag{4.69}$$

All information about possible outcomes of the measurement is encoded in the definition of the function $F : \Gamma \to \mathbb{R}$,

$$F(x) := \sum_\lambda \lambda \chi_\lambda(x), \quad x \in \Gamma. \tag{4.70}$$

The analogy between Eqs. (4.69) and (4.70) and the spectral theorem are obvious. To be precise, there is the following correspondence between classical and quantum notions.

> Self-adjoint operator $\hat{A}$ on $\mathcal{H}$ $\Longleftrightarrow$ function $F : \Gamma \to \mathbb{R}$.
>
> Spectrum of $\hat{A}$ $\Longleftrightarrow$ range of F.
>
> Spectral projectors of $\hat{A}$ $\Longleftrightarrow$ characteristic functions χ_{C_λ} on value sets C_λ.
>
> Subspaces of Hilbert space $\mathcal{H}$ $\Longleftrightarrow$ subsets of state space Γ.

This correspondence underlies the following foundational principle of quantum theory.

> **FP2** A self-adjoint operator $\hat{A}$ on a Hilbert space $\mathcal{H}$ corresponds to a physical quantity of a physical system. Possible measurement outcomes of this quantity correspond to points in the spectrum of $\hat{A}$, and they are associated to spectral projectors of $\hat{A}$.

A measurement that corresponds to one-dimensional projectors is called *fine-grained*. A fine-grained measurement defines an orthonormal basis on the system's Hilbert space. Any measurement that is not fine-grained is called *coarse-grained*.

Consider a self-adjoint operator $\hat{A}_1$ with discrete spectrum and at least one degenerate eigenvalue. The measurements associated to this operator are coarse-grained. We can render the measurements of $\hat{A}_1$ fine-grained, if we also measure other quantities that correspond to operators $\hat{A}_2, \hat{A}_3, \dots, N$, commuting with each other – $[\hat{A}_a, \hat{A}_b] = 0$, for all $a, b = 0, 1, 2, \dots, N$ – so that there exists a unique orthonormal basis $|n\rangle$ satisfying

$$\hat{A}_a|n\rangle = \lambda_{a,n}|n\rangle. \tag{4.71}$$

This way, each vector $|n\rangle$ corresponds to a unique N-plet of eigenvalues

$$(\lambda_{1,n}, \lambda_{2,n}, \dots, \lambda_{N,n}).$$

A set of self-adjoint operators with this property is said to define a *complete set of measurements*.

However, if two observables do not commute, we cannot specify their eigenvalues simultaneously. The observables are then said to be *incompatible*.

Operators with a continuous spectrum such as position or momentum have no one-dimensional spectral projectors; hence, all measurements are coarse-grained. This means that there is no absolute precision in the measurement of the associated physical quantity. In this case, one can partition the spectrum in segments of finite length and use them in order to define exhaustive and mutually exclusive alternatives.

Example 4.11 Consider the position operator $\hat{x}$ on $L^2(\mathbb{R}, dx)$, where $\sigma(\hat{x}) = \mathbb{R}$. We partition the real line to intervals Δ_n of width δ,

$$\Delta_n = \left[\left(n - \frac{1}{2} \right) \delta, \left(n + \frac{1}{2} \right) \delta \right), \tag{4.72}$$

where $n \in Z$. The interval Δ_n is centered around $\bar{x}_n = n\delta$, and δ corresponds to the resolution of the measurement apparatus. The set of projectors

$$\hat{P}_n = \int_{\Delta_n} dx |x\rangle \langle x| \tag{4.73}$$

is both exhaustive

$$\sum_n \hat{P}_n = \int_{\Delta_n} dx |x\rangle \langle x| = \int_{\cup_n \Delta_n} dx |x\rangle \langle x| = \int_{\mathbb{R}} dx |x\rangle \langle x| = \hat{I}, \tag{4.74}$$

and mutually exclusive: $\hat{P}_n \hat{P}_m = \delta_{mn} \hat{P}_n$.

One may wonder whether FP2 provides a full correspondence between measurements and self-adjoint operators. The answer is negative. First, we will see in Chapter 21 that there exist measurements of physical quantities that cannot be described by self-adjoint operators.

Conversely, a self-adjoint operator may represent a physically realizable measurement, as long as it respects superselection rules. For an operator $\hat{A} = \sum_n a_n \hat{P}_n$, the spectral projectors $\hat{P}_n$ must not project to subspaces that contain superpositions of states, not allowed by the superselection rules. For example, suppose we have superselection rules for the electric charge, represented by an operator $\hat{Q}$ with eigenvectors $|q\rangle$. No spectral projector $\hat{P}_n$ of $\hat{A}$ can project to subspaces containing vectors of the form $|q\rangle + |q'\rangle$ for $q \neq q'$. This means that $[\hat{Q}, \hat{P}_n] = 0$; hence, $[\hat{Q}, \hat{A}] = 0$. Only operators that commute with $\hat{Q}$ correspond to physically realizable measurements.

Why should we interpret spectral projectors as corresponding to measurements? Why not interpret them as *properties* of the quantum system, irrespective of whether measurements have been carried out? In classical physics, this interpretation is both common and sensible. When we assign alternatives and assign probabilities to them, we do not have to specify how we will find out which alternative has been realized. Physics is fundamentally about intrinsic properties of physical systems, and measurements are derivative notions.

This is not the case in quantum theory. In Chapter 6, we will see that it is impossible to talk about properties of quantum systems without an explicit reference to the way that these properties are being determined. Properties belong to the complex of the microscopic system and the macroscopic measurement apparatus, and not to the microscopic system alone.

QUESTIONS

4.1 Which of the following operators on $L^2(\mathbb{R})$ are bounded: (i) $\hat{x}$, (ii) $\hat{p}$, (iii) $e^{i\hat{p}a}$, (iv) $\hat{x}^{25}(\hat{x}^{26} + \hat{I})^{-1}$?

4.2 Give an example of an operator, other than the unit, that is both self-adjoint and unitary. Are there any operators that are both positive and unitary?

4.3 If the operator $\hat{A}$ has no degeneracy, then all operators $f(\hat{A})$ have no degeneracy. True or false? If false, what restriction must we place upon f in order to make the proposition true?

4.4 Explain in detail why the following holds: If $[\hat{A}, \hat{B}] = 0$, then $[f(\hat{A}), \hat{B}] = 0$. Does the converse hold?

4.5 Let $\hat{A}$ and $\hat{B}$ be self-adjoint. Which ones of the following operators are self-adjoint: (i) $[\hat{A}, \hat{B}]$, (ii) $\hat{A}\hat{B} + \hat{B}\hat{A}$, (iii) $\hat{B}\hat{A}\hat{B}$, (iv) $\hat{B}\hat{A}\hat{B}\hat{A}$?

4.6 Simplify the following expressions: (i) $\delta(\cosh x)$, (ii) $\delta(3x^2 + 1)$, (iii) $\delta(x^3 + 1)$, and (iv) $\delta(x^4 - 1)$.

4.7 Is the sum of two delta functions well defined? What about their product?

4.8 "To find the possible values of a sum $A + B$ of two physical magnitudes, find the possible values of A and B and add their results." Does this proposal work in quantum theory? If not, give an explicit counterexample.

PROBLEMS

4.1 Find the adjoint of the operator $\hat{D}_s$, defined by $\hat{D}_s\psi(x) := \frac{1}{\sqrt{s}}\psi(x/s)$ on $L^2(\mathbb{R})$.

4.2 Show that, for any $n \times n$ matrix A, $||A|| < n^{3/2}\max_{i,j}|A_{ij}|$.

4.3 Show that any positive operator is self-adjoint.

4.4 Show that if $[\hat{A}, \hat{B}] = 0$ and $\hat{A}$ has no degenerate eigenvalue (or generalized eigenvalue), then $\hat{B}$ is a function of $\hat{A}$.

4.5 Show that if two operators commute, their spectral projectors also commute.

4.6 Show that if two positive operators $\hat{A}$ and $\hat{B}$ commute, then $\hat{A}\hat{B}$ is a positive operator.

4.7 Let a_{max} be the maximum and a_{min} be the minimum eigenvalue of $\hat{A} \geq 0$. Show that $a_{max}\hat{I} \geq \hat{A}$ and that $a_{min}\hat{A} \leq \hat{A}^2$.

4.8 Let $\hat{P}_1$ and $\hat{P}_2$ be projectors with associated subspaces V_1 and V_2, respectively, both of finite dimension. (i) Show that, if $[\hat{P}_1, \hat{P}_2] = 0$, then $\hat{P}_1\hat{P}_2$ is a projector on the intersection of V_1 and V_2. (ii) Show that, if $\hat{P}_1\hat{P}_2 = 0$, then $\hat{P}_1 + \hat{P}_2$ is a projector on $V_1 \oplus V_2$. (iii) Show that, if $\hat{P}_1\hat{P}_2 = \hat{P}_1$, then $V_1 \subset V_2$. (iv) What changes if at least one of the subspaces V_1 and V_2 is infinite-dimensional?

4.9 Evaluate the commutators $[\hat{x}^3\hat{p}, \hat{x}\hat{p}]$ and $[x^2, [\hat{p}^2, \hat{x}^3]]$.

4.10 (i) Show that the most general nontrivial operator $\hat{P}$ on $\mathbb{C}^2$ that satisfies $\hat{P}^2 = \hat{P}$ is
$$\frac{1}{2}\begin{pmatrix} 1+z & \alpha \\ \frac{1-z^2}{\alpha} & 1-z \end{pmatrix},$$ for $z, \alpha \in \mathbb{C}$. (ii) For which values of z, α is $\hat{P}$ normal and for which values is it a projector?

4.11 Show that in a finite-dimensional Hilbert space any self-adjoint operator can be written as a linear combination of two unitary operators.

4.12 (i) Show that the Legendre operator $\hat{\Lambda}$ is self-adjoint. (ii) Show that the nonpolynomial solutions to the eigenvalue equation of $\hat{\Lambda}$ are not square integrable.

4.13 Let $\hat{X}_h$ be an operator on $L^2(\mathbb{R})$, defined as $\hat{X}_h \psi(x) := h(x)\psi(x)$ for some complex-valued function $h(x)$. (i) Show that $\hat{X}_h$ is a normal operator. (ii) Show that if $h(x)$ is real valued, then $\hat{X}_h$ is self-adjoint, and that if $|h(x)| = 1$, then $\hat{X}_h$ is unitary. (iii) Show that if $h(x)$ takes a constant value a in an interval (c, d), then a is an eigenvalue of $\hat{X}_h$ with infinite degeneracy.

4.14 (i) Show that, in a finite-dimensional Hilbert space, a normal operator with only real eigenvalues is self-adjoint. (ii) Find a counterexample to that statement in an infinite-dimensional Hilbert space. (Hint: Use the operators $\hat{X}_h$ of Problem 4.13)

4.15 Let $\hat{H} = \lambda \begin{pmatrix} 1 & 0 & 0 & 0 \\ 0 & 1 & -1 & 0 \\ 0 & -1 & 1 & 0 \\ 0 & 0 & 0 & 1 \end{pmatrix}$, where $\lambda > 0$. (a) Find the spectral projectors of $\hat{H}$.

(b) Evaluate the operators $e^{-i\hat{H}t}$ and $\chi_U(\hat{H})$ for $U = [0, \lambda]$.

4.16 For any distribution ω on $\mathbb{R}$ we define the derivative ω', by $\omega'(f) = -\omega(f')$, where f is a smooth square integrable function. (i) Show that, for distributions that correspond to ordinary functions, this definition reduces to the standard one. (ii) Write the defining equations for the first and second derivatives of the delta function. (iii) Evaluate $\delta'(x^2 - 1)$ and $\delta''(x^2 - 1)$.

4.17 Show that the Weyl operator $\hat{V}(a, b)$ acts as: $\hat{V}(a, b)|x\rangle = e^{\frac{iab}{2}}e^{iax}|x + b\rangle$, and $\hat{V}(a, b)|p\rangle = e^{-\frac{iab}{2}}e^{-ibp}|p + a\rangle$.

4.18 Let $|p\rangle$ and $|x\rangle$ be generalized eigenvectors of momentum $\hat{p}$ and position $\hat{x}$, respectively. Evaluate (i) $\langle p|\hat{p}|x\rangle$, (ii) $\langle p_1|\hat{x}|p_2\rangle$, and (iii) $\langle p|\hat{x}\hat{p}|x\rangle$.

4.19 Show that $\delta^{(3)}(\mathbf{x}) = -\nabla^2 \frac{1}{4\pi r}$, where $r = |\mathbf{x}|$.

4.20 For self-adjoint operators $\hat{A}, \hat{B}, \hat{C}$, where $\hat{B} \geq 0$, show that

$$\text{Tr}(\hat{A}\hat{B}\hat{C}\hat{B}) \leq \frac{1}{2}\text{Tr}\left[(\hat{A}\hat{B})^2 + (\hat{B}\hat{C})^2\right].$$

4.21 We define the *trace norm* of an operator $\hat{A}$ as $||\hat{A}||_{tr} := \text{Tr}|\hat{A}|$, where $|\hat{A}| = \sqrt{\hat{A}^\dagger \hat{A}}$ is the absolute value operator. Show that (i) for a self-adjoint operator the trace norm is the sum of the absolute values of its eigenvalues multiplied with each eigenvalue's degeneracy, (ii) $\text{Tr}|\hat{A}| \geq ||\hat{A}||$, where $||\hat{A}||$ is the usual operator norm, and (iii) $|\text{Tr}(\widehat{AB})| \leq ||\hat{A}||\text{Tr}|\hat{B}|$.

Bibliography

- For a mathematical introduction to matrix theory and spectral theory, see Axler (2004); for a more advanced level, see Roman (2010). For a quick introduction to Hilbert space operators, see chapters 7 and 8 of Young (1988); for a more detailed analysis, see Akhiezer and Glazman (1993).

- For the interpretation of operators in quantum theory, see chapters 3 and 7 in Isham (1995) and chapter 4 in Peres (2002).

5   Operators II

Applications

'Now, let us … see if the ball can get out of the crater without rolling over the top,' and he threw the ball back into the hole. For a while nothing happened, and Mr Tompkins could hear only the slight rumbling of the ball rolling to and fro in the crater. Then, as by a miracle, the ball suddenly appeared in the middle of the outer slope, and quietly rolled down to the table.

Gamow (1965)

5.1   Operators for Qubits

The simplest quantum systems correspond to the smallest nontrivial Hilbert space $\mathbb{C}^2$. They are called *two-level* systems. Some physical magnitudes, such as photon polarization or electron spin, are naturally described by the Hilbert space $\mathbb{C}^2$. However, usually, two-level systems are approximations to more complex systems. Consider, for example, an atom with a lowest-energy state $|0\rangle$. We can excite the atom with a laser pulse, so that it can be found in an excited state $|1\rangle$. With a different pulse we can create a specific superposition of $|0\rangle$ and $|1\rangle$. The set of all such states spans a two-dimensional Hilbert space.

A quantum two-level system is also called a *qubit*, that is, a quantum bit, because it constitutes the quantum analogue of the notion of a bit in information theory. The term "qubit" was first used by Schumacher (1995), as a joke. The word sounds the same as the existing word "cubit" that is the oldest recorded length unit, and probably the most common one in antiquity.

Qubits are the building blocks of quantum computers, that is, devices that perform computation while exploiting the quantum superposition principle. Any two-level quantum system that can be subjected to sufficient control can be viewed as a qubit. This includes atoms, photons, and solid-state systems, whence, one talks about atomic, photonic, and solid-state qubits, respectively.

5.1.1   Pauli Matrices

By definition, the most general self-adjoint operator on $\mathbb{C}^2$ is of the form

$$\hat{A} = \begin{pmatrix} x_1 & z \\ z^* & x_2 \end{pmatrix}, \quad \text{where } x_1, x_2 \in \mathbb{R}, \text{ and } z \in \mathbb{C}. \tag{5.1}$$

This can be written as

$$\hat{A} = a_0 \hat{I} + \sum_{i=1}^{3} a_i \hat{\sigma}_i, \tag{5.2}$$

where σ_i for $i = 1, 2, 3$ are the Pauli matrices,

$$\hat{\sigma}_1 = \begin{pmatrix} 0 & 1 \\ 1 & 0 \end{pmatrix}, \; \hat{\sigma}_2 = \begin{pmatrix} 0 & -i \\ i & 0 \end{pmatrix}, \; \hat{\sigma}_3 = \begin{pmatrix} 1 & 0 \\ 0 & -1 \end{pmatrix}, \tag{5.3}$$

and $a_0, a_i \in \mathbb{R}$.

It is straightforward to prove the following identities for the Pauli matrices.

(i) $(\hat{\sigma}_i)^2 = \hat{I}$.
(ii) $\hat{\sigma}_1 \hat{\sigma}_2 = i\sigma_3, \hat{\sigma}_2 \hat{\sigma}_1 = -i\sigma_3$.
(iii) $\hat{\sigma}_2 \hat{\sigma}_3 = i\sigma_1, \hat{\sigma}_3 \hat{\sigma}_2 = -i\sigma_1$.
(iv) $\hat{\sigma}_3 \hat{\sigma}_1 = i\sigma_2, \hat{\sigma}_1 \hat{\sigma}_3 = -i\sigma_2$.

The identities above can be expressed compactly as

$$\hat{\sigma}_i \hat{\sigma}_j = \hat{I} \delta_{ij} + i \sum_{k=1}^{3} \epsilon_{ijk} \hat{\sigma}_k, \tag{5.4}$$

where ϵ_{ijk} is the *totally antisymmetric symbol*

$$\epsilon_{ijk} = \begin{cases} 1 & \text{if } (i,j,k) = (1,2,3),(3,1,2),(2,3,1) \\ -1 & \text{if } (i,j,k) = (2,1,3),(3,2,1),(1,3,2) \\ 0 & \text{if any two indices } i,j,k \text{ coincide.} \end{cases} \tag{5.5}$$

It is evident that the exchange of any two indices is equivalent to multiplication with -1: $\epsilon_{ijk} = -\epsilon_{jik} = -\epsilon_{kji} = -\epsilon_{ikj}$.

In later chapters, we will often employ the following property of the totally antisymmetric symbol,

$$\sum_{i=1}^{3} \epsilon_{ijk} \epsilon_{ilm} = \delta_{jl} \delta_{km} - \delta_{jm} \delta_{kl}, \tag{5.6}$$

which is fully equivalent to the well-known vector identity

$$(\mathbf{A} \times \mathbf{B})^2 = \mathbf{A}^2 \mathbf{B}^2 - (\mathbf{A} \cdot \mathbf{B})^2 \tag{5.7}$$

for two three-vectors $\mathbf{A}$ and $\mathbf{B}$.

To see this, note that the left-hand side of Eq. (5.7) is $\sum_{ijklm} \epsilon_{ijk} \epsilon_{ilm} A_j B_k A_l B_m$ in coordinate notation, while the right-hand side can be written as $\sum_{jklm} (\delta_{jl}\delta_{km} - \delta_{jm}\delta_{kl}) A_j B_k A_l B_m$. Since Eq. (5.7) holds for all A_i and B_i, Eq. (5.5) follows.

We write $\hat{\boldsymbol{\sigma}} = (\hat{\sigma}_1, \hat{\sigma}_2, \hat{\sigma}_3)$, so that the operator (5.2) is expressed as

$$\hat{A} = a_0 \hat{I} + \mathbf{a} \cdot \hat{\boldsymbol{\sigma}}, \tag{5.8}$$

Equation (5.4) implies that

$$(\mathbf{a} \cdot \hat{\boldsymbol{\sigma}})(\mathbf{b} \cdot \hat{\boldsymbol{\sigma}}) = \mathbf{a} \cdot \mathbf{b} \hat{I} + i(\mathbf{a} \times \mathbf{b}) \cdot \hat{\boldsymbol{\sigma}}, \tag{5.9}$$

since $\mathbf{a} \cdot \mathbf{b} = \sum_{i=1}^{3} a_i b_i$, and

$$(\mathbf{a} \times \mathbf{b})_k = \sum_{i=1}^{3} \sum_{i=1}^{3} \epsilon_{ijk} a_i b_j = (a_2 b_3 - a_3 b_2, a_3 b_1 - a_1 b_3, a_1 b_2 - a_2 b_1).$$

Equation (5.9) implies that

$$[\mathbf{a} \cdot \hat{\sigma}, \mathbf{b} \cdot \hat{\sigma}] = 2i(\mathbf{a} \times \mathbf{b}) \cdot \sigma. \tag{5.10}$$

Of significance is the *completeness identity* of the Pauli matrices,

$$\hat{\sigma}_{ab} \cdot \hat{\sigma}_{cd} := \sum_{i=1}^{3} (\hat{\sigma}_i)_{ab} (\hat{\sigma}_i)_{cd} = 2\delta_{ad}\delta_{bc} - \delta_{ab}\delta_{cd}, \tag{5.11}$$

where the indices $a, b, c, d = 1, 2$ refer to the elements of the Pauli matrices.

To prove Eq. (5.11), we take the trace of both sides in Eq. (5.8), to obtain $a_0 = \frac{1}{2}\mathrm{Tr}\hat{A}$. We also multiply both sides of Eq. (5.8) with $\hat{\sigma}_k$ and then take the trace. By Eq. (5.4), $\mathbf{a} = \frac{1}{2}\mathrm{Tr}(\hat{A}\hat{\sigma})$. Hence, Eq. (5.8) is brought to the form

$$\hat{A} = \frac{1}{2}(\mathrm{Tr}\hat{A})\hat{I} + \frac{1}{2}\mathrm{Tr}(\hat{A}\hat{\sigma}) \cdot \hat{\sigma}. \tag{5.12}$$

We write $\mathrm{Tr}\hat{A} = \sum_{cd} \delta_{cd}\hat{A}_{cd}$ and $\mathrm{Tr}(\hat{A}\hat{\sigma}) = \sum_{cd} \hat{A}_{cd}\hat{\sigma}_{cd}$. Then, Eq. (5.12) becomes

$$\sum_{cd} \hat{A}_{cd} \left(\delta_{ac}\delta_{bd} - \frac{1}{2}\delta_{ab}\delta_{cd} - \frac{1}{2}\hat{\sigma}_{ab} \cdot \hat{\sigma}_{cd} \right) = 0, \tag{5.13}$$

from which Eq. (5.11) follows.

5.1.2 Eigenvalues of Self-Adjoint Operators

The matrix (5.2) reads explicitly

$$\hat{A} = \begin{pmatrix} a_0 + a_3 & a_1 - ia_2 \\ a_1 + ia_2 & a_0 - a_3 \end{pmatrix}. \tag{5.14}$$

It is straightforward to compute its two eigenvalues: $a_+ = a_0 + |\mathbf{a}|$ and $a_- = a_0 - |\mathbf{a}|$, where $|\mathbf{a}| = \sqrt{\mathbf{a} \cdot \mathbf{a}}$. The corresponding eigenvectors are

$$|a_+\rangle = \frac{1}{\sqrt{2|\mathbf{a}|(|\mathbf{a}| - a_3)}} \begin{pmatrix} a_1 - ia_2 \\ |\mathbf{a}| - a_3 \end{pmatrix},$$

$$|a_-\rangle = \frac{1}{\sqrt{2|\mathbf{a}|(|\mathbf{a}| - a_3)}} \begin{pmatrix} -|\mathbf{a}| + a_3 \\ a_1 + ia_2 \end{pmatrix}. \tag{5.15}$$

We evaluate the associated spectral projectors

$$\hat{P}_{\pm} = |a_{\pm}\rangle\langle a_{\pm}| = \frac{1}{2}\left(\hat{I} \pm \mathbf{n} \cdot \sigma \right). \tag{5.16}$$

where $\mathbf{n} = \frac{\mathbf{a}}{|\mathbf{a}|}$ is a unit vector on $\mathbb{R}^3$.

The spectral projectors depend only on $\mathbf{n}$; hence, $\mathbf{n}$ determines the type of measurement performed on a qubit. Without loss of generality, we can always choose $a_0 = 0$ and $|\mathbf{a}| = 1$. Other values of a_0 or $|\mathbf{a}|$ simply effect a change in the eigenvalues by a linear function $f(x) = |\mathbf{a}|x + a_0$. Hence, when we refer to measurement in qubits, we will usually restrict ourselves to self-adjoint operator of the form $\mathbf{n} \cdot \boldsymbol{\sigma}$, where $\mathbf{n}$ is a unit vector.

5.1.3 Raising and Lowering Operators

In qubits, we distinguish the basis consisting of the vectors

$$\begin{pmatrix} 0 \\ 1 \end{pmatrix}, \quad \begin{pmatrix} 1 \\ 0 \end{pmatrix}. \tag{5.17}$$

We refer to these vectors, either as $|0\rangle$ and $|1\rangle$, respectively, or as $|-\rangle$ and $|+\rangle$. The former notation is preferred when the two-level system describes an atom, or it is viewed as an abstract qubit. The latter notation is preferred when describing particles with spin $\frac{1}{2}$ or photon polarization. Here, we proceed with the 0-1 notation.

We define the operators $\hat{\sigma}_+$ and $\hat{\sigma}_-$ by

$$\hat{\sigma}_+ = \frac{1}{2}(\hat{\sigma}_1 + i\hat{\sigma}_2) = \begin{pmatrix} 0 & 1 \\ 0 & 0 \end{pmatrix}, \quad \hat{\sigma}_- = (\hat{\sigma}_+)^\dagger = \begin{pmatrix} 0 & 0 \\ 1 & 0 \end{pmatrix}. \tag{5.18}$$

These operators act on the basis vectors $|0\rangle$ and $|1\rangle$ as

$$\hat{\sigma}_+|0\rangle = |1\rangle, \quad \hat{\sigma}_+|1\rangle = 0, \tag{5.19}$$
$$\hat{\sigma}_-|0\rangle = 0, \quad \hat{\sigma}_-|1\rangle = |0\rangle. \tag{5.20}$$

The operator $\hat{\sigma}_+$ is called *raising operator* because it induces the transition $0 \to 1$. The operator $\hat{\sigma}_-$ is called *lowering operator* because it induces the transition $1 \to 0$.

It is straightforward to show that the operators $\hat{\sigma}_\pm$ satisfy the following identities.

(i) $(\hat{\sigma}_\pm)^2 = 0$.
(ii) $\hat{\sigma}_+\hat{\sigma}_- + \hat{\sigma}_-\hat{\sigma}_+ = \hat{I}$.
(iii) $[\hat{\sigma}_+, \hat{\sigma}_-] = \hat{\sigma}_3$.
(iv) $[\hat{\sigma}_3, \hat{\sigma}_\pm] = \pm 2\hat{\sigma}_\pm$.

5.2   Harmonic Oscillator

In Section 2.6.2, we found the eigenvalues and eigenfunctions of the time-independent Schrödinger equation for a harmonic oscillator potential. Here, we will show that the solutions to the time-independent Schrödinger equation can be determined purely algebraically, without solving a differential equation.

First, we rephrase the results of Section 2.6.2 in the Hilbert space language. The harmonic oscillator Hamiltonian is defined on the Hilbert space $L^2(\mathbb{R}, dx)$ as

$$\hat{H} = \frac{1}{2m}\hat{p}^2 + \frac{1}{2}m\omega^2\hat{x}^2, \tag{5.21}$$

where $\hat{x}$ and $\hat{p}$ are the position and momentum operator, respectively. The operator $\hat{H}$ has purely discrete spectrum with nondegenerate eigenvalues $E_n = (n + \frac{1}{2})\omega$, where $n = 0, 1, 2, \ldots$.

5.2.1 Creation and Annihilation Operators

We define the *annihilation operator*

$$\hat{a} = \sqrt{\frac{m\omega}{2}}\hat{x} + \frac{i}{\sqrt{2m\omega}}\hat{p}; \tag{5.22}$$

its conjugate $\hat{a}^\dagger$ is called *creation operator*. We will see the reason for these names shortly.

The definition (5.22) implies that

$$[\hat{a}, \hat{a}^\dagger] = \hat{I}, \tag{5.23}$$

$$\hat{H} = \omega(\hat{a}^\dagger\hat{a} + \frac{1}{2}\hat{I}). \tag{5.24}$$

Equation (5.23) implies that the operators $\hat{a}$ and $\hat{a}^\dagger$ are not normal. Equation (5.24) asserts that $\hat{H}$ is a linear function of the operator $\hat{N} = \hat{a}^\dagger\hat{a}$; hence, the operators $\hat{H}$ and $\hat{N}$ have the same eigenvectors.

$\hat{N}$ is a positive operator: For any vector $|\psi\rangle$,

$$\langle\psi|\hat{N}|\psi\rangle = \|\hat{a}|\psi\rangle\|^2 \geq 0.$$

It is straightforward to obtain the commutation relations

$$[\hat{a}, \hat{N}] = \hat{a}, \qquad\qquad [\hat{a}^\dagger, \hat{N}] = -\hat{a}^\dagger. \tag{5.25}$$

Let $\hat{N}|n\rangle = \nu_n|n\rangle$ be the eigenvalue equation for $\hat{N}$; n stands for some integer that orders the eigenvalues. Acting the operators on both sides of Eq. (5.25) to $|n\rangle$, we obtain

$$\hat{N}\hat{a}|n\rangle = (\nu_n - 1)\hat{a}|n\rangle, \tag{5.26}$$

$$\hat{N}\hat{a}^\dagger|n\rangle = (\nu_n + 1)\hat{a}^\dagger|n\rangle. \tag{5.27}$$

These equations imply that the vectors $\hat{a}|n\rangle$ and $\hat{a}^\dagger|n\rangle$ are eigenvectors of $\hat{N}$, respectively. The eigenvalue for $\hat{a}|n\rangle$ is $\nu_n - 1$, and the eigenvalue of $\hat{a}^\dagger|n\rangle$ is $\nu_n + 1$.

Equation (5.26) implies that the eigenvalues ν_n must be integers, otherwise the repeated action of $\hat{a}$ on an eigenfunction would eventually lead to a negative eigenvalue of $\hat{N}$. We number the vectors $|n\rangle$ so that $\nu_n = 0$ for $n = 0$. Hence, the vector $|0\rangle$ satisfies $\hat{a}|0\rangle = 0$. We assume that there is no degeneracy in the lowest eigenvalue, so that the vector $|0\rangle$ is unique. Hence, the eigenvalue equation for $\hat{N}$ becomes $\hat{N}|n\rangle = n|n\rangle$, for $n = 0, 1, 2, \ldots$, that is, $\nu_n = n$.

The eigenvalues of the operator $\hat{N}$ are the number of energy quanta in a state. For this reason, we refer to $|n\rangle$ as the *number states* for the harmonic oscillator.

The normalized eigenvectors $|n\rangle$ are obtained by n successive actions of $\hat{a}^\dagger$ on $|0\rangle$. We note that $\langle n|\hat{a}\hat{a}^\dagger|n\rangle = \langle n|[\hat{a},\hat{a}^\dagger] + \hat{a}^\dagger\hat{a}|n\rangle = \langle n|\hat{I} + \hat{N}|n\rangle = (n+1)$. This implies that

$$\hat{a}^\dagger|n\rangle = \sqrt{n+1}|n+1\rangle, \tag{5.28}$$

hence,

$$|n\rangle = \frac{\hat{a}^\dagger}{\sqrt{n}}|n-1\rangle = \frac{(\hat{a}^\dagger)^2}{\sqrt{n(n-1)}}|n-2\rangle = \cdots = \frac{(\hat{a}^\dagger)^n}{\sqrt{n!}}|0\rangle. \tag{5.29}$$

Similarly, we find that $\langle n|\hat{a}^\dagger\hat{a}|n\rangle = n$, leading to

$$\hat{a}|n\rangle = \sqrt{n}|n-1\rangle. \tag{5.30}$$

Thus, we found the eigenvalues and eigenvectors of the operator $\hat{N}$ – hence, of the Hamiltonian (5.24) – without having to solve a differential equation. The only assumption in this calculation is that the lowest-energy state is not degenerate. This requirement is fully justified for the harmonic oscillator in one dimension – see Section 5.3.2. The names of $\hat{a}$ and $\hat{a}^\dagger$ are justified by the fact that the former annihilates an energy quantum and the latter creates an energy quantum.

The preceding analysis is easily adapted to systems with a degenerate ground state. Suppose that the zero eigenvalue of $\hat{N}$ corresponds to a D-dimensional eigenspace V_0. We choose an orthonormal basis on V_0 with vectors $|0,j\rangle$, $j = 1, 2, \ldots D$. The repeated action of $\hat{a}^\dagger$ on $|0,j\rangle$ leads to the formation of D "towers" of eigenvectors of $\hat{N}$,

$$|n,j\rangle = \frac{(\hat{a}^\dagger)^n}{\sqrt{n!}}|0,j\rangle, \tag{5.31}$$

one for each value of j. Every eigenvalue $\hat{N}$ has degeneracy D, since $\hat{N}|n,j\rangle = n|n,j\rangle$, for all $j = 1, 2, \ldots D$.

Example 5.1 It is often necessary to evaluate expressions of the form $\langle n|F(\hat{a},\hat{a}^\dagger)|n\rangle$, where F is a function of $\hat{a}$ and $\hat{a}^\dagger$. In such calculations, we proceed as follows. First, we ignore all terms in F that do not contain the same number of $\hat{a}$ and $\hat{a}^\dagger$, because their contribution vanishes. Second, we use commutation relations to express F as a function of $\hat{N}$ and of $\hat{I}$. In particular, we substitute $\hat{a}^\dagger\hat{a} = \hat{N}$ and $\hat{a}\hat{a}^\dagger = \hat{N} + \hat{I}$.

Let us see how this works for $\langle n|\hat{x}^4|n\rangle$. By definition, $\hat{x} = \frac{1}{\sqrt{2m\omega}}(\hat{a} + \hat{a}^\dagger)$, hence,

$$\langle n|\hat{x}^4|n\rangle = \frac{1}{4m^2\omega^2}\langle n|(\hat{a}+\hat{a}^\dagger)^4|n\rangle = \frac{1}{4m^2\omega^2}\langle n|(\hat{a}^2 + \hat{a}^{\dagger 2} + \hat{a}^\dagger\hat{a} + \hat{a}\hat{a}^\dagger)^2|n\rangle$$

$$= \frac{1}{4m^2\omega^2}\langle n|(\hat{a}^2 + \hat{a}^{\dagger 2} + 2\hat{N} + \hat{I})^2|n\rangle = \frac{1}{4m^2\omega^2}\langle n|\hat{a}^2\hat{a}^{\dagger 2} + \hat{a}^{\dagger 2}\hat{a}^2 + (2\hat{N}+\hat{I})^2|n\rangle.$$

In the last step, we kept only terms with the same number of $\hat{a}$ and $\hat{a}^\dagger$. Next, we use the commutation relations, to obtain

$$\hat{a}^2\hat{a}^{\dagger 2} = \hat{a}\hat{a}^\dagger\hat{a}\hat{a}^\dagger + \hat{a}\hat{a}^\dagger = (\hat{N}+\hat{I})^2 + \hat{N} + \hat{I}, \ \hat{a}^{\dagger 2}\hat{a}^2 = \hat{a}^\dagger\hat{a}\hat{a}^\dagger\hat{a} - \hat{a}^\dagger\hat{a} = \hat{N}^2 - \hat{N}.$$

We conclude that

$$\langle n|\hat{x}^4|n\rangle = \frac{1}{4m^2\omega^2}\langle n|6\hat{N}^2 + 6\hat{N} + 3\hat{I}|n\rangle = \frac{3}{2m^2\omega^2}\left(n^2 + n + \frac{1}{2}\right). \tag{5.32}$$

Example 5.2 We define the operator $\hat{V}$ of Example 4.4 for a harmonic oscillator by its action on the number states, $\hat{V}|n\rangle = |n+1\rangle$ for all $n = 0, 1, 2, \ldots$. By Eq. (5.28), $\hat{V} = \hat{a}^{\dagger}(\hat{N} + \hat{I})^{-1/2}$, hence, $\hat{a}^{\dagger} = \sqrt{\hat{a}\hat{a}^{\dagger}}\,\hat{V}$. The analogy with the polar decomposition of a classical wave amplitude $\alpha = \sqrt{\alpha\alpha^{*}}e^{i\phi}$ – compare with Eq. (1.26) – tempts us to identify $\hat{V}$ with a phase operator, namely, an operator that accounts for the phase of a harmonic oscillator. This is an issue of some importance in quantum optics, where a mode of the EM field corresponds to a harmonic oscillator and phase measurements are common.

The operator $\hat{V}$ represents a phase only if it is unitary. Then, it can be written as $e^{i\hat{\phi}}$, where $\hat{\phi}$ is a self-adjoint operator to represent the phase observables. But, as shown in Example 4.4, $\hat{V}$ is not unitary. It is not even a normal operator as $[\hat{V}, \hat{V}^{\dagger}] = -\hat{P}_0$, where $\hat{P}_0 = |0\rangle\langle 0|$. Hence, a self-adjoint operator for the phase does not exist. A more general argument against the existence of self-adjoint phase operators is given in Section 5.6.3.

Susskind and Glogower (1964) noticed that the operators $\hat{c} = \frac{1}{2}(\hat{V} + \hat{V}^{\dagger})$ and $\hat{s} = \frac{1}{2i}(\hat{V} - \hat{V}^{\dagger})$ are self-adjoint, and argued that they can be viewed as quantum analogues of the cosine and sine of the harmonic oscillator phase. To see this, we calculate

$$[\hat{c}, \hat{s}] = -\frac{i}{2}\hat{P}_0, \qquad \hat{c}^2 + \hat{s}^2 = \hat{I} - \frac{1}{2}\hat{P}_0. \tag{5.33}$$

Hence, outside the subspace of $|0\rangle$, $\hat{c}$ and $\hat{s}$ commute and their squares add up to unity, as would be expected from operators that represent the cosine and the sine of some angle.

We also evaluate the commutators $[\hat{V}, \hat{N}] = [\hat{a}^{\dagger}, \hat{N}](\hat{N} + \hat{I})^{-1/2} = -\hat{V}$, and $[\hat{V}^{\dagger}, \hat{N}] = (\hat{N} + \hat{I})^{-1/2}[\hat{a}, \hat{N}] = \hat{V}^{\dagger}$. They imply that

$$[\hat{c}, \hat{N}] = -i\hat{s}, \qquad [\hat{s}, \hat{N}] = i\hat{c}. \tag{5.34}$$

These are the commutations that would have been obtained, if we could write $\mp \cos\hat{\phi}$ and $\hat{s} = \sin\hat{\phi}$ in terms of a self-adjoint operator $\hat{\phi}$ that satisfies $[\hat{\phi}, \hat{N}] = i\hat{I}$. Since the operator $\hat{\phi}$ does not exist, the Susskind–Glogower (SG) trigonometric operators $\hat{c}$ and $\hat{s}$ are quantum observables that *approximately* represent the notion of the classical phase, for states $|\psi\rangle$ that have small overlap with the ground state, $|\langle 0|\psi\rangle| << 1$.

5.2.2 Coherent States

Next, we study the spectrum of the creation and annihilation operators. The two operators are not normal, hence, the spectrum will be very different from what we have encountered so far.

To solve the eigenvalue equation $\hat{a}^{\dagger}|w\rangle = w|w\rangle$ for the creation operator $\hat{a}^{\dagger}$, we substitute $|w\rangle = \sum_{n=0}^{\infty} c_n|n\rangle$, to obtain $\sum_{n=0}^{\infty} \sqrt{n+1}c_n|n+1\rangle = w\sum_{n=0}^{\infty} c_n|n\rangle$. We change the index of summation of the left-hand side to $n-1$ from n, to obtain

$$\sum_{n=1}^{\infty} (wc_n - \sqrt{n}c_{n-1})|n\rangle + wc_0|0\rangle = 0, \tag{5.35}$$

hence, $c_0 = 0$ and $c_n = c_{n-1}\sqrt{n}/w$. It follows that $c_0 = c_1 = c_2 = \cdots = 0$, so that $|w\rangle = 0$. The creation operator $\hat{a}^{\dagger}$ has no eigenvectors.

To solve the eigenvalue equation $\hat{a}|z\rangle = z|z\rangle$ for the annihilation operator $\hat{a}$, we substitute $|z\rangle = \sum_{n=0}^{\infty} c_n|n\rangle$, to obtain $\sum_{n=0}(\sqrt{n+1}c_{n+1} - zc_n)|n\rangle = 0$. It follows that

$$c_{n+1} = \frac{c_n z}{\sqrt{n+1}}. \tag{5.36}$$

Equation (5.36) yields $c_n = c_0 z^n / \sqrt{n!}$. The constant c_0 is specified by the normalization condition

$$1 = \langle z|z \rangle = c_0^2 \sum_{m=0}^{\infty} \sum_{n=0}^{\infty} \frac{z^{*m} z^n}{\sqrt{n! \, m!}} \langle m|n \rangle = c_0^2 \sum_{n=0}^{\infty} \frac{|z|^{2n}}{n!} = c_0^2 \exp(|z|^2). \tag{5.37}$$

We obtain $c_0 = \exp(-\frac{1}{2}|z|^2)$, modulo a phase factor.

To conclude, the eigenvectors $|z\rangle$ of the annihilation operator are

$$|z\rangle = e^{-\frac{|z|^2}{2}} \sum_{n=0}^{\infty} \frac{z^n}{\sqrt{n!}} |n\rangle = e^{-\frac{|z|^2}{2}} \sum_{n=0}^{\infty} \frac{z^n (\hat{a}^\dagger)^n}{n!} |0\rangle. \tag{5.38}$$

The vectors $|z\rangle$ are known as *coherent states*. By definition, $\langle z|\hat{a}^{\dagger m} \hat{a}^n |z \rangle = (z^*)^m z^n$ for all $m, n = 0, 1, 2, \ldots$. Also, $\langle n|z \rangle = e^{-\frac{1}{2}|z|^2} z^n / \sqrt{n!}$.

There is no restriction on the eigenvalue z; it can be any complex number. To better understand the physical meaning of the coherent states and of the number z, we write the associated wave functions $\psi_z(x) = \langle x|z \rangle$ in the position representation. Equation (5.22) suggests writing $z = \sqrt{\frac{m\omega}{2}} \bar{q} + \frac{i}{\sqrt{2m\omega}} \bar{p}$ for some real parameters $\bar{q}$ and $\bar{p}$, so that we can identify z with the pair $(\bar{q}, \bar{p}) \in \mathbb{R}^2$.

The eigenvalue equation $\hat{a}|z\rangle = z|z\rangle$ becomes $\frac{d\psi_z}{dx} = [-m\omega(x - \bar{q}) + i\bar{p}]\psi_z$, with solution

$$\psi_z(x) = \left(\frac{m\omega}{\pi}\right)^{1/4} e^{i\phi} \exp\left[-\frac{m\omega(x - \bar{q})^2}{2} + i\bar{p}x\right], \tag{5.39}$$

where ϕ is an arbitrary phase factor. In Problem 5.6 it is shown that Eq. (5.39) coincides with Eq. (5.38), if $e^{i\phi} = e^{-\frac{i}{2}\bar{q}\bar{p}}$.

Comparing with Eq. (2.41), we see that the coherent state is described by the wave function of the harmonic oscillator with its center shifted by $\bar{q}$ and with a phase term $e^{i\bar{p}x}$. Also note that, by Eq. (4.27), the Weyl operator $\hat{V}(a, b)$ acts on the coherent state $|z\rangle = |\bar{q}, \bar{p}\rangle$ as

$$\hat{V}(a, b)|\bar{q}, \bar{p}\rangle = e^{\frac{i}{2}(\bar{q}a + \bar{p}b)}|\bar{q} + b, \bar{p} + a\rangle. \tag{5.40}$$

For $\bar{q} = \bar{p} = 0$, Eq. (5.40) implies that the coherent state is obtained from the action of the Weyl operator on the vacuum.

The set of coherent states defines a map $z \to |z\rangle$ from the classical state space $\Gamma = \{(\bar{q}, \bar{p})\}$ to the Hilbert space $\mathcal{H}$ of the quantum theory. In Section 6.4, we will show that Gaussian states of the form (5.39) have minimum spread on Γ, so the map defined by the coherent states provides the optimal resolution of classical microstates afforded by quantum theory.

The coherent state map does not physically distinguish between neighboring state space points. As $\hat{a}$ is not self-adjoint, coherent states with different z are not orthogonal,

$$\langle z|z' \rangle = e^{-\frac{1}{2}|z|^2 - \frac{1}{2}|z'|^2} \sum_{n=0}^{\infty} \sum_{m=0}^{\infty} \frac{z^{*m} z'^n}{\sqrt{n! \, m!}} \langle m|n \rangle = e^{-\frac{1}{2}|z|^2 - \frac{1}{2}|z'|^2} \sum_{n=0}^{\infty} \frac{(z^* z')^n}{n!}$$

$$= e^{-\frac{1}{2}|z|^2 - \frac{1}{2}|z'|^2 + z^* z'}. \tag{5.41}$$

It follows that

$$|\langle z|z' \rangle|^2 = e^{-|z - z'|^2}. \tag{5.42}$$

In Chapter 6, we will see that $|\langle z|z'\rangle|^2$ is the probability that a particle prepared at state $|z\rangle$ will be found at $|z'\rangle$. As such, $|\langle z|z'\rangle|^2$ is a measure of how much quantum theory can distinguish between classical microstates: If $|\langle z|z'\rangle|^2 << 1$, then the two state space "points" z and z' can be efficiently distinguished. For $z = (\bar{q}, \bar{p})$ and $z' = (\bar{q}', \bar{p}')$,

$$|\langle z|z'\rangle|^2 = \exp\left[-\frac{m\omega}{2}(\Delta q)^2 - \frac{1}{2m\omega}(\Delta p)^2\right] < e^{-(\Delta q)(\Delta p)}, \tag{5.43}$$

where $\Delta q = \bar{q} - \bar{q}'$ and $\Delta p = \bar{p} - \bar{p}'$ and we employed the inequality $a^2 + b^2 \geq 2ab$. Hence, the two points z, z' are distinguished if $(\Delta q)(\Delta p) >> 1$, in full agreement with Heisenberg's uncertainty principle.

A crucial property of coherent states is their *overcompleteness*:

$$\int \frac{d^2 z}{\pi} |z\rangle\langle z| = \hat{I}, \tag{5.44}$$

where $d^2 z = dx\, dy$ for $z = x + iy$, $x, y \in \mathbb{R}$. To show this, we evaluate the matrix elements

$$I_{mn} := \int \frac{d^2 z}{\pi} \langle m|z\rangle\langle z|n\rangle = \int \frac{d^2 z}{\pi\sqrt{n!\, m!}} z^{*n} z^m e^{-|z|^2}. \tag{5.45}$$

Setting $z = re^{i\theta}$ for $r \geq 0$ and $\theta \in [0, 2\pi)$, we find

$$I_{mn} = \int_0^{2\pi} d\theta\, e^{i(m-n)\theta} \int_0^\infty \frac{r\, dr}{\pi\sqrt{n!\, m!}} r^{m+n} e^{-r^2}.$$

The integral with respect to θ vanishes unless $m = n$, in which case it equals 2π. Then, $I_{nn} = \int_0^\infty \frac{ds}{n!} s^n e^{-s} = 1$, where $s = r^2$. We conclude that $I_{mn} = \delta_{mn}$, which proves Eq. (5.44).

The overcompleteness relation is a generalization of the usual resolution of the unity for coherent states. It means that there are many more vectors than in a generalized basis such as momentum – hence the set of vectors is more than complete. We can analyze any vector on coherent states,

$$|\psi\rangle = \int \frac{d^2 z}{\pi} |z\rangle\langle z|\psi\rangle = \int \frac{d^2 z}{\pi} \psi(z^*)|z\rangle, \tag{5.46}$$

where $\psi(z^*) = \langle z|\psi\rangle$. We can also express the inner product

$$\langle\psi_1|\psi_2\rangle = \int \frac{d^2 z}{\pi} e^{-|z|^2} \langle\psi_1|z\rangle\langle z|\psi_2\rangle = \int \frac{d^2 z}{\pi} e^{-|z|^2} \psi_1^*(z)\psi_2(z). \tag{5.47}$$

Hence, coherent states provide a new representation for quantum states of a particle. This representation employs wave functions $\psi(z)$ with respect to the complexified state space variable z.

5.3   The Schrödinger Operator in One Dimension

The time-independent Schrödinger equation for a particle of mass m (2.34) in one dimension is an eigenvalue equation for the one-dimensional *Schrödinger operator*

$$\hat{H} = -\frac{1}{2m}\frac{\partial^2}{\partial x^2} + V(x), \tag{5.48}$$

defined on the Hilbert space $L^2(\mathbb{R})$. In what follows, we will examine the most important properties of the operator (5.48) for general potentials $V(x)$.

We have also referred to the operator (5.48) as the Hamiltonian of the system. In Chapter 7, we shall see that the notion of the Hamiltonian is more general, as it refers to the role of the operator in the time evolution of a quantum system, and not to a specific mathematical form. In many systems, the Hamiltonian takes forms very different from Eq. (5.48). The term "Schrödinger operator" is more appropriate when we talk about the specific mathematical form (5.48), that is, the sum of a kinetic energy term and a potential. In any case, we refer to the eigenvalues of (5.48) as energy eigenvalues.

5.3.1   General Properties

The eigenvalue equation for the Schrödinger operator can be written as

$$\psi'' + 2m[E - V(x)]\psi = 0. \tag{5.49}$$

Solutions to Eq. (5.49) are *eigenfunctions* of $\hat{H}$. The square integrable wave functions define the discrete spectrum of $\hat{H}$, as they are vectors of $L^2(\mathbb{R})$. They are called *bound* states, because they correspond to a particle being trapped by the potential. The nonsquare integrable eigenfunctions correspond to generalized eigenvectors and they define the continuous spectrum of $\hat{H}$. They are called *scattering states*, because they correspond to a particle being scattered by the potential and escaping far away.

The imaginary unit does not appear in Eq. (5.48). This means that for any complex solution ψ of Eq. (5.48), both the real and the imaginary part of ψ is also a solution to Eq. (5.48). Hence, we can always choose the eigenfunctions of $\hat{H}$ to be real.

For a symmetric potential, $V(x) = V(-x)$, if ψ is a solution to Eq. (5.49), then so is ψ_-, where $\psi_-(x) := \psi(-x)$. Both ψ and ψ_- correspond to the same eigenvalue E. This is trivially true, if ψ is even or odd. If ψ is neither even nor odd, we can define an even eigenfunction $\psi + \psi_-$ of $\hat{H}$ and an odd eigenfunction $\psi - \psi_-$ of $\hat{H}$. We conclude that in symmetric potentials, we can choose the eigenfunctions of $\hat{H}$ to be either even or odd.

Consider two different eigenfunctions ψ_1 and ψ_2 of $\hat{H}$ with eigenvalues E_1 and E_2, respectively,

$$-\psi_1'' = 2m[E_1 - V(x)]\psi_1, \qquad\qquad -\psi_2'' = 2m[E_2 - V(x)]\psi_2.$$

We multiply both sides of the first equality by ψ_2, and multiply both sides of the second equality by ψ_1. We subtract the resulting equations, to obtain $\psi_2\psi_1'' - \psi_1\psi_2'' = (E_2 - E_1)\psi_1\psi_2$, or equivalently,

$$(\psi_2\psi_1' - \psi_1\psi_2')' = (E_2 - E_1)\psi_1\psi_2. \tag{5.50}$$

The function $W = \psi_2\psi_1' - \psi_1\psi_2'$ is known as the Wronskian of the functions ψ_1, ψ_2.

For $E_2 = E_1$, Eq. (5.50) becomes

$$\psi_2\psi_1' - \psi_1\psi_2' = C \tag{5.51}$$

for some constant C.

A useful corollary follows if we apply Eq. (5.50) for $E_1 = E$, $E_2 = E + \delta E$, and take the limit $\delta E \to 0$. We write $\psi_1 = \psi_E$, $\psi_2 = \psi_{E+\delta E}$, and we integrate Eq. (5.50) from $-\infty$ to x,

$$\psi_{E+\delta E}\psi_E' - \psi_E\psi_{E+\delta E}' = \delta E \int_{-\infty}^{x} dx'\, \psi_E(x')\psi_{E+\delta E}(x'). \tag{5.52}$$

We divide both sides by $\psi_E(x)\psi_{E+\delta E}(x)$, to obtain

$$\frac{1}{\delta E}[b_{E+\delta E}(x) - b_E(x)] = -\frac{1}{\psi_E(x)\psi_{E+\delta E}(x)} \int_{-\infty}^{x} dx'\, \psi_E(x')\psi_{E+\delta E}(x'), \tag{5.53}$$

where $b_E(x) := \frac{d\ln\psi_E(x)}{dx}$ is the *logarithmic derivative* of the wave function at x.

We take the limit $\delta E \to 0$, to obtain

$$\frac{db_E(x)}{dE} = -\frac{1}{\psi_E(x)^2} \int_{-\infty}^{x} dx'\, \psi_E(x')^2 < 0. \tag{5.54}$$

Hence, the logarithmic derivative of a real eigenvalue of the Schrödinger operator is a decreasing function of the energy, for all x.

5.3.2 Discrete Spectrum

First, we present some important results about the discrete spectrum of the one-dimensional Schrödinger operator.

Theorem 5.1 *The operator $\hat{H}$ of Eq. (5.48) has no degenerate eigenvalues.*

Proof For square integrable eigenfunctions, the term on the left-hand side of Eq. (5.51) vanishes at infinity, hence, $C = 0$. It follows that $\psi_1'/\psi_1 = \psi_2'/\psi_2$ or $\frac{d}{dx}\ln(\psi_1/\psi_2) = 0$, hence, $\psi_1 = a\psi_2$ for some constant a. There is no degeneracy. $\qquad\qquad\square$

Since there is no degeneracy, we can label eigenvalues and eigenvectors with a single integer $n = 0, 1, 2, \ldots$, so that $E_0 < E_1 < E_2 < \cdots$. In this notation, $\psi_0(x)$ is the state of lowest energy, and $\psi_n(x)$ the nth excited state.

A *node* of a wave function ψ is any point $x_0 \in \mathbb{R}$ such that $\psi(x_0) = 0$. A crucial result about the nodes of the one-dimensional Schrödinger operator is *Sturm's comparison theorem*.

Theorem 5.2 (Comparison theorem) *The nth excited eigenfunction of the one-dimensional Schrödinger operator has n nodes.*

The comparison theorem is very powerful. It asserts that the nth eigenfunctions for any potential have similar shape, the same number of nodes, and, hence, the same number of extrema. We urge the reader to return to Section 2.6 and verify the comparison theorem for the harmonic oscillator. The proof of the comparison theorem is given in Box 5.1.

Finally, we note that for any eigenstate $|\psi\rangle$ of a Hamiltonian and for any operator $\hat{A}$, $\langle\psi|[\hat{A},\hat{H}]|\psi\rangle = 0$. For a Schrödinger operator $\hat{H}$, $\hat{p} = -im[\hat{x},\hat{H}]$ and $V'(x) = i[\hat{p},\hat{H}]$; hence,

$$\langle\psi|\hat{p}|\psi\rangle = 0, \qquad \langle\psi|V'(\hat{x})|\psi\rangle = 0. \tag{5.55}$$

Box 5.1 Proof of the Comparison Theorem

First, we define the *zero points* of a wave function ψ as its nodes together with the limits $\pm\infty$, given that a square integrable function vanishes at infinity.

The comparison theorem requires the following result.

Theorem 5.3 (Sturm's theorem) *Let ψ_a be an eigenfunction of the one-dimensional Schrödinger operator with eigenvalue E_a, and let x_0 and x_1 be two successive zero points of ψ_a. Then, any eigenfunction ψ_b of the Schrödinger operator with eigenvalue $E_b > E_a$ has one zero point between x_0 and x_1.*

Proof For the eigenfunctions ψ_a and ψ_b, Eq. (5.50) becomes

$$\frac{d}{dx}(\psi_b\psi_a' - \psi_a\psi_b') = 2m(E_b - E_a)\psi_b\psi_a.$$

Let x_0 and x_1 be two successive zero points of ψ_a. We integrate both sides of the above equation in the interval $[x_0, x_1]$. Since $\psi_a(x_0) = \psi_a(x_1) = 0$,

$$\psi_b(x_1)\psi_a'(x_1) - \psi_b(x_0)\psi_a'(x_0) = 2m(E_b - E_a)\int_{x_0}^{x_1} dx\,\psi_b(x)\psi_a(x). \tag{A}$$

Since x_0 and x_1 are successive zero points of ψ_a, ψ_a does not change sign in (x_0, x_1), so it is either positive or negative. In the former case, $\psi_a'(x_0) > 0$ and $\psi_a'(x_1) < 0$, because the function increases from zero at x_0 to a positive value and it decreases from a positive value to zero at x_1. Similarly, in the latter case, $\psi_a'(x_0) < 0$ and $\psi_a'(x_1) > 0$.

We assume that ψ_b does not vanish in (x_0, x_1), and we will show that a contradiction follows. By assumption, ψ_b does not change sign in (x_0, x_1), it is either positive or negative. All possible cases of signs are summarized in the following table.

ψ_a	ψ_b	$\psi_a'(x_0)$	$\psi_b'(x_1)$	lhs	rhs
+	+	+	−	−	+
+	−	+	−	+	−
−	+	−	+	+	−
−	−	−	+	−	+

In the table, we see that the left-hand side (lhs) of Eq. (A) is positive when the right-hand side (rhs) is negative, and vice versa. Hence, we obtained a contradiction. We conclude that ψ_b vanishes at least once in (x_0, x_1).

Square integrable eigenfunctions of the Schrödinger operator have $\pm\infty$ as zero points. For these eigenfunctions, the number of nodes increases with energy. To see this, consider a square integrable eigenfunction $\psi(x)$ with two nodes, at x_1 and at $x_2 > x_1$. This function has four zero points, $-\infty, x_1, x_2, \infty$. Any eigenfunction of higher energy has at least one node in $(-\infty, x_1)$, one in (x_1, x_2), and one in (x_2, ∞), that is, at least three nodes.

The minimum number of nodes (none) is possessed by the eigenfunction of lowest energy, ψ_0. The eigenfunction ψ_1 can only have one node, and so on. Hence, ψ_n has n nodes, as asserted by the comparison theorem.

5.3.3 Special Cases

Potentials that grow unboundedly at infinity: $\lim_{|x|\to\infty} V(x) = \infty$. For large $|x|$, Eq. (5.49) takes the form $\psi'' - k^2(x)\psi = 0$, where $k(x) = \sqrt{2m[V(x) - E]} > 0$. We write $\psi(x) = e^{S(x)}$, and we note that the eigenvalue equation is approximately satisfied for $S(x) = \pm\int^x k(x')dx'$, if

$$k'(x) << k^2(x). \tag{5.56}$$

The condition (5.56) is always satisfied for $|x| \to \infty$. To see this, note that for large $|x|$, $k(x) = \sqrt{2mV(x)}$, and Eq. (5.56) becomes $\frac{V'}{V^{3/2}} < 2\sqrt{2m}\epsilon$, where $\epsilon << 1$. For $x \to \infty$, $V' > 0$; hence, we can integrate both sides of the inequality to obtain $c - \frac{1}{2\sqrt{V}} < 2\sqrt{2m}\epsilon x$, where c is a constant. Since $V \to \infty$, this inequality holds for sufficiently large x. A similar proof applies for the limit $x \to -\infty$.

Therefore, for $|x| \to \infty$, the solutions to Eq. (5.49) take the form $\exp[\pm\int^x k(x')dx']$. There are no solutions with oscillatory behavior at infinity; hence, there are no generalized eigenvectors and no continuous spectrum. Excluding divergent solutions, we obtain an asymptotic behavior $\exp[-\int^x k(x')dx']$ for $x \to \infty$ and $\exp[\int^x k(x')dx']$ for $x \to -\infty$. The eigenfunctions of $\hat{H}$ vanish sufficiently fast at infinity; hence, they are square integrable. Since there is no continuous spectrum, the eigenfunctions define a Hilbert space basis. Hence, they are infinite in number because the Hilbert space $L^2(\mathbb{R})$ is infinite-dimensional. Therefore, we conclude the following.

Theorem 5.4 *If* $\lim_{|x|\to\infty} V(x) = \infty$, *the Schrödinger operator has an infinite number of eigenvectors.*

The theorem implies that potentials that classically never allow a particle to escape continue to do so in quantum theory.

Potentials unbounded from below at infinity: $\lim_{|x|\to\infty} V(x) = -\infty$. In this case, the eigenvalue equation for large $|x|$ becomes $\psi'' + \lambda^2(x)\psi = 0$, where $\lambda(x) = \sqrt{2m[E + |V(x)|]} > 0$. The solutions are oscillatory at infinity, irrespective of E. They correspond to generalized eigenvectors and, hence, to a continuous spectrum for all real values of E. The spectrum has double degeneracy, because there are two solutions for each E, behaving asymptotically as $e^{i\int^x dx'\lambda(x')}$ and $e^{-i\int^x dx'\lambda(x')}$.

For such potentials, the spectrum of the Hamiltonian has no lower bound. In Chapter 7, we will see that this property is physically inadmissible.

Potentials that vanish at infinity: $\lim_{|x| \to \infty} V(x) = 0$. The eigenvalue equation for large $|x|$ takes the form $\psi'' + 2mE\psi = 0$. For $E > 0$, the solution is oscillatory; hence, we have generalized eigenvectors and a continuous spectrum. For $E < 0$, there are solutions that vanish as $e^{-|Ex|}$ at infinity, hence, they are square integrable. This means that the spectrum for $E < 0$ is discrete.

While the continuous spectrum for $E > 0$ is always present, negative eigenvalues may not always exist. For example, if $V(x) > 0$, that is, the potential is everywhere repulsive, then the Hamiltonian $\hat{H}$ is a sum of two positive operators. Therefore, $\hat{H}$ is a positive operator, and it has no negative eigenvalues. Then the spectrum is purely continuous.

In contrast, potentials that are mostly attractive have negative eigenvalues. By "mostly attractive," we mean a potential $V(x)$, such that $v := \int_{-\infty}^{\infty} dx\, V(x) < 0$. We will show that, in this case, the Schrödinger operator is nonpositive, by finding a vector $|\psi\rangle$, such that $\langle\psi|\hat{H}|\psi\rangle < 0$. For the unnormalized Gaussian function $\psi(x) = e^{-\frac{x^2}{4\sigma^2}}$, $\langle\psi|\hat{p}^2|\psi\rangle = \sqrt{\frac{\pi}{2}}\sigma^{-1}$. Furthermore, $\langle\psi|\hat{V}|\psi\rangle = \int_{-\infty}^{\infty} dx\, V(x)e^{-\frac{x^2}{2\sigma^2}}$. At the limit of very large σ, $\langle\psi|\hat{V}|\psi\rangle \to v$. At the same limit, the term, $\sqrt{\frac{\pi}{2}}\sigma^{-1}$ is negligible. Consequently, $\langle\psi|\hat{H}|\psi\rangle \to v < 0$. Therefore, for sufficiently large σ, $\langle\psi|\hat{H}|\psi\rangle < 0$. We have proved the following.

Theorem 5.5 *If* $\lim_{|x| \to \infty} V(x) = 0$ *and* $\int_{-\infty}^{\infty} dx\, V(x) < 0$, *the Schrödinger operator has at least one negative eigenvalue.*

Hence, even a very weak attractive potential can trap a particle. In Chapter 13, we will see that the Schrödinger operator in three dimensions does not share this property.

5.3.4   Scattering States

We continue the analysis of the Schrödinger operator for potentials that vanish at infinity, but we focus on the continuous spectrum.

For simplicity, we assume that the potentials vanish outside a finite interval $U = [a, b]$, such potentials are said to have *finite support*. Outside U, the Schrödinger operator coincides with the kinetic energy operator $\hat{T} = -\frac{1}{2m}\frac{\partial^2}{\partial x^2}$. The generalized eigenvalues of $\hat{T}$ have *double degeneracy*, that is, there are two different eigenfunctions e^{ikx} and e^{-ikx} for each generalized eigenvalue $E = \frac{k^2}{2m}$. This conclusion is generalized to potentials that vanish asymptotically at infinity.

Theorem 5.6 *If the potential* $V(x)$ *vanishes at* $\pm\infty$, *every* $E \in [0, \infty)$ *is a doubly degenerate generalized eigenvalue of the one-dimensional Schrödinger operator* $\hat{H}$.

Consider a potential that is nonvanishing only in the interval $[a, b]$ (region II). In regions I ($x < a$) and III ($x > b$), the eigenvalue equation is $-\psi'' = 2mE\psi$, with general solution $\psi = Ae^{ikx} + Be^{-ikx}$, for $k = \sqrt{2mE}$.

For every value of the energy E, we select two eigenfunctions of $\hat{H}$, in terms of their behavior in regions I and III:

$$f_{k+}(x) = \begin{cases} \frac{1}{\sqrt{2\pi}}(e^{ikx} + R_k e^{-ikx}) & x < a \\ \frac{1}{\sqrt{2\pi}} T_k e^{ikx} & x > b \end{cases}$$

$$f_{k-}(x) = \begin{cases} \frac{1}{\sqrt{2\pi}} \bar{T}_k e^{-ikx} & x < a \\ \frac{1}{\sqrt{2\pi}}(e^{-ikx} + \bar{R}_k e^{ikx}) & x > b. \end{cases} \tag{5.57}$$

The solutions f_{k+} correspond to waves that are outcoming in region III, that is, they move away from region II and they propagate towards $+\infty$. The solutions f_{k-} correspond to outcoming waves in region I, that is, they move away from region II and propagate towards $-\infty$. In an analogy with classical waves, the solutions f_{k+} correspond to wave packets coming from $-\infty$, a part of them reflected on the potential, and the remainder crossing into region III. An analogous description holds for solutions f_{k-}, with the inverse direction of propagation.

The coefficients R_k and $\bar{R}_k$ are called *reflection amplitudes*, and the coefficients T_k and $\bar{T}_k$ are called *transmission amplitudes*. In classical waves, $|T_k|^2$ is the fraction of the wave's energy being transmitted through the barrier. In Section 7.3, we will prove that, in quantum theory, $|T_k|^2$ is the probability that a particle with energy E coming from the left of the barrier is found in region III. For this reason, $|T_k|^2$ is called *transmission probability*.

The coefficients in Eq. (5.57) are not independent. For any pair of the four functions, f_{k+}, f_{k-}, f_{k+}^*, and f_{k-}^*, Eq. (5.51) applies, because all four functions are solutions to the eigenvalue equation for the same value of E. We find that

$$f_{k+} f_{k+}^{*\prime} - f_{k+}^* f_{k+}' = \begin{cases} -\frac{ik(1-|R_k|^2)}{\pi} & x < a \\ -\frac{ik|T_k|^2}{\pi} & x > b \end{cases} \tag{5.58}$$

$$f_{k-} f_{k-}^{*\prime} - f_{k-}^* f_{k-}' = \begin{cases} \frac{ik|\bar{T}_k|^2}{\pi} & x < a \\ \frac{ik(1-|\bar{R}_k|^2)}{\pi} & x > b \end{cases} \tag{5.59}$$

$$f_{k+} f_{k-}' - f_{k-} f_{k+}' = \begin{cases} -\frac{ik\bar{T}_k}{\pi} & x < a \\ -\frac{ik\bar{T}_k}{\pi} & x > b \end{cases} \tag{5.60}$$

$$f_{k+} f_{k-}^{*\prime} - f_{k-}^* f_{k+}' = \begin{cases} \frac{ik\bar{T}_k^* R_k}{\pi} & x < a \\ -\frac{ik T_k \bar{R}_k^*}{\pi} & x > b. \end{cases} \tag{5.61}$$

The above quantities must take the same value in regions I and III. Hence, we obtain four identities,

$$|T_k|^2 + |R_k|^2 = 1, \quad |\bar{T}_k|^2 + |\bar{R}_k|^2 = 1, \quad T_k = \bar{T}_k, \quad \bar{T}_k^* R_k + T_k \bar{R}_k^* = 0. \tag{5.62}$$

The four complex amplitudes can be expressed as a function of only three independent variables. We choose these variables to be the norm $|T_k|$ of the transmission amplitude and the phases $e^{i\phi_k} := R_k/|R_k|$ and $e^{i\bar{\phi}_k} := \bar{R}_k/|\bar{R}_k|$ of the reflection amplitudes. Equations (5.62) imply that $|R_k|^2 = |\bar{R}_k|^2 = 1 - |T_k|^2$, and that the phase of the transmission amplitude, $e^{i\theta_k} := T_k/|T_k|$, is a function of ϕ_k and $\bar{\phi}_k$, $2\theta_k = \pi + \phi_k + \bar{\phi}_k$.

Of particular importance is the case of a symmetric potential, for which $f_{k+}(-x)$ are also eigenfunctions of the Schrödinger operator, and they coincide with $f_{k-}(x)$. This means that, for symmetric potentials, $R_k = \bar{R}_k$. It follows that $\phi_k = \bar{\phi}_k$ and that $\theta_k = \phi_k + \frac{\pi}{2}$.

Suppose we translate the potential by distance a: $V(x) \rightarrow V(x + a)$. Then, the translated eigenfunctions $f_{k\pm}(x + a)$ are eigenfunctions of the Schrödinger operator with the translated potential. However, $f_{k+}(x + a)$ is equal, modulo a phase e^{ika}, to $f_{k+}(x)$ with $R_k \rightarrow R_k e^{-2ika}$ and $T_k \rightarrow T_k$. Similarly, $f_{k-}(x + a)$ is equal, modulo a phase e^{-ika}, to $f_{k-}(x)$ with $\bar{R}_k \rightarrow \bar{R}_k e^{2ika}$ and $\bar{T}_k \rightarrow T_k$. Hence, a translation of the potential by a changes R_k by a phase factor of e^{-2ika} and $\bar{R}_k$ by a phase factor of e^{2ika}.

We will represent the functions f_{k+} and f_{k-} by kets $|k, +\rangle$ and $|k, -\rangle$, respectively. Since these kets are generalized eigenvectors of a self-adjoint operator, the inner product between kets of different k vanish. In fact, we can prove that

$$\langle k, + | k', +\rangle = \delta(k - k'), \quad \langle k, - | k', -\rangle = \delta(k - k'), \quad \langle k, + | k, -\rangle = 0. \tag{5.63}$$

Example 5.3 Consider the potential $V(x) = ax^2 + \frac{b}{x^2+1}$, where a and b are constants. The spectrum of the associated Schrödinger operator and the corresponding degeneracies g are given in the following table.

Parameters	Discrete spectrum	Continuous spectrum
$a > 0$	$E > 0, g = 1$	none
$a = 0, b > 0$	none	$E > 0, g = 2$
$a = 0, b < 0$	$E < 0, g = 1$	$E > 0, g = 2$
$a < 0$	none	$E \in \mathbb{R}, g = 2$

5.3.5 Transfer Matrix

In many cases, we can manipulate particles so that they pass through successive media, each of which can be modeled by a potential $V_i(x)$ of finite support – see Fig. 5.1. We want to evaluate the transmission amplitude for the total system, provided that we know the transmission and reflection amplitudes for each of the potentials.

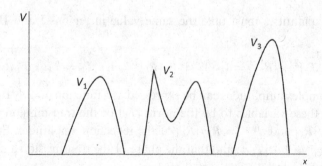

Fig. 5.1 Successive potentials of finite support.

Let the wave function to the left of any potential be of the form $f(x) = Ae^{ikx} + Be^{-ikx}$. As the "wave" passes through a potential V, it changes to $Ce^{ikx} + De^{-ikx}$. The coefficients C and D are obtained from A and B through the action of a 2×2 matrix Σ, the *transfer matrix*,

$$\begin{pmatrix} C \\ D \end{pmatrix} = \Sigma \begin{pmatrix} A \\ B \end{pmatrix}. \tag{5.64}$$

A comparison with (5.57) yields

$$\begin{pmatrix} T \\ 0 \end{pmatrix} = \Sigma \begin{pmatrix} 1 \\ R \end{pmatrix}, \qquad\qquad \begin{pmatrix} \bar{R} \\ 1 \end{pmatrix} = \Sigma \begin{pmatrix} 0 \\ T \end{pmatrix},$$

where we dropped the k dependence from the transmission and reflection amplitudes. We solve these equations for Σ, to obtain

$$\Sigma = \begin{pmatrix} T^{*-1} & \bar{R}T^{-1} \\ -RT^{-1} & T^{-1} \end{pmatrix}. \tag{5.65}$$

It is straightforward to follow the changes of the wave function $f(x) = Ae^{ikx} + Be^{-ikx}$ through any sequence of potentials by multiplying the associated transfer matrices. For two successive potentials with transfer matrices Σ_1 and Σ_2,

$$\Sigma_{tot} = \Sigma_2 \Sigma_1 = \begin{pmatrix} T_2^{*-1} & \bar{R}_2 T_2^{-1} \\ -R_2 T_2^{-1} & T_2^{-1} \end{pmatrix} \begin{pmatrix} T_1^{*-1} & \bar{R}_1 T_1^{-1} \\ -R_1 T_1^{-1} & T_1^{-1} \end{pmatrix}. \tag{5.66}$$

To evaluate the total transmission amplitude T_{tot}, it suffices to compute the 22 element of Σ_{tot}. We find that

$$T_{tot} = \frac{T_1 T_2}{1 - \bar{R}_1 R_2}. \tag{5.67}$$

The physical interpretation of Eq. (5.67) follows from the expansion of the denominators as a geometric series,

$$T_{tot} = T_1 T_2 \sum_{n=0}^{\infty} (\bar{R}_1 R_2)^n.$$

Each term $T_1 T_2 (\bar{R}_1 R_2)^n$ corresponds to the amplitude $T^{(n)}$ that a particle crossed the two potentials in the $(n+1)$th attempt, that is, after n reflections on the second barrier and another n reflections on the first barrier. Then, T_{tot} is the sum of all amplitudes $T^{(n)}$.

Resonances Suppose that the two potentials in the preceding example are identical, but the second one is located at distance a from the first. Then, $T_1 = T_2 = T$ and $\bar{R}_1 = \bar{R}e^{-ika}$ and $R_2 = Re^{-ika}$. It follows that

$$T_{tot} = \frac{T^2}{1 - \bar{R}Re^{-2ika}}. \tag{5.68}$$

Let us write $\bar{R}R = |R|^2 e^{i\theta}$. Then,

$$|T_{tot}|^2 = \frac{|T|^4}{1 + |R|^4 - 2|R|^2 \cos(2ka - \theta)}. \tag{5.69}$$

Fig. 5.2 Plot of $|T_{tot}|^2$ of Eq. (5.69) as a function of ka. In this plot, $|T|^2 = 0.1$ and $\theta = \frac{\pi}{4}$.

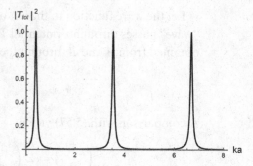

$|T_{tot}|^2$ changes periodically as we vary a – see Fig. 5.2. For $\cos\left[(2ka - \theta)\right] = 1$, $|T_{tot}|^2 = 1$, that is, there is full transmission through the barrier, even if the transmission probability $|T|^2$ of individual barriers is very small. This phenomenon is known as *scattering resonance*. For fixed a, resonance occurs at specific values of energy $E = \frac{k^2}{2m}$, the *resonance* frequencies of the composite potential.

Let E_0 be a resonance frequency. We Taylor-expand the oscillatory term around E_0, $\cos(2ka - \theta) \simeq 1 - \frac{1}{2}b(E - E_0)^2$, for some $b > 0$. The first term in the Taylor expansion vanishes as the resonance occurs at an extremum of the cosine. We assume that the resonance is so sharp that $|T|^2$ is practically a constant. Thus, we obtain

$$|T_{tot}|^2 = \frac{1}{1 + \frac{|R|^2 b}{|T|^4}(E - E_0)^2}. \tag{5.70}$$

The transmission probability near resonance behaves like a Lorentzian. Eq. (5.70) is known as the *Breit–Wigner* formula for resonances. The width of the Lorentzian is $\frac{|T|^2}{|R|\sqrt{b}}$, that is, the peak is sharper when $|T|^2$ is small.

5.4   Schrödinger Operator for Piecewise Constant Potentials

There are very few potentials, for which we can find exact expressions for the eigenvalues and eigenvectors of the Schrödinger operator. However, in many problems we only need a model for the potential that captures its basic features. Often it is a good approximation to describe a potential as piecewise constant, i.e., taking constant values everywhere except for a finite number of points of discontinuity.

In intervals that do not contain the points of discontinuity of a piecewise constant potential, the eigenvalue equation is simply $\psi'' + 2m(E - V)\psi = 0$, where V is a constant. For $E > V$, solutions are of the form $\psi = Ae^{ikx} + Be^{-ikx}$, where $k = \sqrt{2m(E - V)}$, and A and B are constants. For $E < V$, solutions are of the form $\psi = Ce^{-\lambda x} + De^{\lambda x}$, where $\lambda = \sqrt{2m(V - E)}$, and C and D are constants. For $E = V$, solutions are of the form $\psi = Fx + G$, where F and G are constants.

We must combine solutions from intervals with different values of V. To this end, we must find how an eigenfunction $\psi(x)$ of the Schrödinger operator behaves at the points of discontinuity.

Let $x = x_0$ be such a point. Obviously, ψ must be continuous at x_0. Furthermore, if we integrate the eigenvalue equation $\psi'' + 2m[E - V(x)]\psi = 0$ in the interval $[x_0 - \epsilon, x_0 + \epsilon]$, where $\epsilon > 0$ is very small, we obtain

$$\psi'(x_0 + \epsilon) - \psi'(x_0 - \epsilon) = -2m \int_{x_0 - \epsilon}^{x_0 + \epsilon} dx[E - V(x)]\psi(x). \tag{5.71}$$

At the limit $\epsilon \to 0$, the integral at the right-hand side of Eq. (5.71) vanishes, because $\psi(x)$ is continuous. Hence, $\lim_{\epsilon \to 0}[\psi'(x_0 + \epsilon) - \psi'(x_0 - \epsilon)] = 0$, i.e., the first derivative of ψ is continuous at $x = x_0$. We conclude that in the points of discontinuity of the potential, we require continuity of $\psi(x)$ and of its first derivative.

5.4.1 Square-Well Potential

First, we study the Schrödinger operator in a square-well potential

$$V(x) = \begin{cases} -V_0 & -\frac{a}{2} \le x \le \frac{a}{2} \\ 0 & |x| > \frac{a}{2} \end{cases}, \tag{5.72}$$

where $V_0 > 0$ is the well's *depth* and a is its width.

We focus on the discrete spectrum of the Schrödinger operator that corresponds to negative eigenvalues. For $-V_0 < E < 0$, solutions for $|x| > a/2$ are linear combinations of $e^{\lambda x}$ and of $e^{-\lambda x}$, where $\lambda = \sqrt{|E|}$. A square integrable function vanishes as $x \to \pm\infty$. Hence, the acceptable solutions for $|x| > a/2$ are proportional to $e^{-\lambda|x|}$.

Since the potential is symmetric, eigenfunctions have fixed parity: They are either even, denoted by ψ_+, or odd, denoted by ψ_-:

$$\psi_+(x) = \begin{cases} A_+ e^{\lambda x} & x < -\frac{a}{2} \\ B\cos kx & -\frac{a}{2} \le x \le \frac{a}{2} \\ A_+ e^{-\lambda x} & x > \frac{a}{2} \end{cases}, \quad \psi_-(x) = \begin{cases} -A_- e^{\lambda x} & x < -\frac{a}{2} \\ C\sin kx & -\frac{a}{2} \le x \le \frac{a}{2} \\ A_- e^{-\lambda x} & x > \frac{a}{2} \end{cases},$$

where $k = \sqrt{2m(V_0 - |E|)}$.

The requirement of continuity of the eigenfunction and of its first derivative at $x = \pm\frac{a}{2}$ yields four equations for each eigenfunction. Because of symmetry, the conditions at $x = \frac{a}{2}$ are identical to those at $x = -\frac{a}{2}$. Hence, we end up with two equations for each function:

- for ψ_+: $A_+ e^{-\lambda a/2} = B\cos\left(\frac{1}{2}ka\right)$ and $\lambda A_+ e^{-\lambda a/2} = kB\sin\left(\frac{1}{2}ka\right)$,
- for ψ_-: $A_- e^{-\lambda a/2} = C\sin\left(\frac{1}{2}ka\right)$ and $-\lambda A_- e^{-\lambda a/2} = kC\cos\left(\frac{1}{2}ka\right)$.

Dividing the two equations for ψ_+ and the two equations for ψ_-, we obtain

$$\tan\left(\frac{1}{2}ka\right) = \lambda/k \text{ for } \psi_+, \qquad \cot\left(\frac{1}{2}ka\right) = -\lambda/k \text{ for } \psi_-. \tag{5.73}$$

We rewrite Eq. (5.73) using dimensionless variables $x := |E|/V_0 \in [0,1]$ and $b := \sqrt{\frac{1}{2}mV_0 a^2}$,

$$\sqrt{x} = \sqrt{1-x}\tan(b\sqrt{1-x}) \text{ for } \psi_+, \tag{5.74}$$
$$\sqrt{x} = -\sqrt{1-x}\cot(b\sqrt{1-x}) \text{ for } \psi_-. \tag{5.75}$$

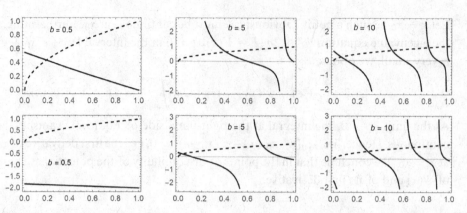

Fig. 5.3 Graphical solution of Eqs. (5.74) and (5.75). In the upper plots, we find the intersection points of the graphs $y_1 = \sqrt{x}$ (dashed line) and $y_2 = \sqrt{1-x}\tan(b\sqrt{1-x})$ for different values of b. In the lower plots, we specify the intersection points of the graphs $y_1 = \sqrt{x}$ (dashed line) and $y_2 = -\sqrt{1-x}\cot(b\sqrt{1-x})$ for different values of b. In both cases, y_2 undergoes more oscillations in $[0,1]$ with increasing b, hence, the number of solutions to Eq. (5.74) and Eq. (5.75) increases.

Equations (5.74) and (5.75) admit one or more solutions depending on the value of b. For $b \ll 1$, we keep only the first term in the Taylor expansion of the tangent. Then, Eq. (5.74) becomes $\sqrt{x} = b(1-x)$, with solution $x \simeq b$. As b increases, more periods of the tangent with values in $(-\infty, \infty)$ become relevant, and Eq. (5.74) admits more solutions. The same holds for solutions to Eq. (5.75), with one difference. For $b \ll 1$, Eq. (5.75) becomes $b\sqrt{x} = -1$, with no admissible solution. Hence, for $b \ll 1$ there is only one even eigenfunction. This behavior is shown in Fig. 5.3.

We conclude that the number N_- of eigenvalues to the Schrödinger operator increases both with the depth and with the width a of the potential.

5.4.2   Square-Barrier Potential

Next, we study the Schrödinger operator in a square-barrier potential,

$$
V(x) = \begin{cases} V_0 & -\frac{a}{2} \le x \le \frac{a}{2} \\ 0 & |x| > \frac{a}{2} \end{cases}, \tag{5.76}
$$

where $V_0 > 0$ is the barrier's *height* and a is the barrier's width.

The Schrödinger operator for this potential is positive; hence, it has only continuous spectrum with eigenfunctions f_{k+} and f_{k-}, as in (5.57). It suffices to consider the solutions f_{k+}, because $f_{k-}(x) = f_{k+}(-x)$ by the symmetry of the potential. We focus on energies energy $E < V_0$. This is the most interesting case for highlighting an important difference between classical and quantum physics. Classically, a particle with energy $E < V_0$ cannot cross the barrier. For $E < V_0$,

$$
f_{k+}(x) = \begin{cases} \frac{1}{\sqrt{2\pi}}(e^{ikx} + R_k e^{-ikx}) & x < -\frac{a}{2} \\ \frac{1}{\sqrt{2\pi}}(Ce^{\lambda x} + De^{-\lambda x}) & -\frac{a}{2} \le x \le \frac{a}{2} , \\ \frac{1}{\sqrt{2\pi}}T_k e^{ikx} & x > \frac{a}{2} \end{cases} \tag{5.77}
$$

where $k = \sqrt{2mE}$, $\lambda = \sqrt{2m(V_0 - E)}$, and C, D are constants.

The continuity of f_{k+} and of its first derivative at $x = \frac{a}{2}$ leads to the equations $Ce^{\lambda a/2} + De^{-\lambda a/2} = T_k e^{ika/2}$, and $Ce^{\lambda a/2} - De^{-\lambda a/2} = (ik/\lambda)T_k e^{ika/2}$, which allow us to express C and D in terms of T_k,

$$C = \frac{1}{2}\left(1 + i\frac{k}{\lambda}\right)e^{-\lambda a/2}e^{ika/2}T_k, \quad D = \frac{1}{2}\left(1 - i\frac{k}{\lambda}\right)e^{\lambda a/2}e^{ika/2}T_k. \tag{5.78}$$

The continuity of f_{k+} and of its first derivative at $x = -\frac{a}{2}$ leads to the equations

$$Ce^{-\frac{1}{2}\lambda a} + De^{\frac{1}{2}\lambda a} = e^{-\frac{i}{2}ka} + R_k e^{\frac{i}{2}ka},$$
$$(-i\lambda/k)\left(Ce^{-\frac{1}{2}\lambda a} - De^{\frac{1}{2}\lambda a}\right) = e^{-\frac{i}{2}ka} - R_k e^{\frac{i}{2}ka}.$$

We add these equations, to obtain $2e^{-\frac{i}{2}ka} = (1 - \frac{i\lambda}{k})e^{-\frac{1}{2}\lambda a}C + (1 + \frac{i\lambda}{k})e^{\frac{1}{2}\lambda a}D$. We substitute C and D from Eq. (5.78), and we solve for the transmission amplitude

$$T_k = \frac{2e^{-ika}}{\left[1 - \frac{i}{2}\left(\frac{k}{\lambda} - \frac{\lambda}{k}\right)\right]e^{\lambda a} + \left[1 + \frac{i}{2}\left(\frac{k}{\lambda} - \frac{\lambda}{k}\right)\right]e^{-\lambda a}}. \tag{5.79}$$

In the *opaque barrier* limit, $\lambda a \gg 1$, the denominator of Eq. (5.79) is dominated by the $e^{\lambda a}$ term, hence,

$$T_k = \frac{e^{-ika}e^{-\lambda a}}{1 - \frac{i}{2}\left(\frac{k}{\lambda} - \frac{\lambda}{k}\right)}. \tag{5.80}$$

The transmission amplitude drops exponentially with the barrier's width.

Nonetheless, the transmission amplitude is nonzero. Hence, there is nonzero probability that the particle crosses the barrier, even if its energy is smaller than the barrier's height. This is a characteristic quantum phenomenon, known as *tunneling*.

By Eq. (5.79), we evaluate the transmission probability,

$$|T_k|^2 = \frac{1}{\cosh^2(\lambda a) + \frac{1}{4}\left(\frac{k}{\lambda} - \frac{\lambda}{k}\right)^2 \sinh^2(\lambda a)} = \frac{1}{1 + \frac{1}{4}\left(\frac{k}{\lambda} + \frac{\lambda}{k}\right)^2 \sinh^2(\lambda a)}$$
$$= \frac{1}{1 + \frac{\sinh^2(b\sqrt{1-x})}{4x(1-x)}}, \tag{5.81}$$

where in the last step, we used the dimensionless variables $x := E/V_0$ and $b := \sqrt{2mV_0}a$. In Fig. 5.4, we plot the logarithm of $|T_k|^2$ as a function of $x = E/V_0$, for different values of b. The transmission probability increases with x for constant b and it decreases with b for constant x.

5.5 Schrödinger Operator for Delta-Function Potentials

At the limit where the height (or the depth) V_0 of a square potential goes to infinity and its width a goes to zero, while the product $\eta = V_0 a$ remains constant, the potential is proportional to a delta function,

$$V(x) = \eta\delta(x). \tag{5.82}$$

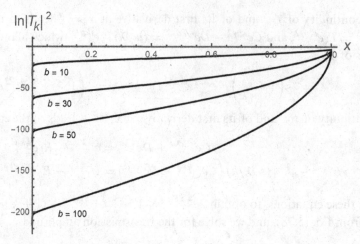

Fig. 5.4 The logarithm of the transmission probability $|T_k|^2$, Eq. (5.81), as a function of $x = E/V_0 \in [0, 1]$, for different values of $\beta = \sqrt{2mV_0}a$.

We can solve the eigenvalue problem for delta-function potentials by taking the appropriate limits to the quantities evaluated for the square well and the square barrier. However, it is more useful to proceed by finding a junction condition for eigenfunctions $\psi(x)$ of the Schrödinger operator around points x_0, where the potential has a delta-function behavior. First, we require that $\psi(x)$ is continuous at x_0. Second, we set $V(x) = \eta\delta(x - x_0)$ in Eq. (5.71), to obtain

$$\lim_{\epsilon \to 0} [\psi'(x_0 + \epsilon) - \psi'(x_0 - \epsilon)] = 2m\eta\psi(x_0), \tag{5.83}$$

that is, the first derivative of ψ is discontinuous.

5.5.1 Single Delta Well

We consider an attractive delta potential, $V(x) = -\eta\delta(x)$, for $\eta > 0$. The negative-energy eigenfunctions are of the form

$$\psi(x) = \begin{cases} A_+ e^{-\lambda x} & x > 0 \\ A_- e^{\lambda x} & x < 0 \end{cases}, \tag{5.84}$$

where A_+ and A_- are constants, and $\lambda = \sqrt{2m|E|}$. Continuity at $x = 0$ implies that $A_+ = A_-$, and Eq. (5.83) implies that $\lambda = m\eta$. Hence, there is only one negative eigenvalue

$$E = -\frac{m\eta^2}{2}, \tag{5.85}$$

with corresponding eigenfunction

$$\psi_0(x) = \sqrt{m\eta}\,e^{-m\eta|x|}. \tag{5.86}$$

For the continuous spectrum, it suffices to consider the eigenfunctions $f_{k+}(x)$, because the potential is symmetric:

$$f_{k+}(x) = \begin{cases} \frac{1}{\sqrt{2\pi}}(e^{ikx} + R_k e^{-ikx}), & x < 0 \\ \frac{1}{\sqrt{2\pi}} T_k e^{ikx}, & x > 0 \end{cases}. \tag{5.87}$$

Continuity at $x = 0$ implies that $1 + R_k = T_k$, and Eq. (5.83) yields $ik(1 - R_k - T_k) = -2m\eta T_k$. We solve the system of equations to obtain

$$T_k = \frac{1}{1 + i\frac{m\eta}{k}}, \quad R_k = -\frac{1}{1 - i\frac{k}{m\eta}}. \tag{5.88}$$

Equation (5.88) applies to a potential $V(x) = -\eta\delta(x)$, irrespective of whether η is positive or negative. In contrast, the ground state (5.86) exists only for positive η.

The relation $E = \frac{k^2}{2m}$ gives an imaginary value for k for negative energies. We note that T_k has a pole at $k = -im\eta$ and R_k has a pole at $k = im\eta$. The corresponding values of energy equal E_0, that is, the energy of the bound states can be read from the poles of T_k and R_k. However, not all poles correspond to bound states. There exist poles even for $\eta < 0$, in which case no bound states exist.

5.5.2 Double Delta Well

We consider the potential $V(x) = -\eta[\delta(x - \frac{a}{2}) + \delta(x + \frac{a}{2})]$, that idealizes a system with two potential wells at distance a. The potential is symmetric, so the eigenfunctions of the Schrödinger operator are either even or odd.

The even eigenfunctions are of the form

$$\psi_+(x) = \begin{cases} A_+ e^{\lambda x} & x < -\frac{a}{2} \\ B_+ \cosh \lambda x & -\frac{a}{2} < x < \frac{a}{2} \\ A_+ e^{-\lambda x} & x > \frac{a}{2} \end{cases}, \tag{5.89}$$

for some constants A_+ and B_+ and $\lambda = \sqrt{2m|E|}$.

The odd eigenfunctions are of the form

$$\psi_-(x) = \begin{cases} -A_- e^{\lambda x} & x < -\frac{a}{2} \\ B_- \sinh \lambda x & -\frac{a}{2} < x < \frac{a}{2} \\ A_- e^{-\lambda x} & x > \frac{a}{2} \end{cases}, \tag{5.90}$$

for some constants A_- and B_-.

The junction conditions for ψ_+ are

$$A_+ e^{-\frac{1}{2}\lambda a} = B_+ \sinh\left(\frac{1}{2}\lambda a\right), \quad (1 - 2m\eta/\lambda)A_+ e^{-\frac{1}{2}\lambda a} = -B_+ \sinh\left(\frac{1}{2}\lambda a\right), \tag{5.91}$$

and, for ψ_-,

$$A_- e^{-\frac{1}{2}\lambda a} = B_- \sinh\left(\frac{1}{2}\lambda\right), \quad (1 - 2m\eta/\lambda)A_- e^{-\frac{1}{2}\lambda a} = -B_- \sinh\left(\frac{1}{2}\lambda a\right). \tag{5.92}$$

It is convenient to define $x := \frac{1}{2}\lambda a$ and $b := ma\eta$. Then, the junction conditions for ψ_+ yield

$$x = \frac{b}{1 + \tanh x}. \tag{5.93}$$

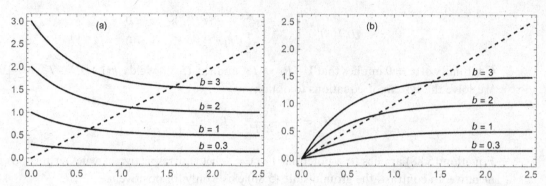

Fig. 5.5 (a) Graphical solution of Eq. (5.93) for different values of b. (b) Graphical solution of Eq. (5.94) for different values of b. A solution exists only for $b > 1$.

The function of x in the right-hand side of Eq. (5.93) is decreasing in $[0, \infty)$; it equals b at $x = 0$, and it converges to $b/2$ as $x \to \infty$. Hence, the graph $y = \frac{b}{1 + \tanh x}$ intersects the line $y = x$ only once. Then, Eq. (5.93) has only one solution, which we will denote as $x_+(b)$ – see Fig. 5.5. For $b \ll 1$, Eq. (5.93) becomes $x \simeq b(1 - x)$, which implies that $x_+(b) \simeq b/(1 + b) \simeq b - b^2$. For b significantly larger than unity, Eq. (5.93) becomes $x = \frac{b}{2}$, hence, $x_+(b) = \frac{b}{2}$.

The junction conditions for ψ_- yield

$$x = \frac{b}{1 + \coth x}. \tag{5.94}$$

The function of x in the right-hand side of Eq. (5.94) is increasing in $[0, \infty)$; it vanishes at $x = 0$ and converges to $\frac{1}{2}b$ as $x \to \infty$. Eq. (5.94) has a solution $x = 0$ that is non admissible, because it leads to $\lambda = 0$, and, hence, to nonsquare integrable $\psi_-(x)$. There is one positive solution only if the slope of the graph $y = \frac{b}{1 + \coth x}$ near zero is greater than the slope of the line $y = x$. Near zero, $\coth x \simeq x^{-1}$, hence, $\frac{b}{1 + \coth x} \simeq bx$. This means that there is one positive solution to Eq. (5.94) for $b > 1$ and none for $b < 1$ – see Fig. 5.5.b. We will denote the solution of Eq. (5.94) as $x_-(b)$.

Since $(1 + \tanh x)^{-1} = \cosh x / (\sinh x + \cosh x) > \sinh x / (\sinh x + \cosh x) = (1 + \coth x)^{-1}$, $x_+(b) > x_-(b)$ for all b. Hence, the lowest eigenvalue of energy always corresponds to the even eigenfunction.

To summarize,

- for $ma\eta \leq 1$, there is a single eigenfunction of the Schrödinger operator, with eigenvalue $E_+ = -\frac{2}{ma^2}[x_+(ma\eta)]^2$;
- for $ma\eta > 1$, there are two eigenfunctions of the Schrödinger operator: an even one with the lowest eigenvalue $E_+ = -\frac{2}{ma^2}[x_+(ma\eta)]^2$, and an odd one with larger eigenvalue $E_- = -\frac{2}{ma^2}[x_-(ma\eta)]^2$.

Numerical values of $x_+(b)$ and $x_-(b)$ are given in the following table.

b	0.01	0.1	0.5	1	1.1	1.5	2	5	10
$x_+(b)$	0.0099	0.092	0.37	0.64	0.69	0.88	1.1	2.51	5.00
$x_-(b)$	–	–	–	–	0.10	0.44	0.80	2.48	5.00

The absence of degeneracy in the lowest-energy state is remarkable. In the corresponding classical system, there are two states of lowest energy, one in which the particle is motionless at $x = \frac{a}{2}$, and one in which the particle is motionless at $x = -\frac{a}{2}$. The quantum state of lowest energy is a superposition of states in which the particle is localized in one of the two wells. We will return to this issue in Section 12.4.4.

5.6 The Role of Boundary Conditions

So far, we have studied several operators defined on the Hilbert space $L^2(\mathbb{R})$, such as the position operator $\hat{x}$, the momentum operator $\hat{p}$, and the Schrödinger operator $\hat{H}$. The Hilbert space $L^2(\mathbb{R})$ describes a point particle moving in line. However, a particle may have constraints in its motion or it may move in an altogether different geometry, for example, in a circle. These cases can be studied by imposing appropriate boundary conditions on the wave functions $\psi(x)$.

5.6.1 Particle at the Half-Line

First, we examine a particle restricted in the half-line $\mathbb{R}^+ = [0, \infty)$. This restriction may originate from a wall located at $x = 0$, upon which the particle is reflected elastically. Alternatively, x may be a particle coordinate that is positive by definition.

In all cases, we require that the associated wave function vanishes at the boundary $x = 0$: $\psi(0) = 0$. Square integrable functions that satisfy this condition define the Hilbert space $L^2_D(\mathbb{R}^+)$.

Any function $\psi(x)$ on $\mathbb{R}^+$ can be Fourier analyzed as

$$\psi(x) = \int_0^\infty \frac{dk}{\sqrt{2\pi}} [a(k) \cos kx + b(k) \sin kx]. \tag{5.95}$$

The boundary condition $\psi(0) = 0$ means that $a(k) = 0$. Hence, an element of $L^2_D(\mathbb{R}^+)$ is a function of the form $\psi(x) = \int_0^\infty \frac{dk}{\sqrt{2\pi}} b(k) \sin kx$. The functions

$$f_k(x) = \frac{1}{\sqrt{2\pi}} \sin kx$$

are not square integrable, but they define a generalized basis on $L^2_D(\mathbb{R}^+)$. We will represent the function f_k by the ket $|k\rangle$.

We evaluate the inner product $\langle k | k' \rangle$ following the reasoning of Section 4.6. That is, we consider the family of functions $f_{k,\epsilon} = \frac{1}{\sqrt{2\pi}} \sin(kx) e^{-\frac{1}{2}\epsilon x^2}$, and we define

$$\langle k | k' \rangle = \lim_{\epsilon \to 0} \int_0^\infty dx f_{k,\epsilon}^*(x) f_{k',\epsilon}(x). \tag{5.96}$$

We calculate $\int_0^\infty dx f_{k,\epsilon}^*(x) f_{k',\epsilon}(x) = \frac{1}{2\pi} \int_0^\infty dx \sin(kx) \sin(k'x) e^{-\epsilon x^2} = \delta_\epsilon(k-k') - \delta_\epsilon(k+k')$, where the function δ_ϵ is given by Eq. (4.38). At the limit $\epsilon \to 0$, $\delta_\epsilon(k + k') = 0$, because $k + k' \neq 0$. Hence, Eq. (5.96) yields

$$\langle k | k' \rangle = \delta(k - k'). \tag{5.97}$$

The kets $|k\rangle$ are not eigenvectors of the momentum operator, because the operator $\hat{p} = -i\frac{\partial}{\partial x}$ is ill defined on $L_D^2(\mathbb{R}^+)$. The derivative of $\sin kx$ is $k\cos kx$, which is nonzero at $x = 0$. However, the second derivative of $\sin kx$ is $-k^2\sin kx$, which vanishes at $x = 0$. Hence, the operator $\widehat{p^2} = -\frac{\partial^2}{\partial x^2}$ is well defined on $L_D^2(\mathbb{R}^+)$, and the kets k are its generalized eigenfunctions,

$$\widehat{p^2}|k\rangle = k^2|k\rangle. \tag{5.98}$$

In Eq. (5.98), the hat is on top of p^2, which means that the operator corresponds to the classical function p^2. The operator $\hat{p}^2 = \hat{p}\hat{p}$ is not defined, because $\hat{p}$ is not defined.

The operator $\widehat{p^2}$ is positive, hence, we can define the operator $\sqrt{\widehat{p^2}}$, which we express as $\widehat{|p|}$. The fact that the absolute value of momentum is an operator but momentum is not means that there is an ambiguity in relation to the sign of momentum in this system. We cannot distinguish between positive and negative momenta because the particle can be reflected at $x = 0$, and hence change its sign, in the absence of any external intervention.

The position operator $\hat{x}$ is well defined by multiplication on wave functions, and so is the Schrödinger operator $\hat{H} = \frac{1}{2m}\widehat{p^2} + V(\hat{x})$.

The results on the discrete spectrum of the Schrödinger operator on $L^2(\mathbb{R})$ are straightforwardly transferred to the Schrödinger operator on $L_D^2(\mathbb{R}^+)$. In particular, the eigenvalues of the Schrödinger operator on $L_D^2(\mathbb{R}^+)$ are nondegenerate, and the nth excited state has n nodes. (The point $x = 0$ does not count as a node.)

Regarding the continuous spectrum, the main difference is that there is no degeneracy on $L_D^2(\mathbb{R}^+)$. This is because there is only one direction of incoming particles in the half-line, from $+\infty$ towards 0. Hence, there is a single eigenfunction $g_k(x)$ for each energy $E = \frac{k^2}{2m}$; it is the single linear combination of the functions f_{k+} and f_{k-} of Eq. (5.57) that vanishes at $x = 0$.

We write $g_k(x) = af_{k+}(x) + bf_{k-}(x)$ for some constants a and b. The condition $g_k(0) = 0$ yields $a(1 + R_k) + bT_k = 0$, hence, $a = -T_k c$ and $b = (1 + R_k)c$ for some constant c. Then, we obtain

$$g_k(x) = \frac{c}{\sqrt{2\pi}} \begin{cases} -2i\sin kx, & 0 \le x < a \\ (1 + R_k)\left[e^{-ikx} - \left(\frac{T_k^2}{1+R_k} - \bar{R}_k\right)e^{ikx}\right], & x > b. \end{cases} \tag{5.99}$$

In Problem 5.19, you are asked to show that the prefactor $\frac{T_k^2}{1+R_k} - \bar{R}_k$ has unit norm – hence it can be written as $e^{i\Theta_k}$. The normalization constant c is found by the requirement that $g_k(x)$ coincides with the kets of $\widehat{p^2}$, for $V(x) = 0$, that is, for $T_k = 1$ and $R_k = \bar{R}_k = 0$. We straightforwardly find that $c = i$.

5.6.2 Particle in a Box

Next, we assume that the values of a particle's position are restricted in a finite interval $I_L = [0, L]$. For example, the particle may be moving in a box with perfectly reflecting walls at $x = 0$ and $x = L$. We use wave functions $\psi(x)$ for $x \in [0, L]$, subject to Dirichlet boundary conditions, that is, they satisfy $\psi(0) = \psi(L) = 0$. The associated Hilbert space is denoted as $L_D^2(I_L)$.

As shown in Section 5.6.1, any function that vanishes at $x = 0$ can be written as $\int \frac{dk}{\sqrt{2\pi}} b(k) \sin(kx)$. The requirement that the function vanishes at $x = L$ implies that $\sin(kL) = 0$. Hence, only discrete values of k are allowed,

$$k_n = n\frac{\pi}{L}, \quad n = 1, 2, 3, \ldots \tag{5.100}$$

We define the functions $f_n(x) = \sqrt{\frac{2}{L}} \sin\left(\frac{n\pi x}{L}\right)$, which satisfy

$$\int_0^L dx f_n(x) f_{n'}(x) = \frac{2}{L} \int_0^a dx \sin\left(\frac{n\pi x}{L}\right) \sin\left(\frac{n'\pi x}{L}\right)$$
$$= \frac{1}{L} \int_0^L dx \left\{ \cos\left[\frac{(n-n')\pi x}{L}\right] - \cos\left[\frac{(n+n')\pi x}{L}\right] \right\} = \delta_{nn'} - \delta_{-nn'} = \delta_{nn'}. \tag{5.101}$$

In the last step, $\delta_{-nn'} = 0$ because $n, n' > 0$. The functions f_n define an orthonormal set on $L_D^2(I_L)$. Since any function subject to these boundary conditions must be written as a linear combination of $\sin(n\pi x/L)$ for different n, the set $\{f_n\}$ forms an orthonormal basis. In Dirac notation, we represent the functions f_n by $|n\rangle$.

Again, we cannot define a momentum operator $\hat{p}$, but we can define the operator $\widehat{p^2} = -\frac{\partial^2}{\partial x^2}$ that satisfies

$$\widehat{p^2}|n\rangle = k_n^2|n\rangle = \frac{n^2\pi^2}{L^2}|n\rangle. \tag{5.102}$$

Since $\widehat{p^2}$ has discrete spectrum, so does the Hamiltonian of a free particle of mass m, $\hat{H} = \frac{\widehat{p^2}}{2m}$. The eigenvalues of $\hat{H}$ are

$$E_n = n^2 \frac{\pi^2}{2mL^2}. \tag{5.103}$$

The presence of the potential does not change the discreteness of the Hamiltonian's spectrum. To see this, consider a potential that is nonzero in an interval $[a, b] \subset [0, L]$. The eigenfunctions of the Hamiltonian with Dirichlet conditions on $x = 0$ are given by Eq. (5.99). The Dirichlet conditions at $x = L$ imply that $e^{2ikL + i\Theta_k} = 1$. Hence, the eigenvalues of the Hamiltonian correspond to solutions of the equation $2kL + \Theta_k = 2\pi n$, for $n = 0, 1, 2, \ldots$. The spectrum is discrete. Furthermore, there are infinite eigenvalues, because $\Theta_k \in [0, 2\pi)$ and for large n the solution to the eigenvalue equation is close to $k = \frac{n\pi}{L}$, that is, the eigenvalues almost coincide with those of the free particle.

The comparison theorem applies for a particle in a box, and it implies that the eigenstates of the Hamiltonian are indexed by the number n of nodes of their wave function. This means that for a particle in a box there is one-to-one correspondence between the eigenstates of *any* two Schrödinger operators.

5.6.3 Periodic Boundary Conditions

Suppose we identify the points $x = 0$ and $x = L$ in the real line. Then, a circle forms, with perimeter equal to L or, equivalently, with radius $R = \frac{L}{2\pi}$. Hence, wave functions that satisfy the periodic boundary conditions $\psi(0) = \psi(L)$ essentially describe the motion of a particle on a ring of radius $R = \frac{L}{2\pi}$. It is often convenient to use the angle variable $\theta = x/R$, with a period of 2π. The inner product of two wave functions ψ and ϕ,

$$\langle \psi | \phi \rangle = \int_0^L \psi^*(x)\phi(x)dx = R \int_0^{2\pi} d\theta \, \psi^*(\theta)\phi(\theta), \tag{5.104}$$

is defined in terms of an integral on the unity circle S^1. The associated Hilbert space consists of all square integrable functions on S^1, and it is denoted as $L^2(S^1, Rd\theta)$.

From Fourier analysis, we know that any function on the circle can be written as

$$\psi(\theta) = \sum_{n=-\infty}^{\infty} a_n e^{in\theta}. \tag{5.105}$$

The functions $f_n(\theta) = \frac{1}{\sqrt{L}} e^{in\theta}$ satisfy $R \int_0^{2\pi} d\theta f_n^*(\theta) f_m(\theta) = \delta_{mn}$ and they define an orthonormal basis on $L^2(S^1, Rd\theta)$. We will represent them by the kets $|n\rangle$.

The angular momentum operator $\hat{\ell} = -i\frac{\partial}{\partial\theta}$ is well defined, and it admits the kets $|n\rangle$ as eigenvectors,

$$\hat{\ell}|n\rangle = n|n\rangle. \tag{5.106}$$

The "momentum" operator $\hat{p} = -i\frac{\partial}{\partial x} = \frac{2\pi}{L}\hat{\ell}$ is proportional to $\hat{\ell}$; its eigenvalues equal $n\frac{2\pi}{L}$.

We cannot define an operator for position x or angle θ. The reason is that it is impossible to implement periodicity in the spectrum of a self-adjoint operator, so that the eigenvalues $\theta = 0$ and $\theta = 2\pi$ coincide. The spectrum of a self-adjoint operator is a subset of the real line, and it is impossible to continuously embed a circle in a line.

However, there is no problem in defining the sine operator $\hat{s}$ and the cosine operator $\hat{c}$,

$$\hat{s}\psi(\theta) = \sin\theta \, \psi(\theta), \quad \hat{c}\psi(\theta) = \cos\theta \, \psi(\theta). \tag{5.107}$$

They are both self-adjoint. Any periodic function $f(\theta)$ can be analysed in sums of powers of sines and cosines of θ. The corresponding operator $\hat{f}$ is defined as a function of $\hat{c}$ and $\hat{s}$.

It is straightforward to prove the commutation relations,

$$[\hat{c}, \hat{\ell}] = -i\hat{s}, \qquad [\hat{s}, \hat{\ell}] = i\hat{c}, \qquad [\hat{c}, \hat{s}] = 0, \tag{5.108}$$

that characterize the main observables of a particle moving in a ring.

The periodic boundary conditions are also used as a mathematical tool in systems with no physical periodicity. The reason is that the imposition of periodic boundary conditions renders the spectrum of the Schrödinger operator discrete; hence, we can avoid calculations that involve a continuous spectrum. In statistical mechanics, it is standard practice to consider particles in a circle of radius R, and taking the limit $R \to \infty$ at the end of any calculation.

In this context, the notion of the *number-of-states function* $\Omega(\epsilon)$ is very useful. Consider a Hamiltonian operator $\hat{H}$ with periodic boundary condition and eigenvalues E_n, indexed by $n = 0, 1, 2, \ldots$. We define

$$\Omega(\epsilon) = \text{number of eigenvalues of } \hat{H} \text{ smaller than } \epsilon = \sum_{n, E_n < \epsilon} 1. \qquad (5.109)$$

The function Ω is discontinuous. However, at the limit where the period L goes to infinity, the distance between successive eigenvalues shrinks to zero, hence, Ω is approximated by a differentiable function. Then, we define the *density of states*,

$$g(\epsilon) = \frac{d}{d\epsilon} \Omega(\epsilon), \qquad (5.110)$$

which contains important information about the generalized eigenvalues of $\hat{H}$ at the continuous limit.

Example 5.4 Consider the Hamiltonian $\hat{H} = \frac{\hat{p}^2}{2m}$ with periodic boundary conditions of period L. The eigenvalues are $E_n = \frac{2\pi^2}{mL^2} n^2$, for $n = 0, \pm 1, \pm 2, \ldots$. The condition $E_n < \epsilon$ becomes $|n| < \frac{L}{\pi} \sqrt{\frac{1}{2} m\epsilon}$, therefore,

$$\Omega(\epsilon) = \text{number of integers } n, \text{ such that } |n| < \frac{L}{\pi} \sqrt{\frac{1}{2} m\epsilon} = 2 \left[\frac{L}{\pi} \sqrt{\frac{1}{2} m\epsilon} \right] + 1,$$

where $[x]$ stands for the integer part of a real number. Since $\frac{x - [x]}{x} \to 0$ as $x \to \infty$, for large L, we substitute $[x]$ for x. Then, the number-of-states function becomes continuous, $\Omega(\epsilon) = \frac{L}{\pi} \sqrt{2m\epsilon}$ (the term $+1$ is negligible). The density of states is

$$g(\epsilon) = \frac{L}{\pi} \sqrt{\frac{m}{2\epsilon}}.$$

5.7 Schrödinger Operator for Periodic Potentials

The Schrödinger operators with periodic potentials $V(x)$ are of particular interest, because they can be used to describe the motion of electrons in a crystalline solid. Such operators can be studied either on the full real line, or with periodic boundary conditions. In the latter case, the period L of the boundary conditions must be much larger than the period a of the potential. The two periodicities are compatible if the ratio L/a is a positive integer. In this section, we will consider the former case, that is, Schrödinger operators defined on the Hilbert space $L^2(\mathbb{R})$.

5.7.1 Bloch's Theorem

We define the unitary translation operator $\hat{T}_a = e^{ia\hat{p}}$. It acts on wave functions as $\hat{T}_a \psi(x) = \psi(x + a)$ – see Eq. (4.26). Then, $\hat{T}_a[V(x)\psi(x)] = V(x + a)\psi(x + a) = V(x)\psi(x + a) = V(x)\hat{T}_a\psi(x)$, hence, $[\hat{T}_a, \hat{V}] = 0$. It follows that the Schrödinger operator $\hat{H} = \frac{1}{2m}\hat{p}^2 + \hat{V}$ satisfies

$$[\hat{H}, \hat{T}_a] = 0. \qquad (5.111)$$

This means that $\hat{H}$ and $\hat{T}_a$ have common eigenfunctions $\psi(x)$,

$$\hat{T}_a\psi(x) = C(a)\psi(x). \tag{5.112}$$

Since $\hat{T}_a$ is unitary, $|C(a)| = 1$. By definition, $\hat{T}_a^n = \hat{T}_{na}$ for all integers n. Hence, $C(a)^n = C(na)$. Writing $C(a) = e^{i\theta(a)}$, we obtain $n\theta(a) = \theta(na)$. This means that $\theta(a) = ka$ for some constant k. Then, Eq. (5.112) becomes

$$\psi(x + a) = e^{ika}\psi(x). \tag{5.113}$$

If we write

$$\psi(x) = e^{ikx}u(x), \tag{5.114}$$

Eq. (5.113) implies that $u(x + a) = u(x)$. Equation (5.113) is known as *Bloch's theorem*.

The constants k are not unique. If we substitute k with $k' = k + \frac{2\pi}{a}r$ for any integer r, Eq. (5.113) still holds. We can always choose $-\frac{\pi}{a} \leq k < \frac{\pi}{a}$, so that k is uniquely defined.

We substitute Eq. (5.114) in the eigenvalue equation $\hat{H}$, $\hat{H}\psi = E\psi$, to obtain

$$-\frac{1}{2m}u'' - i\frac{k}{m}u' + V(x)u = \left(E - \frac{k^2}{2m}\right)u. \tag{5.115}$$

Bloch's theorem implies that we can find the eigenfunctions of $\hat{H}$, by solving Eq. (5.115) for a single period of the potential, and using Eq. (5.114) to extend the solution to the whole real line.

5.7.2 Kronig–Penney Potential

As an example, we study the Kronig–Penney potential

$$V(x) = \eta \sum_{n=-\infty}^{\infty} \delta(x - na). \tag{5.116}$$

By Bloch's theorem, it suffices to solve the eigenvalue equation for a single period of the potential. We consider an interval that contains the single delta function centered on $x = 0$, $V(x) = \eta\delta(x)$.

The general solution to the equation $\psi'' + 2m[E - \eta\delta(x)]\psi = 0$ is

$$\psi_q(x) = \begin{cases} Ae^{-iqx} + Be^{iqx} & 0 < x < a \\ Ce^{-iqx} + De^{iqx} & -a < x < 0 \end{cases}, \tag{5.117}$$

where $q = \sqrt{2mE}$. Equation (5.113) implies that this solution can be extended to the whole real line provided that $C = e^{-i(k+q)a}A$ and $D = e^{-i(k-q)a}B$.

Continuity at $x = 0$ yields $A + B = C + D$, hence,

$$(1 - e^{-i(k+q)a})A = -(1 - e^{-i(k-q)})B. \tag{5.118}$$

Equation (5.83) yields $q(A - B - C + D) = 2m\eta i(A + B)$, hence,

$$\left(1 - e^{-i(k+q)a} - i\frac{2m\eta}{q}\right)A = B\left(1 - e^{-i(k-q)a} + i\frac{2m\eta}{q}\right). \tag{5.119}$$

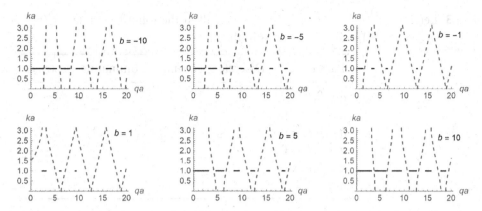

Fig. 5.6 Plots of ka as a function of qa for different values of $b = m\eta a$. The thick horizontal line denotes the presence of gaps.

We divide Eq. (5.119) by Eq. (5.118) to obtain

$$1 - i\frac{2m\eta}{q(1 - e^{-i(k+q)a})} = -1 - i\frac{2m\eta}{q(1 - e^{-i(k-q)a})}, \tag{5.120}$$

which gives

$$\cos(ka) = \cos(qa) - \frac{m\eta}{q}\sin(qa). \tag{5.121}$$

The left-hand side of Eq. (5.121) depends only on k and takes values from -1 to 1. The right-hand side depends only on q and can take values outside $[-1, 1]$. For any value of q, such that the quantity $|\cos qa - \frac{m\eta}{q}\sin qa|$ is greater than 1, Eq. (5.121) has no solution; hence, there are no energy eigenvalues $E = \frac{q^2}{2m}$. The spectrum $\sigma(\hat{H})$ of the Schrödinger operator $\hat{H}$ is characterized by *bands*, that is, disjoint intervals with eigenvalues and *gaps* in between.

For q in a band, Eq. (5.121) can be solved for k,

$$k(q) = \pm\frac{1}{a}\cos^{-1}\left[\cos(qa) - \frac{m\eta}{q}\sin(qa)\right]. \tag{5.122}$$

There are two values of k for any value of q, hence, the energy eigenfunctions have double degeneracy. In Fig. 5.6, we plot k as a function of q for different values of $b := m\eta a$. We see that there is no periodicity in either the location or width of the gaps. As $|b|$ increases, the shift from bands to gaps is more frequent.

QUESTIONS

5.1 Simplify the following operators: (i) $(\hat{\sigma}_2)^{1927}$, (ii) $(\sigma_1\sigma_2\sigma_3)^{2023}$, (iii) $\hat{\sigma}_-^{33}$.

5.2 Describe the spectrum (discrete and/or continuous, degrees of degeneracy) of the Schrödinger operator on the real line with potential $V(x) = $ (i) e^x, (ii) $\cosh x$, (iii) $\sinh x$, and (iv) $\tanh x$. How do your answers change when the Schrödinger operator is defined on the half line?

5.3 Let $V(x) = ae^x + be^{-x}$ for $a, b \in \mathbb{R}$. Fill in the following table for the spectrum of the associated Schrödinger operator in the line.

Parameters	Discrete spectrum	Continuous spectrum
$a \geq 0, b > 0$		
$a > 0, b < 0$		
$a < 0, b \geq 0$		
$a < 0, b < 0$		
$a = 0, b < 0$		
$a < 0, b = 0$		

5.4 Draw a one-dimensional potential that admits a positive-energy bound state in classical physics. Why are such states impossible in quantum theory?

5.5 Explain why we cannot define the momentum operator for a particle in a box and what this means physically.

5.6 Compare the SG trigonometric operators with the corresponding operators for a particle in a ring.

5.7 Plot the number-of-states function $\Omega(\epsilon)$ for the Kronig–Penney model.

PROBLEMS

5.1 Find all operators on $\mathbb{C}^2$ that are both self-adjoint and unitary.

5.2 Show that for any operator $\hat{A}$ on $\mathbb{C}^2$,

$$\sum_{i=1}^{3} \hat{\sigma}_i \hat{A} \hat{\sigma}_i = -\hat{A} + 2(\mathrm{Tr}\hat{A})\hat{I}. \qquad (5.123)$$

5.3 Use Hadamard's identity in order to evaluate the product $e^{i\hat{A}s} \hat{a} e^{-i\hat{A}s}$, where $\hat{A} = \frac{1}{2}(\hat{a}^2 + \hat{a}^{\dagger 2})$.

5.4 (i) Evaluate $\langle n|\hat{p}^4|n \rangle$ and $\langle n|\hat{x}^2\hat{p}^2|n \rangle$ for a number state $|n \rangle$ of a harmonic oscillator. (ii) Evaluate $\langle z|\hat{x}^4|z \rangle$ and $\langle z|\hat{p}^4|z \rangle$ for a coherent state $|z \rangle$.

5.5 Show that $\langle n, j|n', j' \rangle = \delta_{nn'}\delta_{jj'}$ for the vectors $|n, j \rangle$ of Eq. (5.31).

5.6 Show that Eqs. (5.39) and (5.38) coincide if $e^{i\phi} = e^{-\frac{i\hat{q}\hat{p}}{2}}$.

5.7 For a harmonic oscillator, we define the operator $\hat{b} = \lambda\hat{a} + \mu\hat{a}^\dagger$, where $\lambda, \mu \in \mathbb{R}$. (i) Show that $[\hat{b}, \hat{b}^\dagger] = \hat{I}$ if $\lambda^2 - \mu^2 = 1$. (ii) Find the eigenvectors of $\hat{b}$ with zero eigenvalue. **Comment** In quantum optics, the zero-eigenvectors of $\hat{b}$ are known as *squeezed vacua*.

5.8 The most general form of inner product for the generalized eigenvectors $|k, \pm \rangle$ is $\langle k, a|k', b \rangle = w_{ab}(k)\delta(k - k')$, where $a, b = \pm$. Show that $w_{ab} = \delta_{ab}$. To this end, evaluate the integrals of x in the inner product with a regularizing function $e^{-\epsilon|x|}$, and for k = k'. You will obtain $w_{ab}(k)$ by comparing with the behavior of approximate delta functions as $\epsilon \to 0$.

5.9 Evaluate the total reflection amplitude for two successive potential barriers from Eq. (5.66).

5.10 Consider a particle of mass m in the *Morse potential* $V(x) = V_0[e^{-2(x-x_0)/a} - 2e^{-(x-x_0)/a}]$, where λ, V_0 and x_0 are constants. This potential models the relative position of two atoms in a diatomic molecule. The potential becomes very large for negative x and eventually diverges to infinity as $x \to -\infty$, and it vanishes at $x \to \infty$. (i) Show that the minimum of the potential is at x_0, and that the frequency of small oscillations around this minimum is $\omega_0 = \frac{\sqrt{2V_0/m}}{a}$. (ii) Change the variable to $z = \lambda e^{-(x-x_0)/a}$, where $\lambda = \sqrt{2mV_0}a$, and show that the eigenvalue equation for the Schrödinger operator becomes

$$\psi'' + \frac{1}{z}\psi' + \left(\frac{2\lambda}{z} - 1 - \frac{\kappa^2}{z^2}\right)\psi = 0,$$

where $\kappa = \sqrt{2m|E|}a$. (iii) Show that for a bound state $\psi \sim e^{-z}$ as $z \to \infty$, and $\psi \sim z^{\kappa}$ as $z \to 0$. (iv) Use the ansatz $\psi(z) = z^{\kappa}e^{-z}f(z)$, to show that

$$f'' + \left(-2 + \frac{2\kappa + 1}{z}\right) + \frac{2}{z}\left(\lambda - \kappa - \frac{1}{2}\right)f = 0.$$

(v) Show that the requirement that f is a polynomial function, so that ψ is square integrable, allows one to compute the eigenvalues of the Schrödinger operator

$$E_n = -V_0 + \omega_0\left(n + \frac{1}{2}\right) - \frac{\left(n + \frac{1}{2}\right)^2}{2ma^2},$$

where n is a nonnegative integer. (vi) Show that the maximum number of bound states is $[\lambda - \frac{1}{2}]$, where $[x]$ stands for the integer part of x.

5.11 Evaluate the transmission amplitude T_k in the square potential barrier, for $E > V_0$. Show that the transmission probability equals unity, when the resonance condition

$$\sqrt{2m(E - V_0)} = \frac{n\pi}{a},$$

is satisfied, where $n = 1, 2, 3, \ldots$.

5.12 For a symmetric potential that is nonzero only in the interval $[-a, a]$, it is sometimes convenient to use the even $(+)$ and odd $(-)$ eigenstates of the Schrödinger operator, which take the form $\psi_k^+(x) = A_+ \cos(kx + \delta_+)$ and $\psi_k^-(x) = A_- \sin(kx + \delta_-)$ for $x > a$; $A_{\pm}, \delta_{\pm}$ are real constants. Show that the transmission and reflection amplitudes are given by $T_k = \cos(\delta_+ - \delta_-)e^{i\delta_+ + i\delta_-}$ and $R_k = i\sin(\delta_+ - \delta_-)e^{i\delta_+ + i\delta_-}$.

5.13 Consider the phases $\delta_{\pm}$ as defined in Problem 5.12 for a symmetric potential. (i) Use the inequality (5.54) to show that $d\delta_{\pm}/dk \geq -(2k)^{-1} - a$. (ii) If $\theta_k = \arg T_k$, show that $d\theta_k/dk \geq -2a - k^{-1}$.

5.14 The Hamiltonian for a free *relativistic* particle of mass m moving in one dimension is $\hat{H} = \sqrt{\hat{p}^2 + m^2}$. Suppose that the particle moves in a piecewise constant potential. Consider a point x_0 of discontinuity, such that $\psi'' + k_-^2\psi = 0$ for $x < x_0$ and $\psi'' + k_+^2\psi = 0$ for $x > x_0$, for $k_{\pm}$ that are either real or imaginary. (i) Show that the junction conditions at x_0 are: $\psi(x_{0+}) = \psi(x_{0-})$ and

$$\frac{\sqrt{m^2 + k_+^2} - m}{k_+^2}\psi'(x_{0+}) = \frac{\sqrt{m^2 + k_-^2} - m}{k_-^2}\psi'(x_{0-}).$$

(ii) Show that for a square barrier potential of width a and height V_0, the transmission amplitude for $E < V_0$ is

$$T_k = \frac{e^{-ika}}{\cosh(\lambda a) - \frac{i}{2}(\rho - \rho^{-1})\sinh(\lambda a)},$$

where $k = \sqrt{E^2 - m^2}$, $\lambda = \sqrt{m^2 - (E - V_0)^2}$, and $\rho = \frac{\lambda(E-m)}{k(m - \sqrt{m^2 - (E-V_0)^2})}$.

5.15 Evaluate the transmission amplitude and the reflection amplitude for a particle of mass m in a double delta barrier $V(x) = \eta[\delta(x - \frac{a}{2}) + \delta(x + \frac{a}{2})]$. Identify resonance energies in the regime where $k/(m\eta) \ll 1$.

5.16 Find the eigenvalues of the Schrödinger operator for a harmonic oscillator at the half-line.

5.17 Consider a particle moving on the half-line $x > 0$ under a potential

$$V(x) = \begin{cases} -V_0, & 0 < x < a \\ 0 & a \leq x \end{cases}.$$

Show that the negative eigenvalues E of the Schrödinger operator are obtained from solutions of the equation $\sqrt{x}\tan(b\sqrt{1 - x}) = -\sqrt{1 - x}$, where $x = |E|/V_0$ and $b = \sqrt{2mV_0}a$. Show that for b very close to zero, only the solution $x = 1$ exists.

5.18 (i) Show that the operator $\hat{Q} = -ix\frac{\partial}{\partial x} - \frac{i}{2}$ is self-adjoint on $L^2(\mathbb{R}^+)$. (ii) Show that the functions $f_q(x) = \frac{1}{\sqrt{2\pi}}x^{-\frac{1}{2}+iq}$ for $q \in \mathbb{R}$ are generalized eigenvectors of $\hat{Q}$ and they satisfy $\int_0^\infty dx f_q(x)f_{q'}^*(x) = \delta(q - q')$. (iii) Use Hadamard's identity to show that $e^{is\hat{Q}}\hat{x}e^{-is\hat{Q}} = e^s\hat{x}$.

5.19 (i) Show that the quantity $\frac{T_k^2}{1+\bar{R}_k} - \bar{R}_k$ that appears in Eq. (5.99) has unit norm, hence, it can be written as $e^{i\Theta_k}$. (ii) Evaluate Θ_k for a potential $V(x) = \eta\delta(x - a)$.

5.20 Consider an attractive delta function potential $V(x) = -\eta\delta(x - a)$ with $\eta, a > 0$ for a particle of mass m in the half line. Show that a bound state exists only if $2m\eta a > 1$.

5.21 A particle of mass m moves in a box of width L and under a potential $V(x) = \eta\delta(x - \xi L)$, where $0 < \xi < 1$. Show that the eigenvalues of the Schrödinger operator are $E_n = \frac{1}{2mL^2}x_n^2$, where x_n are solutions to the equation $x[\cot(\xi x) + \cot((1 - \xi)x)] = 2m\eta L$.

5.22 Find the eigenvalues and the eigenvectors of the Schrödinger operator for a particle of mass m moving on a ring of radius R and under a potential $V(x) = \eta\delta(x)$. Comment on the result.

Bibliography

- For coherent states and their applications, see the introductory article in Klauder and Skagerstam (1985).
- For mathematical properties of the one-dimensional Schrödinger operator, see chapter 2 of Berezin and Shubin (1991) and also Schechter (1981). For exact solutions to Schrödinger's equation in one dimension, see Flügge (1971). For the early applications of quantum tunneling, see Merzbacher (2007).
- For Bloch's theorem, see chapter 8 of Ashcroft and Mermin (1976).

6 Quantum Probabilities

... present quantum theory can answer only questions of the form: "If this experiment is performed, what are the possible results and their probabilities?" It cannot, as a matter of principle, answer any question of the form: "What is really happening when ....?" Again, the mathematical formalism of present quantum theory, like Orwellian *newspeak*, does not even provide the vocabulary in which one could ask such a question.

Jaynes (2003)

6.1 Generalization of Born's Rule

We saw in Chapter 2 that Born's statistical interpretation of the wave function was one of the building blocks of quantum theory. According to Born's interpretation, the wave function of a particle at a given moment of time defined a probability density $|\psi(x)|^2$ with respect to the position x. This result is generalized to state vectors of an Hilbert space and to general observables through the following procedure.

Consider a wave function $\psi(x)$ on the Hilbert space $L^2(\mathbb{R}, dx)$ that describes a particle at a line. By Born's rule, the probability $\text{Prob}(C)$ that the particle's position lies in $C \subset \mathbb{R}$ is

$$\text{Prob}(C) = \int_C |\langle x|\psi\rangle|^2 = \langle\psi|\left(\int_C dx|x\rangle\langle x|\right)|\psi\rangle = \langle\psi|\hat{P}_C|\psi\rangle, \tag{6.1}$$

where $\hat{P}_C$ is the spectral projector of position that corresponds to the set C.

Having expressed Born's rule in terms of state vectors and spectral projectors, we can generalize it to any physical magnitude.

FP3 Suppose that a physical system is described by the normalized vector $|\psi\rangle$, and that we measure a physical quantity that corresponds to the self-adjoint operator $\hat{A}$. Let C be a subset of $\mathbb{R}$ and $\hat{P}_C = \chi_C(\hat{A})$ the associated projector. The probability $\text{Prob}(C)$ that the measurement outcome lies within C is given by

$$\text{Prob}(C) = \langle\psi|\hat{P}_C|\psi\rangle. \tag{6.2}$$

We observe that the probabilities (6.2) are invariant under the transformation $|\psi\rangle \to e^{i\theta}|\psi\rangle$. A phase change of a state vector does not affect probabilities, hence, the physical predictions.

Hence, the quantum state is best identified with the *ray* $|\psi\rangle\langle\psi|$ rather than the vector $|\psi\rangle$. This is true as long as we do not talk about set-ups where the vector $|\psi\rangle$ is superposed with another vector $|\phi\rangle$. The vector $|\phi\rangle + |\psi\rangle$ does lead to different probabilities from the vector $|\phi\rangle + e^{i\theta}|\psi\rangle$.

For operators $\hat{A}$ with discrete spectrum, $\hat{A} = \sum_n a_n \hat{P}_n$, the probability p_n of measuring the eigenvalue a_n is

$$p_n = \langle\psi|\hat{P}_n|\psi\rangle. \tag{6.3}$$

If no eigenvalue is degenerate, then $\hat{P}_n = |n\rangle\langle n|$, where $|n\rangle$ is the associated eigenvector, and

$$p_n = |\langle n|\psi\rangle|^2. \tag{6.4}$$

The *mean value* $\langle\hat{A}\rangle$ of the physical magnitude associated to $\hat{A} = \sum_n a_n \hat{P}_n$ is $\langle\hat{A}\rangle = \sum_n a_n p_n = \sum_n a_n \langle\psi|\hat{P}_n|\psi\rangle = \langle\psi|\sum_n a_n \hat{P}_n|\psi\rangle$. Hence,

$$\langle\hat{A}\rangle = \langle\psi|\hat{A}|\psi\rangle. \tag{6.5}$$

The *variance* $(\Delta A)^2$ of measurements of $\hat{A}$ is given by

$$(\Delta A)^2 = \langle\hat{A}^2\rangle - \langle\hat{A}\rangle^2. \tag{6.6}$$

The *correlation* $\mathrm{Cor}(\hat{A}, \hat{B})$ or C_{AB} of two measurable quantities that correspond to operators $\hat{A}$ and $\hat{B}$ is defined as

$$\mathrm{Cor}(\hat{A}, \hat{B}) := \frac{1}{2}\langle\psi|\hat{A}\hat{B} + \hat{B}\hat{A}|\psi\rangle - \langle\psi|\hat{A}|\psi\rangle\langle\psi|\hat{B}|\psi\rangle. \tag{6.7}$$

The correlation is real-valued, because the operator $\hat{A}\hat{B} + \hat{B}\hat{A}$ is self-adjoint. By definition, $\mathrm{Cor}(\hat{A}, \hat{B}) = \mathrm{Cor}(\hat{B}, \hat{A})$ and $\mathrm{Cor}(\hat{A}, \hat{A}) = (\Delta A)^2$. The correlation quantifies the statistical dependence between measurements of $\hat{A}$ and $\hat{B}$ when the system has been prepared in the state $|\psi\rangle$.

For an operator with continuous spectrum $\hat{A} = \int d\lambda\, \lambda \hat{P}_\lambda$, probabilities are expressed through a probability density

$$p(\lambda) = \langle\psi|\hat{P}_\lambda|\psi\rangle, \tag{6.8}$$

since λ is a continuous variable. If $\hat{P}_\lambda = |\lambda\rangle\langle\lambda|$ for some ket $|\lambda\rangle$, then $p(\lambda) = |\langle\lambda|\psi\rangle|^2$.

Jensen's inequality, Eq. (1.31), also applies to quantum probabilities

$$\langle F(\hat{A})\rangle \geq F(\langle\hat{A}\rangle) \tag{6.9}$$

for any convex function $F : \mathbb{R} \to \mathbb{R}$ and self-adjoint operator $\hat{A}$. The proof follows directly from the spectral theorem. For $\hat{A} = \sum_n a_n \hat{P}_n$ and $p_n = \langle\psi|\hat{P}|\psi\rangle$, Eq. (6.9) becomes $\sum_n p_n F(a_n) \geq F(\sum_n p_n a_n)$, that is, it coincides with the classical Jensen inequality (1.31).

Example 6.1 Of particular importance are the Gaussian wave functions

$$\psi_{\bar{q},\bar{p}}(x) = \left(2\pi\sigma^2\right)^{-1/4} \exp\left[-\frac{(x - \bar{q})^2}{4\sigma^2} + i\bar{p}x\right], \tag{6.10}$$

where $\bar{q}$ and $\bar{p}$ are the mean values of position and momentum, respectively. For a suitable choice of σ, the Gaussians (6.10) coincide with the coherent states (5.39). The position probability distribution

$$|\psi_{\bar{q},\bar{p}}(x)|^2 = \frac{1}{\sqrt{2\pi\sigma^2}} \exp\left[-\frac{(x-\bar{q})^2}{2\sigma^2}\right] \tag{6.11}$$

is centered around $\bar{q}$ with mean deviation $\Delta x = \sigma$.

In the momentum basis,

$$\tilde{\psi}_{\bar{q},\bar{p}}(p) = \left(\frac{2\sigma^2}{\pi}\right)^{1/4} \exp\left[-\sigma^2(p-\bar{p})^2 - i\bar{q}k\right], \tag{6.12}$$

corresponding to a momentum probability distribution

$$|\tilde{\psi}_{\bar{q},\bar{p}}(p)|^2 = \left(\frac{2\sigma^2}{\pi}\right)^{1/2} \exp\left[-2\sigma^2(p-\bar{p})^2\right] \tag{6.13}$$

that is centered around $\bar{p}$ with mean deviation $\Delta p = (2\sigma)^{-1}$.

Example 6.2 We write a general wave function $\psi \in L^2(\mathbb{R})$ in the polar form $\psi(x) = R(x)e^{iS(x)}$, where $R(x) > 0$. Then, $\hat{p}\psi = -i(R' + iRS')e^{iS}$, and $\langle\hat{p}\rangle = \int dx(-iRR' + R^2 S') = \int dx R^2 S'$, since $\int dx RR' = \frac{1}{2}R^2|_{-\infty}^{\infty} = 0$ for any square integrable function. If ψ is real-valued, then S is a constant and $\langle\hat{p}\rangle = 0$. If $S(x)$ is a linear function of x, $S(x) = \bar{p}x$, then $\langle\hat{p}\rangle = \bar{p}$.

6.2   Statistical Mixing and Density Matrices

In Section 3.3.4, we mentioned that the description of the quantum state in terms of Hilbert space vectors is not the most general possible. The most general description requires the introduction of the notion of a density matrix. To this end, we will first explain the concept of *statistical mixing* and of *pure states* in classical probability theory.

6.2.1   Classical Statistical Mixing

The concepts of statistical mixing and of pure states were bequeathed to probability theory from chemistry, in particular, from the procedures of mixing and separating substances.

Consider a solution inside a container A, consisting of n different substances. Let p_i, $i = 1, 2, \ldots, n$ be the molecular fraction of the substance with index i, that is the ratio of the number of molecules of type i to the total number of molecules. Obviously, $\sum_{i=1}^{n} p_i = 1$. We can describe the state of the container A by a vector $\mathbf{w}_A = (p_1, p_2, \ldots, p_n)$. Similarly, we can assign to a different container B, with a solution from the same substances with molecular fractions q_i, a vector $\mathbf{w}_B = (q_1, q_2, \ldots, q_n)$.

Suppose we mix the solutions from the two containers: we put in a container C x_A units of solution from container A and y_B units of solution from container B. Hence, the fraction of the solution in C originating from A is $\lambda = \frac{x_A}{x_A + y_B}$ and the fraction originating from B is $1 - \lambda$. The resulting solution has molecular fraction for the ith substance equal to $\lambda p_i + (1-\lambda)q_i$. Therefore, it can be described by a vector

$$\mathbf{w}_C = \lambda \mathbf{w}_A + (1 - \lambda)\mathbf{w}_B \tag{6.14}$$

A sum of two vectors as in Eq. (6.14) for $\lambda \in [0, 1]$ is called a *convex combination* of the vectors.

The reverse of mixing is the decomposition of the solution to its substances. Consider the n vectors

$$\mathbf{e}_1 = (1, 0, \dots, 0), \quad \mathbf{e}_2 = (0, 1, \dots, 0), \quad \dots, \quad \mathbf{e}_n = (0, 0, \dots, 1). \tag{6.15}$$

No vector $\mathbf{e}_i$ can be expressed as a convex combination of other vectors. The vectors $\mathbf{e}_i$ correspond to pure substances, that is, substances in which all molecules in a container are of a single type. Obviously, any vector $\mathbf{w} = (p_1, p_2, \dots, p_n)$ can be written as

$$\mathbf{w} = \sum_{i=1}^{n} p_i \mathbf{e}_i, \tag{6.16}$$

that is, a convex combination of vectors that describe pure substances.

The preceding analysis can be directly translated to probability theory, by identifying the set of n different substances in a container with a sample space. A container with a solution corresponds to a statistical ensemble described by a probability distribution p_i for $i = 1, \dots, n$, or, equivalently, by a probability vector $\mathbf{w} = (p_1, p_2, \dots, p_n)$. If we select a molecule randomly from the container, the probability that this molecule is of type i is given by p_i.

The physical mixing of two or more solutions corresponds to the *statistical mixing* of two or more probability distribution. Suppose we have k different statistical ensembles, each characterized by a probability vector

$$\mathbf{w}_r = (p_{r,1}, p_{r,2}, \dots, p_{r,n}), \tag{6.17}$$

for $r = 1, 2, \dots, k$. We define the statistical mixture of these distributions, with relative weights λ_r, with $\sum_{r=1}^{k} \lambda_r = 1$, as the statistical ensemble described by the probability vector

$$\mathbf{w}_{mix} = \sum_{r=1}^{k} \lambda_r \mathbf{w}_r. \tag{6.18}$$

Suppose that the rth statistical ensemble has $N_r \gg 1$ elements, and mixing corresponds to the union of all N_r into a large statistical ensemble of $N = \sum_{r=1}^{k} N_r$ elements. Then, the relative weight of each ensemble is $\lambda_r = N_r/N$.

Within the same analogy between statistical ensembles and solutions, we call any probability distribution that cannot be obtained from mixing of other distributions *pure*. Pure distributions define random experiments that always give the same result, and they are described by the probability vectors (6.15). We emphasize that the decomposition of a probability vector $\mathbf{w}$ to a convex combination of pure distributions by Eq. (6.16) is unique.

6.2.2 Density Matrix

Statistical mixing is an inseparable component of the concept of probability. We can always mix two statistical ensembles characterized by the same alternatives. Hence, statistical mixing applies to the probabilities defined by quantum theory.

Consider a device that prepares a physical system in states described by vectors of a Hilbert space $\mathcal{H}$. In general, the experimentalist does not have absolute control over all parameters of the device. In each run of the experiments, fluctuations of these parameters may lead to a different prepared state. We model this type of randomness as follows. In each run of the experiment, the device may prepare the system in a state vector $|\psi_a\rangle$ with probability w_a.

Suppose we measure a quantity that corresponds to the operator $\hat{A}$ on a statistical ensemble prepared by the aforementioned device. The probability $\mathrm{Prob}(C)$ of obtaining a value in $C \in \sigma(\hat{A})$ follows from mixing the different probabilities $\langle \psi_a | \hat{P}_C | \psi_a \rangle$ corresponding to each possible state vector, with weights w_a,

$$\mathrm{Prob}(C) = \sum_a w_a \langle \psi_a | \hat{P}_C | \psi_a \rangle = \mathrm{Tr}\left(\rho \hat{P}_C \right), \tag{6.19}$$

where $\hat{\rho} = \sum_a w_a |\psi_a\rangle\langle\psi_a|$ is a positive operator that satisfies $\mathrm{Tr}\hat{\rho} = 1$.

Definition 6.1 Any positive Hilbert space operator that satisfies $\mathrm{Tr}\hat{\rho} = 1$ is called a *density matrix*.

Since a density matrix $\hat{\rho}$ has finite trace, its spectrum is discrete. By definition, the eigenvalues of $\hat{\rho}$ are positive, hence, $\hat{\rho}$ can be expressed as

$$\hat{\rho} = \sum_n w_n \hat{P}_n,$$

where $w_n \in [0,1]$ and $\sum_n w_n = \mathrm{Tr}\hat{\rho} = 1$; $\hat{P}_n$ are the associated spectral projectors. A general density matrix on $\mathbb{C}^N$ is specified by $N^2 - 1$ independent real numbers.[1]

A density matrix that has one eigenvalue equal to unity and all others equal to zero is called *pure*. A pure density matrix is of the form $\hat{\rho} = |\psi\rangle\langle\psi|$ for some normalized Hilbert space vector $|\psi\rangle$. Any density matrix that is not pure is called *mixed*.

Of particular importance are the density matrices of *maximum ignorance* that are defined on $\mathbb{C}^N$ as $\hat{\rho} := N^{-1}\hat{I}$. These are the *maximally mixed* states that do not provide any information at all about the system.

Let $|n\rangle$ and $|m\rangle$ be two vectors of an orthonormal basis. The Cauchy–Schwarz inequality for the vectors $\sqrt{\hat{\rho}}|n\rangle$ and $\sqrt{\hat{\rho}}|n\rangle$ yields $|\langle n|\hat{\rho}|m\rangle|^2 \le \langle n|\hat{\rho}|n\rangle\langle m|\hat{\rho}|m\rangle$, namely,

$$|\rho_{mn}| \le \sqrt{\rho_{nn}\rho_{mm}}. \tag{6.20}$$

The following result allows us to distinguish between pure and mixed density matrices.

Theorem 6.2 *Every density matrix $\hat{\rho}$ satisfies $\hat{\rho} \ge \hat{\rho}^2$. Equality holds only if $\hat{\rho}$ is pure.*

Proof Let w_n be the eigenvalues of $\hat{\rho}$ and $\hat{P}_n$ the associated spectral projectors. Since $0 \le w_n \le 1$, $w_n - w_n^2 \ge 0$, hence, $\sum_n (w_n - w_n^2)\hat{P}_n \ge 0$. It follows that $\hat{\rho} - \hat{\rho}^2 \ge 0$. Equality holds if $w_n = w_n^2$ for all n, that is if $w_n = 0$ or $w_n = 1$. Given that $\sum_n w_n = 1$, all but one of w_n vanish except for one, say w_1, that takes value 1. Then, $\mathrm{Tr}\hat{\rho} = \mathrm{Tr}\hat{P}_1 = 1$, hence $\hat{P}_1$ is one-dimensional, and $\hat{\rho}$ is pure. $\qquad\square$

---

[1] This number is obtained as follows. A self-adjoint operator has N diagonal real elements and $\frac{1}{2}(N^2 - N)$ independent off-diagonal complex elements; hence, $N + (N^2 - N) = N^2$ independent real numbers. We subtract one because of the normalization condition to conclude that a general density matrix on $\mathbb{C}^N$ is specified by $N^2 - 1$ real numbers.

Theorem (6.2) implies that $\mathrm{Tr}\hat{\rho} - \mathrm{Tr}\hat{\rho}^2 \geq 0$, or equivalently $\mathrm{Tr}\hat{\rho}^2 \leq 1$, with equality achieved only for pure states. Hence, the quantity

$$\gamma = \mathrm{Tr}\hat{\rho}^2 \tag{6.21}$$

measures how close a density matrix $\hat{\rho}$ is to a pure state. For this reason, it is known as the *purity* of $\hat{\rho}$ – see Problem 1.4 for an analogous definition in classical probability theory. For states on the Hilbert space $\mathbb{C}^N$, γ ranges between a minimum value $\frac{1}{N}$, which is achieved for the maximum ignorance state, and 1 for pure states.

6.2.3   Density Matrix for a Qubit

Equation (5.2) and the fact that $\mathrm{Tr}\hat{\sigma}_i = 0$ imply that the most general self-adjoint operator $\hat{\rho}$ on $\mathbb{C}^2$ that satisfies $\mathrm{Tr}\hat{\rho} = 1$ is of the form

$$\hat{\rho} = \frac{1}{2}(\hat{I} + \mathbf{r} \cdot \hat{\boldsymbol{\sigma}}), \tag{6.22}$$

for some real numbers r_1, r_2, and r_3.

Equation (5.4) allows us to express the constants r_i as

$$\mathrm{Tr}(\hat{\rho}\hat{\sigma}_i) = \frac{1}{2}\sum_j r_j \mathrm{Tr}(\hat{\sigma}_j\hat{\sigma}_i) = \frac{1}{2}\sum_j r_j (\mathrm{Tr}\hat{I})\delta_{ij} = r_i. \tag{6.23}$$

As shown in Section 5.1, the eigenvalues of the operator (6.22) are $\frac{1}{2}(1 \pm |\mathbf{r}|)$. It follows that $\hat{\rho}$ is positive only if $|\mathbf{r}| \leq 1$, so that both eigenvalues are positive. Hence, the space of density matrices on $\mathbb{C}^2$ is identical with the unit ball on $\mathbb{R}^3$,

$$B_2 = \{(r_1, r_2, r_3) | r_1^2 + r_2^2 + r_3^2 \leq 1\}.$$

We calculate $\mathrm{Tr}\hat{\rho}^2 = \frac{1}{4}(1 + |\mathbf{r}|)^2 + \frac{1}{4}(1 - |\mathbf{r}|)^2 = \frac{1}{2}(1 + |\mathbf{r}|^2)$. The purity condition $\mathrm{Tr}\hat{\rho}^2 = 1$ is satisfied only if $|\mathbf{r}| = 1$. This means that pure density matrices correspond to the unit sphere, the boundary of B_2, which is called the *Bloch sphere*. The sphere's center $\mathbf{r} = 0$ corresponds to the density matrix of maximum ignorance $\hat{\rho} = \frac{1}{2}\hat{I}$.

To study the Bloch sphere we express a general normalized vector $|\psi\rangle$ of $\mathbb{C}^2$ as

$$|\psi\rangle = \sin\frac{\theta}{2}|0\rangle + \cos\frac{\theta}{2}e^{-i\phi}|1\rangle = \begin{pmatrix} \cos\frac{\theta}{2}e^{-i\phi} \\ \sin\frac{\theta}{2} \end{pmatrix}, \tag{6.24}$$

where θ and ϕ are the usual spherical coordinates, $\theta \in [0, \pi]$ and $\phi \in [0, 2\pi]$. We use Eq. (6.23) in order to express $\mathbf{r}_\psi$ as a function of θ and ϕ

$$\mathbf{r}_\psi = \langle\psi|\hat{\boldsymbol{\sigma}}|\psi\rangle = (\sin\theta\cos\phi, \sin\theta\sin\phi, \cos\theta). \tag{6.25}$$

Consider two Hilbert space vectors $|\psi_1\rangle$ and $|\psi_2\rangle$. We can always choose a basis so that $\mathbf{r}_{\psi_2} = (0, 0, 1)$, namely, $|\psi_2\rangle$ is on the north pole of the Bloch sphere. For $|\psi_1\rangle$ of the form (6.24),

$$|\langle\psi_1|\psi_2\rangle|^2 = \cos^2\frac{\theta}{2} = \frac{1}{2}(1 + \cos\theta) = \frac{1}{2}(1 + \mathbf{r}_{\psi_1} \cdot \mathbf{r}_{\psi_2}). \tag{6.26}$$

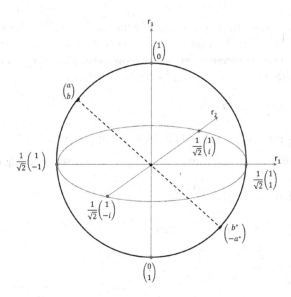

We note that $|\langle\psi_1|\psi_2\rangle| = 0$ only if $\mathbf{r}_{\psi_1} \cdot \mathbf{r}_{\psi_2} = -1$, that is, two orthogonal Hilbert space vectors correspond to two antipodal points of the Bloch sphere – see Fig. 6.1. Hence, an orthonormal basis corresponds to a diameter of the Bloch sphere.

6.2.4   Density Matrix for a Particle in One Dimension

A density matrix on the Hilbert space $L^2(\mathbb{R}, dx)$ corresponds to a function of two variables, $\rho(x, x') := \langle x|\hat{\rho}|x'\rangle$ that satisfies the following properties.

1. $\rho(x, x) \geq 0$, because $\hat{\rho}$ is positive.
2. $\rho^*(x, x') = \rho(x', x)$, because $\hat{\rho}$ is self-adjoint.
3. $\int dx \rho(x, x) = 1$, because $\mathrm{Tr}\hat{\rho} = 1$.
4. $|\rho(x, x')| \leq \sqrt{\rho(x, x)\rho(x', x')}$, that is, the analogue of Eq. (6.20) for continuous spectrum.
5. A density matrix is pure only if $\rho(x, x') = \psi(x)\psi^*(x')$ for some function ψ.

The action of the position operator $\hat{x}$ and of the momentum operator $\hat{p}$ on the function $\rho(x, x')$ is the following. The product $\hat{x}\hat{\rho}$ corresponds to the function $x\rho(x, x')$ and the product $\hat{\rho}\hat{x}$ to the function $x'\rho(x, x')$. This follows from the fact that $\langle x|\hat{x}\hat{\rho}|x'\rangle = x\langle x|\hat{\rho}|x'\rangle = x\rho(x, x')$, and $\langle x|\hat{\rho}\hat{x}|x'\rangle = x'\langle x|\hat{\rho}|x'\rangle = x'\rho(x, x')$.

The product $\hat{p}\hat{\rho}$ corresponds to the function $-i\frac{\partial}{\partial x}\rho(x, x')$ and the product $\hat{\rho}\hat{p}$ to the function $-i\frac{\partial}{\partial x'}\rho(x, x')$. To show this, we use the spectral decomposition $\hat{\rho} = \sum_n w_n|n\rangle\langle n|$, to write,

$$\langle x|\hat{p}\hat{\rho}|x'\rangle = \sum_n w_n\langle x|\hat{p}|n\rangle\langle n|x'\rangle = -i\sum_n w_n\frac{\partial}{\partial x}\langle x|n\rangle\langle n|x'\rangle = -i\frac{\partial}{\partial x}\rho(x, x').$$

Similarly, we show that $\langle x|\hat{\rho}\hat{p}|x'\rangle = -i\frac{\partial}{\partial x'}\rho(x, x')$.

Example 6.3 Consider a Gaussian density matrix $\rho(x, x') = C\exp(-ax^2 - bx'^2 + 2cxx')$ for general complex constants a, b, and c. The density matrix is pure only if $c = 0$, because only then can it be written as $\psi(x)\psi^*(x')$. The requirement of self-adjointness $\rho^*(x, x') = \rho(x', x)$ implies that $a = b^*$ and $c = c^*$. We write $a = \kappa + i\lambda$, for real κ and λ, so that

$$\rho(x, x') = C\exp\left[-\kappa(x^2 + x'^2) - i\lambda(x^2 - x'^2) + 2cxx'\right] \tag{6.27}$$

is expressed in terms of three real constants κ, λ, and c. This means that $\rho(x, x) = Ce^{-2(\kappa-c)x^2}$. The normalization condition implies that $\kappa > c$ and that $C = \sqrt{\frac{2(\kappa-c)}{\pi}}$, where we used Eq. (A.1) for the Gaussian integral.

We also evaluate the purity of the density matrix

$$\gamma = \mathrm{Tr}\hat{\rho}^2 = \int dxdy\rho(x, y)\rho(y, x) = C^2 \int dxdy e^{-2\kappa(x^2+y^2)+4cxy}$$

$$= \frac{2(\kappa - c)}{\pi}\frac{\pi}{2\sqrt{\kappa^2 - c^2}} = \sqrt{\frac{\kappa - c}{\kappa + c}}, \tag{6.28}$$

where we used Eq. (A.5). We confirm that the density matrix is pure ($\gamma = 1$) only if $c = 0$.

6.3   General Description of Quantum States

Given that statistical mixing is inseparable from probabilities, and based on the properties of the density matrix described earlier, we are led to the following generalization of FP1 and FP3.

> **FP1b** Any physical system is described by a Hilbert space. A state of the system at a single moment of time is represented by a density matrix on the Hilbert space.

The principle above is accompanied by the following *rule of statistical mixing* for quantum systems. Suppose we have prepared a statistical ensemble of $N_1 >> 1$ identical systems on a state that corresponds to density matrix $\hat{\rho}_1$, a statistical ensemble of $N_2 >> 1$ identical systems on a state that corresponds to density matrix $\hat{\rho}_2$, and so on. We define the statistical weight of each statistical ensemble as $\lambda_i = N_i/N$, where $N = \sum_i N_i$. Then, the statistical ensemble constructed by the union of the above statistical ensembles is described by the density matrix

$$\hat{\rho} = \sum_i \lambda_i \hat{\rho}_i. \tag{6.29}$$

> **FP3b** Suppose that a physical system is described by the density matrix $\hat{\rho}$, and that we measure a physical quantity that corresponds to the self-adjoint operator $\hat{A}$. Let C be a subset of $\mathbb{R}$ and $\hat{P}_C = \chi_C(\hat{A})$ the associated projector. The probability Prob(C) that the measurement outcome lies within C is given by
>
> $$\mathrm{Prob}(C) = \mathrm{Tr}\left(\hat{\rho}\hat{P}_C\right). \tag{6.30}$$

If the measurement outcomes correspond to a basis on the Hilbert space $|n\rangle$ (an operator with no degenerate eigenvalues), probabilities correspond to the diagonal elements of a density matrix

$$\rho_{nn} = \langle n|\hat{\rho}|n\rangle. \tag{6.31}$$

The average of a physical quantity that corresponds to the self-adjoint operator $\hat{A} = \sum_n a_n \hat{P}_n$ is

$$\langle \hat{A} \rangle = \sum_n \text{Prob}(n)a_n = \sum_n a_n \text{Tr}(\hat{\rho}\hat{P}_n) = \text{Tr}\left(\hat{\rho}\sum_n a_n\hat{P}_n\right) = \text{Tr}(\hat{\rho}\hat{A}). \tag{6.32}$$

It is easy to prove that the Jensen inequality (6.9) also applies to probabilities defined by Eq. (6.30).

The correlation between two operators is defined as

$$\text{Cor}(\hat{A},\hat{B}) = \frac{1}{2}\text{Tr}\left[\hat{\rho}(\hat{A}\hat{B} + \hat{B}\hat{A})\right] - \text{Tr}(\hat{\rho}\hat{A})\text{Tr}(\hat{\rho}\hat{B}), \tag{6.33}$$

and the mean deviation of $\hat{A}$ is $(\Delta A)^2 = \text{Cor}(\hat{A},\hat{A})$.

For measurements that correspond to an operator with continuous spectrum, $\hat{A} = \int d\lambda \hat{P}_\lambda$, probabilities are defined in terms of a density

$$p(\lambda) = \text{Tr}(\hat{\rho}\hat{P}_\lambda). \tag{6.34}$$

If $\hat{P}_\lambda = |\lambda\rangle\langle\lambda|$ for some kets $|\lambda\rangle$, then $p(\lambda) = \langle\lambda|\hat{\rho}|\lambda\rangle$.

Example 6.4 The probability density for position corresponds to the diagonal elements $\rho(x,x)$ of a density matrix $\hat{\rho}$. For the Gaussian density matrix (6.27), we find $\langle \hat{x} \rangle = \int dx x \rho(x,x) = 0$, and $\langle \hat{x}^2 \rangle = \int dx x^2 \rho(x,x) = \frac{1}{4(\kappa-c)}$. By differentiation, we find that $\frac{\partial}{\partial x}\rho(x,x') = 2(-\kappa x - i\lambda x + cx')\rho(x,x')$ and that

$$\frac{\partial^2}{\partial x^2}\rho(x,x') = \left[-2\kappa - 2i\lambda + 4(-\kappa x - i\lambda x + cx')^2\right]\rho(x,x').$$

Hence, we compute

$$\langle \hat{p} \rangle = \int dx \langle x|\hat{p}\hat{\rho}|x\rangle = -2i(-\kappa - i\lambda + c)\int dx x \rho(x,x) = 0$$

$$\langle \hat{p}^2 \rangle = \int dx \langle x|\hat{p}^2\hat{\rho}|x\rangle = 2\kappa + 2i\lambda - (\kappa - c + i\lambda)^2\langle \hat{x}^2 \rangle = \frac{\kappa^2 + \lambda^2 - c^2}{\kappa - c}$$

$$\langle \hat{x}\hat{p} \rangle = \int dx \langle x|\hat{x}\hat{p}\hat{\rho}|x\rangle = -2i(-\kappa - i\lambda + c)\langle \hat{x}^2 \rangle = -\frac{\lambda}{\kappa - c} + \frac{i}{2}.$$

In the calculations above, we used Eq. (A.1) for the Gaussian integrals.

Since the mean values of position and momentum vanish, $C_{xp} = \text{Re}\langle \hat{x}\hat{p} \rangle$. We conclude that

$$(\Delta x)^2 = \frac{1}{4(\kappa - c)}, \quad (\Delta p)^2 = \frac{\kappa^2 + \lambda^2 - c^2}{\kappa - c}, \quad C_{xp} = -\frac{\lambda}{\kappa - c}. \tag{6.35}$$

In Chapter 3, we presented the first fundamental principle of quantum theory, the correspondence of states with Hilbert space vectors. Three chapters later we have generalized this principle, by stating that quantum states correspond to density matrices. The reader has every right to worry about the possibility that, in Chapter 10, he or she will read that the density matrix is not sufficiently general, and that FP3b also needs to be generalized.

Thankfully, this is not going to happen. This is guaranteed by one of the most important mathematical results on quantum theory, *Gleason's theorem* – see Box 6.1. According to this theorem, the most general possible notion of state in quantum theory fully corresponds to a density matrix. Hence, should we ever find ourselves in the position to describe states in terms of objects that are not density matrices, we will have moved beyond quantum theory, or we will have made a mistake.

By Gleason's theorem, we can identify the physical notion of the state of the system at a moment of time with the mathematical object of a density matrix. From now on, we will be using the words "state" and "density matrix" interchangeably. When discussing pure states, we will also refer to the associated Hilbert space vectors as states. Furthermore, we will shorten long phrases of the form "measurement of the physical quantity that corresponds to the self-adjoint operator $\hat{A}$" to "measurement of $\hat{A}$."

Box 6.1 Gleason's Theorem

Gleason's theorem is based on the idea that a measurement on a physical system and the preparation of the system prior to measurement are two distinct and separable processes. For example, suppose we have a Thompson tube that produces electrons and we calibrate it so that it produces a specific statistical ensemble of electrons. If two calibrations are identical, then the corresponding statistical ensembles must be identical. They must be described by the same quantum state.

The state is not affected by our choice of measurement. We can place a Thompson tube with a given calibration either in a set-up where we measure the position of the electrons, or in a set-up where we measure their momentum, or in a set-up where we measure their angular momentum, and so on. A physical theory must provide predictions about the outcomes of an experiment from the knowledge of how the system has been prepared. Hence, the state must provide probabilities for all possible alternatives in all possible measurements on the system.

Measurement outcomes correspond to spectral projectors $\hat{P}_C := \chi_C(\hat{A})$ of self-adjoint operators $\hat{A}$, where $C \subset \sigma(\hat{A})$. The quantum state must provide the probabilities Prob(C) for all C. Suppose that $C = \cup_i C_i$ of some smaller subsets C_i of $\sigma(\hat{A})$, such that $C_i \cap C_j = \emptyset$ for $i \neq j$. The spectral theorem implies that $\hat{P}_C = \sum_i \hat{P}_{C_i}$. Furthermore, by the first Kolmogorov axiom Prob(C) $= \sum_i$ Prob(C_i). Hence, a quantum state must express probabilities as suitable linear functions of the associated projectors.

We are led to the following definition.

A *state* on a Hilbert space $\mathcal{H}$ is a function ω that assigns to each projector $\hat{P}$ on $\mathcal{H}$ a number $\omega(\hat{P}) \in [0, 1]$, such that

(i) $\omega(\hat{0}) = 0$,
(ii) $\omega(\hat{I}) = 1$.

(iii) If the projector $\hat{P}$ is a countable sum $\sum_i \hat{P}_i$ of projectors $\hat{P}_i$, such that $\hat{P}_i \hat{P}_j = 0$ for $i \neq j$, then $\omega(\hat{P}) = \sum_i \omega(\hat{P}_i)$.

With this definition of a state, Gleason proved the following (Gleason, 1957).

Theorem 6.3 (Gleason's theorem) *For any state ω on a Hilbert space $\mathcal{H}$ of dimension $n > 2$, there exists a unique density matrix $\hat{\rho}_\omega$, such that $\omega(\hat{P}) = \mathrm{Tr}\left(\hat{\rho}_\omega \hat{P}\right)$, for all projectors $\hat{P}$ on $\mathcal{H}$.*

The proof of Gleason's theorem is notoriously difficult, in spite of many efforts to simplify it – see Cooke et al. (1985) for one of the simplest proofs. However, the theorem's meaning is clear: Density matrices provide the most general description of quantum states.

6.3.1 The Interpretation of the Quantum State

We return to the explanation of the notion of the state in quantum theory that was first raised in Section 3.2. The state refers to the way that a system has been prepared, either in the lab or naturally, prior to measurement. All information about the preparation is encoded into a mathematical object, the density matrix, from which we can predict probabilities for measurement outcomes by Eq. (6.30).

This interpretation of the quantum state is the *common denominator* for understanding and using quantum theory. It is logically consistent and it suffices for explaining the huge success of quantum theory in predicting experimental results. We will refer to this interpretation as the *minimal interpretation*.

In Section 1.5, we saw that probabilities can be viewed either through a logical interpretation or through a frequentist interpretation. The difference is important, because in the latter case probabilities refer exclusively to a statistical ensemble, while in the former they can be applied to individual systems. This issue is of current practical interest, because it is possible to isolate *individual quantum systems* (e.g., atoms or ions) and perform measurements on them. There are ways to use quantum theory for the description of such systems, and this makes the logical interpretation of quantum probabilities more compelling.

However, the state is most certainly not an *objective property* of *individual* systems. By objective property of a system we mean any property that can be specified by measurements without knowledge of the system's past. In classical physics, states *are* objective properties of individual systems. For example, the state of a single particle is determined by the position x and the momentum p. To identify the state, one has only to measure x and p. In quantum theory, objectivity is precluded by the fact that quantum states are not mutually exclusive – see Section 3.2. Suppose that we measure an operator $\hat{A}$ in an individual system, and we find a value a that corresponds to the eigenvector $|a\rangle$. We cannot conclude that the state of the system (prior to measurement) was $|a\rangle$; this outcome is compatible with all states $|\psi\rangle$ such that $\langle\psi|a\rangle \neq 0$. Since it is impossible to identify a unique quantum state from *any* fine-grained measurement on an individual system, the quantum state cannot be an objective property of the system.

The quantum state must be treated as a semisubjective entity that partly refers to our knowledge about a physical system. This fact contradicts a tradition that originates from Newtonian physics, in which the basic structures of a physical theory correspond to objective features of the world, and not to our knowledge about the world or to the process through which we obtain this knowledge. For this reason, many physicists try to interpret quantum states in such a way that they refer to objective properties of physical systems. Recall that Schrödinger's default position was to interpret wave functions as physical waves.

In later chapters (Chapters 8–10 and 22), we will see that the objective interpretations of the quantum state either fail (because they lead to conflict with fundamental physical principles or to disagreement with experiment), or they are inadequate (they cannot be ruled out, but they require further work before they can become convincing). As a result, they are accepted by a minority of researchers. No interpretation beyond the minimal has universal acceptance.

On the other hand, the minimal interpretation is certainly *incomplete*, because it treats measurement as a fundamental concept. A measurement is itself a physical process that must be described by a physical theory. If this description is part of quantum theory, then we need to go beyond the minimal interpretation, because the latter leads to the vicious circle of using the notion of a measurement in order to define the notion of measurement. If the description of measurements is not subject to quantum theory, then we face a paradox. There are physical systems (the measuring devices and/or the human observer) that are not subject to quantum theory, even though they consist of atoms that are quantum systems. This paradox is known as the *quantum measurement problem*. To quote Bell (1987),

The 'Problem' then is this: how exactly is the world to be divided into speakable apparatus ... that we can talk about ... and unspeakable quantum system that we can not talk about? How many electrons, or atoms, or molecules, make an 'apparatus'? The mathematics of the ordinary theory requires such a division, but says nothing about how it is to be made. In practice the question is resolved by pragmatic recipes which have stood the test of time, applied with discretion and good taste born of experience. But should not fundamental theory permit exact mathematical formulation?

In Box 6.2, we present a sample of different opinions about the interpretation of the quantum state. The reader will immediately notice that the opinions differ significantly. One of the most remarkable facts about quantum theory is that such differences at the fundamental level have never led to ambiguities in physical predictions. This is because the minimal interpretation suffices for almost all predictions.

For this reason, we present quantum theory based on the minimal interpretation. We suggest that the reader adopts this interpretation *temporarily*, in order to familiarize oneself with the ways that quantum theory is used in practice and with the fundamental theoretical and experimental facts that must be respected by any interpretation. In the final chapter of the book (Chapter 22), we will study interpretational issues in some depth, and then the reader will be able to make an informed decision about which interpretation of quantum theory he or she finds most convincing.

Note that taking the minimal interpretation as fundamental, rather than as a provisional or pragmatic point of view, is *not* a neutral position. It implies an adherence to a specific philosophical position, known as *instrumentalism*. One then views quantum theory as

… a set of prescriptions that churns out useful results, but which gives no direct picture of (or assigns any meaning to) the reality that is assumed by most scientists to lie beneath their observations. Many of the conceptual problems are certainly sidestepped by this procedure, but the price paid is an unequivocal anti-realism…. (Isham, 1995)

Box 6.2 Interpretations of the Quantum State

"[The state] combines objective and subjective elements. It contains statements about possibilities or better tendencies ('potentia' in Aristotelian philosophy), and these statements are completely objective, they do not depend on any observer; and it contains statements about our knowledge of the system, which of course are subjective in so far as they may be different for different observers." (Heisenberg, 1999)

"We must think of a wavefunction as one entire thing … Wavefunctions are quite unlike the waves of classical physics in this important respect. The divergent parts of the wave cannot be thought of as local disturbances, each carrying on independently of what is happening in a remote region. Wavefunctions have a strongly non-local character; in this sense they are completely holistic entities." (Penrose, 2005)

"It seems to be clear, therefore, that Born's statistical interpretation of quantum theory is the only possible one. The wave function does not in any way describe a state which could be that of a single system; it relates rather to many systems, to an 'ensemble of systems' in the sense of statistical mechanics." (Einstein, 1936)

"The question of whether the waves are something 'real' or a function to describe and predict phenomena in a convenient way is a matter of taste. I personally like to regard a probability wave, even in 3N-dimensional space, as a real thing, certainly as more than a tool for mathematical calculations. … how could we rely on probability predictions if by this notion we do not refer to something real and objective?" (Born, 1964)

"The state is not an objective property of an individual system but is that information, obtained from a knowledge of how the system was prepared, which can be used for making predictions about future measurements." (Hartle, 1968)

"Quantum states can be given a clear operational definition, based on the notion of test … If these tests are performed many times, after identical preparations, we find that the statistical distribution of outcomes of each test tends to a limit. Each outcome has a definite probability. We can then define a state as follows: A state is characterized by the probabilities of the various outcomes of every conceivable test." (Peres, 2002)

6.3.2   Difference Between Superpositions and Mixtures

Superposition and statistical mixing are two *distinct* procedures for combining pairs of Hilbert space vectors for the definition of a new quantum state. They should not be conflated.

Consider two vectors $|\psi\rangle$ and $|\phi\rangle$ on a Hilbert space $\mathcal{H}$. *Superpositions* of these vectors are constructed through the operation of addition on $\mathcal{H}$, for example, the vector $|\chi\rangle = \frac{1}{\sqrt{2}}(|\psi\rangle + |\phi\rangle)$. The corresponding density matrix is pure,

$$\hat{\rho}_1 = |\chi\rangle\langle\chi|. \tag{6.36}$$

A *mixture* of $|\psi\rangle$ and $|\phi\rangle$ is defined in terms of the addition of density matrices on $\mathcal{H}$, for example,

$$\hat{\rho}_2 = \frac{1}{2}\left(|\psi\rangle\langle\psi| + |\phi\rangle\langle\phi|\right). \tag{6.37}$$

Superpositions and mixtures lead to different probability distributions, hence, different physical predictions. For example, the density matrix $\hat{\rho}_1$ yields the following probability density for position measurements

$$p_1(x) = \langle x|\hat{\rho}_1|x\rangle = \frac{1}{2}|\langle x|\psi\rangle|^2 + \frac{1}{2}|\langle x|\phi\rangle|^2 + \mathrm{Re}\left(\langle\psi|x\rangle\langle x|\phi\rangle\right), \tag{6.38}$$

while the density matrix $\hat{\rho}_2$ leads to the probability density

$$p_2(x) = \langle x|\hat{\rho}_2|x\rangle = \frac{1}{2}|\langle x|\psi\rangle|^2 + \frac{1}{2}|\langle x|\phi\rangle|^2. \tag{6.39}$$

The two probability densities differ with respect to the last term of Eq. (6.38), the *interference* term. The word "interference" is employed because an analogous term is responsible for interference in the two-slit experiment. A comparison of the probability distributions (6.38) and (6.39) is given in Fig. 6.2.

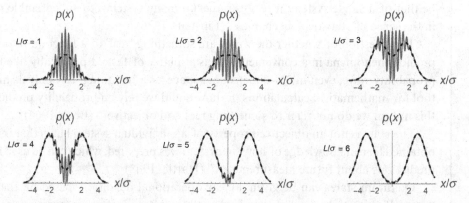

Fig. 6.2 We consider the Gaussian wave functions $\psi(x) = (2\pi\sigma^2)^{-1/4}\exp[-(x - \frac{1}{2}L)/(4\sigma^2) + i\bar{p}x]$ and $\phi(x) = (2\pi\sigma^2)^{-1/4}\exp[-(x + \frac{1}{2}L)/(4\sigma^2)]$. They have the same width σ and their centers are separated by distance L; $\psi(x)$ is complex and has mean momentum equal to $\bar{p}$. We plot the probability densities for position $p_1(x)$ for a superposition (solid line) and $p_2(x)$ for a mixture (dashed line), for different values of L/σ. In these plots, $\bar{p}\sigma = 15$. The two probability distributions become identical for large values of L/σ, whence the interference terms in Eq. (6.38) become negligible.

6.3.3 Difference Between Classical and Quantum Mixing

As shown in Section 6.2.1, mixing is an invertible procedure in classical probability theory. If we mix pure probability vectors $\mathbf{e}_i$ in order to form a probability vector $\mathbf{w}$, the analysis of $\mathbf{w}$ yields uniquely the vectors $\mathbf{e}_i$ and the associated mixing proportions. This is the reason we consider pure probability vectors as more fundamental, and often identify them with the microstates of a physical system.

In quantum theory, statistical mixing is not a reversible procedure. There are infinite ways to express a density matrix $\hat{\rho}$ as a mixture $\sum_i p_i |\psi_i\rangle \langle \psi_i|$ of pure states, as long as we set no restrictions on the vectors $|\psi_i\rangle$. It is obvious that in a qubit there are infinite different ways to express a nonunit Bloch vector as a convex combination of two unit vectors. This means that even if a density matrix $\hat{\rho}$ was formed by a mixture of statistical ensembles with different pure states – as in Section 6.2.2 – the recovery of the initial ensembles is impossible.

We may impose the requirement that the vectors $|\psi_i\rangle$ form an orthonormal set. Again, if a density matrix $\hat{\rho}$ has, say, an eigenvalue with double degeneracy, we cannot find a unique pair of mutually orthogonal eigenvectors: There are infinite such pairs in the corresponding eigenspace.

Assume a device that prepares a qubit in a mixture of states $|0\rangle$ and $|1\rangle$ with equal weight $\frac{1}{2}$. The corresponding density matrix is the one of maximum ignorance $\hat{\rho} = \frac{1}{2}\hat{I}$, since the resolution of the unity implies that $|0\rangle\langle 0| + |1\rangle\langle 1| = \hat{I}$. But the resolution of the unity applies to all bases on the Hilbert space (e.g., one defined by eigenvectors of $\hat{\sigma}_1$); hence, $\hat{\rho}$ can be decomposed to any pair of eigenvectors. We cannot obtain uniquely the vectors $|0\rangle$ and $|1\rangle$ from which $\hat{\rho}$ was formed.

For these reasons, we cannot assert that pure states are more fundamental than mixed states in quantum theory. In contrast to the classical theory, we cannot say that a mixed density matrix describes our ignorance about the pure state that actually describes a system. Box 6.3 shows quantum entropies as measures of mixing.

Box 6.3   Quantum Entropies as Measures of Mixing

Shannon Entropy The most important measure of randomness of a classical probability distribution is *Shannon entropy*. It was first used as a measure of the information lost by a signal that is subject to random perturbations (Shannon, 1948). The name "entropy" originates from the similarity to Gibbs entropy in statistical mechanics.

Definition 6.4 Let Γ be a sample space with N elements and $\mathbf{w} = (p_1, p_2, \ldots, p_N)$ a probability vector on Γ. The Shannon entropy of $\mathbf{w}$ is $I[\mathbf{w}] := -\sum_{i=1}^{N} p_i \ln p_i$, and it is always nonnegative.

The Shannon entropy $I[\mathbf{w}]$ has the following properties.

(i) The minimum of $I[\mathbf{w}]$ is zero, and it is achieved for pure probability vectors.

Proof The coordinates of pure probability vectors are 0 and 1, for which $p \ln p = 0$.

(ii) The maximum of $I[\mathbf{w}]$ is $\ln N$ and it is a achieved by $\vec{w} = N^{-1}(1, 1, \ldots, 1)$.

Proof We maximize $I[\mathbf{w}]$ subject to the condition $\sum_{i=1}^{N} p_i = 1$. The maximum is obtained from the solution of the equation $\frac{\partial}{\partial p_i}[I + c(\sum_{i=1}^{N} p_i - 1)] = 0$, where c is a Lagrange multiplier. We find that $-1 - p_i + c = 0$, that is, all p_i are equal. Normalization implies that $p_i = N^{-1}$.

(iii) $I[\mathbf{w}]$ is *concave*: For any probability vectors $\mathbf{w}_r$ $(r = 1, 2, \ldots k)$ and numbers $0 \le \lambda_r \le 1$ with $\sum_{r=1}^{k} \lambda_r = 1$, $I[\sum_{r=1}^{k} \lambda_r \mathbf{w}_r] \ge \sum_{r=1}^{k} \lambda_r I[\mathbf{w}_r]$.

Proof The concavity of the function $f(x) = -x \ln x$ $(f''(x) < 0)$ implies that $f(\lambda x_1 + (1 - \lambda)x_2) \ge \lambda f(x_1) + (1 - \lambda)f(x_2)$, for all $0 \le \lambda \le 1$. This leads us to the desired inequality for $k = 2$. The proof to general k proceeds by induction.

Property (iii) implies that the Shannon entropy increases with statistical mixing. We always lose information when two statistical ensembles are mixed.

Application to Quantum Probabilities Suppose we prepare a quantum system on a state $\hat{\rho}$ and we measure $\hat{A} = \sum_n a_n \hat{P}_n$. The probability of obtaining outcome a_n is $p_n = \mathrm{Tr}(\hat{\rho}\hat{P}_n)$. We define the Shannon entropy of this measurement as

$$I(\hat{\rho}, \hat{A}) = -\sum_n p_n \ln p_n. \tag{A}$$

For qubits, the density matrix corresponds to a vector $\mathbf{r}$ of Bloch's sphere and $\hat{A} = \mathbf{m} \cdot \hat{\boldsymbol{\sigma}}$, for some unit vector $\mathbf{m}$. The spectral projectors $\hat{P}_\pm$ of $\hat{A}$ are given by Eq. (5.16). Hence, the associated probabilities are $p_\pm = \frac{1}{2}(1 \pm \mathbf{r} \cdot \mathbf{m})$. We substitute into Eq. (A), and we write $\mathbf{r} \cdot \mathbf{m} = rx$, for $r = |\mathbf{r}|$ and $|x| \le 1$. Then,

$$I(r, x) = \ln 2 + \frac{r}{2}x \ln\left[\frac{1-rx}{1+rx}\right] - \frac{1}{2}\ln\left[1 - r^2 x^2\right]. \tag{B}$$

For constant r, the minimum of $I(r, x)$ is obtained for $x = \pm 1$, and it equals $I_{min}(r) = -w_- \ln w_- - w_+ \ln w_+$, where $w_\pm = 1 \pm r$ are the eigenvalues of the density matrix $\hat{\rho}$. The minimum corresponds to an operator $\hat{A}$ that commutes with $\hat{\rho}$.

von Neumann Entropy The preceding analysis is straightforwardly generalized to arbitrary systems. The minimum value (B) of the Shannon entropy depends only on the density matrix $\hat{\rho}$. We write this minimum value as $S[\hat{\rho}]$; it is known as *von Neumann entropy*. For a density matrix $\hat{\rho} = \sum_n w_n |n\rangle\langle n|$,

$$S[\hat{\rho}] = -\sum_n w_n \ln w_n, \tag{C}$$

where it is assumed that an eigenvalue with degeneracy D_n is added D_n times. Equation (C) can also be written as $S[\hat{\rho}] = -\mathrm{Tr}(\hat{\rho} \ln \hat{\rho})$.

The von Neumann entropy also satisfies the following properties.

(i) If $\hat{\rho}$ is pure, then $S[\hat{\rho}] = 0$.
(ii) The maximum value of $S[\hat{\rho}]$ in $\mathbb{C}^N$ is $\ln N$, and it is obtained for $\hat{\rho} = N^{-1}\hat{I}$.
(iii) $S[\hat{U}\hat{\rho}\hat{U}^\dagger] = S[\hat{\rho}]$, for all unitary $\hat{U}$.

6.4   Uncertainty Relations

Quantum states incorporate Heisenberg's uncertainty principle in their definition, albeit with a slightly different physical interpretation. This is a consequence of the following theorem that was first proved by Kennard (1927) and Robertson (1929).

Theorem 6.5 *Let $\hat{A}$ and $\hat{B}$ be self-adjoint operators, with mean deviations ΔA and ΔB, respectively, in a state $\hat{\rho}$. Then,*

$$(\Delta A)(\Delta B) \geq \frac{1}{2} \left| \text{Tr} \left(\hat{\rho}[\hat{A}, \hat{B}] \right) \right|. \tag{6.40}$$

Proof We assume with no loss of generality that $\langle \hat{A} \rangle = \langle \hat{B} \rangle = 0$. If this is not true, we proceed with the operators $\hat{A} - \langle \hat{A} \rangle$ and $\hat{B} - \langle \hat{B} \rangle$, and the proof remains unchanged. Hence, $(\Delta A)^2 = \langle \hat{A}^2 \rangle = \text{Tr}(\hat{\rho}\hat{A}^2)$ and similarly for $\hat{B}$.

We note that $\text{Tr}(\hat{\rho}\hat{A}^2) = \text{Tr}[\sqrt{\hat{\rho}}\hat{A}(\sqrt{\hat{\rho}}\hat{A})^{\dagger}]$. We use Eq. (4.67) for the operators $\sqrt{\hat{\rho}}\hat{A}$ and $\sqrt{\hat{\rho}}\hat{B}$, to obtain

$$(\Delta A)^2(\Delta B)^2 = \text{Tr}\left(\hat{\rho}\hat{A}^2 \right) \text{Tr}\left(\hat{\rho}\hat{B}^2 \right) \geq \left| \text{Tr}\left(\sqrt{\hat{\rho}}\hat{A}(\sqrt{\hat{\rho}}\hat{B})^{\dagger} \right) \right|^2 = \left| \text{Tr}\left(\hat{\rho}\hat{A}\hat{B} \right) \right|^2. \tag{6.41}$$

We write $2\text{Tr}\left(\hat{\rho}\hat{A}\hat{B} \right) = \text{Tr}\left(\hat{\rho}\left(\hat{A}\hat{B} + \hat{B}\hat{A} \right) \right) + \text{Tr}\left(\hat{\rho}[\hat{A}, \hat{B}] \right)$, where the first term in the right-hand side is real and the second is imaginary. Hence,

$$\left| \text{Tr}\left(\hat{\rho}\hat{A}\hat{B} \right) \right|^2 = C_{AB}^2 + \frac{1}{4} \left| \text{Tr}\left(\hat{\rho}[\hat{A}, \hat{B}] \right) \right|^2,$$

where C_{AB} is the correlation between $\hat{A}$ and $\hat{B}$. Then, Eq. (6.41) becomes

$$(\Delta A)^2(\Delta B)^2 - C_{AB}^2 \geq \frac{1}{4} \left| \text{Tr}\left(\hat{\rho}[\hat{A}, \hat{B}] \right) \right|^2, \tag{6.42}$$

from which Eq. (6.40) follows.   □

For the position operator $\hat{x}$ and the momentum operator $\hat{p}$, Eq. (6.40) yields

$$\Delta x \Delta p \geq \frac{1}{2}. \tag{6.43}$$

Equality in Eq. (6.43) is satisfied by the Gaussian wave functions (6.10).

Equation (6.43) resembles Heisenberg's uncertainty relation, Eq. (2.50). Indeed, Kennard and Robertson proved Eq. (6.43) while looking for a mathematically rigorous statement of the uncertainty principle. However, the meaning of Δx and Δp differs. In Eq. (2.50), Δx and Δp are errors in a measurement that determines both position and momentum and they refer to an individual particle. In Eq. (6.43), Δx and Δp are mean deviations that refer to two distinct experiments, one measuring position and one measuring momentum. As such, they are defined at the level of a statistical ensemble and not at the level of individual systems.

Equation (6.42) provides a sharper uncertainty relation that takes into account the correlation between position and momentum. It implies that the *generalized uncertainty* $\mathcal{A} := \sqrt{(\Delta x)^2(\Delta p)^2 - C_{xp}^2}$ satisfies

$$\mathcal{A} \geq \frac{1}{2}. \tag{6.44}$$

Example 6.5   Consider a harmonic oscillator with frequency ω and mass m. For the eigenstates $|n\rangle$ of its Hamiltonian, Eq. (5.55) implies that $\langle n|\hat{x}|n\rangle = 0$ and $\langle n|\hat{p}|n\rangle = 0$. Furthermore,

$$\langle n|\hat{x}^2|n\rangle = \frac{1}{2m\omega}\langle n|(\hat{a}+\hat{a}^\dagger)^2|n\rangle = \frac{1}{2m\omega}\langle n|\hat{a}\hat{a}^\dagger + \hat{a}^\dagger\hat{a}|n\rangle = \frac{2n+1}{2m\omega}$$

$$\langle n|\hat{p}^2|n\rangle = -\frac{m\omega}{2}\langle n|(\hat{a}-\hat{a}^\dagger)^2|n\rangle = \frac{m\omega}{2}\langle n|\hat{a}\hat{a}^\dagger + \hat{a}^\dagger\hat{a}|n\rangle = \frac{m\omega}{2}(2n+1)$$

$$\langle n|\hat{x}\hat{p}|n\rangle = \frac{1}{2i}\langle n|(\hat{a}+\hat{a}^\dagger)(\hat{a}-\hat{a}^\dagger)|n\rangle = \frac{i}{2}\langle n|[\hat{a},\hat{a}^\dagger]|n\rangle = \frac{i}{2}. \tag{6.45}$$

It follows that $(\Delta x)^2 = \frac{2n+1}{2m\omega}$, $(\Delta p)^2 = \frac{m\omega}{2}(2n+1)$, and $C_{xp} = 0$. Hence,

$$(\Delta x)(\Delta p) = \mathcal{A} = n + \frac{1}{2} = \frac{E_n}{\omega}. \tag{6.46}$$

Unlike the coherent states, the uncertainty for number states increases linearly with energy. Hence, for large n, the number states describe a system that is very spread out in the classical state space.

Example 6.6   For the Gaussian density matrix (6.27), Eq. (6.35) yields

$$(\Delta x)(\Delta p) = \frac{1}{2(\kappa - c)}\sqrt{\kappa^2 + \lambda^2 - c^2}. \tag{6.47}$$

The right-hand side of Eq. (6.47) is minimized for $\lambda = c = 0$, which corresponds to coherent states.

The generalized uncertainty $\mathcal{A}$ takes values

$$\mathcal{A} = \frac{1}{2}\sqrt{\frac{\kappa + c}{\kappa - c}} = \frac{1}{2\gamma}, \tag{6.48}$$

where γ is the purity of the state, Eq. (6.28). The minimum value of $\mathcal{A}$ is achieved for Gaussian pure states, that is, for $c = 0$.

Example 6.7   The variance is not always a good measure of uncertainty. To see this, consider a wave function

$$\psi(x) = \frac{\sqrt{a}}{\pi^{1/4}}\frac{\sin\left(\frac{x^2}{a^2}\right)}{x}. \tag{6.49}$$

It is straightforward to verify that in this state $\langle \hat{x}\rangle = 0$ and $\langle \hat{p}\rangle = 0$. Hence,

$$(\Delta x)^2 = \langle \hat{x}^2\rangle = \frac{a}{\sqrt{\pi}}\int_{-\infty}^{\infty} dx \sin^2\left(\frac{x^2}{a^2}\right) = \infty$$

$$(\Delta p)^2 = \langle \hat{p}^2\rangle = \frac{a}{\sqrt{\pi}}\int_{-\infty}^{\infty} dx \left(\frac{2}{a^2}\cos\left(\frac{x^2}{a^2}\right) - \frac{\sin\left(\frac{x^2}{a^2}\right)}{x^2}\right)^2 = \infty.$$

The probability density for position $|\psi(x)|^2$ and the probability density for momentum $|\tilde{\psi}(p)|^2$ are plotted in Fig. 6.3. In spite of divergence of the mean deviations, both probability

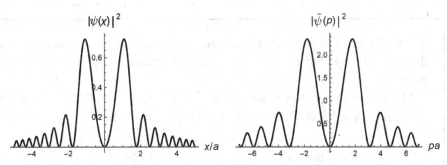

Fig. 6.3 The probability density $|\psi(x)|^2$ for position and the probability density $|\tilde{\psi}(p)|^2$ for momentum, for $\psi(x)$ given by Eq. (6.49).

distributions are well localized and we can define other measures of their spread. For example, we can define as δx the range of values of x that lies under the two central peaks in the plot of $|\psi(x)|^2$, whence $\delta x \simeq 3.6a$. We can similarly define δp as the range of values of p under the two central peaks in the plot of $|\tilde{\psi}(p)|^2$, whence $\delta p = 6.4/a$. For these measures of uncertainty, $\delta x \delta p \simeq 23$.

A measure of uncertainty for position and momentum that remains informative even when the variances are divergent is described in Box 6.4.

Box 6.4 Entropic Uncertainty Relations

We present a reformulation of the Kennard–Robertson uncertainty principle in terms of the Shannon entropy (Hirschman, 1957; Everett, 1956; Bialynicki-Birula and Mycielski, 1975).

A density matrix $\hat{\rho}$ defines a probability distribution $\langle x|\hat{\rho}|x\rangle$ with respect to position and a probability distribution $\langle p|\hat{\rho}|p\rangle$ with respect to momentum. We therefore define Shannon entropies for position and momentum,

$$I_{\hat{x}}(\hat{\rho}) = -\int dx \langle x|\hat{\rho}|x\rangle \ln\langle x|\hat{\rho}|x\rangle \qquad I_{\hat{p}}(\hat{\rho}) = -\int dp \langle p|\hat{\rho}|p\rangle \ln\langle p|\hat{\rho}|p\rangle.$$

The entropic uncertainty relation is the inequality

$$I_{\hat{x}}(\hat{\rho}) + I_{\hat{p}}(\hat{\rho}) \geq \ln(\pi e), \tag{A}$$

The proof of Eq. (A) is highly technical (Beckner, 1975), and will not be given here. The Shannon entropies for continuous probability distributions can become negative and arbitrarily small, as we well see. However, by Eq. (A) the sum $I_{\hat{x}} + I_{\hat{p}}$ has a positive lower bound.

In order to physically interpret Eq. (A), we look for the probability density $p(x)$ that minimizes Shannon entropy for fixed mean deviation σ. It is easy to verify that the Shannon entropy does not change if we shift the mean value of $p(x)$. Hence, we can choose $\langle x\rangle = 0$, so that $\langle x^2\rangle = \sigma^2$. We maximize the quantity $-\int dx p(x)\ln p(x) + \lambda[\int dx x^2 p(x) -$

σ^2], with respect to $p(x)$; λ is a Lagrange multiplier. For ease of calculation, treat x as a discrete index and write a sum rather than an integral; then differentiate with respect to $p_x := p(x)$.

Maximization is achieved for $-\ln p(x) - 1 + \lambda x^2 = 0$, that is, for a Gaussian $p(x)$. Therefore, we use $p(x) = (2\pi\sigma^2)^{-1/2} \exp\left[-\frac{x^2}{2\sigma_x^2}\right]$, to obtain

$$I[p] \leq \frac{1}{2}\ln(2\pi\sigma^2) + \int dx\, p(x)\frac{x^2}{2\sigma^2} = \frac{1}{2}\ln(2\pi\sigma^2) + \frac{1}{2} = \frac{1}{2}\ln(2\pi e\sigma^2).$$

Obviously $I[p]$ becomes negative for sufficiently small σ, and goes to $-\infty$ as $\sigma \to 0$.

We apply Eq. (A) to the entropies $I_{\hat{x}}(\hat{\rho})$ and $I_{\hat{p}}(\hat{\rho})$, to obtain $I_{\hat{x}}(\hat{\rho}) \leq \frac{1}{2}\ln\left[2\pi e\Delta x^2\right]$, and $I_{\hat{p}}(\hat{\rho}) \leq \frac{1}{2}\ln\left[2\pi e\Delta p^2\right]$. Hence,

$$I_{\hat{x}}(\hat{\rho}) + I_{\hat{p}}(\hat{\rho}) \leq \ln\left[2\pi e\Delta x\Delta p\right]. \tag{B}$$

The comparison of Eqs. (A) and (B) confirms the Kennard–Robertson uncertainty relations. Both inequalities are saturated for coherent states.

Let us apply Eq. (A) to the wave function (6.49), with infinite mean deviations Δx and Δp. We find numerically that $I_{\hat{x}} \simeq \ln a + 1.57$ and $I_{\hat{p}} \simeq -\ln a + 5.63$. Hence $I_{\hat{x}} + I_{\hat{p}} \simeq 7.2$, that is, a finite value that allows for a comparison with the lower limit $\ln(\pi e) \simeq 2.14$ of Eq. (A).

6.5   Wigner Function

As explained in Section 1.5, probabilities in a classical system are defined in terms of a probability density on the state space Γ of the system. In quantum theory, probabilities are defined in terms of density matrices on an appropriate Hilbert space $\mathcal{H}$. It would be useful to be able to translate from one language to the other, that is, to compare the classical and quantum probability assignments.

Such a translation is provided by the *Wigner function* (Wigner, 1932). The Wigner function is a *pseudoprobability* distribution in state space, associated to unique density matrices, and giving correct expressions for the quantum probabilities of position and momentum measurements.

The expression "pseudoprobability distribution" signifies that the Wigner function satisfied many properties of a genuine probability distribution, but not all: *It can take negative values.* This is to be expected: If we could describe a quantum system in terms of genuine probability densities on the state space, we would not need quantum theory. A generalization of classical statistical mechanics would suffice.

We consider the Wigner function for a particle at a line. The associated Hilbert space is $L^2(\mathbb{R})$. The classical state space is $\Gamma = \mathbb{R}^2$; it describes the position x and momentum p of a particle.

For any state $\hat{\rho}$, we define the Wigner function $W : \Gamma \to \mathbb{R}$ as

$$W(x,p) := \frac{1}{2\pi} \int dy \left\langle x - \frac{y}{2} | \hat{\rho} | x + \frac{y}{2} \right\rangle e^{ipy}. \tag{6.50}$$

The definition (6.50) is a special case of the *Weyl–Wigner transform*, that assigns a function $F_{\hat{A}}(x,p)$ on Γ to each operator $\hat{A}$ on $L^2(\mathbb{R})$, by

$$F_{\hat{A}}(x,p) = \int dy \left\langle x - \frac{y}{2} | \hat{A} | x + \frac{y}{2} \right\rangle e^{ipy}. \tag{6.51}$$

The inverse Weyl–Wigner transform is given by

$$\langle x | \hat{A} | x' \rangle = \int \frac{dp}{2\pi} F_{\hat{A}} \left(\frac{x + x'}{2}, p \right) e^{ip(x - x')}. \tag{6.52}$$

The proof is straightforward. We substitute Eq. (6.52) into Eq. (6.51), and we obtain an identity.

Example 6.8 For a Weyl operator, $\hat{V}(a,b) = e^{ia\hat{x} - ib\hat{p}}$, Eq. (6.51) yields

$$F_{\hat{V}(a,b)}(x,p) = \int dy e^{ipy} \left\langle x - \frac{y}{2} | \hat{V}(a,b) | x + \frac{y}{2} \right\rangle$$

$$= \int dy e^{ipy} e^{-\frac{i}{2}ab + ia(x - \frac{y}{2})} \delta(y + b) = e^{iax - ibp}. \tag{6.53}$$

Differentiating with respect to a and b, and then taking $a = b = 0$, we obtain $F_{\hat{x}}(x,p) = x$ and $F_{\hat{p}} = p$.

The most important properties of the Wigner function are that (i) it expresses quantum probabilities in a way that *looks like* a classical probabilistic description, and (ii) its marginal distributions for position and momentum coincide with those of quantum theory.

Property (i) is a consequence of the identity

$$\mathrm{Tr}(\hat{\rho}\hat{A}) = \int dx dp \, W(x,p) F_{\hat{A}}(x,p) \tag{6.54}$$

that applies to any operator $\hat{A}$ and density matrix $\hat{\rho}$. To prove Eq. (6.54), we start from the right-hand side and substitute Eqs. (6.50) and (6.51).

We find that $\int dx dp W(x,p) F_{\hat{A}}(x,p) = \int dx dx' \langle x | \hat{\rho} | x' \rangle \langle x' | \hat{A} | x \rangle = \int dx \langle x | \hat{\rho} \hat{A} | x \rangle = \mathrm{Tr}(\hat{\rho}\hat{A})$. Obviously, the right-hand side of Eq. (6.54) *only* looks like the corresponding equation of classical probability theory, since the Wigner function can take negative values.

Property (ii) is expressed by the relations

$$\int dp W(x,p) = \langle x | \hat{\rho} | x \rangle, \qquad \int dx W(x,p) = \langle p | \hat{\rho} | p \rangle. \tag{6.55}$$

The proof of the first Eq. (6.55) is straightforward, since the integration with respect to p in Eq. (6.50) yields a term $\delta(y)$. To prove the second Eq. (6.55), we express the position matrix elements of $\hat{\rho}$

$$\langle x | \hat{\rho} | x' \rangle = \int dk dk' \langle x | k \rangle \langle k | \hat{\rho} | k' \rangle \langle k' | x' \rangle = \int \frac{dk dk'}{2\pi} e^{ikx - ik'x'} \langle k | \hat{\rho} | k' \rangle$$

in terms of momentum kets, here denoted as $|k\rangle$. Then,

$$W(x,p) = \frac{1}{(2\pi)^2} \int dk\, dk' \left(\int dy\, e^{iy(p-\frac{k+k'}{2})} \right) e^{i(k-k')x} \langle k|\hat{\rho}|k'\rangle$$

$$= \frac{1}{2\pi} \int dk\, dk'\, \delta\left(p - \frac{k+k'}{2}\right) e^{i(k-k')x} \langle k|\hat{\rho}|k'\rangle. \tag{6.56}$$

We set $K = \frac{k+k'}{2}$ and $\xi = k' - k$. Then, Eq. (6.56) becomes

$$W(x,p) = \frac{1}{2\pi} \int dK\, d\xi\, \delta(p - K) e^{-i\xi x} \langle K - \xi/2|\hat{\rho}|K + \xi/2\rangle$$

$$= \frac{1}{2\pi} \int d\xi \left\langle p - \frac{1}{2}\xi \Big| \hat{\rho} \Big| p + \frac{1}{2}\xi \right\rangle e^{-i\xi x}. \tag{6.57}$$

We integrate over x. Given that $\int dx\, e^{-i\xi x} = 2\pi\, \delta(\xi)$, we obtain the desired result

With similar calculations, we find that the Weyl–Wigner transform associates (i) any function $f(x)$ of x to the operator $f(\hat{x})$, and (ii) any function $g(x)$ of p to the operator $g(\hat{p})$. We also note that the operator $\frac{1}{2}(\hat{x}\hat{p} + \hat{p}\hat{x})$ corresponds to the function $x \int \frac{dy}{2\pi} \langle x - \frac{y}{2}|\hat{p}|x + \frac{y}{2}\rangle e^{ipy} = xp$.

Example 6.9 We evaluate the Wigner function associated to the Gaussian wave function $\psi_{\bar{q},\bar{p}}$ of Eq. (6.10). We use Eq. (A.1), to obtain

$$W(x,p) = \frac{1}{\pi} e^{-\frac{(x-\bar{q})^2}{2\sigma^2} - 2\sigma^2(p-\bar{p})^2}. \tag{6.58}$$

Note that the Wigner function (6.58) is positive.

Next, we evaluate the Wigner function for a Schrödinger's cat wave function,

$$\psi(x) = \frac{1}{\sqrt{2}} \left[\psi_{\bar{q},\bar{p}}(x) + \psi_{\bar{q}',\bar{p}'}(x) \right], \tag{6.59}$$

that corresponds to the superposition of two Gaussians with different values for mean position and mean momentum. For simplicity, we assume that $\bar{p} = \bar{p}' = 0$ and $\bar{q} = \frac{1}{2}L = -\bar{q}'$, that is, the two wave packets have the same momentum but their centers are separated by distance L. We find that

$$W(x,p) = \frac{1}{\pi} e^{-2\sigma^2 p^2} \left[e^{-\frac{\left(x+\frac{L}{2}\right)^2}{2\sigma^2}} + e^{-\frac{\left(x-\frac{L}{2}\right)^2}{2\sigma^2}} + 2e^{-\frac{x^2}{2\sigma^2}} \cos\left(\frac{pL}{2}\right) \right]. \tag{6.60}$$

For $L \gg \sigma$ and $\cos\left(\frac{1}{2}pL\right)$ close to -1, the Wigner function takes negative values near $x = 0$. This can be seen in the plot of (6.60) in Fig. 6.4. Superpositions of states with different mean position or momentum are characterized by oscillations in the Wigner function.

The Wigner function is directly related to the coherent states $|z\rangle$ that were defined in Section 5.2.1. The Gaussian wave function (6.10) coincides with a coherent state $|z\rangle$ for $z = \frac{1}{\sqrt{2}}(\bar{q}/\sigma + i\sigma\bar{p})$. Substituting $\hat{A} = |z\rangle\langle z|$ into Eq. (6.54), we obtain

$$\langle z|\hat{\rho}|z\rangle = \int dx\, dp\, W_\rho(x,p) W_z(x,p) = \int \frac{dx\, dp}{\pi} e^{-\frac{(\bar{q}-x)^2}{2\sigma^2} - 2\sigma^2(\bar{p}-p)^2} W_\rho(x,p)$$

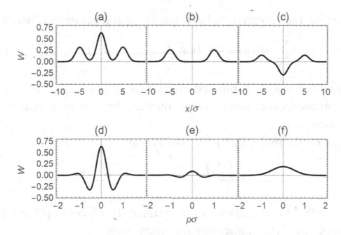

Fig. 6.4 In the upper row, the Wigner function (6.60) is plotted as a function of x, for different values of p: (a) $p = 0$, (b) $p = \pi/L$, and (c) $p = 2\pi/L$. In the lower row, the Wigner function (6.60) is plotted as a function of p, for different values of x: (d) $x = 0$, (e) $x = 2\sigma$, and (c) $x = 4\sigma$. In all plots, $L/\sigma = 10$.

for any density matrix $\hat{\rho}$. Given that $\langle z|\hat{\rho}|z\rangle \geq 0$, we see that the convolution of the Wigner function with a Gaussian always gives a positive function. The function $e^{-\frac{(\bar{q}-x)^2}{2\sigma^2}-2\sigma^2(\bar{p}-p)^2}$ takes values close to unity for $\bar{q},\bar{p}$ around (x,p) with area of order 1, and it is negligible elsewhere. Hence, there is a restriction on the negativity of the Wigner function: The integration of the Wigner function in a region with area of order unity yields a positive number.

6.6   Combining Information from Different Experiments

A position measurement requires a different experimental set-up from that of a momentum measurement. Hence, it is far from obvious that we can combine the probability distribution $\langle x|\hat{\rho}|x\rangle$ for position with the probability $\langle p|\hat{\rho}|p\rangle$ for momentum, to obtain a joint probability distribution. We saw that the Wigner function failed to define joint probabilities of this type, because it can take negative values.

Could some other definition succeed? The answer is an emphatic no. We cannot combine results from different experiments at the level of probability distributions, unless the associated operators commute. The proof proceeds in two steps. First, we show that the combination does not work if the operators do not commute, and then we show that, even for commuting operators, such combinations are inconsistent.

6.6.1   Noncommuting Operators

Consider two experiments that correspond to measurements of two operators, say, $\hat{A} = \sum_n a_n \hat{P}_n$ and $\hat{B} = \sum_m b_m \hat{Q}_n$. Suppose we can combine the results of the two experiments, in the sense that there exists a distribution of joint probabilities for the two magnitudes for each state $\hat{\rho}$.

By definition, the marginal probabilities for a_n and b_n coincide with those of quantum theory,

$$\sum_n p(a_n, b_m) = \mathrm{Tr}(\hat{\rho}\hat{Q}_m), \qquad \sum_m p(a_n, b_m) = \mathrm{Tr}(\hat{\rho}\hat{P}_n). \qquad (6.61)$$

The probabilities $p(a_n, b_m)$ must be linear functions of $\hat{\rho}$, in order to behave properly when statistical ensembles are mixed. Hence, they can be written as $p(a_n, b_m) = \mathrm{Tr}(\hat{\rho}\hat{A}_{nm})$, for some positive operators $\hat{A}_{nm}$.

Since Eqs. (6.61) hold for all $\hat{\rho}$, it follows that

$$\sum_n \hat{A}_{nm} = \hat{Q}_m, \qquad \sum_m \hat{A}_{nm} = \hat{P}_n. \qquad (6.62)$$

We will prove that Eq. (6.62) cannot be satisfied. It suffices to prove it for a qubit, because $\mathbb{C}^2$ is a subspace of any higher-dimensional Hilbert space.

In qubits, the indices n and m take only two values, say, 1 and 2. Equations (6.62) imply that $\hat{A}_{11} + \hat{A}_{12} = \hat{P}_1$. The positivity of $\hat{A}_{1m}$ implies that $\hat{A}_{1m} \leq \hat{P}_1$. Given that the projector $\hat{P}_1$ is one dimensional, the inequality is satisfied only if $\hat{A}_{1m} = 0$ or if $\hat{A}_{1m} = \hat{P}_1$. In both cases, $[\hat{A}_{1m}, \hat{P}_1] = 0$. We similarly show that $[\hat{A}_{2m}, \hat{P}_2] = 0$. Since $\hat{P}_2 = \hat{I} - \hat{P}_1$, we also have $[\hat{A}_{2m}, \hat{P}_1] = 0$. Equation (6.62) yields $\hat{Q}_m = \hat{A}_{1m} + \hat{A}_{2m}$, hence, $[\hat{Q}_m, \hat{P}_n] = 0$. Since spectral projectors commute, so do the corresponding self-adjoint operators. We conclude that $[\hat{A}, \hat{B}] = 0$. Hence, joint probabilities for measurements of $\hat{A}$ and $\hat{B}$ are impossible unless the two operators commute.

6.6.2   Commuting Operators

Next, we show that the assumption that we can define joint probabilities for measurements of commuting operators leads to a contradiction. We follow an analysis by Cabello et al. (2013).

Consider a quantum system described by the Hilbert space $\mathbb{C}^3$. Any vector $|v\rangle \in \mathbb{C}^3$ defines a yes–no measurement, that is, a measurement in which we verify whether the system is one state $|v\rangle$ or not. The associated spectral projectors are $\hat{P} = |v\rangle\langle v|$, and $\hat{I} - \hat{P}$.

We select the following vectors,

$$|v_1\rangle = \frac{1}{\sqrt{3}}\begin{pmatrix} 1 \\ -1 \\ 1 \end{pmatrix}, \quad |v_2\rangle = \frac{1}{\sqrt{2}}\begin{pmatrix} 1 \\ 1 \\ 0 \end{pmatrix}, \quad |v_3\rangle = \begin{pmatrix} 0 \\ 0 \\ 1 \end{pmatrix},$$

$$|v_4\rangle = \begin{pmatrix} 1 \\ 0 \\ 0 \end{pmatrix}, \quad |v_5\rangle = \frac{1}{\sqrt{2}}\begin{pmatrix} 0 \\ 1 \\ 1 \end{pmatrix}. \qquad (6.63)$$

Each vector $|v_n\rangle$ defines one yes–no measurement, with spectral projectors $\hat{P}_n = |v\rangle\langle v|$ and $\hat{I} - \hat{P}_n$. We denote the positive outcome of the measurement associated to $|v_n\rangle$ by n, and the negative outcome by $\bar{n}$.

If $\langle v_n | v_{n'}\rangle = 0$, the associated spectral projectors commute. By assumptions, we can define joint probabilities for the outcomes of the corresponding measurements. Since $|v_n\rangle$ belongs in the complementary subspace of $|v_{n'}\rangle$, $\hat{P}_n \leq \hat{I} - \hat{P}_{n'}$. Hence, if n occurs, then $\bar{n}'$ also occurs, and n' never occurs. This means that $\mathrm{Prob}(n', n) = 0$. Similarly, we show that $\mathrm{Prob}(\bar{n}', \bar{n}) = 0$.

The vectors (6.63) satisfy $\langle v_1|v_2\rangle = \langle v_2|v_3\rangle = \langle v_3|v_4\rangle = \langle v_4|v_5\rangle = \langle v_5|v_1\rangle = 0$. Hence,

$$\text{Prob}(1,2) = \text{Prob}(2,3) = \text{Prob}(3,4) = \text{Prob}(4,5) = \text{Prob}(5,1) = 0$$
$$\text{Prob}(\bar{1},\bar{2}) = \text{Prob}(\bar{2},\bar{3}) = \text{Prob}(\bar{3},\bar{4}) = \text{Prob}(\bar{4},\bar{5}) = \text{Prob}(\bar{5},\bar{1}) = 0.$$

Suppose that 1 is true. Since $\text{Prob}(1,2) = 0$, it follows that 2 is false and $\bar{2}$ is true. But then, since $\text{Prob}(\bar{2},3) = 0$, 3 is true. Since $\text{Prob}(3,4) = 0$, this implies that $\bar{4}$ is true. Finally, by $\text{Prob}(\bar{4},5) = 0$, 5 is true. The result of this chain of inferences is that, if 1 is true, then $\bar{5}$ is false, or $\text{Prob}(\bar{5},1) = 0$. But $\text{Prob}(\bar{5},1) = \text{Prob}(\bar{5},1) + \text{Prob}(5,1) = \text{Prob}(1) \neq 0$ for a general initial state $|\psi\rangle$. Thus, we arrived at a contradiction. The assumption that we can assign joint probabilities to all measurements of commuting observables is wrong.

> It is impossible to combine measurements from different experiments under a joint probability distribution.

This statement is implicit in Bohr's complementarity principle.

Evidence obtained under different experimental conditions cannot be comprehended within a single picture, but must be regarded as complementary in the sense that only the totality of the phenomena exhausts the possible information about the objects. (Bohr, 1998)

This means that we cannot think of the measurement outcomes revealing pre-existing properties of the quantum system. In fact, quantum theory is not compatible with the existence of properties in microscopic systems that make no reference to the way that the property is being measured. This is strongly suggested by the preceding analysis, and it has been rigorously proven by Kochen and Specker – see Box 6.5. We shall come back to this issue in Section 9.4, where we shall see that not only is the assumption of microscopic properties incompatible with quantum theory, but it also leads to predictions that conflict experiment.

> In quantum theory, measurement outcomes are to be thought of as properties of the complex that consists of the microscopic system and the macroscopic measuring apparatus, and not of the microscopic system alone.

Some researchers rephrase this statement by saying that properties of microscopic systems are *contextual*, the context being the macroscopic set-up through which they are measured.

Box 6.5   The Kochen–Specker Theorem

In classical physics, physical quantities possess values that are independent of how those values are eventually measured. The Kochen–Specker theorem proves that this is not the case in quantum theory.

Suppose that a quantum system has properties that are revealed by the measurement. Then, there is a function f, a *valuation*, that assigns to each self-adjoint operator – hence, to each possible measurement – the preexisting value $f(\hat{A})$ of the associated physical quantity. We want valuations to preserve the algebraic relations between physical quantities, that is, to be additive: $f(\hat{A} + \hat{B}) = f(\hat{A}) + f(\hat{B})$.

Obviously, additivity for operators that correspond to incompatible observables is too strong a condition. We should relax additivity to hold only for compatible observables, namely, only if $[\hat{A}, \hat{B}] = 0$. Then, it is reasonable to also require that $f(\hat{A}\hat{B}) = f(\hat{A})f(\hat{B})$ if $[\hat{A}, \hat{B}] = 0$. For a projector $\hat{P}, f(\hat{P}) = f(\hat{P}^2) = f(\hat{P})f(\hat{P})$, hence, $f(\hat{P}) \in \{0, 1\}$.

Consider an orthonormal basis $|n\rangle$ in a N-dimensional Hilbert space, and the associated one-dimensional projectors $\hat{P}_n = |n\rangle\langle n|$. Since $\sum_n \hat{P}_n = \hat{I}$, the additivity of the valuation implies that $\sum_n f(\hat{P}_n) = 1$. It follows that one projector in $\{\hat{P}_n\}$ has value unity; the rest have value zero.

Hence, for any valuation f, we can specify $N - 1$ white vectors (value 0) and one red vector (value 1) in any orthonormal basis. The color of a vector must be the same in all bases in which it appears. Kochen and Specker (1967) proved that this is impossible. They identified a set of bases in $\mathbb{C}^3$ that could not be colored with the above rules. This proved that no valuations are possible in any Hilbert space of dimension greater than or equal to 3: There is no way to assign properties to quantum systems that do not depend on how these properties are measured.

The original proof of Kochen and Specker involved 132 bases that contained 117 distinct vectors. There is a proof with 37 bases in $\mathbb{C}^3$ containing 31 vectors (Conway and Kochen, 1993) and a proof with 9 bases in $\mathbb{C}^4$ containing only 18 vectors (Cabello et al., 1996). The latter allows for the demonstration of the theorem by mere inspection.

Consider the table in this box. Every string of four integers corresponds to an (unnormalized) vector on $\mathbb{C}^4$ – here $\bar{1}$ stands for -1. For example, $10\bar{1}0$ corresponds to the vector $(1, 0, -1, 0)$. It is easy to verify that the vectors in one column are mutually orthogonal. Hence, after normalization, each column defines one orthonormal basis.

We note that each vector appears twice in the table, that is, in two different bases. Since there are nine bases, nine vectors must be colored red, one in each column. However, since each vector appears exactly twice, the number of red vectors must be an even number. Hence, we reach a contradiction.

0001	0001	$1\bar{1}1\bar{1}$	$1\bar{1}1\bar{1}$	0010	$1\bar{1}\bar{1}1$	$11\bar{1}1$	$11\bar{1}1$	$111\bar{1}$
0010	0100	$1\bar{1}\bar{1}1$	1111	0100	1111	$111\bar{1}$	$\bar{1}111$	$\bar{1}111$
1100	1010	1100	$10\bar{1}0$	1001	$100\bar{1}$	$1\bar{1}00$	1010	1001
$1\bar{1}00$	$10\bar{1}0$	0011	$010\bar{1}$	$100\bar{1}$	$01\bar{1}0$	0011	$010\bar{1}$	$01\bar{1}0$

6.6.3   Quantum Tomography

The fact that we cannot combine results of different experiments in a joint probability distribution does not mean that there are no other ways of combining them. We can use information from different experiments in order to reconstruct a system's quantum state. This procedure is called *quantum tomography* or *quantum state estimation*.

The simplest example is the quantum tomography of a qubit. As shown in Section 6.2.3, a qubit's state is fully specified by a vector $\mathbf{r} \in \mathbb{R}^3$ with $|\mathbf{r}| \leq 1$. The coordinates r_i of this vector equal the expectation values of the Pauli matrices $\hat{\sigma}_i$. Hence, it suffices to find the expectation values of $\hat{\sigma}_1, \hat{\sigma}_2$, and $\hat{\sigma}_3$, in order fully to specify the quantum state. We say that the set of the three Pauli matrices is *informationally complete*: The expectation values of the Pauli matrices fully specify a qubit's density matrix. The expectation value of two Pauli matrices is not sufficient. Of course there are other triplets of operators, or even sets with more elements, that are informationally complete.

A quantum state reconstruction is based on expectation values that are known with finite accuracy. This implies that the data do not specify a unique state, but a region in the space of states. This procedure requires new methods of statistical analysis, specially designed for quantum systems.

We have to keep in mind that the quantum state is reconstructed after measurements in a statistical ensemble of identical prepared systems. It is therefore natural to think of the state as an objective property of such statistical ensembles. This is to be contrasted with the nonobjectivity of the state when viewed as a property of individual systems – see Section 6.3.1.

Example 6.10 We present a procedure for quantum tomography in $L^2(\mathbb{R})$ (Vogel and Risken, 1989) that has been used for the reconstruction of photon states in quantum optics.

Consider the operator $\hat{A}_\phi = \hat{x} \cos \phi + \hat{p} \sin \phi$. Its eigenvalue equation,

$$-i \sin \phi \frac{df_{a,\phi}}{dx} + \cos \phi \, x f_{a,\phi} = a f_{a,\phi},$$

has solutions

$$f_{a,\phi}(x) = \frac{1}{\sqrt{2\pi |\sin \phi|}} e^{-\frac{i \cot \phi}{2} x^2 + i \frac{a}{\sin \phi} x},$$

with normalization $\langle a, \phi | a', \phi \rangle = \int dx f_{a,\phi}^*(x) f_{a',\phi}(x) = \delta(a - a')$.

For every state $\hat{\rho}$, the distribution function for measurements of $\hat{A}_\phi$ is

$$p_\phi(a) = \langle a, \phi | \hat{\rho} | a, \phi \rangle = \int dx dx' f_{a,\phi}(x) \langle x | \hat{\rho} | x' \rangle f_{a,\phi}(x'). \tag{6.64}$$

We change variables to $X = \frac{1}{2}(x + x')$ and $y = x' - x$, and we use Eq. (6.50), to obtain

$$p_\phi(a) = \frac{1}{2\pi |\sin \phi|} \int dX d\xi \left\langle X - \frac{1}{2}\xi \Big| \hat{\rho} \Big| X + \frac{1}{2}\xi \right\rangle e^{i\left(\frac{a}{\sin \phi} - X \cot \phi\right) y}$$

$$= \frac{1}{|\sin \phi|} \int dX W \left(X, \frac{a}{\sin \phi} - X \cot \phi \right). \tag{6.65}$$

The Fourier transforms $\tilde{p}_\phi$ and $\tilde{W}$ are defined by $p_\phi(a) = \frac{1}{2\pi} \int dz \tilde{p}_\phi(z) e^{iaz}$, and $W(x, p) = \frac{1}{2\pi} \int du dv e^{iux + ivp} \tilde{W}(u, v)$. Equation (6.65) implies that

$$\tilde{W}(u, v) = \tilde{p}_{\tan^{-1}(v/u)} \left(\sqrt{u^2 + v^2} \right). \tag{6.66}$$

Measurements of $\hat{A}_\phi$ for different values of ϕ allow us to evaluate $p_\phi(a)$, and consequently, use Eq. (6.66) in order to compute the Wigner function.

QUESTIONS

6.1 Let observables A and B be the velocities of two cars in a ten-lane motorway. When do you expect their correlation to be stronger, during rush hours or at midnight?

6.2 What measurement must be performed in a statistical ensemble of qubits in order to distinguish between the pure state $|\psi\rangle = \cos\theta|0\rangle + \sin\theta|1\rangle$ and the mixed state $\hat{\rho} = \cos^2\theta|0\rangle\langle0| + \sin^2\theta|1\rangle\langle1|$.

6.3 Explain why a quantum mixed state cannot, in general, be interpreted in terms of ignorance about pure states. Are there cases where an ignorance interpretation is possible?

6.4 Assuming that we can assign states to individual quantum systems, some authors distinguish two different types of statistical ensemble. In a *proper mixture*, each system in the ensemble is prepared in a pure state, and the density matrix $\hat{\rho}$ for the ensemble is obtained by mixing of the pure states. In an *improper mixture*, all systems in the ensemble are prepared in the state $\hat{\rho}$. Is it possible experimentally to distinguish between those two types of ensemble?

6.5 Explain the difference between the Kennard–Robertson uncertainty relation and Heisenberg's original one.

6.6 The Kennard–Robertson uncertainty relation for position and momentum is trivial for a particle in a box and for a particle in a ring. Why?

6.7 (**B**) The complementarity principle also implies that *the same operator may correspond to different observables*, depending on which other quantities are measured together with it (Peres, 2002). Explain how this conclusion follows from the example of Section 6.6.2.

6.8 (**B**) Why can't we obtain the contradiction in Section 6.6.2 using sets of three or four vectors $|v_n\rangle$?

6.9 (**B**) It is often stated that the quantum tomography is a measurement of the quantum state. Explain the difference between this "measurement" and the measurement of physical quantities that correspond to self-adjoint operators.

6.10 (**B**) A source produces particles described by a density matrix $\hat{\rho}$. In one experiment, we measure the position of particles at time t after emission and construct the probability density $w(x)$ for position. In another experiment, we measure particle momentum at time t after emission, and we construct the probability distribution $\bar{w}(p)$ for momentum. Do we have enough information to reconstruct the quantum state at time t?

PROBLEMS

6.1 A system is prepared in a state $\hat{\rho} = \frac{1}{4}\begin{pmatrix} 1 & 0 & i \\ 0 & 1 & 1 \\ -i & 1 & 2 \end{pmatrix}$. We measure the operator $\hat{A} = \epsilon\begin{pmatrix} 1 & 0 & 0 \\ 0 & 3 & 0 \\ 0 & 0 & 1 \end{pmatrix}$, for $\epsilon > 0$. Find the possible measurement outcomes and the associated probabilities.

6.2 Find the values of α and β for which the density matrix

$$\hat{\rho} = \frac{1}{7} \begin{pmatrix} \alpha & -i & \beta & i \\ i & 1 & 2i & -1 \\ \beta & -2i & 2\beta & 2i \\ -i & -1 & -2i & \alpha \end{pmatrix}$$

describes a pure state.

6.3 We construct a mixture of the state that corresponds to the positive eigenvalue of $\hat{\sigma}_1$ and of the state that corresponds to the negative eigenvalue of $\hat{\sigma}_3$. The analogy is $3 : 1$. What is the Bloch vector of this mixture?

6.4 Let $\hat{N} = \hat{a}^\dagger \hat{a}$ be the number operator for a harmonic oscillator. Calculate the mean deviation ΔN (i) for a coherent state $|z\rangle$ and (ii) for a squeezed vacuum state, as defined in Problem 5.7.

6.5 We have a source of microscopic systems described by the Hilbert space $\mathbb{C}^3$. We measure the operators

$$\hat{J}_1 = \frac{1}{\sqrt{2}} \begin{pmatrix} 0 & 1 & 0 \\ 1 & 0 & 1 \\ 0 & 1 & 0 \end{pmatrix}, \quad \hat{J}_2 = \frac{1}{\sqrt{2}} \begin{pmatrix} 0 & -i & 0 \\ i & 0 & -i \\ 0 & i & 0 \end{pmatrix}, \quad \hat{J}_3 = \begin{pmatrix} 1 & 0 & 0 \\ 0 & 0 & 0 \\ 0 & 0 & -1 \end{pmatrix}.$$

After a series of measurements, we find that $\langle \hat{J}_1 \rangle = \langle \hat{J}_2 \rangle = 0$, and $\langle \hat{J}_3 \rangle = a$. (i) Find all states compatible with the above information for $a = 1$ (ii) Which of those states are pure? (iii) Repeat the preceding questions above for $a = 0$.

6.6 Let $\hat{\rho}_i$ be density matrices with purities γ_i, and let $\hat{\rho} = \sum_i \lambda_i \hat{\rho}_i$ be their mixture, where $0 \leq \lambda_i \leq 1$ and $\sum_i \lambda_i = 1$. Show that the purity γ of $\hat{\rho}$ satisfies the inequality $\gamma \leq \sum_i \lambda_i \gamma_i$. When does equality hold?

6.7 A free particle of mass m is described by the Gaussian wave function

$$\psi(x) = (2\pi\sigma^2)^{-1/4} \exp\left[-\frac{(1 + ir)(x - x_0)^2}{4\sigma^2} + ip_0 x\right],$$

where σ, r, x_0, and p_0 are constants. Evaluate the mean deviation of the Hamiltonian $\hat{H}$ and the correlation C_{xH}.

6.8 The equilibrium states of a system with Hamiltonian $\hat{H}$ in contact with a heat reservoir at temperature β^{-1} is given by the *Gibbs* state $\hat{\rho} = Z^{-1} e^{-\beta\hat{H}}$, where $Z(\beta) = \mathrm{Tr} e^{-\beta\hat{H}}$. Show that $(\Delta H)^2 = (\ln Z)''$ and evaluate this quantity for a harmonic oscillator.

6.9 A particle at the half-line $\mathbb{R}^+$ is prepared in the state $\psi(x) = 2\lambda^{3/2} x e^{-\lambda x}$, for $\lambda > 0$. (i) Calculate the mean deviation Δx. (ii) Find the probability density that corresponds to measurements of the operator $\widehat{|p|}$. (iii) Calculate the mean deviation $\Delta|p|$ and the product $\Delta x \Delta|p|$. (iv) Calculate the probability distribution for the operator $\hat{Q}$ of Problem 5.18.

6.10 A particle in a box of length L is prepared in a state $\psi(x) = \sqrt{\frac{30}{L^5}} x(L - x)$. (i) Calculate the probabilities that correspond to possible values of the operator $\widehat{|p|}$. (ii) Calculate the mean deviations Δx, $\Delta|p|$, and the product $\Delta x \Delta|p|$.

6.11 (i) Find the expectation values of the SG trigonometric operators $\hat{c}$ and $\hat{s}$ (Example 4.2) on a coherent state $|z\rangle$. Show that, for large $|z|$, the expectation values $\langle \hat{c} \rangle$ and $\langle \hat{s} \rangle$

approximate the corresponding trigonometric functions of the phase of z. (ii) In analogy with the classical relations $\delta(\cos\phi) = -\sin\phi\delta\phi$ and $\delta(\sin\phi) = \cos\phi\delta\phi$, we define the phase uncertainty

$$\delta\phi = \min\left\{\frac{\Delta s}{|\langle\hat{c}\rangle|}, \frac{\Delta c}{|\langle\hat{s}\rangle|}\right\}. \tag{6.67}$$

Show that, for the states at which $\delta\phi$ is defined, $\delta\phi\Delta N \geq \frac{1}{2}$.

6.12 A particle on a ring of perimeter L is prepared in a state

$$\psi(x) = \begin{cases} Cx, & x \in \left[0, \frac{L}{2}\right] \\ C(L-x), & x \in \left[\frac{L}{2}, L\right] \end{cases}$$

(i) Find the normalization constant C. (ii) Evaluate the mean deviation of the cosine operator $\hat{c}$ and of the sine operator $\hat{s}$. (iii) Derive the probability distribution for the angular momentum $\hat{\ell}$.

6.13 **(B)** Let $\psi_0(x)$ be the ground state of the Schrödinger operator for a delta potential, Eq. (5.86). (i) Calculate the mean deviations Δx and Δp. (ii) Find the Wigner function.

6.14 **(B)** (i) Find the Wigner function for the first two eigenvectors of the harmonic oscillator Hamiltonian. (ii) Pretending that the Wigner function is a probability distribution, find the associated distributions for energy. Compare with the quantum probability distributions of energy.

6.15 **(B)** We define the Husimi distribution of a state $\hat{\rho}$ on $L^2(\mathbb{R})$ as $f_H(q,p) := \langle qp|\hat{\rho}|qp\rangle$, where $|qp\rangle$ a coherent state that corresponds to the point (q,p) of the classical state space for a particle at a line. (i) Show that f_H is a probability distribution on the state space. (ii) Evaluate the variance of position and of momentum for the Husimi distribution. (iii) Show that (in contrast to the Wigner function) the marginals of the Husimi distribution do not reproduce the prediction of quantum theory for the position and momentum probability distributions.

6.16 **(B)** Let $\hat{P}$ and $\hat{Q}$ be two projectors on a Hilbert space $\mathcal{H}$. We define the self-adjoint operator $\hat{C} = \frac{1}{2}(\hat{P}\hat{Q} + \hat{Q}\hat{P})$. (i) Show that $\hat{C}$ has a negative eigenvalue, unless $[\hat{P}, \hat{Q}] = 0$. (Prove this result for a qubit, and then argue that it generalizes to any Hilbert space.) (ii) Use this result in order to show that, for a general quantum state, the correlation (6.7) of two self-adjoint operators $\hat{A} = \sum_n a_n\hat{P}_n$ and $\hat{B} = \sum_m b_m\hat{Q}_m$ may be obtained from a classical joint probability distribution $p(a_n, b_m)$ only if $[\hat{A}, \hat{B}] = 0$.

Bibliography

- For the mathematical structure of the space of quantum states, see Beltrametti and Cassinelli (2010).
- For entropy in quantum theory, see Wehrl (1978). For the Wigner function, see Balazs (1984). For quantum tomography, see D'Ariano et al. (2003).
- For the Kochen–Specker theorem and its implications, see chapter 9 of Isham (1995) and chapter 7 of Peres (2002).

7 Time Evolution

When you ask what are electrons and protons I ought to answer that this question is not a profitable one to ask and does not really have a meaning. The important thing about electrons and protons is not what they are but how they behave, how they move. I can describe the situation by comparing it to the game of chess. In chess, we have various chessmen, kings, knights, pawns and so on. If you ask what chessman is, the answer would be that it is a piece of wood, or a piece of ivory, or perhaps just a sign written on paper, or anything whatever. It does not matter. Each chessman has a characteristic way of moving and this is all that matters about it. The whole game of chess follows from this way of moving the various chessmen.

Dirac (1955)

7.1 The Quantum Rule of Time Evolution

In Chapter 2, we saw that the nascent quantum theory was divided between two different descriptions of time evolution.

In Heisenberg's description, physical magnitudes correspond to matrices A_{nm} that evolve in time according to the rule

$$A_{mn}(t) = e^{i(E_m - E_n)t} A_{mn}(0), \tag{7.1}$$

where E_n are possible values of energy.

Schrödinger's description emphasizes the wave function $\psi(\mathbf{x}_1, \dots, \mathbf{x}_N)$ for N particles, which evolves by the equation

$$i\frac{d\psi}{dt} = -\sum_{i=1}^{N} \frac{1}{2m_i} \nabla_i^2 \psi + V(\mathbf{x}_1, \dots \mathbf{x}_N)\psi. \tag{7.2}$$

In Section 2.7.1, we showed that these two descriptions of time evolution are mathematically equivalent. In the Hilbert space formalism, they are unified by the notion of *unitary evolution*.

To see this, we note that the operator

$$\hat{H} = -\sum_{i=1}^{N} \frac{1}{2m_i} \nabla_i^2 + V(\mathbf{x}_1, \dots \mathbf{x}_N) \tag{7.3}$$

that appears in Schrödinger's equation for N particles is self-adjoint. For simplicity, we assume that $\hat{H}$ has purely discrete spectrum; hence, it satisfies the eigenvalue equation $\hat{H}|n\rangle = E_n|n\rangle$, and the eigenvectors $|n\rangle$ define an orthonormal basis.

Shrödinger's equation is written as $i\frac{d}{dt}|\psi\rangle = \hat{H}|\psi\rangle$. We left-multiply with the bra $\langle n|$, to obtain the equation $i\frac{d}{dt}\langle n|\psi\rangle = E_n\langle n|\psi\rangle$, with solution $\langle n|\psi(t)\rangle = e^{-iE_n t}\langle n|\psi(0)\rangle$. Using the decomposition of the unity for the basis $|n\rangle$, we obtain

$$|\psi(t)\rangle = \sum_n |n\rangle\langle n|\psi(t)\rangle = \sum_n e^{-iE_n t}|n\rangle\langle n|\psi(0)\rangle = \hat{U}_t|\psi(0)\rangle, \tag{7.4}$$

where

$$\hat{U}_t = \sum_n e^{-iE_n t}|n\rangle\langle n| = e^{-i\hat{H}t} \tag{7.5}$$

is a unitary operator, known as the *time-evolution operator*.

A direct consequence of Eq. (7.5) is the useful identity

$$\hat{H} = i\dot{\hat{U}}_t\hat{U}_t^\dagger = i\hat{U}_t^\dagger\dot{\hat{U}}_t, \tag{7.6}$$

where the dot corresponds to differentiation with respect to t.

Consider an observable that corresponds to the operator $\hat{A}$. Let us assume that it evolves in time as

$$\hat{A}(t) = \hat{U}_t^\dagger\hat{A}(0)\hat{U}_t. \tag{7.7}$$

Then, the matrix element $A_{mn}(t) = \langle m|\hat{A}(t)|n\rangle$, where $|m\rangle$ and $|n\rangle$ are eigenvectors of $\hat{H}$ evolves in time according to Heisenberg's rule. Hence, in both descriptions, time evolution is expressed through the unitary operator (7.5).

Differentiating Eq. (7.7) with respect to time and using Eq. (7.6), we obtain a differential equation for the evolution of an observable $\hat{A}$,

$$\frac{d\hat{A}(t)}{dt} = i\left[\hat{H}, \hat{A}(t)\right]. \tag{7.8}$$

The preceding analysis implies that there are two equivalent ways to describe time evolution in quantum theory.

1. *Schrödinger picture.* The states evolve in time according to the rule $|\psi\rangle \rightarrow \hat{U}_t|\psi\rangle$, while observables remain unchanged.
2. *Heisenberg picture.* Observables evolve in time according to Eq. (7.7), while states remain unchanged.

The two descriptions are equivalent, because all physical predictions of quantum theory are expressed in terms of probabilities. We assume that the system has been prepared in a state $|\psi_0\rangle$ at time $t = 0$, and that the operator $\hat{A} = \sum_n a_n\hat{P}_n$ is measured at time t. Then, the probability that the value a_n will be obtained is

$$\text{Prob}(a_n, t) = \left\langle\psi_0|\hat{U}_t^\dagger\hat{P}_n\hat{U}_t|\psi_0\right\rangle. \tag{7.9}$$

Equation (7.9) can be written as $\text{Prob}(a_n, t) = \left\langle\psi_t|\hat{P}_n|\psi_t\right\rangle$, where $|\psi_t\rangle = \hat{U}_t|\psi_0\rangle$ is the state at time t in the Schrödinger picture.

The spectral analysis of an operator $\hat{A} = \sum_n a_n\hat{P}_n$ remains unchanged under time evolution in the Heisenberg picture. The equation $\hat{A} = \sum_n a_n\hat{P}_n$ implies that $\hat{U}_t^\dagger\hat{A}\hat{U}_t = \sum_n a_n\hat{U}_t^\dagger\hat{P}_n\hat{U}_t$, or, equivalently, that $\hat{A}(t) = \sum_n a_n \hat{P}_n(t)$. Hence, Eq. (7.9) can be expressed as $\text{Prob}(a_n, t) = \langle \psi_0 | \hat{P}_n(t) | \psi_0 \rangle$. Probabilistic predictions are the same in the two pictures.[1]

We express the new foundational principle of quantum theory in Schrödinger's picture, because this picture is more useful for most topics encountered in this book.

FP4 In a closed physical system and in absence of measurements, the quantum state evolves under the action of a family of unitary operators $e^{-i\hat{H}t}$ as $|\psi(t)\rangle = e^{-i\hat{H}t}|\psi(0)\rangle$, where t is the time, and $\hat{H}$ is a self-adjoint operator that is called the *Hamiltonian* of the system.

To see how FP4 translates to mixed states, we analyze every density matrix $\hat{\rho}$ to its eigenvectors $|\psi_i\rangle$,

$$\hat{\rho} = \sum_i w_i |\psi_i\rangle \langle \psi_i|.$$

Every ket $|\psi_i\rangle$ evolves as $|\psi_i\rangle \rightarrow \hat{U}_t|\psi_i\rangle$ and every bra evolves as $\langle \psi_i| \rightarrow \langle \psi_i|\hat{U}_t^\dagger$. Hence, the density matrix evolves as $\hat{\rho} \rightarrow \hat{U}_t \hat{\rho} \hat{U}_t^\dagger$.

Hence, the most general law of time evolution in Schrödinger's picture is

$$\hat{\rho}(t) = \hat{U}_t \hat{\rho}(0) \hat{U}_t^\dagger. \tag{7.10}$$

We differentiate both sides of Eq. (7.10) with respect to time

$$\frac{d\hat{\rho}(t)}{dt} = \dot{\hat{U}}_t \hat{\rho}(0) \hat{U}_t^\dagger + \hat{U}_t \hat{\rho}(0) \dot{\hat{U}}_t^\dagger. \tag{7.11}$$

By Eq. (7.6), we obtain an evolution equation for the density matrix

$$i\frac{d\hat{\rho}(t)}{dt} = \left[\hat{H}, \hat{\rho}(t)\right], \tag{7.12}$$

known as *von Neumann equation*. Note the difference between Eq. (7.12) and Eq. (7.8) by a sign.

For mixed states, Eq. (7.9) can be written as

$$\text{Prob}(a_n, t) = \text{Tr}\left(\hat{U}_t \hat{\rho}_0 \hat{U}_t^\dagger \hat{P}_n\right). \tag{7.13}$$

Example 7.1 For a particle of mass m in one dimension under a general potential V, the Hamiltonian is $\hat{H} = \frac{\hat{p}^2}{2m} + V(\hat{x})$. The Heisenberg equations of motion for $\hat{x}$ and $\hat{p}$ are

$$\frac{d\hat{x}}{dt} = i\left[\hat{H}, \hat{x}\right] = \frac{\hat{p}}{m}, \qquad \frac{d\hat{p}}{dt} = i\left[\hat{H}, \hat{p}\right] = -V'(\hat{x}). \tag{7.14}$$

---

[1] This is not an accident peculiar to quantum theory. Any theory that involves a pairing of states and observables in the definition of probabilities admits two equivalent pictures of time evolution. This includes classical mechanics and classical probability theory.

Equations (7.14) are algebraically identical with the classical Hamilton equations (1.4). They imply an operator's analogue of Newton's equations

$$m\frac{d^2\hat{x}}{dt^2} = -V'(\hat{x}).$$

(7.15)

Taking the expectation value with respect to some state $\hat{\rho}$, we obtain the so-called *Ehrenfest's theorem* (Ehrenfest, 1927),

$$m\frac{d^2\langle\hat{x}\rangle}{dt^2} = -\langle V'(\hat{x})\rangle.$$

(7.16)

We write $\langle V'(\hat{x})\rangle = \langle V'[\langle\hat{x}\rangle + (\hat{x} - \langle\hat{x}\rangle)]\rangle$, and we Taylor-expand V' around $\langle\hat{x}\rangle$,

$$\langle V'(\hat{x})\rangle = V'(\langle\hat{x}\rangle) + \frac{1}{2}V'''(\langle\hat{x}\rangle)(\Delta x)^2 + \cdots.$$

If $|V'''(\langle\hat{x}\rangle)|(\Delta x)^2 << |V'(\langle\hat{x}\rangle)|$, we only keep the first term in the Taylor expansion. Then, $\langle\hat{x}\rangle$ becomes an autonomous variable; it evolves according to the classical equations of motion,

$$m\frac{d^2\langle\hat{x}\rangle}{dt^2} = -V'(\langle\hat{x}\rangle).$$

(7.17)

Equation (7.17) is the simplest example of a classical evolution equation derived from quantum theory as an approximation.

7.2   Implications of the Time-Evolution Rule

In this section, we discuss the implications of the quantum rule for time evolution. In particular, we analyze the relation between the Hamiltonian and energy, we show that time operators do not exist, we present different forms of the time–energy uncertainty principle, and finally, we explain the notion of conserved quantity in quantum theory.

7.2.1   Hamiltonian and Energy

The Hamilton operator $\hat{H}$ plays a double role. On one hand, it determines the time evolution of a closed system; on the other, its eigenvalues correspond to the system's energy. Energy is a conserved quantity that can be measured at the macroscopic level. For this reason, it constitutes the most important link between the microscopic and macroscopic description of a physical system.

However, at the macroscopic level, the rules of thermodynamics apply. The Hamiltonian of a quantum system must have specific properties, in order for the energy to behave in a way compatible with thermodynamics.

In particular, the Hamiltonian $\hat{H}$ must be bounded from below, that is, energy must have a minimum value. If not, the system will emit radiation endlessly, as it will keep transitioning to states of smaller and smaller energy. We will be extracting energy from the system for ever, without doing anything. This would be a violation of the second law of thermodynamics. In the hydrogen atom, the *classical* Hamiltonian function $H = \frac{\mathbf{p}^2}{2m} - \frac{e^2}{4\pi r}$ is not bounded from below, because it goes to $-\infty$ as $r \to 0$. On the contrary, the spectrum of the quantum Hamiltonian is bounded from below, as it has a minimum value – see Chapter 13. The state of minimum energy is called the *ground* state of the system.

In nonrelativistic physics, the minimum energy can be negative, as we saw in many one-dimensional systems. However, in relativity, the total energy includes the particles' rest masses in addition to their kinetic and dynamical energy and, as it turns out, it can never be negative. The reason is that, for systems with negative energy, the energy can become arbitrarily small (i.e., negative with arbitrarily large absolute value) by choosing an appropriate reference frame.

In our study of the Schrödinger operator in one dimension, we saw that Hamiltonians can have both discrete and continuous spectrum. The discrete spectrum corresponds to *bound* states of the particles, that is, states in which the particles are concentrated in a bounded region of space. For example, all eigenfunctions of the square-well Hamiltonian decay exponentially to large distances from the well. The continuous spectrum corresponds to generalized eigenvectors that cannot be localized in a bounded region. They describe particles that have not been trapped by attractive forces, and they can escape to infinity. Generalized eigenvectors of this type are usually referred to as *scattering states*.

A qualitative description of different types of spectrum is given in Fig. 7.1. Cases (a)–(e) are familiar from the study of one-dimensional systems in Chapter 5. Case (f) is novel. It describes systems in which the discrete spectrum lies inside the continuous spectrum, in the sense that a proper eigenvalue is surrounded continuously by generalized eigenvalues. This behavior is common in composite systems. Consider two hydrogen atoms at very large separation. Let us denote by $E_1 = -13.6\,\text{eV}$ the ground state and by $E_2 = \frac{1}{4}E_1$ the first excited state of a *single* atom. Then, the ground state of the two atom systems has energy $2E_1$. The continuous spectrum exists for all $E \geq E_1$, that is, it starts when one atom is in the ground state and the other atom is barely ionized. The state in which both atoms are excited corresponds to energy

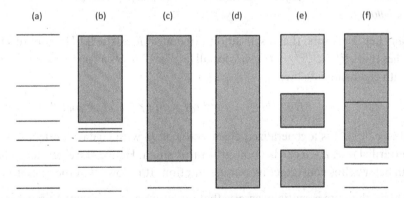

Fig. 7.1 Different types of spectrum for the Hamiltonian $\hat{H}$ of a physical system. (a) Fully discrete spectrum (as in the harmonic oscillator). (b) Discrete spectrum at low energies and continuous spectrum for higher energies. It appears in systems such as a proton–electron pair that can exist either as a hydrogen atom, or as two independent particles. (c) A unique ground state and the continuous spectrum. It characterizes systems with a variable number of particles. (d) Purely continuous spectrum (as in the free particle). (e) Purely continuous spectrum with gaps and bands, as for a particle in a periodic potential. (f) Both discrete and continuous spectrum with part of the former lying within the latter.

$2E_2 = \frac{1}{2}E_1 > E_1$. Hence, $E = E_2$ is a proper energy eigenvalue, but also a point of the continuous spectrum.

There are also spectra that are neither discrete nor continuous, but *anomalous*, that is, they correspond to "weird" subsets of the set of real numbers, for example, subsets with a fractal structure. This type of behavior is rare, but it appears in some Hamiltonians that describe particles within an almost periodic potential (Bellissard and Simon, 1982).

7.2.2 Time in Quantum Theory

In Schrödinger's equation, the time t appears exactly as in classical Newtonian mechanics. It is an external parameter of the system, that is, it corresponds to the time recorded by an external ideal clock that does not interact with the physical system under study.

In Section 2.7.3, we referred to the fundamental asymmetry between space and time in quantum theory: Born's rule defines probabilities with respect to position and not with respect to time. Time is not an observable in quantum theory, in the sense that there is no time operator. To be precise, there is no self-adjoint operator $\hat{T}$ that satisfies

$$e^{i\hat{H}t}\hat{T}e^{-i\hat{H}t} = \hat{T} + t\hat{I}. \tag{7.18}$$

Equation (7.18) is necessary for the operator $\hat{T}$ to be interpreted as a time operator: Its values must be shifted appropriately by Heisenberg time evolution. Equation (7.18) implies that

$$[\hat{T}, \hat{H}] = i\hat{I}. \tag{7.19}$$

Equation (7.19) is obtained by differentiating both sides of Eq. (7.18) with respect to t, and then taking $t = 0$. Pauli (1958) proved the following theorem.

Theorem 7.1 *No self-adjoint operator $\hat{T}$ satisfies Eq. (7.18), if the Hamiltonian $\hat{H}$ is bounded from below.*

Proof Let us assume that a self-adjoint operator $\hat{T}$ exists. By Hadamard's theorem, Eq. (7.19) implies that $e^{ia\hat{T}}\hat{H}e^{-ia\hat{T}} = \hat{H} - a\hat{I}$, for all $a \in \mathbb{R}$. If $|E\rangle$ is a (possibly generalized) eigenvector of $\hat{H}$ with eigenvalue E, then

$$\hat{H}e^{-ia\hat{T}}|E\rangle = e^{-ia\hat{T}}(\hat{H} - a\hat{I})|E\rangle = (E - a)e^{-ia\hat{T}}|E\rangle,$$

that is, $e^{-ia\hat{T}}|E\rangle$ is a generalized eigenvector of $\hat{H}$ with eigenvalue $E - a$. Given that a may be arbitrarily large, $E - a$ can become arbitrarily small. Hence, the eigenvalues of $\hat{H}$ are unbounded from below, thus contradicting our assumption. It follows that the operator $\hat{T}$ does not exist. $\square$

Pauli's theorem essentially asserts that quantum clocks cannot be good clocks. If the variable $\hat{T}$ corresponds to a clock hand, quantum fluctuations force the hand also to go backwards, and not only forward.

The absence of a time operator apparently means that time is not an observable in quantum theory. However, in the laboratory, we often measure the time at which an event happens, for example, we measure the time of detection of a particle that is emitted by a decaying nucleus. These measurements are described by quantum theory, but they require a more elaborate development of quantum measurement theory that will take place in Chapter 21. Still, to describe these measurements we must assume the existence of a background classical time, as in Newtonian mechanics. Here, quantum theory strongly conflicts the other fundamental theory of physics, General Relativity, which describes gravitational phenomena. According to General Relativity, time is fundamentally dynamical, that is, it is determined from the solution of the equations of motion. This fundamental conflict is a major reason why we have not yet been able to construct a unified theory of quantum gravity.

7.2.3   Time–Energy Uncertainty Relation

We saw in Section 6.4 that the Kennard–Robertson inequality expresses Heisenberg's uncertainty principle for position and momentum in terms of operator commutators. In the absence of a time operator, Heisenberg's time–energy uncertainty relation cannot be similarly expressed. Nonetheless, there are several different ways to express analogues of this inequality in terms of operators.

An important version of the time–energy uncertainty relation, due to Mandelshtam and Tamm (1945), is the following. Consider the measurement of an operator $\hat{A}$ in a system with a Hamiltonian $\hat{H}$. By the Kennard–Robertson inequality (6.40),

$$\Delta H \Delta A \geq \frac{1}{2} |\langle [\hat{H}, \hat{A}] \rangle|. \tag{7.20}$$

We use the evolution equation (7.8), in order to express (7.20) as

$$\Delta H \Delta A \geq \frac{1}{2} \left| \left\langle \frac{d\hat{A}}{dt} \right\rangle \right|. \tag{7.21}$$

Equation (7.21) is the *Mandelhstam–Tamm inequality*. The quantity

$$T_A = \frac{\Delta A}{|\langle \frac{d\hat{A}}{dt} \rangle|} \tag{7.22}$$

has dimensions of time. Its physical interpretation depends on the choice of the operator $\hat{A}$. We present some examples.

1. *Clock errors* Let $\hat{A}$ be a degree of freedom that is employed as a clock, that is, the value of $\hat{A}$, as it evolves in time, is a measure of the duration of a phenomenon. Think, for example, of $\hat{A}$ as a quantum version of a clock hand. In an ideal clock, there is a one-to-one correspondence between the background time t and the value of the hand. The presence of quantum fluctuations, as encoded in ΔA, destroys this correspondence, and brings out an error δt in specifying time. We can use T_A as a measure of this error. Hence, Eq. (7.21) becomes

$$\Delta H \delta t \geq \frac{1}{2}. \tag{7.23}$$

If $\Delta H = 0$, that is, a system prepared in an eigenstate of the Hamiltonian cannot be used as a clock, because the error in specifying time will be infinite.

2. *Half-life of unstable systems* We choose $\hat{A} = |\phi\rangle\langle\phi|$ for $|\phi\rangle \in \mathcal{H}$. We should think of $|\phi\rangle$ as describing an excited state of a nucleus or an atom. The system does not stay long in this state, as it decays to the ground state with an emission of a photon. Since $\hat{A} = \hat{A}^2$, $\Delta A = \sqrt{a - a^2}$, where $a = \langle\hat{A}\rangle$. By definition,

$$a(t) = \langle\phi|e^{i\hat{H}t}|\phi\rangle\langle\phi|e^{-i\hat{H}t}|\phi\rangle = |\langle\phi|e^{-i\hat{H}t}|\phi\rangle|^2. \tag{7.24}$$

Equation (7.21) becomes

$$\frac{|\dot{a}|}{\sqrt{a - a^2}} \leq 2\Delta H. \tag{7.25}$$

We assume that $a(0) = 1$: At time $t = 0$ the state of the system is $|\phi\rangle$. Since we describe an unstable state, $\dot{a} \leq 0$. We integrate both sides of Eq. (7.25) from 0 to t, to obtain

$$\arcsin\left(\sqrt{a(t)}\right) + \Delta Ht \geq \frac{\pi}{2}. \tag{7.26}$$

For $t > \frac{\pi}{2\Delta H}$, Eq. (7.26) is satisfied trivially, while for $t < \frac{\pi}{2\Delta H}$,

$$a(t) \geq \sin^2\left(\frac{\pi}{2} - \Delta Ht\right) = \cos^2(\Delta Ht). \tag{7.27}$$

We define the *half-life* $\tau_{\frac{1}{2}}$ of the state $|\phi\rangle$ by the condition $a(\tau_{\frac{1}{2}}) = \frac{1}{2}$. Hence, Eq. (7.27) provides a relation between the energy spread and the half-life of a state

$$\Delta H\tau_{\frac{1}{2}} \geq \arccos\frac{1}{\sqrt{2}} = \frac{\pi}{4}. \tag{7.28}$$

3. *Speed of quantum evolution* Nowadays, we can engineer the Hamiltonian of a quantum system in order to steer the system towards specific quantum states. In this context, we often consider the evolution of an initial state $|\phi\rangle$ to a final state that is normal to $|\phi\rangle$. If this evolution takes time τ, then $\langle\phi|e^{-i\hat{H}\tau}|\phi\rangle = 0$. By Eq. (7.24), this implies that $a(\tau) = 0$. Then, Eq. (7.26) yields

$$\Delta H\tau \geq \frac{\pi}{2}. \tag{7.29}$$

Equation (7.29) is often called a *quantum speed limit*, because it gives a lower bound to the time τ needed for a quantum system to evolve from an initial state to a final one that has no overlap with the initial.

Equation (7.29) apparently implies that there is no speed limit if $\Delta H = \infty$. However, this regime is covered by a different inequality that was derived by Margolus and Levitin (1998),

$$\langle\hat{H}\rangle\tau \geq \frac{\pi}{2}. \tag{7.30}$$

To prove Eq. (7.30), we write $\left\langle\phi|e^{-i\hat{H}\tau}|\phi\right\rangle = x + iy$, where $x, y \in \mathbb{R}$. For simplicity, we assume that the Hamiltonian has purely discrete spectrum, so there is an orthonormal basis $|n\rangle$ of eigenvectors, with corresponding eigenvalues E_n. We expand $|\phi\rangle = \sum_n c_n|n\rangle$, to obtain $\left\langle\phi|e^{-i\hat{H}\tau}|\phi\right\rangle = \sum_n |c_n|^2 e^{-iE_n t}$. Then, $x = \sum_n |c_n|^2 \cos(E_n t)$. We use the trigonometric inequality

$$\cos x \geq 1 - \frac{2}{\pi}(x + \sin x) \tag{7.31}$$

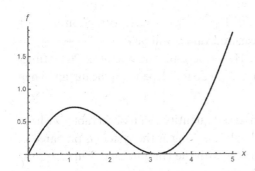

Fig. 7.2 The function $f(x) = \cos(x) + \frac{2}{\pi}(x + \sin x) - 1$ for $x \geq 0$; f is nowhere negative, and $f(0) = f(\pi) = 0$.

that holds for all $x \geq 0$ – see Fig. 7.2. It follows that

$$x \geq 1 - \frac{2}{\pi}\langle \hat{H}\rangle t - \frac{2}{\pi}y. \tag{7.32}$$

For $t = \tau$, $x = y = 0$, hence, Eq. (7.30) follows.

The inequalities (7.29) and (7.30) are combined as (Levitin and Toffoli, 2009)

$$\tau \geq \frac{\pi}{2}\max\left\{\frac{1}{\Delta H}, \frac{1}{\langle \hat{H}\rangle}\right\}. \tag{7.33}$$

Equality in both Eq. (7.29) and Eq. (7.30) is achieved for states $|\psi\rangle = \frac{1}{\sqrt{2}}(|0\rangle + e^{i\theta}|1\rangle)$, where $|0\rangle$ is the ground state, $|1\rangle$ is the first excited state, and $e^{i\theta}$ an arbitrary phase.

7.2.4 Conserved Quantities

In classical physics, a quantity is conserved if its value is constant in time. In Hamiltonian mechanics, we define as a conserved quantity a function F on the state space that satisfies

$$\{F, H\} = 0, \tag{7.34}$$

where H is the classical Hamiltonian of the system. By Eq. (1.6), Eq. (7.34) implies that $\dot{F} = 0$.

In quantum theory, the notion of a conserved quantity splits into two different notions: (i) conservation of a probability distribution, and (ii) conservation of the value of a quantity.

Conservation of probability distribution Motivated by Eq. (7.34), we define as a conserved quantity in a quantum system any operator $\hat{Q} = \sum_q q\hat{P}_q$ that commutes with the Hamiltonian $\hat{H}$,

$$[\hat{Q}, \hat{H}] = 0. \tag{7.35}$$

Equation (7.35) implies that $[\hat{P}_q, \hat{H}] = 0$ for all spectral projectors $\hat{P}_q$ of $\hat{Q}$. Obviously, the Hamiltonian is itself a conserved quantity, as $[\hat{H}, \hat{H}] = 0$.

However, measurements of a conserved quantity, defined this way, do not *give the same value*. For example, if we prepare a system in a superposition of two eigenstates of the Hamiltonian with different energy eigenvalues $c_1|E_1\rangle + c_2|E_2\rangle$, with $|c_1|^2 + |c_2|^2 = 1$, a fraction $|c_1|^2$ of measured values will be E_1 and a fraction $|c_2|^2$ of measured values will be E_2.

In general, the condition $[\hat{H}, \hat{Q}] = 0$ implies that the Heisenberg-picture operator $\hat{Q}(t)$ is time-independent. Hence, the probabilities for measurements of $\hat{Q}$ do not depend on the instant when the measurement takes place.

Preservation of value There is a stronger notion of conserved quantity, as an observable $\hat{Q}$ whose measurements always give a unique time-independent value, as long as the system is prepared in a pure state. This definition presupposes that $\hat{Q}$ commutes with the Hamiltonian. Furthermore, it implies that physically acceptable pure states must be eigenvectors of $\hat{Q}$. No superpositions of states that correspond to different eigenvalues q of $\hat{Q}$ are allowed. In this case, we say that the conserved quantity defines *superselection rules* – see Section 3.2. The conservation law is stronger than the superposition principle.

There are superselection rules for the *electric charge*, and for other conserved charges of particle physics, such as the *lepton number* and the *baryon number*. For the electric charge, the superselection rule implies that there is no superposition of a state of N electrons with a state of $N' \neq N$ electrons. However, a one-electron state can create superpositions with a state of two electrons and one positron.

One may wonder why there are superselection rules for electric charge and not for energy. The difference is that we can measure the electric charge of a microscopic system without changing its value; for example, we can specify the charge by measuring a particle's position after it crossed through a region where an electric field was present. In contrast, the energy of a microscopic system is not preserved during measurements. A particle's interaction with the measurement device requires additional terms in the Hamiltonian.

7.3   Time Evolution in Simple Systems

Next, we solve the evolution equations for some simple quantum systems: the qubit, the free particle, and the harmonic oscillator. We also derive from first principles the formulas for the transmission and reflection probabilities in one-dimensional scattering.

7.3.1   Time Evolution of a Qubit

Consider a qubit with Hamiltonian $\hat{H} = \mathbf{a} \cdot \hat{\boldsymbol{\sigma}}$. The eigenvalues of $\hat{H}$ are $E_+ = |\mathbf{a}|$ and $E_- = -|\mathbf{a}|$, with associated projectors $\hat{P}_\pm = \frac{1}{2}(\hat{I} \pm \frac{\mathbf{a}}{|\mathbf{a}|} \cdot \hat{\boldsymbol{\sigma}})$ – see Section 5.1. The evolution operator $\hat{U}_t = e^{-i\hat{H}t}$ is

$$\hat{U}_t = e^{-i|\mathbf{a}|t}\frac{1}{2}\left(\hat{I} + \frac{\mathbf{a}}{|\mathbf{a}|} \cdot \hat{\boldsymbol{\sigma}}\right) + e^{i|\mathbf{a}|t}\frac{1}{2}\left(\hat{I} - \frac{\mathbf{a}}{|\mathbf{a}|} \cdot \hat{\boldsymbol{\sigma}}\right)$$

$$= \cos(|\mathbf{a}|t)\,\hat{I} - i\sin(|\mathbf{a}|t)\,\frac{\mathbf{a}}{|\mathbf{a}|} \cdot \hat{\boldsymbol{\sigma}}. \tag{7.36}$$

Consider an initial state $|\psi_0\rangle = |0\rangle$. The state $|\psi_t\rangle$ at time t is

$$|\psi_t\rangle = \hat{U}_t|0\rangle = \left(\cos\left(|\mathbf{a}|t\right) + i\frac{a_3}{|\mathbf{a}|}\sin\left(|\mathbf{a}|t\right)\right)|0\rangle - i\frac{a_1 - ia_2}{|\mathbf{a}|}\sin\left(|\mathbf{a}|t\right)|1\rangle. \qquad (7.37)$$

We assume that at time t, the observable $\hat{\sigma}_3$ is measured. The associated probabilities are

$$\text{Prob}(+1, t) = |\langle 1|\psi_t\rangle|^2 = \frac{a_1^2 + a_2^2}{|\mathbf{a}|^2}\sin^2\left(|\mathbf{a}|t\right) \qquad (7.38)$$

$$\text{Prob}(-1, t) = |\langle 0|\psi_t\rangle|^2 = \cos^2\left(|\mathbf{a}|t\right) + \frac{a_3^2}{|\mathbf{a}|^2}\sin^2\left(|\mathbf{a}|t\right). \qquad (7.39)$$

We note that the probability distribution is periodic with frequency $|\mathbf{a}|$, if at least one of a_1 and a_2 is nonzero.

In quantum computing, qubits have zero intrinsic Hamiltonian, and they change through external operations. Such operations are known as quantum *gates*, and they are represented by unitary operators $\hat{U}$. To have full control over a qubit, we must be able to act with all possible unitary operators. We do not need to physically implement one gate for each unitary operator. It suffices to construct a minimal set of gates, so that any unitary operator can be obtained from a successive operation of gates from this set. Such sets of gates are called *universal*.

The most common universal set of gates for a single qubit consists of the so-called Hadamard gate $\hat{H}$, and the phase operations $\hat{R}(\phi)$,

$$\hat{H} = \frac{1}{\sqrt{2}}\begin{pmatrix} -1 & 1 \\ 1 & 1 \end{pmatrix}, \qquad \hat{R}(\phi) = \begin{pmatrix} e^{i\frac{\phi}{2}} & 0 \\ 0 & e^{-i\frac{\phi}{2}} \end{pmatrix} = e^{\frac{i}{2}\hat{\sigma}_3\phi}. \qquad (7.40)$$

Indeed, any quantum state can be obtained by successive action of those gates on $|0\rangle$,

$$\hat{R}\left(\frac{1}{2}\pi + \phi\right)\hat{H}\hat{R}(\theta)\hat{H}|0\rangle = e^{-\frac{i\phi}{2} - i\frac{\pi}{4}}\left(\cos\frac{\theta}{2}|0\rangle + e^{i\phi}\sin\frac{\theta}{2}|1\rangle\right). \qquad (7.41)$$

7.3.2   Free Particle

Consider the time evolution of a free particle of mass m in one dimension. The state vector at time t is $|\psi_t\rangle = e^{-i\hat{H}t}|\psi_0\rangle$, where $|\psi_0\rangle$ is the initial state vector, and $\hat{H} = \frac{1}{2m}\hat{p}^2$.

The wave functions $\psi_t(x) := \langle x|\psi_t\rangle$ and $\psi_0(x) := \langle x|\psi_0\rangle$ are related by

$$\psi_t(x) = \langle x|e^{-i\hat{H}t}|\psi_0\rangle = \int dx'\langle x|e^{-i\hat{H}t}|x'\rangle\langle x'|\psi_0\rangle = \int dx' G_t(x, x')\psi_0(x'), \qquad (7.42)$$

where the function

$$G_t(x, x') := \langle x|e^{-i\hat{H}t}|x'\rangle \qquad (7.43)$$

is the *propagator* for a free particle. Equation (7.42) implies that the propagator allows us to specify the wave function at any time t, provided we know that wave function at time $t = 0$.

We evaluate the propagator, using the spectral decomposition

$$e^{-i\hat{H}t} = \int dp\, e^{-i\frac{p^2}{2m}t} |p\rangle \langle p|, \tag{7.44}$$

where $|p\rangle$ are the generalized eigenvectors of momentum. We obtain,

$$G_t(x, x') = \int dp\, e^{-i\frac{p^2}{2m}t} \langle x|p\rangle \langle p|x'\rangle = \int \frac{dp}{2\pi} e^{-i\frac{p^2}{2m}t + ip(x-x')}. \tag{7.45}$$

The integral in Eq. (7.45) is ill defined, because the integrand oscillates as $p \to \pm\infty$. To evaluate this integral, we substitute $t \to (t - i\epsilon)$ for $\epsilon > 0$. Then, the integrand in Eq. (7.45) gains a term $e^{-\epsilon \frac{p^2}{2m}}$ that allows the integral to converge. By Eq. (A.3),

$$G_{t-i\epsilon}(x, x') = \frac{1}{2\sqrt{\pi \left(\epsilon + i\frac{t}{2m}\right)}} \exp\left[i\frac{m(x-x')^2}{2(t-i\epsilon)}\right]. \tag{7.46}$$

At the limit $\epsilon \to 0$,

$$G_t(x, x') = \sqrt{\frac{m}{2\pi it}} e^{i\frac{m(x-x')^2}{2t}}. \tag{7.47}$$

Example 7.2 Consider an initial Gaussian wave function, centered around $x = 0$, and with mean momentum p_0,

$$\psi_0(x) = (2\pi\sigma^2)^{-1/4} \exp\left[-\frac{x^2}{4\sigma^2} + ip_0 x\right],$$

Substituting into Eq. (7.42) and using the propagator (7.47), we find the wave function ψ_t at time t,

$$\psi_t(x) = (2\pi\sigma^2)^{-1/4} \sqrt{\frac{m}{2\pi it}} e^{\frac{imx^2}{2t}} \int dx' \exp\left[-\left(1 - \frac{2im\sigma^2}{t}\right)\frac{x'^2}{4\sigma^2} + ix'(p_0 - \frac{mx}{t})\right]$$

$$= \frac{1}{(2\pi\sigma^2)^{1/4}\sqrt{1 + \frac{it}{2m\sigma^2}}} \exp\left[\frac{imx^2}{2t} - \frac{\sigma^2(p_0 - \frac{mx}{t})^2}{1 - \frac{2im\sigma^2}{t}}\right]. \tag{7.48}$$

The probability density for position at time t,

$$|\psi_t(x)|^2 = \frac{1}{\sqrt{2\pi\sigma^2}\sqrt{1 + \frac{t^2}{4m^2\sigma^4}}} \exp\left[-\frac{(x - \frac{p_0 t}{m})^2}{2\sigma^2(1 + \frac{t^2}{4m^2\sigma^4})}\right], \tag{7.49}$$

is a Gaussian centered at $p_0 t/m$ and with mean deviation

$$\Delta x(t) = \sigma\sqrt{1 + \frac{t^2}{4m^2\sigma^4}}. \tag{7.50}$$

For $t \ll 2m\sigma^2$, $|\psi_t(x)|^2 = |\psi_0(x - p_0 t/m)|^2$, that is, the center of the initial probability distribution is translated by $p_0 t/m$, as one would expect from the classical equations of motion. However, for $t/(2m\sigma^2)$ of order unity or larger, the probability distribution is deformed as the dispersion of the wave packet becomes increasingly larger. This behavior is plotted in Fig. 7.3. See also Box 7.1.

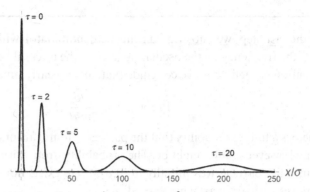

Fig. 7.3 The time-evolved probability distribution $|\psi(x,t)|^2$ of Eq. (7.49) as a function of x, and for different values of $\tau := t/(2m\sigma^2)$. The wave function dispersion is strong for τ of order unity or larger.

Box 7.1   Faster-Than-Light Speeds in Quantum Time Evolution?

Let a particle of mass m be prepared in a state $\psi_0(x) = \frac{1}{\sqrt{a}}\chi_{[0,a]}(x)$. The state is sharply localized in the interval $[0,a]$ and it has zero mean momentum. We evolve ψ_0 with the propagator (7.47),

$$\psi_t(x) = \sqrt{\frac{m}{2\pi iat}} \int_0^a dx' e^{i\frac{m(x-x')^2}{2t}} = \frac{1}{\sqrt{\pi ia}} \int_{-\sqrt{\frac{m}{2t}}x}^{-\sqrt{\frac{m}{2t}}(x-a)} dse^{-is^2},$$

where $s := \sqrt{\frac{m}{2t}}(x'-x)$. Integration over s yields

$$\psi_t(x) = \frac{1}{\sqrt{2ia}}\left[(C-iS)\left(-\sqrt{\frac{m}{2t}}(x-a)\right) - (C-iS)\left(-\sqrt{\frac{m}{2t}}x\right)\right],$$

where $C(x)$ and $S(x)$ are the Fresnel integrals, defined by Eqs. (B.12) and (B.13). By Eq. (B.14), $C(x) - iS(x) = \sqrt{\frac{\pi}{2}}\frac{1+i}{2} - i\frac{e^{ix^2}}{2x}$, for large x. Hence, for $x >> a$ and $t < ma^2$,

$$\psi_t(x) = -\sqrt{\frac{it}{4ma}}\left[\frac{e^{i\frac{m}{2t}(x-a)^2}}{x-a} - \frac{e^{i\frac{m}{2t}x^2}}{x}\right].$$

Since $x >> a$, we substitute $x-a$ in the denominator with x, and we approximate $(x-a)^2 \simeq x^2 - 2ax$ in the exponent. Then, $\psi_t(x) = \sqrt{\frac{it}{4ma}}e^{i\frac{mx^2}{t}}\frac{1-e^{-i\frac{ma}{t}x}}{x}$ and

$$|\psi_t(x)|^2 = \frac{t}{ma}\frac{\sin^2\left(\frac{ma}{2t}x\right)}{x^2}.$$

The probability that the particle is found in $[L-\frac{d}{2}, L+\frac{d}{2}]$ at time t is

$$\text{Prob}(L,t) = \int_{L-\frac{d}{2}}^{L+\frac{d}{2}} dx|\psi_t(x)|^2 \simeq \frac{t}{2maL^2}\int_{L-\frac{d}{2}}^{L+\frac{d}{2}} dx[1 - \cos\left(\frac{ma}{t}x\right)],$$

where in the last step, we substituted x in the denominator with L, since $d \ll L$. For $t \ll mad$, the frequency of the oscillating term in the integral becomes very large, hence, its contribution is negligible. We conclude that, at very early times,

$$\text{Prob}(L, t) = \frac{td}{2maL^2},$$

that is, there is a finite probability that the particle is found around $x = L$, however large L may be and however small t might be. The probability drops with L, but only as an inverse power. Since the particle was initially in $[0, a]$, there exists nonzero probability that the particle moves with arbitrarily high speed.

One might view this result as an artefact of the nonrelativistic description of particles. However, similar results exist even in fully relativistic systems (Hegerfeldt, 1974). They imply that we cannot accept states that are strictly localized in a specific spatial region. Wave functions must have long tails that extend well outside any localization region (Hegerfeldt, 1985).

7.3.3 Harmonic Oscillator

The simplest way to construct the time-evolution operator for a harmonic oscillator is by using coherent states. For a Hamiltonian $\hat{H} = \omega \hat{a}^\dagger \hat{a}$, we find

$$e^{-i\hat{H}t}|z\rangle = e^{-\frac{|z|^2}{2}} \sum_{n=0}^{\infty} \frac{z^n}{n!} e^{-i\hat{H}t}|n\rangle = e^{-\frac{|z|^2}{2}} \sum_{n=0}^{\infty} \frac{z^n}{n!} e^{-in\omega t}|n\rangle$$

$$= e^{-\frac{|z|^2}{2}} \sum_{n=0}^{\infty} \frac{(ze^{-i\omega t})^n}{n!}|n\rangle = |ze^{-i\omega t}\rangle. \tag{7.51}$$

The time evolution of a coherent state remains a coherent state.

We evaluate the propagator, using the resolution of the unity for coherent states,

$$G_t(x, x') = \int \frac{d^2z}{\pi} \left\langle x|e^{-i\hat{H}t}|z\right\rangle \langle z|x'\rangle = \int \frac{d^2z}{\pi} \langle x|ze^{-i\omega t}\rangle \langle z|x'\rangle. \tag{7.52}$$

We substitute $z = \sqrt{\frac{m\omega}{2}} q + \frac{i}{\sqrt{2m\omega}} p$; then, $ze^{-i\omega t} = \sqrt{\frac{m\omega}{2}} q(t) + \frac{i}{\sqrt{2m\omega}} p(t)$, where

$$q(t) = q \cos \omega t + \frac{p}{m\omega} \sin \omega t, \qquad p(t) = p \cos \omega t - m\omega q \sin \omega t,$$

are the mean position and momentum, respectively, evolving according to the classical equations of motion.

We use Eq. (5.39) with the appropriate phase. Then, Eq. (7.52) becomes

$$G_t(x, x') = \sqrt{\frac{m\omega}{\pi}} \int \frac{dqdp}{\pi} \exp\left[-\frac{m\omega}{2}[(x'-q)^2 + (x - q(t))^2] + ip(t)x - ipx' + \frac{i}{2}pq - \frac{i}{2}p(t)q(t)\right]$$

$$= \frac{1}{\pi}\sqrt{\frac{m\omega}{\pi}} e^{-\frac{m\omega(x^2+x'^2)}{2}} \int d^2\xi \exp\left(-\frac{1}{2}\xi^T A\xi + \xi^T B\right), \tag{7.53}$$

where

$$\xi = \begin{pmatrix} q \\ p \end{pmatrix}, \quad B = \begin{pmatrix} m\omega(x' + xe^{i\omega t}) \\ i(xe^{-i\omega t} - x') \end{pmatrix}$$

$$A = \begin{pmatrix} m\omega(1 + e^{-i\omega t}\cos\omega t) & -i\sin^2\omega t \\ -i\sin^2\omega t & -i(m\omega)^{-1}e^{i\omega t}\sin\omega t \end{pmatrix}.$$

By Eq. (A.4), the integral over ξ yields $\frac{\pi}{\sqrt{\det A}}e^{\frac{1}{2}B^T A^{-1}B}$. It is straightforward, if somewhat tedious, to show that

$$G_t(x,x') = \sqrt{\frac{m\omega}{2\pi i \sin\omega t}} e^{\frac{i}{2}\omega t} \exp\left[\frac{im\omega}{2\sin\omega t}\left[(x^2 + x'^2)\cos\omega t - 2xx'\right]\right]. \tag{7.54}$$

The phase $e^{\frac{i}{2}\omega t}$ disappears if we include the ground-state energy $\frac{1}{2}\omega$ in the definition of the Hamiltonian.

7.3.4   Scattering in a General Potential

Consider a particle of mass m moving in one dimension within a potential $V(x)$ that vanishes outside the interval $[a,b]$. As shown in Section 5.3.3, the generalized eigenvalues of the Schrödinger operator correspond to kets $|k,\pm\rangle$, where $k = \sqrt{2mE}$. We assume that the Schrödinger operator has only a continuous spectrum, that is, there are no bound states.

By the spectral theorem, the time-evolution operator reads

$$e^{-i\hat{H}t} = \int_0^\infty dk\, e^{-i\frac{k^2}{2m}t}\left(|k,+\rangle\langle k,+| + |k,-\rangle\langle k,-|\right). \tag{7.55}$$

Outside the interval $[a,b]$, the kets $|k,\pm\rangle$ are of the form (5.57). We assume an initial state $|\psi_0\rangle$, centered around $x_0 < a$; x_0 is far away from a, so that $\psi_0(x)$ is well localized outside the barrier. We also require that the momentum wave function $\tilde{\psi}_0(p) = \frac{1}{\sqrt{2\pi}}\int dx e^{-ipx}\psi_0(x)$ is concentrated around $p_0 > 0$ with spread $\Delta p << p_0$, so that $\tilde{\psi}_0(p) \simeq 0$ for $p < 0$. This condition guarantees that the particle definitely moves to the right.

We evaluate

$$\langle k,+|\psi_0\rangle = \frac{1}{\sqrt{2\pi}}\int dx\left(e^{-ikx} + R_k^* e^{ikx}\right)\psi_0(x) = \tilde{\psi}_0(k) + R_k^*\tilde{\psi}_0(-k) = \tilde{\psi}_0(k)$$

$$\langle k,-|\psi_0\rangle = \frac{1}{\sqrt{2\pi}}T_k^*\int dx e^{ikx}\psi_0(x) = T_k^*\tilde{\psi}_0(-k) = 0,$$

where in the last step we used the fact that $\tilde{\psi}_0(-k) = 0$ for $k > 0$.

The wave function at time t is

$$\psi_t(x) := \langle x|e^{-i\hat{H}t}|\psi_0\rangle = \int_0^\infty dk\langle x|k,+\rangle\tilde{\psi}_0(k)e^{-i\frac{k^2}{2m}t}. \tag{7.56}$$

For $x > b$,

$$\psi_t(x) = \frac{1}{\sqrt{2\pi}}\int_0^\infty dk\, T_k\tilde{\psi}_0(k)e^{ikx-i\frac{k^2}{2m}t}. \tag{7.57}$$

To evaluate Eq. (18.69), we first write $T_k = |T_k|e^{i\phi_k}$. Then, we note that the integrand is strongly peaked around p_0, and we can use the *phase expansion* approximation for the transmission of almost monochromatic waves – see Box 2.2. We proceed as follows.

First, we substitute $|T_k|$ with $|T_{p_0}|$. Then, we expand the phases around p_0, keeping terms up to first order in $k - p_0$. That is, we write $\theta_k = \theta_{p_0} + \theta'_{p_0}(k - p_0)$ and $k^2 = p_0^2 + 2p_0(k - p_0)$. Finally, we extend the lower range of integration to $-\infty$, since the contribution from negative values of k vanishes.

Hence, the integral $\frac{1}{\sqrt{2\pi}} \int_{-\infty}^{\infty} dk e^{ik\left(x - p_0 t/m + \theta'_{p_0}\right)} \tilde{\psi}_0(k) = \psi_0(x - p_0 t/m + \theta'_{p_0})$ appears in Eq. (18.69), leading to

$$\psi_t(x) = T_{p_0} e^{i\frac{p_0^2}{2m}t - ip_0\theta'_{p_0}} \psi_0(x - p_0 t/m + \theta'_{p_0}). \tag{7.58}$$

The probability density for position at time t is

$$|\psi_t(x)|^2 = |T_{p_0}|^2 |\psi_0(x - p_0 t/m + \theta'_{p_0})|^2, \tag{7.59}$$

that is, it is the initial probability distribution with a translated center and reduced height by a factor $|T_{p_0}|^2$. The probability $\text{Prob}(x > b, t)$ that at time t, the particle is found at $x > b$ is

$$\text{Prob}(x > b, t) = \int_b^{\infty} dx |\psi_t(x)|^2 = |T_{p_0}|^2 \int_{b - p_0 t/m + \theta'_{p_0}}^{\infty} dx |\psi_0(x)|^2. \tag{7.60}$$

For sufficiently large t, the lower limit of integration can be taken to $-\infty$, hence, the integral gives unity. In this limit,

$$\text{Prob}(x > b, t) = |T_{p_0}|^2. \tag{7.61}$$

This calculation justifies the interpretation of $|T_p|^2$ as the transmission probability in a potential barrier.

Equation (7.61) holds, as long as Eq. (7.58) is a good approximation for the wave function. Equation (7.58) ignores the spread of the wave packet. We assume that the range $d = b - a$ of the potential is microscopic and that the measurement of position takes place at macroscopic distance from $[a, b]$. This means that the wave packet propagates most of the time as a free particle. Hence, we can estimate the wave packet spread using the free-particle expression. The spread is negligible as long as we restrict to times $t \lesssim m(\Delta x)^2$, where Δx is the mean deviation of the initial wave packet. On the other hand, the long-time limit in Eq. (7.60) presupposes that $p_0 t/m \gg b + \theta'_{p_0}$. The term $b + \theta'_{p_0}$ is of the order of the range d of the potential; hence, $t \gg md/p_0$. Combining both inequalities, we find that a sufficient condition for Eq. (7.61) is

$$\Delta x \gg \sqrt{\frac{d}{p_0}}. \tag{7.62}$$

Furthermore, we have assumed that $\Delta p \ll p_0$. Since $\Delta p > (\Delta x)^{-1}$, we obtain a second condition for Eq. (7.61)

$$p_0 \Delta x \gg 1. \tag{7.63}$$

7.3.5 Tunneling Times

Suppose that we consider energies E for the particles that correspond to quantum tunneling through the potential $V(x)$. Equation (7.58) suggests that the outcoming particles had a time delay

$$\tau_{del} = \frac{m}{p_0}\theta'_{p_0} = \left(\frac{\partial\theta}{\partial E}\right)_{E_0}, \tag{7.64}$$

when compared to particles crossing the same distance in absence of the barrier (Bohm, 1951; Wigner, 1955); here $E_0 = \frac{p_0^2}{2m}$. If we add to τ_{del} the time $m(b-a)/p_0$ that it would take a classical free particle to cross distance $b-a$, we obtain an estimate for the time τ_{tun} that it takes a tunneling particle to cross the interval $[a, b]$

$$\tau_{tun} = \frac{m}{p_0}[\theta'_{p_0} + (b-a)]. \tag{7.65}$$

Both Eq. (7.65) and Eq. (7.64) presuppose the validity of the phase expansion approximation.

Equation (7.59) strongly suggests that the detection time for tunneling particles is not sharply defined, but it is probabilistically distributed. Hence, τ_{del} and τ_{tun} are parameters of probability distributions, and not values of quantum observables. A further caveat in the applicability of Eq. (7.65) is that τ_{tun} can become negative in potentials characterized by a shallow bound state – you will see an example in Problem 7.11. Then, the interpretation of τ_{tun} as the time it takes a particle to cross the barrier is untenable.[2]

There are several different proposals for determining the time a particle spends inside the barrier, which lead to very different estimates for τ_{tun}. This is not a paradox, because different proposals involve different measurement set-ups, and thus, they correlate τ_{tun} with different quantum observables.

We can evaluate Eq. (7.65) for a square-barrier potential of height V_0 and width d. In the opaque barrier limit, the transmission amplitude is given by Eq. (5.80). Then,

$$\tau_{tun} = \frac{2m}{p_0\sqrt{2mV_0 - p_0^2}}. \tag{7.66}$$

We note that τ_{tun} does not depend on the width d of the barrier, but only on energy. Since d can be made arbitrarily large, the expected velocity d/τ_{tun} of the particle during crossing can become arbitrarily large. Apparently, any particle that crossed the barrier has moved faster than light.

This behavior of quantum systems in tunnelling is called the *Hartmann effect* (Hartmann, 1962). The conclusion remains unaffected if the particle is relativistic. However, this faster-than-light transmission is only apparent, and not actual. We should keep in mind that quantum theory deals only with measured quantities. The time τ_{del} is not directly measured, it is defined the mean traversal time in presence of the potential barrier minus the mean traversal time without the barrier. This means that the tunneling time τ_{tun} is statistically inferred by combining the results of two different experiments, one in presence of the potential and one in absence of the potential.

---

[2] In Problem 7.11, it is shown that for a symmetric potential of width d, $\theta'_p > -d - p^{-1}$, which implies that $\tau_{tun} > -m/p_0^2$. The lower bound to τ_{tun} may approach $-\infty$ as $p_0 \to 0$!

In Section 6.6, we saw that the combination of information from different experiments is not allowed by the rules of quantum theory.

Equation (7.59) applies provided that both the particle source and the particle detector are far from the potential barrier. It can be proven that the time that has elapsed between particle emission (in the region $x < a$) and particle detection (in the region $x > b$) is never smaller than the time it takes light to cross the same interval (Anastopoulos and Savvidou, 2013). There is no faster-than-light *signal*. The particle appears to travel faster than light in the forbidden region, because we use reasoning from classical physics – a combination of information from different experiments – that is not appropriate to quantum systems.

7.4   Systems with Time-Dependent Hamiltonian

So far we only studied systems where the evolution operator $\hat{U}_t$ is *time-homogeneous*, that is, it satisfies $\hat{U}_{t+t'} = \hat{U}_t \hat{U}_{t'}$. This means that it can be expressed in terms of a time-independent Hamiltonian. The Hamiltonians for all fundamental closed systems are, indeed, time-independent. However, in many experiments we act through external, time-dependent electric or magnetic fields. To describe the evolution of such systems, we must solve Schrödinger's equation with a time-dependent Hamiltonian

$$i\frac{d}{dt}|\psi\rangle = \hat{H}(t)|\psi\rangle. \tag{7.67}$$

In this section, we study solutions to this equation. First, we present different ways of expressing the solution to Eq. (7.67) as a series. Then, we find solutions to two paradigmatic systems, a harmonic oscillator under an external force, and a qubit under an external force.

7.4.1   Series Expansions of the Evolution Operator

First, we express solutions to Eq. (7.67) as $|\psi(t)\rangle = \hat{U}(t, t_0)|\psi(t_0)\rangle$, for some initial moment of time t_0. The unitary operator $\hat{U}(t, t_0)$ satisfies

$$i\frac{d}{dt}\hat{U}(t, t_0) = \hat{H}(t)\hat{U}(t, t_0), \tag{7.68}$$

with initial condition $\hat{U}(t_0, t_0) = 1$.

Equation (7.68) is solved by the series

$$\hat{U}(t, t_0) = \hat{I} - i\int_{t_0}^{t} ds\hat{H}(s)ds + (-i)^2\int_{t_0}^{t} ds_1 \int_{t_0}^{s_1} ds_2 \hat{H}(s_1)\hat{H}(s_2) + \cdots$$

$$= \hat{I} + \sum_{n=1}^{\infty}(-i)^n \int_{t_0}^{t} ds_1 \int_{t_0}^{s_1} ds_2 \cdots \int_{t_0}^{s_{n-1}} ds_n \hat{H}(s_1)\hat{H}(s_2)\cdots\hat{H}(s_n). \tag{7.69}$$

To see this, note that the time derivative of the nth term in the series equals the $(n-1)$th term, multiplied on the left by $-i\hat{H}(t)$. That is,

$$i\frac{d}{dt}\left(-i\int_{t_0}^{t}ds\hat{H}(s)\right)=\hat{H}(t)\hat{I}$$

$$i\frac{d}{dt}\left[(-i)^2\int_{t_0}^{t}ds_1\int_{t_0}^{s_1}ds_2\hat{H}(s_1)\hat{H}(s_2)\right]=\hat{H}(t)\left(-i\int_{t_0}^{t}ds\hat{H}(s)\right),\tag{7.70}$$

and so on. Hence, when all terms in the series (7.69) are added, Eq. (7.68) is satisfied.

An alternative form to Eq. (7.69) is obtained from the observation that the integral for the second-order term can be expressed as

$$\int_{t_0}^{t}ds_1\int_{t_0}^{s_1}ds_2\hat{H}(s_1)\hat{H}(s_2)=\int_{t_0}^{t}ds_1\int_{t_0}^{t}ds_2\theta(s_1-s_2)\hat{H}(s_1)\hat{H}(s_2)$$

$$=\frac{1}{2}\int_{t_0}^{t}ds_1\int_{t_0}^{t}ds_2\mathcal{T}[\hat{H}(s_1)\hat{H}(s_2)],\tag{7.71}$$

where the object

$$\mathcal{T}[\hat{H}(s_1)\hat{H}(s_2)]:=\theta(s_1-s_2)\hat{H}(s_1)\hat{H}(s_2)+\theta(s_2-s_1)\hat{H}(s_2)\hat{H}(s_1)\tag{7.72}$$

is called the *time-ordered product* of $\hat{H}(s_1)$ and $\hat{H}(s_2)$.

The time-ordered product of n operators $\hat{H}(s_i)$, where $i=1,2,\ldots,n$, is defined as

$$\mathcal{T}[\hat{H}(s_1)\hat{H}(s_2)\cdots\hat{H}(s_n)]=\hat{H}(s_{i_1})\hat{H}(s_{i_2})\cdots\hat{H}(s_{i_n}),\tag{7.73}$$

where $(i_1,i_2,\ldots,i_n)$ is a permutation of $(1,2,\ldots,n)$ such that $s_{i_1}\geq s_{i_2}\geq\cdots\geq s_{i_n}$.

With this definition, Eq. (7.69) becomes

$$\hat{U}(t,t_0)=\hat{I}+\sum_{n=1}^{\infty}\frac{(-i)^n}{n!}\int_{t_0}^{t}ds_1\int_{t_0}^{t}ds_2\cdots\int_{t_0}^{t}ds_n\mathcal{T}[\hat{H}(s_1)\hat{H}(s_2)\cdots\hat{H}(s_n)],\tag{7.74}$$

where the $n!$ in the denominator arises because there are $n!$ different permutations of the n time points $s_1,s_2,\ldots,s_n$.

Because of Eq. (7.74), we write the operator $\hat{U}(t,t_0)$ compactly as $\mathcal{T}e^{-i\int_{t_0}^{t}ds\hat{H}(s)}$, where $\mathcal{T}$ appears in order to designate that we have not the usual exponential, but the series expansion Eq. (7.74). The latter is known as *time-ordered* exponential, or Dyson series.

Yet another useful expression for the evolution operator can be obtained from the fact that $\hat{U}(t,t_0)=\hat{U}(t,t_1)\hat{U}(t_1,t_0)$. We split the interval $[t,t_0]$ into n subintervals $[t_i,t_{i-1}]$, $i=1,2,\ldots,n$, such that $t=t_n$ and $t_i-t_{i-1}=\delta t=\frac{t-t_0}{n}$. For sufficiently large n, we keep only the leading-order term to δt, so that $\hat{U}(t_{i+1},t_i)\simeq 1-\hat{H}(t_i)\delta t\simeq e^{-i\hat{H}(t_i)\delta t}$. Then,

$$\hat{U}(t,t_0)=\lim_{n\to\infty}e^{-i\hat{H}(t_n)\delta t}e^{-i\hat{H}(t_{n-1})\delta t}\cdots e^{-i\hat{H}(t_0)\delta t}.\tag{7.75}$$

We evaluate this product through repeated uses of the BCH identity (Theorem 4.15), and then, we take the limit $\delta t\to 0$. The result is the *Magnus expansion* of the evolution operator (Magnus, 1954),

$$\hat{U}(t,t_0)=e^{-i\sum_{n=1}^{\infty}\hat{\mathcal{M}}_n},\tag{7.76}$$

where $\hat{\mathcal{M}}_n$ is the Magnus term of nth order, involving n nested commutators $[\hat{H}(t_1), [\hat{H}(t_2), \dots$ $[\hat{H}(t_{n-1}), \hat{H}(t_n)]\dots]]$. The first three Magnus terms are

$$\mathcal{M}_1 = \int_{t_0}^{t} dt_1 \hat{H}(t_1),$$

$$\mathcal{M}_2 = \frac{1}{2i} \int_{t_0}^{t} dt_1 \int_{t_0}^{t_1} dt_2 [\hat{H}(t_1), \hat{H}(t_2)],$$

$$\mathcal{M}_3 = \frac{1}{6i^2} \int_{t_0}^{t} dt_1 \int_{t_0}^{t_1} dt_2 \int_{t_0}^{t_2} dt_3 \left([\hat{H}(t_1), [\hat{H}(t_2), \hat{H}(t_3)]] + [\hat{H}(t_3), [\hat{H}(t_2), \hat{H}(t_1)]] \right).$$

The benefit of the Magnus expansion is that if we approximate the evolution operator by terminating the expansion at a finite order N, the resulting operator remains unitary, since it is of the form $e^{i\hat{A}}$ with self-adjoint $\hat{A}$. The Dyson series does not have this property.

7.4.2 Forced Harmonic Oscillator

We consider a harmonic oscillator of frequency ω under a time-dependent external force $F(t)$ that is coupled linearly to position,

$$\hat{H} = \omega \hat{a}^\dagger \hat{a} - F(t)\hat{x} = \omega \hat{a}^\dagger \hat{a} - \frac{F(t)}{\sqrt{2m\omega}}(\hat{a} + \hat{a}^\dagger). \tag{7.77}$$

Our aim is to compute the evolution operator $\hat{U}(t) = \hat{U}(t, 0)$. To this end, we write $\hat{H}_0 = \omega \hat{a}^\dagger \hat{a}$ and $\hat{V} = -F(t)\hat{x}$. Then, we define the operator $\hat{O}(t) := e^{i\hat{H}_0 t} \hat{U}(t)$, in order to separate the influence of the external force. We differentiate $\hat{O}(t)$ with respect to t, to obtain

$$i\frac{d}{dt}\hat{O}(t) = \hat{V}(t)\hat{O}(t), \tag{7.78}$$

where

$$\hat{V}(t) = e^{i\hat{H}_0 t} \hat{V} e^{-i\hat{H}_0 t} = -\frac{F(t)}{\sqrt{2m\omega}}(\hat{a}e^{-i\omega t} + \hat{a}^\dagger e^{i\omega t}). \tag{7.79}$$

We will evaluate $\hat{O}(t)$ using the Magnus expansion, Eq. (7.76), for the solution of Eq. (7.78). The key point is that the commutator $[\hat{V}(t), \hat{V}(t')] = -i\frac{F(t)F(t')}{m\omega}\sin[\omega(t - t')]\hat{I}$ is proportional to unity. Double- and higher-order commutators vanish; hence, only the first two terms in the Magnus expansion contribute. It follows that

$$\hat{O}(t) = \exp\left[i\int_0^t ds \frac{F(s)}{\sqrt{2m\omega}}(\hat{a}e^{-i\omega s} + \hat{a}^\dagger e^{i\omega s})\right.$$
$$\left. + \frac{i}{2m\omega}\int_0^t ds \int_0^s ds' F(s)F(s')\sin[\omega(s - s')]\right] = e^{i\gamma(t)}e^{ia(t)\hat{x} - ib(t)\hat{p}}, \tag{7.80}$$

where $a(t) = \int_0^t dsF(s)\cos\omega s, b(t) = -\frac{1}{m\omega}\int_0^t dsF(s)\sin(\omega s)$ and

$$\gamma(t) = \frac{1}{2m\omega}\int_0^t ds \int_0^s ds' F(s)F(s')\sin[\omega(s - s')].$$

Hence, $\hat{O}(t)$ is a Weyl operator (4.22) with time-dependent coefficients, modulo a phase: $\hat{O}(t) = e^{i\gamma(t)}\hat{V}[a(t), b(t)]$.

The propagator is

$$G_t(x, x') = \langle x|e^{-i\hat{H}_0 t}\hat{O}(t)|x'\rangle = e^{i\gamma(t)}\langle x|e^{-i\hat{H}_0 t}\hat{V}[a(t), b(t)]|x'\rangle$$

$$= e^{i\gamma(t)+\frac{i}{2}a(t)b(t)}e^{ia(t)x'}\langle x|e^{-i\hat{H}_0 t}|x'+b(t)\rangle = e^{i\gamma(t)+\frac{i}{2}a(t)b(t)}e^{ia(t)x'}G_t^{(0)}[x, x'+b(t)],$$

where $G_t^{(0)}$ is the propagator (7.54) of the harmonic oscillator without external force.

Equation (7.54) implies that $G_t^{(0)}(x, x'+b) = G_t^{(0)}(x, x')e^{\frac{im\omega b}{\sin \omega t}\left(x'\cos\omega t - x + \frac{1}{2}b\right)}$. It follows that

$$G_t(x, x') = e^{i\left(x\int_0^t dsF(s)\frac{\sin\omega s}{\sin\omega t} + x'\int_0^t dsF(s)\frac{\sin[\omega(t-s)]}{\sin\omega t}\right)+i\tilde{\gamma}(t)}G_t^{(0)}(x, x'), \tag{7.81}$$

where

$$\tilde{\gamma}(t) = \gamma(t) + \frac{1}{2}a(t)b(t) + \frac{m\omega}{2}b^2(t)\tan\omega t$$

$$= -\frac{1}{2m\omega\sin\omega t}\int_0^t ds\int_0^s ds' F(s)F(s')\sin(\omega s')\,\sin[\omega(t-s)]. \tag{7.82}$$

Example 7.3 Assume that we use the harmonic oscillator in order to measure the force $F(t)$. This problem is relevant to the detection of gravitational waves, where F corresponds to the force exerted by such waves. Suppose that the force has finite duration T. For $t > T$, $\hat{O}(t)$ of Eq. (7.80) is time-independent, and equal to $\hat{S} := e^{i\gamma(T)}e^{ia(T)\hat{x}-ib(T)\hat{p}}$.

Suppose that the probe oscillator is originally in its ground state $|0\rangle$. By Eq. (4.27), $\hat{S}|0\rangle$ is the ground state with the momentum shifted by $a(T)$ and the position by $b(T)$. Then, by Eq. (5.39) $\hat{S}|0\rangle$ is a coherent state $|z\rangle$ modulo a phase, where $z = \sqrt{\frac{m\omega}{2}}b(T) + \frac{i}{\sqrt{2m\omega}}a(T)$.

The force from gravitational waves is very weak, and we want a criterion to assert that it actually has acted on our probe. A simple criterion is that $|z\rangle$ should be distinguished from the vacuum. Given that $|\langle z|0\rangle|^2 = e^{-|z|^2}$, the condition $|z| > 1$ ensures a reasonable degree of distinguishability. It also means that the mean transferred energy is larger than one energy quantum. Hence, we require that

$$\frac{1}{2}m\omega b(T)^2 + \frac{1}{2m\omega}a(T)^2 > 1. \tag{7.83}$$

Assume that the force is impulsive, with duration $T << \omega^{-1}$. We can approximate it by $F(t) = K\delta(t - t_0)$, where K is the transferred momentum (the impulse) and $t_0 \in [0, T]$. Then, $a(T) = K\cos(\omega t_0)$ and $b(T) = -K(m\omega)^{-1}\sin(\omega t_0)$. The distinguishability criterion yields

$$K > \sqrt{2m\omega}, \tag{7.84}$$

that is, a lower bound to the impetus of an impulsive force that can be distinguished by a quantum oscillator probe.

7.4.3 Rabi Oscillations

Next, we study the time evolution of a qubit under the influence of a strong time-dependent external field. Consider the time-dependent Hamiltonian

$$\hat{H}(t) = \frac{\omega}{2}(\hat{I} + \hat{\sigma}_3) + gf(t)\hat{\sigma}_1, \tag{7.85}$$

where $\omega > 0$ and g a constant with dimensions of energy. The Hamiltonian (7.85) describes, for example, transitions between the ground state ($|0\rangle$) and an excited state ($|1\rangle$) of an atom, subject to the influence of an external classical electromagnetic field that varies in time according to the function $f(t)$.

We follow the same steps as in the forced harmonic oscillator. We distinguish the operator $\hat{H}_0$ that describes the intrinsic qubit dynamics from the operator $\hat{V}$ that describes the influence of the external field,

$$\hat{H}_0 = \frac{\omega}{2}(\hat{I} + \hat{\sigma}_3) = \omega \begin{pmatrix} 1 & 0 \\ 0 & 0 \end{pmatrix}$$

$$\hat{V} = gf(t)\hat{\sigma}_1 = g \begin{pmatrix} 0 & f(t) \\ f(t) & 0 \end{pmatrix}. \tag{7.86}$$

Let $\hat{U}(t) = \hat{U}(t,0)$ be the evolution operator corresponding to the Hamiltonian (7.85). Again we define $\hat{O}(t) := e^{i\hat{H}_0 t}\hat{U}(t)$, and we differentiate with respect to t, to obtain

$$i\frac{d}{dt}\hat{O}(t) = \hat{V}(t)\hat{O}(t), \tag{7.87}$$

where

$$\hat{V}(t) = e^{i\hat{H}_0 t}\hat{V}e^{-i\hat{H}_0 t} = \begin{pmatrix} e^{i\omega t} & 0 \\ 0 & 1 \end{pmatrix}\begin{pmatrix} 0 & gf(t) \\ gf(t) & 0 \end{pmatrix}\begin{pmatrix} e^{-i\omega t} & 0 \\ 0 & 1 \end{pmatrix}$$

$$= gf(t)\begin{pmatrix} 0 & e^{i\omega t} \\ e^{-i\omega t} & 0 \end{pmatrix}. \tag{7.88}$$

Unlike the forced harmonic oscillator, there is no exact solution to Eq. (7.87), and we have to rely on approximations. The *rotating wave approximation* (RWA) is useful when the external force is periodic, with frequency close to that of the qubit. Suppose that $f(t) = \cos(\tilde{\omega}t)$, where $\tilde{\omega}$ satisfies $|\tilde{\omega} - \omega| << \omega$. The operator (7.88) becomes

$$\hat{V}(t) = \frac{g}{2}\begin{pmatrix} 0 & e^{i(\omega+\tilde{\omega})t} + e^{i(\omega-\tilde{\omega})t} \\ e^{-i(\omega+\tilde{\omega})t} + e^{-i(\omega-\tilde{\omega})t} & 0 \end{pmatrix}. \tag{7.89}$$

The RWA is based on the fact that, for $|\tilde{\omega}-\omega| << \omega$, oscillations with frequency $\tilde{\omega}+\omega$ are much faster than oscillations with frequency $|\omega - \tilde{\omega}|$. For measurements at a time scale $\tau >> \tilde{\omega}^{-1}$, we can average over the fast oscillations, with no significant error. Hence, $e^{i(\omega+\tilde{\omega})t}$ in Eq. (7.89) can be substituted by its average in an interval of duration τ, which is practically zero. Then,

$$\hat{V}(t) \simeq \frac{g}{2}\begin{pmatrix} 0 & e^{i(\omega-\tilde{\omega})t} \\ e^{-i(\omega-\tilde{\omega})t} & 0 \end{pmatrix} = e^{i\hat{H}_1 t}\hat{V}_1 e^{-i\hat{H}_1 t}, \tag{7.90}$$

where

$$\hat{H}_1 = \frac{1}{2}\Delta\hat{\sigma}_3, \qquad \hat{V}_1 = \frac{g}{2}\sigma_1. \tag{7.91}$$

The difference $\Delta = \omega - \tilde{\omega}$ between the qubit frequency and the external frequency is referred to as the *detuning*.

To solve Eq. (7.87), we define $\hat{O}_1(t) = e^{-i\hat{H}_1 t}\hat{O}(t)$. Then, Eq. (7.90) implies that

$$i\frac{d}{dt}\hat{O}_1(t) = \hat{H}_2\hat{O}_1(t), \tag{7.92}$$

where $\hat{H}_2 = \hat{H}_1 + \hat{V}_1$. The solution of Eq. (7.92) is $\hat{O}_1(t) = e^{-i\hat{H}_2 t}$, and by Eq. (7.36), we obtain

$$\hat{O}_1(t) = \cos(\Omega_R t)\hat{I} + i\frac{\sin(\Omega_R t)}{2\Omega_R}(g\hat{\sigma}_1 + \Delta\hat{\sigma}_3), \tag{7.93}$$

where

$$\Omega_R = \frac{1}{2}\sqrt{\Delta^2 + g^2} \tag{7.94}$$

is called the *Rabi* frequency.

The full evolution operator is $\hat{U}(t) = e^{-i\hat{H}_0 t}e^{-i\hat{H}_1 t}e^{-i\hat{H}_2 t}$. We assume that at time $t = 0$, the system is found on the ground state $|0\rangle$. The probability that the system is found on the excited state $|1\rangle$ at time t is $p(1, t) = |\langle 1|\hat{U}(t)|0\rangle|^2$. Since $\langle 1|\hat{U}(t)|0\rangle = e^{i\omega t + \frac{i\Delta}{2}t}\langle 1|e^{-i\hat{H}_2 t}|0\rangle$, we obtain

$$p(1, t) = \left|\langle 1|e^{-i\hat{H}_2 t}|0\rangle\right|^2 = \frac{g^2}{4\Omega_R^2}\sin^2(\Omega_R t), \tag{7.95}$$

where we used Eq. (7.36) for the last step.

The oscillations (7.95) of the excitation probability are remarkable, because the associated frequency, Ω_R, is close neither to the qubit's frequency nor to the frequency of the external perturbation. In general, Ω_R is much smaller than the former frequencies; the period may be sufficiently large to be macroscopically distinguishable. This phenomenon is known as *Rabi oscillations*. Both the detuning Δ and the coupling g depend on the external perturbation, and they can be controlled by the experimentalist. For an atomic qubit driven by an external laser, g is proportional to the electric field amplitude, that is a tunable parameter of the laser. Rabi oscillations are well attested experimentally; see, for example, Fig. 7.4.

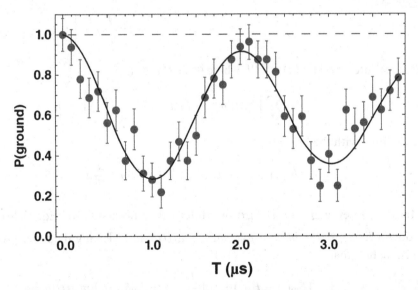

Fig. 7.4 Rabi oscillations for a small number of *Rb* atoms, between the ground state and a highly excited state. The plot shows the probability $p(0, T) = 1 - p(1, T)$ as a function of the duration T of the harmonic pulse. Each data point is the average of 40 single-atom experiments, with the bars showing one standard deviation. [Figure reprinted with permission from Johnson et al. (2008). Copyright 2008 by the American Physical Society.]

7.5   The Adiabatic Approximation

In this section, we analyze evolution under Hamiltonians that change slowly with time. This may happen because the Hamiltonian depends on some external parameters that vary at a time-scale much larger than the microscopic ones of quantum evolution. Examples of this behavior include the time evolution of atoms in a slowly varying external EM field and the time evolution of an EM field mode in a cavity with slowly moving boundaries. We show that such systems are often well described by the *adiabatic approximation*, to be explained shortly. We also present a remarkable topological effect that often appears in this context, Berry's geometric phase.

7.5.1   The Adiabatic Theorem

Suppose that the evolution of a quantum system is determined by the time-dependent Hamiltonian $\hat{H}(t)$. We define the *instantaneous eigenvectors* $|n,t\rangle$ and the instantaneous eigenvalues of $\hat{H}(t)$,

$$\hat{H}|n,t\rangle = E_n(t)|n,t\rangle. \tag{7.96}$$

For simplicity, we assume that $\hat{H}(t)$ has a purely discrete spectrum for all t.

We expand a solution $|\psi(t)\rangle$ to Schrödinger's equation

$$i\frac{d}{dt}|\psi(t)\rangle = \hat{H}|\psi(t)\rangle, \tag{7.97}$$

with respect to the basis of instantaneous eigenvectors, $|\psi(t)\rangle = \sum_n c_n(t)|n,t\rangle$. Equation (7.97) becomes

$$\sum_n \left[(\dot{c}_n(t) - iE_n(t))|n,t\rangle + c_n(t)\frac{d}{dt}|n,t\rangle \right] = 0. \tag{7.98}$$

We substitute $c_n(t) = b_n(t)e^{-i\theta_n(t)}$, where $\theta_n = \int_0^t ds E_n(s)$,

$$\sum_n \left[\dot{b}_n(t)|n,t\rangle + b_n(t)e^{-i\theta_n(t)}\frac{d}{dt}|n,t\rangle \right] = 0. \tag{7.99}$$

We multiply with the bra $\langle m,t|$, to obtain

$$\dot{b}_m(t) + \sum_n b_n(t)e^{-i[\theta_n(t)-\theta_m(t)]}\langle m,t|\frac{d}{dt}|n,t\rangle = 0. \tag{7.100}$$

In order to evaluate $\langle m,t|\frac{d}{dt}|n,t\rangle$, we differentiate both sides of Eq. (7.96) with respect to t, to obtain $\dot{\hat{H}}|n,t\rangle + \hat{H}\frac{d}{dt}|n,t\rangle = \dot{E}_n(t)|n,t\rangle + E_n(t)\frac{d}{dt}|n,t\rangle$. Taking the inner product with the bra $\langle m,t|$, we find that

$$[E_m(t) - E_n(t)]\langle m,t|\frac{d}{dt}|n,t\rangle = \dot{E}_n\delta_{mn} - \langle m,t|\dot{\hat{H}}|n,t\rangle. \tag{7.101}$$

For $m \neq n$, and assuming no degeneracy,

$$\langle m,t|\frac{d}{dt}|n,t\rangle = \frac{\langle m,t|\dot{\hat{H}}|n,t\rangle}{E_n(t) - E_m(t)}. \tag{7.102}$$

We substitute in Eq. (7.100), to obtain

$$\dot{b}_m(t) + \langle m, t | \frac{d}{dt} | m, t \rangle b_m(t) + \sum_{n \neq m} b_n(t) e^{-i[\theta_n(t) - \theta_m(t)]} \frac{\langle m, t | \dot{\hat{H}} | n, t \rangle}{E_n(t) - E_m(t)} = 0. \qquad (7.103)$$

In Box 7.2, we prove that for sufficiently slow change of the Hamiltonian with time, the last term in Eq. (18.6) can be ignored. This result is known as the *adiabatic theorem*. Hence, Eq. (7.103) becomes

$$\dot{b}_m(t) + \langle m, t | \frac{d}{dt} | m, t \rangle b_m(t) = 0. \qquad (7.104)$$

The quantity $\langle m, t | \frac{d}{dt} | m, t \rangle$ is purely imaginary, because the differentiation of the normalization condition $\langle m, t | m, t \rangle = 1$ gives $\text{Re} \langle m, t | \frac{d}{dt} | m, t \rangle = 0$. Hence, we write

$$\langle m, t | \frac{d}{dt} | m, t \rangle = -i R_m(t). \qquad (7.105)$$

It follows that Eq. (7.103) is solved approximately by

$$b_m(t) = b_m(0) e^{i \int_0^t ds R_m(s)}. \qquad (7.106)$$

The general solution to Schrödinger's equation is

$$|\psi(t)\rangle = \sum_n c_n(0) e^{i \int_0^t ds [R_n(s) - E_n(s)]} |n, t\rangle. \qquad (7.107)$$

If the system is initially in an instantaneous eigenstate $|n, t\rangle$ of the Hamiltonian, it will remain on the same instantaneous eigenstate, with an additional phase factor

$$\phi_n(t) = \int ds \langle n, s | i \frac{d}{ds} - \hat{H}(s) | n, s \rangle = \gamma_n(t) - \chi_n(t), \qquad (7.108)$$

where $\gamma_n(t) = \int ds \langle n, s | i \frac{d}{ds} | n, s \rangle$ is the kinematical contribution and $\chi_n(t) = \int ds E_n(s)$ the dynamical contribution to the phase.

Box 7.2 Proof of the Adiabatic Theorem

We implement the idea that the Hamiltonian $\hat{H}$ changes slowly in time by assuming that $\hat{H}$ depends on time t only through the combination $\tau := t/\mu$, where μ is a very large dimensionless number. The instantaneous eigenvectors $|n, t\rangle$ and the instantaneous eigenvalues $E_n(t)$ will also depend on τ. We want to find the limit of Eq. (7.103) as $\mu \to \infty$. To this end, we first express Eq. (7.103) as

$$\frac{d}{d\tau} b_m(\tau) + \langle m, \tau | \frac{d}{d\tau} | m, \tau \rangle b_m(\tau) + \sum_{n \neq m} b_n(\tau) e^{-i\mu[\theta_n(\tau) - \theta_m(\tau)]} \frac{\langle m, \tau | \dot{\hat{H}}(\tau) | n, \tau \rangle}{E_n(\tau) - E_m(\tau)} = 0.$$

We integrate all terms in the above equation, to obtain

$$b_m(\tau) - b_m(0) + \int_0^\tau ds \langle m, s | \frac{d}{ds} | m, s \rangle b_m(s)$$
$$+ \sum_{n \neq m} \int_0^\tau ds \, b_n(s) e^{-i\mu[\theta_n(s) - \theta_m(s)]} \frac{\langle m, s | \dot{\hat{H}}(s) | n, s \rangle}{E_n(s) - E_m(s)} = 0. \qquad (A)$$

We assume that there is a gap ϵ_0 between eigenvalues of $\hat{H}$, $|E_n(t) - E_m(t)| > \epsilon_0$ for all n, m with $n \neq m$, and for all times t. This means that $\theta_n(s) - \theta_m(s)$ is always either positive or negative. Furthermore, $|\theta_n(s) - \theta_m(s)| \geq \epsilon_0 s$, which implies that the phase in the last term of Eq. (A) oscillates with a frequency that is always larger than ϵ_0. Hence, the last term of Eq. (A) is of the form $I(\tau) = \int_0^\tau ds e^{\pm i\mu a(s)} f(s)$, where $a(s)$ is a positive increasing function and $f(s)$ is a continuous function.

We integrate $I(\tau)$ by part, to obtain

$$I(\tau) = \frac{1}{\pm i\mu} \left(e^{\pm i\mu a(s)} \frac{f(s)}{a(s)} \big|_0^\tau - \int_0^\tau ds e^{\pm i\mu a(s)} g(s) \right), \tag{B}$$

where $g(s) = \frac{d}{ds} \frac{f(s)}{a(s)}$. The integral in Eq. (B) is of the same form with $I(\tau)$. Again, integration by parts gives a term proportional to μ^{-1} and another integral of the same type with $I(\tau)$ multiplied by μ^{-1}. We conclude for large μ, $I(\tau)$ is proportional to μ^{-1}, and it vanishes as $\mu \to \infty$.

We proved the adiabatic theorem by showing that the last term in Eq. (A) is negligible for sufficiently weak time variations of the Hamiltonian. In the proof, we assumed a Hamiltonian with a discrete spectrum, no degeneracy, and subject to the existence of an eigenvalue gap ϵ_0, as in the original proof by Born and Fock (1928). The theorem has been generalized to Hamiltonians with degenerate eigenvalues (Kato, 1950), and without assuming the gap condition (Avron and Elgart, 1999).

7.5.2 Geometric Phase

Quantum systems are often characterized by some external parameters $a_1, a_2, \ldots, a_n$ that can be changed in a controlled way. For example, the parameters a_i may correspond to the components of an external electric or magnetic field, or to the dimensions of a box that encloses the system. Any possible value of $a_1, a_2, \ldots, a_n$ defines an element of a set Γ, and it will be denoted by a vector $\mathbf{a}$.

In such a system, the Hamiltonian $\hat{H}(\mathbf{a})$ is a function of $\mathbf{a}$, and so are its eigenvalues $E_n(\mathbf{a})$ and eigenvectors $|n, \mathbf{a}\rangle$. Suppose we change the values of the external parameters in time, thereby defining a continuous path $\mathbf{a}(t)$ on Γ. For a sufficiently slow change, the adiabatic theorem holds. Let the initial state be an eigenvector of the initial Hamiltonian $\hat{H}[\mathbf{a}(0)]$. We express the quantum state at time t as $|\mathbf{a}(t)\rangle$, dropping the index n in order to unclutter the notation.

We also write $R(t) = i\langle \mathbf{a}(t)| \frac{d}{dt} |\mathbf{a}(t)\rangle = \sum_i A_i[\mathbf{a}(t)] da_i/dt$, where

$$A_i(\mathbf{a}) = i\langle \mathbf{a}| \frac{\partial}{\partial a_i} |\mathbf{a}\rangle \tag{7.109}$$

is a vector field on Γ, known as the *Berry connection*.

Let the path $\mathbf{a}(t)$ be closed, that is, $\mathbf{a}(T) = \mathbf{a}(0)$ for some time T. Then, the kinematical contribution to the phase

$$\gamma = \sum_i \int_0^T dt A_i[\mathbf{a}(t)] \frac{da_i(t)}{dt} = \oint \mathbf{A} \cdot d\mathbf{a}, \tag{7.110}$$

depends only on the points of the path, and not on the explicit dependence of the path on time t. For this reason, it is called *geometric phase* (Berry, 1982).

The geometric phase vanishes, if $A_i = \partial f / \partial a_i$ for any scalar function f on Γ. Hence, the gauge transformation,

$$A_i \rightarrow A_i + \frac{\partial f}{\partial a_i}, \tag{7.111}$$

analogous to Eq. (1.24) for the EM field, does not change the geometric phase.

Example 7.4 Consider a qubit with a time-dependent Hamiltonian $\hat{H} = -\mathbf{a}(t) \cdot \boldsymbol{\sigma}$. Since the vector $\mathbf{a}$ has three components, Γ coincides with $\mathbb{R}^3$. By Eq. (5.15), the instantaneous eigenvectors $|\mathbf{a}\rangle$ of minimum energy are of the form

$$|\mathbf{a}\rangle = \frac{1}{\sqrt{2|\mathbf{a}|(|\mathbf{a}| - a_3)}} \begin{pmatrix} a_1 - ia_2 \\ |\mathbf{a}| - a_3 \end{pmatrix}. \tag{7.112}$$

It is convenient to employ the usual polar coordinates (a, θ, ϕ), where $a_1 = a \sin \theta \cos \phi$, $a_2 = a \sin \theta \sin \phi$, $a_3 = a \cos \theta$, in order to write

$$|\mathbf{a}\rangle = \begin{pmatrix} \cos \frac{\theta}{2} e^{-i\phi} \\ \sin \frac{\theta}{2} \end{pmatrix}. \tag{7.113}$$

The vectors $|\mathbf{a}\rangle$ do not depend on $a = |\mathbf{a}|$.

We differentiate,

$$d|\mathbf{a}\rangle = \begin{pmatrix} \left(-\frac{1}{2} \sin \frac{\theta}{2} d\theta - i \cos \frac{\theta}{2} d\phi \right) e^{-i\phi} \\ \frac{1}{2} \cos \frac{\theta}{2} d\theta \end{pmatrix}. \tag{7.114}$$

By Eq. (7.109), the inner product

$$\langle \mathbf{a} | d | \mathbf{a} \rangle = \cos \frac{\theta}{2} \left(-\frac{1}{2} \sin \frac{\theta}{2} d\theta - i \cos \frac{\theta}{2} d\phi \right) + \frac{1}{2} \sin \frac{\theta}{2} \cos \frac{\theta}{2} d\theta$$

$$= -\frac{i}{2}(1 + \cos \theta)d\phi, \tag{7.115}$$

equals $-iA_\theta d\theta - iA_\phi d\phi$; hence,

$$A_\theta = 0, \qquad A_\phi = \frac{1}{2}(1 + \cos \theta). \tag{7.116}$$

We can subtract the constant term $\frac{1}{2}$ from A_ϕ, through a gauge transformation, so that $A_\phi = \frac{1}{2} \cos \theta$.

By Stokes' theorem, the geometric phase $\gamma(C) = \frac{1}{2} \oint_C \cos \theta d\phi$ for a closed path C on the unit sphere equals $-\frac{1}{2} \int_S \sin \theta d\theta d\phi$, where S is the region of the sphere bounded by C. The integral $\int_S \sin \theta d\theta d\phi$ equals the area of S, or equivalently the solid angle $\Omega(C)$ of C with respect to the center of the sphere. We conclude that

$$\gamma(C) = -\frac{1}{2}\Omega(C). \tag{7.117}$$

For the instantaneous eigenvectors of higher energy, the geometric phase in a path C is similarly evaluated to $\gamma(C) = \frac{1}{2}\Omega(C)$.

QUESTIONS

7.1 What is the analogue of the Schrödinger and Heisenberg pictures in classical mechanics?

7.2 "Surely Pauli's theorem is flawed. Consider a particle in a homogeneous gravitational field g with a Hamiltonian $\hat{H} = \frac{\hat{p}^2}{2m} + mg\hat{x}$. If we define $\hat{T} = -\frac{\hat{p}}{mg}$, then $[\hat{T}, \hat{H}] = i\hat{I}$, and $\hat{T}$ is a proper time operator." What is the problem with this argument?

7.3 "The time–energy uncertainty relation implies that, for very small times, energy is not conserved in quantum theory." Is this statement accurate? How would you explain the time–energy uncertainty relation to an audience of nonphysicists?

7.4 Does quantum tunneling violate energy conservation?

7.5 State precisely the definition of the time parameters that appear in the different applications of the time–energy uncertainty relation in Section 7.2.3.

7.6 **(B)** Consider Example 7.3. An alternative criterion for distinguishing the action of a weak external force is that the momentum shift on the oscillator is larger than the mean deviation of momentum of the ground state. What is the analogue of Eq. (7.84) for an impulsive force?

7.7 **(B)** How does the geometric phase change if the Hamiltonian $\hat{H}(t)$ is substituted by $f(t)\hat{H}(t)$ for some function f?

PROBLEMS

7.1 A qubit with Hamiltonian $\hat{H} = \omega\hat{\sigma}_3$, with $\omega > 0$, is prepared ($t = 0$) in an initial state $\hat{\rho}_0$ with Bloch vector $\mathbf{r}_0$. Write the probabilities for the outcomes in a measurement of σ_1 at time t.

7.2 Consider a quantum system described by the Hilbert space $\mathbb{C}^3$. The Hamiltonian has eigenvectors $|1\rangle, |2\rangle, |3\rangle$, with associated eigenvalues $E_1 = 0, E_2 = \omega, E_3 = 3\omega$, where $\omega > 0$. The system is initially prepared ($t = 0$) in a state

$$\hat{\rho}_0 = \frac{2}{3}|\alpha_+\rangle\langle\alpha_+| + \frac{1}{3}|\alpha_-\rangle\langle\alpha_-|,$$

where $|\alpha_\pm\rangle = \frac{1}{2}(|1\rangle \pm \sqrt{2}|2\rangle + |3\rangle)$. After time t, we measure the observable $\hat{A} = k|\alpha_+\rangle\langle a_+| - k|\alpha_-\rangle\langle a_-|$, where $k > 0$. Evaluate the probabilities for all outcomes of the measurement.

7.3 Take the position $\hat{x}$ of a free particle as a "clock" variable. Use the evolved wave function (7.48), in order to evaluate the time uncertainty δt according to Mandelhstam and Tamm. Then, calculate $\Delta H \delta t$.

7.4 A quantum clock is described by the Hilbert space $\mathbb{C}^N$. The clock's Hamiltonian $\hat{H}$ has N eigenstates $|n\rangle$, such that $\hat{H}|n\rangle = n\omega, n = 0, 1, \ldots, N - 1$. (i) Show that the vectors

$$|v\rangle = \frac{1}{\sqrt{N}} \sum_{n=0}^{N-1} e^{-i\frac{2\pi}{N}vn}|n\rangle,$$

with $v = 0, 1, \ldots, N - 1$, form an orthonormal basis. (ii) Show that $e^{-i\hat{H}\tau}|v\rangle = |v + 1\rangle$ for $\tau = \frac{2\pi}{N\omega}$. Hence, the vectors $|v\rangle$ act as discrete hands of a clock with accuracy equal to τ. What happens when the state with $v = N - 1$ evolves by one time step? (iii) We define the time operator $\hat{T} = \tau \sum_{v=0}^{N-1} v|v\rangle\langle v|$. Evaluate $[\hat{H}, \hat{T}]$. (iv) Evaluate $(\Delta H)\tau$ for a hand eigenstate $|v\rangle$. (v) Evolve a hand state $|v\rangle$ with $e^{-i\hat{H}\tau\xi}$, where $\xi \in (0,1)$. What is the probability distribution for time $\hat{T}$ in the resulting state?

7.5 Prove inequality (7.31).

7.6 (i) Prove that the equality in Eq. (7.30) is possible for states $|\psi\rangle = \frac{1}{\sqrt{2}}(|0\rangle + e^{i\theta}|1\rangle)$, where $|0\rangle$ the ground state and $|1\rangle$ the first excited state. (ii) Prove the inequality $-\frac{\pi^2}{4}(1 - \cos x) + x \sin x + 2x^2 \geq 0$, in order to show that $a(t) \geq 1 + \frac{4t}{\pi^2}\dot{a} - \frac{4t^2}{\pi^2}(\Delta H)^2$, where $a(t)$ is given by Eq. (7.24). (iii) Use this result in order to to prove Eq. (7.29), and show that equality in Eq. (7.29) is obtained only for the state $|\psi\rangle$.

7.7 Evaluate the propagator for a free particle of mass m in the half-line.

7.8 Evaluate the propagator for a particle in one dimension, with Hamiltonian $\hat{H} = v|\hat{p}|$, where $v > 0$ is the particle's transmission speed.

7.9 Consider a particle of mass m in one dimension with Hamiltonian $\hat{H} = \frac{\hat{p}^2}{m} + V(\hat{x})$. (i) Show that

$$\frac{d(\Delta x)^2}{dt} = \frac{C_{xp}}{m}, \quad \frac{d(\Delta p)^2}{dt} = -\text{Cor}[\hat{p}, V'(\hat{x})], \quad \frac{dC_{xp}}{dt} = \frac{(\Delta p)^2}{m} - \text{Cor}[\hat{x}, V'(\hat{x})].$$

(ii) Show that, in the regime where Eq. (7.17) applies,

$$\text{Cor}[\hat{p}, V'(\hat{x})] \simeq V''(\langle\bar{x}\rangle)C_{xp}, \quad \text{Cor}[\hat{x}, V'(\hat{x})] \simeq V''(\langle\hat{x}\rangle)(\Delta x)^2.$$

Hence, the quantities $\langle\bar{x}\rangle, \langle\bar{p}\rangle, (\Delta x)^2, (\Delta p)^2$, and C_{xp} satisfy an autonomous set of equations. Using these equations for the study of a quantum system is known as the Gaussian approximation. (iii) Show that these equations are exact for a quadratic potential, and find their general solution.

7.10 Write the wave function (7.56) for $x < a$. For an almost monochromatic initial state with momentum p_0, show that $\text{Prob}(x < a, t) = |R_{p_0}|^2$ as $t \to \infty$.

7.11 (i) Prove Eq. (7.65). (ii) Evaluate the delay time for a particle tunneling through the delta potential.

7.12 Consider scattering by the double barrier potential, described in Section 5.3.5. The phase expansion approximation does not work well for the transmission amplitude T_{tot} of Eq. (5.68), because of the oscillations in the denominator. A different approach is needed. (i) Expand T_{tot} as a geometric series and apply the phase expansion approximation to each term of the series. You will obtain $\psi_t(x)$ as a sum of terms of the form (7.58), each corresponding to the partial amplitude $T_k^{(n)}$. (ii) If the time scale of observation is much larger than the time between successive attempts of the particle to cross the barrier, you can change the sum over n to an integral, and treat the initial wave function $\psi_0(x)$ as proportional to a delta function. Show that the detection probability at x is of the form $Ce^{-\Gamma(t-t_0)}$ for $t > t_0$, and evaluate the constants C, Γ, and t_0. How do you interpret this result?

7.13 **(B)** Derive the first three terms in the Magnus expansion, Eq. (7.76).

7.14 **(B)** (i) Evaluate the probability $P(t) = |\langle 0|\hat{U}(t)|0\rangle|^2$ that a harmonic oscillator of mass m and frequency ω remains in the ground state $|0\rangle$ at time t, while being acted by a periodic force $F(t) = g\sin\omega_0 t$, for some constant g and frequency ω_0. (ii) Suppose that the external force lasts time T, such that $\omega_0 T = 2\pi N$, where $N >> 1$ is an integer. Use the criterion (7.83) in order to identify the smallest value of g that can be distinguished by a harmonic oscillator probe.

7.15 **(B)** Find the time-evolution operator for a qubit with Hamiltonian (7.85), in the adiabatic approximation.

7.16 **(B)** The distance between the walls of a box is a function of time $L(t) = L_0 + a\sin(\omega t)$, $0 < a < L_0$. Consider a particle inside the box, initially prepared in the state $\frac{1}{\sqrt{2}}(|1\rangle + |2\rangle)$, where $|n\rangle$ are instantaneous eigenvectors of the Hamiltonian at time $t = 0$. Find the state at time $t = \frac{2\pi}{\omega}$ in the adiabatic approximation.

7.17 **(B)** A qubit evolves under the unitary operator $\hat{U}(t) = \cos(\omega t)\hat{I} - i\sin(\omega t)\mathbf{n}(t) \cdot \hat{\boldsymbol{\sigma}}$, where $\mathbf{n}(t) = \left(\cos\frac{2\pi t}{\tau}, \sin\frac{2\pi t}{\tau}, 0\right)$. (i) Evaluate the corresponding Hamiltonian and find the instantaneous eigenstates. (ii) Consider the adiabatic regime at $\omega\tau >> 1$. Assuming that the system starts from an instantaneous eigenstate, find the norm distance between the exact and the adiabatic evolving vector at time t. How large can it become?

7.18 **(B)** Assume that a particle in one dimension evolves by a time-dependent Hamiltonian that forces it to evolve along a path of coherent states $|z(t)\rangle$. Compute the associated Berry connection. Identify the adiabatic regime in the evolution of a forced harmonic oscillator, and confirm the value of the geometric phase.

Bibliography

- For time in quantum theory, see Zeh (2009) and Muga et al. (2008). I also recommend the elementary account of Isham and Savvidou (2002).
- For the time–energy uncertainty relation, see Busch (2008). For superselection rules, see Giulini (2009). For tunneling times, see Hauge and Stonveng (1989) and Landauer and Martin (1994).
- For the adiabatic theorem, see Sarandy et al. (2004). See also section 2 of Albash and Lidar (2018) for a summary of its different versions. For the geometric phase and its applications, see Anandan (1992) and Cohen et al. (2019).

8   Quantum State Reduction

… the quantum-mechanical equations of motion do not describe the measurement process; they only help in the calculation of the probabilities of the different outcomes. These probabilities form the real content of quantum-mechanical theory. The formalism of state vectors, equations of motion, etc., are only means to calculate these probabilities. The observation results are the true "reality" which underlie quantum mechanics. The state vector does not represent "reality." It is a calculational tool.

Wigner (1976)

8.1   Classical "State Reduction"

Many experiments require the execution of several measurements in a microscopic system. For example, consider a particle A decaying into a pair of particles B and C that move in different directions. Each product particle is detected by a different apparatus at different moments of time. We need a rule that tells us how to incorporate information obtained from the first measurement in order to predict the outcome of the second.

First, we examine the analogous problem in classical probability theory. Consider a system described by a state space Γ. We prepare an ensemble of such systems, described by the probability density $p(x)$, where $x \in \Gamma$. We perform two measurements in each element of the ensemble. The first measurement corresponds to a function F on Γ and the second measurement to a function G on Γ. Let C_λ be the subset of Γ that corresponds to the value λ of F, $(F(x) = \lambda)$ and let D_μ be the subset of Γ that corresponds to the value μ of G $(G(x) = \mu)$.

When F is measured, a fraction

$$\text{Prob}(\lambda) = \int_{C_\lambda} p(x)dx = \int dx \chi_{C_\lambda}(x) p(x) \tag{8.1}$$

of the statistical ensemble will be found with a value λ. The joint probability $\text{Prob}(\lambda, \mu)$ to find λ in the first measurement and μ in the second measurement is identical to the probability of finding the system in the intersection $C_\lambda \cap D_\mu$, that is,

$$\text{Prob}(\lambda, \mu) = \int_{C_\lambda \cap D_\mu} dx p(x) = \int dx \chi_{C_\lambda \cap D_\mu}(x) = \int dx \chi_{D_\mu}(x) \left[\chi_{C_\lambda}(x) p(x) \right]. \tag{8.2}$$

The conditional probability $\text{Prob}(\mu | \lambda)$ of measuring μ in the second measurement provided we measured λ in the first measurement is

$$\mathrm{Prob}(\mu|\lambda) = \frac{\mathrm{Prob}(\lambda,\mu)}{\mathrm{Prob}(\lambda)} = \int dx \chi_{D_\mu}(x) \frac{\chi_{C_\lambda}(x)p(x)}{\mathrm{Prob}(\lambda)} = \int dx p(x|\lambda) \chi_{D_\mu}(x), \qquad (8.3)$$

where $p(x|\lambda) = \chi_{C_\lambda}(x)p(x)/\mathrm{Prob}(\lambda)$.

We conclude that in classical probability theory, we incorporate information obtained from measurements through the following rule.

Classical State Reduction Rule Suppose that we measure the observable F in a statistical ensemble described by the probability distribution $p(x)$. After the measurement, we describe the subensemble of all systems with $F = \lambda$ by the probability distribution

$$p(x|\lambda) = \frac{\chi_{C_\lambda}(x)p(x)}{\mathrm{Prob}(\lambda)}. \qquad (8.4)$$

8.2   Reduction of Quantum States

In this section, we generalize the classical reduction rule for quantum systems, based on the analogy between classical and quantum measurements that was presented in Section 4.8. We first define reduction for pure states, then consider mixed states. We identify the most pertinent difference between quantum and classical reduction, namely, that in quantum theory we cannot "forget" the fact that a measurement took place; the state of the total ensemble genuinely changes after a measurement. Finally, we discuss quantum state reduction in qubits.

8.2.1   Reduction of Pure States

We proceed to generalize the classical reduction rule for quantum systems. Keeping in mind the analogy between classical and quantum measurements in Section 4.8, we consider the simple case of a statistical ensemble of particles described by a normalized wave function $\psi(x)$. Suppose we measure position, and we want to carry out further measurements in all particles that were found with $x \in C = [a,b]$. The natural thing to do is to describe further measurements in this subensemble by a reduced wave function $\psi(x;C)$ that has shred all information outside C, that is, by $\chi_C(x)\psi(x)$. To normalize, we have to divide by $[\int dx \chi_C(x)|\psi(x)|^2]^{1/2}$; hence,

$$\psi(x;C) = \frac{\chi_C(x)\psi(x)}{\sqrt{\int dx \chi_C(x)|\psi(x)|^2}}. \qquad (8.5)$$

Equivalently, we can use the reduced ket

$$|\psi;C\rangle = \frac{\chi_C(\hat{x})|\psi\rangle}{\sqrt{\langle\psi|\chi_C(\hat{x})|\psi\rangle}} = \frac{\hat{P}_C|\psi\rangle}{\sqrt{\langle\psi|\hat{P}_C|\psi\rangle}}, \qquad (8.6)$$

where $\hat{P}_C = \chi_C(\hat{x})$.

This definition of a reduced ket generalizes to measurements of arbitrary operators, and it defines the *quantum state reduction* rule.

FP5 Let the measurement of $\hat{A} = \sum_n a_n \hat{P}_n$ be carried out in a statistical ensemble that is described by a state vector $|\psi\rangle$. After the measurement, the statistical subensemble in which the value a_n was obtained, is described by the state vector

$$|\psi; a_n\rangle = \frac{\hat{P}_n |\psi\rangle}{\sqrt{\langle\psi|\hat{P}_n|\psi\rangle}}. \tag{8.7}$$

For one-dimensional spectral projectors $\hat{P}_n = |n\rangle\langle n|$, where $|n\rangle$ is an orthonormal basis, the state reduction rule simplifies

$$|\psi\rangle \to |\psi, a_n\rangle = \frac{\langle n|\psi\rangle}{|\langle n|\psi\rangle|}|n\rangle, \tag{8.8}$$

that is, if a value a_n is obtained the subensemble is described by the corresponding eigenvector of $\hat{A}$.

Suppose now we carry a measurement of $\hat{B} = \sum_m b_m \hat{Q}_m$ after the measurement of $\hat{A}$ in an ensemble described by the state $|\psi\rangle$. The *conditional probability* of measuring b_m in the second measurement, provided that a_n was measured in the first, is

$$\text{Prob}(b_m|a_n) = \langle\psi; a_n|\hat{Q}_m|\psi; a_n\rangle = \frac{\langle\psi|\hat{P}_n\hat{Q}_m\hat{P}_n|\psi\rangle}{\langle\psi|\hat{P}_n|\hat{\psi}\rangle}. \tag{8.9}$$

The *joint probability* of finding a_n in the first measurement and then finding b_m in the second measurement is

$$\text{Prob}(a_n, b_m) = \text{Prob}(b_m|a_n) \cdot \text{Prob}(a_n) = \langle\psi|\hat{P}_n\hat{Q}_m\hat{P}_n|\psi\rangle. \tag{8.10}$$

We note that the order of the measurements affects joint probabilities:

$$\text{Prob}(a_n, b_m) \neq \text{Prob}(b_m, a_n) \tag{8.11}$$

unless $[\hat{P}_n, \hat{Q}_m] = 0$ for all n and m, or, equivalently, if $[\hat{A}, \hat{B}] = 0$. In that case,

$$\text{Prob}(a_n, b_m) = \langle\psi|\hat{P}_n\hat{Q}_m|\psi\rangle. \tag{8.12}$$

Equation (8.11) is to be contrasted with the classical joint probability (8.2) that does not depend on the order of the two measurements. The quantum state reduction rule introduces a direction of time in quantum mechanics. To quote Landau and Lifshitz (1977):

The measuring process in quantum mechanics has a "two-faced" character: it plays different parts with respect to the past and future of the electron. With respect to the past, it "verifies" the probabilities of the various possible results predicted from the state brought about by the previous measurement. With respect to the future, it brings about a new state. Thus the very nature of the process of measurement involves a far-reaching principle of irreversibility.

This irreversibility is of fundamental significance... . The basic equations of quantum mechanics are in themselves symmetrical with respect to a change in the sign of the time; here quantum mechanics does not differ from classical mechanics. The irreversibility of the process of measurement, however, causes the two directions of time to be physically non-equivalent, i.e. creates a difference between the future and the past.

8.2.2 Reduction of General States

FP5 easily generalizes for a nonpure initial state $\hat{\rho}$. Since any density matrix $\hat{\rho}$ can be analyzed into eigenvectors $|\psi_i\rangle$, as $\hat{\rho} = \sum_i w_i |\psi_i\rangle\langle\psi_i|$, we simply apply FP5 to each vector $|\psi_i\rangle$. We obtain the following.

FP5b If a statistical ensemble is described by a density matrix $\hat{\rho}$, the statistical subensemble in which the measurement of $\hat{A}$ led to value a_n is described by the density matrix

$$\hat{\rho}(a_n) = \frac{\hat{P}_n \hat{\rho} \hat{P}_n}{\text{Tr}(\hat{\rho}\hat{P}_n)}.$$

The conditional probability $\text{Prob}(a_n|b_m)$ of finding b_m in the second measurement, provided that a_n was found in the first measurement, is

$$\text{Prob}(b_m|a_n) = \frac{\text{Tr}(\hat{Q}_m \hat{P}_n \hat{\rho} \hat{P}_n)}{\text{Tr}(\hat{\rho}\hat{P}_n)}, \tag{8.13}$$

while the joint probability $\text{Prob}(a_n, b_m)$ takes the form

$$\text{Prob}(a_n, b_m) = \text{Tr}(\hat{Q}_m \hat{P}_n \hat{\rho} \hat{P}_n) = \text{Tr}(\hat{P}_n \hat{Q}_m \hat{P}_n \hat{\rho}). \tag{8.14}$$

Again, probabilities depend on the ordering of the measurements, unless $[\hat{A}, \hat{B}] = 0$, in which case,

$$\text{Prob}(a_n, b_m) = \text{Tr}(\hat{P}_n \hat{Q}_m \hat{\rho}) \tag{8.15}$$

Suppose that the system is described by a Hamiltonian $\hat{H}$. Let the first measurement take place at time t_1 and the second measurement at time t_2. To incorporate time evolution into Eq. (8.15), we evolve each state $\hat{\rho}(a_n)$ that is obtained after the first measurement as $e^{-i\hat{H}(t_2-t_1)} \hat{\rho}(a_n) e^{i\hat{H}(t_2-t_1)}$. Then,

$$\text{Prob}(a_n, t_1; b_m, t_2) = \text{Tr}\left(\hat{Q}_m e^{-i\hat{H}(t_2-t_1)} \hat{P}_n e^{-i\hat{H}t_1} \hat{\rho} e^{i\hat{H}t_1} \hat{P}_n e^{i\hat{H}(t_2-t_1)} \right). \tag{8.16}$$

The strong dependence of Eq. (8.16) at the time t_1 of the first measurement suggests that measurement is viewed as an almost instantaneous process in quantum theory. However, there are many measurements that involve a long interaction of the quantum system with the apparatus. For example, particles may spend long times in the arms of a very large interferometer before they are detected. When should we apply the reduction rule?

Following a suggestion of Wheeler (1978), many experiments have been carried out that involve a *delayed choice* of the observable that is being measured, that is, the observable is selected long after the particle has entered an interferometer. The result is unambiguous – see, for example, Jacques et al. (2007) – and it is also conceptually simple. *A measurement occurs when a macroscopic record appears on the detector.* What happened before the emergence of this record should not be viewed as part of the measurement *per se*, but part of the time evolution of the quantum system. Hence, the state reduction rule is applied at the moment that we obtain information from the quantum system.

Finally, we note that FP5 associates states to statistical ensembles. If we assume that the state refers to individual systems and not only to ensembles, we should rephrase as follows.

> **FP5c** If a system is described by a density matrix $\hat{\rho}$, and the measurement of $\hat{A}$ led to value a_n, then the system after measurement is described by the density matrix
>
> $$\hat{\rho}(a_n) = \frac{\hat{P}_n \hat{\rho} \hat{P}_n}{\mathrm{Tr}(\hat{\rho} \hat{P}_n)}.$$

8.2.3   The Difference Between Classical and Quantum State Reduction

Based on the preceding analysis, one might claim that quantum state reduction defines the notion of conditional probability in quantum theory. However, there is a crucial difference between classical and quantum state reduction, regarding the description of a statistical ensemble after a measurement.

Consider a measurement of the observable F in a classical statistical ensemble described by a probability density $p(x)$. Possible measurement outcomes correspond to a statistical subensemble described by the probability density $p(x|\lambda)$ given by Eq. (8.4). Suppose we mix back all subensembles created by the measurement in order to reconstruct the original ensemble; the subensemble of outcome λ will have a weight equal to $\mathrm{Prob}(\lambda)$ in the mixture. The resulting ensemble is described by a probability density

$$p'(x) = \sum_\lambda \mathrm{Prob}(\lambda) p_\lambda(x) = \sum_\lambda \chi_{C_\lambda}(x) p(x) = 1 \cdot p(x) = p(x). \tag{8.17}$$

Hence, the probability density of the statistical ensemble is unaffected.

Let us apply the same rationale in a statistical ensemble of quantum systems, described by a density matrix $\hat{\rho}$. Any possible outcome a_n of a measurement corresponds to a statistical subensemble with statistical weight $\mathrm{Prob}(a_n)$, described by a density matrix

$$\hat{\rho}(a_n) = \frac{\hat{P}_n \hat{\rho} \hat{P}_n}{\mathrm{Prob}(a_n)}. \tag{8.18}$$

After the measurement, the full statistical ensemble is described by the density matrix

$$\hat{\rho}' = \sum_n \mathrm{Prob}(a_n) \hat{\rho}(a_n) = \sum_n \hat{P}_n \hat{\rho} \hat{P}_n, \tag{8.19}$$

which in general differs from the initial density matrix $\hat{\rho}$. In contrast to classical probability theory, the incorporation of information about measurement outcomes changes the state.

This means the following. In classical probability theory, we cannot distinguish between our ignorance for the outcome of a measurement and the absence of a measurement. In quantum theory this is not the case: Even if we have full ignorance about the outcome of a measurement, the results of any future measurement *depend on the fact that a measurement took place.*

> The quantum state does not describe only our knowledge about a physical system, because it changes in measurements whether we know the measurement outcomes or not.

One might wonder whether the change of the quantum state is due to the interaction of the measured system with the apparatus. The answer is negative. An interaction with the apparatus would lead to unitary evolution of the quantum state according to Schrödinger's equation: $\hat{\rho} \to \hat{\rho}' = \hat{U}\hat{\rho}\hat{U}^\dagger$, for some unitary operator $\hat{U}$, which is very different from the change of the state described by Eq. (8.19).

8.2.4   Successive Measurements on a Qubit

Suppose we perform two successive measurements on a qubit. Let the first correspond to a self-adjoint operator $\hat{A} = \hat{\sigma}_1$ and the second to a self-adjoint operator $\hat{B} = \hat{\sigma}_3$. According to Eq. (5.16), the spectral projectors $\hat{P}_\pm$ of $\hat{A}$ and $\hat{Q}_\pm$ of $\hat{B}$ are

$$\hat{P}_+ = \frac{1}{2}\begin{pmatrix} 1 & 1 \\ 1 & 1 \end{pmatrix}, \quad \hat{P}_- = \frac{1}{2}\begin{pmatrix} 1 & -1 \\ -1 & 1 \end{pmatrix}, \quad \hat{Q}_+ = \begin{pmatrix} 1 & 0 \\ 0 & 0 \end{pmatrix}, \quad \hat{Q}_- = \begin{pmatrix} 0 & 0 \\ 0 & 1 \end{pmatrix}.$$

Let us assume an initial state $|\psi = |1\rangle$. By Eq. (8.14), the joint probabilities $p_{ab} := \text{Prob}(a, b)$ to those measurement outcomes are

$$p_{++} = \langle 1|\hat{P}_+\hat{Q}_+\hat{P}_+|1\rangle = \frac{1}{4}, \quad p_{+-} = \langle 1|\hat{P}_+\hat{Q}_-\hat{P}_+|1\rangle = \frac{1}{4},$$

$$p_{-+} = \langle 1|\hat{P}_-\hat{Q}_+\hat{P}_-|1\rangle = \frac{1}{4}, \quad p_{--} = \langle 1|\hat{P}_-\hat{Q}_-\hat{P}_-|1\rangle = \frac{1}{4}.$$

The probability of finding $+1$ in the second measurement, irrespective of what happened in the first, is $p_{++} + p_{-+} = \frac{1}{4} + \frac{1}{4} = \frac{1}{2}$. In contrast, if there was no first measurement, the probability of measuring $b = +1$ would be $\langle 1|\hat{Q}_+|1\rangle = 1$. We confirm our earlier statement, namely, that ignorance about the outcome of a measurement is not equivalent to the absence of a measurement.

It is elementary to show that the joint probabilities change if the order of the measurement changes. If we measure first $\hat{B}$ and then $\hat{A}$, we find probabilities $\bar{p}_{ba} := \text{Prob}(b, a)$

$$\bar{p}_{++} = \langle 1|\hat{Q}_+\hat{P}_+\hat{Q}_+|1\rangle = \frac{1}{2}, \quad \bar{p}_{+-} = \langle 1|\hat{Q}_+\hat{P}_-\hat{Q}_+|1\rangle = \frac{1}{2},$$

$$\bar{p}_{-+} = \langle 1|\hat{Q}_-\hat{P}_+\hat{Q}_-|1\rangle = 0, \quad \bar{p}_{--} = \langle 1|\hat{Q}_-\hat{P}_-\hat{Q}_-|1\rangle = 0.$$

8.3   Interpretation of Quantum State Reduction

We proceed to discuss the meaning of the quantum state reduction rule, and the challenges that it poses for the interpretation of quantum theory. The key point is that in the minimal interpretation, state reduction is a mathematical procedure for updating probabilities after measurements and *not* a physical process. The treatment of reduction as a physical process faces significant problems, and the only way it may conceivably work is by postulating physics beyond quantum theory.

8.3.1 The Quantum Demarcation Problem

In Eq. (8.16), quantum state reduction appears as an almost instantaneous process that alternates with unitary time evolution. This seems to imply that there are two distinct rules of quantum evolution in quantum theory:

- a deterministic rule of time evolution according to Schrödinger equation, as long as no measurement takes place;
- the reduction of the quantum state at measurements, where the quantum system instantaneously "selects" the realized value of a physical magnitude.

Hence, the randomness of quantum theory appears only at the instant of the measurement. This interpretation is appealing because it provides a simple physical interpretation of mathematical terms. Unfortunately, it is wrong, or, at best, highly problematic. The reasons are the following.

Consider the measurement of an observable A of a microscopic system M. Suppose that we include the measuring apparatus A into the quantum description. Hence, we associate a quantum state with the composite system M + A. The pointer P_A of the apparatus is a variable that correlates the value of A (think of a needle in an analogue scale). Since the apparatus is quantum, the pointer is also a quantum observable represented by a self-adjoint operator. Hence, working at the level of M + A we can substitute the measurement of A with the measurement of P_A. This requires implementing quantum state reduction at the level of the system M + A.

We can also use quantum theory to describe a yet larger quantum system that consists of M, A, and a camera B that records the pointer. Then, there is a quantum observable P_{P_A} on the camera that correlates with the reading of the pointer P_A. To describe the measurement at this level, we need to implement quantum state reduction at the system M + A + B. We can continue this procedure to increasingly larger systems, M + A + B + C, and so on.

The key point here is the following. If reduction occurs at the level of M, then the degrees of freedom of A are not affected. If reduction occurs at the level of M + A, then the degrees of freedom of A are affected, but the degrees of freedom of B are not affected, and so on. The physics depends on the level at which the reduction occurs. Quantum theory does not reveal this level, because it does not tell us which physical processes count as measurements. Hence, the theory cannot uniquely specify the state of the system M + A + B after measurement. This ambiguity is unacceptable in a physical theory, and it defines the quantum *demarcation* problem.[1] Quantum theory requires the separation of physical systems into measured and measuring ones, but it does not explain where to set the boundary between them.

---

[1] In Section 6.3.1, we identified the quantum measurement problem as our inability to describe measurement as a physical quantum process, and the consequent treatment of measurement as a fundamental notion. The demarcation problem is the form that the quantum measurement problem takes when we attempt to describe reduction as a physical process.

There is no demarcation problem in the minimal interpretation. As long as we view the quantum state as an informational object, state reduction is a mere update of the system's description after measurement. It does not matter where we set the boundary between measuring and measured system, as long as the resulting probabilities are compatible (Heisenberg, 1935).

In contrast, the demarcation problem is a huge burden for any interpretation of the quantum state as an objective element of the physical description, and the reduction process as a physical change. Such interpretations can be consistent only if they specify uniquely and unambiguously which physical processes count as measurements.

There *is* a resolution to the demarcation problem, which, however, raises more questions than it answers. Eventually, the pointer, or the photo of the pointer, is seen by a human. This means that we must eventually refer to the records in a person's retina, and beyond that to neurons and the electrochemical signal from the retina to the brain. But now this process has an endpoint: the record of the measurement outcome in the observer's consciousness. Hence, one is led to the thought that quantum state reduction occurs at the level where the measurement outcome enters the stream of consciousness or, less precisely, that quantum state reduction is caused by the observer's consciousness (Wigner, 1967). This solution cannot be ruled out on the basis of our current knowledge, and some researchers find it attractive. However, it does not rise beyond the level of speculation, as it provides no separate mathematical formalism and it proposes no experimental tests. It suggests, however, that the quantum measurement problem might be solvable within a larger theory than quantum mechanics that would also describe mind–body interactions. While the existence of such a theory is hardly implausible (Anastopoulos, 2021), it lies well beyond our current state of knowledge.

8.3.2 State Reduction as Part of the Quantum Probability Rule

Relativity places great constraints on our understanding of quantum state reduction. These problems have little to do with the detailed physics of relativistic quantum systems, to be studied in Chapter 16. They originate from elementary properties of relativity; to understand them, we need only the following information.

1. Physical events correspond to spacetime points. A spacetime point A is described in a reference frame Σ by one time coordinate $t(A)$ and three space coordinates $x_1(A), x_2(A), x_3(A)$. The same point has different coordinates $t'(A)$ and $x'_1(A), x'_2(A), x'_3(A)$ in a different reference frame Σ'.
2. There are spacetime points, say A and B, with the following property. A occurs before B in a reference frame Σ ($t(A) < t(B)$), and B occurs before A in a reference frame Σ'. Then the events A and B are called spacelike separated.

We describe a quantum system in the reference frame Σ. We assume that the system is prepared in the state $|\psi\rangle$. Let event A corresponds to the measurement of observable $\hat{X} = \sum_i a_i \hat{P}_i$, and event B to the measurement of observable $\hat{Y} = \sum_j b_j \hat{Q}_j$. We assume that the two events are spacelike separated. Let the outcome of the two measurements be a_n and b_m, respectively. According to the quantum state reduction rule, the quantum state evolves as follows:

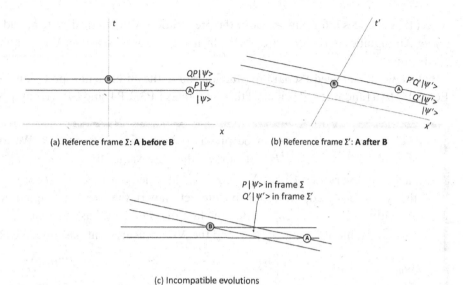

(a) Reference frame Σ: **A before B**

(b) Reference frame Σ': **A after B**

(c) Incompatible evolutions

Fig. 8.1 Evolution of the quantum state under successive measurements in a relativistic system. (a) Reference frame Σ. (b) Reference frame Σ'. (c) The spacetime region in which the two evolutions are incompatible.

$$
\begin{aligned}
|\psi\rangle && t < t(\mathrm{A}) \\
c_1 \hat{P}_n |\psi\rangle && t(\mathrm{A}) < t < t(\mathrm{B}) \\
c_2 \hat{Q}_m \hat{P}_n |\psi\rangle && t > t(\mathrm{B}),
\end{aligned}
\tag{8.20}
$$

where c_1, c_2 are constants. This evolution of the state is depicted in Fig. 8.1a.

Now let us describe the same process in a reference frame Σ'. The initial state, the observables and the associated spectral projectors must be transformed to this frame; we denote the transformed quantities by a prime. In Chapter 16, we will see that this transformation is implemented by a unitary operator, but this information is irrelevant to present purposes. In Σ', the quantum state evolves as follows:

$$
\begin{aligned}
|\psi'\rangle && t' < t'(\mathrm{A}) \\
c_1' \hat{Q}_n' |\psi'\rangle && t'(\mathrm{A}) < t' < t'(\mathrm{B}) \\
c_2' \hat{Q}_m' \hat{P}_n' |\psi'\rangle && t' > t'(\mathrm{B}),
\end{aligned}
\tag{8.21}
$$

where c_1', c_2' are constants. This evolution is depicted in Fig. 8.1b.

The two evolutions give incompatible results in the parallelogram that is indicated in Fig. 8.1c. In this spacetime region, the quantum state is $c_1 \hat{P}_n |\psi\rangle$ in the reference frame Σ, and $c_1' \hat{Q}_n' |\psi'\rangle$ in the reference system Σ'. Since $\hat{X}$ and $\hat{Y}$ are arbitrary, there is no transformation that depends only on the reference frames that take one state to the other. We obtain genuinely different evolutions for the quantum state in the two reference frames.

Nonetheless, this ambiguity in the quantum state does not lead to an ambiguity in physical predictions, which are expressed in terms for probabilities. Probabilities are uniquely defined by Eq. (8.14), provided that $[\hat{X}, \hat{Y}] = 0$. A paradox exists only if we think of the state as an objective attribute of the quantum system. In the minimal interpretation, the state is the mathematical object that encodes information about the preparation of a physical system, and not a physical entity. Ambiguities in the state are physically irrelevant as long as they do not affect the theory's predictions.

In fact, we do not need a separate reduction rule in the minimal interpretation. We can upgrade Eq. (8.14) to a fundamental principle that integrates FP3, FP4, and FP5 (Houtappel et al., 1965).

FP345 Let a quantum system be prepared in a state $\hat{\rho}$ at time t_0. We measure the observable $\hat{A}^{(1)} = \sum_{n_1} a_{n_1}^{(1)} \hat{P}_{n_1}^{(1)}$ at time t_1, the observable $\hat{A}^{(2)} = \sum_{n_2} a_{n_2}^{(2)} \hat{P}_{n_2}^{(2)}$ at time t_2, ..., and the observable $\hat{A}^{(N)} = \sum_{n_N} a_{n_N}^{(N)} \hat{P}_{n_N}^{(N)}$ at time t_N, where $t_0 < t_1 < t_2 < \cdots < t_N$. If the system is closed, except for the interaction with the measuring apparatus, then the probability $p(a_{n_1}^{(1)}, a_{n_2}^{(2)}, \ldots, a_{n_N}^{(N)})$, that the first measurement has outcome $a_{n_1}^{(1)}$, the second measurement has outcome $a_{n_2}^{(2)}$, ..., and the Nth measurement has outcome $a_{n_N}^{(N)}$ is

$$p(a_{n_1}^{(1)}, a_{n_2}^{(2)}, \ldots, a_{n_N}^{(N)}) = \mathrm{Tr}\left[\hat{P}_{n_N}^{(N)} e^{-i\hat{H}(t_N - t_{N-1})} \cdots e^{-i\hat{H}(t_2 - t_1)} \hat{P}_{n_1}^{(1)} e^{i\hat{H}(t_1 - t_0)} \hat{\rho} \right.$$
$$\left. e^{i\hat{H}(t_1 - t_0)} \hat{P}_{n_1}^{(1)} e^{i\hat{H}(t_2 - t_1)} \cdots e^{i\hat{H}(t_N - t_{N-1})} \right], \tag{8.22}$$

where $\hat{H}$ is the system's Hamiltonian.

It is easy to show that FP3, FP4, and FP5, are consequences of FP345. In this formulation, the fundamental elements of quantum theory are the histories of measurement outcomes $\alpha := (a_{n_1}^{(1)}, a_{n_2}^{(2)}, \ldots, a_{n_N}^{(N)})$, and the probability distribution $p(\alpha)$ of those histories, given by Eq. (8.22). The rule of unitary evolution and the rule of quantum state reduction are components of this fundamental probability rule, and not separate physical processes.[2] Hence, FP345 leaves no room for ambiguities similar to those of our preceding discussion, or for an interpretation of state reduction as a physical process.

Equation (8.22) can be written in a more compact form. We define the *history operator*

$$\hat{C}_\alpha = \hat{P}_{n_N}^{(N)}(t_N) \cdots \hat{P}_{n_2}^{(2)}(t_2) \hat{P}_{n_1}^{(1)}(t_1), \tag{8.23}$$

where $\hat{P}_{n_i}^{(i)}(t_i) := e^{i\hat{H}(t_i - t_0)} \hat{P}_{n_i}^{(i)} e^{-i\hat{H}(t_i - t_0)}$ is the Heisenberg-picture evolution of $\hat{P}_{n_i}^{(i)}$. Then, the probability distribution for histories becomes

$$p(\alpha) = \mathrm{Tr}\left[\hat{C}_\alpha \hat{\rho}_0 \hat{C}_\alpha^\dagger \right]. \tag{8.24}$$

8.3.3 State Reduction as a Physical Process

As mentioned in Chapter 6, the minimal interpretation is incomplete, because it uses the notion of the measurement as fundamental, leading to the quantum measurement problem. Perhaps the problem can be solved through the addition of a physical mechanism for quantum

---

[2] There is an obvious objection: Isn't FP345 too unwieldy for a foundational principle? The answer is that if one demands the existence of a probability rule for histories, subject to some simple conditions, FP345 follows uniquely (Anastopoulos, 2003). Hence, FP345 can be obtained as a consequence of mathematically more-intuitive axioms (Isham, 1994).

state reduction. We have already mentioned the proposal that state reduction is a real process that originates from the observers' consciousness. Consciousness has two crucial features that enable it to play this role: It is not described by quantum theory, and it is ever-present in all experiments. Any entity with those two characteristics can also play this role. For example, the gravitational field is ever-present and it interacts with all matter. Hence, if the gravitational field is not fundamentally quantum, then it could serve as the agent of quantum state reduction (Karolyhazy, 1966; Diosi, 1987; Penrose, 1986, 1996). Crucially, models of gravity-induced reduction make predictions that can be tested experimentally.

Alternatively one could postulate state reduction as a physical process of microscopic origins that is *universal*, in the sense that it is *intrinsic to all physical systems*, and it does not appear only in measurements. This process is called *dynamical state reduction* (Ghirardi et al., 1986; Pearle, 1989). The probability of a single microscopic particle undergoing reduction is tiny; however, these tiny probabilities add up to a significant change in the dynamics for systems with large numbers of particles. This is the reason why measurement apparatuses behave classically, and appear as though they, themselves, cause reduction to microscopic systems.

The notion of dynamical reduction is attractive to many researchers, because it may lead to a formulation of quantum theory that refers to the evolution of objective entities, with no special role for measurement. However, it does not cure the incompatibility of the reduction process with relativity. When applied to relativistic systems, it leads to faster-than-light transmission of information.

Furthermore, no fundamental theory of dynamical reduction has been developed yet, just a collection of different models. The form of the reduction process in each model is postulated ad hoc, and typically it involves some free parameters. The good thing is that these models lead to different predictions from quantum theory, and thus, they are amenable to experimental tests. So far, experiments have failed to confirm any model, and they have excluded some ranges of values for the parameters of those models. More experiments are planned for the near future. We will come back to these issues in Section 21.5 and in Section 22.2.

8.4 Applications

The state reduction rule has significant observational consequences. Here, we present three of them: postselected measurements, quantum jump experiments, and the quantum Zeno effect.

8.4.1 Postselected Measurements

When several measurements are performed on a quantum system, we can define *postselected* probabilities, in which we ignore all outcomes of the first measurement unless the second measurement has a specific outcome.

Suppose that we first measure the observable $\hat{A} = \sum_n a_n \hat{P}_n$ and then the observable $\hat{B} = \sum_m b_m \hat{Q}_m$. We decide to ignore any run of the experiment, in which the second measurement does not give b_1 for $\hat{B}$. The relevant probability is the conditional probability $\text{Prob}(a_n|b_1)$ for the outcomes of the first measurements, given b_1 for the second measurement. It is different from the conditional probability (8.13), because the alternative b_1 corresponds to a later measurement.

The conditional probability rule gives

$$\text{Prob}(a_n|b_1) = \frac{\text{Prob}(a_n, b_1)}{\sum_n \text{Prob}(a_n, b_1)} = \frac{\text{Tr}(\hat{\rho}\hat{P}_n\hat{Q}_1\hat{P}_n)}{\sum_n \text{Tr}(\hat{\rho}\hat{P}_n\hat{Q}_1\hat{P}_n)}. \tag{8.25}$$

The denominator Eq. (8.25) is the probability to obtain b_1 in the second measurement irrespective of what happened in the first.

For $\hat{Q}_1 = |\phi\rangle\langle\phi|$ and pure initial state, $\hat{\rho} = |\psi\rangle\langle\psi|$,

$$\text{Prob}(a_n|b_1) = \frac{|\langle\psi|\hat{P}_n|\phi\rangle|^2}{\sum_n |\langle\psi|\hat{P}_n|\phi\rangle|^2}. \tag{8.26}$$

In Eq. (8.26), we cannot distinguish whether $|\psi\rangle$ or $|\phi\rangle$ is the initial state. In measurements with postselection, it looks as though the system is described by two different states, the initial state $|\psi\rangle$ that defines preselection (i.e., the preparation of the system) and the final state $|\phi\rangle$ that defines postselection.

There is no reason why we cannot take Eq. (8.26) as a fundamental principle of quantum theory in place of FP3, that is, to postulate that the fundamental quantum mechanical probabilities involve both preselection and postselection (Aharonov et al., 1964). In most cases, the experimenter does not postselect, so we sum over all possible final states, and recover the usual probability rule. A modified probability rule with both a final and an initial state makes little difference in the context of the minimal interpretation that focuses solely on measurement outcomes.

However, the difference is profound for any interpretation of quantum theory that purports to make no fundamental reference to measurement. Equation (8.26) implies that final conditions are at the same footing with initial conditions, as far as quantum probabilities are concerned. Such interpretations would make it possible to talk scientifically not only about the initial conditions of the Universe – which has been done ever since the time of Newton – but also about the final conditions of the Universe, which is completely novel in physics.

8.4.2   Quantum Jumps

As mentioned in Section 2.3, in the days of the old quantum theory, Bohr had hypothesized that the emission of photons from atoms is a discontinuous process, a *quantum jump*. Nowadays, we have verified the existence of apparently discontinuous transition processes that partly fit with Bohr's early ideas about jumps (Nagourney et al., 1986). This has been made possible by our technological ability, ever since the 1980s, to isolate individual atoms or ions and to operate upon them. Apparatuses that isolate individual atoms or ions are called *traps*. Traps employ combinations of electric and magnetic fields that restrict the possible motions of an individual particle, eventually binding in a specific region.

The experimental confirmation of quantum jumps relies on the phenomenon of *resonant fluorescence*, described in Fig. 8.2. Laser light of frequency ω is directed towards an ion at the center of a trap. This excites the ion and causes a transition from the ground state $|0\rangle$ to

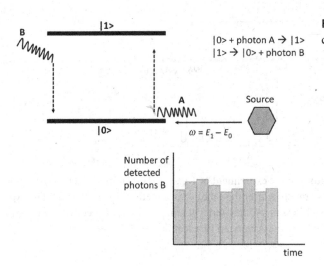

$$|0\rangle + \text{photon A} \rightarrow |1\rangle$$
$$|1\rangle \rightarrow |0\rangle + \text{photon B}$$

Fig. 8.2 Key principles of resonant fluorescence.

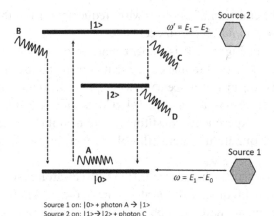

Source 1 on: $|0\rangle$ + photon A $\rightarrow$ $|1\rangle$
Source 2 on: $|1\rangle \rightarrow |2\rangle$ + photon C
Spontaneous emission: $|1\rangle \rightarrow |0\rangle$ + photon B
Rare spontaneous emission: $|2\rangle \rightarrow |1\rangle$ + photon D

Fig. 8.3 An experimental set-up for observing quantum jumps in the transitions between levels $|2\rangle$ and $|0\rangle$. Source 2 induces transitions from $|1\rangle$ to $|2\rangle$. As long as the atom is found in state $|2\rangle$, no fluorsescence (photons A) is observed. After it transitions to $|0\rangle$ through a quantum jump, fluorescence restarts.

an excited state $|1\rangle$. The state $|1\rangle$ then decays, emitting a photon of the same frequency as the incoming laser photons, but traveling in a different direction. Outcoming photons are detected. If we expose a photographic plate long enough, a photo of the trapped atom will be formed.

We observe quantum jumps in trapped ions by exploiting resonant fluorescence in a system with three levels $|0\rangle$, $|1\rangle$ and $|2\rangle$, as in Fig. 8.3. The energy levels are selected so that (i) spontaneous transitions $|1\rangle \rightarrow |2\rangle$ are impossible, and (ii) spontaneous transitions $|2\rangle \rightarrow |0\rangle$ are much rarer than spontaneous transitions $|1\rangle \rightarrow |0\rangle$. This means that when the ion is found in state $|2\rangle$, it will stay there for a long time (of the order of a second).

We direct two laser beams to the trapped ion, one with frequency $\omega = E_2 - E_0$ and the other with frequency $\omega' = E_1 - E_2$. Photons of frequency ω cause resonant fluorescence between level $|1\rangle$ and $|0\rangle$. Photons of frequency ω' cause transitions $|1\rangle \rightarrow |2\rangle$, which cannot happen spontaneously. If the ion is found in state $|2\rangle$, resonant fluorescence stops, and photons of frequency ω are not emitted until the transition $|2\rangle \rightarrow |0\rangle$ occurs. Then, fluorescence restarts.

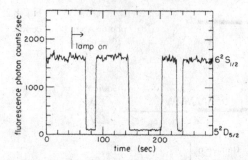

Fig. 8.4 Number of fluorescence photons in the experiment of Fig. 8.3. Transitions to state $|1\rangle$ are practically instantaneous, they occur at a timescale much shorter than the average lifetime of the $|2\rangle$ state. The experiment was carried out on a Ba^+ ion with two lasers at 493 nm and 650 nm. [Figure reprinted with permission from Nagourney et al. (1986). Copyright 1985 by the American Physical Society.]

This means that by monitoring the emission of photons with frequency ω from the ion, we can specify when the transitions $|1\rangle \to |2\rangle$ and $|2\rangle \to |0\rangle$ occurred.

Experimental results are shown in Fig. 8.4. It is evident that the transition from $|1\rangle$ to $|2\rangle$ occurs at a timescale much smaller than the lifetime of $|2\rangle$, so it appears almost instantaneous.

The results of the experiments are in full agreement with quantum theory that correctly predicts the statistical behavior of results as in Fig. 8.4 (Cook and Kimble, 1985; Zoller et al., 1987). Hence, quantum jump experiments constitute an important example of applying quantum theory to an individual quantum system (the photo-emitting ion), and not to a statistical ensemble.

Quantum jump experiments demonstrate that the transition between atomic states takes place in a timescale much smaller than the lifetime of the excited state. Now, we know that the jump is not instantaneous, recent experiments have been able to resolve its characteristic timescale (Minev et al., 2019). These experiments clarified that the quantum jump should not be identified with the transition between the energy levels. The instantaneous, noncausal process that Bohr envisioned takes place *before* the transition between the energy levels $|2\rangle$ and $|0\rangle$ of Fig. 8.3, and it leaves an imprint in the fluorescent photons. We can detect this imprint before the transition takes place, and from this detection we can predict that a transition will happen, and even hinder it.

Many physicists are tempted to identify the physical concept of the quantum jump with the mathematical operation of quantum state reduction. This identification makes sense only in interpretations that treat quantum state reduction as a physical process. It is inadmissible in the minimal interpretation, unless one *reinterprets* the quantum jumps as jumps in our knowledge of the measured system, and not as physical processes. Heisenberg (1999) was of this opinion.

The observation itself changes the probability function discontinuously; it selects of all possible events the actual one that has taken place. Since through the observation our knowledge of the system has changed discontinuously, its mathematical representation also has undergone the discontinuous change and we speak of a 'quantum jump.'

Obviously, Heisenberg's quantum jumps are not the quantum jumps of quantum-jump experiments. The latter are physical processes and not abrupt changes in our knowledge. Their quantum description is not mathematically or physically equivalent to quantum state reduction, namely, FP5. The mathematical description of quantum jumps involves a complex combination of FP3, FP4, and FP5.

8.4.3   The Quantum Zeno Effect

Zeno of Elea was a Greek philosopher, famous for his paradoxes that attempted to prove the impossibility of motion. One of his paradoxes referred to a moving arrow. At any instant of time, the arrow cannot move towards where it is not, because it cannot move there instantaneously. But it also cannot move towards where it is now, because it is already there. Hence, at any given instant of time, the arrow does not move. If everything is motionless at all instants, and time consists entirely of instants, then motion is impossible.

It turns out that the quantum state of a system does not change if one subjects it continuously to measurements (Degasperis et al., 1974). This effect is reminiscent of the arrow paradox, hence, it is called the *quantum Zeno* effect (Misra and Sudarshan, 1977). Consider a system characterized by a Hamiltonian $\hat{H}$. We have a measuring device that corresponds to a projection operator $\hat{P}$, that is, it records whether a defining property of $\hat{P}$ is true (eigenvalue +1) or false (eigenvalue 0). We prepare the system in an eigenstate $|\psi\rangle$ of $\hat{P}$, $\hat{P}|\psi\rangle = |\psi\rangle$.

We measure $\hat{P}$ N times, with a time interval δt between successive measurements. The probability $\mathrm{Prob}(\hat{P}, N)$ that $\hat{P}$ is true in all N measurements

$$\mathrm{Prob}(\hat{P}, N) = \langle\psi|\hat{P}\hat{U}_{\delta t}^{\dagger}\hat{P}\hat{U}_{\delta t}^{\dagger}\cdots\hat{U}_{\delta t}\hat{P}\hat{U}_{\delta t}\hat{P}|\psi\rangle = \langle\psi|\hat{S}_{N,t}^{\dagger}\hat{S}_{N,t}|\psi\rangle, \tag{8.27}$$

where $\hat{S}_{N,t} = (\hat{P}\hat{U}_{\delta t}\hat{P})^N$, and $t = N\delta t$. At the limit of continuous measurement, $N \to \infty$ with t constant, $e^{-i\hat{H}\delta t} \simeq 1 - i\hat{H}t/N$, hence,

$$\hat{S}_t := \lim_{N\to\infty}(\hat{P} - i\hat{P}\hat{H}\hat{P}t/N)^N. \tag{8.28}$$

To evaluate this limit, we first note that $\hat{S}_t$ vanishes outside the defining subspace V of $\hat{P}$. Inside V,

$$\hat{S}_t = \lim_{N\to\infty}(\hat{I} - i\hat{P}\hat{H}\hat{P}t/N)^N = e^{-i\hat{P}\hat{H}\hat{P}t}. \tag{8.29}$$

Hence, $\hat{S}_t$ is unitary in V. For any $|\psi\rangle \in V$, $\hat{S}_t^{\dagger}\hat{S}_t|\psi\rangle = |\psi\rangle$.

From Eq. (8.27), we evaluate the probability $\mathrm{Prob}(\hat{P}, t) := \lim_{N\to\infty}\mathrm{Prob}(\hat{P}, N)$ that $\hat{P}$ is found true for continuous measurements up to time t,

$$\mathrm{Prob}(\hat{P}, t) = \langle\psi|\hat{S}_t^{\dagger}\hat{S}_t|\psi\rangle = \langle\psi|\psi\rangle = 1. \tag{8.30}$$

Hence, the system cannot leave V if it is measured continuously. For example, the system may be an unstable particle, and V containing the states that describe a particle that has not decayed. Then, the quantum Zeno effect implies that if we test continuously whether the particle decayed, the particle will never decay.

The first observation of the quantum Zeno effect was reported by Itano et al. (1990). The experiment is similar to the quantum jump experiments, in that it exploits the behavior of a three-level system under the action of two lasers of different frequency. The theoretical description of this experiment is given in Box 8.1.

Box 8.1 Quantum Zeno Effect in Trapped Ions

Consider an ion with three eigenstates of the Hamiltonian $|0\rangle$, $|1\rangle$ and $|2\rangle$, such that

- $|0\rangle$ is the ground state with energy $E_0 = 0$;
- $|1\rangle$ is a stable excited state with energy $E_1 = \omega_1$;
- $|2\rangle$ is an unstable excited state with energy $E_2 = \omega_2$. It decays spontaneously to $|0\rangle$ with the emission of a photon of frequency ω_2.
- There is no spontaneous transition from $|2\rangle$ to $|1\rangle$.

If we direct a laser of frequency ω_1 to the ion, we will observe Rabi oscillations between the states $|0\rangle$ and $|1\rangle$, with Rabi frequency Ω_R, given by Eq. (7.94) for $\Delta = 0$. In the subspace spanned by $|0\rangle$ and $|1\rangle$, time evolution is given by the unitary operator

$$\hat{U}(t) = \begin{pmatrix} e^{-i\omega_1 t}\cos(\Omega_R t) & i e^{-i\omega_1 t}\sin(\Omega_R t) \\ i\sin(\Omega_R t) & \cos(\Omega_R t) \end{pmatrix}.$$

Under a short pulse of frequency ω_2, an ion in the ground state $|0\rangle$ is excited to $|2\rangle$, it stays there for a short time, and returns to $|0\rangle$ with the emission of a photon. In contrast, if the ion is in $|1\rangle$, it is not affected by the pulse. Hence, if we detect a photon of frequency ω_2 after the pulse, we confirm that the ion was in the state $|0\rangle$. If not, the ion was in $|1\rangle$. Hence, the ω_2-pulse acts as a measurement with spectral projectors $\hat{P}_0 = |0\rangle\langle 0|$ and $\hat{P}_1 = |1\rangle\langle 1|$.

Let the laser of frequency ω_1 be switched on at $t = 0$. Then, an initial state $|0\rangle$ evolves to $|\psi(t)\rangle = \hat{U}(t)|0\rangle$ at time t. For $t = T := \frac{\pi}{2\Omega_R}$, the system ends up in the state $|1\rangle$. Suppose that during the time interval $[0, T]$, we emit n pulses of frequency ω_2, with a time step $\tau = T/n$ between one pulse and the next. Let us denote $\hat{U}_\tau = \hat{U}(T/n)$; the state $\hat{\rho}_t$ becomes after one time step

$$\hat{\rho}_{t+\tau} = \hat{P}_0 \hat{U}_\tau \hat{\rho}_t \hat{U}_\tau^\dagger \hat{P}_0 + \hat{P}_1 \hat{U}_\tau \hat{\rho}_t \hat{U}_\tau^\dagger \hat{P}_1. \tag{A}$$

The off-diagonal elements of a density matrix vanish after one time step. Hence, we need only to consider diagonal density matrices. Their Bloch vectors are of the form $(0, 0, r)$. Then, Eq. (A) implies that $r(t + \tau) = r(t)\cos(2\Omega_R \tau) = r(t)\cos(\pi/n)$. For an initial state $|0\rangle$, with $r(0) = -1$, we find that for n steps, $r(T) = -\cos^n(\pi/n)$. For large n, $\cos(\pi/n) \simeq 1 - \frac{\pi^2}{2n^2}$, hence, $r(T) = -(1 - \frac{\pi^2}{2n^2})^n = -\exp\left[n\ln(1 - \frac{\pi^2}{2n^2})\right] \simeq -e^{-\frac{\pi^2}{2n}}$.

The probability of finding the ion on $|1\rangle$ at time $t = T$ is

$$p_1(T) = \frac{1}{2}\left(1 - e^{-\frac{\pi^2}{2n}}\right) \simeq \frac{\pi^2}{4n},$$

that is, it drops to zero with the number n of ω_2-pulses.

We conclude that if we act only with a laser of frequency ω_1 for time T, the ion will be certainly found on $|1\rangle$. This can be verified by detecting the photon of frequency ω_1 that will be emitted after the ion has decayed. However, if during this time T, we send n pulses of frequency ω_2, the transition to $|1\rangle$ and the subsequent photon emission is strongly suppressed, being inversely proportional to n.

QUESTIONS

8.1 Argue that quantum state reduction should not be viewed as a physical process. Then, give arguments for the opposite. Which case do you find more convincing?

8.2 Einstein (1948) wrote about quantum state reduction:

> The change in the psi function through observation then does not correspond to the change in a real matter of fact but rather to an alteration in *our knowledge* of this matter of fact.

Explain why this claim is wrong.

8.3 Explain Wigner's quote at the beginning of this chapter, in light of the discussion of Section 8.3.2.

8.4 **(B)** How would you describe the notion of a quantum jump, given the results of the quantum jump experiments?

8.5 **(B)** Suppose that the time δt between successive measurements cannot be smaller than τ. What is the leading correction to the probability $\mathrm{Prob}(\hat{P}, t)$ of Eq. (8.30)?

PROBLEMS

8.1 How does the Bloch vector of an ensemble of qubits change after the measurement of an observable $\hat{A} = \mathbf{n} \cdot \hat{\boldsymbol{\sigma}}$?

8.2 A qubit is prepared in a state $\frac{1}{\sqrt{2}}(|0\rangle + i|1\rangle)$, and it is subjected to measurements of $\hat{\sigma}_1$ and then $\hat{\sigma}_2$. Evaluate the joint probabilities for all measurement outcomes. Repeat the calculation for the converse order of measurements.

8.3 Let γ be the purity of a state $\hat{\rho}$ prior to the measurement of $\hat{A} = \sum_n a_n \hat{P}_n$ and let γ' be the purity of the state $\hat{\rho}' = \sum_n \hat{P}_n \hat{\rho} \hat{P}_n$ after measurement. Show that $\gamma' \leq \gamma$. When does equality hold? (ii) Use this result to show that there is no unitary operator $\hat{U}$, such that $\hat{U} \hat{\rho} \hat{U}^\dagger = \sum_n \hat{P}_n \hat{\rho} \hat{P}_n$.

8.4 A qubit with a Hamiltonian $\hat{H} = \omega \hat{\sigma}_1$ is prepared on a pure state $|\psi\rangle$ that corresponds to the positive eigenvalue of $\hat{H}$. At $t = 0$, we measure $\hat{A} = \hat{\sigma}_3$. At time t, we measure $\hat{A}$ again. (i) Evaluate the joint probabilities for the outcomes of the two measurements. (ii) Compute the probabilities for the outcomes of the first measurement after we postselect to keep only outcomes $+1$ in the second measurement.

8.5 Consider two successive measurements on a qubit. The first apparatus has a random number generator that gives 60 percent probability to measure $\hat{\sigma}_1$ and 40 percent probability that no measurement takes place. The second apparatus measures $\hat{\sigma}_3$. Evaluate $\langle \hat{\sigma}_3 \rangle$ in the second measurement for the initial state $|1\rangle$.

8.6 Consider two successive measurements on a qubit. The first apparatus measures $\mathbf{n} \cdot \hat{\boldsymbol{\sigma}}$, where $\mathbf{n}$ is a random variable, subject to the homogenous probability distribution on the unit sphere. Given an initial pure state $|\psi\rangle$, what is the state of the statistical ensemble prior to the second measurement?

8.7 Any vector $|c\rangle \in \mathbb{C}^3$ defines an observable C with two values: 1, associated to the projector $\hat{P}_c = |c\rangle\langle c|$; and 0, associated to the projector $\hat{I} - \hat{P}_c$. Consider a system prepared in a state $|0\rangle$, subject to the successive measurement of two such observables, first B and then C, with corresponding vectors $|b\rangle$ and $|c\rangle$. Suppose that we only measure C, and attempt to infer what we would have observed had a measurement of B taken place. (i) Show that the conditional probability equation $\mathrm{Prob}(B = 1 | C = 1) = 1$ implies $|\langle 0|b\rangle\langle b|c\rangle| = |\langle 0|c\rangle|$. (ii) Choose

$$|0\rangle = \begin{pmatrix} 1 \\ 0 \\ 0 \end{pmatrix}, |c\rangle = \begin{pmatrix} \cos\theta \\ \sin\theta \\ 0 \end{pmatrix},$$

for $\theta \in [0, \pi/2]$. Show that the equation $\mathrm{Prob}(B = 1 | C = 1) = 1$ admits two families of solutions $|b\rangle = |y, \pm\rangle$, parameterized by $y \in [0, 1]$:

$$|y, \pm\rangle = \frac{1}{\sqrt{\cos^2\theta + y\sin^2\theta}} \begin{pmatrix} \cos\theta \\ y\sin\theta \\ \pm\sin\theta\sqrt{y - y^2} \end{pmatrix}.$$

(iii) Show that for $\tan\theta \geq \frac{1}{2\sqrt{2}}$ there are values of y, for which $|y, +\rangle$ is orthogonal to $|y, -\rangle$. **Comment** In classical probability, the condition $\mathrm{Prob}(B = 1 | C = 1) = 1$ implies that if we measure $C = 1$, we can infer that $B = 1$, even if we had carried out no measurement. If we attempt to do the same in quantum theory, we will obtain two different conclusions that are not compatible with each other, since the corresponding vectors $|b\rangle$ are normal to each other (Kent, 1997). This is the *retrodiction problem* of quantum theory. There is no problem for predictions, because they refer to measurements that have not been carried out yet; hence, the observable to be measured is indeterminate.

8.8 (**B**) Consider a family of states $|a\rangle$ specified by a parameter $a \in \Gamma$. We want to measure whether the system follows a continuous path $a(t)$ on Γ, where $t \in [0, T]$. This means that we measure the projector $|a(t)\rangle\langle a(t)|$ at all times t. To construct the associated probabilities, consider N successive measurements with time step $\delta t = T/N$, and take the limit $N \to 0$ at the end. Show that the probability $\mathrm{Prob}[a(\cdot)]$ of measuring $a(t)$ at the continuum limit equals $\langle a(0)|\hat{\rho}_0|a(0)\rangle$. Hence, if $\hat{\rho}_0 = |a(0)\rangle\langle a(0)|$, then $\mathrm{Prob}[a(\cdot)] = 1$.

Comment This result was first obtained by Vardi and Aharonov (1980). It is the opposite of the quantum Zeno effect: if we measure continuously whether the system follows a given trajectory, then this trajectory is realized with probability one.

Bibliography

- Many classic discussions about quantum state reduction are reprinted in Wheeler and Zurek (1983); Wigner's Princeton lectures are highly recommended. I also recommend chapter 8 of Isham (1995) and chapter 14 of d'Espagnat (1999). See also section 8.3 of the latter reference for the incompatibility of state reduction with relativity. Part I of Ghirardi and Bassi (2003) provides a road map of different approaches towards the quantum measurement problem.

- On interpretations of quantum state reduction: See Stapp (2001) for consciousness-induced reduction; see Gisin (2018) for dynamical reduction; see Penrose (2014) for gravity-induced reduction.
- For postselected measurements and their implications, see Aharonov and Vaidman (2008). For quantum jump experiments, see Carmichael (2009). For the quantum Zeno effect, see Joos (2009).

9   Composite Quantum Systems and Entanglement

Anybody who's not bothered by Bell's theorem has to have rocks in his head.

Unnamed physicist, quoted by Mermin (1985)

9.1   Tensor Product

The composition of subsystems in quantum theory is defined in terms of a mathematical operation known as the *tensor product*. We proceed to explain this concept, and to show how it fits in the Hilbert space calculus.

9.1.1   Definition of the Tensor Product

We temporarily abandon the Dirac notation for the definitions of Section 9.1.1.

The notion of the tensor product originates from the following property of matrices. Let

$$\phi = \begin{pmatrix} \phi_1 \\ \phi_2 \\ \dots \\ \phi_n \end{pmatrix} \in \mathbb{C}^n, \qquad \psi = \begin{pmatrix} \psi_1 \\ \psi_2 \\ \dots \\ \psi_m \end{pmatrix} \in \mathbb{C}^m. \tag{9.1}$$

Using these vectors we can construct a $n \times m$ matrix, denoted by $\phi \otimes \psi$, with elements $(\phi \otimes \psi)_{ij} := \phi_i \psi_j$, where $i = 1, 2, \dots, n$ in $j = 1, 2, \dots, m$, that is,

$$\phi \otimes \psi = \begin{pmatrix} \phi_1\psi_1 & \phi_2\psi_1 & \dots & \phi_n\psi_1 \\ \phi_1\psi_2 & \phi_2\psi_2 & \dots & \phi_n\psi_2 \\ \dots & \dots & \dots & \dots \\ \phi_1\psi_m & \phi_2\psi_m & \dots & \phi_n\psi_m \end{pmatrix}. \tag{9.2}$$

The tensor product Hilbert space $\mathbb{C}^n \otimes \mathbb{C}^m$ consists of all linear combinations of the form

$$\sum_{k=1}^{r} \phi^{(k)} \otimes \psi^{(k)},$$

for any r-tuples of vectors $\phi^{(k)} \in \mathbb{C}^n$ and $\psi^{(k)} \in \mathbb{C}^m$, where $k = 1, 2, \dots, r$. Note that the index k is inside a parenthesis because it labels the selected vectors and not the components of individual vectors. The inner product on $\mathbb{C}^n \otimes \mathbb{C}^m$ is first defined for vectors $\phi \otimes \psi$ as

$$(\phi \otimes \psi, \phi' \otimes \psi')_{\mathbb{C}^n \otimes \mathbb{C}^m} = (\phi, \phi')_{\mathbb{C}^n} \times (\psi, \psi')_{\mathbb{C}^n}, \tag{9.3}$$

and then extended to any linear combination of the form (9.3) by linearity. For example,

$$(\phi^{(1)} \otimes \psi^{(1)} + \phi^{(2)} \otimes \psi^{(2)}, \phi' \otimes \psi') = (\phi^{(1)}, \phi')(\psi^{(1)}, \psi') + (\phi^{(2)}, \phi')(\psi^{(2)}, \psi').$$

Any $n \times m$ matrix can be written as a sum (9.3); hence, the Hilbert space $\mathbb{C}^n \otimes \mathbb{C}^m$ is isometric to the Hilbert space $\mathbb{C}^{nm}$ of $n \times m$ matrices,

$$\mathbb{C}^n \otimes \mathbb{C}^m = \mathbb{C}^{nm}. \tag{9.4}$$

The preceding description suffices for the definition of the tensor product $\mathcal{H}_1 \otimes \mathcal{H}_2$, for finite-dimensional Hilbert space $\mathcal{H}_1$ and $\mathcal{H}_2$. If at least one of the Hilbert spaces is infinite-dimensional, then a more intricate definition is required. It is described in Box 9.1. The definition of a tensor product of n Hilbert spaces $\mathcal{H}_1 \otimes \mathcal{H}_2 \otimes \ldots \otimes \mathcal{H}_n$ is a straightforward generalization.

Box 9.1 Definition of the Tensor Product

First, we define the *free vector space $F(S)$* associated to a set S: $F(S)$ is a complex vector space, characterized by a basis $|s\rangle$, where s runs over all elements of S. Obviously, if S has N elements then $F(S) = \mathbb{C}^n$.

Consider two Hilbert spaces $\mathcal{H}_1$ and $\mathcal{H}_2$, with vectors expressed as $|\phi\rangle_1$ and $|\psi\rangle_2$, respectively. The Cartesian product $\mathcal{H}_1 \times \mathcal{H}_2$ consists of pairs $(|\phi\rangle, |\psi\rangle)$. The free vector space $F(\mathcal{H}_1 \times \mathcal{H}_2)$ consists of elements of the form

$$\alpha = \sum_{i=1}^{r} c_i(|\phi_i\rangle, |\psi_i\rangle) = c_1(|\phi_1\rangle, |\psi_1\rangle) + c_2(|\phi_2\rangle, |\psi_2\rangle) + \ldots + c_r(|\phi_r\rangle, |\psi_r\rangle), \text{ for } c_i \in \mathbb{C}.$$

We define a norm on $F(\mathcal{H}_1 \times \mathcal{H}_2)$ by

$$\| \textstyle\sum_{i=1}^{r} c_i(|\phi_i\rangle, |\psi_i\rangle) \|^2 = \sum_{i=1}^{r} \sum_{j=1}^{r} c_i c_j^* \langle \phi_j | \phi_i \rangle \langle \psi_j | \psi_i \rangle.$$

This definition is improper because there are vectors $\alpha \in F(\mathcal{H}_1 \times \mathcal{H}_2)$, such that $\|\alpha\| = 0$. We obtain a proper norm by identifying any two vectors $\alpha, \beta \in F(\mathcal{H}_1 \times \mathcal{H}_2)$ if $\|\alpha - \beta\| = 0$. To be precise, we define the equivalence class $[\alpha]$ as the set of all vectors $\beta \in F(V_1 \times V_2)$ that satisfy $\|\alpha - \beta\| = 0$. By the triangle inequality, $|\|\alpha - \beta\| - \|\alpha\|| \leq \|\beta\| \leq \|\alpha - \beta\| + \|\alpha\|$. It follows that $\|\beta\| = \|\alpha\|$, that is, all vectors in the equivalence class $[\alpha]$ have the same norm. Hence, the norm $\|[\alpha]\|$ of the equivalence class $[\alpha]$ is well defined.

The tensor product $\mathcal{H}_1 \otimes \mathcal{H}_2$ is the vector space of all equivalence classes $[\alpha]$. The norm $\|[\alpha]\|$ defines an inner product through the polarization identity. We denote the equivalence class $[(|\phi\rangle, |\psi\rangle)]$ as $|\psi\rangle \otimes |\phi\rangle \in \mathcal{H}_1 \otimes \mathcal{H}_2$.

9.1.2 Dirac Notation

We switch back to Dirac notation.

Dirac notation is very convenient for describing the tensor product. The Hilbert space $\mathcal{H}_1 \otimes \mathcal{H}_2$ consists of vectors $\sum_{k=1}^{r} |\phi_k\rangle \otimes |\psi_k\rangle$, where $|\phi_k\rangle \in \mathcal{H}_1$ and $|\psi_k\rangle \in \mathcal{H}_2$, while the inner product is expressed as

$$\left(\sum_{l=1}^{r'} \langle \phi_l | \otimes \langle \psi_l | \right) \left(\sum_{k=1}^{r} |\phi_k\rangle \otimes |\psi_k\rangle \right) = \sum_{k=1}^{r} \sum_{l=1}^{r'} \langle \phi_l | \phi_k \rangle \langle \psi_l | \psi_k \rangle. \tag{9.5}$$

Vectors of the form $|\phi\rangle \otimes |\psi\rangle \in \mathcal{H}_1 \otimes \mathcal{H}_2$ are called *separable*. We will often write the vectors $|\phi\rangle \otimes |\psi\rangle$ as $|\phi, \psi\rangle$, but this notation requires prior explanation in order to avoid ambiguities.

If the vectors $|n\rangle$ define an orthonormal basis on $\mathcal{H}_1$ and the vectors $|i\rangle$ define an orthonormal basis on $\mathcal{H}_2$, then the vectors $|n, i\rangle := |n\rangle \otimes |i\rangle$ define the *induced* orthonormal basis on $\mathcal{H}_1 \otimes \mathcal{H}_2$. Any vector $|\Psi\rangle \in \mathcal{H}_1 \otimes \mathcal{H}_2$ can be expressed as $|\Psi\rangle = \sum_{n,i} a_{n,i} |n, i\rangle$, where $a_{n,i} = \langle n, i|\Psi\rangle$. Hence, the resolution of the unity in $\mathcal{H}_1 \otimes \mathcal{H}_2$ takes the form

$$\hat{I} = \sum_{n,i} |n, i\rangle \langle n, i|. \tag{9.6}$$

The tensor product and the direct sum satisfy the distributive property,

$$\mathcal{H}_1 \otimes (\mathcal{H}_2 \oplus \mathcal{H}_3) = (\mathcal{H}_1 \otimes \mathcal{H}_2) \oplus (\mathcal{H}_1 \otimes \mathcal{H}_2). \tag{9.7}$$

To show this, one selects orthonormal basis in each Hilbert space $\mathcal{H}_1$, $\mathcal{H}_2$, and $\mathcal{H}_3$. Then, it is straightforward to show that the induced basis on $\mathcal{H}_1 \otimes (\mathcal{H}_2 \oplus \mathcal{H}_3)$ coincides with the induced basis on $(\mathcal{H}_1 \otimes \mathcal{H}_2) \oplus (\mathcal{H}_1 \otimes \mathcal{H}_2)$.

9.1.3 Tensor Product of Operators

Let $\hat{A}$ be an operator on the Hilbert space $\mathcal{H}_1$ and let $\hat{B}$ be an operator on the Hilbert space $\mathcal{H}_2$. We define the operator $\hat{A} \otimes \hat{B}$ on the Hilbert space $\mathcal{H}_1 \otimes \mathcal{H}_2$ by

$$(\hat{A} \otimes \hat{B}) \sum_i |\phi_i\rangle \otimes |\psi_i\rangle = \sum_i \hat{A}|\phi_i\rangle \otimes \hat{B}|\psi_i\rangle. \tag{9.8}$$

The definition implies that

$$(\hat{A} \otimes \hat{B})(\hat{C} \otimes \hat{D}) = \hat{A}\hat{C} \otimes \hat{B}\hat{D}. \tag{9.9}$$

Note: We use the same symbol ($\otimes$) for the tensor product of Hilbert spaces, for the factorized vectors, and for the tensor product of operators. Strictly speaking, these are distinct mathematical operations, and the use of the same symbol for all three of them is an abuse of notation.

We will also use the same symbol ($\hat{I}$) for the unit operator in $\mathcal{H}_1$, in $\mathcal{H}_2$ and in $\mathcal{H}_1 \otimes \mathcal{H}_2$. With this convention, equations of the form $\hat{I} = \hat{I} \otimes \hat{I}$ are valid.

Operators of the form $\hat{A} \otimes \hat{B}$ are called *factorizable*. Any operator $\hat{X}$ on $\mathcal{H}_1 \otimes \mathcal{H}_2$ can be expressed as a sum of factorizable operators $\hat{X} = \sum_i \hat{A}_i \otimes \hat{B}_i$, where $\hat{A}_i$ are operators on $\mathcal{H}_1$ and $\hat{B}_i$ are operators on $\mathcal{H}_2$.

Let $|n\rangle$ and $|i\rangle$ be orthonormal bases in the Hilbert spaces $\mathcal{H}_1$ and $\mathcal{H}_2$, respectively. Writing $|n, i\rangle = |n\rangle \otimes |i\rangle$, we find that

$$\mathrm{Tr}(\hat{A} \otimes \hat{B}) = \sum_{n,i} \langle n, i|\hat{A} \otimes \hat{B}|n, i\rangle = \sum_n \langle n|\hat{A}|n\rangle \cdot \sum_i \langle i|\hat{B}|i\rangle = \mathrm{Tr}(\hat{A})\mathrm{Tr}(\hat{B}).$$

9.2 Quantum Combination of Subsystems

First, we examine the combination of subsystems in classical physics. Consider two particles in a line, each described by one position and one momentum coordinate. The pairs (q_1, p_1) span the state space Γ_1 of the first particle and the pairs (q_2, p_2) span the state space Γ_2 of the second particle. We have full information about the two-particle system if we know the positions and momenta of both particles, that is, if we specify the quadruplet (q_1, q_2, p_1, p_2). Hence, the state space for the composite system is $\Gamma_1 \times \Gamma_2$.

We conclude that, in classical physics, the state space of a composite system is the *Cartesian product* $\Gamma_1 \times \Gamma_2 \times \cdots \times \Gamma_n$ of the state spaces Γ_i of its components.

To find the corresponding law in quantum theory, we consider two quantum systems described by the Hilbert spaces $\mathcal{H}_1$ and $\mathcal{H}_2$, and with associated Hamiltonians $\hat{H}_1$ and $\hat{H}_2$. Let E_n^1 and $|n\rangle_1$ be the eigenvalues and eigenvectors of $\hat{H}_1$, and let E_n^2 and $|n\rangle_2$ be the eigenvalues and eigenvectors of $\hat{H}_2$. If the two systems do not interact, we expect that the composite system will be described by a Hilbert space $\mathcal{H}$, and a Hamiltonian $\hat{H}$. The eigenvalues of $\hat{H}$ must be of the form $E_n^1 + E_{n'}^2$, for different values of n and n', because energy is an additive quantity.

The only way to satisfy this condition is by choosing the Hamiltonian

$$\hat{H} = \hat{H}_1 \otimes \hat{I} + \hat{I} \otimes \hat{H}_2 \tag{9.10}$$

on the Hilbert space $\mathcal{H}_1 \otimes \mathcal{H}_2$. The eigenvectors of $\hat{H}$ are $|n, n'\rangle = |n\rangle \otimes |n'\rangle$, since

$$\hat{H}|n, n'\rangle = \hat{H}_1 \otimes \hat{I}|n, n'\rangle + \hat{I} \otimes \hat{H}_2|n, n'\rangle = \hat{H}_1|n\rangle \otimes \hat{I}|n'\rangle + \hat{I}|n\rangle \otimes \hat{H}_2|n'\rangle$$
$$= E_n^1|n, n'\rangle + E_{n'}^2|n, n'\rangle = (E_n^1 + E_{n'}^2)|n, n'\rangle. \tag{9.11}$$

The Hamiltonian (9.10) can also be obtained by assuming a factorizable evolution operator $\hat{U}_t = e^{-it\hat{H}_1} \otimes e^{-it\hat{H}_2}$. By Leibniz's rule,

$$i\frac{d}{dt}\hat{U}_t = \hat{H}_1 e^{-it\hat{H}_1} \otimes e^{-it\hat{H}_2} + e^{-it\hat{H}_1} \otimes \hat{H}_2 e^{-it\hat{H}_2}$$
$$= (\hat{H}_1 \otimes \hat{I} + \hat{I} \otimes \hat{H}_2)\hat{U}_t. \tag{9.12}$$

Since time evolution operators satisfy $i\frac{d}{dt}\hat{U}_t = \hat{H}\hat{U}_t$, we confirm Eq. (9.10) for the Hamiltonian of the composite system.

We conclude that subsystems in quantum theory must be composed through the tensor product. We will refer to composite systems with n components as n-partite systems.

FP6 An n-partite system is described by the tensor product $\mathcal{H}_1 \otimes \mathcal{H}_2 \otimes \cdots \otimes \mathcal{H}_n$ of the Hilbert spaces $\mathcal{H}_i$, $i = 1, 2, \ldots, n$ of its subsystems.

The tensor product dramatically increases the number of available states in a composite system. Consider two subsystems described by the Hilbert space $\mathbb{C}^n$. The space of pure states for each system has real dimension $2(n-1)$. In classical physics, the state space for the composite system would have dimension $4(n-1)$. The Hilbert space of the composite system $\mathbb{C}^n \otimes \mathbb{C}^n = \mathbb{C}^{n^2}$

corresponds to a space of pure states of dimension $2(n^2 - 1)$, that is, of $2n^2 - 2 - 4(n-1) = 2(n-1)^2$ more dimensions than the space obtained through the classical rule for combining subsystems!

If a particle is described by wave functions $\phi_1(x)$, $(x \in \mathbb{R})$, and another particle by wave functions $\phi_2(y)$ $(y \in \mathbb{R})$, then the composite system of the two particles is described by a wave function $\psi(x, y)$, that is, a function that takes pairs $(x, y) \in \mathbb{R}^2$ to the complex numbers. This implies a natural identification of Hilbert spaces $L^2(\mathbb{R}, dx) \otimes L^2(\mathbb{R}, dy) = L^2(\mathbb{R}^2, dxdy)$. In general,

$$\mathcal{L}^2(\mathbb{R}^n, d^n x) \otimes \mathcal{L}^2(\mathbb{R}^m, d^m y) = \mathcal{L}^2(\mathbb{R}^{n+m}, d^n x d^m y). \tag{9.13}$$

In a bipartite system with two identical subsystems, each described by the Hilbert space $\mathcal{H}$, it is convenient to define the operators $\hat{A}_1 := \hat{A} \otimes \hat{I}$ and $\hat{A}_2 := \hat{I} \otimes \hat{A}$ on $\mathcal{H} \otimes \mathcal{H}$, for any operator $\hat{A}$ in $\mathcal{H}$. Hence, for two particles, we write the position operator of the first particle as $\hat{x}_1 = \hat{x} \otimes \hat{I}$ and of the second particle as $\hat{x}_2 = \hat{I} \otimes \hat{x}$. Furthermore, if the context is clear, we will occasionally write operators of the form (9.10) as $\hat{H} = \hat{H}_1 + \hat{H}_2$.

Example 9.1 Consider a free particle of mass m in a three-dimensional box, with sides L_1, L_2, and L_3. Since motion in each direction is an independent degree of freedom, it can be treated as a composite system of three degrees of freedom. Hence, it is described by the Hilbert space $L^2(I_{L_1}) \times L^2(I_{L_2}) \times L^2(I_{L_3}) = L^2(I_{L_1} \times I_{L_2} \times I_{L_3})$, with wave functions $\psi(\mathbf{x}) = \psi(x_1, x_2, x_2)$, where $x_i \in I_{L_i} = [0, L_i]$ for $i = 1, 2, 3$.

Its Hamiltonian $\hat{H} = \frac{1}{2m}(\hat{p}^2 \otimes \hat{I} \otimes \hat{I} + \hat{I} \otimes \hat{p}^2 \otimes \hat{I} + \hat{I} \otimes \hat{I} \otimes \hat{p}^2$ is written simply as $\hat{H} = \frac{1}{2m}(\hat{p}_1^2 + \hat{p}_2^2 + \hat{p}_3^2)$, or compactly as $\hat{H} = \frac{\hat{\mathbf{p}}^2}{2m}$. The eigenvalues are additive, that is, they are of the form

$$E_{n_1, n_2, n_3} = \frac{\pi^2}{2m} \left(\frac{n_1^2}{L_1^2} + \frac{n_2^2}{L_2^2} + \frac{n_3^2}{L_3^2} \right), \tag{9.14}$$

for $n_1, n_2, n_3 = 1, 2, \ldots$.

9.3   Entanglement

Consider a bipartite system with Hilbert space $\mathcal{H} = \mathcal{H}_1 \otimes \mathcal{H}_2$. Pure states on $\mathcal{H}$ of the form $|\phi\rangle \otimes |\psi\rangle$, where $|\phi\rangle \in \mathcal{H}_1$ and $|\psi\rangle \in \mathcal{H}_2$, are called *separable*. Any pure state that is not separable is called *entangled*. Entanglement of states is a unique characteristic of quantum systems, with no analogue in classical physics.

To understand entanglement it is useful first to introduce the concept of the *partial trace*. Let $|i\rangle$ be an orthonormal basis on $\mathcal{H}_1$ and let $|n\rangle$ be an orthonormal basis on $\mathcal{H}_2$. The vectors $|i, n\rangle = |i\rangle \otimes |n\rangle$ define an orthonormal basis on $\mathcal{H}_1 \otimes \mathcal{H}_2$. For any density matrix $\hat{\rho}$ on $\mathcal{H}_1 \otimes \mathcal{H}_2$, measurements on the first subsystem $\mathcal{H}_1$ are described by projectors of the form $\hat{P} \otimes \hat{I}$ and probabilities

$$\mathrm{Tr}\left[\hat{\rho}(\hat{P} \otimes \hat{I}) \right] = \sum_{i,n} \langle n, i | \hat{\rho} \hat{P} \otimes \hat{I} | n, i \rangle = \sum_{i,n} \sum_{j,m} \langle n, i | \hat{\rho} | m, j \rangle \langle m, j | \hat{P} \otimes \hat{I} | n, i \rangle$$

$$= \sum_{i,n} \sum_{j,m} \langle n, i | \hat{\rho} | m, j \rangle \langle m | \hat{P} | n \rangle \langle i | j \rangle = \sum_{n,m} \langle n | \hat{\rho}_1 | m \rangle \langle m | \hat{P} | n \rangle = \mathrm{Tr}(\hat{\rho}_1 \hat{P}). \tag{9.15}$$

In Eq. (9.15) we defined the *first partial trace* of $\hat{\rho}$ as a density matrix ρ_1 on $\mathcal{H}_1$ with elements

$$\langle n|\hat{\rho}_1|m\rangle = \sum_i \langle n,i|\hat{\rho}|m,i\rangle. \qquad (9.16)$$

Here, $\hat{\rho}_1$ contains all information of the composite system that is accessible by measurements on the first subsystem.

Similarly, we define the *second partial trace* of $\hat{\rho}$ as the density matrix $\hat{\rho}_2$ on $\mathcal{H}_2$, with elements

$$\langle i|\hat{\rho}_2|j\rangle = \sum_n \langle n,i|\hat{\rho}|n,j\rangle. \qquad (9.17)$$

We will often write $\hat{\rho}_1 = \mathrm{Tr}_{\mathcal{H}_2}\hat{\rho}$ and $\hat{\rho}_2 = \mathrm{Tr}_{\mathcal{H}_1}\hat{\rho}$. The states $\hat{\rho}_1$ and $\hat{\rho}_2$ are also called *reduced* density matrices.

9.3.1 Entanglement of Pure States

For pure states $\hat{\rho} = |\Psi\rangle\langle\Psi|$ on $\mathcal{H}$, the following property holds. If $|\Psi\rangle$ is separable (i.e., of the form $|\phi\rangle \otimes |\psi\rangle$), then the two partial traces of $\hat{\rho}$ are pure states: $\hat{\rho}_1 = |\phi\rangle\langle\phi|$ and $\hat{\rho}_2 = |\psi\rangle\langle\psi|$. However, if $|\Psi\rangle$ is entangled, then the partial traces are mixed states.

This remark is made precise by the following theorem – see Box 9.2 for proof.

Theorem 9.1 (Schmidt decomposition) *All vectors $|\Psi\rangle \in \mathcal{H}_1 \otimes \mathcal{H}_2$ can be expressed as*

$$|\Psi\rangle = \sum_{i=1}^{N} c_i|\phi_i\rangle \otimes |\psi_i\rangle, \qquad (9.18)$$

where $c_i > 0$, the vectors $|\phi_i\rangle$, $i = 1,2,\ldots,N$ define an orthonormal set on $\mathcal{H}_1$, and the vectors $|\psi_i\rangle$, $i = 1,2,\ldots,N$ define an orthonormal set on $\mathcal{H}_2$. The integer N is smaller than or equal to the smallest among the dimensions of $\mathcal{H}_1$ and $\mathcal{H}_2$; it is called the Schmidt order of the vector $|\Psi\rangle$. The numbers c_i are unique, the orthonormal sets $|\phi_i\rangle$ and $|\psi_i\rangle$ are not.

Obviously, if $N = 1$, then $|\Psi\rangle$ is separable, and if $N > 1$, $|\Psi\rangle$ is entangled. The partial traces of the density matrix $|\Psi\rangle\langle\Psi|$ are

$$\hat{\rho}_1 = \sum_{i=1}^{N} c_i^2|\phi_i\rangle\langle\phi_i|, \quad \hat{\rho}_2 = \sum_{i=1}^{N} c_i^2|\psi_i\rangle\langle\psi_i|. \qquad (9.19)$$

The two partial traces have the same purity $\gamma = \sum_{i=1}^{N} c_i^4$. The quantity

$$E := 1 - \gamma \qquad (9.20)$$

takes values in the interval $(0,1]$, and it provides a measure of entanglement, as it vanishes for all separable states.[1] We will refer to E as *quadratic entanglement*, as it is quadratic with respect to the reduced state.

---

[1] A better measure of entanglement for pure states is the von Neumann entropy of $\hat{\rho}_1$: $S[\hat{\rho}_1] = -\mathrm{Tr}(\hat{\rho}_1 \ln \hat{\rho}_1) = -\sum_{i=1}^{N} c_i^2 \ln c_i^2$. For properties of the von Neumann entropy, see Box 6.3.

If $N = \dim \mathcal{H}_1 = \dim \mathcal{H}_2$, and a state $|\Psi\rangle$ is characterized by $c_1 = c_2 = \cdots = c_n = \frac{1}{N}$, the partial traces are states of maximal ignorance, that is, the most mixed states. Then, the vector $|\Psi\rangle$ is a *maximal-entanglement* state. For such states, $E = 1 - N^{-1}$.

Note that any unitary transformation of the form $\hat{U}_1 \otimes \hat{U}_2$ only changes the orthonormal sets $|\phi_i\rangle$ and $|\psi_i\rangle$ in the Schmidt decomposition. The values of the c_i remain unchanged. Hence, any entanglement measure defined solely from the c_i, like the quadratic entanglement E of Eq. (9.20) remains invariant under the transformations $|\Psi\rangle \rightarrow \hat{U}_1 \otimes \hat{U}_2 |\Psi\rangle$.

The state of an entangled system cannot be reconstructed from the knowledge of the states of its constituents. The reduced density matrices of the latter are mixed, hence, they have less information than maximal. For this reason,

...the phenomenon of quantum entanglement suggests a holistic structure for the physical world that contrasts strongly with the predominantly reductionist views of Western philosophy whereby composite systems may be analysed in terms of their constituent subsystems. (Isham, 1995)

Box 9.2   Proof of the Schmidt Decomposition

The Schmidt decomposition is a special case of a key result of matrix theory, the *singular value expansion*.

Singular Value Expansion Any $n \times m$ complex matrix A can be expressed as $A = U\Sigma V^\dagger$, where (i) U is a unitary $n \times n$ matrix, (ii) Σ is a diagonal $n \times m$ matrix, whose elements are either positive numbers or zero, and (iii) V is a unitary $m \times m$ matrix.

The unitary matrices U and V are not unique, the diagonal elements of Σ are unique up to rearrangement. The latter are called *singular values* of A.

Consider two orthonormal bases $|i\rangle$ and $|k\rangle$ in $\mathbb{C}^n$ and in $\mathbb{C}^n$, respectively. A vector $|\Psi\rangle \in \mathbb{C}^n \otimes \mathbb{C}^m$ can be expressed as $|\Psi\rangle = \sum_{i=1}^{n} \sum_{k=1}^{m} C_{ik} |i\rangle \otimes |k\rangle$, in terms of a $n \times m$ matrix C_{ik}. We use the singular value expansion for C, to obtain

$$C_{ik} = \sum_{j=1}^{n} \sum_{l=1}^{m} U_{ij} \Sigma_{jl} V_{kl}^*. \tag{A}$$

Without loss of generality, we take $n \leq m$. Then, the matrix Σ is of the form

$$\Sigma = \begin{pmatrix} c_1 & 0 & 0 & \ldots & 0 & 0 & 0 & \ldots & 0 \\ 0 & c_2 & 0 & \ldots & 0 & 0 & 0 & \ldots & 0 \\ \ldots & \ldots & \ldots & \ldots & \ldots & \ldots & \ldots & \ldots & \ldots \\ 0 & 0 & 0 & \ldots & c_n & 0 & 0 & \ldots & 0 \end{pmatrix},$$

where $c_1, c_2, \ldots, c_n$ are the singular values of C. We order c_i so that they appear in decreasing order in the diagonal. Summation in Eq. (A) runs for $j, l = 1, 2, \ldots N$, where N is the number of nonzero singular values c_i. Hence, $C_{ik} = \sum_{j=1}^{N} c_j U_{ij} V_{kj}^*$.

We conclude that

$$|\Psi\rangle = \sum_{j=1}^{N} \sum_{i=1}^{n} \sum_{k=1}^{m} c_j U_{ij} V_{kj}^* |i\rangle \otimes |k\rangle = \sum_{i=1}^{N} c_i |\phi_i\rangle \otimes |\psi_i\rangle,$$

where $|\phi_j\rangle = \sum_{i=1}^{n} U_{ij} |i\rangle$ and $|\psi_j\rangle = \sum_{k=1}^{m} V_{kj}^* |k\rangle$ define orthonormal sets on $\mathbb{C}^n$ and on $\mathbb{C}^m$ respectively, since the matrices U and V are unitary.

9.3.2   Entanglement of Mixed States

The notion of entanglement also applies to mixed states. Factorized density matrices obviously correspond to separable states. Since mixing is a classical operation, we expect that mixtures of factorized states will also behave classically. With this reasoning, we call a density matrix $\hat{\rho}$ on the Hilbert space $\mathcal{H}_1 \otimes \mathcal{H}_2$ separable, if it can be expressed as a mixture of factorized density matrices, that is, if

$$\hat{\rho} = \sum_{k=1}^{r} \lambda_k \hat{\rho}_1^{(k)} \otimes \hat{\rho}_2^{(k)}, \tag{9.21}$$

where $\lambda_k \geq 0$ and $\sum_{k=1}^{r} \lambda_k = 1$. The maximal-ignorance density matrix on $\mathcal{H}_1 \otimes \mathcal{H}_2$ is obviously separable. Any nonseparable density matrix is entangled.

The first partial trace of the separable state (9.21) is $\hat{\rho}_1 = \sum_k \lambda_k \hat{\rho}_1^{(k)}$; hence,

$$\hat{\rho}_1 \otimes \hat{I} - \hat{\rho} = \sum_{k=1}^{r} \lambda_k \lambda_k \hat{\rho}_1^{(k)} \otimes (\hat{I} - \hat{\rho}_2^{(k)}) \geq 0, \tag{9.22}$$

where the inequality in the last steps follows because $\hat{\rho}_1^{(k)} \geq 0$ and $\hat{I} - \hat{\rho}_2^{(k)} \geq 0$. It follows that a density matrix $\hat{\rho}$ that fails the inequality (9.22) is entangled.

The violation of inequality (9.22) is a sufficient but not necessary condition for entanglement, that is, any state that violates it is entangled, but not all entangled states violate it.

Another widely used entanglement criterion is due to Peres (1996) and Horodecki et al. (1996). It is usually called *PPT criterion* where PPT stands for positive partial transpose.

Let $|n\rangle$ be an orthonormal basis on the Hilbert space $\mathcal{H}_1$ and let $|i\rangle$ be an orthonormal basis on the Hilbert space $\mathcal{H}_2$. The partial transpose transformation T_2 swaps the matrix elements of a density matrix $\hat{\rho}$ on $\mathcal{H}_1 \otimes \mathcal{H}_2$ with respect to the basis $|i\rangle$,

$$\langle m, j | T_2[\hat{\rho}] | n, i \rangle = \langle m, i | \hat{\rho} | n, j \rangle. \tag{9.23}$$

For a separable density matrix (9.21),

$$T_2[\hat{\rho}] = \sum_{k=1}^{r} \lambda_k \hat{\rho}_1^{(k)} \otimes [\hat{\rho}_2^{(k)}]^T, \tag{9.24}$$

where by T we denote the transpose of an operator. The transpose of a density matrix is still a density matrix. Hence, the action of T_2 on a separable density matrix $\hat{\rho}$ gives another (separable) density matrix $T_2[\hat{\rho}]$.

In general, entangled density matrices are not mapped to density matrices by T_2. To see this, consider a pure entangled state $\hat{\rho} = |\Psi\rangle\langle\Psi|$. We choose $|\Psi\rangle$ to be of Schmidt order 2, so that $|\Psi\rangle = \cos\theta|1\rangle \otimes |a\rangle + \sin\theta|2\rangle \otimes |b\rangle$, where $|1\rangle$ and $|2\rangle$ form an orthonormal set in $\mathcal{H}_1$, $|a\rangle$ and $|b\rangle$ form an orthonormal set in $\mathcal{H}_2$, and $\theta \in [0, \frac{\pi}{2}]$. Then,

$$\begin{aligned}
\hat{\rho} = {} & \cos^2\theta|1\rangle\langle 1| \otimes |a\rangle\langle a| + \sin^2\theta|2\rangle\langle 2| \otimes |b\rangle\langle b| \\
& + \sin\theta\cos\theta \left(|1\rangle\langle 2| \otimes |a\rangle\langle b| + |2\rangle\langle 1| \otimes |b\rangle\langle a|\right).
\end{aligned} \tag{9.25}$$

Hence,

$$T_2[\hat{\rho}] = \cos^2\theta|1\rangle\langle 1| \otimes |a\rangle\langle a| + \sin^2\theta|2\rangle\langle 2| \otimes |b\rangle\langle b|$$
$$+ \sin\theta\cos\theta\left(|1\rangle\langle 2| \otimes |b\rangle\langle a| + |2\rangle\langle 1| \otimes |a\rangle\langle b|\right). \tag{9.26}$$

It is straightforward to verify that the vector $|\Phi\rangle = \frac{1}{\sqrt{2}}\left(|1\rangle \otimes |b\rangle - |2\rangle \otimes |a\rangle\right)$ is an eigenvector of $T_2[\hat{\rho}]$ with negative eigenvalue,

$$T_2[\hat{\rho}]|\Phi\rangle = -\sin\theta\cos\theta|\Phi\rangle. \tag{9.27}$$

Hence, $T_2[\hat{\rho}]$ is not positive.

The PPT criterion asserts that any density matrix $\hat{\rho}$ on $\mathcal{H}_1 \otimes \mathcal{H}_2$, with a negative eigenvalue for $T_2[\hat{\rho}]$ is entangled. In general, the PPT criterion is sufficient but not necessary for entanglement. It can be proven that for bipartite systems with Hilbert spaces $\mathbb{C}^2 \otimes \mathbb{C}^2$ and $\mathbb{C}^2 \otimes \mathbb{C}^3$, PPT is both sufficient and necessary.

The absolute value of the smallest negative eigenvalue λ_{min} of $T_2[\hat{\rho}]$ can be used to quantify entanglement. For example, the reduced density matrix associated to $\hat{\rho}$ of Eq. (9.25) has matrix elements $\langle 1|\hat{\rho}_1|1\rangle = \cos^2\theta$, $\langle 2|\hat{\rho}_1|2\rangle = \sin^2\theta$, and $\langle 1|\hat{\rho}_1|2\rangle = 0$. Hence, the quadratic entanglement (9.20) is $E = 1 - \cos^4\theta - \sin^4\theta = 2\sin^2\theta\cos^2\theta = 2|\lambda_{min}|^2$.

There is no general mathematical measure of entanglement for mixed states. There are several different proposals, but they are all hard to compute. All such measures must satisfy two crucial properties. First, they must be invariant under factorized unitary transformations $\hat{\rho} \to \left(\hat{U}_1 \otimes \hat{U}_2\right)\hat{\rho}\left(\hat{U}_1^\dagger \otimes \hat{U}_2^\dagger\right)$, similar to the measure (9.20) for pure states. Second, entanglement must not increase when we extract local information, that is, information from only one of the subsystems. This means that the state $\hat{\rho}'$ of the statistical ensemble after the measurement of an operator $\hat{A} \otimes \hat{I}$ must have entanglement equal or less than the entanglement of the state $\hat{\rho}$ prior to measurement.

9.4   The Two-Qubit System

The two-qubit system is the simplest system in which to study entanglement. It is described by the Hilbert space $\mathcal{H} = \mathbb{C}^2 \otimes \mathbb{C}^2$. In this system, Eq. (9.16) for the reduced density matrix becomes

$$\langle 1|\hat{\rho}_1|1\rangle = 1 - \langle 0|\hat{\rho}_1|0\rangle = \langle 1,1|\Psi\rangle\langle\Psi|1,1\rangle + \langle 1,0|\Psi\rangle\langle\Psi|1,0\rangle$$
$$\langle 1|\hat{\rho}_1|0\rangle = \langle 1,1|\Psi\rangle\langle\Psi|0,1\rangle + \langle 1,0|\Psi\rangle\langle\Psi|0,0\rangle. \tag{9.28}$$

The following four vectors are known as *Bell vectors*, and they define an orthonormal basis on $\mathcal{H}$, the *Bell basis*.

$$|\Psi_\pm\rangle = \frac{1}{\sqrt{2}}\left(|1,0\rangle \pm |0,1\rangle\right) \tag{9.29}$$

$$|\Phi_\pm\rangle = \frac{1}{\sqrt{2}}\left(|1,1\rangle \pm |0,0\rangle\right). \tag{9.30}$$

By Eq. (9.28), the Bell vectors satisfy

$$\langle 1|\hat{\rho}_1|1\rangle = \langle 0|\hat{\rho}_1|0\rangle = \frac{1}{2} \qquad \langle 1|\hat{\rho}_1|0\rangle = 0. \qquad (9.31)$$

$$\langle 1|\hat{\rho}_2|1\rangle = \langle 0|\hat{\rho}_2|0\rangle = \frac{1}{2} \qquad \langle 1|\hat{\rho}_2|0\rangle = 0. \qquad (9.32)$$

Hence, $\hat{\rho}_1 = \hat{\rho}_2 = \frac{1}{2}\hat{I}$. The partial trace of all Bell vectors is the state of maximum ignorance for a single qubit. Hence, the Bell vectors are maximum-entanglement states.

9.4.1 Probabilities and Correlations

Next, we construct the probabilities for the measurement of an operator $\hat{A} = \mathbf{m} \cdot \boldsymbol{\sigma}$ on the first qubit and an operator $\hat{B} = \mathbf{n} \cdot \boldsymbol{\sigma}$ on the second qubit, where $\mathbf{m}$ and $\mathbf{n}$ are unit vectors. The associated operators on $\mathcal{H}$ are $\hat{A} \otimes \hat{I}$ and $\hat{I} \otimes \hat{B}$.

The spectral projectors of $\hat{A}$ are $\hat{P}_\pm = \frac{1}{2}(\hat{I} \pm \mathbf{m} \cdot \boldsymbol{\sigma})$ and the spectral projectors of $\hat{B}$ are $\hat{Q}_\pm = \frac{1}{2}(\hat{I} \pm \mathbf{n} \cdot \boldsymbol{\sigma})$. The spectral projectors of $\hat{A} \otimes \hat{I}$ are $\hat{P}_a \otimes \hat{I}$, where $a = \pm 1$; the spectral projectors of $\hat{I} \otimes \hat{B}$ are $\hat{I} \otimes \hat{Q}_b$, where $b = \pm 1$.

The operators $\hat{A} \otimes \hat{I}$ and $\hat{I} \otimes \hat{B}$ commute. Hence, the joint probability of finding a in the first qubit and b in the second qubit does not depend on the order of the measurements, and it is given by Eq. (8.15),

$$\text{Prob}(a,b) = \text{Tr}[\hat{\rho}(\hat{P}_a \otimes \hat{I})(\hat{I} \otimes \hat{Q}_b)] = \text{Tr}(\hat{\rho}\hat{P}_a \otimes \hat{Q}_b), \qquad (9.33)$$

for all initial states $\hat{\rho}$.

Example 9.2 We evaluate the probabilities $\text{Prob}(a,b)$ for the initial state $|\Psi_-\rangle$ of Eq. (9.29),

$$\text{Prob}(a,b) = \langle \Psi_-|\hat{P}_a \otimes \hat{Q}_b|\Psi_-\rangle = \frac{1}{2}\Big(\langle 1|\hat{P}_a|1\rangle\langle 0|\hat{Q}_b|0\rangle + \langle 0|\hat{P}_a|0\rangle\langle 1|\hat{Q}_b|1\rangle$$
$$-\langle 1|\hat{P}_a|0\rangle\langle 0|\hat{Q}_b|1\rangle - \langle 0|\hat{P}_a|1\rangle\langle 1|\hat{Q}_b|0\rangle \Big). \quad (9.34)$$

By Eq. (5.16),

$$\langle 0|\hat{P}_\pm|0\rangle = \frac{1}{2}(1 \mp m_3), \quad \langle 1|\hat{P}_\pm|1\rangle = \frac{1}{2}(1 \pm m_3), \quad \langle 0|\hat{P}_\pm|1\rangle = \pm\frac{1}{2}(m_1 + im_2),$$

$$\langle 0|\hat{Q}_\pm|0\rangle = \frac{1}{2}(1 \mp n_3), \quad \langle 1|\hat{Q}_\pm|1\rangle = \frac{1}{2}(1 \pm n_3), \quad \langle 0|\hat{Q}_\pm|1\rangle = \pm\frac{1}{2}(n_1 + in_2).$$

We substitute into Eq. (9.34), to obtain

$$\text{Prob}(+,+) = \text{Prob}(-,-) = \frac{1}{4}(1 - \mathbf{m} \cdot \mathbf{n}) \qquad (9.35)$$

$$\text{Prob}(+,-) = \text{Prob}(-,+) = \frac{1}{4}(1 + \mathbf{m} \cdot \mathbf{n}). \qquad (9.36)$$

We define the bipartite correlation $C(\mathbf{m}, \mathbf{n})$ between a measurement of $\hat{A} = \mathbf{m} \cdot \boldsymbol{\sigma}$ on the first qubit and a measurements of $\hat{B} = \mathbf{n} \cdot \boldsymbol{\sigma}$ on the second qubit, as

$$C(\mathbf{m}, \mathbf{n}) = \text{Tr}\left[\hat{\rho}(\hat{A} \otimes \hat{B})\right]. \qquad (9.37)$$

For maximum entanglement states $\text{Tr}\left[\hat{\rho}(\hat{A} \otimes \hat{I})\right] = \text{Tr}(\hat{\rho}_1 \hat{A}) = \frac{1}{2}\text{Tr}\hat{A} = 0$, and, similarly, $\text{Tr}\left[\hat{\rho}(\hat{I} \otimes \hat{B})\right] = 0$. Then, by Eq. (6.33),

$$C(\mathbf{m}, \mathbf{n}) := \text{Cor}(\hat{A} \otimes \hat{I}, \hat{I} \otimes \hat{B}). \tag{9.38}$$

Example 9.3 For $\hat{\rho} = |\Psi_-\rangle\langle\Psi_-|$,

$$C(\mathbf{m}, \mathbf{n}) = \frac{1}{2}\left(\langle 1|\hat{A}|1\rangle\langle 0|\hat{B}|0\rangle + \langle 0|\hat{A}|0\rangle\langle 1|\hat{B}|1\rangle - \langle 1|\hat{A}|0\rangle\langle 0|\hat{B}|1\rangle - \langle 0|\hat{A}|1\rangle\langle 1|\hat{B}|0\rangle\right).$$

We insert the values of the matrix elements A_{ij} and B_{ij}, to obtain

$$C(\mathbf{m}, \mathbf{n}) = -\mathbf{m} \cdot \mathbf{n}. \tag{9.39}$$

We can also derive Eq. (9.39) as follows. By the spectral theorem $\hat{A} = \hat{P}_+ - \hat{P}_-$ and $\hat{B} = \hat{Q}_+ - \hat{Q}_-$. Hence,

$$\hat{A} \otimes \hat{B} = \hat{P}_+ \otimes \hat{Q}_+ + \hat{P}_- \otimes \hat{Q}_- - \hat{P}_+ \otimes \hat{Q}_- - \hat{P}_- \otimes \hat{Q}_+. \tag{9.40}$$

It follows that

$$C(\mathbf{m}, \mathbf{n}) = \text{Prob}(+, +) + \text{Prob}(-, -) - \text{Prob}(+, -) - \text{Prob}(-, +). \tag{9.41}$$

Then, Eq. (9.36) leads to Eq. (9.39).

9.4.2   Mixed States

An interesting family of entangled mixed states are the Werner states (Werner, 1989)

$$\hat{\rho}_W = \alpha|\Psi_-\rangle\langle\Psi_-| + \frac{1-\alpha}{4}\hat{I}. \tag{9.42}$$

Ostensibly, they are obtained as a mixture of the maximal entanglement state $|\Psi_-\rangle$ and the maximal ignorance state $\frac{1}{4}\hat{I}$, which is separable.

We express the Werner state as a matrix in the basis defined by the vectors $|0,0\rangle, |0,1\rangle, |1,0\rangle, |1,1\rangle$,

$$\hat{\rho}_W = \frac{1}{4}\begin{pmatrix} 1-\alpha & 0 & 0 & 0 \\ 0 & 1+\alpha & -2\alpha & 0 \\ 0 & -2\alpha & 1+\alpha & 0 \\ 0 & 0 & 0 & 1-\alpha \end{pmatrix}. \tag{9.43}$$

The matrix (9.42) has a single eigenvalue $\frac{1}{4}(1 + 3\alpha)$ and a triple eigenvalue $\frac{1}{4}(1 - \alpha)$. The Werner states are well defined (the eigenvalues are positive) for $-\frac{1}{3} \le \alpha \le 1$. For $-\frac{1}{3} \le \alpha \le 0$, the Werner states cannot be interpreted as a mixture of $|\Psi_-\rangle$ and the maximal ignorance state.

The first partial trace is $\hat{\rho}_1 = \frac{1}{2}\hat{I}$. Hence,

$$\hat{\rho}_1 \otimes \hat{I} - \hat{\rho}_W = \frac{1}{4}\begin{pmatrix} 1+\alpha & 0 & 0 & 0 \\ 0 & 1-\alpha & +2\alpha & 0 \\ 0 & +2\alpha & 1+\alpha & 0 \\ 0 & 0 & 0 & 1-\alpha \end{pmatrix}. \tag{9.44}$$

The matrix $\hat{\rho}_1 \otimes \hat{I} - \hat{\rho}_W$ is identical to the matrix (9.43) for $\alpha \to -\alpha$. Hence, its eigenvalues are $\frac{1}{4}(1 - 3\alpha)$ and $\frac{1}{4}(1 + \alpha)$ (triple). The eigenvalues are both nonnegative only for $\alpha \leq \frac{1}{3}$. By the separability criterion (9.22), Werner states with $\alpha > \frac{1}{3}$ are entangled.

9.5 Bell's Theorem

Bell's theorem (Bell, 1964) is one of the most important results of quantum theory. It shows that the assumption that measurements reveal preexistent properties of quantum systems leads to predictions that are incompatible with quantum theory.

Bell's theorem originates from a paper by Einstein, Podolsky, and Rosen (EPR), who used quantum entanglement in order to argue that quantum theory is incomplete (Einstein et al., 1935). The term "entanglement" did not exist then; in some sense, the EPR paper is the first analysis of entanglement.

In modern terminology, the argument is the following. Consider a system of two qubits[2] prepared in the state $|\Psi_-\rangle$. By Eq. (9.36), the joint probabilities for the measurement of the same observable ($\mathbf{m} = \mathbf{n}$) are

$$\text{Prob}(+, +) = \text{Prob}(-, -) = 0, \quad \text{Prob}(+, -) = \text{Prob}(-, +) = \frac{1}{2}. \tag{9.45}$$

The associated conditional probabilities are $\text{Prob}(+|-) = \text{Prob}(-|+) = 1$.

Suppose then that we measure $\hat{\sigma}_3$ on the first qubit. If we find ± 1, we can infer that a measurement of $\hat{\sigma}_3$ on the second qubit will give with certainty ∓ 1. EPR then invoked their "criterion of reality":

If, without in any way disturbing a system, we can predict with certainty (i.e., with probability equal to unity) the value of a physical quantity, then there exists an element of physical reality corresponding to this physical quantity.

According to this criterion, the value of $\hat{\sigma}_3$ on the second qubit is an element of reality after the measurement of $\hat{\sigma}_3$ on the first qubit has been carried out, even if we do not carry out a measurement on the second qubit.[3] On the other hand, had we chosen to measure $\hat{\sigma}_1$ on the first qubit, the element of reality of the second qubit would be its value for $\hat{\sigma}_1$.

However, if the qubits are far enough apart, the choice of measurement on the first qubit should not affect the elements of reality in the second qubit. Hence, the qubit should have a definite value of both $\hat{\sigma}_3$ and $\hat{\sigma}_1$, which is impossible in quantum theory.

At this point in the book, the reader can easily identify the problem with the EPR argument. The quantum conditional probabilities refer solely to measurement outcomes, and not to any properties possessed by the system. Quantum theory admits no elements of reality in the EPR

---

[2] The original EPR paper involved measurements of position and momentum in a pair of particles. The simpler argument with qubits first appeared in sections 22.15–22.18 of Bohm (1951).

[3] The inference of the value for $\hat{\sigma}_3$ on the second qubit does not involve any transfer of information. Since we know the initial state, we know that the outcomes of the two measurements are fully correlated. The same applies to classical systems. For example, when two particles scatter, the measurement of the momentum of one particle in the center-of-mass frame fully specifies the momentum of the other particle.

sense. If quantum elements of reality exist, they refer to the complex that consists of the microscopic system and the measuring apparatus, and not solely to the microscopic system (Bohr, 1935).

9.5.1 Bell's Inequality

Bell realized that the EPR's view of "elements of reality" leads to concrete experimental predictions that differ from those of quantum theory. This statement is known as Bell's theorem.

Again, we consider a two-qubit system, subject to measurements of observables $\mathbf{m} \cdot \boldsymbol{\sigma}$ for each qubit. The unit vectors $\mathbf{m}$ correspond to macroscopic parameters that are specified in advance. The possible measurement outcomes are $+1$ and -1.

Suppose that the measurements reveal preexisting properties of the quantum system, "elements of reality" at a level of description that is deeper, and more fundamental than quantum theory. These properties must be expressed in terms of some variables ξ that define a state space Γ for the two-qubit system. As shown in Section 1.5.3, the measurement of any observable corresponds to a function $f : \Gamma \to \mathbb{R}$. For a dichotomic observable with values ± 1, we can restrict to functions $f : \Gamma \to \{+1, -1\}$.

Hence, the measurement of $\mathbf{m} \cdot \hat{\boldsymbol{\sigma}} \otimes \hat{I}$ corresponds to a function $f_{\mathbf{m}}^1 : \Gamma \to \{+1, -1\}$, and the measurement of $\hat{I} \otimes \mathbf{m} \cdot \hat{\boldsymbol{\sigma}}$ corresponds to a function $f_{\mathbf{m}}^2 : \Gamma \to \{+1, -1\}$.

Then, we consider the following experiment. A source emits entangled qubit pairs, and each qubit is driven towards a different apparatus. In qubit 1, we measure $\mathbf{m}_1 \cdot \hat{\boldsymbol{\sigma}}$, and in qubit 2, we measure $\mathbf{m}_2 \cdot \hat{\boldsymbol{\sigma}}$. We carry out the measurement in a statistical ensemble of $N \gg 1$ qubit pairs. For the kth pair, we denote by $a_k \in \{-1, 1\}$ the outcome of the measurement of the first qubit, and by $b_k \in \{-1, 1\}$ the outcome of the measurement of the second qubit. From these measurement outcomes, we construct the correlation

$$C(\mathbf{m}_1, \mathbf{m}_2) = \frac{1}{N} \sum_{k=1}^{N} a_k b_k, \tag{9.46}$$

which depends on the type of measurement carried out on the two qubits, as described by the unit vectors $\mathbf{m}_1$ and $\mathbf{m}_2$.

If $C(\mathbf{m}_1, \mathbf{m}_2)$ converges to a constant at the limit $N \to \infty$, then we can attempt to reproduce it from a classical probability distribution ρ on the state Γ,

$$C(\mathbf{m}_1, \mathbf{m}_2) = \int d\xi \rho(\xi) f_{\mathbf{m}_1}^1(\xi) f_{\mathbf{m}_2}^2(\xi). \tag{9.47}$$

Using the same source of qubit pairs, we perform measurements in four different experiments, each characterized by a different pair of unit vectors for the two qubit measurements: (i) $(\mathbf{m}_1, \mathbf{m}_2)$, (ii) $(\mathbf{m}_1, \mathbf{n}_2)$, (iii) $(\mathbf{n}_1, \mathbf{m}_2)$, and (iv) $(\mathbf{n}_1, \mathbf{n}_2)$.

We evaluate the measured correlation function in each experiment, and then we construct the quantity

$$\Delta = C(\mathbf{m}_1, \mathbf{m}_2) + C(\mathbf{m}_1, \mathbf{n}_2) + C(\mathbf{n}_1, \mathbf{m}_2) - C(\mathbf{n}_1, \dot{\mathbf{n}}_2). \tag{9.48}$$

Since the four experiments were carried out with the same preparation of the system, they must be described by the same probability distribution $\rho(\xi)$. Then,

$$\Delta = \int d\xi \rho(\xi) \Phi(\xi), \tag{9.49}$$

where

$$\Phi := f^1_{\mathbf{m}_1} f^2_{\mathbf{m}_2} + f^1_{\mathbf{m}_1} f^2_{\mathbf{n}_2} + f^1_{\mathbf{n}_1} f^2_{\mathbf{m}_2} - f^1_{\mathbf{n}_1} f^2_{\mathbf{n}_2} \tag{9.50}$$

is a function of Γ. We write Φ as

$$\Phi = f^1_{\mathbf{m}_1}(f^2_{\mathbf{m}_2} + f^2_{\mathbf{n}_2}) + f^1_{\mathbf{n}_1}(f^2_{\mathbf{m}_2} - f^2_{\mathbf{n}_2}). \tag{9.51}$$

The possible values of the functions $f^2_{\mathbf{m}_2} \pm f^2_{\mathbf{n}_2}$ are $-2, 0$ and $+2$. We note that if $f^2_{\mathbf{m}_2} + f^2_{\mathbf{n}_2} = \pm 2$, then $f^2_{\mathbf{m}_2} - f^2_{\mathbf{n}_2} = 0$, and if $f^2_{\mathbf{m}_2} + f^2_{\mathbf{n}_2} = 0$, then $f^2_{\mathbf{m}_2} - f^2_{\mathbf{n}_2} = \pm 2$. Hence, if $|f^2_{\mathbf{m}_2} \pm f^2_{\mathbf{n}_2}| = 2$, then $|f^2_{\mathbf{m}_2} \mp f^2_{\mathbf{n}_2}| = 0$.

Therefore,

$$|\Phi| \leq |f^1_{\mathbf{m}_1}||f^2_{\mathbf{m}_2} + f^2_{\mathbf{n}_2}| + |f^1_{\mathbf{n}_1}||f^2_{\mathbf{m}_2} - f^2_{\mathbf{n}_2}| = |f^2_{\mathbf{m}_2} + f^2_{\mathbf{n}_2}| + |f^2_{\mathbf{m}_2} - f^2_{\mathbf{n}_2}| = 2.$$

We conclude that the measurable quantity Δ, Eq. (9.48), always satisfies

$$|\Delta| \leq 2; \tag{9.52}$$

Eq. (9.52) is *Bell's inequality*.[4]

Bell's inequality is a consequence of describing measurements through classical probability theory. Quantum theory violates it. If the system has been prepared in a pure state $|\Psi\rangle \in \mathcal{H}$, $\Delta = \langle\Psi|\hat{D}|\Psi\rangle$, where

$$\hat{D} = \mathbf{m}_1 \cdot \hat{\boldsymbol{\sigma}} \otimes \mathbf{m}_2 \cdot \hat{\boldsymbol{\sigma}} + \mathbf{m}_1 \cdot \hat{\boldsymbol{\sigma}} \otimes \mathbf{n}_2 \cdot \hat{\boldsymbol{\sigma}} + \mathbf{n}_1 \cdot \hat{\boldsymbol{\sigma}} \otimes \mathbf{m}_2 \cdot \hat{\boldsymbol{\sigma}} - \mathbf{n}_1 \cdot \hat{\boldsymbol{\sigma}} \otimes \mathbf{n}_2 \cdot \hat{\boldsymbol{\sigma}}. \tag{9.53}$$

For the initial state $|\Psi_-\rangle$ of Eq. (9.29), Eq. (9.39) implies that

$$\Delta = -\mathbf{m}_1 \cdot \mathbf{m}_2 - \mathbf{m}_1 \cdot \mathbf{n}_2 - \mathbf{n}_1 \cdot \mathbf{m}_2 + \mathbf{n}_1 \cdot \mathbf{n}_2. \tag{9.54}$$

We select the following unit vectors

$$\mathbf{m}_1 = (1,0,0) \quad \mathbf{m}_2 = (\cos\theta, 0, \sin\theta)$$
$$\mathbf{n}_1 = (0,0,1) \quad \mathbf{n}_2 = (\cos\theta, 0, -\sin\theta), \tag{9.55}$$

for some angle θ. Equation (9.54) yields $|\Delta| = 2|\sin\theta + \cos\theta|$. The maximum value

$$|\Delta|_{\max} = 2\sqrt{2} > 2, \tag{9.56}$$

is achieved for $\theta = \frac{\pi}{4}$. Hence, *some quantum states violate Bell's inequality*.

The problem is that Bell's inequality follows from elementary, ostensibly obvious, assumptions. The Bell-inequality assumptions (BIA) are the following.

---

[4] In fact, Eq. (9.52) is not due to Bell, but to Clauser, Horne, Shimony, and Holt (Clauser et al., 1969), and it is often called CHSH inequality. This inequality both simplifies and generalizes the original inequality by John Bell.

BIA1 Measurements reveal preexisting properties of quantum systems. Finding $+1$ in a qubit measurement asserts that the qubit has a specific property, say a_+, while finding -1 implies that the qubit has a specific property a_- that is not compatible with a_+. This is simply described by the introduction of a state space Γ for the qubit. The property a_+ corresponds to the subset $C_+ = f_{\mathbf{m}}^{-1}(+1)$, and a_- corresponds to the subset $C_- = f_{\mathbf{m}}^{-1}(-1)$. Obviously C_+ and C_- are disjoint.

This assumption is called *microscopic-property realism*, *microrealism* for short, or simply *realism*. The word "realism" is used as a technical philosophical term, meaning that something is objectively real and not dependent on human activities, beliefs, or narratives. For example, "mathematical realism" stands for the statement that the objects of mathematics exist independently of our mathematical activity, and that a mathematical proposition is true or false, irrespective of whether we can prove it or not. In the context of physics, microscopic-property realism means that a microscopic system has objective properties, that is, values for physical variables, that do not depend on how we choose to measure them.

Traditionally, experimental sciences have been treating property realism as a truism, that is, they affirm that measurements reveal properties of physical systems, and that the goal of the experimentalist is to come as close to the "true value" of physical variables as possible.

BIA2 We assumed that the probability density ρ on Γ is the same in all four experimental configurations. Again, this is an obvious assumption, given that the procedure of preparing the qubit system and the procedure of configuring the apparatuses (i.e., choosing the observables to be measured) are completely independent from each other. BIA2 is often referred to as *measurement independence*. Measurement independence has been traditionally assumed in all experiments, in fact it is a precondition for a meaningful experimental science.

Nonetheless, it is conceivable that the apparatuses exert some influence on the detectors, by a process yet unknown to us, so that the choice of observables in the apparatus affects the probability density ρ. Of course, we can always place the two measurement apparatuses very far from each other and from the source. We can also make the choice about what observable to measure in each apparatus so late that no signal traveling slower-than-light can reach qubit 2 from the configuration of the apparatus for qubit 1.

Hence, BIA2 is guaranteed by the *principle of locality*, that there is no instantaneous transmission of information. The locality principle is a consequence of relativity, and, to our present state of knowledge, it holds with no exceptions. Note that the locality principle is a more stringent condition than BIA2. BIA2 is eminently reasonable even in a world with instantaneous transmission of information.

The violation of Bell's inequality implies that either BIA1 or BIA2 is false. If BIA2 is false then the principle of locality is also false. Hence the violation of Bell's inequality contradicts the conjunction of property realism (BIA1) and the locality principle. This conjunction is known as *local realism*.

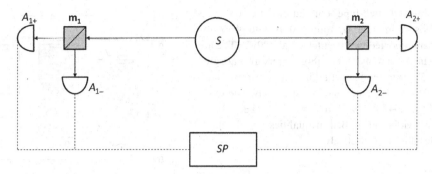

Fig. 9.1 Experiment for measuring Bell-type correlations in photonic qubits. S: source of entangled photon pairs. $\mathbf{m}_1, \mathbf{m}_2$: analyzers that split a photon beam with respect to the polarization in the chosen direction. $A_{1\pm}, A_{2\pm}$: photodetectors. SP: signal processor for counting coincidences.

9.5.2 Experimental Tests

Since both locality and realism are foundational to our understanding of the physical world, experiments that test Bell's inequality are of the highest importance. The first series of experiments to confirm the violation of Bell's inequality was undertaken by Alain Aspect in the early 1980s (Aspect et al., 1981, 1982). Aspect used photonic qubits, in which the vectors $\mathbf{m}$ correspond to the direction of the polarization. One of the experimental configurations is described in Fig. 9.1. Each photon in a pair passes through an analyzer of polarizations in the $\mathbf{m}_i$ direction ($i = 1, 2$), and it is recorded by detector A_{i+} if its polarization is positive and by detector A_{i-} if its polarization is negative.

For each choice of unit vectors $\mathbf{m}$ and $\mathbf{n}$, we repeat the experiment with N qubit pairs, and we record the following numbers of coincidences.

1. N_{++}: coincidences in A_{1+} and A_{2+}.
2. N_{+-}, coincidences in A_{1+} and A_{2-}.
3. N_{-+}, coincidences in A_{1-} and A_{2+}.
4. N_{--}, coincidences in A_{1-} and A_{2-}.

The correlation for this experiment

$$C(\mathbf{m}_1, \mathbf{m}_2) = \frac{N_{++} + N_{--} - N_{+-} - N_{-+}}{N_{++} + N_{--} + N_{+-} + N_{-+}}. \tag{9.57}$$

In an ideal measurement, $N = N_{++} + N_{--} + N_{+-} + N_{-+}$. In an actual experiment, the coincidences of many qubit pairs are not recorded, and these are not taken into account in determining the correlation.

Repeating this procedure for four different pairs of polarization directions $(\mathbf{m}_1, \mathbf{m}_2)$, one can evaluate the quantity Δ of Eq. (9.48). The results of Aspect's experiments were unambiguous. *Bell's inequality is violated.* This result has been confirmed in a large number of subsequent experiments – see Box 9.3. As an illustration, Fig. 9.2 presents the results of a Bell experiment with atomic qubits separated by 398 m that was carried out in Munich (Rosenfeld et al., 2017).

Fig. 9.2 Observed Bell-type correlations for entangled atomic qubits (a) in state $|\Psi_-\rangle$ and (b) in state $|\Psi_+\rangle$ from an experiment reported by Rosenfeld et al. (2017). The four vectors $\mathbf{m}_1, \mathbf{m}_2, \mathbf{n}_1$, and $\mathbf{n}_2$ for the qubit observables are given by Eq. (9.55) for $\theta = \frac{\pi}{4}$. In the plot, they correspond to the angles $0°$, $45°$, $90°$, and $-45°$, respectively. The results $|\Delta| = 2.240 \pm 0.047$ for $|\Psi_-\rangle$ and $|\Delta| = 2.204 \pm 0.047$ for $|\Psi_+\rangle$ imply a violation of Bell inequalities by 5.1 and 4.3 standard deviations, respectively.

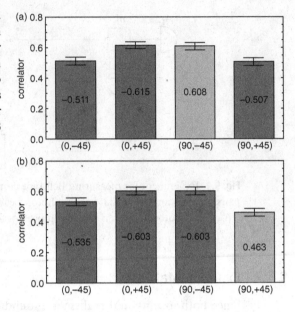

Box 9.3   Bell-Inequality Experiments

The greatest challenge in Bell-inequality tests is to exclude any reason of doubt about the result, no matter how implausible. All experiments involve practical assumptions in the preparation of the system or in the statistics of the outcomes, in addition to the hypothesis that is being tested. These assumptions are usually thought to be benign and to not affect the conclusions. However, the stakes of Bell-type experiments are enormous, and these additional assumptions may work as *loopholes*. It is therefore necessary to devise experiments that close such loopholes.

The most important loopholes are the following.

- *Locality loophole* We have to account for the possibility that the apparatuses somehow communicate the choice of measurement to the qubits, so that BIA2 is violated. This can be avoided in set-ups (i) where the vectors $\mathbf{m}$ and $\mathbf{n}$ are selected randomly in each run of the experiment when the qubits are already far from each other, and (ii) the duration of the measurement is small, so that any such influence can only occur with faster-than-light propagation.
- *Detection loophole* As explained in Section 9.5.2, in Bell-test experiments with photons, one measures coincidences of photodetection events. Since there are no perfect photodetectors, there will be undetected photons. If one photon of a pair is undetected, then the pair does not count towards the evaluation of $C(\mathbf{m}, \mathbf{n})$. The obvious assumption is that detected photon pairs are a representative sample of all emitted photon pairs. It is highly implausible that this assumption is false, but it is logically possible, so one has to find ways to carry out the experiment without this assumption.

Hence, to remove all doubts about the failure of local realism, Bell-type experiments must involve qubits that are well separated before measurement, and detectors that are both fast and highly efficient. These conditions are difficult to achieve. It took 35 years from the first observed violation of Bell's inequality to the first observed violation without the two loopholes above (Hensen et al., 2015). Of course, a huge number of experiments on entangled systems had been giving excellent agreement with the predictions of quantum theory in the intervening years.

Some important Bell-test experiments are presented in the following table.

Year	Location	Qubits	Loophole closed	Distance	Reference
1998	Geneva	photonic	–	10 km	Tittel et al. (1998)
1998	Vienna	photonic	locality	400 m	Weihs et al. (1998)
2001	Boulder	atomic	detection	3 μm	Rowe et al. (2001)
2009	Sta. Barbara	solid-state	detection	3 mm	Ansmann et al. (2009)
2013	Vienna	photonic	detection	–	Giustina et al. (2013)
2015	Delft	electron spin	both	1.3 km	Hensen et al. (2015)
2015	Vienna	photonic	both	58 m	Giustina et al. (2015)
2015	Boulder	photonic	both	180 m	Shalm et al. (2015)
2017	Munich	atomic	both	400 m	Rosenfeld et al. (2017)

9.5.3 Meaning of the Violation of Bell's Inequality

Since experiments confirm the violation of Bell's inequality unambiguously, local realism has been proven false. Some loopholes are still logically possible, but they strain credulity. For example, BIA2 presupposes that the experimentalist is free to choose which observable to measure for one qubit; hence, if the detectors are sufficiently far apart this choice cannot affect the other qubit. What if the experimentalist does not have this freedom of choice? What if her choice of measurement in each particular instance has been predetermined long ago, and the qubits have also been predetermined to behave in such a way as to violate Bell's inequality? This type of predetermination may very well be an outcome of processes that respect locality. But then it invokes a cosmic conspiracy, apparently aiming to make humanity misunderstand physics. We would have to postulate an entity like Descartes's *malicious demon*, that is, "a being of utmost power and cunning that employs all his energies in order to deceive us" (Descartes, 1641/2017).

Evil demons aside, the violation of Bell's inequality implies that either BIA1 or BIA2 is false. Which one? BIA2 appears stronger. First, if BIA2 is false, then so is the locality principle. We have to abandon the highly successful theory of relativity, without any inkling of what to substitute it with. Second, the failure of BIA2 implies the existence of *ghostly* interactions between apparatuses and microscopic systems. These interactions are ghostly,[5] because they are not generated by any term in a Hamiltonian; hence, they have no relation to the fundamental forces (electromagnetic, strong and weak) of microscopic physics. Their only function is to change the qubits in accordance with our choice of configuration for the apparatus, but for some reason, they can never be used in order to send faster-than-light signals (this is not allowed in quantum theory, see Section 9.6.1).

Dropping BIA2 creates more problems than it solves. In contrast, the abandonment of realism is consistent with many other properties of quantum theory, as seen in Chapters 7 and 8. Still, it is not entirely trouble-free. If microscopic systems have no properties, and macroscopic systems consist of microscopic ones, then how do macroscopic systems have properties?[6] Compared with the chaos that is unleashed by dropping BIA2, the denial of BIA1 appears as the sane choice, at least for the present state of knowledge.[7]

The rejection of realism means that the notion of a "property" has no meaning beyond the classical macroscopic level that corresponds to measurements. A microscopic system has no properties in itself; properties arise after its interaction with an apparatus, and they refer to the joint system of microscopic system and apparatus, not to the microscopic system itself. Hence, words like "property" and "fact" cannot be used for microscopic entities the same way they are used for objects of our everyday experience.

9.5.4 Cirel'son's Inequality

The quantity $|\Delta|$ of Eq. (9.48) has not only an upper bound from classical probability, but also one from quantum theory. By definition, the maximum value of $|\Delta|$ is 4. We showed that systems described by classical probability theory satisfy $|\Delta| \leq 2$. Cirel'son (1980) proved that systems described by quantum theory satisfy $|\Delta| \leq 2\sqrt{2}$. Hence, experiments that test Bell's inequality can also check for violations of quantum theory. So far, none has been observed.

Proposition 9.2 *(Cirel'son's inequality)* $|\Delta| \leq 2\sqrt{2}$.

Proof The operator $\hat{D}$ of Eq. (9.53) satisfies

$$\hat{D}^2 = 4\hat{I} + (\mathbf{m}_1 \cdot \hat{\boldsymbol{\sigma}}, \mathbf{n}_1 \cdot \hat{\boldsymbol{\sigma}}) \otimes (\mathbf{m}_2 \cdot \hat{\boldsymbol{\sigma}}, \mathbf{n}_2 \cdot \hat{\boldsymbol{\sigma}}).$$

The proof is straightforward, only recall that $(\mathbf{m} \cdot \hat{\boldsymbol{\sigma}})^2 = |\mathbf{m}|^2 \hat{I}$.   □

By Eq. (5.10),

$$\hat{D}^2 = 4\hat{I} - 4(\mathbf{m}_1 \times \mathbf{n}_1) \cdot \hat{\boldsymbol{\sigma}} \otimes (\mathbf{m}_2 \times \mathbf{n}_2) \cdot \hat{\boldsymbol{\sigma}}.$$

---

[5] "Ghostly" is a translation of the German word "spukhaft," used by Einstein to describe the supposed action at a distance in the EPR paradox (Born, 1955). Another common translation is "spooky."

[6] The emergence of the classical macroscopic world in which objects have definite properties is an important foundational problem, which we will take up in Section 10.5. But dropping BIA2 does not provide a solution; in fact, it makes the problem worse. One would have to also explain how locality emerges in the macroscopic world from fundamentally nonlocal physics with ghostly interactions at the microscopic scale.

[7] There is at least one theory that keeps microscopic realism and drops BIA2, namely, *Bohmian mechanics*. We discuss it in Section 10.3 and in Section 22.2.2.

Then,

$$|\Delta|^2 = (\langle\psi|\hat{D}|\psi\rangle)^2 \leq \langle\psi|\hat{D}^2|\psi\rangle = 4 - 4\langle\psi|(\mathbf{m}_1 \times \mathbf{n}_1)\cdot\hat{\boldsymbol{\sigma}}\otimes(\mathbf{m}_2 \times \mathbf{n}_2)\cdot\hat{\boldsymbol{\sigma}}|\psi\rangle. \qquad (9.58)$$

For any self-adjoint operators $\hat{A}$ and $\hat{B}$, the Cauchy–Schwarz inequality gives

$$|\langle\psi|\hat{A}\otimes\hat{B}|\psi\rangle| \leq \sqrt{\langle\psi|\hat{A}^2\otimes\hat{I}|\psi\rangle\langle\psi|\hat{I}\otimes\hat{B}^2|\psi\rangle},$$

hence,

$$|\langle\psi|(\mathbf{m}_1 \times \mathbf{n}_1)\cdot\hat{\boldsymbol{\sigma}}\otimes(\mathbf{m}_2 \times \mathbf{n}_2)\cdot\hat{\boldsymbol{\sigma}}|\psi\rangle| \leq |\mathbf{m}_1 \times \mathbf{n}_1||\mathbf{m}_2 \times \mathbf{n}_2|. \qquad (9.59)$$

Equation (9.58) becomes

$$|\Delta|^2 \leq 4 + 4|\mathbf{m}_1 \times \mathbf{n}_1||\mathbf{m}_2 \times \mathbf{n}_2|. \qquad (9.60)$$

The maximum of $|\mathbf{m}_1 \times \mathbf{n}_1||\mathbf{m}_2 \times \mathbf{n}_2|$ is 1 (for $\mathbf{m}_1$ normal to $\mathbf{n}_1$ and $\mathbf{m}_2$ normal to $\mathbf{n}_2$). Hence, $|\Delta|^2 \leq 8$, or $|\Delta| \leq 2\sqrt{2}$.

9.5.5 Bell's Theorem without Inequalities

Bell's inequality involves correlations defined from measurements in a statistical ensemble of quantum systems. Nonetheless, Bell's theorem can be proved without inequalities, and without any reference to a statistical ensemble, ideally solely in terms of measurements on a individual quantum system. This proof is due to Greenberger, Horne and Zeilinger, and it is known as the *GHZ theorem* (Greenberger et al., 1989, 1990). Here, we present a simpler proof by Mermin (1990).

Consider a three-qubit system, described by the Hilbert space $\mathcal{H} = \mathbb{C}^2 \otimes \mathbb{C}^2 \otimes \mathbb{C}^2$. We define four operators

$$\hat{A} = \hat{\sigma}_1 \otimes \hat{\sigma}_2 \otimes \hat{\sigma}_2, \quad \hat{B} = \hat{\sigma}_2 \otimes \hat{\sigma}_1 \otimes \hat{\sigma}_2, \quad \hat{C} = \hat{\sigma}_2 \otimes \hat{\sigma}_2 \otimes \hat{\sigma}_1, \quad \hat{D} = \hat{\sigma}_1 \otimes \hat{\sigma}_1 \otimes \hat{\sigma}_1.$$

These operators commute with each other. To see this,

$$\hat{A}\hat{B} = \hat{\sigma}_1\hat{\sigma}_2 \otimes \hat{\sigma}_2\hat{\sigma}_1 \otimes \hat{\sigma}_2^2 = \hat{\sigma}_3 \otimes \hat{\sigma}_3 \otimes \hat{I}, \qquad \hat{B}\hat{A} = \hat{\sigma}_2\hat{\sigma}_1 \otimes \hat{\sigma}_1\hat{\sigma}_2 \otimes \hat{\sigma}_2^2 = \hat{\sigma}_3 \otimes \hat{\sigma}_3 \otimes \hat{I},$$

hence, $[\hat{A},\hat{B}] = 0$. Similarly, we show that $[\hat{A},\hat{C}] = [\hat{B},\hat{C}] = 0$. Furthermore,

$$\hat{A}\hat{D} = \hat{\sigma}_1^2 \otimes \hat{\sigma}_2\hat{\sigma}_1 \otimes \hat{\sigma}_2\hat{\sigma}_1 = -\hat{I} \otimes \hat{\sigma}_3 \otimes \hat{\sigma}_3, \qquad \hat{D}\hat{A} = \hat{\sigma}_1^2 \otimes \hat{\sigma}_1\hat{\sigma}_2 \otimes \hat{\sigma}_1\hat{\sigma}_2 = -\hat{I} \otimes \hat{\sigma}_3 \otimes \hat{\sigma}_3,$$

hence, $[\hat{A},\hat{D}] = 0$. Similarly, $[\hat{B},\hat{D}] = [\hat{C},\hat{D}] = 0$.

The key point is that

$$\hat{A}\hat{B}\hat{C}\hat{D} = \hat{\sigma}_1\hat{\sigma}_2^2\hat{\sigma}_1 \otimes \hat{\sigma}_2\hat{\sigma}_1\hat{\sigma}_2\hat{\sigma}_1 \otimes \hat{\sigma}_2^2\hat{\sigma}_1^2 = -\hat{I}. \qquad (9.61)$$

The GHZ theorem is essentially Eq. (9.61). Let us denote by $f_1^{(i)}$ and $f_2^{(i)}$ the functions that correspond to the measurements of $\hat{\sigma}_1$ and $\hat{\sigma}_2$ at the ith qubit ($i = 1, 2, 3$), as in Section 9.4.1; their values are ± 1. Then, the product $\hat{A}\hat{B}\hat{C}\hat{D}$ corresponds to the function

$$[f_1^{(1)}f_2^{(2)}f_2^{(3)}][f_2^{(1)}f_1^{(2)}f_2^{(3)}][f_2^{(1)}f_2^{(2)}f_1^{(3)}][f_1^{(1)}f_1^{(2)}f_1^{(3)}] = \left(f_1^{(1)}f_1^{(2)}f_1^{(3)}f_2^{(1)}f_2^{(2)}f_2^{(3)}\right)^2 = 1,$$

obviously contradicting Eq. (9.61).

Ideally, the predictions of quantum theory can be distinguished from those of a local realist theory by a *single* measurement of the observables $\hat{A}$, $\hat{B}$, $\hat{C}$, and $\hat{D}$. We only need to prepare the system in a common eigenstate of $\hat{A}$, $\hat{B}$, and $\hat{C}$ – by Eq. (9.61) it will also be an eigenstate of $\hat{D}$.

Let us denote by $a = \pm 1$ the eigenvalues of $\hat{A}$, by $b = \pm 1$, the eigenvalues of $\hat{B}$ and by $c = \pm 1$ the eigenvalues of $\hat{C}$. We denote by $|a, b, c\rangle$ the common eigenvectors of $\hat{A}$, $\hat{B}$, $\hat{C}$ and $\hat{D}$. There are 2^3 such eigenvectors, known as the *GHZ states*, they define an orthonormal basis on the Hilbert space $\mathcal{H} = \mathbb{C}^8$. They are the following:

$$|---\rangle = \frac{1}{\sqrt{2}}\left(|1,1,1\rangle + |0,0,0\rangle\right), \quad |++\,+\rangle = \frac{1}{\sqrt{2}}\left(|1,1,1\rangle - |0,0,0\rangle\right),$$

$$|+\,+\,-\rangle = \frac{1}{\sqrt{2}}\left(|1,1,0\rangle + |0,0,1\rangle\right), \quad |-\,-\,+\rangle = \frac{1}{\sqrt{2}}\left(|1,1,0\rangle - |0,0,1\rangle\right),$$

$$|+\,-\,+\rangle = \frac{1}{\sqrt{2}}\left(|1,0,1\rangle + |0,1,0\rangle\right), \quad |-\,-\,+\rangle = \frac{1}{\sqrt{2}}\left(|1,0,1\rangle - |0,1,0\rangle\right),$$

$$|-\,+\,+\rangle = \frac{1}{\sqrt{2}}\left(|0,1,1\rangle + |1,0,0\rangle\right), \quad |-\,-\,+\rangle = \frac{1}{\sqrt{2}}\left(|0,1,1\rangle - |1,0,0\rangle\right). \tag{9.62}$$

Suppose we measure $\hat{A}$, $\hat{B}$, and $\hat{C}$ successively on one of these states, and we find values a, b, and c, respectively. We can predict that a subsequent measurement of $\hat{D}$ will give value $-1/(abc)$, while local realism predicts $1/(abc)$. Hence, we have a sharp differentiation of predictions even in a single quantum system. We do not need to perform measurements on a statistical ensemble.

Experiments, carried out with three entangled photons (Pan et al., 2000), do not involve four successive measurements on the same system, but $\hat{A}$, $\hat{B}$, $\hat{C}$, and $\hat{D}$ are measured on different systems, prepared in the same state. The low efficiency of the detectors makes it necessary to take a statistical ensemble of such measurements. So far, the results of such experiments are fully compatible with quantum theory.

9.6   Applications

Next, we present four crucial results that elaborate on the quantum rule for the composition of subsystems and the notion of entanglement.

9.6.1   No-Communication Theorem

The no-communication theorem asserts that it is impossible to use entangled states to transmit information instantaneously.

Let us assume two observers, Alice and Bob. Alice can act on quantum systems described by the Hilbert space $\mathcal{H}_A$, and Bob can act on quantum systems described by the Hilbert space $\mathcal{H}_B$. Suppose that a source produces entangled pairs of systems: One element of the pair is accessed by Alice and the other by Bob. The state ρ of these pairs can always be expressed as

$$\hat{\rho} = \sum_i \hat{S}_i \otimes \hat{T}_i, \tag{9.63}$$

for some operators $\hat{S}_i$ and $\hat{T}_i$ on $\mathcal{H}_A$ and $\mathcal{H}_B$, respectively.

Bob picks up his system and moves far away. Then Alice acts on her own system. The most general action on a state $\hat{\rho}$ on a Hilbert space $\mathcal{H}$ is of the form

$$\hat{\rho} \rightarrow \sum_a \hat{K}_a \hat{\rho} \hat{K}_a^\dagger, \tag{9.64}$$

where $\hat{K}_a$ are operators on $\mathcal{H}$ that satisfy $\sum_a \hat{K}_a^\dagger \hat{K}_a = \hat{I}$; they are called *Kraus operators*. This result is proven in Section 20.1.2. However, the form (9.64) should not be a surprise: Unitary evolution corresponds to a single unitary Kraus operator, and the rule (8.19) for measurements has spectral projectors as Kraus operators. If we combine a measurement of an operator $\hat{F} = \sum_a f_a \hat{P}_a$ and time evolution through a unitary operator $\hat{U}$, the Kraus operators are $\hat{K}_a = \hat{U} \hat{P}_a$.

Suppose that Alice's action on her system is described by the Kraus operators $\hat{K}_a$ on $\mathcal{H}_A$. When describing the entangled pair on the Hilbert space $\mathcal{H}_A \otimes \mathcal{H}_B$, Alice's Kraus operators are expressed as $\hat{K}_a \otimes \hat{I}$. After Alice's action, the state of the composite system is

$$\hat{\rho}' = \sum_a (\hat{K}_a \otimes \hat{I}) \hat{\rho} (\hat{K}_a^\dagger \otimes \hat{I}) = \sum_a \sum_i (\hat{K}_a \otimes \hat{I})(\hat{S}_i \otimes \hat{T}_i)(\hat{K}_a^\dagger \otimes \hat{I})$$

$$= \sum_a \sum_i \hat{K}_a \hat{S}_i \hat{K}_a^\dagger \otimes \hat{T}_i. \tag{9.65}$$

Then, Bob measures the operator $\hat{Y} = \sum y_n \hat{P}_n$ on $\mathcal{H}_B$. This is represented by $\hat{I} \otimes \hat{Y} = \sum_n y_n (\hat{I} \otimes \hat{P}_n)$ on $\mathcal{H}_A \otimes \mathcal{H}_B$. The associated probabilities are

$$\text{Prob}(n) = \text{Tr}[\hat{\rho}'(\hat{I} \otimes \hat{P}_n)] = \sum_i \left(\sum_a \text{Tr}[\hat{K}_a \hat{S}_i \hat{K}_a^\dagger] \cdot \text{Tr}(\hat{T}_i \hat{P}_n) \right)$$

$$= \sum_i \left[\text{Tr}\left(\hat{S}_i \sum_a \hat{K}_a^\dagger \hat{K}_a \right) \cdot \text{Tr}(\hat{T}_i \hat{P}_n) \right] = \sum_i \left[\text{Tr}(\hat{S}_i) \cdot \text{Tr}(\hat{T}_i \hat{P}_n) \right]. \tag{9.66}$$

We conclude that the probabilities determined by Bob do not depend on the Kraus operators $\hat{K}_a$, that is, on the way Alice chose to act on her system. Hence, entanglement cannot be used for instantaneous transfer of information from Alice to Bob.

Quantum theory does not allow entanglement to be used for instantaneous transfer of information.

9.6.2 No-Cloning Theorem

Copying machines are ubiquitous in the modern world. Their inputs are physical systems in a given state, and their outputs are exact copies of the input, while the input is preserved. For example, a photocopier records *all information* on a piece of paper and prints it on a blank page.

In a classical world, there is no limit to the accuracy of copiers. The quantum world is different (Park, 1970; Wooters and Zurek, 1982; Dieks, 1982). Suppose we want to copy a system that is found on the state $|\psi\rangle$ of a Hilbert space $\mathcal{H}$. To this end, we need the quantum equivalent of a blank page of a photocopier. This is another system, prepared in a specific state $|e\rangle$ of the same

Hilbert space $\mathcal{H}$. The quantum copier is an apparatus that acts with a unitary operator $\hat{U}$ on the Hilbert space $\mathcal{H} \otimes \mathcal{H}$, so that

$$\hat{U}(|\psi\rangle \otimes |e\rangle) = |\psi\rangle \otimes |\psi\rangle, \tag{9.67}$$

for all $|\psi\rangle \in \mathcal{H}$. Eq. (9.67) means that the initial state is preserved, and its content is fully copied into the "blank."

Consider two vectors $|\phi_1\rangle$ and $|\phi_2\rangle$ that both satisfy Eq. (9.67): $\hat{U}(|\phi_1\rangle \otimes |e\rangle) = |\phi_1\rangle \otimes |\phi_1\rangle$ and $\hat{U}(|\phi_2\rangle \otimes |e\rangle) = |\phi_2\rangle \otimes |\phi_2\rangle$. We take the inner product of the left-hand sides of these equations to obtain

$$\langle \phi_1 | \phi_2 \rangle = (\langle \phi_1 | \phi_2 \rangle)^2. \tag{9.68}$$

Hence, either $\langle \phi_1 | \phi_2 \rangle = 0$ or $\langle \phi_1 | \phi_2 \rangle = 1$. These conditions do not hold for generic vectors $|\phi_1\rangle$ and $|\phi_2\rangle$, but only for ones in an orthonormal set. Hence, there can be no copying machine (i.e., unitary operator $\hat{U}$) that works for an arbitrary set of initial states. This statement is the *no-cloning theorem*. We can only construct copiers of limited purposes, that is, apparatuses that can copy a specific orthonormal set of states.

The no-cloning theorem is a foundational result for quantum cryptography. Suppose that the state $|\psi\rangle$ contains a message between Alice and Bob and that Eve attempts to eavesdrop the conservation by copying the quantum state. It is impossible to do so without affecting the message, that is, by changing the transmitted state $|\psi\rangle$, hence, alerting Alice and Bob.

9.6.3 Quantum Teleportation

Quantum teleportation is the transfer of the quantum state from one point in space to another, using the properties of entanglement.

Let Alice and Bob be two observers that share an entangled pair of qubits in the state $|\Phi_+\rangle$, of Eq. (9.30). Alice's qubit is described by the Hilbert space $\mathcal{H}_A$, and Bob's qubit is described by the Hilbert space $\mathcal{H}_B$. We will keep track of the Hilbert spaces by writing the state of the shared qubit as $|\Phi_+\rangle_{AB}$.

Alice has another qubit C, described by the Hilbert space $\mathcal{H}_C$. C has been prepared in a state $|\psi\rangle_C = \alpha|0\rangle_C + \beta|1\rangle_C$, where $\alpha, \beta \in \mathbb{C}$. The state of the system of three qubits is $|\Phi_+\rangle_{AB} \otimes |\psi\rangle_C$. This can be written as

$$|\Phi_+\rangle_{AB} \otimes |\psi\rangle_C = \frac{1}{2} \big[|\Phi_+\rangle_{AC} \otimes (\alpha|0\rangle_B + \beta|1\rangle_B) + |\Phi_-\rangle_{AC} \otimes (\alpha|0\rangle_B - \beta|1\rangle_B)$$
$$+ |\Psi_+\rangle_{AC} \otimes (\beta|0\rangle_B + \alpha|1\rangle_B) + |\Psi_-\rangle_{AC} \otimes (\beta|0\rangle_B - \alpha|1\rangle_B) \big]. \tag{9.69}$$

To prove Eq. (9.69), start from the right-hand side; 12 out of the 16 terms cancel, and the remaining ones give the left-hand side.

Next, Alice performs a measurement on her two qubits (A and C) that corresponds to the orthonormal basis defined by the four vectors $|\Phi_\pm\rangle_{AC}, |\Psi_\pm\rangle_{AC}$. Equation (9.69) implies that the probability of each of the four outcomes equals $\frac{1}{4}$. After the measurement, the three-qubit system is found in one of the following four states:

$$|\Phi_+\rangle_{AC} \otimes (\alpha|0\rangle_B + \beta|1\rangle_B), \quad |\Phi_-\rangle_{AC} \otimes (\alpha|0\rangle_B - \beta|1\rangle_B),$$
$$|\Psi_+\rangle_{AC} \otimes (\beta|0\rangle_B + \alpha|1\rangle_B), \quad |\Psi_-\rangle_{AC} \otimes (\beta|0\rangle_B - \alpha|1\rangle_B). \tag{9.70}$$

Then, Alice informs Bob about the outcome of her measurement. Bob executes that following protocol.

1. If the measurement outcome corresponds to $|\Phi_+\rangle_{AC}$, Bob does nothing.
2. If the measurement outcome corresponds to $|\Phi_-\rangle_{AC}$, Bob acts on his qubit with the operator $\hat{U} = \hat{\sigma}_3$.
3. If the measurement outcome corresponds to $|\Psi_+\rangle_{AC}$, Bob acts on his qubit with the operator $\hat{U} = \hat{\sigma}_1$.
4. If the measurement outcome corresponds to $|\Psi_+\rangle_{AC}$, Bob acts on his qubit with the operator $\hat{U} = \hat{\sigma}_2$.

In all cases, the state of the three-qubit system is separable between the Hilbert space $\mathcal{H}_A \otimes \mathcal{H}_C$ that describes Alice's two qubits and the Hilbert space $\mathcal{H}_B$ that describes Bob's qubit. The state of Bob's qubit is identical to the initial state of qubit C. Hence, the information contained in C was transferred to Bob without any propagation of matter. For this reason, we call this phenomenon *quantum teleportation*.

We note the following.

1. Teleportation is possible for any initial state $|\psi\rangle_C$, even if it is unknown.
2. Teleportation requires Alice communicating the outcome of her measurement to Bob. This communication takes place at slower-than-light speeds. Hence, quantum teleportation does not allow faster-than-light transmission of information.
3. The state of qubit C is fully transferred to B, but C does not remain in its initial state. The no-cloning theorem is not violated.
4. Teleportation requires an entangled pair of qubits. For this reason, *entanglement* constitutes a *resource* for quantum teleportation.
5. Alice sends to Bob only two bits of information (specifying one out of four outcomes) and manages to teleport a qubit state. In general, the exact information contained in the qubit state depends on our accuracy in preparing a quantum state. For example, if we can prepare a qubit with accuracy of one degree with respect to the Bloch sphere angles, we would be able to distinguish between $180 \times 360 = 64,800$ quantum states. The specification of one of those quantum states requires $\log_2(64,800) \simeq 16$ bits. Hence, Alice transmitted information for 16 bits using only 2 bits. Teleportation is a protocol for *superdense* encoding of information.

Quantum teleportation was first proposed by Bennett et al. (1993). The first successful teleportation experiment with photonic qubits was reported by Boschi et al. (1998); teleportation experiments with atomic qubits were first reported by Riebe et al. (2004) and Barrett et al. (2004). The current record on teleportation distance is 1400 km and it was achieved with photonic qubits (Yin et al., 2017).

9.6.4 Entanglement Distillation

We saw that quantum teleportation requires the preparation of a pair of qubits in a maximal entanglement state. How can we have a steady supply of such states, for use in teleportation or any other application? This is made possible by the process known as *entanglement distillation*, the preparation of maximum entanglement pairs from qubit pairs of less-than-maximum entanglement.

Here, we present a simple example of entanglement distillation (Bennett et al., 1996). Consider two pairs of qubits, each prepared in the state $|\Psi_\beta\rangle = \cos\frac{\beta}{2}|0,0\rangle + \sin\frac{\beta}{2}|1,1\rangle$, where $0 < \beta < \pi$. The first qubit in each pair belongs to Alice, and the second one to Bob.

By the analysis of Section 9.3.1, the entanglement E of $|\Psi_\beta\rangle$ is proportional to $|\sin\beta|$, and it is maximized for $\beta = \frac{\pi}{2}$, whence $|\Psi_\beta\rangle$ becomes a maximal entanglement state.

The joint system of the two qubit pairs is described by the state $|\Psi_\beta\rangle \otimes |\Psi_\beta\rangle$. We measure the operator

$$\hat{A} = \hat{\sigma}_3 \otimes \hat{I} \otimes \hat{\sigma}_2 \otimes \hat{I}. \tag{9.71}$$

This is a measurement on Alice's two qubits. We substitute $\hat{\sigma}_3 = |1\rangle\langle 1| - |0\rangle\langle 0|$, to obtain the spectral analysis of $\hat{A}$,

$$\hat{A} = (1)\hat{P}_1 + (-1)\hat{P}_{-1}, \tag{9.72}$$

where

$$\hat{P}_1 = |1\rangle\langle 1| \otimes \hat{I} \otimes |1\rangle\langle 1| \otimes \hat{I} + |0\rangle\langle 0| \otimes \hat{I} \otimes |0\rangle\langle 0| \otimes \hat{I}$$
$$\hat{P}_{-1} = |0\rangle\langle 0| \otimes \hat{I} \otimes |1\rangle\langle 1| \otimes \hat{I} + |1\rangle\langle 1| \otimes \hat{I} \otimes |0\rangle\langle 0| \otimes \hat{I}. \tag{9.73}$$

The probability of outcome -1 is

$$\text{Prob}(-1) = \langle \Psi_\beta, \Psi_\beta | \hat{P}_{-1} | \Psi_\beta, \Psi_\beta \rangle = 2\langle \Psi_\beta | \left(|0\rangle\langle 0| \otimes \hat{I} \right) |\Psi_\beta\rangle \langle \Psi_\beta | \left(|1\rangle\langle 1| \otimes \hat{I} \right) |\Psi_\beta\rangle$$
$$= 2\cos^2\frac{\beta}{2}\sin^2\frac{\beta}{2} = \frac{1}{2}\sin^2\beta, \tag{9.74}$$

while the state of the four-qubit system after a measurement of -1 is

$$|\Psi; -1\rangle = \frac{\hat{P}_{-1}|\Psi_\beta, \Psi_\beta\rangle}{\sqrt{\text{Prob}(-1)}} = \frac{1}{\sqrt{2}} \left(|0\rangle \otimes |0\rangle \otimes |1\rangle \otimes |1\rangle + |1\rangle \otimes |1\rangle \otimes |0\rangle \otimes |0\rangle \right). \tag{9.75}$$

In Eq. (9.75), Alice has the first and third qubit, while Bob has the second and fourth qubit. We rewrite this state, by placing Alice's and Bob's qubits in separate kets, labeled by A and B respectively,

$$|\Psi; -1\rangle = \frac{1}{\sqrt{2}} \left(|1,0\rangle_A \otimes |1,0\rangle_B + |0,1\rangle_A \otimes |0,1\rangle_B \right). \tag{9.76}$$

Alice's two qubits involve only the two-dimensional subspace $\mathcal{V}_A$ generated by $|+\rangle_A := |0,1\rangle_A$ and $|-\rangle_A := |1,0\rangle_A$. Similarly, Bob's qubits involves only the two-dimensional subspace $\mathcal{V}_B$ generated by $|+\rangle_B := |0,1\rangle_B$ and $|-\rangle_B := |1,0\rangle_B$. Hence, Eq. (9.76) is the maximum entanglement state

$$|\Psi; -1\rangle = \frac{1}{\sqrt{2}} \left(|-\rangle_A \otimes |-\rangle_B + |+\rangle_A \otimes |+\rangle_B \right) \tag{9.77}$$

in the Hilbert space $\mathcal{V}_A \otimes \mathcal{V}_B$.

With this protocol, two qubit pairs in the state $|\Psi_\beta\rangle$ are expended in order to obtain a single maximum-entanglement qubit pair with probability $\frac{1}{2}\sin^2\beta$. On average we distill one maximum-entanglement qubit pair from $\frac{4}{\sin^2\beta}$ qubit pairs in the state $|\Psi_\beta\rangle$. This protocol can be made more efficient by performing measurements on n qubit pairs.

9.7 Entanglement in Continuous-Variable Systems

Finally, we consider entanglement in continuous-variable systems. A particle on the line is described by a wave function $\psi(x)$ that is an element of the Hilbert space $L^2(\mathbb{R})$. N such particles are described by wave functions $\psi(x_1, x_2, \ldots, x_n)$ in the Hilbert space $L^2(\mathbb{R}) \otimes L^2(\mathbb{R}) \ldots L^2(\mathbb{R}) = L^2(\mathbb{R}^n)$, or by density matrices $\rho(x_1, x_2, \ldots, x_n; x_1', x_2', \ldots, x_n') := \langle x_1, x_2, \ldots, x_n | \hat{\rho} | x_1', x_2', \ldots, x_n' \rangle$.

Interestingly, in these systems, entanglement can be quantified in terms of a variation of the Kennard–Robertson uncertainty relation (Simon, 2000; Duan et al., 2000). We will show this for $N = 2$; the generalization to arbitrary N is straightforward.

For $N = 2$, we define position operators $\hat{x}_1 := \hat{x} \otimes \hat{I}$, $\hat{x}_2 = \hat{I} \otimes \hat{x}$, and momentum operators $\hat{p}_1 := \hat{p} \otimes \hat{I}$, $\hat{p}_2 = \hat{I} \otimes \hat{p}$. The only nonzero commutators between these operators are $[\hat{x}_1, \hat{p}_1] = [\hat{x}_2, \hat{p}_2] = i\hat{I}$.

We define the operators $\hat{x}_\pm := \frac{1}{2}(\hat{x}_1 \pm \hat{x}_2)$ and $\hat{p}_\pm := \hat{p}_1 \pm \hat{p}_2$. They satisfy the commutation relations $[\hat{x}_+, \hat{p}_+] = [\hat{x}_-, \hat{p}_-] = i\hat{I}$ and $[\hat{x}_+, \hat{p}_-] = [\hat{x}_-, \hat{p}_+] = 0$. The Kennard–Robertson uncertainty relation implies that

$$(\Delta x_+)(\Delta p_+) \geq \frac{1}{2}, \qquad (\Delta x_-)(\Delta p_-) \geq \frac{1}{2}. \tag{9.78}$$

Next, we apply the PPT criterion to a general density matrix $\rho(x_1, x_2; x_1', x_2')$. We work with the associated Wigner function,

$$W(x_1, x_2, p_1, p_2) = \frac{1}{(2\pi)^2} \int dy_1 dy_2 \rho(x_1 - \frac{y_1}{2}, x_2 - \frac{y_2}{2}; x_1 + \frac{y_1}{2}, x_2 + \frac{y_2}{2}) e^{ip_1 y_1 + ip_2 y_2}.$$

The PPT transformation takes y_2 to $-y_2$ in this integral; hence, it acts upon the Wigner function as

$$PPT : W(x_1, x_2, p_1, p_2) \to W(x_1, x_2, p_1, -p_2). \tag{9.79}$$

This means that under PPT (x_1, x_2, p_1, p_2) transforms to $(x_1, x_2, p_1, -p_2)$, or equivalently, (x_+, x_-, p_+, p_-) transforms to (x_+, x_-, p_-, p_+).

By the PPT criterion, if $\hat{\rho}$ is separable, then Eqs. (9.78) apply also for the PPT-transformed state. Hence, for separable states

$$(\Delta x_+)(\Delta p_-) \geq \frac{1}{2}, \qquad (\Delta x_-)(\Delta p_+) \geq \frac{1}{2}. \tag{9.80}$$

Thus, we obtained a simple entanglement criterion: if either of Eqs. (9.80) is violated, then $\hat{\rho}$ is entangled. However, the converse does not hold: There exist entangled states that satisfy Eq. (9.80). The uncertainties $(\Delta x_+)(\Delta p_-)$ and $(\Delta x_-)(\Delta p_+)$ define *entanglement witnesses*.

Example 9.4 Consider a Gaussian state with Wigner function

$$W(x_1, x_2, p_1, p_2) = Ce^{-a(x_1^2 + x_2^2 - 2rx_1x_2) - b(p_1^2 + p_2^2 - 2sp_1p_2)}, \tag{9.81}$$

where $a, b > 0$ and $|r|, |s| < 1$, and $C = ab\sqrt{1 - r^2}\sqrt{1 - s^2}/\pi^2$. We change variables to $x_\pm, p_\pm$,

$$W(x_+, x_-, p_+, p_-) = Ce^{-2a(1-r)x_+^2 - 2a(1+r)x_-^2 - \frac{1}{2}b(1-s)p_+^2 - \frac{1}{2}b(1+s)p_-^2}. \tag{9.82}$$

We straightforwardly obtain

$$(\Delta x_\pm)^2 = \frac{1}{4a(1 \mp r)}, \qquad (\Delta p_\pm)^2 = \frac{1}{b(1 \mp s)}. \tag{9.83}$$

The uncertainty relations $(\Delta x_+)(\Delta p_+) \geq \frac{1}{2}$ and $(\Delta x_-)(\Delta p_-) \geq \frac{1}{2}$ imply that $ab(1-r)(1-s) \leq 1$ and that $ab(1+r)(1+s) \leq 1$ for all states.

The conditions $(\Delta x_+)(\Delta p_-) \geq \frac{1}{2}$ and $(\Delta x_-)(\Delta p_+) \geq \frac{1}{2}$, satisfied by all separable states, imply that $ab(1-r)(1+s) \leq 1$ and $ab(1+r)(1-s) \leq 1$. Hence, a state with either $ab(1-r)(1+s) > 1$ or $ab(1+r)(1-s) > 1$ is entangled. This is possible only if r and s have opposite signs. Then, $(1+r)(1+s)$ and $(1-r)(1-s)$ are both smaller than either $(1-r)(1+s)$ (if $r < 0, s > 0$) or $(1+r)(1-s)$ (if $r > 0, s < 0$).

QUESTIONS

9.1 State the difference between $\mathcal{H}_1 \otimes \mathcal{H}_2$ and $\mathcal{H}_1 \oplus \mathcal{H}_2$ using logical connectives.

9.2 In what sense do we treat entanglement as a quantity?

9.3 Is the successive measurement of $\hat{A} \otimes \hat{I}$ and $\hat{I} \otimes \hat{B}$ equivalent to the measurement of $\hat{A} \otimes \hat{B}$?

9.4 Explain why all self-adjoint operators of the form $\hat{A} \otimes \hat{I}$ have degenerate spectrum. What is their smallest degeneracy?

9.5 You are told that "probabilities in quantum theory imply ignorance about the true values of physical quantities, as these are determined by measurement." Explain how Bell's theorem contradicts this claim.

9.6 Explain why the violation of Bell's inequality does not imply "action from a distance" or faster-than-light transmission of information.

9.7 Identify the errors in the following narrative (Chopra, 2015).

> The commonsense idea of local reality is true only at a certain level. The whole of reality, as explained by quantum physics, lies deeper. A famous mathematical formula, known as Bell's theorem (after its author, the Irish physicist John Bell), holds that the reality of the universe must be nonlocal; in other words, all objects and events in the cosmos are inter-connected with one another and respond to one another's changes of state.[...]
>
> What kind of explanation would satisfy Bell's requirement for a totally inter-connected, nonlocal reality? It would have to be a quantum explanation, because if [...] a change of spin in one particle causes an equal but opposite change instantly in its partner somewhere in outer space, it is obvious that the information going from one place to the other is traveling faster than the speed of light. This is not permitted in ordinary reality, either by Newton or Einstein.

9.8 How would you go on explaining Bell's theorem and its implications to an audience of nonphysicists?

9.9 What would it mean if a Bell-type experiment found $\Delta = 3.1 \pm 0.1$?

9.10 **(B)** Suppose that Alice and Bob are in different planetary systems. Why does Bob have to wait years for Alice's message in the teleportation protocol? Why can't he just measure his qubit?

9.11 **(B)** Consider the following comment on quantum teleportation.

> The transmission of observable effects is subject to the same restrictions as classical communication, and there is no action at a distance—Einstein, Podolski and Rosen's bugbear—unless one believes that the state vector has real physical existence. It is not universally accepted that it does ... Nevertheless, the teleportation device might be taken as a reason to believe in the reality of the quantum state vector; if something can be transmitted from one object to another, it might be argued, it must be real. The debate will continue. (Sudbery, 1993)

State explicitly the arguments in support of and against the reality of the quantum state on the basis of the teleportation experiments.

PROBLEMS

9.1 A two-qubit system is prepared in the state $\cos\theta|1,1\rangle + \sin\theta e^{i\phi}|0,0\rangle$. Evaluate the average and the variance of the observable $\hat{A} = \frac{1}{2}(\hat{\sigma}_1 \otimes \hat{\sigma}_2 - \hat{\sigma}_2 \otimes \hat{\sigma}_1)$.

9.2 The swap operator for a bipartite system is defined by $\hat{S}(|\phi\rangle \otimes |\psi\rangle) := |\psi\rangle \otimes |\phi\rangle$. **(a)** Prove that for a pair of qubits $\hat{S} = \frac{1}{2}(\hat{I} + \sum_i \hat{\sigma}_i \otimes \hat{\sigma}_i)$. **(b)** Find the associated eigenvalues and eigenvectors.

9.3 Find the eigenvalues and the spectral projectors of the operators $\hat{K}_\pm = \hat{\sigma}_1 \otimes \hat{\sigma}_2 \pm \hat{\sigma}_2 \otimes \hat{\sigma}_1$.

9.4 A two-qubit system is characterized by a Hamiltonian $\hat{H} = \frac{1}{2}\omega_1 \hat{\sigma}_3 \otimes \hat{I} + \frac{1}{2}\omega_2 \hat{I} \otimes \hat{\sigma}_3$, and it is prepared in a state $|\theta\rangle = \frac{1}{\sqrt{2}}(|1,0\rangle + e^{i\theta}|0,1\rangle)$. At time t_1 we measure $\hat{\sigma}_1$ in the first qubit, and at time t_2, we measure $\hat{\sigma}_1$ in the second qubit. Evaluate the probabilities for all possible measurement outcomes.

9.5 **(i)** Show that we can always choose a basis on $\mathbb{C}^2$ so that a pure two-qubit state can be expressed as $\cos\frac{\beta}{2}|0,0\rangle + \sin\frac{\beta}{2}|1,1\rangle$, for $0 < \beta < \pi$. **(ii)** Show that for this state, the maximum value of $|\Delta|$ is $2\sqrt{1 + \sin^2\beta}$. It follows that any entangled pure state violates Bell's inequality.

9.6 Evaluate Δ of Eq. (9.48) for the Werner state (9.42). Which values of α violate Bell's inequality?

9.7 **(i)** Prove *Bell's original inequality* for the two-qubit system,

$$|C(\mathbf{m}, \mathbf{n}_1) - C(\mathbf{m}, \mathbf{n}_2)| \leq 1 - C(\mathbf{n}_1, \mathbf{n}_2),$$

for all unit vectors $\mathbf{m}, \mathbf{n}_1$ and $\mathbf{n}_2$. **(ii)** Show that the inequality is violated for the state $|\Psi_-\rangle$.

9.8 The most general pure state for a two-qubit system is $|\psi\rangle = a|0,0\rangle + b|0,1\rangle + c|1,0\rangle + d|1,1\rangle$. **(i)** Show that the quadratic entanglement E of Eq. (9.20) is a increasing function of $|ad - bc|$. This result suggests the use of the *concurrence* $C(\psi) = 2|ad - bc|$ as an entanglement measure. **(ii)** Show that $C(\psi) = |\langle\psi|\tilde{\psi}\rangle|^2$, where $|\tilde{\psi}\rangle = \hat{\sigma}_2 \otimes \hat{\sigma}_2|\psi^*\rangle$ and $|\psi^*\rangle$ is the complex conjugate of $|\psi\rangle$ in the standard basis.

9.9 The concurrence defined in Problem 9.8 generalizes to an entanglement measure for mixed states. For any density matrix $\hat{\rho} = \sum_i p_w |\psi_i\rangle \langle \psi_i|$, $C(\hat{\rho}) := \min \sum_i p_i C(\psi_i)$, where the minimum is over all orthonormal sets that diagonalize $\hat{\rho}$. Evaluate the concurrence for the Werner state.

9.10 A state $\hat{\rho}$ that is diagonal in the Bell basis is called *Bell diagonal*. Show that a Bell-diagonal state is separable only if its largest eigenvalue $p_{max} \leq \frac{1}{2}$.

9.11 Consider the GHZ state $|\Psi\rangle = \frac{1}{\sqrt{2}} (|0,0,0\rangle + |1,1,1\rangle)$ of a three-qubit system. (i) Show that the reduced density matrix for the first two qubits is an unentangled mixed state. (ii) Show that, after the measurement of $\hat{\sigma}_1$ on the third qubit, the reduced density matrix for the first two qubits is maximally entangled.

9.12 For any operator $\hat{A}$ on a Hilbert space $\mathcal{H}$, we can define a vector $|\hat{A}\rangle = \sum_{ij} \langle i|\hat{A}|j\rangle |i,j\rangle \in \mathcal{H} \otimes \mathcal{H}$, where $|i\rangle$ is a basis on $\mathcal{H}$. (i) Show that $\langle \hat{A}|\hat{B}\rangle = \text{Tr}(\hat{A}^\dagger \hat{B})$, and, hence that $|\hat{A}\rangle$ is well defined only if $\text{Tr}(\hat{A}^\dagger \hat{A}) < \infty$. (ii) Show that for any unitary $\hat{U}$, $|\hat{U}\rangle$ is a maximum-entanglement vector. (iii) Show that in a normalized state $|\hat{C}\rangle$, $\langle \hat{C}|\hat{A} \otimes \hat{B}|\hat{C}\rangle = \text{Tr}(\hat{C}^\dagger \hat{A} \hat{C} \hat{B}^T)$.

9.13 A three-qubit system is prepared in the W-state, $|W\rangle = \frac{1}{\sqrt{3}}(|1,0,0\rangle + |0,1,0\rangle + |0,0,1\rangle)$. (i) Show that the reduced density matrix for the first two qubits is an entangled mixed state. (ii) Calculate the quadratic entanglement of the first two qubits after a measurement of $\hat{\sigma}_1$ in the third qubit with outcome $+1$.

9.14 In the Hilbert space $\mathbb{C}^n \otimes \mathbb{C}^n$, we define the completely symmetric vector $|\Omega\rangle = \frac{1}{\sqrt{n}} \sum_{i=1}^{n} |i\rangle \otimes |i\rangle$, and the isotropic states

$$\hat{\rho}_\lambda = \frac{1-\lambda}{n^2} \hat{I} + \lambda |\Omega\rangle\langle\Omega|. \tag{9.84}$$

(i) Show that the reduced density matrix is $\hat{\rho}_1 = \frac{1}{n}\hat{I}$. (ii) Show that $\hat{\rho}_\lambda$ is entangled for $\lambda > \frac{1}{n+1}$.

9.15 Consider the Hilbert space $\mathbb{C}^n \otimes \mathbb{C}^n$, with the totally symmetric vector $|\Omega\rangle$ as in Problem 9.14. We define $\hat{W} = \hat{I} - n|\Omega\rangle\langle\Omega|$. (i) Show that all separable density matrices $\hat{\rho}$ satisfy $\text{Tr}(\hat{\rho}\hat{W}) \geq 0$. This implies that any state $\hat{\rho}$ that satisfies $\text{Tr}(\hat{\rho}\hat{W}) < 0$ is entangled. (ii) Find the values of λ for which the states (9.84) satisfy $\text{Tr}(\hat{\rho}_\lambda \hat{W}) < 0$. (iii) Take $n = 2$. Can this criterion identify $|\Psi_-\rangle$ as an entangled state?

9.16 The *Jaynes–Cummings* model (Jaynes and Cummings, 1963) describes the interaction of a two-level atom of frequency ω with a mode of the quantum EM field of the same frequency. The Hilbert space is $\mathbb{C}^2 \otimes L^2(\mathbb{R})$, and the Hamiltonian is

$$\hat{H} = \frac{1}{2}\omega\hat{\sigma}_3 \otimes \hat{I} + \omega \hat{I} \otimes \hat{a}^\dagger \hat{a} + \frac{g}{2}(\hat{\sigma}_- \otimes \hat{a}^\dagger + \hat{\sigma}_+ \otimes \hat{a}).$$

(i) Express the Hamiltonian as a sum $\hat{H}_1 + \hat{H}_2$, where $\hat{H}_1 = \omega(\frac{1}{2}\hat{\sigma}_3 \otimes \hat{I} + \hat{I} \otimes \hat{a}^\dagger \hat{a})$, and show that $[\hat{H}_1, \hat{H}] = 0$. (ii) Show that $\hat{H}_1$ has eigenvalues $(n - \frac{1}{2})\omega$, with no degeneracy for $n = 0$, but with double degeneracy for $n = 1, 2, \ldots$. Show also that, for $n > 0$, the corresponding eigenspace V_n is spanned by $|0\rangle \otimes |n\rangle$ and $|1\rangle \otimes |n-1\rangle$. (iii) Since $\hat{H}$ commutes with $\hat{H}_1$, it suffices to diagonalize $\hat{H}$ in each V_n. Find the associated eigenvalues and eigenvectors, and evaluate $e^{-i\hat{H}t}$ on each subspace. (iv) Evaluate the probability that the atom is found in an excited state at time t, for an initial state $|0\rangle \otimes |n_0\rangle$. What is the associated Rabi frequency? (v) Evaluate the reduced density matrix for the first qubit as a function of time, and the associated value of the quadratic entanglement.

9.17 (B) An improved quantum copier uses two copies of the "blank" state $|e\rangle$, in order to achieve better results. Let a general input state $|\psi\rangle$ be copied through the unitary transformation $|\psi\rangle \otimes |e\rangle \otimes |e\rangle \to |\psi\rangle \otimes |\psi\rangle \otimes |f\rangle$, where $|f\rangle$ is a trash-state that occurs at the end of the procedure and depends on $|\psi\rangle$. Show that the no-cloning theorem still holds.

9.18 (B) Consider two qubit pairs prepared in the state $|\Psi\rangle = \cos\theta |0,0\rangle + \sin\theta |1,1\rangle$. Alice and Bob share the first pair (qubits 1 and 2), while Bob owns the second pair (qubits 3 and 4). Bob measures qubits 2 and 3 with respect to the Bell basis. (i) Show that there is a probability $2\sin^2\theta\cos^2\theta$ that the final state for the qubits 1 and 4 is maximally entangled. (ii) Show that the other produced states for qubits 1 and 4 have less entanglement than $|\Psi\rangle$.

9.19 (B) In an N-particle system, define $\hat{\xi}_a$, $a = 1,\ldots,2N$ by $(\hat{\xi}_1,\hat{\xi}_2,\ldots,\hat{\xi}_{2N-1},\hat{\xi}_{2N}) := (\hat{x}_1,\hat{p}_1,\ldots,\hat{x}_N,\hat{p}_N)$. Then the commutation relations become $[\xi_a,\xi_b] = i\Omega_{ab}\hat{I}$ for some antisymmetric matrix Ω. (i) Write the matrix Ω explicitly. (ii) Show that the correlation matrix $C_{ab} := \text{Cor}(\hat{\xi}_a,\hat{\xi}_b)$ satisfies $C + \frac{i}{2}\Omega \geq 0$. (iii) Show that, for $N = 1$, this inequality reduces to the uncertainty relation (6.44), and, hence, to the Kennard–Robertson uncertainty relation.

9.20 (B) (i) Show that for a factorized state, the correlation matrix C of the previous problem satisfies $C + \frac{i}{2}\tilde{\Omega} \geq 0$, where $\tilde{\Omega} = \Lambda\Omega\Lambda$, $\Lambda = \text{diag}\{1,1,1,-1\}$. (ii) Use this matrix inequality in order to rederive Eq. (9.80).

Bibliography

- For further details on quantum entanglement, see chapter 6 of Jaeger (2007) and for quantum communication, including teleportation, see chapter 8 of the same book. Horodecki et al. (2009) is a thorough review of entanglement theory with emphasis in quantum-information applications.
- For Bell's theorem, its background, related experiments and their implications, see chapter 6 of Peres (2002), Goldstein et al. (2011), and Wayne et al. (2020). For the various "loopholes" in Bell-type experiments, see Larsson (2014). I highly recommend the semipopular articles of Mermin (1981, 1985), and papers 13 and 15 in Bell (1987).

10 Quantum–Classical Correspondence

Quantum mechanics occupies a very unusual place among physical theories: it contains classical mechanics as a limiting case, yet at the same time it requires this limiting case for its own formulation.

Landau and Lifshitz (1977)

10.1 The Relation between Quantum and Classical Physics

In Chapter 1, we presented the fundamental principles of classical physics, and then we motivated and presented the fundamental principles of quantum physics. The two sets of principles are summarized and compared in Table 10.1.

Both classical and quantum theories are highly successful in their respective domains. Classical physics is not supposed to be fundamental, but it works excellently in the macroworld. Indeed, natural sciences outside physics rarely employ quantum physics. Quantum physics is believed to be a fundamental theory. It works in the microcosm, but it also predicts macroscopic quantum phenomena such as Schrödinger's cats. The natural expectation is that classical physics emerges from quantum theory as a limit, probably by taking $\hbar \to 0$ in an appropriate way. However, this expectation does not work, because of the central role played by measurements in quantum theory. Measurements have definite outcomes, and these outcomes must be expressed in terms of classical concepts.

For this reason, the founders of quantum theory, insisted on the coexistence of the classical and quantum description of the world, and of the importance of finding the regime where both descriptions are compatible. Bohr (1934) wrote that

...it would be a misconception to believe that ... the fundamental concepts of the classical theories will ever become superfluous for the description of physical experience ... [Quantum theory] not only depends on an analysis of measurements based on classical concepts, but it continues to be the application of these concepts alone that makes it possible to relate the symbolism of the quantum theory to the data of experience.

Bohr's viewpoint is reasonable, but it runs counter to a basic logical intuition that a fundamental theory should make no reference to a different theory in its formulation. There are many attempts either to explain the emergence of classical physics from quantum physics, or to justify their coexistence. The key point is that our understanding of quantum theory is incomplete, unless we explicitly identify its connections to classical theory. This is the aim of the present chapter.

Table 10.1 Fundamental principles: classical vs. quantum of quantum theory

Physical concepts	Classical physics	Quantum physics
Pure states	Points on **state space** Γ	Unit vectors on **Hilbert space** $\mathcal{H}$
Mixed states	Probability distributions ρ on Γ	Density matrices $\hat{\rho}$ on $\mathcal{H}$
Observables	Real functions F on Γ	Self-adjoint operators $\hat{A}$ on $\mathcal{H}$
Measurement outcomes	Numbers λ in the range of F	Points λ on the spectrum of $\hat{A}$,
Represented by	Value sets C_λ of F	Spectral projectors $\hat{P}_\lambda$
Probabilities	$\text{Prob}(\lambda) = \int dx \rho(x) \chi_{C_\lambda}(x)$	$\text{Prob}(\lambda) = \text{Tr}(\hat{\rho}\hat{P}_\lambda)$
Time evolution	Hamilton's eqn/Liouville eqn	Schrödinger's eqn (unitary evolution)
Incorporation of information	Classical conditional probability	Quantum state reduction
Composite systems	Cartesian product of state spaces	Tensor product of Hilbert spaces

10.2 The Semiclassical Approximation

In optics, we can describe light propagation by using rays, that is, lines that are perpendicular to the wave fronts of an EM wave, and point in the direction of energy flow. The ray description is approximate because it cannot account for diffraction phenomena; it corresponds to the older theory of *geometric optics*. However, it is very useful in describing complex optical systems, as a ray analysis is much easier to implement than finding a full solution to the wave equation. The semiclassical approximation is the analogue of the ray approximation for Schrödinger's equation, rays corresponding to particle motion according to classical mechanics.

10.2.1 GO and WKB Approximations

We will analyze the semiclassical approximation for a particle of mass m in a potential $V(x)$ in one dimension. We express the time-independent Schrödinger equation (5.49) as

$$\psi''(x) + k^2(x)\psi(x) = 0, \tag{10.1}$$

where $k(x) = \sqrt{2m[E - V(x)]} \geq 0$ for $E \geq V(x)$, and $k(x) = i\kappa(x)$, where $\kappa(x) = \sqrt{2m[V(x) - E]} > 0$, for $E < V(x)$. In the latter case, there are no classical paths, hence, no direct quantum-to-classical correspondence.

We write $\psi = Re^{iS}$, for $R > 0$ and real S. Then $\psi' = (R' + iRS')e^{iS}$, and $\psi'' = [R'' + 2iR'S' + iRS'' - R(S')^2]e^{iS}$. Then, Eq. (10.1) yields

$$RS'' + 2S'R' = 0 \tag{10.2}$$
$$(S')^2 = R^{-1}R'' + k^2. \tag{10.3}$$

Equation (10.2) implies that $(R^2 S')' = 0$, hence, $S'R^2 = c$, where c is a constant.

The semiclassical approximation applies when $R^{-1}R''$ in Eq. (10.3) is much smaller than k^2, and consequently much smaller than $(S')^2$. In this regime, $S' = \pm k$, hence, $S(x) = \pm \int_a^x dx' k(x')$, for some reference point a. Then, $R(x) \sim 1/\sqrt{k(x)}$ and the solution for $E \geq V(x)$ is

$$\psi(x) = \frac{1}{\sqrt{k(x)}}\left(Ae^{i\int_a^x dx' k(x')} + Be^{-i\int_a^x dx' k(x')}\right), \tag{10.4}$$

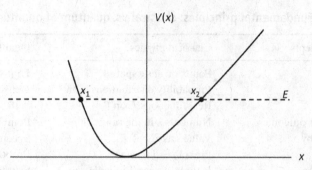

Fig. 10.1 Potential with a periodic orbit between points x_1 and x_2 at energy E.

for some constants A and B. Note that in the classically allowed region, $\frac{1}{2m}(S')^2 + V(x) = E$, and S' can be identified with the classical momentum p. In classical mechanics, the function S is known as the Hamilton–Jacobi action.

In the classically forbidden region $E < V(x)$, there are two versions of the semiclassical approximation.

1. In the *geometric optics* (GO) approximation, the wave function vanishes: $\psi = 0$. Clearly, the GO approximation cannot account for tunneling.
2. In the *Wentzel–Kramers–Brillouin* (WKB) approximation, $S' = \pm\, i\kappa$; hence, $S(x) = \pm i \int_a^x dx'\kappa(x')$ for some reference point a. Then, $R(x) \sim 1/\sqrt{\kappa(x)}$, and the solution is

$$\psi(x) = \frac{1}{\sqrt{\kappa(x)}} \left(Ce^{\int_a^x dx'\kappa(x')} + De^{-\int_a^x dx'\kappa(x')} \right) \tag{10.5}$$

for constants C and D.

A classical turning point x_0 corresponds to $k(x_0) = 0$. In the GO approximation, $\psi(x_0) = 0$. Equation (10.4) implies that $B = -Ae^{2i\int_a^{x_0} dx'k(x')}$; hence,

$$\psi(x) = \frac{A'}{\sqrt{k(x)}} \sin\left(\int_{x_0}^x dx'k(x') \right), \tag{10.6}$$

where A' is a constant.

Suppose now that, classically, there exists a periodic orbit between two turning points x_1 and x_2, such that $x_1 < x_2$ – see Fig. 10.1. Then, the condition $\psi(x_1) = \psi(x_2) = 0$, implies that $\sin\left(\int_{x_1}^{x_2} dx'k(x')\right) = 0$, or equivalently that $\int_{x_1}^{x_2} dx'k(x') = (n+1)\pi$, where $n = 0, 1, 2, 3, \ldots$. Since the path from x_1 to x_2 is half of the full periodic orbit, we can write

$$\oint p\,dx = 2\pi(n+1), \tag{10.7}$$

where the integral is over one periodic orbit. Equation (10.7) is known as the Wilson–Sommerfeld quantization condition, and it had been proposed as an ad hoc procedure for energy quantization prior to the development of quantum mechanics (Wilson, 1915; Sommerfeld, 1916).

10.2.2 The WKB Matching Formula

Naively, the WKB approximation near a turning point x_0 would lead to a connection between the coefficients A, B of Eq. (10.4) and C, D of Eq. (10.5) by requiring continuity of the wave function of its first derivative. However, since $k(x_0) = 0$, the semiclassical condition $R''/R << k^2$ fails near the turning point.

This problem is easily corrected. Suppose that x_0 is a simple solution to the equation $E - V(x) = 0$. Then, in the vicinity of x_0, $2m[E - V(x)] \simeq b(x - x_0)$; $b > 0$ if $x > x_0$ in the classically allowed region, and $b < 0$ if $x < x_0$ in the classically allowed region. Then, Eq. (10.1) becomes $\psi'' - b(x - x_0) = 0$. We change the dependent variable to $z := b^{1/3}(x - x_0)$, to obtain

$$\frac{d^2\psi}{dz^2} - z\psi = 0. \tag{10.8}$$

Equation (10.8) is the Airy equation. It admits two solutions, the *Airy functions Ai(z)* and *Bi(z)*. Their properties are described in Appendix B (Section B.4).

Near the turning point

$$\psi(x) = a_1 Ai\left[b^{1/3}(x - x_0)\right] + a_2 Bi\left[b^{1/3}(x - x_0)\right], \tag{10.9}$$

for some constants a_1 and a_2.

We use the asymptotic expressions (B.17) and (B.18) of the Airy functions, to write

$$\psi(x) = \begin{cases} \frac{1}{\sqrt{\pi}|b(x-x_0)^3|^{1/4}}(a_1 \cos(\zeta - \frac{\pi}{4}) - a_2 \sin(\zeta - \frac{\pi}{4})), & b(x - x_0) < 0 \\ \frac{1}{\sqrt{\pi}|b(x-x_0)^3|^{1/12}}(\frac{1}{2}a_1 e^{-\zeta} + a_2 e^{\zeta}), & b(x - x_0) > 0 \end{cases}, \tag{10.10}$$

where $\zeta = \frac{2}{3}|b^{1/3}(x - x_0)|^{3/2}$. On the other hand, for x close to x_0,

$$\int_{x_0}^x dx' k(x') = \int_0^z dz' \sqrt{z'} = \frac{2}{3}z^{\frac{3}{2}} = \begin{cases} i\zeta & z < 0 \\ \zeta & z > 0 \end{cases}. \tag{10.11}$$

Combining Eqs. (10.10) and (10.11), we obtain the following recipe for matching WKB solutions. A semiclassical solution of the form

$$\psi(x) = \frac{1}{\sqrt{k(x)}}\left[a_1 \cos\left(|\int_{x_0}^x dx' k(x')| - \frac{\pi}{4}\right) - a_2 \sin\left(|\int_{x_0}^x dx' k(x')| - \frac{\pi}{4}\right)\right] \tag{10.12}$$

in the classically allowed region joins with a solution of the form

$$\psi(x) = \frac{1}{\sqrt{\kappa(x)}}\left[\frac{1}{2}a_1 \exp\left(-|\int_{x_0}^x dx' \kappa(x')|\right) + a_2 \exp\left(|\int_{x_0}^x dx' \kappa(x')|\right)\right] \tag{10.13}$$

in the classically forbidden region.

Let us return now to the periodic orbit of Fig. 10.1. In the forbidden region $x < x_1$, only the term with $\exp[-\int_x^{x_1} dx'\kappa(x')]$ exists, because the other term diverges at infinity. By the matching formula, only the term with a_1 persists for $x \in (x_1, x_2)$. The integral in this term can be written as $\int_{x_1}^{x} dx' k(x') = \Phi - \int_{x}^{x_2} dx' k(x')$, where $\Phi = \int_{x_1}^{x_2} dx' \kappa(x')$. Then,

$$
\psi(x) = \frac{a_1}{\sqrt{k(x)}} \cos \left(\Phi - \int_{x}^{x_2} dx' k(x') - \frac{\pi}{4} \right)
$$

$$
= \frac{a_1}{\sqrt{k(x)}} \left[\sin \Phi \cos \left(\int_{x}^{x_2} dx' k(x') - \frac{\pi}{4} \right) - \cos \Phi \sin \left(\int_{x}^{x_2} dx' k(x') - \frac{\pi}{4} \right) \right].
$$

By the matching formula, the second term above gives rise to a term proportional to $\cos \Phi \exp \left(\int_{x_2}^{x} dx' \kappa(x') \right)$ in the forbidden region. This term diverges at infinity; hence, its coefficient must vanish. Thus, we obtain $\cos \Phi = 0$, which implies that

$$
\int_{x_1}^{x_2} dx' \kappa(x') = (n + \frac{1}{2})\pi,
$$

where $n = 0, 1, 2, \ldots$. Equivalently, when integrating over a closed trajectory,

$$
\oint p dx = 2\pi \left(n + \frac{1}{2} \right). \tag{10.14}
$$

The integral $\oint p dx$ is smaller by a factor of $\frac{\pi}{2}$ in the WKB approximation than in the GO approximation, Eq. (10.7). The difference is negligible at $n \gg 1$, where the semiclassical approximation works best.

Suppose that the wave function satisfies Dirichlet boundary conditions at $x = x_2$. The condition $\psi(x_2) = 0$, implies that $\sin (\Phi + \frac{\pi}{4}) = 0$, hence, $\Phi = \pi(n + \frac{3}{4})$, for $n = 0, 1, 2, \ldots$. Equivalently,

$$
\oint p dx = 2\pi \left(n + \frac{3}{4} \right). \tag{10.15}
$$

Example 10.1 For a harmonic oscillator of mass m and frequency ω, we use Eq. (A.10) to evaluate $\oint p dx = 2\pi \frac{E}{\omega}$. Hence, the WKB approximation gives the exact eigenvalues $E_n = \left(n + \frac{1}{2} \right) \omega$, and the GO approximation gives $E_n = (n + 1) \omega$, that is, an error of $\frac{1}{2}\omega$.

Example 10.2 Consider the *quantum bouncing ball*, that is, a particle of mass m in a homogeneous gravitational field g that is reflected elastically by the ground at $x = 0$. The system is described by the Hilbert space $L_D^2(\mathbb{R}^+)$, and the Hamiltonian is $\hat{H} = \frac{\hat{p}^2}{2m} + mg\hat{x} \geq 0$. The energy eigenvalues are positive, and they are obtained by solving the Schrödinger equation $\psi'' + 2m(E - mgx)\psi = 0$ subject to Dirichlet boundary conditions at $x = 0$.

We define $z = (2m^2 g)^{1/3} \left(x - \frac{E}{mg} \right)$. Then, the Schrödinger equation becomes $\frac{d^2\psi}{dz^2} - z\psi = 0$, that is, it coincides with the Airy equation. The general solution is of the form $CAi(z) + DBi(z)$ for some constants C and D. Since $Bi(z)$ diverges as $x \to \infty$, square integrability of ψ implies that $D = 0$. Hence,

$$
\psi(x) = CAi \left[(2m^2 g)^{1/3} \left(x - \frac{E}{mg} \right) \right]. \tag{10.16}
$$

The boundary condition $\psi(0) = 0$, implies that $Ai\left[-\left(\frac{2}{mg^2}\right)^{1/3} E\right] = 0$. Hence, the energy eigenvalues are

$$E_n = \left(\frac{mg^2}{2}\right)^{1/3} s_n, \qquad (10.17)$$

where $-s_n$ is the $(n+1)$th root of the Airy function, $n = 0, 1, 2, \ldots$. By the analysis of Section B.4 (in Appendix B), an excellent approximation to s_n is $s_n = \left[\frac{3\pi}{2}(n+\frac{3}{4})\right]^{2/3}$.

Next, we evaluate the eigenvalues in the semiclassical approximation. The turning points for energy E are $x = 0$ and $x = \frac{E}{mg}$. It follows that

$$\oint p\,dx = 2\sqrt{2m}\int_0^{\frac{E}{mg}} dx\sqrt{E - mgx} = \frac{4\sqrt{2}}{3}\frac{E^{3/2}}{\sqrt{mg}}. \qquad (10.18)$$

Then, we use Eqs. (10.7) and (10.15). They lead to an equation of the form (10.17), but with different expressions for s_n. In the GO approximation, $s_n^{(GO)} = \left[\frac{3\pi}{2}(n+1)\right]^{2/3}$, and in the WKB approximation, $s_n^{(WKB)} = \left[\frac{3\pi}{2}(n+\frac{3}{4})\right]^{2/3}$. The WKB approximation is remarkably accurate.

Looking at the plot of the Airy function, we see that its value at $z = 2$ is about 10 percent of its maximum value. Hence, the position probability distribution associated to Eq. (10.16) is well concentrated to values less than $x = h$, where $h = \frac{E}{mg} + \left(\frac{4}{m^2 g}\right)^{\frac{1}{3}}$. Comparing with Eq. (10.17), we see that the second term is negligible for $n \gg 1$, hence in this regime the maximum height is well approximated by the classical expression $h = \frac{E}{mg}$.

To summarize, the GO and the WKB approximations allow us to determine quantum properties using solely calculations from classical mechanics. Some physical predictions coincide with those of classical physics at the limit $n \to \infty$.

10.2.3 Tunneling in the WKB Approximation

Next, we employ the WKB matching conditions to the problem of particle tunneling through a general potential $V(x)$ that vanishes outside an interval $[a, b]$. In particular, we will identify the transmission amplitude T_k and the reflection amplitude R_k for the eigenfunctions $f_{k\pm}$ of Eq. (5.57).

To this end, we write the eigenfunction $f_{k+}(x) = \frac{T_k}{\sqrt{2\pi}}e^{ikx}$ for $x > b$ as

$$\frac{T_k}{\sqrt{2\pi}}e^{i\pi/4}\left[\cos\left(kx - \frac{\pi}{4}\right) + i\sin\left(kx - \frac{\pi}{4}\right)\right].$$

The WKB matching conditions at x_2 imply that in the interval $[x_1, x_2]$,

$$f_{k+}(x) = \frac{\sqrt{k}T_k e^{i\pi/4}}{\sqrt{2\pi\kappa(x)}}\left[\frac{1}{2}\exp\left[-\int_x^{x_2}\kappa(x')dx'\right] - i\exp\left[\int_x^{x_2}\kappa(x')dx'\right]\right]$$

$$= \frac{\sqrt{k}T_k e^{i\pi/4}}{\sqrt{2\pi\kappa(x)}}\left[\frac{1}{2}\Theta^{-1}\exp\left[\int_{x_1}^{x}\kappa(x')dx'\right] - i\Theta\exp\left[-\int_{x_1}^{x}\kappa(x')dx'\right]\right],$$

where $\Theta = \exp\left[\int_{x_1}^{x_2}\kappa(x')dx'\right]$.

Using the WKB conditions at x_1, we find that for $x < a$,

$$f_{k+}(x) = -\frac{T_k e^{i\pi/4}}{\sqrt{2\pi}} \left[2i\Theta \cos\left(-kx - \frac{\pi}{4}\right) + \frac{1}{2}\Theta^{-1} \sin\left(-kx - \frac{\pi}{4}\right) \right]$$

$$= \frac{T_k}{2\sqrt{2\pi}} \left[\left(2\Theta + \frac{1}{2}\Theta^{-1}\right) e^{ikx} - i\left(2\Theta - \frac{1}{2}\Theta^{-1}\right) e^{-ikx} \right]. \qquad (10.19)$$

Comparing with Eq. (5.57), we conclude that

$$T_k = \frac{2}{2\Theta + \frac{1}{2}\Theta^{-1}}, \qquad R_k = -i\frac{2\Theta - \frac{1}{2}\Theta^{-1}}{2\Theta + \frac{1}{2}\Theta^{-1}}. \qquad (10.20)$$

In the WKB approximation, $\arg T_k = 0$. Hence, the effective time that it takes the particle to cross the barrier according to Eq. (7.65) is negative, and it does not depend on the details for the potential. This means that *the WKB approximation greatly misrepresents the phases of T_k and R_k*. However, it provides a reasonable estimate for the transmission probability $|T_k|^2$.

We can see this in the square-barrier potential of Section 5.4.3. In this case, $\Theta = e^{\lambda a}$, where $\lambda = \sqrt{2m(V_0 - E)}$ and V_0 is the height of the potential. Hence,

$$|T_k|^2 = \frac{1}{\cosh^2(\lambda a + \ln 2)}.$$

At the limit of the opaque barrier ($\lambda a \gg 1$) $|T_k|^2 = e^{-2\lambda a}$. The exact solution (5.81) gives $|T_k|^2 = [1 + \frac{1}{4}(\frac{k}{\lambda} - \frac{\lambda}{k})^2]^{-1} e^{-2\lambda a}$. The WKB approximate captures correctly the exponential dependence of $|T_k|^2$ but it overestimates the prefactor.

10.3 Bohmian Paths

The WKB approximation allows for the derivation of quantum properties using notions from classical physics. In fact, it is possible to describe quantum phenomena using classical concepts without any approximation. To see this, note that Eq. (10.3) can be written as $\frac{1}{2m}(S')^2 + V + V_Q = E$, where

$$V_Q(x, t) = -\frac{R''(x, t)}{2mR(x, t)}, \qquad (10.21)$$

is the so-called *quantum potential*, that is, a "correction" to the classical potential $V(x)$ that is dependent on the wave function. The identification $p = S'$ leads to the energy conservation equation $\frac{p^2}{2m} + V_{tot} = E$ for a particle in the potential $V_{tot} = V + V_Q$.

This property generalizes straightforward to the time-dependent Schrödinger equation. Substituting $\psi = Re^{iS}$ into Eq. (2.30), we obtain

$$\dot{R} = -\frac{1}{2mR}(R^2 S') \qquad (10.22)$$

$$\dot{S} = -\frac{1}{2m}(S')^2 - V_{tot}, \qquad (10.23)$$

where the dot denotes a partial time derivative.

Equation (10.22) is simply the continuity equation $\dot{\rho} + J' = 0$, where $\rho = R^2$ is the position probability density and $J = \frac{1}{m}R^2 S'$ is the probability current. To interpret Eq. (10.23), we first identify the momentum function $p(x, t) = S'(x, t)$. Then, we identify particle trajectories $x(t)$ by the condition that their momentum equals $p(x, t)$, that is, by the differential equation

$$m\frac{dx(t)}{dt} = S'[x(t), t].$$ (10.24)

We differentiate both sides in Eq. (10.23) with respect to x, to obtain $\dot{S}' = -\frac{1}{m}S'S'' - V'_{tot}$, or equivalently, $\dot{p} + \frac{p}{m}p' = V'_{tot}$. For particle trajectories that satisfy Eq. (10.24), we obtain $\dot{p} + \frac{dx(t)}{dt}p' = V_{tot}$. For any function $F(x, t)$, the total rate of change along a trajectory $x(t)$ is $\frac{dF}{dt} := \frac{\partial F}{\partial t} + \frac{dx(t)}{dt}\frac{\partial F}{\partial x}$. Hence, the momentum $p(x, t)$ satisfies

$$\frac{d}{dt}p = -V'_{tot}.$$ (10.25)

We conclude that particle trajectories satisfy Newton's equation with respect to the quantum-corrected potential.

This awkward rewriting of Schrödinger equation achieves the introduction of classical particle trajectories into quantum theory. The wave of Schrödinger's equation and the particle coexist: the wave guides the motion of the particle by affecting the potential in which the particle moves. This description is the key idea of Bohmian mechanics, an alternative description of quantum theory that was developed by Bohm (1952), drawing on earlier ideas by de Broglie.

Bohmian mechanics is a deterministic theory. Schrödinger's equation is a deterministic wave equation, and the particle's trajectory follows from a deterministic equation of motion. Nonetheless, physical predictions involve probabilities. However, in Bohmian mechanics,

... quantum-mechanical probabilities are regarded (like their counterparts in classical statistical mechanics) as only a practical necessity and not as an inherent lack of complete determination in the properties of matter at the quantum level. (Bohm, 1952)

Bohmian mechanics requires that the particle positions in statistical ensembles are distributed according to Born's rule, that is, with a probability density for position given by $|\psi(x, t)|^2$. This probabilistic assumption is rather ad hoc in an otherwise fully deterministic theory, but it is crucial in order to reproduce the statistical predictions of quantum theory. There are some arguments that Born's rule arises naturally as an equilibrium distribution for statistical ensembles, much as the Maxwellian velocity distribution emerges from an analysis of classical thermodynamic equilibrium (Bohm, 1953; Dürr et al., 1992).

Assuming Born's rule, Bohmian mechanics fully reproduces the predictions of quantum theory for nonrelativistic systems. It also describes quantum measurements through the inclusion of the degrees of freedom of the apparatus into the quantum description. However, the behavior of Bohmian paths runs counter to all expectations from classical physics. This can be seen in the following two examples.

1. Let $\psi(x)$ describe a bound state of the Hamiltonian. The wave function is real-valued; hence, S vanishes and so does the momentum p. The particle is motionless. This means, for example, that in each individual hydrogen atom, the electron is at rest with respect to the nucleus. The

position of rest is different in each atom, and it is distributed by the probability distribution $|\psi(x)|^2$ in any statistical ensemble.

2. Consider a free particle with an initial Gaussian state of mean position $\langle \hat{x} \rangle = 0$, position spread σ and mean momentum $\langle \hat{p} \rangle = p_0$. The time-evolved wave function is given by Eq. (7.48). It is straightforward to evaluate $p(x, t) = \mathrm{Im}\frac{\partial}{\partial x}\psi(x, t)$ and thus, to write the equation of motion for the Bohmian paths. We obtain

$$\frac{dx}{dt} = \frac{\frac{x}{t}}{1 + \left(\frac{2m\sigma^2}{t}\right)^2} + \frac{\frac{p_0}{m}}{1 + \left(\frac{t}{2m\sigma^2}\right)^2}.$$

For $t \ll m\sigma^2$, Bohmian paths approximately satisfy the classical equations of motion $\frac{dx}{dt} = \frac{p_0}{m}$. However, they strongly diverge from classical paths at later times. For example, if $p_0 = 0$, the associated classical paths are given by $x(t) = x(0)$. The Bohmian paths are $x(t) = x(0)\sqrt{1 + \left(\frac{t}{2m\sigma^2}\right)^2}$. At long times $x(t) = x(0)t/(2m\sigma^2)$, that is, Bohmian paths diverge from the classical paths linearly in time.

In Bohmian mechanics, particle positions have specific values prior to measurement, and measurement determines these values. Bell's theorem then implies that Bohmian mechanics violates locality. Indeed, the guiding field leads to faster-than-light interactions between particles. However, as the theory leads to the same statistical predictions with quantum theory, these faster-than-light interactions are never manifested in experiments.

Besides the problem of nonlocality, the main drawback of Bohmian mechanics is its restricted domain of applicability. Its generalizations to relativistic systems cannot, as yet, reproduce the predictions of standard quantum theory. Still, Bohmian mechanics is of great importance, because it is a realist theory that reproduces a significant subset of the predictions of quantum theory. Hence, it allows us to understand the structure of theories that violate the measurement independence assumption in Bell inequalities – see Section 9.4.1.

10.4 Feynman Paths

In this section, we will describe the evolution of quantum states in terms of paths on the configuration space. These paths are not physical, however, they allow for a formulation of quantum dynamics with an simple and intuitive connection to classical physics, first developed by Richard Feynman. This formulation has found a myriad of applications in quantum phenomena.

10.4.1 The Path Integral

We will present Feynman's formulation of quantum dynamics in the simplest system, a particle of mass m in one dimension, with a Hamiltonian of the Schrödinger type, $\hat{H} = \frac{\hat{p}^2}{2m} + V(\hat{x})$. We want to compute the propagator $G_t(x_f, x_i) = \langle x_f | e^{-i\hat{H}t} | x_i \rangle$. To this end, we take a sequence of N time points $t_n = n\delta t$, where $\delta t = \frac{t}{N}$, $n = 1, 2, \ldots N$. Then, we express the propagator as

$$G_t(x_f, x_i) = \int \prod_{n=1}^{N-1} dx_n \langle x_f | e^{-i\hat{H}\delta t} | x_{N-1} \rangle \dots \langle x_2 | e^{-i\hat{H}\delta t} | x_1 \rangle \langle x_1 | e^{-i\hat{H}\delta t} | x_i \rangle,$$

where we split the evolution operator into $N-1$ time steps $e^{-i\hat{H}t} = (e^{-i\hat{H}\delta t})^n$, and introduced a resolution of the unity at each time step t_n, for $n = 1, 2, \dots N-1$.

We evaluate the matrix elements

$$\langle x_n | e^{-i\hat{H}\delta t} | x_{n-1} \rangle = \langle x_n | \hat{I} - i\hat{H}\delta t | x_{n-1} \rangle = \langle x_n | \hat{I} - iV(\hat{x})\delta t - \frac{i\delta t}{2m}\hat{p}^2 | x_{n-1} \rangle$$

$$= \int dp_n \left[1 - iV(x_n)\delta t - \frac{ip_n^2 \delta t}{2m} \right] \langle x_n | p_n \rangle \langle p_n | x_{n-1} \rangle$$

$$= \int \frac{dp_n}{2\pi} e^{-i\left(\frac{p_n^2}{2m} + V(x_n)\right)\delta t + ip_n(x_n - x_{n-1})} = \sqrt{\frac{2m\pi}{i\delta t}} e^{\frac{im(x_n - x_{n-1})^2}{2\delta t} - iV(x_n)\delta t},$$

where we inserted a decomposition of the unity in the momentum basis, and performed the Gaussian integral over p_n as in Section 7.3.2.

Then, the propagator becomes

$$G_t(x_f, x_i) = \left(\frac{m}{2\pi i \delta t}\right)^{\frac{N}{2}} \int \prod_{n=1}^{N-1} dx_n e^{i\sum_{n=1}^{N}\left(\frac{m(x_n - x_{n-1})^2}{2\delta t} - V(x_{n-1})\delta t\right)}. \tag{10.26}$$

Suppose that we take the limit $N \to \infty$ for the right-hand side of Eq. (10.26), and that the limit can be interchanged with the integral. In fact, we cannot do this, but let us for the moment pretend differently. Then, the term in the exponential becomes equal to $iS[x(\cdot)]$, where $S[x(\cdot)] = \int_0^t ds \left[\frac{1}{2}m\dot{x}^2 - V(x)\right]$ is the classical action (1.16). Then, we can write formally

$$G_t(x_f, x_i) = \int Dx(\cdot) e^{iS[x(\cdot)]}, \tag{10.27}$$

where $Dx(\cdot)$ is a short-hand for the integration measure $\sqrt{\frac{m}{2\pi i \delta t}} \prod_{n=1}^{N-1} \sqrt{\frac{m}{2\pi}} dx_n$ at the limit where $N \to \infty$. The integration in Eq. (10.27) is over all paths $x(\cdot)$, such that $x(0) = x_i$ and $x(t) = x_f$.

The heuristic integral in Eq. (10.27) is known as a *path integral*. Equation (10.27) expresses quantum dynamics in terms of a purely classical object, the action. Feynman (1942) derived this expression for the propagator following earlier ideas of Dirac (1933) about the role of the Lagrangian in quantum theory. He then extended this formulation to more general quantum systems, including relativistic quantum fields (Feynman, 1948), and used it to construct computational procedures that work even when a rigorous analysis is impossible. Since then, heuristic Feynman path integrals have been widely used in many fields of physics. They provide a good quantization procedure, that is, a fast, if not completely rigorous, way of expressing the quantum dynamics from the action $S[x(\cdot)]$ of an associated classical theory. For rigorous definitions of the path integral, see Box 10.1.

The success of the path integral formalism has often led to strong statements that path integrals provide a new formulation of quantum theory, distinct from the Hilbert space formulation. This is definitely not true: path integrals provide only a useful expression for the evolution operator. This may be sufficient for a large class of physical problems, but not for all. The Hilbert space structure is indispensable for defining general observables, for analysing the structure of quantum states and for describing quantum measurements.

Box 10.1   Rigorous Definition of the Path Integral

Integrals over paths had been defined well before Feynman, in the context of classical probability theory (Wiener, 1921, 1923). To understand the connection, note that the free-particle Schrödinger's equation with imaginary time coincides with the diffusion equation $\dot{\rho} = \frac{1}{2}\nu\rho''$, which is the continuum limit of a random walk – see Problem 1.8.

The diffusion equation admits a propagator $g_t(x,x') = \frac{1}{\sqrt{2\pi\nu t}}e^{-\frac{(x-x')^2}{2\nu t}}$ that is similar to Eq. (7.47). The propagator $g_t(x,x')$ is interpreted as the conditional probability density that a particle initially around x' will be found around x after time t.

Using the propagator, we construct the probability density $p(x_1, x_2, \ldots, x_n)$ for paths starting at x_0 to be found at points $x_1, x_2, \ldots, x_N$ at times $t_1, t_2, \ldots, t_N$, respectively, where the time instants are given by $t_n = n\delta t$, for $n = 1, 2, \ldots, N$:

$$p(x_1, x_2, \ldots, x_N) = \prod_{n=1}^{N} g_{\delta t}(x_n - x_{n-1}) = (2\pi\nu\delta t)^{-\frac{N}{2}}e^{-\frac{\sum_{n=1}^{N}(x_n-x_{n-1})^2}{2\nu\delta t}}. \tag{A}$$

When taking the limit $\delta t \to 0$ and $N \to \infty$ for constant $N\delta t$, we obtain a heuristic expression for a probability density on paths $P[x(\cdot)] \sim e^{-\frac{1}{2\nu}\int dt\dot{x}^2}$. While this formal expression is mathematically meaningless, it turns out that there *is* a rigorous way of taking the limit $N \to 0$ in Eq. (A). We can define the *Wiener measure*,

$$d\mu_W^\nu[x(\cdot)] = e^{-\frac{1}{2\nu}\int dt\dot{x}^2}Dx(\cdot),$$

even though neither the exponential factor nor $Dx(\cdot)$ are separately well defined.

We pay a price for this definition: The Wiener measure can be defined only on a space of paths Π that is too large. Almost all paths in Π (i) are continuous but nowhere differentiable and (ii) they have an infinite number of local maxima and minima in any finite interval $[t_1, t_2]$. A pair of typical paths in Π are plotted in the figure in this box. Certainly these paths are very different from the usual paths of classical mechanics.

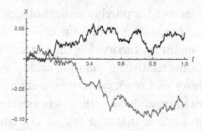

Equipped with the Wiener measure, we can prove that the propagator $G_\tau(x,x')$ for the Schrödinger equation with imaginary time $\frac{\partial\psi}{\partial\tau} = \frac{1}{2m}\psi'' - V(x)\psi$, can be expressed as a path integral

$$G_\tau(x,x') = \int d\mu_W^{m^{-1}}[x(\cdot)]e^{-\int_0^\tau d\tau' V[x(\tau')]}\delta[x(0) - x']\delta[x(\tau) - x]. \tag{B}$$

> Equation (B) was proved by Kac (1949), and it is known as the Feynman–Kac formula. It demonstrates that the Feynman path integral is well defined for imaginary time $\tau = it$. In most cases, obtain the correct quantum result by transforming back to real time $t = -i\tau$ at the end of the calculation.

10.4.2 Evaluating Path Integrals

Path integrals are analogous to ordinary integrals of the form (A.18), in the sense that they involve integration over a rapidly oscillating phase. As shown in Section A.3, such integrals are dominated by the contributions near the minima of the phase function. In a path integral, the phase function is the action S, which, by Hamilton's principle, is minimized for solutions to the classical equations of motion. Let us denote by $x_{cl}(s)$ the solution to the classical equations of motion $m\ddot{x} + V'(x) = 0$, subject to the boundary conditions $x(0) = x_i$ and $x(t) = x_f$; here, the dot refers to derivative with respect to s. We define $\xi(s) := x(s) - x_{cl}(s)$. For paths in the path integral (10.27), $\xi(0) = 0$ and $\xi(t) = 0$.

We expand the action $S[x(\cdot)]$ to second order in $\xi(\cdot)$,

$$
\begin{aligned}
S[x(\cdot)] &\simeq S_{cl} + \frac{1}{2}\int_0^t ds\left[m\dot{\xi}^2(s) - V''[x_{cl}(s)]\xi^2(s)\right] \\
&= S_{cl} - \frac{1}{2}\int_0^t ds\,\xi(s)\left[m\ddot{\xi}(s) + V''[x_{cl}(s)]\xi(s)\right],
\end{aligned}
\tag{10.28}
$$

where $S_{cl} := S[x_{cl}(\cdot)]$. Equation (10.28) is exact for quadratic potentials.

Equation (10.27) becomes

$$
G_t(x_f, x_i) = e^{iS_{cl}(x, x')}\int D\xi(\cdot)e^{-\frac{i}{2}\int_0^t ds\,\xi(s)\left[m\ddot{\xi}(s) + V''[x_{cl}(s)]\xi(s)\right]}.
\tag{10.29}
$$

We evaluate the path integral above for imaginary time $\tau = it$. We also use the imaginary variable $\sigma = is$ for the paths. We introduce the (real) Hilbert space $L_D^2([0, \tau], d\sigma)$ with inner product given by $(f, g) = C^{-1}\int_0^\tau d\sigma f(\sigma)g(\sigma)$, for some constant $C > 0$. The operator $\mathcal{E}_\tau := \frac{C}{2\pi}\left[-m\frac{d^2}{d\sigma^2} + \frac{1}{2}V[x_{cl}(\sigma)]\right]$ is self-adjoint and positive. We introduced the constant C in order to compensate for the multiplicative ambiguity in the definition of the measure $D\xi(\cdot)$; we will determine C later.

The integrand reads $\int D\xi(\cdot)e^{-\pi(\mathcal{E}_\tau\xi, \xi)}$. In a finite-dimensional vector space, Eq. (A.4) implies that this integral $[\det\mathcal{E}_\tau]^{-1/2}$. We take the same expression to apply in $L_D^2([0, \tau], d\sigma)$, provided we make an appropriate definition of the determinant for the operator $\mathcal{E}_\tau$. Then, we obtain the *semiclassical propagator*

$$
G_t(x_f, x_i) = \frac{1}{\sqrt{\det\mathcal{E}_{it}}}e^{iS_{cl}(x_f, x_i)}.
\tag{10.30}
$$

The determinant $\det\mathcal{E}_{it}$ depends on x_i and x_f.

The usual definition of an operator determinant is given by Eq. (4.68). With this definition, $\det\mathcal{E}_\tau$ diverges. For example, consider the operator $\mathcal{E}_\tau^0 = -\frac{mC}{2\pi}\frac{d^2}{d\sigma^2}$ for a free particle. It coincides with the Schrödinger operator for a particle in a box (Section 5.6.2), hence, its eigenvalues are $\epsilon_n = \frac{n^2 \pi mC}{2\tau^2}$, with $n = 1, 2, \ldots$. Then, $\det \mathcal{E}_\tau^0 = \prod_{n=1}^\infty \epsilon_n = \exp\left[\sum_{n=1}^\infty \ln\left(\frac{n^2 \pi mC}{2\tau^2}\right)\right] \to \infty$.

Several methods exist for the construction of finite operator determinants. The most powerful is the *zeta-function regularization*. This works as follows. For any self-adjoint operator $\hat{A}$ with eigenvalues a_n, $\ln \det \hat{A} = -\sum_n a_n = -\frac{d}{ds} \sum_n a_i^{-s}|_{s=0}$. This motivates the introduction of the *zeta function* associated to $\hat{A}$,

$$\zeta_A(s) := \sum_n \frac{1}{a_n^s}, \tag{10.31}$$

a generalization of Riemann's zeta function $\zeta(s) = \sum_{n=1}^\infty n^{-s}$.

If $a_n \to \infty$ as $n \to 0$, then $\zeta_A(s)$ is well defined for all complex s with sufficiently large real part. We analytically continue $\zeta_A(s)$ to a neighbourhood of $s = 0$, and thus, *define* the determinant as

$$\det \hat{A} := e^{-\zeta_A'(0)}. \tag{10.32}$$

As an example, we apply this procedure to the operator $\mathcal{E}_\tau^0$,

$$\zeta_{\mathcal{E}_\tau^0}(s) = \left(\frac{2\tau^2}{m\pi^2}\right)^s \sum_{n=1}^\infty n^{-2s} = \left(\frac{2\tau^2}{m\pi^2}\right)^s \zeta(2s). \tag{10.33}$$

We obtain $\zeta_{\mathcal{E}_\tau^0}'(0) = 2\zeta'(0) + \zeta(0) \ln\left(\frac{2\tau^2}{\pi mC}\right)$. From Table B.1, we see that $\zeta'(0) = -\frac{1}{2} \ln(2\pi)$ and $\zeta(0) = -\frac{1}{2}$. Hence, $\zeta_{\mathcal{E}_\tau^0}'(0) = -\frac{1}{2} \ln\left(\frac{8\pi\tau^2}{mC}\right)$, and by Eq. (10.32), $\det \mathcal{E}_\tau^0 = 2\tau\sqrt{\frac{2\pi}{mC}}$. Equation (7.47) implies that $\det \mathcal{E}_\tau^0 = \frac{2\pi\tau}{m}$, which determines $C = \sqrt{\frac{2m}{\pi}}$. Alternatively, we can determine C by the requirement that the free particle propagator satisfies $G_0(x_f, x_i) = \delta(x_f - x_i)$. The multiplicative constant C fixes the integration measure, so its value is the same for all potentials.

For a harmonic oscillator of frequency ω, the classical solution to the equations of motion with $x(0) = x_i$ and $x(t) = x_f$ is

$$x_{cl}(s) = \frac{x_i \sin \omega(t - s) + x_f \sin \omega s}{\sin \omega t}. \tag{10.34}$$

Therefore,

$$S_{cl} = \int_0^t ds\left(\frac{1}{2}m\dot{x}_{cl}^2 - \frac{1}{2}m\omega^2 x_{cl}^2\right) = \int_0^t ds \frac{d}{ds}(mx_{cl}\dot{x}_{cl}) - m\int_0^t ds\, x_{cl}(s)(\ddot{x}_{cl} + \omega^2 x_{cl})$$

$$= m\dot{x}_{cl}(t)x_{cl}(t) - m\dot{x}_{cl}(0)x_{cl}(0) = \frac{m\omega}{\sin \omega t}[(x_i^2 + x_f^2)\cos \omega t - 2x_i x_f]. \tag{10.35}$$

The operator $\mathcal{E}_\tau = \frac{mC}{2\pi}(-\frac{d^2}{d\sigma^2} + \omega^2)$ has eigenvalues $\epsilon_n = \frac{mC}{2\pi}\left(\frac{n^2\pi^2}{2\tau^2} + \omega^2\right)$, hence,

$$\det \mathcal{E}_\tau = \prod_{n=1}^\infty \frac{mC}{2\pi}\left(\frac{n^2\pi^2}{\tau^2} + \omega^2\right) = \left(\prod_{n=1}^\infty \frac{n^2\pi mC}{2\tau^2}\right) \prod_{n=1}^\infty \left(1 + \frac{\omega^2\tau^2}{n^2\pi^2}\right)$$

$$= \det \mathcal{E}_\tau^0 \prod_{n=1}^\infty \left(1 + \frac{\omega^2\tau^2}{\pi^2 n^2}\right).$$

We use the identity

$$\prod_{n=1}^{\infty} \left(1 + \frac{x^2}{\pi^2 n^2}\right) = \frac{\sinh x}{x}, \tag{10.36}$$

obtained by setting $z = ix$ in Euler's product formula, Eq. (A.15). Then,

$$\det \mathcal{E}_\tau = \det \mathcal{E}_\tau^0 \frac{\sinh \omega\tau}{\omega\tau} = \frac{2\pi \sinh \omega\tau}{m\omega}, \tag{10.37}$$

and we recover Eq. (7.54) for the propagator.

10.5 Quasiclassical Paths

In Chapter 2, we saw the importance of Heisenberg's idea that the notion of a particle trajectory must be abandoned in quantum theory. In this chapter, we saw that we can rewrite quantum theory, in ways that introduce some notion of particle trajectory, like Bohmian paths or Feynman paths. We also saw that classical trajectories can be employed in order to construct meaningful approximations to quantum dynamics. Nonetheless, Heisenberg's idea remains valid; the trajectories that we introduced in this chapter *do not correspond to physical observables*.

To see this, recall that, in classical physics, trajectories are themselves physical observables. For example, the trajectory of a planet in the Solar System is determined by many successive measurements of its position. We will distinguish the measurable trajectories of a physical system from paths that refer solely to a mathematical description (e.g., Feynman, Bohmian, or WKB paths) by referring to the former as *histories*.[1] In this section, we will use histories in order to analyse how classical behavior emerges in quantum systems.

10.5.1 The Frailness of Quantum Histories

The probabilities associated to classical histories are *robust*, in the sense that they are largely insensitive to the properties of a recording apparatus. We can use these probabilities in order to predict a physical system's future or to retrodict its past without any reference to measurement. In general, this is not possible for histories of quantum systems. To show this, we consider a two-time history for an observable $\hat{A}$, that is, we measure $\hat{A}$ at time $t = 0$ and then again at time t. Despite its simplicity, this example demonstrates all relevant features of quantum histories.

We assume an accuracy δ in the measurement of $\hat{A}$. To describe such measurements, we split the real line into subsets $\Delta_n = [(n-\frac{1}{2})\delta, (n+\frac{1}{2})\delta)$, for $n = 0, \pm 1, \pm 2, \ldots$. The sets Δ_n are mutually exclusive and exhaustive. Hence, the projectors $\hat{P}_n = \chi_{\Delta_n}(\hat{A})$ are also mutually exclusive and exhaustive.

---

[1] The concept of a history does not apply solely to particle trajectories, but also to fields. The totality of the EM waves received by my radio while I listen to a program defines a history for the electromagnetic field.

For an initial state $\hat{\rho}_0$, the probability that the particle is found first in cell Δ_n and then in Δ_k is given by Eq. (8.16),

$$\text{Prob}(\Delta_n, 0; \Delta_k, t) = \text{Tr}\left(\hat{P}_k \hat{U}_t \hat{P}_n \hat{\rho}_0 \hat{P}_n \hat{U}_t^\dagger\right),\tag{10.38}$$

where $\hat{U}_t = e^{-i\hat{H}t}$ is the evolution operator of the system.

Consider now the same set-up, but with double accuracy in the first measurement, that is, the real line is split into subsets $\bar{\Delta}_n = [(n - \frac{1}{2})\bar{\delta}, (n + \frac{1}{2})\bar{\delta}]$, where $\bar{\delta} = \frac{1}{2}\delta$. Then, we define the projectors $\hat{P}'_n = \chi_{\bar{\Delta}_n}(\hat{A})$, and evaluate the probability

$$\text{Prob}(\bar{\Delta}_n, 0; \Delta_k, t) = \text{Tr}\left(\hat{P}_k \hat{U}_t \hat{P}'_n \hat{\rho}_0 \hat{P}'_n \hat{U}_t^\dagger\right).\tag{10.39}$$

The key point is that the two probabilities (10.38) and (10.39) do not follow from the same probability measure. To see this, we note that $\Delta_n = \bar{\Delta}_{2n} \cup \bar{\Delta}_{2n-1}$; hence, $\hat{P}_n = \hat{P}'_{2n} + \hat{P}'_{2n-1}$. If a common probability measure exists, the Kolmogorov additivity property – see Section 1.4.2 – implies that

$$\mathcal{D}_{n,k} := \text{Prob}(\Delta_n, 0; \Delta_k, t) - \text{Prob}(\bar{\Delta}_{2n}, 0; \Delta_k, t) - \text{Prob}(\bar{\Delta}_{2n-1}, 0; \Delta_k, t) = 0.$$

This condition fails. We straightforwardly evaluate

$$\mathcal{D}_{n,k} = 2\text{Re}\,\text{Tr}\left(\hat{P}_k \hat{U}_t \hat{P}'_{2n-1} \hat{\rho}_0 \hat{P}'_{2n} \hat{U}_t^\dagger\right) \neq 0.\tag{10.40}$$

In general, $|\mathcal{D}_{n,k}|$ is of the order of $\text{Prob}(\Delta_n, 0; \Delta_k, t)$, so the Kolmogorov additivity condition does not hold even approximately – see Box 10.2 for an example.

In classical physics, we get more accurate results when *fine-graining* our description, that is, when we increase the resolution of the measuring apparatus. The probabilities converge to an ideal value. This is also true for quantum measurements at a single moment of time: The probabilities for a position measurement with accuracy δ converge to the Born probability distribution $|\psi_0(x)|^2$ as $\delta \to 0$.

This (seemingly obvious) property fails for quantum sequential measurements. Fine-graining does not lead to more accurate results, it leads to *different results*. There is no convergence to some ideal probability distribution, because measurements with different resolution are not described by a common probability measure. We will refer to this counterintuitive feature as the *frailness* of quantum histories.

Box 10.2 The Frailness of Quantum Histories: An Example

We will evaluate $\mathcal{D}_{n,k}$ of Eq. (10.40) for measurements of position on a free particle of mass m. This means that $\hat{P}_n = \int_{\delta(n-\frac{1}{2})}^{\delta(n+\frac{1}{2})} dx |x\rangle\langle x|$, and $\hat{P}'_n = \int_{\bar{\delta}(n-\frac{1}{2})}^{\bar{\delta}(n+\frac{1}{2})} dx |x\rangle\langle x|$, where $\bar{\delta} = \frac{1}{2}\delta$. For simplicity, we assume an initial state with real wave function $\psi_0(x)$ that varies at scales much larger than δ, so that $\hat{P}'_{2n}|\psi_0\rangle \simeq \psi_0(x_n) \int_{x_n - \frac{\delta}{4}}^{x_n + \frac{\delta}{4}} dx |x\rangle$, and $\hat{P}'_{2n-1}|\psi_0\rangle \simeq \psi_0(x_n) \int_{x_n - \frac{3\delta}{4}}^{x_n - \frac{\delta}{4}} dx |x\rangle$, where $x_n = n\delta$.

Then, $\mathcal{D}_{n,k} = |\psi_0(x_n)|^2 \int_{-\frac{\delta}{2}}^{\frac{\delta}{2}} dy \, \text{Re} \left[L(y + x_k - x_n + \frac{\delta}{2}) L^*(y - x_k + x_n) \right]$,

where $L(x) := \int_{-\frac{\delta}{4}}^{\frac{\delta}{4}} dy \, G_t(x,y)$ is defined in terms of the free-particle propagator $G_t(x,y)$, given by Eq. (7.47). In this example, $\mathcal{D}_{n,k}$ depends only on the difference $n - k$, and so do the probabilities $\text{Prob}(\bar{\Delta}_n, 0; \Delta_k, t)$.

We carry out the integration over y in $\mathcal{D}_{n,k}$, and we express $L(x)$ in terms of the function $E(x) := C(x) + iS(x)$, where $C(x)$ and $S(x)$ are the Fresnel integrals, given by Eqs. (B.12) and (B.13),

$$ L(x) = \frac{1}{\sqrt{2i}} \left[E \left(\sqrt{\frac{m}{\pi t}} (x + \frac{\delta}{4}) \right) - E \left(\sqrt{\frac{m}{\pi t}} (x - \frac{\delta}{4}) \right) \right]. $$

The ratio $\epsilon_{n-k} := \frac{|\mathcal{D}_{n,k}|}{\text{Prob}(\bar{\Delta}_n, 0; \Delta_k, t)}$ is a measure of the frailness of quantum probabilities that does not depend on the initial state, It depends on the accuracy δ of the position measurements through the combination $\sigma = \sqrt{\frac{\pi m}{4t}} \delta$. If $\epsilon_{n,k} << 1$ for all n and k, then the probabilities for the two-time histories would be robust.

It turns out that this is not the case. The plot of ϵ_{n-k} as a function of σ for different values of $n - k$ shows that $\epsilon_{n,k}$ takes values around $\frac{1}{2}$. This class of histories are frail.

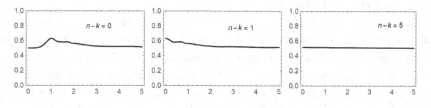

10.5.2 Quasiclassical Histories

We revisit the case of the two-time measurements of an observable $\hat{A}$ that it is described by the probabilities (10.38). We saw that the probabilities for histories are frail because, in general, no underlying probability measure exists. However, in specific cases, such a measure may exist, at least as an approximation.

A probability measure exists, if Kolmogorov's additivity condition is satisfied,

$$ \text{Prob}(\Delta_n \cup \Delta_{n'}, 0; \Delta_k, t) = \text{Prob}(\Delta_n, 0; \Delta_k, t) + \text{Prob}(\Delta_{n'}, 0; \Delta_k, t) \quad (10.41) $$

$$ \text{Prob}(\Delta_n, 0; \Delta_k \cup \Delta_{k'}, t) = \text{Prob}(\Delta_n, 0; \Delta_k, t) + \text{Prob}(\Delta_n, 0; \Delta_{k'}, t), \quad (10.42) $$

for all $n, n' \neq n, k$ and $k' \neq k$. Note that the projector associated to $\Delta_n \cup \Delta_{n'}$ is $\chi_{\Delta_n \cup \Delta_{n'}}(\hat{A}) = \chi_{\Delta_n}(\hat{A}) + \chi_{\Delta_{n'}}(\hat{A}) = \hat{P}_n + \hat{P}_{n'}$. Since Eq. (10.38) is linear with respect to the projector $\hat{P}_k$ of the second measurement, Eq. (10.42) is trivially satisfied. In contrast, Eq. (10.41) is not satisfied, unless the quantity

$$ \mathcal{D}_{n,n',k}(\delta) := \text{Re} \, \text{Tr} \left(\hat{P}_k \hat{U}_t \hat{P}_{n'} \hat{\rho}_0 \hat{P}_n \hat{U}_t^\dagger \right) \quad (10.43) $$

vanishes, for all k, n and n'.

It is not necessary that $\mathcal{D}_{n,n',k}(\delta)$ be exactly zero; it suffices that the Kolmogorov rule holds up to an error $\epsilon \ll 1$. This motivates the introduction of the quantity

$$\mathcal{E}(\delta) = \sum_{n,n' \neq n,k} |\mathcal{D}_{n,n',k}(\delta)|. \tag{10.44}$$

Equation (10.41) is satisfied with an accuracy less than $\mathcal{E}(\delta)$ for all $n, n' \neq n$ and k. Hence, a probability measure is approximately defined if $\mathcal{E}(\delta) \ll 1$.

As shown earlier, the operators $\hat{P}'_n = \hat{P}_{2n} + \hat{P}_{2n-1}$ form an exhaustive and exclusive set of width 2δ. Using these projectors we can evaluate $\mathcal{D}_{n,n',k}(2\delta)$, and, hence, $\mathcal{E}(2\delta)$. It is straightforward to show that $\mathcal{E}(2\delta) \leq \mathcal{E}(\delta)$, and then to show that $\mathcal{E}(N\delta) \leq \mathcal{E}(\delta)$ for all integers $N > 1$. Hence, if $\mathcal{E}(\delta) \ll 1$, then $\mathcal{E}(N\delta) \ll 1$. We expect that this condition will hold for all δ that interpolate between the integer multiples $N\delta$. This suggests the following definition.

> An observable $\hat{A}$ defines a *quasiclassical history* for an initial state $|\psi_0\rangle$, if there exists a scale δ_0, such that $\mathcal{E}(\delta) \ll 1$ for all $\delta \geq \delta_0$.

Hence, in a quasiclassical history, all two-time measurements of $\hat{A}$ with accuracy $\delta > \delta_0$ are described by a common probability measure up to an error that is smaller than $\mathcal{E}(\delta_0)$. We refer to δ_0 as the *classicalization scale* for $\hat{A}$.

To understand the meaning of a classicalization scale, think of a nanoparticle that consists of a large number of atoms, and let $\hat{A}$ be the position of its center of mass (CoM). If the CoM is monitored at an accuracy worse than δ_0, say of $5\,\mathrm{nm}$, its evolution can be reasonably well expressed in terms of classical trajectories, subject to a robust probability distribution. We can use this probability distribution in order to predict the future or to retrodict the past from present knowledge. However, if we attempt to measure the CoM with higher accuracy than δ_0, the "disturbance" will be so great as to render a description in terms of classical trajectories impossible. This is why we use the prefix "quasi" for histories of this type. Their classicality is conditional: Some measurements will reveal their fundamentally quantum nature.

Nonetheless, quasiclassical histories for N measurements are the closest representation to the notion of a classical trajectory that we can achieve in quantum theory (Gell-Mann and Hartle, 1993). As such, they are indispensable for understanding the quantum-to-classical correspondence.

The two most important mechanisms that lead to the emergence of quasiclassical histories are the following.

1. *Environment-induced decoherence* (See Zeh, 1970; Zurek, 1982; Joos and Zeh 1985.) Assume that the measured system is in contact with an environment. The total Hilbert space is of the form $\mathcal{H}_{sys} \otimes \mathcal{H}_{env}$, and the measured observable is $\hat{A} \otimes \hat{I}$. For simplicity, we assume that $\hat{A}$ is nondegenerate on $\mathcal{H}_{sys}$ so that its eigenvectors $|a\rangle$ define a basis on $\mathcal{H}_{sys}$. We also assume that the initial state of the system (at $t = 0$) factorizes as $\hat{\rho}_{sys} \otimes \hat{\rho}_{env}$, but the time evolution operator is not factorizable.

The environment is said to *decohere* the observable $\hat{A}$ if the reduced density matrix $\hat{\rho}_{sys}$ of the system at time $t > t_0$ satisfies

$$|\langle a|\hat{\rho}_{sys}(t)|a'\rangle| < |\langle a|\hat{\rho}_{sys}(t_0)|a'\rangle| \, F(|a - a'|, t), \tag{10.45}$$

where $F(a, t)$ is a decreasing function of both a and t that vanishes as $a \to \infty$ and as $t \to \infty$. This means that the environment tends to diagonalize the reduced density matrix of the system in the $|a\rangle$ basis. In the simplest examples of environment-induced decoherence, $F(a, t) \sim e^{-Da^2 t}$, where $D > 0$ is a constant that depends on properties of the environment – for examples, see Chapter 20.

In a decohering environment, $\mathcal{D}_{n, n', k}(\delta)$ in Eq. (10.43) is suppressed by a factor $F[|n-n'|\delta, |(t-t_0)|] \sim \exp[-D\delta^2(n - n')^2(t - t_0)]$. This means that, for sufficiently large δ, $\mathcal{D}_{n, n', k}(\delta)$ is negligible for all $n \neq n'$. Hence, we can find a classicalization scale δ_0, such that the definition of the quasiclassical history is satisfied. It follows that decohering environments generate quasiclassical histories.

It is important to emphasise that the decoherence condition (10.45) is *special*, and not generic. It presupposes a natural system–environment splitting, specific initial states $\hat{\rho}_{env}$ of the environment and specific forms of the system–environment interaction – see Section 20.5.2 for a system that demonstrates this point.

2. *Approximate determinism* Let us assume that, for each n, $\hat{U}_t^\dagger \hat{P}_n \hat{U}_t$ approximately equals another projector of the partition, say, $\hat{P}_{\sigma(n)}$. For example, the following condition may hold

$$||\hat{U}_t^\dagger \hat{P}_n \hat{U}_t - \hat{P}_{\sigma(n)}|| < \epsilon << 1 \tag{10.46}$$

for all n and for all $\delta > \delta_0$. It implies that

$$\mathcal{D}_{n, n', k} \simeq \mathrm{ReTr}\left(\hat{P}_{\sigma(k)}\hat{P}_{n'}\hat{\rho}_0\hat{P}_n\right) = \delta_{n, \sigma(k)}\delta_{nn'}\mathrm{Tr}(\hat{\rho}_0\hat{P}_n) + O(\epsilon), \tag{10.47}$$

Then, $\mathcal{E}(\delta)$ is of order ϵ, and the histories are quasiclassical.

The key point here is that the conditional probability

$$\mathrm{Prob}(\Delta_{\sigma(n)}, t | \Delta_n, 0) = \frac{\mathrm{Prob}(\Delta_n, 0; \Delta_{\sigma(n)}, t)}{\mathrm{Tr}(\hat{\rho}_0\hat{P}_n)} = 1 - O(\epsilon),$$

which implies that the projectors $\hat{P}_n$ behave approximately as alternatives of a deterministic system, that evolve according to the evolution law $n \to \sigma(n)$ with time step t.

In this case, the existence of quasiclassical histories is a consequence of approximate determinism. In particular, this is the case for quasiclassical histories that correspond to time evolution according to classical mechanics (Omnés, 1989b) – you can see how this works by solving Problem 10.12.

10.6 The Emergence of the Classical World

The concept of quasiclassical histories explains how classical behavior can arise as a special case in quantum systems. Still, it relies on the notion of measurement, and as such it presupposes an external apparatus to record these histories. Again, the apparatus must be postulated to be classical, so we cannot avoid the coexistence of classical and quantum concepts pointed out by Bohr. Indeed, in the minimal interpretation this coexistence must be taken for granted.

Can we use the notion of quasiclassical histories without reference to measurements? If this were possible, we would be able to derive classical physics from quantum physics without any reference to measurements – perhaps, this could also lead to a solution of the measurement problem. Unfortunately, this does not work.

To understand why, consider a history that corresponds to one physical magnitude $\hat{A}_1$ at time t_1 and another magnitude $\hat{A}_2$ at time t_2. Let us consider $\hat{A}_1$ and $\hat{A}_2$ to be the unknowns. Then, we can view Eqs. (10.41) and (10.42) as equalities that can be solved in order to determine $\hat{A}_1$ and $\hat{A}_2$. There is an infinite number of solutions, and, hence, an infinite number of sets of quasiclassical histories. These sets may be mutually incompatible, and they typically behave very differently from the histories of classical physics (Dowker and Kent, 1996).

In the minimal interpretation, the set of histories is specified by the measurement apparatus. In the absence of the apparatus, there is an infinity of possible ways that the quantum system may behave. If we want to recover our macroscopic experience, that is, a world described by classical physics, we must add axioms that select the appropriate sets of histories. A set of such axioms with universal validity has not been identified yet; however, even if it is discovered, it would mean that the specification of the classical world should enter the axiomatics of quantum theory, a result that would not be far from Bohr's conception.

Another issue that arises in this context is the absence of nonclassical states, like superpositions of macroscopically distinct states for big objects such as cats, or the Burj Khalifa, or the Moon. For any pair of macroscopically distinct states, say $|1\rangle$ and $|2\rangle$, there is an infinity of superpositions $c_1|1\rangle + c_2|2\rangle$, so, by any measure, the vast majority of states in a system's Hilbert space is nonclassical.

We know that the construction of nonclassical states in the laboratory is obstructed by the environment, which acts uncontrollably upon the prepared system and can destroy macroscopic superpositions – we will see explicit examples in Chapter 20. It is therefore tempting to postulate that the environment is responsible for the absence of such states from everyday experience. For the case of the cat, the environment can be taken as the air molecules in the atmosphere. But what if we include the totality of air molecules on Earth in the system? As we explained earlier, the Hilbert space of all air molecules has many more nonclassical states than classical ones; for example, states involving strong entanglement between all air molecules. If the air molecules were in a nonclassical state, they would not, in general, suppress the superpositions of a Schrödinger's cat. To explain why air molecules are found in a classical state, we have to postulate an environment for them, the whole Earth perhaps. But then, we can continue this procedure *ad infinitum*, or rather, until our system is the Universe, so that no environment is possible. Again, we will ask, how come we observe a universe that behaves classically at a macroscopic scale? Why do we not observe a quantum superposition of expanding and collapsing universes, or quantum correlations between separated clusters of galaxies? We have no choice but to postulate a very special state of the Universe, but of course, we do not have a quantum theory that works at this scale, to even know if such states are possible.

Hence, when we try to explain how the classical world emerges from quantum theory by invoking the action of an environment, we must ground our answers in a highly speculative quantum theory of the Universe. This is too large a leap from our usual applications of quantum theory, and the majority of physicists feel uncomfortable with it. Most importantly, such leaps have not progressed beyond the level of conjecture. There is simply no theory about the initial state of the Universe that explains the emergence of the classical world.

Some modifications of quantum theory could explain the emergence of the macroscopic world without reference to measurement. In Section 8.3.3, we mentioned the dynamical state reduction models that have been proposed as solutions to the measurement problems. Dynamical reduction can explain the emergence of the classical world, at least in nonrelativistic physics, where the problems of dynamical reduction with faster-than-light signals are not manifested. Furthermore, predictions from dynamical reduction models are, in principle, testable.

Finally, we note that the coexistence of quantum and classical physics may be due to the existence of *irreducibly classical entities* (ICEs) which interact with quantum systems. These interactions lead to the suppression of nonclassical states in matter and to the emergence of the classical world. The ICEs cannot be made of ordinary matter, because ordinary matter consists of atoms, and atoms are quantum systems. Two candidate ICEs appeared in our discussion of quantum measurements in Section 8.3, namely, mind/consciousness and the gravitational field. While there is as yet no concrete theory, simple models that treat the gravitational field as an ICE lead to testable predictions (Bassi et al., 2017). It is to be noted, however, that existing models are rather ad hoc and the predictions are strongly model-dependent. If ICEs exist, their fundamental description likely requires new physics.

We will come back to the issue of the emergent classical world in Chapter 22.

QUESTIONS

10.1 Reintroduce $\hbar$ in Schrödinger's equation, and confirm that the semiclassical evolution equations correspond to the leading-order terms in an expansion with respect to $\hbar$.

10.2 Dirac (1930a) wrote that "classical mechanics may be regarded as the limiting case of quantum mechanics when $\hbar$ tends to zero." This statement is often asserted as a general fact, even if it originally referred to the very specific context of explaining the relation between commutators and Poisson brackets (see Section 12.2). Argue that this statement is not in general true, by considering highly nonclassical states like cat states or entangled states.

10.3 A popular account of Feynman's path integral asserts that "an electron takes all possible paths from emitter to detector at the same time." Explain why this statement is problematic.

10.4 In Bohmian mechanics, a system in a quantum state ψ behaves classically if the quantum potential V_Q is negligible in comparison with the classical potential V at all times. Is the emergence of a classical world from quantum theory a problem in Bohmian mechanics?

10.5 Explain Landau and Lifschitz's quote at the beginning of this chapter. Does it make literal sense?

10.6 Which account of the relation between quantum and classical physics do you find more convincing? Explain its advantages and its problems.

PROBLEMS

10.1 Find the energy eigenvalues in the WKB approximation for a particle of mass m in a potential $V(x) = kx^{2n}$, where $k > 0$ and n integer.

10.2 Find the energy eigenvalues in the WKB approximation for the bouncing ball in a box, that is, for a particle of mass m with Hamiltonian $\hat{H} = \frac{\hat{p}^2}{2m} + mg\hat{x} \geq 0$ with $x \in [0, L]$, and Dirichlet boundary conditions at both boundaries.

10.3 A particle of mass m lies in a box of length L with a potential $V(x) = a|x - \frac{L}{2}|$, $a > 0$. Evaluate the energy eigenvalues in the WKB approximation.

10.4 Consider a pair of particles, described by the wave function $\psi(x_1, x_2) = C \exp[-ax_1^2 - ax_2^2 + 2bx_1 x_2]$, where $a > 0$, b, and $C > 0$ are constants. (i) Calculate the quadratic entanglement E for ψ and show that the state is entangled if $b \neq 0$. (ii) Evaluate the quantum potential as

$$V_Q(x_1, x_2) = -\frac{\frac{\partial^2 R}{\partial x_1^2} + \frac{\partial^2 R}{\partial x_2^2}}{2mR},$$

where $R = |\psi|$. (iii) Show that if ψ is entangled then the quantum potential causes instantaneous interaction at a distance for the two particles.

10.5 Derive a path-integral expression for the propagator of the Hamiltonian $\hat{H} = \frac{1}{2m}[\hat{p} - a(\hat{x})]^2 + V(\hat{x})$, where $a(x)$ is a general function.

10.6 (i) Express the propagator in the coherent state basis as a path integral,

$$\langle z_f, t|z_i, 0\rangle = e^{-\frac{1}{2}(|z_f|^2 + |z_i|^2)} \int Dz(\cdot)Dz^*(\cdot) e^{iS[z(\cdot), z^*(\cdot)]},$$

where integration extends over all paths such that $z(0) = z_i$, $z^*(t) = z_f^*$, and the action at the continuous limit is

$$iS[z(\cdot), z^*(\cdot)] = z^*(t)z(t) + \int_0^t ds\left[-z^*(s)\dot{z}(s) - iH(z^*(s), z(s))\right],$$

where $H(z^*, z') = \langle z|\hat{H}|z'\rangle$. (ii) Write the semiclassical approximation for this path integral. (iii) Calculate the propagator for a harmonic oscillator.

10.7 Prove that

$$\int Dx(\cdot)x(t_1)e^{iS[x(\cdot)]} = \langle x_f, t|\hat{x}(t_1)|x_i, 0\rangle$$
$$\int Dx(\cdot)x(t_1)x(t_2)e^{iS[x(\cdot)]} = \langle x_f, t|\mathcal{T}[\hat{x}(t_1)\hat{x}(t_2)]|x_i, 0\rangle,$$

where integration is over paths such that $x(0) = x_i$ and $x(t) = x_f$, $\hat{x}(t_i)$ is the Heisenberg evolved operator, $0 < t_1, t_2 < t$ and $\mathcal{T}$ stands for time ordering.

10.8 In calculating $\det \mathcal{E}_\tau^0$ with zeta-function regularization, we used the following values for Riemann's zeta function: $\zeta(0) = -\frac{1}{2}$ and $\zeta'(0) = -\frac{1}{2}\ln(2\pi)$. Prove these identities from the functional equation (B.10).

10.9 For a harmonic oscillator, the path integral over $\xi(\cdot)$ in Eq. (10.29) is identical with the path integral of Eq. (10.26) for $x_i = x_f = 0$, namely, with $\langle 0|e^{-i\hat{H}t}|0\rangle$. (i) Use Eq. (10.26) for N time steps, in order to show that

$$\langle 0|e^{-i\hat{H}t}|0\rangle = \frac{m}{2\pi i\delta t}(\det A_N)^{-1/2},$$

where

$$A_N = \begin{pmatrix} 2 - \omega^2 \delta t^2 & -1 & 0 & 0 & \cdots & 0 \\ -1 & 2 - \omega^2 \delta t^2 & -1 & 0 & \cdots & 0 \\ 0 & -1 & 2 - \omega^2 \delta t^2 & -1 & \cdots & 0 \\ \cdots & \cdots & \cdots & \cdots & \cdots \\ 0 & 0 & 0 & \cdots & 2 - \omega^2 \delta t^2 \end{pmatrix}.$$

(ii) Let $K_N = \det A_N$. Show that $K_N = (2 - \omega^2 \delta t^2) K_{N-1} - K_{N-2}$. (iii) Show that the general solution to this difference equation is $K_N = c_1 (1 - i\omega \delta t)^N + c_2 (1 + i\omega \delta t)^N$ for constants c_1 and c_2. (iv) Specify c_1 and c_2 by the requirements that $K_1 = 1$ and $K_2 = 2 - \omega^2 \delta t^2$. (v) Take the limit $N \to \infty$, and confirm that $\langle 0 | e^{-i\hat{H}t} | 0 \rangle = \sqrt{\frac{m\omega}{2\pi i \sin \omega t}}$.

10.10 Evaluate the path integral for a harmonic oscillator with an external force $F(t)$, that is, with a term $-F(t)x$ added to the Hamiltonian.

10.11 Prove that $\mathcal{E}(2\delta) \le \mathcal{E}(\delta)$, for $\mathcal{E}$ defined by Eq. (10.44).

10.12 Consider a harmonic oscillator of mass m and frequency ω in units where $m = \omega = 1$. Let $|z\rangle$ be the associated coherent states. Define the operator $\hat{F}_C := \int_C \frac{d^2 z}{\pi} |z\rangle \langle z|$, where C is a bounded set on the complex plane $\mathbb{C}$. (i) Show that $\hat{F}_C$ is a positive operator with discrete spectrum, and that $\mathrm{Tr} \hat{F}_C = A(C)$, where $A(C)$ is the area of C. (ii) Show that all eigenvalues of $\hat{F}_C$ are less than one. (iii) Let z_0 be a point "well inside" C, in the sense that the distance D of z_0 from the boundary of C is much larger than unity. Show that $|z_0\rangle$ is an approximate eigenvector of $\hat{F}_C$ with eigenvalue one, modulo an error of order $e^{-\frac{1}{2}D^2}$. Prove the analogous property for z_0 "well outside" z_0. (iv) Argue that

$$\frac{\mathrm{Tr}|\hat{F}_C - \hat{F}_C^2|}{\mathrm{Tr}\hat{F}_C} \sim \frac{L(C)}{A(C)},$$

where $L(C)$ is the length of the boundary of C. We take C to be a disk of radius $R \gg 1$. Then, $\hat{F}_C$ approximates a projector with an error of order R^{-1}, that is, $\hat{F}_C$ approximately expresses the classical property that the oscillator is found in the subset C of the state space. (iv) Show that Eq. (10.46) for approximate determinism holds for $\hat{F}_C$ with an error of order R^{-1}.

Bibliography

- For the semiclassical approximations in quantum theory, see the review by Berry and Mount (1972). For Bohmian mechanics, see Bohm and Hiley (1995) and Dürr and Teufel (2010).
- For path integrals and their applications see Feynman and Hibbs (2010) and Zinn-Justin (2005).
- Bohr's explanation of the coexistence of classical and quantum descriptions is notoriously opaque. The first essay in Bohr (1963) is one of his clearest accounts.
- For the frailness of quantum histories, see Anastopoulos (2004).

- For the emergence of classical behavior from quantum theory, see Gell-Mann and Hartle (1993), chapters 16–20 from Omnés (1999), Zurek (2003), and Joos (2006). The outlook of these references is very optimistic about the prospects of explaining the emergence of the classical world through decoherence. They should be balanced with the more cautious account of Anastopoulos (2002) and critical viewpoints such as those of Leggett (2002b), Adler (2003), and Kastner (2014).

Part III

Elementary Systems and Their Symmetries

11 Symmetries I

Rotations

Until then we believed in the ancient saying of Democritus "in the beginning was the particle"... But perhaps that philosophy was totally wrong. In the beginning, there was a natural law, mathematics, symmetry? Perhaps we should approach Plato and accept that "in the beginning was the symmetry."

Heisenberg (1969)

11.1 Symmetries in Quantum Theory

In previous chapters, we encountered the fundamental principles of quantum theory and we saw how the representation of physical magnitudes by Hilbert space operators allows us to construct probabilities for the outcome of any experiment. However, these principles do not tell us what the fundamental physical systems are, how to construct their associated Hilbert spaces, and which operators correspond to physical magnitudes.

This information is encoded in the *symmetries* that characterize a system. Recall the importance in the development of quantum theory of Heisenberg's commutation relations (2.28). The fundamental observables of a physical system are defined in terms of an algebraic relation, which, as we will see, reflects a symmetry of the system. To clarify this point, it is necessary first to explain the three distinct ways that symmetry appears in physics.

First, there are the symmetries of states or of histories. A state or a history of a system is characterized by a symmetry if it remains invariant under a specific class of transformation. For example, the ground state of the harmonic oscillator is invariant under the transformation $x \rightarrow -x$. In classical physics, a circular orbit of a satellite is unchanged if we rotate the satellite (mentally) by any angle around an axis vertical at its center. Historically, this was the first type of symmetry that was discovered or postulated in physics.

Second, there are dynamical symmetries. Any transformation that maps solutions to the equations of motion to other solutions is a dynamical symmetry. The existence of a dynamical symmetry in a system does not imply that the states or histories of this system manifest this symmetry. A snapshot of the Solar System would show planets dispersed chaotically in space, and would not reveal the symmetric simplicity of Newton's law of gravity.

Third, there are kinematical symmetries. These symmetries *define* the fundamental observables of the system. One example is Heisenberg's commutation relations. We will see that the most important kinematical symmetries express the invariance of a system under changes of reference frames in space and in time.

273

The three types of symmetry we have mentioned are not exclusive. The presence of dynamical symmetries typically implies the existence of *some* symmetric states, and there are symmetries that are both dynamical and kinematical. There is also a fourth type of symmetry, gauge symmetry, which combines features from all three forms above. We will encounter the simplest forms of gauge symmetry in Section 13.5.

The proper mathematical framework for describing symmetries in quantum theory is the theory of groups and their representations. In this chapter, we focus on one of the most important symmetries of quantum theory, the symmetry of spatial rotations, without explicitly introducing concepts from group theory. This way, we shall understand the motivation for many of these concepts, mirroring the historical development of the topic: Rotations were well understood prior to the formal development of group theory. Applications of group theory to quantum mechanics will be presented in Chapter 12.

11.1.1 Wigner's Theorem

The three types of symmetry we have described refer to the constancy of some physical entity under a class of transformations. Hence, in order to describe symmetries in quantum theory, we must first define the notion of a symmetry transformation in the Hilbert space of a quantum system. The obvious idea is to identify transformations with operators, since any operator $\hat{U}$ transforms a vector $|\psi\rangle$ into a vector $|\psi'\rangle = \hat{U}|\psi\rangle$. These operators must preserve the inner product, so that if $\hat{U}$ acts on two vectors $|\psi\rangle$ and $|\phi\rangle$,

$$\langle\phi'|\psi'\rangle = \langle\phi|\psi\rangle, \tag{11.1}$$

which means that $\hat{U}^\dagger\hat{U} = \hat{I}$. We also assume that if $\hat{U}$ is an admissible transformation, then so is $\hat{U}^{-1}$. This implies that $\hat{U}\hat{U}^\dagger = \hat{I}$, hence, $\hat{U}$ is unitary.

Unitary operators transform other operators $\hat{A}$ as $\hat{A} \rightarrow \hat{U}\hat{A}\hat{U}^\dagger$, so that the combined transformation of observables and states leaves probabilities invariant,

$$\langle\psi'|\hat{A}'|\psi'\rangle = \langle\psi|\hat{A}|\psi\rangle. \tag{11.2}$$

The description of symmetry transformations in terms of unitary operators turns out to be too restrictive. The requirement (11.1) is stronger than necessary.

We can also represent symmetries with *antilinear* operators. An antilinear operator $\hat{S}$ is a function that assigns to each Hilbert space vector $|\psi\rangle$ another vector $\hat{S}|\psi\rangle$, such that

$$\hat{S}(c|\psi\rangle + |\phi\rangle) = c^*\hat{S}|\psi\rangle + \hat{S}|\phi\rangle, \tag{11.3}$$

for all $c \in \mathbb{C}$ and $|\phi\rangle, |\psi\rangle \in \mathcal{H}$. (Recall that a linear operator satisfies $\hat{S}(c|\psi\rangle + |\phi\rangle) = c\hat{S}|\psi\rangle + \hat{S}|\phi\rangle$.) It is elementary to show that the product of two antilinear operators is a linear operator.

The conjugate operator $\hat{S}^\dagger$ of an antilinear operator $\hat{S}$ is defined by

$$\langle\phi|\hat{S}|\psi\rangle = \langle\psi|\hat{S}^\dagger|\phi\rangle, \tag{11.4}$$

for all $|\phi\rangle, |\psi\rangle \in \mathcal{H}$. An *antiunitary* operator is an antilinear operator that satisfies $\hat{U}\hat{U}^\dagger = \hat{I}$. If $|\phi'\rangle = \hat{U}|\phi\rangle$ and $|\psi'\rangle = \hat{U}|\psi\rangle$, then $\langle\phi'|\psi'\rangle = \langle\phi'|\hat{U}|\psi\rangle = \langle\psi|\hat{U}^\dagger|\phi'\rangle = \langle\psi|\hat{U}^\dagger\hat{U}|\phi\rangle = \langle\psi|\phi\rangle = (\langle\phi|\psi\rangle)^*$. Antiunitary operators do not preserve the inner product of two vectors, but they do preserve the inner product's absolute value. This suffices for physical predictions, since quantum probabilities involve the inner product norm square $|\langle \phi | \psi \rangle|^2$.

A crucial theorem by Wigner (1932) confirms that symmetry transformations can only be expressed by unitary and antiunitary operators.

Theorem 11.1 (Wigner's) *Let $\mathcal{H}$ be a complex Hilbert space, and $S : |\phi\rangle \to S(|\phi\rangle) = |S(\phi)\rangle$ an onto function from $\mathcal{H}$ to $\mathcal{H}$, such that*

$$|\langle S(\phi) | S(\psi) \rangle| = |\langle \phi, |\psi \rangle| \tag{11.5}$$

for all $|\phi\rangle, |\psi\rangle \in \mathcal{H}$. Then $(|S(\psi)\rangle) = \hat{U}|\psi\rangle$, where $\hat{U}$ is a unitary or antiunitary operator.

Wigner's theorem is remarkable in that it does not require that the function S be linear (or antilinear). Linearity follows as a consequence of the theorem. In Box 11.1, we present an elementary proof of Wigner's theorem that is due to Peres (2002).

Box 11.1   Proof of Wigner's Theorem

Let $\mathcal{H} = \mathbb{C}^N$ and $|n\rangle$ be an orthonormal basis on $\mathcal{H}$, where $n = 1, 2, \dots, N$. By assumption, $|S(n)\rangle$ also defines an orthonormal basis. The multiplication of each vector $|S(n)\rangle$ by a different phase is equivalent to a unitary transformation $\hat{V}$ acting on any vector $|S(\phi)\rangle$. As such it can be absorbed into a redefinition of S that does not affect the assumptions or conclusions of the theorem

We define $|k_n\rangle = |1\rangle + |n\rangle$, for $n = 2, 3, \dots, N$. By assumption, $|S(k_n)\rangle$ satisfies $\langle S(k_n)|S(1)\rangle = \langle k_n|1\rangle = 1$ and $\langle S(k_n)|S(n')\rangle = \langle k_n|n'\rangle = \delta_{nn'}$.

This implies that $|S(k_n)\rangle = a|S(1)\rangle + b_n|S(n)\rangle$ for $|a| = |b_n| = 1$. We absorb the phase a into a redefinition of $|S(1)\rangle$ and the phase b_n into a redefinition of $|S(n)\rangle$, so that

$$|S(k_n)\rangle = |S(1)\rangle + |S(n)\rangle. \tag{A}$$

Let $|\phi\rangle = \sum_n c_n|n\rangle$. We express $|S(\phi)\rangle$ with respect to the basis $|S(n)\rangle$ as $\sum_n d_n|S(n)\rangle$. By assumption, $|d_n| = |\langle S(\phi)|S(n)\rangle| = |\langle \phi|n\rangle| = |c_n|$. Furthermore, by Eq. (A), $|d_1 + d_n| = |\langle S(\phi)|S(k_n)\rangle| = |\langle \phi|k_n\rangle| = |c_1 + c_n|$. Therefore we obtain, $c_1^* c_n + c_1 c_n^* = d_1^* d_n + d_1 d_n^*$. Dividing both sides by $|c_1 c_n| = |d_1 d_n|$, we obtain

$$\sqrt{\frac{c_1^* c_n}{c_1 c_n^*}} + \sqrt{\frac{c_1 c_n^*}{c_1^* c_n}} = \sqrt{\frac{d_1^* d_n}{d_1 d_n^*}} + \sqrt{\frac{d_1 d_n^*}{d_1^* d_n}}. \tag{B}$$

Equation (B) is of the form $e^{i\theta} + e^{-i\theta} = e^{i\theta'} + e^{-i\theta'}$, which admits two solutions, $\theta = \theta'$ and $\theta = -\theta'$.

If $\theta = \theta'$, then $\frac{c_1^* c_n}{c_1 c_n^*} = \frac{d_1^* d_n}{d_1 d_n^*}$. We choose the overall phase factor of $|\phi\rangle$, so that $c_1 = d_1$. Then, $c_n/c_n^* = d_n/d_n^*$, and since $|c_n| = |d_n|$, we obtain $c_n = d_n$. Hence, $|S(\phi)\rangle = \sum_n c_n|S(n)\rangle$, that is, the map S corresponds to a change of basis, namely, it is a unitary operator.

If $\theta = -\theta'$, then $\frac{c_1^* c_n}{c_1 c_n^*} = \frac{d_1 d_n^*}{d_1^* d_n}$. We choose the overall phase factor of $|\phi\rangle$, so that $c_1 = d_1^*$. Then, $c_n/c_n^* = d_n^*/d_n$. Since $|c_n| = |d_n|$, we obtain $c_n = d_n^*$. Hence,

$|S(\phi)\rangle = \sum_n c_n^* |S(n)\rangle$, that is, the map S is an antiunitary operator. This concludes the proof.

11.1.2 Examples

In Section 4.4.3, we encountered the unitary Weyl operators $\hat{V}(a,b)$, which describe translations of the position and of the momentum of a particle. We have also encountered the unitary operator of time evolution $e^{-i\hat{H}t}$, which implements time translations.

In the Hilbert space $L^2(\mathbb{R})$, we define the *space inversion* operator, or parity operator, $\hat{\mathbb{P}}$ as

$$\hat{\mathbb{P}}\psi(x) = \psi(-x). \tag{11.6}$$

It is straightforward to show that $\hat{\mathbb{P}}$ is self-adjoint. Furthermore, $\hat{\mathbb{P}}^2\psi(x) = \hat{\mathbb{P}}\psi(-x) = \psi(x)$, namely, $\hat{\mathbb{P}}^2 = \hat{I}$. This implies that $\hat{\mathbb{P}}$ is also unitary.

In the Hilbert space $L^2(\mathbb{R})$, we also define the *time reversal* operator $\hat{\mathbb{T}}$, as $\hat{\mathbb{T}}\psi(x) = \psi^*(x)$. It is easy to show that $\hat{\mathbb{T}}$ is antilinear, and that $\hat{\mathbb{T}}^2 = \hat{I}$. The name time reversal follows from the fact that, for any solution $\psi(x,t)$ to Schrödinger's equation $i\frac{\partial \psi}{\partial t} = -\frac{1}{2m}\frac{\partial^2 \psi}{\partial x^2} + V(x)\psi$, $\psi^*(x,t)$ is a solution to

$$i\frac{\partial \psi^*}{\partial(-t)} = -\frac{1}{2m}\frac{\partial^2 \psi^*}{\partial x^2} + V(x)\psi^*,$$

that is, to Schrödinger's equation with time t transformed to $-t$.

We note that $\int dx \psi^*(x)\hat{\mathbb{T}}\phi(x) = \int dx \psi^*(x)\phi^*(x) = \int dx(\hat{\mathbb{T}}\psi)(x)\phi^*(x)$. Hence, by Eq. (11.4), $\hat{\mathbb{T}}^\dagger = \hat{\mathbb{T}}$. This means that the equation $\hat{\mathbb{T}}^2 = \hat{I}$ can be written as $\hat{\mathbb{T}}\hat{\mathbb{T}}^\dagger = \hat{I}$. It follows that $\hat{\mathbb{T}}$ is antiunitary. As it turns out, $\hat{\mathbb{T}}$ is antiunitary in all physical systems. As shown in Section 14.5 and in Section 16.5, this is a consequence of the positivity of energy.

11.2 Rotations

Matrices in this section are not quantum observables, so they do not carry a hat. One of the most important symmetries that characterizes quantum systems is the rotation symmetry. In this section, we analyze rotations in two and three dimensions purely geometrically, with no reference to quantum theory.

11.2.1 Planar Rotations

We consider the rotation of a vector $\mathbf{r} = \begin{pmatrix} x_1 \\ x_2 \end{pmatrix}$ by an angle θ around the coordinate origin. This rotation is implemented by a 2×2 matrix,

$$O(\theta) = \begin{pmatrix} \cos\theta & \sin\theta \\ -\sin\theta & \cos\theta \end{pmatrix}, \tag{11.7}$$

acting on $\mathbf{r}$ as $\mathbf{r} \to O(\theta)\mathbf{r}$. The matrix $O(\theta)$ has two eigenvalues $e^{i\theta}$ and $e^{-i\theta}$.

It is easy to show that

$$O(\theta_1)O(\theta_2) = O(\theta_1 + \theta_2), \tag{11.8}$$

and that $O(0) = 1$.

For θ very close to 0, $O(\theta) \simeq 1 + \theta J$, where

$$J = \begin{pmatrix} 0 & 1 \\ -1 & 0 \end{pmatrix} = i\sigma_2. \tag{11.9}$$

We find that $e^{\theta J} = e^{i\theta\sigma_2} = \cos\theta I + i\sigma_2 \sin\theta = O(\theta)$. By exponentiating J, we recover all elements of the rotation matrix $O(\theta)$. For this reason, we refer to J as the *generator* of the rotation matrices $O(\theta)$.

In $\mathbb{R}^2$, it is common to use complex coordinates $z = x_1 + ix_2$ and $z^* = x_1 - ix_2$. A rotation transforms these coordinates as $z \to ze^{-i\theta}$ and $z^* \to z^*e^{i\theta}$.

11.2.2 Spatial Rotations

Next, we study rotations in three-dimensional space. The rotation of a vector $\mathbf{r} \in \mathbb{R}^3$ leaves its length $\sqrt{\mathbf{r}^T\mathbf{r}}$ invariant. In this notation, we treat $\mathbf{r}$ as a column vector, and $\mathbf{r}^T$ is its associated row vector; obviously $\mathbf{r}^T\mathbf{r} = \mathbf{r}^2$.

If rotation is implemented by 3×3 matrix, O, as

$$\mathbf{r} \to \mathbf{r}' = O\mathbf{r}, \tag{11.10}$$

then $\mathbf{r}'^T\mathbf{r}' = (\mathbf{r}O)^T O\mathbf{r} = \mathbf{r}^T\mathbf{r}$. It follows that $O^T O = I$. Since the inverse of a rotation always exists, then the matrix O must be orthogonal.

The relation $O^T O = I$, implies $\det O = \pm 1$. For the moment, we will restrict ourselves to rotation matrices with $\det O = 1$. Later, we will study matrices with $\det O = -1$.

We can build a general rotation matrix from successive rotations around the three axes of a Cartesian reference frame. Each orthogonal matrix

$$O_1(\theta_1) = \begin{pmatrix} 1 & 0 & 0 \\ 0 & \cos\theta_1 & \sin\theta_1 \\ 0 & -\sin\theta_1 & \cos\theta_1 \end{pmatrix}, O_2(\theta_2) = \begin{pmatrix} \cos\theta_2 & 0 & -\sin\theta_2 \\ 0 & 1 & 0 \\ \sin\theta_2 & 0 & \cos\theta_2 \end{pmatrix},$$

$$O_3(\theta_3) = \begin{pmatrix} \cos\theta_3 & \sin\theta_3 & 0 \\ -\sin\theta_3 & \cos\theta_3 & 0 \\ 0 & 0 & 1 \end{pmatrix}, \tag{11.11}$$

contains a planar rotation submatrix of form (11.7) around the axes 1, 2, and 3, respectively. A general rotation matrix is the product $O_1(\theta_1)O_2(\theta_2)O_3(\theta_3)$ for some values of the angles θ_1, θ_2, and θ_3.

The generators of the matrices O_1, O_2, and O_3 are the antisymmetric matrices

$$J_1 = \begin{pmatrix} 0 & 0 & 0 \\ 0 & 0 & 1 \\ 0 & -1 & 0 \end{pmatrix}, J_2 = \begin{pmatrix} 0 & 0 & -1 \\ 0 & 0 & 0 \\ 1 & 0 & 0 \end{pmatrix}, J_3 = \begin{pmatrix} 0 & 1 & 0 \\ -1 & 0 & 0 \\ 0 & 0 & 0 \end{pmatrix}. \tag{11.12}$$

It is straightforward to confirm that $O_1(\theta_1) = e^{\theta_1 J_1}, O_2(\theta_2) = e^{\theta_2 J_2}$ and $O_3(\theta_3) = e^{\theta_3 J_3}$.

The generators J_i of Eq. (11.12) can be written compactly using the totally antisymmetric symbol ϵ_{ijk}, Eq. (5.5),

$$(J_i)_{jk} = \epsilon_{ijk}, \quad \text{for } i,j,k = 1,2,3. \tag{11.13}$$

From Eq. (11.12), we straightforwardly derive the commutation relations $[J_1,J_2] = J_3, [J_3,J_1] = J_2$ and $[J_2,J_3] = J_1$. These are expressed as

$$[J_i,J_j] = \sum_k \epsilon_{ijk} J_k. \tag{11.14}$$

Equation (11.14) defines the *rotation algebra*, also known as the *angular momentum algebra*.

11.2.3 Rotation around an Arbitrary Axis

The eigenvalues λ of an orthogonal matrix O have unit absolute value, because an orthogonal matrix on $\mathbb{R}^3$ is also a unitary matrix on $\mathbb{C}^3$. The three eigenvalues of O are the roots $\lambda_1, \lambda_2, \lambda_3$ of the third-order characteristic polynomial $\det(O - \lambda I)$. Since $\det O = \lambda_1 \lambda_2 \lambda_3 = 1$, there are two roots $e^{\pm i\theta}$ and one root equal to 1.

Let $\mathbf{n} = (n_1, n_2, n_3)$ be the normalized eigenvector of O with eigenvalue 1. The eigenvalue equation

$$O\mathbf{n} = \mathbf{n} \tag{11.15}$$

means that $\mathbf{n}$ defines the axis of rotation. The other two eigenvalues $e^{\pm i\theta}$ of $O\mathbf{n}$ correspond to the rotation angle θ around $\mathbf{n}$.

For small θ, $O = I + \theta J_\mathbf{n}$, and Eq. (11.15) yields

$$J_\mathbf{n}\mathbf{n} = 0. \tag{11.16}$$

The relation $O^T O = I$ implies that $I + \theta(J_\mathbf{n} + J_\mathbf{n}^T) = I$. Hence, $J_\mathbf{n} = -J_\mathbf{n}^T$, that is, $J_\mathbf{n}$ is an antisymmetric matrix. As such it is specified by three independent parameters. We can express $J_\mathbf{n}$ as $\sum_{i=1}^3 n_i J_i$, where the generators J_i are given by Eq. (11.12). We substitute into Eq. (11.16), to obtain $a_1 = n_1, a_2 = n_2, a_3 = n_3$. Hence,

$$J_\mathbf{n} = \begin{pmatrix} 0 & n_3 & -n_2 \\ -n_3 & 0 & n_1 \\ n_2 & -n_1 & 0 \end{pmatrix}. \tag{11.17}$$

The matrix $O_\mathbf{n}(\theta)$ implementing a rotation of angle θ around the axis $\mathbf{n}$ is $e^{\theta J_\mathbf{n}}$.

Example 11.1 Let $\mathbf{n} = \frac{1}{\sqrt{2}}(1, 0 - 1)$. By Eq. (11.17),

$$J_\mathbf{n} = \frac{1}{\sqrt{2}} \begin{pmatrix} 0 & -1 & 0 \\ 1 & 0 & 1 \\ 0 & -1 & 0 \end{pmatrix}.$$

To evaluate $e^{\theta J_\mathbf{n}}$, we work with the self-adjoint matrix $iJ_\mathbf{n}$, for which we can use the spectral theorem. The eigenvalues of $iJ_\mathbf{n}$ are $0, 1$ and -1. We also evaluate the associated spectral projectors, to obtain the spectral decomposition

$$iJ_{\mathbf{n}} = (0) \begin{pmatrix} 1 & 0 & -1 \\ 0 & 0 & 0 \\ -1 & 0 & 1 \end{pmatrix} + (1)\frac{1}{4} \begin{pmatrix} 1 & -\sqrt{2}i & 1 \\ \sqrt{2}i & 2 & \sqrt{2}i \\ 1 & -\sqrt{2}i & 1 \end{pmatrix}$$

$$+(-1)\frac{1}{4} \begin{pmatrix} 1 & \sqrt{2}i & 1 \\ -\sqrt{2}i & 2 & -\sqrt{2}i \\ 1 & \sqrt{2}i & 1 \end{pmatrix}.$$

Then, the matrix $e^{\theta J_{\mathbf{n}}} = e^{-i\theta(iJ_{\mathbf{n}})}$ is

$$e^{\theta J_{\mathbf{n}}} = \begin{pmatrix} 1 & 0 & -1 \\ 0 & 0 & 0 \\ -1 & 0 & 1 \end{pmatrix} + e^{-i\theta}\frac{1}{4} \begin{pmatrix} 1 & -\sqrt{2}i & 1 \\ \sqrt{2}i & 2 & \sqrt{2}i \\ 1 & -\sqrt{2}i & 1 \end{pmatrix}$$

$$+e^{i\theta}\frac{1}{4} \begin{pmatrix} 1 & \sqrt{2}i & 1 \\ -\sqrt{2}i & 2 & -\sqrt{2}i \\ 1 & \sqrt{2}i & 1 \end{pmatrix}.$$

11.2.4 Euler Angles

We can also parameterize the rotation matrices O with the *Euler angles* (ψ, θ, ϕ). Consider a general unit vector $\mathbf{n}$, expressed in terms of the sphere coordinates (θ, ϕ), $\mathbf{n} = (\sin\theta\cos\phi, \sin\theta\sin\phi, \cos\theta)$. Let $\mathbf{n}_0 = (0, 0, 1)$. A rotation by $-\theta$ around the 2-axis, implemented by $O_2(-\theta)$, takes $\mathbf{n}_0$ to the vector $(\sin\theta, 0, \cos\theta)$. We further rotate this vector by ϕ around the 3-axis using $O_3(\phi)$. The end result is $\mathbf{n}$.

Hence, we can write any unit vector $\mathbf{n}$ as $O_3(\phi)O_2(-\theta)\mathbf{n}_0$. We further note that any $O_3(\psi)\mathbf{n}_0 = \mathbf{n}_0$, for any angle ψ, so we can write

$$\mathbf{n} = O_3(\phi)O_2(-\theta)O_3(\psi)\mathbf{n}_0. \tag{11.18}$$

As ψ, θ, ϕ vary, the matrices

$$O(\psi, \theta, \phi) := O_3(\phi)O_2(-\theta)O_3(\psi) \tag{11.19}$$

range over all rotation matrices. The angles (ψ, θ, ϕ) are the Euler angles.

11.2.5 Space Inversion

We define the space-inversion matrix

$$\mathcal{P} = \begin{pmatrix} -1 & 0 & 0 \\ 0 & -1 & 0 \\ 0 & 0 & -1 \end{pmatrix}, \tag{11.20}$$

which implements the inversion of vectors in relation to the origin: $\mathcal{P}\mathbf{r} = -\mathbf{r}$.

The matrix $\mathcal{P}$ satisfies $\det\mathcal{P} = -1$ and $\mathcal{P}^2 = 1$. For any orthogonal matrix O with $\det O = 1$, the matrix $O' = \mathcal{P}O = -O$ satisfies $\det O' = -1$.

The action of $\mathcal{P}$ on the generators J_i is trivially

$$\mathcal{P}J_i\mathcal{P} = J_i, \tag{11.21}$$

is important. We will see that Eq. (11.21) is nontrivial when employed in a quantum context.

11.2.6 Spatial Rotations and the Bloch Sphere

In Section 6.2.3, we showed that normalized vectors $|\psi\rangle$ on $\mathbb{C}^2$ corresponds to points of the Bloch sphere, with associated unit vector $\hat{\mathbf{r}} = \langle\psi|\hat{\boldsymbol{\sigma}}|\psi\rangle$. A unitary transformation $|\psi\rangle \to \hat{U}|\psi\rangle$ takes the Bloch vector $\mathbf{r}$ to $\mathbf{r}' = \langle\psi|\hat{U}^\dagger\hat{\boldsymbol{\sigma}}\hat{U}|\psi\rangle$.

The operator $\hat{U}^\dagger\hat{\sigma}_i\hat{U}$ is self adjoint for $i = 1, 2, 3$ and has zero trace. Hence, it can be expressed as a linear combination of Pauli matrices,

$$\hat{U}^\dagger\hat{\sigma}_i\hat{U} = \sum_j \Lambda_{ij}\hat{\sigma}_j, \tag{11.22}$$

so that $r'_i = \sum_j \Lambda_{ij}r_j$. Since $|\mathbf{r}| = |\mathbf{r}'| = 1$, the matrix Λ_{ij} is orthogonal. We evaluate Λ, by multiplying both sides of Eq. (11.22) with $\hat{\sigma}_k$, and then taking the trace,

$$\Lambda_{ij} = \frac{1}{2}\mathrm{Tr}\left(\hat{U}^\dagger\hat{\sigma}_i\hat{U}\hat{\sigma}_j\right). \tag{11.23}$$

Equation (11.23) implies that the multiplication of $\hat{U}$ by a phase $e^{i\phi}$ does not affect Λ_{ij}. We can also prove that the product $\hat{U}_1\hat{U}_2$ of two unitary operators is mapped by Eq. (11.23) to the product of the associated matrices, Λ_1 and Λ_2 – see Problem 11.1.

By Eq. (11.23), $\mathrm{Tr}\Lambda = \frac{1}{2}\sum_i \mathrm{Tr}\left(\hat{U}^\dagger\hat{\sigma}_i\hat{U}\hat{\sigma}_i\right)$. We use Eq. (5.123) to obtain

$$\mathrm{Tr}\Lambda = -1 + 2|\mathrm{Tr}\hat{U}|^2 \geq -1. \tag{11.24}$$

The trace of the spatial inversion matrix $\mathcal{P}$ of Eq. (11.20) is -3. Hence, no unitary operator is mapped to $\mathcal{P}$ by Eq. (11.23). The orthogonal matrices Λ_{ij} of Eq. (11.23) have positive determinant.

The inversion transformation $\mathbf{r} \to -\mathbf{r}$ in Bloch's sphere is implemented through the antiunitary map (see Problem 11.2)

$$\hat{\mathbb{T}}\begin{pmatrix} c_1 \\ c_2 \end{pmatrix} = \begin{pmatrix} -c_2^* \\ c_1^* \end{pmatrix}. \tag{11.25}$$

The orthogonal matrices with negative determinant correspond to antiunitary operators of the form $\hat{\mathbb{T}}\hat{U}$ where $\hat{U}$ is a unitary operator.

Orthogonal matrices of either positive or negative determinant define the most general transformations that preserve the inner product of vectors on $\mathbb{R}^3$. By Eq. (6.26), they are in one to one correspondence with maps on $\mathbb{C}^2$ that preserve the absolute value of the inner product. Hence, the latter are either unitary operators or antiunitary operators of the form $\hat{\mathbb{T}}\hat{U}$. This is Wigner's theorem in $\mathbb{C}^2$.

11.3 Angular Momentum Operators

We want to implement transformations that correspond to spatial rotations in a quantum system described by a Hilbert space $\mathcal{H}$. By Wigner's theorem, we must find unitary (or antiunitary) operators that correspond to the rotation matrices.

This is achieved through the following procedure.[1] To each generator J_i of rotations, we associate an operator $\frac{1}{i}\hat{J}_i$ on $\mathcal{H}$, where $\hat{J}_i$ is self-adjoint. This way, the fundamental commutation relation (11.14) of angular momentum operators becomes

$$[\hat{J}_i, \hat{J}_j] = i \sum_k \epsilon_{ijk} \hat{J}_k. \tag{11.26}$$

The idea is that we will then proceed to construct the unitary operators $e^{-i\theta \mathbf{n}\cdot\hat{\mathbf{J}}}$ that correspond to a rotation by angle θ around the axis $\mathbf{n} = (n_1, n_2, n_3)$. Hence, we will recover an algebraic description of rotations for quantum systems.

The operators $\hat{J}_i$ are called *angular momentum* operators, because they implement the classical notion of angular momentum in quantum theory. In what follows we will provide the most general definition of angular momentum operators in quantum theory, and we will analyze their spectrum.

11.3.1 The Operator $\hat{\mathbf{J}}^2$

First, we define the positive operator

$$\hat{\mathbf{J}}^2 = \hat{J}_1^2 + \hat{J}_2^2 + \hat{J}_3^2, \tag{11.27}$$

which is the quantum analogue of the length of the "vector" $(\hat{J}_1, \hat{J}_2, \hat{J}_3)$. The operator $\hat{\mathbf{J}}^2$ is positive and it satisfies the commutation relations

$$[\hat{\mathbf{J}}^2, \hat{J}_i] = 0, \quad i = 1, 2, 3. \tag{11.28}$$

In general, we expect that the eigenvalues λ of $\hat{\mathbf{J}}^2$ are degenerate. Since $\hat{J}_3$ commutes with $\hat{\mathbf{J}}^2$, they have common eigenvectors $|\lambda, m\rangle$, such that

$$\hat{\mathbf{J}}^2|\lambda, m\rangle = \lambda|\lambda, m\rangle \tag{11.29}$$

$$\hat{J}_3|\lambda, m\rangle = m|\lambda, m\rangle. \tag{11.30}$$

We assume that the operators $\hat{J}_3$ and $\hat{\mathbf{J}}^2$ form a complete set of measurements. This implies that the eigenvectors $|\lambda, m\rangle$ are unique, up to a change of phase. This case involves no loss of generality, as all other cases can be reduced to it. If $\hat{J}_3$ and $\hat{\mathbf{J}}^2$ do not form a complete set of measurements, the kets of the Hilbert space will carry additional indices, but the angular momentum operators will be defined solely with respect to its action on λ and m.

11.3.2 Ladder Operators

We define the *ladder operators* $\hat{J}_+$ and $\hat{J}_-$ by

$$\hat{J}_\pm = \hat{J}_1 \pm i\hat{J}_2. \tag{11.31}$$

---

[1] The procedure presented in this section is physically straightforward, even if here it is presented as an ad hoc recipe. In fact, it is a unique consequence of the principles of quantum theory, but for the full mathematical justification in terms of group theory, the reader will have to wait for Chapter 12.

The ladder operators are adjoint, $\hat{J}_+^\dagger = \hat{J}_-$, and they satisfy the commutation relations

$$[\hat{J}_+, \hat{J}_-] = 2\hat{J}_3 \tag{11.32}$$

$$[\hat{J}_+, \hat{J}_3] = -\hat{J}_+ \tag{11.33}$$

$$[\hat{J}_-, \hat{J}_3] = \hat{J}_- \tag{11.34}$$

$$[\hat{J}_\pm, \hat{\mathbf{J}}^2] = 0. \tag{11.35}$$

Moreover, $\hat{J}_+\hat{J}_- = \hat{J}_1^2 + \hat{J}_2^2 - i[\hat{J}_1, \hat{J}_2]$, and by Eq. (11.27),

$$\hat{J}_+\hat{J}_- = \hat{\mathbf{J}}^2 - \hat{J}_3^2 + \hat{J}_3. \tag{11.36}$$

Similarly, we obtain

$$\hat{J}_-\hat{J}_+ = \hat{\mathbf{J}}^2 - \hat{J}_3^2 - \hat{J}_3. \tag{11.37}$$

We act the operators from both sides of Eq. (11.33) to the vector $|\lambda, m\rangle$. We obtain $\hat{J}_+\hat{J}_3|\lambda, m\rangle - \hat{J}_3\hat{J}_+|\lambda, m\rangle = -\hat{J}_+|\lambda, m\rangle$. By Eq. (11.30),

$$\hat{J}_3\hat{J}_+|\lambda, m\rangle = (m+1)\hat{J}_+|\lambda, m\rangle.$$

Equation (11.35) implies that

$$\hat{\mathbf{J}}^2\hat{J}_+|\lambda, m\rangle = \hat{J}_+\hat{\mathbf{J}}^2|j, m\rangle = \lambda\hat{J}_+|\lambda, m\rangle.$$

Comparing with Eqs. (11.29) and (11.30), we see that the ket $\hat{J}_+|\lambda, m\rangle$ is an eigenvector of $\hat{\mathbf{J}}^2$ with eigenvalue λ and of $\hat{J}_3$ with eigenvalue $m+1$, that is,

$$\hat{J}_+|\lambda, m\rangle = c_{\lambda m}|\lambda, m+1\rangle, \tag{11.38}$$

for some constant $c_{\lambda m}$. We evaluate $c_{\lambda m}$ by the normalization of $|\lambda, m+1\rangle$:

$$|c_{\lambda m}|^2 = \langle\lambda, m|\hat{J}_-\hat{J}_+|\lambda, m\rangle = \langle\lambda, m|\hat{\mathbf{J}}^2 - \hat{J}_3^2 - \hat{J}_3|\lambda, m\rangle$$
$$= (\lambda - m^2 - m) \tag{11.39}$$

where we used Eq. (11.37). We conclude that

$$\hat{J}_+|\lambda, m\rangle = \sqrt{\lambda - m(m+1)}|\lambda, m+1\rangle. \tag{11.40}$$

Similarly, we obtain

$$\hat{J}_-|\lambda, m\rangle = \sqrt{\lambda - m(m-1)}|\lambda, m-1\rangle. \tag{11.41}$$

Because of Eqs. (11.40) and (11.41), $\hat{J}_+$ is called a *raising operator*, and $\hat{J}_-$ a *lowering operator*.

11.3.3 The Spectrum of Angular Momentum Operators

For fixed λ, $|c_{\lambda m}|^2$ in Eq. (11.39) becomes negative if m is allowed to take arbitrarily large values. Hence, there is a maximum value of m for each λ. We denote this value by j. For $m = j$, we must have $\hat{J}_+|\lambda, j\rangle = 0$, otherwise the action of $\hat{J}_+$ will go on giving larger values of m. Hence, we conclude that

$$\lambda = j(j + 1). \tag{11.42}$$

Since $\lambda \geq 0$, Eq. (11.42) implies that $j \geq 0$.

With a similar argument, Eq. (11.41) implies that there is a minimum value m_{min} of m, so that $\hat{J}_-|\lambda, m_{min}\rangle = 0$. This implies that $\lambda = m_{min}(m_{min} - 1)$; by Eq. (11.42), $m_{min} = -j$.

Hence, m takes values from $-j$ to j. Given that $\hat{J}_+$ increases the value of m by 1, there is an integer number of steps N from $m = -j$ to $m = j$. It follows that $-j + N = j$, that is,

$$j = \frac{N}{2}. \tag{11.43}$$

We conclude that j takes both integer and half-integer values.

There are $2j + 1$ different values of m for each j: $-j, -j + 1, \ldots, j - 1, j$. Evidently, if j is an integer, then m is also an integer, and if j is a half-integer, then m is also a half-integer. The eigenspace of constant j has degeneracy $2j + 1$, that is, it coincides with $\mathbb{C}^{2j+1}$.

We summarize the basic formulas, using j rather than λ as ket label, that is, writing $|j, m\rangle$ rather than $|\lambda, m\rangle$.

$$\hat{\mathbf{J}}^2|j, m\rangle = j(j + 1)|j, m\rangle, \tag{11.44}$$

$$\hat{J}_3|j, m\rangle = m|j, m\rangle, \tag{11.45}$$

$$\hat{J}_+|j, m\rangle = \sqrt{j(j + 1) - m(m + 1)}|j, m + 1\rangle, \tag{11.46}$$

$$\hat{J}_-|j, m\rangle = \sqrt{j(j + 1) - m(m - 1)}|j, m - 1\rangle. \tag{11.47}$$

We will refer to the kets $|j, m\rangle$ as the angular momentum basis.

11.3.4 Matrix Description

It is often convenient to express the angular momentum operators, defined by Eqs. (11.44)–(11.47), as matrices. For constant j, the vectors $|j, m\rangle$ form an orthonormal basis on the Hilbert space $\mathcal{H}_j = \mathbb{C}^{2j+1}$. Therefore, we write

$$|j, j\rangle = \begin{pmatrix} 1 \\ 0 \\ \cdots \\ 0 \\ 0 \end{pmatrix}, \quad |j, j-1\rangle = \begin{pmatrix} 0 \\ 1 \\ \cdots \\ 0 \\ 0 \end{pmatrix}, \quad \cdots |j, -j\rangle = \begin{pmatrix} 0 \\ 0 \\ \cdots \\ 0 \\ 1 \end{pmatrix}. \tag{11.48}$$

Equation (11.45) implies that $\hat{J}_3$ is diagonal, so that

$$\hat{J}_3 = \begin{pmatrix} j & 0 & \cdots & 0 \\ 0 & j - 1 & \cdots & 0 \\ \cdots & \cdots & \cdots & \cdots \\ 0 & 0 & \cdots & -j \end{pmatrix}. \tag{11.49}$$

Equation (11.46) implies that all elements of the matrix $\hat{J}_+$ vanish, except for those in the first line above and parallel to the diagonal, which take values $c_{j,m} = \sqrt{j(j+1) - m(m+1)}$. Similarly, Eq. (11.47) implies that all elements of the matrix $\hat{J}_-$ vanish, except for those in the first line below and parallel to the diagonal, which take values $c_{j,-m} = \sqrt{j(j+1) - m(m-1)}$.

For $j = \frac{1}{2}$,

$$\hat{J}_+ = \begin{pmatrix} 0 & 1 \\ 0 & 0 \end{pmatrix}, \ \hat{J}_- = \begin{pmatrix} 0 & 0 \\ 1 & 0 \end{pmatrix}. \tag{11.50}$$

Since $\hat{J}_1 = \frac{1}{2}(\hat{J}_+ + \hat{J}_-)$ and $\hat{J}_2 = \frac{1}{2i}(\hat{J}_+ - \hat{J}_-)$, we find

$$\hat{J}_i = \frac{1}{2}\hat{\sigma}_i, \tag{11.51}$$

where $\hat{\sigma}_i$ are the Pauli matrices.

Then, the unitary operators that implement a rotation of angle θ around the axis $\mathbf{n}$ are

$$e^{-i\frac{\theta}{2}\mathbf{n}\cdot\sigma} = \cos\frac{\theta}{2}\hat{I} - i\sin\frac{\theta}{2}\mathbf{n}\cdot\sigma, \tag{11.52}$$

where we followed the same steps as in the derivation of Eq. (7.36).

For $j = 1$,

$$\hat{J}_+ = \begin{pmatrix} 0 & \sqrt{2} & 0 \\ 0 & 0 & \sqrt{2} \\ 0 & 0 & 0 \end{pmatrix} \ \hat{J}_- = \begin{pmatrix} 0 & 0 & 0 \\ \sqrt{2} & 0 & 0 \\ 0 & \sqrt{2} & 0 \end{pmatrix}, \tag{11.53}$$

hence,

$$\hat{J}_1 = \frac{1}{\sqrt{2}}\begin{pmatrix} 0 & 1 & 0 \\ 1 & 0 & 1 \\ 0 & 1 & 0 \end{pmatrix} \ \hat{J}_2 = \frac{1}{\sqrt{2}}\begin{pmatrix} 0 & -i & 0 \\ i & 0 & -i \\ 0 & i & 0 \end{pmatrix} \ \hat{J}_3 = \begin{pmatrix} 1 & 0 & 0 \\ 0 & 0 & 0 \\ 0 & 0 & -1 \end{pmatrix}. \tag{11.54}$$

Consider now the case of arbitrary j, and let $\hat{J}_i$ be the angular momentum operators in the Hilbert space $\mathbb{C}^{2j+1}$. For any unit vector $\mathbf{n}$, the unitary operator $\hat{U}^{(j)}(\mathbf{n}, \theta) = e^{-i\theta\mathbf{n}\cdot\hat{\mathbf{J}}}$ represents the rotation matrix $O(\mathbf{n}, \theta) = e^{\theta J_\mathbf{n}}$. Hence, for each j, we have a different map that assigns a rotation matrix O to a unitary operator $\hat{U}^{(j)}(O)$ in $\mathbb{C}^{2j+1}$. The fact that the operators $\hat{J}_i$ satisfy the angular momentum algebra guarantees that the map $\hat{U}^{(j)}$ preserves matrix multiplication, namely, $\hat{U}^{(j)}(O_1 O_2) = \hat{U}^{(j)}(O_1)\hat{U}^{(j)}(O_2)$ – for more details, see Section 12.1. This construction works for rotations with matrices O such that $\det O = 1$. We will deal with space inversions later.

We express the matrix elements of $\hat{U}^{(j)}(O)$ in the angular momentum basis as

$$\mathcal{U}^{(j)}_{mm'}(O) = \langle j, m| \hat{U}^{(j)}(O)|j, m'\rangle. \tag{11.55}$$

If we express the matrix O in terms of the Euler angles as in Eq. (11.19), then $\hat{U}^{(j)}(O) = e^{-i\phi\hat{J}_3}e^{i\theta\hat{J}_2}e^{i\psi\hat{J}_3}$, and the matrix elements (11.55) simplify,

$$\mathcal{U}^{(j)}_{mm'}(O) = \langle j, m|e^{-i\phi\hat{J}_3}e^{i\theta\hat{J}_2}e^{-i\psi\hat{J}_3}|j, m'\rangle = e^{-im'\psi + im\phi}\mathcal{V}^{(j)}_{mm'}(\theta), \tag{11.56}$$

where

$$V^{(j)}_{mm'}(\theta) = \langle j,m|e^{i\theta \hat{J}_2}|j,m'\rangle.$$ (11.57)

Hence, it suffices to calculate the unitary operator $e^{i\theta \hat{J}_2}$, in order to write the matrix elements $\mathcal{U}^{(j)}_{mm'}(O)$ explicitly.

11.3.5 2π-Rotations

Since the matrix $\hat{J}_3$ is diagonal, the unitary operator $e^{-i\theta \hat{J}_3}$ that implements rotations around axis 3 by an angle θ takes the simple form

$$e^{-i\theta \hat{J}_3} = \begin{pmatrix} e^{-ij\theta} & 0 & \cdots & 0 \\ 0 & e^{-i(j-1)\theta} & \cdots & 0 \\ \cdots & \cdots & \cdots & \cdots \\ 0 & 0 & \cdots & e^{ij\theta} \end{pmatrix}.$$ (11.58)

Let us set $\theta = 2\pi$. If j is an integer, then all diagonal elements in Eq. (11.58) are of the form $e^{i2\pi k}$ for integer k, hence, they equal unity. Then, $e^{-2\pi i \hat{J}_3} = \hat{I}$; rotation by 2π causes no change. If j is a half-integer, then all diagonal elements in Eq. (11.58) are of the form $e^{i\pi k}$ for odd k, that is, they equal -1. It follows that $e^{-2\pi i \hat{J}_3} = -\hat{I}$; rotation by 2π causes a multiplication of a Hilbert space vector by -1. A further 2π-rotation is needed to return to the initial vector. Overall, we write

$$e^{-i\theta \hat{J}_3} = (-1)^{2j}\hat{I}.$$ (11.59)

The traditional notion of angular momentum that corresponds to rotating motions of particles in space is incompatible with a transformation that changes the state when the system is rotated by 2π. In Chapter 14, we will see that half-integer j corresponds to rotations that cannot be interpreted in terms of a rotating motion in space, but they define an intrinsic feature of particles, named *spin*.

Nonetheless, we expect that probabilities are unaffected by a 2π-rotation. Let $|\psi\rangle = c_1|j_1,m_1\rangle + c_2|j_2,m_2\rangle$, where j_1 is an integer and j_2 a half-integer. Then, the state $e^{2\pi i \hat{J}_3}|\psi\rangle = c_1|j_1,m_1\rangle - c_2|j_2,m_2\rangle$ is distinct from $|\psi\rangle$, with different probability distributions for physical observables. This strongly suggests that the states $|\psi\rangle$ are not physically realizable, that is, there is a superselection rule for the values of $e^{2\pi i \hat{J}_3}$ (Wick et al., 1952).

> There exist superpositions only between integer values of j or between half-integer values of j.

11.3.6 Space Inversion

In order to describe space inversion in conjunction with spatial rotations, we must define a unitary operator $\hat{\mathbb{P}}$ to represent the matrix $\mathcal{P}$ of Eq. (11.20). The operator $\hat{\mathbb{P}}$ must implement the defining properties of $\mathcal{P}$, that is, to satisfy

$$\hat{P}^2 = \hat{I}, \tag{11.60}$$

$$\hat{P}\hat{J}_i\hat{P}^\dagger = \hat{J}_i. \tag{11.61}$$

Equation (11.60) implies that the eigenvalue of $\hat{P}$ are ± 1, hence, $\hat{P}$ is also self-adjoint: $\hat{P} = \hat{P}^\dagger = \hat{P}^{-1}$. Equation (11.61) follows from Eq. (11.21) and it implies that $[\hat{P}, \hat{J}_i] = 0$. Hence, the kets $|j, m\rangle$ are eigenvectors of $\hat{P}$,

$$\hat{P}|j, m\rangle = a_{j,m}|j, m\rangle, \tag{11.62}$$

where $a_{j,m} = \pm 1$. Given that $[\hat{P}, \hat{J}_\pm] = 0$, $\hat{P}\hat{J}_+|j, m\rangle = \hat{J}_+\hat{P}|j, m\rangle$, hence, $a_{j,m} = a_{j,m+1}$. We conclude that $a_{j,m}$ is m-independent,

$$\hat{P}|j, m\rangle = a_j|j, m\rangle, \tag{11.63}$$

where $a_j = \pm 1$. The precise form of a_j depends on the physical system under consideration.

11.3.7 Time Reversal

So far, we have said nothing about the relation between rotations and time. Intuitively, we expect that time reversal also reverses rotation. Hence, we expect that an antiunitary time-reversal operator $\hat{\mathbb{T}}$ should act on the angular momentum operators as

$$\hat{\mathbb{T}}\hat{J}_i\hat{\mathbb{T}}^\dagger = -\hat{J}_i. \tag{11.64}$$

We will prove Eq. (11.64) in Section 14.5. Here, we follow its consequences.

Equation (11.64) implies that

$$\hat{\mathbb{T}}\hat{\mathbf{J}}^2\hat{\mathbb{T}}^\dagger = \sum_{i=1}^3 \hat{\mathbb{T}}\hat{J}_i\hat{\mathbb{T}}^\dagger\hat{\mathbb{T}}\hat{J}_i\hat{\mathbb{T}}^\dagger = \sum_{i=1}^3 (-\hat{J}_i)(-\hat{J}_i) = \hat{\mathbf{J}}^2. \tag{11.65}$$

Equations (11.64) and (11.65) imply that $\hat{\mathbb{T}}\hat{J}_3 = -\hat{J}_3\hat{\mathbb{T}}$ and $\hat{\mathbb{T}}\hat{\mathbf{J}}^2 = \hat{\mathbf{J}}^2\hat{\mathbb{T}}$. Hence,

$$\hat{J}_3\hat{\mathbb{T}}|j, m\rangle = -m\hat{\mathbb{T}}|j, m\rangle, \qquad \hat{\mathbf{J}}^2\hat{\mathbb{T}}|j, m\rangle = j(j+1)\hat{\mathbf{J}}^2\hat{\mathbb{T}}|j, m\rangle.$$

We conclude that $\hat{\mathbb{T}}|j, m\rangle$ is a common eigenvector of $\hat{J}_3$ and of $\hat{\mathbf{J}}^2$, with eigenvalues $-m$ and $j(j+1)$, respectively. It follows that

$$\hat{\mathbb{T}}|j, m\rangle = b_{j,m}|j, -m\rangle \tag{11.66}$$

for some phases $b_{j,m}$.

Since $\hat{\mathbb{T}}$ is antiunitary, Eq. (11.64) implies that $\hat{\mathbb{T}}\hat{J}_+\hat{\mathbb{T}}^\dagger = -\hat{J}_-$, or, equivalently, $\hat{\mathbb{T}}\hat{J}_+ = -\hat{J}_-\hat{\mathbb{T}}$. By Eq. (11.66),

$$\hat{\mathbb{T}}\hat{J}_+|j, m\rangle = \sqrt{j(j+1) - m(m+1)}b_{j,m+1}|j, -m-1\rangle,$$
$$\hat{J}_-\hat{\mathbb{T}}|j, m\rangle = \sqrt{j(j+1) - m(m+1)}b_{j,m}|j, -m-1\rangle.$$

This is possible only if $b_{j,m+1} = -b_{j,m}$. Hence, $b_{j,m} = (-1)^{j+m}b_{j,-j}$, where $b_{j,-j}$ is an m-independent phase factor.

We can choose any value of $b_{j,-j}$ by redefining the basis vectors $|j, m\rangle$. Set $|j, m\rangle' = e^{i\phi_j}|j, m\rangle$ for some phase $e^{i\phi_j}$. Then,

$$\hat{\mathbb{T}}|j, m\rangle' = e^{-i\phi_j}\hat{\mathbb{T}}|j, m\rangle = e^{-i\phi_j}b_{j,-j}(-1)^{j+m}|j, -m\rangle = e^{-2i\phi_j}b_{j,-j}(-1)^{j+m}|j, -m\rangle'.$$

There are two common choices for $b_{j,-j}$. First, we can set $e^{-2i\phi_j}b_{j,-j} = 1$, so the phase change under time reversal is ± 1,

$$\hat{\mathbb{T}}|j,m\rangle = (-1)^{j+m}|j,-m\rangle. \tag{11.67}$$

Second, we can choose $e^{-2i\phi_j}b_{j,-j}(-1)^j = 1$, so that the phase under time reversal depends solely on m,

$$\hat{\mathbb{T}}|j,m\rangle = i^{2m}|j,-m\rangle. \tag{11.68}$$

Either way, $\hat{\mathbb{T}}^2|j,m\rangle = (-1)^{2j}|j,m\rangle$, that is,

$$\hat{\mathbb{T}}^2 = (-1)^{2j}\hat{I}. \tag{11.69}$$

For half-integer values of j, $\hat{\mathbb{T}}^2 = -\hat{I}$. Hence, double time reversal does not return systems with half-integer j to their initial state, but it creates a phase difference π, just like rotation by 2π. As long as no superpositions of even and odd j exist, probabilities remain unchanged under the action of $\hat{\mathbb{T}}^2$.

11.4 Orbital Angular Momentum

So far, our analysis of angular momentum has been abstract, as it is applied to any system where spatial rotations can be implemented. In this section, we analyze a physical system with a well-known classical analogue, the angular momentum of a particle that moves in three dimensions.

11.4.1 Definitions and Properties

A particle in three dimensions is described by position operators $\hat{x}_i$ and momentum operators $\hat{p}_i$, for $i = 1, 2, 3$, subject to Heisenberg's commutation relations

$$[\hat{x}_i, \hat{x}_j] = 0, \qquad [\hat{p}_i, \hat{p}_j] = 0, \qquad [\hat{x}_i, \hat{p}_j] = i\delta_{ij}\hat{I}. \tag{11.70}$$

Motivated by the classical formula $\mathbf{L} = \mathbf{r} \times \mathbf{p}$ for the angular momentum $\mathbf{L}$, in terms of the position and momentum vectors $\mathbf{r}$ and $\mathbf{p}$, we define the *orbital angular momentum* operators as

$$\hat{\ell}_i = \sum_{jk} \epsilon_{ijk}\hat{x}_j\hat{p}_k. \tag{11.71}$$

It is easy to show that $[\hat{\ell}_1, \hat{\ell}_2] = i\hat{\ell}_3$ and so on, that is, that the operators $\hat{\ell}_i$ satisfy the angular momentum algebra

$$[\hat{\ell}_i, \hat{\ell}_j] = i\sum_{k=1}^{3} \epsilon_{ijk}\hat{\ell}_k. \tag{11.72}$$

Since Eq. (11.71) is satisfied, the eigenvalues of $\hat{\ell}^2 := \hat{\ell}_1^2 + \hat{\ell}_2^2 + \hat{\ell}_3^2$ are of the form $\ell(\ell + 1)$, and the eigenvalues of $\hat{\ell}_3$ are $m = -\ell, -\ell + 1, \ldots, \ell$. For orbital angular momentum, we use the symbol ℓ rather than j.

Rotations of a particle in physical space by 2π must *by definition* leave the system invariant. This means that ℓ cannot be a half-integer. Hence, we expect that ℓ *takes only integer values*.

The position and momentum operators can be expressed in terms of their action on wave functions $\psi(x_1, x_2, x_3)$, that is, on elements of the Hilbert space $L^2(\mathbb{R}^3, d^3x)$: $\hat{x}_i \psi(x_1, x_2, x_3) = x_i \psi(x_1, x_2, x_3)$ and $\hat{p}_i \psi(x_1, x_2, x_3) = -i\frac{\partial}{\partial x_i} \psi(x_1, x_2, x_3)$, so that

$$\hat{\ell}_i = -i \sum_{jk} \epsilon_{ijk} x_j \frac{\partial}{\partial x_k}. \tag{11.73}$$

We note that a rescaling of position $x_i \to \mu x_i$ for all $\mu > 0$ leaves the operators $\hat{\ell}_i$ invariant. This means that the operators $\hat{\ell}_i$ do not depend on the length $|\mathbf{x}|$ of the vector $\mathbf{x} = (x_1, x_2, x_3)$, but only on the unit vectors $\mathbf{n} = \mathbf{x}/|\mathbf{x}|$. We use the coordinates $\theta \in [0, \pi]$ and $\phi \in [0, 2\pi)$ the unit sphere, $(x_1, x_2, x_3) = (\sin\theta\cos\phi, \sin\theta\sin\phi, \cos\theta)$. Then,

$$\frac{\partial}{\partial x_1} = \frac{1}{\cos\theta\cos\phi}\frac{\partial}{\partial\theta} - \frac{1}{\sin\theta\sin\phi}\frac{\partial}{\partial\phi}, \quad \frac{\partial}{\partial x_2} = \frac{1}{\cos\theta\sin\phi}\frac{\partial}{\partial\theta} + \frac{1}{\sin\theta\cos\phi}\frac{\partial}{\partial\phi},$$
$$\frac{\partial}{\partial x_3} = -\frac{1}{\sin\theta}\frac{\partial}{\partial\theta}.$$

We obtain

$$\hat{\ell}_1 = -i\left(x_2\frac{\partial}{\partial x_3} - x_3\frac{\partial}{\partial x_2}\right) = i(\sin\phi\frac{\partial}{\partial\theta} + \cot\theta\cos\phi\frac{\partial}{\partial\phi}) \tag{11.74}$$

$$\hat{\ell}_2 = -i\left(x_3\frac{\partial}{\partial x_1} - x_1\frac{\partial}{\partial x_3}\right) = i(-\cos\phi\frac{\partial}{\partial\theta} + \cot\theta\sin\phi\frac{\partial}{\partial\phi}) \tag{11.75}$$

$$\hat{\ell}_3 = -i\left(x_1\frac{\partial}{\partial x_2} - x_2\frac{\partial}{\partial x_1}\right) = -i\frac{\partial}{\partial\phi} \tag{11.76}$$

$$\hat{\ell}^2 = \frac{1}{\sin\theta}\frac{\partial}{\partial\theta}\left(\sin\theta\frac{\partial}{\partial\theta}\right) + \frac{1}{\sin^2\theta}\frac{\partial^2}{\partial\phi^2}. \tag{11.77}$$

This means that $\hat{\ell}_i$ act on functions only of θ and ϕ, that is, they act on wave functions on the unit sphere S^2. Indeed, $\hat{\ell}^2$ coincides with the Laplace operator on the sphere S^2; see Eq. (B.52).

We therefore consider $\hat{\ell}_i$ as operators on the Hilbert space $\mathcal{H}_{S^2} = L^2(S^2, \sin\theta d\theta d\phi)$ of square integrable functions on the unit sphere S^2, with the usual integration measure $\sin\theta d\theta d\phi$. We express the vectors of $\mathcal{H}_{S^2}$ as wave functions $\psi(\mathbf{n})$, where $\mathbf{n} = (\sin\theta\cos\phi, \sin\theta\sin\phi, \cos\theta)$, and we define kets $|\mathbf{n}\rangle = |\theta, \phi\rangle$ on $\mathcal{H}_{S^2}$ such that $\psi(\mathbf{n}) = \langle\mathbf{n}|\psi\rangle$.

Given that all functions $\psi(x_1, x_2, x_3)$ can be written using polar coordinates as $\psi(r, \theta, \phi)$ and that $d^3x = r^2\sin\theta dr d\theta d\phi$, we can split

$$L^2(\mathbb{R}^3, d^3x) = \mathcal{H}_{rad} \otimes \mathcal{H}_{S^2}, \tag{11.78}$$

where the Hilbert space $\mathcal{H}_{rad}$ consists of functions of the form $\psi(r)$ that satisfy $\int_0^\infty dr|r\psi(r)|^2 < \infty$ and appropriate boundary conditions as $r \to 0$. It is straightforward to show that

$$\delta^3(\mathbf{x} - \mathbf{x}') = \frac{1}{r^2}\delta(r - r')\delta(\cos\theta - \cos\theta')\delta(\phi - \phi'), \tag{11.79}$$

so that we can split the position kets $|\mathbf{x}\rangle = |r\rangle \otimes |\mathbf{n}\rangle$, where $|r\rangle$ are kets on $\mathcal{H}_{rad}$ satisfying $\langle r|r'\rangle = r^{-2}\delta(r - r')$ and

$$\langle\mathbf{n}|\mathbf{n}'\rangle = \delta^2(\mathbf{n}, \mathbf{n}'). \tag{11.80}$$

The delta function on the sphere $\delta^2(\mathbf{n}, \mathbf{n}')$ is a shorthand for $\delta(\cos\theta - \cos\theta')\delta(\phi - \phi')$.

The operators (11.74)–(11.76) act on $\mathcal{H}_{rad} \otimes \mathcal{H}_{S^2}$ as $\hat{I} \otimes \hat{\ell}_i$. We will abuse notation and use the symbol $\hat{\ell}_i$ in both Hilbert spaces. The spectral analysis that we present next makes sense only if $\hat{\ell}_i$ is defined on $\mathcal{H}_{S^2}$.

11.4.2 Spherical Harmonics

The ladder operators for orbital angular momentum are

$$\hat{\ell}_\pm = e^{\pm i\phi} \left(\pm \frac{\partial}{\partial\theta} + i\cot\theta \frac{\partial}{\partial\phi} \right). \tag{11.81}$$

Let us denote the common eigenvectors of $\hat{\ell}_3$ and $\hat{\ell}^2$ by $|\ell, m\rangle$ and the associated functions of θ and ϕ by $Y_{\ell m}(\theta, \phi)$. The latter are known as *spherical harmonics* in the theory of differential equations. Occasionally, we will also denote the spherical harmonics as $Y_{\ell m}(\mathbf{n})$, in terms of a unit vector $\mathbf{n}$.

By the general theory of angular momentum, $\hat{\ell}_+|\ell, \ell\rangle = 0$, or, equivalently,

$$\frac{\partial Y_{\ell\ell}}{\partial\theta} + i\cot\theta \frac{\partial Y_{\ell\ell}}{\partial\phi} = 0. \tag{11.82}$$

Since $|\ell, \ell\rangle$ is an eigenvector of $\hat{\ell}_3$ with eigenvalue ℓ, we have $-i\frac{\partial}{\partial\phi} Y_{\ell\ell} = \ell Y_{\ell\ell}$. Hence, $Y_{\ell\ell}(\theta, \phi) = e^{i\ell\phi} F_\ell(\theta)$, for some function F_ℓ of θ. We substitute in Eq. (11.82), to obtain

$$\frac{dF_\ell}{d\theta} - \ell\cot\theta F_\ell = 0, \tag{11.83}$$

with solution $F_\ell(\theta) = C_\ell \sin^\ell\theta$, for some constant C_ℓ. We conclude that

$$Y_{\ell\ell}(\theta, \phi) = C_\ell \sin^\ell\theta e^{i\ell\phi}. \tag{11.84}$$

The requirement that $Y_{\ell\ell}(\theta, \phi + 2\pi) = Y_{\ell\ell}(\theta, \phi)$ leads to the conclusion that $e^{2\pi i\ell} = 1$. Hence, ℓ is an integer, as we expected.

The constant C_ℓ is computed by the normalization requirement $\int \sin\theta d\theta d\phi |Y_{\ell\ell}|^2 = 1$. Integration with respect to ϕ gives a factor 2π, while with a change of variables $\xi = \cos\theta$, we obtain $2\pi |C_\ell|^2 \int_{-1}^1 d\xi (1 - \xi^2)^\ell = 1$. The integral over ξ is evaluated by substituting $\xi = y^2$,

$$\int_{-1}^1 d\xi (1 - \xi^2)^\ell = \int_0^1 dy\, y^{-\frac{1}{2}}(1 - y)^\ell = B(\frac{1}{2}, \ell + 1) = \frac{\Gamma(\frac{1}{2})\Gamma(\ell + 1)}{\Gamma(\ell + \frac{3}{2})} = \frac{2^{2\ell+1}(\ell!)^2}{(2\ell + 1)!},$$

where we used properties of the beta and the gamma function listed in Appendix B; in particular, we evaluated $\Gamma(\ell + \frac{3}{2})$ using Eq. (B.5). We conclude that

$$C_\ell = \frac{1}{2^\ell \ell!} \sqrt{\frac{(2\ell + 1)!}{4\pi}}. \tag{11.85}$$

The vector $|\ell, m\rangle$ is obtained by successive actions of $\hat{\ell}_-$ on $|\ell, \ell\rangle$, according to Eq. (11.47),

$$|\ell, m\rangle = \frac{1}{\sqrt{1 \cdot 2\ell \cdots [\ell(\ell + 1) - m(m + 1)]}} \hat{\ell}_-^{\ell-m} |\ell, \ell\rangle. \tag{11.86}$$

A

Table 11.1 Spherical harmonics

ℓ	m	$Y_{\ell m}(\theta,\phi)$	ℓ	m	$Y_{\ell m}(\theta,\phi)$
0	0	$\frac{1}{2\sqrt{\pi}}$	2	± 2	$\frac{1}{4}\sqrt{\frac{15}{2\pi}}\sin^2\theta\, e^{\pm 2i\phi}$
1	0	$\sqrt{\frac{3}{4\pi}}\cos\theta$	3	0	$\frac{1}{4}\sqrt{\frac{7}{\pi}}(5\cos^3\theta - 3\cos\theta)$
1	± 1	$\mp\sqrt{\frac{3}{8\pi}}\sin\theta\, e^{\pm i\phi}$	3	± 1	$\mp\frac{1}{8}\sqrt{\frac{21}{\pi}}\sin\theta(5\cos^2\theta - 1)e^{\pm i\phi}$
2	0	$\frac{1}{4}\sqrt{\frac{5}{\pi}}(3\cos^2\theta - 1)$	3	± 2	$\frac{1}{4}\sqrt{\frac{105}{2\pi}}\sin^2\theta\cos\theta\, e^{\pm 2i\phi}$
2	± 1	$\mp\sqrt{\frac{15}{8\pi}}\sin\theta\cos\theta\, e^{\pm i\phi}$	3	± 3	$\mp\frac{1}{8}\sqrt{\frac{35}{\pi}}\sin^3\theta\, e^{\pm 3i\phi}$

Substituting into Eq. (11.84), we obtain a general expression for spherical harmonics:

$$Y_{\ell m}(\theta,\phi) = \frac{C_\ell e^{im\phi}\left(-\frac{\partial}{\partial\theta} - \ell\cot\theta\right)^{\ell-m}\sin^\ell\theta}{\sqrt{1\cdot 2\ell\cdots\cdot[\ell(\ell+1)-m(m+1)]}}. \tag{11.87}$$

As expected from the solutions of the eigenvalue equation $-i\frac{\partial}{\partial\phi}Y_{\ell m} = mY_{\ell m}$, the spherical harmonics (11.87) consist of a θ-dependent term times $e^{im\phi}$.

The case of $m = 0$ is of particular interest. The eigenvalue equation $\hat{\ell}_3|\ell, 0\rangle = 0$ implies that $\frac{\partial}{\partial\phi}Y_{\ell 0} = 0$, therefore, $Y_{\ell 0}$ depends only on θ. By Eq. (11.77), the eigenvalue equation $\hat{\ell}^2|\ell, 0\rangle = \ell(\ell+1)|\ell, 0\rangle$ is a differential equation only with respect to θ. We set $\xi = \cos\theta$, to find that for zero eigenvalue of $\hat{\ell}_3$, $\hat{\ell}^2 = -\frac{d}{d\xi}\left[(1-\xi^2)\frac{d}{d\xi}\right]$. It follows that $\hat{\ell}^2$ coincides with the Legendre operator $\hat{\Lambda}$. Hence, the eigenvectors of $\hat{\ell}^2$ are the Legendre polynomials, normalized to unity according to Eq. (3.13),

$$Y_{\ell 0}(\theta,\phi) = \sqrt{\frac{2\ell+1}{4\pi}}P_\ell(\cos\theta). \tag{11.88}$$

Equation (11.87) determines the spherical harmonics, normalized to unity, up to a phase. We usually choose a phase so that the time reversal operator $\hat{\mathbb{T}}$ satisfies the convention (11.68). For orbital angular momentum, time reversal is defined in terms of complex conjugation of the wave function, the same way that it is defined for particles,

$$\hat{\mathbb{T}}Y_{\ell m}(\theta,\phi) = Y_{\ell m}^*(\theta,\phi). \tag{11.89}$$

Indeed, Eq. (11.89) implies that $\hat{\mathbb{T}}$ takes $Y_{\ell m}$ to $Y_{\ell(-m)}$, as in Eq. (11.68). The phase is fixed by multiplying the spherical harmonics (11.87) of positive m with a phase factor $(-1)^m$, so that $Y_{\ell m}^* = Y_{\ell(-m)}(-1)^m$. A list of the spherical harmonics up to $\ell = 3$ with this phase convention is given in Table 11.1.

Example 11.2 The parity transformation $\mathbf{r} \to -\mathbf{r}$ in spherical coordinates becomes $r \to r, \theta \to \pi - \theta$ and $\phi \to \pi + \phi$. By Eq. (11.63), the eigenvalues of the space inversion operator $\hat{\mathbb{P}}$ do not depend on m, so it suffices to evaluate them on a single vector for a given ℓ. We choose $|\ell, \ell\rangle$. By Eq. (11.84), $Y_{\ell\ell}(\pi - \theta, \pi + \phi) = e^{i\ell\pi}Y_{\ell\ell}(\theta,\phi) = (-1)^\ell Y_{\ell\ell}(\theta,\phi)$. We conclude that

$$\hat{\mathbb{P}}|\ell, m\rangle = (-1)^\ell|\ell, m\rangle. \tag{11.90}$$

11.5 Addition of Angular Momenta

Consider a composite system with Hilbert space $\mathcal{H}_1 \otimes \mathcal{H}_2$. Let $^1\hat{J}_i$ be the angular momentum operators defined in the Hilbert space $\mathcal{H}_1$, and let $^2\hat{J}_i$ be the angular momentum operators defined in the Hilbert space $\mathcal{H}_2$. In $\mathcal{H}_1 \otimes \mathcal{H}_2$, we define the operators

$$\hat{J}_i = {}^1\hat{J}_i \otimes \hat{I} + \hat{I} \otimes {}^2\hat{J}_i. \tag{11.91}$$

In this notation, operators of the composite system carry no labels, while operators in each subsystem carry a label 1 or 2 that specifies the subsystem.

The operators $\hat{J}_i$ satisfy the angular momentum algebra, since

$$
\begin{aligned}
[\hat{J}_i, \hat{J}_j] &= [{}^1\hat{J}_i \otimes \hat{I} + \hat{I} \otimes {}^2\hat{J}_i, {}^1\hat{J}_j \otimes \hat{I} + \hat{I} \otimes {}^2\hat{J}_j] \\
&= [{}^1\hat{J}_i, {}^1\hat{J}_j] \otimes \hat{I} + \hat{I} \otimes [{}^2\hat{J}_i, {}^2\hat{J}_j] = i\sum_k \epsilon_{ijk} {}^1\hat{J}_k \otimes \hat{I} + \sum_k \epsilon_{ijk} \hat{I} \otimes {}^2\hat{J}_k \\
&= i\sum_k \epsilon_{ijk}({}^1\hat{J}_k \otimes \hat{I} + \hat{I} \otimes {}^2\hat{J}_k) = i\sum_k \epsilon_{ijk}\hat{J}_k.
\end{aligned}
$$

This means that the joint eigenvectors $|J,M\rangle$ of the operators $\hat{\mathbf{J}}^2$ and $\hat{J}_3$ define a basis on $\mathcal{H}_1 \otimes \mathcal{H}_2$. By definition the vectors $|J,M\rangle$ satisfy

$$\hat{\mathbf{J}}^2|J,M\rangle = J(J+1)|J,M\rangle, \qquad \hat{J}_3|J,M\rangle = M|J,M\rangle. \tag{11.92}$$

Another orthonormal basis on $\mathcal{H}_1 \otimes \mathcal{H}_2$ is formed by the vectors $|j_1,j_2;m_1,m_2\rangle := |j_1,m_1\rangle_1 \otimes |j_2,m_2\rangle_2$. The aim of this section is to find the relation between the two bases $|J,M\rangle$ and $|j_1,j_2;m_1,m_2\rangle$. This will allow us to express rotations in the composite system in terms of rotations in the components.

11.5.1 Triangle Inequality

Let us denote by $\mathcal{H}_j$ a Hilbert space spanned by vectors $|j,m\rangle$, for fixed j and with no degeneracy. Any Hilbert space $\mathcal{H}$ in which the angular momentum algebra is realized can be expressed as $\mathcal{H} = \oplus_{j \in C}\mathcal{H}_j$, where C is a set with several different values of j. If we write $\mathcal{H}_1 = \oplus_{j \in C_1}\mathcal{H}_j$ and $\mathcal{H}_2 = \oplus_{j \in C_2}\mathcal{H}_j$ for two different sets C_1 and C_2, then $\mathcal{H}_1 \otimes \mathcal{H}_2 = \oplus_{j_1 \in C_1, j_2 \in C_2}\mathcal{H}_{j_1} \otimes \mathcal{H}_{j_2}$. Hence, it suffices to consider the simplest case of tensor products of the form $\mathcal{H}_{j_1} \otimes \mathcal{H}_{j_2}$, as the more complex cases can be built from combinations thereof.

For such tensor products, $\mathcal{H}_1 = \mathbb{C}^{2j_1+1}$ and $\mathcal{H}_2 = \mathbb{C}^{2j_2+1}$, hence, $\mathcal{H}_1 \otimes \mathcal{H}_2 = \mathbb{C}^{(2j_1+1)(2j_2+1)}$. We obtain

$$
\begin{aligned}
\hat{J}_\pm(|j_1,m_1\rangle_1 \otimes |j_2,m_2\rangle_2) &= {}^1\hat{J}_\pm|j_1,m_1\rangle_1 \otimes |j_2,m_2\rangle_2 + |j_1,m_1\rangle_1 \otimes {}^2\hat{J}_\pm|j_2,m_2\rangle_2 \\
&= \sqrt{j_1(j_1+1) - m_1(m_1 \pm 1)}|j_1,m_1 \pm 1\rangle_1 \otimes |j_2,m_2\rangle_2 \\
&\quad + \sqrt{j_2(j_2+1) - m_2(m_2 \pm 1)}|j_1,m_1\rangle_1 \otimes |j_2,m_2 \pm 1\rangle_2. \tag{11.93}
\end{aligned}
$$

Furthermore,

$$
\begin{aligned}
\hat{J}_3(|j_1,m_1\rangle_1 \otimes |j_2,m_2\rangle_2) &= {}^1\hat{J}_3|j_1,m_1\rangle_1 \otimes |j_2,m_2\rangle_2 + |j_1,m_1\rangle_1 \otimes {}^2\hat{J}_3|j_2,m_2\rangle_2 \\
&= (m_1 + m_2)|j_1,m_1\rangle_1 \otimes |j_2,m_2\rangle_2. \tag{11.94}
\end{aligned}
$$

Hence, m_1 and m_2 combine additively to M,

$$M = m_1 + m_2. \tag{11.95}$$

This means that the maximum value of M is $j_1 + j_2$, which implies that the maximum value J is also $j_1 + j_2$. Given that M changes by adding or subtracting integers, the possible values of J are $j_1 + j_2 - 1, j_1 + j_2 - 2$, and so on, up to a minimum value of J,

$$J_{min} = j_1 + j_2 - N, \tag{11.96}$$

where N is a nonnegative integer. We proceed to prove that $J_{min} = |j_1 - j_2|$.

There are $2J + 1$ basis vectors for each J; hence, the dimension of $\mathcal{H}_1 \otimes \mathcal{H}_2$ equals

$$\sum_{J=J_{min}}^{J_{max}} (2J + 1) = N + 1 + 2 \sum_{n=0}^{N} (J_{max} - n) = 2J_{max} + 1 + 2J_{max}N - N^2,$$

where we used the identity $\sum_{k=1}^{n} k = \frac{1}{2}n(n + 1)$. But the dimension of $\mathcal{H}_1 \otimes \mathcal{H}_2$ is also $(2j_1 + 1)(2j_2 + 1)$, hence,

$$N^2 - 2(j_1 + j_2)N + 4j_1 j_2 = 0. \tag{11.97}$$

This binomial equation has solution $N_{\pm} = j_1 + j_2 \pm |j_1 - j_2|$. The solution N_+ is unacceptable as it implies that $J_{min} < 0$. The solution N_- implies that $J_{min} = |j_1 - j_2|$. We conclude that J satisfies the inequality

$$|j_1 - j_2| \leq J \leq j_1 + j_2, \tag{11.98}$$

and that J increases with integer steps from the minimum to the maximum value. Equation (11.98) is the *triangle identity* for the addition of angular momenta.

Some examples are as follows.

(i) For $j_1 = j_2 = \frac{1}{2}$, possible values of J are 0 and 1.
(ii) For $j_1 = 1$ and $j_2 = \frac{1}{2}$, possible values of J are $\frac{1}{2}$ and $\frac{3}{2}$.
(iii) For $j_1 = 2$ and $j_2 = 1$, possible values of J are 1, 2, and 3.

At the Hilbert space level the triangle inequality implies that

$$\mathbb{C}^{2j_1+1} \otimes \mathbb{C}^{2j_1+1} = \mathbb{C}^{2|j_1-j_2|+1} \oplus \mathbb{C}^{2|j_1-j_2|+3} \oplus \cdots \oplus \mathbb{C}^{2(j_1+j_2)+1}. \tag{11.99}$$

It is useful to denote the Hilbert space $\mathbb{C}^{2j+1}$ that corresponds to angular momentum j as $\underline{2j + 1}$. Then, Eq. (11.99) becomes

$$\underline{2j_1 + 1} \otimes \underline{2j_2 + 1} = \underline{2|j_1 - j_2| + 1} \oplus \underline{2|j_1 - j_2| + 3} \oplus \cdots \oplus \underline{2(j_1 + j_2) + 1}.$$

In this notation, example (i) means that $\underline{2} \otimes \underline{2} = \underline{1} \oplus \underline{3}$, example (ii) means that $\underline{3} \otimes \underline{2} = \underline{2} \oplus \underline{4}$, and example (iii) means that $\underline{5} \otimes \underline{3} = \underline{3} \oplus \underline{5} \oplus \underline{7}$.

We can also use the triangle inequality for the addition of more than two angular momenta. For three subsystems with angular momentum j_1, j_2, and j_3, we first identify the possible values of the angular momentum $j_{1,2}$ of the first two subsystems. Then we apply the triangle inequality for the total angular momentum for J: $|j_{1,2} - j_3| \leq J \leq j_{1,2} + j_3$, for *all* $j_{1,2}$.

Example 11.3 Let $j_1 = j_2 = \frac{1}{2}$ and $j_3 = \frac{3}{2}$. By the triangle inequality, $j_{1,2} = 0, 1$. For $j_{1,2} = 0$, $J = \frac{3}{2}$. For $j_{1,2} = 1$, $J = \frac{1}{2}, \frac{3}{2}, \frac{5}{2}$. Overall, possible values of J are $J = \frac{1}{2}, \frac{3}{2}, \frac{5}{2}$, with $J = \frac{3}{2}$ appearing in two different ways. Hence, $\underline{2} \otimes \underline{2} \otimes \underline{4} = \underline{2} \oplus \underline{4} \oplus \underline{4} \oplus \underline{6}$.

11.5.2 Case: $j_1 = j_2 = \frac{1}{2}$

Next, we find the transformation that relates the bases $|J, M\rangle$, and $|j_1, j_2; m_1, m_2\rangle$ for special values of $j_1 = j_2 = \frac{1}{2}$. This is a prelude to a general algorithm that works for all j_1 and j_2.

By the triangle inequality, possible values of J are 0 and 1. Given Eq. (11.95), the state $J = 1$, $M = 1$ can only be obtained for $m_1 = m_2 = \frac{1}{2}$. Hence,

$$|1, -1\rangle = \left|\frac{1}{2}, \frac{1}{2}\right\rangle_1 \otimes \left|\frac{1}{2}, \frac{1}{2}\right\rangle_2. \qquad (11.100)$$

Next, we act on $|1, 1\rangle$ with the lowering operator $\hat{J}_-$ of the composite system,

$$\hat{J}_- = {}^1\hat{J}_- \otimes \hat{I} + \hat{I} \otimes {}^2\hat{J}_-. \qquad (11.101)$$

We note that

$$\hat{J}_-|1, 1\rangle = \sqrt{2}|1, 0\rangle. \qquad (11.102)$$

By Eq. (11.93),

$$\hat{J}_-|1, 1\rangle = \left|\frac{1}{2}, -\frac{1}{2}\right\rangle_1 \otimes \left|\frac{1}{2}, \frac{1}{2}\right\rangle_2 + \left|\frac{1}{2}, \frac{1}{2}\right\rangle_1 \otimes \left|\frac{1}{2}, -\frac{1}{2}\right\rangle_2. \qquad (11.103)$$

We equate Eqs. (11.102) and (11.103), to obtain

$$|1, 0\rangle = \frac{1}{\sqrt{2}}\left(\left|\frac{1}{2}, \frac{1}{2}\right\rangle_1 \otimes \left|\frac{1}{2}, -\frac{1}{2}\right\rangle_2 + \left|\frac{1}{2}, -\frac{1}{2}\right\rangle_1 \otimes \left|\frac{1}{2}, +\frac{1}{2}\right\rangle_2\right). \qquad (11.104)$$

Then, we act with $\hat{J}_-$ on $|1, 0\rangle$, to obtain $|1, -1\rangle$. However, we can exploit the fact that $|J, M\rangle$ is obtained (modulo an overall phase) by substituting M, m_1, m_2 with $-M, -m_1, -m_2$ in the expression for $|J, M\rangle$. Hence,

$$|1, 1\rangle = \left|\frac{1}{2}, -\frac{1}{2}\right\rangle_1 \otimes \left|\frac{1}{2}, -\frac{1}{2}\right\rangle_2. \qquad (11.105)$$

We constructed all vectors $|J, M\rangle$ with $J = 1$. We must also construct $|0, 0\rangle$, which is the only vector with $J = 0$. This vector must be of the form $a|\frac{1}{2}, \frac{1}{2}\rangle_1 \otimes |\frac{1}{2}, -\frac{1}{2}\rangle_2 + b|\frac{1}{2}, -\frac{1}{2}\rangle_1 \otimes |\frac{1}{2}, +\frac{1}{2}\rangle_2$, so that Eq. (11.95) applies. Given that $|0, 0\rangle$ and $|1, 0\rangle$ correspond to different eigenvalues of $\hat{J}^2$, $\langle 0, 0|1, 0\rangle = 0$; hence, $a + b = 0$. Together with the normalization condition, we obtain $a = -b = \frac{1}{\sqrt{2}}$. Then,

$$|0, 0\rangle = \frac{1}{\sqrt{2}}\left(\left|\frac{1}{2}, \frac{1}{2}\right\rangle_1 \otimes \left|\frac{1}{2}, -\frac{1}{2}\right\rangle_2 - \left|\frac{1}{2}, -\frac{1}{2}\right\rangle_1 \otimes \left|\frac{1}{2}, +\frac{1}{2}\right\rangle_2\right). \qquad (11.106)$$

Note that if we exchange the kets of the two subsystems, the $|1, M\rangle$ kets remain unchanged, while $|0, 0\rangle$ goes to $-|0, 0\rangle$.

In the qubit notation, $|\frac{1}{2},\frac{1}{2}\rangle = |1\rangle, |\frac{1}{2}, -\frac{1}{2}\rangle = |0\rangle$, the vector (11.104) corresponds to $|\Psi_+\rangle$ and the vector (11.106) corresponds to $|\Psi_-\rangle$ of Eq. (9.29). Both are maximal-entanglement states. In general, most $|J,M\rangle$ vectors are entangled. The only exception are vectors with extremal values of J and M such as the previous vectors $|1, \pm 1\rangle$.

11.5.3 An Algorithm for Addition of Angular Momenta

The previous example illustrates the key ideas of an algorithm for the construction of the vectors $|J,M\rangle$ for any j_1 and j_2.

1. Start from the extremal state with $J_{max} = j_1 + j_2$ and $M = -J_{max}$. This is always separable and equal to $|j_1, -j_1\rangle_1 \otimes_1 |j_2, -j_2\rangle_2$.
2. Act successively with $\hat{J}_-$ from Eq. (11.101) to obtain all vectors with $J = J_{max}$. Use the symmetry between positive and negative values of M, m_1, m_2 to expedite calculations.
3. Construct the states with $J = J_{max} - 1$. First, we construct the state with $|J_{max}-1, J_{max}-1\rangle$. This is identified uniquely by the requirement that it is normal to $|J_{max}, J_{max} - 1\rangle$, which is the only other state with the same M.
4. Act successively with $\hat{J}_-$ to obtain all vectors with $J = J_{max} - 1$.
5. Repeat this procedure by decreasing J by one at each step. The first vector for each J is $|J,J\rangle$, which is determined by requiring that it is normal to all vectors with $M = J$ that were identified in previous steps.
6. The procedure terminates after you complete the calculation for $J = J_{min} = |j_1 - j_2|$.

At the end of this procedure, we have expressed all vectors $|J,M\rangle$ as

$$|J,M\rangle = \sum_m C_{j_1,j_2,m_1,m_2}^{J,M} |j_1,m_1\rangle_1 \otimes |j_2,m_2\rangle_2, \qquad (11.107)$$

by specifying the constants $C_{j_1,j_2,m}^{J,M}$ for all allowed J, M, and m given specific j_1 and j_2. The constants $C_{j_1,j_2,m}^{J,M}$ are real by construction and they are known as *Clebsch–Gordan coefficients* for rotations.

Writing $|j_1,j_2;m_1,m_2\rangle = |j_1,m_1\rangle_1 \otimes |j_2,m_2\rangle_2$, we express the Clebsch–Gordan coefficients as

$$C_{j_1,j_2,m_1,m_2}^{J,M} = \langle J,M|j_1,j_2;m_1,m_2\rangle = \langle j_1,j_2;m_1,m_2|J,M\rangle. \qquad (11.108)$$

Hence,

$$|j_1,m_1\rangle_1 \otimes |j_2,m_2\rangle_2 = \sum_{J,M} C_{j_1,j_2,m_1,m_2}^{J,M}|J,M\rangle. \qquad (11.109)$$

11.5.4 Case: $j_1 = \ell, j_2 = \frac{1}{2}$

Next, we proceed to apply the algorithm above to the case where j_1 corresponds to an orbital angular momentum ℓ and $j_2 = \frac{1}{2}$. In Chapter 14, we will use these results in calculating the total angular momentum of electrons.

There are only two values of J for each ℓ: $J = \ell \pm \frac{1}{2}$. The extremal state corresponds to $M = \ell + \frac{1}{2}$,

$$\left| \ell + \frac{1}{2}, \ell + \frac{1}{2} \right\rangle = |\ell, \ell\rangle_1 \otimes \left| \frac{1}{2}, \frac{1}{2} \right\rangle_2. \tag{11.110}$$

The action of $\hat{J}_-$ on both sides of Eq. (11.110) yields,

$$\left| \ell + \frac{1}{2}, \ell - \frac{1}{2} \right\rangle = \sqrt{\frac{2\ell}{2\ell + 1}} |\ell, \ell - 1\rangle_1 \otimes \left| \frac{1}{2}, \frac{1}{2} \right\rangle_2 + \frac{1}{\sqrt{2\ell + 1}} |\ell, \ell\rangle_1 \otimes \left| \frac{1}{2}, -\frac{1}{2} \right\rangle_2.$$

The key point here is that the next action of $\hat{J}_-$ again leads to two terms in the right-hand side, because $^2\hat{J}_-$ annihilates $|\frac{1}{2}, -\frac{1}{2}\rangle_2$. We calculate

$$\left| \ell + \frac{1}{2}, \ell - \frac{3}{2} \right\rangle = \sqrt{\frac{2\ell - 1}{2\ell + 1}} |\ell, \ell - 2\rangle_1 \otimes \left| \frac{1}{2}, \frac{1}{2} \right\rangle_2 + \sqrt{\frac{2}{2\ell + 1}} |\ell, \ell - 1\rangle_1 \otimes \left| \frac{1}{2}, -\frac{1}{2} \right\rangle_2.$$

The pattern is now obvious. It is easy to prove by induction that k actions of $\hat{J}_-$ yield

$$\left| \ell + \frac{1}{2}, \ell + \frac{1}{2} - k \right\rangle = \sqrt{\frac{2\ell + 1 - k}{2\ell + 1}} |\ell, \ell - k\rangle_1 \otimes \left| \frac{1}{2}, \frac{1}{2} \right\rangle_2$$

$$+ \sqrt{\frac{k}{2\ell + 1}} |\ell, \ell - k + 1\rangle_1 \otimes \left| \frac{1}{2}, -\frac{1}{2} \right\rangle_2.$$

Substituting $k = \ell - m$, we find

$$\left| \ell + \frac{1}{2}, m + \frac{1}{2} \right\rangle = \sqrt{\frac{\ell + m + 1}{2\ell + 1}} |\ell, m\rangle_1 \otimes \left| \frac{1}{2}, \frac{1}{2} \right\rangle_2$$

$$+ \sqrt{\frac{\ell - m}{2\ell + 1}} |\ell, m + 1\rangle_1 \otimes \left| \frac{1}{2}, -\frac{1}{2} \right\rangle_2. \tag{11.111}$$

The vectors $|\ell - \frac{1}{2}, m - \frac{1}{2}\rangle$ are of the form $a|\ell, m - 1\rangle_1 \otimes |\frac{1}{2}, \frac{1}{2}\rangle_2 + b|\ell, m\rangle_1 \otimes |\frac{1}{2}, -\frac{1}{2}\rangle_2$, for some constants a and b, and they are also orthogonal to $|\ell + \frac{1}{2}, m - \frac{1}{2}\rangle$. The latter are read from Eq. (11.111) by substituting m with $m - 1$. The orthogonality condition implies that $a\sqrt{\frac{\ell + m}{2\ell + 1}} + b\sqrt{\frac{\ell - m + 1}{2\ell + 1}} = 0$. Hence, after normalization, we find

$$\left| \ell - \frac{1}{2}, m - \frac{1}{2} \right\rangle = \sqrt{\frac{\ell - m + 1}{2\ell + 1}} |\ell, m - 1\rangle_1 \otimes \left| \frac{1}{2}, \frac{1}{2} \right\rangle_2$$

$$- \sqrt{\frac{\ell + m}{2\ell + 1}} |\ell, m\rangle_1 \otimes \left| \frac{1}{2}, -\frac{1}{2} \right\rangle_2. \tag{11.112}$$

Thus, we have provided an explicit relation between the two bases. The Clebsch–Gordan coefficients are

$$C^{\ell + \frac{1}{2}, m + \frac{1}{2}}_{\ell, \frac{1}{2}, m, \frac{1}{2}} = \sqrt{\frac{\ell + m + 1}{2\ell + 1}} \qquad C^{\ell + \frac{1}{2}, m + \frac{1}{2}}_{\ell, \frac{1}{2}, m+1, -\frac{1}{2}} = \sqrt{\frac{\ell - m}{2\ell + 1}}$$

$$C^{\ell - \frac{1}{2}, m - \frac{1}{2}}_{\ell, \frac{1}{2}, m-1, \frac{1}{2}} = \sqrt{\frac{\ell - m + 1}{2\ell + 1}} \qquad C^{\ell - \frac{1}{2}, m - \frac{1}{2}}_{\ell, \frac{1}{2}, m, -\frac{1}{2}} = -\sqrt{\frac{\ell + m}{2\ell + 1}}. \tag{11.113}$$

11.5.5 The Inner Product Operator

In many problems, an analogue of the inner product between angular momentum vectors of two subsystems appears. This inner product operator is defined as

$$\hat{\mathbf{J}}_1 \cdot \hat{\mathbf{J}}_2 := \sum_{i=1}^{3} {}^1\hat{J}_i \otimes {}^2\hat{J}_i. \tag{11.114}$$

We evaluate $\hat{\mathbf{J}}^2$ for the composite system. By Eq. (11.91),

$$\hat{\mathbf{J}}^2 = {}^1\hat{\mathbf{J}}^2 \otimes \hat{I} + \hat{I} \otimes {}^2\hat{\mathbf{J}}^2 + 2\sum_{i=1}^{3} {}^1\hat{J}_i \otimes {}^2\hat{J}_i, \tag{11.115}$$

hence,

$$\hat{\mathbf{J}}_1 \cdot \hat{\mathbf{J}}_2 = \frac{1}{2}\left(\hat{\mathbf{J}}^2 - {}^1\hat{\mathbf{J}}^2 \otimes \hat{I} - \hat{I} \otimes {}^2\hat{\mathbf{J}}^2\right). \tag{11.116}$$

Suppose j_1 and j_2 are fixed. Then, the action of $\hat{\mathbf{J}}_1 \cdot \hat{\mathbf{J}}_2$ on $|J, M\rangle$ yields

$$\hat{\mathbf{J}}_1 \cdot \hat{\mathbf{J}}_2 |J, M\rangle = \frac{1}{2}[J(J+1) - j_1(j_1+1) - j_2(j_2+1)]|J, M\rangle. \tag{11.117}$$

Hence, $\hat{\mathbf{J}}_1 \cdot \hat{\mathbf{J}}_2$ has eigenvalues $\frac{1}{2}[J(J+1) - j_1(j_1+1) - j_2(j_2+1)]$ and eigenvectors $|J, M\rangle$. Since the eigenvalues do not depend on M, each eigenvalue has degeneracy equal to $(2J+1)$.

11.6 Matrix Elements of Tensor Operators

Some operators represent observables that have a simple transformation rule under rotations. For example, scalars in $\mathbb{R}^3$ are invariant under rotations, while vectors transform according to Eq. (11.10). The matrix elements of such operators in the $|j, m\rangle$ basis are highly constrained. In this section, we identify the general form of those matrix elements.

11.6.1 Scalar Operators

Consider a self-adjoint operator $\hat{A}$ that represents a scalar quantity. This means that it is invariant under rotations, $\hat{U}(O)\hat{A}\hat{U}^\dagger(O) = \hat{A}$ for all rotation matrices O. Since a general $\hat{U}(O)$ can be expressed as $e^{i\theta \mathbf{n}\cdot\hat{\mathbf{J}}}$, $\hat{A}$ must commute with all generators $\hat{J}_i$, and also with $\hat{\mathbf{J}}^2$.

The condition $\hat{A}\hat{J}_3 - \hat{J}_3\hat{A} = 0$ implies that $\langle j', m'|\hat{A}\hat{J}_3 - \hat{J}_3\hat{A}|j, m\rangle = 0$; hence, $(m - m')\langle j', m'|\hat{A}|j, m\rangle = 0$. We conclude that $\langle j', m'|\hat{A}|j, m\rangle$ vanishes unless $m = m'$. Similarly, the condition $\hat{A}\hat{\mathbf{J}}^2 - \hat{\mathbf{J}}^2\hat{A} = 0$ implies that $\langle j', m'|\hat{A}|j, m\rangle$ vanishes unless $j = j'$.

The condition $\hat{A}\hat{J}_+ - \hat{J}_+\hat{A} = 0$ implies that $\langle j, m'|\hat{A}\hat{J}_+ - \hat{J}_+\hat{A}|j, m\rangle = 0$, or, equivalently,

$$\sqrt{j(j+1) - m(m+1)}\langle j, m'|\hat{A}|j, m+1\rangle = \sqrt{j(j+1) - m'(m'-1)}\langle j, m'-1|\hat{A}|j, m\rangle.$$

These matrix elements vanish unless $m = m' - 1$. Then, $\langle j, m+1|\hat{A}|j, m+1\rangle = \langle j, m|\hat{A}|j, m\rangle$, that is, $\langle j, m|\hat{A}|j, m\rangle$ does not depend on m.

We conclude that, for any scalar quantity,

$$\langle j', m' | \hat{A} | j, m \rangle = (j | \hat{A} | j) \delta_{jj'} \delta_{mm'}, \tag{11.118}$$

where $(j | \hat{A} | j)$ is the *reduced matrix element* of $\hat{A}$. This is nothing but a function of j, but this more complex notation conforms with our subsequent analysis of general tensor operators.

Equation (11.118) implies that any scalar operator can be expressed as

$$\hat{A} = \sum_j (j | \hat{A} | j) \hat{E}_j, \tag{11.119}$$

where $\hat{E}_j = \sum_{m=-j}^{j} |j, m\rangle \langle j, m|$ is a projector to a subspace of constant j.

Example 11.4 We will prove that the projector $\hat{E}_\ell = \sum_{m=-\ell}^{\ell} |\ell, m\rangle \langle \ell, m|$ has matrix elements

$$\langle \mathbf{n}' | \hat{E}_\ell | \mathbf{n} \rangle = \frac{2\ell + 1}{4\pi} P_\ell(\mathbf{n} \cdot \mathbf{n}'), \tag{11.120}$$

where P_ℓ are the Legendre polynomials. To prove Eq. (11.120), we start from an expansion of the function $P_\ell(\mathbf{n} \cdot \mathbf{n}')$ in spherical harmonics

$$P_\ell(\mathbf{n} \cdot \mathbf{n}') = \sum_{m, m'} a_{mm'} Y_{\ell m}(\mathbf{n}) Y_{\ell m'}^*(\mathbf{n}').$$

The right-hand side depends on ϕ through the combination

$$e^{im\phi - im'\phi'} = e^{\frac{i}{2}(m+m')(\phi - \phi')} e^{\frac{i}{2}(m-m')(\phi + \phi')}.$$

The inner product $\mathbf{n} \cdot \mathbf{n}' = \cos\theta \cos\theta' + \sin\theta \sin\theta' \cos(\phi - \phi')$ does not depend on $\phi + \phi'$. This is possible only if all terms with $m \neq m'$ vanish. Hence,

$$P_\ell(\mathbf{n} \cdot \mathbf{n}') = \sum_{m=-\ell}^{\ell} a_m Y_{\ell m}(\mathbf{n}) Y_{\ell m}^*(\mathbf{n}'), \tag{11.121}$$

for some constants a_m. These constants must be real, because complex conjugation is equivalent to the exchange of $\mathbf{n}$ with $\mathbf{n}'$ which leaves $\mathbf{n} \cdot \mathbf{n}'$ invariant. To find a_m, we first set $\mathbf{n} = \mathbf{n}'$ in Eq. (11.121) and we integrate over $\mathbf{n}$. Since the spherical harmonics are normalized to unity,

$$\sum_{m=-\ell}^{\ell} a_m = 4\pi P_\ell(1) = 4\pi. \tag{11.122}$$

Next, we take the square of Eq. (11.121),

$$P_\ell(\mathbf{n} \cdot \mathbf{n}')^2 = \sum_{m=-\ell}^{\ell} \sum_{m'=-\ell}^{\ell} a_m a_{m'} Y_{\ell m}(\mathbf{n}) Y_{\ell m'}^*(\mathbf{n}) Y_{\ell m}^*(\mathbf{n}') Y_{\ell m'}(\mathbf{n}'),$$

and we integrate over $\mathbf{n}$. We can always choose coordinates so that $\mathbf{n}' = (0, 0, 1)$, and $P_\ell(\mathbf{n} \cdot \mathbf{n}')^2 = P_\ell(\cos\theta)$. Then,

$$\int_0^\pi \sin\theta \, d\theta \int_0^{2\pi} d\phi P_\ell(\cos\theta)^2 = 2\pi \int_{-1}^1 d\xi P_\ell(\xi)^2 = \frac{4\pi}{2\ell + 1},$$

where $\xi = \cos\theta$ and we used Eq. (3.13). Hence, we obtain $\sum_{m=-\ell}^{\ell} a_m^2 |Y_{\ell m}(\mathbf{n}')|^2 = \frac{4\pi}{2\ell+1}$. Further integration over $\mathbf{n}'$ yields

$$\sum_{m=-\ell}^{\ell} a_m^2 = \frac{16\pi^2}{2\ell+1}. \tag{11.123}$$

Equation (11.122) implies that the mean $\langle a \rangle$ of the $2\ell + 1$ numbers a_m is $4\pi/(2\ell+1)$. Equation (11.123) implies that the mean square $\langle a^2 \rangle = \langle a \rangle^2$. This is possible only if all a_m are equal, and by Eq. (11.123), $a_m = 4\pi/(2\ell+1)$. Equation (11.120) then follows.

11.6.2 Vector Operators

Consider now a vector operator $\hat{\mathbf{X}}$, namely, a set of three operators $\hat{X}_i$, $i = 1, 2, 3$ that transforms under rotations as

$$\hat{U}(O)\hat{X}_i\hat{U}^\dagger(O) = \sum_j O_{ij}\hat{X}_j. \tag{11.124}$$

For an infinitesimal rotation $O = 1 + \theta J_{\mathbf{n}}$, $\hat{U}(O) = 1 + i\theta\mathbf{n} \cdot \hat{\mathbf{J}}$, and we obtain

$$[\hat{X}_i, \hat{J}_j] = i\sum_k (J_j)_{jk}\hat{X}_k. \tag{11.125}$$

By Eq. (11.13),

$$[\hat{X}_i, \hat{J}_j] = i\sum_k \epsilon_{ijk}\hat{X}_k. \tag{11.126}$$

We introduce the operators $\hat{X}_\pm = \mp\frac{1}{\sqrt{2}}(\hat{X}_1 \pm \hat{X}_2)$. Then, Eq. (11.126) becomes

$$[\hat{X}_+, \hat{J}_+] = 0, \qquad [\hat{X}_+, \hat{J}_-] = -\sqrt{2}\hat{X}_3, \qquad [\hat{X}_+, \hat{J}_3] = -\hat{X}_+,$$
$$[\hat{X}_-, \hat{J}_+] = -\sqrt{2}\hat{X}_3, \quad [\hat{X}_-, \hat{J}_-] = 0, \qquad [\hat{X}_-, \hat{J}_3] = \hat{X}_3,$$
$$[\hat{X}_3, \hat{J}_+] = \sqrt{2}\hat{X}_+, \quad [\hat{X}_3, \hat{J}_-] = -\sqrt{2}\hat{X}_-, \quad [\hat{X}_3, \hat{J}_3] = 0.$$

These equations can be simplified by introducing the index $q = -1, 0, 1$, and defining operators $\hat{X}_q$ as: $\hat{X}_{\pm 1} := \hat{X}_\pm$ and $\hat{X}_0 := \hat{X}_3$. Then,

$$[\hat{J}_\pm, \hat{X}_q] = \sqrt{2 - q(q \pm 1)}\hat{X}_{q\pm 1}, \tag{11.127}$$
$$[\hat{J}_3, \hat{X}_q] = q\hat{X}_q. \tag{11.128}$$

We will use these equations in order to construct the matrix elements of the operators $\hat{X}_q$ with respect to the eigenvectors $|j, m\rangle$ of angular momentum. If the operators from Eqs. (11.127) and (11.128) act on kets $|j, m\rangle$, we obtain

$$\hat{J}_\pm\left(\hat{X}_q|j,m\rangle\right) = \sqrt{j(j+1) - m(m \pm 1)}\hat{X}_q|j, m \pm 1\rangle + \sqrt{2 - q(q \pm 1)}\hat{X}_{q\pm 1}|j,m\rangle,$$
$$\hat{J}_3\left(\hat{X}_q|j,m\rangle\right) = (m + q)\hat{X}_q|j,m\rangle. \tag{11.129}$$

Comparing with Eqs. (11.93) and (11.94), we see that $\hat{X}_q|j,m\rangle$ behaves like a ket $|1, q\rangle \otimes |j, m\rangle$, that is, it corresponds to the addition of an angular momentum $\underline{3}$ for $\hat{X}_q$ and an angular momentum

$\underline{2j+1}$ for $|j,m\rangle$. By the triangle inequality, the possible values of J are $j-1$, j, and $j+1$, while $M = m + q$.

We analyze $\hat{X}_q|j,m\rangle$ following the algorithm of Section 11.5.4. We start with the extremal vector $\hat{X}_{+1}|j,j\rangle$, which must be of the form $\alpha_{+1}(j)|J,J\rangle$ for some constant $\alpha_{+1}(j)$ and for $J = j+1$. Then, we act with the lowering operator on $\hat{X}_{+1}|j,j\rangle$ to obtain all vectors with $J = j + 1$ – all these vectors are proportional to $\alpha_{+1}(j)$.

Then, we construct the vectors with $J = j$, for which we have to introduce a new proportionality constant $\alpha_0(j)$. Finally, we construct the vectors with $J = j - 1$, with the proportionality constant $\alpha_{-1}(j)$. At the end of this procedure, we will write

$$\alpha_{J-j}(j)|J,M\rangle = \sum_{q,m} C^{J,M}_{1,j,q,m}\hat{X}_q|j,m\rangle. \tag{11.130}$$

Then,

$$\hat{X}_q|j,m\rangle = \sum_{J,M} C^{J,M}_{1,j,q,m}\alpha_{J-j}(j)|J,M\rangle. \tag{11.131}$$

Hence, we obtain the matrix elements

$$\langle j',m'|\hat{X}_q|j,m\rangle = (j'|\hat{X}|j)C^{j',m'}_{1,j,q,m}, \tag{11.132}$$

where we wrote $\alpha_{j'-j}(j)$ as a reduced matrix element $(j'|\hat{X}|j)$. Equation (11.132) is a special case of the *Wigner–Eckart theorem*, to be presented shortly. We will encounter several cases where Eq. (11.132) greatly simplifies calculations in later chapters. A key point to remember is that the matrix elements of vector operators vanish if $|j - j'| > 1$.

Equation (11.132) also holds when the eigenvalues of $\hat{\mathbf{J}}^2$ and $\hat{J}_3$ are degenerate. In this case, we consider an orthonormal basis of the form $|\nu,j,m\rangle$, for some label ν of the degeneracy subspaces. Then, Eq. (11.132) becomes

$$\langle \nu',j',m'|\hat{X}_q|\nu,j,m\rangle = (\nu',j'|\hat{X}|\nu,j)C^{j',m'}_{1,j,q,m}, \tag{11.133}$$

that is, the only difference is that the reduced matrix elements also depend on the degeneracy indices.

For two vector operators $\hat{X}_i$ and $\hat{Y}_i$, Eq. (11.132) implies that

$$\langle j',m'|\hat{X}_q|j,m\rangle = A(j,j')\langle j',m'|\hat{Y}_q|j,m\rangle, \tag{11.134}$$

where A depends only on j and j'. Equation (11.134) applies only if $(j'|\hat{Y}|j)$ is nonzero for $j \neq j'$. For example, it does not apply if $\hat{Y}_i$ is identified with the angular momentum $\hat{J}_i$, because the matrix elements of $\hat{J}_i$ vanish for $j \neq j'$.

The components $\hat{X}_i$ and $\hat{Y}_i$ are linear combinations of $\hat{X}_q$ and $\hat{Y}_q$, respectively. Hence, we can express Eq. (11.134) as

$$\langle j',m'|\hat{X}_i|j,m\rangle = A(j,j')\langle j',m'|\hat{Y}_i|j,m\rangle.$$

Consider the special case $\hat{Y}_i = \hat{J}_i$ and $j = j'$. We evaluate the matrix element

$$\langle j,m'|\hat{\mathbf{X}} \cdot \hat{\mathbf{J}}|j,m\rangle = \sum_{i=1}^{3} \sum_{m_1} \langle j,m'|\hat{X}_i|j,m_1\rangle \langle j,m_1|\hat{J}_i|j,m\rangle$$

$$= A(j,j) \sum_{i=1}^{3} \sum_{m_1} \langle j,m'|\hat{J}_i|j,m_1\rangle \langle j,m_1|\hat{J}_i|j,m\rangle$$

$$= A(j,j)\langle j,m'|\hat{\mathbf{J}}^2|j,m\rangle = A(j,j)j(j+1)\delta_{mm'}. \qquad (11.135)$$

The inner product $\hat{\mathbf{X}} \cdot \hat{\mathbf{J}}$ is a scalar. Hence, by Eq. (11.118), $\langle j,m'|\hat{\mathbf{X}} \cdot \hat{\mathbf{J}}|j,m\rangle = \langle j|\hat{\mathbf{X}} \cdot \hat{\mathbf{J}}|j\rangle\delta_{mm'}$. Then, Eq. (11.135) implies that

$$A(j,j) = \frac{\langle j|\hat{\mathbf{X}} \cdot \hat{\mathbf{J}}|j\rangle}{j(j+1)}.$$

Thus, we obtained the *projection theorem*,

$$\langle j,m'|\hat{X}_i|j,m\rangle = \frac{\langle j|\hat{\mathbf{X}} \cdot \hat{\mathbf{J}}|j\rangle}{j(j+1)} \langle j,m'|\hat{J}_i|j,m\rangle. \qquad (11.136)$$

The matrix elements of any vector operator in a subspace of constant j are proportional to the matrix elements of the angular momentum operator.

11.6.3 Spherical Tensor Operators

The transformation of a vector operator $\hat{X}_i$ to $\hat{X}_q$ can be written as

$$\begin{pmatrix} \hat{X}_+ \\ \hat{X}_0 \\ \hat{X}_- \end{pmatrix} = S \begin{pmatrix} \hat{X}_1 \\ \hat{X}_2 \\ \hat{X}_3 \end{pmatrix}, \quad \text{where } S = \begin{pmatrix} -\frac{1}{\sqrt{2}} & -\frac{i}{\sqrt{2}} & 0 \\ 0 & 0 & 1 \\ \frac{1}{\sqrt{2}} & -\frac{i}{\sqrt{2}} & 0 \end{pmatrix}.$$

It is straightforward to show that the matrix S is unitary. It is also straightforward to show that the generators $J_{\mathbf{n}}$ of Eq. (11.17) transform under S as

$$SJ_{\mathbf{n}}S^{-1} = \frac{1}{i}\mathbf{n} \cdot \hat{\mathbf{J}}, \qquad (11.137)$$

where here $\hat{J}_i$ are the angular momentum operators (11.54) for $j = 1$. This means that $e^{\theta J_{\mathbf{n}}} = S^\dagger e^{-i\theta\mathbf{n}\cdot\hat{\mathbf{J}}} S$. Setting $O = e^{\theta J_{\mathbf{n}}}$, we obtain $O = S^\dagger \hat{U}(O)\hat{S}$, and Eq. (11.124) becomes

$$\hat{U}(O)\hat{X}_q\hat{U}^\dagger(O) = \sum_{q'} \mathcal{U}_{qq'}^{(1)}(O)\hat{X}_{q'}, \qquad (11.138)$$

where $\mathcal{U}_{qq'}(O)$ are the matrix elements of $\hat{U}(O)$ for $j = 1$, Eq. (11.55). Equation (11.138) explains the mathematical correspondence between vector operators and angular momentum with $j = 1$.

Furthermore, Eq. (11.138) suggests a generalization of the notion of a vector operator.

Definition 11.2 For any integer k, a set of $2k+1$ operators $\hat{X}_q$, where $q = -k, -k+1, \ldots, k-1, k$ defines a *spherical k-tensor operator* (or a tensor operator of *rank k*), if it transforms under rotations as

$$\hat{U}(O)\hat{X}_q\hat{U}^\dagger(O) = \sum_{q'} \mathcal{U}^{(k)}_{qq'}(O)\hat{X}_{q'}, \tag{11.139}$$

where $\mathcal{U}_{qq'}(O)$ are the matrix elements of $\hat{U}(O)$ for $j = k$, Eq. (11.55).

Scalar operators are spherical 0-tensor operators and vector operators are spherical 1-tensor operators.

By Eq. (11.139), we obtain the commutation relations

$$[\hat{J}_\pm, \hat{X}_q] = \sqrt{k(k+1) - q(q \pm 1)}\hat{X}_{q\pm1}, \tag{11.140}$$

$$[\hat{J}_3, \hat{X}_q] = q\hat{X}_q. \tag{11.141}$$

Then, it is straightforward to prove the generalization of Eq. (11.132) for spherical k-tensors – see Problem 11.16.

Theorem 11.3 (Wigner–Eckart theorem) *The matrix elements of a spherical k-tensor operator $\hat{X}_q$ in the angular momentum basis are*

$$\langle j', m' | \hat{X}_q | j, m \rangle = (j' \| \hat{X} \| j) C^{j', m'}_{k, j, q, m}, \tag{11.142}$$

where $C^{j', m'}_{k, j, q, m}$ are Clebsch–Gordan coefficients, and the reduced matrix element $(j' \| \hat{X} \| j)$ is a function only of j and j'.

The Wigner–Eckart theorem implies that all matrix elements $\langle j', m' | \hat{X}_q | j, m \rangle$ vanish if $|j - j'| > k$ or if $m' \neq m + q$. Furthermore, the matrix elements of two different spherical k-tensors with $j = j'$ differ only by a multiplicative constant.

Spherical tensors are *different* from the more familiar *Cartesian tensors*. A Cartesian n-tensor is an element of $\otimes^n_{i=1} \mathbb{R}^3$, and it is represented by $3n$ numbers $T_{a_1 a_2 \dots a_n}$, $a_i = 1, 2, 3$, $i = 1, 2, \dots, n$, that transforms under rotations as

$$T_{a_1 a_2 \cdots a_n} \to \sum_{b_1, b_2, \dots, b_n} O_{a_1 b_1} O_{a_2 b_2} \cdots O_{a_n b_n} T_{b_1 b_2 \cdots b_n}, \tag{11.143}$$

where O_{ab} is an orthogonal matrix. Similarly, a Cartesian n-tensor operator is a set of $3n$ self-adjoint operators $\hat{T}_{a_1 a_2 \cdots a_n}$ that transform as

$$\hat{U}(O)\hat{T}_{a_1 a_2 \cdots a_n}\hat{U}^\dagger(O) = \sum_{b_1, b_2, \dots, b_n} O_{a_1 b_1} O_{a_2 b_2} \cdots O_{a_n b_n} \hat{T}_{b_1 b_2 \cdots b_n} \tag{11.144}$$

for all orthogonal matrices O. Scalar operators and vector operators are Cartesian 0-tensor operators and 1-tensor operators respectively, but, in general, higher-order Cartesian tensor operators do not coincide with higher-order spherical tensor operators. The elements of a Cartesian n-tensor operator typically define several spherical k-tensor operators with $k < 3n$, since a Cartesian n-tensor corresponds to the addition of n angular momenta with $j = 1$.

Example 11.5 Consider two vector operators $\hat{Y}_i$ and $\hat{Z}_i$, $i = 1, 2, 3$, from which we construct the Cartesian 2-tensor $\hat{T}_{ij} := \frac{1}{2}(\hat{Y}_i\hat{Z}_j + \hat{Z}_i\hat{Y}_j)$. We switch to the q-basis and write $\hat{T}_{qq'} := \frac{1}{2}(\hat{Y}_q\hat{Z}_{q'} + \hat{Z}_q\hat{Y}_{q'})$. Using Eqs. (11.127) and (11.128), we obtain

$$[\hat{J}_\pm, \hat{T}_{qq'}] = \sqrt{2 - q(q \pm 1)}T_{q\pm1, q'} + \sqrt{2 - q'(q' \pm 1)}T_{q, q'\pm1}, \tag{11.145}$$

$$[\hat{J}_3, \hat{T}_{qq'}] = (q + q')\hat{T}_{qq'}. \tag{11.146}$$

The maximum value of $q + q'$ is $+2$, and it is obtained only for $q = q' = +$. This means that $\hat{T}_{++}$ is the $q = 2$ element of a spherical 2-tensor $\hat{T}_q^{(2)}$, that is, $\hat{T}_{+2}^{(2)} = \hat{T}_{++}$. Equation (11.140) implies that $\left[\hat{J}_-, \hat{T}_{+2}^{(2)}\right] = 2\hat{T}_{+1}^{(2)}$, while Eq. (11.145) implies that $\left[\hat{J}_-, \hat{T}_{+2}^{(2)}\right] = \sqrt{2}(\hat{T}_{0+} + \hat{T}_{+0})$. Hence,

$$\hat{T}_{+1}^{(2)} = \frac{1}{\sqrt{2}}\left(\hat{T}_{0+} + \hat{T}_{+0}\right). \tag{11.147}$$

The reader must have realized that the procedure is similar to the algorithm of Section 11.5.4. Equation (11.140) implies that $\left[\hat{J}_-, \hat{T}_{+1}^{(2)}\right] = \sqrt{6}\hat{T}_0^{(2)}$, while Eqs. (11.145) and (11.147) imply that $\left[\hat{J}_-, \hat{T}_{+1}^{(2)}\right] = \hat{T}_{-+} + 2\hat{T}_{00} + \hat{T}_{+-}$. Hence,

$$\hat{T}_0^{(2)} = \frac{1}{\sqrt{6}}(\hat{T}_{-+} + 2\hat{T}_{00} + \hat{T}_{+-}). \tag{11.148}$$

The elements $\hat{T}_{-1}^{(2)}$ and $\hat{T}_{-2}^{(2)}$ are straightforwardly computed from $\hat{T}_{+1}^{(2)}$ and $\hat{T}_{+2}^{(2)}$ by the exchange $\pm \to \mp$.

The element $T_{+1}^{(1)}$ of a spherical 1-tensor $T^{(1)}$ is the unique (modulo a multiplicative factor) linear combination of $\hat{T}_{+0}$ and $\hat{T}_{0+}$ that satisfies $[\hat{J}_+, \hat{T}_{+1}^{(1)}] = 0$. We find straightforwardly,

$$\hat{T}_{+1}^{(1)} = \hat{T}_{0+} - \hat{T}_{+0}, \tag{11.149}$$

where the multiplicative factor is chosen unity for convenience. By Eq. (11.140), $\left[\hat{J}_-, \hat{T}_{+1}^{(1)}\right] = \sqrt{2}\hat{T}_0^{(1)}$, while Eqs. (11.145) and (11.149) imply that $\left[\hat{J}_-, \hat{T}_{+1}^{(1)}\right] = \hat{T}_{-+} - \hat{T}_{+-}$. Hence, $\hat{T}_0^{(1)} = \frac{1}{\sqrt{2}}(\hat{T}_{-+} - \hat{T}_{+-})$. Similarly, we find that $T_{-1}^{(1)} = \frac{1}{2}(\hat{T}_{-0} - \hat{T}_{0-})$.

A scalar tensor $T^{(0)}$ is the unique linear combination of $\hat{T}_{+-}, \hat{T}_{-+}$ and $\hat{T}_{00}$ that satisfies $[\hat{J}_\pm, T^{(0)}] = 0$. We find that

$$\hat{T}^{(0)} = \hat{T}_{00} - \hat{T}_{+-} - \hat{T}_{-+}, \tag{11.150}$$

modulo an overall multiplicative factor.

Switching back to Cartesian coordinates, we obtain

$$\hat{T}^{(0)} = \hat{T}_{11} + \hat{T}_{22} + \hat{T}_{33}, \qquad \hat{T}^{(1)} = \begin{pmatrix} \hat{T}_{13} - \hat{T}_{31} + i(\hat{T}_{23} - \hat{T}_{32}) \\ i(\hat{T}_{21} - \hat{T}_{12}) \\ \hat{T}_{13} - \hat{T}_{31} + i(\hat{T}_{23} - \hat{T}_{32}) \end{pmatrix}$$

$$\hat{T}^{(2)} = \begin{pmatrix} \frac{1}{2}(\hat{T}_{11} - \hat{T}_{22}) + \frac{i}{2}(\hat{T}_{12} + \hat{T}_{21}) \\ -\frac{1}{2}(\hat{T}_{13} + \hat{T}_{31}) + \frac{i}{2}(\hat{T}_{32} + \hat{T}_{23}) \\ \frac{1}{\sqrt{6}}(2\hat{T}_{33} - \hat{T}_{11} - \hat{T}_{22}) \\ \frac{1}{2}(\hat{T}_{13} + \hat{T}_{31}) - \frac{i}{2}(\hat{T}_{32} + \hat{T}_{23}) \\ \frac{1}{2}(\hat{T}_{11} - \hat{T}_{22}) - \frac{i}{2}(\hat{T}_{12} + \hat{T}_{21}) \end{pmatrix}. \tag{11.151}$$

We see that $\hat{T}^{(0)}$ is the trace of $\hat{T}_{ij}$, $\hat{T}^{(1)}$ contains the antisymmetric elements of $\hat{T}_{ij}$, and $\hat{T}^{(2)}$ the remaining elements of $\hat{T}_{ij}$.

QUESTIONS

11.1 How do the transmission and reflection amplitudes of a particle in a potential $V(x)$ transform under the time reversal and space inversion operators?

11.2 Suppose that in some system we observe a superposition of integer and half-integer j. What would this imply?

11.3 Show that the attempt to construct spherical harmonics with $\ell = \frac{1}{2}$ through the procedure of Section 11.4.2 leads to a contradiction.

11.4 (**B**) Specify the type of the spherical tensors that are defined by the 27 elements of a Cartesian 3-tensor.

11.5 (**B**) The electric quadrupole moment operator of a nucleus is defined as $\hat{Q}_{ij} = e \sum_a (3\hat{x}_{ai}\hat{x}_{aj} - \sum_a \hat{r}_a^2 \delta_{ij})$, where $\hat{x}_{ai}$ is the ith position coordinate of the ath proton. Let $|\psi\rangle$ be an eigenstate of the total angular momentum of the nucleus. What is the most general form of the expectation value $\langle\psi|\hat{Q}_{ij}|\psi\rangle$ according to the Wigner–Eckart theorem?

PROBLEMS

11.1 Let the unitary operators $\hat{U}_1$ and $\hat{U}_2$ in $\mathbb{C}^2$ correspond to the 3×3 orthogonal matrices Λ_1 and Λ_2 on $\mathbb{R}^3$ through Eq. (11.23). Show that $\hat{U}_1\hat{U}_2$ corresponds to $\Lambda_1\Lambda_2$. (Hint: Use Eq. (5.11).)

11.2 Find the eigenvalues and the eigenvectors for the following operators on $L^2(\mathbb{R})$: (i) the time reversal operator $\hat{\mathbb{T}}$, (ii) the space inversion operator $\hat{P}$, and (iii) their product $\hat{P}\hat{\mathbb{T}}$.

11.3 We define the time reversal operator for a qubit by Eq. (11.25). (i) Show that $\hat{\mathbb{T}}^\dagger = -\hat{\mathbb{T}}$ and that $\hat{\mathbb{T}}^2 = -\hat{I}$. (ii) What transformation on the Bloch sphere corresponds to the action of $\hat{\mathbb{T}}$? (iii) Show that $\hat{\mathbb{T}}^\dagger \hat{\sigma}_i \hat{\mathbb{T}} = -\hat{\sigma}_i$.

11.4 Show that an operator that commutes with two components of the angular momentum $\hat{J}_i$ also commutes with the third.

11.5 Show that $\text{Tr}\hat{J}_i = 0$ and that $\text{Tr}(\hat{J}_i\hat{J}_j) = \frac{1}{3}j(j+1)(2j+1)\delta_{ij}$.

11.6 Compute all spherical harmonics with $\ell \leq 2$ from Eq. (11.87).

11.7 The quantum rigid rotator is described by the Hamiltonian

$$\hat{H} = \sum_{i=1}^{3} \frac{1}{2I_i} \hat{\ell}_i^2,$$

where $\hat{\ell}_i$ are the operators of orbital angular momentum, and I_i the moments of inertia along the three axes. (i) Find the eigenvalues and eigenvectors of $\hat{H}$ for a symmetric rotator with $I_1 = I_2$. (ii) Find the eigenvalues and eigenvectors of a generic rotator with $\ell = 1$.

11.8 Evaluate the matrix $\mathcal{V}^{(j)}(\theta)$ of Eq. (11.57) for $j = 1$.

11.9 Show that (i) $\underline{2} \otimes \underline{2} \otimes \underline{2} = \underline{2} \oplus \underline{2} \oplus \underline{4}$, (ii) $\underline{2} \otimes \underline{2} \otimes \underline{3} = \underline{1} \oplus \underline{3} \oplus \underline{3} \oplus \underline{5}$, and (iii) $\underline{2} \otimes \underline{2} \otimes \underline{2} \otimes \underline{2} = \underline{1} \oplus \underline{1} \oplus \underline{3} \oplus \underline{3} \oplus \underline{3} \oplus \underline{5}$.

11.10 Evaluate the Clebsch–Gordan coefficients for the addition of the angular momenta $j_1 = j_2 = 1$.

11.11 A composite system is found on a state with $j_1 = 7, m_1 = -3$ and $j_2 = \frac{1}{2}, m_2 = \frac{1}{2}$. Find the possible values of the total angular momentum and the associated probabilities.

11.12 Consider the addition of two angular momenta $j_1 = j_2 = j$. Show that the totally antisymmetric vector

$$|\Psi\rangle = \frac{1}{\sqrt{2j+1}} \sum_{m=-j}^{j} (-1)^{j-m} |j,m\rangle \otimes |j,-m\rangle$$

corresponds to total angular momentum $J = 0$.

11.13 Construct the eigenvectors $|J,M\rangle$ for the total angular momentum of a system composed of three subsystems with $j_1 = j_2 = j_3 = \frac{1}{2}$.

11.14 Consider a system of three angular momenta $\hat{\mathbf{J}}_1, \hat{\mathbf{J}}_2,$ and $\hat{\mathbf{J}}_3$ with quantum numbers $j_1, j_2,$ and j_3, respectively. (i) Express the operator $\hat{A} = \hat{\mathbf{J}}_1 \cdot \hat{\mathbf{J}}_2 + \hat{\mathbf{J}}_2 \cdot \hat{\mathbf{J}}_3 + \hat{\mathbf{J}}_3 \cdot \hat{\mathbf{J}}_1$ in terms of squares of spin operators. (ii) Find the eigenvalues of $\hat{A}$ and their degeneracy. (iii) Write the eigenvector of $\hat{A}$ with the minimum eigenvalue for $j_1 = j_2 = j_3 = 1$.

11.15 (**B**) Prove Eqs. (11.140) and (11.141). Start from Eq. (11.139) and consider rotations very close to unity, $O = I + \theta J_{\mathbf{n}}$ with $\theta << 1$.

11.16 (**B**) Prove the Wigner–Eckart theorem for spherical k-tensors, starting from Eqs. (11.140) and (11.141).

11.17 (**B**) For any operator $\hat{A}$ define $\hat{\mathbf{J}}^2[\hat{A}] := \sum_{i=1}^{3} [\hat{J}_i, [\hat{J}_i, \hat{A}]]$. Show that for a spherical k-tensor $\hat{X}_q$, $\hat{\mathbf{J}}^2[\hat{X}_q] = k(k+1)\hat{X}_q$.

11.18 (**B**) Let $\hat{x}_i$ be the position operator, and $|v, \ell, m\rangle$ the eigenstates of orbital angular momentum, where we included a degeneracy parameter v. (i) Identify the matrix elements $\langle v, \ell, m | \hat{x}_i | v', \ell', m' \rangle$ that do not vanish identically. (ii) Identify all matrix elements $\langle v, \ell, m | \hat{x}_i \hat{x}_j | v', \ell', m' \rangle$ that do not vanish identically.

Bibliography

- For more details about angular momentum in quantum theory, including tables of Clebsch–Gordan coefficients and related quantities, see the classic monograph of Edmonds (1996).
- For other elementary proofs of Wigner's theorem, see Simon et al. (2007). For time reversal, see chapter 26 of Wigner (1959).

12 Symmetries II

Group Theory in Quantum Mechanics

> We need a super-mathematics in which the operations are as unknown as the quantities they operate on, and a super-mathematician who does not know what he is doing when he performs these operations. Such a super-mathematics is the Theory of Groups.
>
> *Eddington (1934)*

12.1 Group Representations

In Chapter 11, we saw that by Wigner's theorem, symmetry transformations in quantum theory are represented by unitary or antiunitary operators. Then, we focused exclusively in the symmetry of space rotations. Since our focus was so narrow, we did not have to introduce the most appropriate language for the description of symmetries, namely group theory.

Some elements of group theory are crucial for understanding the role of symmetries in quantum theory. Appendix C presents a self-contained introduction to group theory, with an emphasis on the theory of Lie groups and Lie algebras. The material presented there is sufficient background for the use of group theory in this book. Readers with no background in group theory should go through Appendix C in full detail. Readers familiar with group theory may find it useful to skim through Appendix C before proceeding with this chapter.

Regarding notation, we will write the elements of a general group G as g, h and so on. The group operation between elements g_1 and g_2 will be denoted as $g_1 g_2$, and the unit element of the group as e.

We will treat all Lie groups that appear in this book as matrix groups, that is, each element of a Lie group G corresponds to some $n \times n$ matrix A, where n is the same for all group elements. The group multiplication $A_1 A_2$ coincides with matrix multiplication, and the unit element of the group is the unit matrix I. When we use calculus on a Lie group G, we will treat G as a subset of either $\mathbb{R}^{n^2}$ or $\mathbb{C}^{n^2}$. Not all Lie groups are matrix groups, but this identification suffices for present purposes.

Elements of the Lie algebra $\mathfrak{g}$ of a Lie group G also correspond to $n \times n$ matrices, and they will be represented by letters T, R, S, and so on. Each $T \in \mathfrak{g}$ defines a one-parameter subgroup of G that consists of elements $A(s) = e^{Ts}$, for all $s \in \mathbb{R}$.

12.1.1 Unitary Representations

Groups are first defined *abstractly*, as mathematical entities characterized by specific algebraic relations. An abstract group appears *concretely* in a physical system, if its algebraic structure is mirrored by the mathematical objects that implement transformations in the theory. In quantum theory, the latter are unitary (and antiunitary) operators. Thus, we are led to the notions of unitary representations of groups that are central to the description of symmetries in quantum theory.

Definition 12.1 A unitary representation of a group G on a Hilbert space $\mathcal{H}$ is a map that assigns to each $g \in G$ a unitary $\hat{U}(g)$, such that for any $g_1, g_2 \in G$,

$$\hat{U}(g_1)\hat{U}(g_2) = \hat{U}(g_1 g_2). \tag{12.1}$$

If G is a Lie group, then we also require that the map $g \to \hat{U}(g)$ is differentiable.

It is straightforward to show that in a unitary representation $\hat{U}(g^{-1}) = \hat{U}^{\dagger}(g)$, and that $\hat{U}(e) = \hat{I}$, where e is the unit element of G. Note that every group has a *trivial representation* on any Hilbert space, in which $\hat{U}(g) = \hat{I}$ for all $g \in G$.

By Wigner's theorem, we would expect antiunitary representations also to be physically meaningful. A key difference is that only discrete groups have antiunitary representations. To see this, consider a Lie group G that consists of $n \times n$ matrices. G contains elements of the form e^{sT}, where T is an element of the group's Lie algebra $\mathfrak{g}$ and $s \in \mathbb{R}$. Obviously, $\hat{U}(e^{2sT}) = \hat{U}(e^{sT})\hat{U}(e^{sT})$, hence, if the operator $\hat{U}(e^{sT})$ is antiunitary, the operator $\hat{U}(e^{2sT})$ is unitary. Hence, there is no $s \neq 0$, such that $\hat{U}(s)$ is antiunitary.

In this chapter, we will consider only group representations via unitary operators. We will encounter antiunitary representations in our next discussion of time reversal, in Chapter 14.

Let G be a Lie group of dimension D, and $\mathfrak{g}$ its associated Lie algebra. Consider a unitary representation $\hat{U}(A)$ of G on a Hilbert space $\mathcal{H}$. For each $T \in \mathfrak{g}$, we define a self-adjoint operator

$$\hat{T} := i\frac{\partial}{\partial s}\hat{U}\left[\exp(sT)\right]_{s=0}, \tag{12.2}$$

on $\mathcal{H}$. We call $\hat{T}$ a *generator of the representation* of G in $\mathcal{H}$. Compare this name "generators of the group G" for the elements of the Lie algebra $\mathfrak{g}$. The similarity of the names is due to the fact that the operator $\frac{1}{i}\hat{T}$ represents the element T on the Hilbert space $\mathcal{H}$.

An immediate consequence of the definition in (12.2) is the following identity,

$$e^{-is\hat{T}} = \hat{U}\left(e^{sT}\right). \tag{12.3}$$

To prove Eq. (12.3), we first define $\hat{U}(s) := \hat{U}\left[\exp(sT)\right]$, and then we evaluate the ratio $\frac{1}{\delta s}\left[\hat{U}(s+\delta s) - \hat{U}(s)\right] = \frac{1}{\delta s}\left[\hat{U}(s+\delta s)\hat{U}^{-1}(s) - \hat{I}\right]\hat{U}(s)\hat{I} = \frac{1}{\delta s}\left[\hat{U}(e^{\delta s T}) - \hat{I}\right]\hat{U}(s)$. Taking the limit $\delta s \to 0$, and using Eq. (12.2), we find that $i\frac{\partial}{\partial s}\hat{U}(s) = \hat{T}\hat{U}(s)$. Then, Eq. (12.3) follows.

Equation (12.3) allows us to define the action of $\hat{U}(A)$ on the generator $\hat{T}$. By definition, $\hat{U}(A)e^{-is\hat{T}}\hat{U}^{\dagger}(A) = \hat{U}(Ae^{sT}A^{-1}) = \hat{U}(e^{sAd_A(T)})$, where $Ad_A(T) = ATA^{-1}$. Differentiating both sides with respect to s and taking $s = 0$, we obtain

$$\hat{U}(A)\hat{T}\hat{U}^{\dagger}(A) = \widehat{Ad_A(T)}, \tag{12.4}$$

where $\widehat{Ad_A(T)}$ stands for the self-adjoint operator that represents $Ad_A(T) \in \mathfrak{g}$ – see Eq. (C.8).

Next, we substitute e^{sS} for A in Eq. (12.4), where $s \in \mathbb{R}$ and $S \in \mathfrak{g}$. By Eq. (12.3), $e^{-is\hat{S}}\hat{T}e^{is\hat{S}} = \widehat{Ad_{e^{sS}}(T)}$. Then, we differentiate with respect to s at $s = 0$, to obtain

$$\left[\hat{S}, \hat{T}\right] = i\widehat{[S, T]}, \tag{12.5}$$

where $\widehat{[S, T]}$ is the self-adjoint operator that represents the element $[S, T] \in \mathfrak{g}$. Hence, a unitary representation on a Hilbert space $\mathcal{H}$ maps the commutator of the Lie algebra $\mathfrak{g}$ to the commutator of operators on $\mathcal{H}$. Equation (12.5) justifies the use of the commutator (11.26) for analyzing spatial rotations in quantum theory.

Suppose we choose a basis T_a in the Lie algebra $\mathfrak{g}$, where $a = 1, \ldots, D$. The elements of the basis satisfy the commutation relations $[T_a, T_b] = \sum_{i=1}^{D} f_{abc} T_c$, where f_{abc} are the structure constants of $\mathfrak{g}$. By Eq. (12.5), the associated operator $\hat{T}_a$ will satisfy

$$\left[\hat{T}_a, \hat{T}_b\right] = i\sum_{c=1}^{D} f_{abc}\hat{T}_c. \tag{12.6}$$

12.1.2 Projective Representations

For some physical applications, the notion of the unitary representation turns out to be stronger than necessary. We can represent group elements by unitary operators *up to a phase*.

Definition 12.2 A projective representation of a group G on a Hilbert space $\mathcal{H}$ is a map that assigns to each $g \in G$ a unitary $\hat{U}(g)$, such that for any $g_1, g_2 \in G$,

$$\hat{U}(g_1)\hat{U}(g_2) = \hat{U}(g_1 g_2)e^{i\phi(g_1, g_2)}, \tag{12.7}$$

for some phase $e^{i\phi}$ that depends on g_1 and g_2. If G is a Lie group, then we also require that the map $g \to \hat{U}(g)$ is differentiable.

If $\phi(g_1, g_2) = 0$ for all g_1, g_2, then the projective representation reduces to a unitary representation. We can always take $\phi(e, e) = 0$ in a projective representation, where e is the unit element of G. If $\phi(e, e) \neq 0$, we redefine $\hat{U}'(g) = \hat{U}(g)e^{-i\phi(e, e)}$, and the redefined operators satisfy Eq. (12.7) with the phase $\phi'(g_1, g_2) = \phi(g_1, g_2) - \phi(e, e)$; hence, $\phi'(e, e) = 0$.

We use the term "projective," because the phase $e^{i\phi}$ cancels when the operators $\hat{U}(g)$ act on projectors $|\psi\rangle\langle\psi|$,

$$\hat{U}(g_1)\hat{U}(g_2)|\psi\rangle\langle\psi|\hat{U}^\dagger(g_2)\hat{U}^\dagger(g_1) = \hat{U}(g_1 g_2)|\psi\rangle\langle\psi|\hat{U}^\dagger(g_1 g_2). \tag{12.8}$$

We have already encountered a projective representation in Chapter 11, even though we did not call it by that name. In Section 11.3, we found the operators $\hat{J}_i$ that represent the angular momentum algebra, and we implemented rotations through the unitary operators $e^{-i\theta \mathbf{n}\cdot\hat{\mathbf{J}}}$. Take for $g_1 = g_2$ the rotation around the 3-axis by π. Then, $U(g_1)U(g_2) = e^{-i\pi\hat{J}_3}e^{-i\pi\hat{J}_3} = e^{-2i\pi\hat{J}_3} = (-1)^j\hat{I}$, by Eq. (11.59). Since $g_1 g_2$ is a rotation by 2π, it equals unity, so $\hat{U}(g_1 g_2) = \hat{I}$. We see that Eq. (12.7) is satisfied with a phase term $e^{i\phi} = (-1)^{2j}$. For integer j, $e^{i\phi} = 1$, and the representation is unitary. However, for half-integer j, $e^{i\phi} = -1$, and the representation of $SO(3)$ is only projective.

Consider the group $SU(2)$ of 2×2 unitary matrices with unit determinant. As shown in Appendix C.3, $SU(2)$ has the same Lie algebra as $SO(3)$. Hence, the representation of the angular momentum algebra of Chapter 11 also defines a representation of $SU(2)$. However, the groups $SU(2)$ and $SO(3)$ are not isomorphic. As shown by Eq. (11.23), the homomorphism between elements of $SU(2)$ and $SO(3)$ is two-to-one, that is, one element of $SO(3)$ corresponds to two elements of $SU(2)$ with the opposite sign. We say that $SU(2)$ is a *double cover* of $SO(3)$ and, since $SU(2)$ is simply connected, it is also the universal cover of $SO(3)$ – see Section C.1.3 for the definition of these terms.

The 2×2 matrices that define $SU(2)$ coincide with the matrices of the $j = \frac{1}{2}$ representation of angular momentum. Hence, in $SU(2)$, a rotation by 2π corresponds to $-I$ and not I. It is then straightforward to show that the matrices of the form $e^{i\theta \mathbf{n} \cdot \hat{J}}$ define a unitary representation of $SU(2)$. This is an instant of a more general fact: *A projective representation of a group G can often be lifted to a unitary representation of its universal cover $\bar{G}$.*

By *lifting* a projective representation $\hat{U}(g)$ to a unitary one, we mean the following. We define $\hat{V}(g) = \hat{U}(g)e^{i\theta(g)}$ for some phase function $\theta(g)$. Then,

$$\hat{V}(g_1)\hat{V}(g_2) = \hat{V}(g_1 g_2)\exp[i\phi(g_1, g_2) - \theta(g_1 g_2) + \theta(g_1) + \theta(g_2)]. \tag{12.9}$$

If there exists a function $\theta(g)$, such that $\phi(g_1, g_2) = \theta(g_1 g_2) - \theta(g_1) - \theta(g_2)$, then the operators $\hat{V}(g)$ define a unitary representation. Whether such a function $\theta(g)$ can be found or not depends primarily on the algebraic structure of the group.

A crucial theorem by Bargmann (1954) identifies the groups, whose projective representations can be lifted to unitary representations. These groups include

- all simply connected groups G with semisimple Lie algebras;
- the inhomogeneous versions IG of such groups.

Projective representations that cannot be lifted to unitary ones do not, in general, lead to faithful representations of the associated Lie algebras. The phase factor $\phi(g_1, g_2)$ in Eq. (12.7) leads to the appearance of an additional term, proportional to unity, in the commutation relations,

$$\left[\hat{T}_a, \hat{T}_b\right] = i\sum_{c=1}^{D} f_{abc}\hat{T}_c + ic_{ab}\hat{I}, \tag{12.10}$$

where c_{ab} is an antisymmetric matrix.

Equation (12.10) defines a *central extension* of the Lie algebra $\mathfrak{g}$; we call it central because the added elements belong to the center of the new algebra, that is, they commute with all other elements. The nonzero elements of the matrix c_{ab} are the central charges of the representation.

12.1.3 Irreducible Representations

Let $\hat{U}_1(g)$ and $\hat{U}_2(g)$ be two representations of the group G on the Hilbert spaces $\mathcal{H}_1$ and $\mathcal{H}_2$, respectively. We call the representations *unitarily equivalent*, if (i) the two Hilbert spaces are isometric, $\mathcal{H}_1 = \mathcal{H}_2 = \mathcal{H}$, and (ii) there exists a unitary map $\hat{V}$ on H, such that

$$\hat{V}\hat{U}_1(g)\hat{V}^\dagger = \hat{U}_2(g) \tag{12.11}$$

for all $g \in G$.

Let $\hat{T}_a^{(1)}$ and $\hat{T}_a^{(2)}$ be a basis of generators in the two representations. If the representations are unitarily equivalent, then $\hat{V}\hat{T}_a^{(1)}\hat{V}^\dagger = \hat{T}_a^{(2)}$. This means that the operators $\hat{T}_a^{(1)}$ and $\hat{T}_a^{(2)}$ have the same spectrum. This provides us with a simple criterion for verifying the inequivalence of two representations; it suffices to show that the spectrum of at least one generator is different in the two representations.

A unitary representation $\hat{U}(g)$ of a group G on a Hilbert space $\mathcal{H}$ is called *reducible*, if $\mathcal{H} = \mathcal{H}_1 \oplus \mathcal{H}_2$, and it is possible to define a unitary representation of G in each subspace $\mathcal{H}_1$ and $\mathcal{H}_2$. An *irreducible* representation cannot be analyzed into simpler representations via the direct sum. Hence, irreducible representations describe the simplest possible quantum system that carries the symmetry corresponding to the group G; they are the *atoms* of that symmetry.

Theorem 12.3 (Schur's lemma) *Let $\hat{U}(g)$ be a unitary representation of a group G in a Hilbert space $\mathcal{H}$. The representation is irreducible if and only if the only self-adjoint operators $\hat{A}$ that satisfies $\left[\hat{A}, \hat{U}(g)\right] = 0$ for all $g \in G$ is a multiple of the unit operator $\hat{I}$.*

Proof Let $\hat{A}$ be a self-adjoint operator that commutes with all $\hat{U}(g)$ and is not a multiple of $\hat{I}$. Then $\hat{A}$ has at least one spectral projector $\hat{P} \neq \hat{I}$. Since $\hat{P}$ is a function of $\hat{A}$, $\hat{P}$ also commutes with all $\hat{U}(g)$. Then, the eigenspace corresponding to $\hat{P}$ carries a unitary representation of G by the operators $\hat{P}\hat{U}(g)\hat{P}$. Indeed, $\left(\hat{P}\hat{U}(g_1)\hat{P}\right)\left(\hat{P}\hat{U}(g_2)\hat{P}\right) = \hat{P}\hat{U}(g_1)\hat{U}(g_2)\hat{P} = \hat{P}\hat{U}(g_1 g_2)\hat{P}$. Hence, the representation $\hat{U}(g)$ on $\mathcal{H}$ is reducible.   □

For a Lie group G, Schur's lemma implies that the only operator to commute with all generators $\hat{T}$ is a multiple of unity.

We define the *Casimir operator*, or simply the *Casimir*, of a unitary representation $\hat{U}(g)$ of a group G as any self-adjoint operator $\hat{C}$ that commutes with all $\hat{U}(g)$ and is not a multiple of the unity. In Lie groups, Casimirs commute with all generators $\hat{T}_a$.

Clearly, if $\hat{C}$ is a Casimir, any function $f(\hat{C})$ is also a Casimir. Furthermore, if $\hat{C}_1$ and $\hat{C}_2$ are Casimirs, all linear combinations $a\hat{C}_1 + b\hat{C}_2$ with real a and b are also Casimirs. When we say that a group has n Casimirs, we mean n linearly and functionally independent Casimirs.

Schur's lemma implies that an irreducible representation is the common eigenspace of all Casimir operators. Hence, to identify all irreducible representations of a Lie group, we have to construct a very large – that is, highly reducible – representation of G, find the Casimirs there, and then analyze their spectrum.

For example, we can construct the Hilbert space $L^2(G)$ that consists of square integrable wave functions on the group G. Then, we define $\hat{U}(g)\psi(g') := \psi(g^{-1}g')$ for all $g, g' \in G$. With an appropriate choice of the inner product, $\hat{U}(g)$ defines a unitary representation of G that is highly reducible. For compact groups, this construction suffices, as shown by a theorem of Peter and Weyl (1927).

Theorem 12.4 (Peter–Weyl) *All unitary irreducible representations of a compact group G correspond to subspaces of $L^2(G)$ and they are finite-dimensional.*

The explicit construction of the representations proceeds through an analysis that generalizes that of Chapter 11 for the representations of $SU(2)$. That is, we first identify a maximal commuting set of linearly independent operators among the generators, say, $\hat{L}_1, \hat{L}_2, \ldots, \hat{L}_q$. The number q of such commuting generators is known as the Lie algebra's *rank*. Second, we identify the Casimirs $\hat{C}_i$ of the group. If there are n independent Casimirs, we denote them by $\hat{C}_i$, where $i = 1, 2, \ldots, n$. Then, a representation of the Hilbert space is characterized by the common eigenvectors (or generalized eigenvectors) $|c_1, c_2, \ldots, c_n, \lambda_1, \lambda_2, \ldots, \lambda_q\rangle$, which satisfy

$$\hat{L}_k|c_1, c_2, \ldots, c_n, \lambda_1, \lambda_2, \ldots, \lambda_q\rangle = \lambda_k|c_1, c_2, \ldots, c_n, \lambda_1, \lambda_2, \ldots, \lambda_q\rangle \qquad (12.12)$$
$$\hat{C}_i|c_1, c_2, \ldots, c_n, \lambda_1, \lambda_2, \ldots, \lambda_q\rangle = c_i|c_1, c_2, \ldots, c_n, \lambda_1, \lambda_2, \ldots, \lambda_q\rangle.$$

The eigenspace with fixed values of all c_i carries an irreducible representation of G. The remaining generators of the Lie algebra can be combined to create the analogue of the $\hat{J}_\pm$ in the representations of $SO(3)$, that is, operators that act on the kets $|c_1, c_2, \ldots, c_n, \lambda_1, \lambda_2, \ldots, \lambda_q\rangle$ by changing the values of λ_k.

The kets $|c_1, c_2, \ldots, c_n, \lambda_1, \lambda_2, \ldots, \lambda_q\rangle$ generalize the kets $|j, m\rangle$ that we encountered in our study of rotations. In the latter case, m had a maximal value equal to j, and the extremal vectors $|j, j\rangle$ allow us to construct all other vectors through successive operations of $\hat{J}_-$. It turns out that the vectors $|c_1, c_2, \ldots, c_n, \lambda_1, \lambda_2, \ldots, \lambda_q\rangle$ share this property, as long as the Lie algebra $\mathfrak{g}$ is semisimple. That is, there is a unique set of maximal values $\lambda_1(\mathbf{c}), \ldots, \lambda_q(\mathbf{c})$ for each $\mathbf{c} = (c_1, c_2, \ldots, c_n)$. Since there are q independent maximal values of λ_i, there must also be q independent Casimirs. This fact is known as *Racah's theorem*.

Theorem 12.5 (Racah's) *A group G with a semisimple Lie algebra $\mathfrak{g}$ of rank q has q independent Casimirs.*

A Casimir operator that is constructed as a monomial of order p with respect to the generators $\hat{T}_a$ is denoted by $\hat{C}^{(p)}$. In semisimple algebras, where the Cartan metric w_{ab}, Eq. (C.14), can be inverted, there is a simple expression for the quadratic Casimir.

Theorem 12.6 *The operator $\hat{C}^{(2)} = -\sum_{a,b} \bar{w}_{ab} \hat{T}_a \hat{T}_b$ is a Casimir, where $\bar{w}_{ab}$ is the inverse matrix of the Cartan metric w_{ab}, Eq. (C.14).*

Proof By Eq. (12.6), $[\hat{C}^{(2)}, \hat{T}_a] = -i \sum_{bcd} [\bar{w}_{cd} f_{dab} + \bar{w}_{db} f_{dac}] \hat{T}_c \hat{T}_b$. We evaluate the first term in the square bracket,

$$\sum_d \bar{w}_{cd} f_{dab} = -\sum_d f_{adb} \bar{w}_{cd} = -\sum_{def} f_{ade} w_{ef} \bar{w}_{fb} \bar{w}_{cd} = -\sum_{df} \bar{f}_{adf} \bar{w}_{fb} \bar{w}_{cd}$$

$$= \sum_{df} \bar{f}_{afd} \bar{w}_{fb} \bar{w}_{cd} = \sum_{df} \bar{f}_{afd} \bar{w}_{dc} \bar{w}_{fb} = \sum_f f_{afc} \bar{w}_{fb} = -\sum_d f_{dac} \bar{w}_{db},$$

where $\bar{f}_{abc}$ are the modified structure constants that are given by Eq. (C.15). We conclude that $[\hat{C}^{(2)}, \hat{T}_a] = 0$. $\qquad\square$

For $SO(3)$ and $SU(2)$, $w_{ab} = -2\delta_{ab}$, hence, $\hat{C}^{(2)} = \frac{1}{2}\hat{J}^2$, in agreement with our analysis of Chapter 11.

12.2   The Heisenberg–Weyl Group

Group theory lies at the core of quantum mechanics; it even appears in the pioneering paper of Heisenberg (1925). The fundamental commutation relation for a particle in one dimension $[\hat{x}, \hat{p}] = i\hat{I}$, together with the trivial relations $\left[\hat{x}, \hat{I}\right] = \left[\hat{p}, \hat{I}\right] = 0$ define the representation of a Lie group H_1 with generators $\hat{x}$, $\hat{p}$, and $\hat{I}$. This group is known as the *Heisenberg–Weyl* group. Each element of H_1 corresponds to the operator

$$\hat{U}(a,b,c) = \exp\left[ia\hat{x} - ib\hat{p} + ic\hat{I}\right], \tag{12.13}$$

for $a, b, c \in \mathbb{R}$.

By the multiplication rule (4.23),

$$\hat{U}(a_1,b_1,c_1)\hat{U}(a_2,b_2,c_2) = \hat{U}[a_1 + a_2, b_1 + b_2, c_1 + c_2 + \tfrac{1}{2}(a_1 b_2 - a_2 b_1)]. \tag{12.14}$$

Hence, the group H_1 has elements $(a,b,c) \in \mathbb{R}^3$ and operation

$$(a_1,b_1,c_1)(a_2,b_2,c_2) = (a_1 + a_2, b_1 + b_2, c_1 + c_2 + \tfrac{1}{2}(a_1 b_2 - a_2 b_1)). \tag{12.15}$$

Interestingly, the operators $\hat{U}(a,b,0)$ satisfy the condition (12.7) for a projective representation of the Abelian group $\mathbb{R}^2$ with group law $(a_1,b_1)(a_2,b_2) = (a_1 + a_2, b_1 + b_2)$. The associated phase factor is $\phi = \tfrac{1}{2}(a_2 b_1 - a_1 b_2)$. This means that Heisenberg's commutation relation can also be interpreted as a central extension of the Lie algebra $\mathfrak{r}^2$. This point of view will prove important when discussing the notion of quantization in Section 12.3.

Given a positive parameter σ, we define the annihilation operators $\hat{a} = \frac{1}{\sqrt{2\sigma}}\hat{x} + i\sqrt{\frac{\sigma}{2}}\hat{p}$ and their conjugates (creation operators) $\hat{a}^\dagger = \frac{1}{\sqrt{2\sigma}}\hat{x} - i\sqrt{\frac{\sigma}{2}}\hat{p}$. Then, Heisenberg's commutation relations become $[\hat{a}, \hat{a}^\dagger] = \hat{I}$. We express the operator $\hat{U}(a,b,0)$ as

$$\hat{U}(z) = \exp[z\hat{a}^\dagger - z^*\hat{a}], \tag{12.16}$$

where $z = \frac{1}{\sqrt{2\sigma}}b + i\sqrt{\frac{\sigma}{2}}a \in \mathbb{C}$. Then, the multiplication law (4.23) becomes

$$\hat{U}(z_1)\hat{U}(z_2) = \hat{U}(z_1 + z_2)e^{\frac{1}{2}(z_1^* z_2 - z_2^* z_1)}. \tag{12.17}$$

Furthermore, we can use the BCH identity to prove that

$$\hat{U}(z) = e^{-\frac{|z|^2}{2}} e^{z\hat{a}^\dagger} e^{-z^*\hat{a}}. \tag{12.18}$$

In Section 5.2, we saw that the operator $\hat{a}$ has at least one eigenvector with zero eigenvalue. Using Schur's lemma, we can show that in an irreducible representation of H_1, $\hat{a}$ has no degeneracies – see Problem 12.1. Hence, there is a unique vector $|0\rangle$ such that $\hat{a}|0\rangle = 0$. For $\sigma = (m\omega)^{-1}$, $|0\rangle$ is the ground state of a harmonic oscillator of mass m and frequency ω. Since $e^{-z^*\hat{a}}|0\rangle = |0\rangle$, we find that

$$\hat{U}(z)|0\rangle = e^{-\frac{|z|^2}{2}} \sum_{n=0}^{\infty} \frac{z^n (\hat{a}^\dagger)^n}{\sqrt{n!}}|0\rangle = |z\rangle, \tag{12.19}$$

that is, the coherent states $|z\rangle$ are obtained by the action of the operator $\hat{U}(z)$ on the state $|0\rangle$. This remark leads to the generalization of the notion of coherent states to Hilbert spaces with representations of Lie groups other than H_1.

The Heisenberg–Weyl group H_n in n dimensions has $2n+1$ generators: n position operators $\hat{x}_i$, n momentum operators $\hat{p}_i$, and the unity $\hat{I}$; here, $i = 1, 2, \ldots, n$. The fundamental commutation relations are

$$[\hat{x}_i, \hat{x}_j] = 0, \qquad [\hat{p}_i, \hat{p}_j] = 0, \qquad [\hat{x}_i, \hat{p}_j] = i\delta_{ij}\hat{I}, \qquad (12.20)$$

and the associated unitary operators $\hat{U}(\mathbf{a}, \mathbf{b}, c) = \exp\left[i\mathbf{a} \cdot \hat{\mathbf{x}} - i\mathbf{b} \cdot \hat{\mathbf{p}} + ic\hat{I}\right]$ satisfy

$$\hat{U}(\mathbf{a}_1, \mathbf{b}_1, c_1)\hat{U}(\mathbf{a}_2, \mathbf{b}_2, c_2) = \hat{U}(\mathbf{a}_1 + \mathbf{a}_2, \mathbf{b}_1 + \mathbf{b}_2, c_1 + c_2 + \tfrac{1}{2}(\mathbf{a}_1 \cdot \mathbf{b}_2 - \mathbf{a}_2 \cdot \mathbf{b}_1)). \qquad (12.21)$$

Equation (12.21) is straightforwardly proved by the BCH identity.

We conclude that the group H_n has elements $(\mathbf{a}, \mathbf{b}, c) \in \mathbb{R}^{2n+1}$ and group law

$$(\mathbf{a}_1, \mathbf{b}_1, c_1)(\mathbf{a}_2, \mathbf{b}_2, c_2) = (\mathbf{a}_1 + \mathbf{a}_2, \mathbf{b}_1 + \mathbf{b}_2, c_1 + c_2 + \tfrac{1}{2}(\mathbf{a}_1 \cdot \mathbf{b}_2 - \mathbf{a}_2 \cdot \mathbf{b}_1)). \qquad (12.22)$$

The most common representation of H_n is the familiar *Schrödinger representation*, on the Hilbert space $L^2(\mathbb{R}^n, d^n x)$, where $\hat{x}_i \psi(\mathbf{x}) = x_i \psi(\mathbf{x})$ and $\hat{p}_i \psi(\mathbf{x}) = -i\frac{\partial}{\partial x_i}\psi(\mathbf{x})$.

The Schrödinger representation is irreducible. We show this for $n = 1$; the generalization is straightforward. The generalized eigenvectors of the position operator $\hat{x}$ on $L^2(\mathbb{R})$ are not degenerate. Hence, any operator $\hat{A}$ that commutes with $\hat{x}$, can be written as $f(\hat{x})$ for some function f – see Problem 4.4. Then, $\left[\hat{A}, \hat{p}\right] = [f(\hat{x}), p] = if'(\hat{x})$; the commutator $[\hat{A}, \hat{p}]$ vanishes only if f is constant. Hence, the only operator that commutes with all generators of H_1 is a multiple of the identity. By Schur's lemma, the Schrödinger representation is irreducible.

The Schrödinger representation is not unique. For example, the momentum representation is obtained by defining $\hat{x}_i \psi(\mathbf{p}) = i\frac{\partial}{\partial p_i}\psi(\mathbf{p})$ and $\hat{p}\psi(\mathbf{p}) = p_i \psi(\mathbf{p})$ on the Hilbert space $L^2(\mathbb{R}^n, d^n p)$. In fact, there is an infinity of unitary representation of H_n, and the question arises whether physical predictions depend on the choice of representation.

Thankfully, the answer is negative, as shown by a major theorem from Stone (1930, 1932) and von Neumann (1931, 1932b).

Theorem 12.7 (Stone–von Neumann) *All unitary representations of the Heisenberg–Weyl group H_n are unitarily equivalent to the Schrödinger representation.*

The Stone–von Neumann theorem is a rigorous confirmation of the complete equivalence between Heisenberg's matrix mechanics and Schrödinger's wave mechanics. It implies that the Schrödinger representation is not special; we can use any representation of H_n that is convenient for a given calculation.

We proceed to a sketch of the proof of the theorem for $n = 1$. The generalization to arbitrary n is relatively straightforward. The key point is the demonstration of Section 5.2.2 that the annihilation operator $\hat{a} = \frac{1}{\sqrt{2\sigma}}\hat{x} + i\sqrt{\frac{\sigma}{2}}\hat{p}$, defined for arbitrary $\sigma > 0$, has at least one zero eigenvalue. Irreducible representations are those with only one zero eigenvector $|0\rangle$.

Given $|0\rangle$ we define the coherent states $|z\rangle$ of Eq. (12.19). From the coherent states, we obtain the Schrödinger representation by defining wave functions $\psi(x) = \int \frac{d^2z}{\pi} \langle z|\psi\rangle\langle x|z\rangle$, where $\langle x|z\rangle$ is given by Eq. (5.39). The only thing that remains is to show that representations for different σ are unitarily equivalent. The unitary operator that implements this transformation is determined in Problem 12.3.

There is also an infinite-dimensional Weyl–Heisenberg group H_∞. Heuristically, it can be defined by taking the number of generators n to infinity. A better definition is provided by representing the vectors **a** and **b** in Eq. (12.21) by elements a and b of a real vector space V, and writing the inner product $\mathbf{a} \cdot \mathbf{b}$ as (a, b). Then, Eq. (12.21) becomes

$$\hat{U}(a_1, b_1, c_1)\hat{U}(a_2, b_2, c_2) = \hat{U}[a_1 + a_2, b_1 + b_2, c_1 + c_2 + \tfrac{1}{2}(a_1, b_2) - \tfrac{1}{2}(a_2, b_1)], \qquad (12.23)$$

which is well defined even for infinite-dimensional V. The Stone–von Neumann theorem does not apply to H_∞. There is an infinity of unitarily inequivalent representations of H_∞ – see Example 15.5 for an explicit construction. This fact leads to grave problems in the quantum theory of systems with an infinite number of degrees of freedom, namely, quantum fields.

12.3 Kinematical Symmetries

The Heisenberg–Weyl group is a special case of a kinematical symmetry, that is, a group of transformations whose generators define fundamental observables of the system. In this section, we analyze how this notion can be generalized to other systems, and how one can use kinematical symmetries in order to obtain a consistent correspondence rule between classical and quantum observables.

12.3.1 Quantization Rules

Heisenberg's postulate of the fundamental commutation relations raised many questions about its origins. Position and momentum are observables in classical physics; they correspond to functions on the state space. Why do we focus on an algebraic relation of these observables in quantum theory? Is this relation somehow contained in classical physics?

Dirac (1926, 1930a) was the first to attempt an answer. He noted the similarity of the operator commutator to the Poisson bracket of Hamiltonian mechanics. In any Hamiltonian system, the Poisson bracket takes two functions F and G on the classical state space Γ and maps them to another function $\{F, G\}$ on Γ, defined by Eq. (1.7).

In the state space $\Gamma = \mathbb{R}^{2n}$, with elements (x_a, p_a), $a = 1, 2, \ldots n$, we have the following Poisson brackets

$$\{x_a, x_b\} = 0, \qquad \{p_a, p_b\} = 0, \qquad \{x_a, p_b\} = \delta_{ab}. \qquad (12.24)$$

Dirac suggested that the similarity of the Poisson bracket to the operator commutator provides the basis for a *quantization* procedure, i.e., a set of rules for associating quantum observables to classical ones. He proposed that we associate functions on the classical state space Γ to self-adjoint operators on the Hilbert space $\mathcal{H}$ according to the following rules.

1. The functions x_a and p_a on Γ respectively correspond to the operators $\hat{x}_a$ and $\hat{p}_a$ on $\mathcal{H}$.
2. Any function $f(x_a)$ on Γ corresponds to the operator $f(\hat{x}_a)$ and any function $g(p_a)$ on Γ corresponds to the operator $g(\hat{p}_a)$.
3. If the functions F and G on Γ correspond to the operators $\hat{F}$ and $\hat{G}$, respectively, then the function $F + G$ corresponds to the operator $\hat{F} + \hat{G}$.
4. The Poisson bracket $\{\cdot, \cdot\}$ on Γ corresponds to $\frac{1}{i}[\cdot, \cdot]$, where $[\cdot, \cdot]$ is the operator commutator on $\mathcal{H}$.

Dirac's rules have proved very useful. For example, they allow us to specify the Hamiltonian of a particle as a Schrödinger operator $\hat{H} = \frac{\hat{p}^2}{2m} + V(\hat{x})$, from the knowledge of the classical Hamiltonian $H(x, p) = \frac{p^2}{2m} + V(x)$.

Still, these rules say nothing about how to write the quantum mechanical analogues of functions that involve products of x and p. In this case, there is no unique function-to-operator correspondence. To see this, consider the classical function $F(x) = xp^2$. We can map it to the operator $\hat{F}_1 = \frac{1}{2}(\hat{x}\hat{p}^2 + \hat{p}^2\hat{x})$ or to the operator $\hat{F}_2 = \hat{p}\hat{x}\hat{p}$. We can even map it to any linear combination $c_1\hat{F}_1 + c_2\hat{F}_2$, with $c_1 + c_2 = 1$. The operators $\hat{F}_1$ and $\hat{F}_2$ differ by a term that is proportional to the commutator $[\hat{x}, \hat{p}]$ and hence by a factor of $\hbar$. We would expect that in some semiclassical regime, measurements of $\hat{F}_1$ and $\hat{F}_2$ would lead to the same predictions. This ambiguity is known as the *factor-ordering* problem.

A more serious problem is that the set of quantization rules is not consistent. As shown by Groenewold (1946), they lead to contradictions. To see this, consider the following identity for classical observables

$$\left\{x^3, p^3\right\} + \frac{1}{12}\left\{\left\{p^2, x^3\right\}, \left\{x^2, p^3\right\}\right\} = 0. \tag{12.25}$$

If we use the quantization rules to express the left-hand side of Eq. (12.25) quantum mechanically, we obtain

$$\frac{1}{i}\left[\hat{x}^3, \hat{p}^3\right] + \frac{1}{12i}\left[\frac{1}{i}\left[\hat{p}^2, \hat{x}^3\right], \frac{1}{i}\left[\hat{x}^2, \hat{p}^3\right]\right] = -3. \tag{12.26}$$

This means that one of the four quantization rules is problematic. Rules 1 and 4 are inviolable, as they are the essence of Heisenberg's proposal. But rules 2 and 3 are simple and practical, and they have also proved useful in a variety of cases. If we drop one of them, the very idea of quantization as a rules-based procedure may turn out to be problematic.

The problem is made worse by the realization that the Heisenberg commutation relations do not apply to all systems. We saw that the momentum operator is not defined for particles in the half-line and for particles in a box, and that the position operator is not defined for particles in a ring. Hence, there are realistic systems in which Heisenberg's commutation relations are not defined. The underlying reason in all cases is that the state space is not a linear space.

In recent years, research in many fields of physics (condensed matter, high-energy physics, gravity theory) has uncovered many systems with nontrivial classical state spaces. For such systems, the Heisenberg commutation relations are a nonstarter; we must find a new set of quantization rules. Two complementary approaches have been developed. The first approach aims to construct the quantum theory geometrically, that is, through the introduction of additional geometric structures on the classical state space. The result is a formalism known as *geometric quantization*. The second approach is closer to Heisenberg's logic; it tries to identify an analogue of the Heisenberg–Weyl group, the *canonical group*, in each classical system. The associated quantum theory is obtained from the study of the unitary representations of the canonical group. In what follows, we present the main ideas of quantization via the canonical group.

12.3.2 Canonical Transformations

In Hamiltonian mechanics, symmetries are implemented through canonical transformations. These are defined as follows. Consider the standard classical state space $\Gamma = \mathbb{R}^{2n}$ with elements $\xi = (x_1, p_1, x_2, p_2, \ldots, x_n, p_n)$ that is equipped with the Poisson bracket (1.7). The canonical variables (x_i, p_i), $i = 1, 2, \ldots, n$, satisfy the fundamental Poisson brackets (12.20).

Every function $F : \Gamma \to \mathbb{R}$ generates a one-parameter family of transformations $(x_a, p_a) \to (x_a, p_a)(s)$ for all $s \in \mathbb{R}$. These transformations are obtained from the solution of the differential equations

$$\frac{dx_a(s)}{ds} = \{x_a(s), F\} = \frac{\partial F}{\partial p_a}, \qquad \frac{dp_a(s)}{ds} = \{p_a(s), F\} = -\frac{\partial F}{\partial x_a}, \qquad (12.27)$$

with initial conditions $(x_a, p_a)(0) = (x_a, p_a)$. Any transformation of this form is a *canonical transformation*, and the function F is called its *generator*.

For fixed s, the transformation (12.27) takes a function G on Γ to a different function $G_s(x_a, p_a) = G[x_a(s), p_a(s)]$. Then,

$$\frac{dG_s}{ds} = \sum_a \left(\frac{\partial G_s}{\partial x_a} \frac{dx_a}{ds} + \frac{\partial G_s}{\partial p_a} \frac{dp_a}{ds} \right) = \sum_a \left(\frac{\partial G_s}{\partial x_a} \frac{\partial F}{\partial p_a} - \frac{\partial G_s}{\partial p_a} \frac{\partial F}{\partial x_a} \right) = \{G_s, F\}. \qquad (12.28)$$

It is convenient to express the canonical transformations in terms of linear operators acting on functions on Γ. Hence, we define the operator

$$X_F = \sum_a \left(\frac{\partial F}{\partial x_a} \frac{\partial}{\partial p_a} - \frac{\partial F}{\partial p_a} \frac{\partial}{\partial x_a} \right), \qquad (12.29)$$

which acts on functions G as $X_F(G) = \{F, G\}$. Then, Eq. (12.28) becomes $\frac{dG_s}{ds} = -X_F(G_s)$, with formal solution

$$G_s = e^{-X_F s}(G). \qquad (12.30)$$

Hence, the canonical transformations can be expressed in terms of the operator X_F. Note that if we add a constant c to F, the operator X_F is not affected: $X_{F+c} = X_F$. It is straightforward to show that the operator commutator reflects the Poisson bracket

$$[X_F, X_G] = X_{\{F, G\}}. \qquad (12.31)$$

The key property of canonical transformations is that they preserve the Poisson bracket, that is, for any functions G and H on Γ,

$$\{e^{-X_{FS}}(G), e^{-X_{FS}}(H)\} = e^{-X_{FS}}(\{G, H\}). \tag{12.32}$$

To prove Eq. (12.32), we evaluate

$$\frac{d}{ds}\{G_s, H_s\} = \{\frac{dG_s}{ds}, H_s\} + \{G_s, \frac{dH_s}{ds}\} = \{\{G_s, F\}, H_s\} + \{G_s, \{H_s, F\}\}$$
$$= \{\{G_s, F\}, H_s\} + \{\{F, H_s\}, G_s\} = \{\{G_s, H_s\}, F\},$$

where in the last step we used Jacobi's identity. Then, Eq. (12.32) follows.

Hence, when a canonical transformation takes (x_a, p_a) to (x'_a, p'_a), (x'_a, p'_a) is also a canonical pair

$$\{x'_a, x'_b\} = 0, \qquad \{p'_a, p'_b\} = 0, \qquad \{x'_a, p'_b\} = \delta_{ab}. \tag{12.33}$$

It follows that we can characterize canonical transformations as maps $f : \Gamma \to \Gamma$ that preserve the Poisson bracket.

12.3.3 The Canonical Group

Canonical transformations in classical mechanics are analogous to unitary transformations in quantum mechanics. For historical reasons, mathematicians use the word "action" rather than "representation" when talking about the realization of a group on a classical state space, even if the meaning is the same. A *canonical group action* is a map that assigns to each $g \in G$, a canonical transformation $f_g : \Gamma \to \Gamma$, such that $f_{g_2}[f_{g_1}(\xi)] = f_{g_2 g_1}(\xi)$ for all $g_1, g_2 \in G$ and $\xi \in \Gamma$.

As an example, let us consider the following transformation on $\Gamma = \mathbb{R}^2$:

$$(x, p) \to (x + x_0, p + p_0), \tag{12.34}$$

where x_0 and p_0 are constants. The action obviously preserves the Poisson brackets (12.24). As x_0 and p_0 vary over $\mathbb{R}^2$, the transformation defines an action of the Abelian group $\mathbb{R}^2$ on Γ. An element of the Lie algebra $\mathfrak{r}^2$ is a pair of reals $T = (a, b)$, which generates the one-parameter group of transformations

$$(x, p) \to (x - sb, p - sa). \tag{12.35}$$

The transformations (12.35) are generated by the function $F_{a,b} = ax - bp$ through the Poisson bracket. Hence, the function $F_{a,b}$ represents $(a, b) \in \mathfrak{r}^2$.

The Poisson bracket between a pair of such functions is

$$\{F_{a_1, b_1}, F_{a_2, b_2}\} = a_2 b_1 - a_1 b_2. \tag{12.36}$$

It *does not* represent accurately the Lie algebra of $\mathfrak{r}^2$, which is Abelian.

The reason is that the transformations generated by a function F are unchanged if F is shifted by a constant. Hence, when attempting to represent a Lie algebra $\mathfrak{g}$ by functions F_T on a state space Γ, where $T \in \mathfrak{g}$, we obtain relations of the form

$$\{F_{T_1}, F_{T_2}\} = F_{[T_1, T_2]} + z(T_1, T_2), \tag{12.37}$$

where $T_1, T_2 \in \mathfrak{g}$ and z is a constant on Γ that depends on T_1 and T_2. Hence, Lie algebras $\mathfrak{g}$ are represented only modulo a central extension. If $z(T_1, T_2)$ depends on T_1 and T_2 through their commutator, that is, if $z(T_1, T_2) = \zeta([T_1, T_2])$ for some function ζ on $\mathfrak{g}$, then we can redefine $F_T \to F_T - \zeta(T)$, so that

$$\{F_{T_1}, F_{T_2}\} = F_{[T_1, T_2]}, \tag{12.38}$$

with respect to the redefined F_T.

Such a redefinition is not always possible. In particular, the central charge in Eq. (12.36) cannot be written as a function of a commutator, because the Lie algebra is commutative. Hence, Eq. (12.36) can only be viewed as a representation of an extended Lie algebra, with one additional generator that corresponds to the unit function on Γ. This is nothing but the Lie algebra of the Heisenberg–Weyl group.

Therefore, the Heisenberg–Weyl group H_1 can be identified classically as a minimal extension of the group $\mathbb{R}^2$ that implements *position and momentum translations*. The symmetry of position and momentum translations is common in both the classical description and the quantum description. It describes the most important kinematical features of the system, namely, that position and momentum are independent variables that take values in the whole of the real line. In effect, this symmetry asserts that the classical state space is a vector space.

There exist many important physical systems with state spaces that are not vector spaces. For such systems, we need to find the appropriate group that captures their kinematical symmetry, the *canonical group*. The quantum theory will then be constructed by identifying the appropriate unitary representations of the canonical group. The overall procedure is the following (Isham, 1983).

(i) First, we specify the classical state space Γ, a manifold of even dimension $2n$. We can always find coordinates on Γ, such that its elements can be expressed as n canonical pairs (x_a, p_a), $a = 1, 2, \ldots, n$. However, such coordinates may not be well defined everywhere. The state space Γ is equipped with a Poisson bracket $\{\cdot, \cdot\}$, expressed in terms of the local coordinates (x_a, p_a) by Eq. (1.7).

Hence, the state space Γ looks locally like the vector space $\mathbb{R}^{2n}$, but globally it may be very different.

(ii) We identify the canonical group of the classical system. This is the *smallest* Lie group G that acts through canonical transformations f_g on Γ, in a way that captures the *global structure* of Γ.

By "smallest" Lie group, we mean that there is no "inactive" group element, that is, there is no $g \in G$ such that $f_g(\xi) = \xi$ for all $\xi \in \Gamma$.

To capture the global structure of Γ, the group action must be *transitive*: For any pair of points $\xi_1, \xi_2 \in \Gamma$, there is at least one $g \in G$, such that $f_g(\xi_1) = \xi_2$.

(iii) Any element T of the Lie algebra $\mathfrak{g}$ of G generates a one-parameter group of canonical transformations on Γ through its correspondence with a function F_T on Γ. The map $T \to F_T$ is unique only up to a constant. In some cases, it is possible to choose the functions F_T as to represent the Lie algebra via the Poisson bracket, Eq. (12.38). If not, we employ a Lie algebra with a central extension; the canonical group is the associated extended Lie group.

(iv) We identify all inequivalent unitary irreducible of the canonical group G, or of its universal cover, if necessary. Each irreducible representation describes a different type of quantum system with symmetry.

(vi) The generators $\hat{T}$ of a unitary representation of the canonical group G on $\mathcal{H}$ correspond to the classical functions F_T for all $T \in \mathfrak{g}$. The map $F_T \to \hat{T}$ implements the quantization of a set of classical observables, and it satisfies the following properties.

- Additivity: $F_{T_1} + F_{T_2} \to \hat{T}_1 + \hat{T}_2$, for all $T_1, T_2 \in \mathfrak{g}$.
- Scalar multiplication: $\lambda F_T \to \lambda \hat{T}$, for all $\lambda \in \mathbb{R}$ and $T \in \mathfrak{g}$.
- Preservation of Poisson bracket: $\{F_{T_1}, F_{T_2}\} \to i[\hat{T}_1, \hat{T}_2]$.

Quantization via the canonical group provides a clear answer to an important foundational question. Why do we refer to microscopic quantum systems as particles when they are so different from the macroscopic particles that we are familiar with? Quantum systems have no sharp position and momentum; in fact, they have no properties at all unless they are being measured; they do not follow trajectories, so aren't we stretching the language thin when we call them particles? The answer is that *quantum particles have the same kinematical symmetry as their classical counterparts*. We can relate specific quantum observables with classical observables, because they are generators of the same symmetry transformations – even if these transformations are implemented in physical theories that are radically different.

12.3.4 Examples of Canonical Groups

Example 12.1 A spinning top is a simple Hamiltonian system with a state space $\Gamma = S^2$ that contains all vectors **s** with fixed norm $s = |\mathbf{s}|$. We describe the sphere in terms of the standard coordinates θ and ϕ. This system is Hamiltonian, as ϕ and $\cos\theta$ form a canonical pair, with Poisson bracket

$$\{\phi, \cos\theta\} = s^{-1}. \tag{12.39}$$

It is straightforward to show that the Cartesian coordinates of **s**, $s_1 = s\sin\theta\cos\phi, s_2 = s\sin\theta\sin\phi$ and $s_3 = s\cos\theta$, satisfy the angular momentum algebra $\{s_i, s_j\} = \sum_k \epsilon_{ijk} s_k$. They generate an action of the $SO(3)$ group on Γ through canonical transformations.

Hence, $SO(3)$, or rather its double cover $SU(2)$, is the canonical group of the spinning top. Any irreducible representation of $SU(2)$ defines a quantization map that takes the classical observables **s** to the quantum operators that represent angular momentum. The difference is that classically the norm of the angular momentum vectors **s** takes any positive value, while in quantum theory it is quantized.

Example 12.2 In Section 5.6.1, we showed that a particle moving in the half-line is described by the Hilbert space $L_D^2(\mathbb{R}^+)$ of square integrable functions subject to Dirichlet boundary conditions at $x = 0$. We also showed that the operator $\hat{p} = -i\frac{\partial}{\partial x}$ is not well defined on $L_D^2(\mathbb{R}^+)$, because its action does not preserve the boundary conditions. However, the operator

$$\hat{Q}\psi(x) = -\frac{i}{2}[x\frac{\partial\psi}{\partial x} + \frac{\partial(x\psi)}{\partial x}] = -ix\frac{\partial\psi}{\partial x} - \frac{i}{2}\psi \tag{12.40}$$

is self-adjoint, and its action does preserve the boundary conditions – see Problem 5.18 for other properties of $\hat{Q}$. The commutator

$$[\hat{x}, \hat{Q}] = i\hat{x} \tag{12.41}$$

implies the existence of a unitary representation of a Lie group with generators $\hat{x}$ and $\hat{Q}$.

We represent the group elements with the unitary operators $\hat{U}(a,b) = e^{ia\hat{x}+ib\hat{Q}}$. The function $\phi(x,a,b) := \hat{U}(a,b)\psi(x)$ satisfies the differential equations

$$\frac{\partial\phi}{\partial a} = iax\phi, \qquad \frac{\partial\phi}{\partial b} = x\frac{\partial}{\partial x}\phi + \frac{1}{2}\phi, \qquad (12.42)$$

subject to the initial condition $\phi(x,0,0) = \psi(x)$. The solutions to the first equation is $\phi(x,a,b) = \phi(x,0,b)e^{iax}$, and the solution to the second equation is $\phi(x,a,b) = \phi(e^b x, a, 0)e^{\frac{b}{2}}$. Hence,

$$\hat{U}(a,b)\psi(x) = e^{\frac{b}{2}+iax}\psi(e^b x). \qquad (12.43)$$

We evaluate the product of two such operators

$$\hat{U}(a_1,b_1)\hat{U}(a_2,b_2) = \hat{U}(a_2 + a_1 e^{b_2}, b_1 + b_2). \qquad (12.44)$$

Hence, the canonical group of this system consists of pairs $(a,b) \in \mathbb{R}^2$ with operation

$$(a_1,b_1)(a_2,b_2) = (a_2 + a_1 e^{b_2}, b_1 + b_2), \qquad (12.45)$$

that is, it coincides with the affine group Af_1.

The corresponding classical system has state space $\Gamma = \mathbb{R}^+ \times \mathbb{R}$ with a conjugate pair (x,p). The functions $F_{a,b} = ax + bxp$ generate the canonical action of the affine group on Γ

$$(x,p) \rightarrow \left(e^{sb}x, \left(p + \frac{a}{b}\right)e^{-bs} - \frac{a}{b}\right), \qquad (12.46)$$

for all $s \in \mathbb{R}$. The group Af_1 defines a quantization map that takes the functions $ax + bxp$ on Γ to the operators $a\hat{x} + b\hat{Q}$ on the Hilbert space $L_D^2(\mathbb{R}^+)$.

Example 12.3 In Section 5.6.3, we showed that a particle moving in a ring is described by the Hilbert space $L^2(S^1)$. For simplicity, we assume a circle with unit radius. The self-adjoint operators

$$\hat{c}\psi(\theta) = \cos\theta\,\psi(\theta), \quad \hat{s}\psi(\theta) = \sin\theta\,\psi(\theta), \quad \hat{\ell}\psi(\theta) = -i\frac{\partial\psi}{\partial\theta}, \qquad (12.47)$$

are defined on $L^2(S^1)$. Their commutators are

$$[\hat{c},\hat{\ell}] = -i\hat{s}, \quad [\hat{s},\hat{\ell}] = i\hat{c}, \quad [\hat{c},\hat{s}] = 0, \qquad (12.48)$$

and they represent the algebra $\mathfrak{iso}(2)$ of the Euclidean group, Eq. (C.29).

It is convenient to work with the unitary operators $\hat{U}(a,b,\phi) = e^{ia\hat{c}+ib\hat{s}}e^{i\phi\hat{\ell}}$ that act on wave functions as

$$\hat{U}(a,b,\phi)\psi(\theta) = e^{ia\cos\theta+ib\sin\theta}\psi[(\theta+\phi)\mathrm{mod}\,2\pi]. \qquad (12.49)$$

Then, we evaluate the group law $\hat{U}(a_1,b_1,\phi_1)\hat{U}(a_2,b_2,\phi_2) = \hat{U}(a',b',(\phi_1+\phi_2)\mathrm{mod}\,2\pi)$, where

$$\begin{pmatrix} a' \\ b' \end{pmatrix} = \begin{pmatrix} a_1 \\ b_1 \end{pmatrix} + \begin{pmatrix} \cos\phi_1 & \sin\phi_1 \\ -\sin\phi_1 & \cos\phi_1 \end{pmatrix}\begin{pmatrix} a_2 \\ b_2 \end{pmatrix}, \qquad (12.50)$$

in accordance with the group law of the Euclidean group $ISO(2)$, as defined by Eq. (C.4).

The corresponding classical system has phase space $S^1 \times \mathbb{R}$, with canonical pairs (θ,ℓ), with an angular coordinate $\theta \in [0,2\pi)$ and angular momentum $\ell \in \mathbb{R}$. The fundamental Poisson bracket is $\{\theta, \ell\} = 1$. The functions $F_{a,b,\phi} = a\cos\theta + b\sin\theta + \omega\ell$ generate the canonical action of $ISO(2)$ on Γ,

$$(\theta, \ell) \to ((\theta + \omega s)\mathrm{mod}\, 2\pi, \ell - \frac{a}{\omega}\sin(\theta + \omega s) - \frac{b}{\omega}\cos(\theta + \omega s)). \tag{12.51}$$

The group $ISO(2)$ provides the quantization map for a particle moving in a ring.

12.4 Dynamical Symmetries

Dynamical symmetries are symmetries of time evolution, that is, of the Hamiltonian. In this section, we analyse their properties, and we undertake a first investigation of related concepts, like internal symmetries and spontaneous symmetry breaking, that come to full fruition in the quantum theory of fields.

12.4.1 Noether's Theorem

A dynamical symmetry leaves the dynamics of the system invariant. Hence, the unitary representation $\hat{U}(g)$ of a group G on a Hilbert space $\mathcal{H}$ defines a dynamical symmetry if it leaves the evolution equation invariant. This means that, for any solution $|\psi(t)\rangle$ of Schrödinger's equation with Hamiltonian $\hat{H}$, the vectors $\hat{U}(g)|\psi(t)\rangle$ are also solutions,

$$i\frac{d}{dt}\hat{U}(g)|\psi(t)\rangle = \hat{H}\hat{U}(g)|\psi(t)\rangle, \tag{12.52}$$

for all $g \in G$.

Equation (12.52) implies that $i\frac{d}{dt}|\psi(t)\rangle = \hat{U}^\dagger(g)\hat{H}\hat{U}(g)|\psi(t)\rangle$. Since this condition applies for all initial states $|\psi(0)\rangle \in \mathcal{H}$, we conclude that the representation $\hat{U}(g)$ defines a dynamical symmetry, if

$$\hat{U}^\dagger(g)\hat{H}\hat{U}(g) = \hat{H}, \tag{12.53}$$

or equivalently if,

$$[\hat{U}(g), \hat{H}] = 0, \tag{12.54}$$

for all $g \in G$. This means that $\hat{U}(g)$ defines a dynamical symmetry, if the Hamiltonian is a Casimir of the group G.

Equation (12.54) implies that $\hat{U}(g)$ commutes with all functions of the Hamiltonian $\hat{H} = \sum_n E_n\hat{P}_n$ and, hence, with all its spectral projectors $\hat{P}_n$. Thus, each eigenspace of $\hat{H}$ carries a unitary representation of the group G defined by the operators $\hat{P}_n\hat{U}(g)\hat{P}_n$.

For Lie groups, Eq. (12.53 implies) that

$$[\hat{H}, \hat{T}] = 0, \tag{12.55}$$

for all generators $\hat{T}$ of the representation. The generators are self-adjoint operators, hence, physical observables. As explained in Section 7.2.4, Eq. (12.55) implies that $\hat{T}$ are conserved quantities, in the sense that their probability distributions are preserved in time. Hence, there is a direct connection between dynamical symmetries and conserved quantities, a statement known as *Noether's theorem*.

> **Theorem 12.8 (Noether's theorem)** *If a unitary representation $\hat{U}(g)$ of the Lie group G is a dynamical symmetry, then every generator of the representation defines a conserved quantity.*

In some cases, we can define conserved quantities associated to discrete groups. For example, this is possible if the unitary operators $\hat{U}(g)$ of the representation are also self-adjoint, like the parity operators $\hat{\mathbb{P}}$.

12.4.2 Energy Symmetry

A weaker notion from dynamical symmetry is that of an *energy symmetry*. A representation $\hat{U}(g)$ of the group G defines an energy symmetry, if the transformed Hamiltonian $\hat{H}_g := \hat{U}^\dagger(g)\hat{H}\hat{U}(g)$ is a conserved quantity,

$$[\hat{H}_g, \hat{H}] = 0, \tag{12.56}$$

for all $g \in G$.

Clearly, a dynamical symmetry is always an energy symmetry, but the converse does not hold. It turns out that some of the most important symmetries in physics, namely, spacetime symmetries, are energy symmetries rather than dynamical symmetries – see Section 14.4 for the Galilei group and Chapter 16 for the Poincaré group.

The two Hamiltonians $\hat{H}_g$ and $\hat{H}$ are related by a unitary transformation, so they share the same eigenvalues. Since they commute, they also share a common set of eigenvectors; that is, we can find eigenvectors $|E, r\rangle$ of $\hat{H}$, such that

$$\hat{H}_g|E, r\rangle = E_g|E, r\rangle, \tag{12.57}$$

where E_g is the eigenvalue of $\hat{H}_g$; r labels the corresponding eigenspace. It follows that $\hat{H}_g$ and $\hat{H}$ share the same eigenspaces.

For a Lie group, we choose $\hat{U}(g) = e^{i\hat{T}s}$ for some generator $\hat{T}$. Substituting into Eq. (12.56) and differentiating at $s = 0$, we obtain

$$[\hat{H}, [\hat{H}, \hat{T}]] = 0. \tag{12.58}$$

The generators evolve in the Heisenberg picture as $\hat{T}(t) = e^{i\hat{H}t}\hat{T}e^{-i\hat{H}t}$. Using the Hadamard identity and Eq. (12.58), we find

$$\hat{T}(t) = \hat{T} + i[\hat{H}, \hat{T}]t, \tag{12.59}$$

that is, $\hat{T}(t)$ increases linearly with time.

From Eq. (12.58), it is straightforward to show that the Heisenberg–Weyl group defines an energy symmetry for the free particle Hamiltonian $\hat{H} = \frac{\hat{p}^2}{2m}$, or for any Hamiltonian that is a function solely of $\hat{p}$.

12.4.3 The Symmetry of the Ground State

Let $\hat{U}(g)$ be a unitary representation of a Lie group G that defines an energy symmetry of the Hamiltonian $\hat{H}$. Let $|E\rangle$ be an eigenstate of $\hat{H}$ with a nondegenerate eigenvalue E. By Eq. (12.58), this implies that $(\hat{H}^2\hat{T} + \hat{T}\hat{H}^2 - 2\hat{H}\hat{T}\hat{H})|E\rangle = 0$ for all generators $\hat{T}$. It follows that

$(\hat{H} - E\hat{I})^2 \hat{T}|E\rangle = 0$. This is possible only if $\hat{T}|E\rangle$ is an eigenvector of $\hat{H}$ with eigenvalue E. Since E is a nondegenerate eigenvalue, it follows that

$$\hat{T}|E\rangle = \tau|E\rangle, \tag{12.60}$$

for some $\tau \in \mathbb{R}$.

We select a set of operators $\hat{T}_a$ that represent a basis of the Lie algebra $\mathfrak{g}$, and denote by τ_a their eigenvalues on $|E\rangle$. Then, the unitary operators $\exp\left(-i\sum_a s_a \tau_a\right)\hat{I}$ define a one-dimensional unitary representation of the Lie group G, on the subspace defined by $|E\rangle$. We obtain the same conclusion for a discrete group G that is a dynamical symmetry.

Let $\hat{T}$ be a generator that can be expressed as a commutator of two other generators $\hat{T} = i\left[\hat{S}_1, \hat{S}_2\right]$ of the representation.[1] Then, $\mathrm{Tr}\hat{T} = 0$. In a one-dimensional representation, this is only possible if $\hat{T}$ vanishes. It follows that

$$\hat{T}|E\rangle = 0, \tag{12.61}$$

for all nondegenerate eigenstates of $\hat{H}$. As a result, the associated subgroup of G leaves $|E\rangle$ invariant.

In semisimple Lie algebras, $[\mathfrak{g}, \mathfrak{g}] = \mathfrak{g}$, and all generators satisfy Eq. (12.61). Hence, the associated group representation satisfies $\hat{U}(g)|E\rangle = |E\rangle$ for all $g \in G$.

This conclusion is particularly important when employed for the ground state of the Hamiltonian $\hat{H}$.[2] Most physical systems have a unique ground state $|0\rangle$, even in the presence of an energy symmetry, or a dynamical symmetry. Consider, for example, the Schrödinger operator $\hat{H} = \frac{1}{2m}\hat{p}^2 + V(\hat{x})$, with a potential that has several minima at the same value V_{min}. The corresponding classical system has one state of minimum energy for each minimum of the potential. In contrast, by Theorem 5.1, the quantum ground state is unique. We saw that in the double delta potential system, where the ground state is a superposition of states concentrated at the different minima. A similar property holds for particles in three dimensions (Theorem 13.2).

12.4.4 Spontaneous Symmetry Breaking

We saw that any quantum system with an energy symmetry that corresponds to a semisimple algebra has an invariant ground state, if this state is unique.

In contrast, if there is more than one ground state, then each such state does not remain invariant under the action of $\hat{U}(g)$. If we denote the different ground states by $|0, r\rangle$ for some degeneracy index r, then there exist group elements $g \in G$ such that $\hat{U}(g)|0, r\rangle = |0, r'\rangle$ for some $r \neq r'$. Then, the ground state does not manifest the symmetry of the Hamiltonian, and we have a phenomenon known as *spontaneous symmetry breaking*.[3]

Let us explain the notion of spontaneous symmetry breaking with a concrete example. Consider a macroscopic sphere of some metal. Viewed as a quantum system, the sphere is

---

[1] $\hat{T}$ represents an element of the derived algebra $[\mathfrak{g}, \mathfrak{g}]$; see Section C.2.2 for definitions and properties.

[2] For relativistic systems, the Hamiltonian is an element of the derived algebra of the Poincaré group – see Section 16.1.

[3] The term "symmetry breaking" has been well established in the bibliography. However, it is rather misleading: The symmetry is not broken, it is merely hidden. The group G still defines a dynamical symmetry.

described by a Hilbert space $\mathcal{H}$ and a Hamiltonian $\hat{H}$. The Hamiltonian is invariant under spatial rotations, that is, $\hat{H}$ is invariant under the $SO(3)$ group that is represented by unitary operators $\hat{U}(O)$ on $\mathcal{H}$ with generators $\hat{J}_i$, for $i = 1, 2, 3$. Suppose we are interested in the magnetic properties of the ball, so that we study the vector operator $\hat{\mathbf{M}}$ of magnetization.

Since $\mathfrak{so}(3)$ is semisimple, if $\hat{H}$ has a unique ground state $|0\rangle$, then $\hat{J}_i$ will satisfy $\hat{J}_i|0\rangle = 0$. By the projection theorem, Eq. (11.136), the expectation value of any vector operator on $|0\rangle$ vanishes. In particular, this is the case for the magnetization vector $\hat{\mathbf{M}}$: $\langle 0|\hat{\mathbf{M}}|0\rangle = 0$. This is a property characteristic of *paramagnetic metals*, such as aluminum or sodium.

However, there also exist *ferromagnetic materials*, such as iron or nickel, that manifest spontaneous magnetization at sufficiently low temperatures (i.e., when the system's state is close to the ground state): $\langle \hat{\mathbf{M}} \rangle \neq 0$. This is only possible if the ground state is degenerate. There is an infinity of ground states, labeled by different directions of the magnetization vector, that is, by unit vectors $\mathbf{n} = \langle \hat{\mathbf{M}} \rangle / |\langle \hat{\mathbf{M}} \rangle|$. We denote such states by $|\mathbf{n}\rangle$. Certainly, these states are not invariant under rotations, as $\hat{U}(O)|\mathbf{n}\rangle = |O\mathbf{n}\rangle$.

Suppose that the sphere is initially at a high temperature with no net magnetization, and then we gradually lower the temperature towards zero. The sphere will settle in one of its ground states. The "choice" of the ground state is due to nonreproducible factors (the history of the metal, the presence of very weak uncontrolled external fields, and so on), and in this sense it is *spontaneous*.

Mathematically, spontaneous symmetry breaking appears only in systems with an infinite number of degrees of freedom. Such systems include quantum fields, and many-particle systems in the thermodynamic limit, where the volume V and the number of particles N tend to infinity, with their ratio remaining constant. This may appear paradoxical, since, after all, ferromagnetic materials in nature have a finite, if very large, number of atoms. But as we will show in a simple model, the large number of particles leads to *approximate degeneracy* of the ground state for large N, which is as good as exact degeneracy when the systems are subjected to even the tiniest of external perturbations.

Consider N particles of mass m in a one-dimensional ring of radius R, characterized by the classical Hamiltonian

$$H = \sum_{i=1}^{N} \frac{\ell_i^2}{8\pi^2 mR^2} + \frac{1}{2}m\omega^2 R^2 \sum_{i=1}^{N+1} (\theta_i - \theta_{i-1})^2, \qquad (12.62)$$

where we identify $\theta_{N+1} = \theta_1$. The Poisson brackets for this system are $\{\theta_i, \theta_j\} = 0, \{\ell_i, \ell_j\} = 0$, and $\{\theta_i, \ell_j\} = \delta_{ij}$.

It is straightforward to show that the total angular momentum $L = \sum_{i=1} \ell_i$ is a conserved quantity. It generates canonical transformations that shift each angle θ_i by a constant a. The corresponding symmetry group is $SO(2)$. The conjugate variable to L is $\Theta := \frac{1}{N}\sum_{i=1}^{N} \theta_i$, that is, the center-of-mass coordinate.

It is convenient to employ the $N-1$ coordinate variables $\chi_i = \theta_{i+1} - \theta_i$, $i = 1, 2, \ldots N-1$, with their associated conjugate momenta λ_i. It is a standard result in classical mechanics that the Hamiltonian splits as

$$H = \frac{L^2}{8\pi^2 NmR^2} + H_{rel}(\chi_i, \lambda_i), \qquad (12.63)$$

that is, as a sum of a free-particle Hamiltonian for the center of mass and a Hamiltonian term for the relative degrees of freedom. The latter correspond to oscillations with frequencies $\omega > 0$; their explicit form is not needed for present purposes. The lowest energy state of the system is obtained when there is zero energy on oscillations, so that $H_{rel} = 0$. Zero energy for oscillations implies that $\chi_i = 0$ and $\lambda_i = 0$.

Then, we have to deal only with the kinetic term of the center of mass. In classical physics, minimum energy corresponds to $L = 0$. Hence, there is an infinity of classical states of minimum energy, each corresponding to a different value of Θ. For each such state, the center of mass of the system is localized.

Consider now the corresponding quantum system. The internal degrees of freedom must be frozen in the ground state, so that the only relevant contribution to the Hamiltonian is the kinetic term for the center of mass, $\hat{H} = \frac{\hat{L}^2}{8\pi^2 NmR^2}$. As shown in Section 5.6.3, the energy eigenvalues are $E_n = \frac{1}{2NmR^2}n^2$, where $n \in \mathbb{Z}$. There is a unique ground state for $n = 0$, with wave function $\psi_0(\Theta) = \frac{1}{\sqrt{2\pi}}$. Indeed, the ground state is *invariant under SO(2) transformations*. The center of mass of the system is completely delocalized; it is described by a homogeneous probability density.

In the thermodynamic limit, we take N and R going to infinity with linear density $\nu = \frac{N}{2\pi R}$ held constant. Then, $E_n = \frac{2\pi^2\nu^2}{mN^3}n^2$. When $N \to \infty$ all states with finite n have the same energy, so all eigenstates $\psi_n(\Theta) = \frac{1}{\sqrt{2\pi}}e^{in\Theta}$ are degenerate ground states. Spontaneous symmetry breaking occurs.

Consider now the case of large but finite N. The difference between successive energy levels is of the order of $\Delta E = \frac{2\pi^2\nu^2}{mN^3}$. For linear density $\nu = 10\text{nm}^{-1}$, mass $m = 1$ amu, we obtain $\Delta E \sim (0.1eV)/N^3$. A single photon of the cosmic microwave background radiation carries energy $E \sim 0.6 \cdot 10^{-3}$ eV. Even for N as small as 10^5, the energy of such a photon is higher than the energy difference between the lowest $400,000$ energy levels for the center of mass.

This means that, even for relatively small N, the ground state is unstable with respect to even the tiniest perturbations. Such perturbations are usually local. For example, an external photon will be absorbed by a specific atom in the chain. The system is driven towards an effective ground state that is well localized around some point Θ_0. The exact point of localization is impossible to predict, as it depends on uncontrollable parameters of the environment. Each time we cool a ring, in order to obtain its ground state, we will find a different value of Θ_0, since this value is determined "spontaneously."

12.4.5 Internal Symmetries

Let us assume that there are N particle species that are completely identical in their fundamental characteristics, such as mass m and electric charge q, but we can still distinguish between them from the way they interact with other particles. We will call these particles A, and we will label the different species by the index $a = 1, 2, \dots, N$. Each particle is described by a Hilbert space $\mathcal{V}_a$ and a Hamiltonian $\hat{h}$. Since the A particles are identical with respect to everything except for the index a, we can write $\mathcal{V}_a = \mathcal{V}$ and $\hat{h}_a = \hat{h}$ for all a.

Suppose now that a process creates one A particle, but we have no way of knowing which one. We will describe this system by the Hilbert space $\mathcal{H} = \oplus_{a=1}^{N} \mathcal{V}_a$, so that $\mathcal{H}$ has all Hilbert spaces $\mathcal{V}_a$ as subspaces. Similarly, the Hamiltonian is $\hat{H} = \oplus_{a=1}^{N} \hat{h}_a$. Any orthonormal basis $|n\rangle$ on $\mathcal{V}$

Table 12.1 Internal symmetries

Name	Group	Particles involved	Status
isospin	$SU(2)$	proton, neutron	approximate
flavor	$SU(3)$	u, d, s quarks	approximate
color	$SU(3)$	three quark colors	exact
weak isospin	$SU(2)$	electron, neutrino	exact (broken)

defines an orthonormal basis $|n, a\rangle$ on $\mathcal{H}$. Since any vector on $\mathcal{H}$ can be written as $\sum_{n,a} c_{n,a} |n, a\rangle$, for some complex numbers $c_{n,a}$, $\mathcal{H}$ coincides with $\mathcal{V} \otimes \mathbb{C}^N$. With this identification, $\hat{H} = \hat{h} \otimes \hat{I}$. We say that the particles A form an N-plet.

There is a natural unitary representation of the group $U(N)$ on $\mathcal{H}$, by the unitary operators $\hat{U}(u) := \hat{I} \otimes \hat{u}$, where $\hat{u} \in U(N)$. This representation defines a dynamical symmetry of the Hamiltonian, since $(\hat{I} \otimes \hat{u}^\dagger)(\hat{h} \otimes \hat{I})(\hat{I} \otimes \hat{u}) = \hat{h} \otimes \hat{I}$.

This symmetry might appear rather contrived, and perhaps trivial. However, it becomes of the utmost importance if the terms in the Hamiltonian that couple those particles to the rest of the world remain invariant under this $U(N)$ symmetry, or a subgroup G thereof. Then, G defines a nontrivial dynamical symmetry, an *internal symmetry* for the A particles.

Internal symmetries are important, because our fundamental interactions (electromagnetic, and strong and weak nuclear forces) are defined in terms of such symmetries. Furthermore, many systems are well described by *approximate internal symmetries*. This means that the Hamiltonian $\hat{H}$ of the A particles is not invariant under the unitary operators $\hat{U}(u)$, but that it can be written as $\hat{H}_0 + \hat{H}_1$, such that $\hat{H}_0$ is invariant, $\hat{H}_1$ is not, and $\hat{H}_1$ is in some sense small in comparison to $\hat{H}_0$. For example, this may be the case if the masses m_a of the A particles differ but the ratio $|m_a - m_b|/m_a$ is very small, for all $a \neq b$.

The neutron and the proton satisfy this condition, as the neutron mass is 0.14 percent higher than that of the proton. The strong nuclear forces do not distinguish between protons and neutrons. Hence, it is useful to view the proton and neutron as forming a *doublet*, and that they are characterized by an $SU(2)$ internal symmetry. For historical reasons, this symmetry was called *isospin*.[4] Isospin is an exact symmetry for the strong interactions, but it is broken by the EM interactions – obviously, proton and neutron have different charges; hence, they interact very differently with other charged particles. What makes the notion of isospin useful is that the strong interactions are about 100 times stronger than the EM interactions; hence, even an approximate symmetry suffices for making good estimates.

Some of the internal symmetries found in nature are listed in Table 12.1.

12.5 Quantization of Systems with Constraints

Our fundamental theory of gravity (General Relativity) and our fundamental model for microscopic interactions (the Standard Model) share one characteristic. They describe *constrained systems*. It is therefore imperative to understand how to quantize such systems. Here, we undertake

---

[4] Because it has the same symmetry group with spin; see Chapter 14.

a preliminary investigation, explaining the meaning of constraints in classical mechanics and analyzing strategies for implementing the constraints in quantum systems.

12.5.1 Classical Constrained Systems

Consider a Hamiltonian system with a state space Γ and a Hamiltonian H. The system is *constrained* if its evolution takes place only in a subset C of Γ, the *constraint surface*. The constraint surface is typically determined by n independent functions $\phi_a : \Gamma \rightarrow \mathbb{R}$, $a = 1, 2, \ldots, n$, such that $\phi_a = 0$ for all a. Time evolution is compatible with the constraints, in the sense that it cannot take the system out of the constraint surface if the Hamiltonian satisfies $\{H, \phi_a\} = 0$ on C for all a.

A constrained system is called *first-class*, if $\{\phi_a, \phi_b\}$ vanishes on the constraint surface C. Since any function commutes with itself, all systems with a single constraint are first-class.

First-class constrained systems are of the greatest importance in physics: The fundamental microscopic forces (electromagnetism, weak and strong nuclear forces) and gravity are first-class constrained systems. Their full description requires a field-theoretic description that is, mostly, beyond the scope of this book. Here, we will present only the main ideas in the quantization of systems with first-class constraints, and their relation to the concept of *gauge symmetry*.

The constraints do not only specify the constraint surface C; they also generate canonical transformations. Any function $G = \sum r_a \phi_a$, where r_a is a function on Γ, generates a one-parameter family of canonical transformations on functions F on Γ, through the solution of the differential equation

$$\frac{dF_s}{ds} = \{F_s, G\} = \sum_a r_a \{F_s, \phi_a\} + \sum_a \{F_s, r_a\} \phi_a = \sum_a r_a \{F_s, \phi_a\} \quad \text{on} \quad C. \tag{12.64}$$

The maps $F \rightarrow F_s$ are known as *gauge transformations*. It can be shown that the familiar gauge transformations of the EM field, Eq. (1.24), is a special case of the transformations (12.64).

Only gauge-invariant functions define physical observables, that is, the functions that satisfy

$$\{F, \phi_a\} = 0 \quad \text{on} \quad C, \tag{12.65}$$

for all a. To understand this point, we must keep in mind that, in Hamiltonian mechanics, all physical observables generate canonical transformations. In the presence of constraints, the canonical transformations generated by an observable must not leave the constraint surface. This means that $X_F(\phi_a) = 0$ on C, where the operator X_F is given by Eq. (12.29). Hence, a function F is a physical observable if Eq. (12.65) holds. The Hamiltonian H is obviously a physical observable.

Consider a set of coordinate functions ξ_α that specify a point on the constraint surface. The change of these coordinate functions by the canonical transformation (12.64) corresponds to a solution curve $\xi_\alpha(s)$ of the differential equation

$$\frac{d\xi_\alpha(s)}{ds} = \sum_a r_a \{\xi_\alpha(s), \phi_a\} = -\sum_a r_a X_{\phi_a}[\xi_\alpha(s)]. \tag{12.66}$$

The key point is that all observables are constant along $\xi(s)$,

$$\frac{dF[\xi(s)]}{ds} = \sum_\alpha \frac{\partial F}{\partial \xi_\alpha} \frac{d\xi_\alpha(s)}{ds} = -\sum_a r_a \sum_\alpha \frac{\partial F}{\partial \xi_\alpha} X_{\phi_a}[\xi_\alpha(s)]$$

$$= -\sum_a r_a X_{\phi_a}[F[\xi(s)]] = \sum_a r_a \{F, \phi_a\}[\xi(s)] = 0.$$

Two points of C are *gauge-equivalent* if there is a curve $\xi_\alpha(s)$ generated by a gauge transformation that connects them. The set of all points in C that are gauge-equivalent to a point ξ defines a *gauge orbit* through ξ, denoted by O_ξ. Thus the constraint surface splits into a collection of orbits, such that each point of C belongs to only one orbit. The set of all orbits is the *reduced state space* Γ_{red}, that is, the state space of the true degrees of freedom of the system. Any function on C that can be used as a coordinate along the gauge orbits is called a *pure-gauge* variable.[5]

Let the dimension of Γ be $2N$. In presence of n constraints, the constraint surface has dimension $2N - n$. Each X_{ϕ_a} generates a transformation along a different direction in a gauge orbit, so the dimension of a gauge orbit is n. It follows that the dimension of the reduced state space is $2N - 2n$. The reduction procedure is sketched in Fig. 12.1.

Each observable F takes a single value in each orbit. Hence, it projects to a function $\tilde{F}$ on Γ_{red}, defined by

$$\tilde{F}(O_\xi) := F(\xi). \tag{12.67}$$

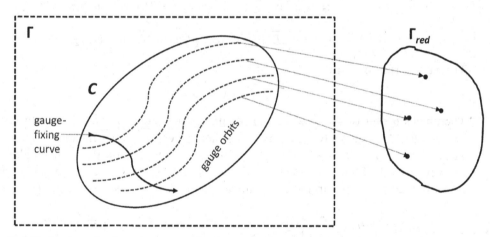

Fig. 12.1 Reduction of a first-class constrained system. The constraints restrict from the full space Γ to the constraint surface C, and they generate gauge orbits. Each gauge orbit defines a point of the reduced state space. We fix the gauge by selecting a curve that interesects each orbit once. Then, we treat the points of intersection as a representative of the orbits.

---

[5] Since pure-gauge variables are not observables, it is often convenient to give them arbitrary values. This procedure is called *gauge-fixing*. The reader may be familiar with this notion in classical electromagnetism, when we often impose auxiliary conditions on the magnetic potential $\mathbf{A}$, for example, the Coulomb-gauge condition $\nabla \cdot \mathbf{A} = 0$.

The Poisson bracket of two observables is also an observable – see Problem (12.9) – hence, the reduced state space also accepts a Poisson bracket, defined by

$$\{\tilde{F}, \tilde{G}\} = \widetilde{\{F, G\}}. \tag{12.68}$$

Thus, we obtain a well-defined Hamiltonian system on the reduced state space, with time evolution given by the projection $\tilde{H}$ of the Hamiltonian on the original state space Γ.

Example 12.4 We consider two harmonic oscillators of unit mass and frequency constrained so that their total energy is constant. The phase space $\Gamma = \{(x_1, p_1, x_2, p_2)\} = \mathbb{R}^4$ carries the standard Poisson bracket (12.24). The constraint function is $\phi = \frac{1}{2}(p_1^2 + p_2^2 + x_1^2 + x_2^2) - E$.

The constraint function generates the canonical transformations

$$\begin{pmatrix} x_1 \\ p_1 \\ x_2 \\ p_2 \end{pmatrix} \to \begin{pmatrix} x_1 \cos s + p_1 \sin s \\ p_1 \cos s - x_1 \sin s \\ x_2 \cos s + p_2 \sin s \\ p_2 \cos s - x_2 \sin s \end{pmatrix}.$$

On the constraint surface, we use the coordinates $\theta \in [0, \frac{\pi}{2}]$, $\phi_1 \in [0, 2\pi)$ and $\phi_2 \in [0, 2\pi)$, defined by

$$x_1 := \sqrt{2E} \cos \frac{\theta}{2} \cos \phi_1, \quad p_1 := \sqrt{2E} \cos \frac{\theta}{2} \sin \phi_1,$$
$$x_2 := \sqrt{2E} \sin \frac{\theta}{2} \cos \phi_2, \quad p_2 := \sqrt{2E} \sin \frac{\theta}{2} \sin \phi_2.$$

The gauge orbits on C are

$$(\theta, \phi_1, \phi_2) \to (\theta, \phi_1 - s, \phi_2 - s).$$

The angles θ and $\phi := \frac{1}{2}(\phi_2 - \phi_1) \in [0, 2\pi)$ are gauge-independent, hence, they define observables. They span a two-dimensional sphere S^2 which defines the reduced state space Γ_{red}. The coordinate $\chi := \frac{1}{2}(\phi_2 + \phi_1)$ transforms to $\chi - s$, so it is a natural parameter along the orbits.

To evaluate the Poisson bracket on Γ_{red}, we use the fact that χ is a pure-gauge variable, so it can be treated as a mere parameter and not a variable in the Poisson brackets. Then, we calculate,

$$1 = \{x_1, p_1\} = 2E \left\{ \cos \frac{\theta}{2} \cos(\chi - \phi), \cos \frac{\theta}{2} \sin(\chi - \phi) \right\} = \frac{E}{2} \sin \theta \{\phi, \theta\},$$

hence, $\{\phi, \cos \theta\} = \frac{2}{E}$. The Poisson bracket is the same as that of the spinning top in Example 12.1, with angular momentum norm equal to $\frac{1}{2}E$.

12.5.2 Quantization Methods

The key idea in the quantization of systems with constraints is that only the true degrees of freedom are physically relevant. We need not assign probabilities to pure-gauge variables. There are two distinct quantization procedures, depending on whether one enforces the constraints before or after quantization.

1. Solve constraints before quantization. If we solve the constraints classically, and construct the reduced state space, we have a proper Hamiltonian system that can be quantized with the methods described in Section 12.3. This method is known as *reduced–state-space quantization.*

In general, reduced–state-space quantization is the physically most transparent method, but also the most difficult to implement technically. It is transparent, because it is not concerned with the quantum representation of pure-gauge degrees of freedom. It is technically demanding because the reduced state space may not be a continuous and differentiable surface. There are some procedures for quantizing such systems; however, the degree of difficulty is such as to discourage the attempt of all but the simplest cases.

2. Solve constraints after quantization. In this method, we first quantize the initial system with the state space Γ prior to the imposition of the constraints. In the resulting Hilbert space $\mathcal{H}$, we define self-adjoint operators $\hat{\phi}_a$ to represent the constraints. We also define an operator $\hat{H}$ to represent the Hamiltonian, and we require that $[\hat{H}, \hat{\phi}_a] = 0$ for all a.

Then, we define the *physical Hilbert space* $\mathcal{H}_{phys}$ as the zero eigenspace of all the constraints, that is, the subspace that consists of all vectors $|\psi\rangle$, such that

$$\hat{\phi}_a |\psi\rangle = 0. \tag{12.69}$$

Physical states are elements of $\mathcal{H}_{phys}$, and probabilities are defined using the structure of $\mathcal{H}_{phys}$. Furthermore, any self-adjoint operator $\hat{X}$ on $\mathcal{H}$ that satisfies $[\hat{X}, \hat{\phi}_a] = 0$ is a physical observable. When $\hat{X}$ acts on $|\psi\rangle \in \mathcal{H}_{phys}$, we find that $\hat{\phi}_a \hat{X} |\psi\rangle = 0$, that is, $\hat{X}$ projects to a well-defined operator on $\mathcal{H}_{phys}$. This procedure is known as *Dirac quantization.*

There are many technical complications in Dirac quantization; for example, the definition of $\mathcal{H}_{phys}$ is rather intricate when the constraint operators have continuous spectrum at zero. Nonetheless, it is the most widely used method for the quantization of constrained systems. The reason is that the constraint equation (12.69) is amenable to many approximation schemes (for example, semiclassical approximations, perturbation theory), and it can also be implemented in path integrals.

Example 12.5 We will consider the quantization of the system of two oscillators with fixed total energy E that was analyzed in Example 12.4. Reduced–state-space quantization is straightforward since we showed that the reduced state space coincides with a two-sphere of radius $\frac{1}{2}E$.

The canonical group of the sphere is $SO(3)$, hence, the quantum systems are identified by the unitary irreducible representations of its double cover $SU(2)$. There is one such representation on $\mathbb{C}^N$ for each integer $N = 1, 2, \ldots$, and it corresponds to angular momentum $j = \frac{1}{2}(N-1)$. Since j is to be identified with the radius of the two-sphere, we obtain that $E = N - 1$, that is, we can only quantize pairs of oscillators if the energy is an integer. No quantization is possible for other values of energy.

For Dirac quantization, the Hilbert space for the two-oscillator system is $L^2(\mathbb{R}^2)$. We use the standard creation and annihilation operators in order to express the constraint operator as

$$\hat{\phi} = \hat{a}_1^\dagger \hat{a}_1 + \hat{a}_2^\dagger \hat{a}_2 - \tilde{E}\hat{I}, \tag{12.70}$$

where we wrote $\tilde{E} = E - 1$, that is, we subtracted from E the vacuum energy of the oscillators. We use the number basis $|n_1, n_2\rangle$ with $n_1, n_2 = 0, 1, 2, \ldots$ for the two oscillators. The constraint equation $\hat{\phi}|n_1, n_2\rangle = 0$ is solved for $\tilde{E} = n_1 + n_2$. Hence, $\tilde{E} = 0, 1, 2, \ldots$. For fixed $\tilde{E}$, there are

$\tilde{E} + 1$ different pairs (n_1, n_2) that add up to $\tilde{E}$. Hence, the dimension N of the Hilbert space for energy $\tilde{E}$ is $N = \tilde{E} + 1 = E$. We obtain the same result with reduced–state-space quantization, modulo the difference in the vacuum energy ($N = E$ versus $N = E + 1$).

We also note that the operators

$$\hat{J}_1 := \frac{1}{2}(\hat{a}_1^\dagger \hat{a}_2 + \hat{a}_2^\dagger \hat{a}_1), \quad \hat{J}_2 := \frac{i}{2}(\hat{a}_2^\dagger \hat{a}_1 - \hat{a}_1^\dagger \hat{a}_2), \quad \hat{J}_3 := \frac{1}{2}(\hat{a}_1^\dagger \hat{a}_1 - \hat{a}_2^\dagger \hat{a}_2)$$

satisfy the angular momentum algebra. It is straightforward to evaluate the Casimir $\mathbf{J}^2 = \frac{1}{4}\hat{N}^2 + \hat{N}$, where $\hat{N} = \hat{a}_1^\dagger \hat{a}_1 + \hat{a}_2^\dagger \hat{a}_2 = \hat{\phi} + \tilde{E}\hat{I}$. Hence, in the zero subspace of $\hat{\phi}$, the Casimir $\mathbf{J}^2$ is proportional to unity, in accordance with Schur's lemma.

In Example 12.5, the two methods lead to almost the same results. There is a difference in the energy associated to the quantum states, because in Dirac quantization we must incorporate the contribution of the vacuum energy. The two methods usually agree in simple problems. However, there is no mathematical reason why they should always produce the same results, and they often disagree (Romano and Tate, 1989). There has so far been no system in which different predictions by these methods could be tested by experiment. Hence, at the moment, we must think of both methods as equally reliable for the quantization of constrained systems, differing only in their practicality for specific problems.

QUESTIONS

12.1 The consensus among specialists is that *quantization is an art, not an algorithm*. What aspects of quantization have you encountered that cannot fit into general rules, and have to proceed on a case-by-case basis?

12.2 Suppose that the canonical group of a classical system has many inequivalent unitary representations. How should we proceed to select the correct one?

12.3 Can the canonical group of a quantum system define a dynamical symmetry?

12.4 Consider the particle ring of Section 12.4.4. Assuming that the atoms in the ring almost touch each other, what is the largest number of atoms N for which we can create a stable delocalized ground state at temperature $T = 1\,K$? What is the corresponding ring radius R?

12.5 In Dirac quantization, pure-gauge quantities are described by operators on the Hilbert space $\mathcal{H}$. In reduced–state-space quantization, pure-gauge quantities are not quantized. Are they classical or quantum variables?

PROBLEMS

12.1 Show that in an irreducible representation of the Heisenberg–Weyl group, the annihilation operator has a single zero eigenstate $|0\rangle$.

12.2 Consider a representation $\hat{U}(a, b)\psi(x) = e^{\frac{b}{2} + iatx}\psi(e^b x)$ for the affine group, where $t \in \mathbb{R}$. Show that if $t > 0$, all representations are unitarily equivalent to the representation (12.43). Show that, for $t < 0$, all representations are unitarily equivalent to the representation for $t = -1$ and unitarily inequivalent to the representation (12.43). What is the physical meaning of the two inequivalent representations?

12.3 The linear transformation $(x, p) \to (ax + bp, cx + dp)$ on $\mathbb{R}^2$ is canonical for $ad - bc = 1$. (i) Specify the generators of these canonical transformations in the classical state space. (ii) Identify the unitary operators $\hat{U}$ on $L^2(\mathbb{R})$ that implement those transformations quantum mechanically: $\hat{U}\hat{x}\hat{U}^\dagger = a\hat{x} + b\hat{p}$ and $\hat{U}\hat{p}\hat{U}^\dagger = c\hat{x} + d\hat{p}$.

12.4 In the Hilbert space $\mathcal{H}$ of a single harmonic oscillator, we define the operators $\hat{K}_1 = \frac{1}{4}(\hat{a}^{\dagger 2} + \hat{a}^2)$, $\hat{K}_2 = \frac{i}{4}(\hat{a}^2 - \hat{a}^{\dagger 2})$, and $\hat{K}_3 = \frac{1}{2}(\hat{a}^\dagger \hat{a} + \frac{1}{2}\hat{I})$. (i) Show that these operators form a representation of the Lie algebra $\mathfrak{sl}(2, \mathbb{R})$. (ii) Show that the Casimir $\hat{C}^{(2)}$ is a multiple of the unity. (iii) Construct creation and annihilation operators $\hat{K}_\pm = \hat{K}_1 \pm i\hat{K}_2$ and show that they raise/lower the number states by two. (iv) Show that there are two distinct irreducible representations, one corresponding to even and one to odd number states.

12.5 Consider a system of n oscillators with the same frequency labeled by $i = 1, 2, \ldots, n$ described on the Hilbert space $\mathcal{H}^n = L^2(\mathbb{R}^n)$. For any $n \times n$ matrix A_{ij}, define the operator $\hat{T}(A) = \sum_{i,j} A_{ij} \hat{a}_i^\dagger \hat{a}_j$, where $\hat{a}_i$ are the annihilation operators of the ith oscillator. (i) Show that $[\hat{T}(A), \hat{T}(B)] = \hat{T}([A, B])$. (ii) Show that the Hilbert space carries a representation of the algebra $\mathfrak{su}(n)$. (iii) Show that the Hamiltonian is a Casimir for $\mathfrak{su}(n)$.

12.6 A particle moving on a sphere is described classically by the standard spherical coordinates θ and ϕ, and their conjugate momenta p_θ and p_ϕ. (i) Express p_θ and p_ϕ as components of a three-vector $\mathbf{p}$ tangent to the sphere. (ii) The canonical group G for this system consists of transformations $(\mathbf{n}, \mathbf{p}) \to (O\mathbf{n}, O\mathbf{p} + \mathbf{a})$ where $O \in SO(3)$ and $a \in \mathbb{R}^3$. Write the associated generators. (iii) In quantum theory, the system is described by square integrable wave functions $\psi(\theta, \phi)$ on S^2. Write the natural representation of G on these wave functions. (iv) Identify the corresponding generators and find their spectrum.

12.7 Write all possible self-adjoint operators that correspond to the classical function $F(x) = x^3 p^3$.

12.8 Prove Eqs. (12.25) and (12.26).

12.9 Show that the Poisson bracket of two observables in a first-class constrained system is also an observable.

12.10 Consider the system of two harmonic oscillators with constant energy difference, that is, with the constraint

$$\phi = \frac{1}{2}(p_1^2 + x_1^2 - p_2^2 - x_2^2) - \delta.$$

(i) Show that the reduced state space is a cylinder $\mathbb{R} \times S^1$, and identify the fundamental Poisson brackets. (ii) Show that the functions $K_1 = \frac{1}{2}(x_1 p_2 + x_2 p_1)$, $K_2 = \frac{1}{2}(x_1 x_2 - p_1 p_2)$ and $K_3 = \frac{1}{4}(p_1^2 + x_1^2 + p_2^2 + x_2^2)$ represent the algebra $\mathfrak{sl}(2, \mathbb{R})$. (iii) Show that K_1, K_2, and K_3 are observables and express them in terms of coordinates on the reduced state space.

12.11 Quantize the system of Problem 12.10 using the Dirac method. (i) Show that quantization is possible only if δ is an integer. (ii) Identify the quantum operators $\hat{K}_1, \hat{K}_2$, and $\hat{K}_3$ that correspond to the generators of $\mathfrak{sl}(2, \mathbb{R})$ in the previous problem and verify that they still satisfy the $\mathfrak{sl}(2, \mathbb{R})$ algebra. (iii) Evaluate the corresponding Casimir operator. (iv) Define the ladder operators $\hat{K}_\pm = \hat{K}_1 \pm i\hat{K}_2$, and identify the unitary irreducible representations of $SL(2, \mathbb{R})$ for each value of δ.

Bibliography

- For an introduction to group theory with an emphasis on applications in physics, I recommend Jones (1998) or Ramond (2010). For a more mathematical introduction, see Hall (2004). For an account of quantum theory that strongly emphasizes the role of symmetries, see Mackey (1963a).
- The full development of quantization techniques requires a background in differential geometry. For quantization with the canonical group, see Isham (1983). For geometric quantization, see Souriau (1997). For a review of different quantization procedures, see Twareque Ali and Englis (2005).
- For constrained systems and their quantization, see Sundermeyer (1982).

13 Particles in Three Dimensions

There would be a very real difficulty in supposing that the (force) law $1/r^2$ held down to zero values of r. For the force between two charges at zero distance would be infinite; we should have charges of opposite sign continually rushing together and, when once together, no force would be adequate to separate them ... Thus the matter in the universe would tend to shrink into nothing or to diminish indefinitely in size.

Jeans (1915)

13.1 The Schrödinger Operator in Three Dimensions

The study of composite systems typically requires their analysis into simpler systems. In classical physics, the simplest systems are *particles*, that is, pointlike bodies that move in space. A particle is traditionally described by three position coordinates and three momenta, so the associated state space is $\mathbb{R}^6$. In Chapter 14, we will see that this description is incomplete: Particles also have an additional degree of freedom called *spin*.

In many cases, we can ignore spin, or we can study it in separation from the other degrees of freedom. Then, the six degrees of freedom or position and momentum suffice for describing a pointlike particle. We call a particle "pointlike" when we are not interested in its internal structure. This does not mean that the particle is elementary, but that the existence or not of an internal structure is not relevant to the issue under study. For example, when we study the electronic structure of an atom, we can ignore the internal structure of the nucleus; when we study the motion of an ion in an EM field, we focus only on the ion's center of mass.

Ignoring spin, a particle is described by a Hilbert space $\mathcal{H}$ with operators $\hat{x}_i$ and $\hat{p}_i$ subject to the commutation relations

$$[\hat{x}_i, \hat{x}_j] = 0, \qquad [\hat{p}_i, \hat{p}_j] = 0, \qquad [\hat{x}_i, \hat{p}_j] = i\delta_{ij}\hat{I}, \tag{13.1}$$

where $i, j = 1, 2, 3$. We saw in Section 12.2 that all representations of the canonical commutation relations are equivalent to the Schrödinger representation, in which: (i) $\mathcal{H} = L^2(\mathbb{R}^3, d^3x)$, (ii) the position operators act by multiplication, $\hat{x}_i\psi(\mathbf{x}) = x_i\psi(\mathbf{x})$, and (iii) the momentum operators act by differentiation: $\hat{p}_i\psi(\mathbf{x}) = -i\frac{\partial}{\partial x_i}\psi(\mathbf{x})$.

For a free particle of mass m, the Hamiltonian is $\hat{H} = \frac{1}{2m}\hat{\mathbf{p}}^2$. In the presence of an external potential $V(\mathbf{x})$, the Hamiltonian is

$$\hat{H} = \frac{1}{2m}\hat{\mathbf{p}}^2 + V(\hat{\mathbf{x}}). \tag{13.2}$$

13.1.1 The Quantum Two-Body Problem

The Schrödinger operator (13.2) also describes pairs of interacting particles. To see this, consider a pair of particles with masses m_1 and m_2 that interact through the potential $V(\hat{\mathbf{x}}_1 - \hat{\mathbf{x}}_2)$. The Hamiltonian is of the form

$$\hat{H} = \frac{\hat{\mathbf{p}}_1^2}{2m_1} + \frac{\hat{\mathbf{p}}_2^2}{2m_2} + V(\hat{\mathbf{x}}_1 - \hat{\mathbf{x}}_2), \tag{13.3}$$

where $\hat{\mathbf{x}}_a, \hat{\mathbf{p}}_a$, $a = 1, 2$, are vector operators for the particle's position and momentum. We define the center-of-mass position vector $\hat{\mathbf{R}} := (m_1\hat{\mathbf{x}}_1 + m_2\hat{\mathbf{x}}_2)/(m_1 + m_2)$, the center-of-mass momentum $\hat{\mathbf{P}} := \hat{\mathbf{p}}_1 + \hat{\mathbf{p}}_2$, the relative position $\hat{\mathbf{r}} := \hat{\mathbf{x}}_1 - \hat{\mathbf{x}}_2$, and the relative momentum $\hat{\mathbf{p}} := (m_1\hat{\mathbf{p}}_1 - m_2\hat{\mathbf{p}}_2)/(m_1 + m_2)$. It is straightforward to verify that each pair $(\hat{\mathbf{R}}, \hat{\mathbf{P}})$ and $(\hat{\mathbf{r}}, \hat{\mathbf{p}})$ satisfies the canonical commutation relations.

The Hamiltonian (13.3) can be expressed as $\hat{H} = \hat{H}_{CM} + \hat{H}_{rel}$, where $\hat{H}_{CM} = \frac{1}{2M}\hat{\mathbf{P}}^2$ is the Hamiltonian of the center of mass, describing a free particle of mass $M = m_1 + m_2$, and $\hat{H}_{rel} = \frac{\hat{\mathbf{p}}^2}{2\mu} + V(\hat{\mathbf{r}})$ is the Hamiltonian for the relative degrees of freedom. It coincides with the Schrödinger operator for a particle of mass $\mu = m_1 m_2/(m_1 + m_2)$. The quantity μ is the *reduced mass* of the two-particle system.

13.1.2 Energy Eigenstates

If $V(\mathbf{x}) \geq 0$ for all $\mathbf{x}$, then the Schrödinger operator $\hat{H}$ of Eq. (13.2) is positive. Furthermore, if $V(\mathbf{x})$ has a minimum value V_{min}, then $\hat{H} \geq V_{min}\hat{I}$.

Let $V(\mathbf{x})$ be a potential that vanishes at infinity: $\lim_{r \to \infty} V(\mathbf{x}) = 0$, where $r = |\mathbf{x}|$. In Chapter 5, we saw that, for particles in one dimension, positive energy solutions correspond to continuous spectrum and scattering states and negative energy solutions correspond to discrete spectrum and bound states.

This pattern holds in three dimensions if the potential is everywhere attractive. There are no bound states of positive energy E, because any positive energy eigenfunction satisfies $\nabla^2 \psi + k^2(\mathbf{x})\psi = 0$, where $k^2(\mathbf{x}) = E - V(\mathbf{x}) > 0$, and it is oscillatory at infinity. However, if the potential oscillates between positive and negative values at large distances, *bound states of positive energy are possible*. We present an example in Box 13.1.

An important property of the eigenstates of the Schrödinger operator follows from the simple commutator identity $[\hat{\mathbf{x}} \cdot \hat{\mathbf{p}}, \hat{H}] = i(2\hat{T} - \hat{\mathbf{x}} \cdot \nabla \hat{V})$, where $\hat{T} = \frac{\hat{\mathbf{p}}^2}{2m}$ is the kinetic energy operator. Since $\langle \psi | [\hat{\mathbf{x}} \cdot \hat{\mathbf{p}}, \hat{H}] | \psi \rangle = 0$ for any eigenstate $|\psi\rangle$ of $\hat{H}$, we obtain the so-called *virial theorem*,

$$\langle \psi | \hat{T} | \psi \rangle = \frac{1}{2} \langle \psi | \hat{\mathbf{x}} \cdot \nabla \hat{V} | \psi \rangle. \tag{13.4}$$

Box 13.1   Bound States of Positive Energy

We will construct a potential that admits bound states of positive energy, following von Neumann and Wigner (1929). To this end, we write the eigenvalue equation $\hat{H}\psi = E\psi$ for the Hamiltonian (13.2) as

$$V = E + \frac{\nabla^2 \psi}{2m\psi} \tag{A}$$

and we look for solutions of the form $\psi(\mathbf{x}) = \frac{\sin kr}{kr} f(r)$, where $r = |\mathbf{x}|$ and $E = \frac{k^2}{2m}$. The wave function ψ is square integrable if $f(r)$ vanishes as r^{-a} at infinity, for $a > \frac{1}{2}$. For $f(r) = 1$, $\psi(\mathbf{x})$ is a solution of the eigenvalue equation for $V = 0$. Substituting this trial wave function for ψ in Eq. (A), we obtain

$$V(r) = \frac{1}{m} k \cot(kr) \frac{f'}{f} + \frac{1}{2m} \frac{f''}{f}. \tag{B}$$

The potential $V(r)$ is bounded only if f'/f vanishes on the poles of $\cot(kr)$, that is, for $\sin(kr) = 0$. We choose the function $f(r) = \frac{1}{1+[2kr-\sin(2kr)]^2}$ that, indeed, satisfies this property. Since $f(r)$ decays with r^{-2} at large r, ψ is square integrable. Substituting in Eq. (B), we find

$$V(r) = -\frac{64k^2 \sin^4(kr)}{m[1+[2kr-\sin(2kr)]^2]^2} + \frac{48k^2 \sin^4(kr) - 8k^2[2kr-\sin(2kr)]\sin(2kr)}{m[1+[2kr-\sin(2kr)]^2]}.$$

Hence, we found a potential that admits the trial function ψ as a positive-energy bound state. This potential vanishes with r^4 as $r \to 0$, and decays as $\sin(kr)/r$ for large r. It is a long-range potential that decays with alternating sign for large r.

13.1.3 Existence of a Ground State

As explained in Section 7.2.1, the spectrum of a physical Hamiltonian must be bounded from below. In one-dimensional systems, a potential that is unbounded from below makes the spectrum of the associated Schrödinger operator also unbounded from below. However, in three dimensions, a minimum-energy state exists even in potentials that are unbounded from below. One such example is the Coulomb potential $V(r) = -\frac{\kappa}{r}$, for some constant κ. In classical physics, there is no state of minimum energy, because the energy goes to $-\infty$ as $r \to 0$. Indeed, this was the main problem in Rutherford's model for the atom.

The following theorem provides a solid criterion for the existence of a ground state in a Schrödinger operator.

Theorem 13.1 (Hardy's inequality) $\langle \psi | \hat{\mathbf{p}}^2 | \psi \rangle \geq \frac{1}{4} \langle \psi | \hat{r}^{-2} | \psi \rangle$, for all $|\psi\rangle \in L^2(\mathbb{R}^3)$.

Proof First, we write $\langle \psi | \hat{\mathbf{p}}^2 | \psi \rangle = -\int d^3x \psi^* \nabla^2 \psi = -\int d^3x \nabla \cdot (\psi^* \nabla \psi) + \int d^3x |\nabla \psi|^2$. By Gauss' theorem, the first term equals the surface integral $-\int_{S_\infty} d^2\sigma \cdot (\psi^* \nabla \psi)$ on a sphere S_∞ at infinity; it vanishes for square integrable functions. Hence, $\langle \psi | \hat{\mathbf{p}}^2 | \psi \rangle = \int d^3x |\nabla \psi|^2$.

We write $\psi(\mathbf{x}) = g(\mathbf{x})/\sqrt{r}$. Square integrability of ψ implies that $g(0) = g(\infty) = 0$. Then,

$$|\nabla \psi|^2 = \frac{1}{4} \frac{|\psi|^2}{r^2} + \frac{|\nabla g|^2}{r} - \frac{\frac{\partial}{\partial r}|g|^2}{2r^2}.$$

In spherical coordinates, $d^3x = r^2 \sin\theta \, dr d\theta d\phi$. Integration of the last term in the equation above leads $\int_0^\infty dr \frac{\partial |g|^2}{\partial r} = g(\infty) - g(0) = 0$. Hence, $\int d^3x [|\nabla \psi|^2 - \frac{1}{4} \frac{|\psi|^2}{r^2}] \geq \int d^3x \frac{|\nabla g|^2}{r} \geq 0$. $\qquad \square$

Hardy's inequality implies that $\langle\psi|\hat{H}|\psi\rangle \geq \langle\psi|\frac{1}{8m}\hat{r}^{-2} + V(\hat{\mathbf{x}})|\psi\rangle$. Hence, the Hamiltonian is bounded from below, if $V(\mathbf{x}) + \frac{1}{8mr^2}$ has a global minimum, that is, if there exists a constant V_0, such that

$$V(\mathbf{x}) \geq -\frac{1}{8mr^2} + V_0. \tag{13.5}$$

For r approaching zero, V_0 is negligible in comparison with the other term in the right-hand side of Eq. (13.5). This implies that the Hamiltonian is bounded from below if the potential diverges to $-\infty$ slower than $-r^{-2}$ as $r \to 0$. The Coulomb potential satisfies this criterion.

If Eq. (13.5) is satisfied, then $\hat{H} \geq V_0\hat{I}$, and the ground-state energy E_0 always satisfies $E_0 \geq V_0$. For example, for the Coulomb potential $V(r) = -\frac{\kappa}{r}$, the minimum of $V(r) + \frac{1}{8mr^2}$ is achieved for $r = (4\kappa m)^{-1}$ and it equals $V_0 = -2\kappa^2 m$. This value is smaller than the exact expression for the ground state energy $E_0 = -\frac{1}{2}\kappa^2 m$ (to be derived in Section 13.3) by a factor of four.

Unlike the one-dimensional case, the eigenvalues of the Schrödinger operators in three dimensions can be degenerate. An exception is the ground state.

Theorem 13.2 *If the Schrödinger operator has a ground state, then it is unique, and the associated wave function $\psi_0(\mathbf{x})$ can be chosen to be real.*

In Section 13.2, we give a proof of this theorem for central potentials. For the general case, see section 10.5 of Teschl (2014).

13.2 Central Potentials: General Properties

A potential $V(r)$ is called *central* if it depends only on the radius $r = |\mathbf{x}|$. Schrödinger operators with central potential are important, because the symmetry allows for the derivation of many properties of their spectrum. For this reason, they are often used as a first approximation in modeling systems such as atoms and nuclei.

In this section, we analyze the spectrum of the Schrödinger operator for such potentials.

13.2.1 The Hamiltonian

Since the operators $\hat{\mathbf{p}}^2$ and $\hat{r} = \sqrt{\hat{\mathbf{r}}^2}$ are scalars, they commute with the angular momentum generators $\hat{\ell}_i$, defined by Eq. (11.73). Hence, the Schrödinger operator $\hat{H} = \frac{1}{2m}\hat{\mathbf{p}}^2 + V(\hat{r})$ satisfies $[\hat{H}, \hat{\ell}_i] = 0$. The Hamiltonian is a scalar operator with respect to rotations, and angular momentum is a conserved quantity.

In what follows, we will use the following identity,

$$\hat{\boldsymbol{\ell}}^2 = \hat{\mathbf{x}}^2\hat{\mathbf{p}}^2 - (\hat{\mathbf{x}}\cdot\hat{\mathbf{p}})^2 + i(\hat{\mathbf{x}}\cdot\hat{\mathbf{p}}). \tag{13.6}$$

Equation (13.6) is similar to the vector identity (5.7), the only difference being one additional term in the right-hand side, due to the noncommutativity of position and momentum.

To prove Eq. (13.6), we write $\hat{\boldsymbol{\ell}}^2 = \sum_{ijklm} \epsilon_{ijk}\epsilon_{ilm}\hat{x}_j\hat{p}_k\hat{x}_l\hat{p}_m$. By Eq. (5.6),

$$\hat{\boldsymbol{\ell}}^2 = \sum_{jklm} (\delta_{jl}\delta_{km} - \delta_{jm}\delta_{kl})\hat{x}_j\hat{p}_k\hat{x}_l\hat{p}_m.$$

The first term becomes $\sum_{jklm} \delta_{jl}\delta_{km}(\hat{x}_j\hat{x}_l\hat{p}_k\hat{p}_m - i\delta_{kl}\hat{x}_j\hat{p}_m) = \hat{\mathbf{x}}^2\hat{\mathbf{p}}^2 - i\hat{\mathbf{x}}\cdot\hat{\mathbf{p}}$ after a single exchange of $\hat{p}_k$ with $\hat{x}_l$. The term $-\sum_{jklm} \delta_{jm}\delta_{kl}\hat{x}_j\hat{p}_k\hat{x}_l\hat{p}_m$ simplifies after two exchanges. The first exchange between $\hat{p}_m$ and $\hat{x}_l$ yields a term $-i\hat{\mathbf{x}}\cdot\hat{\mathbf{p}}$; the second one, between $\hat{p}_k$ and $\hat{x}_l$, yields a term $+3i\hat{\mathbf{x}}\cdot\hat{\mathbf{p}}$. The end result is $-(\hat{\mathbf{x}}\cdot\hat{\mathbf{p}})^2 + 2i\hat{\mathbf{x}}\cdot\hat{\mathbf{p}}$. We add the two terms to obtain Eq. (13.6).

We write Eq. (13.6) as

$$\hat{\mathbf{p}}^2 = \hat{r}^{-2}[(\hat{\mathbf{x}}\cdot\hat{\mathbf{p}})^2 - i(\hat{\mathbf{x}}\cdot\hat{\mathbf{p}})] + \hat{r}^{-2}\hat{\boldsymbol{\ell}}^2. \tag{13.7}$$

The operator $\hat{\mathbf{x}}\cdot\hat{\mathbf{p}} = -ir\sum_i \frac{x_i}{r}\frac{\partial}{\partial x_i} = -ir\sum_i \frac{\partial x_i}{\partial r}\frac{\partial}{\partial x_i} = -ir\frac{\partial}{\partial r}$ is purely radial, and so is the operator,

$$\hat{r}^{-2}[(\hat{\mathbf{x}}\cdot\hat{\mathbf{p}})^2 - i(\hat{\mathbf{x}}\cdot\hat{\mathbf{p}})] = -\frac{\partial}{\partial r^2} - \frac{2}{r}\frac{\partial}{\partial r}. \tag{13.8}$$

Recall from Section 11.4.1 that the Hilbert space $L^2(\mathbb{R}^3)$ can be expressed as a tensor product $L^2(\mathbb{R}^3, d^3x) = \mathcal{H}_{rad}\otimes\mathcal{H}_{S^2}$. This implies that $\hat{r}^{-2}[(\hat{\mathbf{x}}\cdot\hat{\mathbf{p}})^2 + i(\hat{\mathbf{x}}\cdot\hat{\mathbf{p}})] = \hat{\Pi}_r\otimes\hat{I}$, where $\hat{\Pi}_r := -\frac{\partial}{\partial r^2} - \frac{2}{r}\frac{\partial}{\partial r}$ acts on the Hilbert space $\mathcal{H}_{rad}$.

The elements of $\mathcal{H}_{rad}$ are functions $R(r)$ that satisfy $\int_0^\infty dr r^2|R(r)|^2 < \infty$. Equivalently, we can use functions $u(r) = rR(r)$ that satisfy $\int_0^\infty dr|u(r)|^2 < \infty$. Then, $\hat{\Pi}_r u(r) = -\frac{d^2 u}{dr^2}$. The operator $\hat{\Pi}_r$ is self-adjoint if u satisfies Dirichlet boundary conditions, $u(0) = 0$. Then, $\mathcal{H}_{rad} = L_D^2(\mathbb{R}^+, dr)$, and $\hat{\Pi}_r$ coincides with the operator $\widehat{p^2}$ of Section 5.6.2.

Since the operator $\hat{\boldsymbol{\ell}}^2$ acts on $\mathcal{H}_{S^2}$, we can write the Schrödinger operator as

$$\hat{H} = \left[\frac{1}{2m}\hat{\Pi}_r + V(\hat{r})\right]\otimes\hat{I} + \frac{1}{2m}\hat{r}^{-2}\otimes\hat{\boldsymbol{\ell}}^2. \tag{13.9}$$

For vectors of the form $|\Psi\rangle\otimes|\ell, m_\ell\rangle$,[1] the eigenvalue equation for the Hamiltonian (13.9) implies that $\hat{H}_\ell|\Psi\rangle = E|\Psi\rangle$, where

$$\hat{H}_\ell = \frac{1}{2m}\hat{\Pi}_r + V(\hat{r}) + \frac{\ell(\ell+1)}{2m}\hat{r}^{-2} \tag{13.10}$$

is a self-adjoint operator on $\mathcal{H}_{rad}$.

We denote the eigenvectors of $\hat{H}_\ell$ as $|E, \ell\rangle$. Keep in mind that ℓ in Eq. (13.10) is an index that distinguishes between the different operators $\hat{H}_\ell$ on $\mathcal{H}_{rad}$ and is not a label of operator eigenvalues.

When representing $|E, \ell\rangle$ by wave functions $R_{E,\ell}(r)$, the eigenvalue equation for $\hat{H}_\ell$ becomes

$$-\frac{1}{2m}\frac{\partial^2 R_{E,\ell}}{\partial r^2} - \frac{1}{mr}\frac{\partial R_{E,\ell}}{\partial r} + \left[\frac{\ell(\ell+1)}{2mr^2} + V(r)\right]R_{E,\ell} = ER_{E,\ell}. \tag{13.11}$$

---

[1] We denote the eigenvalues of $\hat{\ell}_3$ as m_ℓ, in order to avoid confusion with the mass m.

We obtain a simpler expression when we use $u_{E,\ell}(r) = r R_{E,\ell}(r)$ to represent $|E, \ell\rangle$,

$$-\frac{1}{2m}\frac{d^2 u_{E,\ell}}{dr^2} + \left[\frac{\ell(\ell+1)}{2mr^2} + V(r)\right] u_{E,\ell}(r) = E u_{E,\ell}(r). \tag{13.12}$$

Equation (13.12) is a Schrödinger equation in the half-line with an effective potential $V_{eff}(r) = \frac{\ell(\ell+1)}{2mr^2} + V(r)$ and Dirichlet boundary conditions at $r = 0$.

We conclude that, for any energy E that solves Eq. (13.12), the wave function

$$\Psi_{E,\ell,m_\ell}(r,\theta,\phi) = \frac{1}{r} u_{E,\ell}(r) Y_{\ell m_\ell}(\theta,\phi) \tag{13.13}$$

is an eigenfunction of the Schrödinger operator with eigenvalue E.

For potentials that diverge more slowly than $\frac{1}{r^2}$ near $r = 0$ (or take finite values), the contribution of the potential can be neglected. Then, setting $u_{E,\ell} \sim r^\alpha$, for $\alpha > 0$, we find $\alpha(\alpha - 1) = \ell(\ell+1)$. The only positive solution to this equation is $\alpha = \ell + 1$. Hence, solutions to Eq. (13.12) behave as $r^{\ell+1}$ as $r \to 0$.

13.2.2 Discrete Spectrum

Next, we study square integrable solutions to Eq. (13.12). By Theorem 5.1, the eigenvalues of Eq. (13.12) are not degenerate. The comparison theorem (Theorem 5.2 adapted for the half-line) implies that we can label by using the *node number* $n_r = 0, 1, 2, \ldots$, so that the n_rth excited eigenfunction of $\hat{H}_\ell$ has n_r nodes.

The usual convention is to label the eigenstates with the *principal quantum number* $n := n_r + \ell + 1 > \ell$, rather than n_r. Hence, we write the eigenvalues of the Hamiltonian as $E_{n,\ell}$. Since energy increases with n_r for constant ℓ,

$$E_{n+1,\ell} > E_{n,\ell}. \tag{13.14}$$

The eigenvalues $E_{n,\ell}$ do not depend on the label m_ℓ of the spherical harmonics. Since there are $2\ell + 1$ different values of m for each ℓ, each eigenvalue $E_{n,\ell}$ has degeneracy $2\ell + 1$. This degeneracy is caused by the invariance of the Hamiltonian under rotations; the actual degeneracy may be larger if the Hamiltonian is invariant under a larger group of symmetries.

The operators $\hat{H}_\ell$ of Eq. (13.10) satisfy $\hat{H}_{\ell+1} - \hat{H}_\ell = \frac{\ell+1}{m}\hat{r}^{-2} \geq 0$. By Weyl's ordering theorem (Theorem 10.18), the nth eigenvalue of $\hat{H}_{\ell+1}$ is larger than the nth eigenvalue of $\hat{H}_\ell$. Hence, of two energy eigenvalues with the same node number n_r, the one with larger ℓ is larger. It follows that

$$E_{n+1,\ell+1} > E_{n,\ell}. \tag{13.15}$$

Equation (13.15) implies that the ground state *always has the lowest value of ℓ, namely, $\ell = 0$.* There is no rotational degeneracy; hence, the ground state has energy $E_{1,0}$ and it is unique. The eigenfunction $u_{1,0}$ for $\hat{H}_\ell$ has zero nodes. Since it is real, it can be chosen to be everywhere positive. By Eq. (11.84), $Y_{00}(\theta,\phi)$ equals a positive constant. We conclude that the wave function $\Psi_{1,0,0}(r,\theta,\phi) = \frac{1}{r} u_{1,0}(r) Y_{00}(\theta,\phi)$ of the ground state of the Schrödinger operator is positive. Thus, we proved Theorem 13.2 for the special case of a central potential.

An important theorem provides a relation between the sign of $\nabla^2 V(r)$ and the order of the eigenvalues $E_{n,\ell}$ (Baumgartner et al., 1984).

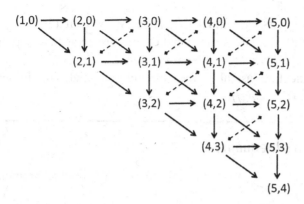

Fig. 13.1 The order of the energy eigenvalues for a central potential that satisfies $\nabla^2 V(r) < 0$. An arrow from (n, ℓ) to (n', ℓ') signifies that $E_{n, \ell} < E_{n', \ell'}$. The dashed lines represent the gaps in the ordering, that is, pairs of eigenvalues whose order depends on the potential.

Theorem 13.3 *(i) If $\nabla^2 V(r) < 0$ for all r, then $E_{n, \ell} \leq E_{n, \ell+1}$. (ii) If $\nabla^2 V(r) > 0$ for all r, then $E_{n, \ell} \geq E_{n, \ell+1}$.*

The following attractive potentials are of particular interest.

- $V(r) = -\frac{C}{r^a}$ for $a, C > 0$. Then, $\nabla^2 V = a(a - 1)V(r)/r^2$. Hence, $a > 1$ corresponds to case (i) and $a < 1$ corresponds to case (ii).

- $V(r) = Kr^a$ for $a, K > 0$. Then, $\nabla^2 V = a(a + 1)Kr^{a-2} > 0$. Case (ii) applies always.

- The *Yukawa potential*, $V(r) = -g\frac{e^{-\mu r}}{r}$, $\mu, g > 0$, describes the strong interaction between nucleons. It satisfies $\nabla^2 V = \mu^2 V < 0$, hence case (i) applies.

In atomic systems, case (i) of Theorem 13.3 applies. Then, the energy eigenvalues are ordered as follows:

$$E_{n, \ell} < E_{n+1, \ell} < E_{n+1, \ell+1}, \qquad E_{n, \ell} < E_{n, \ell+1} < E_{n+1, \ell+1}.$$

The inequalities above do not order all eigenvalues. In particular, there is no general relation between $E_{n+1, \ell}$ and $E_{n, \ell+1}$; their relation depends on the potential. Figure 13.1 shows the order of the first few eigenvalues. For an explicit calculation of the ordering of eigenvalues, see Example 13.1.

The number N_ℓ of bound state solutions to Eq. (13.12) with angular momentum ℓ is estimated by Bargmann's inequality (Bargmann, 1952; Schwinger, 1960). In Box 13.2, we present an elementary proof, from Schmidt (2002).

Theorem 13.4 *For a central potential $V(r)$, such that $\lim_{r \to \infty} V(r) = 0$,*

$$N_\ell < \frac{I_V}{2\ell + 1}, \tag{13.16}$$

where $I_V := 2m \int_0^\infty dr r V_-(r)$ and $V_-(r) := \max\{0, -V(r)\}$.

For sufficiently large ℓ, $N_\ell < 1$, hence, $N_\ell = 0$. Let ℓ_{max} be the largest value of ℓ for which the right-hand side of Eq. (13.16) is greater than unity, $\ell_{max} = \frac{1}{2}(I_V - 1)$. Hence, the total number N_- of bound states for the Schrödinger operator is

$$N_- := \sum_{\ell=0}^{\ell_{max}}(2\ell + 1)N_\ell \leq \sum_{\ell=0}^{\ell_{max}} I_V = (\ell_{max} + 1)I_V = \frac{1}{2}I_V(I_V + 1). \tag{13.17}$$

For a sufficiently weak potential, it is possible that $I_V < 1$. Then, $N_- < 1$, that is, $N_- = 0$; no bound states exist. This contrasts with the one-dimensional case where an attractive potential has at least one bound state.

Bargmann's inequality is not particularly useful for the Coulomb potential, since in this case $I_V = \infty$. For the Yukawa potential, $I_V = 2mg/\mu$.

Box 13.2 Proof of Bargmann's Inequality

Consider the operator $\hat{H}_\ell = -\frac{1}{2m}\frac{\partial^2}{\partial r^2} + \frac{\ell(\ell+1)}{2mr^2} + V(r)$ of Eq. (13.10) with Dirichlet boundary conditions at $r = 0$.

If $\hat{H}_\ell$ has N negative eigenvalues, then, by Sturm's theorem, Box 5.1, the eigenfunction ψ_0 of $\hat{H}_\ell$ with zero eigenvalue has N nodes, or $N+1$ zeros if we include the zero at $r = 0$.

Consider two successive zeros, r_1 and r_2. We define the function $f(r) := r\frac{\psi_0'(r)}{\psi_0(r)}$ for $r \in (r_1, r_2)$. Since r_1 and r_2 are successive zeros, $\psi(r)$ is either only positive or only negative for all $r \in (r_1, r_2)$. If $\psi(r) > 0$, then $\psi'(r_1) > 0$ and $\psi'(r_2) < 0$; if $\psi(r) < 0$, then $\psi'(r_1) < 0$ and $\psi'(r_2) > 0$. In both cases, $\lim_{r \to r_{1+}} f(r) = \infty$ and $\lim_{r \to r_2} f(r) = -\infty$. By continuity, f passes through all real values as r varies in (r_1, r_2).

The eigenvalue equation $\hat{H}_\ell \psi_0 = 0$, implies that $f' = -\frac{1}{r}(f+\ell)(f-\ell-1) + 2mrV$. We note that $f' \geq 2mrV$ if $-\ell \leq f \leq \ell+1$. Since f passes through all real values in (r_1, r_2), there is an interval $(r_+, r_-) \subset (r_1, r_2)$, in which $f \in [-\ell, \ell+1]$, with $f(r_+) = \ell+1$ and $f(r_-) = -\ell$. Then,

$$\int_{r_1}^{r_2} dr\, r V_-(r) \geq \int_{r_+}^{r_-} dr\, r V_-(r) \geq -\int_{r_+}^{r_-} dr\, r V(r)$$
$$\geq -\frac{1}{2m}\int_{r_1}^{r_-} dr f' = -\frac{1}{2m}[f(r_-) - f(r_+)] = \frac{2\ell+1}{2m}.$$

We conclude that each interval (r_1, r_2) between two successive zeros of ψ_0 contributes a term larger than $\frac{2\ell+1}{2m}$ to the integral $\int_0^\infty dr\, r V(r)$. Since there are at least N such intervals, $\int_0^\infty dr\, r V_-(r) > N\frac{2\ell+1}{2m}$, that is, we obtained Bargmann's inequality.

We also note two useful identities that follow from the form of the Hamiltonian (13.10). First, differentiating Eq. (13.10) with respect to ℓ (keeping n_r fixed), and using the Feynman–Hellmann theorem, we find that the expectation $\langle \hat{r}^{-2} \rangle$ on any eigenstate $|n, \ell, m_\ell\rangle$ of the Hamiltonian is

$$\langle \hat{r}^{-2} \rangle = \frac{m}{\ell + \frac{1}{2}} \left(\frac{\partial E_{n,\ell}}{\partial \ell} \right)_{n_r}. \tag{13.18}$$

Second, Eq. (5.55) applied to V_{eff} implies that

$$\langle \hat{r}^{-3} \rangle = \frac{m}{\ell(\ell+1)} \langle V'(r) \rangle. \tag{13.19}$$

13.2.3 The Free Particle

The case of a free particle $V(r) = 0$ can be solved straightforwardly by noting that the Hamiltonian $\hat{H}$ is a function of the momentum $\hat{\mathbf{p}}$, so the generalized eigenstates $|\mathbf{p}\rangle$ of the latter are also generalized eigenstates of the Hamiltonian, $\hat{H}|\mathbf{p}\rangle = \frac{\mathbf{p}^2}{2m}|\mathbf{p}\rangle$. Each generalized energy eigenvalue $E = \frac{\mathbf{p}^2}{2m}$ has infinite degeneracy, corresponding to all vectors $\mathbf{p}$ with the same norm $|\mathbf{p}|$.

A different eigenbasis for the Hamiltonian is obtained by treating the free particle Hamiltonian as a trivial case of a Schrödinger operator with vanishing central potential. This basis is particularly useful when studying the Schrödinger operator with potentials that vanish at infinity.

For $V(r) = 0$, Eq. (13.11) becomes

$$-\frac{1}{2m}\frac{\partial^2 R_{E,\ell}}{\partial r^2} - \frac{1}{mr}\frac{\partial R_{E,\ell}}{\partial r} + \frac{\ell(\ell+1)}{2mr^2}R_{E,\ell} = ER_{E,\ell}. \tag{13.20}$$

It is convenient to label eigenfunctions with $k = \sqrt{2mE}$, and to use a dimensionless coordinate $x = kr$. Then, Eq. (13.20) becomes

$$x^2\frac{d^2 R_{k,\ell}}{dx^2} + 2x\frac{dR_{k,\ell}}{dx} + \left[x^2 - \ell(\ell+1)\right]R_{k,\ell} = 0, \tag{13.21}$$

Equation (13.21) admits as solutions the spherical Bessel functions of the first kind $j_\ell(x)$ and of the second kind $\eta_\ell(x)$. The definition and properties of these functions are given in Appendix B.6.

The functions η_ℓ diverge with $x^{-(\ell+1)}$ at $x = 0$. They are not admissible solutions because $xR_{k,\ell}(x)$ must vanish at $x = 0$. Hence,

$$R_{k,\ell}(r) = C_\ell(k)j_\ell(kr), \tag{13.22}$$

where $C_\ell(k)$ is a normalization constant. We identify $R_{k,\ell}(r)$ with a ket $|k,\ell\rangle$ on $\mathcal{H}_{rad}$. By Eq. (B.40), the normalization condition $\langle k,\ell|k',\ell\rangle = \delta(k - k')$ determines

$$C_\ell(k) = k\sqrt{\frac{2}{\pi}}.$$

Hence, we represent generalized eigenvectors $|k,\ell,m_\ell\rangle = |k,\ell\rangle \otimes |\ell,m_\ell\rangle$ of the free-particle Hamiltonian by the wave functions

$$\psi_{k,\ell,m_\ell}(r,\theta,\phi) = R_{k,\ell}(r)Y_{\ell,m_\ell}(\theta,\phi) = k\sqrt{\frac{2}{\pi}}j_\ell(kr)Y_{\ell,m_\ell}(\theta,\phi). \tag{13.23}$$

Example 13.1 Assume that a free particle is enclosed in a spherical box of radius r_B. This means that $R_{k,\ell}(r_B) = 0$. By Eq. (13.22), this is possible only if $x := ka/\pi$ is a positive solution to the equation $j_\ell(x\pi) = 0$. By the comparison theorem, the energy eigenvalues (for fixed ℓ) are ordered by the number of nodes n_r of the wave functions. Therefore, we denote the n_rth positive solution to equation $j_\ell(x\pi) = 0$ by $s_{n_r,\ell}$. The energy eigenvalues, expressed in terms of the principal quantum number $n = n_r + \ell + 1$, are

$$E_{n,\ell} = \frac{\pi^2}{2mR^2}s_{n-\ell-1,\ell}^2. \tag{13.24}$$

Since $j_0(x) = \frac{\sin x}{x}$, $s_{n_r,0} = n_r + 1$. For $\ell \geq 1$, the numbers $s_{n_r,\ell}$ must be computed numerically. Specific values are given in the following table.

	$n_r = 0$	$n_r = 1$	$n_r = 2$	$n_r = 3$	$n_r = 4$
$\ell = 1$	1.43	2.46	3.47	4.48	5.48
$\ell = 2$	1.83	2.90	3.92	4.94	5.95
$\ell = 3$	2.22	3.32	4.36	5.39	6.40
$\ell = 4$	2.60	3.72	4.78	5.82	6.85
$\ell = 5$	2.98	4.13	5.20	6.26	7.29

It is then straightforward to find the order of eigenvalues,

$$E_{1,0} < E_{2,1} < E_{3,2} < E_{2,0} < E_{4,3} < E_{3,1} < E_{5,4} < E_{4,2} < E_{3,0} < E_{5,3} < \cdots .$$

Theorem 13.3 does not apply here, because the bounding box is equivalent to an infinite discontinuity in the potential. Nonetheless, we find the ordering $E_{n,\ell} > E_{n,\ell+1}$ predicted by the theorem's case (ii).

13.2.4 Scattering States for Short-Range Potentials

A short-range potential is characterized by a length parameter r_0, the potential's range, such that $V(r) = 0$ for $r > r_0$. This means that, for $r > r_0$, the eigenfunctions $R_{k,\ell}(r)$ satisfy Eq. (13.21) with solutions

$$R_{k,\ell}(r) = A_\ell(k)j_\ell(kr) + B_\ell(k)\eta_\ell(kr), \tag{13.25}$$

for some constants $A_\ell(k)$ and $B_\ell(k)$. As shown in Appendix B.6, the spherical Bessel functions behave asymptotically as

$$j_\ell(x) \simeq \frac{\sin\left(x - \frac{\ell\pi}{2}\right)}{x}, \quad \eta_\ell(x) \simeq -\frac{\cos\left(x - \frac{\ell\pi}{2}\right)}{x}. \tag{13.26}$$

Hence, for sufficiently large r, $R_{k,\ell}(r)$ is proportional to

$$x^{-1}\left[A_\ell(k)\sin\left(x - \frac{\ell\pi}{2}\right) - B_\ell(k)\cos\left(x - \frac{\ell\pi}{2}\right)\right]$$

$$= \frac{C_\ell(k)}{x}\left[\cos\delta_\ell \sin\left(x - \frac{\ell\pi}{2}\right) + \sin\delta_\ell \cos\left(x - \frac{\ell\pi}{2}\right)\right]$$

$$= \frac{C_\ell(k)}{x}\sin\left(x - \frac{\ell\pi}{2} + \delta_\ell(k)\right),$$

where we defined

$$C_\ell(k) = \sqrt{A_\ell(k)^2 + B_\ell(k)^2}, \qquad \delta_\ell(k) = -\tan^{-1}[B_\ell(k)/A_\ell(k)], \tag{13.27}$$

so that $A_\ell(k) = C_\ell(k)\cos\delta_\ell(k)$ and $B_\ell(k) = -C_\ell(k)\sin\delta_\ell(k)$.

The quantities $\delta_\ell(k)$ are called *phase shifts*, because they express how the short-range potential modifies the asymptotic oscillating behavior of the free-particle eigenfunctions.[2] In Chapter 19, we will see that they are crucial for the derivation of probabilities in scattering experiments.

---

[2] Even though phase shifts typically appear inside trigonometric functions, we do not define them modulo 2π, but we let them take values in the full real axis. This way we can count the number of 2π circles that a phase shift goes through as the energy increases.

Hence, for $r > r_0$,

$$R_{k,\ell}(r) = C_\ell(k)[\cos \delta_\ell j_\ell(kr) - \sin \delta_\ell \eta_\ell(kr)]. \qquad (13.28)$$

The value of $C_\ell(k)$ are fixed by the normalization condition $\langle k, \ell | k', \ell' \rangle = \delta(k - k')$, where the ket $|k, \ell\rangle$ represented $R_{k,\ell}$ in $\mathcal{H}_{rad}$. Again, we construct the generalized eigenvectors $|k, \ell, m\rangle = |k, \ell\rangle \otimes |\ell, m\rangle$ of the Hamiltonian. These kets are represented by the wave functions $\psi_{k,\ell,m}(r, \theta, \phi) = R_{k,\ell}(r) Y_{\ell,m}(\theta, \phi)$. We will come back to these eigenfunctions when studying scattering theory for central potentials, in Section 19.3.

Example 13.2 The simplest potential that allows an analytic calculation of the phase shifts is the hard sphere potential,

$$V(r) = \begin{cases} \infty, & r < a \\ 0, & r \geq 0 \end{cases}, \qquad (13.29)$$

where $a > 0$. In this potential, $R_{k,\ell}(r) = 0$ for $r < a$, while Eq. (13.25) applies for $r \geq a$. Continuity of the eigenfunctions at $r = a$ implies that $\cos \delta_\ell j_\ell(ka) - \sin \delta_\ell \eta_\ell(ka) = 0$. Hence,

$$\tan \delta_\ell(k) = \frac{j_\ell(ka)}{\eta_\ell(ka)}. \qquad (13.30)$$

Since $j_0(x) = \frac{\sin x}{x}$ and $\eta_0(x) = -\frac{\cos x}{x}$, Eq. (13.30) implies that $\delta_0(k) = -ka$.

Example 13.3 There is a remarkable relation between the number of bound states in a potential and phase shifts. To show this, we assume that the system is enclosed in a spherical box of radius r_B. We take r_B to be so large that the asymptotic form of the solution $R_{k,\ell}(r) \sim r^{-1} \sin\left[kr - \frac{1}{2}\ell\pi + \delta_\ell(k)\right]$ applies near the box. With Dirichlet boundary conditions at $r = r_B$, the bound states correspond to solutions of the equation

$$kr_B - \frac{1}{2}\ell\pi + \delta_\ell(k) = n\pi, \qquad (13.31)$$

with respect to k, where n can be any positive integer.

We can always take r_B so large that the rate of change of $\delta_\ell(k)$ is much smaller than r_B. This means that the left-hand side of Eq. (13.31) is an increasing function of k. Hence, the values of k that solve Eq. (13.31) increase with n. Equivalently, this means that the left-hand side of Eq. (13.31) equals π times the number $\Omega_\ell^{sc}(k)$ of scattering states with energy less than $\frac{k^2}{2m}$.

By definition, $\Omega_\ell^{sc}(0) = 0$, which implies that $\delta_\ell(0) = \frac{1}{2}\ell\pi$, hence,

$$\Omega_\ell^{sc}(k) = \frac{1}{\pi}[kr_B + \delta_\ell(k) - \delta_\ell(0)]. \qquad (13.32)$$

The first term, kr_B/π, is the number-of-states function for a free particle in the box. As shown in Section 5.6.2, there is one-to-one correspondence between the eigenstates of any two Schrödinger operators in a box, and their eigenvalues coincide in the limit of large k. Hence, in this limit, kr_B/π converges to the number-of-states function $\Omega(k)$ for the potential V as $k \to \infty$. Since $\Omega(k)$ also counts the bound states, the total number N_ℓ of the latter equals $\Omega(\infty) - \Omega_\ell^{sc}(\infty)$, that is,

$$N_\ell = \frac{1}{\pi}[\delta_\ell(\infty) - \delta_\ell(0)]. \qquad (13.33)$$

Equation (13.33) is known as *Levinson's theorem* (Levinson, 1949). It applies to all potentials as long as there is no bound state for $k = 0$ and $\ell = 0$. Otherwise, we must subtract a factor of $\frac{1}{2}\pi$ from N_0.

13.3 Central Potentials: Exact Solutions

In this section, we study two central potentials for which the Schrödinger operator admits exact solutions: the isotropic harmonic oscillator and the Coulomb potential. The latter solution essentially describes the hydrogen atom.

These two systems can be solved exactly because the Hamiltonian is invariant under a larger group of transformations, and not only rotations. For the isotropic oscillator, the corresponding symmetry has been identified in Problem 12.5. The case of the Coulomb potential is analyzed in Section 13.3.3, where the discrete spectrum of the Hamiltonian is rederived through the theory of group representations.

13.3.1 Isotropic Harmonic Oscillator

The isotropic harmonic oscillator is characterized by the potential

$$V(r) = \frac{1}{2}m\omega^2 r^2 - \frac{3}{2}\omega. \tag{13.34}$$

We subtracted the constant $\frac{3}{2}\omega$, so that the ground state has zero energy.

The isotropic oscillator corresponds to three one-dimensional oscillators with the same frequency, one for each of the spatial directions. Hence, the eigenvectors of the Hamiltonian can be expressed as $|n_1, n_2, n_3\rangle$, where $n_i = 0, 1, 2, \ldots$, and the eigenvalues are

$$E_N = N\omega, \tag{13.35}$$

where $N = n_1 + n_2 + n_3$. The degeneracy $g(E_N)$ of the energy E_N is the number of all triples of integers (n_1, n_2, n_3) that sum up to N. This means that, for each n_1, $n_2 + n_3 = N - n_1$. This equation is satisfied by $N - n_1 + 1$ pairs (n_2, n_3), because n_2 takes values $0, 1, \ldots, N - n_1$ and n_3 is fully specified by n_2. Hence,

$$g(E_N) = \sum_{n_1=0}^{N}(N + 1 - n_1) = (N + 1)^2 - \sum_{n_1=0}^{N} n_1$$

$$= (N + 1)^2 - \frac{1}{2}N(N + 1) = \frac{1}{2}(N + 1)(N + 2). \tag{13.36}$$

Alternatively, we find the eigenvalues of energy by solving the differential equation (13.12). The natural length scale of the systems is $1/\sqrt{m\omega}$ and the natural energy scale is ω. We use the dimensionless variables $\xi = \sqrt{m\omega}\, r$ and $\epsilon = E/\omega$, in order to bring Eq. (13.12) into the form

$$\frac{d^2u}{d\xi^2} + \left(2\epsilon + 3 - \xi^2 - \frac{\ell(\ell + 1)}{\xi^2}\right)u = 0, \tag{13.37}$$

where the prime denotes a differentiation with respect to ξ. For simplicity, we dropped the indices E and ℓ from u.

For large ξ, Eq. (13.37) simplifies to $\frac{d^2u}{d\xi^2} - \xi^2 u = 0$, which accepts an approximate solution $u \simeq e^{-\xi^2/2}$. In Section 13.2.1, we showed that, for small ξ, $u \sim \xi^{\ell+1}$. We consider the ansatz $u(\xi) = \xi^{\ell+1} e^{-\frac{\xi^2}{2}} f(\xi)$, so that $f(\xi)$ interpolates between the asymptotic forms of the solution at $\xi = 0$ and $\xi \to \infty$. Hence, f must be a constant at $\xi = 0$ and grow slower than $e^{\frac{\xi^2}{2}}$ as $\xi \to \infty$.

With the above ansatz, we obtain

$$\frac{d^2f}{d\xi^2} + 2\left(\frac{\ell+1}{\xi} - \xi\right)\frac{df}{d\xi} + 2(\epsilon - \ell)f = 0. \tag{13.38}$$

A change of variables to $x = \frac{1}{2}\xi^2$ brings Eq. (13.38) into the form

$$xf'' + \left(\ell + \frac{3}{2} - 2x\right)f' + (\epsilon - \ell)f = 0, \tag{13.39}$$

where the prime stands for derivative with respect to x.

Equation (13.38) is a special case of the differential equation

$$xf'' + (1 + \alpha + \beta x)f' + \beta\gamma f = 0, \tag{13.40}$$

for $\alpha = \ell + \frac{1}{2}, \beta = -2$ and $\gamma = \frac{1}{2}(\ell - \epsilon)$. In Box 13.3, it is shown that the solutions to Eq. (13.40) satisfy the following properties.

Theorem 13.5 *For $\alpha > 0$, Eq. (13.40) admits two solutions, f_1 that is constant at $x = 0$ (regular) and f_2 that diverges with $x^{-\alpha}$ at $x = 0$ (singular). For $\gamma = -j$, where $j = 0, 1, 2, \ldots, f_1(x) = L_j^{(\alpha)}(-\beta x)$, where L_j^α are polynomials of order j, known as the* generalized Laguerre polynomials.[3] *For $\gamma \neq -j, f_1(x) \sim e^{-\beta x}$ at large x.*

Since $\alpha > 0$, only the solution f_1 is admissible by the boundary conditions at $x = 0$. The asymptotic behavior $e^{-\beta x}$ for solutions with $\gamma \neq -j$ corresponds to $e^{2x} = e^{\xi^2}$, and it is unacceptable as it leads to nonsquare integrable function u. Hence, we obtain bound-state solutions only if $\gamma = -j$. Except for $x = 0$, $u(x) = 0$ only if $f(x) = 0$. If $f(x)$ is a polynomial of degree j, then j coincides with the number n_r of nodes for $u(r)$. It follows that $2n_r - \epsilon + \ell = 0$, and we recover Eq. (13.35) for $N = 2n_r + \ell$.

In terms of the principal quantum number $n = n_r + \ell + 1$,

$$E_{n,\ell} = \omega[2(n-1) - \ell]. \tag{13.41}$$

As predicted by Theorem 13.3, $E_{n,\ell} > E_{n,\ell+1}$. We also note that N is even for even ℓ and odd for odd ℓ.

The degeneracy of the isotropic harmonic oscillator is much larger than $2\ell + 1$. This is due to the fact that the Hamiltonian is invariant under a larger group of transformations, namely $SU(3)$ – see Problem 12.5.

The eigenfunctions $u_{n,\ell}(r)$ of the Hamiltonian $\hat{H}_\ell$ are

$$u_{n,\ell}(r) = C_{n,\ell}(\sqrt{m\omega}r)^{\ell+1} e^{-\frac{m\omega r^2}{2}} L_{n-\ell-1}^{\left(\ell+\frac{1}{2}\right)}\left(m\omega r^2\right), \tag{13.42}$$

---

[3] See Appendix B.9.

where $C_{n,\ell}$ is a normalization constant. By Eq. (B.66), the normalization condition $\int_0^\infty dr |u_{n,\ell}(r)|^2 = 1$ implies that

$$C_{n,\ell} = (m\omega)^{1/4} \sqrt{\frac{2(n-\ell-1)!}{\Gamma(n+\ell+\frac{3}{2})}}. \tag{13.43}$$

Box 13.3 Solutions to the Differential Equation (13.40)

We express solutions to the differential equation $xf'' + (1 + \alpha + \beta x)f' + \beta\gamma f = 0$ as a power series in x: $f(x) = \sum_{k=0}^\infty c_k x^{k+\lambda}$ for some real λ. Then, $xf''(x) = (\lambda - 1)\lambda c_0 + \sum_{k=0}^\infty (k+\lambda+1)(k+\lambda)c_{k+1}x^{k+\lambda}$, $f'(x) = \lambda c_0 + \sum_{k=0}^\infty (k+\lambda+1)c_{k+1}x^{k+\lambda}$, and $xf'(x) = \sum_{k=0}^\infty (k+\lambda)c_k x^{k+\lambda}$. Substituting into the differential equation, we obtain

$$\lambda(\lambda+\alpha)c_0 + \sum_{k=0}^\infty \left[(k+\lambda+1)(k+\lambda+\alpha+1)c_{k+1} + \beta(k+\lambda+\gamma)c_k\right] = 0. \tag{A}$$

This means that there are two solutions, f_1 with $\lambda = 0$ and f_2 with $\lambda = -\alpha$. For f_1, Eq. (A) implies the recurrence relation

$$c_{k+1} = -\beta \frac{k+\gamma}{(k+1)(k+\alpha+1)} c_k. \tag{B}$$

This recursion relation is well defined for any α that is not a negative integer smaller than -1. The presence of the multiplication factor $(-\beta)$ means that the function depends on x through the combination $-\beta x$. If $-\gamma$ is a nonnegative integer j, then the series terminates at $k = n$. Hence, $f_1(x)$ is a polynomial of order j with respect to $-\beta x$ that depends only on α. These polynomials are known as generalized Laguerre polynomials, they are expressed as L_j^α. Their properties are analyzed in Appendix B.9.

If $\gamma \neq -j$, then the recursion relation (B) at large k becomes $c_{k+1} = -\frac{\beta}{k+1} c_k$, which is the same with the recurrence relation for the coefficients of the exponential. Hence for large x, $f_1(x) \sim e^{-\beta x}$.

For the solution f_2, the recursion relation becomes $c_{k+1} = (-\beta)\frac{k-\alpha+\gamma}{(k+1)(k+2-\alpha)} c_k$.

The power series terminates at finite k if $\alpha - \gamma = n$, where $n = 0, 1, 2, \ldots$. Then, the solution is of the form $(\beta x)^{-\alpha} G_j^{(\alpha)}(\beta x)$, where $G_j^{(\alpha)}$ is a polynomial of order j that depends on α. Hence, $f_2 \sim x^{-\alpha}$ at $x = 0$, and it diverges for $\alpha > 0$.

If $\alpha - \gamma \neq n$, the recursion relation implies that $f_2 \sim e^{-\beta x}$ for large x.

13.3.2 The Coulomb Potential

Next, we specialize to the case of an attractive Coulomb potential,

$$V(r) = -\frac{\kappa}{r}, \tag{13.44}$$

where $\kappa > 0$. The most common case, $\kappa = \frac{Ze^2}{4\pi}$, corresponds to a hydrogen-like atom, that is, an electron moving around a nucleus of atomic number Z. This system corresponds to the hydrogen atom for $Z = 1$, ionized helium for $Z = 2$, doubly ionized lithium for $Z = 3$ and so on.

In this system, the particle mass m corresponds to the reduced mass $\mu = \frac{m_e M}{m_e + M}$, where m_e is the electron mass and M is the mass of the nucleus. The natural length scale is $a := (m\kappa)^{-1}$. For $\kappa = \frac{Ze^2}{4\pi}$, $a = a_0/Z$, where a_0 is the Bohr radius, Eq. (2.16). The natural energy scale is $(2m_e a^2)^{-1}$.

We define the dimensionless variables $x = r/a$ and $\epsilon = 2ma^2 E$, and we bring Eq. (13.12) into the form

$$u'' + \left(\epsilon + \frac{2}{x} - \frac{\ell(\ell+1)}{x^2} \right) u = 0, \tag{13.45}$$

where the prime denotes differentiation with respect to x. For large x, Eq. (13.45) becomes $u'' + \epsilon u = 0$. For $\epsilon < 0$, it admits only one solution that vanishes at infinity, namely, $u = e^{-\sqrt{|\epsilon|}x}$. We employ the ansatz

$$u(x) = x^{\ell+1} e^{-\sqrt{|\epsilon|}x} f(x), \tag{13.46}$$

where f interpolates between the behavior of u near $x = 0$ and for $x \to \infty$. We substitute into Eq. (13.46) to obtain

$$xf'' + 2(\ell + 1 - \sqrt{|\epsilon|}x)f' + 2[1 - (\ell+1)\sqrt{|\epsilon|}]f = 0. \tag{13.47}$$

Equation (13.47) is a special case of Eq. (13.40) for $\alpha = 2\ell + 1$, $\beta = -2\sqrt{|\epsilon|}$ and $\gamma = \ell + 1 - |\epsilon|^{-1/2}$.

By Theorem 13.5, only the solution f_1 that vanishes at the origin is acceptable, because all other solutions diverge at $x = 0$. Square integrability is incompatible with the asymptotic behavior $f_1 \sim e^{-\beta x} \sim e^{2\sqrt{|\epsilon|}x}$. Hence, the only square integrable solutions exist for $\gamma = -n_r$, for $n_r = 0, 1, 2, \ldots$, and they correspond to generalized Laguerre polynomials $L_{n_r}^{(\alpha)}(-\beta x)$. As in the isotropic harmonic oscillator, n_r is the number of nodes of u. In terms of the principal quantum number $n = n_r + \ell + 1$, the condition $\gamma = -n_r$ implies that $|\epsilon| = n^{-2}$. Hence, the energy eigenvalues are

$$E_n = -\frac{1}{2ma^2 n^2} = -\frac{m\kappa^2}{2n^2}. \tag{13.48}$$

Remarkably, for $\kappa = \frac{Ze^2}{4\pi}$ and $m = m_e$, the energy eigenvalues coincide with those of Bohr's model.

The energy eigenvalues do not depend on ℓ. Hence, their degeneracy is

$$g(E_n) = \sum_{\ell=0}^{n-1} (2\ell + 1) = 2\frac{n(n-1)}{2} + n = n^2. \tag{13.49}$$

In Section 13.3.3, we will identify the symmetry of the Coulomb potentials that leads to degeneracies much larger than the $2\ell + 1$ of spherical symmetry.

The eigenvectors $|n, \ell\rangle$ of the Hamiltonian $\hat{H}_\ell$ on $\mathcal{H}_{rad}$ correspond to the functions

$$u_{n,\ell}(r) = C_{n,\ell} \left(\frac{r}{a} \right)^{\ell+1} e^{-\frac{r}{na}} L_{n-\ell-1}^{(2\ell+1)} \left(\frac{2r}{na} \right), \tag{13.50}$$

where $C_{n,\ell}$ is determined by the normalization condition $\int_0^\infty dr |u_{n,\ell}|^2(r) = 1$. Using Eq. (B.67), we find that

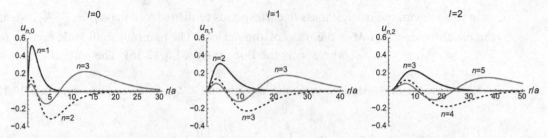

Fig. 13.2 Plots of the dimensionless radial wave function $\sqrt{a}u_{n,\ell}$ as a function of r/a. The figures plot the wave function for fixed ℓ and different n. Each wave function has $n_r = n - \ell - 1$ nodes.

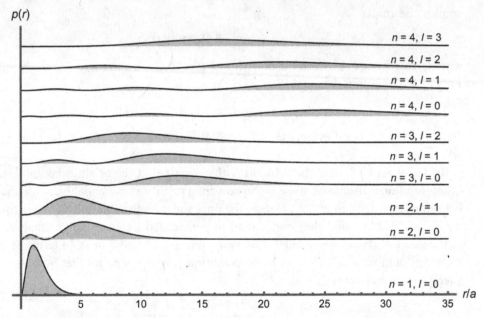

Fig. 13.3 Plots of the probability density $p(r) = r^2 R_{n,\ell}^2(r)$ for the 10 lowest-energy eigenstates of the Schrödinger operator for the Coulomb potential. Both the mean distance of the particle from the origin and the spread of the distribution increase with n.

$$C_{n,\ell} = \sqrt{\frac{(n-\ell-1)!}{(2a)2n(n+\ell)!}} \left(\frac{2}{n}\right)^{\ell+\frac{3}{2}}. \tag{13.51}$$

In Fig. 13.2, the function $u_{n,\ell}$ is plotted for different n and ℓ. In Fig. 13.3, the probability density $p(r) = r^2 R_{n,\ell}^2(r) = u_{n,\ell}^2(r)$ of radial distance is plotted as a function of r.

The eigenfunctions of the Hamiltonian $|n,\ell,m\rangle$ are represented by the wave functions

$$\psi_{n\ell m_\ell}(r,\theta,\phi) = \sqrt{\left(\frac{2}{na}\right)^3 \frac{(n-\ell-1)!}{2n(n+\ell)!}} e^{-\frac{r}{na}} \left(\frac{2r}{na}\right)^\ell L_{n-\ell-1}^{2\ell+1}\left(\frac{2r}{na}\right) Y_{\ell m_\ell}(\theta,\phi).$$

The first few eigenfunctions of the Schrödinger operator for the Coulomb potential are given in Table 13.1.

Table 13.1 Eigenfunctions of the Schrödinger operator for the Coulomb potential

n	ℓ	m_ℓ	ψ_{n,ℓ,m_ℓ}
1	0	0	$\frac{1}{\sqrt{\pi}\,a^{3/2}}e^{-r/a}$
2	0	0	$\frac{1}{4\sqrt{2\pi}\,a^{3/2}}e^{-\frac{r}{2a}}\left(2-\frac{r}{a}\right)$
2	1	0	$\frac{1}{4\sqrt{2\pi}\,a^{5/2}}e^{-\frac{r}{2a}}r\cos\theta$
2	1	± 1	$\frac{1}{8\sqrt{\pi}\,a^{5/2}}e^{-\frac{r}{2a}}r\sin\theta\,e^{\pm i\phi}$
3	0	0	$\frac{1}{81\sqrt{3\pi}\,a^{3/2}}e^{-\frac{r}{3a}}\left(27-\frac{18r}{a}+\frac{2r^2}{a^2}\right)$
3	1	0	$\frac{\sqrt{2}}{81\sqrt{\pi}\,a^{5/2}}e^{-\frac{r}{3a}}r\left(6-\frac{r}{a}\right)\cos\theta$
3	1	± 1	$\frac{1}{81\sqrt{\pi}\,a^{5/2}}e^{-\frac{r}{3a}}r\left(6-\frac{r}{a}\right)\sin\theta\,e^{\pm i\phi}$
3	2	0	$\frac{1}{81\sqrt{6\pi}\,a^{7/2}}e^{-\frac{r}{3a}}r^2(3\cos^2\theta-1)$
3	2	± 1	$\frac{1}{81\sqrt{\pi}\,a^{7/2}}e^{-\frac{r}{3a}}r^2\sin\theta\cos\theta\,e^{\pm i\phi}$
3	2	± 2	$\frac{1}{162\sqrt{\pi}\,a^{7/2}}e^{-\frac{r}{3a}}r^2\sin^2\theta\,e^{\pm 2i\phi}$

Example 13.4 We evaluate some expectation values of the form $\langle \hat{r}^{-k}\rangle$ on energy eigenstates $|n,\ell,m_\ell\rangle$ that will be useful in later chapters. First, we differentiate the eigenvalue equation (13.48) with respect to κ to obtain, $\partial E_n/\partial \kappa = -\frac{m\kappa}{n^2}$. By the Feynman–Hellmann theorem, Eq. (4.10), we obtain

$$\langle \hat{r}^{-1}\rangle = \frac{\kappa m}{n^2}. \tag{13.52}$$

Equation (13.18) yields

$$\langle \hat{r}^{-2}\rangle = \frac{\kappa^2 m^2}{(\ell+\frac{1}{2})n^3}. \tag{13.53}$$

Equation (13.19) implies that

$$\langle \hat{r}^{-3}\rangle = \frac{m\kappa}{\ell(\ell+1)}\langle \hat{r}^{-2}\rangle = \frac{\kappa^3 m^3}{\ell(\ell+\frac{1}{2})(\ell+1)n^3}. \tag{13.54}$$

13.3.3 The Coulomb Potential: Group Theoretic Treatment

A few weeks before Schrödinger published his treatment of the hydrogen atom, Pauli (1926) calculated the spectrum of the hydrogen atom Hamiltonian using Heisenberg's matrix mechanics. Pauli exploited the fact that the Coulomb potential is characterized by a symmetry larger than the rotational symmetry.

The analogue of this symmetry exists already in classical physics. The classical Hamiltonian for an attractive Coulomb potential is $H = \frac{\mathbf{p}^2}{2m} - \frac{\kappa}{r}$. The additional symmetry of the system is related to the Runge–Lenz vector.

$$\mathbf{A} = \frac{\mathbf{p}\times\boldsymbol{\ell}}{m} - \kappa\frac{\mathbf{x}}{r}. \tag{13.55}$$

It is straightforward to show that the Runge–Lenz vector is normal to the angular momentum ℓ and that it is conserved: $\{H, \mathbf{A}\} = 0$.

In quantum theory, we define the Runge–Lenz vector operator as

$$\hat{A}_i = \frac{1}{2m} \sum_{j,k} \epsilon_{ijk} \left(\hat{p}_j \hat{\ell}_k + + \hat{\ell}_k \hat{p}_j \right) - \kappa \hat{x}_i \hat{r}^{-1}. \tag{13.56}$$

It is straightforward to show that

$$\hat{\mathbf{A}} \cdot \hat{\ell} = 0. \tag{13.57}$$

The Hamiltonian operator $\hat{H} = \frac{\hat{\mathbf{p}}^2}{2m} - \kappa \hat{r}^{-1}$ satisfies the commutation relations $[\hat{x}_i, \hat{H}] = i\hat{p}_i/m$, $[\hat{p}_i, \hat{H}] = -i\hat{x}_i \hat{r}^{-3}$, and $[\hat{\ell}_i, \hat{H}] = 0$. Using these relations, we find that $\hat{A}_i$ is a conserved quantity also in quantum theory,

$$[\hat{H}, \hat{A}_i] = 0. \tag{13.58}$$

Hence, the group generated by $\hat{A}_i$ leaves the eigenspaces V_E of $\hat{H}$ invariant. It is simpler to identify the algebraic properties of $\hat{A}_i$ by restricting to V_E, where we can substitute $\hat{H}$ with $E\hat{I}$.

We substitute Eq. (11.71) into Eq. (13.56), to obtain

$$\hat{A}_i = \left(\hat{x}_i \frac{\hat{\mathbf{p}}^2}{2m} + \frac{\hat{\mathbf{p}}^2}{2m} \hat{x}_i \right) - \frac{\hat{\mathbf{x}} \cdot \hat{\mathbf{p}} + \hat{\mathbf{p}} \cdot \hat{\mathbf{x}}}{2m} \hat{p}_i - \kappa \hat{x}_i \hat{r}^{-1}$$

$$= (\hat{x}_i \hat{H} + \hat{H} \hat{x}_i) - \frac{\hat{\mathbf{x}} \cdot \hat{\mathbf{p}} + \hat{\mathbf{p}} \cdot \hat{\mathbf{x}}}{2m} \hat{p}_i + \kappa \hat{x}_i \hat{r}^{-1}$$

$$= -\frac{\hat{\mathbf{x}} \cdot \hat{\mathbf{p}} + \hat{\mathbf{p}} \cdot \hat{\mathbf{x}}}{2m} \hat{p}_i + \left(2E\hat{I} + \kappa \hat{r}^{-1} \right) \hat{x}_i. \tag{13.59}$$

Equation (13.59) leads to the commutation relation

$$[\hat{A}_i, \hat{\ell}_j] = i \sum_k \epsilon_{ijk} \hat{A}_k, \tag{13.60}$$

which confirms that $\hat{A}_i$ is a vector operator. We also find that

$$[\hat{A}_i, \hat{A}_j] = -\frac{2iE}{m} \sum_k \epsilon_{ijk} \hat{\ell}_k. \tag{13.61}$$

Equation (13.59) also implies that

$$\hat{\mathbf{A}}^2 = \kappa^2 \hat{I} + \frac{2E}{m} \left(\hat{\ell}^2 + \hat{I} \right). \tag{13.62}$$

For the study of bound states, we take $E < 0$. We define the operators

$$\hat{\mathbf{S}} = \frac{1}{2} \left(\hat{\ell} - \sqrt{\frac{m}{2|E|}} \hat{\mathbf{A}} \right), \quad \hat{\mathbf{T}} = \frac{1}{2} \left(\hat{\ell} + \sqrt{\frac{m}{2|E|}} \hat{\mathbf{A}} \right). \tag{13.63}$$

Equations (13.60) and (13.61) imply that

$$[\hat{S}_i, \hat{T}_j] = 0, \quad [\hat{S}_i, \hat{S}_j] = i \sum_k \epsilon_{ijk} \hat{S}_k, \quad [\hat{T}_i, \hat{T}_j] = i \sum_k \epsilon_{ijk} \hat{T}_k. \tag{13.64}$$

Hence, in the eigenspace V_E, there are two independent representations of the angular momentum algebra,[4] one with generators $\hat{S}_i$ and one with generators $\hat{T}_i$.

Equation (13.57) implies that

$$\hat{\mathbf{S}}^2 = \hat{\mathbf{T}}^2 = \frac{1}{4}(\hat{\ell}^2 + \frac{m}{2|E|}\hat{\mathbf{A}}^2), \tag{13.65}$$

hence, the two representations have the same dimension. Therefore, we write $\hat{\mathbf{S}}^2 = \hat{\mathbf{T}}^2 = s(s+1)\hat{I}$, where s can take any values $0, \frac{1}{2}, 1, \ldots$. Hence, the dimension of V_E is $(2s+1)^2$.

Combining Eq. (13.62) with Eq. (13.65), we obtain

$$\hat{\mathbf{S}}^2 = \frac{1}{4}\left(-1 + \frac{\kappa^2 m}{2|E|}\right)\hat{I}. \tag{13.66}$$

It follows that $E = -\frac{\kappa^2 m}{2n^2}$, where $n = 2s + 1$ takes values $1, 2, 3, \ldots$. We obtained the usual expression for the eigenvalues and the degeneracy of the attractive Coulomb potential.

13.4 Particles Interacting with the Electromagnetic Field

In classical mechanics, the motion of a particle under the action of an EM field $(\mathbf{E}, \mathbf{B})$ is determined by the Lorentz force, Eq. (1.18). In quantum theory, force is not a fundamental notion; time evolution is generated by the Hamiltonian. Hence, the interaction of a particle with an EM field is implemented through the addition of terms in the Hamiltonian.

In this section, we present the quantum Hamiltonian that describes the interaction of point particles with a background EM field, and then we specialize to some elementary cases.

13.4.1 The Minimal Substitution

For a point particle without spin, the classical Hamiltonian formalism provides a rule for incorporating EM interaction into a particle's Hamiltonian. This rule is known as *minimal substitution*, and it works as follows. Consider a particle of mass m and charge q moving in an EM field $(\mathbf{E}(\mathbf{x}), \mathbf{B}(\mathbf{x}))$. In the equation $\hat{H} = \frac{1}{2m}\hat{\mathbf{p}}^2$ for the free particle Hamiltonian, we substitute

- the Hamiltonian $\hat{H}$ with $\hat{H} - q\phi(\hat{\mathbf{x}}, t)$, where $\phi(\mathbf{x}, t)$ is the electric potential,
- the momentum $\hat{\mathbf{p}}$ with $\hat{\mathbf{p}} - q\mathbf{A}(\hat{\mathbf{x}}, t)$, where $\mathbf{A}(\mathbf{x}, t)$ is the magnetic potential.

We remind the reader that the EM potentials are related to the EM fields through Eq. (1.23),

$$\mathbf{E} = -\frac{\partial \mathbf{A}}{\partial t} - \nabla\phi \qquad \mathbf{B} = \nabla \times \mathbf{A}.$$

Then, the Hamiltonian for a particle in an EM field reads

$$\hat{H} = \frac{1}{2m}\left[\hat{\mathbf{p}} - q\mathbf{A}(\hat{\mathbf{x}}, t)\right]^2 + q\phi(\hat{\mathbf{x}}, t). \tag{13.67}$$

---

[4] We have a representation of the Lie algebra $\mathfrak{so}(3) \oplus \mathfrak{so}(3)$, which is identical to $\mathfrak{so}(4)$.

The commutation relation $[\hat{p}_i, \hat{A}_j] = -i\frac{\partial \hat{A}_j}{\partial x_i}$ means that $\hat{\mathbf{p}} \cdot \hat{\mathbf{A}} = \hat{\mathbf{A}} \cdot \hat{\mathbf{p}} - i\nabla \cdot \hat{\mathbf{A}}$. Hence, the quadratic term in Eq. (13.67) becomes

$$\left(\hat{\mathbf{p}} - q\hat{\mathbf{A}}\right)^2 = \hat{\mathbf{p}}^2 - 2q\hat{\mathbf{A}} \cdot \hat{\mathbf{p}} + iq\nabla \cdot \hat{\mathbf{A}} + q^2\hat{\mathbf{A}}^2. \tag{13.68}$$

Two different EM potentials give the same values for the electric and magnetic fields, if they are related by a gauge transformation, Eq. (1.24). This property allows us to impose an arbitrary condition on the potentials – a *gauge choice* – that drastically restrict their possible form. We choose the so-called *Coulomb gauge*, in which $\nabla \cdot \mathbf{A} = 0$. Then, the Hamiltonian simplifies to

$$\hat{H} = \frac{1}{2m}\hat{\mathbf{p}}^2 + q\hat{\phi} - \frac{q}{m}\hat{\mathbf{A}} \cdot \hat{\mathbf{p}} + \frac{q^2}{2m}\hat{\mathbf{A}}^2. \tag{13.69}$$

Excepting the term $-\frac{q}{m}\hat{\mathbf{A}} \cdot \hat{\mathbf{p}}$ that is linear in momentum, the terms in Eq. (13.69) define a Schrödinger operator with potential $V_{eff} = q\phi + \frac{q^2}{2m}\mathbf{A}^2$. Note that $\hat{\phi}$ and $\mathbf{A}$ denotes that they are functions of position, which is a quantum observable, and not that the EM field is itself a quantum observable.

To understand the physical meaning of the term $-\frac{q}{m}\hat{\mathbf{A}} \cdot \hat{\mathbf{p}}$, we choose the magnetic potential $\mathbf{A} = \frac{1}{2}\mathbf{B} \times \mathbf{x}$ that corresponds to a constant and homogeneous magnetic field $\mathbf{B}$. We find that $\hat{\mathbf{A}} \cdot \hat{\mathbf{p}} = \frac{1}{2}(\mathbf{B} \times \hat{\mathbf{x}}) \cdot \hat{\mathbf{p}} = \frac{1}{2}\mathbf{B} \cdot (\hat{\mathbf{x}} \times \hat{\mathbf{p}}) = \frac{1}{2}\mathbf{B} \cdot \hat{\boldsymbol{\ell}}$. Here, we used the identity $(\mathbf{A} \times \mathbf{B}) \cdot \mathbf{C} = \mathbf{A} \cdot (\mathbf{B} \times \mathbf{C})$, which also applies to operators, as long as their ordering remains unchanged.

Hence, for a constant and homogeneous magnetic field, the Hamiltonian (13.69) contains a term

$$\hat{V} = -\frac{q}{2m}\mathbf{B} \cdot \hat{\boldsymbol{\ell}}, \tag{13.70}$$

that describes a coupling of the magnetic field to the orbital angular momentum. This term is analogous to the classical interaction energy for a magnetic dipole in a magnetic field $-\boldsymbol{\mu} \cdot \mathbf{B}$, where $\boldsymbol{\mu}$ is the dipole's magnetic moment. Hence, we define a magnetic moment operator as $\hat{\boldsymbol{\mu}} = \frac{q}{2m}\hat{\boldsymbol{\ell}}$.

13.4.2 A Particle in a Homogeneous Electric Field

Consider a homogeneous electric field of the form $\mathbf{E} = (\mathcal{E}, 0, 0)$, which corresponds to an electric potential $\phi = -\mathcal{E}x_1$ and magnetic potential $\mathbf{A} = 0$. Directions 2 and 3 contribute trivially to the Hamiltonian, so we can restrict ourselves to one dimension. The Hamiltonian is

$$\hat{H} = \frac{\hat{p}^2}{2m} - q\mathcal{E}\hat{x}. \tag{13.71}$$

The Hamiltonian (13.71) is not bounded from below, since x can become arbitrarily large. We have to assume implicitly that the electric field extends to a finite region of space. Equation (13.71) is meaningful provided the particle does not move out of that finite region.

It is simpler to work in the momentum representation. We use momentum wave functions $\phi(p)$, upon which the momentum operator acts by multiplication $\hat{p}\phi(p) = p\phi(p)$, and the position operator by differentiation, $\hat{x} = i\frac{\partial}{\partial p}$. Then, the eigenvalue equation for $\hat{H}$ becomes

$$-iq\mathcal{E}\frac{d}{dp}\phi_E = \left(E - \frac{p^2}{2m}\right)\phi_E, \tag{13.72}$$

with solutions $\phi_E(p) = Ce^{\frac{i}{q\mathcal{E}}\left(Ep - \frac{p^3}{6m}\right)}$.

The functions ϕ_E define kets $|E\rangle$, such that $\langle p|E\rangle = \phi_E(p)$. The constant C is determined by the normalization condition $\langle E'|E\rangle = \delta(E - E')$. We calculate $\langle E'|E\rangle = \int dp\phi^*_{E'}(p)\phi_E(p) = |C|^2 \int dpe^{\frac{ip}{q\mathcal{E}}(E-E')} = |C|^2 q\mathcal{E}2\pi\,\delta(E - E')$. It follows that $C = 1/\sqrt{2\pi q\mathcal{E}}$, hence,

$$\phi_E(p) = \frac{1}{\sqrt{2\pi q\mathcal{E}}}e^{\frac{i}{q\mathcal{E}}\left(Ep - \frac{p^3}{6m}\right)}. \tag{13.73}$$

The matrix elements of the time evolution operator in the momentum basis are

$$\langle p'|e^{-i\hat{H}t}|p\rangle = \int dE\langle p'|E\rangle\langle E|p\rangle e^{-iEt} = \frac{1}{2\pi q\mathcal{E}}e^{i\frac{p^3-p'^3}{6m\mathcal{E}}}\int dEe^{iE\left(\frac{p'-p}{q\mathcal{E}}-t\right)}$$

$$= \frac{1}{q\mathcal{E}}\delta\left(\frac{p'-p}{q\mathcal{E}}-t\right)e^{i\frac{p^3-p'^3}{6mq\mathcal{E}}} = \delta(p'-p-q\mathcal{E}t)e^{i\frac{p^3-p'^3}{6mq\mathcal{E}}}. \tag{13.74}$$

The time evolution operator adds to the initial momentum p a term $q\mathcal{E}t$, just like in the analogous classical system. It also induces rapid phase changes that have no classical analogue.

In the momentum representation, the time evolution of an initial state ψ_0 is

$$\langle p|\psi_t\rangle = \int dp'\langle p|e^{-i\hat{H}t}|p'\rangle\langle p'|\psi_0\rangle = e^{i\frac{(p-q\mathcal{E}t)^3-p^3}{6mq\mathcal{E}}}\langle p - q\mathcal{E}t|\psi_0\rangle. \tag{13.75}$$

The probability distribution for momentum at time t,

$$|\langle p|\psi_t\rangle|^2 = |\langle p - q\mathcal{E}t|\psi_0\rangle|^2, \tag{13.76}$$

coincides with the initial probability distribution shifted by $q\mathcal{E}t$.

13.4.3 A Particle in a Homogeneous Magnetic Field

Consider a magnetic field $\mathbf{B} = (0, 0, B)$. The field does not affect a particle's motion along direction 3, so we can restrict ourselves to the study of particle motion at the plane $x_1 - x_2$. Therefore, we consider a two-dimensional system described by the Hilbert space $L^2(\mathbb{R}^2)$. It is convenient to choose a magnetic potential $\mathbf{A} = (0, Bx_1, 0)$.

The Hamiltonian (13.69) becomes

$$\hat{H} = \frac{1}{2m}(\hat{p}_1^2 + \hat{p}_2^2) - \frac{qB}{m}\hat{x}_1\hat{p}_2 + \frac{q^2}{2m}B^2\hat{x}_1^2 = \frac{1}{2m}\hat{p}_1^2 + \frac{1}{2}m\omega_c^2\left(\hat{x}_1 - \frac{1}{m\omega_c}\hat{p}_2\right)^2,$$

where $\omega_c = \frac{qB}{m}$. We note that, along direction 1, the motion is that of a harmonic oscillator with frequency ω_c and a center that depends on $\hat{p}_2$. The only role of $\hat{p}_2$ in the Hamiltonian is to determine the center of the oscillations in direction 1.

This structure of the Hamiltonian is compatible with the classical motion of particles in a magnetic field: the particles follow a circular orbit of frequency ω_c on a plane normal to the field's direction. The frequency ω_c is known as the *cyclotron frequency*.

The reader might worry that, with a different choice of the magnetic potential, say $\mathbf{A} = (-Bx_2, 0, 0)$, the oscillations will take place in direction 2, and the center of the oscillations will be determined by $\hat{p}_1$. Won't this lead to different physical predictions? The answer – to be elaborated on in Section 13.5.1 – is that neither $\hat{p}_1$ nor $\hat{p}_2$ are physical observables in this system. The physical observables do not depend on the arbitrary choice of $\mathbf{A}$.

Suppose the Hamiltonian acts on a ket $|\psi\rangle \otimes |k\rangle$, where $|k\rangle$ is a generalized eigenvector of $\hat{p}_2$. We obtain

$$\hat{H}(|\psi\rangle \otimes |k\rangle) = \left[\frac{1}{2m}\hat{p}_1^2 + \frac{1}{2}m\omega_c^2(\hat{x}_1 - \frac{k}{m\omega_c})^2 \right] |\psi\rangle \otimes |k\rangle. \tag{13.77}$$

The kets $|\psi\rangle \otimes |k\rangle$ are generalized eigenvectors of $\hat{H}$, if $|\psi\rangle$ is a number state $|n\rangle$ of a harmonic oscillator, translated in position by $\frac{k}{m\omega_c}$. The unitary operator that translates the position by a is $e^{ia\hat{p}}$, whence $|\psi\rangle = e^{-i\frac{k}{m\omega_c}\hat{p}}|n\rangle$.

We conclude that the eigenvalues of the Hamiltonian (13.77) are $E_n = \omega_c(n + \frac{1}{2})$, where $n = 0, 1, 2, \ldots$, and they correspond to generalized eigenvectors

$$|n, k\rangle = e^{-i\frac{k}{m\omega_c}\hat{p}}|n\rangle \otimes |k\rangle. \tag{13.78}$$

Since the energy does not depend on k, and k has an infinity of possible values, the system is characterized by infinite degeneracy. The infinite-dimensional eigenspaces of the Hamiltonian are known as *Landau levels*.

The spectral theorem for $\hat{H}$ yields

$$\hat{H} = \sum_{n=0}^{\infty} \int dk \omega_c \left(n + \frac{1}{2} \right) e^{-i\frac{k}{m\omega_c}\hat{p}} |n\rangle\langle n| e^{i\frac{k}{m\omega_c}\hat{p}} \otimes |k\rangle\langle k| = \sum_{n=0}^{\infty} \omega_c \left(n + \frac{1}{2} \right) \hat{P}_n,$$

where $\hat{P}_n = \int dk e^{-i\frac{k}{m\omega_c}\hat{p}} |n\rangle\langle n| e^{i\frac{k}{m\omega_c}\hat{p}} \otimes |k\rangle\langle k|$ are the spectral projectors to the Landau levels.

The infinite degeneracy of the Landau levels originates from the fact that the magnetic field extends to the whole plane. We can render this degeneracy finite, by taking periodic boundary conditions for x_1 and x_2, so that the magnetic field acts in a finite region. Let L be the associated period. Then, $k = j\frac{2\pi}{L}$, where $j = 0, \pm 1, \pm 2, \ldots$. The shift $\frac{k}{m\omega_c}$ of the center of the oscillations cannot be larger than L in absolute value. Hence,

$$|j| \leq \frac{m\omega_c}{2\pi} A := j_{max}, \tag{13.79}$$

where $A = L^2$ is the total area. The degeneracy g is the number of all values of j that satisfy Eq. (13.79), that is, $g = 2j_{max} + 1$. For macroscopic values of the area A, $j_{max} \gg 1$, hence

$$g \simeq 2j_{max} = \frac{m\omega_c}{\pi} A = \frac{q\Phi}{\pi}, \tag{13.80}$$

where $\Phi = BA$ is the total magnetic flux.

13.4.4 Motion in a Ring That Surrounds a Magnetic Flux

Next, we study a particle of mass m and charge q that is constrained to move in a ring that surrounds magnetic flux, see Fig. 13.4.

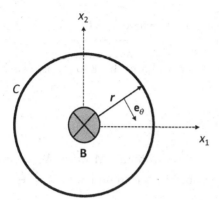

Fig. 13.4 A ring of radius R around a magnetic flux.

Let C be a ring of radius R on the x_1–x_2 plane. We assume an inhomogeneous magnetic field along direction 3, which depends only on the radial coordinate $r = \sqrt{x_1^2 + x_2^2}$ on the plane. We choose the magnetic potential to be of the form $\mathbf{A} = A_\theta(r)\mathbf{e}_\theta$, where $\mathbf{e}_\theta$ is a unit vector on the x_1–x_2 plane that is everywhere normal to the position vector $\mathbf{x} = (x_1, x_2, 0)$. If we define the angular variable θ by $\tan\theta = x_2/x_1$, then $\mathbf{e}_\theta = (-\sin\theta, \cos\theta, 0)$.

We evaluate the magnetic flux $\Phi = \int \mathbf{B} \cdot d\mathbf{S}$ on the disk bounded by the ring C. By Stokes' theorem, $\Phi = \int \mathbf{\nabla} \times \mathbf{A} \cdot d\mathbf{S} = \oint_C \mathbf{A} \cdot d\boldsymbol{\ell} = 2\pi R A_\theta(R)$; hence, $A_\theta(R) = \frac{\Phi}{2\pi R}$.

The particle's position is fully specified by the angle θ, so the system is effectively one dimensional. The Hamiltonian takes the form

$$\hat{H} = \frac{1}{2m}\left(-\frac{i}{R}\frac{\partial}{\partial\theta} - qA_\theta\right)^2 = -\frac{1}{2mR^2}\left(\frac{\partial}{\partial\theta} - i\frac{q\Phi}{2\pi}\right)^2. \tag{13.81}$$

To solve the eigenvalue equation $\hat{H}\psi(\theta) = E\psi(\theta)$, we write $\psi(\theta) = e^{i\frac{q\Phi}{2\pi}\theta}\chi(\theta)$. Then,

$$\frac{d^2\chi}{d\theta^2} + 2mR^2 E\chi = 0. \tag{13.82}$$

The solutions to Eq. (13.82) are $\chi(\theta) = e^{\pm i\sqrt{2mR^2 E}\theta}$. Hence, the eigenfunctions of the Hamiltonian are

$$\psi_E(\theta) = C\exp\left[i\left(\pm\sqrt{2mR^2 E} + \frac{q\Phi}{2\pi}\right)\theta\right]. \tag{13.83}$$

The requirement that $\psi(2\pi) = \psi(0)$ implies that $\pm\sqrt{2mR^2 E} + \frac{q\Phi}{2\pi} = n$ for some integer n. It follows that the Hamiltonian has a discrete spectrum, with eigenvalues

$$E_n = \frac{1}{2mR^2}\left(n - \frac{q\Phi}{2\pi}\right)^2, \tag{13.84}$$

for $n = 0, \pm 1, \pm 2, \dots$. The associated eigenfunctions are $\psi(\theta) = \frac{1}{\sqrt{2\pi}}e^{in\theta}$. The ground state corresponds to the integer n_0 that is closest to the quantity $\frac{q\Phi}{2\pi}$. The energy eigenvalues depend only *on the total magnetic flux across the disk*. A magnetic field that is generated by a solenoid of radius $r_s < R$ vanishes at the ring, but the flux across the disk is still nonzero. We arrive at a remarkable conclusion.

> The EM field may affect quantum systems even in regions where $\mathbf{E} = \mathbf{B} = 0$.

At the moment, we refrain from further explanation. We will come back to this issue in Section 13.5.2.

13.4.5 The Dipole Approximation

Of particular interest is the case of a particle's interaction with an EM wave. We assume that in the absence of this interaction, the particle is described by a Hamiltonian $\hat{H}_0 = \frac{\hat{\mathbf{p}}^2}{2m} + V(\hat{\mathbf{x}})$, with eigenvalues E_a and eigenvectors $|a\rangle$.

We describe the EM wave in the Coulomb gauge, in which the electric potential vanishes, and the magnetic potential $\mathbf{A}$ satisfies $\nabla \cdot \mathbf{A} = 0$. We analyze the EM wave to plane waves as

$$\mathbf{A}(\mathbf{x}, t) = \sum_{\mathbf{k}} \tilde{\mathbf{A}}(\mathbf{k}) e^{i\mathbf{k}\cdot\mathbf{x} - ikt}, \tag{13.85}$$

where $k = |\mathbf{k}|$ and where $\mathbf{k} \cdot \tilde{\mathbf{A}}(\mathbf{k}) = 0$.

For a weak field, the quadratic term $\mathbf{A}^2$ in Eq. (13.69) is negligible. Hence, the Hamiltonian for the particle interacting with the EM field is of the form $\hat{H}_0 + \hat{H}_I(t)$, $\hat{H}_I(t) = -\frac{q}{m}\mathbf{A}(\mathbf{x}) \cdot \hat{\mathbf{p}}$. The interaction term simplifies if the wavelength λ of the EM wave is much larger than the typical spatial extent of the bound states $|a\rangle$. For example, if the particle is an electron bound to an atom, the typical extent of the quantum state is of the order of the Bohr radius $a_0 \simeq 0.05$ nm. Hence, radiation at wavelengths of the order of hundreds of nanometers, as in the optical spectrum, satisfies $\lambda >> a_0$. If we assume that the bounding potential is nonzero in the vicinity of $\mathbf{x} = 0$, then for the particle's motion $|\mathbf{k} \cdot \mathbf{x}| << 1$. Therefore the *dipole approximation* $e^{i\mathbf{k}\cdot\mathbf{x}} \sim 1$ is justified.

In the dipole approximation, $\hat{H}_I(t) = -\frac{q}{m}\mathbf{A}(t) \cdot \hat{\mathbf{p}}$, where we wrote $\mathbf{A}(t) = \mathbf{A}(0, t)$. The matrix elements of $\hat{H}_I(t)$ in the basis $|a\rangle$ are

$$\langle a|\hat{H}_I(t)|a'\rangle = -\frac{q}{m}\mathbf{A}(t) \cdot \langle a|\hat{\mathbf{p}}|a'\rangle. \tag{13.86}$$

We use the fact that $\hat{\mathbf{p}} = -im[\hat{\mathbf{x}}, \hat{H}_0]$, in order to evaluate $\langle a|\hat{\mathbf{p}}|a'\rangle = -im\langle a|\hat{\mathbf{x}}\hat{H}_0 - \hat{H}_0\hat{\mathbf{x}}|a'\rangle = im(E_a - E_{a'})\langle a|\hat{\mathbf{x}}|a'\rangle$. Hence,

$$\langle a|\hat{H}_I(t)|a'\rangle = -i(E_a - E_{a'})\mathbf{A}(t) \cdot \mathbf{d}_{aa'}, \tag{13.87}$$

where $\mathbf{d}_{aa'} = q\langle a|\hat{\mathbf{x}}|a'\rangle$.

If all frequencies of the EM wave are close to a transition frequency $\omega = E_a - E_{a'} > 0$, and the eigenstates $|a\rangle$ and $|a'\rangle$ are nondegenerate, then we can use the two-level approximation. That is, we identify $|a'\rangle = |0\rangle$ and $|a\rangle = |1\rangle$. Then, $\langle 0|\hat{H}_I(t)|0\rangle = \langle 1|\hat{H}_I(t)|1\rangle = 0$ and

$$\langle 1|\hat{H}_I(t)|0\rangle = -i\omega\mathbf{A}(t) \cdot \mathbf{d}, \tag{13.88}$$

where $\mathbf{d} = \mathbf{d}_{10}$. It follows that

$$\hat{H}_I(t) = -i\omega\mathbf{A}(t) \cdot \left(\mathbf{d}\,\hat{\sigma}_+ + \mathbf{d}^*\,\hat{\sigma}_-\right), \tag{13.89}$$

in the two-level-system Hilbert space. If $\mathbf{d}$ is real, or can be made real by a phase change of $|0\rangle$ and $|1\rangle$, then $\hat{H}_I(t) = -2i\omega\mathbf{A}(0,t)\cdot\mathbf{d}\hat{\sigma}_1$, which is of the same form with the interaction term (7.85) that leads to Rabi oscillations.

For a particle in a central potential, relevant transitions take place between a state $|a\rangle = |n,\ell,m\rangle$ and a state $|a'\rangle = |n',\ell',m'\rangle$. Hence, $\mathbf{d} = q\langle n,\ell,m|\hat{\mathbf{x}}|n',\ell',m'\rangle$. Since $\hat{\mathbf{x}}$ is a vector operator, the Wigner–Eckart theorem (Section 11.6.2) implies that $\mathbf{d}$ vanishes unless $|\ell - \ell'| \leq 1$. We also note that if $\ell = \ell'$, the product $\psi_{n,\ell,m}(r,\theta,\phi)\psi_{n',\ell,m'}(r,\theta,\phi)$ is an even function, because the parity of each eigenfunction is proportional to $(-1)^\ell$. Then, $\langle n,\ell,m|\hat{\mathbf{x}}|n',\ell,m'\rangle$ vanishes, because it involves the integral of an odd function over all space.

We conclude that in the dipole approximation, an EM wave leads to transitions that are subject to the *selection rule*,

$$|\ell - \ell'| = 1. \tag{13.90}$$

Example 13.5   We evaluate the vector $\mathbf{d}$ for transitions between the ground state $|1,0,0\rangle$ and the first excited states $|2,\ell,m_\ell\rangle$ of the hydrogen atom. As shown earlier, $\mathbf{d}$ vanishes for $\ell = 0$, so we consider only the matrix elements $\mathbf{d}^\pm = -e\langle 1,0,0|\hat{\mathbf{x}}|2,1,\pm 1\rangle$ and $\mathbf{d}^0 = -e\langle 1,0,0|\hat{\mathbf{x}}|2,1,\pm 0\rangle$. We evaluate

$$d_1^\pm = -e\int d^3x \left(\frac{e^{-r/a_0}}{\sqrt{\pi}a_0^{3/2}}\right)(r\sin\theta\cos\phi)\left(\frac{e^{-\frac{r}{2a}}r\sin\theta e^{\pm i\phi}}{8\sqrt{\pi}a_0^{5/2}}\right)$$

$$= -\frac{e}{8\pi a_0^4}\left(\int_0^\infty dr r^4 e^{-\frac{3r}{2a_0}}\right)\left(\int_0^\pi d\theta\sin^3\theta\right)\left(\int_0^{2\pi}d\phi\cos\phi e^{\pm i\phi}\right)$$

$$= -\frac{e}{8\pi a_0^4}\left(\frac{256a_0^5}{81}\right)\left(\frac{4}{3}\right)(\pi) = \frac{128i}{243}ea_0,$$

$$d_2^\pm = -\frac{e}{8\pi a_0^4}\left(\int_0^\infty dr r^4 e^{-\frac{3r}{2a_0}}\right)\left(\int_0^\pi d\theta\sin^3\theta\right)\left(\int_0^{2\pi}d\phi\sin\phi e^{\pm i\phi}\right)$$

$$= -\frac{e}{8\pi a_0^4}\left(\frac{256a_0^5}{81}\right)\left(\frac{4}{3}\right)(\pm i\pi) = \mp\frac{128i}{243}ea_0,$$

$$d_3^\pm = -\frac{e}{8\pi a_0^4}\left(\int_0^\infty dr r^4 e^{-\frac{3r}{2a_0}}\right)\left(\int_0^\pi d\theta\sin^2\theta\cos\theta\right)\left(\int_0^{2\pi}d\phi e^{\pm i\phi}\right) = 0.$$

Similarly, we find that $\mathbf{d}^0 = 0$. Hence, transitions due to the dipole interaction are nonzero only in the three-dimensional subspace spanned by $|1,0,0\rangle$, $|2,1,1\rangle$, and $|2,1,-1\rangle$. By Eq. (13.86),

$$\hat{H}_I = -2i\mathbf{A}(t)\cdot\left(\mathbf{d}^+|1,0,0\rangle\langle 2,1,1| + \mathbf{d}^{+*}|2,1,1\rangle\langle 1,0,0|\right.$$
$$\left. + \mathbf{d}^-|1,0,0\rangle\langle 2,1,-1| + \mathbf{d}^{-*}|1,0,0\rangle\langle 2,1,-1|\right).$$

13.5 Gauge Symmetry in Quantum Theory

In classical electromagnetic theory, the electric field $\mathbf{E}$ and the magnetic field $\mathbf{B}$ are invariant under the so-called *gauge transformations* of the potentials ϕ and $\mathbf{A}$,

$$\phi \to \phi' = \phi - \frac{\partial f}{\partial t} \qquad \mathbf{A} \to \mathbf{A}' = \mathbf{A} + \nabla f, \tag{13.91}$$

where f is a differentiable function of $\mathbf{x}$ and t. Gauge transformations constitute a symmetry of classical electromagnetism, in the sense that they leave physical quantities unchanged. In particular, the forces acting upon charged particles depend only on the fields, and they are invariant under gauge transformations. We proceed to the description of gauge transformations in quantum theory.

13.5.1 Gauge Transformations

In quantum theory, particle dynamics are implemented by the Hamiltonian operator (13.67), where the electric potentials appear explicitly. So, we must take gauge transformations into account. To make the dynamics invariant under a gauge transformation, we must supplement Eq. (13.91) with a gauge transformation for the quantum state $|\psi\rangle$,

$$|\psi\rangle \to |\psi'\rangle = e^{iqf(\hat{\mathbf{x}}, t)}|\psi\rangle. \tag{13.92}$$

The commutation relation $[\hat{\mathbf{p}}, e^{iqf}] = qe^{iqf}\nabla f$ implies that

$$e^{-iqf(\hat{\mathbf{x}}, t)}\hat{\mathbf{p}}e^{iqf(\hat{\mathbf{x}}, t)} = \hat{\mathbf{p}} - q\nabla f(\hat{\mathbf{x}}, t), \tag{13.93}$$

hence,

$$e^{-iqf(\hat{\mathbf{x}}, t)}\left[\hat{\mathbf{p}} - q\mathbf{A}(\hat{\mathbf{x}})\right]e^{iqf(\hat{\mathbf{x}}, t)} = \hat{\mathbf{p}} - q\mathbf{A}'(\hat{\mathbf{x}}), \tag{13.94}$$

where $\mathbf{A}'$ is given by Eq. (1.24).

The position operator is unchanged under the action of $e^{-iqf(\hat{\mathbf{x}}, t)}$,

$$e^{-iqf(\hat{\mathbf{x}}, t)}\hat{\mathbf{x}}e^{iqf(\hat{\mathbf{x}}, t)} = \hat{\mathbf{x}}. \tag{13.95}$$

Let $|\psi(t)\rangle = e^{-i\hat{H}t}|\psi\rangle$, where $\hat{H}$ is the Hamiltonian (13.67). We write $|\psi'(t)\rangle = e^{iqf(\hat{\mathbf{x}}, t)}|\psi(t)\rangle$. Then,

$$i\frac{d}{dt}|\psi'(t)\rangle = e^{iqf(\hat{\mathbf{r}}, t)}(i\frac{\partial}{\partial t} - q\frac{\partial f}{\partial t})|\psi(t)\rangle. \tag{13.96}$$

Equations (13.91) and (13.96) imply that

$$i\frac{d}{dt}|\psi'(t)\rangle = \hat{H}'|\psi'(t)\rangle, \tag{13.97}$$

where $\hat{H}'$ is given by Eq. (13.67), but expressed in terms of the transformed EM potentials $\mathbf{A}'$ and ϕ'. We conclude that gauge transformations leave the dynamics of a particle in an EM field invariant.

By Eq. (13.93), the momentum $\hat{\pi}$ is gauge-transformed in a way that affects its spectrum and, hence, the associated probabilities. In contrast, the *generalized momentum* $\hat{\boldsymbol{\pi}} := \hat{\mathbf{p}} - q\hat{\mathbf{A}}$ remains unaffected under a gauge transformation. Hence, in the presence of an external magnetic field, the measurable quantity is not the momentum $\hat{\mathbf{p}}$, but the generalized momentum $\hat{\boldsymbol{\pi}}$. While it may be convenient to employ the ordinary momentum operators $\hat{p}_i$ for calculations, as in Section 13.4.3, the only meaningful predictions are the ones expressed in terms of the gauge-invariant observables $\hat{x}_i$ and $\hat{\pi}_i$.

The components of the generalized momentum $\hat{\pi}$ do not commute with each other, $[\hat{\pi}_i, \hat{\pi}_j] = iq\left(\frac{\partial \hat{A}_j}{\partial x_i} - \frac{\partial \hat{A}_i}{\partial x_j}\right)$. Hence, $[\hat{\pi}_1, \hat{\pi}_2] = iq\hat{B}_3$, and so on.

The commutation relations for the gauge-invariant particle observables are

$$[\hat{x}_i, \hat{x}_j] = 0, \quad [\hat{x}_i, \hat{\pi}_j] = i\delta_{ij}, \quad [\hat{\pi}_i, \hat{\pi}_j] = iq\sum_k \epsilon_{ijk}\hat{B}_k. \tag{13.98}$$

The Hamiltonian is $\hat{H} = \frac{\hat{\pi}^2}{2m}$. The Heisenberg equations of motion $m\frac{d\hat{x}_i}{dt} = \hat{\pi}_i$ imply that the generalized momentum $\hat{\pi}$ can still be interpreted as mass times velocity.

Example 13.6   Consider a particle of mass m and charge q in an external EM field with position-independent magnetic potential $\mathbf{A}(t)$. The external magnetic field vanishes and the external electric field is $\mathbf{E}(t) = -\dot{\mathbf{A}}(t)$. This case is relevant to atomic electrons under the influence of an external EM field in the dipole approximation. The particle state $\psi(\mathbf{x}, t)$ evolves under the Schödinger equation

$$i\frac{\partial}{\partial t}\psi(\mathbf{x}, t) = -\frac{1}{2m}[\nabla - iq\mathbf{A}(t)]^2 \psi(\mathbf{x}, t) + V(\mathbf{x})\psi(\mathbf{x}, t). \tag{13.99}$$

We gauge-transform the quantum state to $\psi_g(\mathbf{x}, t) := e^{-i\mathbf{A}(t)\cdot\mathbf{x}}\psi(\mathbf{x}, t)$. The evolution equation becomes

$$i\frac{\partial}{\partial t}\psi_g(\mathbf{x}, t) = \left[-\frac{1}{2m}\nabla^2 + V(x) - q\mathbf{E}(t)\cdot\mathbf{x}\right]\psi_g(\mathbf{x}, t). \tag{13.100}$$

Hence, the Hamiltonian for $\psi_g(\mathbf{x}, t)$ is a Schrödinger operator with potential V plus a dipole-interaction term

$$\hat{H}_I = -q\mathbf{E}(t)\cdot\hat{\mathbf{x}}. \tag{13.101}$$

In the atom optics literature, the interaction Hamiltonian (13.101) is referred to as the *dipole Hamiltonian*, while the Hamiltonian (13.87) is referred to as the *minimal coupling* Hamiltonian. Note that the Hamiltonian (13.101) is well defined only in the dipole approximation.

For a two-level atom, we follow the reasoning of Section 13.4.5, to obtain an interaction Hamiltonian

$$\hat{H}_I(t) = -\mathbf{E}(t)\cdot\left(\mathbf{d}\,\hat{\sigma}_+ + \mathbf{d}^*\,\hat{\sigma}_-\right). \tag{13.102}$$

13.5.2 The Aharonov–Bohm effect

In Section 13.4.4, we showed that quantum systems may be affected by the EM field even if they are in a region of space where the electric and the magnetic fields vanish. Here, we will analyse an experiment proposed by Aharonov and Bohm (1959), that allows for a measurement of this effect, while also highlighting its relation to gauge transformations.

Consider the set-up of Fig. 13.5, consisting of a source of charged particles, and an interferometer with two beam splitters and two mirrors. A solenoid with an electric current is placed normal to the plane of the interferometer.

Suppose that the current in the solenoid is switched on. For any loop γ that surrounds the solenoid $\oint_\gamma d\mathbf{r}\cdot\mathbf{A} = \int\nabla\times\mathbf{A}\cdot d\mathbf{S} = \int\mathbf{B}\cdot d\mathbf{S} = \Phi$, where Φ is the magnetic flux through

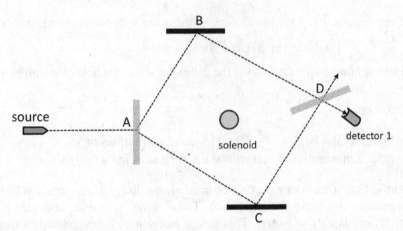

Fig. 13.5 The Aharonov–Bohm experiment. An interferometer consists of two beam splitters (A and D) and two mirrors (B and C). A solenoid is placed along the vertical direction of the interferometer. The presence of a magnetic flux through the solenoid affects the interference pattern, even if particles pass only through regions where the magnetic field vanishes.

the solenoid. Since $\Phi \neq 0$, **A** must take nonvanishing values on *any loop* that surrounds the solenoid.[5]

Let $\Psi^{(0)}(\mathbf{x}, t)$ be the solution to Schrödinger's equation for this system, when no current passes through the solenoid. We denote by $\Psi_1^{(0)}(\mathbf{x}, t)$ the solution that corresponds to propagation along the path ABD in the interferometer and by $\Psi_2^{(0)}(\mathbf{x}, t)$ the solution that corresponds to propagation along the path ACD. We can choose the parameters of the interferometer so that wave function recorded at detector 1 is $\Psi_1^{(0)}(\mathbf{x}, t) + \Psi_2^{(0)}(\mathbf{x}, t)$.

We assume that the paths ABD and ACD are of macroscopic dimension, much larger than the position spread of the states $\Psi_1^{(0)}(\mathbf{x}, t)$ and $\Psi_2^{(0)}(\mathbf{x}, t)$, so that the wave functions are well localized around the paths. Hence, when the current in the solenoid is switched on, the particles will encounter values of the magnetic potential along the two paths ABD and ACD.

This means that the wave function $\Psi_1(\mathbf{x}, t)$ that propagates along the arm is approximately $e^{iq \int_{ABD} d\mathbf{r} \cdot \mathbf{A}} \Psi_1^{(0)}(\mathbf{x}, t)$ at D, and that the wave function $\Psi_2(\mathbf{x}, t)$ that propagates along the arm ACD is approximately $e^{iq \int_{ACD} d\mathbf{r} \cdot \mathbf{A}} \Psi_2^{(0)}(\mathbf{x}, t)$ at D. Hence, the wave function that enters detector 1 is $e^{iq \int_{ABD} d\mathbf{r} \cdot \mathbf{A}} \Psi_1^{(0)}(\mathbf{x}, t) + e^{iq \int_{ACD} d\mathbf{r} \cdot \mathbf{A}} \Psi_2^{(0)}(\mathbf{x}, t)$.

The probability distribution for the position at detector 1 is

$$p(\mathbf{x}, t) = |\Psi_1^{(0)}(\mathbf{x}, t)|^2 + |\Psi_1^{(0)}(\mathbf{x}, t)|^2 + 2\text{Re}\left[e^{i\chi} \Psi_1^{(0)}(\mathbf{x}, t) \Psi_2^{(0)*}(\mathbf{x}, t) \right], \tag{13.103}$$

where $\chi = q \int_{ABD} d\mathbf{r} \cdot \mathbf{A} - q \int_{ACD} d\mathbf{r} \cdot \mathbf{A} = q \oint d\mathbf{r} \cdot \mathbf{A} = q\Phi$.

---

[5] The reason that we cannot choose $\mathbf{A} = 0$ everywhere outside the solenoid, is that there is no differentiable function f, such that $A_i = \partial f / \partial x_i$ everywhere outside the solenoid. In any Cartesian system of coordinates x_1 and x_2 with origin at the center of the solenoid, we can write $\mathbf{A} = \frac{\Phi}{2\pi}(-\frac{x_2}{r^2}, \frac{x_1}{r^2}, 0)$, where $r = \sqrt{x_1^2 + x_2^2}$. Then $\mathbf{A} = \nabla f$, where $f = \frac{\Phi}{2\pi} \tan^{-1}\left(\frac{x_2}{x_1}\right)$. The function f is not everywhere differentiable, as it diverges at $x_1 = 0$. We can make $\mathbf{A}$ vanish by a gauge transformation $\hat{\mathbf{A}} \to \hat{\mathbf{A}} - \nabla f$ only in a region that does not contain the x_2 axis. Since any loop around the solenoid crosses the x_2 axis, it encounters a region with nonzero $\mathbf{A}$.

Hence, the interference pattern depends on the magnetic flux through the solenoid, despite the fact that the particles cross only regions where the magnetic field vanishes. This prediction has been verified experimentally (Chambers, 1960; Tonomura et al., 1986).

In classical mechanics, the equations of motion for charged particles involve only the EM fields **E** and **B** and, for this reason, only the EM fields were thought to have direct physical meaning. Potentials were regarded as auxiliary mathematical constructs. In quantum theory, the evolution equation for charged particles explicitly involves the potentials, and the Aharonov–Bohm effect seems to imply that, at least in some cases, the potentials are physically active even when the fields vanish. The alternative is to keep the classical notion that only the EM fields are physically meaningful, but accept that they can act nonlocally. After all, the Aharonov–Bohm effect depends on the flux, which is defined in terms of the magnetic field. But this contradicts the widely held belief that the fields act only locally, that is, that charged particles are affected only by the field in its immediate vicinity.

Hence, the Aharonov–Bohm effect manifests a duality that is characteristic of gauge systems. Interactions can be described locally in terms of gauge-dependent objects such as the potentials, or nonlocally in terms of gauge-invariant quantities (true degrees of freedom in the language of constraint systems). This nonlocal interaction is not problematic at the fundamental level, because it cannot be used to transmit faster-than-light signals. However, its exact nature and its relation to other aspects of quantum theory, such as entanglement, is an open problem of current research.

QUESTIONS

13.1 Find the range of values of the quantity I_V in Theorem 13.4, for which the Schrödinger operator has (i) only one eigenvalue, and (ii) only two eigenvalues.

13.2 **(B)** Draw the classical Runge–Lenz vector for a particle that follows an elliptical orbit in an attractive Coulomb potential.

13.3 Is the dipole approximation compatible with Maxwell's equations for the external EM field? If not, how can it be justified?

13.4 **(B)** Sketch a set-up to measure an electric analogue of the Aharonov–Bohm effect, that is, appropriate time-dependent electric potentials $\phi(t)$ that generate a phase $q \oint \phi dt$ in interference experiments.

PROBLEMS

13.1 (i) Show that, for any eigenvector $|\psi\rangle$ of the Schrödinger operator, $\langle \psi | \hat{\mathbf{p}} | \psi \rangle = 0$. (ii) Show that, for a central potential, $\langle \psi | \hat{\mathbf{x}} | \psi \rangle = 0$. (ii) Show that, for a central potential, Hardy's inequality implies that $\Delta p \Delta r \geq \frac{1}{2}$, where $(\Delta p)^2 = \sum_{i=1}^{3}(\Delta p_i)^2$ and $(\Delta r)^2 = \sum_{i=1}^{3}(\Delta x_i)^2$.

13.2 Let $V_1(r) = V_0 \left[\left(\frac{r_0}{r} \right)^{12} - \left(\frac{r_0}{r} \right)^6 \right]$ with $V_0, r_0 > 0$ be the intermolecular Lennard-Jones potential, and let $V_2(r) = - \frac{V_0}{1 + e^{\frac{r - r_0}{a}}}$ with $V_0, r_0, a > 0$ be the nuclear Woods–Saxon potential. (i) Find an upper bound to the total number of states for the Schrödinger operators associated to these potentials. (ii) For what range of parameters do those potentials admit a single bound state?

13.3 Show that the Schrödinger operator with an attractive delta potential $V(\mathbf{x}) = -\lambda \delta^3(\mathbf{x})$ $\lambda > 0$, is not bounded from below in three dimensions. (Hint: Evaluate the expectation value $\langle \psi | \hat{H} | \psi \rangle$ for Gaussians centered around $\mathbf{x} = 0$.)

13.4 Assume that the spherical Bessel functions j_ℓ and η_ℓ are defined by Eq. (B.31). (i) Prove the recursion relations

$$j_{\ell+1}(x) + j_{\ell-1}(x) = \frac{2\ell+1}{x} j_\ell(x), \quad j'_\ell(x) = \frac{\ell}{x} j_\ell(x) - j_{\ell+1}(x),$$

and the analogous ones for $\eta_\ell(x)$. (ii) Using these relations, prove that the spherical Bessel functions are solutions to Eq. (13.21). (iii) Show that $j_\ell(x) \simeq \frac{2^\ell \ell!}{(2\ell+1)!} x^\ell$ and $\eta_\ell(x) \simeq -\frac{(2\ell)!}{2^\ell \ell!} x^{-(\ell+1)}$, as $x \to 0$. (iv) Prove Eq. (B.32).

13.5 Consider a particle in a central potential

$$V(r) = \begin{cases} V_0 & r < a \\ 0 & r \geq a \end{cases},$$

where $a, V_0 > 0$. (i) Show that, for $E < V_0$, the phase shift $\delta_0(k)$ satisfies the algebraic equation $k \tanh(\lambda a) = \lambda \tan(ka + \delta_0)$, where $k = \sqrt{2mE}$ and $\lambda = \sqrt{2m(V_0 - E)}$. (ii) Show that, for $ka << 1$, $\delta_0(k) \simeq ka \left(\frac{\tanh(\lambda_0 a)}{\lambda_0 a} - 1 \right)$, where $\lambda_0 = \sqrt{2mV_0}$.

13.6 Evaluate the probability distribution for the momentum norm $p = |\mathbf{p}|$ of the electron in the ground state of the hydrogen atom.

13.7 The Kratzer potential, $V(r) = V_0 \left(\frac{a^2}{r^2} - \frac{2a}{r} \right)$, for $V_0, a > 0$ is often employed in the study of diatomic molecules. (i) Show that the potential has minimum value $-V_0$ at $r = a$, and calculate the frequency of small oscillations around this minimum. (ii) Show that the energy eigenvalues of the associated Schrödinger operator are

$$E_{n_r, \ell} = -\frac{2V_0^2 a^2 m}{(n_r + \lambda_\ell + 1)^2},$$

with degeneracy $2\ell + 1$. Here, $\lambda_\ell = \sqrt{2mV_0 a^2 + (\ell + \frac{1}{2})^2} - \frac{1}{2}$ and n_r is the node number of the radial wave functions.

13.8 The attractive potential $V(r) = -V_0 e^{-r/a}$, where $V_0, a > 0$, provides a toy model for the interaction between nucleons. Show that the ground-state energy for the Schrödinger operator with this potential is $E_{1,0,0} = -\frac{v^2(b)}{8ma^2}$, where $b = \sqrt{8mV_0 a^2}$, and $v(b)$ is the real solution of the equation $J_v(b) = 0$ with the smallest absolute value; $J_v(x)$ is the Bessel function of order v – see Appendix B.5. (Hint: Change variables to $x = e^{-\frac{r}{2a}}$.)

13.9 Consider the Schrödinger operator $\hat{H}$ in two dimensions with a central potential $V(r)$, where $r = \sqrt{x_1^2 + x_2^2}$. (i) Show that the angular momentum operator $\hat{\ell} := \hat{x}_1 \hat{p}_2 - \hat{x}_2 \hat{p}_1$ commutes with $\hat{H}$. (ii) Show that the eigenfunctions of $\hat{H}$ in polar coordinates are $\psi_{E,\ell}(r, \theta) = \frac{C_{E,\ell}}{\sqrt{r}} u_{E,\ell}(r) e^{i\ell\theta}$, where $\ell = 0, \pm 1, \pm 2, \ldots$, and $u_{E,\ell}(r)$ satisfies the Schrödinger equation in $\mathbb{R}^+$ with an effective potential

$$V_{eff}(r) = \frac{\ell^2 - \frac{1}{4}}{2mr^2} + V(r)$$

and Dirichlet boundary conditions at $r = 0$. $C_{E,\ell}$ is a constant. (iii) Evaluate the eigenvalues of $\hat{H}$ and their degeneracy for the attractive Coulomb potential $V(r) = -\frac{\kappa}{r}$, $\kappa > 0$.

13.10 Prove Eqs. (13.60), (13.61), and (13.62).

13.11 Show that the Heisenberg equations of motion for the position and momentum operators with respect to the Hamiltonian (13.67) correspond to the classical motion of a particle under the Lorentz force (1.18).

13.12 A particle in a homogeneous electric field is prepared in a Gaussian state $\psi_0(x) = (2\pi\sigma^2)^{-1/4} \exp\left(-\frac{x^2}{4\sigma^2}\right)$. (i) Find the probability distribution for position at time t. (ii) Evaluate the variance $(\Delta x)^2$ as a function of time.

13.13 Consider a homogeneous magnetic field $\mathbf{B} = (0, 0, B)$, and particles restricted on the plane $x_1 - x_2$. Since the operators $\hat{\pi}_1$ and $\hat{x}_2$ commute they have a common set of joint (generalized) eigenvectors $|k_1, x_2\rangle$, such that $\hat{\pi}_1 |k_1, x_2\rangle = k_1 |k_1, x_2\rangle$ and $\hat{x}_2 |k_1, x_2\rangle = x_2 |k_1, x_2\rangle$. (i) Write $|k_1, x_2\rangle$ in the position representation. (ii) Do the same for the joint generalized eigenvectors $|x_1, k_2\rangle$ of $\hat{x}_1$ and $\hat{\pi}_2$. (iii) Compute the inner product $\langle k_1, x_2 | x_1, k_2 \rangle$.

13.14 Find the dipole matrix element $\mathbf{d}$ for transitions between the ground state and the first excited states of the isotropic harmonic oscillator.

13.15 (**B**) If the dominant term in the dipole approximation vanishes, then transitions are guided by the term $i\mathbf{k} \cdot \mathbf{x}$ in the expansion of $e^{i\mathbf{k}\cdot\mathbf{x}}$. (i) Show that the interaction Hamiltonian is $\hat{H}_I = -iq/2m \sum_{\mathbf{k}} \sum_{ij} k_i A_j(\mathbf{k}) e^{-ikt} \hat{R}_{ij}$, where $\hat{R}_{ij} = \hat{x}_i \hat{p}_j + \hat{p}_j \hat{x}_i$. (ii) Consider a particle in a central potential. Use the Wigner–Eckart theorem, and arguments from parity, in order to identify the selection rules for the transitions induced by $\hat{H}_I$.

13.16 (**B**) (i) Show that the solutions $\psi(\mathbf{x}, t)$ of Schrödinger's equation with the Hamiltonian (13.67) satisfy the continuity equation $\frac{\partial |\psi|^2}{\partial t} + \nabla \cdot \mathbf{J} = 0$, where

$$\mathbf{J} = \frac{1}{2mi} \left(\psi^* \nabla \psi - \psi \nabla \psi^* - 2iq\mathbf{A}|\psi|^2 \right)$$

is the probability current in presence of a magnetic field. (ii) Show that the probability current is invariant under gauge transformations.

Bibliography

- For further properties of Schrödinger operators in three dimensions, see chapter 6 of Galindo and Pascual (1990). For a mathematical treatment, see chapter 3 of Berezin and Shubin (1991).
- For the Aharonov–Bohm effect, see chapter 4 of Aharonov and Rohrlich (2005).

14 Particles with Spin

Today I read your new work, and it is certain that I am the one who rejoices most of it, because you push the swindle to unimagined height, by introducing electrons with four degrees of freedom, thus going beyond any idea of mine you have made fun of.

Heisenberg, in a card addressed to Pauli

14.1 The Discovery of Spin

Spin was introduced as part of the effort to understand the structure of atoms prior to the development of mature quantum theory. Variations of Bohr's model described atoms in terms of three quantum numbers, roughly similar to n, ℓ, and m that appear when solving the Schrödinger equation in central potentials. In this context, Pauli proposed that, in each atom, there exists at most one electron for each triplet of quantum numbers. This proposal is Pauli's famous "exclusion principle," which we will analyze in Chapter 15.

The exclusion principles did not fit the data perfectly; there appeared to be twice as many electrons as triplets of quantum numbers. Then, Pauli (1925) proposed a fourth quantum number. But there can be only three quantum numbers to account for the motion of an electron in space. The new quantum number does not refer to those motions, but to some "internal" property of the electron, whatever this might mean.

It was soon realized that Pauli's fourth quantum number had been observed a few years before, in an experiment carried out by Otto Stern and Walther Gerlach. Stern and Gerlach (1922) had directed a beam of silver atoms to cross an *inhomogeneous* magnetic field – see Fig. 14.1. They tried to verify a prediction of Bohr's theory, called quantization of space. They expected the beam to split into two smaller subbeams, and this is exactly what they observed. In retrospect, we can say that their interpretation of the experimental findings was wrong; the phenomenon they were looking for does not exist.

The correct interpretation of the Stern–Gerlach experiment was made possible by the work of Uhlenbeck and Goudsmit (1925). They proposed that the electron has an additional degree of freedom that can take only two values, like Pauli's fourth quantum number. This degree of freedom corresponds to an *intrinsic magnetic moment* of the electron that persists even when the electron is at rest. In the presence of a magnetic field, the two values of the new degree of freedom correspond to the magnetic moment being either parallel or antiparallel to the field. The main motive of Goudsmit and Uhlenbeck was to explain some inconsistencies in the spectroscopy

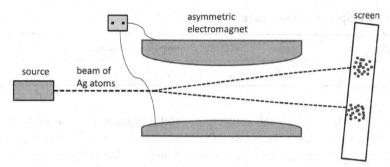

Fig. 14.1 The Stern–Gerlach experiment. A beam of Ag atoms passes through an inhomogeneous magnetic field and splits into two distinguishable subbeams.

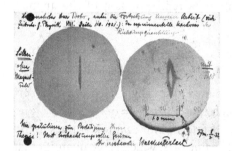

Fig. 14.2 A card sent by Stern to Bohr with a photo of the beam without the magnetic field (left) and the split beam in presence of the magnetic field (right). (Photo from the Bohr Archive in . Copenhagen.)

of the hydrogen atom, and they did not address the Stern–Gerlach experiment. However, their results explained it. Silver has an odd number of electrons, so the magnetic moments of electrons cannot cancel each other. Hence, an Ag atom has a nonzero magnetic moment which turns out to be due to a single electron in the outer shell. Since there are only two values for the magnetic moment, there are two different trajectories for the atoms inside the magnetic field, that is, two subbeams (see Fig. 14.2).

An obvious interpretation for the intrinsic magnetic moment of the electron is that it originates from self-rotation. This interpretation would imply that the electron has a finite radius r_0, because only rigid bodies self-rotate, and rigid bodies have finite extension. Since the electron is smaller than the atomic nucleus, $r_0 < 10^{-15}$ m. The angular momentum of the self-rotation would be of the order of $\hbar$, if the electron rotated with a speed $\frac{\hbar}{m_e r_0} > 10^{12}$ m/s, that is, if the rotation were faster than light. The interpretation of spin as self-rotation does not work.

Today, we understand spin as a degree of freedom that has the same symmetries as classical self-rotation without being an actual self-rotation. As explained in Section 12.3, classical and quantum concepts can be related only through the sharing of common symmetries. Spin is a degree of freedom that originates from the kinematical symmetry that defines the very notion of a particle. We will explain this point in more detail in Section 14.4.

14.2 The Quantum Description of Spin

Spin is related to the mini paradox in the quantum description of angular momentum that we encountered in Chapter 11. What is the physical significance of the representation of the angular

Table 14.1 Spin of various particle species

particle	electron	proton	neutron	quark	^{4}He nucleus	^{39}K nucleus	^{3}He atom	^{4}He atom
spin s	$\frac{1}{2}$	$\frac{1}{2}$	$\frac{1}{2}$	$\frac{1}{2}$	0	$\frac{3}{2}$	$\frac{1}{2}$	0

momentum algebra with half-integer j, if orbital angular momentum accepts only integer values of j? Now, we can state the answer: Representations with half-integer j involve spin.

According to Pauli, spin is a degree of freedom for particles, in addition to position and momentum, which describe a particle's motion in space. The latter degrees of freedom are fully described by the Hilbert space $L^2(\mathbb{R}^3)$. The additional degree of freedom will be described by a Hilbert space $\mathcal{V}_s$, so that a particle will be described by the Hilbert space $L^2(\mathbb{R}^3) \otimes \mathcal{V}_s$.

The relation of spin to rotation implies that $\mathcal{V}_s$ contains a representation of the angular momentum algebra,

$$[\hat{s}_i, \hat{s}_j] = i \sum_k \epsilon_{ijk} \hat{s}_k, \tag{14.1}$$

where we have represented the generators of rotations associated to spin by $\hat{s}_i$. Hence, the Hilbert space $\mathcal{V}_s$ has an orthonormal basis of vectors $|s, m_s\rangle$, such that $\hat{\mathbf{s}}^2|s, m_s\rangle = s(s+1)|s, m_s\rangle$ and $\hat{s}_3|s, m_s\rangle = m_s|s, m_s\rangle$. The integer s can be an integer or half-integer, while m_s take values $-s, -s+1, \ldots, s-1, s$. A particle has a unique and immutable value of s, which is the quantity that we refer to as the *particle's spin*. Spin is a defining characteristic of a particle species, like mass.

In an abuse of terminology, one often refers to the eigenvalue m_s also as spin because, with s constant, only m_s is a variable quantity; hence, it coincides with the fourth quantum number postulated by Pauli.

Table 14.1 lists the values of s for different particles; elementary and composite particles are treated at the same footing. The atomic constituents, electrons, protons and neutrons, all have spin $s = \frac{1}{2}$.

Given that s is constant for a given particle, the associated Hilbert space $\mathcal{V}_s = \mathbb{C}^{2s+1}$. Hence, the Hilbert space that describes a particle with spin s is $\mathcal{H}_s = L^2(\mathbb{R}^3) \otimes \mathbb{C}^{2s+1}$. On this Hilbert space, we define the *total angular momentum* operator,

$$\mathbf{\hat{J}} = \hat{\boldsymbol{\ell}} \otimes \hat{I} + \hat{I} \otimes \hat{\mathbf{s}}. \tag{14.2}$$

To avoid confusion, we will refer to the quantum numbers of orbital angular momentum as (ℓ, m_ℓ) and to those of total angular momentum as (j, m_j).

We express Eq. (14.2) as

$$\mathbf{\hat{J}} = \mathbf{\hat{L}} + \mathbf{\hat{S}}, \tag{14.3}$$

where $\hat{L}_i := \hat{\ell}_i \otimes \hat{I}$ and $\hat{S}_i := \hat{I} \otimes \hat{s}_i$. In the same way, we write $\hat{P}_i = \hat{p}_i \otimes \hat{I}$ and $\hat{X}_i = \hat{x}_i \otimes \hat{I}$, to denote the position and momentum operators in $\mathcal{H}_s$.

To summarize, in nonrelativistic physics, we describe a particle of mass m and spin s by the Hilbert space $\mathcal{H}_s = L^2(\mathbb{R}^3) \otimes \mathbb{C}^{2s+1}$, where the following commutation relations hold:

$$[\hat{X}_i, \hat{X}_j] = 0, \quad [\hat{P}_i, \hat{P}_j] = 0, \quad [\hat{X}_i, \hat{P}_j] = i\delta_{ij}, \quad [\hat{J}_i, \hat{J}_j] = i \sum_k \epsilon_{ijk} \hat{J}_k. \tag{14.4}$$

Equation (14.4) together with Eq. (14.3) are *the fundamental relations of nonrelativistic quantum mechanics.*

14.2.1 Bases on $\mathcal{H}_s$

For any orthonormal basis $|n\rangle$ on $L^2(\mathbb{R}^3)$, we define the orthonormal basis $|n, m_s\rangle := |n\rangle \otimes |m_s\rangle$ on $\mathcal{H}_s$. We do not use s as a label in the ket because it is constant. The same holds for generalized bases such as position $|\mathbf{r}\rangle$ and momentum $|\mathbf{k}\rangle$ on $L^2(\mathbb{R}^3)$, that is, we can define the kets $|\mathbf{r}, m_s\rangle$ and $|\mathbf{k}, m_s\rangle$ on $\mathcal{H}_s$.

If we employ the basis $|n, \ell, m_\ell\rangle$ for the eigenfunctions of the Schrödinger operator with central potential on $L^2(\mathbb{R}^3)$, then the basis $\mathcal{H}_s$ is $|n, \ell, m_\ell, m_s\rangle$, which consists of eigenvectors to the orbital angular momentum operators $\hat{\mathbf{L}}^2$ and $\hat{L}_3$.

Alternatively, we can define a basis $|n, \ell, j, m_j\rangle$ of eigenvectors to the total angular momentum operators $\hat{\mathbf{J}}^2$ and $\hat{J}_3$. To this end, we use the analysis of Section 11.5.5. There are two values of ℓ for each j, as $j = \ell \pm \frac{1}{2}$. By Eqs. (11.111) and (11.112), we define the basis vectors $|n, j, \ell, m_j\rangle$ as

$$\left| n, j, j + \frac{1}{2}, m_j \right\rangle = \sqrt{\frac{j - m_j + 1}{2(j+1)}} \left| n, j + \frac{1}{2}, m_j - \frac{1}{2}, \frac{1}{2} \right\rangle$$

$$- \sqrt{\frac{j + m_j + 1}{2(j+1)}} \left| n, j + \frac{1}{2}, m_j + \frac{1}{2}, -\frac{1}{2} \right\rangle, \tag{14.5}$$

$$\left| n, j, j - \frac{1}{2}, m_j \right\rangle = \sqrt{\frac{j + m_j}{2j}} \left| n, j - \frac{1}{2}, m_j - \frac{1}{2}, \frac{1}{2} \right\rangle$$

$$+ \sqrt{\frac{j - m_j}{2j}} \left| n, j - \frac{1}{2}, m_j + \frac{1}{2}, -\frac{1}{2} \right\rangle, \tag{14.6}$$

where the vectors in the right-hand side of the equations are the $|n, \ell, m_\ell, m_s\rangle$ eigenvectors.

In the Schrödinger representation, we express the elements of $\mathcal{H}_s$ as wave functions indexed by m_s; we write $\psi_{m_s}(\mathbf{x})$. We can also represent elements of $\mathcal{H}_s$ as column vectors,

$$\begin{pmatrix} \psi_s(\mathbf{r}) \\ \psi_{s-1}(\mathbf{r}) \\ \cdots \\ \psi_{-s}(\mathbf{r}) \end{pmatrix}. \tag{14.7}$$

Wave functions of the form $\psi(\mathbf{x})c_{m_s}$ are factorized states of $\mathcal{H}_s$. For these states, the correlations $\mathrm{Cor}(P_i, S_j)$ between momentum and spin vanish. These correlations are nonzero in entangled vectors of $\mathcal{H}_s$.

14.2.2 Spin of Composite Particles

So far, we made no distinction between elementary and composite particles in the definition of spin. There are some differences, which now we proceed to explain.

Consider two particles with masses m_1 and m_2, and spins s_1 and s_2, which interact through a central potential $V(|\hat{\mathbf{x}}_1 - \hat{\mathbf{x}}_2|)$. In Section 13.1, we showed that the Hamiltonian of this system can be expressed as $\hat{H} = \hat{H}_{CM} + \hat{H}_{rel}$, where $\hat{H}_{CM}$ is the Hamiltonian of the center of mass describing a free particle of mass $M = m_1 + m_2$, and $\hat{H}_{rel}$ is the Hamiltonian for the relative degrees of freedom.

When we treat the composite system as a particle, we focus on the degrees of freedom of the center of mass, that is, we only employ the Hamiltonian $\hat{H}_{CM}$. We treat the remaining degrees of freedom as frozen, that is, we assume that they are found in an eigenstate of $\hat{H}_{rel}$, usually the ground state.

The total angular momentum of the system is

$$\hat{\mathbf{J}} = \hat{\mathbf{L}}_{CM} + \hat{\mathbf{L}}_{rel} + \hat{\mathbf{S}}_1 + \hat{\mathbf{S}}_2, \tag{14.8}$$

where $\hat{\mathbf{L}}_{CM} = \mathbf{X} \times \mathbf{P}$ is the orbital angular momentum for the center of mass, $\hat{\mathbf{L}}_{rel} = \hat{\mathbf{r}} \times \hat{\mathbf{p}}$ is the orbital angular momentum of the internal degrees of freedom, and $\hat{\mathbf{S}}_1$, $\hat{\mathbf{S}}_2$ are the spins of the constituent particles.

Since we treat the total system as a single particle, its orbital angular momentum is $\hat{\mathbf{L}}_{CM}$. Then, Eq. (14.8) implies that the spin of the composite particle is

$$\hat{\mathbf{S}} = \hat{\mathbf{L}}_{rel} + \hat{\mathbf{S}}_1 + \hat{\mathbf{S}}_2. \tag{14.9}$$

Hence, the spin of a composite particle depends not only on the spins of its constituents, but also on their relative orbital angular momentum. If the internal degrees of freedom are in their ground state, then $\ell_{rel} = 0$; hence, $\hat{\mathbf{S}} = \hat{\mathbf{S}}_1 + \hat{\mathbf{S}}_2$. However, an excitation of the internal degrees of freedom may bring them in a state with $\ell_{rel} > 0$; hence, $\hat{\mathbf{S}}^2$ may change. Usually, when we refer to the spin of a composite particle, we refer to the value of s obtained when the internal degrees of freedom are in a ground state.[1]

Equation (14.9) means that if we know the spins s_1 and s_2 of the components of a composite particle, we can decide whether the spin of the composite particle is an integer or a half-integer. Since ℓ_{rel} is always an integer, the spin s of the composite system is an integer if $s_1 + s_2$ is an integer and a half-integer if $s_1 + s_2$ is a half-integer. This generalizes to composite particles of N constituents, each of spin s_i, with $i = 1, 2, \ldots, N$. The spin of the composite particle is an integer or a half-integer if $\sum_{i=1}^{N} s_i$ is an integer or a half-integer, respectively.

14.3 Electromagnetic Interactions of Particles with Spin

A particle with spin has an additional term in the Hamiltonian (13.67) that describes its interaction with an EM field, as the spin couples directly to the magnetic field. A particle of charge q, mass m, and spin s inside an EM field $(\mathbf{E}, \mathbf{B})$ evolves according to the Hamiltonian,

---

[1] Whether we consider an excited state of the internal degrees of freedom as defining a different particle is a matter of convention of particular research fields. In atomic and nuclear physics, we do not view an excited state of an atom or of a nucleus as a different particle. We talk about an excited atom or an excited nucleus of the same element. In contrast, in particle physics we often view the excited states of composite particles as different particles that merit their own name. For example, the ground state for the internal degrees of freedom in a pair of a u quark and a d antiquark is called a π meson; the first excited state is called the ρ meson.

Table 14.2 The g-factors of elementary and composite particles

particle	electron	muon	proton	neutron	^{2}H nucleus	^{7}Li nucleus
g-factor	−2.00232	−2.00233	5.58569	−3.83136	1.72697	15.23157

$$\hat{H} = \frac{1}{2m}\left(\hat{\mathbf{P}} - q\hat{\mathbf{A}}\right)^2 + q\phi - \gamma\mathbf{B}\cdot\hat{\mathbf{S}}. \tag{14.10}$$

The additional term $-\gamma\mathbf{B}\cdot\hat{\mathbf{S}}$ is similar to the term (13.70) that describes the coupling of a homogeneous magnetic field with orbital angular momentum. The coefficient is different, though. The constant γ is the *gyromagnetic ratio*, and has a fixed value for each particle. The gyromagnetic ratio can be nonzero even for electrically neutral particles.

It is useful to define the *g-factor*,

$$g = \frac{2m}{e}\gamma, \tag{14.11}$$

where e is the unit of the electric charge. Then, for a homogeneous magnetic field, Eq. (14.10) contains a dipole interaction term $-\frac{e}{2m}\mathbf{B}\cdot(N_q\hat{\mathbf{L}} + g\hat{\mathbf{S}})$, where N_q is the integer q/e. From this term, we identify the magnetic moment of the particle as

$$\hat{\mu} = \frac{e}{2m}(N_q\hat{\mathbf{L}} + g\hat{\mathbf{S}}). \tag{14.12}$$

For $m = m_e$, the electron mass, the ratio $\mu_B = \frac{e}{2m}$ is known as the *Bohr magneton*, the characteristic scale for the magnetic moment in atoms. For $m = m_p$, the proton mass, the ratio $\mu_N = \frac{e}{2m_p}$ is the *nuclear magneton*, the characteristic scale for the magnetic moment in nuclei. Since $\mu_B/\mu_N = m_p/m_e \gg 1$, the magnetic moment of the electrons dominates over the magnetic moment of the nuclei in atoms.

The values of the g-factor for some particles are given in Table 14.2.

The gyromagnetic ratio of composite particles can be calculated if one knows the gyromagnetic ratio of their components. Consider, for example, a system of two particles with spin s_1 and s_2, and gyromagnetic ratios γ_1 and γ_2, respectively. Let $s > 0$ be the spin of the composite particle, and let γ be its gyromagnetic ratio. The total Hamiltonian of the system involves a term $-\gamma_1\mathbf{B}\cdot\hat{\mathbf{S}}_1 - \gamma_2\mathbf{B}\cdot\hat{\mathbf{S}}_2$. If we treat the composite system as a particle, this term must equal $-\gamma\mathbf{B}\cdot\hat{\mathbf{S}}$. Hence,

$$\gamma\hat{\mathbf{S}} = \gamma_1\hat{\mathbf{S}}_1 + \gamma_2\hat{\mathbf{S}}_2. \tag{14.13}$$

Assume that the internal degrees of freedom are in the ground state so that $\mathbf{S} = \hat{\mathbf{S}}_1 + \hat{\mathbf{S}}_2$. Then,

$$\gamma\hat{\mathbf{S}}^2 = (\gamma_1\hat{\mathbf{S}}_1 + \gamma_2\hat{\mathbf{S}}_2)\cdot(\hat{\mathbf{S}}_1 + \hat{\mathbf{S}}_2) = \gamma_1\hat{\mathbf{S}}_1^2 + \gamma_2\hat{\mathbf{S}}_2^2 + (\gamma_1 + \gamma_2)\hat{\mathbf{S}}_1\cdot\hat{\mathbf{S}}_2. \tag{14.14}$$

We use Eq. (11.117), to obtain

$$\gamma = \frac{1}{2}(\gamma_1 + \gamma_2) + \frac{1}{2}(\gamma_1 - \gamma_2)\frac{s_1(s_1 + 1) - s_2(s_2 + 1)}{s(s + 1)}. \tag{14.15}$$

Example 14.1 The deuterium nucleus D^+ consists of one proton and one neutron. Its ground state corresponds to $\ell = 0$ and $s = 1$ – for an explanation, see Section 15.1. Then, Eq. (14.15) applies, with $s_1 = s_2 = \frac{1}{2}$ and $s = 1$, from which we obtain $\gamma = \frac{1}{2}(\gamma_1 + \gamma_2)$. Taking the masses of the nucleons to be approximately equal, and the mass of the deuterium about twice the mass of the proton, we obtain $g_{D^+} = g_n + g_p$. Comparing with the values from Table 14.2, we find an error of 2 percent, which is remarkably small given that we made the drastic assumption of an interaction through a central potential (so that $\ell = 0$ in the ground state).

The gyromagnetic ratio can be calculated from first principles for elementary particles, but this requires a relativistic treatment in the context of quantum field theory. For an electron, the value of g (written as g_e) is very close to -2; we will see shortly why. Experiment has given $g_e = -2.0023193043615(5)$, in agreement with the theoretical prediction in 13 significant digits. This renders the prediction of g_e the most precise confirmed prediction in the history of science. The measurement of g for the muon, $g_\mu = -2.0023318412(8)$ is accurate up to 10 significant digits, but there seems to be a deviation from the current best theoretical calculation at the sixth significant digit. At the time of this writing (early 2023), it is not clear whether this discrepancy is due to a calculational error or if it is a sign of new physics, that is, of effects beyond the Standard Model of particle physics.

14.3.1 Explanation of the Stern–Gerlach Experiment

Equation (14.10) explains the beam splitting in the Stern–Gerlach experiment. We recall that the experiment was about a beam of Ag atoms propagating in an inhomogeneous field, which can be taken as $\mathbf{B}(\mathbf{x}) = (0, 0, B(\mathbf{x}))$. The Ag atoms are neutral ($q = 0$), they have spin $s = \frac{1}{2}$, and nonzero gyromagnetic ratio γ. The Hamiltonian (14.10) reads

$$\hat{H} = \frac{1}{2m}\hat{\mathbf{P}}^2 - \gamma B(\hat{X})\hat{S}_3. \tag{14.16}$$

The momentum operator evolves in the Heisenberg picture as follows

$$\frac{d}{dt}\hat{P}_i = i\gamma[B(\hat{X}), \hat{P}_i]\hat{S}_3 = -\gamma\frac{\partial B}{\partial x_i}(\hat{\mathbf{X}})\hat{S}_3. \tag{14.17}$$

In subspaces $\mathcal{H}_\pm$ of constant $m_s = \pm\frac{1}{2}$, Eq. (14.17) becomes

$$\frac{d}{dt}\hat{P}_i = \mp\frac{\gamma}{2}\frac{\partial B}{\partial x_i}(\hat{X}), \tag{14.18}$$

that is, momentum evolves in two distinct ways, one for each value of m_s. As a result, the beam breaks into two subbeams that differ in the moment along the directions of inhomogeneity $\left(\frac{\partial B}{\partial x_i} \neq 0\right)$ of the magnetic field.

14.3.2 Larmor Precession

The physical meaning of the gyromagnetic ratio is best understood when examining an electrically neutral particle moving in a homogeneous magnetic field. In this case, the degrees of

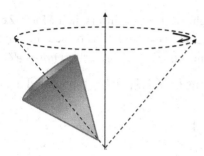

Fig. 14.3 Precession of a self-rotating rigid body in classical mechanics. The axis of rotation of a top spans a cone around the vertical axis during its motion.

freedom for position and momenta do not interact with the field, so we can separate the action of the magnetic field on spin. Hence, we restrict to the Hilbert space $\mathbb{C}^{2s+1}$ for spin, and we consider the Hamiltonian $\hat{H} = -\gamma \mathbf{B} \cdot \hat{\mathbf{s}}$ for a constant magnetic field $\mathbf{B}$. In the Heisenberg picture, the spin operators evolve as

$$\dot{\hat{s}}_i = i[\hat{H}, \hat{s}_i] = -i\gamma \sum_j B_j[\hat{s}_j, \hat{s}_i] = -\gamma \sum_{jk} \epsilon_{ijk} B_j \hat{s}_k, \tag{14.19}$$

or, equivalently,

$$\dot{\hat{\mathbf{s}}} = -\gamma \mathbf{B} \times \hat{\mathbf{s}}. \tag{14.20}$$

We choose $\mathbf{B} = (0, 0, B)$. Then, Eq. (14.20) yields $\dot{\hat{s}}_1 = \gamma B \hat{s}_2, \dot{\hat{s}}_2 = -\gamma B \hat{s}_1, \dot{\hat{s}}_3 = 0$. In terms of the operators $\hat{s}_\pm = \hat{s}_1 \pm i\hat{s}_2$, we obtain

$$\dot{\hat{s}}_\pm = \mp i\gamma B \hat{s}_\pm \qquad \dot{\hat{s}}_3 = 0. \tag{14.21}$$

The solution is

$$\hat{s}_\pm(t) = \hat{s}_\pm(0)e^{\mp i\gamma Bt}, \quad \hat{s}_3(t) = \hat{s}_3(0). \tag{14.22}$$

Equivalently,

$$\hat{s}_1(t) = \hat{s}_1(0)\cos(\gamma Bt) + \hat{s}_2(0)\sin(\gamma Bt), \tag{14.23}$$

$$\hat{s}_2(t) = -\hat{s}_1(0)\sin(\gamma Bt) + \hat{s}_2(0)\cos(\gamma Bt), \tag{14.24}$$

$$\hat{s}_3(t) = \hat{s}_3(0). \tag{14.25}$$

Hence, the spin vector rotates around the axis of the magnetic field with angular frequency $\omega = \gamma B$. This phenomenon is known as *Larmor precession*. The classical analogue of the Larmor precession is the rotation of a top (see Fig. 14.3). The name "gyromagnetic ratio" is explained by the expression $\gamma = \omega/B$; "gyro" meaning rotation in Greek.

By choosing the direction of the magnetic field, and the time t that the particle spends inside the magnetic field, we can rotate the spin around any axis and by any angle $\chi = \gamma Bt$. To determine the time t, we assume that the particles' momentum is well concentrated around a momentum p_0, and that the magnetic field extends for a distance L. For particles of mass m, this implies that $t = Lm/p_0$, so that $\chi = \gamma BLm/p_0$.

Let us assume that the particle beam passes through two beam splitters, in a Mach–Zehnder interferometer similar to that of Fig. 3.2. A magnetic field $\mathbf{B} = (0, 0, B)$ is present in one of the two interferometric paths. For an input state $|in\rangle$, the output state is $|out\rangle = \frac{1}{\sqrt{2}} \left(|in\rangle + e^{i\chi \hat{s}_3} |in\rangle \right)$.

For concreteness, we take the particles to be neutrons $\left(s = \frac{1}{2}, \hat{s}_3 = \frac{1}{2}\hat{\sigma}_3 \right)$, prepared in a state

$$|in\rangle = \frac{1}{\sqrt{2}} \begin{pmatrix} 1 \\ 1 \end{pmatrix}.$$

Then, the output state is

$$|out\rangle = \frac{1}{2} \begin{pmatrix} 1 + e^{\frac{i}{2}\chi} \\ 1 + e^{-\frac{i}{2}\chi} \end{pmatrix}.$$

The probability Prob(+) that a measurement of $\hat{s}_1$ on $|out\rangle$ gives $+1$ depends parametrically on χ,

$$\text{Prob}(+) = |\langle in|out\rangle|^2 = \frac{1}{2} \left(1 + \cos\frac{\chi}{2} \right)^2. \tag{14.26}$$

Suppose we carry out this experiment for different values of the magnetic field B, and, thereby of the angle χ. The probabilities Prob(+) will then be periodic functions of χ. The key point is that the distance between two successive maxima of Prob(+) corresponds to $\Delta\chi = 4\pi$ and not $\Delta\chi = 2\pi$. A rotation of 2π will take us from the maximum value of Prob(+) = 1 to the minimum value Prob(+) = 0. This provides direct experimental proof that the state of particles with half-integer spin changes by a factor of -1 under a rotation of 2π, as pointed out in Section 11.3.5. Such experiments were first carried out by Rauch et al. (1975) and Werner et al. (1975). They confirmed the 4π symmetry of the neutron's spin. In Fig. 14.4, we describe the setup and results of the first experiment.

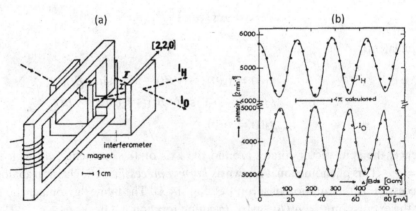

Fig. 14.4 (a) Sketch of the experimental setup of Rauch et al. (1975). A neutron beam is split by a perfect crystal. One of the subbeam traverses distance L within a magnetic field B. The two subbeams are recombined by a second crystal, leading to two output beams, I_H and I_0. (b) Plot of the intensity of the I_H and I_0 beams as a function of BL. The angular distance between the peaks was found to be 704 ± 34 degrees, clearly establishing the 4π symmetry of neutron's spin. [Reprinted from Rauch et al. (1975). Copyright 1975, with permission from Elsevier.]

14.3.3 Pauli's Equation

Pauli explained the fact that the gyromagnetic ratio of the electron is almost exactly -2 in terms of a new, more fundamental, Hamiltonian for particles with spin $\frac{1}{2}$ interacting with the EM'field.

For $s = \frac{1}{2}$, a particle's spin can be described as a qubit. Using Pauli matrices, we define the operator

$$\hat{\mathbf{K}} \cdot \hat{\boldsymbol{\sigma}} = \sum_{i=1}^{3} \hat{K}_i \otimes \hat{\sigma}_i, \tag{14.27}$$

on $\mathcal{H}_{\frac{1}{2}}$, where $\hat{K}_i$, $i = 1, 2, 3$ is a vector operator on $L^2(\mathbb{R}^3)$. We use Eq. (5.4), to obtain

$$(\hat{\mathbf{K}} \cdot \hat{\boldsymbol{\sigma}})^2 = \sum_{ij} \hat{K}_i \hat{K}_j \otimes \hat{\sigma}_i \hat{\sigma}_j = \sum_{ij} \hat{K}_i \hat{K}_j \otimes \left(\delta_{ij} \hat{I} + i \sum_k \epsilon_{ijk} \hat{\sigma}_k \right)$$

$$= \hat{\mathbf{K}}^2 \otimes \hat{I} + \frac{i}{2} \sum_{ijk} \epsilon_{ijk} [\hat{K}_i, \hat{K}_j] \otimes \hat{\sigma}_k. \tag{14.28}$$

Pauli's Hamiltonian is

$$\hat{H} = \frac{1}{2m} [(\hat{\mathbf{p}} - q\hat{\mathbf{A}}) \cdot \hat{\boldsymbol{\sigma}}]^2 + q\hat{\phi}. \tag{14.29}$$

We use Eqs. (13.98) and (14.28) to obtain

$$[(\hat{\mathbf{p}} - q\hat{\mathbf{A}}) \cdot \hat{\boldsymbol{\sigma}}]^2 = (\hat{\mathbf{p}} - q\hat{\mathbf{A}})^2 \otimes \hat{I} - 2q \sum_i \hat{B}_i \otimes \hat{s}_i, \tag{14.30}$$

where $\hat{s}_i = \frac{1}{2}\hat{\sigma}_i$. Hence, Pauli's Hamiltonian becomes

$$\hat{H} = \frac{1}{2m} \left(\hat{\mathbf{P}} - q\mathbf{A} \right)^2 + q\hat{\phi} - \frac{q}{m} \hat{\mathbf{B}} \cdot \hat{\mathbf{S}}. \tag{14.31}$$

Comparing with Eq. (14.10), we find that $\gamma = q/m = -e/m$; hence, $g = -2$.

Pauli's Hamiltonian applies only to charged elementary particles with spin $\frac{1}{2}$. It does not apply to protons or any other charged composite particles. In Chapter 16, we will see that Eq. (14.29) is the nonrelativistic limit of Dirac's Hamiltonian, which provides a relativistic description of the electron.

14.3.4 Spin Interactions

The magnetic field that interacts with a spin may be generated by some other microscopic magnetic dipole, resulting in induced interaction terms between angular momenta. Such terms can be evaluated from first principles only in a fully relativistic description. In particular, the Hamiltonian of a charged particle of mass m, spin s, charge q, and gyromagnetic ratio in a central electric potential $\phi(r)$ must include a term

$$\hat{V}_{LS} = \frac{\gamma}{2mq} \hat{r}^{-1} \phi'(\hat{r}) \hat{\mathbf{L}} \cdot \hat{\mathbf{S}}, \tag{14.32}$$

known as *LS coupling* or *spin–orbit coupling*. This expression cannot be derived from any argument within classical electromagnetic theory, it requires the quantum theory of relativistic electrons.

In classical electromagnetic theory, the interaction energy of two dipoles with magnetic moments $\boldsymbol{\mu}_1$ and $\boldsymbol{\mu}_2$ is

$$E = -\frac{1}{4\pi r^3}\left[3(\boldsymbol{\mu}_1 \cdot \mathbf{n})(\boldsymbol{\mu}_2 \cdot \mathbf{n}) - \boldsymbol{\mu}_1 \cdot \boldsymbol{\mu}_2\right], \tag{14.33}$$

where r is the distance between the dipoles and $\boldsymbol{n}$ the unit vector along the axis that passes through the dipoles. The quantum analogue of Eq. (14.33) is obtained by expressing the magnetic moments in terms of operators (14.12).

14.4 The Fundamental Symmetry of Nonrelativistic Physics

The fundamental symmetry of nonrelativistic physics follows from Newton's notions of absolute time and space. It is based on the identification of *inertial reference frames*, namely, a special class of reference frames with respect to which Newton's law holds without modification. The set of transformations between inertial reference frames defines the *Galilei group*.[2] In this section, we analyze the properties of this group, and we show that it defines the fundamental symmetry of nonrelativistic quantum physics.

14.4.1 The Generators of the Galilei Group

A reference frame Σ describes a physical event by four numbers: the time t, and a vector $\mathbf{x} = (x_1, x_2, x_3)$ for spatial coordinates with respect to an orthogonal system of axes. A different reference frame Σ' assigns to the same event a different time t' and a different vector $\mathbf{x}' = (x_1', x_2', x_3')$. If both frames are inertial, they are related by the following transformations.

1. Space translation: $(t', \mathbf{x}') = (t, \mathbf{x} + \mathbf{a})$, where $\mathbf{a} \in \mathbb{R}^3$. (The origin of the coordinates is shifted by a constant vector.)
2. Time translation: $(t', \mathbf{x}') = (t + \tau, \mathbf{x})$, where $\tau \in \mathbb{R}$.
3. Rotation: $(t', \mathbf{x}') = (t, (O\mathbf{x}))$, where $O \in SO(3)$ and $(O\mathbf{x})_i = \sum_j O_{ij} x_j$. (The coordinate axes are rotated but the coordinate origin remains the same.)
4. Boost: $(t', \mathbf{x}') = (t, \mathbf{x} + \mathbf{v}t)$, where $\mathbf{v} \in \mathbb{R}^3$. (Relative motion at constant velocity.)

These transformations define the Galilei group Gal. The elements of Gal can be expressed as a 5×5 matrices, with 10 independent variables

$$G(O, \mathbf{v}, \mathbf{a}, \tau) = \begin{pmatrix} O_{11} & O_{12} & O_{13} & v_1 & a_1 \\ O_{21} & O_{22} & O_{23} & v_2 & a_2 \\ O_{31} & O_{32} & O_{33} & v_3 & a_3 \\ 0 & 0 & 0 & 1 & \tau \\ 0 & 0 & 0 & 0 & 1 \end{pmatrix}. \tag{14.34}$$

---

[2] The idea that the laws of physics do not depend on the reference frame originates from "Galileo's ship," a famous thought experiment described by Galileo in his book *Dialogue on the Two Chief World Systems*. In this experiment, a person in a closed cabin in a ship cannot tell whether the ship is still or moving with constant speed.

It is straightforward to show that the action of $G(O, \mathbf{v}, \mathbf{a}, \tau)$ on a column vector reproduces the Galilei transformations

$$G(O, \mathbf{v}, \mathbf{a}, \tau) \begin{pmatrix} x_1 \\ x_2 \\ x_3 \\ t \\ 1 \end{pmatrix} = \begin{pmatrix} (Ox)_1 + v_1 t + a_1 \\ (Ox)_2 + v_2 t + a_2 \\ (Ox)_3 + v_3 t + a_3 \\ t + \tau \\ 1 \end{pmatrix}. \tag{14.35}$$

The group law is determined by matrix multiplication

$$G(O_2, \mathbf{v}_2, \mathbf{a}_2, \tau_2) G(O_1, \mathbf{v}_1, \mathbf{a}_1, \tau_1) = G(O_2 O_1, \mathbf{v}_2 + \mathbf{v}_1, O_2 \mathbf{a}_1 + \mathbf{a}_2 + \mathbf{v}_2 \tau_1, \tau_1 + \tau_2).$$

There are three generators of rotations J_i, three generators of boosts K_i, three generators of space translations P_i and one generator of time translations H. The corresponding 5×5 matrices are straightforwardly obtained from Eq. (14.34),

$$\boldsymbol{\theta} \cdot \mathbf{J} = \begin{pmatrix} 0 & \theta_3 & -\theta_2 & 0 & 0 \\ -\theta_3 & 0 & \theta_1 & 0 & 0 \\ \theta_2 & -\theta_1 & 0 & 0 & 0 \\ 0 & 0 & 0 & 0 & 0 \\ 0 & 0 & 0 & 0 & 0 \end{pmatrix}, \quad \mathbf{v} \cdot \mathbf{K} = \begin{pmatrix} 0 & 0 & 0 & v_1 & 0 \\ 0 & 0 & 0 & v_2 & 0 \\ 0 & 0 & 0 & v_3 & 0 \\ 0 & 0 & 0 & 0 & 0 \\ 0 & 0 & 0 & 0 & 0 \end{pmatrix},$$

$$\mathbf{a} \cdot \mathbf{P} = \begin{pmatrix} 0 & 0 & 0 & 0 & a_1 \\ 0 & 0 & 0 & 0 & a_2 \\ 0 & 0 & 0 & 0 & a_3 \\ 0 & 0 & 0 & 0 & 0 \\ 0 & 0 & 0 & 0 & 0 \end{pmatrix}, \quad H = \begin{pmatrix} 0 & 0 & 0 & 0 & 0 \\ 0 & 0 & 0 & 0 & 0 \\ 0 & 0 & 0 & 0 & 0 \\ 0 & 0 & 0 & 0 & 0 \\ 0 & 0 & 0 & 1 & 0 \end{pmatrix}, \tag{14.36}$$

for arbitrary vectors $\boldsymbol{\theta}, \mathbf{v}$, and $\mathbf{a}$. Then, we find the commutation relations for the associated Lie algebra $\mathfrak{gal}$. It is convenient to write these relations in the form of a *commutator table*, as follows, in which the entry for column A and row B is the commutator $[A, B]$.

	J_j	K_j	P_j	H
J_i	$\sum_k \epsilon_{ijk} J_k$	$\sum_k \epsilon_{ijk} K_k$	$\sum_k \epsilon_{ijk} P_k$	0
K_i	$\sum_k \epsilon_{ijk} K_k$	0	0	P_i
P_i	$\sum_k \epsilon_{ijk} P_k$	0	0	0
H	0	$-P_j$	0	0

14.4.2 Representations of the Galilei Group

In a unitary representation of Gal, we identify the generators with self-adjoint operators $\hat{J}_i, \hat{K}_i, \hat{P}_i$ and $\hat{H}$ on a Hilbert space $\mathcal{H}$. The commutator table is as follows.

	$\hat{J}_j$	$\hat{K}_j$	$\hat{P}_j$	$\hat{H}$
$\hat{J}_i$	$i\sum_k \epsilon_{ijk}\hat{J}_k$	$i\sum_k \epsilon_{ijk}\hat{K}_k$	$i\sum_k \epsilon_{ijk}\hat{P}_k$	0
$\hat{K}_i$	$i\sum_k \epsilon_{ijk}\hat{K}_k$	0	0	$i\hat{P}_i$
$\hat{P}_i$	$i\sum_k \epsilon_{ijk}\hat{P}_k$	0	0	0
$\hat{H}$	0	$-i\hat{P}_j$	0	0

We attempt to represent Gal in the Hilbert space $\mathcal{H}_s = L^2(\mathbb{R}^3)\otimes\mathbb{C}^{2s+1}$ of a particle with spin s and mass m. There are two evident identifications, of $\hat{P}_i$ with the momentum operator and of $\hat{J}_i$ with the angular momentum operator. The obvious choice for the generator of time translations is the free-particle Hamiltonian $\hat{H} = \frac{1}{2m}\hat{\mathbf{P}}^2$. If we select $\hat{K}_i = m\hat{X}_i$, then all commutation relations are satisfied except for $[\hat{K}_i, \hat{P}_j] = 0$. Instead, we have Heisenberg's commutator,

$$[\hat{K}_i, \hat{P}_j] = im\delta_{ij}\hat{I}. \tag{14.37}$$

Equation (14.37) defines a central extension of $\mathfrak{gal}$, with the mass m playing the role of the central charge. Hence, the Hilbert space $\mathcal{H}_s$ carries a projective representation of the Galilei group.

In principle, we can do better than a mere projective representation. We can construct unitary representations of the Galilei group, but they turn out to be unphysical.[3] The only representations that correspond to physical particles are the projective ones. For these representations, the commutator table becomes the following.

	$\hat{J}_j$	$\hat{K}_j$	$\hat{P}_j$	$\hat{H}$	$\hat{I}$
$\hat{J}_i$	$i\sum_k \epsilon_{ijk}\hat{J}_k$	$i\sum_k \epsilon_{ijk}\hat{K}_k$	$i\sum_k \epsilon_{ijk}\hat{P}_k$	0	0
$\hat{K}_i$	$i\sum_k \epsilon_{ijk}\hat{K}_k$	0	$im\delta_{ij}\hat{I}$	$i\hat{P}_i$	0
$\hat{P}_i$	$i\sum_k \epsilon_{ijk}\hat{P}_k$	$-im\delta_{ij}\hat{I}$	0	0	0
$\hat{H}$	0	$-i\hat{P}_j$	0	0	0
$\hat{I}$	0	0	0	0	0

This commutator table defines the *Bargmann algebra* with 11 generators, the 10 generators of the Galilei group, and the unity $\hat{I}$. The associated Lie group is the *Bargmann group*.

The Bargmann group has two Casimirs, both involving second-order polynomials of the generators. The first Casimir is

$$\hat{C}_1 = m\hat{H} - \frac{\hat{\mathbf{P}}^2}{2}. \tag{14.38}$$

In an irreducible representation, $\hat{H} = \frac{\hat{\mathbf{P}}^2}{2m} + E_0\hat{I}$, where E_0 is a constant that corresponds to minimum energy. Hence, the Casimir $\hat{C}_1$ implements the well-known fact that there is unique

---

[3] The unitary irreducible representations of the Galilei group are defined in a Hilbert space where no position kets can be defined, and the momentum kets are too few to define a generalized basis (Inonü and Wigner, 1952). Any attempt to identify the generator of spatial translations with physical momentum leads to absurdities.

zero of energy in nonrelativistic physics. In this context, the positivity of the mass m arises a consequence of the Hamiltonian being bounded from below.

The second Casimir is

$$\hat{C}_2 = \hat{\mathbf{W}}^2, \tag{14.39}$$

where $\hat{\mathbf{W}} = -m\hat{\mathbf{J}} + \hat{\mathbf{K}} \times \hat{\mathbf{P}}$. It is straightforward to compute $[\hat{W}_i, \hat{W}_j] = im \sum_k \epsilon_{ijk} \hat{W}_k$, which implies that the operators $m^{-1}\hat{W}_i$ satisfy the angular momentum algebra, and they coincide with the spin vector. Hence, the Casimir $\hat{\mathbf{W}}^2$ takes the constant value $m^2 s(s+1)$ in an irreducible representation, where s can be $0, \frac{1}{2}, 1, \frac{3}{2}, \ldots$. The Stone–von Neumann theorem (Theorem 12.7) guarantees that the subgroup H_3 generated by $\hat{C}_i, \hat{P}_i$, and $\hat{I}$ has a unique representation for a given value of m. We conclude that the most general unitary irreducible representation of the Bargmann group for $m > 0$ is characterized by the triplet (m, E_0, s), and it describes a particle with spin s in the Hilbert space $\mathcal{H}_s$.

This means that the requirement that a physical system carries a representation of the Galilei group leads to the introduction of spin as an intrinsic property of a particle.

> In nonrelativistic physics, we define a particle as an irreducible (projective) representation of the Galilei group. Hence, a particle is the simplest system that manifests the characteristic symmetry of inertial reference frames.

The definition of particles, or of other elementary quantum systems, in terms of symmetries originates from Wigner (1939), in his analysis of relativistic symmetries in quantum theory (Chapter 16), and from Mackey (1963a,b) who developed the theory for the most general case. These ideas underlie the quantization through the canonical group that was presented in Section 12.3.

14.4.3 Space Inversion and Time Reversal

Coming back to the formulation of Section 14.4.1, we note that the transformation of time reversal $(t, \mathbf{x}) \rightarrow (-t, \mathbf{x})$ corresponds to the 5×5 diagonal matrix

$$\mathcal{T} = \text{diag}\{1, 1, 1, -1, 1\}. \tag{14.40}$$

$\mathcal{T}$ acts on the generators (14.36) of the Galilei group as

$$\mathcal{T}J_i\mathcal{T}^{-1} = J_i, \quad \mathcal{T}K_i\mathcal{T}^{-1} = -K_i, \quad \mathcal{T}P_i\mathcal{T}^{-1} = P_i, \quad \mathcal{T}H\mathcal{T}^{-1} = H. \tag{14.41}$$

The space inversion transformation $(t, \mathbf{x}) \rightarrow (t, -\mathbf{x})$ corresponds to the 5×5 diagonal matrix

$$\mathcal{P} = \text{diag}\{-1, -1, -1, 1, 1\}, \tag{14.42}$$

which acts on the generators (14.36) as

$$\mathcal{P}J_i\mathcal{P}^{-1} = J_i, \quad \mathcal{P}K_i\mathcal{P}^{-1} = -K_i \quad \mathcal{P}P_i\mathcal{P}^{-1} = -P_i, \quad \mathcal{P}H\mathcal{P}^{-1} = H. \tag{14.43}$$

In quantum theory, we represent the transformations $\mathcal{T}$ and $\mathcal{P}$ by unitary or antiunitary operators on the Hilbert space $\mathcal{H}_s$ of a particle with spin s.

Consider first the case of time reversal. Suppose we can represent it by a unitary operator $\hat{\mathbb{T}}$. Equation (14.41) implies that, for the generator $\hat{H}$ of time translations,

$$\hat{\mathbb{T}}\frac{1}{i}\hat{H}\hat{\mathbb{T}}^\dagger = -\frac{1}{i}\hat{H}. \tag{14.44}$$

If $\hat{\mathbb{T}}$ is unitary, then $\hat{\mathbb{T}}\hat{H}\hat{\mathbb{T}}^\dagger = -\hat{H}$. However, such a transformation takes any positive point of the Hamiltonian's spectrum to its opposite. It is incompatible with a Hamiltonian that is bounded from below but not from above, like the Hamiltonian of a free particle. This implies that we cannot represent time reversal with a unitary operator.

In contrast, if $\hat{\mathbb{T}}$ is antiunitary, Eq. (14.44) implies that

$$\hat{\mathbb{T}}\hat{H}\hat{\mathbb{T}}^\dagger = \hat{H}, \tag{14.45}$$

and there is no problem with the spectrum of the Hamiltonian. Thus, we proved the necessity of describing time reversal by an antiunitary operator, a fact that we took for granted in Chapter 11.

Since $\hat{\mathbb{T}}$ is antiunitary, Eq. (14.41) implies that

$$\hat{\mathbb{T}}\hat{J}_i\hat{\mathbb{T}}^\dagger = -\hat{J}_i, \qquad \hat{\mathbb{T}}\hat{K}_i\hat{\mathbb{T}}^\dagger = \hat{K}_i, \qquad \hat{\mathbb{T}}\hat{P}_i\hat{\mathbb{T}}^\dagger = -\hat{P}_i. \tag{14.46}$$

Space inversion is implemented by a unitary operator on $\mathcal{H}_s$, so Eq. (14.43) leads to

$$\hat{\mathbb{P}}\hat{J}_i\hat{\mathbb{P}}^\dagger = \hat{J}_i, \quad \hat{\mathbb{P}}\hat{K}_i\hat{\mathbb{P}}^\dagger = -\hat{K}_i, \quad \hat{\mathbb{P}}\hat{P}_i\hat{\mathbb{P}}^\dagger = -\hat{P}_i, \quad \hat{\mathbb{P}}\hat{H}\hat{\mathbb{P}}^\dagger = \hat{H}. \tag{14.47}$$

In Chapter 11, we showed how the these two operators act on position wave function (Section 11.1.3) and how they act on angular momenta (Sections 11.3.6 and 11.3.7). Combining, we write their action in the Schrödinger representation for $\mathcal{H}_s$,

$$\hat{\mathbb{T}}\psi_{m_s}(\mathbf{x}) = (-1)^{s-m_s}\psi^*_{s-m_s}(\mathbf{x}), \tag{14.48}$$

$$\hat{\mathbb{P}}\psi_{m_s}(\mathbf{x}) = \varpi\,\psi_{m_s}(-\mathbf{x}), \tag{14.49}$$

where we used the convention (11.67) for the action of $\hat{\mathbb{T}}$ on spin. The quantity ϖ takes values ± 1, and originates from the action of $\hat{\mathbb{P}}$ on the spin eigenvectors.[4] It is an intrinsic property of a particle, known as the particle's *intrinsic parity*.

We can also define the operators $\hat{\mathbb{T}}$ and $\hat{\mathbb{P}}$ by their actions on the kets $|\mathbf{p}, m_s\rangle$

$$\hat{\mathbb{T}}|\mathbf{p}, m_s\rangle = (-1)^{s-m_s}|-\mathbf{p}, s-m_s\rangle, \tag{14.50}$$

$$\hat{\mathbb{P}}|\mathbf{p}, m_s\rangle = \varpi|-\mathbf{p}, m_s\rangle. \tag{14.51}$$

We note that $[\hat{\mathbb{T}}, \hat{H}] = [\hat{\mathbb{P}}, \hat{H}] = 0$; hence, both time reversal and space inversion are dynamical symmetries for free particles with spin.

14.4.4 Time Reversal and Space Inversion as Dynamical Symmetries

Next, we consider the more general case of particles interacting with an external EM field. Let the external field be described by the potentials $\mathbf{A}$ and ϕ. We write the Hamiltonian (14.10) for a particle interacting with the field as $\hat{H}_{\phi, \mathbf{A}}$. Then, using Eqs. (14.46)

---

[4] Note that ϖ is the phase a_j of Eq. (11.63) for $j = s$.

$$\hat{\mathbb{T}}\hat{H}_{\phi,\mathbf{A}}\hat{\mathbb{T}}^\dagger = \hat{H}_{\phi,-\mathbf{A}}.\tag{14.52}$$

In absence of external magnetic fields, time reversal is a dynamical symmetry. This result straightforwardly generalizes to many-particle systems interacting with the EM field.

If time reversal is a dynamical symmetry, then $\hat{\mathbb{T}}$ takes eigenstates of the Hamiltonian $\hat{H}$ to other eigenstates with the same energy. By Eq. (14.48), $\hat{\mathbb{T}}$ can leave no eigen state invariant, unless $m_s = 0$. But this is possible only if the total spin of the system is an integer. Thus, we arrive at the following theorem (Kramers, 1930).

Theorem 14.1 (Kramers') *The energy eigenvalues of a system with half-integer total spin are degenerate.*

In particular, Kramers' theorem implies that the ground state for the internal degrees of freedom of a composite particle with half-integer spin is degenerate. For example, the deuterium atom consists of one proton, one neutron, and one electron, so its spin is half-integer. Its ground state is doubly degenerate. In contrast, the hydrogen atom has integer spin, and its ground state turns out to be unique. This is not evident on the basis of Section 13.2. The degeneracy is broken by an interaction term between the spins of the electron and the proton that is described in Problem 17.8.

For space inversion, Eq. (14.47) implies that

$$\hat{\mathbb{P}}\hat{H}_{\phi,\mathbf{A}}\hat{\mathbb{P}}^\dagger = \hat{H}_{\phi',\mathbf{A}'},\tag{14.53}$$

where $\phi'(\mathbf{x}) = \phi(-\mathbf{x})$ and $\mathbf{A}'(\mathbf{x}) = -\mathbf{A}(-\mathbf{x})$. Note that the magnetic field $\mathbf{B}'(\mathbf{x})$ associated to $\mathbf{A}'$ is $\mathbf{B}'(\mathbf{x}) = \mathbf{B}(-\mathbf{x})$. Space inversion is a dynamical symmetry, if $\phi' = \phi$ and $\mathbf{A}' = \mathbf{A}$. This is the case, for example, if ϕ corresponds to a central potential and $\mathbf{A}$ to a homogeneous magnetic field $\mathbf{B}$.

Example 14.2 Consider a system of two particles 1 and 2, with spins s_1 and s_2, and intrinsic parities ϖ_1 and ϖ_2, respectively. The system is described by the Hilbert space $\mathcal{H}_{s_1} \otimes \mathcal{H}_{s_2}$, with elements wave functions $\psi_{m_1 m_2}(\mathbf{x}_1, \mathbf{x}_2)$. Space inversion is implemented by the operator $\hat{\mathbb{P}}_{12} = \hat{\mathbb{P}} \otimes \hat{\mathbb{P}}$, where $\hat{\mathbb{P}}$ is given by Eq. (14.49). Hence,

$$\hat{\mathbb{P}}_{12}\psi_{m_1 m_2}(\mathbf{x}_1, \mathbf{x}_2) = \varpi_1 \varpi_2 \psi_{m_1 m_2}(-\mathbf{x}_1, -\mathbf{x}_2).\tag{14.54}$$

Suppose that the system is described by a Hamiltonian $\hat{H}$ of the form (13.3), with a central potential $V(r)$. Then, the generalized eigenstates of $\hat{H}$ are of the form $\Psi_{\bar{m}_s}(\mathbf{X})\phi_{E,\ell,m}(\mathbf{r})$, where $\mathbf{X}$ is the center-of-mass coordinate and $\mathbf{r} = \mathbf{x}_1 - \mathbf{x}_2$ is the relative position vector; $\bar{m}_s$ is the spin coordinate for the center-of-mass degrees of freedom; $\phi_{E,\ell,m}$ the eigenstates for the Hamiltonian of the relative degrees of freedom.

Space inversion inverts both $\mathbf{x}_1$ and $\mathbf{x}_2$. Hence, it also inverts $\mathbf{X}$ and $\mathbf{r}$. Since $\phi_{E,\ell,m}(-\mathbf{r}) = (-1)^\ell \phi_{E,\ell,m}(\mathbf{r})$, Eq. (14.54) implies that

$$\hat{\mathbb{P}}_{12}[\Psi_{\bar{m}_s}(\mathbf{X})\phi_{E,\ell,m}(\mathbf{r})] = \varpi_1 \varpi_2 (-1)^\ell \Psi_{\bar{m}_s}(-\mathbf{X})\phi_{E,\ell,m}(\mathbf{r}).\tag{14.55}$$

We can view space inversion as acting only on the center-of-mass degrees of freedom. The center of mass behaves like a composite particle with intrinsic parity

$$\varpi_{12} = \varpi_1 \varpi_2 (-1)^\ell.\tag{14.56}$$

Hence, the intrinsic parity of a composite particle changes with the excitation of the angular momentum for the internal degrees of freedom.

QUESTIONS

14.1 Can there be a superposition of two particle states with different values of the spin s?

14.2 Which of the following can have nonzero gyromagnetic ratio: (i) particles with zero charge, (ii) particles with zero spin?

14.3 How would you define the angle of precession for a spin? Which values can this angle take?

14.4 What magnetic fields are required so that the dipole interaction for an electron is of the same order of magnitude as the energy eigenvalues of the hydrogen atom? Compare with the strongest magnetic fields in the laboratory, which are of the order of $100\,\text{T}$.

14.5 **(B)** Is the Heisenberg–Weyl group a subgroup of the Galilei group?

14.6 **(B)** How are composite particles described in terms of representations of the Galilei group?

PROBLEMS

14.1 A free neutron is prepared in a state $\dfrac{1}{\sqrt{2}}\begin{pmatrix} \psi_1(x) \\ -\psi_2(x) \end{pmatrix}$, with $\psi_1(x) = \phi(x + \tfrac{1}{2}L), \psi_2(x) = \phi(x - \tfrac{1}{2}L)$, and $\phi(x) = (2\pi\sigma^2)^{1/4}e^{-\frac{x^2}{4\sigma^2}}$, for some constants σ and L. (i) Calculate the correlation $\text{Cor}(\hat{X}, \hat{S}_i)$. (ii) Calculate $\text{Cor}(\hat{P}, \hat{S}_i)$ and $\text{Cor}(\hat{H}, \hat{S}_i)$ if $\psi_1(x) = \phi(x)$ and $\psi_2(x) = \phi(x)e^{ipx}$.

14.2 Find the solution to Eq. (14.20) for a magnetic pulse $\mathbf{B} = (0, 0, \beta\delta(t - t_0))$. Show that with an appropriate choice of β we can rotate the spin vector by any angle around the 3-axis.

14.3 An homogeneous magnetic field $\mathbf{B}$ acts on two particles with the same spin $s = \tfrac{1}{2}$ and gyromagnetic ratio γ. The Hamiltonian describing the particles is $\hat{H} = -\gamma\mathbf{B}\cdot\hat{\mathbf{S}}_1 - \gamma\mathbf{B}\cdot\hat{\mathbf{S}}_2 + a\mathbf{S}_1\cdot\mathbf{S}_2$, where a is a constant. Find the eigenvalues and the eigenvectors of $\hat{H}$. Evaluate the probability that the total spin of the ground state is $s = 1$.

14.4 We model the interaction of the proton with the neutron in the deuterium nucleus by a potential $\hat{V} = V_0(\hat{r}) + \lambda V_0(\hat{r})\hat{\mathbf{S}}_p\cdot\hat{\mathbf{S}}_n$, where λ is a constant, $\hat{r}$ is the relative position vector of the two particles, and $V_0(r)$ a central potential. Assume that both particles have the same mass m. Write the eigenvalues of $\hat{H}$ as a function of the eigenvalues $E_{n,\ell}$ of the Schrödinger operator for the potential $V_0(r)$.

14.5 A particle with spin $\tfrac{1}{2}$ and gyromagnetic ratio $\gamma > 0$ is acted by a time-dependent magnetic field $\mathbf{B} = (B_1\cos\omega t, -B_1\sin\omega t, B_0)$, where B_0 and B_1 are constants. (i) Show that the eigenvalues of the time-dependent Hamiltonian are $\pm\gamma\sqrt{B_0^2 + B_1^2}$, and then evaluate the corresponding instantaneous eigenvectors $|\pm, t\rangle$. (ii) Express the quantum state as a linear combination $c_+(t)|+, t\rangle + c_-(t)|-, t\rangle$, and show that the coefficients $c_\pm$ satisfy a system of linear differential equations with constant coefficients. Find the general solution, and confirm the appearance of a frequency $\Omega = \sqrt{\left(\gamma B_0 - \tfrac{1}{2}\omega\right)^2 + \gamma^2 B_1^2}$. (iii)

For an initial state $|+\rangle$ at $t = 0$ (i.e., eigenvector of $\hat{s}_3$ for eigenvalue $\frac{1}{2}$), show that the probability of $\hat{s}_3$ to take value $-\frac{1}{2}$ at time t is

$$\text{Prob}(-, t) = \left(\frac{\gamma B_1}{\Omega}\right)^2 \sin^2 \Omega t.$$

(iv) Show that the time average of the probability $\text{Prob}(-, t)$ is maximized for $\omega = 2\gamma B_0$.

Comment This phenomenon is known as *spin resonance*. It allows us to measure the gyromagnetic ratio γ of a particle by determining its resonance frequency. Compare with the Rabi oscillations in Section 8.4.4.

14.6 Consider a sample of a material that contains $N \gg 1$ particles with spin s. We assume that the position of the particles is approximately fixed so that they interact with magnetic fields only through their spin. For example, this is the case for the atomic nuclei in a solid. Suppose that all particles are initially prepared in the state $|s, s\rangle$. The total magnetization $\langle \mathbf{M} \rangle = (0, 0, N\gamma s)$ is macroscopically large and it can be measured. An electric field $\mathbf{B} = (B, 0, 0)$ is switched on at time t. The key point here is that the magnetic field $\mathbf{B}$ is not homogeneous within the material, so each spin i precesses with a slightly different value B_i. (i) Suppose that the distribution of the field values B_i is a Gaussian with mean value B_0. Show that the total magnetization quickly becomes vanishingly small. (i) At $t = \tau$, a pulse is applied that inverts $\hat{S}_2$ for all spins. Show that, at $t = 2\tau$, the total magnetization becomes again $(0, 0, N\gamma s)$.

Comment This phenomenon is known as a *spin echo* and it was first observed by Hahn (1950). Typically, the recovered magnetization is smaller than the initial one, because of dissipation effects. The dissipation rate depends on the properties of the material; hence, it allows for a characterization of the sample. The theoretical importance of the echo effects originates from the fact that, after the inversion of spins by the action of external pulses, a macroscopic system seems to evolve spontaneously from a disordered state into an ordered one. This would appear to contradict the nonequilibrium version of the second law of thermodynamics. However, this contradiction is only apparent. Evolution is truly irreversible: The decay of magnetization during an echo provides a measure of entropy change. In the time interval $[0, \tau]$, the information of the initial spin state is encoded into the correlations between spin and particle positions, and this information is retrieved reversibly after the spin inversion (Anastopoulos and Savvidou, 2011).

14.7 **(B)** Prove that the operators (14.38) and (14.39) are Casimir.

14.8 **(B)** (i) Show that Maxwell's equations are invariant under the time-reversal transformation $T : (t, \mathbf{E}, \mathbf{B}) \to (-t, \mathbf{E}, -\mathbf{B})$. (ii) How do the potentials ϕ and $\mathbf{A}$ transform under time reversal? (iii) Show that the Hamiltonian (14.10) is invariant under time reversal if we include the transformation T for the fields. (iv) Show that an electric dipole interaction term $d\mathbf{E} \cdot \mathbf{S}$ is not invariant under time reversal. Hence, time-reversal invariance implies that $d = 0$ for all particles. (Current experiments give $d/e < 10^{-28}$ m for the neutron.)

14.9 **(B)** (i) Identify the space-inversion transformation P that leaves Maxwell's equations invariant. (ii) Show that the Hamiltonian (14.10) is invariant under space inversion if we include the P transformation for the fields. (iii) Show that the electric dipole interaction term is not invariant under space inversion.

14.10 (B) The Galilean symmetry allows us to write the most general Hamiltonian for a nonrelativistic particle of spin s that interacts with external fields. Let us denote by $\hat{U}(\mathbf{v})$ the representation of the subgroup of the Galilei group that consists of boosts. A Hamiltonian $\hat{H}$ is Galilei-compatible, if $\hat{U}(\mathbf{v})\frac{d\hat{\mathbf{X}}}{dt}\hat{U}^{\dagger}(\mathbf{v}) = \frac{d\hat{\mathbf{X}}}{dt} - \mathbf{v}\hat{I}$, for $\frac{d\hat{\mathbf{X}}}{dt} = i[\hat{H}, \hat{\mathbf{X}}]$. (i) Show that, for $s = 0$, the most general Galilei-compatible Hamiltonian is

$$\hat{H} = \frac{1}{2m}\left[\hat{\mathbf{P}} - \mathbf{A}(\hat{\mathbf{X}}, t)\right]^2 + V(\hat{\mathbf{X}}, t),$$

for some scalar function V and vector function $\mathbf{A}$. (ii) Show that, for $s = \frac{1}{2}$, the most general Hamiltonian is

$$\hat{H} = \frac{1}{2m}\hat{\Pi}^2 + V(\hat{\mathbf{X}}, t) + \mathbf{B}(\hat{\mathbf{X}}, t)\cdot\hat{\mathbf{S}},$$

where $\mathbf{B}$ is a vector function, $\hat{\Pi}_i = \hat{P}_i - A_i(\hat{\mathbf{X}}, t) + \sum_j C_{ij}(\hat{\mathbf{X}}, t)\hat{S}_j$, and C_{ij} is a matrix-valued function. (iii) Identify all terms linear to $\hat{\mathbf{S}}$ in the Hamiltonian for the special case of an antisymmetric C_{ij}, in which case we can write $C_{ij} = \sum_k \epsilon_{ijk}E_k$, for some vector function $\mathbf{E}$. (iv) Compare with the Hamiltonian for a particle in an EM field, Eq. (14.10). When do we obtain a spin–orbit coupling, that is, a term that is proportional to $\hat{\mathbf{L}}\cdot\hat{\mathbf{S}}$, where $\hat{\mathbf{L}}$ is the orbital angular momentum?

Bibliography

- For the history of the concept of spin, see chapters 1–3 of Tomonaga (1998).
- For the Galilei group in quantum theory, see chapters 3 and 4 of Ballentine (1998). For unitary representations of the Galilei group, see Inonü and Wigner (1952).

15 Particle Statistics and Field–Particle Duality

Even in principle, one cannot demand an alibi from an electron.

Weyl (1950)

15.1 Identical Particles

In Chapter 9, we presented the quantum rule of combination of subsystems through the tensor product. In this chapter, we will discuss a key elaboration of this rule that applies to composite systems with a specific symmetry, namely, invariance under *exchange of identical particles*.

First, we explain the notion of identical particles. Particles of the same species have the same fundamental characteristics, mass m, spin s, and charge q. How can we distinguish one from the other? In deterministic theories, such as classical mechanics, we can distinguish particles by their history. If we identify one particle at a given moment of time, we can follow its time evolution and identify it again at a later time. The identity of a particle is never lost, because we can follow its evolution with arbitrary accuracy, at least in principle.

Quantum mechanics is not deterministic. Suppose we identify two electrons at some initial moment of time,[1] and then we allow them to move freely inside a box. After some time, no measurement can identify which electron is which. All "tags" of their initial identification are lost. We conclude that particles of the same species cannot, in general, be distinguished, so we call them *identical*. In what follows, we analyze the implications of particle identity to quantum phenomena.

15.1.1 Exchange Symmetry

Particle identity defines a symmetry: Physical predictions are unchanged by the exchange of two identical particles. Let a single particle be described by the Hilbert space $\mathcal{H}$. A system of two identical particles is described by the Hilbert space $\mathcal{H}^2 = \mathcal{H} \otimes \mathcal{H}$. For any orthonormal basis $|a\rangle$ on $\mathcal{H}$; the vectors $|a,b\rangle = |a\rangle \otimes |b\rangle$ define an orthonormal basis on $\mathcal{H}^2$. We define the unitary *exchange operator* $\hat{V}$ by

$$\hat{V}|a,b\rangle = |b,a\rangle. \tag{15.1}$$

---

[1] For example, each electron may originate from a different source.

By definition, $\hat{V}^2|a,b\rangle = \hat{V}|b,a\rangle = |a,b\rangle$, that is, $\hat{V}^2 = \hat{I}$. It follows that the eigenvalues of $\hat{V}$ are $+1$ and -1. Hence, $\hat{V}$ is both unitary and self-adjoint; together with the unit $\hat{I}$, they define a unitary representation of the permutation group S_2 on $\mathcal{H}^2$.

We define the symmetric vectors $|a,b\rangle_S$ and the antisymmetric basis vectors $|a,b\rangle_A$ by

$$|a,b\rangle_S := \frac{1}{\sqrt{2}}(|a,b\rangle + |b,a\rangle), \qquad |a,b\rangle_A := \frac{1}{\sqrt{2}}(|a,b\rangle - |b,a\rangle). \tag{15.2}$$

These vectors satisfy

$$\hat{V}|a,b\rangle_S = |a,b\rangle_S, \qquad \hat{V}|a,b\rangle_A = -|a,b\rangle_A. \tag{15.3}$$

The symmetric vectors $|a,b\rangle_S$ define the *symmetric subspace* $\mathcal{H}_S^2$, namely, the eigenspace of $\hat{V}$ for the eigenvalue $+1$. The antisymmetric vectors $|a,b\rangle_A$ define the *antisymmetric subspace* $\mathcal{H}_A^2$, namely, the eigenspace of $\hat{V}$ for the eigenvalue -1. We write the projectors associated to these subspaces as $\hat{P}_S$ and $\hat{P}_A$, respectively. By the spectral theorem, $\hat{P}_S + \hat{P}_A = \hat{I}$, and $\hat{V} = \hat{P}_S - \hat{P}_A$.

If we cannot distinguish between two particles, then the states $|\psi\rangle$ and $\hat{V}|\psi\rangle$ on $\mathcal{H}^2$ should give the same probabilities for all measurements. Hence, physical quantities correspond to self-adjoint operators $\hat{A}$ on $\mathcal{H}^2$ that satisfy $\langle\psi|\hat{A}|\psi\rangle = \langle\psi|\hat{V}^\dagger\hat{A}\hat{V}|\psi\rangle$ for all $|\psi\rangle$. It follows that $\hat{V}^\dagger\hat{A}\hat{V} = \hat{A}$, or, equivalently,

$$[\hat{A}, \hat{V}] = 0. \tag{15.4}$$

The condition (15.4) for all physical observables is called the *indistinguishability postulate*. In particular, the Hamiltonian $\hat{H}$ of the two-particle system satisfies Eq. (15.4).

Any operator $\hat{A}$ on $\mathcal{H}^2$ satisfies the trivial identity $\hat{A} = (\hat{P}_S + \hat{P}_A)\hat{A}(\hat{P}_S + \hat{P}_A) = \hat{P}_S\hat{A}\hat{P}_S + \hat{P}_S\hat{A}\hat{P}_A + \hat{P}_A\hat{A}\hat{P}_S + \hat{P}_A\hat{A}\hat{P}_A$. We note that $\hat{P}_S\hat{A}\hat{P}_S\hat{V} = \hat{P}_S\hat{A}\hat{P}_S(\hat{P}_S - \hat{P}_A) = \hat{P}_S\hat{A}\hat{P}_S = \hat{V}\hat{P}_S\hat{A}\hat{P}_S$; hence, $[\hat{P}_S\hat{A}\hat{P}_S, \hat{V}] = 0$. Similarly, we find that $[\hat{P}_A\hat{A}\hat{P}_A, \hat{V}] = 0$, and that $[\hat{P}_S\hat{A}\hat{P}_A, \hat{V}] = -2\hat{P}_S\hat{A}\hat{P}_A$. Then, Eq. (15.4) implies that $\hat{P}_S\hat{A}\hat{P}_A + \hat{P}_S\hat{A}\hat{P}_A = 0$, or, equivalently,

$$\hat{A} = \hat{P}_S\hat{A}\hat{P}_S + \hat{P}_A\hat{A}\hat{P}_A. \tag{15.5}$$

Let $|\phi\rangle \in \mathcal{H}_S^2$, $|\chi\rangle \in \mathcal{H}_A^2$ and $|\psi\rangle = |\phi\rangle + |\chi\rangle$ be a superposition of vectors from the two eigenspaces of $\hat{V}$. Then, $\langle\psi|\hat{A}|\psi\rangle = \langle\phi|\hat{A}|\phi\rangle + \langle\chi|\hat{A}|\chi\rangle$; all terms that mix contributions from the two eigenspaces vanish.

Hence, the exchange symmetry implies a superselection rule for systems: superpositions between $\mathcal{H}_S^2$ and $\mathcal{H}_a^2$ are not observable.

15.1.2 The Symmetrization Postulate and the Spin-Statistics Theorem

The description of identical particles through superselection rules turned out to be too weak a condition for accurately describing physical systems. A much stronger assumption, the *symmetrization postulate*, is needed. According to this postulate, a pair of two particles is described exclusively either from vectors of $\mathcal{H}_S^2$ or from vectors of $\mathcal{H}_A^2$. The symmetrization postulate cannot be justified from the fundamental principles of quantum theory, and it has to be added as an independent axiom (Messiah and Greenberg, 1964).

> **Symmetrization postulate (for two particles)** The Hilbert space that describes a system of two identical particles is either the symmetric subspace $\mathcal{H}_S^2$ or the antisymmetric subspace $\mathcal{H}_A^2$. The choice of the subspace is the same for all particles of the same species.

The symmetrization postulate rules out, for example, the possibility that the state of two identical particles is a mixture $(1 - \lambda)\hat{\rho}_S + \lambda\hat{\rho}_A$, where $\lambda \in (0, 1)$, $\hat{\rho}_S$ is a density matrix on $\mathcal{H}_S^2$, and $\hat{\rho}_A$ is a density matrix on $\mathcal{H}_A^2$. The possibility that electrons are described by such a state was tested experimentally by Ramberg and Snow (1975), the results constraining λ to be smaller than 10^{-26}. The symmetrization postulate holds with remarkable accuracy.

Particles described by the antisymmetric subspace are called *fermions* (from Enrico Fermi) and particles described by the symmetric subspace are called *bosons* (from Satyendra Bose). Equivalently, one talks about the fermionic or bosonic *statistics* of a particle species.

The statistics of any particle species is determined by the following fundamental theorem.

> **Theorem 15.1** (Spin-statistics theorem) *Particles with integer spin s are bosons, and particles with half-integer spin s are fermions.*

The validity of the spin-statistics theorem was already evident in the 1920s, because it would guarantee the stability of atoms. However, it was first proven much later by Pauli (1940). Pauli's proof relies on the fundamental principle of relativity theory, that there can be no signal transmission at faster-than-light speeds, and also on the description of particles in terms of quantum fields. In Section 15.4.4, we present this proof for the simplest case of $s = 0$.

The spin-statistics theorem cannot be proven in nonrelativistic quantum theory without introducing additional postulates. These postulates are related to the property that a 2π rotation of a particle with spin s correspond to the multiplication of the state vector with $(-1)^{2s}$. The spin-statistics theorem is identical with the statement that *an exchange of two particles is equivalent to the rotation of one of the particles by 2π*. This statement is far from obvious, and this is the reason why additional postulates, supposedly more intuitive, are needed in order to prove it as a theorem.

Example 15.1 Consider a system of two identical particles of spin s. The associated Hilbert space is $\mathcal{H}_s \otimes \mathcal{H}_s$. In the Schrödinger representation, we use the wave functions $\psi_{m_1 m_2}(\mathbf{x}_1, \mathbf{x}_2)$, where m_1 and m_2 are spin indices. For bosons, $\psi_{m_2 m_1}(\mathbf{x}_2, \mathbf{x}_1) = \psi_{m_1 m_2}(\mathbf{x}_1, \mathbf{x}_2)$, and for fermions, $\psi_{m_2 m_1}(\mathbf{x}_2, \mathbf{x}_1) = -\psi_{m_1 m_2}(\mathbf{x}_1, \mathbf{x}_2)$. Note that the exchange applies both to position and the spin label, not each one separately. For example, a fermion may have a wave function that is symmetric with respect to the exchange of $\mathbf{x}_1$ and $\mathbf{x}_2$, as long as it is antisymmetric in the exchange of m_1 and m_2.

Take $s = \frac{1}{2}$. If the total spin of the two particles $S = 1$, then the spin part of the state is symmetric under exchange; see Section 11.5.3. If $S = 0$, then the spin part of the state is antisymmetric under exchange. Hence, for $S = 0$, the state is symmetric with respect to the exchange of positions: $\psi_{m_1 m_2}(\mathbf{x}_1, \mathbf{x}_2) = \psi_{m_1 m_2}(\mathbf{x}_2, \mathbf{x}_1)$. The relative position vector $\mathbf{r} = \mathbf{x}_1 - \mathbf{x}_2$ becomes $-\mathbf{r}$ under the exchange of the two particles. This means that the eigenstates of the

Hamiltonian for the internal degrees of freedom are parity-even. Hence, only even values for the relative orbital angular momentum ℓ are allowed.

If $S = 1$, the state is antisymmetric with respect to the exchange of positions: $\psi_{m_1 m_2}(\mathbf{x}_1, \mathbf{x}_2) = -\psi_{m_1 m_2}(\mathbf{x}_2, \mathbf{x}_1)$. Hence, the eigenstates of the Hamiltonian for the internal degrees of freedom are parity-odd. The relative angular momentum ℓ is also odd.

In contrast, if the two particles are *not* identical, there exist eigenstates of the Hamiltonian with $S = 0$ and odd ℓ, and $S = 1$ and even ℓ.

15.1.3 N-Particle Systems

Next, we generalize our analysis of the exchange symmetry to N-particle systems. We denote the Hilbert space of a single particle by $\mathcal{H}$, and we define

$$\mathcal{H}^N = \underbrace{\mathcal{H} \otimes \mathcal{H} \otimes \cdots \otimes \mathcal{H}}_{N \text{ times}}. \tag{15.6}$$

For any orthonormal basis $|a\rangle$ on $\mathcal{H}$, we define an orthonormal basis on $\mathcal{H}^N$, with vectors

$$|a_1, a_2, \ldots, a_N\rangle := |a_1\rangle \otimes |a_2\rangle \otimes \cdots \otimes |a_N\rangle.$$

We denote permutations of the N labels $a_1, a_2, \ldots, a_N$ as $\sigma(a_1, a_2, \ldots, a_N)$. We define the *sign* $\text{sgn}(\sigma)$ of a permutation σ to be 1 if σ is even and -1 if σ is odd. There is a unitary representation of the permutation group S_N on $\mathcal{H}^N$ by the unitary operators $\hat{V}(\sigma)$, defined by

$$\hat{V}(\sigma)|a_1, a_2, \ldots, a_N\rangle = |\sigma(a_1, a_2, \ldots, a_N)\rangle. \tag{15.7}$$

We define the symmetrized and antisymmetrized basis vectors

$$|a_1, a_2, \ldots, a_N\rangle_S = \frac{1}{\sqrt{N!}} \sum_\sigma |\sigma(a_1, a_2, \ldots, a_n)\rangle \tag{15.8}$$

$$|a_1, a_2, \ldots, a_N\rangle_A = \frac{1}{\sqrt{N!}} \sum_\sigma \text{sgn}(\sigma)|\sigma(a_1, a_2, \ldots, a_n)\rangle, \tag{15.9}$$

respectively. The sum extends to all $(N!)$ possible permutations of $a_1, a_2, \ldots, a_N$. By definition, an exchange of a_i with a_j gives a sign $+1$ to $|a_1, a_2, \ldots, a_N\rangle_S$ and a sign -1 to $|a_1, a_2, \ldots, a_N\rangle_A$. Note that $|a_1, a_2, \ldots, a_N\rangle_A$ vanishes if any two a_i and a_j are equal, since their exchange gives $|a_1, a_2, \ldots, a_N\rangle_A = -|a_1, a_2, \ldots, a_N\rangle_A$.

It is straightforward to show that for any permutation $\sigma \in S_N$,

$$\hat{V}(\sigma)|a_1, a_2, \ldots, a_N\rangle_S = |a_1, a_2, \ldots, a_N\rangle_S,$$
$$\hat{V}(\sigma)|a_1, a_2, \ldots, a_N\rangle_A = \text{sgn}(\sigma)|a_1, a_2, \ldots, a_N\rangle_A.$$

The subspace of $\mathcal{H}^N$ spanned by the vectors $|a_1, a_2, \ldots, a_N\rangle_S$ is called the *completely symmetric subspace*, and it is denoted $\mathcal{H}_S^N$. The subspace of $\mathcal{H}^N$ spanned by the vectors $|a_1, a_2, \ldots, a_N\rangle_A$ is called the *completely antisymmetric subspace*, and it is denoted by $\mathcal{H}_A^N$. The subspaces $\mathcal{H}_S^N$ and $\mathcal{H}_A^N$ do not depend on the choice of basis $|a\rangle$. For $N > 2$, $\mathcal{H}_S^N \oplus \mathcal{H}_A^N$ is a proper subspace of $\mathcal{H}^N$, that is, there are vectors in $\mathcal{H}^N$ that cannot be split into a totally symmetric and a totally antisymmetric component.

Let $\hat{A}$ be an operator on $\mathcal{H}^N$ that commutes with all $\hat{V}(\sigma)$. It is then straightforward to show that $\hat{A}$ preserves $\mathcal{H}_S^N$ and $\mathcal{H}_A^N$, that is, it takes vectors of $\mathcal{H}_S^N$ to vectors of $\mathcal{H}_S^N$ and vectors of $\mathcal{H}_A^N$ to vectors of $\mathcal{H}_A^N$.

Two operators of this type are of particular interest. For each $\hat{f}$ on $\mathcal{H}$, we define

$$\hat{f}^{\otimes N} = \hat{f} \otimes \hat{f} \otimes \cdots \otimes \hat{f} \tag{15.10}$$

$$\hat{f}^{\oplus N} = \hat{f} \otimes \hat{I} \otimes \hat{I} \otimes \cdots \hat{I} + \hat{I} \otimes \hat{f} \otimes \hat{I} \otimes \cdots \hat{I} + \cdots + \hat{I} \otimes \hat{I} \otimes \hat{I} \otimes \cdots \hat{f}. \tag{15.11}$$

Both operators $\hat{f}^{\otimes N}$ and $\hat{f}^{\oplus N}$ commute with all $\hat{V}(\sigma)$. Hence, their restriction to $\mathcal{H}_S^N$ and $\mathcal{H}_A^N$ is well defined.

Slater determinant Let $|a\rangle$ and $|n\rangle$ denote two different orthonormal bases in the Hilbert space $\mathcal{H}$. They induce two different bases $|a_1, \ldots a_N\rangle$ and $|n_1, \ldots n_N\rangle$ in the Hilbert space $\mathcal{H}^N$, and also the associated bases (15.8) and (15.9) in the subspaces $\mathcal{H}_S^N$ and $\mathcal{H}_A^N$, respectively.

Of particular usefulness is the inner product of the different basis vectors in $\mathcal{H}_A^N$,

$$\langle n_1, \ldots, n_N | a_1, \ldots, a_N \rangle_A = \frac{1}{\sqrt{N!}} \begin{vmatrix} \langle n_1|a_1\rangle & \langle n_1|a_2\rangle & \ldots & \langle n_1|a_N\rangle \\ \langle n_2|a_1\rangle & \langle n_2|a_2\rangle & \ldots & \langle n_2|a_N\rangle \\ \ldots & \ldots & \ldots & \ldots \\ \langle n_N|a_1\rangle & \langle n_N|a_2\rangle & \ldots & \langle n_N|a_N\rangle \end{vmatrix}. \tag{15.12}$$

The determinant in Eq. (15.12) is called the *Slater determinant*; it arises because of the negative contributions from the odd permutations in Eq. (15.9). Equation (15.12) is straightforward to prove for $N = 2$, and the proof for general N proceeds by induction. Note that Eq. (15.12) also applies for generalized bases $|a\rangle$ and $|n\rangle$.

Dimension of the subspaces While the Hilbert space for a single particle is infinite-dimensional, it is sometimes convenient to work with a finite-dimensional subspace, so that $\mathcal{H} = \mathbb{C}^d$, for some integer d. Then, the labels a_i in Eq. (15.7) take values $1, 2, \ldots, d$. By definitions, $\dim \mathcal{H}^N = d^N$.

The dimension of $\mathcal{H}_S^N$ equals with the number of linearly independent vectors $|a_1, a_2, \ldots, a_N\rangle_S$. There is one such vector for each N-plet of numbers a_i, each with d different values. Hence, $\dim \mathcal{H}_S^N$ is the number of ways that we can distribute N indistinguishable balls in d boxes, known from elementary combinatorics to equal

$$\dim \mathcal{H}_S^N = \frac{(N + d - 1)!}{N!(d - 1)!}. \tag{15.13}$$

Similarly the dimension of $\mathcal{H}_A^N$ equals the number of ways that we can place N balls into d boxes, so that there is at most one ball in each box. For $N > d$, this task is impossible, hence, $\dim \mathcal{H}_A^N = 0$. For $N \leq d$,

$$\dim \mathcal{H}_A^N = \frac{d!}{N!(d - N)!}. \tag{15.14}$$

Given the definitions above, the symmetrization postulate for N particles takes the following form.

> **Symmetrization postulate (for N particles)** The Hilbert space that describes a system of N identical particles is either the totally symmetric subspace $\mathcal{H}_S^N$ or the totally antisymmetric subspace $\mathcal{H}_A^N$. The choice of the subspace is the same for all particles of the same type.

The spin-statistics theorem is evidently unchanged for N particles.

General case In the most general case, we have composite systems that consist of r different species of fermions and s different species of bosons. We use the index $a = 1, 2, \ldots, r$ for fermion species,[2] and the index $b = 1, 2, \ldots, s$ for boson species. We assume that a fermion of type a is described by the Hilbert space $\mathcal{H}_a$ and a boson of type b is described by the Hilbert space $\mathcal{K}_b$.

Suppose that the system consists of N_a fermions of type a, for all $a = 1, 2, \ldots, r$, and M_b bosons of type b for all $b = 1, 2, \ldots, s$. The Hilbert space for the total system is

$$(\mathcal{H}_1)_A^{N_1} \otimes (\mathcal{H}_2)_A^{N_2} \otimes (\mathcal{H}_r)_A^{N_r} \otimes (\mathcal{K}_1)_S^{M_1} \otimes (\mathcal{K}_2)_S^{M_2} \otimes \cdots (\mathcal{K}_s)_S^{M_s}. \tag{15.15}$$

That is, we use the combination of subsystems through the tensor product, and we take symmetric or antisymmetric combinations of state vector *only for particles of the same species*, according to the spin-statistics theorem.

This structure is rather complicated; the good thing is that it simplifies dramatically when many-particle systems are expressed in terms of Fock spaces, as we will see in Section 15.3.

15.1.4 The Stability of Matter

The range of the nuclear force is of the order of 10^{-15} m; hence, at the atomic scale ($\sim 10^{-11}$ m), we can ignore the internal structure of nuclei. We treat a nucleus of atomic number Z as a pointlike particle with spin, and electric charge equal to $+Ze$. Hence, to describe matter from the atomic scale upwards, we consider systems that consist of N electrons and K nuclei. We label the electrons by an index $i = 1, 2, \ldots, N$. The K nuclei may belong to different species, so we use an index $a = 1, 2, \ldots, K$, and we assume that the atomic numbers Z_a and the masses M_a depend on a.

For particles that move with nonrelativistic velocities, the Coulomb electrostatic force dominates all other components of the electromagnetic interaction. Hence, a system of N electrons and K nuclei[3] is well described by the Hamiltonian

$$\hat{H} = \sum_{i=1}^{N} \frac{\hat{\mathbf{p}}_i^2}{2m_e} + \sum_{a=1}^{K} \frac{\hat{\mathbf{P}}_a^2}{2M_a} + V\left(\hat{\mathbf{x}}_i, \hat{\mathbf{X}}_a\right), \tag{15.16}$$

where $\hat{\mathbf{x}}_i$ are the position vectors of the electrons, $\hat{\mathbf{X}}_a$ are the positions of the nuclei, $\hat{\mathbf{p}}_i$ and $\hat{\mathbf{P}}_a$ are the associated momenta, m_e is the electron mass, and V is the potential:

$$V(\mathbf{x}_i, \mathbf{X}_a) = \alpha \left(-\sum_{i=1}^{N}\sum_{a=1}^{K} \frac{Z_a}{|\mathbf{x}_i - \mathbf{X}_a|} + \sum_{1 \le a < b \le K} \frac{Z_a Z_b}{|\mathbf{X}_a - \mathbf{X}_b|} + \sum_{1 \le i < j \le N} \frac{1}{|\mathbf{x}_i - \mathbf{x}_j|} \right),$$

---

[2] For example, we can set $a = 1$ for electrons, $a = 2$ for protons, and so on.

[3] $K = 1$ corresponds to atoms; K and N up to a few thousands correspond to molecules; K and N of the order of 10^{23} correspond to electrically neutral ($N = \sum_{a=1}^{K} Z_a$) macroscopic matter.

where $\alpha = \frac{e^2}{4\pi} \simeq \frac{1}{137}$. The first term in the potential describes the attractive electron–nuclei interaction, the second term describes the repulsion between nuclei, and the third term describes the repulsion between electrons.

The Hamiltonian (15.16) is defined on a Hilbert space of the form (15.15), where care must be taken to symmetrize or antisymmetrize the state vectors for identical nuclei, according to their spin.

Let the ground state energy of the Hamiltonian be denoted by E_0. As explained in Section 8.2, the stability of matter presupposes that $E_0 > -\infty$. There is also the *strong stability condition*

$$E_0 > -C(Z)(N + K), \tag{15.17}$$

for all N and K. In Eq. (15.17), $C(Z)$ is a positive constant that depends on the maximal atomic number of the nuclei, $Z = \max_a Z_a$. Since $N + K$ is the total number of particles, Equation (15.17) expresses the fact that the energy per particle is bounded from below, as the number of particles goes to infinity. For example, if $E_0 \sim -(N + K)^a$, for $a > 1$, the total energy is indeed bounded from below, but Eq. (15.17) fails.

The strong stability condition is essential for the large-scale organization of matter and its description by thermodynamic principles. In equilibrium, the energy per particle is not only finite, but it is also constant. Consider, for example, a gas in a box. If we double the number of particles and the volume of the box while keeping the temperature fixed, then the internal energy and the entropy of the box will also double. As long as gravitational effects are negligible, energy is proportional to the number of particles; this property is known as *extensivity*.

To better understand the meaning of extensivity, consider a box that contains K atoms of the same type kept at a temperature very close to zero, so that its energy approximately equals E_0. (We assume $N = ZK$ so that the system is electrically neutral.) Suppose that $E_0 = -CK^a$, for some $a > 1$, so that Eq (15.17) is violated. Two identical boxes have energy $2E_0 = -2CK^a$. Suppose that we bring those boxes in contact, so that they can exchange particles. Then, the state where all particles are in one box has total energy $-C(2K)^a = 2^a E_0 < 2E_0$, that is, it is energetically favorable. Hence, particles will tend to concentrate in one of the two boxes, leaving the other empty, and releasing in the process energy equal to $(2^a - 2)E_0$. The mere act of bringing two boxes of gas in contact liberates a huge amount of energy! Certainly, we do not observe that. Macroscopic matter satisfies the extensivity principle, otherwise it would have been wildly unstable.

Strong stability guarantees extensivity. The proof that strong stability holds in quantum theory was given by Dyson and Lenard (1967, 1968), in one of the most celebrated theorems of mathematical physics in the twentieth century. The proof has been subsequently improved and simplified. The following version is due to Lieb and Seiringer (2010).

Theorem 15.2 (Stability of matter) *The ground-state energy* (15.16) *satisfies Eq.* (15.17) *for* $C(Z) \simeq 1.7\alpha^2 Z^2$.

Strong stability is possible because electrons are fermions. If the electron were a boson, the energy E_0 would decrease with $-N^{7/5}$ (Dyson, 1967; Conlon et al., 1988), and Eq. (15.17) would be violated. The world would be radically different.

Finally, we note that the Dyson–Lenard theorem was proven in the context of nonrelativistic quantum mechanics. If relativity is taken into account, the stability of matter becomes a more complicated issue that has not yet been settled.

15.2   Noninteracting Identical Particles

In this section, we will study systems of N noninteracting identical particles. By "noninteracting," we mean that the Hamiltonian on the Hilbert space $\mathcal{H}^N$ of N particles is of the form

$$\hat{H} = \hat{h}^{\oplus N} \tag{15.18}$$

where $\hat{h}$ is the Hamiltonian in the Hilbert space $\mathcal{H}$ of a single particle – see also Eq. (15.11). The restriction to noninteracting particles is hardly unphysical, we employ the same restriction when we postulate gases to be ideal in statistical mechanics. It is a good first approximation. We will show that, in the mean-field approximation, even systems of interacting particles can be approximated by Hamiltonians of the form (15.18).

15.2.1   The Spectrum of the Many-Particle Hamiltonian

For simplicity, we assume that the single-particle Hamiltonian $\hat{h}$ has a discrete spectrum, and so that its eigenvectors $|n\rangle$ form an orthonormal basis on $\mathcal{H}$. We choose the index $n = 0, 1, 2, \ldots$ so that it orders the eigenvalues of $\hat{h}$ as $\epsilon_0 \leq \epsilon_1 \leq \epsilon_2 \leq \ldots$. In this ordering, an eigenvalue of degeneracy g appears g successive times in the sequence ϵ_n of the eigenvalues.

By Eqs. (15.8) and (15.9), the vectors $|n_1, n_2, \ldots, n_N\rangle_S \in \mathcal{H}_S^N$ and $|n_1, n_2, \ldots, n_N\rangle_A \in$ are eigenvectors of the Hamiltonian (15.18),

$$\hat{H}|n_1, n_2, \ldots, n_N\rangle_S = \left(\sum_{i=1}^{N} \epsilon_{n_i}\right) |n_1, n_2, \ldots, n_N\rangle_S \tag{15.19}$$

$$\hat{H}|n_1, n_2, \ldots, n_N\rangle_A = \left(\sum_{i=1}^{N} \epsilon_{n_i}\right) |n_1, n_2, \ldots, n_N\rangle_A, \tag{15.20}$$

with eigenvalue $\left(\sum_{i=1}^{N} \epsilon_{n_i}\right)$.

First, we consider bosons. Assume that the ground state $|0\rangle$ for one particle is nondegenerate. Then, the lowest energy state for N bosons is $|0, 0, \ldots, 0\rangle_S$ with energy $E_0 = N\epsilon_0$. If the ground state of a single boson has degeneracy g_0, then by Eq. (15.13), the ground state of the N boson system $\frac{(N+g_0-1)!}{N!(g_0-1)!}$.

For fermions, no integer n_i can be repeated, because of antisymmetry. This statement is known as *Pauli's exclusion principle*. Pauli first stated it in the context of early atomic models, and it provided the template for building the general theory of identical particles. As discussed in Section 14.1, it also led to the discovery of spin.

By the exclusion principle, the lowest energy state in a system of N fermions is $|0, 1, 2, \ldots N - 1\rangle_A$, that is, it corresponds to the N smallest values of the integers n_i. The ground-state energy for the system is

$$E_0 = \sum_{n=0}^{N-1} \epsilon_n. \tag{15.21}$$

The term ϵ_{N-1}, which is the largest in the sum (15.21), is the *Fermi energy* of the system; it is denoted by ϵ_F.

Example 15.2 Consider five fermionic oscillators of frequency ω and spin $s = \frac{1}{2}$. We write the single-particle eigenstates as $|n\pm\rangle$, where $\pm$ refers to the sign of the eigenvalues of $\hat{S}_3$. By Kramers' theorem, we expect that the lowest energy has double degeneracy. Indeed, the two ground states are $|0+,0-,1+,1-,2+\rangle_A$ and $|0+,0-,1+,1-,2-\rangle_A$. They have energy $E_0 = 4\omega$, Fermi energy $\epsilon_F = 2\omega$, and they differ on the spin direction of the oscillator at the Fermi energy.

The energy eigenvalues of the systems are of the form $E_0 + K\omega$ for $K = 0, 1, 2, \ldots$. The first few energy eigenstates are given in the following table.

K	Energy	Eigenvectors	Degeneracy
0	4ω	$\lvert 0+,0-,1+,1-,2\pm\rangle_A$	2
1	5ω	$\lvert 0+,0-,1+,1-,3\pm\rangle_A, \lvert 0+,0-,1\pm,2+,2-\rangle_A$	4
2	6ω	$\lvert 0+,0-,1+,1-,4\pm\rangle_A, \lvert 0+,0-,1\pm,2\pm,3\pm\rangle_A,$	12
		$\lvert 0\pm,1+,1-,2+,2-\rangle_A$	
3	7ω	$\lvert 0+,0-,1+,1-,5\pm\rangle_A, \lvert 0+,0-,1\pm,2\pm,4\pm\rangle_A,$	20
		$\lvert 0+,0-,2+,2-,3\pm\rangle_A, \lvert 0\pm,1+,1-,2\pm,3\pm\rangle_A$	

15.2.2 Mean-Field Theory and Atomic Orbitals

Mean-field theory is a widely used approximation in the study of many-particle systems. The key idea is to assume that each particle moves within a mean potential that is generated by the other particles. This approximation transforms the study of N interacting particles into a problem of N noninteracting particles, where the single-particle Hamiltonian incorporates the mean potential.

Mean-field theory fails to account for the statistical correlations and entanglement between particles. Still, it provides a good description for many systems, at least as a first approximation.

Here, we apply the mean-field approximation to an atom of atomic number Z. We assume that the nucleus is pointlike and motionless, so the only degrees of freedom are of the electrons. In mean-field theory, each electron is described by a Hamiltonian

$$\hat{h}_Z = \frac{\hat{p}^2}{2m} + V_Z(\hat{r}), \tag{15.22}$$

where the effective potential $V_Z(r)$ is central and it incorporates both the attraction from the nucleus and the average repulsion from the remaining $Z - 1$ electrons. In general, the effective potential differs for different atomic numbers. It is possible to calculate a reasonable form for $V_Z(r)$ from first principles, for example, through a generalization of the methods presented in

Fig. 15.1 Memorization rule for the Aufbau principle.

Section 17.4. Here, we will explore only the implications of the mean-field approximation to the general properties of atoms.

We express the effective potential as $V_Z(r) = -\frac{\alpha}{r}\chi_Z(r)$. The function χ_Z satisfies $\chi_Z(0) = Z$, because the electron sees only the Coulomb potential of the nucleus at small r; also $\chi_Z(\infty) = 1$, because at large r the nucleus and the remaining $Z - 1$ electrons behave like an ion of charge $+e$.

The eigenvalues $E_{n,\ell}$ of the Hamiltonian (15.22) have degeneracy equal to $2(2\ell + 1)$; the additional factor of 2 is due to spin. The ground state of the atom is obtained by placing one electron in each of the first Z energy eigenvalues. In atomic physics, the eigenspaces of $\hat{h}_Z$ are called *orbitals*, whence the procedure of identifying the ground state of the atom is referred to as *filling the orbitals*.

To do so, we need to know the ordering of the eigenvalues $\epsilon_{n,\ell}$ of the Hamiltonian. In Section 13.2.2, we showed that the ordering is only partially fixed by spherical symmetry. In fact, the Hamiltonian $\hat{h}_Z$ depends on Z, hence, the ordering may vary from atom to atom.

The correct ordering of the eigenvalues can be ascertained with input from spectroscopic data. We present it by using the traditional spectroscopic notation, where different values of ℓ are represented by letters, as in the following table.

symbol	s	p	d	f	g	h
ℓ	0	1	2	3	4	5
$2(2\ell + 1)$	2	6	10	14	18	22

The correct ordering of eigenvalues is the following:

$$1s, 2s, 2p, 3s, 3p, (4s, 3d), 4p, (5s, 4d), 5p, (6s, 4f, 5d), 6p, (7s, 5f, 6d), \dots . \tag{15.23}$$

The parentheses contain orbitals whose ordering changes from atom to atom.

We note that, for fixed n, the s and p orbitals have the lowest energy, and then we pass to orbitals with the next value of the principal quantum number, $n + 1$. Hence, in a given atom, the outer shell, that is, the orbitals with the highest value of n, contains at most $2 + 6 = 8$ electrons. The electrons of the outer shell are to a large extent responsible for an atom's chemical behavior. The periodicity in the chemical properties of the elements reflects the periodicity of the number of electrons in the outer shell as a function of Z.

The ordering (15.23) without the parentheses is the most common in atoms, and it has a simple memorization rule, shown in Fig. 15.1. It has become known as the *Aufbau principle* for atoms. The promotion of this rule to the status of a principle is rather arbitrary, because about 20 percent of the atoms do not follow this rule, and they are treated as exceptions. These exceptions are given in Table 15.1.

Table 15.1 Exceptions to the Aufbau principle

Z	atom	Aufbau principle	correct order
24	Cr	$\ldots 4s^2 3d^4$	$\ldots 4s^1 3d^5$
29	Cu	$\ldots 4s^2 3d^9$	$\ldots 4s^1 3d^{10}$
$41 + n, n = 0, 1$	Nb, Mo	$\ldots 5s^2 4d^{3+n}$	$\ldots 5s^1 4d^{4+n}$
$44 + n, n = 0, 1, 2, 3$	Ru, Rh, Pd, Ag	$\ldots 5s^2 4d^{6+n}$	$\ldots 5s^1 4d^{7+n}$
57	La	$\ldots 6s^2 4f^1$	$\ldots 6s^2 5d^1$
58	Ce	$\ldots 6s^2 4f^2$	$\ldots 6s^2 4f^1 55d^1$
64	Gd	$\ldots 6s^2 4f^8$	$\ldots 6s^2 4f^7 5d^1$
$78 + n, n = 0, 1$	Pt, Au	$\ldots 6s^2 4f^{14} 5d^{8+n}$	$\ldots 6s^1 4f^{14} 5d^{9+n}$
$89 + n, n = 0, 1$	Ac, Th	$\ldots 7s^2 5f^{1+n}$	$\ldots 7s^2 6d^{1+n}$
$91 + n, n = 0, 1, 2$	Pa, U, Np	$\ldots 7s^2 5f^{3+n}$	$\ldots 7s^2 5f^{2+n} 6d^1$
96	Cm	$\ldots 7s^2 5f^8$	$\ldots 7s^2 5f^7 6d^1$
103	Lr	$\ldots 7s^2 5f^{14} 6d^1$	$\ldots 7s^2 5f^{14} 7p^1$

Many more details can be added into the orbital description. However, even the elementary account given here shows that this approximation provides a good preliminary explanation of the periodicity in the chemical properties of the elements, as recognized by Mendeleyev in the nineteenth century. There are three main elements of this explanation:

1. Pauli's exclusion principle;
2. the approximate spherical symmetry of the atoms, leading to a degeneracy of $2\ell + 1$;
3. the spin of the electrons that doubles the degeneracy to $2(2\ell + 1)$.

Obviously, the description of atoms in terms of orbitals is not fundamental. The fundamental description involves a state vector in the Hilbert space that describes the Z electrons of an atom. The orbitals, the shells, and the procedure of filling is nothing but a visualization of mathematical terms that appear in the mean-field approximation.

15.2.3 Many-Particle Systems

When the number N of identical particles becomes very large, the energy E_0 of the ground state of the N-particle system is much larger than the separation $\epsilon_n - \epsilon_{n-1}$ of successive energy eigenvalues of a single particle. In this case, we can take the continuum limit. We do this by employing the number-of-states function $\Omega(\epsilon)$, which was defined in Section 5.6.3. The function $\Omega(\epsilon)$ gives the number of eigenvectors of $\hat{h}$ with energy less than ϵ. Then, the Fermi energy ϵ_F is obtained from the solution of the equation $\Omega(\epsilon_F) = N$.

When Ω is a continuous function of ϵ, we define the density of states $g(\epsilon) = \frac{d\Omega}{d\epsilon}$, and we substitute sums over energy eigenvalues with an integral over energies ϵ, weighted by $g(\epsilon)$: $\int d\epsilon g(\epsilon)$. Then, Eq. (15.21) becomes

$$E_0 = \int_{\epsilon_0}^{\epsilon_F} g(\epsilon) \epsilon d\epsilon. \tag{15.24}$$

Example 15.3 Consider a gas of N free fermions of mass m and spin s. The single-particle Hamiltonian is just the kinetic energy $\hat{h} = \frac{\hat{p}^2}{2m}$. To make the spectrum of the Hamiltonian discrete, we impose periodic boundary conditions in the three directions of space. Hence, the energy eigenvalues of a single particle are determined by three integers $n_1, n_2, n_3 = 0, \pm 1, \pm 2, \pm 3, \ldots$, as

$$\epsilon_{\mathbf{n}} = \frac{2\pi^2}{mL^2}\left(n_1^2 + n_2^2 + n_3^2\right) = \frac{2\pi^2}{mL^2}\mathbf{n}^2, \tag{15.25}$$

where we write $\mathbf{n} = (n_1, n_2, n_3)$. Each energy eigenvalue has at least a $2s + 1$-fold degeneracy because of spin.

We evaluate the number-of-states function $\Omega(\epsilon) = g_s \sum_{\mathbf{n}, |\mathbf{n}|<a} 1$, where $a^2 = m\epsilon L^2/(2\pi^2)$, and we write $g_s = 2s + 1$. For $m\epsilon L^2/\pi^2 >> 1$, we can substitute the sum by an integral, since the distance between two successive energy eigenvalues is much smaller than ϵ. Then,

$$\Omega(\epsilon) = g_s \int_{|\mathbf{n}|<a} d^3n = \frac{4\pi}{3}g_s a^3 = \frac{g_s V}{6\pi^2}(2m\epsilon)^{3/2}, \tag{15.26}$$

where $V = L^3$ is the volume occupied by the gas. In the derivation, we have used the fact that the equation $|\mathbf{n}| < a$ describes the interior of a sphere of radius a. Solving the equation $\Omega(\epsilon_F) = N$ for the Fermi energy ϵ_F, we obtain

$$\epsilon_F = \frac{1}{2m}\left(\frac{6\pi^2 N}{g_s V}\right)^{2/3}. \tag{15.27}$$

The density of states is

$$g(\epsilon) = \frac{g_s V}{4\pi^2}(2m)^{3/2}\sqrt{\epsilon}. \tag{15.28}$$

By Eq. (15.24),

$$E_0 = \frac{g_s V}{4\pi^2}(2m)^{3/2}\int_0^{\epsilon_F}\epsilon^{3/2}d\epsilon = \frac{g_s V}{10\pi^2}(2m)^{3/2}\epsilon_F^{5/2} = \frac{C_F}{mg_s^{2/3}}\frac{N^{5/3}}{V^{2/3}}, \tag{15.29}$$

where $C_F = \frac{3}{10}(6\pi^2)^{2/3} \simeq 4.56$. Equivalently, we can express the energy density $\rho = E_0/V$ as a function of the particle density $n = N/V$,

$$\rho = \frac{C_F}{mg_s^{2/3}}n^{5/3}. \tag{15.30}$$

The pressure P exercised by the fermions is $P = -\frac{\partial U}{\partial V}$; by Eq. (15.29), we find that $PV = \frac{2}{3}$. This expression is to be contrasted with the classical equation of state of ideal gases $PV = nT$, which implies that the pressure vanishes at zero absolute temperature T. The zero-temperature pressure of fermions is known as the *degeneracy pressure*, and it balances the gravitational pull in compact astrophysical bodies such as white dwarves and neutron stars, which consist of fermions at very low temperature.

15.3 Fock Spaces

So far, we have studied only systems with a fixed number of particles of each species: In Eq. (15.15), the particle numbers N_a and M_b are constants. This restriction prevents us from describing many important phenomena, such as chemical processes or nuclear reactions. In chemical reactions, molecules may split, combine, or change their species through interactions. The same holds also in nuclear reactions. Such processes do not occur only in composite particles but also in elementary ones.[4] For example, the scattering of one electron and one positron leads to their annihilation and to the appearance of two photons; the μ particle decays to one electron and two neutrinos.

To account for these phenomena, we must make the number of particles of a given species an observable that can change dynamically. Such observables are defined as operators on a specific type of Hilbert space, known as the *Fock space*. In this section, we define Fock spaces, we analyze their properties and we explain how the Fock space calculus works.

15.3.1 Defining Properties

Let $\mathcal{H}$ be the Hilbert space associated with a single particle. The Hilbert space that accommodates any number of such particles is obtained by "joining" the Hilbert space for zero particles with the Hilbert space for one particle, with the Hilbert space for two particles and so on. Joining is implemented by the direct sum. Since the Hilbert spaces for $N > 1$ particles depend on whether these particles are bosons or fermions, there are two different Fock spaces.

- The bosonic Fock space:

$$\mathcal{F}_B(\mathcal{H}) = \mathbb{C} \oplus \mathcal{H} \oplus \mathcal{H}_S^2 \oplus \mathcal{H}_S^3 \oplus \cdots . \tag{15.31}$$

- The fermionic Fock space:

$$\mathcal{F}_F(\mathcal{H}) = \mathbb{C} \oplus \mathcal{H} \oplus \mathcal{H}_A^2 \oplus \mathcal{H}_A^3 \oplus \cdots . \tag{15.32}$$

Let $|a\rangle$ be an orthonormal basis on $\mathcal{H}$, so that $\langle a'|a\rangle = \delta_{a'a}$. The label a takes values in some set L. L may be a continuous set, in which case the kets $|a\rangle$ correspond to a generalized basis and the symbol $\delta_{a'a}$ stands for a delta function.

For bosons, the orthonormal basis $|a\rangle$ on $\mathcal{H}$ induces an orthonormal basis $|a_1, a_2\rangle$ on H_S^2, an orthonormal basis $|a_1, a_2, a_3\rangle_S$ on H_S^3, and so on. Given that vectors with different number of particles n belong to orthogonal subspaces of $\mathcal{F}_B(\mathcal{H})$, basis vectors with different values of n are orthogonal. It follows that the set of all vectors of the form

$$|0\rangle, |a\rangle, |a_1, a_2\rangle_S, |a_1, a_2, a_3\rangle_S, \ldots, |a_1, a_2, \ldots, a_n\rangle_S, \ldots, \tag{15.33}$$

defines an orthonormal basis on the Fock space $\mathcal{F}_B(\mathcal{H})$; $|0\rangle$ is the unique vector that describes a state with zero particles. We will refer to $|0\rangle$ as the Fock vacuum, or simply vacuum.

---

[4] Elementary particles are neither eternal nor immutable as they were supposed to be in pre–twentieth-century physics.

It is convenient to parameterize this basis using a function $n : L \to \mathbb{N}$, that assigns a nonnegative integer n_a to each $a \in L$. Every vector of the basis (15.33) defines such a function: n_a is the number of particles in the state a. For example, if $L = \mathbb{N}$, the vector $|0, 1, 1, 2, 4, 4\rangle_S$ corresponds to the function with elements $n_0 = 1, n_1 = 2, n_2 = 1, n_3 = 0, n_4 = 2$, and $n_i = 0$, for $i \geq 5$.

Since any basis vector corresponds to a finite number of particles, n_a is nonzero for a finite number of labels a, hence, $\sum_a n_a < \infty$. Conversely, any function $n : L \to \mathbb{N}$ with a finite number of nonzero values corresponds to one of the vectors of the basis (15.33). Therefore, we can write the vectors of (15.33) as $|\{n_a\}\rangle$ or as $|n_{a_1}, n_{a_2}, n_{a_3}, \ldots\rangle$.

Similarly, in the fermionic Fock space $\mathcal{F}_B(\mathcal{H})$, there is an orthonormal basis of the form

$$|0\rangle, |a\rangle, |a_1, a_2\rangle_A, |a_1, a_2, a_3\rangle_A, \ldots, |a_1, a_2, \ldots, a_n\rangle_A, \ldots, \tag{15.34}$$

which can be described in terms of functions $n : L \to \{0, 1\}$, such that $\sum_a n_a < \infty$. By Pauli's principle, n_a can only take values 0 and 1. Hence, we will write the basis (15.34) of $\mathcal{F}_B(\mathcal{H})$ as $|\{n_a\}\rangle$.

15.3.2 Operators in the Bosonic Fock Space

In the bosonic Fock space $\mathcal{F}_B(\mathcal{H})$, we define the annihilation operator $\hat{a}^c$ and the creation operator $\hat{a}_c^\dagger$ for any $c \in L$,

$$\hat{a}^c |n_1, n_2, \ldots, n_c, \ldots\rangle = \sqrt{n_c} |n_1, n_2, \ldots, n_c - 1, \ldots\rangle, \tag{15.35}$$

$$\hat{a}_c^\dagger |n_1, n_2, \ldots, n_c, \ldots\rangle = \sqrt{n_c + 1} |n_1, n_2, \ldots, n_c + 1, \ldots\rangle. \tag{15.36}$$

There is an obvious similarity here with the annihilation and creation operator for a harmonic oscillator. The reason why the index c is up on $\hat{a}$ and down on $\hat{a}^\dagger$ will be explained shortly.

We straightforwardly obtain the *canonical* commutation relations

$$\left[\hat{a}^c, \hat{a}^d\right] = 0, \quad \left[\hat{a}_c^\dagger, \hat{a}_d^\dagger\right] = 0, \quad \left[\hat{a}^c, \hat{a}_d^\dagger\right] = \delta_d^c \hat{I}. \tag{15.37}$$

As in the harmonic oscillator, we can express the basis vectors $|\{n_a\}\rangle$ in terms of successive actions of annihilation operators on the vacuum,

$$|n_{a_1}, n_{a_2}, \ldots\rangle = \frac{1}{\sqrt{n_{a_1}! \, n_{a_2}! \cdots}} \left(\hat{a}_{a_1}^\dagger\right)^{n_{a_1}} \left(\hat{a}_{a_2}^\dagger\right)^{n_{a_2}} \cdots |0\rangle. \tag{15.38}$$

Let us denote by ξ^a the coefficients in the analysis of a vector $|\xi\rangle \in \mathcal{H}$ with respect to the basis $|a\rangle$. We denote by $\bar{\xi}_a$ the complex conjugates of ξ^a; they correspond to the decomposition of $\langle\xi|$ in the basis $\langle a|$. Then, we define the operators

$$\hat{a}^\dagger(\xi) = \hat{a}_c^\dagger \xi^c, \quad \hat{a}(\bar{\xi}) = \hat{a}^c \bar{\xi}_c. \tag{15.39}$$

Here, we used the so-called *Einstein summation convention*: Any index that appears twice, one up and one down, is summed over.

The benefit of using the operators $\hat{a}^\dagger(\xi)$ and $\hat{a}(\bar{\xi})$ is that they do not depend on the choice of basis $\mathcal{H}$. They satisfy the commutation relations

$$\left[\hat{a}(\bar{\xi}),\hat{a}(\bar{\eta})\right] = 0, \quad \left[\hat{a}^\dagger(\xi),\hat{a}^\dagger(\eta)\right] = 0, \quad \left[\hat{a}(\bar{\eta}),\hat{a}^\dagger(\xi)\right] = (\bar{\eta}_c\xi^c)\hat{I}. \tag{15.40}$$

Any operator $\hat{A}$ on the Fock space $\mathcal{F}_B(\mathcal{H})$ can be decomposed as a sum of maps A_{mn} from the subspace $\mathcal{H}_S^n$ of n particles to the subspace $\mathcal{H}_S^m$ of m particles. To construct the map A_{mn}, we act with n annihilation operators on a state with n particles to obtain the vacuum. Then, we act on the vacuum with m creation operators, ending up with a state of m particles. Hence, we can express any Fock space operator as a sum of terms that involve n annihilation and m creation operators for all $n, m = 0, 1, 2, \ldots$,

$$\hat{A} = \sum_{n=0}^\infty \sum_{m=0}^\infty K_{a_1 a_2 \cdots a_n}{}^{b_1 b_2 \cdots b_m} \hat{a}^\dagger_{b_1} \cdots \hat{a}^\dagger_{b_m} \hat{a}^{a_1} \cdots \hat{a}^{a_n}, \tag{15.41}$$

for some constant coefficients $K_{a_1 a_2 \cdots a_n}{}^{b_1 b_2 \cdots b_m}$. In Eq. (15.41), all annihilation operators are to the right of all creation operators. This expression for an operator $\hat{A}$ is known as the *normal ordered* form of $\hat{A}$.

Given an operator $\hat{f}$ on $\mathcal{H}$, we can define its *linear extension* $\hat{\sigma}(\hat{f})$ to $\mathcal{F}_B(\mathcal{H})$ as

$$\hat{\sigma}(\hat{f}) := \hat{a}^\dagger_a f^a{}_b \hat{a}^b, \tag{15.42}$$

where $f^a{}_b := \langle a|\hat{f}|b\rangle$ are the matrix elements of $\hat{f}$. By definition, $\hat{\sigma}(\lambda\hat{f} + \hat{g}) = \lambda\hat{\sigma}(\hat{f}) + \hat{\sigma}(\hat{g})$, for all operators $\hat{f}$ and $\hat{g}$ on $\mathcal{H}$, and $\lambda \in \mathbb{C}$.

Note that we can define $\hat{\sigma}(\hat{f})$ without making use of a specific basis,

$$\hat{\sigma}(\hat{f}) = 0 \oplus \hat{f} \oplus \hat{f}^{\oplus 2} \oplus \hat{f}^{\oplus 3} \oplus \cdots . \tag{15.43}$$

To show the equivalence of the two definitions, we first note that the operator (15.42) does not change the number of particles; hence, it preserves the subspaces $\mathcal{H}_S^n$. For $n = 2$, $\hat{\sigma}(\hat{f})|a_1, a_2\rangle_S = \sum_{b_1} f^{a_1}{}_{b_1}|b_1, a_2\rangle_S + \sum_{b_2} f^{a_2}{}_{b_2}|a_1, b_2\rangle_S = \hat{f}^{(2)}|a_1, a_2\rangle$. We can generalize this result by induction to all n, thus confirming Eq. (15.43).

It is straightforward to show that the map $\hat{\sigma}$ preserves commutation relations,

$$\left[\hat{\sigma}\left(\hat{f}\right), \hat{\sigma}(\hat{g})\right] = \hat{\sigma}\left(\left[\hat{f}, \hat{g}\right]\right), \tag{15.44}$$

for all operators $\hat{f}$ and $\hat{g}$ on $\mathcal{H}$. This means that any representation of a Lie algebra on $\mathcal{H}$ induces a representation of the same Lie algebra on $\mathcal{F}_B(\mathcal{H})$. For example, if $\hat{h}$ is the generator of time translations for a representation of the Galilei group on $\mathcal{H}$, $\hat{\sigma}(\hat{h})$ is the generator of time translations on $\mathcal{F}_B(\mathcal{H})$.

If $\hat{f}$ is self-adjoint on $\mathcal{H}$, then $\hat{\sigma}(\hat{f})$ is self-adjoint on $\mathcal{F}_B(\mathcal{H})$. Let $\hat{f}|a\rangle = f_a|a\rangle$ be the eigenvalue equation for $\hat{f}$. Suppose that the eigenvectors $|a\rangle$ of $\hat{f}$ define an orthonormal basis on $\mathcal{H}$. Then, the associated basis vectors $|\{n_a\}\rangle$ on $\mathcal{F}_B(\mathcal{H})$ are eigenvectors of $\hat{\sigma}(\hat{f})$,

$$\hat{\sigma}\left(\hat{f}\right)|\{n_a\}\rangle = \left(\sum_a n_a f_a\right)|\{n_a\}\rangle. \tag{15.45}$$

If $\hat{f}$ is a positive operator, then $f_a \geq 0$; hence, $\hat{\sigma}(\hat{f})$ is also positive.

For $\hat{f} = \hat{I}, f_a = 1$, hence, the eigenvalues of $\hat{\sigma}(\hat{I})$ coincide with the total number of particles $\sum_a n_a$. We call $\hat{\sigma}(\hat{I})$, the *number operator*, and we denote it by $\hat{N}$,

$$\hat{N} = \hat{\sigma}(\hat{I}) = \hat{a}_c^\dagger \hat{a}^c. \tag{15.46}$$

For an operator $\hat{u}$ on $\mathcal{H}$, we define the operator $\exp_\otimes(\hat{u})$ on $\mathcal{F}(\mathcal{H}_B)$ as

$$\exp_\otimes(\hat{u}) = \hat{I} \oplus \hat{u} \oplus \hat{u}^{\otimes 2} \oplus \hat{u}^{\otimes 3} \oplus \cdots. \tag{15.47}$$

It is straightforward to show that the map $\exp_\otimes$ preserves operator multiplication, namely, $\exp_\otimes(\hat{u}_1)\exp_\otimes(\hat{u}_2) = \exp_\otimes(\hat{u}_1 \hat{u}_2)$. Furthermore, if $\hat{u}$ is unitary on $\mathcal{H}$, $\exp_\otimes(\hat{u})$ is unitary on $\mathcal{F}(\mathcal{H}_B)$. Hence, any unitary representation of a group G on $\mathcal{H}$, induces a unitary representation of G on $\mathcal{F}(\mathcal{H}_B)$. For further properties of the map $\exp_\otimes$, including its relation to the map σ, see Problem 15.10.

Example 15.4 If $\mathcal{H} = \mathbb{C}^n$, the creation and annihilation operators satisfy the commutation relations $[\hat{a}^a, \hat{a}^b] = [\hat{a}_a^\dagger, \hat{a}_b^\dagger] = 0$ and $[\hat{a}^a, \hat{a}_b^\dagger] = \delta^a{}_b \hat{I}$, for $a, b = 1, 2, \ldots, n$. Hence, the Fock space $\mathcal{F}_B(\mathbb{C}^n)$ describes n harmonic oscillators.

We define $\hat{x}_a = \frac{1}{\sqrt{2}}(\hat{a}^a + \hat{a}_a^\dagger)$ and $\hat{p}_a = \frac{1}{\sqrt{2}i}(\hat{a}^a - \hat{a}_a^\dagger)$. Then,

$$[\hat{x}_a, \hat{x}_b] = 0, \quad [\hat{p}_a, \hat{p}_b] = 0, \quad [\hat{x}_a, \hat{p}_b] = i\delta_{ab}\hat{I}, \tag{15.48}$$

that is, the Fock space carries a representation of the Heisenberg–Weyl group H_n. This representation is irreducible, because the only operators that commute with all creation and annihilation operators are multiples of the unity; see the discussion about the Stone–von Neumann theorem in Section 12.2.

Example 15.5 Consider the operators $\hat{b}^a$ and $\hat{b}_a^\dagger$, defined by the linear transformation of the creation and annihilation operator on the Fock space

$$\hat{b}^a = A^a{}_b \hat{a}^b + B^{ab} \hat{a}_b^\dagger, \quad \hat{b}_a^\dagger = \bar{A}_a{}^b \hat{b}_b^\dagger + \bar{B}_{ab} \hat{b}^b. \tag{15.49}$$

Here $\bar{A}$ and $\bar{B}$ stand for the complex conjugate matrices of A and B, respectively; the indices of these matrices are raised with a complex conjugation, in order to preserve the Einstein summation convention.

Suppose that the operators $\hat{b}^a$ and $\hat{b}_a^\dagger$ satisfy the canonical commutation relations (15.37). The condition $[\hat{b}^a, \hat{b}_b^\dagger] = \delta_b^a \hat{I}$ implies that $A^a{}_c \bar{A}_b{}^c - B^{ac} \bar{B}_{bc} = \delta_b^a$. The conditions $[\hat{b}^a, \hat{b}^b] = [\hat{b}_a^\dagger, \hat{b}_b^\dagger] = 0$ imply that $A^a{}_c B^{bc} = B^{ac} A^b{}_c$. In matrix form, these conditions read

$$AA^\dagger - BB^\dagger = I, \quad AB^T = BA^T. \tag{15.50}$$

Transformations of the form (15.49), where A and B satisfy Eq. (15.50), are called *Bogoliubov transformations*. Given Eqs. (15.50), the inverse of the transformations (15.49) are

$$\hat{a}^a = \bar{A}^b{}_a \hat{b}^b + \bar{B}^{ba} \hat{b}_b^\dagger, \quad \hat{a}_a^\dagger = A_b{}^a \hat{b}_b^\dagger + B_{ba} \hat{b}^b. \tag{15.51}$$

Let us denote by $|\Omega\rangle$ the Bogoliubov vacuum,[5] that is, the state annihilated by all annihilation operators b^a, $\hat{b}^a|\Omega\rangle = 0$. The existence of this state is guaranteed by the algebraic analysis of the spectrum of the annihilation operators – see Section 5.2.2. Then, the expectation value of any operator $\hat{\sigma}(\hat{f})$ on $|\Omega\rangle$ is

$$\langle\Omega|\hat{\sigma}(\hat{f})|\Omega\rangle = f^a{}_b\langle\Omega|\hat{a}^\dagger_a\hat{a}^b|\Omega\rangle = B_{ca}f^a{}_b\bar{B}^{db}\langle\Omega|\hat{b}^c\hat{b}^\dagger_d|\Omega\rangle = B_{ca}f^a{}_b\bar{B}^{db}\delta^c_d = \mathrm{Tr}(B^\dagger B\hat{f}).$$

For $\hat{f} = \hat{I}$, we find the expected number of particles on $|\Omega\rangle$, $\langle\Omega|\hat{N}|\Omega\rangle = \mathrm{Tr}(B^\dagger B)$.

If $\mathcal{H}$ is finite-dimensional, then the matrix B is also finite-dimensional and $\langle\Omega|\hat{N}|\Omega\rangle$ is finite. If $\mathcal{H}$ is infinite-dimensional, then $\mathrm{Tr}(D^\dagger D)$ may fail to be finite. But then $|\Omega\rangle$ is not a vector on the Fock space $\mathcal{F}_B(\mathcal{H})$, since $\hat{N}$ takes finite values on all basis vectors (15.33) and its expectation values are always finite. This means that the Bogoliubov transformation (15.49) cannot be implemented by a unitary operator $\hat{U}$ as $\hat{b}_a = \hat{U}\hat{a}_a\hat{U}^\dagger$, because then $|\Omega\rangle = \hat{U}|0\rangle$ is a vector of $\mathcal{F}_B(\mathcal{H})$ that satisfies $\hat{b}_a|\Omega\rangle = 0$.

Hence, if $\mathrm{Tr}(B^\dagger B) = \infty$, the operators $\hat{b}^a$ and $\hat{b}^\dagger_a$ define a representation of the infinite-dimensional Weyl group, which is unitarily inequivalent to the representation defined by $\hat{a}^a$ and $\hat{a}^\dagger_a$.

15.3.3 Operators on the Fermionic Fock Space

In the fermionic Fock space $\mathcal{F}_F(\mathcal{H})$, we define the annihilation operators $\hat{c}^b$ and the creation operators $\hat{c}^\dagger_b$ for all $b \in L$ through their action on the basis vectors (15.34)

$$\hat{c}^b|a_1, a_2, \ldots, a_n\rangle_A = \sum_i \delta^b{}_{a_i}|a_1, a_2, \ldots, \not{a}_i, \ldots, a_n\rangle_A, \tag{15.52}$$

$$\hat{c}^\dagger_b|a_1, a_2, \ldots, a_n\rangle_A = |b, a_1, a_2, \ldots, a_n\rangle_A. \tag{15.53}$$

By $|a_1, a_2, \ldots, \not{a}_i, \ldots, a_n\rangle_A$ we denote the vector $|a_1, a_2, \ldots, a_{i-1}, a_{i+1}, \ldots, a_n\rangle_A$, from which we removed the index a_i.

We find that

$$\hat{c}^a\hat{c}^b|a_1, a_2, \ldots, a_n\rangle_A = \sum_{i,j} \delta^a{}_{a_i}\delta^b{}_{a_j}|a_1, a_2, \ldots, \not{a}_i, \ldots, \not{a}_j, \ldots, a_n\rangle_A$$

$$= -\sum_{i,j} \delta^a{}_{a_i}\delta^b{}_{a_j}|a_1, a_2, \ldots, \not{a}_j, \ldots, \not{a}_i, \ldots, a_n\rangle_A$$

$$= -\sum_{i,j} \delta^b{}_{a_i}\delta^a{}_{a_j}|a_1, a_2, \ldots, \not{a}_i, \ldots, \not{a}_j, \ldots, a_n\rangle_A = -\hat{c}^b\hat{c}^a|a_1, a_2, \ldots, a_n\rangle_A,$$

hence,

$$[\hat{c}^a, \hat{c}^b]_+ = 0, \tag{15.54}$$

where we defined the *anticommutator* of two operators as

$$[\hat{A}, \hat{B}]_+ = \hat{A}\hat{B} + \hat{B}\hat{A}. \tag{15.55}$$

---

[5] In the quantum optics literature, Bogoliubov transformations are known as *squeezing transformations*, and the state $|\Omega\rangle$ is called the *squeezed vacuum*.

Similarly, we find that

$$\hat{c}_a^\dagger \hat{c}_b^\dagger |a_1, a_2, \ldots, a_n\rangle_A = |a, b, a_1, a_2, \ldots, a_n\rangle_A =$$
$$-|b, a, a_1, a_2, \ldots, a_n\rangle_A = -\hat{c}_b^\dagger \hat{c}_a^\dagger |a_1, a_2, \ldots, a_n\rangle_A,$$

hence,

$$[\hat{c}_a^\dagger, \hat{c}_b^\dagger]_+ = 0. \tag{15.56}$$

Furthermore,

$$\hat{c}^a \hat{c}_b^\dagger |a_1, a_2, \ldots, a_n\rangle_A = \delta^a{}_b |a_1, a_2, \ldots, a_n\rangle_A + \sum_i \delta^a{}_{a_i} |b, a_1, a_2, \ldots, \phi_i, \ldots, a_n\rangle_A$$

$$= \delta^a{}_b |a_1, a_2, \ldots, a_n\rangle_A + \hat{c}_b^\dagger \hat{c}^a |a_1, a_2, \ldots, a_n\rangle_A.$$

Hence,

$$[\hat{c}^a, \hat{c}_b^\dagger]_+ = \delta^a{}_b \hat{I}. \tag{15.57}$$

Eqs. (15.54)–(15.57) are called *canonical anticommutation relations*.

The vectors $|a_1, a_2, \ldots, a_n\rangle_A$ can be expressed as

$$|a_1, a_2, \ldots, a_n\rangle_A = \hat{c}_{a_n}^\dagger \cdots \hat{c}_{a_2}^\dagger \hat{c}_{a_1}^\dagger |0\rangle. \tag{15.58}$$

In analogy with the bosonic case, we define the operators $\hat{c}(\bar{\xi}) = \hat{c}^a \bar{\xi}_a$ and $\hat{c}^\dagger(\xi) = \hat{c}_a^\dagger \xi^a$, which do not depend on the choice of basis on $\mathcal{H}$. The anticommutation relations (15.54) – (15.57) become

$$[\hat{c}(\bar{\xi}), \hat{c}(\bar{\eta})]_+ = 0, \quad [\hat{c}^\dagger(\xi), \hat{c}^\dagger(\eta)]_+ = 0, \quad [\hat{c}(\bar{\eta}), \hat{c}^\dagger(\xi)]_+ = (\bar{\eta}_c \xi^c) \hat{I}. \tag{15.59}$$

Every operator $\hat{A}$ on $\mathcal{F}_F(\mathcal{H})$ can be written in a normal-ordered form

$$\hat{A} = \sum_{n=0}^\infty \sum_{m=0}^\infty K_{a_1 a_2 \cdots a_n}{}^{b_1 b_2 \cdots b_m} \hat{c}_{b_1}^\dagger \cdots \hat{c}_{b_m}^\dagger \hat{c}^{a_1} \cdots \hat{c}^{a_n}, \tag{15.60}$$

where $K_{a_1 a_2 \cdots a_n}{}^{b_1 b_2 \cdots b_m}$ are constants.

For any operator $\hat{f}$ on $\mathcal{H}$, we define its *linear extension* $\hat{\sigma}(\hat{f})$ on $\mathcal{F}_F(\mathcal{H})$ by

$$\hat{\sigma}(\hat{f}) := \hat{c}_a^\dagger f^a{}_b \hat{c}^b. \tag{15.61}$$

The linear extension map $\hat{\sigma}$ satisfies the same properties with the corresponding map in the bosonic Fock space. In particular, the number operator is defined by $\hat{N} = \hat{\sigma}(\hat{I})$ and the commutation relations (15.44) still apply. The latter are straightforwardly proven by using the operator identity

$$[\hat{A}, \hat{B}\hat{C}] = [\hat{A}, \hat{B}]_+ \hat{C} - \hat{B}[\hat{A}, \hat{C}]_+. \tag{15.62}$$

The map $\exp_\otimes$ is also well defined on $\mathcal{F}_F(\mathcal{H})$ and satisfies properties analogous to its bosonic counterpart.

Example 15.6 If $\mathcal{H} = \mathbb{C}^d$, the series of direct sums in the definition of the fermionic Fock space (15.32) has a finite number of terms, because $\mathcal{H}_A^n = \emptyset$ for $n > d$. Given that $\dim\mathcal{H}_A^n = \frac{d!}{(d-n)!n!}$, $\dim\mathcal{F}_A(\mathbb{C}^d) = \sum_{n=0}^d \frac{d!}{(d-n)!n!} = 2^d$. Hence, $\mathcal{F}_A(\mathbb{C}^d) = \mathbb{C}^{2^d}$, that is, the fermionic Fock space is identical to the Hilbert space of d qubits. For $d = 1$, $\mathcal{F}_A(\mathbb{C}) = \mathbb{C}^2$, the creation operator $\hat{c}^\dagger$ coincides with $\hat{\sigma}_+$ and the annihilation operator $\hat{c}$ with $\hat{\sigma}_-$.

15.3.4 The Symmetrization Postulate Revisited

The introduction of the Fock spaces allows us to rewrite the symmetrization postulate as follows.

> **Symmetrization postulate** Particles are either bosons with creation and annihilation operators satisfying the commutation relations (15.40), or fermions with creation and annihilation operators satisfying the anticommutation relations (15.59).

This version of the symmetrization postulate is fundamentally algebraic, and it is more useful when passing to the theory of quantum fields.

A system that involves particles of two types, say, C and D, is described by the tensor product $\mathcal{F}_C \otimes \mathcal{F}_D$ of the Fock spaces $\mathcal{F}_C$ and $\mathcal{F}_D$, associated to C and D, respectively. Suppose C and D are fermions. Hence, there are operators $\hat{c}^a$ and $\hat{c}_a^\dagger$ on $\mathcal{F}_C$ that satisfy the canonical anticommutation relations and the same holds for operators $\hat{d}^a$ and $\hat{d}_a^\dagger$ on $\mathcal{F}_D$. Since these operators are defined on different Hilbert spaces of a tensor product,[6] they commute, that is,

$$[\hat{c}^a, \hat{d}^b] = [\hat{c}^a, \hat{d}_b^\dagger] = 0. \tag{15.63}$$

However, it is possible to redefine the operators $\hat{d}_b$ and $\hat{d}_b^\dagger$, so that the commutators in Eq. (15.63) are substituted by anticommutators.

To see this, let $\hat{N} = \sum_a \hat{c}_a^\dagger \hat{c}^a + \sum_a \hat{d}_a^\dagger \hat{d}^a$ be the operator for the total number of particles. Then, we define $\hat{\eta} := e^{i\pi\hat{N}}$. Since $\hat{N}$ takes integer eigenvalues, the eigenvalues of $\hat{\eta}$ are ± 1; hence, $\hat{\eta}^\dagger = \hat{\eta}$. Since the creation and the annihilation operator change the number of particles by one, $\hat{d}^a e^{i\pi\hat{N}} = e^{i\pi(\hat{N}-\hat{I})}\hat{d}^a$, which implies that $\hat{d}^a\hat{\eta} = -\hat{\eta}\hat{d}^a$. Similarly, $\hat{c}^a\hat{\eta} = -\hat{\eta}\hat{c}^a$, $\hat{d}_a^\dagger\hat{\eta} = -\hat{\eta}\hat{d}_a^\dagger$, and $\hat{c}_a^\dagger\hat{\eta} = -\hat{\eta}\hat{c}_a^\dagger$.

Then, we define new creation and annihilation operators through the *Klein transformation*,

$$\hat{d}^a \to \hat{d}'^a = \hat{d}^a\hat{\eta}, \qquad \hat{d}_a^\dagger \to \hat{d}_a'^\dagger = \hat{\eta}\hat{d}_a^\dagger. \tag{15.64}$$

Then, $[\hat{c}^a, \hat{d}'^b]_+ = \hat{c}^a\hat{d}^b\eta + \hat{d}^b\eta\hat{c}^a = \hat{\eta}[\hat{c}^a, \hat{d}^b] = 0$. Similarly, we find that all anticommutators between $\hat{c}, \hat{c}^\dagger \hat{d}', \hat{d}'^\dagger$ vanish,

$$[\hat{c}^a, \hat{d}'^b]_+ = [\hat{c}_a, \hat{d}_b'^\dagger]_+ = 0. \tag{15.65}$$

---

[6] Equation (15.63) is an abbreviated notation of the $[\hat{c}_a \otimes \hat{I}, \hat{I} \otimes \hat{d}_b] = 0$, and so on.

We conclude that it is always possible to make the creation and annihilation operators of different particles to anticommute with each other. However, the operators $\hat{d}'^b$ and $\hat{d}'^{\dagger}_b$ do not involve only degrees of freedom of the D particles, but they also have a contribution from the C particles through the operator $\hat{\eta}$.

15.4 Quantum Fields

The Fock space description of many-particle systems leads naturally to the definition of quantum fields. Here, we focus on the nonrelativistic domain, but this is sufficient in order to establish one of the most fundamental properties of quantum theory, the duality between particles and fields. We also introduce the simplest relativistic quantum field, which leads us to a proof of the spin-statistics theorem for $s = 0$. Alternative statistics are shown in Box 15.1.

Box 15.1 Alternative Statistics

The fact that the symmetrization postulate cannot be accounted for by the axioms of quantum mechanics has led to the search of statistics beyond bosonic and fermionic.

The first example is *parastatistics* (Green, 1952). Parabose statistics allow up to p identical particles in an antisymmetric state, but any n particles with $n > p$ must be in a symmetric state. In analogy, parafermi statistics allows up to p identical particles in a symmetric state, but any n particles with $n > p$ must be in an antisymmetric state. The integer $p > 1$ is the *order* of the parastatistics.

We can represent parastatistics on a Fock space, in terms of p creation and annihilation operators $\hat{a}^a_\lambda$ and $\hat{a}^{\dagger}_{a\lambda}$, where $\lambda = 1, 2, \ldots, p$, and a labels a basis on the single-particle Hilbert space. For parabose statistics,

$$[\hat{a}^a_\lambda, \hat{a}^{\dagger}_{b\lambda}] = \delta_{ab}, \text{ and } [\hat{a}^a_\lambda, \hat{a}^b_{\lambda'}]_+ = [\hat{a}^{\dagger}_{a\lambda}, \hat{a}^{\dagger}_{b\lambda'}]_+ = [\hat{a}^a_\lambda, \hat{a}^{\dagger}_{b\lambda'}]_+ = 0 \text{ for } \lambda \neq \lambda'.$$

For parafermi statistics,

$$[\hat{a}^a_\lambda, \hat{a}^{\dagger}_{b\lambda}]_+ = \delta_{ab}, \text{ and } [\hat{a}^a_\lambda, \hat{a}^b_{\lambda'}] = [\hat{a}^{\dagger}_{a\lambda}, \hat{a}^{\dagger}_{b\lambda'}] = [\hat{a}^a_\lambda, \hat{a}^{\dagger}_{b\lambda'}] = 0 \text{ for } \lambda \neq \lambda'.$$

Parastatistics give gross violations of Bose or Fermi statistics, and they have never been observed. To parameterize small deviations from Bose or Fermi statistics, one often employs *quon statistics*, defined by the commutation relation,

$$\hat{a}_a \hat{a}^{\dagger}_b - q \hat{a}^{\dagger}_b \hat{a}_a = \delta_{ab}, \text{ where } q \in [-1, 1].$$

There are no commutation relations with two $\hat{a}$s or two $\hat{a}^{\dagger}$s for quons. Experiments constrain q to be -1 with an accuracy of 10^{-26} for electrons and with an accuracy of order 10^{-12} for quarks (Greenberg and Hilborn, 1999).

A different approach to quantum statistics, originating from Leinass and Myrheim (1977), analyzes the quantum states in the position representation. Let Q be the configuration space of a single particle. States correspond to functions $\psi(x)$ on Q. We implement a particle exchange by the requirement that $\psi(x_1, x_2) = e^{i\theta} \psi(x_2, x_1)$.

The key difference from the analysis of Section 15.1, is that we require that the exchange from x_1 to x_2 and x_2 to x_1 is implemented *continuously*, that is, that there are paths taking x_1 to x_2 and x_2 to x_1. The phase $e^{i\theta}$ depends only on the topology of these paths: Paths that can be continuously deformed to each other have the same θ. A double exchange transformation takes (x_1, x_2) to (x_1, x_2), so it defines a loop. In three dimensions, it is always possible to contract a loop to a single point, so that there is no exchange. Hence, $e^{2i\theta} = 1$, or $e^{i\theta} = \pm 1$. Only fermionic and bosonic statistics are possible.

This is not the case in two spatial dimensions. Not all loops can be deformed to each other. Furthermore, any loop has an invariant orientation: A trajectory is either left handed or right handed, the former corresponding to phase $e^{i\theta}$ and the latter to phase $e^{-i\theta}$. Orientation plays no role in three dimensions, because a right-handed path can be continuously changed to a left-handed one by a rotation in the extra dimension.

Values of $e^{i\theta}$ other than ± 1 define *anyonic* statistics. The corresponding particles are called *anyons*. Anyons exist only in two dimensions, so the known elementary particles cannot be anyons. However, in condensed matter physics the creation of effectively two-dimensional systems is possible, and in such systems anyons may emerge as quasiparticles, that is, propagating excitations. Anyons were first proposed in Wilczek (1982) and their signature was first identified in Bartolomei et al. (2020).

15.4.1 Noninteracting Particles

We consider a system of particles of mass M and spin s. The single-particle Hilbert space is $\mathcal{H}_s = L^2(\mathbb{R}^3) \otimes \mathbb{C}^{2s+1}$. For concreteness, we consider fermions; the bosonic case is treated analogously.

We employ the generalized basis $|a\rangle = |\mathbf{p}, m\rangle$ on $\mathcal{H}_s$, where $\mathbf{p}$ is a momentum vector and m is the spin projection along direction 3. We write the corresponding annihilation and creation operators as $\hat{c}_m(\mathbf{p})$ and $\hat{c}_m^\dagger(\mathbf{p})$, respectively. The anticommutation relations read

$$[\hat{c}_m(\mathbf{p}), \hat{c}_{m'}(\mathbf{p}')]_+ = [\hat{c}_m^\dagger(\mathbf{p}), \hat{c}_{m'}^\dagger(\mathbf{p}')]_+ = 0, \quad [\hat{c}_m(\mathbf{p}), \hat{c}_{m'}^\dagger(\mathbf{p}')]_+ = \delta_{mm'}\delta^3(\mathbf{p} - \mathbf{p}')\hat{I}.$$

We assume that the particles do not interact with each other. The Fock space Hamiltonian is $\hat{H} = \hat{\sigma}(\hat{h})$, where $\hat{h}$ is the Hamiltonian for a single particle. For free particles, $\hat{h} = \frac{\hat{\mathbf{p}}^2}{2M} \otimes \hat{I}$. It follows that

$$\hat{H} = \hat{\sigma}(\hat{h}) = \sum_m \int d^3p \frac{\mathbf{p}^2}{2M} \hat{c}_m^\dagger(\mathbf{p})\hat{c}_m(\mathbf{p}). \tag{15.66}$$

Similarly, we define the total momentum of the system:

$$\hat{\mathbf{P}} = \hat{\sigma}(\hat{\mathbf{p}}) = \sum_m \int d^3p \, \mathbf{p}\hat{c}_m^\dagger(\mathbf{p})\hat{c}_m(\mathbf{p}). \tag{15.67}$$

We define the corresponding *quantum fields* as

$$\hat{\psi}_m(\mathbf{x}) = \int \frac{d^3p}{(2\pi)^{3/2}} \hat{c}_m(\mathbf{p})e^{i\mathbf{p}\cdot\mathbf{x}}, \qquad \hat{\psi}_m^\dagger(\mathbf{x}) = \int \frac{d^3p}{(2\pi)^{3/2}} \hat{c}_m^\dagger(\mathbf{p})e^{-i\mathbf{p}\cdot\mathbf{x}}. \tag{15.68}$$

It is straightforward to show that the fields satisfy anticommutation relations

$$[\hat{\psi}_m(\mathbf{x}), \hat{\psi}_{m'}(\mathbf{x}')]_+ = [\hat{\psi}_m^\dagger(\mathbf{x}), \hat{\psi}_{m'}^\dagger(\mathbf{x}')]_+ = 0, \quad [\hat{\psi}_m(\mathbf{x}), \hat{\psi}_{m'}^\dagger(\mathbf{x}')]_+ = \delta_{mm'}\delta^3(\mathbf{x} - \mathbf{x}')\hat{I}.$$

Then, we express the following operators in terms of the quantum fields:

- the Hamiltonian $\hat{H} = -\frac{1}{2M} \sum_m \int d^3x \hat{\psi}_m^\dagger(\mathbf{x}) \nabla^2 \hat{\psi}_m(\mathbf{x})$;
- the momentum $\hat{\mathbf{P}} = -i \sum_m \int d^3x \hat{\psi}_m^\dagger(\mathbf{x}) \nabla \hat{\psi}_m(\mathbf{x})$;
- the particle number $\hat{N} = \sum_m \int d^3x \hat{\psi}_m^\dagger(\mathbf{x}) \hat{\psi}_m(\mathbf{x})$.

The three operators $\hat{H}, \hat{\mathbf{P}}$ and $\hat{N}$ are *local functions* of the field, that is, they can be expressed as $\hat{H} = \int d^3x \hat{\epsilon}(\mathbf{x})$, $\hat{\mathbf{P}} = \int d^3x \hat{\boldsymbol{\pi}}(\mathbf{x})$ and $\hat{N} = \int d^3x \hat{v}(\mathbf{x})$, in terms of

- the *energy density* operator: $\hat{\epsilon}(\mathbf{x}) = -\frac{1}{2M} \sum_m \hat{\psi}_m^\dagger(\mathbf{x}) \nabla^2 \hat{\psi}_m(\mathbf{x})$,
- the *momentum density* operator: $\hat{\boldsymbol{\pi}}(\mathbf{x}) = -i \sum_m \hat{\psi}_m^\dagger(\mathbf{x}) \nabla \hat{\psi}_m(\mathbf{x})$,
- and the *particle density* operator: $\hat{v}(\mathbf{x}) = \sum_m \hat{\psi}_m^\dagger(\mathbf{x}) \hat{\psi}_m(\mathbf{x})$.

The density operators depend only on the values of the field and of a finite number of derivatives at each point.

The Heisenberg evolution equations for the field operators yield

$$i\frac{\partial}{\partial t} \hat{\psi}_m(\mathbf{x}) = \hat{h} \hat{\psi}_m(\mathbf{x}), \tag{15.69}$$

with solutions

$$\hat{\psi}_m(\mathbf{x}, t) = \int \frac{d^3p}{(2\pi)^{3/2}} \hat{c}_m(\mathbf{p}) e^{i\mathbf{p}\cdot\mathbf{x} - i\frac{\mathbf{p}^2}{2M}t}, \quad \hat{\psi}_m^\dagger(\mathbf{x}) = \int \frac{d^3p}{(2\pi)^{3/2}} \hat{c}_m^\dagger(\mathbf{p}) e^{-i\mathbf{p}\cdot\mathbf{x} + i\frac{\mathbf{p}^2}{2M}t}.$$

By Eq. (15.69), the quantum field evolves under the same rule as a single-particle wave function, except for the fact that it is an operator. For this reason, the description of quantum systems with fields used to be called *second quantization*, meaning that the first quantization is the construction of Schrödinger's equation for particles, and the second quantization was the substitution of the single-particle wave function with a field operator. Today, we view this fact as an aspect of the more general principle of *field–particle duality*.

> *Field–particle duality* We can define field observables on the Hilbert space that describes an indefinite number of particles, so that the Hamiltonian is a local functional of the fields. The evolution law of the fields is expressed in terms of a partial differential equation with respect to the time and space coordinate.

The field–particle duality is not a separate quantum principle. It is a statement of a pattern in our fundamental quantum theories. It asserts that the elementary quantum systems that appear in nature, namely, particles, can also be described in terms of local quantum fields.

Next, consider a particle of mass M, spin s, charge q, and gyromagnetic ratio γ in an external EM field with potentials $(\phi, \mathbf{A})$. We denote the associated Hamiltonian (14.10) by $\hat{h}$. The Fock space Hamiltonian is $\hat{H} = \hat{\sigma}(\hat{h})$; for particles that do not interact with each other the Hamiltonian is

$$\hat{H} = \sum_m \int d^3x \hat{\psi}_m^\dagger(\mathbf{x}) \left(-\frac{1}{2M}\nabla^2 + q\phi + \frac{iq}{m}\mathbf{A}\cdot\nabla + \frac{q^2}{2m}\mathbf{A}^2 \right) \hat{\psi}_m(\mathbf{x})$$
$$+ \gamma \sum_{mm'} \int d^3x \hat{\psi}_m^\dagger(\mathbf{x})\mathbf{B}\cdot\hat{\mathbf{S}}_{mm'}\hat{\psi}_{m'}(\mathbf{x}), \tag{15.70}$$

where $\hat{\mathbf{S}}_{mm'}$ are the matrix elements of the spin operator. The Hamiltonian (15.70) is local with respect to the fields $\hat{\psi}_m$, but also with respect to the EM potentials.

15.4.2 Interacting Particles

For interacting particles, the Hamiltonian includes a term $\hat{U}$ that describes the interaction,

$$\hat{H} = \hat{\sigma}(\hat{h}) + \hat{U}. \tag{15.71}$$

The usual case involves pairwise interaction, that is, each pair of particles contributes a separate term in $\hat{U}$. Let U_{ab} be the term that corresponds to the interaction of one particle in state $|a\rangle$ with one particle in state $|b\rangle$. For n_a particles in state $|a\rangle$ and n_b particles in state $|b\rangle$, the total contribution to the energy is $U_{ab}n_an_b$. We substitute n_a with the number operator for particles in state $|a\rangle$, $\hat{N}_a = \hat{c}_a^\dagger\hat{c}_a$ (no summation in a); we obtain the most general form of $\hat{U}$ for pairwise interaction,

$$\hat{U} = \frac{1}{2}\sum_{ab} U_{ab}\hat{c}_a^\dagger\hat{c}_a\hat{c}_b^\dagger\hat{c}_b. \tag{15.72}$$

The term $\frac{1}{2}$ is introduced so that the terms U_{ab} and U_{ba} do not contribute twice.

In Eq. (15.72), we do not use the Einstein summation condition, and for this reason we have lowered the index of $\hat{c}$. We can rewrite $\hat{U}$ so that the summation condition is preserved,

$$\hat{U} = W^{ab}{}_{cd}\hat{c}_a^\dagger\hat{c}^c\hat{c}_b^\dagger\hat{c}^d, \tag{15.73}$$

where $W^{ab}_{cd} = \frac{1}{2}U_{ab}\delta_c^a\delta_d^b$.

Suppose that the particles interact through a potential, so that the interaction does not depend on spin. Then, U_{ab} can be written as

$$U_{mm'}(\mathbf{x},\mathbf{x}') = \delta_{mm'} U(\mathbf{x} - \mathbf{x}'), \tag{15.74}$$

with the basis correspondence $a \rightarrow (\mathbf{r},m)$ and $n \rightarrow (\mathbf{r}',m')$. The interaction term (15.72) is expressed in terms of the particle-density operator $\hat{v}(\mathbf{x})$ as

$$\hat{U} = \frac{1}{2}\int d^3x d^3x' \, V(\mathbf{x} - \mathbf{x}')\hat{v}(\mathbf{x})\hat{v}(\mathbf{x}'). \tag{15.75}$$

The full Hamiltonian for particles interacting through the potential is

$$\hat{H} = -\frac{1}{2M}\sum_m \int d^3x \hat{\psi}_m^\dagger(\mathbf{x})\nabla^2\hat{\psi}_m(\mathbf{x}) + \frac{1}{2}\int d^3x d^3y \, U(\mathbf{x},\mathbf{y})\hat{v}(\mathbf{x})\hat{v}(\mathbf{y}). \tag{15.76}$$

The Hamiltonian (15.76) is not a local functional of the fields, because it involves a double integral with respect to position. The field–particle duality is not fully implemented in nonrelativistic quantum mechanics, because particle interactions involve nonlocal terms in the Hamiltonian.

In contrast, the locality of the Hamiltonian is crucial in relativistic physics, because nonlocal Hamiltonians typically lead to faster-than-light transmission of information. Since quantum fields are indispensable in relativistic quantum theory, the issue is often raised whether fields or particles are more fundamental. In one viewpoint, fields are nothing but mathematical objects introduced in order to construct interactions compatible with relativistic symmetries and unitarity. In another, fields are fundamental, and particles are just field excitations. Interestingly, both approaches work equally well in special relativity. One can start from particles and introduce fields as derivative concepts, and one can start from fields and derive particles from the eigenstates of the field Hamiltonian. The equivalence between these approaches is guaranteed by the kinematical symmetry of relativistic quantum systems, the Poincaré symmetry that is described in Chapter 16. We need to go beyond special relativity, and consider gravitational phenomena, to see a difference. At the moment, it seems that treating fields as fundamental and particles as derivative is better suited for describing quantum effects in the presence of strong gravitational fields.

15.4.3 Chemical Reactions

The Fock space formalism allows us to describe processes in which particles of one type disappear and particles of another type appear. Let us consider, for example, a chemical reaction of the form $A + C \to D$, in which one particle of type A and one particle of type C interact and give a particle of type D. By conservation of angular momentum, the particles A, C, and D cannot all be fermions. We will assume that C and D are fermions with the same spin and that A is a boson with zero spin.

We write the annihilation and creation operators: of A as $\hat{a}^a$ and $\hat{a}_a^\dagger$, respectively; of C as $\hat{c}^a$ and $\hat{c}_a^\dagger$, respectively; and of D as $\hat{d}^a$ and $\hat{d}_a^\dagger$, respectively. Obviously, $\hat{c}_a$, $\hat{d}_a$ and their conjugates are fermionic operators, while $\hat{a}_a$ and their conjugates are bosonic operators.

This reaction would occur if time evolution involves a term that destroys a term that destroys one A particle and one C particle, while creating a D particle. This can be achieved by a term $\hat{c}^a \hat{a}^b \hat{d}_c^\dagger$ in the Hamiltonian. Self-adjointness of the Hamiltonian implies that the term $\hat{c}_a^\dagger \hat{a}_b^\dagger \hat{d}_c$ must also be present. Hence, the most general term that describes this reaction is of the form

$$\hat{V} = K_{abc}\hat{c}^a \hat{a}^b \hat{d}_c^\dagger + \bar{K}_{ab}{}^c \hat{c}_a^\dagger \hat{a}_b^\dagger \hat{d}^c, \tag{15.77}$$

where $\bar{K}_{ab}{}^c$ stands for the complex conjugate of K_{abc}.

If particle interactions are of small range, we can use the *ultralocal approximation*, according to which, the term $\hat{V}$ is a local function of the field operator. Let $\hat{\psi}_m(\mathbf{x})$ be the field operator for C-type particles, $\hat{\chi}_m(\mathbf{x})$ the field operators for D-type particles, and $\hat{\phi}(\mathbf{x})$ the field operators for A-type particles.

In the ultralocal approximation, the interaction term is of the form

$$\hat{V} = g \sum_m \int d^3x \left[\hat{\psi}_m^\dagger(\mathbf{x})\hat{\phi}^\dagger(\mathbf{x})\hat{\chi}_m(\mathbf{x}) + \hat{\chi}_m^\dagger(\mathbf{x})\hat{\psi}_m(\mathbf{x})\hat{\phi}(\mathbf{x}) \right], \tag{15.78}$$

where g is the coupling constant of the interaction. This approximation applies to many problems, ranging from the description of Bose–Einstein condensates to the (approximate) description of weak interactions in nuclear and high-energy physics.

15.4.4 Relativistic Scalar Field

We conclude this chapter with a simple example of a relativistic quantum field. We construct the field theory associated to free relativistic particles with spin $s = 0$, and then we prove the spin-statistics theorem for these particles.

For $s = 0$, the momentum kets $|\mathbf{p}\rangle$ define a generalized basis on the single-particle Hilbert space. Hence, we write the annihilation and creation operators as $\hat{c}(\mathbf{p})$ and $\hat{c}^\dagger(\mathbf{p})$, respectively. We will not yet specify whether these operators are fermionic or bosonic.

The key difference in the relativistic case is that energy is not a polynomial function of momentum. Rather, it involves a square root, $\epsilon_\mathbf{p} = \sqrt{\mathbf{p}^2 + m^2}$. This means that the definition (15.68) for the field operators does not work. We cannot write the Hamiltonian $\hat{\sigma}(\sqrt{\hat{\mathbf{p}}^2 + m^2})$ as a local function of the fields.

However, since the square of the energy is a polynomial, we can modify the definition of the fields by the insertion of a factor $\sqrt{\epsilon_\mathbf{p}}$ into the momentum integral,

$$\hat{\psi}(\mathbf{x}, t) = \int \frac{d^3p}{(2\pi)^{3/2}\sqrt{\epsilon_\mathbf{p}}} \hat{a}(\mathbf{p}) e^{i\mathbf{p}\cdot\mathbf{x} - i\epsilon_\mathbf{p} t},$$

$$\hat{\psi}^\dagger(\mathbf{x}, t) = \int \frac{d^3p}{(2\pi)^{3/2}\sqrt{\epsilon_\mathbf{p}}} \hat{a}^\dagger(\mathbf{p}) e^{-i\mathbf{p}\cdot\mathbf{x} + i\epsilon_\mathbf{p} t}. \tag{15.79}$$

These fields satisfy the *Klein–Gordon* equation,

$$\frac{\partial^2}{\partial t^2}\hat{\psi}(\mathbf{x}, t) - \nabla^2\hat{\psi}(\mathbf{x}, t) = 0. \tag{15.80}$$

We can also write the Hamiltonian as a local function of the fields,

$$\hat{H} = \int d^3x \left[(\nabla\hat{\psi}^\dagger) \cdot (\nabla\hat{\psi}) + m^2\hat{\psi}^\dagger\hat{\psi} \right]. \tag{15.81}$$

We pay a price: The fields $\hat{\psi}(\mathbf{x}, t)$ and $\hat{\psi}^\dagger(\mathbf{x}, t)$ no longer satisfy natural (anti)commutation relations at equal times. We evaluate the commutator for bosons, and the anticommutator for fermions, to obtain

$$[\hat{\psi}(\mathbf{x}, t), \hat{\psi}^\dagger(\mathbf{x}, t)]_\pm = \Delta_0(\mathbf{x} - \mathbf{x}'), \tag{15.82}$$

where

$$\Delta_0(\mathbf{x}) = \int \frac{d^3k}{(2\pi)^3\epsilon_\mathbf{k}} e^{i\mathbf{k}\cdot\mathbf{x}} = \int \frac{d^3k}{(2\pi)^3\epsilon_\mathbf{k}} \cos(\mathbf{k}\cdot\mathbf{x}), \tag{15.83}$$

is an even function of $\mathbf{x}$. $\Delta_0(\mathbf{x})$ peaks around zero with width of order m^{-1} (see Fig. 15.2) but it is not a delta function.

Nonetheless, we can find meaningful commutation relations for bosons. We define the new fields

$$\hat{\phi}(\mathbf{x}, t) := \frac{1}{\sqrt{2}}[\hat{\psi}(\mathbf{x}, t) + \hat{\psi}^\dagger(\mathbf{x}, t)] = \int \frac{d^3p}{(2\pi)^3\sqrt{2\epsilon_\mathbf{p}}} \left[\hat{a}(\mathbf{p}) e^{i\mathbf{p}\cdot\mathbf{x} - i\epsilon_\mathbf{p} t} + \hat{a}^\dagger(\mathbf{p}) e^{-i\mathbf{p}\cdot\mathbf{x} + i\epsilon_\mathbf{p} t} \right]$$

$$\hat{\pi}(\mathbf{x}, t) := \frac{d}{dt}\hat{\phi}(\mathbf{x}, t), \tag{15.84}$$

Fig. 15.2 The function $\Delta_0(r)/m$ of Eq. (15.83) plotted as a function of mr.

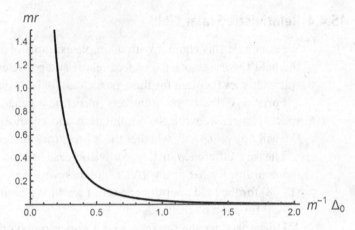

which satisfy

$$[\hat{\phi}(\mathbf{x},t),\hat{\phi}(\mathbf{x}',t)] = [\hat{\pi}(\mathbf{x},t),\hat{\pi}(\mathbf{x}',t)] = 0, \quad [\hat{\phi}(\mathbf{x},t),\hat{\pi}(\mathbf{x}',t)] = \delta^3(\mathbf{x}-\mathbf{x}').$$

The operators $\hat{\phi}(\mathbf{x},t)$ and $\hat{\pi}(\mathbf{x},t)$ define a representation of the infinite-dimensional Heisenberg–Weyl group.[7] Analogous anti-commutation relations are not possible for fermions. For example, if the creation and annihilation operators in Eq. (15.84) are fermionic, then

$$\left[\hat{\phi}(\mathbf{x},t),\hat{\phi}(\mathbf{x}',t)\right]_+ = \Delta_0(\mathbf{x}-\mathbf{x}'). \tag{15.85}$$

This strongly suggests that spin-zero particles cannot be fermions. The reason is the following. We expect that self-adjoint field operators localized at different spatial points at the same time commute, so that they can be measured simultaneously. If they do not commute, then one field measurement will affect the other, even if they are spacelike separated. This requires a faster-than-light signal.

Suppose that $\phi(\mathbf{x},t)$ is quantized with fermionic statistics and that the anti-commutation relation $[\hat{\phi}(\mathbf{x},t),\hat{\phi}(\mathbf{x}',t)]_+ = 0$ is valid. Since the field operators anti-commute, they cannot commute. This implies that *a fermion field operator cannot be a physical observable*. This is not a problem; we show that interesting physical observables like energy density are constructed from operators that are quadratic to the fields.

Consider, for example, the quadratic operator $\hat{\phi}^2(\mathbf{x},t)$. By Eq. (15.62), the commutator $[\hat{\phi}^2(\mathbf{x},t),\hat{\phi}^2(\mathbf{x}',t)]$ can be expressed as a sum of four terms, each of which contains the *anticommutator* $[\hat{\phi}(\mathbf{x},t),\hat{\phi}(\mathbf{x}',t)]_+$. Hence, if the latter vanishes, then the commutator $[\hat{\phi}^2(\mathbf{x},t),\hat{\phi}^2(\mathbf{x}',t)]$ also vanishes, and the operators $\hat{\phi}^2(\mathbf{x},t)$ are acceptable physical observables.

However, if $[\hat{\phi}^2(\mathbf{x},t),\hat{\phi}^2(\mathbf{x}',t)] \neq 0$, quadratic operators are not acceptable physical observables. The theory cannot describe consistently the most important physical quantities. For this reason, Eq. (15.85) implies that fermionic quantization is not appropriate for particles with spin $s = 0$. Assuming that no other alternative is possible, spin-zero particles can only be bosons. Thus, we proved the spin-statistics theorem for $s = 0$.

---

[7] To be precise, the generators of the Heisenberg–Weyl group H_∞ are the smeared fields $\hat{\phi}(f) := \int d^3x\,\hat{\phi}(\mathbf{x},0)f(\mathbf{x})$ and $\hat{\pi}(f) := \int d^3x\,\hat{\pi}(\mathbf{x},0)g(\mathbf{x})$, where f and g are functions that vanish rapidly at infinity. Smearing is necessary for the proper mathematical definition of the field operators.

QUESTIONS

15.1 Describe the atoms in a world where electrons have spin $s = 0$. Then, describe the atoms in a world where electrons have spin $s = \frac{3}{2}$.

15.2 A physicist proposes a variation of Pauli's principle according to which at most two electrons can be found in one state. How would you disprove it?

15.3 Which values of the relative angular momentum are excluded in a system of two particles with spin $s = 0$?

15.4 The eigenvalues of the Hamiltonian of a central potential that models an atom in the mean-field approximation are of the form $\epsilon_{n,\ell} = -E_0(n + a_n\ell)^2$, where $E_0 > 0$, and $0 < a_n < 1$. Identify values for a_n that guarantee that the Aufbau principle holds up to the orbital $6s$.

15.5 **(B)** Use Eq. (15.40), in order to argue that $\mathcal{F}_B(\mathcal{H}_1 \oplus \mathcal{H}_2) = \mathcal{F}_B(\mathcal{H}_1) \otimes \mathcal{F}_B(\mathcal{H}_2)$. What about $\mathcal{F}_F(\mathcal{H}_1 \oplus \mathcal{H}_2)$?

15.6 **(B)** What is the general form of an interaction term $\hat{V}$ that describes reactions of the form $A \to B + C + D$?

15.7 **(B)** Do you find the demonstration of the spin-statistics theorem for $s = 0$ in Section 15.4.4 convincing? If not, what do you think is missing?

PROBLEMS

15.1 Consider a systems of two particles in one dimension. (i) Show that the operator $(\hat{A} \otimes \hat{I} - \hat{I} \otimes \hat{A})^2$ is compatible with the exchange symmetry, and write its expectation values on the states $|a, b\rangle, |a, b\rangle_S, |a, b\rangle_A$. (ii) Evaluate these expectation values if $\hat{A}$ is the position operator and $|a\rangle, |b\rangle$ are, respectively, the ground state and first excited state of a harmonic oscillator of frequency ω.

15.2 Find the five lowest energy states and their degeneracy in a system of four oscillators of equal frequency ω, (i) if the oscillators are fermionic, and (ii) if the oscillators are bosonic.

15.3 Twenty-eight noninteracting electrons are bound by a central potential with energy levels $\epsilon_{n,\ell} = n^2 - 2\ell$. Evaluate the ground-state energy of the system.

15.4 Find the energy of the ground state and of the first excited state for six electrons in a ring of radius a.

Comment This is a crude model for the six outer electrons in the benzene molecule C_6H_6. The electrons are delocalized, in the sense that, in the molecular ground state, they are spread over all six carbon atoms.

15.5 In the simplest mean-field model for the nuclei, protons and neutrons move in an effective potential $V(r) = -V_0 + \frac{1}{2}m\omega^2 r^2$, where $V_0 > 0$. The frequency ω is the same for both protons and neutrons and it decreases with the mass number A as $\omega = \omega_0 A^{-1/3}$, for some constant ω_0. Ignoring the effects of the electromagnetic repulsion between protons, evaluate the ground-state energy per nucleon E_0/A for the isotopes of argon with $A = 36, 37, \ldots, 42$. Which isotope has the smallest energy per particle E_0/A? Why? Which other nuclei do you expect to correspond to local minima of E_0/A?

15.6 Find the ground-state energy for a free fermion gas of N particles in one and in two spatial dimensions.

15.7 (i) Show that for high energies the density of states for an isotropic harmonic oscillator of frequency ω in three dimensions is $g(\epsilon) = \frac{\epsilon^2}{2\omega^3}$. (ii) Find the ground-state energy of N such oscillators of spin $\frac{1}{2}$.

15.8 Consider an atom with Z electrons in an approximation where the repulsion between electrons is ignored. Show that, in the limit of large Z, the ground state energy is $E_0 = -(12)^{1/3} (2ma_0^2)^{-1} Z^{7/3}$.

15.9 **(B)** Confirm that, in the fermionic Fock space, $[\hat{c}(\bar{\xi})]^{\dagger} = \hat{c}^{\dagger}(\xi)$.

15.10 **(B)** This problem explores the properties of the map $\exp^{\otimes}$ of Eq. (15.47). (i) Show that, if $\hat{u}$ is unitary, then $\exp^{\otimes}(\hat{u})$ is unitary and that if $\hat{q}$ is a projector, then $\exp^{\otimes}(\hat{q})$ is a projector. (ii) Show that $\exp^{\otimes}(\hat{u})\hat{a}^{\dagger}(\xi)\exp^{\otimes}(\hat{u}^{\dagger}) = \hat{a}^{\dagger}(\hat{u}\xi)$, and that the same holds for fermionic annihilation operators. (iii) Show that

$$\exp^{\otimes}(e^{-i\hat{f}t}) = e^{-i\hat{\sigma}(\hat{f})t}. \tag{15.86}$$

15.11 **(B)** On the bosonic Fock space $\mathcal{F}_B(\mathcal{H})$, we define the (unnormalized) coherent states by $|z\rangle := \sum_{n=0}^{\infty} \frac{\hat{a}^{\dagger}(z)^n}{n!}|0\rangle$, where $z \in \mathcal{H}_B$. (i) Show that $\langle w|z\rangle = \exp(\bar{w}_a z^a)$. (ii) Show that $\hat{a}^a|z\rangle = z^a|z\rangle$. (iii) Show that $\exp^{\otimes}(\hat{u})|z\rangle = |\hat{u}z\rangle$ for any operator $\hat{u}$ on $\mathcal{F}_B(\mathcal{H})$.

15.12 **(B)** Show that the linear transformation $\hat{c}^a \rightarrow \hat{d}^a = A^a{}_b \hat{c}^b + B^{ab} \hat{c}_b$ for fermions preserves the canonical anticommutation relations if $AA^{\dagger} + BB^{\dagger} = I$ and $AB^T + BA^T = 0$.

15.13 **(B)** Identify fermionic creation and annihilation operators in a two-qubit system.

15.14 **(B)** Evaluate the expectation values $\langle\Psi|\hat{v}(\mathbf{x}, t)|\Psi\rangle$ and $\langle\Psi|\hat{v}(\mathbf{x}, t)\hat{v}(\mathbf{x}', t')|\Psi\rangle$ for free particles of spin zero. Take as $|\Psi\rangle$, (i) the vacuum, and (ii) a general one-particle state.

15.15 **(B)** (i) Show that the Hamiltonian (15.81) can be expressed as

$$\hat{H} = \frac{1}{2} \int d^3x \left[\hat{\pi}^2 + (\nabla\hat{\phi})^2 + m^2\hat{\phi}^2 \right]$$

modulo a constant infinite term $E_0 = \frac{1}{2} \int d^3 p\, \varepsilon_p$. (ii) Show that the Klein–Gordon equation for $\hat{\phi}$ arises from the Heisenberg equations of motion.

15.16 **(B)** (i) Show that, for the bosonic relativistic particle, $[\hat{\psi}(\mathbf{x}, t), \psi^{\dagger}(\mathbf{x}, t')] = \Delta(\mathbf{x} - \mathbf{x}', t - t')$, where

$$\Delta(\mathbf{x}, t) = \int \frac{d^3k}{(2\pi)^3 \epsilon_{\mathbf{k}}} e^{i\mathbf{k}\cdot\mathbf{x} - i\epsilon_{\mathbf{k}}t},$$

reduces to $\Delta_0(\mathbf{x})$, Eq. (15.83) for $t = 0$. (ii) Show that $\Delta(\mathbf{x}, t)$ depends on $\mathbf{x}$ only through $r = |\mathbf{x}|$ and that

$$\Delta(r, t) = \frac{1}{2\pi^2 r} \int_0^{\infty} \frac{k\, dk}{\sqrt{k^2 + m^2}} \sin(kr) e^{-i\sqrt{k^2 + m^2}t}.$$

(iii) Show that $\Delta_0(r) = \frac{m}{2\pi^2 r} K_1(mr)$, where K_1 is the modified Bessel function. Find the asymptotic form of $\Delta_0(r)$ as $r \rightarrow \infty$. (iv) Show that for $m = 0$,

$$\Delta(r, t) = -\frac{1}{2\pi^2} \lim_{\epsilon \to 0^+} \frac{1}{(t - i\epsilon)^2 - r^2}. \tag{15.87}$$

Bibliography

- For a history of Pauli's exclusion principle, see Pauli's Nobel lecture (Pauli, 1945) and Straumann (2004). For the spin-statistics theorem, see Duck and Sudarshan (1998). For the stability of matter, see Lieb and Seiringer (2010).
- For further properties of Fock spaces, see Berezin (1966).
- For the implications of field–particle duality, and the debate on whether particles or fields are more fundamental entities, see chapter 10 of Anastopoulos (2008), and Kuhlmann (2013, 2020).

16 Relativistic Systems

Within the regime of special relativity we may distinguish two different types of information in the description of an instrument. On the one hand there is the intrinsic structure of the apparatus, given essentially by a workshop drawing. On the other hand, we have to specify where and how this equipment is to be placed within some established space-time reference system in the laboratory. We can put its center of mass in different positions, orient its axes, trigger the apparatus at a time of our choice, let it rest in the laboratory or move it with a constant velocity. There are altogether 10 parameters which specify the "placement". They correspond to the 10 parameters of the Poincaré group. Poincaré invariance of the laws of nature means that the result of a complete experiment does not depend on the placement.

Haag (1996)

16.1 The Poincaré Group

The fundamental kinematical symmetry is the invariance under transformations between inertial reference frames. In the regime of small velocities, this symmetry corresponds to the Galilei group. However, the Galilei symmetry is only approximate. The exact symmetry, in the absence of gravity, is defined by the Poincaré group, which we analyse here.

16.1.1 Definitions and Key Properties

Any physical event is characterized by one time coordinate t and three Cartesian spatial coordinates x^1, x^2, x^3. In relativity, time is treated as a fourth coordinate $x^0 = t$ (in units where the speed of light is unity), and an event corresponds to a four-vector $x = (x^0, x^1, x^2, x^3)$ of *Minkowski spacetime* $M = \mathbb{R}^4$. In what follows, we will write four-vectors such as x as x^μ, where the Greek indices μ take values $0, 1, 2, 3$. We will use Latin indices i, j, k, and so on, with values $1, 2, 3$ for spatial variables.

Let x and x' describe two events, and let $\delta x^\mu := x^\mu - x'^\mu$. The fundamental principle of relativity asserts that the laws of physics are invariant under coordinate transformations that leave invariant the quantity

$$(\Delta x^0)^2 - (\Delta x^1)^2 - (\Delta x^2)^2 - (\Delta x^3)^2 = \Delta x^T \eta \Delta x = \eta_{\mu\nu} \Delta x^\mu \Delta x^\nu, \qquad (16.1)$$

where in the last formula we use Einstein's summation convention.

The matrix $\eta = \text{diag}\{1, -1, -1, -1\}$ in Eq. (16.1) is the *Minkowski metric*. We use this in order to raise and lower indices. To this end, we first express the inverse matrix η^{-1} with upper indices as $\eta^{\mu\nu}$. Then, we define $A_\mu := \eta_{\mu\nu} A^\nu$ for any four-vector A^μ. Conversely, we can raise indices by $A^\mu = \eta^{\mu\nu} A_\nu$. In this convention, $A_0 = A^0$ and $A^i = -A_i$. Furthermore, $A_i B^i = -\sum_i A_i B_i = -\mathbf{A} \cdot \mathbf{B}$.

The most general transformation that leaves (16.1) invariant is of the form

$$x \rightarrow \Lambda x + c, \tag{16.2}$$

where $c \in \mathbb{R}^4$ and Λ is a 4×4 matrix that satisfies $\Lambda^T \eta \Lambda = \eta$. Matrices Λ of this form are called *Lorentz transformations*. By definition, $\Lambda^T \eta \Lambda = \eta$, that is, Λ is an element of the indefinite orthogonal group $O(3,1)$. We will restrict ourselves to matrices $\Lambda \in SO(3,1)$, that is, to matrices with determinant equal to one. Matrices of the form $\text{diag}\{-1,1,1,1\}$ that implement time reversal, and matrices of the form $\text{diag}\{1, -1, -1, -1\}$ that implement space inversion, will be dealt with separately.

Using Greek indices, we will express Lorentz transformations as $x^\mu \rightarrow \Lambda^\mu{}_\nu x^\nu$. The inverse Lorentz transformation $\Lambda^{-1} = \eta \Lambda^T \eta^{-1}$ reads $(\Lambda^{-1})^\mu{}_\nu = \Lambda_\nu{}^\mu$.

The scalar product $A \cdot B := \eta_{\mu\nu} A^\mu B^\nu$ of two four-vectors is invariant under Lorentz transformations. This product is not a genuine inner product, because $A \cdot A$ is not always positive. A four-vector A^μ is called *timelike* if $A \cdot A > 0$, *spacelike* if $A \cdot A < 0$, and *null* if $A \cdot A = 0$. We will be writing A^2 for $A \cdot A$, if the context is unambiguous, for example, if the exponent 2 cannot be confused with a coordinate.

A path $x^\mu(\lambda)$ in Minkowski spacetime with an everywhere timelike tangent vector $\dot{x}^\mu(\lambda)$ is called timelike. We can always parameterize a timelike path by the *proper time* $\tau = \int d\lambda \sqrt{\eta_{\mu\nu} \dot{x}^\mu \dot{x}^\nu}$; hence, we can always express the paths as $x^\mu(\tau)$.

In classical physics, particles follow timelike paths $x^\mu(\tau)$. The four-velocity u^μ of a path is the unit timelike four-vector $\dot{x}^\mu(\tau)$, where the dot denotes the derivative with respect to proper time. The four-velocity of a particle in inertial motion is constant, and it can always be expressed as $(\frac{1}{\sqrt{1-\mathbf{v}^2}}, \frac{\mathbf{v}}{\sqrt{1-\mathbf{v}^2}})$, where $\mathbf{v} = \frac{d\mathbf{x}}{dt} = \dot{\mathbf{x}}/\dot{x}^0$ is its ordinary velocity with respect to the $(t, \mathbf{x})$ coordinates. Any unit timelike vector ξ describes a family of inertial observers with four-velocity ξ^μ. Hence, it defines an inertial reference frame, to which we will refer as the ξ-frame.

The four-momentum $p^\mu := mu^\mu$ of a particle of mass m satisfies

$$p \cdot p = m^2. \tag{16.3}$$

The zeroth component p^0 coincides with the energy E and it must be positive, while $p_i = -p^i$ is the particle's momentum. The relativistic relation between energy and momentum is $E = \sqrt{\mathbf{p}^2 + m^2}$. In the limit $|\mathbf{p}| << m$, $E \simeq m + \frac{\mathbf{p}^2}{2m}$, that is, we obtain the nonrelativistic kinetic energy plus the mass m. Note that we use the term "mass" for the invariant quantity m of Eq. (16.3). Many authors refer to the same object as "rest mass."

The transformations (16.2) are defined by a pair $(\Lambda, c) \in SO(3,1) \times \mathbb{R}^4$, and they identify the Poincaré group as $ISO(3,1)$, the ten-dimensional inhomogeneous group associated to $SO(3,1)$.

Two successive transformations (Λ_1, c_1) and (Λ_2, c_2) act as $x \to \Lambda_2\Lambda_1 x + \Lambda_2 c_1 + c_2$. Hence, the multiplication law of the Poincaré group is

$$(\Lambda_2, c_2)(\Lambda_1, c_1) = (\Lambda_2\Lambda_1, \Lambda_2 c_1 + c_2). \tag{16.4}$$

The unit of the Poincaré group is $(I, 0)$, where I is the unit matrix and 0 is the zero vector. The inverse of (Λ, c) is $(\Lambda^{-1}, -\Lambda^{-1}c - c)$.

16.1.2 The Group $SO(3, 1)$

The most general matrix $\Lambda \in SO(3, 1)$ is of the form

$$\Lambda(u_1, u_2, u_3, \theta_1, \theta_2, \theta_3) = R_3(\theta_3)R_2(\theta_2)R_1(\theta_1)B_3(u_3)B_2(u_2)B_1(u_1), \tag{16.5}$$

where $\theta_i \in [0, 2\pi]$, $u_i \in \mathbb{R}$ and

$$B_1(u_1) = \begin{pmatrix} \cosh u_1 & \sinh u_1 & 0 & 0 \\ \sinh u_1 & \cosh u_1 & 0 & 0 \\ 0 & 0 & 1 & 0 \\ 0 & 0 & 0 & 1 \end{pmatrix}, B_2(u_2) = \begin{pmatrix} \cosh u_2 & 0 & \sinh u_2 & 0 \\ 1 & 0 & 0 & 0 \\ \sinh u_2 & 0 & \cosh u_2 & 0 \\ 0 & 0 & 0 & 1 \end{pmatrix},$$

$$B_3(u_3) = \begin{pmatrix} \cosh u_3 & 0 & 0 & \sinh u_3 \\ 0 & 1 & 0 & 0 \\ 0 & 0 & 1 & 0 \\ \sinh u_3 & 0 & 0 & \cosh u_3 \end{pmatrix}, R_1(\theta_1) = \begin{pmatrix} 1 & 0 & 0 & 0 \\ 0 & 1 & 0 & 0 \\ 0 & 0 & \cos\theta_1 & \sin\theta_1 \\ 0 & 0 & -\sin\theta_1 & \cos\theta_1 \end{pmatrix},$$

$$R_2(\theta_2) = \begin{pmatrix} 1 & 0 & 0 & 0 \\ 0 & \cos\theta_2 & 0 & -\sin\theta_2 \\ 0 & 0 & 1 & 0 \\ 0 & \sin\theta_2 & 0 & \cos\theta_2 \end{pmatrix}, R_3(\theta_3) = \begin{pmatrix} 1 & 0 & 0 & 0 \\ 0 & \cos\theta_3 & \sin\theta_3 & 0 \\ 0 & -\sin\theta_3 & \cos\theta_3 & 0 \\ 0 & 0 & 0 & 1 \end{pmatrix}.$$

Each matrix B_i corresponds to an $SO(1, 1)$ transformation in the x^0–x^i plane. These transformations are called *boosts*; they mix spatial and temporal coordinates and they describe a change to a reference frame that moves with constant velocity in relation to the original one. Each matrix R_i corresponds to an $SO(2)$ rotation around the axis x^i.

We denote by K_i the generators for the matrices B_i, and by J_i the generators for the matrices R_i. The generators are compactly expressed as

$$\mathbf{n} \cdot \mathbf{K} = \begin{pmatrix} 0 & n_1 & n_2 & n_3 \\ n_1 & 0 & 0 & 0 \\ n_2 & 0 & 0 & 0 \\ n_3 & 0 & 0 & 0 \end{pmatrix}, \quad \mathbf{m} \cdot \mathbf{J} = \begin{pmatrix} 0 & 0 & 0 & 0 \\ 0 & 0 & m_3 & -m_2 \\ 0 & -m_3 & 0 & m_1 \\ 0 & m_2 & -m_1 & 0 \end{pmatrix},$$

in terms of vectors $\mathbf{n} = (n_1, n_2, n_3)$ and $\mathbf{m} = (m_1, m_2, m_3)$.

The six matrices K_i and J_i define a basis on the Lie algebra $\mathfrak{so}(3,1)$. We straightforwardly calculate the commutation relations

$$[K_i, K_j] = \sum_k \epsilon_{ijk} J_k, \quad [J_i, J_j] = \sum_k \epsilon_{ijk} J_k, \quad [K_i, J_j] = \sum_k \epsilon_{ijk} K_k. \tag{16.6}$$

The matrices J_i define the subalgebra $\mathfrak{so}(3)$ of spatial rotations.

We can also parameterize the basis of $\mathfrak{so}(3,1)$ using an antisymmetric pair of spacetime indices $\mu\nu$. Any matrix $S \in \mathfrak{so}(3,1)$ satisfies the condition $S^T \eta + \eta S = 0$. Writing the elements of S explicitly as $S^\alpha{}_\beta$, this condition becomes

$$\eta_{\alpha\gamma} S^\gamma{}_\beta + \eta_{\beta\gamma} S^\gamma{}_\alpha = 0. \tag{16.7}$$

Equation (16.7) is solved by matrices of the form $S^\alpha{}_\beta = \epsilon_{\mu\nu}(\delta^\alpha{}_\nu \eta_{\mu\beta} - \delta^\alpha{}_\mu \eta_{\nu\beta})$ for some parameters $\epsilon^{\mu\nu}$. Hence, the matrices $M_{\mu\nu}$ with elements

$$(M_{\mu\nu})^\alpha{}_\beta = \delta^\alpha{}_\nu \eta_{\mu\beta} - \delta^\alpha{}_\mu \eta_{\nu\beta} \tag{16.8}$$

define a basis of $\mathfrak{so}(3,1)$. It is straightforward to show that

$$M_{0i} = -M^{0i} = K_i, \quad M_{ij} = \sum_k \epsilon_{ijk} J_k, \quad J_i = \frac{1}{2}\sum_{jk} \epsilon_{ijk} M_{jk}.$$

The commutation relations (16.6) become

$$[M_{\mu\nu}, M_{\rho\sigma}] = \eta_{\mu\rho} M_{\nu\sigma} + \eta_{\nu\sigma} M_{\mu\rho} - \eta_{\nu\rho} M_{\mu\sigma} - \eta_{\mu\sigma} M_{\nu\rho}. \tag{16.9}$$

To prove Eq. (16.9), we write the matrix product $(M_{\mu\nu} M_{\rho\sigma})^\alpha{}_\beta$ as $(M_{\mu\nu})^\alpha{}_\gamma (M_{\rho\sigma})^\gamma{}_\beta$, and we use Eq. (16.8).

16.1.3 The Lie Algebra $\mathfrak{iso}(3,1)$

Like the Galilei group, the Poincaré group can be expressed as a matrix group by identifying each element (Λ, c) with the 5×5 matrix

$$(\Lambda, c) = \begin{pmatrix} \Lambda^0{}_0 & \Lambda^0{}_1 & \Lambda^0{}_2 & \Lambda^0{}_3 & c^0 \\ \Lambda^1{}_0 & \Lambda^1{}_1 & \Lambda^1{}_2 & \Lambda^1{}_3 & c^1 \\ \Lambda^2{}_0 & \Lambda^2{}_1 & \Lambda^2{}_2 & \Lambda^2{}_3 & c^2 \\ \Lambda^3{}_0 & \Lambda^3{}_1 & \Lambda^3{}_2 & \Lambda^3{}_3 & c^3 \\ 0 & 0 & 0 & 0 & 1 \end{pmatrix}.$$

The generators $M^{\mu\nu}$ of the $\mathfrak{so}(3,1)$ subgroup were identified in Section 16.1.2. The generators of spacetime translations P^μ are given by

$$b^\mu P_\mu = \begin{pmatrix} 0 & 0 & 0 & 0 & b^0 \\ 0 & 0 & 0 & 0 & b^1 \\ 0 & 0 & 0 & 0 & b^2 \\ 0 & 0 & 0 & 0 & b^3 \\ 0 & 0 & 0 & 0 & 1 \end{pmatrix},$$

in terms of the four-vector $b^\mu = (b^0, b^1, b^2, b^3)$. Clearly,

$$[P_\mu, P_\nu] = 0. \tag{16.10}$$

It is also straightforward to show that $(\Lambda, 0)P^\mu(\Lambda^{-1}, 0) = \Lambda^\mu{}_\nu P^\nu$. For an infinitesimal Lorentz transformation $(\Lambda, 0) = I + \epsilon_{\mu\nu}M^{\mu\nu}$, we obtain

$$[M_{\mu\nu}, P_\rho] = \eta_{\mu\rho}P_\nu - \eta_{\nu\rho}P_\mu. \tag{16.11}$$

The Lie algebra of the Poincaré group is defined by Eqs. (16.9), (16.10), and (16.11).

Using spatial indices, we can express the Lie algebra of the Poincaré group in terms of three generators K_i of boosts, three generators J_i of spatial rotations, three generators P_i of space translations, and one generator $H := P_0$ of time translations. The corresponding commutation relations are given in the following table.

	J_j	K_j	P_j	H
J_i	$\sum_k \epsilon_{ijk} J_k$	$\sum_k \epsilon_{ijk} K_k$	$\sum_k \epsilon_{ijk} P_k$	0
K_i	$\sum_k \epsilon_{ijk} J_k$	$\sum_k \epsilon_{ijk} J_k$	$\delta_{ij} H$	P_i
P_i	$\sum_k \epsilon_{ijk} P_k$	$-\delta_{ij} H$	0	0
H	0	$-P_j$	0	0

16.2   The Group $SL(2, \mathbb{C})$

The group $SL(2, \mathbb{C})$ is the double cover of $SO(3, 1)$, and, for this reason, it plays a crucial role in the description of relativistic systems. It allows us to describe Lorentz transformations by 2×2 complex matrices, and it leads to the introduction of relativistic spinors.

Note that, in this section, Pauli matrices are unhatted, because they represent geometric objects in spacetime and not quantum observables.

16.2.1   Properties of $SL(2, \mathbb{C})$

In Section 6.1, we saw that all the most general 2×2 self-adjoint matrix A can be expressed as $A = a_0 I + \mathbf{a} \cdot \boldsymbol{\sigma}$, in terms of the unit matrix I and the Pauli matrices σ_i. The coefficients $(a_0, \mathbf{a})$ define a four-vector. Conversely, we can map each four-vector ξ^μ to a self-adjoint 2×2 matrix

$$\tilde{\xi} := \xi^\mu \sigma_\mu = \begin{pmatrix} \xi^0 + \xi^3 & \xi^1 - i\xi^2 \\ \xi^1 + i\xi^2 & \xi^0 - \xi^3 \end{pmatrix}, \tag{16.12}$$

where we have defined the matrix four-vector $\sigma_\mu := (I, \sigma_1, \sigma_2, \sigma_3)$. We will refer to $\tilde{\xi}$ as the *spin matrix* associated to the four-vector ξ.

We also define the matrix four-vector $\bar{\sigma}_\mu := (I, -\sigma_1, -\sigma_2, -\sigma_3)$. Using Eq. (5.4), we find that $[\sigma_\mu, \bar{\sigma}_\nu]_+ = 2\eta_{\mu\nu} I$; hence,

$$\text{Tr}(\sigma_\mu \bar{\sigma}_\nu) = 2\eta_{\mu\nu}. \tag{16.13}$$

Equation (16.13) allows us to invert the map (16.12), and to express ξ^μ in terms of its spin matrix $\tilde{\xi}$,

$$\xi_\mu = \frac{1}{2}\mathrm{Tr}(\tilde{\xi}\bar{\sigma}_\mu). \tag{16.14}$$

We also define the inverse spin matrix

$$\underset{\sim}{\xi} := \xi^\mu \bar{\sigma}_\mu = \begin{pmatrix} \xi^0 - \xi^3 & -\xi^1 + i\xi^2 \\ -\xi^1 - i\xi^2 & \xi^0 + \xi^3 \end{pmatrix}. \tag{16.15}$$

We find that $\tilde{\xi}\underset{\sim}{\xi} = \underset{\sim}{\xi}\tilde{\xi} = \xi^\mu \xi^\nu \sigma_\mu \bar{\sigma}_\nu = \xi^\mu \xi^\nu \eta_{\mu\nu} I$. If ξ^μ is a unit timelike vector, then $\underset{\sim}{\xi} = (\tilde{\xi})^{-1}$.
Equation (16.12) implies the crucial identity

$$\det \tilde{\xi} = (\xi^0)^2 - (\xi^1)^2 - (\xi^2)^2 - (\xi^3)^2 = \xi^\mu \xi_\mu. \tag{16.16}$$

The eigenvalues of the spin matrix $\tilde{\xi}$ are $\xi^0 \pm |\xi|$. If $\xi^0 \geq 0$, then $\tilde{\xi}$ is a positive matrix, for all timelike vectors ξ^μ. The converse also holds.

Consider the transformation $\tilde{\xi} \to \tilde{\xi}' = \alpha\tilde{\xi}\alpha^\dagger$, where $\alpha \in SL(2,\mathbb{C})$. We use Eq. (16.14) in order to specify the four-vector ξ'^μ that corresponds to $\tilde{\xi}'$,

$$\xi'^\mu = \frac{1}{2}\mathrm{Tr}(\tilde{\xi}'\bar{\sigma}^\mu) = \frac{1}{2}\mathrm{Tr}(\alpha\tilde{\xi}\alpha^\dagger\bar{\sigma}^\mu) = \frac{1}{2}\mathrm{Tr}(\alpha\sigma_\nu\alpha^\dagger\bar{\sigma}^\mu)\xi^\nu = \Lambda^\mu{}_\nu(\alpha)\xi^\nu, \tag{16.17}$$

where

$$\Lambda^\mu{}_\nu(\alpha) := \frac{1}{2}\mathrm{Tr}(\alpha\sigma_\nu\alpha^\dagger\bar{\sigma}^\mu). \tag{16.18}$$

Since $\det \tilde{\xi}' = \det \tilde{\xi} |\det \alpha|^2 = \det \tilde{\xi}$, Eq. (16.16) implies that $\xi'^\mu \xi'_\mu = \xi^\mu \xi_\mu$, that is, $\Lambda^\mu{}_\nu(\alpha) \in SO(3,1)$.

Hence, Eq. (16.18) defines a function from $SL(2,\mathbb{C})$ to $O(3,1)$ that maps α to $\Lambda(\alpha)$. This function is two-to-one, because the matrices α and $-\alpha$ are mapped to the same element $\Lambda(\alpha)$. It follows that $SL(2,\mathbb{C})$ is the double cover of $SO(3,1)$, in a similar way that $SU(2)$ is the double cover of $SO(3)$. We will be using the notation $\alpha \cdot \xi$ for the vector $\Lambda^\mu{}_\nu(\alpha)\xi^\nu$.

A natural basis on the Lie algebra $\mathfrak{sl}(2,\mathbb{C})$ consists of the six matrices

$$K_i = \frac{1}{2}\sigma_i, \quad J_i = \frac{i}{2}\sigma_i, \quad i = 1,2,3, \tag{16.19}$$

which satisfy the relations (16.6). Note that all matrices of the form $e^{\mathbf{n}\cdot\mathbf{K}}$ for real vectors $\mathbf{n}$ are positive. We will refer to such matrices as *pure boosts*.

The antisymmetric matrix $\epsilon = i\sigma_2$ deserves a symbol of its own, because it satisfies $\epsilon\alpha\epsilon^{-1} = \alpha^{-1}$, for all $\alpha \in SL(2,\mathbb{C})$. As a result $\epsilon\sigma_i\epsilon^{-1} = -\sigma_i$, and $\epsilon\tilde{\xi}\epsilon^{-1} = \underset{\sim}{\xi}$ for all four-vectors ξ.

Example 16.1 The action of an infinitesimal boost $\alpha = I + \frac{1}{2}\lambda \cdot \sigma$ on $\tilde{\xi}$ yields $\tilde{\xi} + \delta\tilde{\xi}$, where

$$\delta\tilde{\xi} = \frac{1}{2}\left[\lambda \cdot \sigma(\xi^0 I + \xi \cdot \sigma) + (\xi^0 I + \xi \cdot \sigma)\lambda \cdot \sigma\right] = (\lambda \cdot \xi)I + \xi^0\lambda \cdot \sigma.$$

Hence, an infinitesimal boost changes ξ^0 by $\delta\xi^0 = \lambda \cdot \xi$ and ξ^i by $\delta\xi = \lambda\xi^0$.

The action of an infinitesimal rotation $\alpha = I - \frac{i}{2}\lambda \cdot \sigma$ on $\tilde{\xi}$ yields $\tilde{\xi} + \delta\tilde{\xi}$, where

$$\delta\tilde{\xi} = -\frac{i}{2}\left[\lambda \cdot \sigma(\xi^0 I + \xi \cdot \sigma) - (\xi^0 I + \xi \cdot \sigma)\lambda \cdot \sigma\right] = (\lambda \times \xi) \cdot \sigma.$$

Hence, for an infinitesimal rotation $\delta\xi^0 = 0$ and $\delta\xi = \lambda \times \xi$.

16.2.2 Boosts

Consider the reference four-vector $n = (1,0,0,0)$. Its associated spin matrix is the unit operator, $\tilde{n} = I$. Let ξ^μ be another unit timelike four-vector with $\xi^0 > 0$, and let $\Lambda(\xi)$ be a Lorentz matrix that transforms n to ξ: $\xi^\mu = \Lambda^\mu{}_\nu(\xi)n^\nu$. For α an $SL(2,\mathbb{C})$ matrix that corresponds to $\Lambda(\xi)$,

$$\alpha\alpha^\dagger = \tilde{\xi}. \tag{16.20}$$

This equation admits infinite solutions for α. We restrict ourselves to solutions that are pure boosts, that is, positive matrices, and we denote them by $b(\xi)$. Equation (16.20) becomes $b(\xi)^2 = \tilde{\xi}$, whence,

$$b(\xi) = \sqrt{\tilde{\xi}} = \frac{1}{\sqrt{2(\xi^0 + 1)}}\left[(\xi^0 + 1)I + \xi \cdot \sigma\right]. \tag{16.21}$$

The most general solution to Eq. (16.20) can be expressed as

$$\alpha = \sqrt{\tilde{\xi}}u, \tag{16.22}$$

where $u \in SU(2)$. We conclude that all elements of $SL(2,\mathbb{C})$ can be expressed as a product of a positive matrix that describes a boost and a unitary operator that describes rotations and leaves the reference vector invariant.

Example 16.2 The pure boosts generated by $K_3 = \frac{1}{2}\sigma_3$ define the family of matrices

$$b_3(u) = e^{\frac{u}{2}\sigma_3} = \begin{pmatrix} e^{u/2} & 0 \\ 0 & e^{-u/2} \end{pmatrix}. \tag{16.23}$$

The action of $b_3(u)$ on $\tilde{n} = I$ yields

$$\tilde{\xi} = b_3(u)b_3^\dagger(u) = \begin{pmatrix} e^u & 0 \\ 0 & e^{-u} \end{pmatrix},$$

By Eq. (16.14), $\tilde{\xi}$ is the spin-matrix of the four-velocity $\xi^\mu = (\cosh u, 0, 0, \sinh u)$ for a particle moving along the 3-axis.

Consider a particle A with four-velocity $\beta = (\sqrt{1 + \beta^2}, \beta, 0, 0)$ in the n-frame. We will derive the four-vector $\beta' = b_3(u) \cdot \beta$ that describes the particle's velocity in the ξ-frame, using spin matrices. The spin matrix of β is

$$\tilde{\beta} = \begin{pmatrix} \sqrt{1 + \beta^2} & \beta \\ \beta & \sqrt{1 + \beta^2} \end{pmatrix}.$$

We obtain the spin matrix of β' by acting the boosts $b_3(u)$ on $\tilde{\beta}$,

$$\tilde{\beta}' = b_3(u)\tilde{\beta}b_3^\dagger(u) = \begin{pmatrix} e^u\sqrt{1+\beta^2} & \beta \\ \beta & e^{-u}\sqrt{1+\beta^2} \end{pmatrix}.$$

We use Eq. (16.13) to obtain β',

$$\beta^\mu \to \beta'^\mu = \left(\cosh u\sqrt{1+\beta^2}, \beta, 0, \sinh u\sqrt{1+\beta^2} \right). \tag{16.24}$$

16.2.3 Spinors

By Eq. (16.16), a null four-vector k corresponds to a self-adjoint matrix $\tilde{k}$, with $\det \tilde{k} = 0$. This means that $\tilde{k}$ has a single nonzero eigenvalue. Therefore it can be written as

$$\tilde{k} = c(k)c^\dagger(k), \tag{16.25}$$

where $c(k) \in \mathbb{C}^2$ is a column vector, which we call an $SL(2, \mathbb{C})$ *spinor*.[1] Two spinors that differ by a phase correspond to the same null vector k. Equation (16.25) implies that the transformation $c(k) \to \alpha c(k)$ for $\alpha \in SL(2, \mathbb{C})$ corresponds to the Lorentz transformation $k^\mu \to \Lambda^\mu{}_\nu(\alpha)k^\nu$.

The most general spinor $c(k)$ that satisfies Eq. (16.25) is

$$c(k) = \frac{1}{\sqrt{k^0 + k^3}} \begin{pmatrix} k^0 + k^3 \\ k^1 + ik^2 \end{pmatrix}, \tag{16.26}$$

up to a phase. The choice of the phase in the spinor (16.26) is conventional. This convention is well defined for all null vectors except for $(k, 0, 0, -k)$.

Let us denote by $d(k)$ the eigenvector for the zero eigenvalue of $\tilde{k}$, that is, the solution to the equation $\tilde{k}d(k) = 0$. This spinor is orthogonal to $c(k)$, that is, it satisfies $d^\dagger(k)c(k) = 0$. It is straightforward to show that $d(k) = \epsilon c^*(k)$. Since $\underline{k} = \epsilon \tilde{k}\epsilon^{-1}$, Eq. (16.25) implies that $\underline{k} = c'(k)c'^\dagger(k)$, where $c'(k) = \epsilon c(k)$. Let $d'(k)$ be the solution to the zero eigenvalue equation $\underline{k}d'(k) = 0$. Then $d'(k) = \epsilon c'^*(k) = -c^*(k)$.

There is an analogue of the decomposition (16.22) for null vectors. Every $\alpha \in SL(2, \mathbb{C})$ can be expressed as a product $\omega(k)\chi$, where $\omega(k)$ is a "null boost" defined by a null vector k, and χ is an element of a subgroup G_0 of $SL(2, \mathbb{C})$, to be determined shortly. To derive this decomposition, we choose a reference null four-vector $l = \kappa(1, 0, 0, 1)$, where κ is an arbitrary constant with the dimension of mass. By Eq. (16.26), the corresponding spinor is $c(l) = \sqrt{\kappa} \begin{pmatrix} 1 \\ 0 \end{pmatrix}$.

The group G_0 is specified by all matrices that leave $c(l)$ invariant up to a phase. It is straightforward to identify its elements as

$$\chi(z, \phi) = \begin{pmatrix} e^{i\phi} & z \\ 0 & e^{-i\phi} \end{pmatrix}, \tag{16.27}$$

where $z \in \mathbb{C}$ and ϕ a phase. These matrices form an E_2 subgroup of $SL(2, \mathbb{C})$, hence, $G_0 = E_2$.

---

[1] Spinors are not quantum states of some qubit. They are fundamentally classical objects. For example, spinors describe the four-momentum of *classical* massless particles.

We define $\omega(k)$ as the lower triangular solution to the equation $\omega(k)c(l) = c(k)$. This solution is unique,

$$\omega(k) = \begin{pmatrix} \sqrt{\frac{k^0+k^3}{\kappa}} & 0 \\ \frac{k^1+ik^2}{\sqrt{\kappa(k^0+k^3)}} & \sqrt{\frac{\kappa}{k^0+k^3}} \end{pmatrix}. \tag{16.28}$$

We will refer to the matrices (16.28) as *null boosts*. The end result is that any $\alpha \in SL(2,\mathbb{C})$ can be expressed as the product $\omega(k)\chi(z,\phi)$, where the null boost $\omega(k)$ is given by (16.28) and the E_2 transformation $\chi(z,\phi)$ is given by Eq. (16.27).

16.2.4 Wigner Rotations

The product of two self-adjoint operators is not, in general, a self-adjoint operator. Hence, the product $b(\xi_2)b(\xi_1)$ of two pure boosts $b(\xi_1)$ and $b(\xi_2)$ is not a pure boost. The two boosts take the vector n to $b(\xi_2) \cdot \xi_1$, and, by Eq. (16.22),

$$b(\xi_2)b(\xi_1) = b[b(\xi_2) \cdot \xi_1]u_W(\xi_1, \xi_2), \tag{16.29}$$

where $u_W(\xi_1, \xi_2)$ is a unitary operator that is called a *Wigner rotation*.

An alternative characterization of the Wigner rotation is the following. Let us set $\xi_2 = \eta$, and $\xi_1 = v \cdot \xi$ for some unitary v. By Eq. (16.21), $b(\xi_1) = vb(\xi)v^\dagger$. Hence, $b(\xi_2)b(\xi_1) = b(\eta)vb(\xi)v^\dagger$. As η and v vary over all possible values, their product varies over all elements of $SL(2,\mathbb{C})$. Hence, we can substitute $b(\eta)v$ with a generic element $\alpha \in SL(2,\mathbb{C})$. Then, $b(\xi_2) \cdot \xi_1 = \alpha \cdot \xi$, and Eq. (16.29) becomes

$$w(\alpha, \xi) = b(\alpha \cdot \xi)^{-1}\alpha b(\xi), \tag{16.30}$$

where we redefined the Wigner rotation as $w(\alpha, \xi) := u_W(\xi_1, \xi_2)v$. Equation (16.30) defines the Wigner rotation for a general element of $SL(2,\mathbb{C})$ (and not only for pure boosts) with respect to reference frame defined by the unit null vector ξ. Note that if α is unitary, then $w(\alpha, \xi) = \alpha$.

Equation (16.30) implies that for any $\alpha_1, \alpha_2 \in SL(2,\mathbb{C})$,

$$w(\alpha_2, \alpha_1 \cdot \xi)w(\alpha_1, \xi) = w(\alpha_2\alpha_1, \xi). \tag{16.31}$$

The physical interpretation of the Wigner rotation will have to wait until Section 16.4. There, we will show that the $w(\alpha, \xi)$ is the rotation of the spin vector of a particle with four-velocity ξ under the action of the $SL(2,\mathbb{C})$ transformation α.

Example 16.3 We evaluate the Wigner rotation for an infinitesimal boost $\alpha = I + \frac{1}{2}\lambda \cdot \sigma$. As shown in Example 16.1, $\alpha \cdot \xi = \xi + \delta\xi$, where $\delta\xi = (\lambda \cdot \xi, \lambda\xi^0)$. We straightforwardly calculate $b(\alpha \cdot \xi) = b(\xi) + \delta b(\xi)$, where

$$\delta b(\xi) = -\frac{\lambda \cdot \xi}{2(1+\xi^0)}b(\xi) + \frac{1}{\sqrt{2(1+\xi^0)}}\left(\lambda \cdot \xi I + \xi^0\lambda \cdot \sigma\right).$$

Then $b(\alpha \cdot \xi)^{-1} = b(\xi)^{-1} - b(\xi)^{-1}\delta b(\xi)b(\xi)^{-1}$, and we obtain $w(\alpha, \xi) = I + \frac{1}{2}\lambda \cdot \sigma - b(\xi)^{-1}\delta b(\xi)$. A straightforward substitution yields

$$w(\alpha, \xi) = I - \frac{i\xi^0}{2(1+\xi^0)}(\lambda \times \xi) \cdot \sigma.$$

Example 16.4 We can also define Wigner rotations with respect to null vectors k, using the null boosts of Eq. (16.28),

$$w(\alpha, k) := \omega^{-1}(\alpha \cdot k)\alpha\omega(k), \tag{16.32}$$

for all $\alpha \in SL(2, \mathbb{C}$. Strictly speaking, the matrices $w(\alpha, k)$ are not rotations, that is, elements of $SU(2)$; they belong in the E_2 subgroup with matrices of the form (16.27).

To evaluate $w(\alpha, k)$, we first express the null boost $\omega(k)$ as $\begin{pmatrix} x & 0 \\ z & x^{-1} \end{pmatrix}$ with $x \in \mathbb{R}$ and $z \in \mathbb{C}$.

The corresponding spinor $c(k)$ is $\begin{pmatrix} x \\ z \end{pmatrix}$. Hence, for $\alpha = \begin{pmatrix} a & b \\ c & d \end{pmatrix} \in SL(2, \mathbb{C})$, with $ad - bc = 1$,

$$\alpha c(k) = \begin{pmatrix} ax + bz \\ cx + dz \end{pmatrix} = \begin{pmatrix} |ax + bz| \\ (cx + dz)^{-i\phi(\alpha,k)} \end{pmatrix} e^{i\phi(\alpha,k)}, \tag{16.33}$$

where in the second step $\alpha c(k)$, we brought the spinor in the form (16.26) modulo the phase $e^{i\phi(\alpha,k)} := \frac{ax+bz}{|ax+bz|}$. Substituting x and z from Eq. (16.28), we obtain

$$\phi(\alpha, k) = \arg\left[a + b\frac{k^1 + ik^2}{k^0 + k^3} \right]. \tag{16.34}$$

The pure boost $\omega(\alpha \cdot k)$ associated to $\alpha c(k)$ is $\begin{pmatrix} |ax + bz| & 0 \\ (cx + dz)^{-i\phi} & |ax + bz|^{-1} \end{pmatrix}$. Substituting into Eq. (16.32), we obtain the Wigner rotation

$$w(\alpha, k) = \begin{pmatrix} |ax + bz|^{-1} & 0 \\ -(cx + dz)e^{-i\phi} & |ax + bz| \end{pmatrix} \begin{pmatrix} a & b \\ c & d \end{pmatrix} \begin{pmatrix} x & 0 \\ z & \frac{1}{x} \end{pmatrix}$$
$$= \begin{pmatrix} e^{i\phi} & \frac{b}{x|ax+bz|} \\ 0 & e^{-i\phi} \end{pmatrix}. \tag{16.35}$$

16.3 The Casimirs of the Poincaré Group

Consider a unitary representation $\hat{U}(\Lambda, c)$ of the Poincaré group on a Hilbert space $\mathcal{H}$. The associated generators $\hat{M}_{\mu\nu}$ and $\hat{P}_\mu$ satisfy the commutation relations

$$[\hat{M}_{\mu\nu}, \hat{M}_{\rho\sigma}] = i(\eta_{\mu\rho}\hat{M}_{\nu\sigma} + \eta_{\nu\sigma}\hat{M}_{\mu\rho} - \eta_{\nu\rho}\hat{M}_{\mu\sigma} - \eta_{\mu\sigma}\hat{M}_{\nu\rho}), \tag{16.36}$$
$$[\hat{M}_{\mu\nu}, \hat{P}_\rho] = i(\eta_{\mu\rho}\hat{P}_\nu - \eta_{\nu\rho}\hat{P}_\mu), \tag{16.37}$$
$$[\hat{P}_\mu, \hat{P}_\nu] = 0. \tag{16.38}$$

The Poincaré group has two Casimirs. The first Casimir is quadratic to the generators, and is known as the *mass operator*,

$$\hat{C}_2 := \eta^{\mu\nu}\hat{P}_\mu\hat{P}_\nu. \tag{16.39}$$

$\hat{C}_2$ commutes trivially with $\hat{P}_\mu$. Furthermore,

$$[\hat{C}_2, \hat{M}_{\mu\nu}] = 2\eta^{\rho\sigma}[\hat{P}_\rho, \hat{M}_{\mu\nu}]\hat{P}_\sigma = -2i\eta^{\rho\sigma}(\eta_{\mu\rho}\hat{P}_\nu - \eta_{\nu\rho}\hat{P}_\mu)\hat{P}_\sigma = 2i(\hat{P}_\mu\hat{P}_\nu - \hat{P}_\nu\hat{P}_\mu) = 0.$$

For a single particle of mass m, $\hat{P}_\mu$ is the particle's four-momentum, and $\hat{C}_2 = m^2\hat{I}$. However, $\hat{C}_2$ is not constant in systems of multiple particles. For example, in a system of two identical particles with masses m, the particle's four-momenta $\hat{p}_1$ and $\hat{p}_2$ satisfy $\hat{p}_1 \cdot \hat{p}_1 = \hat{p}_2 \cdot \hat{p}_2 = m^2\hat{I}$. The total four-momentum is $\hat{P}^\mu = \hat{p}_1^\mu + \hat{p}_2^\mu$. Then, the mass operator $\hat{C}_2 = 2m^2\hat{I} + 2\hat{p}_1 \cdot \hat{p}_2$ is momentum dependent. Since Casimirs are constant in irreducible representations, we conclude that *irreducible representations of the Poincaré group correspond to particles*.

The second Casimir of the Poincaré group is defined in terms of the *Pauli–Lubanski vector*

$$\hat{W}_\mu := \frac{1}{2}\epsilon_{\mu\nu\rho\sigma}\hat{M}^{\nu\rho}\hat{P}^\sigma, \tag{16.40}$$

where $\epsilon_{\mu\nu\rho\sigma}$ is the totally antisymmetric symbol in four dimensions:

$$\epsilon_{\mu\nu\rho\sigma} = \begin{cases} 1 & \text{if } \mu\nu\rho\sigma \text{ is an even permutation of } 0123 \\ -1 & \text{if } \mu\nu\rho\sigma \text{ is an odd permutation of } 0123 \\ 0 & \text{if any index appears twice.} \end{cases} \tag{16.41}$$

By definition, $\epsilon_{0ijk} = \epsilon_{ijk}$, where ϵ_{ijk} is the totally antisymmetric symbol in three dimensions, defined by Eq. (5.5). The following identity is a straightforward generalization of Eq. (5.6),

$$\epsilon_{\mu\nu\rho\sigma}\epsilon^{\mu\alpha\beta\gamma} = \delta_\nu^\alpha\delta_\rho^\beta\delta_\sigma^\gamma + \delta_\sigma^\alpha\delta_\nu^\beta\delta_\rho^\gamma + \delta_\rho^\alpha\delta_\sigma^\beta\delta_\nu^\gamma - \delta_\rho^\alpha\delta_\nu^\beta\delta_\sigma^\gamma - \delta_\sigma^\alpha\delta_\rho^\beta\delta_\nu^\gamma - \delta_\nu^\alpha\delta_\sigma^\beta\delta_\rho^\gamma.$$

The components of the Pauli–Lubanski vector are

$$\hat{W}_0 = \frac{1}{2}\epsilon_{0ijk}\hat{M}^{ij}\hat{P}^k = \hat{J}_k\hat{P}^k = -\hat{\mathbf{J}} \cdot \hat{\mathbf{P}}, \tag{16.42}$$

$$\hat{W}_i = \frac{1}{2}(\epsilon_{i0jk}\hat{M}^{0j}\hat{P}^k + \epsilon_{ij0k}\hat{M}^{j0}\hat{P}^k + \epsilon_{ijk0}\hat{M}^{jk}\hat{P}^0)$$

$$= -\epsilon_{ijk}\hat{M}^{0j}\hat{P}^k + \frac{1}{2}\epsilon_{ijk}\hat{M}^{jk}\hat{P}_0 = \hat{H}\hat{J}_i - \epsilon_{ijk}\hat{K}^j\hat{P}^k. \tag{16.43}$$

Equation (16.40) implies that $\hat{W}_\mu\hat{P}^\mu = 0$. We readily calculate

$$[\hat{P}_\mu, \hat{W}_\nu] = \frac{1}{2}\epsilon_{\nu\rho\sigma\tau}[\hat{P}_\mu, \hat{M}^{\rho\sigma}]\hat{P}^\tau = \frac{i}{2}\epsilon_{\nu\rho\sigma\tau}(\delta_\mu^\rho\hat{P}^\sigma - \delta_\mu^\sigma\hat{P}^\rho)\hat{P}^\tau = 0. \tag{16.44}$$

The commutation relation

$$[\hat{M}_{\mu\nu}, \hat{W}_\rho] = i(\eta_{\nu\rho}\hat{W}_\mu - \eta_{\mu\rho}\hat{W}_\nu) \tag{16.45}$$

is also straightforward, even if the proof is somewhat more lengthy.

The second Casimir of the Poincaré group is

$$\hat{C}_4 = \hat{W}_\mu\hat{W}^\mu. \tag{16.46}$$

Equation (16.44) implies that $[\hat{W}_\rho, \hat{P}_\mu] = 0$. We also find that

$$
\begin{aligned}
[\hat{C}_4, \hat{M}_{\mu\nu}] &= \hat{W}^\rho[\hat{W}_\rho, \hat{M}_{\mu\nu}] + [\hat{W}_\rho, \hat{M}_{\mu\nu}]\hat{W}^\rho \\
&= i\hat{W}^\rho(\eta_{\mu\rho}\hat{W}_\nu - \eta_{\nu\rho}\hat{W}_\mu + i(\eta_{\mu\rho}\hat{W}_\nu - \eta_{\nu\rho}\hat{W}_\mu)\hat{W}^\rho = 0.
\end{aligned}
\tag{16.47}
$$

To interpret the Casimir $\hat{C}_4$, we consider a particle of mass m in its rest frame. Then, $\mathbf{P} = 0$, whence $\hat{W}_0 = 0$ and $\hat{W}_i = m\hat{J}_i$. It follows that $\hat{C}_4 = -\hat{\mathbf{W}} \cdot \hat{\mathbf{W}} = -m^2\hat{\mathbf{J}}^2$. In the particle's rest frame, the only contribution to angular momentum comes from spin. Hence, $\hat{\mathbf{J}}^2 = s(s+1)\hat{I}$, where s is the particle's spin. It follows that $\hat{C}_4 = -m^2 s(s+1)\hat{I}$. Since $\hat{C}_4$ is a scalar, its value is the same in all reference frames. The constancy of $\hat{C}_4$ in an irreducible representation implies the constancy of s. Hence, we conclude the following.

> Unitary irreducible representations of the Poincaré group correspond to relativistic particles with constant rest mass m and spin s.

The full classification of the unitary irreducible representations of the Poincaré group was achieved by Wigner (1939). Each representation corresponds to fixed values for mass m and spin s. There are five classes of representations: (i) with $m^2 > 0$ and discrete spin, (ii) with $m^2 = 0$ and discrete spin, (iii) with $m^2 < 0$, (iv) with $m^2 = 0$ and continuous spin, and (v) with $m^2 = 0$ and trivial space translations. Only the first two classes correspond to particles observed in nature.

16.4   Massive Representations of the Poincaré Group

In this section, we construct the massive representations of the Poincaré group, or rather of its double cover, that is obtained by the substitution of $SO(3,1)$ with $SL(2,\mathbb{C})$. We will abusingly use the name "Poincaré group" also for this double cover, and we will denote its elements by (α, c), where $\alpha \in SL(2,\mathbb{C})$ and $c \in \mathbb{R}^4$.

16.4.1   Construction of the Representation

In a massive representation, the four-momentum $\hat{P}^\mu$ is timelike. Its generalized eigenvalues are of the form $m\xi^\mu$, where ξ^μ is a unit timelike vector.

The space V of all unit timelike vectors splits as $V_+ \cup V_-$, where V_+ describes particles with positive energy ($\xi^0 > 0$) and V_- describes particles with negative energy ($\xi^0 < 0$). Here, we consider only positive-energy representations.

First, we define the Hilbert space $\mathcal{H}_{m,0} = L^2[V_+, d\mu_+(\xi)]$, where

$$
d\mu_+(\xi) = m^2 d^4\xi\,\delta(\xi^\mu\xi_\mu - 1) = m^2 d^4\xi\,\delta[(\xi^0)^2 - \boldsymbol{\xi}^2 - 1] = m^2\frac{d^3\xi}{2\sqrt{1 + \boldsymbol{\xi}^2}}
$$

is an integration measure on V_+ that remains invariant under Lorentz transformations. The Hilbert space $\mathcal{H}_{m,0}$ carries a unitary irreducible representation of the Poincaré group for particles with zero spin,

$$\hat{U}(\alpha, c)\psi(\xi) = e^{-imc\cdot\xi}\psi(\alpha^{-1} \cdot \xi). \tag{16.48}$$

To incorporate spin, we consider the unitary irreducible representations of $SU(2)$ on $\mathbb{C}^{2s+1}$. Each element of $u \in SU(2)$ is represented by a unitary matrix $\mathcal{U}_{\sigma\sigma'}^{(s)}(u)$ on $\mathbb{C}^{2s+1}$. We write the corresponding generators as $(S_i)_{\sigma\sigma'}^i$. The indices σ, σ' refer to the eigenvalues of $\hat{S}_3$, and they take values $-s, -s+1, \ldots, s-1, s$.

The key point is to identify the $SU(2)$ group of spin with the $SU(2)$ subset of $SL(2, \mathbb{C})$. To this end, we define the Hilbert space $\mathcal{H}_{m,s} = \mathcal{H}_{m,0} \otimes \mathbb{C}^{2s+1}$. Then, we represent each element (α, c) of the Poincaré group by the unitary operator $\hat{U}(\alpha, c)$,

$$\hat{U}(\alpha, c)\psi_\sigma(\xi) = e^{-imc\cdot\xi}\sum_{\sigma'}\mathcal{U}_{\sigma\sigma'}^{(s)}[w(\alpha, \alpha^{-1} \cdot \xi)]\psi_{\sigma'}(\alpha^{-1} \cdot \xi), \tag{16.49}$$

expressed in terms of the Wigner rotation $w(\alpha, \alpha^{-1} \cdot \xi)$. Using the identity (16.31), it is straightforward to show that Eq. (16.49) satisfies the Poincaré group law, Eq. (16.4). Furthermore, if α is a unitary matrix u, $\psi_{\sigma'}$ is rotated by $\mathcal{U}_{\sigma\sigma'}^{(s)}(u)$.

The derivation of the generators is straightforward, if tedious,

$$\hat{H}\psi_\sigma(\xi) = m\xi^0\psi_\sigma(\xi), \tag{16.50}$$

$$\hat{\mathbf{P}}\psi_\sigma(\xi) = m\boldsymbol{\xi}\psi_\sigma(\xi), \tag{16.51}$$

$$\hat{\mathbf{J}}\psi_\sigma(\xi) = -i\boldsymbol{\xi} \times \frac{\partial}{\partial\boldsymbol{\xi}}\psi_\sigma(\xi) + \sum_{\sigma'}\mathbf{S}_{\sigma\sigma'}\psi_{\sigma'}(\xi), \tag{16.52}$$

$$\hat{\mathbf{K}}\psi_\sigma(\xi) = -i\left(\xi^0\frac{\partial}{\partial\boldsymbol{\xi}} - \boldsymbol{\xi}\frac{\partial}{\partial\xi^0}\right)\psi_\sigma(\xi) + \frac{1}{1+\xi^0}\boldsymbol{\xi} \times \sum_{\sigma'}\mathbf{S}_{\sigma\sigma'}\psi_{\sigma'}(\xi). \tag{16.53}$$

Equation (16.52) suggests that we identify $-i\boldsymbol{\xi} \times \frac{\partial}{\partial\boldsymbol{\xi}}$ with the orbital angular momentum $\hat{\mathbf{L}}$ so that the total angular momentum is the sum of $\hat{\mathbf{L}}$ and the spin operator $\hat{\mathbf{S}}$, defined by $\hat{\mathbf{S}}\psi_\sigma(\xi) = \sum_{\sigma'}\mathbf{S}_{\sigma\sigma'}\psi_\sigma(\xi)$. This split is meaningful if $\hat{\mathbf{L}} = \hat{\mathbf{X}} \times \hat{\mathbf{P}}$ for some position operator $\hat{\mathbf{X}}$. The operators

$$\hat{X}_i = \frac{1}{m}\left[i\frac{\partial}{\partial\xi^i} + f(\xi^0)\xi^i\right] \tag{16.54}$$

satisfy this property. They also satisfy the canonical commutation relations $[\hat{X}_i, \hat{P}^j] = i\delta_i^j$ and $[\hat{X}_i, \hat{X}_j] = 0$ for all functions f. They are therefore candidates for position operators. Unfortunately, such identifications turn out to be problematic. First, these operators do not transform like a position coordinate under boosts[2] – see Problem 16.5. More importantly, a theorem by Malament (1996) asserts that a position operator for nonrelativistic particles cannot exist, as its existence is incompatible with relativistic causality.

In the absence of a position operator, there is no unique split of $\hat{\mathbf{J}}$ into orbital angular momentum and spin. This has resulted into a large number of proposals for the relativistic spin operator. Here, we will take $\hat{\mathbf{S}}$ for the spin operator, solely for reasons of mathematical simplicity. By construction, $\hat{\mathbf{S}}$ is independent of ξ, so it commutes with $\hat{H}$, $\hat{\mathbf{P}}$ and $\hat{\mathbf{L}}$.

---

[2] In the representations of the Galilei group, the position operator is essentially the boost generator. Such an identification is not possible in relativistic systems, because the boost generators do not commute with each other.

The operator $\hat{\mathbf{R}} = (m\hat{I} + \hat{H})^{-1}\hat{\mathbf{P}} \times \hat{\mathbf{S}}$ generates boosts for spin, so we will refer to it as the *spin boost* operator. Note that $\hat{\mathbf{R}} \cdot \hat{\mathbf{S}} = \hat{\mathbf{R}} \cdot \hat{\mathbf{P}} = 0$, so $\hat{\mathbf{R}}$ is normal to the plane defined by the four momentum and the spin operator.

We also define the *helicity*

$$\hat{\boldsymbol{\Sigma}} := |\hat{\mathbf{P}}|^{-1}\hat{\mathbf{P}} \cdot \hat{\mathbf{S}}, \tag{16.55}$$

that describes the spin projection along the particle's momentum. Helicity, being a spin projection along an axis, has $2s + 1$ eigenvalues: $-s, -(s-1), \ldots, s-1, s$.

For many calculations in the Hilbert space $\mathcal{H}_{m,s}$, it is convenient to use the generalized basis $|p, \sigma\rangle$ on $\mathcal{H}_{m,s}$, labeled by a four-vector p and the spin projection σ. The kets $|p, \sigma\rangle$ are common generalized eigenstates of both $\hat{P}^\mu$ and $\hat{S}_3$. The Poincaré group acts on the kets as

$$\hat{U}(\alpha, c)|p, \sigma\rangle = e^{-ic\cdot p} \sum_{\sigma'} \mathcal{U}_{\sigma\sigma'}^{(s)}\left[w\left(\alpha, \frac{p}{m}\right)\right]|\alpha \cdot p, \sigma'\rangle. \tag{16.56}$$

Equation (16.56) provides the physical interpretation of Wigner rotations. They record the rotation of a spin vector under a Lorentz transformation.

16.4.2 Time Reversal and Space Inversion

The time reversal and space inversion transformations are also defined for relativistic systems. They correspond to the transformations $\mathcal{T} : x^0 \to -x^0$ and $\mathcal{P} : \mathbf{x} \to -\mathbf{x}$, respectively, which leave invariant the quantity (16.1).

These transformations are implemented through an analysis similar to that of Section 14.4.3. Time reversal must be represented by an antiunitary operator $\hat{\mathbb{T}}$, in order to preserve the positivity of the Hamiltonian. For the same reason, space inversion must be represented by a unitary operator $\hat{\mathbb{P}}$. The two operators act on the generators of the Poincaré group, as

$$\hat{\mathbb{T}}(\hat{H}, \hat{\mathbf{P}}, \hat{\mathbf{K}}, \hat{\mathbf{J}})\hat{\mathbb{T}}^\dagger = (\hat{H}, -\hat{\mathbf{P}}, \hat{\mathbf{K}}, -\hat{\mathbf{J}}), \tag{16.57}$$

$$\hat{\mathbb{P}}(\hat{H}, \hat{\mathbf{P}}, \hat{\mathbf{K}}, \hat{\mathbf{J}})\hat{\mathbb{P}}^\dagger = (\hat{H}, -\hat{\mathbf{P}}, -\hat{\mathbf{K}}, \hat{\mathbf{J}}). \tag{16.58}$$

Since the transformations are identical to the corresponding transformations of the Galilei group, the analysis of Section 14.4.3 passes unchanged. Hence, the space inversion operator acts as

$$\hat{\mathbb{P}}\Psi_\sigma(\xi) = \varpi \Psi_\sigma(\mathcal{P}p), \tag{16.59}$$

where ϖ is the particle's intrinsic parity. The time reversal operator acts as

$$\hat{\mathbb{T}}|p, \sigma\rangle = \eta(-1)^{s-\sigma}\Psi_{-\sigma}^*(\mathcal{P}\xi), \tag{16.60}$$

where the phase η has no physical significance and it can be chosen equal to unity. As in the nonrelativistic case, $\hat{\mathbb{T}}^2 = (-1)^{2s}\hat{I}$.

16.5 Massless Representations of the Poincaré Group

In this section, we analyze massless ($m^2 = 0$) irreducible unitary representations of the Poincaré group. We show that they correspond to massless particles with spin and fixed values of helicity. We also show that space inversion can be implemented only in reducible representations of the Poincaré group, because it inverts helicity.

16.5.1 Construction of the Representation

In massless representations, the generalized eigenvalues of the four-momentum $\hat{P}^\mu$ are null vectors. Let N be the space of all nonzero null vectors k^μ. N splits as $N_+ \cup N_-$, where N_+ describes particles with positive energy ($k^0 > 0$), and N_- describes particles with negative energy. Here, we consider only the first case.

We define the Hilbert space $\mathcal{H}_{m=0} = L^2[N_+, d\mu_+(k)]$, where

$$d\mu_+(k) = d^4\xi\,\delta(k^\mu k_\mu) = d^4k\,\delta\left[\left(k^0\right)^2 - \mathbf{k}^2\right] = \frac{d^3k}{2|\mathbf{k}|} \tag{16.61}$$

is an integration measure on N_+ that remains invariant under Lorentz transformations. $\mathcal{H}_{m=0}$ carries a representation of the Poincaré group

$$\hat{U}(\alpha, c)\psi(k) = e^{-ic \cdot k}\psi(\alpha^{-1} \cdot k), \tag{16.62}$$

which describes massless particles with spin zero.

For nonzero spin, we first identify the unitary irreducible representation of the group $G_0 = E_2$, which describes the analogue of rotations for null vectors. The representations that correspond to known particles are one-dimensional. They map all $\chi(\phi, z) \in E_2$ to a phase $e^{ir\phi}$, where $r \in \mathbb{Z}$, that is, they are of the form

$$\hat{V}_r\left[\begin{pmatrix} e^{i\phi} & z \\ 0 & e^{-i\phi} \end{pmatrix}\right] := e^{ir\phi}. \tag{16.63}$$

This means that the information contained in the parameter z of Eq. (16.27) is lost.

Then, we define a unitary representation of the Poincaré group on $\mathcal{H}_{m=0}$ by

$$\hat{U}(\alpha, c)\psi(k) = e^{-ik \cdot c}e^{ir\phi(\alpha, \alpha^{-1} \cdot k)}\psi(\alpha^{-1} \cdot k), \tag{16.64}$$

where the phase $\phi(\alpha, k)$ is given by Eq. (16.34). We will refer to $\phi(\alpha, k)$ as the *Wigner phase* of the massless representations.

The associated generators are (Problem 16.6)

$$\hat{H}\psi(k) = k^0\psi(k) \tag{16.65}$$

$$\hat{\mathbf{P}}\psi(k) = \mathbf{k}\psi(k) \tag{16.66}$$

$$\hat{\mathbf{J}}\psi(k) = -i\mathbf{k} \times \frac{\partial}{\partial \mathbf{k}}\psi(k) + \mathbf{S}(k)\psi(k) \tag{16.67}$$

$$\hat{\mathbf{K}}\psi(k) = -i\left(k^0\frac{\partial}{\partial \mathbf{k}} - \mathbf{k}\frac{\partial}{\partial k^0}\right)\psi(k) + \mathbf{R}(k)\psi(k), \tag{16.68}$$

where the spin vector S_i and the spin-boost operator R_i are defined as

$$\mathbf{S} = \frac{r}{2}(\frac{k^1}{k^0 + k^3}, \frac{k^2}{k^0 + k^3}, 1), \quad \mathbf{R} = \frac{r}{2}(\frac{k^2}{k^0 + k^3}, -\frac{k^1}{k^0 + k^3}, 0). \tag{16.69}$$

The spin-boost vector is normal to both $\mathbf{S}$ and to $\mathbf{k}$. Spin is not an independent degree of freedom, but a function of momentum. Furthermore, the helicity operator $\hat{\Sigma} = |\hat{\mathbf{P}}|^{-1}\hat{\mathbf{P}} \cdot \mathbf{S}$ is constant, $\hat{\Sigma} = \frac{r}{2}\hat{I}$. Hence, each massless irreducible representation is characterized by the value of helicity, which is invariant under the Poincaré group.

It is conventional to redefine the meaning of terms, and to call $s := \frac{1}{2}|r|$ the spin of the representation, and to reserve *helicity* for the sign $\sigma = \pm$ of r. With this terminology, the spin takes integer and half-integer values, so that the spin-statistic theorem may be relevant, and we say that massless particles of spin $s > 0$ come in two helicities. We write the associated Hilbert spaces as $\mathcal{H}_{0,s,\sigma}$.

By Eqs. (16.42) and (16.43) we find the components of the Pauli–Lubanski vector:

$$\hat{W}_0\psi(k) = -\frac{r}{2}k_0\psi(k), \qquad \hat{W}_i\psi(k) = -\frac{r}{2}k_i\psi(k); \tag{16.70}$$

hence, $\hat{W}_\mu = -\frac{r}{2}\hat{P}_\mu$, and $\hat{C}_4 = \frac{r^2}{4}\hat{C}_2 = 0$.

16.5.2 Photon Polarization

For $s > 0$, we define the Hilbert space $\mathcal{H}_{0,s} = \mathcal{H}_{0,s,+} \oplus \mathcal{H}_{0,s,-}$ that carries a *reducible* representation of the Poincaré group, with both helicities σ for the same spin s. Hence, the elements of $\mathcal{H}_{0,s}$ are wave functions $\psi_\sigma(k)$, and there exists a generalized basis $|k,\sigma\rangle$ of eigenstates of $\hat{P}^\mu$, labeled by a null four-momentum k and the helicity σ. Then, we write Eq. (16.64) as

$$\hat{U}(\alpha,c)|p,\sigma\rangle = e^{2i\sigma s\phi(\alpha,k)-ik\cdot c}|\alpha \cdot k,\sigma\rangle. \tag{16.71}$$

The only massless particles we currently know are the photons (for $s = 1$), and it is an experimental fact that they can exist in superpositions of helicities, that is, we can describe photons of a given momentum by kets of the form $c_+|k,+\rangle + c_-|k,-\rangle$, with $|c_+|^2 + |c_-|^2 = 1$. According to the conventions of optics, a photon in this state is called (i) circularly polarized if either c_+ or c_- vanishes, (ii) linearly polarized if $|c_+| = |c_-|$, and (iii) elliptically polarized for any other value of c_+ and c_-.

A Lorentz transformation takes the state $c_1|k,+\rangle + c_2|k,-\rangle$ to $c_1 e^{2i\phi(\alpha,k)}|k,+\rangle + c_2 e^{-2i\phi(\alpha,k)}|k,-\rangle$, where $\phi(\alpha,k)$ is the Wigner phase. As an example, consider a photon with momentum given $(k,0,0,k)$. The $SU(2)$ matrix that describes a rotation of angle θ around the axis $\mathbf{n}$ is given by Eq. (11.52). The Wigner phase is $\phi = \arg(\cos\frac{\theta}{2} + in_3\sin\frac{\theta}{2})$. It vanishes for $n_3 = 0$, i.e., for the rotation axis in the 1–2 plane. For $\mathbf{n}$ along the 3-axis, $\phi = \frac{\theta}{2}$, and $c_1|k,+\rangle + c_2|k,-\rangle$ transforms to $c_1 e^{i\theta}|k,+\rangle + c_2 e^{-i\theta}|k,-\rangle$.

This means that the coefficients c_1 and c_2 transform under rotations like the complex coordinates $z = x_1 + ix_2$ and $z^* = x_1 - ix_2$ of a two-vector $\begin{pmatrix} x_1 \\ x_2 \end{pmatrix}$ on the 1–2 plane. This vector is the *polarization vector*, known from classical EM theory.

Example 16.5 Consider an inertial observer A and an observer B that moves with velocity v along direction 1 in relation to A. The boost

$$b_1(u) = e^{\frac{u}{2}\sigma_1} = \begin{pmatrix} \cosh\frac{u}{2} & \sinh\frac{u}{2} \\ \sinh\frac{u}{2} & \cosh\frac{u}{2} \end{pmatrix}, \tag{16.72}$$

where $u = \tanh v$, transforms from the rest frame of A to the rest frame of B.

The observer A prepares a photon state $\frac{1}{\sqrt{2}}(|k,+\rangle + |k,-\rangle)$, where $k^\mu = (k,0,k,0)$, for $k > 0$. The associated spinor $c(k)$ is $\sqrt{\frac{k}{2}}\begin{pmatrix} 1 \\ i \end{pmatrix}$. The boosted spinor

$$b_1(u)c(k) = \sqrt{\frac{k}{2}}\begin{pmatrix} \cosh\frac{u}{2} + i\sinh\frac{u}{2} \\ \sinh\frac{u}{2} + i\cosh\frac{u}{2} \end{pmatrix},$$

corresponds to the null vector $k'^\mu = c(k')^\dagger \sigma^\mu c(k') = (k\cosh u, k\sinh u, k, 0)$.

By Eq. (16.34), the Wigner phase for $b_1(u)$ is $\phi = \arg\left[\cosh\frac{u}{2} + i\sinh\frac{u}{2}\right]$, which implies that $\tan\phi = \tanh\frac{u}{2}$, or equivalently that $e^{2i\phi} = \frac{1+i\tanh\frac{u}{2}}{1-i\tanh\frac{u}{2}}$. Hence, the observer B describes the photon by the ket $\frac{1}{\sqrt{2}}\left(\frac{1+i\tanh\frac{u}{2}}{1-i\tanh\frac{u}{2}}|k',+\rangle + \frac{1-i\tanh\frac{u}{2}}{1+i\tanh\frac{u}{2}}|k',-\rangle\right)$.

16.5.3 Time Reversal and Space Inversion

The space inversion transformation $\hat{\mathbb{P}}$ inverts momentum but preserves angular momentum. In a massless particle, this implies that $\hat{\mathbb{P}}$ inverts helicity. This is impossible in an irreducible representation of the Poincaré group, where the helicity is constant, so we can only represent $\hat{\mathbb{P}}$ in reducible representations of the Poincaré group.

Space inversion can be represented in the Hilbert space $\mathcal{H}_{0,s} = \mathcal{H}_{0,s,+} \oplus \mathcal{H}_{0,s,-}$. Since $\hat{\mathbb{P}}$ inverts momentum, $\mathbf{k} \to \mathbf{k}' = -\mathbf{k}$. Then, the spin vector $\mathbf{S}$ transforms to $\mathbf{S}' = -\frac{r}{2}(\frac{-k^1}{k^0-k^3}, \frac{-k^2}{k^0-k^3}, 1)$, which is normal to $\mathbf{S}$. We straightforwardly verify that $\mathbf{k}\cdot\mathbf{R} = \mathbf{k}'\cdot\mathbf{R} = 0$ and that $\mathbf{S}'\cdot\mathbf{R} = \mathbf{S}\cdot\mathbf{R} = 0$, where $\mathbf{R}$ is the spin boost. This means that all vectors $\mathbf{k}, \mathbf{k}', \mathbf{S}$, and $\mathbf{S}'$ lie on the plane normal to $\mathbf{R}$.

The state $\hat{\mathbb{P}}|k,\sigma\rangle$ is not quite equal to $|\mathcal{P}k, -\sigma\rangle$, because the vector $\mathbf{S}'$ is not in the same direction as $\mathbf{S}$. We need to rotate $\mathbf{S}'$ back to $\mathbf{S}$ while leaving $\mathbf{k}$ invariant. This requires a rotation of angle $-\pi$ around the $\mathbf{k}$-axis if $k_3 > 0$, or of angle π if $k_3 < 0$; see Fig. 16.1.

Let u represent the matrix that implements the desired rotation. Then,

$$\hat{U}(u,0)\hat{\mathbb{P}}\psi_\sigma(k) = \varpi\,\psi_{-\sigma}(\mathcal{P}k),$$

where $\varpi = \pm 1$ is the particle's intrinsic parity. Hence,

$$\hat{\mathbb{P}}\psi_\sigma(k) = \varphi\hat{U}(u^{-1},0)\psi_{-\sigma}(\mathcal{P}k) = \varpi e^{-2is\sigma\phi(u^{-1},\mathcal{P}k)}\psi_{-\sigma}(\mathcal{P}k).$$

By definition, $u = \exp\left[-i\,\text{sgn}(k_3)\pi\frac{\mathbf{k}}{k^0}\cdot\frac{\sigma}{2}\right] = -i\,\text{sgn}(k_3)\frac{\mathbf{k}\cdot\sigma}{k^0}$. We straightforwardly evaluate the Wigner rotation $\phi(u^{-1},\mathcal{P}k) = \text{sgn}(k_3)\frac{\pi}{2}$. Hence, we obtain

$$\hat{\mathbb{P}}\psi_\sigma(k) = \varpi e^{-is\sigma\,\text{sgn}(k_3)\pi}\psi_{-\sigma}(\mathcal{P}k). \tag{16.73}$$

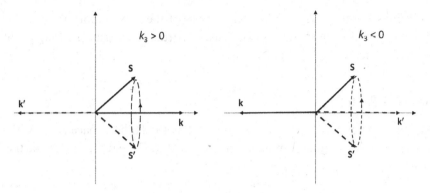

Fig. 16.1 The momentum vector **k** and the spin vector **S** transform to **k′** and **S′** under parity. All four vectors are in the same plane. We must rotate **S′** by π around the **k**-axis to return it to **S**, so that spin is well defined after parity. For $k_3 > 0$, the rotation angle is negative (right-handed rotation) and for $k_3 < 0$, it is positive (left-handed rotation).

The discontinuity due to $\mathrm{sgn}(k_3)$ is the result of our convention (16.26) for the parameterization of spinors, which has a singularity at $k_3 = -k^0$. A discontinuity is unavoidable because when space inversion is appended to the Poincaré group, the total group is not connected.

The time reversal operator $\hat{\mathbb{T}}$ on $|k,\sigma\rangle$ reverses momentum, but also spin, so it preserves helicity. Its action is calculated similarly. There is only a sign difference in the phase we calculated for $\hat{\mathbb{P}}$ that originates from the fact that helicity is preserved. The final result is

$$\hat{\mathbb{T}}\psi_\sigma(k) = \eta e^{is\sigma\,\mathrm{sgn}(k_3)\pi}\,\psi_\sigma^*(\mathcal{P}k), \tag{16.74}$$

where η is phase that can be absorbed in a redefinition of the phases of the kets $|k,\sigma\rangle$ – see Section 11.3.7. Again, we see that $\hat{\mathbb{T}}^2 = (-1)^{2s}\hat{I}$.

16.6   Relativistic Wave Equations

The irreducible representations of the Poincaré group provide a quantum description of relativistic particles with no physical or mathematical ambiguity. One can also describe relativistic particles with differential equations for wave functions in spacetime. The latter description came historically first, from the effort to generalize Schrödinger's equation to the relativistic regime. This effort was quite successful, as it led to *Dirac's equation* for electrons (Dirac, 1928) and to the prediction of antiparticles (Dirac, 1930b). Later, Bargmann and Wigner (1948) provided a general rule for constructing wave equations from the representations of the Poincaré group.

Nowadays, we know that the description of particles through relativistic wave equations is not sufficient as a foundation of a relativistic quantum theory. The reason is that we cannot associate all solutions of a wave equation to physically acceptable state vectors, because the solutions of relativistic wave equations come in pairs: there is one solution with energy $-E$ for each solution with energy $E > 0$. The associated Hamiltonian would have to be unbounded from below.

In this section, we derive and analyze some of the most important relativistic wave equations. Then, we will describe how they can be used in order to describe the coupling of charged particles to the EM field.

16.6.1 Spinless Particles

Massive particles with no spin are described by the Hilbert space $\mathcal{H}_{m,0}$, with element functions $\psi(\xi)$ of the dimensionless momentum $\xi^\mu = p^\mu/m$. For each ψ, we define the wave function

$$\phi(X) = \int d\mu_+(\xi)e^{-im\xi \cdot X}\psi(\xi), \tag{16.75}$$

where $X^\mu = (t,\mathbf{x})$ is a point of Minkowski spacetime. The spacetime translations $\psi(\xi) \to e^{-imc\cdot\xi}\psi(\xi)$ correspond to the transformation $\phi(X) \to \phi(X + a)$.

Since $\xi \cdot \xi = 1$, the wave function $\phi(X)$ satisfies the *Klein–Gordon equation*

$$\Box\phi(X) + m^2\phi(X) = 0, \tag{16.76}$$

where $\Box = \eta^{\mu\nu}\frac{\partial^2}{\partial X^\mu}\frac{\partial}{\partial X^\nu} = \frac{\partial^2}{\partial t^2} - \mathbf{\nabla}^2$ is the d'Alembert operator.

Wave functions of the form (16.75) are not the only solutions of Eq. (16.76). The complex conjugate $\psi^*(\xi)$ of $\psi(\xi)$ evolves in time as $\psi^*(\xi)e^{i\xi^0 t}$, and the corresponding wave function also solves Eq. (16.76). However, the generator for time translations of $\psi^*(\xi)$ equals $-\xi^0$, that is, the energy is negative. Hence, the Klein–Gordon equation admits one negative-energy solution for each positive-energy one. Such solutions are unphysical, hence, Eq. (16.76) cannot serve as the starting point for the dynamics of a relativistic quantum theory.

The most general solution to the Klein–Gordon equation is of the form

$$\phi(X) = \int d\mu_+(\xi)\left[e^{-im\xi \cdot X}\psi_1(\xi) + e^{im\xi \cdot X}\psi_2^*(\xi)\right], \tag{16.77}$$

where $\psi_1, \psi_2 \in \mathcal{H}_{m,0}$. For real $\phi(X)$, $\psi_1 = \psi_2$.

16.6.2 Massive Particles of Spin $\frac{1}{2}$

Massive particles of spin $s = \frac{1}{2}$ are described by elements of $\mathcal{H}_{m,\frac{1}{2}} = \mathbb{C}^2 \otimes \mathcal{H}_{m,0}$,

$$\psi(\xi) = \begin{pmatrix} \psi_{\frac{1}{2}}(\xi) \\ \psi_{-\frac{1}{2}}(\xi) \end{pmatrix}. \tag{16.78}$$

Since $\mathbb{C}^2$ carries the defining representation of $SU(2)$, Eq. (16.49) leads to a representation of the Lorentz group,

$$\hat{U}(\alpha)\psi(\xi) = b^{-1}(\xi)\alpha b(\alpha^{-1} \cdot \xi)\psi(\alpha^{-1} \cdot \xi), \tag{16.79}$$

in terms of the $SL(2,\mathbb{C})$ boosts $b(\xi) = \sqrt{\tilde{\xi}}$.

We construct a wave function associated to $\psi \in \mathcal{H}_{m,\frac{1}{2}}$ through the following procedure. First, we define the 4×4 *Dirac matrices*

$$\gamma_\mu := \begin{pmatrix} 0 & \sigma_\mu \\ \bar{\sigma}_\mu & 0 \end{pmatrix}. \tag{16.80}$$

Equation (16.13) implies that

$$[\gamma_\mu, \gamma_\nu]_+ = 2\eta_{\mu\nu}I. \tag{16.81}$$

We commonly write $\not{\xi} = \xi^\mu \gamma_\mu$ for all four-vectors ξ^μ. Then,

$$\not{\xi} = \begin{pmatrix} 0 & \tilde{\xi} \\ \xi & 0 \end{pmatrix}. \tag{16.82}$$

For any $\psi \in \mathcal{H}_{m,\frac{1}{2}}$, we define the *Dirac spinor*

$$\tilde{\Psi}(\xi) = \begin{pmatrix} b(\xi)\psi(\xi) \\ b^{-1}(\xi)\psi(\xi) \end{pmatrix} = \begin{pmatrix} \sqrt{\tilde{\xi}}\psi(\xi) \\ \sqrt{\xi}\psi(\xi) \end{pmatrix} \in \mathbb{C}^4 \otimes \mathcal{H}_{m,0}. \tag{16.83}$$

If we transform the upper copy of ψ with $\hat{U}(\alpha)$ and the lower copy with $\epsilon \hat{U}(\alpha)\epsilon^{-1}$, the Dirac spinor transforms as

$$\tilde{\Psi}(\xi) \rightarrow \begin{pmatrix} \alpha & 0 \\ 0 & \alpha^{-1} \end{pmatrix} \tilde{\Psi}(\alpha^{-1} \cdot \xi). \tag{16.84}$$

The spacetime translations $\psi(\xi) \rightarrow e^{-im\xi \cdot a}\psi(\xi)$ correspond to the transformation $\Psi(X) \rightarrow \Psi(X + a)$. Hence, Dirac spinors define a nonunitary, reducible representation of the Poincaré group.

Using Eq. (16.78), we can write the Dirac spinor as

$$\tilde{\Psi}(\xi) = \psi_{\frac{1}{2}}(\xi)u_+(\xi) + \psi_{-\frac{1}{2}}(\xi)u_-(\xi), \tag{16.85}$$

where

$$u_r(\xi) = \begin{pmatrix} \sqrt{\tilde{\xi}}\chi_r \\ \sqrt{\xi}\chi_r \end{pmatrix}, \text{ with } r = \pm, \text{ and } \chi_+ = \begin{pmatrix} 1 \\ 0 \end{pmatrix}, \chi_- = \begin{pmatrix} 0 \\ 1 \end{pmatrix}. \tag{16.86}$$

The action of $\not{\xi}$ on a Dirac spinor yields

$$\not{\xi}\tilde{\Psi}(\xi) = \begin{pmatrix} 0 & \tilde{\xi} \\ \xi & 0 \end{pmatrix} \begin{pmatrix} \sqrt{\tilde{\xi}}\psi(\xi) \\ \sqrt{\xi}\psi(\xi) \end{pmatrix} = \begin{pmatrix} \tilde{\xi}\sqrt{\xi}\psi(\xi) \\ \xi\sqrt{\tilde{\xi}}\psi(\xi) \end{pmatrix} = \begin{pmatrix} \sqrt{\tilde{\xi}}\psi(\xi) \\ \sqrt{\xi}\psi(\xi) \end{pmatrix} = \tilde{\Psi}(\xi).$$

Hence, the wave function

$$\Psi(X) := \int d\mu_+(\xi)e^{-im\xi \cdot X}\tilde{\Psi}(\xi) \tag{16.87}$$

satisfies *Dirac's equation*

$$i\gamma^\mu \frac{\partial}{\partial X^\mu}\Psi(X) - m\Psi(X) = 0. \tag{16.88}$$

Every $\psi \in \mathcal{H}_{m,\frac{1}{2}}$ defines a solution to Dirac's equation, but the converse does not hold. To see this, we associate a different Dirac spinor $\tilde{\Psi}(\xi)$ to $\psi(\xi)$,

$$\tilde{\Psi}'(\xi) = \begin{pmatrix} \sqrt{\xi}\,\psi^*(\xi) \\ \sqrt{\tilde{\xi}}\,\psi^*(\xi) \end{pmatrix}. \tag{16.89}$$

The spinor $\tilde{\Psi}'$ transforms the same way with $\tilde{\Psi}$ under the Lorentz transformations and it satisfies the Dirac equation. However, it has the opposite behavior with respect to spacetime translations: The transformation $\Psi(X) \to \Psi(X + a)$ is generated by a negative Hamiltonian. Hence, $\tilde{\Psi}'$ is a negative-energy solution. Using Eq. (16.78), we can write

$$\tilde{\Psi}'(\xi) = \psi^*_{\frac{1}{2}}(\xi)v_+(\xi) + \psi^*_{-\frac{1}{2}}(\xi)v_-(\xi), \tag{16.90}$$

where

$$v_r(\xi) = \begin{pmatrix} \sqrt{\xi}\,\chi_r \\ \sqrt{\tilde{\xi}}\,\chi_r \end{pmatrix}, \tag{16.91}$$

with $r = \pm$. The general solution to Dirac's equation is

$$\Psi(X) = \int d\mu_+(\xi) \left[e^{-im\xi \cdot X}\tilde{\Psi}_1(\xi) + e^{im\xi \cdot X}\tilde{\Psi}'_2(\xi) \right], \tag{16.92}$$

where $\tilde{\Psi}_1$ and $\tilde{\Psi}'_2$ are defined in terms of two separated vectors $\psi_1, \psi_2 \in \mathcal{H}_{m,\frac{1}{2}}$.

16.6.3 Massless Particles with Spin $\frac{1}{2}$

Consider massless particles with spin $s = \frac{1}{2}$ and positive helicity. Let $\psi_- \in \mathcal{H}_{0,\frac{1}{2},-}$. We define the *left-handed Weyl spinor*

$$\tilde{\psi}_L(k) = \omega^*(k) \begin{pmatrix} \psi_-(k) \\ 0 \end{pmatrix} = c^*(k)\psi_-(k), \tag{16.93}$$

where $\omega(k)$ is the null boost (16.28), and $c(k)$ is the spinor (16.26).

A Lorentz transformation $\hat{U}(\alpha, 0)$ acts as $\psi_-(k) \to \psi_-(\alpha^{-1} \cdot k)e^{-i\phi}$, where ϕ is the Wigner angle. Recall that the factor $e^{i\phi}$ is generated by the action of the null Wigner rotation $w(\alpha, k)$ on $\begin{pmatrix} 1 \\ 0 \end{pmatrix}$ – see Eq. (16.35). Then, using Eq. (16.32), we find that under a Lorentz transformation

$$\tilde{\psi}_L(k) \to \omega^*(k)w^*(\alpha, \alpha^{-1} \cdot k) \begin{pmatrix} \psi_-(\alpha^{-1} \cdot k) \\ 0 \end{pmatrix}$$

$$= \alpha^*\omega^*(\alpha^{-1} \cdot k) \begin{pmatrix} \psi_-(\alpha^{-1} \cdot k) \\ 0 \end{pmatrix} = \alpha^*\tilde{\psi}_L(\alpha^{-1} \cdot k),$$

where we used Eq. (16.32). Hence, a Lorentz transformation on ψ_- transforms the left-handed Weyl spinor under a nonunitary representation of $SL(2,\mathbb{C})$,

As seen in Section 16.2.3, $\underline{k}c^*(k) = 0$. Hence, $\underline{k}\tilde{\psi}_L(k) = 0$. Therefore, the left-handed wave function

$$\psi_L(X) = \int d\mu(k)e^{-ik\cdot X}\tilde{\psi}_L(k) \tag{16.94}$$

satisfies the left-handed Weyl equation

$$i\bar{\sigma}^\mu \frac{\partial}{\partial X^\mu}\psi_L(X) = 0. \tag{16.95}$$

It is straightforward to show that $\psi_L(X) \to \psi_L(X+a)$ under spacetime translations.

Equation (16.95) admits negative-energy solutions. They correspond to vectors $\psi_+ \in \mathcal{H}_{0,\frac{1}{2},+}$ that define left-handed spinors

$$\psi'_L(k) = \omega(k)\begin{pmatrix} \psi_+^*(k) \\ 0 \end{pmatrix}. \tag{16.96}$$

These spinors transform like $\psi_L(k)$ under Lorentz transformations, but inversely under space-time translations. Hence, the general solution of the left-handed Weyl equation is

$$\psi_L(X) = \int d\mu(k)\left[e^{-ik\cdot X}\tilde{\psi}_{L-}(k) + e^{ik\cdot X}\tilde{\psi}'_{L+}(k)\right], \tag{16.97}$$

where the Weyl spinors $\tilde{\psi}_{L+}(k)$ and $\tilde{\psi}'_{L-}(k)$ are defined with respect to one vector $\psi_- \in \mathcal{H}_{0,\frac{1}{2},-}$ and one vector $\psi_+ \in \mathcal{H}_{0,\frac{1}{2},+}$, respectively.

Similarly, we find that a vector $\psi_+ \in \mathcal{H}_{0,\frac{1}{2},+}$ defines a right-handed Weyl spinor $\psi_R(k) = d(k)\psi_+(k)$ that satisfies $\tilde{k}\tilde{\psi}_R(k) = 0$. The corresponding wave function $\psi_R(X) = \int d\mu(k)e^{-ik\cdot X}\tilde{\psi}_R(k)$ satisfies the right-handed Weyl equation

$$i\sigma^\mu \frac{\partial}{\partial X^\mu}\psi_R(X) = 0. \tag{16.98}$$

16.6.4 Coupling to External Fields

The description of relativistic particles in terms of the wave equation has the benefit of a direct analogy with the Schrödinger representation of nonrelativistic quantum theory. For the reasons we mentioned earlier, this analogy is mostly illusory. It is nonetheless useful. It allows us to derive the evolution equation for charged relativistic particles coupled to an external EM field, by appropriately generalizing the minimal substitution rule.

Recall that the minimal substitution for a particle of charge q consists in substituting its momentum $\hat{\mathbf{p}}$ with $\hat{\mathbf{p}} - q\mathbf{A}(\hat{\mathbf{x}},t)$, where $\mathbf{A}(\mathbf{x},t)$ is an external magnetic potential and its Hamiltonian with $\hat{H} - q\phi(\mathbf{x},t)$, where $\phi(\hat{\mathbf{x}},t)$ is an external electric potential. This is equivalent to substituting the spatial derivative ∇ with $\nabla - iq\mathbf{A}$ and the time derivative $\frac{\partial}{\partial t}$ with $\frac{\partial}{\partial t} + iq\phi$.

The minimal substitution applies to all three wave equations that we derived. For the Dirac equation, $i\gamma_0\frac{\partial\Psi}{\partial t} - (i\sum_i \gamma_i\nabla_i + m)\Psi = 0$, we obtain

$$i\frac{\partial}{\partial t}\Psi = -i\boldsymbol{\beta}\cdot(\nabla - iq\mathbf{A})\Psi + q\phi\Psi + m\gamma_0\Psi, \tag{16.99}$$

where we use the fact that $(\gamma^0)^2 = I$, and we defined $\hat{\mathbf{p}} = -i\nabla$ and

$$\beta_i = -\gamma_0\gamma_i = \begin{pmatrix} \sigma_i & 0 \\ 0 & -\sigma_i \end{pmatrix}. \tag{16.100}$$

This looks like a one-particle Schrödinger equation with Hamiltonian

$$\hat{H} = -i\boldsymbol{\beta} \cdot (\nabla - iq\mathbf{A}) + q\phi I + m\gamma_0, \tag{16.101}$$

but the formal analogy is deceiving as $\hat{H}$ is unbounded from below and, therefore, unphysical. Nonetheless, the "Hamiltonian" (16.101), when restricted to its positive-energy subspace, turns out to be quite reliable in many calculations.

To solve the eigenvalue equation $\hat{H}\Psi = E\Psi$, we write $\hat{\boldsymbol{\pi}} := -i(\nabla - q\mathbf{A})$, and we set $\Psi = \begin{pmatrix} \psi_1 \\ \psi_2 \end{pmatrix}$. Then, we obtain the set of equations

$$\boldsymbol{\sigma} \cdot \hat{\boldsymbol{\pi}}\psi_1 + q\phi\psi_1 + m\psi_2 = E\psi_1, \quad -\boldsymbol{\sigma} \cdot \hat{\boldsymbol{\pi}}\psi_2 + q\phi\psi_2 + m\psi_1 = E\psi_2.$$

We solve the first equation for ψ_2 and substitute in the second, to obtain

$$\frac{(\boldsymbol{\sigma} \cdot \hat{\boldsymbol{\pi}})^2}{m}\psi_1 - i\frac{q}{m}(\boldsymbol{\sigma} \cdot \nabla\phi)\psi_1 + \frac{2qE}{m}\phi\psi_1 - \frac{q^2\phi^2}{m} = \frac{E^2 - m^2}{m}\psi_1. \tag{16.102}$$

In the nonrelativistic limit, we set $E = m + \epsilon$, where $\epsilon << m$ is the nonrelativistic kinetic energy. We also take the dynamical energy $q\phi$ to be of the same order as the kinetic energy ϵ. To leading order in ϵ/m, Eq. (16.102) yields

$$\frac{(\boldsymbol{\sigma} \cdot \boldsymbol{\pi})^2}{2m}\psi_1 + q\phi\psi_1 = \epsilon\psi_1, \tag{16.103}$$

that is, we obtain the eigenvalue equation for the Pauli Hamiltonian (14.29). Hence, Dirac's theory predicts the value -2 for the g-factor of the electron. A more systematic procedure of taking the nonrelativistic limit (Foldy and Wouthuysen, 1950) provides corrections to Pauli's equation, including the L–S interaction term (14.32).

The Dirac equation for the Coulomb potential ($q\phi = -\frac{\kappa}{r}, \mathbf{A} = 0$) can be solved exactly. The corresponding energy eigenvalues

$$E_{n,j} = \frac{m}{\sqrt{1 + \frac{\kappa^2}{(n-\epsilon_j)^2}}}, \quad \text{where} \quad \epsilon_j = j + \frac{1}{2} - \sqrt{\left(j + \frac{1}{2}\right) - \kappa^2}, \tag{16.104}$$

are well defined for $\kappa < 1$. The energy eigenvalues depend only on the principal quantum number n and the total angular momentum j. Hence, the degeneracy equals $2(2j + 1)$.

For hydrogen-like atoms $\kappa = Z\alpha \simeq Z/137$, hence, if Z is small enough that $\kappa^2 << 1$ we can approximate $\epsilon_j \simeq \frac{\kappa^2}{2j+1}$. In this approximation,

$$E_{n,j} = m - \frac{\kappa^2 m}{2\left(n - \frac{\kappa^2}{2j+1}\right)^2} + \frac{3\kappa^4 m}{8n^4} + O(\kappa^6)$$

$$= m - \frac{\kappa^2 m}{2n^2} + \frac{\kappa^4 m}{2n^3}\left(\frac{3}{4n} - \frac{1}{2j+1}\right) + O(\kappa^6). \tag{16.105}$$

The first term is the contribution of the mass, the second term is the standard non-relativistic expression and the third term is the lowest-order correction, known as the *Sommerfeld term*. In Section 17.3.2, we derive this term using perturbation theory.

16.7 Towards Relativistic Quantum Field Theory

The most important role of relativistic wave equations is that they provide a "scaffolding" for the construction of relativistic quantum fields. To see this, compare the relativistic scalar field of Eq. (15.84) with the solutions to the Klein–Gordon equation (16.77). To facilitate the comparison, we redefine the annihilation and creation operators by multiplying them both with $\sqrt{2\epsilon_{\mathbf{p}}}$, so that they satisfy the modified commutation relations

$$[\hat{a}(p), \hat{a}^\dagger(p')] = 2\epsilon_{\mathbf{p}}\delta^3(\mathbf{p} - \mathbf{p}'). \tag{16.106}$$

Note that the argument of the creation and annihilation operators is the four-momentum $p = (\epsilon_{\mathbf{p}}, \mathbf{p})$. Then, the field operator becomes

$$\hat{\phi}(X) = \int d\mu_+(p)\left[\hat{a}(p)e^{-ip\cdot X} + \hat{a}^\dagger(p)e^{ip\cdot X}\right], \tag{16.107}$$

where $d\mu_+(p) = \frac{d^3p}{2\epsilon_{\mathbf{p}}}$. Then, Eq. (16.106) is formally identical to Eq. (16.77), modulo the correspondence of positive-energy solution $\psi_1(\xi)$ with the annihilation operator $\hat{a}(p)$ for $p^\mu = m\xi^\mu$, and of the negative-energy solution $\psi_2(\xi)$ with the creation operator $\hat{a}^\dagger(p)$.

The problem of negative energies disappears in the field description. The negative-energy solutions in the relativistic wave equation simply define the transformation rule for $\hat{a}^\dagger$ under spacetime translations. Furthermore, the creation and annihilation operators are defined on the Fock space $\mathcal{F}_B(\mathcal{H}_{m,0})$. So if $\psi_{1,2}$ transform under a representation $\hat{U}(g)$ of a group G, then the operators $\exp_\otimes[\hat{U}(g)]$ define a representation of G on the Fock space. As shown in Problem 15.10, the annihilation operators behave similarly to ψ_1 under the action of $\hat{U}(g)$.

Hence, the substitution of positive-energy solutions with annihilation operators and of negative-energy solutions with creation operators allows us to employ the relativistic wave equations for constructing quantum fields. In fact, the scalar quantum field of Eq. (15.84) is not the most general one. Since we can take ψ_1 different from ψ_2 in Eq. (16.77), we have the freedom of using creation and annihilation operators for two types of particles, say $\hat{a}(p)$ and $\hat{a}^\dagger(p)$ for the first type and $\hat{b}(p)$ and $\hat{b}^\dagger(p)$ for the second type. They are both subject to the commutation relation (16.106). This means that we can define complex fields

$$\hat{\phi}(X) = \int d\mu_+(p)\left[\hat{a}(p)e^{-ip\cdot X} + \hat{b}^\dagger(p)e^{ip\cdot X}\right], \tag{16.108}$$

$$\hat{\phi}^\dagger(X) = \int d\mu_+(p)\left[\hat{b}(p)e^{-ip\cdot X} + \hat{a}^\dagger(p)e^{ip\cdot X}\right]. \tag{16.109}$$

Since the scalar field involves two types of particle, it is defined on the Fock space $\mathcal{F}_B(\mathcal{H}^a_{m,0}) \otimes \mathcal{F}_B(\mathcal{H}^b_{m,0}) = \mathcal{F}_B(\mathcal{H}^a_{m,0} \oplus \mathcal{H}^b_{m,0})$. In this notation, we placed indices a and b on $\mathcal{H}_{m,0}$ in order to distinguish between the two types of particle. It is straightforward to show that, under a Lorentz transformation Λ, $\hat{\phi}(X) \to \hat{\phi}(\Lambda^{-1}X)$ and $\hat{\phi}^{\dagger}(X) \to \hat{\phi}^{\dagger}(\Lambda^{-1}X)$.

By construction, the two types of particle have the same mass. To see how they differ, we must couple the particles to an external field. If the field $\hat{\phi}$ couples with a magnetic field – via minimal substitution – through terms of the form $(\nabla - iq\mathbf{A})\hat{\phi}$, $\hat{\phi}^{\dagger}$ couples through terms of the form $(\nabla + iq\mathbf{A})\hat{\phi}^{\dagger}$, that is, the two types of particle have opposite charge. We say that one is the *antiparticle* to the other.

The exchange operator on $\mathcal{F}_B(\mathcal{H}_{m,0}) \otimes \mathcal{F}_B(\mathcal{H}_{m,0})$ is customarily referred to as *charge conjugation*, and it is denoted by $\hat{\mathcal{C}}$. By definition, $\hat{\mathcal{C}}(|a\rangle \otimes |b\rangle) = |b\rangle \otimes |a\rangle$ for all $|a\rangle \in \mathcal{F}_B(\mathcal{H}_{m,0}^a)$ and $|b\rangle \in \mathcal{F}_B(\mathcal{H}_{m,0}^b)$. As it implements an exchange, $\hat{\mathcal{C}}$ is unitary and it satisfies $\hat{\mathcal{C}}^2 = \hat{I}$. Under charge conjugation, the fields transform as

$$\hat{\mathcal{C}}\hat{\phi}(X)\hat{\mathcal{C}}^{\dagger} = \hat{\phi}^{\dagger}(X), \qquad \hat{\mathcal{C}}\hat{\phi}^{\dagger}(X)\hat{\mathcal{C}}^{\dagger} = \hat{\phi}(X). \tag{16.110}$$

We compare charge conjugation with the other two fundamental discrete transformations, namely, space inversion $\hat{\mathbb{P}}$ and time reversal $\hat{\mathbb{T}}$. Equation (16.59) implies that the annihilation operators $\hat{a}(p)$ and $\hat{b}(p)$ transform under $\hat{\mathbb{P}}$ to $\varpi\hat{a}(\mathcal{P}p)$ and $\varpi\hat{b}(\mathcal{P}p)$, respectively. It is straightforward to show that

$$\hat{\mathbb{P}}\hat{\phi}(t, \mathbf{x})\hat{\mathbb{P}}^{\dagger} = \varpi\hat{\phi}(t, -\mathbf{x}), \qquad \hat{\mathbb{P}}\hat{\phi}^{\dagger}(t, \mathbf{x})\hat{\mathbb{P}}^{\dagger} = \varpi\hat{\phi}^{\dagger}(t, -\mathbf{x}). \tag{16.111}$$

Equation (16.60) implies that $\hat{a}(p)$ and $\hat{b}(p)$ transform under $\hat{\mathbb{T}}$ to $\hat{a}^{\dagger}(\mathcal{P}p)$ and $\hat{b}^{\dagger}(\mathcal{P}p)$, respectively. It follows that

$$\hat{\mathbb{T}}\hat{\phi}(t, \mathbf{x})\hat{\mathbb{T}}^{\dagger} = \hat{\phi}^{\dagger}(-t, \mathbf{x}), \tag{16.112}$$

that is, $\hat{\mathbb{T}}$ switches particles with antiparticles while reversing the time direction.

We can also define the CTP operator, $\hat{\Theta} = \hat{\mathcal{C}}\hat{\mathbb{P}}\hat{\mathbb{T}}$, which transforms as follows:

$$\hat{\Theta}\hat{\phi}(X)\hat{\Theta}^{\dagger} = \varpi\hat{\phi}(-X), \qquad \hat{\Theta}\hat{\phi}^{\dagger}(X)\hat{\Theta}^{\dagger} = \varpi\hat{\phi}^{\dagger}(-X). \tag{16.113}$$

The transformation $\hat{\Theta}$ is subject to one of the most important theorems of quantum field theory, the *CTP theorem* (Schwinger, 1951). This theorem asserts that $\hat{\Theta}$ is a dynamical symmetry for any quantum field theory with a unique ground state.

The construction of the field operators presented here can be straightforwardly generalized to all particles. Here, we study particles with spin $s = \frac{1}{2}$. Again, we consider a particle–antiparticle pair, described by creation and annihilation operators $\hat{c}_r(\mathbf{p})$, $\hat{c}_r^{\dagger}(\mathbf{p})$ for the particle, and $\hat{d}_r(\mathbf{p})$, $\hat{d}_r^{\dagger}(\mathbf{p})$ for its antiparticle. The index r takes values $\pm$: $\hat{c}_+$ corresponds to $\psi_{\frac{1}{2}}$ and $\hat{c}_-$ corresponds to $\psi_{-\frac{1}{2}}$ in Eq. (16.85). Similarly $\hat{d}_+^{\dagger}$ corresponds to $\psi_{\frac{1}{2}}^*$ and $\hat{d}_-^{\dagger}$ corresponds to $\psi_{-\frac{1}{2}}^*$ in Eq. (16.90).

These operators must be fermionic due to the spin-statistics theorem. By the analysis of Section 15.3.4, we can always choose the $\hat{c}, \hat{c}^{\dagger}$ operators to anticommute with the $\hat{d}, \hat{d}^{\dagger}$ operators. In analogy to Eq. (16.106), we write the anticommutation relations as

$$[\hat{c}_r(p), \hat{c}_{r'}^{\dagger}(p')]_+ = [\hat{d}_r(p), \hat{d}_{r'}^{\dagger}(p')]_+ = 2\epsilon_{\mathbf{p}}\delta^3(\mathbf{p} - \mathbf{p}'). \tag{16.114}$$

Then, we define the Dirac field

$$\hat{\psi}(X) = \int d\mu_+(p) \sum_{r=\pm} \left[\hat{c}_r(p) u_r(p) e^{-ip \cdot X} + \hat{d}_r^\dagger(p) v_r(p) e^{-ip \cdot X} \right] \tag{16.115}$$

$$\hat{\bar{\psi}}(X) = \int d\mu_+(p) \sum_{r=\pm} \left[\hat{d}_r(p) \bar{v}_r(p) e^{-ip \cdot X} + \hat{c}_r^\dagger(p) \bar{u}_r(p) e^{-ip \cdot X} \right], \tag{16.116}$$

where $u_r(p)$ are given by Eq. (16.86), $v_r(p)$ are given by Eq. (16.91), $\bar{u}_r(p) = u_r^\dagger(p)\gamma_0$ and $\bar{v}_r(p) = v_r^\dagger(p)\gamma_0$.

With the definition of the Dirac field, we stop – somewhat abruptly – our analysis of relativistic quantum systems. Further elaboration would take us deeply into the formulation of quantum field theory, which is outside the scope of this volume. The key point of this chapter was to show how relativistic quantum physics is founded on the representation theory of the Poincaré group. This theory determines the structure of elementary relativistic systems, namely, of particles, but also implements the field–particle correspondence through the definition of quantum fields.

QUESTIONS

16.1 Equation (16.26) is ill defined for null vectors $(k, 0, 0, -k)$. Can you find a parameterization that is everywhere well defined? If not, why not?

16.2 What does Theorem 7.1 imply about the possibility of defining a relativistic position operator?

16.3 Is there an analogue of the Wigner rotation in nonrelativistic physics?

16.4 Verify Eq. (16.74).

16.5 Consider the quote by Haag in the beginning of this chapter. If invariance under the Poincaré group refers to the placing of detectors, how would you interpret the fact that the unitary irreducible representations of the Poincaré group correspond to particles?

PROBLEMS

16.1 Show that $[\hat{W}_\mu, \hat{W}_\nu] = -i\epsilon_{\mu\nu\rho\sigma} \hat{W}^\rho \hat{P}^\sigma$, where $\hat{W}_\mu$ is the Pauli–Lubanski vector.

16.2 Calculate the Wigner rotation $w[b_3(u), \beta]$, where $b_3(u)$ is the boost operator in Example 16.2 and $\beta = (\sqrt{1 + \beta^2}, \beta, 0, 0)$ is a particle's four-velocity.

16.3 Verify Eqs. (16.50)–(16.53).

16.4 Show that the phase $\phi(\alpha, k)$ of the null Wigner rotation equals (i) $\frac{\epsilon_1 k^2 - \epsilon_2 k^2}{2(k^0 + k^3)}$ for an infinitesimal boost $\alpha = I + \frac{1}{2}\lambda \cdot \sigma$ and (ii) $\frac{\epsilon_3}{2} + \frac{\epsilon_1 k^1 + \epsilon_2 k^2}{2(k^0 + k^3)}$ for an infinitesimal rotation $\alpha = I + \frac{i}{2}\lambda \cdot \sigma$.

16.5 Evaluate the commutator $[\hat{X}_i, \hat{K}_j]$, where $\hat{X}_i$ is given by Eq. (16.54), and show that $\hat{X}_i$ does not transform like a position coordinate under boosts.

16.6 Derive Eqs. (16.65)–(16.68).

16.7 Observer A prepared two stationary electrons in an entangled spin state, $|\Psi\rangle = \frac{1}{\sqrt{2}}\left[|\frac{1}{2}, -\frac{1}{2}\rangle - |-\frac{1}{2}, \frac{1}{2}\rangle \right]$, where the spin basis is defined with respect to the 3-axis.

An observer B describes moves with velocity $\mathbf{v}$ along the 1-axis with respect to the observer A. (i) How does B describe the state $|\Psi\rangle$? (ii) What probabilities does B assign to spin measurements along the 3-axis? (iii) Consider the same problem, but with one electron in the pair moving with momentum $\mathbf{p} = (0,0,p)$ and the other with momentum $-\mathbf{p}$. How does B describe this state? What probabilities does B assign?

16.8 Observer A has constructed qubits with photon polarization. She prepares a pair of photon qubits, moving with the same momentum $\mathbf{k} = (k,0,0)$, in an entangled $|\Psi\rangle = \frac{1}{\sqrt{2}}(|+,-\rangle - |-,+\rangle)$, where $+$ and $-$ refer to the photon's helicity. Observer B moves along the 1-axis with velocity $\mathbf{v}$ in relation to A. (i) What state does B assign to the photon helicities? (ii) Answer the same question if the first photon has momentum $\mathbf{k}$ and the second has momentum $-\mathbf{k}$.

16.9 Write the action of the $\hat{\mathbb{P}}$, $\hat{\mathbb{T}}$, and $\hat{\mathcal{C}}$ transformations on the Dirac field.

Bibliography

- Even if this section is self-contained, a familiarity with special relativity is useful; see, for example, chapters 2–7 of Rindler (2001).
- For the representations of the Poincaré group, I recommend the introductory discussion in Jones (1998), and also the monograph of Sexl and Urbantke (2001) for the analysis of unitary irreducible representations. The analysis presented here is inspired by the more mathematical perspective of Simms (1968). For relativistic particles with spin in classical mechanics, see Souriau (1997).
- For properties of Dirac's equation and its applications, see Greiner (2000).
- There are literally dozens of textbooks on quantum field theory and these are written from several different perspectives. Most modern textbooks start with fields, and downplay the importance of particles and their symmetries in building the field description. Weinberg (1996) is an exception, but I would suggest that the reader goes first through a more introductory text such as Mandl and Shaw (2010) before tackling it. For a foundational perspective to quantum field theory, I strongly recommend the classic text by Bogoliubov et al. (1976).

Part IV

Techniques

17 Energy Spectra and the Structure of Composite Systems

To try to make a model of an atom by studying its spectrum is like trying to make a model of a grand piano by listening to the noise it makes when thrown downstairs.

Anonymous

17.1 The Problem under Study

One of the most important motives for the development of quantum theory was the need to understand the structure of atoms and molecules, and to account for their emission spectra. Later on, analogous issues were raised for other composite systems, such as nuclei and hadrons. In all cases, the answer requires finding the *discrete spectrum of the Hamiltonian $\hat{H}$* that describes the composite system, that is, solving the eigenvalue equation $\hat{H}|n\rangle = E_n|n\rangle$ for the Hamiltonian.

If we solve the eigenvalue equation, we can determine the structure of a composite system from the ground state $|0\rangle$. The ground state defines the probability distribution for important physical magnitudes, such as the positions of nuclei and electrons in a molecule. Thus, it determines not only the shape and size of the molecule, but also quantities such as the electric dipole moments and moments of inertia. The emission frequencies are also determined by the solution of the eigenvalue problem, since they coincide with the differences $E_n - E_m$ between energy eigenvalues.

In realistic systems, we cannot find explicit solutions for the eigenvalue equation for the Hamiltonian, so we have to rely on numerical methods. But even with the fastest supercomputers available, numerical methods do not suffice. We must find *physical approximations* that will simplify the problem to the extent that numerical methods will work. The reason is that the quantum rule of composition of subsystems through the tensor products increases computational requirements exponentially with respect to the number of particles that constitute the systems.

The Hilbert spaces that describe realistic physical systems are infinite-dimensional, so the eigenvalue equation for the Hamiltonian corresponds to the diagonalization of an infinite-dimensional matrix. We can handle this problem computationally, only if we make the matrix finite-dimensional. To this end, we restrict to a N-dimensional subspace V, which we assume to contain the first $k < N$ eigenvectors of $\hat{H}$. In effect, we substitute $\hat{H}$ with $\hat{P}\hat{H}\hat{P}$, where $\hat{P}$ is the projector to V, and we assume a small error in the estimation of the eigenvalues and eigenvectors, $|\langle n|\hat{H} - \hat{P}\hat{H}\hat{P}|n\rangle| << |E_n|$, for $0 \leq n \leq k$.

Consider an atom with atomic number Z. A single electron is described by the Hilbert space $\mathcal{H}_s = L^2(\mathbb{R}^3) \otimes \mathbb{C}^2$. Even if we ignore spin (except for its implications for the symmetry of the spatial wave function), we would need at least a 10-dimensional subspace of $L^2(\mathbb{R}^3)$ for a decent accuracy in any computation of the electron's behavior. This means that we need a 10^Z-dimensional Hilbert space in order to describe an atom. For a light atom such as Na with $Z = 11$, we will use vectors of dimension 10^{11}. Assuming 8 bytes of memory for each component of the state vector, we need a 800 GB hard disk *only for storing* a state vector.

For a medium-weight atom such as Ti with $Z = 22$, a state vector requires about 10^{11} TB, that is, about the same amount of information that is currently (2023) stored in the Internet. For a heavy atom such as Hg, we get an outrageous number like 10^{79} TB for the storage of the state vector. It is obvious that we need approximation methods that will greatly simplify computations, at least for some predictions of direct physical interest. In this chapter, we will present the main ideas of such methods.

17.2   Perturbation Theory

The simplest approximation method is *perturbation theory*. It works for Hamiltonian operators $\hat{H}$ of the form $\hat{H}_0 + \lambda \hat{V}$, where $\hat{H}_0$ is an operator with a solved eigenvalue problem,

$$\hat{H}_0|n\rangle_0 = E_n^{(0)}|n\rangle_0, \tag{17.1}$$

and $\lambda \hat{V}$ is an operator that is "small," in the sense that it is proportional of some dimensionless constant $\lambda \ll 1$ that is much smaller than unity.

In perturbation theory, we treat the eigenvalues E_n and the eigenvectors $|n\rangle$ of $\hat{H}$ as small corrections to the eigenvalues and eigenvectors of $\hat{H}_0$. To this end, we express E_n and $|n\rangle$ as a series with respect to λ,

$$E_n = E_n^{(0)} + \lambda E_n^{(1)} + \lambda^2 E_n^{(2)} + \cdots \tag{17.2}$$

$$|n\rangle = |n\rangle_0 + \lambda|n\rangle_1 + \lambda^2|n\rangle_2 + \cdots . \tag{17.3}$$

The terms $|n\rangle_i$ and $E_n^{(i)}$, for $i > 0$, are called *ith-order corrections* to the eigenfunctions and the eigenvalues, respectively.

17.2.1 Perturbative Corrections

The calculation of perturbative corrections proceeds as follows. First, we substitute Eqs. (17.2) and (17.3) into the eigenvalue equation $\hat{H}|n\rangle = E_n|n\rangle$, to obtain

$$(\hat{H}_0 + \lambda \hat{V})(|n\rangle_0 + \lambda|n\rangle_1 + \lambda^2|n\rangle_2 + \cdots)$$
$$= (E_n^{(0)} + \lambda E_n^{(1)} + \lambda^2 E_n^{(2)} + \cdots)(|n\rangle_0 + \lambda|n\rangle_1 + \lambda^2|n\rangle_2 + \cdots). \tag{17.4}$$

The normalization condition $\langle n|n\rangle = 1$ yields

$$\lambda \left({}_0\langle n|\langle n\rangle_1 + {}_1\langle n|n\rangle_0 \right) + \lambda^2 \left({}_1\langle n|n\rangle_1 + {}_2\langle n|n\rangle_0 + {}_0\langle n|n\rangle_2 \right) + \cdots = 0 \tag{17.5}$$

Every power of λ in Eq. (17.5) must vanish independently. This means that

$$\mathrm{Re}\,_0\langle n|n\rangle_1 = 0 \tag{17.6}$$

$$\mathrm{Re}\,_0\langle n|n\rangle_2 + {}_1\langle n|n\rangle_1 = 0. \tag{17.7}$$

We can always choose the phase of the eigenvectors $|n\rangle_0$ so that the product ${}_1\langle n|n\rangle_0$ is a real number. Then, Eq. (17.6) becomes

$$_0\langle n|n\rangle_1 = 0. \tag{17.8}$$

First-order corrections Next, we equate the powers of λ in Eq. (17.4). To order λ^0, we obtain Eq. (17.1). To order λ^1,

$$\hat{H}_0|n\rangle_1 + \hat{V}|n\rangle_0 = E_n^{(0)}|n\rangle_1 + E_n^{(1)}|n\rangle_0. \tag{17.9}$$

We multiply Eq. (17.9) with $_0\langle n|$ from the left, to obtain

$$E_n^{(1)} = {}_0\langle n|\hat{V}|n\rangle_0. \tag{17.10}$$

Then, we multiply Eq. (17.9) from the left with $_0\langle k|$ for $k \neq n$. This yields

$$_0\langle k|n\rangle_1 = \frac{V_{kn}}{E_n^{(0)} - E_k^{(0)}}, \tag{17.11}$$

where $V_{kn} = {}_0\langle k|\hat{V}|n\rangle_0$ and we assumed that $E_n^{(0)} \neq E_k^{(0)}$.

Combining Eq. (17.11) with Eq. (17.8), we find the first-order correction to the eigenvectors of $\hat{H}$,

$$|n\rangle_1 = \sum_{k\neq n} \frac{V_{kn}}{E_n^{(0)} - E_k^{(0)}}|k\rangle_0. \tag{17.12}$$

Degeneracy The expressions above are problematic for degenerate eigenvalues of $\hat{H}_0$. Let S_E be the N-dimensional eigenspace of an eigenvalue E of $\hat{H}_0$. Then, any normalized vector $|n\rangle_0 \in S_E$ can be used in Eq. (17.10). As a result, the first-order correction $E^{(1)}$ is not uniquely defined. Moreover, the denominator in Eq. (17.12) vanishes for all $|k\rangle \in S_E$.

The very statement of the problem contains its solution: We simply choose a basis on S_E, in which $|n\rangle_1$ does not diverge. Consider N vectors $|k\rangle_0$ that define an orthonormal basis on S_E. We multiply Eq. (17.9) with $_0\langle k|$ on the left, to obtain

$$V_{kn} = E_n^{(1)}\delta_{kn}. \tag{17.13}$$

This means that the vectors $|k\rangle_0$ must be eigenvectors of the $N \times N$ matrix V_{kn}, which corresponds to the projection of $\hat{V}$ in the eigenspace S_E. The eigenvalues of V_{kn} are the desired first-order corrections $E_n^{(1)}$.

The same problem appears when there is *approximate degeneracy* in $\hat{H}_0$, that is, if there are N eigenvectors $|r\rangle$ of $\hat{H}_0$, with eigenvalues very close to each other, in the sense that $|E_r - E_{r'}| \sim \lambda|E_r|$, for all $r, r' = 1, \dots, N$. In this case, we choose arbitrarily an energy $\bar{E}$ such that $\Delta E_r = E_r - \bar{E}$ is of order $\lambda|E_r|$ for all r. For example, we can choose $\bar{E}$ to be the mean value or the median of the eigenvalues E_r.

Then, we redefine the unperturbed Hamiltonian as $\hat{H}'_0 := \hat{H}_0 - \sum_r \Delta E_r |r\rangle\langle r|$ and the perturbation as $\hat{V}' := \hat{V} + \sum_r \Delta E_r |r\rangle\langle r|$. We still have $\hat{H} = \hat{H}'_0 + \hat{V}'$, and $\hat{V}'$ remains of the order λ, so that perturbation theory applies. But now, $\hat{H}'_0$ has exact degeneracy, and we recover the previous case.

Second-order corrections Equating terms of order λ^2 in Eq. (17.4), we obtain

$$\hat{V}|n\rangle_1 + \hat{H}_0|n\rangle_2 = E_n^{(2)}|n\rangle_0 + E_n^{(1)}|n\rangle_1 + E_n^{(0)}|n\rangle_2. \tag{17.14}$$

We multiply by $_0\langle n|$ on the left, and we employ Eq. (17.8), to obtain the second-order correction for nondegenerate eigenvalues,

$$E_n^{(2)} = {}_0\langle n|\hat{V}|n\rangle_1 = \sum_{k \neq n} \frac{|V_{kn}|^2}{E_n^{(0)} - E_k^{(0)}}. \tag{17.15}$$

Second-order corrections to the eigenvectors, as well as higher-order corrections, can be calculated with a similar procedure. Expressions become increasingly complex. We will not need them in this book.

Example 17.1 Consider the Schrödinger operator for a unharmonic oscillator potential, $\hat{H} = \frac{\hat{p}^2}{2m} + \frac{1}{2}m\omega^2\hat{x}^2 + g\hat{x}^4$. Here, $\hat{H}_0$ is the Schrödinger operator for the harmonic oscillator.

The constant g has dimensions [energy]/[length]^4. Hence, if we multiply g with the harmonic oscillator length scale $1/\sqrt{m\omega}$ to the fourth power and divide by the energy scale ω, we obtain a dimensionless quantity $\lambda = \frac{g}{m^2\omega^3}$.

Perturbation theory applies for $\lambda \ll 1$. The first-order correction to energy is $E_n^{(1)} = g\langle n|\hat{x}^4|n\rangle$. By Eq. (5.32), we obtain

$$E_n^{(1)} = \frac{3g}{2m^2\omega^2}\left(n^2 + n + \frac{1}{2}\right) = \lambda\frac{3\omega}{2}\left(n^2 + n + \frac{1}{2}\right).$$

For large n, the correction increases proportionally to n^2 and eventually becomes larger than $E_n^{(0)}$. Obviously there is no meaning to perturbation theory for highly excited energy eigenvalues.

Example 17.2 Consider the Hamiltonian $\hat{H} = \hat{\sigma}_3 + \lambda\hat{\sigma}_1$ for a qubit. As shown in Section 5.1.3, the eigenvalues are $E_\pm = \pm\sqrt{1 + \lambda^2}$. Here, we will approximate $E_\pm$ by perturbing the eigenvalues $E_\pm^{(0)} = \pm 1$ of the operator $\hat{H}_0 = \hat{\sigma}_3$.

The first-order correction is $E_\pm^{(0)} = \lambda\langle\pm|\hat{\sigma}_1|\pm\rangle = 0$. For the second-order correction, we use the fact that $\langle +|\hat{V}|-\rangle = \langle -|\hat{V}|+\rangle = \lambda$. Then, Eq. (17.15) yields

$$E_+^{(2)} = \lambda^2\frac{|\langle +|\hat{V}|-\rangle|^2}{E_+^{(0)} - E_-^{(0)}} = \frac{1}{2}\lambda^2, \qquad E_-^{(2)} = \lambda^2\frac{|\langle -|\hat{V}|+\rangle|^2}{E_-^{(0)} - E_+^{(0)}} = -\frac{1}{2}\lambda^2. \tag{17.16}$$

Hence, to second order in λ, $E_\pm = \pm\left(1 + \frac{1}{2}\lambda^2\right)$, that is, we obtain the first two terms in the Taylor expansion of the exact eigenvalues with respect to λ^2. This property persists in higher perturbative orders: The perturbative series coincides with the Taylor series of the exact eigenvalues with respect to λ.

17.2.2 Convergence of the Perturbative Series

In general, the perturbative series is *not* a Taylor series; it is not convergent, no matter how small λ may be. We can see this in Example 17.1. If a series $\sum_{n=0}^{\infty} a_n \lambda^n$ converges to the exact energy of the ground state, it must do so for all λ in a small neighborhood $(-\epsilon, \epsilon)$ around $\lambda = 0$, where $\epsilon > 0$. But then, the series would also converge for some negative values of λ. But this is impossible: A negative value of λ renders the operator $\hat{H}$ unbounded from below, so that no ground state exists.

The perturbative series for the unharmonic oscillator is not convergent. Rather, it is an *asymptotic series*. To see the difference, compare the following definitions.

- A series $\sum_{n=0}^{\infty} a_n \lambda^n$ *converges* to a function $f(\lambda)$ if, for all λ, the first N terms in the series $f_N(\lambda) := \sum_{n=0}^{N} a_n \lambda^n$ approximate $f(\lambda)$ with arbitrary accuracy, as long as we choose N sufficiently large.
- A series $\sum_{n=0}^{\infty} a_n \lambda^n$ *is asymptotic* to the function $f(\lambda)$ if, for all positive integers N, the relative error in the approximation of $f(\lambda)$ by $f_N(\lambda)$ can be made arbitrarily small for sufficiently small λ, that is, if $\lim_{\lambda \to 0} \lambda^{-N} \left[f_N(\lambda) - f(\lambda) \right] = 0$.

An asymptotic series is not convergent. For example, the series $\sum_{n=0}^{\infty} (-1)^n n! \, \lambda^n$ is divergent for $\lambda \neq 0$, because no matter how small λ may be, the terms in the series become arbitrarily large for large values of n – this is easily seen by using Stirling's approximation $\ln n! \simeq n \ln n - n$ for large n. Nonetheless, this series is asymptotic to the function $f(\lambda) = E_1(\lambda^{-1}) e^{\lambda^{-1}} / \lambda$, where $E_1(x) = \int_x^{\infty} \frac{e^{-t}}{t} dt$.

Suppose that we approximate a function f with the first N terms in an asymptotic series. If we want relative accuracy $\epsilon \ll 1$, then the ratio of the $(N+1)$th term to the Nth term must be smaller than ϵ, that is, $|a_{N+1}| \lambda^{N+1} < \epsilon |a_N| \lambda^N$. Hence,

$$\lambda < \epsilon \left| \frac{a_N}{a_{N+1}} \right|. \tag{17.17}$$

The desired accuracy is possible only if λ is smaller than the critical value $\lambda_c(N, \epsilon) := \epsilon^{-1} \left| \frac{a_N}{a_{N+1}} \right|$.

In physical problems, the value of λ is fixed; it cannot be made arbitrarily small. Hence, there is a specific value of N that achieves the optimal approximation to the function f, and the approximation fails for larger N – see Fig. 17.1. In other words, *taking more terms in the*

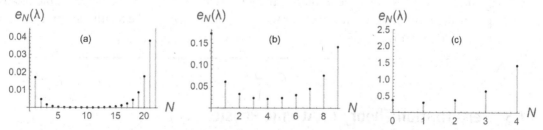

Fig. 17.1 The relative accuracy $e_N(\lambda) = \frac{|f_N(\lambda) - f(\lambda)|}{|f(\lambda)|}$ in the approximation of $f_N(\lambda) = \sum_{n=0}^{N} (-1)^n n! \, \lambda^n$ to the function $f(\lambda) = E_1(\lambda^{-1}) e^{\lambda^{-1}} / \lambda$, as a function of N, for (a) $\lambda = 0.1$, (b) $\lambda = 0.2$, and (c) $\lambda = 0.3$. In case (a), the series gives a reasonably good accuracy for a range of values of N, but eventually fails. In case (b), we obtain a 3 percent accuracy for $N = 4$, and then the series starts diverging. In case (c), there is gross error even in the first term.

perturbation series does not necessarily lead to a better result. We cannot bludgeon our way to a good approximation by calculating increasingly high orders of perturbation theory. Box 17.1 discusses the perturbative series of the unharmonic oscillator as an example.

Box 17.1　Perturbative Series for the Unharmonic Oscillator

Bender and Wu (1971) analyzed the high-order behavior of the perturbation series for the unharmonic oscillator (Example 17.1). They evaluated the coefficients a_n in the perturbative series $E_0(\lambda) = \frac{1}{2}\omega \left(1 + \sum_{n=0}^{\infty} a_n\lambda^n\right)$ for the ground state energy. The first 12 terms in the series are given in the following table.

n	1	2	3	4	5	6
a_n	0.75	-2.6	11	-240	$3.6 \cdot 10^3$	$-6.4 \cdot 10^4$

n	7	8	9	10	11	12
a_n	$1.3 \cdot 10^6$	$-3.1 \cdot 10^7$	$8.3 \cdot 10^8$	$-2.4 \cdot 10^{10}$	$7.9 \cdot 10^{11}$	$-2.8 \cdot 10^{13}$

Suppose we want to approximate the ground-state energy of the unharmonic oscillator with a relative accuracy $\epsilon = 10^{-3}$ to sixth order in perturbation theory. By Eq. (17.17), this is possible only if $\lambda < 10^{-3}|a_6/a_7| = 0.00005$.

Suppose λ is fixed in a given system and it takes the value 0.05, which is significantly smaller than unity. At sixth order in perturbation theory, Eq. (17.17) is satisfied only for $\epsilon > 2.5$. This means that, for this value of λ, perturbation theory to the sixth order gives meaningless results. With the same reasoning, we can show that relative accuracy to first order is 17 percent, it degenerates to 21 percent to second order, and we obtain worse than 100 percent error to third order. Hence, $\lambda = 0.05$ is not small enough for reliable perturbative estimates.

If we decrease λ by a factor of ten, we increase the relative accuracy by the same factor. This implies that $\lambda = 0.005$ allows perturbative approximations accurate up to a few percent.

Bender and Wu also showed that, for large n, $a_n \simeq -\frac{(-3)^n \sqrt{6}}{\pi^{3/2}}\Gamma(n + \frac{1}{2})$. This means that $a_n\lambda^n \sim n!\,(3\lambda)^n$ for large n, that is, the perturbative series for the unharmonic oscillator has an asymptotic behavior similar to the series $\sum_{n=0}^{\infty} n!\,\lambda^n$, the relative accuracy of which is plotted in Fig. 17.1.

17.3　Perturbation Theory in Atomic Physics

We proceed to applications of perturbation theory to problems in atomic physics.

17.3.1 Helium-Like Atoms

A helium-like atom is described by the Hamiltonian (15.16) for $K = 1$ and $N = 2$, while the atomic number Z may vary. We can treat the nucleus as static, since it is much heavier than the electrons. Hence, we study a two-electron system, that is described by the antisymmetric subspace of $\mathcal{H}_{\frac{1}{2}} \otimes \mathcal{H}_{\frac{1}{2}}$.

We express the Hamiltonian as $\hat{H} = \hat{H}_0 + \hat{V}$. Here, $\hat{H}_0 = \hat{h}_Z \otimes \hat{I} + \hat{I} \otimes \hat{h}_Z$, where $\hat{h}_Z = \frac{\hat{\mathbf{p}}^2}{2m_e} - Z\alpha\hat{r}^{-1}$ is the Hamiltonian of a single electron in a Coulomb potential that is generated by a nucleus of atomic number Z. The eigenvectors of $\hat{h}_Z$ are $|n, \ell, m_\ell\rangle \otimes |m_s\rangle$ with eigenvalues $\epsilon_n = -\frac{Z^2}{2m_e a_0^2 n^2}$, where $a_0 = (m_e\alpha)^{-1}$ is the Bohr radius. We treat the Coulomb interaction of the two electrons as perturbation,

$$\hat{V} = \alpha|\hat{\mathbf{r}}_1 - \hat{\mathbf{r}}_2|^{-1}, \tag{17.18}$$

where $\hat{\mathbf{r}}_{1,2}$ are the position vectors of the two electrons.

We should not expect perturbation theory to be particularly accurate in this system, because the interaction term is of the same order of magnitude as the potential term in the unperturbed Hamiltonian.

The ground state of $\hat{H}_0$ splits into a symmetric spatial component $|1, 0, 0\rangle \otimes |1, 0, 0\rangle$ and an antisymmetric spin component (that corresponds to $s = 0$). The first-order correction to the ground-state energy is

$$E_0^{(1)} = \alpha \left(\langle 1, 0, 0| \otimes \langle 1, 0, 0| \right) |\hat{\mathbf{r}}_1 - \hat{\mathbf{r}}_2|^{-1} \left(|1, 0, 0\rangle \otimes |1, 0, 0\rangle \right)$$

$$= \alpha \int d^3r_1 d^3r_2 \frac{|\langle \mathbf{r}_1|1, 0, 0\rangle|^2 |\langle \mathbf{r}_2|1, 0, 0\rangle|^2}{|\mathbf{r}_1 - \mathbf{r}_2|}. \tag{17.19}$$

We substitute $\langle \mathbf{r}|1, 0, 0\rangle = \frac{1}{\sqrt{\pi}}(Z/a_0)^{3/2}e^{-Zr/a_0}$, to obtain

$$E_0^{(1)} = \frac{\alpha}{\pi^2}(Z/a_0)^6 \int d^3r_1 d^3r_2 \frac{e^{-2Zr_1/a_0 - 2Zr_2/a_0}}{|\mathbf{r}_1 - \mathbf{r}_2|} = \frac{2Z}{ma_0^2}K, \tag{17.20}$$

where we set $\rho_{1,2} = 2Z\mathbf{r}_{1,2}/a_0$, and we write

$$K = \frac{1}{2^6\pi^2} \int d^3\rho_1 d^3\rho_2 \frac{e^{-\rho_1 - \rho_2}}{|\rho_1 - \rho_2|}. \tag{17.21}$$

In Box 17.2, we show that $K = \frac{5}{16}$, hence,

$$E_0 = -2\frac{Z^2}{2m_e a_0^2} + \frac{5Z}{8m_e a_0^2} = -\frac{1}{m_e a_0^2}\left(Z^2 - \frac{5}{8}Z\right). \tag{17.22}$$

The experimental value for E_0 in the helium atom is -79 eV, which implies that the electron interaction term is 30 eV. Our calculation yields 34 eV for this term, that is, an error of 13 percent. The error is reduced with increasing Z: For $Z = 6$ (C^{4+}), the perturbative calculation yields 102 eV for the electron interaction term, while the experimental value is about 97 eV, that is, the error is only 5 percent.

Box 17.2   Evaluation of K in Eq. (17.21)

Equation (17.21) involves integration with respect to the spherical coordinates $(\rho_1, \theta_1, \phi_1)$ of ρ_1 and with respect to the spherical coordinates $(\rho_2, \theta_2, \phi_2)$ of ρ_2. We can choose θ_2 to be the angle between ρ_1 and ρ_2, so that the integrand is independent of θ_1, ϕ_1 and ϕ_2. Since $|\rho_1 - \rho_2| = \sqrt{\rho_1^2 + \rho_2^2 - 2\rho_1\rho_2\xi}$, where $\xi = \cos\theta_2$, we obtain

$$K = \frac{1}{64\pi^2}(4\pi)(2\pi)\int_0^\infty d\rho_1 \rho_1^2 \int_0^\infty d\rho_2 \rho_2^2 e^{-\rho_1-\rho_2} \int_0^\pi \frac{d\theta_2 \sin\theta_2}{\sqrt{\rho_1^2+\rho_2^2-2\rho_1\rho_2\cos\theta_2}}$$

$$= \frac{1}{8}\int_0^\infty d\rho_1\rho_1^2 \int_0^\infty d\rho_2\rho_2^2 e^{-\rho_1-\rho_2}\int_{-1}^1 \frac{d\xi}{\sqrt{\rho_1^2+\rho_2^2-2\rho_1\rho_2\xi}}.$$

We use the indefinite integral $\int \frac{dx}{\sqrt{a-x}} = -2\sqrt{a-x}$, to evaluate

$$\int_{-1}^1 \frac{d\xi}{\sqrt{\rho_1^2+\rho_2^2-2\rho_1\rho_2\xi}} = \frac{1}{\rho_1\rho_2}(\rho_1+\rho_2-|\rho_1-\rho_2|) = \frac{2}{\rho_1\rho_2}\min\{\rho_1,\rho_2\}.$$

This means that the integral in K splits into two pieces, one for $\rho_1 < \rho_2$ and one for $\rho_1 > \rho_2$, but owing to symmetry between ρ_1 and ρ_2 the two contributions are equal. It follows that

$$K = \frac{1}{2}\int_0^\infty d\rho_1\rho_1 e^{-\rho_1}\int_0^{\rho_1} d\rho_2\rho_2 e^{-\rho_2} = \frac{1}{2}\int_0^\infty d\rho_1 e^{-\rho_1}\rho_1(2-2e^{-\rho_1}-2\rho_1 e^{-\rho_1}-\rho_1^2 e^{-\rho_1}) =$$
$$\frac{1}{2}\int_0^\infty d\rho_1(2\rho_1 e^{-\rho_1}-2\rho_1 e^{-2\rho_1}-2\rho_1^2 e^{-2\rho_1}-2\rho_1^3 e^{-2\rho_1}) = \frac{1}{2}(2-\frac{1}{2}-\frac{1}{2}-\frac{3}{8}) = \frac{5}{16}.$$

17.3.2 Fine Structure of the Hydrogen Atom's Spectrum

The eigenvalue problem for hydrogen-like atoms, with Hamiltonian $\hat{H}_0 = \frac{\hat{\mathbf{p}}^2}{2m_e} - \kappa\hat{r}^{-1}$, is exactly solvable; $\kappa = Z\alpha$. However, the Hamiltonian $\hat{H}_0$ already involves an approximation, as it is defined in nonrelativistic physics. Since the typical size of the atom is of the order of the Bohr radius a_0, the typical momentum is of order $1/a_0$, and the typical speed of the order of $(m_e a_0)^{-1} \sim 10^6\,m/s$, that is, about 1 percent of the speed of light. This means that the nonrelativistic approximation is well justified, and that relativistic phenomena can be treated with perturbation theory.

Two relativistic corrections are important. First, there is a correction in the kinetic energy. The relativistic expression for the energy $E = \sqrt{\mathbf{p}^2 + m^2} = m\sqrt{1+(\mathbf{p}/m)^2}$ for a particle of mass m and momentum $\mathbf{p}$ implies that for $|\mathbf{p}|/m \ll 1$, $E = m + \frac{\mathbf{p}^2}{2m} - \frac{\mathbf{p}^4}{8m^3} + \cdots$. Hence, the first relativistic correction to $\hat{H}_0$ is the operator

$$\hat{V}_1 = -\frac{\hat{\mathbf{p}}^4}{8m_e^3} = -\frac{1}{2m_e}\left(\frac{\hat{\mathbf{p}}^2}{2m_e}\right)^2 = -\frac{1}{2m_e}\left(\hat{H}_0 + \kappa\hat{r}^{-1}\right)^2.$$

The associated first-order correction to energy eigenvalues in the $|n,j,m_j,\ell\rangle$ basis is

$$\langle n,j,m_j,\ell|\hat{V}_1|n,j,m_j,\ell\rangle = -\frac{1}{2m_e}\langle n,j,m_j,\ell|\hat{H}_0^2 + \kappa\hat{H}_0\hat{r}^{-1} + \kappa\hat{r}^{-1}\hat{H}_0 + \kappa^2\hat{r}^{-2}|n,j,m_j,\ell\rangle$$

$$= -\frac{1}{2m_e}\left\{[E_n^{(0)}]^2 + 2\kappa E_n^{(0)}\langle\hat{r}^{-1}\rangle + \kappa^2\langle\hat{r}^{-2}\rangle.\right\} \tag{17.23}$$

We use Eq. (13.52) for $\langle \hat{r}^{-1} \rangle$ and Eq. (13.53) for $\langle \hat{r}^{-2} \rangle$, to obtain

$$\langle n,j,m_j,\ell|\hat{V}_1|n,j,m_j,\ell\rangle = \frac{\kappa^4 m_e}{2}\left(\frac{3}{4n^4} - \frac{1}{(\ell + \frac{1}{2})n^3}\right). \tag{17.24}$$

The second relativistic correction is the spin–orbit coupling, given by Eq. (14.32). For an electron in the hydrogen atom, where $\phi(r) = -\kappa/r$, Eq. (14.32) yields the correction term $\hat{V}_2 = \frac{\kappa}{2m_e^2}\hat{r}^{-3}\hat{\mathbf{L}}\cdot\hat{\mathbf{S}}$. By Eq. (11.117),

$$\langle n,j,m_j,\ell|\hat{V}_2|n,j,m_j,\ell\rangle = \frac{\kappa}{4m_e^2}\langle\hat{r}^{-3}\rangle\left(j(j+1) - \ell(\ell+1) - \frac{3}{4}\right). \tag{17.25}$$

We substitute $\langle\hat{r}^{-3}\rangle$ from Eq. (13.19), to obtain

$$\langle n,j,m_j,\ell|\hat{V}_2|n,j,m_j,\ell\rangle = \frac{\kappa^4 m_e}{4}\frac{\left(j(j+1) - \ell(\ell+1) - \frac{3}{4}\right)}{n^3\ell(\ell+\frac{1}{2})(\ell+1)}. \tag{17.26}$$

For $j = \ell + \frac{1}{2}$, $j(j+1) - \ell(\ell+1) - \frac{3}{4} = j - \frac{1}{2}$, whence

$$\langle\hat{V}_2\rangle = \frac{\kappa^4 m_e}{4n^3 j(j+\frac{1}{2})}.$$

For $j = \ell - \frac{1}{2}$, $j(j+1) - \ell(\ell+1) - \frac{3}{4} = -(j+\frac{3}{2})$, whence

$$\langle\hat{V}_2\rangle = -\frac{\kappa^4 m_e}{4n^3(j+\frac{1}{2})(j+1)}.$$

In both cases, the sum of the contributions from $\hat{V}_1$ and $\hat{V}_2$ depends to leading order only on n and j, and it is given by *Sommerfeld's formula*

$$E_{n,j}^{(1)} = \frac{\kappa^4 m_e}{2n^4}\left(\frac{3}{4} - \frac{n}{j+\frac{1}{2}}\right). \tag{17.27}$$

Energy eigenstates with the same j but different m_j and ℓ have the same energy. Hence, the degeneracy of an energy eigenvalues $E_{n,j}$ is $2(2j+1)$. Equation (17.27) is in agreement with the results of the relativistic analysis that presented in Section 16.6.4.

We conclude that relativistic effects partially break the degeneracy of the hydrogen-atom Hamiltonian. The corrections from these effects are the *fine structure* of the atom's spectrum. The ratio $E_{n,j}^{(1)}/E_n^{(0)}$ is of the order of κ^2; hence, the frequencies of the fine spectrum are about $1 : 10^4$ of the frequencies of the unperturbed spectrum.

17.3.3 The Stark Effect

Consider a system that consists of N particles of charges q_i and masses m_i, where $i = 1, 2, \ldots, N$. We denote the eigenvalues of the Hamiltonian $\hat{H}_0$ by $E_n^{(0)}$ and the corresponding eigenvectors by $|n\rangle$.

Suppose that we switch on a homogeneous electric field $\mathbf{E}$. The Hamiltonian becomes

$$\hat{H} = \hat{H}_0 - \sum_i q_i \mathbf{E} \cdot \hat{\mathbf{x}}_i, \tag{17.28}$$

where $\hat{\mathbf{x}}_i$ is the position operator for the ith particle. For a weak electric field, the first-order corrections to the energy are

$$E_n^{(1)} = -\sum_i q_i \mathbf{E} \cdot \langle n|\mathbf{x}_i|n\rangle. \tag{17.29}$$

If the unperturbed Hamiltonian $\hat{H}_0$ happen to be degenerate, the states $|n\rangle$ in Eq. (17.29) are chosen so that the matrix V_{kn} is diagonal in each eigenspace. Then, degeneracy is lifted, at least partially.

The interaction energy of an electric dipole with an electric field equals $-\mathbf{E} \cdot \mathbf{d}$, where $\mathbf{d}$ is the electric dipole moment. Equation (17.29) suggests that the electric dipole moment of the state $|n\rangle$ is $\mathbf{d}_n = \sum_i q_i \langle n|\hat{\mathbf{x}}_i|n\rangle$. Note that when we talk about the electric dipole moment of an atom or a molecule, we mean its value for the ground state ($n = 0$).

The shift of energy levels in a system that is placed in an electric field is known as the *Stark effect*.

Example 17.3 We study the Stark effect in the hydrogen atom. The electric field does not couple to the spin, so we can ignore the spin degrees of freedom. The contribution of the nucleus to the electron dipole moment is negligible. Hence, we consider only electrons in the Coulomb potential.

We employ the basis $|n, \ell, m_\ell\rangle$. We can always choose a coordinate system so that $\mathbf{E} = (0, 0, \mathcal{E})$. Then, given the degeneracy of the hydrogen-atom Hamiltonian, we need to evaluate matrix elements of the form $\langle n, \ell', m'_\ell | \hat{x}_3 | n, \ell, m_\ell \rangle$. By the Wigner–Eckart theorem, this matrix element is proportional to the Clebsch–Gordan coefficient $C_{1,\ell,0,m}^{\ell',m'}$. This means that $\langle n, \ell', m'_\ell | \hat{x}_3 | n, \ell, m_\ell \rangle$ vanishes unless $\ell' - \ell = 1, 0, -1$ and $m = m'$.

Furthermore, all diagonal matrix elements $\langle n, \ell, m_\ell | \hat{x}_3 | n, \ell, m_\ell \rangle$ vanish. The reason is that all eigenfunctions $\langle \mathbf{x} | n, \ell, m_\ell \rangle$ have definite parity, so $|\langle \mathbf{x} | n, \ell, m_\ell \rangle|^2 x_3$ is an odd function, and its integral over all space is zero. We conclude that the only nonvanishing matrix elements of the perturbation satisfy $\ell' = \ell \pm 1$ and $m' = m$.

The $n = 1$ eigenspace is spanned by the eigenvector $|1, 0, 0\rangle$, hence, $E_{1,0,0}^{(1)} = 0$.

The $n = 2$ eigenspace is spanned by the eigenvectors $|2, 0, 0\rangle$, $|2, 1, -1\rangle$, $|2, 1, 0\rangle$, and $|2, 1, 1\rangle$. The only nonvanishing matrix elements are those between $|2, 0, 0\rangle$ and $|2, 1, 0\rangle$. Hence, the eigenvectors $|2, 0, \pm 1\rangle$ are unaffected. In the subspace spanned by $|2, 0, 0\rangle$ and $|2, 1, 0\rangle$, the interaction corresponds to a matrix $\gamma \begin{pmatrix} 0 & 1 \\ 1 & 0 \end{pmatrix}$, where $\gamma = e\mathcal{E}\langle 2, 0, 0 | \hat{x}_3 | 2, 1, 0 \rangle$ is a real number since the wave functions for $|2, 0, 0\rangle$ and $|2, 1, 0\rangle$ are real. The perturbation matrix has the eigenvector $\frac{1}{\sqrt{2}}(|2, 0, 0\rangle + |2, 1, 0\rangle)$ with eigenvalue γ and the eigenvector $\frac{1}{\sqrt{2}}(|2, 0, 0\rangle - |2, 1, 0\rangle)$ with eigenvalue $-\gamma$.

We conclude that, to first-order in perturbation theory, the $n = 2$ eigenspace splits into (i) a two-dimensional subspace with energy $E_2^{(0)}$ that is spanned by the vectors $|2, 1, -1\rangle$ and $|2, 1, 1\rangle$, with energy $E_2^{(0)}$; (ii) the eigenvector $\frac{1}{\sqrt{2}}(|2,0,0\rangle + |2,1,0\rangle)$ with energy $E_2^{(0)} + \gamma$; and (iii) the eigenvector $\frac{1}{\sqrt{2}}(|2,0,0\rangle - |2,1,0\rangle)$ with energy $E_2^{(0)} - \gamma$.

Finally, we evaluate γ using the wave functions from Table 13.1,

$$\gamma = \frac{e\mathcal{E}}{32\pi a_0^4} \left(\int_0^{2\pi} d\phi \right) \left(\int_0^\pi d\theta \sin\theta \cos^2\theta \right) \left(\int_0^\infty dr^4 (2 - r/a_0) e^{-r/a_0} \right)$$
$$= \frac{e\mathcal{E}}{32\pi a_0^4} (2\pi)(\frac{2}{3})(-72a_0^5) = -3e\mathcal{E}a_0. \tag{17.30}$$

17.3.4 The Zeeman Effect

Suppose we place the system of N charged particles that was considered at the beginning of Section 17.3.3 inside a weak homogeneous magnetic field $\mathbf{B}$. The Hamiltonian (14.10) involves the addition of a term $-\sum_i \frac{q_i}{2m_i} \mathbf{B} \cdot (\mathbf{L}_i + g_i \mathbf{S}_i)$ to the unperturbed Hamiltonian $\hat{H}_0$. As long as we restrict ourselves only to the first-order correction in energy, we can ignore the term in Eq. (14.10) that is quadratic with respect to $\mathbf{B}$. Then,

$$E_n^{(1)} = -\sum_i \frac{q_i}{2m_i} \mathbf{B} \cdot \langle n|\mathbf{L}_i + g_i \mathbf{S}_i|n\rangle. \tag{17.31}$$

The energy of the interaction of a dipole with magnetic moment $\boldsymbol{\mu}$ with a magnetic field $\mathbf{B}$ is $-\mathbf{B} \cdot \boldsymbol{\mu}$. Hence, the magnetic moment $\boldsymbol{\mu}_n$ of the system in the state $|n\rangle$ is $\boldsymbol{\mu}_n = \sum_i \frac{q_i}{2m_i} \langle n|\mathbf{L}_i + g_i \mathbf{S}_i|n\rangle$.

In the hydrogen atom, the nucleus is much heavier than the electron; hence, the electron's contribution dominates in Eq. (17.31). For $\mathbf{B} = (0, 0, B)$, the interaction term is $\frac{e}{2m_e}(\hat{L}_3 + 2\hat{S}_3)$, where we set $g_e = 2$. The perturbation is diagonal in the basis $|n, \ell, m_\ell, m_s\rangle$, so we readily evaluate the eigenvalues

$$E_{n,\ell,m_\ell,m_s}^{(1)} = \frac{eB}{2m_e} \langle n, \ell, m_\ell, m_s|\hat{L}_3 + 2\hat{S}_3|n, \ell, m_\ell, m_s\rangle = \frac{eB}{2m_e}(m_\ell + 2m_s). \tag{17.32}$$

The $n = 1$ eigenspace splits into two one-dimensional subspaces ($m_s = \pm\frac{1}{2}$) with energies $E_{1,0,0} \pm \frac{eB}{2m_e}$. The $n = 2$ eigenspace splits into two one-dimensional subspaces ($m_\ell = \pm 1, m_s = \pm\frac{1}{2}$) with energy $E_{2,0,0} \pm \frac{eB}{m_e}$, two two-dimensional subspaces ($m_\ell = 0, m_s = \pm\frac{1}{2}$ for $\ell = 0, 1$) with energy $E_{2,0,0} \pm \frac{eB}{2m_e}$, and one two-dimensional subspace ($m_\ell = \pm 1, m_s = \mp\frac{1}{2}$) with energy $E_{2,0,0}$. It is straightforward to continue this analysis to eigenspaces of higher n. We see that the magnetic field partially lifts the degeneracy of the energy levels of the system, a phenomenon known as the *Zeeman effect*.

The validity of perturbation theory requires that eB/m_e is much smaller than the unperturbed energies of the hydrogen atom. However, if eB/m_e is of the order of the fine structure energies (17.27) of the hydrogen atom, we are in the domain of almost degenerate perturbation theory that was described in Section 17.2.1. In order to treat this case, we split the total Hamiltonian, including both the fine-structure corrections and the coupling to the magnetic field, as $\hat{H}_0' + \hat{V}'$, where the term

$$\hat{H}'_0 = \hat{H}_0 - \frac{3}{8}\alpha^4 m_e \sum_{n,j,m_j,m_s} n^{-4} |n,j,m_j,\ell\rangle\langle n,j,m_j,\ell|$$

depends only on n, and $\tilde{V}'$ is the perturbation. The latter can be expressed as $\sum_{n=1}^{\infty} \hat{V}'_n$, where

$$\hat{V}'_n = \frac{e}{2m_e}\left(\hat{L}_3 + 2\hat{S}_3\right) - \frac{\alpha^4 m_e}{2n^3} \sum_{j,m_j,\ell} \frac{1}{j+\frac{1}{2}} |n,j,m_j,\ell\rangle\langle n,j,m_j,\ell| \qquad (17.33)$$

is the component of the interaction in the n-eigenspace of $\hat{H}'_0$. Then, we obtain the first-order corrections to the energy by diagonalizing the matrix elements of $\hat{V}'_n$ for each n. In the $|\ell, m_\ell, m_s\rangle$ basis, we find

$$\langle \ell', m'_\ell, m'_s | \hat{V}'_n | \ell, m_\ell, m_s\rangle = \frac{eB}{2m_e}(m_\ell + 2m_s)\delta_{m_\ell m'_\ell}\delta_{m_s m'_s}$$
$$-\frac{\alpha^4 m_e}{2n^3}\sum_j (j+\frac{1}{2})^{-1} C^{j,m'_\ell + m'_s}_{\ell',\frac{1}{2},m'_\ell,m'_s} C^{j,m_\ell + m_s}_{\ell,\frac{1}{2},m_\ell,m_s}. \qquad (17.34)$$

in terms of the Clebsch–Gordan coefficients (11.113).

The case $n = 1$ is straightforward, since $\ell = 0$ and $j = \frac{1}{2}$. The two states differ only on the value of $m_s = m_j = \pm\frac{1}{2}$ and the corrections are

$$\langle \hat{V}'\rangle = -\frac{\alpha^4 m_e}{2} \pm \frac{eB}{m_e}.$$

The case of $n = 2$ is taken up in Problem 17.9.

17.4 The Variational Method

In this section, we present the main ideas of the variational method, which provides one of the most powerful sets of techniques for evaluating the ground states of composite systems.

17.4.1 The Variational Principle

Let $\hat{H}$ be a Hamiltonian on a Hilbert space $\mathcal{H}$ with ground state $|0\rangle$ and corresponding eigenvalue E_0. The variational method starts off from the elementary observation that for any normalized $|\psi\rangle \in \mathcal{H}$,

$$\langle\psi|\hat{H}|\psi\rangle \geq E_0. \qquad (17.35)$$

Consider a set S and a map that assigns each $b \in S$ to a normalized vector $|\psi(b)\rangle$ of $\mathcal{H}$. We refer to the vectors $|\psi(b)\rangle$ as variational vectors or test (wave) functions.

We define the *energy function* $E : S \to \mathbb{R}$ by

$$E(b) := \langle\psi(b)|\hat{H}|\psi(b)\rangle. \qquad (17.36)$$

For b varying smoothly, E is minimized at $b = b_{min}$, such that $\frac{\partial}{\partial b}E(b_{min}) = 0$. Let $E_{min} := \min_{b \in S} E(b)$. Obviously $E_{min} \geq E_0$, and the equality is achieved only if $|0\rangle \in S$.

The key point is that, for an appropriate choice of variational vectors, E_{min} may approximate E_0 with high precision. To see this, let us write $|b_{min}\rangle = |\psi(b_{min})\rangle$, and assume that the distance between $|b_{min}\rangle$ and $|0\rangle$ is $\epsilon << 1$. This means that there exists a unit vector $|\chi\rangle$, such that $|b_{min}\rangle = |0\rangle + \epsilon|\chi\rangle$. Since $\langle b_{min}|b_{min}\rangle = \langle 0|0\rangle = 1$,

$$\epsilon\langle\chi|0\rangle + \epsilon\langle 0|\chi\rangle + \epsilon^2 = 0. \tag{17.37}$$

On the other hand, $E_{min} = \langle 0|\hat{H}|0\rangle + \epsilon\langle\chi|\hat{H}|0\rangle + \epsilon\langle 0|\hat{H}|\chi\rangle + \epsilon^2\langle\chi|\hat{H}|\chi\rangle = E_0(1 + \epsilon\langle\chi|0\rangle + \epsilon\langle 0|\chi\rangle) + \epsilon^2\langle\chi|\hat{H}|\chi\rangle$. Using Eq. (17.37), we find that

$$|E_{min} - E_0| = \epsilon^2|\langle\chi|\hat{H}|\chi\rangle - E_0|. \tag{17.38}$$

This means that the relative error in energy is of order ϵ^2, hence, much smaller than the relative error in the quantum state. Even a mediocre approximation to the quantum state with, say, $\epsilon = 0.1$, will allow for an estimation of the ground-state energy with an accuracy of the order of a few percent.

A good choice of the test functions $|\psi(b)\rangle$ allows us to substitute the problem of diagonalizing a Hamiltonian in an infinite-dimensional Hilbert space with the problem of minimizing a function on the set S, a problem that is much easier computationally. Of course, the converse also holds: A bad choice of test functions will give an unreliable estimate for energy.

There is no general rule for choosing test functions, except for the fact that they should include states with properties that the true ground state is known to possess. For example, since the ground-state wave function of the Schrödinger operator in one dimension has no nodes, the set of test functions should include only wave functions with no nodes.

We can also evaluate the energy of excited states with a variational method. The first excited state of the Hamiltonian $\hat{H}$ coincides with the first excited energy of the Hamiltonian $\hat{H}' = \hat{H} - E_0|0\rangle\langle 0|$. Hence, if we have estimated E_0 and $|0\rangle$ through the variational method, we can again use a variational method for a different set of test functions, to find the ground-state energy of $\hat{H}'$. However, $\hat{H}'$ already contains the error of evaluating the ground state, so the error in the evaluation of the excited state will be larger. For this reason, variational methods work reliably only for the first few energy levels of a system.

Example 17.4   In Example 10.2, we studied the quantum bouncing ball, which is characterized by a Hamiltonian $\hat{H} = \frac{p^2}{2m} + mg\hat{x}$ and Dirichlet boundary conditions at $x = 0$. We found that the ground-state energy is $E_0 = \left(\frac{mg^2}{2}\right)^{1/3} s_0$, where $s_0 \simeq 2.338$ is the absolute value of the first root of the Airy function. Here, we will evaluate the ground-state energy with a variational method. We will consider different families of test functions, in order to compare the results.

We employ test functions $|b\rangle$ that depend only on a single parameter b of dimensions inverse length. Purely on dimensional grounds, $\langle b|\widehat{p^2}|b\rangle = c_1 b^2$ and $\langle b|\hat{x}|b\rangle = c_2 b^{-1}$, where c_1 and c_2 are pure numbers. Hence $E(b) = \frac{c_1}{2m}b^2 + c_2 mg b^{-1}$. The minimum is $E_{min} = \left(\frac{mg^2}{2}\right)^{1/3} s$, where $s = \frac{3}{2}c_2^{2/3}(2c_1)^{1/3}$. In the following table, we evaluate s for four families of test functions. All families satisfy the boundary conditions, except for (iv). In Fig. 17.2, we plot the function $E(b)$ for cases (i), (ii), and (iii).

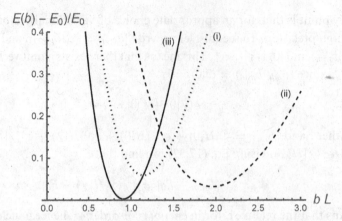

Fig. 17.2 Variational estimation of the ground state of the quantum bouncing ball. We plot the normalized difference $[E(b) - E_0]/E_0$ of the energy function $E(b)$ from the ground-state energy, for three families of test functions (i), (ii), and (iii). The variational parameter b is made dimensionless by multiplying it with the natural length scale of the problem $L = (m^2 g)^{-1/3}$.

Case	$\psi_b(x)$	c_1	c_2	s	Error
(i)	$2b^{3/2} x e^{-bx}$	1	$\frac{3}{2}$	2.476	6%
(ii)	$\sqrt{\frac{4b^5}{3}} x^2 e^{-bx}$	$\frac{1}{3}$	$\frac{5}{2}$	2.414	3%
(iii)	$\frac{\sqrt{2b^3}}{\pi^{1/4}} x e^{-\frac{1}{2} b^2 x^2}$	$\frac{3}{2}$	$\frac{2}{\sqrt{\pi}}$	2.345	0.3%
(iv)	$\sqrt{2b} e^{-bx}$	1	$\frac{1}{2}$	1.190	52%

Cases (i), (ii), and (iii) show variable success in evaluating the ground-state energy, case (iii) being by far the best, with an accuracy of 0.3 percent. Case (iv) is completely off-mark. What is worse, it leads to energy smaller than E_0! This is not a failure of the variational method. As the test function (iv) does not satisfy Dirichlet boundary conditions at $x = 0$, it does not belong in the Hilbert space of the system, so it is not subject to Eq. (17.35). Failing to respect boundary conditions is the worst mistake one can make in selecting test functions.

17.4.2 Application to Helium-Like Atoms

We return to the problem of helium-like atoms that was studied in Section 17.3.1 through perturbation theory. Assuming that the total spin of the two electrons is zero in the ground state, the spatial part of the ground state is symmetric. Hence, we choose symmetric test functions of the form $|\psi_b\rangle \otimes |\psi_b\rangle$, where

$$\langle \mathbf{r} | \psi_b \rangle = \frac{1}{\sqrt{\pi}} (b/a_0)^{3/2} e^{-br/a_0} \tag{17.39}$$

is the ground state of a hydrogen-like atom, but for an atomic number equal to b.

Let $\hat{h}_Z = \frac{\mathbf{p}^2}{2m_e} - Z\alpha\hat{r}^{-1}$. Then, $\hat{h}_b|\psi_b\rangle = -\frac{b^2}{2m_e a_0^2}|\psi_b\rangle$. Since $\hat{h}_Z - \hat{h}_b = -(Z-b)\alpha\hat{r}^{-1}$, $\langle\psi_b|\hat{h}_Z|\psi_b\rangle = -\frac{b^2}{2m_e a_0^2} - (Z-b)\alpha\langle\psi_b|\hat{r}^{-1}|\psi_b\rangle$. We straightforwardly evaluate

$$\langle\psi_b|\hat{r}^{-1}|\psi_b\rangle = (b/a_0)^3 4\int_0^\infty dr\, r e^{-2br/a_0} = b/a_0.$$

It follows that $\langle\psi_b|\hat{h}_Z|\psi_b\rangle = -\frac{b^2}{2m_e a_0^2} - \frac{(Z-b)b}{m_e a_0^2}$.

The Hamiltonian of a helium-like atom is $\hat{H} = \hat{h}_Z \otimes \hat{I} + \hat{I} \otimes \hat{h}_Z + \hat{V}$, where $\hat{V}$ is given by Eq. (17.18). Equation (17.20) implies that $\langle\psi_b, \psi_b|\hat{V}|\psi_b, \psi_b\rangle = \frac{5b}{8m_e a_0^2}$. We conclude that

$$E(b) := \langle\psi_b, \psi_b|\hat{H}|\psi_b, \psi_b\rangle = -\frac{1}{m_e a_0^2}\left(-b^2 + 2Zb - \frac{5b}{8}\right). \tag{17.40}$$

The energy function $E(b)$ is minimized for $b_{min} = Z - \frac{5}{16}$:

$$E(b_{min}) = -\frac{1}{m_e a_0^2}\left(Z - \frac{5}{16}\right)^2. \tag{17.41}$$

For the helium atom ($Z = 2$), $E(b_{min}) = -77.4\,\text{eV}$. Hence, the electron repulsion equals $31.4\,\text{eV}$, a divergence of 8.5 percent from the experimental value. This is a significant improvement over the perturbative calculation with the same intensity of calculations. We can further improve the calculations if we include further variational parameters.

17.4.3 The Reyleigh–Ritz Method

Suppose we choose our test vectors $|\psi(b)\rangle$ to be linear combinations of a finite orthonormal set $|n\rangle$, where $n = 1, 2, \ldots, N$, that is, $|\psi(b)\rangle = \sum_{n=1}^N b_n|n\rangle$. Then, the set S is spanned by all complex N-tuples $(b_1, b_2, \ldots, b_n)$ such that $\sum_{n=1}^N |b_n|^2 = 1$.

In effect, the test vectors define a subspace $\mathcal{V}$ of the Hilbert space with associated projector $\hat{P} = \sum_{n=1}^N |n\rangle\langle n|$, and we look for the element of the subspace that best approximates the ground state. This approach is known as the *Reyleigh–Ritz variational method*.

We evaluate the energy function $E(b) = \sum_{n,m=1}^N \sum_{m=1}^N b_m^* b_n H_{mn}$, and we minimize it subject to the normalization condition. To this end, we use the Lagrange multiplier method, that is, we introduce a Lagrange multiplier λ and we minimize the quantity

$$E(b) - \lambda\left(\sum_{n=1}^N |b_n|^2 - 1\right) = \sum_{n=1}^N \sum_{m=1}^N b_m^* b_n(H_{mn} - \lambda\delta_{mn}) + \lambda,$$

with respect to b_n and b_n^*. Differentiation with respect to b_m^* yields $\sum_{n=1}^N H_{mn}b_n = \lambda b_m$, that is, the coefficients b_n define an eigenvector of the $N \times N$ matrix H_{mn}. Thus, we obtain the eigenvalue equation for $\hat{H}$ projected into the subspace $\mathcal{V}$. This means that we are justified in replacing the Hamiltonian $\hat{H}$ with $\hat{P}\hat{H}\hat{P}$.

17.5 Mean-Field Approximation

The key idea in the mean-field approximation is to treat an individual particle in a many-particle system as being subject to a mean potential generated by all other particles. We already encountered the mean-field approximation in Section 15.2.2, but we left open the question of how the mean field is determined. In this section, we answer this question by implementing the mean-field approximation through the variational principle. We will present the method for a many-boson system, in order to avoid inessential complications that appear in fermionic systems.

The mean-field approximation is best expressed in the Fock-space formalism of Section 15.3. Consider a system of N bosons that is described by a Hamiltonian of the form (15.71),

$$\hat{H} = \hat{a}_a^\dagger h^a{}_b \hat{a}^b + W^{ab}{}_{cd}\hat{a}_a^\dagger \hat{a}^c \hat{a}_b^\dagger \hat{a}^d. \tag{17.42}$$

For pairwise interactions, $W^{ab}{}_{cd} = \frac{1}{2}U_{ab}\delta_c^a\delta_d^b$.

In the mean-field approximation, we assume that the correlations between particles are negligible, so that the ground state can be expressed as $|\phi\rangle = \frac{1}{\sqrt{N!}}\hat{a}^\dagger(\phi)^N|0\rangle$, where ϕ_a is a normalized vector of the single-particle Hilbert space $\mathcal{H}$. Hence, we take $|\phi\rangle$ as variational test functions, where the variational parameter ϕ_a takes values in an infinite-dimensional Hilbert space.

Next, we evaluate

$$\langle\phi|\hat{a}_a^\dagger \hat{a}^b|\phi\rangle = N\phi_a^*\phi^b \tag{17.43}$$

$$\langle\phi|\hat{a}_a^\dagger \hat{a}^c \hat{a}_b^\dagger \hat{a}^d|\phi\rangle = \langle\phi|\hat{a}_a^\dagger \hat{a}_b^\dagger \hat{a}^c \hat{a}^d|\phi\rangle \phi_a\phi_b^* + \delta_b^c\langle\phi|\hat{a}_a^\dagger \hat{a}^d|\phi\rangle$$

$$= N(N-1)\phi_a^*\phi_b^*\phi^c\phi^d + N\delta_b^c\phi_a^*\phi^d. \tag{17.44}$$

In proving the equations above, we brought the annihilation operators to act on the vacuum, using the commutator $[\hat{a}_a, \hat{a}^\dagger(\phi)^N] = N\phi_a\hat{a}^\dagger(\phi)^{N-1}$.

The energy function is

$$E(\phi) := \langle\phi|\hat{H}|\phi\rangle = N\phi_a^*\phi^b\bar{h}^a{}_b + N(N-1)W^{ab}{}_{cd}\phi_a^*\phi_b^*\phi^c\phi^d, \tag{17.45}$$

where $\bar{h}^a{}_b = h^a{}_b + W^{ac}{}_{cb}$.

We minimize the energy function $E(\phi)$ with respect to ϕ_a, subject to the condition $\phi_a^*\phi^a = 1$. To this end, we introduce a Lagrange multiplier λ and we minimize $E(\phi) - \lambda(\phi_a^*\phi^a - 1)$ with respect to ϕ and ϕ^*. The condition $\partial E/\partial\phi_a^* = 0$ yields the *Hartree equation*,

$$\bar{h}^a{}_b\phi^b + 2(N-1)W^{ab}{}_{cd}\phi_b^*\phi^c\phi^d = \frac{\lambda}{N}\phi_a. \tag{17.46}$$

For pairwise interactions

$$(h\phi)_a + \frac{1}{2}U_{aa}\phi_a + U_a^H(\phi)\phi_a = \frac{\lambda}{N}\phi_a, \tag{17.47}$$

where U_a^H is the Hartree mean-field potential

$$U_a^H(\phi) = (N-1)\sum_b U_{ab}\phi_b^*\phi_b. \tag{17.48}$$

In Eq. (17.47), ϕ has a lower index, because we abandoned the Einstein summation convention.

For particles with zero spin, we work in the position representation: $\phi^a \to \phi(\mathbf{x})$. We identify h with the Schrödinger operator for a particle of mass m in a potential $V(\mathbf{x})$. The particle interaction term U_{ab} becomes $U(\mathbf{x} - \mathbf{x}')$. Then, the Hartree equation becomes

$$\left[-\frac{1}{2m}\nabla^2 + V(\mathbf{x}) + (N-1)\left(\int d^3x' \, U(\mathbf{x} - \mathbf{x}')|\phi(\mathbf{x}')|^2 \right) \right] \phi(\mathbf{x}) = \mu\phi(\mathbf{x}), \qquad (17.49)$$

where $\mu := \frac{\lambda}{N} - U(0)$.

Equation (17.49) is a nonlinear integrodifferential equation, which must be solved numerically. Still, it refers to a single particle, and, as such, it is much easier to solve than a linear differential equation for N particles. We have to keep in mind that the "wave function" $\phi(\mathbf{x})$ does *not* describe the quantum state of any specific particle. It is a variational parameter, like the parameters b of Section 17.4.

Equation (17.49) can be employed for the study of a large number ($N \gg 1$) of bosonic atoms at low temperatures, for example, in superfluids or Bose–Einstein condensates. In this case, it is useful to remove the prefactor $N-1$ from the nonlinear term by rescaling $\phi(\mathbf{x})$ so that $\int d^3x |\phi(\mathbf{x})|^2 = N-1$. When we analyze the behavior of the system at scales much larger than the range of the potential $U(\mathbf{x})$, we can approximate the latter with a delta function, $U(\mathbf{x}) = a\delta^3(\mathbf{x})$, for some constant a. Then, Eq. (17.49) yields the *Gross–Pitaevskii equation*,

$$-\frac{1}{2m}\nabla^2\phi(\mathbf{x}) + V(\mathbf{x})\phi(\mathbf{x}) + a|\phi(\mathbf{x})|^2\phi(\mathbf{x}) = \mu\phi(\mathbf{x}). \qquad (17.50)$$

Example 17.5 The Gross–Pitaevskii equation is nonlinear, and, hence, very difficult to solve. However, it admits a simple solution for a constant potential V_0. Taking periodic boundary conditions with period L, we see that Eq. (17.49) admits plane-wave solutions $\phi_\mathbf{k}(\mathbf{x}) = \frac{1}{L^{3/2}}e^{i\mathbf{k}\cdot\mathbf{x}}$ with $k_i = \frac{2\pi}{L}n_i$, for $n_i = 0, \pm 1, \pm 2, \ldots$ and $i = 1, 2, 3$. The corresponding values of μ are

$$\mu_\mathbf{k} = \frac{\mathbf{k}^2}{2m} + V_0 + \frac{a}{L^3}. \qquad (17.51)$$

However, superpositions of $\phi_\mathbf{k}(\mathbf{x})$ are not solutions, because the Gross–Pitaevskii equation is nonlinear. Hence, we cannot solve the equation by approximating the potential with a piecewise constant one, because the continuity conditions admit solutions only if superpositions of plane waves are considered.

17.6 Density-Functional Methods

Perhaps surprisingly, all information about the ground state of the electrons in atoms or molecules can be encoded into single function, the electron density $n(\mathbf{x})$. This fact has led to the development of a family of approximation methods which aim to identify $n(\mathbf{x})$. In this section, we present the key ideas of these methods and we present one important example, the Thomas–Fermi model.

17.6.1 The Role of Electron Densities

Consider an atom or a molecule with N electrons. The Hilbert space of the electrons $\mathcal{H}_N$ is spanned by the generalized basis $|\mathbf{x}_1, m_1; \mathbf{x}_2, m_2; \ldots; \mathbf{x}_N, m_N\rangle$ with respect to the coordinate vectors $\mathbf{x}_i$ and spin components m_i of each electron. We define the electron density $n_\psi(\mathbf{x})$ associated to any state vector $|\psi\rangle$ as

$$n_\psi(\mathbf{x}) := N \sum_{m_1, m_2, \cdots, m_Z} \int d^3x_2 \ldots d^3x_N \langle \mathbf{x}, m_1; \mathbf{x}_2, m_2; \ldots; \mathbf{x}_N, m_N | \psi \rangle$$
$$\times \langle \psi | \mathbf{x}, m_1; \mathbf{x}_2, m_2; \ldots; \mathbf{x}_N, m_N \rangle, \tag{17.52}$$

normalized so that $\int d^3x\, n = N$. The electron density is nothing but the expectation value of the number density operator $\hat{v}(\mathbf{x}, t)$ that was defined in Section 15.4.1.

It is straightforward to show that for any function $f(\mathbf{x})$

$$\langle \psi | \sum_{i=1}^{N} f(\hat{\mathbf{x}}_i) | \psi \rangle = \int d^3x\, n_\psi(\mathbf{x}) f(\mathbf{x}). \tag{17.53}$$

The density $n_\psi(\mathbf{x})$ contains only a small fraction of the information encoded in the quantum state $|\psi\rangle$, as it ignores all correlations between electrons. For this reason, the map $|\psi\rangle \to n_\psi$ is many-to-one, that is, the different $|\psi\rangle$ map to the same density $n_\psi(\mathbf{x})$. We will denote the subset of vectors that map to the same density n by S_n.

The Hamiltonian for N electrons is $\hat{H} = \hat{T} + \hat{U}_e + \hat{V}_{en}$, where $\hat{T}$ is the kinetic term, $\hat{U}_e$ contains the repulsive interactions between electrons, and $\hat{V}_{en} = \sum_i V(\mathbf{x}_i)$, where $V(\mathbf{x}_i)$ sums over all interactions of the ith electron with the nuclei. The terms $\hat{T}$ and $\hat{U}_e$ are *universal*, that is, they are the same for all systems of N electrons, either atomic or molecular. In contrast, the term $\hat{V}_{en}$ is system-specific: For an atom or an ion, the potential $V = -\frac{Z\alpha}{r}$ where Z is the atomic number; for molecules with several nuclei V is a sum of several Coulomb terms.

By the variational principle, the ground state E_0 of the Hamiltonian is obtained from the minimization of $\langle \psi | \hat{T} + \hat{U}_e + \hat{V}_{en} | \psi \rangle$ over all normalized $|\psi\rangle \in \mathcal{H}_Z$. We can perform this minimization in two steps. First, we fix an electron density n and we minimize over all $|\psi\rangle \in S_n$. Then, we minimize over all possible n. Hence,

$$E_0 = \min_n \left[\min_{|\psi\rangle \in S_n} \langle \psi | \hat{T} + \hat{U}_e + \hat{V}_{en} | \psi \rangle \right]. \tag{17.54}$$

By Eq. (17.53), $\langle \psi | \hat{V}_{en} | \psi \rangle = \int d^3x\, n(\mathbf{x}) V(\mathbf{x})$, where n is the density associated to $|\psi\rangle$. We also define the functionals

$$T[n] := \min_{|\psi\rangle \in S_n} \langle \psi | \hat{T} | \psi \rangle, \quad U[n] := \min_{|\psi\rangle \in S_n} \langle \psi | \hat{U}_e | \psi \rangle.$$

Then, Eq. (17.54) becomes $E_0 = \min_n \left(T[n] + U[n] + \int d^3x\, n(\mathbf{x}) V(\mathbf{x}) \right)$, that is, the ground-state energy is obtained by minimizing the energy functional $E[n] := T[n] + U[n] + \int d^3x\, n(\mathbf{x}) V(\mathbf{x})$ with respect to the particle density $n(\mathbf{x})$.

Thus, we have replaced minimization with respect to all states with minimization with respect to all electron densities, which is a much simpler task computationally. Furthermore, the functionals $T[n]$ and $U[n]$ are universal; they do not vary from system to system. Unfortunately,

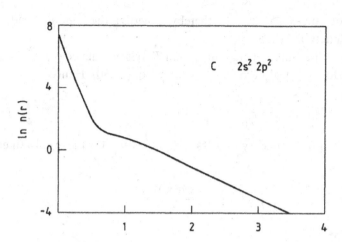

Fig. 17.3 A logarithmic plot of the spherically averaged density $n(r)$ as a function of r for the carbon atom. [Reprinted figure with permission from Jones and Gunnarsson (1989). Copyright 1989 by the American Physical Society.]

we do not know the explicit form of $T[n]$ and $U[n]$, and we have no way of computing it. Nonetheless, we can make some physically intuitive "guesses" about the form of these functionals, and these guesses lead to approximate solutions through the minimization of $E[n]$. This idea is the basis of the *density functional theory*, which has many applications in condensed matter physics and quantum chemistry.

In general, the electron density depends on the three spherical coordinates (r, θ, ϕ). For some applications, it is useful to employ, the spherically averaged density $\bar{n}(r) = (4n)^{-1} \int_0^\pi \sin\theta\, d\theta \int_0^{2\pi} d\phi\, n(r, \theta, \phi)$. A typical plot of $\bar{n}(r)$ for an atom is given in Fig. 17.3.

We can make some statements about the asymptotic behavior of $\bar{n}(r)$ for small and large r. As $r \to 0$, we expect that the nuclear Coulomb term dominates over electron repulsion, so $\bar{n}(r)$ should be proportional to the position probability density of the ground state of a hydrogen-like atom with a nuclear charge Ze. Hence, by Table 13.1,

$$\bar{n}(r) \sim (1 - 2Zr/a_0). \tag{17.55}$$

Indeed, Eq. (17.55) is known as *Kato's theorem*, and it can be proved rigorously for the many-body Coulomb Hamiltonian (Kato, 1957).

In the opposite limit, $r \to \infty$, an individual electron will see the other $Z - 1$ electrons and the nucleus forming a roughly spherical body of charge e. So at large r, we expect $\bar{n}(r)$ to behave like the position probability density of solutions of the hydrogen-atom Hamiltonian, that is, $\bar{n}(r) \sim \exp[-2\sqrt{2m_e|E|}r]$. The relevant energy $|E|$ for a remote electron is the atom's ionization energy E_{ion}, hence, we expect an asymptotic behavior,

$$\bar{n}(r) \sim e^{-2\sqrt{2m_e E_{ion}}r}. \tag{17.56}$$

17.6.2 The Thomas–Fermi Model

The preceding ideas originate from a model that was developed in the early days of quantum mechanics by Thomas (1927) and Fermi (1927). Thomas and Fermi were the first to describe the structure of atoms in terms of an electron density, thereby providing the impetus for the later development of the density-functional theory.

In the Thomas–Fermi model, the kinetic energy functional $T[n]$ is postulated equal to $\int d^3x\rho$, where ρ is the energy density for the ideal Fermi gas, given by Eq. (15.30). Hence,

$$T[n] = \frac{3(3\pi^2)^{2/3}}{10m_e} \int n^{5/3} d^3x. \tag{17.57}$$

The interaction functional is approximated by the classical expression for the repulsion energy of the charge distribution $en(\mathbf{x})$,

$$U[n] = \frac{\alpha}{2} \int d^3x d^3x' \frac{n(\mathbf{x})n(\mathbf{x}')}{|\mathbf{x} - \mathbf{x}'|}. \tag{17.58}$$

Hence, the energy functional is

$$E[n] = \int d^3x \left[\frac{3(3\pi^2)^{2/3}}{10m_e} n^{5/3} + n(\mathbf{x}) \left(V(\mathbf{x}) + \frac{\alpha}{2} \int d^3x' \frac{n(\mathbf{x}')}{|\mathbf{x} - \mathbf{x}'|} \right) \right]. \tag{17.59}$$

We minimize $E[n]$ subject to the condition that $\int d^3xn = N$, where N is the total number of electrons. To this end, we introduce a Lagrange multiplier μ, and minimize the functional $E[n] - \mu \left(\int d^3xn - N \right)$. The minimum is achieved for $n = n_0$, where

$$\frac{(3\pi^2)^{2/3}}{2m_e} n_0^{2/3} + V(\mathbf{x}) + \alpha \int d^3x' \frac{n(\mathbf{x}')}{|\mathbf{x} - \mathbf{x}'|} = \mu. \tag{17.60}$$

The constant μ plays the role of the chemical potential, that is, of the energy that is needed in order to increase the number of electrons by one. For a neutral system, no energy is expended in bringing in a new electron, so $\mu = 0$. In this case, Eq. (17.60) implies that

$$n_0 = \frac{1}{3\pi^2}(-2m_e V_{tot})^{3/2}, \tag{17.61}$$

where $V_{tot}(\mathbf{x}) = V(\mathbf{x}) + \alpha \int d^3x' \frac{n(\mathbf{x}')}{|\mathbf{x}-\mathbf{x}'|}$ is the total electrostatic potential.

For a single nucleus of charge Ze at $\mathbf{x} = 0$, $N = Z$, $V = -Z\alpha/r$ and V_{tot} satisfies the Poisson equation

$$\nabla^2 V_{tot}(\mathbf{x}) = 4\pi Z\alpha\delta^3(\mathbf{x}) - 4\pi\alpha n_0(\mathbf{x}). \tag{17.62}$$

We use Eq. (17.61) and the fact that $\nabla^2 r^{-1} = -4\pi\delta^3(\mathbf{r})$ to write Eq. (17.62) as

$$\nabla^2(V_{tot} + \frac{Z\alpha}{r}) = -\frac{4\alpha}{3\pi}(-2m_e V_{tot})^{3/2}. \tag{17.63}$$

For atoms, we assume rotational invariance; hence, V_{tot} depends only on r. We substitute $V_{tot}(r) = -\frac{Z\alpha}{r}\phi(r)$ in terms of some function $\phi(r)$, to obtain the *Thomas–Fermi equation*,

$$\sqrt{x}\frac{d^2\phi}{dx^2} = \phi^{3/2}, \tag{17.64}$$

where we set $x = Z^{1/3}br/a_0$ with $b = 2(4\pi/3)^{2/3} \simeq 5.2$.

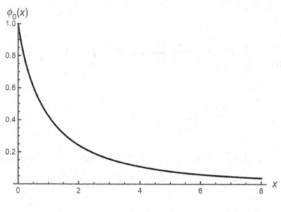

Fig. 17.4 The Thomas–Fermi function.

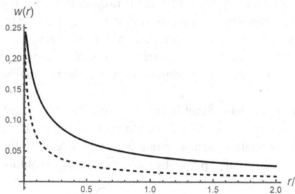

Fig. 17.5 The function $w(r) = 4\pi r^2 n_0(r)/Z$ is plotted as a function of r/a_0 for $Z = 20$ (solid line) and $Z = 80$ (dashed line).

As $r \to 0$, V_{tot} contains only the nuclear contribution, so $\phi(0) = 1$, while for a neutral atom $\phi(\infty) = 0$. Equation (17.64) admits a unique solution ϕ_0 that satisfies these boundary conditions; ϕ_0 is known as the *Thomas–Fermi function*.

In Problem 17.16, the asymptotic behavior of ϕ_0 is derived: For $x << 1$, $\phi_0(x) = 1 + cx + \frac{4}{3}x^{3/2} + \frac{2c}{5}x^{5/2} + \cdots$, where $c = -1.588\ldots$. For $x >> 1$, $\phi(x) \simeq 144x^{-3}$. We plot ϕ_0 in Fig. 17.4.

By Eq. (17.61),

$$n_0(r) = \frac{1}{3\pi^2}\left(\frac{2Z}{a_0 r}\right)^{3/2}\phi_0(Zbr/a_0)^{3/2}. \tag{17.65}$$

The density n_0 diverges as $r^{-3/2}$ near the nucleus, and decays with r^{-6} as $r \to \infty$. It does not have the asymptotic behavior expected from a physical electron density.

The fraction of the electron charge in a thin-shell of width δr at r equals $w(r)\delta r$ where $w(r) = 4\pi r^2 n_0(r)/Z$. This quantity is everywhere finite in the Thomas–Fermi model. In Fig. 17.5, we plot w_0 as a function of r.

The virial theorem, Eq. (17.61), is straightforwardly generalized to systems of many particles that interact through Coulomb forces – see Problem 17.17. The theorem asserts that for eigenstates of the Hamiltonian, the expectation of the kinetic energy is $-\frac{1}{2}$ times the expectation of the potential. It implies that

$$E[n_0] = -T[n_0] = -\frac{3(3\pi^2)^{2/3}}{10m_e} \int d^3x n_0^{5/3} = -\frac{6\pi^2}{5m_e}(9\pi)^{1/3} \int_0^\infty dr r^2 n_0^{5/3}.$$

Substituting n_0 from Eq. (17.65), we obtain

$$E[n_0] = -\lambda Z^{7/3}\frac{1}{2m_e a_0^2}, \tag{17.66}$$

where $\lambda = \frac{16}{5\pi}(3\pi/4)^{1/3} \int_0^\infty \frac{dx}{\sqrt{x}}\phi_0(x)^{5/2} \simeq 1.537$. We note that the energy of Z noninteracting electrons in a neutral atom also scales with $Z^{7/3}$ – see Problem 15.8 – but with a prefactor larger than λ (~ 2.28).

The error of the Thomas–Fermi model is 54 percent for the hydrogen atom, but it drops to 15 percent for large Z. This is a rather large error, but this is to be expected given the crudity of the model. The model also extends to molecules, but it has been proven that it is not compatible with stable chemical bonds (Teller, 1962). Chemical bonds are related to finer properties of the state vector of an atom which are better described by orbital theory than by the analysis of the average behavior of all electrons.

It turns out that the Thomas–Fermi model is exact in the $Z \to \infty$ limit (Lieb and Simon, 1973). This suggests that the Thomas–Fermi model is a cornerstone of atomic physics, an exact solution that can serve as a first approximation for understanding some aspects of the atomic structure. In this sense, its role is analogous to that played by the hydrogen-atom solutions in the regime of small Z.

17.7 The Born–Oppenheimer Approximation

Many physical systems are characterized by very large differences in the characteristic scales of their degrees of freedom. For example, in atoms and molecules, the nuclei are much heavier than electrons. In such systems, the Born–Oppenheimer approximation greatly simplifies all calculations of the Hamiltonian's spectrum. In this section, we will describe the most general form of the Born–Oppenheimer approximation, and then we will employ it for molecules.

Consider a system described by a Hilbert space $\mathcal{H} = \mathcal{H}_l \otimes \mathcal{H}_h$, where $\mathcal{H}_l$ hosts light degrees of freedom and $\mathcal{H}_h$ hosts heavy degrees of freedom. The Hamiltonian is

$$\hat{H} = \hat{H}_l \otimes \hat{I} + \hat{I} \otimes \hat{H}_h + \hat{H}_{int}, \tag{17.67}$$

where $\hat{H}_l$ describes the light degrees of freedom, $\hat{H}_h$ describes the heavy degrees of freedom and $\hat{H}_{int}$ is an interaction term.

Our first step is to find a (generalized) basis $|\xi\rangle$ on $\mathcal{H}_h$ that diagonalizes the interaction term, so that for every $|\psi_l\rangle \in \hat{H}_l$,

$$\hat{H}_{int}(|\psi_l\rangle \otimes |\xi\rangle) = \hat{V}(\xi)|\psi_l\rangle \otimes |\xi\rangle. \tag{17.68}$$

Here $\hat{V}(\xi)$ is an operator on $\hat{H}_l$ that depends parametrically on ξ, in the sense that ξ plays the role of a constant parameter on $\hat{V}$.

Next, we solve the eigenvalue equation for the operator $\hat{H}_l + \hat{V}(\xi)$ on the Hilbert space $\hat{H}_l$,

$$[\hat{H}_l + \hat{V}(\xi)]|\nu,\xi\rangle = \epsilon_\nu(\xi)|\nu,\xi\rangle, \tag{17.69}$$

where ν labels the eigenvalues. The eigenvalues $\epsilon_\nu(\xi)$ and the eigenvectors $|\nu,\xi\rangle$ depend parametrically on ξ. We assume that the spectrum is discrete, or we make it discrete by taking appropriate boundary conditions. Hence, for any given ξ, the eigenvectors $|\nu,\xi\rangle$ define a basis on $\mathcal{H}_l$, and the vectors $|\nu,\xi\rangle \otimes |\xi\rangle$ define a generalized basis on $\mathcal{H}$. We can express any vector $|\Psi\rangle \in \mathcal{H}$ as $\sum_\nu \int d\xi f_\nu(\xi)|\nu,\xi\rangle \otimes |\xi\rangle$.

The eigenvalue equation for the full Hamiltonian, $\hat{H}|\Psi\rangle = E|\Psi\rangle$, becomes

$$\sum_{\nu'} \int d\xi' f_{\nu'}(\xi')|\nu',\xi'\rangle \otimes [\hat{H}_h - E + \epsilon_{\nu'}(\xi')]|\xi'\rangle = 0.$$

We take the inner product with $\langle \nu,\xi| \otimes \langle\xi|$, to obtain

$$\sum_{\nu'} \int d\xi' f_{\nu'}(\xi')\langle \nu,\xi|\nu',\xi'\rangle\langle\xi|\hat{H}_h|\xi'\rangle + [\epsilon_\nu(\xi) - E]f_\nu(\xi) = 0. \tag{17.70}$$

Equation (17.70) is exact. It is simply a rewriting of the eigenvalue equation.

The Born–Oppenheimer approximation applies when $\langle \nu,\xi|\nu',\xi'\rangle \simeq 0$ if ξ substantially differs from ξ'. To make this statement more precise, we assume that some notion of norm $|\cdot|$ can be defined for ξ. Then, there exists a scale σ, such that $\langle \nu,\xi|\nu',\xi'\rangle \simeq 0$ for all ν and ν' if $|\xi-\xi'| > \sigma$. Assuming that $f_\nu(\xi)$ varies with ξ at scales much larger than σ, we can approximate

$$\langle \nu,\xi|\nu',\xi'\rangle \simeq \langle \nu,\xi|\nu',\xi\rangle = \delta_{\nu\nu'}, \tag{17.71}$$

so that Eq. (17.70) becomes

$$\int d\xi' \langle\xi|\hat{H}_h|\xi'\rangle f_\nu(\xi') + [\epsilon_\nu(\xi) - E]f_\nu(\xi) = 0. \tag{17.72}$$

We define $|f_\nu\rangle := \int d\xi f_\nu(\xi)|\xi\rangle$ and $\hat{\epsilon}_\nu := \int d\xi \epsilon_\nu(\xi)|\xi\rangle\langle\xi|$. Then, Eq. (17.72) becomes an eigenvalue equation on $\mathcal{H}_h$,

$$\left(\hat{H}_h + \hat{\epsilon}_\nu\right)|f_\nu\rangle = E|f_\nu\rangle. \tag{17.73}$$

Hence, with the Born–Oppenheimer approximation we replace an eigenvalue problem in $\mathcal{H}_l \otimes \mathcal{H}_h$ with eigenvalue problems on the smaller Hilbert spaces $\mathcal{H}_l$ and $\mathcal{H}_h$. The accuracy in the Born–Oppenheimer approximation is of the order of σ/L, where L is the scale of variation of $f_\nu(\xi)$.

As an application, consider the Hamiltonian (15.16) for N electrons and K nuclei that interact through Coulomb forces. The light degrees of freedom are the electrons, hence,

$$\hat{H}_l = \sum_{i=1}^{N} \frac{\hat{\mathbf{p}}^2}{2m_e} + \sum_{1\leq i<j\leq N} \frac{\alpha}{|\hat{\mathbf{x}}_i - \hat{\mathbf{x}}_j|}, \tag{17.74}$$

where $\hat{\mathbf{x}}_i$ and $\hat{\mathbf{p}}_i$ are the position and momentum operators, respectively, of the electron with index i. The heavy degrees of freedom correspond to nuclei; hence,

$$\hat{H}_h = \sum_{a=1}^{K} \frac{\hat{\mathbf{P}}_a^2}{2M_a} + \sum_{1\leq a<b\leq K} \frac{Z_a Z_b \alpha}{|\hat{\mathbf{X}}_a - \hat{\mathbf{X}}_b|}, \tag{17.75}$$

where the nucleus with index a has atomic number Z_a, mass M_a, and it is described by the position operator $\hat{\mathbf{X}}_a$ and the momentum operator $\hat{\mathbf{P}}_a$.

The interaction Hamiltonian is

$$\hat{H}_{int} = -\sum_{i=1}^{N} \sum_{a=1}^{K} \frac{Z_a \alpha}{|\hat{\mathbf{x}}_i - \hat{\mathbf{X}}_a|}. \tag{17.76}$$

The heavy degrees of freedom appear in $\hat{H}_{int}$ through the position operators $\hat{\mathbf{X}}_a$ of the nuclei. Hence, the generalized basis $|\mathbf{X}_a\rangle$ plays the role of $|\xi\rangle$. It follows that

$$\hat{V}(\mathbf{X}_a) = -\sum_{i=1}^{N} \sum_{a=1}^{K} \frac{Z_a \alpha}{|\hat{\mathbf{x}}_i - \mathbf{X}_a|}, \tag{17.77}$$

so that the positions of the nuclei are treated as classical parameters in the eigenvalue equation (17.69).

Equation (17.73) is the eigenvalue equation for a Schrödinger operator $\sum_{a=1}^{K} \frac{\hat{\mathbf{P}}_a^2}{2M_a} + V_{eff}(\hat{\mathbf{X}}_a)$ with effective potential

$$V_{eff}(\mathbf{X}_a) = \sum_{1 \le a < b \le K} \frac{Z_a Z_b \alpha}{|\hat{\mathbf{X}}_a - \hat{\mathbf{X}}_b|} + \epsilon_\nu(\mathbf{X}_a). \tag{17.78}$$

In the rest-mass frame for the nuclei, and for $\nu = 0$, the ground state for the electrons, the effective potential typically has a sharp minimum at $\mathbf{X}_a = \mathbf{X}_a^{(0)}$. We expand the effective potential as a Taylor series around the minimum,

$$V_{eff}(\mathbf{X}_a) = V_{eff}(\mathbf{X}_a^{(0)}) + \frac{1}{2} \sum_{1 \le a, b \le K} \frac{\partial^2 V}{\partial \mathbf{X}_a \partial \mathbf{X}_b}(\mathbf{X}_a - \mathbf{X}_a^{(0)})(\mathbf{X}_b - \mathbf{X}_b^{(0)}) \tag{17.79}$$

and find the $3K - 3$ eigenfrequencies ω_s for oscillations around equilibrium, where $s = 1, 2, \ldots, 3K - 3$. (Three degrees of freedom correspond to the motion of the center of mass.) The energy for the electrons is then well approximated by $\epsilon_0(\mathbf{X}_a^{(0)})$.

Example 17.6 We analyze the domain of validity of the Born–Oppenheimer approximation in a simple model of two harmonic oscillators. The light oscillator has Hamiltonian $\hat{H}_l = \frac{\hat{p}^2}{2m} + \frac{1}{2}m\omega^2 \hat{x}^2$ and the heavy oscillator has Hamiltonian $\hat{H}_h = \frac{\hat{P}^2}{2M} + \frac{1}{2}M\Omega^2 \hat{X}^2$. The interaction term is $\hat{H}_{int} = \lambda \hat{x}\hat{X}$.

The effective potential for the light oscillator is $V(x, X) = \frac{1}{2}m\omega^2 x^2 + \lambda X x = \frac{1}{2}m\omega^2(x + \frac{\lambda X}{m\omega^2})^2 - \frac{\lambda^2}{m^2\omega^4}X^2$. This potential corresponds to a harmonic oscillator centered around $x_0 = -\frac{\lambda X}{m\omega^2}$. Hence, the energy eigenvectors are $|n, X\rangle = e^{i\frac{\lambda X}{m\omega^2}\hat{p}}|n\rangle$, where $|n\rangle$ are the eigenvectors of $\hat{H}_l$. It follows that

$$\langle n', X'|n, X\rangle = \langle n'|e^{i\frac{\lambda}{m\omega^2}(X - X')\hat{p}}|n\rangle. \tag{17.80}$$

The characteristic length scale of $|n\rangle$ is $(m\omega)^{-1/2}$; all eigenfunctions of the harmonic oscillators decay asymptotically as $e^{-\frac{1}{2}m\omega x^2}$. Hence, the overlap $\langle n', X'|n, X\rangle$ decays with $\exp\left[-\frac{\lambda^2}{2m\omega^3}(X - X')^2\right]$, and it is negligible if

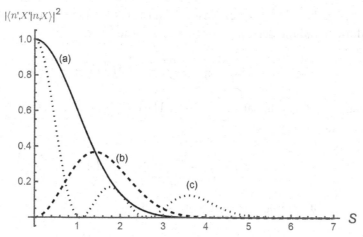

Fig. 17.6 The overlap $|\langle n', X'|n, X\rangle|^2$ of the light oscillator eigenstates as a function of $S = \frac{\lambda}{\sqrt{m\omega^3}}|X - X'|$, for different values of n and n': (a) $n = n' = 0$, (b) $n = 0, n' = 1$, (c) $n = 2, n' = 2$. The overlap becomes negligible for $S \gtrsim 5$.

$$\frac{\lambda}{\sqrt{m\omega^3}}|X - X'| \gg 1. \tag{17.81}$$

Indeed, this can be seen in Fig. 17.6, where we plot the overlap $|\langle n', X'|n, X\rangle|^2$ as a function of $\frac{\lambda}{\sqrt{m\omega^3}}|X - X'|$.

The difference $|X - X'|$ varies on the characteristic scale $(M\Omega)^{-1/2}$ of the heavy oscillator. Hence, the necessary condition for the validity of the Born–Oppenheimer approximation is that

$$\frac{\lambda}{\sqrt{mM\Omega\omega^3}} \gg 1. \tag{17.82}$$

The Born–Oppenheimer approximation requires a sufficiently strong coupling between heavy and light degrees of freedom. It works in the opposite regime from perturbation theory.

Example 17.7 Optomechanical systems are characterized by one or more modes of the EM field interacting with a mechanical degree of freedom. Consider the EM field in an ideal cavity in which one mirror is movable, so that the length L of the cavity oscillates around an equilibrium value L_0. The moving mirror has mass M, and it oscillates with a frequency ω. We focus on a single mode of the field at the resonance frequency π/L, and we treat the wall as a quantum system so that $x := L - L_0$ is promoted to a self-adjoint operator. The Hamiltonian is

$$\hat{H} = \frac{\pi}{L_0 + \hat{x}}\hat{a}^\dagger\hat{a} + \frac{\hat{p}^2}{2M} + \frac{1}{2}M\omega^2\hat{x}^2, \tag{17.83}$$

where $\hat{a}$ is the annihilation operator for the EM field mode, and $\hat{p}$ is the conjugate momentum to $\hat{x}$. Taking the EM mode as the light degree of freedom, we find that its energy eigenvalues parameterized by x are $\epsilon_n(x) = n\frac{\pi}{L_0+x}$, where $n = 0, 1, 2, \ldots$.

For fixed n, the effective potential for the mirror is $V_{eff}(x) = \frac{1}{2}M\omega^2 x^2 + n\frac{\pi}{(L_0+x)}$. For $x \ll L_0$, we Taylor-expand the denominator to second order in x,

$$V_{eff}(x) = \frac{\pi}{L_0} - \frac{n\pi}{L_0^2}x + \frac{1}{2}M\left(\omega^2 + \frac{2n\pi}{ML_0^3}\right)x^2. \tag{17.84}$$

The minimum of V_{eff} lies at $x_0 = \frac{n\pi}{ML_0^2\omega^2}$. The energy eigenvalues for the EM field are obtained by evaluating $\epsilon_n(x)$ at $x = x_0$,

$$\epsilon_n(x_0) = \frac{\pi}{(L_0 + \frac{n\pi}{ML_0^2\omega^2})} \simeq \frac{\pi}{L_0}\left(1 - \frac{n\pi}{ML_0^3\omega^2}\right).$$

The approximation holds for sufficiently small n, so that $\frac{n\pi}{ML_0^3\omega^2} \ll 1$.

QUESTIONS

17.1 A theorist claims to have defined the Hamiltonian of Everything, not by explicit construction, but by providing an algorithm that generates all terms in a perturbation expansion around a known Hamiltonian. Suppose the algorithm works and gives finite results for each individual term. Would you accept the theorist's claims?

17.2 Find the dimensionless constant that must be much smaller than unity for the perturbative treatment of the Stark effect in the hydrogen atom to be meaningful. Do the same for the Zeeman effect.

17.3 Do you expect the relative size of the fine-structure corrections to increase with the atomic number Z?

17.4 Why is the contribution of the nucleus to the dipole electric moment of an atom much smaller than the contribution of the electrons? What about dipole magnetic moments?

17.5 Two different variational calculations of the ground-state energy of a system give values E_1 and E_2 with $E_1 < E_2$. Is there a reason to prefer one over the other?

17.6 What test functions would you use for the Schrödinger operator in three dimensions with central potential $V(r) = -b/r^{3/2}$ for $b > 0$?

17.7 **(B)** Which physical assumption in the Thomas–Fermi theory leads to the bad behavior of the electron density at small and large r?

17.8 **(B)** Estimate the accuracy of the Born–Oppenheimer approximation for molecules. To this end, take the scale L of variation of the electronic state to equal a_0, and identify σ as the natural length scale of oscillations for nuclei. Take typical electronic energies to be of the order of $1/(ma_0^2)$ as in the hydrogen atom.

PROBLEMS

17.1 A harmonic oscillator of frequency ω is perturbed by the potential $g\hat{x}^3$. Calculate the first- and second-order corrections to the energy eigenvalues.

17.2 A particle in a one-dimensional box of length L is perturbed by a weak potential $V(x) = V_0\cos(\pi x/L)$. (i) Show that the first-order corrections to the energy vanish. (ii) Calculate the correction in the ground state. (iii) Find the second-order correction to the energy for the ground state and the first excited state of the system.

17.3 Consider two fermionic oscillators of spin $s = \frac{1}{2}$ of frequency ω interacting through a potential $\hat{V} = \lambda \hat{x}_1^2 \otimes \hat{x}_2^2$. Find the corrections to energy for the three lowest energy levels of the unperturbed system.

17.4 Show that, to second order in perturbation theory, the correction ΔE of the ground-state energy of the hydrogen atom due to an external homogeneous electric field $\mathcal{E}$ is $\Delta E = -\frac{1}{2}b\mathcal{E}^2$, where $b = 2e^2 \sum_{n=2}^{\infty} \frac{|\langle 1,0,0|\hat{x}_3|n,1,0\rangle|^2}{E_n - E_1}$. Use the identity $E_n \geq E_2$ for $n \geq 2$ to show that $b < \frac{16}{3}4\pi a_0^3$.

17.5 Consider two hydrogen atoms in the ground state at distance $R \gg a_0$. In the dipole approximation, the classical interaction energy between the two atoms is

$$W = \frac{e^2}{4\pi R^3}[\mathbf{r}_1 \cdot \mathbf{r}_2 - 3(\mathbf{r}_1 \cdot \mathbf{n})(\mathbf{r}_2 \cdot \mathbf{n})],$$

where $\mathbf{r}_i$ is the position vector of each electron with respect to the corresponding nucleus, and $\mathbf{n}$ is the unit vector along the line that connects the two nuclei. Assume that the quantum version of W defines an interaction term. (ii) Show that the first-order correction vanishes. (ii) Write the second-order correction, and then, take the (rather rough) approximation $E_n \ll E_1$ for $n > 1$. The resulting force is proportional to R^{-6}, like the van der Waals forces.

17.6 Suppose that the proton in the hydrogen atom is a sphere of radius R, where the charge $+e$ is homogeneously distributed. Use perturbation theory in order to calculate the change in the ground-state energy of the hydrogen atom.

17.7 Find the first-order correction to the energy eigenvalues for the $n = 3$ eigenspace of the hydrogen atom in presence of a weak external electric field.

17.8 In classical electrodynamics, the potential energy for two interacting dipoles of magnetic moments $\boldsymbol{\mu}_1$ and $\boldsymbol{\mu}_2$ is

$$V = -\frac{2}{3}\boldsymbol{\mu}_1 \cdot \boldsymbol{\mu}_2 \delta(\mathbf{r}) - \frac{(3\boldsymbol{\mu}_1 \cdot \mathbf{r})(\boldsymbol{\mu}_2 \cdot \mathbf{r}) - \boldsymbol{\mu}_1 \cdot \boldsymbol{\mu}_2 r^2}{4\pi r^5},$$

where $\mathbf{r}$ is the vector that connects the two dipoles. We use the quantum version of this term, in order to describe the interaction between the dipole moments of the nucleus and the electron in the hydrogen atom. We write $\hat{\mu}_1 = -\frac{e}{m_e}\hat{\mathbf{S}}_e$ and $\hat{\mu}_2 = \frac{e}{2m_p}g_p\hat{\mathbf{S}}_p$. We consider the first-order correction to the ground state of the hydrogen atom, for which the contribution from the electron's orbital angular momentum vanishes. (i) Show that the contribution from the second term in V vanishes. (ii) Show that the total first-order correction is

$$\Delta E = \frac{e^2 g_p}{3m_e m_p}|\psi_0(0)|^2 \hat{\mathbf{S}}_e \cdot \hat{\mathbf{S}}_p,$$

where ψ_0 is the wave function for the ground state of the hydrogen atom. (iii) Show that the degenerate ground state of the hydrogen atom splits into two levels that depend on the total angular momentum F of the atom. In particular, show that

$$\Delta E = \begin{cases} -\dfrac{a^2 g_p}{m_p a_0^2}, & F = 0 \\[2mm] \dfrac{a^2 g_p}{3 m_p a_0^2} & F = 1. \end{cases}$$

Comment This split defines the *hyperfine structure* of the spectrum of the hydrogen atom. The photon energy in the hyperfine transitions is $6 \cdot 10^{-6}$ eV and the associated wavelength is 21 cm.

17.9 Use Eq. (17.34) in order to identify the Zeeman-effect corrections to the energy of the hydrogen atom for $n = 2$.

17.10 (i) Use a variational method to estimate the ground-state energy of a particle of mass m that moves in the real line under an unharmonic potential $V = \lambda \hat{x}^4$, where $\lambda > 0$. (ii) Do the same assuming that the particle moves in the half-line.

17.11 Let $E(\psi) = \langle \psi | \hat{H} | \psi \rangle$ be the energy functional for a Hamiltonian $\hat{H}$; the states $|\psi\rangle$ are normalized to unity. (i) Show that for a harmonic oscillator of mass m and frequency ω,

$$E(\psi) = \frac{1}{2m} \int dx \left| \psi' + m\omega x \psi \right|^2 + \frac{1}{2}\omega,$$

and thus verify the standard expressions for the ground-state energy and wave function. (ii) Follow a similar procedure for a particle of mass m in the Pöschl–Teller potential $V(x) = -\dfrac{V_0}{\cosh^2(ax)}$, where $V_0, a > 0$, to show that

$$E(\psi) = \frac{1}{2m} \int dx \left| \psi' + \lambda \tanh(ax)\psi \right|^2 - \frac{\lambda^2}{2m},$$

where λ satisfies $\lambda^2 + a\lambda - 2mV_0 = 0$. Find the ground-state energy and wave function.

17.12 Evaluate the energy of the first excited state for the quantum bouncing ball using the variational method.

17.13 Let $\hat{H}$ be a Hamiltonian with eigenvectors $|\psi_n\rangle$ and eigenvalues E_n, $n = 0, 1, 2, \ldots$, with nondegenerate ground state. Let $|\psi\rangle$ be a test function. Show that we can always choose the phase of $|\psi\rangle$, so that

$$\frac{1}{2}||\psi - \psi_0||^2 \le 1 - \sqrt{\frac{E_1 - \langle \hat{H} \rangle}{E_1 - E_0}}.$$

Show that to the lowest order with respect to the error $\Delta := \langle \hat{H} \rangle - E_0$ of the variational calculation, $||\psi - \psi_0|| \le \sqrt{\frac{\Delta}{E_1 - E_0}}$.

17.14 (B) Show that the Gross–Pitaevskii equation in one dimension with no external potential admits solutions $\phi(x) = C \tanh(x/L)$ for $a > 0$ and $\phi(x) = C'[\cosh(x/L')]^{-1}$ for $a < 0$. Specify the values of L and L'.

17.15 (B) Let $n_0(\mathbf{x})$ be the density associated to the ground state $|0\rangle$ of a Hamiltonian $\hat{H} = \hat{T} + \hat{U}_e + \hat{V}_{en}$, as in Section 17.6.1. Prove that n_0 uniquely specifies the potential V in V_{en}, as long as the ground state is nondegenerate. Assume that there are two potentials V and V' with the same $n_0(\mathbf{x})$; hence, two different Hamiltonians $\hat{H}$ and $\hat{H}'$ and two different ground states $|0\rangle$ and $|0'\rangle$ and reach a contradiction.

Comment This result is known as the first Hohenberg–Kohn theorem (Hohenberg and Kohn, 1964). It implies that n_0 uniquely specifies the Hamiltonian $\hat{H}$ and, hence, the ground state $|0\rangle$.

17.16 (B) (i) Show that all solutions to the Thomas–Fermi equation (TFE) with $\phi(0) = 1$ behave as $\phi(x) = 1 + cx + \frac{4}{3}x^{3/2} + \cdots$ for x near zero, where c is an unspecified constant. (ii) Show that the function $\phi_s(x) = \frac{144}{x^3}$ solves the TFE. (iii) Show that all solutions of the TFE that vanish at infinity, have an asymptotic behavior $\phi(x) = \frac{144}{x^3} + \frac{c'}{x^\sigma} + \cdots$, where c' is a constant and $\sigma = \frac{1}{2}(1 + \sqrt{73}) \simeq 4.8$.

17.17 (B) Generalize the virial theorem to many particles interacting through Coulomb forces. Show that, in any eigenstate of the Hamiltonian, $2\langle \hat{T} \rangle = \langle \hat{V} \rangle$, where $\hat{T}$ is the total kinetic energy and $\hat{V}$ is the total potential energy.

17.18 (B) Given an electron density $n(r)$, the fraction w of the electric charge within a sphere of radius R around the nucleus is given by $w = \frac{4\pi}{Z} \int_0^R dr\, r^2 n(r)$. (i) Use the electron density (17.65) of the Thomas–Fermi theory to show that $w = \frac{1}{4\pi^3} \int_0^{bZR/a_0} dx\, x \phi_0''$. (ii) Evaluate the integral for sufficiently large R, so that $\phi_0(x) \simeq 144/x^3$, and show that $R = C_w a_0 Z^{-1/3}$, with $C_w = \frac{3^{1/3}}{4\pi^{5/3}(1-w)^{1/3}}$. Compute C_w for $w = 0.9$.

Comment The Thomas–Fermi theory suggests that the size of the atom *decreases* with the atomic number. In real atoms, this is true only along the rows of the periodic system. As we move from one row to another, the effective size of the atom increases because the outer electrons have a higher value of the quantum number n (in the orbital picture); hence, they are further away from the nucleus. This behavior cannot be seen in the Thomas–Fermi theory, which considers only the average behavior of electrons.

17.19 (B) A quantum system is described by the Hamiltonian $\hat{H} = \frac{\hat{p}^2}{2m} + \frac{\hat{P}^2}{2M} + a|\hat{Q}|^{-1} + \lambda \hat{x}^2 \hat{Q}^2$, where $\lambda, a > 0$, $\hat{x}$ and $\hat{p}$ are conjugate variables of position and momentum, and the same applies to $\hat{Q}$ and $\hat{P}$. The masses satisfy $m << M$, so you can use the Born–Oppenheimer approximation. (i) Find the equilibrium configuration of the heavy particle, and calculate the frequency of small oscillations around equilibrium. (i) Evaluate the energy of the ground state for the light particle.

Bibliography

- For more details about perturbation theory in quantum mechanics, see Fernandez (2001).
- For calculational techniques of the spectrum and structure of atoms, see the quantum chemistry book by Levine (2013). For the mean-field approximation for fermions, see chapter 18 of Ballentine (1998). For a brief introduction to the density functional theory, see Argaman and Makov (2000). For the Thomas–Fermi theory, see the review article by March (1957).
- For the accuracy of the Born–Oppenheimer approximation, see Jecko (2014).

18 Transitions and Decays

After the exponential law in radioactive decay had been discovered in 1902, it soon became clear that the time of disintegration of an atom was as independent of the previous history of the atom as it was of its physical condition. One could not for example suppose that an atom at its birth begins to lose energy by radiation and that its instability is the result of the drain of energy from the nucleus. On such a view it would be expected that the rate of decay would increase with the age of the atoms. [Eventually] ... it became clear that the disintegrating depended solely on chance. This has been very puzzling so long as we have accepted a dynamics by which the behaviour of particles is definitely fixed by the conditions ... Now, however, we throw the whole responsibility on to the laws of quantum mechanics, recognizing that the behaviour of particles everywhere is equally governed by probability. *

Gurney and Condon (1927)

18.1 Perturbative Treatment of Transitions

In this section, we will analyze transitions between energy eigenstates caused by a transient external force, for example, an EM pulse. We will assume that the external force is weak so that we can describe these decays by a version of perturbation theory. This results into very simple expressions for the transition probabilities and rates with universal validity, that is, they apply to all kinds of phenomena from atomic to nuclear and high-energy physics.

18.1.1 Time Evolution in the Presence of an External Interaction

Consider a system described by a Hamiltonian $\hat{H}_0$ that is disturbed by an external agent. The disturbance takes the form of an additional term $\hat{V}$ in the Hamiltonian that is time-dependent; it becomes nonzero at time $t = 0$ and it has finite duration T. In order to describe the dynamics of this system, we must construct the evolution operator $\hat{U}(t)$ associated to the Hamiltonian $\hat{H} = \hat{H}_0 + \hat{V}$. This operator satisfies Eq. (7.68), subject to the initial condition $\hat{U}(0) = \hat{I}$.

We isolate the effects of the perturbation in the dynamics, by defining the unitary operator $\hat{O}_t := e^{i\hat{H}_0 t}\hat{U}(t)$, which satisfies

$$i\frac{d}{dt}\hat{O}_t = -e^{i\hat{H}_0 t}\hat{H}_0\hat{U}(t) + e^{i\hat{H}_0 t}\hat{H}\hat{U}(t)$$

$$= e^{i\hat{H}_0 t}\hat{V}\hat{U}(t) = e^{i\hat{H}_0 t}\hat{V}e^{-i\hat{H}_0 t}\hat{U}(t) = \hat{V}(t)\hat{O}_t, \tag{18.1}$$

where $\hat{V}(t) = e^{i\hat{H}_0 t}\hat{V}e^{-i\hat{H}_0 t}$ is the Heisenberg-type evolution of $\hat{V}$ with the unperturbed Hamiltonian.

Equation (18.1) is of the form (7.68). Hence, we can write the operator $\hat{O}_t$ as a time-ordered exponential

$$\hat{O}_t = \mathcal{T}e^{-i\int_0^t ds\hat{V}(s)}. \tag{18.2}$$

Therefore, we write $\hat{U}(t) = e^{-i\hat{H}_0 t}\hat{O}_t$. We note that $\hat{O}_t^\dagger$ satisfies $-i\frac{d}{dt}\hat{O}_t^\dagger = \hat{O}_t^\dagger\hat{V}(t)$, which implies that $\hat{U}(t) = \hat{O}_t^\dagger e^{i\hat{H}_0 t}$.

If the disturbance is nonzero only in an interval $[0, T]$, then, $\hat{O}_t = \mathcal{T}e^{-i\int_0^T ds\hat{V}(s)}$ for all $t > T$, that is, $\hat{O}_t$ is time-independent, and we often denote it by $\hat{S}$. Then, we can extend the range of time integration in Eq. (18.2) to the whole of the real axis,

$$\hat{S} = \mathcal{T}e^{-i\int_{-\infty}^\infty ds\hat{V}(s)}, \tag{18.3}$$

and we can write $\hat{U}(t) = e^{-i\hat{H}_0 t}\hat{S} = \hat{S}^\dagger e^{i\hat{H}_0 t}$. The operator $\hat{S}$ is related to the so-called *scattering operator* that is analyzed in Section 19.6.

Example 18.1   Consider an instantaneous interaction at time t_0, $\hat{V} = \delta(t - t_0)\hat{A}$ for some self-adjoint operator $\hat{A}$. By Eq. (7.74), the operator $\hat{S}$ involves integrals of the form

$$\int_{-\infty}^\infty dt_1 \cdots \int_{-\infty}^\infty dt_n \delta(t_1 - t_0) \cdots \delta(t_n - t_0)\mathcal{T}[\hat{A}(t_1) \cdots \hat{A}(t_n)] = [\hat{A}(t_0)]^n,$$

where $\hat{A}(t) = e^{i\hat{H}_0 t}\hat{A}e^{-i\hat{H}_0 t}$. It follows that

$$\hat{S} = \sum_{n=0}^\infty \frac{[-i\hat{A}(t_0)]^n}{n!} = e^{-i\hat{A}(t_0)} = e^{i\hat{H}_0 t_0}e^{-i\hat{A}}e^{-i\hat{H}_0 t_0}. \tag{18.4}$$

Hence, the evolution operator is

$$\hat{U}(t) = e^{-i\hat{H}_0(t-t_0)}e^{-i\hat{A}}e^{-i\hat{H}_0 t_0} = e^{i\hat{H}_0 t_0}e^{i\hat{A}}e^{i\hat{H}_0(t-t_0)}. \tag{18.5}$$

18.1.2 Transition Probabilities

If the term $\hat{V}$ is small, we obtain approximate solutions to Eq. (18.2) by keeping a finite number of terms from the expansion (7.69). To first order in $\hat{V}$,

$$\hat{O}_t \simeq \hat{I} - i\int_0^t ds\hat{V}(s). \tag{18.6}$$

Let $\hat{V} = f(t)\hat{A}$, for some operator $\hat{A}$ and a function $f(t)$ that differs from zero only in a time interval $[0, T]$. Let $\hat{H}_0|n\rangle = E_n|n\rangle$, be the eigenvalue equation for $\hat{H}_0$, for some index n.

We take the initial state of the system to be an eigenvector $|n_0\rangle$ of $\hat{H}_0$. The probability $\text{Prob}(n, t)$ that the system is found in the state $|n\rangle$ at time t is

$$\text{Prob}(n, t) = |\langle n|e^{-i\hat{H}t}|n_0\rangle|^2 = |\langle n|e^{-i\hat{H}_0 t}\hat{O}_t|n_0\rangle|^2 = |\langle n|\hat{O}_t|n_0\rangle|^2.$$

For $|n\rangle \neq |n_0\rangle$, the approximation (18.6) yields

$$\langle n|\hat{O}_t|n_0\rangle \simeq -i \int_0^t ds f(s)\langle n|e^{i\hat{H}_0 s}\hat{A}e^{-i\hat{H}_0 s}|n_0\rangle = -iA_{nn_0}\int_0^t ds f(s)e^{-i(E_{n_0}-E_n)s}.$$

Assume that $t > T$; then we can extend the limits of the integral to $(-\infty, \infty)$. Hence, $\langle n|e^{-i\hat{H}t}|n_0\rangle = -iA_{nn_0}\tilde{f}(E_{n_0} - E_n)$, where $\tilde{f}$ is the Fourier transform of f. The probability $\text{Prob}(n, t)$ is time-independent, so we write

$$\text{Prob}(n) = |\tilde{f}(E_{n_0} - E_n)|^2 |A_{nn_0}|^2. \tag{18.7}$$

Equation (18.7) is valid, as long as $\text{Prob}(n)$ remains much smaller than unity.

Suppose that the external perturbation is harmonic, $f(t) = \cos(\Omega t)$ for some frequency Ω, and that its duration T is much larger than Ω^{-1}. It is convenient to assume that T is a large integer N times the period $2\pi/\Omega$. Then, $\tilde{f}(E) = \int_0^{2\pi N/\Omega} dt \cos(\Omega t)e^{-iEt} = i\frac{E}{E^2-\Omega^2}(e^{-2iN\pi E/\Omega} - 1)$, and

$$|\tilde{f}(E)|^2 = \frac{2E^2}{E^2 + \Omega^2}\left[\frac{1}{(E+\Omega)^2} + \frac{1}{(E-\Omega)^2}\right]\sin^2\frac{N\pi E}{\Omega}. \tag{18.8}$$

We introduce the function $\delta_\epsilon(x) := \frac{\epsilon \sin^2(x/\epsilon)}{\pi x^2}$, which, by Eq. (4.42), converges to a delta function as $\epsilon \to 0$. Noting that $\sin^2\frac{N\pi E}{\Omega} = \sin^2\frac{N\pi(E\pm\Omega)}{\Omega} = \sin^2\frac{(E\pm\Omega)T}{2}$, we write

$$|\tilde{f}(E)|^2 = \frac{\pi E^2 T}{E^2 + \Omega^2}[\delta_\epsilon(E - \Omega) + \delta_\epsilon(E + \Omega)],$$

where $\epsilon = 2/T$. It follows that

$$\frac{\text{Prob}(n_0 \to n)}{T} = \frac{\pi\Delta^2}{\Delta^2 + \Omega^2}|A_{nn_0}|^2[\delta_\epsilon(\Delta - \Omega) + \delta_\epsilon(\Delta + \Omega)],$$

where $\Delta = E_n - E_{n_0}$. For $T \to \infty$, the right-hand side of the equation above converges to a sum of delta functions. This means that the left-hand side must also converge as $T \to \infty$. We write this limit as $w_{n_0 \to n}$. This quantity has dimensions of inverse time and it defines a constant transition rate from $|n_0\rangle$ to $|n\rangle$. Thus, we have obtained *Fermi's golden rule*[1] for harmonically driven transitions:

$$w_{n_0 \to n} = \frac{\pi}{2}|A_{nn_0}|^2\left[\delta(E_n - E_{n_0} - \Omega) + \delta(E_n - E_{n_0} + \Omega)\right]. \tag{18.9}$$

For a harmonic perturbation of infinite duration, the transition rate is constant and it is nonzero only for transitions at exact resonance, where $\Omega = |E_n - E_{n_0}|$. For finite but large T, where we still use the approximate delta function, the transition rate is nonzero for a thin strip of frequencies of width T^{-1} around $|E_n - E_{n_0}|$.

We must point out that the derivation of Eq. (18.9) that we gave here is highly problematic. This is why we talk about Fermi's rule and not Fermi's theorem. The ratio $\frac{p_{ion}}{T}$ converges to a constant, only if $p_{ion} \to \infty$ as $T \to \infty$. This is incompatible with the definition of p_{ion} as a probability ($p_{ion} \leq 1$), and even more so with the perturbative approximation that requires $p_{ion} \ll 1$. Fermi's golden rule does work, but its proper derivation requires more sophisticated methods. In Section 18.3, we will provide a proper proof of Fermi's golden rule for decays.

---

[1] Fermi's golden rule is actually due to Dirac (1927), but receives the name of Fermi, who highlighted its importance.

18.1.3 Ionization Probability

Next, we discuss transitions from the initial bound state to the continuous spectrum of $\hat{H}_0$. One example is the ionization of an atom by an external EM pulse: The electron transitions from a bound state to a scattering state.

Let $|E, r\rangle$ be the generalized eigenvectors of $\hat{H}_0$ for energy E, where r is a degeneracy index. We follow the same reasoning as in the derivation of Eq. (18.7), and we evaluate the probability density of ionization $p(E)$ with final energy E, by summing over r,

$$p(E) = |\tilde{f}(E_{n_0} - E)|^2 A(E),\tag{18.10}$$

where

$$A(E) := \sum_r |\langle E, r|\hat{A}|n_0\rangle|^2 \tag{18.11}$$

is the *transition function*. Assuming that the continuous spectrum starts at $E = E_c$, $A(E) = 0$ for $E < E_c$.

We obtain the total ionization probability p_{ion} by integrating over all energies in the continuous spectrum $\sigma_c(\hat{H}_0)$ of $\hat{H}_0$:

$$p_{ion} = \int_{\sigma_c(\hat{H}_0)} dE|\tilde{f}(E_{n_0} - E)|^2 A(E).\tag{18.12}$$

For a harmonic perturbation $f(t) = \cos(\Omega t)$ of long duration T, the (not very rigorous) arguments of the previous section apply. They lead to a constant ionization rate $w_{ion} = p_{ion}/T$, given by Fermi's golden rule

$$w_{ion} = \pi \int dE \frac{(E - E_0)^2}{(E - E_0)^2 + \Omega^2} A(E)[\delta(E - E_0 - \Omega) + \delta(E - E_0 + \Omega)].\tag{18.13}$$

For $\Omega \neq 0$,

$$w_{ion} = \frac{\pi}{2}[A(E_0 + \Omega) + A(E_0 - \Omega)].\tag{18.14}$$

The ionization rate is nonzero only if $\Omega > E_c - E_{n_0}$.

To obtain the ionization rate for a constant perturbation, we set $\Omega = 0$ in Eq. (18.13) prior to the integration. Then, we obtain

$$w_{ion} = 2\pi A(E_0).\tag{18.15}$$

This version of Fermi's golden rule allows us to evaluate the *decay rate* of unstable quantum systems – see Section 18.2. We note that w_{ion} is nonzero only if there are states in the continuous spectrum with the same energy as the initial state $|n_0\rangle$. This means that the spectrum of the Hamiltonian must be of type (f) in Fig. 7.1, and that $|n_0\rangle$ cannot be the ground state of the system (which is usually assumed to be unique).

Example 18.2 Let $\hat{H}_0$ be the Schrödinger operator for a particle of mass m in the delta potential $V(x) = -\eta\delta(x)$, that was studied in Section 5.1. This Hamiltonian has a single eigenvector $\psi_0(x) = \sqrt{m\eta}e^{-m\eta|x|}$ with negative energy $E_0 = -\frac{m\eta^2}{2}$, and infinite generalized eigenvectors $|k, \pm\rangle$, given by Eq. (5.87), for positive energies $E = \frac{k^2}{2m}$ in the continuous spectrum. Here we will use the kets $|E, \pm\rangle = \sqrt{\frac{m}{k}}|k, \pm\rangle$, that satisfy the normalization condition $\langle E, \pm | E', \pm \rangle = \delta(E - E')$.

We consider an external pulse $\hat{V}(t) = f(t)\hat{x}$, where $f(t)$ is nonzero only in an interval $[0, T]$. By Eq. (5.88) and the identity $\int_0^\infty dx x e^{-ax} = a^{-2}$, we find that

$$\langle 0|\hat{x}|E, \pm\rangle = \sqrt{\frac{\eta}{2\pi}} \frac{i\eta(2mE)^{1/4}}{(E - E_0)^2}. \tag{18.16}$$

This implies that

$$A(E) = \begin{cases} \frac{\eta^3}{\pi} \frac{\sqrt{2mE}}{(E-E_0)^4}, & E \geq 0 \\ 0, & E < 0. \end{cases} \tag{18.17}$$

The ionization probability is

$$p_{ion} = \frac{\eta^3}{\pi} \int_0^\infty dE |\tilde{f}(E_0 - E)|^2 \frac{\sqrt{2mE}}{(E - E_0)^4}. \tag{18.18}$$

For a very brief pulse, we write $f(t) = \lambda \delta(t - t_0)$ for $\lambda > 0$ and $t_0 \in [0, T]$. Then, $\tilde{f}(E) = \lambda e^{-iEt_0}$, and

$$p_{ion} = \frac{\eta^3 \lambda^2}{\pi} \int_0^\infty dE \frac{\sqrt{2mE}}{(E - E_0)^4} = \frac{\lambda^2}{2m^2 \eta^2}, \tag{18.19}$$

where we used the integral (A.8).

By Eq. (18.15), the ionization rate for a constant pulse of long duration vanishes, since $A(E_0) = 0$. For a periodic perturbation of long duration, Eq. (18.14) implies that the ionization rate is nonzero for $\Omega > |E_0|$, and it equals $\frac{\pi}{2} A(\Omega - |E_0|)$.

18.2 Quantum Decays

A decay of an unstable particle A is a process of the type $A \rightarrow B_1 + B_2 + \cdots B_n$, where the particles B_i are the *decay products*. Decays are ubiquitous in physics. Examples include the emission of photons from excited atoms or nuclei, alpha and beta emission, decays of composite subatomic particles, such as neutrons or pions, and decays of elementary particles (for example, a muon decaying to one electron and two neutrinos).

Most decays follow an exponential law. The probability that a decay takes place within the time interval $[t, t + \delta t]$, for $t > 0$, equals $p(t)\delta t$, where the probability density $p(t)$ is given by

$$p(t) = \Gamma e^{-\Gamma t}. \tag{18.20}$$

The *decay constant* Γ is positive and has dimensions of inverse time.

Experiments typically involve a large number of decaying particles with identical preparation and the detection of decay products. Recording the number of detection events within given time intervals $[t, t + \delta t]$, we can reconstruct the probability density $p(t)$ associated to the decay. Hence, $p(t)$ is directly observable.

The exponential decay law was conceptually puzzling from the perspective of classical physics, for the reasons given by Gurney and Condon (1927) at the beginning of this chapter. There is no such problem in the quantum description of decays. Ideally, quantum theory should allow us to construct the observable probability density $p(t)$ from first principles. Note, however, that this probability density is different from those we have treated so far, because the random variable t is temporal. We cannot use Born's rule, because there is no self-adjoint operator for time – see Section 7.2.2.

A direct construction of $p(t)$ requires a better understanding of quantum measurement theory, which we will present in Chapter 21. Here, we will focus on a slightly different issue, namely, on finding the probability that a quantum system, described by a Hamiltonian $\hat{H}$, persists in its initial configuration. By "configuration" we mean an appropriate subspace of the system's Hilbert space. In some unstable systems, the subspace of the initial configuration is one-dimensional, and it coincides with the initial state $|\psi\rangle$. For such systems, it is convenient to employ the *persistence amplitude* (or *survival amplitude*)

$$\mathcal{A}_\psi(t) = \langle\psi|e^{-i\hat{H}t}|\psi\rangle. \tag{18.21}$$

The modulus square of $\mathcal{A}_\psi(t)$ is the *persistence probability* (or *survival probability*), that is, the probability a system prepared in the state $|\psi\rangle$ at $t = 0$ will still be found in $|\psi\rangle$ at a later time t. We assume that the decay happens when the system leaves $|\psi\rangle$. Then, it is natural to define the decay probability density

$$p(t) = -\frac{d}{dt}|\mathcal{A}_\psi(t)|^2. \tag{18.22}$$

The definition (18.22) is meaningful as long as (i) all states normal to $|\psi\rangle$ that can be reached via Hamiltonian evolution correspond to decay products, and (ii) the probability of the reverse process to the decay is negligible.

If condition (i) is not satisfied, then the propositions A = "the system left $|\psi\rangle$" and B = "the decay happened" are not identical: B implies A, but A does not imply B. Hence, Eq. (18.22) cannot be identified with the decay probability. For example, if $|\psi\rangle$ evolves to a different state $|\phi\rangle$ for the initial particle A, a part of the persistence amplitude will describe oscillations between $|\psi\rangle$ and $|\phi\rangle$. An oscillating persistence probability will lead to negative values for the function $p(t)$ of Eq. (18.22). Obviously, in this case, $p(t)$ is not a probability density.

Condition (ii) is satisfied if there are many more states available to the decay products than to the initial particle. Furthermore, the configuration of the experiment must allow the decay products to "explore" their available states, by leaving the locus of their production. Consider, for example, an excited atom in a cavity that decays with the emission of the photon. For some cavity geometries, the emitted photon does not exit the cavity immediately, and it may be reabsorbed by the atom at a later time. Again, the persistence amplitude of the excited atomic state will have an oscillating component, and the probability density $p(t)$, Eq. (18.22), will be ill-defined.

18.3 Perturbative Evaluation of the Persistence Amplitude

In this section, we study decays in which the persistence amplitude can be evaluated perturbatively. It turns out that all physical properties of the decays are encoded into a function that is defined on the complex energy plane, the *self-energy function*. We identify the conditions for the emergence of exponential decay, and we identify regimes for nonexponential decays.

By definition, the initial state $|\psi\rangle$ of an unstable system is not an eigenstate of the Hamiltonian $\hat{H}$. However, in many problems of physical interest, $|\psi\rangle$ is close to an eigenstate of $\hat{H}$, in the sense that $\hat{H}$ is of the form $\hat{H} = \hat{H}_0 + \hat{V}$, where $|\psi\rangle$ is an eigenstate of $\hat{H}_0$ and $\hat{V}$ is a small perturbation. In this case, a perturbative evaluation works well.

We denote the eigenstates of $\hat{H}_0$ by $|b\rangle$ and the corresponding eigenvalues by E_b. We assume that the initial state is one of the eigenstates of $\hat{H}_0$, say $|a\rangle$.

Without loss of generality, we assume that $V_{bb} = 0$ for all b. If $V_{bb} \neq 0$, we redefine $\hat{H}'_0 = \hat{H}_0 + \sum_b V_{bb}|b\rangle\langle b|$ and $\hat{V}' = \hat{V} - \sum_b V_{bb}|b\rangle\langle b|$, so that $|b\rangle$ remain eigenstates of $\hat{H}'_0$, but now $V'_{bb} = 0$.

18.3.1 The Resolvent of the Hamiltonian

The perturbative calculation of the persistence amplitude is simplified by using the *resolvent* $\hat{G}(z)$ associated to the Hamiltonian operator:

$$\hat{G}(z) = (z - \hat{H})^{-1}, \tag{18.23}$$

for $z \in \mathbb{C}$. We will often write the resolvent as $\frac{1}{z-\hat{H}}$.

For $t > 0$, the resolvent is related to the evolution operator $e^{-i\hat{H}t}$ by

$$e^{-i\hat{H}t} = \lim_{\epsilon \to 0^+} \frac{i}{2\pi} \int_{-\infty}^{\infty} \frac{dE e^{-iEt}}{E + i\epsilon - \hat{H}}. \tag{18.24}$$

Equation (18.24) follows from the identity

$$e^{-i\omega t} = \lim_{\epsilon \to 0^+} \frac{i}{2\pi} \int_{-\infty}^{\infty} \frac{dE e^{-iEt}}{E + i\epsilon - \omega}. \tag{18.25}$$

To prove Eq. (18.25) we evaluate the line integral

$$I(t) = \oint_{C_+} dz \frac{e^{-izt}}{z - \omega}, \tag{18.26}$$

along the negative-oriented contour C_+ of Fig. 18.1. At the lower half of the imaginary plane, $\mathrm{Im}z = -y$, for $y > 0$. Therefore, the integrand along the semicircle is proportional to e^{-yt} and vanishes as the radius of the semicircle goes to infinity. Hence,

$$I(t) = \lim_{\epsilon \to 0^+} \int_{-\infty}^{\infty} \frac{dE e^{-iEt}}{E + i\epsilon - \omega}. \tag{18.27}$$

The contour C_+ includes the single pole of the integrand (18.26), at $z = \omega$. Using Cauchy's residue theorem, we arrive at Eq. (18.25).

By integrating along the contour C_-, we can similarly prove that

$$\lim_{\epsilon \to 0^+} \int_{-\infty}^{\infty} \frac{dE e^{-iEt}}{E - i\epsilon - \hat{H}} = 0, \tag{18.28}$$

for $t > 0$. The integral vanishes because the contour C_- does not enclose any pole.

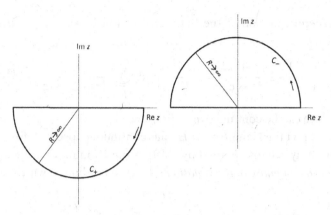

Fig. 18.1 Integration contours $C_\pm$ for calculating the integrals (18.25) and (18.28). The integrals are evaluated at the limit where the radius R of the semicircle goes to infinity.

A crucial property of the resolvent is that it can be expanded in a perturbative series. For a Hamiltonian $\hat{H}$ of the form $\hat{H}_0 + \hat{V}$,

$$(z - \hat{H})^{-1} = (z - \hat{H}_0 - \hat{V})^{-1} = [(z - \hat{H}_0)(1 - (z - \hat{H}_0)^{-1}\hat{V})]^{-1}$$
$$= (1 - (z - \hat{H}_0)^{-1}\hat{V})^{-1}(z - \hat{H}_0)^{-1}.$$

Using the geometric series formula $(\hat{I} - \hat{A})^{-1} = \sum_{n=0}^{\infty} \hat{A}^n$, we obtain

$$\frac{1}{z - \hat{H}} = \frac{1}{z - \hat{H}_0} + \frac{1}{z - \hat{H}_0}\hat{V}\frac{1}{z - \hat{H}_0} + \frac{1}{z - \hat{H}_0}\hat{V}\frac{1}{z - \hat{H}_0}\hat{V}\frac{1}{z - \hat{H}_0} + \cdots. \quad (18.29)$$

18.3.2 The Random Phase Approximation

We construct the persistence amplitude $\mathcal{A}_a(t)$ for an initial eigenstate $|a\rangle$ of $\hat{H}_0$. By Eq. (18.24),

$$\mathcal{A}_a(t) = \lim_{\epsilon \to 0^+} \frac{i}{2\pi} \int_{-\infty}^{\infty} dE\, e^{-iEt} G_a(E + i\epsilon), \quad (18.30)$$

where $G_a(z) := \langle a|(z - \hat{H})^{-1}|a\rangle$.

We evaluate $G_a(z)$ using the perturbative series (18.29). We assume that $\hat{V}$ is proportional to a dimensionless parameter $\lambda \ll 1$. To zeroth-order in λ, $G_a(z)$ coincides with $\frac{1}{z - E_a}$. The first-order contribution to $G_a(z)$ is $\frac{1}{(z - E_a)^2}\langle a|\hat{V}|a\rangle = 0$.

The second-order contribution is $\frac{1}{(z - E_a)^2}\langle a|\hat{V}\frac{1}{z - \hat{H}_0}\hat{V}|a\rangle$. Writing $\frac{1}{z - \hat{H}_0} = \sum_b \frac{1}{z - E_b}|b\rangle\langle b|$, this term becomes $\frac{1}{(z - E_a)^2}\Sigma_a(z)$, where

$$\Sigma_a(z) := \sum_b \frac{|V_{ab}|^2}{z - E_b}, \quad (18.31)$$

is the *self-energy* function of the state $|a\rangle$. The third-order and fourth-order terms are, respectively,

$$\sum_{bc} \frac{1}{(z-E_a)^2} \frac{V_{ab}V_{bc}V_{ca}}{(z-E_b)(z-E_c)} \quad \text{and} \quad \sum_{bcd} \frac{1}{(z-E_a)^2} \frac{V_{ab}V_{bc}V_{cd}V_{da}}{(z-E_b)(z-E_c)(z-E_d)}.$$

The procedure can be continued ad infinitum.

We assume that the Hamiltonian $\hat{H}_0$ has continuous spectrum for $E > \mu$, for some parameter μ. We can always choose $\mu = 0$ by shifting the Hamiltonian by a constant factor. Then, we invoke the *random phase approximation* (RPA), according to which

$$\sum_c \frac{V_{ac}V_{cb}}{z-E_c} \simeq \delta_{ab}\Sigma_a(z). \tag{18.32}$$

The reasoning for Eq. (18.32) is the following. Suppose that the system is contained in a box of volume V with periodic boundary conditions and that it has a large number N of degrees of freedom. The RPA is the assumption that the phases of the matrix elements $\langle a|\hat{V}|b\rangle$, for $b \neq a$ are *randomized* in the continuous limit, that is, in the limit where N and V goes to infinity, with N/V constant. By "randomized," we mean that the phases of $\langle a|\hat{V}|b\rangle$ do not exhibit any periodicity or quasiperiodicity as b varies. Hence, the summation over b is a sum of many random phases, and is expected to be much smaller than the term for $a = b$ that involves no such phases. Equation (18.32) then follows.

It is straightforward to show that, by the RPA, odd-order terms in the expansion of $G_a(z)$ vanish, and that even terms of even order $2n$ equal $\Sigma_a(z)^n/(z-E_a)^{n+1}$. Hence, $G_a(z)$ is given by a geometric series,

$$G_a(z) = \frac{1}{z-E_a}\sum_{n=0}^{\infty} \frac{\Sigma_a(z)^n}{(z-E_a)^n} = \frac{1}{z-E_a}\frac{1}{1-\frac{\Sigma_a(z)}{z-E_a}} = \frac{1}{z-E_a-\Sigma_a(z)}. \tag{18.33}$$

Note that the RPA is exact to second order in perturbation theory, that is, it differs from the exact solution by a term of order $O(\lambda^4)$. However, the RPA is fundamentally not a perturbative approximation; it may work even in regimes where perturbation theory is untrustworthy.

18.3.3 Structure of the Self-Energy Function

Equations (18.33) and (18.30) imply that

$$\mathcal{A}_a(t) = \frac{i}{2\pi}\lim_{\epsilon \to 0^+}\int_{-\infty}^{\infty}\frac{dE e^{-iEt}}{E+i\epsilon-E_a-\Sigma_a(E+i\epsilon)}. \tag{18.34}$$

If the self-energy function $\Sigma_a(z)$ were analytic, the integral (18.34) could be evaluated by integrating along the contour C_+ of Fig. 18.1, and using Cauchy's theorem. However, this is not the case; the self-energy function is discontinuous and it may contain poles or branch points.

To see this, consider the defining equation (18.31) for $\Sigma_a(z)$. Since all $E > 0$ lie in the spectrum of $\hat{H}_0$, $\Sigma_a(z)$ diverges along the half-line $\mathbb{R}^+$. To analyze the behavior of $\Sigma_a(z)$ near $\mathbb{R}^+$, we define $\Sigma_a(E^{\pm}) := \lim_{\eta \to 0^+}\Sigma_a(E \pm i\eta)$, for $\eta > 0$. We find that

$$\text{Re } \Sigma_a(E \pm i\eta) = \sum_b \frac{|V_{ab}|^2(E - E_b)}{(E - E_b)^2 + \eta^2}. \tag{18.35}$$

Hence the real parts of $\Sigma_a(E^+)$ and $\Sigma_a(E^-)$ coincide, a fact that allows us to define the *level-shift function*

$$F_a(E) := \text{Re } \Sigma_a(E^\pm). \tag{18.36}$$

On the other hand, the imaginary part of $\Sigma_a(z)$,

$$\text{Im } \Sigma_a(E \pm i\eta) = \text{Im} \sum_b \frac{|V_{ab}|^2}{E - E_b + i\eta} = \mp\eta \sum_b \frac{|V_{ab}|^2}{(E - E_b)^2 + \eta^2}, \tag{18.37}$$

is discontinuous at $\mathbb{R}^+$. Equation (18.37) implies that $\text{Im } \Sigma(E^-) = -\text{Im}\Sigma(E^+) \geq 0$.
We define the *decay function*

$$\Gamma_a(E) := 2\text{Im } \Sigma(E^-) > 0, \tag{18.38}$$

so that $\Sigma_a(E^\pm) = F_a(E) \mp \frac{i}{2}\Gamma_a(E)$. Then, Eq. (18.34) becomes

$$\mathcal{A}_a(t) = \frac{i}{2\pi} \int_{-\infty}^{\infty} \frac{dE e^{-iEt}}{E - E_a - F_a(E) + \frac{i}{2}\Gamma_a(E)}. \tag{18.39}$$

Since $\Gamma(E) = 0$ for $E < 0$,

$$\mathcal{A}_a(t) = \frac{i}{2\pi} \left(\int_{-\infty}^{0} \frac{dE e^{-iEt}}{E - E_a - F_a(E)} + \int_{0}^{\infty} \frac{dE e^{-iEt}}{E - E_a - F_a(E) + \frac{i}{2}\Gamma_a(E)} \right). \tag{18.40}$$

On the other hand, Eq. (18.28) implies that

$$\lim_{\epsilon \to 0} \int_{-\infty}^{\infty} \frac{dE e^{-iEt}}{E - i\epsilon - E_a - \Sigma_a(E^-)} = 0, \tag{18.41}$$

hence,

$$\int_{-\infty}^{0^+} \frac{dE e^{-iEt}}{E - E_a - F_a(E)} = -\int_{0}^{\infty} \frac{dE e^{-iEt}}{E - E_a - F_a(E) - \frac{i}{2}\Gamma_a(E)}.$$

Substituting into Eq. (18.39), we obtain

$$\mathcal{A}_a(t) = \frac{i}{2\pi} \int_{0}^{\infty} dE e^{-iEt} \left[\frac{1}{E - E_a - F_a(E) + \frac{i}{2}\Gamma_a(E)} - \frac{1}{E - E_a - F_a(E) - \frac{i}{2}\Gamma_a(E)} \right]$$

$$= \frac{1}{2\pi} \int_{0}^{\infty} \frac{dE \, \Gamma_a(E) \, e^{-iEt}}{[E - E_a - F_a(E)]^2 + \frac{1}{4}[\Gamma_a(E)]^2}. \tag{18.42}$$

Equation (18.42) relates the persistence amplitude to the components of the self-energy function.

18.3.4 The Wigner–Weisskopf Approximation

The integral (18.42) involves the functions $F_a(E)$ and $\Gamma_a(E)$ in the denominator. These functions are second-order with respect to the perturbation parameter $\lambda \ll 1$. If $|F_a(E)| \ll E_a$ and $\Gamma(E) \ll E_a$ for E in the vicinity of E_a, we can evaluate the persistence amplitude using the *Wigner–Weisskopf approximation* (WWA) (Weisskopf and Wigner, 1930).

The WWA essentially postulates the substitution of the Lorentzian-like function of E in Eq. (18.42) with an actual Lorentzian. The justification is the following. The integral (18.42) is dominated by values of E within distance of order λ^2 from $E \simeq E_a$. For these values, the integrand is of order λ^{-2}, otherwise it is of order λ^0. Hence, with an error of order λ^2, we can substitute the energy-shift function $F_a(E)$ with the constant $\delta E := F_a(E_a)$, and the decay function $\Gamma_a(E)$ with the constant $\Gamma := \Gamma_a(E_a)$.

Within an error of the same order of magnitude, we extend the range of integration to $(-\infty, \infty)$. Thus,

$$\mathcal{A}_a(t) = \frac{\Gamma}{2\pi} \int_{-\infty}^{\infty} \frac{dE e^{-iEt}}{(E - E_a - \delta E)^2 + \frac{1}{4}\Gamma^2} = e^{-i(E_a + \delta E)t - \frac{\Gamma}{2}t}, \tag{18.43}$$

where we used Eq. (A.9) for the evaluation of the integral.

Substituting Eq. (18.43) in Eq. (18.22), we conclude that $p(t) = \Gamma e^{-\Gamma t}$. Hence, the WWA leads to an exponential decay law with a decay constant Γ that originates from the imaginary part of the self-energy function. The real part of the self-energy function generates a shift δE of the energy level E_a, usually referred to as the *Lamb shift*.[2]

The WWA leads to the same decay constant Γ as Fermi's decay rule. To see this, we note that

$$\Gamma = \lim_{\eta \to 0^+} \sum_b \frac{2\eta |V_{ab}|^2}{(E_a - E_b)^2 + \eta^2} = 2\pi \sum_b |V_{ab}|^2 \delta(E_b - E_a). \tag{18.44}$$

This expression is identical to Eq. (18.15). The analysis of this section provides a proper mathematical justification for Fermi's golden rule for decays.

18.3.5 Beyond Exponential Decay

We derived exponential decay as a consequence of two approximations, the RPA and the WWA. Here, we will consider only very weak interactions $\hat{V}$, for which the RPA is reliable, because it gives an expression that is exact to second order in perturbation theory. In this regime, we need only consider the domain of validity of the WWA.

First, we note that exponential decay cannot be valid at very early times. This is a general statement that originates from the definition (18.22). We Taylor-expand the persistence amplitude around $t = 0$, to obtain $\mathcal{A}_\psi = 1 - it\langle\hat{H}\rangle - \frac{t^2}{2}\langle\hat{H}^2\rangle + \cdots$. Keeping terms up to order t^2, the probability density becomes $|\mathcal{A}_\psi|^2 = 1 - (\Delta H)^2 t^2 + \cdots$. Equation (18.22) implies that $p(t) = (\Delta H)^2 t$.

---

[2] The original Lamb shift referred to a small energy difference between energy levels of the hydrogen atom that are degenerate even in the relativistic treatment. The shift originates from the interaction of the atom with the quantum electromagnetic field, and it was first measured by Lamb and Retherford (1947).

It follows that $p(0) = 0$ while, in exponential decays, $p(0) = \Gamma$. This behavior is reminiscent of the quantum Zeno effect, as for times very close to zero the decay rate almost vanishes. For this reason, the early-time regime where the exponential decay law fails is known as the *quantum Zeno regime*.

Second, we consider the case of very long times. We change the integration from E to $y = Et$ in Eq. (18.42),

$$\mathcal{A}_a(t) = \frac{1}{2\pi t} \int_0^\infty \frac{dy\, \Gamma_a(y/t)\, e^{-iy}}{[y/t - E_a - F_a(y/t)]^2 + \frac{1}{4}[\Gamma_a(y/t)]^2}. \tag{18.45}$$

As $t \to \infty$, $F_a(y/t) \to F_a(0)$ and $\Gamma_a(y/t) \to \Gamma_a(0)$. Both terms are of order λ^2, hence, the term E_a dominates in the denominator. As a result,

$$\mathcal{A}_a(t) \simeq \frac{1}{2\pi E_a^2} \int_0^\infty dE\, \Gamma_a(E)\, e^{-iEt}. \tag{18.46}$$

The long-time limit of $\mathcal{A}_a(t)$ is determined by the behavior of $\Gamma_a(E)$ near zero. By continuity, $\Gamma_a(0) = 0$, since $\Gamma_a(E) = 0$ for $E < 0$. Let $\Gamma_a(E) \simeq \kappa E^n$ as $x \to 0$, for some positive constant κ and integer n. We evaluate the integral $\int_0^\infty dx e^{-iEt} E^n$ for imaginary time $t = -i\tau$, with $\tau > 0$, to obtain $n!\,/\tau^{1+n}$. We analytically continue back to t and substitute into Eq. (18.46) to obtain

$$\mathcal{A}_a(t) = \frac{(-i)^{n+1} n!\, \kappa}{2\pi E_a^2 t^{n+1}}. \tag{18.47}$$

The persistence amplitude drops as an inverse power of t. Hence, in the long-time limit, decays are characterized by an inverse-power law and not an exponential one. In most systems, the inverse-power behavior appears in timescales too large to be measurable. Nonetheless, it has been observed in luminescence decays of dissolved organic material (Rothe et al., 2006).

The decay probability exhibits nonexponential behavior at both very early and very late times *in all systems*. There is no quantum system with pure exponential decay, even if exponential decay dominates the timescales that are most easily accessible to experiments. A representative plot of the persistence probability, showing the deviations from purely exponential behavior, is shown in Fig. 18.2

Some decays are nonexponential at all times. This happens near resonances, that is, when E_a is close to a value of E where $\Sigma(z)$ blows up. For example, suppose that the self-energy function is of the form $\Sigma_0(z) + \frac{\alpha}{z - \omega_0}$, where $\alpha > 0$ is of order λ^2, and $\Sigma_0(z)$ is a term for which the WWA works. Then, we can approximate the amplitude (18.39) with

$$\mathcal{A}_a(t) = \frac{i}{2\pi} \int_{-\infty}^\infty \frac{dE(E - \omega_0)e^{-iEt}}{(E - \bar{E})(E - \omega_0) - \alpha + \frac{i}{2}\Gamma(E - \omega_0)}, \tag{18.48}$$

where we wrote $\bar{E} = E_a + F_0(E_a)$ and $\Gamma = \Gamma_0(E_a)$, with $F_0(E)$ and $\Gamma_0(E)$ defined with respect to $\Sigma_0(z)$. For simplicity, we consider the case of exact resonance: $\bar{E} = \omega_0$. Then, the integrand of Eq. (18.48) has poles at $E = \omega_0 - \frac{i}{4}\Gamma \pm \sqrt{\alpha - \frac{\Gamma^2}{16}}$. Both α and Γ are of order λ^2, hence, $\alpha \gg \Gamma^2$. Then, the poles are approximately at $E = \omega_0 - \frac{i}{4}\Gamma \pm \sqrt{\alpha}$.

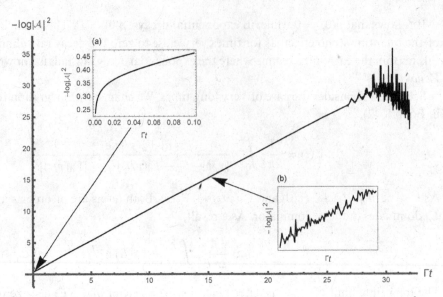

Fig. 18.2 A representative plot of the negative logarithm of the persistence probability $|\mathcal{A}(t)|^2$ – determined through Eq. (18.42) – as a function of Γt. A straight line denotes exponential decay. For Γt about 25 or higher, strong deviations from exponential decay appear in the form of oscillations. In this regime, the persistence probability decays on average with an inverse power law. (This behavior appears when all but a fraction of 10^{-10} of the original unstable systems have decayed.) In inset (a), we see the deviation from exponential decay at early times, namely, the quantum Zeno regime. In inset (b), we zoom deeply in to the plotted curve, and see that the exponential decay is a mean behavior over fluctuations at very short timescales.

We evaluate the persistence amplitude (18.48), by integrating along the curve C_+ of Fig. 18.1,

$$\mathcal{A}_a(t) = -e^{-i\omega_0 t - \frac{\Gamma}{4}t}\left[\cos(\sqrt{\alpha}t) - \frac{\Gamma}{\sqrt{\alpha}}\sin(\sqrt{\alpha}t)\right] \simeq -e^{-i\omega_0 t - \frac{\Gamma}{4}t}\cos(\sqrt{\alpha}t).$$

Hence, the persistence probability $|\mathcal{A}_a(t)|^2 = e^{-\frac{\Gamma}{2}t}\cos^2(\sqrt{\alpha}t)$ has an oscillating component due to resonance, and it does not describe exponential decay. The very definition (18.22) of a decay probability density is invalidated, as $p(t)$ takes negative values.

18.4 Lee's Model

In this section, we present a general model for decays that applies to many different physical situations. This model originates from Lee (1954). We shall consider two versions of the model, one where the decay is accompanied by the emission of a bosonic particle and one where the decay is accompanied by the emission of two fermions. The former describes spontaneous emission of photons, the latter describes beta decay.

18.4.1 Bosonic Emission

We consider decays of the form $A' \to A + B$, in which the emitted bosonic particle B is much lighter than the particles A' and A. Examples of this type of decay are as follows.

- A' is the excited state of a nucleus, or of an atom, or of a molecule, A is the corresponding ground state and B is a photon.
- A' is a heavy nucleus that decays to the nucleus A and an alpha particle.
- A' and A are baryons and B is a light meson.

The key idea in Lee's model is to ignore all degrees of freedom pertaining to the motion of the heavy particles. The heavy particles are then treated as a qubit, with the ground state $|g\rangle$ corresponding to the particle A and the excited state $|e\rangle$ to the particle A'. Note that here we write the usual qubit states $|0\rangle$ and $|1\rangle$ as $|g\rangle$ and $|e\rangle$, respectively.

The B particle is described by a bosonic Fock space $\mathcal{F}_B$. We denote the vacuum of $\mathcal{F}_B$ by $|0\rangle$ and denote the creation and annihilation operators by $\hat{a}_r$ and $\hat{a}_r^\dagger$. The index r labels the (generalized) basis on the Hilbert space of a single B particle that diagonalizes the single-particle Hamiltonian; we denote its eigenvalues by ω_r.

The Hilbert space of the total system is $\mathbb{C}^2 \otimes \mathcal{F}_B$. The Hamiltonian $\hat{H}$ of Lee's model consists of three terms $\hat{H} = \hat{H}_A + \hat{H}_B + \hat{V}$, where $\hat{H}_A = \frac{1}{2}\Omega(\hat{I} + \hat{\sigma}_3)$ is the qubit Hamiltonian, and Ω stands for the energy difference of the two levels; $\hat{H}_B = \sum_r \omega_r \hat{a}_r^\dagger \hat{a}_r$ is the Hamiltonian for noninteracting particles B particles; and

$$\hat{V} = \sum_r \left(g_r \hat{\sigma}_+ \hat{a}_r + g_r^* \hat{\sigma}_- \hat{a}_r^\dagger \right) \tag{18.49}$$

is the interaction term. It describes the excitation of the qubit accompanied by the absorption of a B particle, and the decay of the qubit accompanied by the emission of a B particle. The coefficients g_r are different in different systems. The initial state for Lee's model is $|A'\rangle = |e\rangle \otimes |0\rangle$, that is, the two-level system is excited and no B particle is present.

We evaluate the survival amplitude $\mathcal{A}(t) := \langle A'|e^{-i\hat{H}t}|A'\rangle$ using the series (18.29) for $\hat{H}_0 = \hat{H}_A + \hat{H}_B$. We find that

$$(z - \hat{H}_0)^{-1}\hat{V}(z - \hat{H}_0)^{-1}|A'\rangle = \frac{1}{z - \Omega}|g\rangle \otimes \sum_r \frac{g_r}{z - \omega_r}\hat{a}_r^\dagger|0\rangle;$$

hence, the matrix elements $\langle \psi|\hat{V}|A'\rangle$ are nonzero only when the $|\psi\rangle$ are of the form $|g\rangle \otimes \hat{a}_r^\dagger|0\rangle$. In particular, $\langle A'|\hat{V}|A'\rangle = 0$.

The next term in the perturbative series is

$$(z - \hat{H}_0)^{-1}\hat{V}(z - \hat{H}_0)^{-1}\hat{V}(z - \hat{H}_0)^{-1}|A'\rangle = \frac{\Sigma(z)}{(z - \Omega)^2}|A'\rangle, \tag{18.50}$$

where

$$\Sigma(z) = \sum_r \frac{|g_r|^2}{z - \omega_r} \tag{18.51}$$

is the self-energy function for the initial state $|A'\rangle$.

Since the second-order term is proportional to the zeroth-order one, the above expressions are reproduced to all orders of perturbation. We obtain

$$[(z - \hat{H}_0)^{-1}\hat{V}]^{2n}(z - \hat{H}_0)^{-1}|A'\rangle = \frac{1}{z - \Omega}\left(\frac{\Sigma(z)}{z - \Omega}\right)^n|A'\rangle,$$

$$[(z - \hat{H}_0)^{-1}\hat{V}]^{2n+1}(z - \hat{H}_0)^{-1}|A'\rangle = \frac{1}{z - \Omega}\left(\frac{\Sigma(z)}{z - \Omega}\right)^n |g\rangle \otimes \sum_r \frac{g_r \hat{a}_r^\dagger}{z - \omega_r}|0\rangle.$$

Hence,

$$\frac{1}{z - \hat{H}}|A'\rangle = \frac{1}{z - \Omega}\sum_{n=0}^{\infty}\frac{\Sigma(z)^n}{(z - \Omega)^n}\left(|A'\rangle + |g\rangle \otimes \sum_r \frac{g_r \hat{a}_r^\dagger}{z - \omega_r}|0\rangle\right)$$

$$= \frac{1}{z - \Omega - \Sigma(z)}\left(|A'\rangle + |g\rangle \otimes \sum_r \frac{g_r \hat{a}_r^\dagger}{z - \omega_r}|0\rangle\right). \tag{18.52}$$

We conclude that

$$G(z) = \frac{1}{z - \Omega - \Sigma(z)}. \tag{18.53}$$

We obtained Eq. (18.33) by resumming the full perturbative series (18.29), without any approximation. This means that Lee's model incorporates the RPA in its definition, in particular in the choice of the coupling term (18.49). When taking the two-level approximation in realistic systems, one typically encounters a different coupling term

$$\hat{V}_1 = \hat{\sigma}_1 \sum_a \left(g_r \hat{a}_r + g_r^* \hat{a}_r^\dagger\right). \tag{18.54}$$

Since $\hat{\sigma}_1 = \hat{\sigma}_+ + \hat{\sigma}_-$, the interaction term $\hat{V}_1$ includes terms $\hat{\sigma}_+ \hat{a}_r^\dagger$ and $\hat{\sigma}_- \hat{a}_r$, in addition to the ones of $\hat{V}$ of Eq. (18.49).

The corresponding Hamiltonian $\hat{H}_1 := \hat{H}_A + \hat{H}_B + \hat{V}_1$ is known as the *spin–boson* Hamiltonian. It is straightforward to show that the spin–boson Hamiltonian subject to the RPA and Lee's Hamiltonian lead to the same predictions for decays,

$$\langle A'|(z - \hat{H}_1)^{-1}|A'\rangle = \langle A'|(z - \hat{H})^{-1}|A'\rangle = \frac{1}{z - \Omega - \Sigma(z)}.$$

18.4.2 Spontaneous Emission from Atoms

We will evaluate the self-energy function in a simple model where the emitting particles have zero spin and zero rest mass, namely, scalar photons. This model ignores the effects of polarization, but otherwise it describes well the emission of photons by excited atoms. The inclusion of polarization changes $\Sigma(z)$ only by a multiplicative factor that can be absorbed in a redefinition of the coupling constant.

In this model, the basis r corresponds to photon momenta $\mathbf{k}$, ω_r corresponds to $\omega_{\mathbf{k}} = |\mathbf{k}|$, and the summation over r corresponds to integration with measure $\frac{d^3k}{(2\pi)^3}$. If the size a_0 of the emitting atom is much smaller than the wavelengths of the emitted radiation, we can describe the atom–radiation interaction in the dipole approximation – see Section 13.4.5. Then, the coupling coefficients are $g_{\mathbf{k}} = \frac{\lambda}{\sqrt{2\omega_{\mathbf{k}}}}e^{i\mathbf{k}\cdot\mathbf{x}_0}$, where λ is a constant and $\mathbf{x}_0$ is the position vector of the atom. This can be seen from the interaction Hamiltonian (13.89) if we substitute the EM potential $\mathbf{A}(\mathbf{x}_0)$ with a scalar field $\hat{\phi}(\mathbf{x}_0)$ and employ a scalar constant in place of the dipole-moment matrix element. Then, the interaction takes the form $\lambda \hat{\sigma}_1 \hat{\phi}(\mathbf{x}_0)$. When expanding the scalar field in modes, we obtain an interaction term of the form (18.54) with the aforementioned values for $g_\mathbf{k}$. Note that the constant λ is proportional to the atomic frequency Ω.

The self-energy function (18.51) becomes

$$\Sigma(z) = \frac{\lambda^2}{4\pi^2} \int_0^\infty \frac{kdk}{z-k}. \tag{18.55}$$

The integral in Eq. (18.55) diverges as $k \to \infty$. However, arbitrarily large photon energies are not physically relevant. Hence, we *regularize* the integral (18.55) (that is, we make it finite) by introducing a high-frequency cut-off $\Lambda >> \Omega$,

$$\Sigma(z) = \frac{\lambda^2}{4\pi^2} \int_0^\Lambda \frac{kdk}{z-k} = -\frac{\lambda^2}{4\pi^2} \left[\Lambda + z(\ln(\Lambda - z) - \ln(-z)) \right]. \tag{18.56}$$

There are many other ways to regularize the integral (18.55), for example, by inserting an exponential cut-off function $e^{-k/\Lambda}$. The choice of regularization does not affect the form of $\Sigma(z)$ for the physical range of values of z, namely, $|z| << \Lambda$. However, it introduces an arbitrariness in the behavior of $\Sigma(z)$ for z of the order of Λ. In particular, the apparent branch point at $z = \Lambda$ in Eq. (18.56) is a artefact of the regularization. As far as the physically relevant values of z are concerned, there is no error in substituting $\ln(\Lambda - z) \simeq \ln \Lambda$ in Eq. (18.56). Since $\Sigma(z)$ is added to Ω in Eq. (18.53) for $G(z)$, it is convenient to absorb the constant term in $\Sigma(z)$ into a redefinition of the frequency $\tilde{\Omega} = \Omega - \frac{\lambda^2 \Lambda}{4\pi^2}$. With these modifications, Eq. (18.56) becomes

$$\Sigma(z) = \frac{\lambda^2}{4\pi^2} \left[-(\ln \Lambda)z + z \ln(-z) \right]. \tag{18.57}$$

The logarithm in Eq. (18.57) is defined in the principal branch, that is, its argument lies in $(-\pi, \pi]$. When evaluating $\Sigma(E^-)$, we substitute $z = E - i\eta$, for $\eta > 0$. Hence, $\ln(-z) = \ln E + \ln[-(1 - i\eta/E)]$. As $\eta \to 0$, $1 - i\eta/E \simeq e^{-i\eta/E}$. We have two options for the -1 term in the logarithm; we can express it either as $e^{i\pi}$ or as $e^{-i\pi}$. The first choice gives $\ln(e^{i(\pi - \eta/E)})$, hence, the argument lies in the principal branch. The second choice gives $\ln(e^{i(-\pi - \eta/E)})$, and the argument lies outside the principal branch. Only the first choice is acceptable. Hence, $\ln[-(1 - i\eta/E)] = i(\pi - \eta/E)$, and $\lim_{\eta \to 0} \ln[-(E - i\eta)] = \ln E + i\pi$. It follows that

$$\Sigma(E^-) = -\frac{\lambda^2}{4\pi^2} \left[E \ln(\Lambda/E) - i\pi E \right]. \tag{18.58}$$

By Eqs. (18.36) and (18.38),

$$F(E) = -\frac{\lambda^2}{4\pi^2} E \ln(\Lambda/E), \text{ and } \Gamma(E) = \frac{\lambda^2}{2\pi} E. \tag{18.59}$$

Hence, the decay constant in the WWA is $\Gamma = \Gamma(\tilde{\Omega}) = \frac{\lambda^2}{2\pi} \tilde{\Omega}$. Since λ is proportional to Ω, the decay constant is proportional to Ω^3 (See also Box 18.1).

Box 18.1 Validity of Exponential Decay

The Lee model for spontaneous emission allows us to demonstrate explicitly the domain of validity of the WWA. First, we consider the possible contribution to the persistence amplitude from poles other than the one determined by the WWA. The equation for the poles,

$$z - \tilde{\Omega} - \frac{\lambda^2}{4\pi^2} z \ln(1 - \Lambda/z) = 0,$$

admits another solution for z near Λ. To see this, we substitute $z = \Lambda(1 + x)$ for $|x| << 1$, and we obtain $x \simeq \exp(-\pi^2\lambda^2) << 1$ to leading order in λ^2. This solution is possibly an artefact of regularization, but in any case it corresponds to oscillations much more rapid than any physically relevant timescale. It averages to zero in any measurement with temporal resolution much larger than Λ^{-1}. Hence, the contribution of non-WWA poles to the decay probability is negligible.

Next, we consider the asymptotic term, Eq. (18.47). For spontaneous emission, $n = 1$, $E_a = \tilde{\Omega}$ and $\kappa = \lambda^2/(2\pi) = \Gamma/\tilde{\Omega}$. For sufficiently large times, this term dominates over the exponential decay term $e^{-\Gamma t/2}$ found by the WWA. The relevant timescale τ is found by equating the norm of $\mathcal{A}(t)$ of Eq. (18.47) with $e^{-\Gamma t/2}$. Setting $\Gamma\tau/2 = x$ and $\alpha = \frac{1}{8\pi}(\Gamma/\tilde{\Omega})^3$, we obtain the condition: $2\ln x - x = \ln\alpha$. Solutions to this equation for different values of $\Gamma/\tilde{\Omega}$ are given in the following able.

$\Gamma/\tilde{\Omega}$	10^{-2}	10^{-3}	10^{-4}	10^{-5}	10^{-6}	10^{-7}
x	23.3	30.8	38.1	45.4	52.5	59.8
$e^{-\Gamma\tau}$	$5 \cdot 10^{-21}$	$2 \cdot 10^{-27}$	$8 \cdot 10^{-34}$	$4 \cdot 10^{-40}$	$2 \cdot 10^{-46}$	10^{-52}

Even for values of $\Gamma/\tilde{\Omega}$ as large as 0.01, the exponential decay law breaks down at a time where fewer than $1{:}10^{20}$ of the initial atoms remain in the excited state. Deviations from exponential decay in photoemission are negligible outside the quantum Zeno regime.

18.4.3 Fermionic Emission

Next, we consider decays of the form $A' \to A + B_1 + B_2$, in which B_1 and B_2 are fermionic particles, much lighter than A' and A. The most important example of this type is beta decay, where A' and A are nuclei, B_1 is an electron (or positron), and B_2 is an antineutrino (or a neutrino). Again the nucleus is described as a two-level system, with a ground state $|g\rangle$ and an excited state $|e\rangle$.

A fermionic Fock space $\mathcal{F}_1$ is associated to the particle B_1 and a fermionic Fock space $\mathcal{F}_2$ is associated to the particle B_2. The corresponding ground states are $|0\rangle_1$ and $|0\rangle_2$, respectively. The creation and annihilation operators on $\mathcal{F}_1$ will be denoted as $\hat{c}_r$ and $\hat{c}_r^\dagger$, and the creation and annihilation operators on $\mathcal{F}_2$ as $\hat{d}_l$ and $\hat{d}_l^\dagger$. They satisfy the canonical anticommutation relations.

In the above, r (l) is a label of the basis on the Hilbert space of a single B_1 (B_2) particle that diagonalizes the Hamiltonian; we denote the corresponding eigenvalues by ω_r $(\tilde{\omega}_l)$. The Hilbert space of the total system is $\mathbb{C}^2 \otimes \mathcal{F}_1 \otimes \mathcal{F}_2$.

The Hamiltonian for this system again consists of three parts $\hat{H} = \hat{H}_A + \hat{H}_B + \hat{V}$, where $\hat{H}_A = \frac{1}{2}\Omega(\hat{I} + \hat{\sigma}_3)$, $\hat{H}_B = \sum_r \omega_r \hat{c}_r^\dagger \hat{c}_r + \sum_l \tilde{\omega}_l \hat{d}_l^\dagger \hat{d}_l$, and

$$\hat{V} = \sum_{r,l} \left(g_{r,l} \hat{\sigma}_+ \hat{c}_r \hat{d}_l + g_{r,l}^* \hat{\sigma}_- \hat{c}_r^\dagger \hat{d}_l^\dagger \right) \tag{18.60}$$

is an interaction term, with model-dependent coefficients $g_{r,l}$.

We straightforwardly calculate the self-energy function for an initial state $|A'\rangle = |e\rangle \otimes |0\rangle_1 \otimes |0\rangle_2$

$$\Sigma(z) = \sum_{r,l} \frac{|g_{r,l}|^2}{z - \omega_r - \tilde{\omega}_l}. \tag{18.61}$$

Example 18.3 We consider a simplified model for beta decay, in which both emitted particles have zero spin and zero mass. This model is similar to the earliest theories of weak interactions. The zero-mass approximation is reasonable, if the energy Ω is much larger than the masses of the emitted particles, as is often the case in beta decay. The zero-spin approximation is bad; spin is important in the weak interactions.

The basis r corresponds to momenta $\mathbf{p}$, the basis l corresponds to momenta $\mathbf{q}$, ω_r corresponds to $\omega_{\mathbf{k}} = |\mathbf{p}|$, and $\tilde{\omega}_l$ corresponds to $\tilde{\omega}_{\mathbf{q}} = |\mathbf{q}|$. The summation over r corresponds to integration over $\frac{d^3p}{(2\pi)^3}$ and the summation over s corresponds to integration over $\frac{d^3q}{(2\pi)^3}$. The appropriate constants $g_{r,l}$ for beta decay are $g_{\mathbf{p},\mathbf{q}} = \frac{1}{\mu^2} e^{i(\mathbf{p}+\mathbf{q})\cdot\mathbf{x}_0}$, where μ has dimensions of mass and $\mathbf{x}_0$ is the position vector of the nucleus.

We substitute in to Eq. (18.61), to obtain

$$\Sigma(z) = \frac{1}{64\pi^6\mu^4} \int \frac{d^3p\,d^3q}{z - |\mathbf{p}| - |\mathbf{q}|} = \frac{1}{16\pi^4\mu^4} \int_0^\infty p^2 dp \int_0^\infty q^2 dq \frac{1}{z - p - q}.$$

We change the integration variables to $y = p + q, \xi = p - q$. Then,

$$\Sigma(z) = \frac{1}{256\pi^4\mu^4} \int_0^\infty \frac{dy}{z - y} \int_0^y d\xi (y^2 - \xi^2)^2 = \frac{1}{480\pi^4\mu^4} \int_0^\infty \frac{dy\,y^5}{z - y}, \tag{18.62}$$

The integral in Eq. (18.62) diverges, so we introduce a high-frequency cut-off $\Lambda \ll \mu$ and restrict the integration over y to the interval $[0, \Lambda]$. Thus, we obtain

$$\Sigma(z) = -\frac{1}{240\pi^4\mu^4} \left[\Lambda^5 \left(\frac{1}{5} + \frac{z}{4\Lambda} + \frac{z^2}{3\Lambda^2} + \frac{z^3}{2\Lambda^3} + \frac{z^4}{\Lambda^4} \right) + z^5 \ln \Lambda - z^5 \ln(-z) \right],$$

where we approximated $\ln(\Lambda - z) \simeq \ln \Lambda$.

As in Section 18.4.2, the branch point $z = 0$ is logarithmic. It is then straightforward to evaluate the level-shift and decay functions. In particular,

$$\Gamma(E) = \frac{E^5}{120\pi^3\mu^4}. \tag{18.63}$$

18.5   Decay through Barrier Tunneling

The methodology developed in Section 18.3 applies to decays that originate from a small perturbation in the Hamiltonian. In this section, we consider nonperturbative decays that can be understood in terms of tunneling. Examples of such decays are the alpha emission of nuclei, tunneling ionization of atoms due to an external field, and leakage of particles from a trap. We will consider a simple model of a particle in one dimension that manifests the main properties of tunneling-induced decays.

18.5.1   Set-Up

We consider a particle in the half-line $\mathbb{R}^+ = [0, \infty)$ in the presence of a potential $V(x)$. The potential vanishes outside $[a, b]$, where a and b are microscopic lengths. For simplicity, we assume that the Schrödinger operator has no bound states. Then, it has nondegenerate continuous spectrum for all positive energies $E = \frac{k^2}{2m}$. The corresponding eigenfunctions $g_k(x)$ are expressed in terms of the transmission and reflection amplitudes T_k and R_k associated to the potential by Eq. (5.99), or equivalently by

$$g_k(x) = \begin{cases} -\frac{2i}{\sqrt{2\pi}} \frac{T_k}{1+R_k} \sin kx, & x < a \\ \frac{1}{\sqrt{2\pi}} \left[e^{-ikx} - e^{iS_k} e^{ikx} \right] & x > b \end{cases}, \tag{18.64}$$

where

$$e^{iS_k} = \frac{T_k^2}{1+R_k} - \tilde{R}_k = \frac{1+R_k^*}{1+R_k} \left(\frac{T_k^2}{|T_k|^2} \right). \tag{18.65}$$

The eigenfunctions $g_k(x)$ are normalized so that $\int_0^\infty g_k(x)^* g_{k'}(x) = \delta(k - k')$. We will represent them by kets $|k\rangle$.

We consider an initial state $\psi_0(x)$ that is localized within $[0, a]$ with the following three properties. First, $\psi_0(x)$ belongs to the Hilbert space of square integrable harmonic functions on $\mathbb{R}^+$ subject to Dirichlet boundary conditions. Hence, it can be expressed as

$$\psi_0(x) = \sqrt{\frac{2}{\pi}} \int_0^\infty \sin(kx) \tilde{\psi}_0(k). \tag{18.66}$$

The function $\tilde{\psi}_0(k)$ is defined for $k > 0$. Extending to $k < 0$ by $\tilde{\psi}_0(-k) = -\tilde{\psi}(k)$, we can write Eq. (18.66) as $\psi_0(x) = -\frac{i}{\sqrt{2\pi}} \int_{-\infty}^\infty dk e^{ikx} \tilde{\psi}_0(k)$. By Eq. (18.64), $\langle k|\psi_0\rangle = i\frac{T_k^*}{1+R_k^*} \tilde{\psi}_0(k)$.

Second, we assume that $\psi_0(x)$ is real-valued. For example, ψ_0 may be an eigenstate of a Schrödinger operator $\hat{H}'$ with a different potential $U(x)$. The physical interpretation of this condition is that we prepare the system in an eigenstate of $\hat{H}'$ and at time $t = 0$ we change the potential to $V(x)$, for example, by switching on an external electric field. Third, we assume that ψ_0 has a sharp energy distribution with respect to the Hamiltonian $\hat{H}$: The energy spread for ψ_0 is much smaller than the mean energy.

Given an initial state ψ_0, we find the state at time t,

$$\psi_t(x) = \int_0^\infty dk \langle k|\psi_0\rangle g_k(x) e^{-i\frac{k^2}{2m}t}. \tag{18.67}$$

Equations (18.64) and (18.67) imply that

$$\psi_t(x) = -\frac{i}{\sqrt{2\pi}} \int_0^\infty dk \frac{T_k^*}{1+R_k^*} \left[e^{iS_k+ikx} - e^{-ikx} \right] e^{-i\frac{k^2}{2m}t} \tilde{\psi}_0(k). \tag{18.68}$$

For $x \gg b$, there is no stationary phase in the integral involving $e^{-ikx-i\frac{k^2}{2m}t}$. Hence, its contribution to $\psi_t(x)$ is much smaller than that of the integral involving $e^{ikx-i\frac{k^2}{2m}t}$, which has a stationary phase. By Eq. (18.65),

$$\psi_t(x) = -\frac{i}{\sqrt{2\pi}} \int_0^\infty dk \frac{T_k}{1+R_k} e^{ikx-i\frac{k^2}{2m}t} \tilde{\psi}_0(k). \tag{18.69}$$

We expand $(1+R_k)^{-1} = \sum_{n=0}^\infty (-R_k)^n$, to write Eq. (18.69) as

$$\psi_t(x) = -\frac{i}{\sqrt{2\pi}} \sum_{n=0}^\infty \int_0^\infty dk\, T_k (-R_k)^n e^{ikx-i\frac{k^2}{2m}t} \tilde{\psi}_0(k). \tag{18.70}$$

18.5.2 Exponential Decay

Equation (18.69) is accurate for the asymptotic behavior of the wave function at $x \gg b$. To proceed further, we exploit the fact that $\psi_0(k)$ is strongly peaked about a specific value k_0, and we evaluate Eq. (18.69) in the phase-expansion approximation as in Section 7.3.4. To this end, we write $R_k = -|R_k|e^{i\phi_k}$ and $T_k = |T_k|e^{i\chi_k}$, so that Eq. (18.70) becomes

$$\psi_t(x) = -\frac{i}{\sqrt{2\pi}} \sum_{n=0}^\infty \int_0^\infty dk |T_k R_k^n| e^{ikx-i\frac{k^2}{2m}t+i\chi_k+in\phi_k} \tilde{\psi}_0(k). \tag{18.71}$$

Then, we extend the range of integration of k to $(-\infty, \infty)$ setting $\tilde{\psi}_0(-k) = -\tilde{\psi}_0(k)$ for negative k. The integral changes little, because the added range involves a term $e^{-i|k|x-i\frac{k^2}{2m}t}$, with a rapidly oscillating phase. Then, we approximate $|T_k| \simeq |T_{k_0}|, |R_k| \simeq |R_{k_0}|, k^2 \simeq k_0^2 + 2k_0(k-k_0)$, $\phi_k \simeq \phi_{k_0} + \phi_{k_0}'(k-k_0), \chi_k \simeq \chi_{k_0} + \chi_{k_0}'(k-k_0)$. The resulting integral is the inverse Fourier transform of $\tilde{\psi}_0(k)$. Hence,

$$\psi_t(x) = T_{k_0} e^{ik_0x-i\frac{k_0^2}{2m}t} \sum_{n=0}^\infty (-R_{k_0})^n \psi_0 \left(x - \frac{k_0 t}{m} + \chi_{k_0}' + n\phi_{k_0}' \right). \tag{18.72}$$

Equation (18.72) has a natural interpretation in terms of classical concepts. The particle makes successive attempts to cross the barrier at $x = a$. On failure, it is reflected back, it is reflected again at $x = 0$, and then it makes a new attempt. The nth term in the sum of Eq. (18.72) is the amplitude associated to a particle that succeeded in crossing the barrier at its $(n+1)$th attempt: It is proportional to T_{k_0} (one success) and to $R_{k_0}^n$ (after n failures).

Since $\psi_0(x)$ has support only on $[0, a]$, $\psi_t(x)$ vanishes for $t < t_0 := m(x - a + \chi'_{k_0})/k_0$. The timescale t_0 has an obvious classical interpretation: It is the time it takes a particle inside the barrier region to traverse the distance to point x. The term χ'_{k_0} is proportional to the time delay of the particle crossing the classically forbidden region – see Section 7.3.5. We rewrite Eq. (18.72) as

$$\psi_t(x) = T_{k_0} e^{ik_0 x - i\frac{k_0^2}{2m}t} \theta(t - t_0) \sum_{n=0}^{\infty} (-R_{k_0})^n \psi_0 \left(a - \frac{k_0(t - t_0)}{m} + n\Delta x \right), \tag{18.73}$$

where we defined as $\Delta x = \phi'_{k_0}$ the position shift between successive terms in the series (18.73). If $|\Delta x| > a$, the partial amplitudes at different n do not overlap. Hence, there is no quantum interference between different attempts of the particle to cross the barrier. We will assume that this is the case. We will see later that there is no difference in the conclusions as long as the ratio $|\Delta x|/a$ does not become too small.

The wave function $\psi_t(x)$ depends on x only through the combination $x - \frac{k_0 t}{m}$. This means that we can take a constant x as the location of the detector, and treat $\frac{k_0}{m}|\psi_t(x)|^2$ as a normalized probability density $p(t)$ with respect to t. Then,

$$p(t) = \frac{k_0}{m}|T_{k_0}|^2 \theta(t - t_0) \sum_{n=0}^{\infty} \sum_{\ell=0}^{\infty} (-R_{k_0})^n (-R_{k_0}^*)^\ell$$
$$\times \psi_0 \left(a - \frac{k_0(t - t_0)}{m} + \ell\Delta x \right) \psi_0 \left(a - \frac{k_0(t - t_0)}{m} + n\Delta x \right). \tag{18.74}$$

Terms in the summation with $n \neq \ell$ vanish because the corresponding wave functions do not overlap. Then, we write

$$p(t) = \frac{k_0}{m}|T_{k_0}|^2 \theta(t - t_0) \sum_{n=0}^{\infty} |R_{k_0}|^{2n} \rho \left(\frac{k_0(t - t_0)}{m} - n\Delta x \right), \tag{18.75}$$

where $\rho(x) = |\psi_0(a - x)|^2$ is a probability distribution that is localized within a width of order a.

In order to connect Eq. (18.75) with experiments, we have to treat both x and t as macroscopic variables. This means that they are measured with accuracies of order σ_X and σ_T, respectively, that are much larger than the microscopic scales that characterize the system. Hence, $\sigma_X >> a$ and $\sigma_T >> ma/k_0$. At such scales, the width of $\rho(x)$ is negligible, and we can substitute it with a delta function, $\rho(x) \simeq \delta(x)$. In this regime, we can also approximate the sum over n with an integral, so that

$$p(t) = \frac{k_0}{m}|T_{k_0}|^2 \theta(t - t_0) \int_0^{\infty} dn |R_{k_0}|^{2n} \delta \left(\frac{k_0(t - t_0)}{m} - n\Delta x \right)$$
$$= \frac{|T_{k_0}|^2}{\Delta t} \theta(t - t_0) e^{\ln |R_{p_0}|^2 \frac{(t - t_0)}{\Delta t}}, \tag{18.76}$$

where $\Delta t = m\Delta x/k_0$ has the classical interpretation as the time between two successive attempts of the particle to cross the barrier. In Fig. 18.3, we plot the probability distribution given by $p(t)$ versus the one obtained from Eq. (18.76). The difference vanishes if we sample time at scales much larger than Δt.

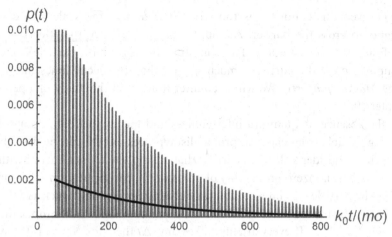

Fig. 18.3 The probability density of Eq. (18.75) as a function of the dimensionless time parameter $\frac{k_0 t}{m\sigma}$. The probability density is characterized by rapid oscillations at the scale of Δt. If we sample at timescales $\sigma_T \gg \Delta t$, we obtain an average decay behavior given by the exponential curve. The separation of scales is crucial for the validity of exponential decay. Note that here, we used the unphysically large value 0.02 for $|T_{k_0}|^2$, in order to show both the oscillations and the overall decay of the probability in the same graph.

For $|T_{k_0}| \ll 1$, $\ln|R_{p_0}|^2 \simeq -|T_{p_0}|^2$. Then, Eq. (18.76) describes exponential decay,

$$p(t) = \Gamma e^{-\Gamma(t-t_0)}\theta(t - t_0), \tag{18.77}$$

with a decay constant

$$\Gamma = \frac{|T_{k_0}|^2}{\Delta t}, \tag{18.78}$$

that does not depend on the detailed properties of the initial state.

Note that exponential decay fails at early times; the derivation of Eq. (18.77) requires that $|t - t_0| \gg \Delta t$. The exponential decay law also fails at very long times, when wave-function dispersion becomes important. To see this, we change variables to $y = \frac{k^2 t}{2m}$ in Eq. (18.69) for $\psi_t(x)$. The dominant term at $t \to \infty$ is

$$\psi_t(x) = -i\sqrt{\frac{m}{4\pi t}}A_0 \int_0^\infty \frac{dy}{y}e^{-iy}\tilde{\psi}_0(\sqrt{2my/t}). \tag{18.79}$$

where A_0 stands for $\lim_{k\to\infty} T_k/(1 + R_k)$. In general, $A_0 \neq 0$, as can be readily checked in elementary systems. Hence, the asymptotic behavior of $\psi_t(x)$ depends on the infrared behavior of $\tilde{\psi}_0$. For a power-law dependence, $\psi_0 \sim k^n$ near $k = 0$; Eq. (18.79) gives $\psi_t(x) \sim t^{-\frac{n+1}{2}}$. It follows that $p(t) \sim t^{-(n+1)}$, that is, the detection probability decays with an inverse power law.

The key property in deriving the exponential decay law is negligible interference between different attempts of the particle to cross the barrier. In this derivation, we assumed no interference (the condition $\Delta x > a$), but this assumption is too strong. To see this, let N be the number of successive attempts that interfere in the probability amplitude. If the position accuracy σ_X remains much larger than Na, the arguments in the derivation pass through.

In fact, even this condition is too strict. The decay time scale Γ^{-1} corresponds to $|T_{k_0}|^{-2}$ attempts to cross the barrier. As long as $|T_{k_0}|^{-2} >> N$ the effects of interference are not manifested at the decay scales. This condition implies that the particle's "memory" about its past attempts to cross the barrier is much shorter than the decay timescale. This feature is known as the *Markov property*. We will encounter it again in the study of open quantum systems in Chapter 20.

In the absence of quantum interferences and memory effects, decays due to tunneling are indistinguishable from classical probabilistic processes that can be described using elementary arguments. Consider a classical particle that attempts to cross a barrier with probability $w << 1$ of success. The nonzero probability of crossing the barrier needs not be quantum mechanical in origin. The particle may interact with a stochastic environment, such as a thermal bath. Then, the crossing of the barrier may be due to a random force. After N attempts, the survival probability is $(1 - w)^N \simeq e^{-Nw}$. If every attempt takes time Δt then, for $N >> 1$, the system is described by an exponential decay law with constant $\Gamma = w/\Delta t$, in full agreement with Eq. (18.78).

QUESTIONS

18.1 The restriction $\Omega = |E_n - E_{n_0}|$ for transitions in Fermi's golden rule applies only for pulses of long duration. For short pulses, there are transitions that violate this condition. Does this imply a failure of energy conservation?

18.2 **(B)** Why was it difficult to reconcile exponential decay of nuclei or atoms with classical mechanics?

18.3 **(B)** It is often stated that the exponential decay law in radioactivity is a prediction of quantum mechanics. Would you agree with this statement? If not, how would you qualify it?

18.4 **(B)** An energy threshold is a value of energy E_{th} at which the density of states $g(E)$ of the Hamiltonian exhibits a large discontinuity. Argue that for initial states with energies near a threshold of the Hamiltonian $\hat{H}_0$, the WWA should be expected to fail.

18.5 **(B)** Identify the approximations involved in the derivation of the exponential law for tunneling decays. Can you think of a scenario in which those approximations fail?

PROBLEMS

18.1 A qubit of frequency ω is prepared in its ground state $|0\rangle$. At time $t = 0$, it is acted upon by a weak pulse of duration T that corresponds to the addition of a term $f(t)\hat{\sigma}_1$ in the Hamiltonian. Calculate the probability that the qubit is found in the excited state for $t > T$ with (i) $f(t) = g \sin\left(\frac{\pi t}{T}\right)$ and (ii) $f(t) = \lambda\delta(t - t_0)$, for constants g and λ.

18.2 In the dipole approximation, the interaction of the electron in the hydrogen atom with an external EM field is given by the dipole Hamiltonian $\hat{H}_I = e\mathbf{E}(t) \cdot \mathbf{x}$. In order to evaluate the transition function $A(E)$ (18.11) for the ground state $|1, 0, 0, \rangle$, we approximate the generalized eigenfunctions of the hydrogen atom Hamiltonian with those of the free particle, $\psi_{\mathbf{k}}(\mathbf{x}) = \frac{1}{(2\pi)^{3/2}} e^{i\mathbf{k}\cdot\mathbf{x}}$.

(i) Show that

$$\langle 1,0,0|\mathbf{x}|\mathbf{k}\rangle = \frac{8\sqrt{2}ia_0^{7/2}}{\pi[1+(ka_0)^2]^3}\mathbf{k}.$$

(ii) Show that the transition function is

$$A(E) = \frac{1024 m_e k^3 a_0^7}{\pi[1+(ka_0)^2]^6}.$$

18.3 **(B)** Let $\hat{H}$ be the Hamiltonian of an unstable quantum system, and let $\hat{P}_E$ be its spectral projectors. The persistence amplitude $\mathcal{A}(t)$ can be expressed as $\int_{E_0}^{\infty} dE \rho(E)e^{-iEt}$, where $\rho(E) := \langle \psi_0|\hat{P}_E|\psi_0\rangle$ is the energy density of the initial state $|\psi_0\rangle$ and E_0 is the energy of the ground state. (i) Show that a purely exponential persistence amplitude $\mathcal{A}(t) = e^{-\frac{1}{2}\Gamma t + i\omega t}$, for $\Gamma > 0$ and $\omega \in \mathbb{R}$ is impossible for a Hamiltonian that is bounded from below. (ii) Suppose that E is peaked around $E = E_1$. Show that a Gaussian $\rho(E)$ with this property is not compatible, even approximately, with exponential decay. (iii) Evaluate the decay probability for $\rho(E)$ given by a Boltzmann-type distribution $\rho(E) = \beta e^{-\beta(E-E_0)}$, where $\beta > 0$.

18.4 **(B)** Consider the spontaneous decay of an atom within an optical fiber of diameter much smaller than the wavelength of the emitted radiation. In this case, we treat the EM field as one-dimensional. Hence, the (scalar) photon states in Lee's model are specified by a wavenumber $k \in \mathbb{R}$, they have energy $\omega_k = |k|$ and couplings $g_k = \lambda/\sqrt{2\omega_k}$. (i) Show that the self-energy function is

$$\Sigma(z) = \frac{\lambda^2}{2\pi}\int_0^{\infty} \frac{dk}{k(z-k)}.$$

This integral converges for large k, but it diverges at $k = 0$. We change the integration range to $[\mu, \infty)$, where μ is a low-frequency cut-off. (ii) Evaluate $\Sigma(z)$ and show that $\Gamma(E) = \lambda^2/E$.

18.5 **(B)** We can mimic the dipole coupling (13.102) by an interaction term $\frac{\partial\hat{\phi}(\mathbf{x}_0,t)}{\partial t}\hat{\sigma}_1$ that is proportional to the time derivative of the quantum field. (i) Show that the Lee-model coefficients are of the form $\kappa\sqrt{\omega_\mathbf{k}}$ for some constant κ. (ii) Evaluate the self-energy function. (iii) Show that the decay constant is proportional to Ω^3, where Ω is the transition frequency, in agreement with the analysis of Section 18.4.2.

18.6 **(B)** A two-level atom of frequency Ω is contained within a cavity that allows only field frequencies close to an eigenfrequency ω_0. Photoemission is described by Lee's model but with $g_\mathbf{k} = \frac{\lambda}{\sqrt{2\omega_k}}\sqrt{\omega_0\chi(\omega_\mathbf{k},\omega_0)}$, where $\chi(\omega,\omega_0)$ is a probability density for $\omega \in [0,\infty)$, centered around ω_0 with width $\gamma << \omega_0$. (i) Show that the self-energy function is

$$\Sigma(z) = -\frac{\lambda^2\omega_0}{4\pi^2} + \frac{\lambda^2\omega_0}{4\pi^2}z\int_0^{\infty} \frac{dk\,\chi(k,\omega_0)}{z-k}.$$

Since the integral is narrowly concentrated around ω_0, we extend the integration range $(-\infty,\infty)$. In this approximation, we choose a Lorentzian $\chi(\omega,\omega_0) = \frac{1}{\pi}\frac{\gamma}{(\omega-\omega_0)^2+\gamma^2}$. (ii) Show that

$$\Sigma(z) = \Gamma \frac{\omega_0 - i\gamma}{z - \omega_0 + i\gamma}, \tag{18.80}$$

where $\Gamma = \frac{\lambda^2 \omega_0}{4\pi^2}$. (iii) Show that the persistence amplitude is determined by the roots of the binomial equation $x^2 - (\delta - i\gamma)x - \Gamma(\omega_0 - i\gamma) = 0$, where $x = z - \Omega$ and $\delta := \omega_0 - \Omega$. (iv) Show that for exact resonance $\delta = 0$, the roots are approximately $\Omega \pm \sqrt{\Gamma\omega_0 - \frac{1}{4}\gamma^2} - i\frac{\gamma}{2}$. Evaluate the survival probability (v) Show that in an ideal cavity with $\gamma \to 0$, both roots are real, so there is no decay. Evaluate the persistence probability in this case.

18.7   (B) We use Lee's model to study the decay of a heavy particle with the emission of a slowly moving boson of mass μ, such as, for example, alpha decay or some heavy baryon decays ($\Omega^- \to \Lambda^0 + K^-$). The appropriate couplings are $g_{\mathbf{k}} = \lambda/\sqrt{2\omega_{\mathbf{k}}}$, while for nonrelativistic velocities, $\omega_{\mathbf{k}} = \mu + \frac{\mathbf{k}^2}{2\mu}$. (i) Show that the self-energy function is

$$\Sigma(z) = -\frac{\lambda^2 \mu}{2\sqrt{2}\pi} \frac{1}{1 + \sqrt{-\frac{z-\mu}{\mu}}}.$$

(ii)Assume that $|z - \mu| << \mu$ to show that $\Sigma(z) = \frac{\lambda^2}{2\pi}\sqrt{-\frac{\mu z}{2}}$ modulo a constant and a redefinition of energy. (iii) Show that for $E > 0$, $F(E) = 0$ and $\Gamma(E) = \frac{\lambda^2}{2\pi}\sqrt{2\mu E}$.

18.8   (B) We generalize Lee's model in order to describe the decay of an entangled pair of qubits. The Hamiltonian is of the form $\hat{H}_0 + \hat{V}$ with

$$\hat{H}_0 = \frac{1}{2}\Omega(\hat{I} + \hat{\sigma}_3^{(1)}) + \frac{1}{2}\Omega(\hat{I} + \hat{\sigma}_3^{(2)}) + \sum_r \omega_r \hat{a}_r^\dagger \hat{a}_r,$$

$$\hat{V} = \sum_r \left(g_r^{(1)} \hat{\sigma}_+^{(1} \hat{a}_r + g_r^{(1)\star} \hat{\sigma}_-^{(1)} \hat{a}_r^\dagger \right) + \sum_r \left(g_r^{(2)} \hat{\sigma}_+^{(2)} \hat{a}_r + g_r^{*(2)} \hat{\sigma}_-^{(2)} \hat{a}_r^\dagger \right),$$

where the upper indices (1) and (2) label the qubit. For identical qubits, the absolute values of the coupling constants $|g_r^{(i)}|$ are the same $|g_r^{(1)}| = |g_r^{(2)}| = g_r$. Hence, we write $g_r^{(i)} = g_r e^{i\Theta_r^{(i)}}$.

We consider an initial state $|B\rangle \otimes |0\rangle$, where $|0\rangle$ is the field vacuum and $|B_\pm\rangle$ is a Bell-type state $|B_\pm\rangle = \frac{1}{\sqrt{2}}(|e, g\rangle \pm |g, e\rangle)$. (i) Show that the self-energy function $\Sigma_{B_\pm}(z) = \Sigma_0(z) \pm \Sigma_1(z)$ where $\Sigma_0(z)$ is the self-energy function for a single two-level system and $\Sigma_1(z) = \sum_r \frac{g_r^2 \cos[\Theta_r^{(1)} - \Theta_r^{(2)}]}{z - \omega_r}$. (ii) Show that, for scalar photons at distance r from each other,

$$\Sigma_1(z) = \frac{\lambda^2}{4\pi^2 r}\left[[\gamma + \ln(-rz) + \text{Cin}(rz)]\sin rz - \text{si}(rz)\cos rz \right],$$

where

$$\text{Cin}(z) := \int_0^z dt \frac{1 - \cos t}{t} \qquad \text{si}(z) := \int_z^\infty dt \frac{\sin t}{t},$$

with γ the Euler–Mascheroni constant. (iii) Show that the associated decay constants are $\Gamma_\pm = \frac{\lambda^2 \tilde{\Omega}}{2\pi}(1 \pm \frac{\sin(\tilde{\Omega}r)}{\tilde{\Omega}r})$.

18.9   (B) We consider tunneling decays of a particle of mass m through a delta-function potential barrier $V(x) = \eta\delta(x - a)$. Show that, in the opaque barrier limit, $|T_k| << 1$, the dominant contribution to the decay rate of an almost monochromatic initial state at energy $\frac{k_0^2}{2m}$ is

$$\Gamma = \frac{k_0^3}{2m^3\eta^2(a - 1/2mn)}.$$

18.10 (B) In the simplest model of alpha decay, we consider an alpha particle with mass m and charge $2e$ interacting with a potential generated by the remaining $Z - 2$ nucleons, in a nucleus with atomic number Z,

$$V(r) = \begin{cases} \frac{2(Z-2)\alpha}{r}, & r > R \\ 0, & r \leq R. \end{cases}$$

Here R is a phenomenological length parameter that models the range of nuclear interactions that cancel the Coulomb barrier. Evaluate the transition amplitude in the WKB approximation for vanishing angular momentum of the outcoming particle. Show that, in the opaque barrier limit, the decay rate for an alpha particle of energy E is

$$\Gamma = \sqrt{\frac{2E}{m}}(8R)^{-1}\exp\left[-2\sqrt{2mE}r_1\left(\tan^{-1}\left(\sqrt{\frac{r_1}{r} - 1}\right) - \frac{R}{r_1}\sqrt{\frac{r_1}{r} - 1}\right)\right],$$

where $r_1 = 2(Z - 2)\alpha/E$. (Assume that the time delay between successive attempts to cross the barrier is given by the classical time it takes to cross the distance $2R$.)

Bibliography

- For a presentation of the RPA and its application to perturbative decays, see Nakazato et al. (1996). For a review of quantum decays that elaborates on the topics presented here, see Anastopoulos (2019).

19 Scattering Theory

To understand the importance of scattering theory, consider the variety of ways in which it arises. First, there are various phenomena in nature (like the blue of the sky) which are the result of scattering. In order to understand the phenomenon ... one must understand the underlying dynamics and its scattering theory. Second, one often wants to use the scattering of waves or particles whose dynamics one knows to determine the structure and position of small or inaccessible objects. For example, in x-ray crystallography (which led to the discovery of DNA), tomography, and the detection of underwater objects by sonar, the underlying dynamics is well understood. What one would like to construct are correspondences that link, via the dynamics, the position, shape, and internal structure of the object to the scattering data ... A third use of scattering theory is as a probe of dynamics itself. In elementary particle physics, the underlying dynamics is not well understood and essentially all the experimental data are scattering data. The main test of any proposed particle dynamics is whether one can construct for the dynamics a scattering theory that predicts the observed experimental data.

Reed and Simon (1979)

19.1 Scattering by a Potential

Scattering experiments are one of our most important tools for extracting information about the structure and interactions of microscopic systems. In these experiments, we prepare a beam of particles of a given type and we direct it towards a target. The interaction of the particles in the beam with those of the target may lead to various phenomena: changes in the direction and the energy of incoming particles, absorption of incoming particles, the appearance of new species of particles, and so on. The target is surrounded by particle detectors that identify the particles that exit the interaction region and measure their momenta.

The statistical distribution of target particles with respect to their species and observables such as momentum, energy, and angular momentum are the scattering data. The aim of the theoretical description is to explain this data from the knowledge of the initial conditions of the incoming particles and of the target and from their mutual interactions.

A great simplification occurs in the theoretical description of scattering experiments due to large separation of their characteristic scales. The source of the incoming beam and the particle detectors are placed at macroscopically large distances from the target. This implies that both incoming and outgoing particles behave like free particles most of the time, because the timescale of the interaction between incoming and target particle is much smaller than the time it takes the particles to propagate from the source to the target and from the target to the detector. Also, the width of the initial beam is much larger than the typical dimensions of the target particles. This means that the minimum distance between an incoming particle and a target particle is a random variable, and that the observed probabilities involve a summation over all possible values of this variable.

19.1.1 Detection Probability

In Section 8.3.4, we studied particle scattering off a potential in one dimension. In this section, we will consider the corresponding problem in three dimensions. The set-up is sketched in Fig. 19.1.

We assume that the target corresponds to a potential $V(\mathbf{x})$ that takes nonzero values only in a region of linear dimension D around the coordinate origin; we call D the *range* of the potential. The potential describes the interaction between one particle from the beam and one particle from the target, in a reference frame where the center of mass of the two particles is stationary. We assume *elastic scattering*: No internal states of the particles can be excited in scattering, so that kinetic energy is preserved.

The incoming particles are prepared in a state $\psi_0(\mathbf{x})$, which is concentrated around $\mathbf{x}_0 = (0, 0, -x_0)$ with width Δx, such that $x_0 >> \Delta x >> D$. The Fourier transform $\tilde{\psi}_0$ of ψ_0 is peaked around momentum $\mathbf{k}_0 = (0, 0, k_0)$ with width Δp, so that $k_0 >> \Delta k$. This means that $\psi(\mathbf{x}) \simeq 0$ if $|\mathbf{x} - \mathbf{x}_0| >> \Delta x$, and $\tilde{\psi}_0(\mathbf{p}) \simeq 0$ if $|\mathbf{p} - \mathbf{k}_0| >> \Delta k$.

The wave packet moves a macroscopic distance along axis 3. In contrast, it extends to only a small distance from the 3-axis along the transverse directions 1 and 2. The only physically relevant property of the wave packet in the transverse directions is its width, because it defines the cross-section A of the incoming beam. It is convenient to take $\psi_0(\mathbf{x})$ to differ from zero only in a cylinder of radius a around the 3-axis, so that $A = \pi a^2$, and

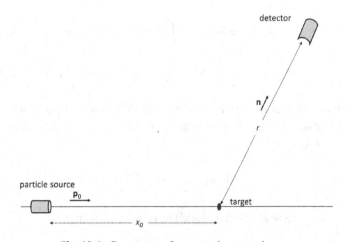

Fig. 19.1 Geometry of a scattering experiment.

$$\psi_0(x_1, x_2, x_3) = \begin{cases} \psi(0,0,x_3), & \sqrt{x_1^2 + x_2^2} \leq a \\ 0, & \sqrt{x_1^2 + x_2^2} > a \end{cases} \tag{19.1}$$

This implies that

$$\int_{-\infty}^{\infty} dx_3 |\psi(0,0,x_3)|^2 = \frac{1}{A}. \tag{19.2}$$

The Hamiltonian that describes particle scattering is $\hat{H} = \frac{\mathbf{p}^2}{2m} + V(\mathbf{x})$, where the potential vanishes everywhere except for a small region around the target. Outside this region, the eigenfunctions of $\hat{H}$ coincide with those of the free-particle Hamiltonian. The degeneracy of the latter implies that no unique generalized basis of eigenvectors exists.

We select a specific basis $|\mathbf{k}\rangle$ of generalized eigenvectors of $\hat{H}$ with corresponding eigenvalues $E_{\mathbf{k}} = \frac{\mathbf{k}^2}{2m}$. The eigenvectors $|\mathbf{k}\rangle$ are defined by their behavior in the region where the potential vanishes,

$$\langle \mathbf{x}|\mathbf{k}\rangle = \frac{1}{(2\pi)^{3/2}} \left[e^{i\mathbf{k}\cdot\mathbf{x}} + f_{\mathbf{k}}(\mathbf{n})\frac{e^{ikr}}{r} \right], \tag{19.3}$$

where $r = |\mathbf{x}|, k = |\mathbf{k}|$, and $\mathbf{n} = \mathbf{x}/r$. The function $f_{\mathbf{k}}(\mathbf{n})$ is called *scattering amplitude*. We will justify this choice in terms of a general theory of scattering in Section 19.5. For the moment, we note that, at the level of classical waves, Eq. (19.3) is a linear combination of a plane wave (that corresponds to the incoming beam) and of a spherical wave (that corresponds to a scattered wave). This choice of basis leads to a very simple expression for the wave function $\psi_t(\mathbf{x})$ after scattering.

To see this, we use the spectral theorem for the time evolution operator to write

$$\psi_t(\mathbf{x}) = \int d^3k \langle \mathbf{x}|\mathbf{k}\rangle e^{-iE_{\mathbf{k}}t} \langle \mathbf{k}|\psi_0\rangle. \tag{19.4}$$

We calculate

$$\langle \mathbf{k}|\psi_0\rangle = \frac{1}{(2\pi)^{3/2}} \int d^3r \left[e^{-i\mathbf{k}\cdot\mathbf{x}} + f_{\mathbf{k}}^*(\mathbf{n})\frac{e^{-ikr}}{r} \right] \psi_0(\mathbf{r})$$
$$\simeq \tilde{\psi}_0(\mathbf{k}) + \frac{1}{(2\pi)^{3/2}r_0} f_{\mathbf{k}}^*(-\mathbf{n}_0) \int d^3r e^{-ikr} \psi_0(\mathbf{r}), \tag{19.5}$$

where we used the fact that ψ_0 is narrowly focused around a point with radial coordinate $r = r_0$ and unit vector $\mathbf{n} = -\mathbf{k}_0/|\mathbf{k}_0| := -\mathbf{n}_0$. For the same reason, we can substitute kr with $\mathbf{k}' \cdot \mathbf{r}$, where $\mathbf{k}' = -k\mathbf{n}_0$. Then,

$$\langle \mathbf{k}|\psi_0\rangle \simeq \tilde{\psi}_0(\mathbf{k}) + \frac{1}{r_0} f_{\mathbf{k}}^*(-\mathbf{n}_0)\tilde{\psi}_0(\mathbf{k}'). \tag{19.6}$$

Since $|\mathbf{k}' - \mathbf{k}_0| = k_0 + k >> \Delta p$, $\tilde{\psi}_0(\mathbf{k}') \simeq 0$. It follows that

$$\langle \mathbf{k}|\psi_0\rangle \simeq \tilde{\psi}_0(\mathbf{k}). \tag{19.7}$$

We substitute Eq. (19.7) into Eq. (19.4), and we take $\mathbf{x}$ in the region where the potential vanishes. Writing $f_{\mathbf{k}}(\mathbf{n}) = |f_{\mathbf{k}}(\mathbf{n})|e^{i\chi_{\mathbf{k}}(\mathbf{n})}$, we obtain

$$\psi_t(\mathbf{x}) = \psi_t^{(0)}(\mathbf{x}) + \frac{1}{(2\pi)^{3/2}r} \int d^3k |f_{\mathbf{k}}(\mathbf{n})|e^{ikr - iE_k t + i\chi_{\mathbf{k}}(\mathbf{n})} \tilde{\psi}_0(\mathbf{k}), \qquad (19.8)$$

where $\psi_t^{(0)}(\mathbf{x})$ is the time evolution of the initial state under the free-particle Hamiltonian. This term corresponds to the fraction of the initial beam that has not been scattered.

We approximate the integral in Eq. (19.8) using the phase expansion method for almost monochromatic wave packets – see Box 2.2. We write $|f_{\mathbf{k}}(\mathbf{n})| \simeq |f_{\mathbf{k}_0}(\mathbf{n})|$, and we expand the oscillating terms in Eq. (19.8) to leading order in $\mathbf{k} - \mathbf{k}_0$,

$$E_{\mathbf{k}} \simeq E_{\mathbf{k}_0} + \frac{\mathbf{k}_0}{m} \cdot (\mathbf{k} - \mathbf{k}_0), \quad k \simeq k_0 + \mathbf{n}_0 \cdot (\mathbf{k} - \mathbf{k}_0),$$

$$\chi_{\mathbf{k}}(\mathbf{n}) = \chi_{\mathbf{k}_0}(\mathbf{n}) + \mathbf{b}_{\mathbf{k}_0}(\mathbf{n}) \cdot (\mathbf{k} - \mathbf{k}_0),$$

where $\mathbf{b}_{\mathbf{k}_0}(\mathbf{n}) = \nabla_{\mathbf{k}}\chi_{\mathbf{k}}(\mathbf{n})|_{\mathbf{k}=\mathbf{k}_0}$.

Integration over $\mathbf{k}$ gives the inverse Fourier transform of $\tilde{\psi}_0$,

$$\psi_t(\mathbf{x}) = \psi_t^{(0)}(\mathbf{x}) + \frac{f_{\mathbf{k}_0}(\mathbf{n})}{r} e^{iE_{\mathbf{k}_0}t - i\mathbf{b}_{\mathbf{k}_0}(\mathbf{n})\cdot\mathbf{k}_0} \psi_0[\mathbf{n}_0 r - \frac{\mathbf{k}_0}{m}t + \mathbf{b}_{\mathbf{k}_0}(\mathbf{n})]. \qquad (19.9)$$

We will see that the quantity $\mathbf{b}_{\mathbf{k}_0}(\mathbf{n})$ is an effective position shift of the particle due to interaction, typically of the order of the range D. It is therefore much smaller than r, and it can be safely ignored. Then,

$$\psi_t(\mathbf{x}) = \psi_t^{(0)}(\mathbf{x}) + \frac{f_{\mathbf{k}_0}(\mathbf{n})}{r} e^{iE_{\mathbf{k}_0}t} \psi_0\left[\mathbf{n}_0\left(r - \frac{\mathbf{k}_0}{m}t\right)\right]. \qquad (19.10)$$

Assuming that the wave packet has not significantly dispersed at time t, $\psi_t^{(0)}(\mathbf{x})$ differs appreciably from zero only for small angles around $\mathbf{n}_0$. For any other direction $\mathbf{n} = \mathbf{x}/r$, the probability density for $\mathbf{r}$ at constant t is

$$|\psi_t(\mathbf{x})|^2 = \frac{|f_{\mathbf{k}_0}(\mathbf{n})|^2}{r^2} \left|\psi_0\left[\mathbf{n}_0\left(r - \frac{\mathbf{k}_0}{m}t\right)\right]\right|^2. \qquad (19.11)$$

We want to compute the probability that the particle will be detected at some moment of time by an elementary detector located at $\mathbf{x}$. This type of probability is difficult to compute; Born's rule gives the probability for position at a fixed moment of time, and we cannot consistently add probabilities from different moments of time. However, the probability density $|\psi_t(\mathbf{x})|^2$ depends on $\mathbf{n}$ only through $|f_{\mathbf{k}_0}(\mathbf{n})|^2$, while t appears only through the combination $r - \frac{k_0}{m}t$. This means that integration over r, allowed by Born's rule, effectively implements integration over time.

Therefore, we calculate the probability $\mathrm{Prob}(\mathbf{n}, t)\delta\Omega$ that, at time t, the particle has "passed through" the detector and it is found within a solid angle $\delta\Omega$ around the direction $\mathbf{n}$,

$$\mathrm{Prob}(\mathbf{n}, t)\delta\Omega := \delta\Omega \int_r^\infty r'^2 dr' |\psi_t(r', \mathbf{n})|^2 = \delta\Omega |f_{\mathbf{k}_0}(\mathbf{n})|^2 \int_r^\infty dr' |\psi_0[\mathbf{n}_0(r' - \frac{k_0}{m}t)]|^2$$

$$= \delta\Omega |f_{\mathbf{k}_0}(\mathbf{n})|^2 \int_{r - \frac{k_0}{m}t}^\infty dx_3 |\psi_0(0, 0, x_3)|^2, \qquad (19.12)$$

where we set $x_3 = r' - \frac{k_0}{m}t$.

For $t >> m(r_0 + r)/k_0$, we can substitute the lower limit of integration with $-\infty$. Then, by Eq. (19.2), we obtain a time-independent probability density $p(\mathbf{n})$ for the directions of the outgoing particles,

$$p(\mathbf{n}) = \frac{|f_{\mathbf{k}_0}(\mathbf{n})|^2}{A}. \tag{19.13}$$

19.1.2 Scattering Cross-Section

We showed that the probability density for particles moving in the direction $\mathbf{n}$ after scattering is the ratio of a quantity $\sigma(\mathbf{n})$ with dimensions of area to the cross-section area A of the initial beam

$$p(\mathbf{n}) = \frac{\sigma(\mathbf{n})}{A}. \tag{19.14}$$

This conclusion is generic for scattering experiments, and it constitutes a nontrivial prediction of quantum theory. The quantity $\sigma(\mathbf{n})$ is called the *differential cross-section*.[1] It is experimentally determined from the fraction of particles measured along the direction $\mathbf{n}$ and the width of the initial beam.

The analysis performed here presupposes *a single incoming particle* and *a single target particle*. In real experiments, we send many-particle beams towards the target. Our analysis is applicable only if the beam particles are uncorrelated. In this case, each particle of the beam scatters independently of the others, and the particle beam defines a statistical ensemble of individual scattering events. The probability density $p(\mathbf{n})$ refers to this ensemble. If the beam consists of N_i particles, then the expected number of particles scattering in a solid angle $\delta\Omega$ around $\mathbf{n}$ is $N_i p(\mathbf{n})\delta\Omega$.

Real targets consist of a large number of particles. Suppose that there are N_t target particles within the incoming beam's cross-section. This is possible only if the width a of the wave packet is much larger than the range D of the interaction. Assume that (i) scattering terms from different target particles do not interfere and (ii) the scattering probability is so small that multiple scattering of incoming particles is negligible. Then, scattering events with different target particles are statistically independent. We can enlarge the statistical ensemble to consist of the $N_i N_t$ possible *pairs* of one incoming and one target particle. The expected number of particles scattering in a solid angle $\delta\Omega$ around $\mathbf{n}$ is $N_i N_t p(\mathbf{n})\delta\Omega$.

However, if conditions (i) and (ii) do not apply, we must view the totality of the N_t particles as a single target. For example, incoming particles with large de Broglie wavelength scatter off the whole crystal in a solid, not off individual atoms – see Problem 19.7.

Suppose that the conditions that we assumed in the derivation of Eq. (19.13) apply. Then, by Eq. (19.14),

$$\sigma(\mathbf{n}) = |f_{\mathbf{k}_0}(\mathbf{n})|^2. \tag{19.15}$$

---

[1] The term "differential" means that $\sigma(\mathbf{n})$ is a density with respect to $\mathbf{n}$ or, equivalently, with respect to angles θ and ϕ. For this reason, it is often represented as $d\sigma/d\Omega$, where $d\Omega = \sin\theta d\theta d\phi$ stands for an infinitesimal solid angle.

The probability density (19.14) is not normalized to unity, because we ignored the fraction of particles that did not scatter. If we integrate over all **n**, we obtain the total scattering probability $p_{sc} = \sigma_{tot}/A$, where

$$\sigma_{tot} = \int d^2\mathbf{n} \,\sigma(\mathbf{n}) \tag{19.16}$$

is the *total cross-section*. The total scattering probability is smaller than unity, as there is always a nonzero probability of no scattering,[2] equal to $1 - p_{sc}$.

This analysis presupposes that the width of the wave packet does not change appreciably during time evolution, either before or after scattering. This means that the propagation time t must be smaller than ma^2 – see Section 7.3.2. The propagation time is $x_0 m/k_0$ prior to scattering and rm/k_0 after scattering. Hence, the inequalities $x_0 > a^2 k_0$ and $r > a^2 k_0$ are essential constraining conditions for the validity of Eq. (19.15).

19.1.3 Center-of-Mass versus Laboratory Frame

The description of two-particle scattering by a potential presupposes that we work in the center-of-mass frame. This coincides with the laboratory frame, where the target is motionless, only in the limit where the target mass M is much larger than the incoming particle's mass m. This means that the differential cross-section $\sigma(\mathbf{n})$ defined here refers to the center-of-mass frame.

The differential cross-section $\sigma_L(\mathbf{n}_L)$ in the laboratory frame depends on the outgoing momentum vector $\mathbf{n}_L$ as determined in the laboratory frame. Since the number of recorded events must be the same, $\sigma(\mathbf{n})d^2\mathbf{n} = \sigma_L(\mathbf{n}_L)d^2\mathbf{n}_L$ or, equivalently,

$$\sigma(\theta,\phi)\sin\theta \, d\theta \, d\phi = \sigma_L(\theta_L,\phi_L)\sin\theta_L \, d\theta_L \, d\phi_L, \tag{19.17}$$

in terms of the angles θ_L and ϕ_L, defined on the laboratory frame.

The scattering of two particles in the two frames is described in Fig. 19.2. In the laboratory frame, the initial momenta of the incoming and target particle are **k** and 0, respectively. Their momenta after collision are $\mathbf{k}_1'$ and $\mathbf{k}_2'$.

In the center-of-mass frame, the initial momenta of the incoming and target particle are **p** and $-\mathbf{p}$, respectively. Their momenta after collision are $\mathbf{p}'$ and $-\mathbf{p}'$, where by energy conservation, $p = p'$.

We note that all momenta lie in the same plane, so that we can always choose coordinates with $\phi_L = \phi$. The center-of-mass velocity in the laboratory frame is $\frac{\mathbf{k}}{m+M}$. Hence, by a Galilei boost, $\mathbf{p} = \mathbf{k} - \frac{m\mathbf{k}}{m+M} = \frac{M}{m+M}\mathbf{k}$ and $\mathbf{p}' = \mathbf{k}_1' - \frac{m\mathbf{k}}{m+M} = \mathbf{k}_1' - \frac{m}{M}\mathbf{p}$. We project the latter equation to the two axes, to obtain

$$p\sin\theta = k_1'\sin\theta_L, \qquad p\cos\theta + \frac{m}{M}p = k_1'\cos\theta_L. \tag{19.18}$$

Dividing those equations by parts, we obtain the relation between θ_L and θ,

$$\tan\theta_L = \frac{\sin\theta}{\cos\theta + \frac{m}{M}}. \tag{19.19}$$

---

[2] This means that the sample space of a scattering experiment is $\Gamma = S^2 \cup \{NO\}$, that is, the union of the unit sphere S^2 of directions with the singlet for the alternative NO of no-scattering.

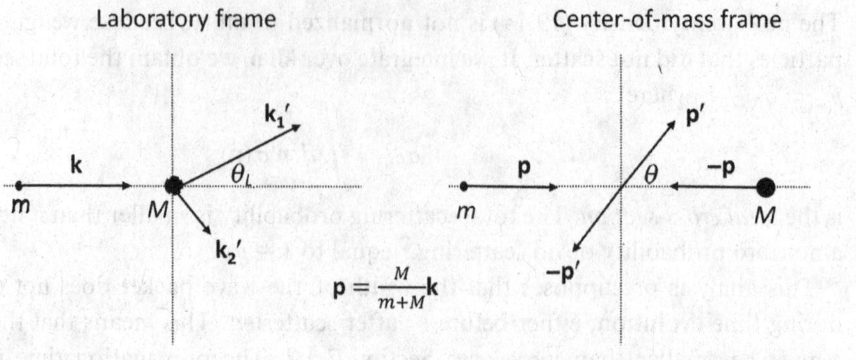

Laboratory frame   Center-of-mass frame

Fig. 19.2 Scattering in the laboratory frame versus scattering in the center-of-mass frame.

If $m <<< M, \theta_L = \theta$, while, for $m = M$, $\tan\theta_L = \frac{2\sin\frac{\theta}{2}\cos\frac{\theta}{2}}{2\cos^2\frac{\theta}{2}} = \tan\frac{\theta}{2}$, that is, $\theta_L = \frac{1}{2}\theta$.

From Eq. (19.19), we calculate

$$\sigma_L(\theta_L, \phi_L) = \sigma(\theta, \phi)\frac{\left(1 + \frac{m^2}{M^2} + 2\frac{m}{M}\cos\theta\right)^{3/2}}{|1 + \frac{m}{M}\cos\theta|}. \tag{19.20}$$

It is straightforward to find the relation between the energy E_L in the laboratory frame and the energy E in the center-of-mass frame: $E_L = \left(1 + \frac{m}{M}\right)E$.

Having established the relation between the two frames, in the rest of this chapter we will work exclusively in the center-of-mass frame.

19.1.4 The Optical Theorem

Equation (19.15) applies for all directions $\mathbf{n}$ except for those that form a very small angle, of the order of $\Delta x/r$, with $\mathbf{n}_0$. To deal with the latter case, we must employ the full Eq. (19.10) for calculating the relevant probabilities. We obtain

$$|\psi_t(\mathbf{x})|^2 = \left|\psi_t^{(0)}(\mathbf{x})\right|^2 + \frac{|f_{\mathbf{k}_0}(\mathbf{n})|^2}{r^2}\left|\psi_0\left[\mathbf{n}_0(r - \frac{k_0}{m}t)\right]\right|^2$$
$$+ \frac{2}{r}\mathrm{Re}\left[f_{\mathbf{k}_0}(\mathbf{n})e^{iE_{\mathbf{k}_0}t}\psi_t^{*(0)}(\mathbf{x})\psi_0\left[\mathbf{n}_0\left(r - \frac{k_0 t}{m}\right)\right]\right]. \tag{19.21}$$

The probability density $p(\mathbf{n})$ for a direction $\mathbf{n}$ is given by $\int_r^\infty r'^2 dr' |\psi_t(r'\mathbf{n})|^2$ for large t. The corresponding quantity $p^{(0)}(\mathbf{n})$ in the absence of a target ($V = 0$) is $\int_r^\infty r'^2 dr'|\psi_t^{(0)}(r'\mathbf{n})|^2$, again at large t. From Eq. (19.21), we obtain an expression that relates these probability densities

$$p(\mathbf{n}) - p^{(0)}(\mathbf{n}) = \frac{|f_{\mathbf{k}_0}(\mathbf{n})|^2}{A} + 2\mathrm{Re}\left[e^{iE_{\mathbf{k}_0}t}\int_r^\infty r'dr'f_{\mathbf{k}_0}(\mathbf{n})\psi_t^{*(0)}(r'\mathbf{n})\psi_0\left[\mathbf{n}_0\left(r' - \frac{k_0}{m}t\right)\right]\right].$$

The phase-expansion approximation in the free-particle evolution yields

$$\psi_t^{(0)}(\mathbf{x}) \simeq e^{-iE_{\mathbf{k}_0}t}\psi_0(\mathbf{x} - \mathbf{k}_0 t/m), \tag{19.22}$$

hence,

$$p(\mathbf{n}) - p^{(0)}(\mathbf{n}) = \frac{|f_{\mathbf{k}_0}(\mathbf{n})|^2}{A} + 2\mathrm{Re}\left[\int_r^\infty r'dr' f_{\mathbf{k}_0}(\mathbf{n})\psi_0^*\left(r'\mathbf{n} - \frac{\mathbf{k}_0}{mt}\right)\psi_0\left[\mathbf{n}_0\left(r' - \frac{k_0}{m}t\right)\right]\right].$$

(19.23)

The product $\psi_0^*(r'\mathbf{n} - \mathbf{k}_0 t/m)\psi_0[\mathbf{n}_0(r' - \frac{k_0}{m}t)]$ differs from zero only if $r'|\mathbf{n} - \mathbf{n}_0| < \Delta x$. Since, $r' \geq r$ in the integral, the condition becomes $|\mathbf{n} - \mathbf{n}_0| < \Delta x/r$ or, equivalently, $4\sin^2\frac{\theta}{2} < \Delta x/r$, where θ is the angle between $\mathbf{n}$ and $\mathbf{n}_0$. The angle θ is very small for sufficiently large r, hence, $4\sin^2\frac{\theta}{2} \simeq \theta^2$. The condition then becomes $|\theta| < \sqrt{\Delta x/r}$. This means that the integral in Eq. (19.23) is proportional to an approximate delta function on the unit-sphere, $\delta_\epsilon^2(\mathbf{n}, \mathbf{n}_0)$. The exact delta function is defined by the condition $\lim_{\epsilon \to 0} \int d^2\mathbf{n} \delta_\epsilon(\mathbf{n}, \mathbf{n}_0)f(\mathbf{n}) = f(\mathbf{n}_0)$. In our case, ϵ is finite and of the order of $\sqrt{\frac{\Delta x}{r}}$.

We write Eq. (19.23) as

$$p(\mathbf{n}) - p^{(0)}(\mathbf{n}) = \frac{|f_{\mathbf{k}_0}(\mathbf{n})|^2}{A} + C\delta_\epsilon^2(\mathbf{n}, \mathbf{n}_0),$$

(19.24)

for some constant C. We evaluate C by integrating both sides of Eq. (19.24) with respect to $\mathbf{n}$. For large times ($t >> m(r_0 + r)/k_0$), the probability densities $p(\mathbf{n})$ and $p^{(0)}(\mathbf{n})$ are normalized to unity, so we obtain $\sigma_{tot}/A + C = 0$. We conclude that

$$p(\mathbf{n}) - p^{(0)}(\mathbf{n}) = \frac{|f_{\mathbf{k}_0}(\mathbf{n})|^2 - \sigma_{tot}\delta_\epsilon^2(\mathbf{n}, \mathbf{n}_0)}{A}.$$

(19.25)

Equation (19.25) implies that the differential cross-section $\mathbf{n}_0$ is smaller than predicted by Eq. (19.14).

An important result allows us to express the total cross-section σ_{tot} in terms of the scattering amplitude,

$$\sigma_{tot} = \frac{4\pi}{k_0}\mathrm{Im}f_{\mathbf{k}_0}(\mathbf{n}_0).$$

(19.26)

Equation (19.26) is known as the *optical theorem*. The name stems from an analogous formula in the scattering of classical EM waves. For the proof, see Box 19.1.

Equation (19.26) allows us to express Eq. (19.24) as a local equation with respect to the scattering direction $\mathbf{n}$, since the integrated quantity σ_{tot} does not appear:

$$p(\mathbf{n}) - p^{(0)}(\mathbf{n}) = \frac{|f_{\mathbf{k}_0}(\mathbf{n})|^2 - \frac{4\pi}{k_0}\mathrm{Im}f_{\mathbf{k}_0}(\mathbf{n})\delta_\epsilon^2(\mathbf{n}, \mathbf{n}_0)}{A}.$$

(19.27)

Box 19.1   Proof of the Optical Theorem

We integrate both sides of Eq. (19.21) over x. Wave functions are normalized: $\int d^3x|\psi_t(\mathbf{x})|^2 = \int d^3x|\psi_t^{(0)}(\mathbf{x})|^2 = 1$. Furthermore,

$$\int d^3x \frac{|f_{\mathbf{k}_0}(\mathbf{n})|^2}{r^2}|\psi_0[\mathbf{n}_0(r - \frac{k_0}{m}t)]|^2 = \int d^2\mathbf{n}p(\mathbf{n}) = \sigma_{tot}/A.$$

We combine with Eq. (19.22), to obtain

$$\sigma_{tot} = -2A \operatorname{Re} \left[\int \frac{d^3x}{r} f_{\mathbf{k}_0}(\mathbf{n}) \psi_0^* \left(r\mathbf{n} - \frac{\mathbf{k}_0 t}{m} \right) \psi_0 \left(r\mathbf{n}_0 - \frac{\mathbf{k}_0 t}{m} \right) \right]. \tag{A}$$

The product $\psi_0^* \left(r\mathbf{n} - \frac{\mathbf{k}_0 t}{m} \right) \psi_0 \left(r\mathbf{n}_0 - \frac{\mathbf{k}_0 t}{m} \right)$ differs from zero only for negligibly small angles between $\mathbf{n}$ and $\mathbf{n}_0$. Setting $f_{\mathbf{k}_0}(\mathbf{n}) \simeq f_{\mathbf{k}_0}(\mathbf{n}_0)$, Eq. (A) becomes

$$\sigma_{tot} = -2A \operatorname{Re} \left[f_{\mathbf{k}_0}(\mathbf{n}_0) Q \right], \tag{B}$$

where $Q := \int \frac{d^3x}{r} \psi_0^* \left(r\mathbf{n} - \frac{\mathbf{k}_0}{m} t \right) \psi_0 \left(r\mathbf{n}_0 - \frac{\mathbf{k}_0}{m} t \right)$.

We evaluate Q using spherical coordinates, with angle θ given by $\cos\theta = \mathbf{n} \cdot \mathbf{n}_0 = \xi$. Then, $Q = 4\pi \int_0^\infty r dr \psi_0(0, 0, r - k_0 t/m) \Psi^*(r)$, where $\Psi(r) := \frac{1}{2} \int_{-1}^1 d\xi \psi_0(0, 0, r\xi - k_0 t/m)$ is the angle-averaged wave function.

We write $\psi_0(0, 0, x) = \int dk e^{ikx} \tilde{\psi}(k)$ in terms of its Fourier components $\tilde{\psi}(k)$, to obtain

$$\Psi(r) = \frac{1}{2} \int dk \tilde{\psi}(k) e^{-ikk_0 t/m} \int_{-1}^1 d\xi e^{ikr\xi} = \int dk \tilde{\psi}(k) \frac{\sin kr}{kr} e^{-ikk_0 t/m}.$$

By assumption, $\tilde{\psi}_0$ is narrowly focused around k_0, hence, we can substitute k in the denominator with k_0. It follows that

$$\Psi(r) = \frac{1}{2ik_0 r} \left[\psi_0(0, 0, r - k_0 t/m) - \psi_0(0, 0, -r - k_0 t/m) \right].$$

Since $\psi_0^*(0, 0, -r - k_0 t/m) \psi_0(0, 0, r - k_0 t/m) \simeq 0$, we find $Q = \frac{2\pi}{ik_0} \int_0^\infty dr |\psi_0(0, 0, r - k_0 t/m)|^2$.

For sufficiently large t, Eq. (19.2) applies, hence, $Q = \frac{2\pi}{ik_0 A}$. Then Eq. (B) yields the optical theorem.

Given that $|z| \geq \operatorname{Im} z$ for all $z \in \mathbb{C}$, the optical theorem implies that

$$\sigma(\mathbf{n}_0) \geq \frac{k_0^2 \sigma_{tot}^2}{16\pi^2}. \tag{19.28}$$

We define the average cross-section $\bar{\sigma} := \sigma_{tot}/4\pi$, in order to express Eq. (19.28) as

$$\frac{\sigma(\mathbf{n}_0)}{\bar{\sigma}} \geq \frac{\pi \sigma_{tot}}{\lambda^2}, \tag{19.29}$$

where λ is the incoming particle's de Broglie wavelength. Suppose that the total cross-section as a function of energy has a nonzero lower bound. Then, at high energies, the ratio σ_{tot}/λ^2 becomes much larger than unity. Hence, the differential cross-section in the direction $\mathbf{n}_0$ will be much larger than $\bar{\sigma}$. This means that most particles will scatter at small angles to $\mathbf{n}_0$. The optical theorem then predicts that forward scattering dominates at high energies.

19.1.5 Identical Particles

In Section 13.1, we showed that the relative motion of two particles in their center of mass can be described by a single-particle Schrödinger operator. Hence, the scattering of two particles in the center of mass is equivalent to the scattering by a potential. However, when describing the scattering of two identical particles, we must take into account their bosonic or fermionic nature.

If the interaction of the particles is spin-independent, then we can study the effects of statistics by symmetrizing or antisymmetrizing the initial state. The exchange of two particles is equivalent to the transformation $\mathbf{x} \to -\mathbf{x}$ for their relative position, that is, it is equivalent to a parity transformation. The mean position and momentum of the initial state transform as $(\mathbf{r}_0, \mathbf{k}_0) \to (-\mathbf{r}_0, -\mathbf{k}_0)$.

(Anti)symmetrization is equivalent with the substitution of the initial state $\psi_0(\mathbf{x})$ with $\frac{1}{\sqrt{2}}[\psi_0(\mathbf{x}) \pm \psi_0(-\mathbf{x})]$. The time-evolved wave function (19.10) is substituted by

$$\psi_t(\mathbf{x}) = \frac{1}{\sqrt{2}}\left[\psi_t^{(0)}(\mathbf{x}) \pm \psi_t^{(0)}(-\mathbf{x})\right] + \frac{1}{\sqrt{2}}\left[f_{\mathbf{k}_0}(\mathbf{n}) \pm f_{-\mathbf{k}_0}(\mathbf{n})\right]\frac{e^{iE_{\mathbf{k}_0}t}}{r}\psi_0\left[\mathbf{n}_0\left(r - \frac{k_0}{m}t\right)\right].$$

Following the same procedure as in Section 19.1.2, we find that, for direction $\mathbf{n} \neq \pm\mathbf{n}_0$,

$$\sigma(\mathbf{n}) = \frac{1}{2}|f_{\mathbf{k}_0}(\mathbf{n}) \pm f_{-\mathbf{k}_0}(\mathbf{n})|^2. \tag{19.30}$$

Suppose that the scattered particles are electrons. If their total spin $S = 0$, the spatial wave function is symmetric; hence, the sign $+$ applies in Eq. (19.30). If $S = 1$, the spatial wave function is antisymmetric; hence, the $-$ sign applies. In a state of complete ignorance for spins, there is probability $\frac{3}{4}$ for $S = 1$ and $\frac{1}{4}$ for $S = 0$. Then,

$$\begin{aligned}\sigma(\mathbf{n}) &= \frac{3}{8}|f_{\mathbf{k}_0}(\mathbf{n}) - f_{-\mathbf{k}_0}(\mathbf{n})|^2 + \frac{1}{8}|f_{\mathbf{k}_0}(\mathbf{n}) + f_{-\mathbf{k}_0}(\mathbf{n})|^2 \\ &= \frac{1}{2}|f_{\mathbf{k}_0}(\mathbf{n})|^2 + \frac{1}{2}|f_{-\mathbf{k}_0}(\mathbf{n})|^2 - \frac{1}{2}\text{Re}\left[f_{\mathbf{k}_0}^*(\mathbf{n})f_{-\mathbf{k}_0}(\mathbf{n})\right].\end{aligned} \tag{19.31}$$

19.1.6 Inelastic Scattering

Suppose that the target particle is composite. Then, scattering can excite its internal degrees of freedom. The kinetic energy of the outgoing particles is less than the kinetic energy of incoming particles: Scattering is *inelastic*.

We denote by $\mathcal{H}_{int}$ the Hilbert space of the internal degrees of freedom of the target particle. The associated Hamiltonian is $\hat{H}_{int}$, with eigenvalues E_a and eigenvectors $|a\rangle$, for a discrete label a. The ground state corresponds to $a = 0$. The total system is described by the Hilbert space $L^2(\mathbb{R}^3, d^3x) \otimes \mathcal{H}_{int}$ and the Hamiltonian

$$\hat{H} = \frac{\hat{\mathbf{p}}^2}{2m} + \hat{H}_{int} + U(\hat{\mathbf{x}}, \hat{Q}). \tag{19.32}$$

The term U incorporates the interactions of the incoming particle with the target particle's internal degrees of freedom, collectively denoted by $\hat{Q}$. It vanishes far from the scattering region, where the Hamiltonian is simply $\hat{H}_0 = \frac{\hat{\mathbf{p}}^2}{2m} + \hat{H}_{int}$.

We select a basis of eigenstates $|k, a\rangle$ of $\hat{H}$ that generalizes Eq. (19.3),

$$\langle \mathbf{x}, b|\mathbf{k}, a\rangle = \frac{1}{(2\pi)^{3/2}}\left[e^{i\mathbf{k}\cdot\mathbf{x}} + \sum_b f_{\mathbf{k},a}(\mathbf{n}, b)\frac{e^{ik_{ab}r}}{r}\right], \tag{19.33}$$

where $\mathbf{n} = \mathbf{x}/r$, and we defined $k_{ab} := \sqrt{k^2 - 2m(E_b - E_a)}$ so that the kinetic energy of the outcoming particle is $\frac{k_{ab}^2}{2m}$. The quantity $f_{\mathbf{k},a}(\mathbf{n}, b)$ is the scattering amplitude. It incorporates both the transitions $\mathbf{k} \to k_{a,b}\mathbf{n}$ of incoming particle momentum and internal transitions $a \to b$.

We compute the detection probabilities following the same procedure as in Section 19.1.1. We only sketch the main steps. We assume an initial state $|\psi_0\rangle \otimes |0\rangle$, where the wave function $\psi_0(\mathbf{x})$ is the same as in Section 19.1.1; in particular, it has mean momentum $\mathbf{k}_0$ and mean position $\mathbf{x}_0$. We also assume that the target starts in the ground state. Hence, $\langle \mathbf{k}, a|(|\psi_0\rangle \otimes |0\rangle) = \tilde{\psi}_0(\mathbf{k})\delta_{a0}$.

We calculate the wave function $\psi_t(\mathbf{x}, a)$ at time t. The analogue of Eq. (19.11) for the probability density $|\psi_t(\mathbf{x}, a)|^2$ with respect to $\mathbf{x}$ for given a is

$$|\psi_t(\mathbf{x}, a)|^2 = \frac{|f_{\mathbf{k}_0, 0}(\mathbf{n}, a)|^2}{r^2} \left|\psi_0\left[\frac{\mathbf{k}_0}{k_a'}\left(r - \frac{k_a' t}{m}\right)\right]\right|^2, \tag{19.34}$$

where $k_a' = \sqrt{k_0^2 - 2m(E_a - E_0)}$ is the momentum of the outgoing particles. Note that for $k_0 < \sqrt{2m(E_a - E_0)}$, inelastic scattering cannot occur.

Then, the probability density $p_a(\mathbf{n})$ for the directions $\mathbf{n}$ of the outgoing particle and the target excited to state $|a\rangle$ is

$$p_a(\mathbf{n}) = \frac{k_a'}{k_0} \frac{|f_{\mathbf{k}_0, 0}(\mathbf{n}, a)|^2}{A}. \tag{19.35}$$

The corresponding differential cross-section is

$$\sigma_a(\mathbf{n}) = \frac{k_a'}{k_0} |f_{\mathbf{k}_0, 0}(\mathbf{n}, a)|^2. \tag{19.36}$$

The total cross-section is obtained by summing over all directions $\mathbf{n}$ and all internal states a,

$$\sigma_{tot} = \sum_a \left(1 - \frac{E_a - E_0}{E_{k_0}}\right)^{1/2} \int d^2n |f_{\mathbf{k}_0, 0}(\mathbf{n}, a)|^2, \tag{19.37}$$

where $E_{k_0} = \frac{k_0^2}{2m}$ is the initial kinetic energy.

19.2 The Born Approximation

If the potential is weak so that its contribution to the Hamiltonian is much smaller than the particle's kinetic energy, we can use the *Born approximation*, which provides a fast and simple calculation of the scattering amplitude.

In particular, we assume that the potential is proportional to a dimensionless constant $\lambda << 1$. Since the scattering amplitude vanishes for zero potential, we expect that it will be proportional to λ. The Born approximation consists in keeping the terms of lowest order in λ in the eigenvalue equation $\hat{H}|\mathbf{k}\rangle = \frac{k^2}{2m}|\mathbf{k}\rangle$, for the generalized eigenvectors $|\mathbf{k}\rangle$ of Eq. (19.3). For $\hat{H} = -\frac{1}{2m}\nabla^2 + V(\mathbf{x})$, the eigenvalue equation yields

$$(\nabla^2 + k^2)\left(\frac{e^{ikr}}{r} f_{\mathbf{k}}(\mathbf{n})\right) = 2mV(\mathbf{x})\left[e^{i\mathbf{k}\cdot\mathbf{x}} + \frac{e^{ikr}}{r} f_{\mathbf{k}}(\mathbf{n})\right]. \tag{19.38}$$

All terms are of order λ, except for the term that involves the product $V(\mathbf{x})f(\mathbf{n})$, which is of order λ^2. In the Born approximation, we ignore this term. Equation (19.38) becomes

$$(\nabla^2 + k^2)\left(\frac{e^{ikr}}{r}f_{\mathbf{k}}(\mathbf{n})\right) = 2mV(\mathbf{x})e^{i\mathbf{k}\cdot\mathbf{x}}. \tag{19.39}$$

This equation is solved by

$$\frac{e^{ikr}}{r}f(\mathbf{n}) = 2m\int d^3r'\, G(\mathbf{x}-\mathbf{x}')V(\mathbf{x}')e^{i\mathbf{k}\cdot\mathbf{x}'}, \tag{19.40}$$

where $G(\mathbf{x})$ is a solution of the differential equation

$$(\nabla^2 + k^2)G(\mathbf{x}) = \delta^{(3)}(\mathbf{x}). \tag{19.41}$$

Keeping in mind that $\nabla^2 r^{-1} = -4\pi\delta^{(3)}(\mathbf{x})$ – see Problem 4.19 – we readily find that the most general solution to Eq. (19.41) is of the form $-[ae^{ikr} + (1-a)e^{-ikr}]/(4\pi r)$ for some constant a. Equation (19.40) implies that the asymptotic dependence of the solutions on r must be of the form $\frac{e^{ikr}}{r}$. This means that we must choose $G(\mathbf{x})$ with $a = 1$. It follows that

$$\frac{e^{ikr}}{r}f_{\mathbf{k}}(\mathbf{n}) = -\frac{m}{2\pi}\int d^3x'\,\frac{e^{ik|\mathbf{x}-\mathbf{x}'|}}{|\mathbf{x}-\mathbf{x}'|}e^{i\mathbf{k}\cdot\mathbf{x}'}V(\mathbf{x}'). \tag{19.42}$$

Equation (19.42) holds for macroscopically large values of r, while the potential differs from zero only for $|\mathbf{r}'|$ smaller than the potential's range D, which is a macroscopic quantity. Therefore, we can approximate $|\mathbf{x} - \mathbf{x}'| = \sqrt{r^2 - 2\mathbf{x}\cdot\mathbf{x}' + r'^2} \simeq \sqrt{r^2 - 2\mathbf{x}\cdot\mathbf{x}'} = r\sqrt{1 - 2\mathbf{x}\cdot\mathbf{x}'/r^2} \simeq r(1 - \mathbf{x}\cdot\mathbf{x}'/r^2) = r - \mathbf{n}\cdot\mathbf{x}'$. In the denominator, the approximation $|\mathbf{x} - \mathbf{x}'| \simeq r$ suffices. We conclude that

$$f_{\mathbf{k}}(\mathbf{n}) = -\frac{m}{2\pi}\int d^3x\, e^{-i\mathbf{q}\cdot\mathbf{x}}V(\mathbf{x}), \tag{19.43}$$

where $\mathbf{q} = k\mathbf{n} - \mathbf{k}$ is a particle's change in momentum due to scattering. In the Born approximation, the scattering amplitude depends on the vectors $\mathbf{k}$ and $\mathbf{n}$ only through their combination $\mathbf{q}$. We will therefore write $f_{\mathbf{k}}(\mathbf{n})$ as $f(\mathbf{q})$.

For a central potential, we choose coordinates so that $\mathbf{q}$ is along the 3-axis. Then, we evaluate the integral (19.43) in spherical coordinates, to obtain

$$f(\mathbf{q}) = -m\int_0^\infty dr\, r^2 V(r)\int_0^\pi d\theta \sin\theta\, e^{-iqr\cos\theta} = -\frac{2m}{q}\int_0^\infty dr\, rV(r)\sin(qr). \tag{19.44}$$

The scattering amplitude depends only on the norm of $\mathbf{q}$. We find that $q^2 = 2k^2(1 + \cos\theta) = 4k^2\sin^2\frac{\theta}{2}$, where $\cos\theta = \mathbf{k}\cdot\mathbf{n}/k$. Hence,

$$q = 2k\sin\frac{\theta}{2}. \tag{19.45}$$

Born's formula is straightforwardly generalized to inelastic scattering. For a composite target, where scattering is described by a Hamiltonian of the form (19.32), the scattering amplitude is

$$f_{\mathbf{k},0}(\mathbf{n}, a) = -\frac{m}{2\pi}\int d^3x\, e^{-i\mathbf{q}_a\cdot\mathbf{x}}\langle a|U(\mathbf{x},\hat{Q})|0\rangle, \tag{19.46}$$

where $\mathbf{q}_a = \mathbf{k}_a' - \mathbf{k}$ is the difference between the outgoing momentum $\mathbf{k}_a' = k_a'\mathbf{n}$ and the incoming momentum $\mathbf{k}$. By energy conservation, $k_a' = \sqrt{k^2 - 2m\Delta E_a}$, where $\Delta E_a = E_a - E_0$. The minimum of $q_a = |\mathbf{q}_a|$ with respect to the scattering angle θ is $k - k'_a = k - \sqrt{k^2 - 2m\Delta E_a}$. This quantity equals $\sqrt{2m\Delta E_a}$ at the onset of inelastic scattering ($k'_a = 0$), and it decays to zero with $m\Delta E_a/k$ at large k.

Consider a target that is composed of N particles (electrons and nuclei in atoms or molecules, neutrons and protons in nuclei, and so on) of masses m_ν, $\nu = 1, 2, \ldots, N$. Let us denote by $\mathbf{r}_\nu$ the coordinate vector of the νth constituent particle. We choose the coordinate origin to coincide with the target's center of mass. For pairwise interactions between particles, $\hat{U}(\mathbf{x}, \mathbf{r}_\nu) = \sum_{\nu=1}^{N} \upsilon_\nu(\mathbf{x} - \mathbf{r}_\nu)$, Eq. (19.46) becomes

$$f_{\mathbf{k},0}(\mathbf{n}, a) = -\frac{1}{2\pi} \sum_{\nu=1}^{N} m_\nu \tilde{\upsilon}_\nu(\mathbf{q}_a) \langle a | e^{-i\mathbf{q}_a \cdot \hat{\mathbf{r}}_\nu} | 0 \rangle. \tag{19.47}$$

Example 19.1 For the Yukawa potential $V(r) = ge^{-\mu r}/r$, Eq. (19.44) yields

$$f(q) = -\frac{2mg}{q} \int_0^\infty dr e^{-\mu r} \sin(qr) = -\frac{2mg}{q^2 + \mu^2}. \tag{19.48}$$

The differential cross-section is

$$\sigma(\theta) = \frac{4m^2 g^2}{\left(\mu^2 + 4k^2 \sin^2 \frac{\theta}{2}\right)^2}. \tag{19.49}$$

The total cross-section is

$$\begin{aligned}
\sigma_{tot} &= 4m^2 g^2 (2\pi) \int_0^\pi \frac{d\theta \sin\theta}{\left[\mu^2 + 2k^2(1 + \cos\theta)\right]^2} \\
&= \frac{8\pi m^2 g^2}{\mu^4} \int_0^2 \frac{dy}{\left(1 + \frac{2k^2}{\mu^2} y\right)^2} = \frac{16\pi m^2 g^2}{\mu^2(\mu^2 + 4k^2)},
\end{aligned} \tag{19.50}$$

where we changed variables to $y := 1 + \cos\theta$. The total cross-section decreases with energy.

At the limit $\mu \to 0$, the Yukawa potential coincides with the Coulomb potential. Equation (19.49) becomes

$$\sigma(\theta) = \frac{m^2 g^2}{4k^4 \sin^4 \frac{\theta}{2}} = \frac{g^2}{16E^2 \sin^4 \frac{\theta}{2}}. \tag{19.51}$$

Equation (19.51) is known as *Rutherford's formula*, as it had been derived within classical physics for Rutherford's famous experiment – see Problem 2.4. It also coincides with the result of the exact quantum mechanical calculation – see Section 19.4.

The total cross-section diverges for $\mu \to 0$. This is because the Coulomb potential is long range, so Eq. (19.3) does not describe accurately the asymptotic behavior of the eigenfunctions of the corresponding Schrödinger operator. As shown in Section 19.4, the form (19.3) for the mode functions fails near $\theta = 0$. For this reason, the scattering cross-section diverges at $\theta = 0$.

Example 19.2 Let the target be an atom with atomic number Z. It consists of one nucleus with charge Ze and Z electrons with charge $-e$. We invoke the Born–Oppenheimer approximation, and treat the nucleus as a motionless point charge located at the coordinate origin. For an incoming particle of charge λe, the interaction is given by the Coulomb potential

$$U(\mathbf{x}, \mathbf{r}_\nu) = \lambda Z \alpha |\mathbf{x}|^{-1} - \lambda \alpha \sum_{\nu=1}^{Z} |\mathbf{x} - \mathbf{r}_\nu|^{-1}, \tag{19.52}$$

where now the coordinate vectors $\mathbf{r}_\nu$ refer solely to the electrons. The scattering amplitude (19.47) is

$$f_{\mathbf{k},0}(\mathbf{n}, a) = \frac{m_e \lambda Z \alpha}{q_a^2} \left[\delta_{0a} - F_a(\mathbf{q}_a) \right], \tag{19.53}$$

where $F_a(\mathbf{q}) := Z^{-1} \sum_{\nu=1}^{Z} \langle a | e^{-i\mathbf{q} \cdot \hat{\mathbf{r}}_\nu} | 0 \rangle$ is the atomic *form-factor*.

We take the limit of low momentum, $k r_0 \ll 1$, where r_0 is the atomic size. Then, we evaluate the form factor $F_0(\mathbf{q})$ for elastic scattering by Taylor-expanding the exponential. The ground state for atoms is approximately spherically symmetric. It follows that $\langle 0 | \hat{\mathbf{r}}_\nu | 0 \rangle \simeq 0$ and $\langle 0 | \hat{r}_{\nu i} \hat{r}_{\nu j} | 0 \rangle \simeq \frac{1}{3} \delta_{ij} \langle 0 | \hat{\mathbf{r}}_\nu^2 | 0 \rangle$. Hence, $F_0(\mathbf{q}) = 1 - \frac{1}{6} R^2 q^2$, where $R^2 := Z^{-1} \sum_{\nu=1}^{Z} \langle 0 | \mathbf{r}_\nu^2 | 0 \rangle$ is the average position variance of the electrons. (R is a good measure for the atomic size r_0.) By Eq. (19.53), the divergence of the scattering amplitude of the Coulomb potential at $q = 0$ is removed,

$$f_{\mathbf{k},0}(\mathbf{n}, 0) = \frac{m_e \lambda Z \alpha R^2}{6}.$$

The total elastic cross-section is therefore finite, $\sigma_{tot}^{(el)} = \frac{1}{9} \pi m_e^2 \lambda^2 Z^2 \alpha^2 R^4$.

In the same regime, we approximate the form factor $F_a(\mathbf{q}) \simeq Z^{-1}$. Then, $f_{\mathbf{k},0}(\mathbf{n}, a) = -m_e \lambda \alpha q_a^{-2}$ is Z-independent. This means that, unlike elastic cross-sections, inelastic cross-sections do not increase with the atom's size.

We use Eq. (19.36), in order to evaluate the total inelastic cross-section $\sigma_{tot}^{(a)}$ for a final internal state a,

$$\sigma_{tot}^{(a)} = \frac{m_e^2 \lambda^2 \alpha^2 k_a'}{k} (2\pi) \int_0^\pi \frac{d\theta \sin\theta}{(k^2 + k_a'^2 - 2kk_a' \cos\theta)^2} = \frac{\pi \lambda^2 \alpha^2 k_a'}{k (\Delta E_a)^2}, \tag{19.54}$$

where we have carried out the integral by setting $\xi = \cos\theta$ and using Eq. (A.11). Hence, the ratio between the inelastic and elastic cross-sections is

$$\frac{\sigma_{tot}^{(a)}}{\sigma_{tot}^{(el)}} = \frac{9}{Z^2 m_e^2 R^4} \frac{k_a'}{k (\Delta E_a)^2}. \tag{19.55}$$

This ratio is typically much smaller than unity. It starts from 0 at $k = \sqrt{2 m_e \Delta E_a}$ and it increases up towards an asymptotic value $\frac{9}{Z^2 m_e^2 R^4 (\Delta E_a)^2}$ for large k. Elastic scattering is predominant at low momenta.

19.3   Scattering in a Central Potential

We proceed with the analysis of scattering in central potentials. As shown in Section 13.2.4, the scattering eigenfunctions of the Schrödinger operator for a central potential are specified by the phase shifts $\delta_\ell(k)$ that characterize the asymptotic behavior of energy eigenfunctions at large distances from the center. We will see that the scattering amplitude can be solely expressed in terms of the phase shifts.

19.3.1 The Scattering Amplitude

The Hamiltonian for a particle in a central potential $V(r)$ is invariant under rotations. This implies that the associated scattering amplitude must be invariant under the transformations $\mathbf{k} \to O\mathbf{k}$ and $\mathbf{n} \to O\mathbf{n}$, for any rotation matrix O. Hence, $f_{\mathbf{k}}(\mathbf{n})$ depends only on the scalar quantities that can be formed by the vectors $\mathbf{k}$ and $\mathbf{n}$, namely, the norm $k = |\mathbf{k}|$, and the angle θ between the two vectors, given by $\cos\theta = \mathbf{k} \cdot \mathbf{n}/k$. Therefore, we will write the scattering amplitude as $f_k(\theta)$.

By the analysis of Section 13.2.4, a general scattering eigenfunction of the Hamiltonian with energy $E = \frac{k^2}{2m}$ has the following asymptotic behavior:

$$\psi_k(r,\theta,\phi) = \frac{1}{kr} \sum_{\ell=0}^{\infty} \sum_{m=-\ell}^{\ell} c_{\ell,m} \sin\left(kr - \frac{\ell\pi}{2} + \delta_\ell(k)\right) Y_{\ell m}(\theta,\phi), \qquad (19.56)$$

where $c_{\ell,m}$ are arbitrary constants. We want to choose appropriate $c_{\ell,m}$, so that the eigenfunction (19.57) takes the form (19.3). Owing to spherical symmetry, we can always orient the vector $\mathbf{k}$ of Eq. (19.3) along the 3-axis, so that the eigenfunctions are ϕ-independent. Then, only the spherical harmonics $Y_{\ell 0}(\theta)$ appear in Eq. (19.56). These are proportional to the Legendre polynomials $P_\ell(\cos\theta)$ – see Eq. (11.88). Hence, we write Eq. (19.56) as

$$\psi_k(r,\theta) = \frac{1}{kr} \sum_{\ell=0}^{\infty} d_\ell(2\ell+1) \sin\left(kr - \frac{\ell\pi}{2} + \delta_\ell(k)\right) P_\ell(\cos\theta), \qquad (19.57)$$

for some constants d_ℓ. We will specify d_ℓ be equating this expression with $e^{ikr\cos\theta} + f_k(\theta)e^{ikr}/r$. To this end, we expand the scattering amplitude in the basis of Legendre polynomials: $f_k(\theta) = \sum_{\ell=0}^{\infty}(2\ell+1)f_\ell P_\ell(\cos\theta)$. We also use the identity

$$\int_{-1}^{1} d\xi\, e^{i\rho\xi} P_\ell(\xi) = 2i^\ell j_\ell(\rho), \qquad (19.58)$$

the proof of which is found in Box 19.2. Equation (19.3) becomes

$$\psi_k(r,\theta) = \sum_{\ell=0}^{\infty}(2\ell+1)\left[i^\ell j_\ell(kr) + f_\ell \frac{e^{ikr}}{r}\right] P_\ell(\cos\theta). \qquad (19.59)$$

Using Eq. (13.26) for large r, we find that

$$\psi_k(r,\theta) = \frac{1}{2ikr} \sum_{\ell=0}^{\infty}(2\ell+1)\left\{e^{ikr}\left(i^\ell e^{-i\ell\pi/2} + 2ikf_\ell\right) - e^{-ikr}\left[i^\ell e^{i\ell\pi/2}\right]\right\} P_\ell(\cos\theta).$$

We equate with Eq. (19.57), to obtain

$$\sum_{\ell=0}^{\infty}(2\ell+1)\left[e^{ikr}\left(1 + 2ikf_\ell - d_\ell(-i)^\ell e^{i\delta_\ell(k)}\right) - e^{-ikr} i^\ell\left(i^\ell - d_\ell e^{-i\delta_\ell(k)}\right)\right] P_\ell(\cos\theta) = 0.$$

The vanishing of the coefficient of e^{-ikr} yields $d_\ell = i^\ell e^{i\delta_\ell(k)}$. The vanishing of the coefficient of e^{ikr} yields $2ikf_\ell = d_\ell(-i)^\ell e^{i\delta_\ell(k)} - 1$. We substitute d_ℓ, to obtain

$$f_\ell = \frac{e^{2i\delta_\ell} - 1}{2ik} = \frac{\sin\delta_\ell e^{i\delta_\ell}}{k}. \qquad (19.60)$$

Hence, the scattering amplitude is

$$f_k(\theta) = \frac{1}{k} \sum_{\ell=0}^{\infty} (2\ell + 1) \sin \delta_\ell e^{i\delta_\ell} P_\ell(\cos \theta). \tag{19.61}$$

Box 19.2 Proof of Eq. (19.58)

We use Rodriguez's formula, Eq. (3.12), to obtain

$$\int_{-1}^{1} d\xi\, e^{i\rho\xi} P_\ell(\xi) = \frac{1}{2^\ell \ell!} \int_{-1}^{1} d\xi\, e^{i\rho\xi} \frac{d^\ell}{d\xi^\ell}(x^2 - 1)^\ell = \frac{(-1)^\ell}{2^\ell \ell!} \int_{-1}^{1} d\xi (\xi^2 - 1)^\ell \frac{d^\ell e^{i\rho\xi}}{d\xi^\ell}$$

$$= \frac{(-i\rho)^\ell}{2^\ell \ell!} \int_{-1}^{1} d\xi (\xi^2 - 1)^\ell e^{i\rho\xi}. \tag{A}$$

In Eq. (A), we integrated by parts ℓ times and we used the fact that $\frac{d^n}{d\xi^n}(\xi^2 - 1)^\ell|_{\xi=\pm 1} = 0$, for $n < \ell$.

We note that

$$\frac{1}{\rho}\frac{\partial}{\partial\rho} \int_{-1}^{1} d\xi\, e^{i\rho\xi} = \frac{i}{\rho} \int_{-1}^{1} d\xi\, \xi e^{i\rho\xi} = \frac{i}{2\rho} \int_{-1}^{1} d(\xi^2 - 1) e^{i\rho\xi} = \frac{1}{2} \int_{-1}^{1} d\xi (\xi^2 - 1) e^{i\rho\xi}.$$

We repeat this operation ℓ times, to obtain

$$\left(\frac{1}{\rho}\frac{\partial}{\partial\rho}\right)^\ell \int_{-1}^{1} d\xi\, e^{i\rho\xi} = \frac{1}{2^\ell \ell!} \int_{-1}^{1} d\xi (\xi^2 - 1)^\ell e^{i\rho\xi}.$$

We substitute into Eq. (A), and we employ Eq. (B.31) to conclude the proof:

$$\int_{-1}^{1} dx\, e^{i\rho\xi} P_\ell(\xi) = (-i\rho)^\ell \left(\frac{1}{\rho}\frac{\partial}{\partial\rho}\right)^\ell \int_{-1}^{1} dx\, e^{i\rho\xi} = 2i^\ell(-\rho)^\ell \left(\frac{1}{\rho}\frac{\partial}{\partial\rho}\right)^\ell \frac{\sin\rho\xi}{\rho} = 2i^\ell j_\ell(\xi).$$

We evaluate the total cross-section $\sigma_{tot} = 2\pi \int_0^\pi d\theta \sin\theta |f_k(\theta)|^2$ using Eq. (3.13). We find that

$$\sigma_{tot} = 4\pi \sum_{\ell=0}^{\infty} (2\ell + 1)|f_\ell|^2 = \frac{4\pi}{k^2} \sum_{\ell=0}^{\infty} (2\ell + 1) \sin^2 \delta_\ell. \tag{19.62}$$

Since $P_\ell(1) = 1$, we find that $\mathrm{Im} f_k(0) = \frac{1}{k} \sum_{\ell=0}^{\infty} (2\ell + 1) \sin^2 \delta_\ell = \frac{4\pi}{k}\sigma_{tot}$, thereby confirming the optical theorem.

In the scattering of identical particles, a symmetrized spatial eigenfunction implies that only even values of ℓ contribute to the scattering amplitude. An antisymmetrized spatial eigenfunction implies the contribution of only odd values of ℓ.

The phase shift decomposition of the scattering amplitude provides a direct connection between experiment and theory. The phase shifts can be directly measured in experiments – an example is shown in Fig. 19.3 – and thus, they provide direct information about the properties of the potential.

19.3.2 Hard-Sphere Potential

The hard-sphere potential is important, because it provides an accurate model for scattering in many regimes. In Example 13.1, we had shown that, for the hard-sphere potential, the

Fig. 19.3 Phase shifts measured in alpha–alpha scattering at energies between 5 and 9 MeV. Since alpha particles are bosons, odd phase shifts vanish. In these energies, only phase shifts for $\ell = 0$ (S-wave) and for $\ell = 2$ (D-wave) contribute. [Figure reprinted with permission from Jones et al. (1960). Copyright 1960 by the American Physical Society.]

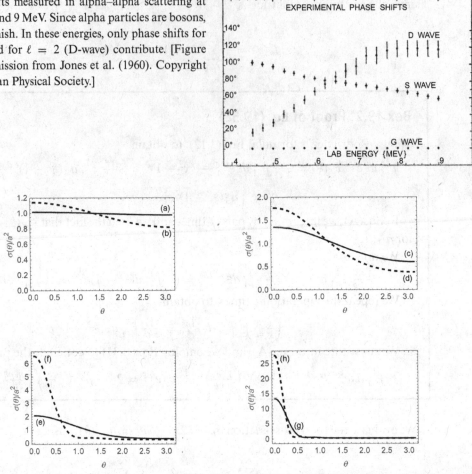

Fig. 19.4 The dimensionless differential cross-section $\sigma(\theta)/a^2$ for a hard sphere of radius a as a function of the scattering angle θ, for different energies: (a) $ka = 0.1$, (b) $ka = 0.3$, (c) $ka = 0.5$, (d) $ka = 0.8$, (e) $ka = 1$, (f) $ka = 3$, (g) $ka = 5$, and (h) $ka = 8$. The differential cross-section is almost angle-independent at low energies, while forward scattering strongly dominates at high energies.

phase shifts are given by Eq. (13.30), $\tan\delta_\ell = \frac{j_\ell(ka)}{\eta_\ell(ka)}$, where a is the radius of the sphere. This means that

$$e^{2i\delta_\ell} = -\frac{j_\ell(ka) - i\eta_\ell(ka)}{j_\ell(ka) + i\eta_\ell(ka)}. \tag{19.63}$$

By Eq. (19.60), $f_\ell = -\frac{1}{k}\frac{j_\ell(ka)}{j_\ell(ka) + i\eta_\ell(ka)}$. We plot the differential cross-section as a function of θ in Fig. 19.4. We confirm that forward scattering is stronger at high energies.

The total cross-section is

$$\sigma_{tot} = \frac{4\pi}{k^2}\sum_{\ell=0}^{\infty}(2\ell + 1)\zeta_\ell(ka), \qquad \bullet \tag{19.64}$$

where

$$\zeta_\ell(x) := \frac{j_\ell(x)^2}{j_\ell(x)^2 + n_\ell(x)^2}. \tag{19.65}$$

For $ka << \ell$, we use Eqs. (B.38) and (B.39) for the behavior of the spherical Bessel functions for small values of their arguments. We obtain

$$\tan \delta_\ell = -\frac{2^{2\ell}(\ell!)^2}{(2\ell+1)!\,(2\ell)!}(ka)^{2\ell+1} << 1.$$

This means that angular momenta with $\ell >> ka$ contribute negligibly to scattering. In classical physics, the ratio ℓ/k is the smallest distance of a particle's orbit from the scattering center. If this is greater than the potential's range, then the particle does not scatter, that is, the scattering probability vanishes. Of course, in quantum theory there is not such a thing as particle orbits, but the ratio ℓ/k still retains some significance.

For $ka << 1$, $\delta_\ell \sim (ka)^{2\ell+1}$, and the dominant contribution to scattering comes from $\ell = 0$. Hence, within a good approximation, the differential cross-section is θ-independent at very small energies. In this regime,

$$\delta_0 = -ka, \tag{19.66}$$

and Eq. (19.62) yields $\sigma_{tot} = 4\pi a^2$. This result differs by a factor of four from the classical prediction. Classically, only particles that approach the center at distances smaller than a scatter; hence, $\sigma_{tot} = \pi a^2$ is the geometric cross-section of the sphere.

For identical particles, with symmetrized spatial eigenfunctions, Eq. (19.30) yields $\sigma_{tot} = 8\pi a^2$. For antisymmetrized spatial eigenfunctions, the dominant contribution at low energies comes form $\ell = 1$. Since $\delta_1 = -\frac{2}{3}(ka)^3$, Eq. (19.30) yields $\sigma_{tot} = \frac{32}{3}a^3 k$.

For high energies ($ka >> \ell$), we analyze the function $\zeta_\ell(x)$ of Eq. (19.65). For large ℓ, $\zeta_\ell(x)$ is negligible for $x < \ell$, it increases sharply at $x \simeq \ell$, and then it oscillates for $x > \ell$. An indicative plot is given in Fig. 19.5.

Hence, for large ka, only values of $\ell < ka$ contribute appreciably into the sum of Eq. (19.64). For these terms, we employ the asymptotic behavior of the spherical Bessel functions for large arguments, Eqs. (B.36) and (B.37), in Eq. (13.30). We obtain $\tan \delta_\ell = -\tan(ka - \frac{\ell\pi}{2})$, which implies that

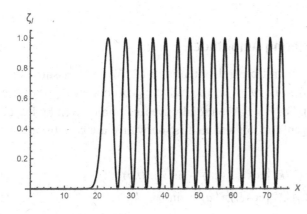

Fig. 19.5 The function $\zeta_\ell(x)$ of Eq. (19.65) for $\ell = 20$.

Fig. 19.6 The total cross-section σ_{tot} for the hard-sphere potential divided by the classical cross-section πa^2, as a function of ka.

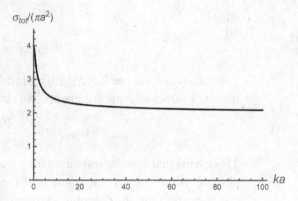

$$\sin^2 \delta_\ell = \begin{cases} \sin^2 ka, & \ell = 2n \\ \cos^2 ka, & \ell = 2n + 1. \end{cases}$$

Two successive terms with $\ell = 2n$ and $\ell = 2n + 1$ contribute a term $(4n + 1) \sin^2 ka + (4n + 3) \cos^2 ka = 2(2n + 1) + \cos(2ka)$ to σ_{tot}. Hence,

$$\sigma_{tot} = \frac{8\pi}{k^2} \sum_{n=0}^{n_{max}} \left[(2n + 1) + \frac{1}{2} \cos(2ka) \right],$$

where $n_{max} = \frac{1}{2}ka \gg 1$ the largest value of n with a substantial contribution to the sum. Keeping the leading order terms with respect to ka, we find that

$$\sigma_{tot} \simeq \frac{8\pi}{k^2} n_{max}^2 = 2\pi a^2. \tag{19.67}$$

The result differs by a factor of two from the classical expression. Naively, one would expect to obtain the classical expression for the total cross-section at high energies, because the de Broglie factor of the particles is too small; hence, particles are supposed to behave like classical particles. Instead, we get a different result, by a factor of two. This failure of quantum-to-classical correspondence in scattering is due to the optical theorem – see Question 19.2.

In Fig. 19.6, we plot the total cross-section σ_{tot} as a function of ka. Indeed, σ_{tot} decreases with energy, and it interpolates between the value $4\pi a^2$ at $k = 0$ and the asymptotic values $2\pi a^2$ as $k \to \infty$.

19.3.3 General Properties of the Scattering Amplitude

Next, we analyze scattering by a general central potential $V(r)$. We only assume that $V(r)$ vanishes for $r > a$ for some radius a. For $r < a$, the radial eigenfunction $R_{k,\ell}(r)$ for energy $E = \frac{k^2}{2m}$ is a solution of Eq. (13.11). For $r > a$, the radial wave function is given by Eq. (13.28). The requirement that the radial eigenfunction and its first derivative are continuous at $r = a$ yields the equations

$$R_{k,\ell}(a) = C_\ell(k) \left[\cos \delta_\ell j_\ell(ka) - \sin \delta_\ell \eta_\ell(ka) \right], \tag{19.68}$$

$$R'_{k,\ell}(a) = C_\ell(k)k \left[\cos \delta_\ell j'_\ell(ka) - \sin \delta_\ell \eta'_\ell(ka) \right]. \tag{19.69}$$

Let $b_\ell(k) := a\frac{d \ln R_{k,\ell}}{dr}(a)$ be the logarithmic derivative of $R_{k,\ell}$ at $r = a$. In general, the logarithmic derivative is positive for a repulsive potential and negative for an attractive potential.

By Eq. (5.54), $b_\ell(k)$ is a decreasing function of k. Equations (19.68) and (19.69) imply that

$$b_\ell(k) = ka\frac{\cos\delta_\ell j'_\ell(ka) - \sin\delta_\ell \eta'_\ell(ka)}{\cos\delta_\ell j_\ell(ka) - \sin\delta_\ell \eta_\ell(ka)}$$

or, equivalently,

$$\tan\delta_\ell = \frac{b_\ell(k)j_\ell(ka) - kaj'_\ell(ka)}{b_\ell(k)\eta_\ell(ka) - ka\eta'_\ell(ka)}. \tag{19.70}$$

Equation (19.70) implies that the phase shifts are fully specified from the logarithmic derivatives $b_\ell(k)$. For $b_\ell(k) \to \infty$, $\tan\delta_\ell \to j_\ell(ka)/\eta_\ell(ka)$, that is, we obtain the phase shifts of the hard-sphere potential. We will denote the latter by $\delta_\ell^{(0)}$.

We make the relation between δ_ℓ and $\delta_\ell^{(0)}$ explicit, by rewriting Eq. (19.70) as

$$e^{2i\delta_\ell} = -\frac{j_\ell(ka) - i\eta_\ell(ka)}{j_\ell(ka) + i\eta_\ell(ka)}\frac{b_\ell(k) - ka\frac{j'_\ell(ka) - i\eta'_\ell(ka)}{j_\ell(ka) - i\eta_\ell(ka)}}{b_\ell(k) - ka\frac{j'_\ell(ka) + i\eta'_\ell(ka)}{j_\ell(ka) + i\eta_\ell(ka)}}. \tag{19.71}$$

By Eq. (19.63), the first term in the product is $e^{2i\delta_\ell^{(0)}}$. We define the dimensionless variables $Q_\ell(k)$ and $S_\ell(k)$ by

$$Q_\ell(k) + iS_\ell(k) := ka[j'_\ell(ka) + i\eta'_\ell(ka)]/[j_\ell(ka) + i\eta_\ell(ka)]. \tag{19.72}$$

Then, we write Eq. (19.71) as

$$e^{2i\delta_\ell - 2i\delta_\ell^{(0)}} = \frac{b_\ell(k) - Q_\ell(k) + iS_\ell(k)}{b_\ell(k) - Q_\ell(k) - iS_\ell(k)}, \tag{19.73}$$

from which we obtain

$$\tan[\delta_\ell - \delta_\ell^{(0)}] = \frac{S_\ell(k)}{b_\ell(k) - Q_\ell(k)}. \tag{19.74}$$

By Eq. (B.32), the quantity

$$S_\ell(k) = ka\frac{j_\ell(ka)\eta'_\ell(ka) - j'_\ell(ka)\eta_\ell(ka)}{j_\ell^2(ka) + \eta_\ell^2(ka)} = \frac{1}{ka[j_\ell^2(ka) + \eta_\ell^2(ka)]}$$

is always positive, a fact of importance in our later study of resonances.

For $ka \ll 1$, we use the expansion of (B.38) and (B.39) for the spherical Bessel function, while keeping the value $b_\ell(0)$ of the logarithmic derivative. Equation (19.70) becomes

$$\tan\delta_\ell = -\frac{b_\ell(0) - \ell}{b_\ell(0) + \ell + 1}\frac{2^{2\ell}(\ell!)^2}{(2\ell + 1)!(2\ell)!}(ka)^{2\ell+1}.$$

Hence, at low energies, $\delta_\ell \sim (ka)^{2\ell+1}$. The dominant contribution comes from $\ell = 0$,

$$\delta_0 \simeq \tan\delta_0 = -\frac{b_\ell(0)}{b_\ell(0) + 1}ka = -ka_s. \tag{19.75}$$

where $a_s = \frac{b_\ell(0)}{b_\ell(0)+1}a$ is the *scattering length* of the potential.[3] As in hard-sphere scattering, $\sigma_{tot} = 4\pi a_s^2$. Note that, despite its name, the scattering length may be negative if $b_\ell(0)$ is negative.

For scattering between nuclei, where the short-range nuclear force dominates over the repulsive Coulomb force, a is of the order of 10^{-15}m. The condition $ka \ll 1$ implies that $E \ll (ma^2)^{-1} \sim 100$ MeV. Indeed, the phase shift δ_0 dominates for energies up to 10–20 MeV, depending on the types of particles that are being scattered. Experiments in this energy typically also determine the next term in the expansion of $\delta_0(k)$ with respect to k. To calculate this term, we assume that the logarithmic derivative $b_0(k)$ is an analytic function of energy, so that, for low energies, $b_0(k) \simeq b_0(0) + ha^2k^2 + \cdots$, for some dimensionless number h. Expanding also $j_0(x) = 1 - \frac{1}{6}x^2$ and $\eta_0(x) = -x^{-1} + \frac{1}{2}x$, we bring Eq. (19.70) into the form

$$k \cot \delta_0 = -\frac{1}{a_s} + r_0 k^2, \tag{19.76}$$

where r_0 is the *effective range* of the potential. Experiments at low energies only determine a_s and r_0 as parameters of the nuclear potential.

19.3.4 Resonances

In general, the total scattering cross-section σ_{tot} varies slowly with energy. However, in some cases, σ_{tot} changes abruptly within a short interval of energies, and then returns to its earlier values. This phenomenon is called a scattering *resonance*.

We first encountered resonances in Section 5.3.5 for particles scattering in one dimension. Resonances appeared in potentials that may trap a particle for some finite time, such as, for example, potentials with two barriers at finite separation. The same idea applies roughly to three-dimensional scattering. The simplest system that exhibits scattering is the spherical well potential

$$V(r) = \begin{cases} -V_0, & r \le a \\ 0, & r > a \end{cases}, \tag{19.77}$$

where $a, V_0 > 0$.

For $r < a$, Eq. (13.11) is like Schrödinger's equation of a free particle with energy $E + V_0$. Hence, solutions are of the form $j_\ell(Kr)$, where we defined $K := \sqrt{2m(E + V_0)} = \sqrt{K_0^2 + k^2}$ with $K_0 = \sqrt{2mV_0}$. The logarithmic derivative at $r = a$ is $b_\ell(k) = Kaj_\ell'(Ka)/j_\ell(Ka)$.

We consider a deep well, namely, $K_0a \gg 1$. For small ℓ, the approximation (13.26) applies, so that

$$b_\ell(k) = Ka \cot\left(Ka - \frac{\ell\pi}{2}\right). \tag{19.78}$$

We are interested in energies such that ka is at most of the order of unity, which means that $Ka = K_0a$ up to corrections of order k/K_0. For generic values of K_0a, $|b_\ell(k)| \simeq K_0a \gg 1$; hence, by Eq. (19.70), $\tan \delta_\ell \simeq \frac{j_\ell(ka)}{\eta_\ell(ka)}$. However, if K_0a is close to a point where the cotangent

---

[3] By Eq. (19.75), δ_0 may not vanish at low energies if $b_\ell(0) = -1$. For $k = 0$ and $r > a$, Eq. (13.11) becomes $(r^2 R_{0,0}')' = 0$, with solution $R_{0,0}(r) = c_1 + c_2 a/r$ for constants c_1, c_2. Then, $b_0(0) = -c_2/(c_1 + c_2)$, hence, $b_\ell(0) = -1$ implies that $c_1 = 0$. But then, $R_{0,0}(r)$ decays with $1/r$; hence, it describes a spatially localized state, even if it is not square integrable.

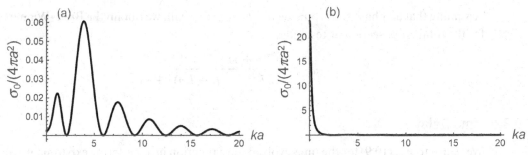

Fig. 19.7 The two plots give the contribution σ_0 to the total cross-section divided by $4\pi a^2$, as a function of the product ka. (a) $K_0 a = 111.518$; there is a resonance at $ka \simeq 4$, and a series of smaller peaks. (b) $K_0 a = 111.525$; resonances have effectively disappeared.

vanishes, then there is a value of Ka near $K_0 a$ for which $b_\ell(k) \simeq 0$, so that $\tan \delta_\ell \simeq \frac{j'_\ell(ka)}{\eta'_\ell(ka)}$. Hence, there is an abrupt change of phase shift, and consequently of the scattering cross-section.

We give an example in Fig. 19.7, where we plot the $\ell = 0$ contribution to the scattering cross-section, $\sigma_0 := 4\pi k^{-2} \sin^2 \delta_0$. Indeed, we see a strong peak with very small width in the cross-section. We also notice a strong sensitivity of resonance on $K_0 a$. The low-energy resonance disappears with a tiny change of $K_0 a$, and new resonances appear at higher energies.

Our analysis shows that resonances appear, when, for some value ℓ_0 of angular momentum, (i) the logarithmic derivative b_{ℓ_0} vanishes for some energy E_0, and (ii) $|b_{\ell_0}| >> 1$ for all energies except for a small interval around E_0. Hence, there is an interval of a small width ΔE around E_0 where

$$b_{\ell_0}(E) = \alpha(E_0 - E). \tag{19.79}$$

The parameter α is necessarily positive, because logarithmic derivatives are decreasing functions of energy – see Section 5.3.1. It is necessary to take $\alpha \Delta E >> 1$ so that $|b_{\ell_0}|$ rapidly becomes very large away from E_0.

We assume that Q_{ℓ_0} and S_{ℓ_0} are approximately constant near resonance. Then, we define the constants $\Gamma := 2S_{\ell_0}(k_0)/\alpha > 0$ and $B := Q_{\ell_0}(k_0)/\alpha$, where $k_0 = \sqrt{2mE_0}$. By definition, $0 < \Gamma << \Delta E$ and $|B| << \Delta E$. Then, Eq. (19.74) yields

$$\tan[\delta_{\ell_0} - \delta_{\ell_0}^{(0)}] = \frac{\Gamma}{2(E_R - E)}, \tag{19.80}$$

where we defined the resonance energy $E_R := E_0 - B$. Since $|B| << \Delta E$, E_R is within distance ΔE from E_0, so Eq. (19.79) applies.

By Eq. (19.80), $\delta_{\ell_0} - \delta_{\ell_0}^{(0)} = \frac{\pi}{2}$ for $E = E_R$. In general, $\delta_{\ell_0}^{(0)} << \frac{\pi}{2}$, so we can ignore it near resonance. Then,

$$\tan \delta_{\ell_0} = \frac{\Gamma}{2(E_R - E)}. \tag{19.81}$$

Assuming that all phase shifts outside resonance are small, we obtain the Breit–Wigner equation for the total cross-section at resonance

$$\sigma_{tot} = \frac{\pi(2\ell_0 + 1)}{k^2} \frac{\Gamma^2}{(E - E_R)^2 + \frac{1}{4}\Gamma^2}. \tag{19.82}$$

19.3.5 Time Delay

We return to Eq. (19.9) for the time-evolved wave function in a scattering experiment. We observe that the quantity

$$\tau(\mathbf{n}) = -\frac{m}{k}\mathbf{n}_0 \cdot \mathbf{b}_{\mathbf{k}_0}(\mathbf{n}) = -\frac{m}{k}\mathbf{n}_0 \cdot \mathrm{Im}\,(\nabla_{\mathbf{k}} \ln f_{\mathbf{k}}) \tag{19.83}$$

can be interpreted as an effective time delay for a particle that was detected in the direction $\mathbf{n}$ after scattering. This quantity is analogous to the time delay (7.64) for scattering in one dimension. As in its one-dimensional counterpart, Eq. (7.64), the interpretation of (19.83) as a time delay must be taken with a grain of salt, because $\tau(\mathbf{n})$ is fundamentally a parameter of a probability distribution and not the value of an observable.

For a central potential, the scattering amplitude depends on $\mathbf{k}$ only through k; hence, Eq. (19.83) becomes $\tau(\theta) = -\mathrm{Im}[\partial \ln f_E(\theta)/\partial E]$, where $E = \frac{k^2}{2m}$.

We can evaluate $\tau(\theta)$ using Eq. (19.61) for the scattering amplitude. Calculations are simplified significantly if the scattering amplitude is dominated by a single term with angular momentum $\ell = \ell_0$. Then, τ is θ-independent, as it takes the simple form $\tau = -\partial\delta_{\ell_0}/\partial E$.

Two cases are of immediate interest. At very low energies, the contribution from the phase shift δ_0 is dominant, and, by Eq. (19.75),

$$\tau = \frac{b_\ell(0)}{b_\ell(0) + 1} \frac{ma}{k}, \tag{19.84}$$

that is, the time delay is of the order of the time it takes a free particle to cross the potential's range.

The second case corresponds to resonances, where the dominant phase shift is given by Eq. (19.81). Then,

$$\tau = \frac{\frac{1}{2}\Gamma}{(E - E_R)^2 + \frac{1}{4}\Gamma^2}. \tag{19.85}$$

For $E = E_R$, $\tau = 2\Gamma^{-1}$, that is, the time delay can be very large, much larger than the time needed to traverse the potential's range. For this reason, we say that a resonance has a "lifetime" of the order of Γ^{-1}.

19.4 Scattering from a Coulomb Potential

Our treatment of scattering relies strongly on the assumption that the potential is short-range. This assumption excludes the Coulomb potential $V(r) = -\frac{\kappa}{r}$. In this section, we analyze scattering in this potential, and we identify the differences from the case of a short-range potential.

The eigenvalue equation for the Schrödinger operator of a particle with mass m in the Coulomb potential is

$$\left(\nabla^2 + k^2 + \frac{2m\kappa}{r}\right)\psi(\mathbf{x}) = 0, \tag{19.86}$$

where $E = \frac{\mathbf{k}^2}{2m}$. Because of spherical symmetry, we can take $\mathbf{k} = (0,0,k)$. Suppose that we write $\psi(\mathbf{x}) = e^{ikx_3} G(\mathbf{x})$. Our boundary conditions imply that, for large r and for $\theta \neq 0$, $\psi_k \sim e^{ikr}/r$; hence, $G(\mathbf{x}) \sim e^{ik(r-x_3)}/(r-x^3)$, where a multiplicative term $(1-\cos\theta)$ is absorbed in the angular dependence. This suggests looking for solutions with G a function of $\xi = k(r-x_3)$.

It is a long but straightforward calculation to show that, for $\psi(\mathbf{x}) = e^{ikx_3} G(\xi)$,

$$\nabla^2 \psi = e^{ikx_3}\left[2G''k\frac{\xi}{r} + 2G'\frac{k}{r}(1-i\xi) - k^2 G\right],$$

where the prime denotes derivative with respect to ξ. We substitute into Eq. (19.86), to obtain

$$\xi G'' + (1 - i\xi)G' - i\gamma G = 0, \tag{19.87}$$

where $\gamma = m\kappa/k$. Let $\tilde{G}(s) := \int_0^\infty d\xi e^{-s\xi} G(\xi)$ be the Laplace transform of $G(\xi)$. With repeated integration by parts, we evaluate the Laplace transform of $\xi G''(\xi)$ as $G(0) - 2s\tilde{G}(s) - s^2 d\tilde{G}(s)/ds$, and the Laplace transform of $(1 - i\xi)G'(\xi)$ as $-G(0) + (s+i)\tilde{G}(s) + isd\tilde{G}(s)/ds$. Then, Eq. (19.87) becomes

$$s(s - i)\frac{d\tilde{G}}{ds} = (-s + i + \gamma)\tilde{G}(s).$$

We solve this differential equation through separation of variables, to obtain

$$\tilde{G}(s) = Cs^{-1+i\gamma}(s - i)^{-i\gamma} = Cs^{-1}\exp\left[i\gamma\left(\ln s - \ln(s - i)\right)\right], \tag{19.88}$$

for some constant C. The function $\tilde{G}(s)$ has two branch-cuts, at $s = 0$ and $s = i$.

The inverse Laplace transform of $\tilde{G}$ is given by the Bromwitch integral

$$G(\xi) = \frac{1}{2\pi i}\int_{c-i\infty}^{c+i\infty} ds e^{s\xi}\tilde{G}(s), \tag{19.89}$$

where c is greater than the real part of the singularities of $\tilde{G}(s)$.

To evaluate $G(\xi)$, we consider the line integral $\oint_{C_\epsilon} dsF(s)$ for $F(s) = \tilde{G}(s)e^{\xi s}$ over the curve C_ϵ that is shown in Fig. 19.8. This curve has two deformations of width ϵ, in order to exclude the half-lines $(\mathrm{Re}\, s \leq 0, \mathrm{Im}\, s = 0)$ and $(\mathrm{Re}\, s \leq 0, \mathrm{Im}\, s = 1)$. Then, the singular points of $F(s)$ lie in the exterior of C_ϵ. By Cauchy's residue theorem, $\oint_{C_\epsilon} dsF(s) = 0$. Taking the limit $\epsilon \to 0$, we obtain

$$G(\xi) + \frac{1}{2\pi i}\lim_{\epsilon \to 0^+}\int_{-\infty}^{\infty} dx[F(i - x + i\epsilon) - F(i - x - i\epsilon)]$$

$$+ \frac{1}{2\pi i}\lim_{\epsilon \to 0^+}\int_{-\infty}^{\infty} dx[F(-x + i\epsilon) - F(-x - i\epsilon)] = 0.$$

Fig. 19.8 The curve C_ϵ on the complex plane that is employed for the evaluation of the integral (19.89). The curve consists of the horizontal axis at $x = c$, a semicircle at infinity that covers all the complex plane to the right of the $\mathrm{Re}\,s = c$ line, with two deformations of width ϵ to exclude the half-lines ($\mathrm{Re}\,s \leq 0, \mathrm{Im}\,s = 0$) and ($\mathrm{Re}\,s \leq 0, \mathrm{Im}\,s = 1$). Hence, the branch-cuts of $\tilde{G}$ at $s = 0$ and $s = i$ are in the exterior of C_ϵ.

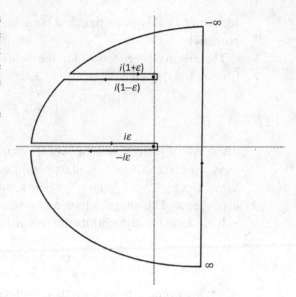

In Section 18.4.2, we showed that $\lim_{\epsilon \to 0^+} \ln(-x \pm i\epsilon) = \ln x \pm i\pi$, for $x > 0$. Using this identity, we obtain

$$G(\xi) = \frac{iC}{\pi} \sinh(\pi\gamma) \left[Q_1(\xi) + Q_2(\xi) e^{i\xi} \right], \qquad (19.90)$$

where

$$Q_1(\xi) = \int_0^\infty dx\, e^{-x\xi} (-i - x)^{-i\gamma} x^{-1+i\gamma}, \qquad Q_2(\xi) = \int_0^\infty dx\, e^{-x\xi} (i - x)^{-1+i\gamma} x^{-i\gamma}.$$

For large ξ, the term $e^{-x\xi}$ suppresses the contribution from nonzero values of x. Then, the two integrals are dominated by the values of x near zero:

$$Q_1(\xi) \simeq (-i)^{-i\gamma} \int_0^\infty dx\, x^{-1+i\gamma} e^{-x\xi} = e^{-\frac{1}{2}\pi\gamma} \Gamma(i\gamma) e^{-i\gamma \ln \xi},$$

$$Q_2(\xi) \simeq i^{-1+i\gamma} \int_0^\infty dx\, x^{-i\gamma} e^{-x\xi} = -i\xi^{-1} e^{-\frac{1}{2}\pi\gamma} \Gamma(1 - i\gamma) e^{i\gamma \ln \xi}.$$

Hence,

$$G(\xi) = \frac{iC}{\pi} \sinh(\pi\gamma) \Gamma(i\gamma) e^{-\frac{1}{2}\pi\gamma} \left[e^{-i\gamma \ln \xi} + \gamma \frac{\Gamma(1 - i\gamma)}{\Gamma(1 + i\gamma)} \frac{e^{i\xi + i\gamma \ln \xi}}{\xi} \right],$$

where we used the identity (B.2), $\Gamma(1 + i\gamma) = i\gamma \Gamma(i\gamma)$.

The corresponding mode functions are

$$\psi(\mathbf{x}) = \frac{iC}{\pi} \sinh(\pi\gamma) \Gamma(i\gamma) e^{-\frac{1}{2}\pi\gamma} \left[e^{ikx_3 - i\gamma \ln(k(r-x_3))} + \frac{\gamma}{k} \frac{\Gamma(1 - i\gamma)}{\Gamma(1 + i\gamma)} \frac{e^{ikr + i\gamma \ln(k(r-x_3))}}{r - x_3} \right].$$

The first term in the square bracket corresponds to incoming particles and the second term corresponds to scattered ones. The logarithmic components $\gamma \ln(k(r - x_3))$ of the oscillatory term signify the divergence from the mode functions in Eq. (19.3). We choose C to mirror the prefactor of Eq. (19.3), that is, we set $\frac{iC}{\pi} \sinh(\pi\gamma)\Gamma(i\gamma)e^{-\frac{1}{2}\pi\gamma} = (2\pi)^{-3/2}$. Then,

$$\psi(r,\theta) = \frac{1}{(2\pi)^{3/2}} \left[e^{ikr\cos\theta - i\frac{m\kappa}{k}\ln[kr(1-\cos\theta)]} \right.$$
$$\left. + \frac{m\kappa}{k^2(1-\cos\theta)}\frac{\Gamma(1-i\gamma)}{\Gamma(1+i\gamma)}e^{ikr + i\frac{m\kappa}{k}\ln[kr(1-\cos\theta)]} \right]. \qquad (19.91)$$

We recover the mode functions (19.3) for $kr >> \frac{m\kappa}{k}|\ln(1-\cos\theta)|$. For sufficiently large r, this condition can only be violated by small angles θ so that $1 - \cos\theta \simeq \frac{\theta^2}{2}$. Hence, the condition for recovering Eq. (19.3) becomes

$$|\theta| >> e^{-\frac{k^2 r}{2m\kappa}}. \qquad (19.92)$$

If Eq. (19.92) is satisfied, we straightforwardly identify the scattering amplitude as

$$f(\theta) = \frac{m\kappa}{k^2(1-\cos\theta)}\frac{\Gamma(1-i\gamma)}{\Gamma(1+i\gamma)} = \frac{m\kappa}{2k^2\sin^2\frac{\theta}{2}}\frac{\Gamma(1-i\gamma)}{\Gamma(1+i\gamma)}. \qquad (19.93)$$

Since $\Gamma(1-i\gamma) = [\Gamma(1+i\gamma)]^*$, the ratio $\frac{\Gamma(1-i\gamma)}{\Gamma(1+i\gamma)}$ is just a phase. It follows that

$$|f(\theta)|^2 = \frac{\kappa^2 m^2}{4k^4 \sin^4\frac{\theta}{2}}, \qquad (19.94)$$

that is, we recover the scattering cross-section that was obtained by the Born approximation, Eq. (19.51). The Born approximation happens to be exact for the modulus of the scattering amplitude $f(\theta)$, but it misrepresents the phase.

We are now in a position to account for the divergence of the total cross-section for the Coulomb potential. This divergence originates from the singular behavior of $f(\theta)$ at $\theta = 0$. For $|\theta|$ violating condition (19.92), the mode functions are not of the form (19.3), hence, the analysis of Section 19.1 does not pass through. The detection probability in the forward direction is not proportional to $|f(\theta)|^2$ of Eq. (19.51). In fact, Eq. (19.91) for the mode functions does not apply, because it follows from the assumption that ξ is large.

19.5   General Theory of Scattering: In and Out States

In the previous sections, we developed a theory for the scattering of a particle from a potential, or, equivalently, for the scattering of two particles that interact through a potential. However, there are other types of interactions: spin-dependent interactions, interactions that change particle species, or even multiparticle scattering. At a fundamental level, interactions are defined through quantum fields, and it is highly nontrivial to write a wave-packet analysis similar to the one of Section 19.1 for the scattering cross-section.

In the remainder of this chapter, we present a general theory of scattering that broadly generalizes the main concepts of the previous sections. The aim is to construct a general expression for probabilities in scattering experiments, without solving explicitly the equations for time evolution. These probabilities are determined solely from the properties of the spectrum of the Hamiltonian. In this process, we derive a rule for selecting a preferred basis for scattering experiments that also accounts for the choice of the basis (19.3) in potential scattering.

19.5.1 Definition of In and Out States

Consider a Hamiltonian $\hat{H} = \hat{H}_0 + \hat{V}$, where $\hat{H}_0$ describes free particles and $\hat{V}$ describes interactions between those particles. We specify neither the number nor the species of particles. The Hilbert space is a tensor product of Fock spaces, one Fock space for each particle species. The Hamiltonian $\hat{H}_0$ is highly degenerate, and it satisfies the eigenvalue equation $\hat{H}_0|\alpha\rangle_0 = E_\alpha|\alpha\rangle_0$. The generalized eigenvectors $|\alpha\rangle_0$ define a (generalized) orthonormal basis for the particles' momenta and spins.

For continuous spectrum, the spectral theorem reads

$$f(\hat{H}_0) = \int d\alpha f(E_\alpha)|\alpha\rangle_0 {}_0\langle\alpha|. \tag{19.95}$$

The crucial step in our study of scattering by a potential was the choice of the generalized eigenvectors (19.3) of $\hat{H}$. These eigenvectors define a basis, and they satisfy Eq. (19.7). The latter is crucial for the derivation of Eq. (19.11), and, hence, for identifying the scattering cross-section. For $t < 0$, Eq. (19.11) yields

$$\psi_t(\mathbf{x}) \simeq \psi_t^{(0)}(\mathbf{x}), \tag{19.96}$$

since the scattering term is proportional to $\psi_0[\mathbf{n}_0(r - \frac{k_0}{m}t)]$ and it is negligible for $t < 0$. We will employ a condition analogous to Eq. (19.96), in order to define a useful basis of eigenvectors of $\hat{H}$, analogous to the vectors (19.3).

We assume that the interaction term does not radically modify the spectrum of the Hamiltonian, in the sense that $\hat{H}$ also has a continuous spectrum in $[0, \infty)$, and for some eigenbasis $|\alpha\rangle_0$ of $\hat{H}_0$, there exists an eigenbasis $|\alpha\rangle$ of $\hat{H}$, $\hat{H}|\alpha\rangle = E_\alpha|\alpha\rangle$, such that $|\alpha\rangle$ converges to $|\alpha\rangle_0$ at the limit of vanishing $\hat{V}$.

We select a specific eigenbasis $|\alpha\rangle_+$ of $\hat{H}$ that satisfies the following condition. Let $|\Psi_0\rangle$ be an initial state appropriate for a scattering experiment. We analyze $|\Psi_0\rangle$ in the $|\alpha\rangle_0$ basis as $|\Psi_0\rangle = \int d\alpha\, \psi(\alpha)|\alpha\rangle_0$. We define the basis $|\alpha\rangle_+$ by the requirement that

$$e^{-i\hat{H}_0 t}\int d\alpha\, \psi(\alpha)|\alpha\rangle_0 = e^{-i\hat{H}t}\int d\alpha\, \psi(\alpha)|\alpha\rangle_+, \text{ for } t \overset{sc}{\to} -\infty, \tag{19.97}$$

for all $\psi(\alpha)$ that correspond to initial conditions appropriate for scattering. Hence, for all appropriate initial states, the free-particle Hamiltonian $\hat{H}_0$ gives the same evolution as the total Hamiltonian for $t \to -\infty$.

The limit $t \to -\infty$ in Eq. (19.97) is not literal, because in this limit free-particle wave packets have dispersed over all space by the time they scatter. Rather, we mean that t is much earlier than the time that the incoming particles approach each other, but it is not so large in absolute value that dispersion is significant. We write sc on top of the arrow for the limit, in order to clarify that we do not mean the mathematical notion of a limit. The eigenvectors $|\alpha\rangle_+$ generalize the wave functions (19.3) for potential scattering. They are known as *in states*.

We also define the *out states* $|\alpha\rangle_-$ by the requirement that

$$e^{-i\hat{H}_0 t}\int d\alpha\,\psi(\alpha)|\alpha\rangle_0 = e^{-i\hat{H}t}\int d\alpha\,\psi(\alpha)|\alpha\rangle_-,\ \text{for}\ t \overset{sc}{\to} \infty, \tag{19.98}$$

for any $\psi(\alpha)$ that corresponds to a scattering set-up.

19.5.2 The Lippman–Schwinger Equation

Next, we identify the in and out states solely from the spectrum of the Hamiltonian. First, we express the eigenvalue equation $\hat{H}|\alpha\rangle_+ = E_\alpha|\alpha\rangle_+$ as

$$(E_\alpha - \hat{H}_0)|\alpha\rangle_+ = \hat{V}|\alpha\rangle_+. \tag{19.99}$$

Equation (19.99) can be viewed as a linear equation with a non-homogeneous term $|\phi\rangle :=$ $\hat{V}|\alpha\rangle_+$. Hence, the general solution is a sum of $|\alpha\rangle_0$ that solves the corresponding homogeneous equation with a partial solution. A large class of partial solutions is of the form $\lim_{z\to E_\alpha}\hat{G}_0(z)|\phi\rangle$, where $G_0(z) = (z - \hat{H}_0)^{-1}$ is the resolvent associated to the Hamiltonian $\hat{H}_0$. The explicit form of the solution depends on the way we take the limit $z \to E_\alpha$, namely, whether we approach E_α from the upper complex half-plane or from the lower half-plane. For a general complex function $F(z)$, we denote these limits as $F(E^\pm) := \lim_{\epsilon\to 0^+} F(E \pm i\epsilon)$.

For the E^+ limit, we select solutions to Eq. (19.99) of the form

$$|\alpha\rangle_+ = |\alpha\rangle_0 + \hat{G}_0(E_a^+)\hat{V}|\alpha\rangle_+. \tag{19.100}$$

The physical interpretation of this solution follows from the time evolution of the vector $|\Psi_0\rangle = \int d\alpha\,\psi(\alpha)|\alpha\rangle_+$,

$$\begin{aligned}
e^{-i\hat{H}t}|\Psi_0\rangle :&= \int d\alpha\,\psi(\alpha)e^{-iE_\alpha t}|\alpha\rangle_+ \\
&= \int d\alpha\,\psi(\alpha)e^{-iE_\alpha t}|\alpha\rangle_0 + \int d\alpha\,\psi(\alpha)e^{-iE_\alpha t}\hat{G}(E_\alpha^+)\hat{V}|\alpha\rangle_+ \\
&= e^{-i\hat{H}_0 t}|\Psi_0\rangle + \int d\alpha \int d\beta\,\psi(\alpha)\frac{e^{-iE_\alpha t}}{E_\alpha - E_\beta + i\epsilon}{}_0\langle\beta|\hat{V}|\alpha\rangle_+|\beta\rangle_0,
\end{aligned} \tag{19.101}$$

where, in the last step, we inserted a resolution of the unity with respect to $|\beta\rangle_0$.

Given that the vectors $|\alpha\rangle_0$ define a basis of energy eigenvectors, integration over α involves integration with respect to energy. Hence, the second term in the right-hand side of Eq. (19.101) contains an integral

$$J = \int dE f(E)\frac{e^{-iEt}}{E - E_\beta + i\epsilon},$$

where the function $f(E)$ accounts for the energy dependence of $\psi(\alpha)\,{}_0\langle\beta|\hat{V}|\alpha\rangle_+$. For $t < 0$, we evaluate J by integrating along the curve C_- of Fig. 18.1. If $f(E)$ has no singularities in the upper half-plane, then $J = 0$. This means that the kets $|\alpha\rangle_+$ satisfy Eq. (19.97), that is, they are in states. This justifies the use of the index $+$ for the in states.

With similar arguments, we find that the out states satisfy

$$|\alpha\rangle_- = |\alpha\rangle_0 + \hat{G}_0(E_a^-)\hat{V}|\alpha\rangle_-. \tag{19.102}$$

Together, Eqs. (19.100) and (19.102) are known as the *Lippman–Schwinger equation* (LSE).

The LSE is solved solely from the knowledge of the resolvent $\hat{G}(z) = (z - \hat{H})^{-1}$ of the full Hamiltonian. To see this, we recall that the resolvents $\hat{G}(z)$ and $\hat{G}_0(z)$ are related by the perturbative series (18.29). Explicitly,

$$\hat{G}(z) = \hat{G}_0(z) + \hat{G}_0(z)\hat{V}\hat{G}_0(z) + \hat{G}_0(z)\hat{V}\hat{G}_0(z)\hat{V}\hat{G}_0(z) + \cdots. \tag{19.103}$$

Suppose that we multiply both sides of the equation either with $\hat{G}_0(z)\hat{V}$ on the right or with $\hat{V}\hat{G}_0(z)$ on the left. In both cases, all terms of the series (19.103) are reproduced except for the first one. The right-hand side of Eq. (19.103) becomes $\hat{G}(z) - \hat{G}_0(z)$. Hence, Eq. (19.103) implies a pair of equations

$$\hat{G}(z) = \hat{G}_0(z) + \hat{G}(z)\hat{V}\hat{G}_0(z), \qquad \hat{G}(z) = \hat{G}_0(z) + \hat{G}_0(z)\hat{V}\hat{G}(z), \tag{19.104}$$

or, equivalently, $[\hat{I} - \hat{G}_0(z)\hat{V}]\hat{G}(z) = \hat{G}_0(z)$ and $[\hat{I} + \hat{G}(z)\hat{V}]\hat{G}_0(z) = \hat{G}(z)$. Then,

$$[\hat{I} - \hat{G}_0(z)\hat{V}][\hat{I} + \hat{G}(z)\hat{V}] = [\hat{I} + \hat{G}(z)\hat{V}][\hat{I} - \hat{G}_0(z)\hat{V}] = \hat{I};$$

hence, $[\hat{I} - \hat{G}_0(z)\hat{V}]^{-1} = \hat{I} + \hat{G}(z)\hat{V}$. The LSE can be written as $[\hat{I} - \hat{G}_0(E_\alpha^\pm)\hat{V}]|\alpha\rangle_\pm = |\alpha\rangle_0$, and solving for $|\alpha\rangle_\pm$, we find

$$|\alpha\rangle_\pm = [\hat{I} - \hat{G}_0(E_\alpha^\pm)\hat{V}]^{-1}|\alpha\rangle_0 = [\hat{I} + \hat{G}(E_\alpha^\pm)\hat{V}]|\alpha\rangle_0. \tag{19.105}$$

19.5.3 Examples

Next, we solve the LSE for elastic scattering of a particle from a potential in one dimension and in three dimensions.

Example 19.3 Consider scattering in one dimension, where $\hat{H}_0 = \frac{\hat{p}^2}{2m}$ and $\hat{V} = V(\hat{x})$. The eigenvectors $|p\rangle$ of $\hat{H}_0$ are specified by the momenta p. The matrix elements of $\hat{G}_0(E^+)$ in the position basis are

$$\langle x|\hat{G}_0(E^+)|x'\rangle = \int dp\langle x|p\rangle \frac{1}{E + i\epsilon - \frac{p^2}{2m}}\langle p|x'\rangle = \int \frac{dp}{2\pi}\frac{e^{ip(x-x')}}{E + i\epsilon - \frac{p^2}{2m}}.$$

The integrand has poles at $p = \pm\sqrt{2mE(1 \pm i\epsilon/E)} = \pm\sqrt{2mE}(1 \pm i\frac{\epsilon}{2E})$. For $x - x' > 0$, we integrate along the curve C_- of Fig. 18.1. Only the pole with positive imaginary part contributes. We obtain

$$\langle x|\hat{G}_0(E^+)|x'\rangle = -\frac{im}{\sqrt{2mE}}e^{i\sqrt{2mE}(x-x')}. \tag{19.106}$$

For $x - x' < 0$, we integrate along the curve C_+ of Fig. 18.1. Only the pole with negative imaginary part contributes. We obtain the same result, modulo the exchange of x with x'. We conclude that

$$\langle x|\hat{G}_0(E^+)|x'\rangle = -\frac{im}{\sqrt{2mE}}e^{i\sqrt{2mE}|x-x'|}. \tag{19.107}$$

The LSE for $\psi_p^+(x) := \langle x|p\rangle_+$ at energy $E = \frac{p^2}{2m}$ takes the form

$$\psi_p^+(x) = \frac{1}{\sqrt{2\pi}}e^{ipx} - \frac{im}{|p|}\int dx'\, e^{i|p||x-x'|}V(x')\psi_p^+(x').\tag{19.108}$$

Suppose that the potential is nonzero only in an interval $[a,b]$. For $p > 0$,

$$\psi_p^+(x) = \begin{cases} \frac{1}{\sqrt{2\pi}}e^{ipx} - \frac{im}{|p|}e^{-ipx}\int dx'\, e^{ipx'}V(x')\psi_p^+(x'), & x < a \\ \frac{1}{\sqrt{2\pi}}e^{ipx} - \frac{im}{|p|}e^{ipx}\int dx'\, e^{-ipx'}V(x')\psi_p^+(x'), & x > b, \end{cases}$$

while for $p < 0$,

$$\psi_p^+(x) = \begin{cases} \frac{1}{\sqrt{2\pi}}e^{ipx} - \frac{im}{|p|}e^{ipx}\int dx'\, e^{-ipx'}V(x')\psi_p^+(x'), & x < a \\ \frac{1}{\sqrt{2\pi}}e^{ipx} - \frac{im}{|p|}e^{-ipx}\int dx'\, e^{ipx'}V(x')\psi_p^+(x'), & x > b. \end{cases}$$

We see that for $p > 0$, the solutions ψ_p^+ coincide with the solutions $f_{|p|,+}$ of Eq. (5.57); for $p < 0$ they coincide with the solutions $f_{|p|,-}$ of Eq. (5.57). We also derive the following formulas for the transmission and reflection amplitudes

$$T_k = 1 + \frac{\sqrt{2\pi}\,m}{ik}\int dx'\, e^{-ikx'}V(x')\psi_k^+(x'),\quad \bar{T}_k = 1 + \frac{\sqrt{2\pi}\,m}{ik}\int dx'\, e^{ikx'}V(x')\psi_{-k}^+(x')$$

$$R_k = \frac{\sqrt{2\pi}\,m}{ik}\int dx'\, e^{ikx'}V(x')\psi_k^+(x'),\quad \bar{R}_k = \frac{\sqrt{2\pi}\,m}{ik}\int dx'\, e^{-ikx'}V(x')\psi_{-k}^+(x'),\tag{19.109}$$

where $k = |p| = \sqrt{2mE}$.

Example 19.4 For a particle in three dimensions $\hat{H}_0 = \frac{\hat{\mathbf{p}}^2}{2m}$ and $\hat{V} = V(\hat{\mathbf{x}})$. The eigenvectors $|\mathbf{p}\rangle$ of $\hat{H}_0$ are specified by the momentum vectors $\mathbf{p}$. The matrix elements of the resolvent $\hat{G}_0(E^+)$ in the position basis are

$$\langle\mathbf{x}|\hat{G}_0(E^+)|\mathbf{x}'\rangle = \int d^3p\langle\mathbf{x}|\mathbf{p}\rangle\frac{1}{E + i\epsilon - \frac{p^2}{2m}}\langle\mathbf{p}|\mathbf{x}'\rangle = \int\frac{d^3p}{(2\pi)^3}\frac{e^{i\mathbf{p}\cdot\mathbf{R}}}{E + i\epsilon - \frac{p^2}{2m}},$$

where $\mathbf{R} = \mathbf{x} - \mathbf{x}'$. We express $\mathbf{p}$ in polar coordinates taking the north pole in the direction of $\mathbf{R}$, so that $\mathbf{p}\cdot\mathbf{R} = pR\cos\theta$. We obtain,

$$\langle\mathbf{x}|\hat{G}_0(E^+)|\mathbf{x}'\rangle = \frac{1}{4\pi^2}\int_0^\infty dp\frac{p^2}{E + i\epsilon - \frac{p^2}{2m}}\int_0^\pi d\theta\,\sin\theta\, e^{ipR\cos\theta}$$

$$= \frac{1}{4\pi^2 iR}\int_0^\infty dp\frac{p(e^{ipR} - e^{-ipR})}{E + i\epsilon - \frac{p^2}{2m}} = \frac{1}{4\pi^2 iR}\int_{-\infty}^\infty dp\frac{pe^{ipR}}{E + i\epsilon - \frac{p^2}{2m}}.\tag{19.110}$$

The integral is similar to that of the one-dimensional case. We evaluate it by integrating along the curve C_- of Fig. 19.1. Only the pole with a positive imaginary part contributes, hence,

$$\langle\mathbf{x}|\hat{G}_0(E^+)|\mathbf{x}'\rangle = -\frac{m}{2\pi|\mathbf{x} - \mathbf{x}'|}e^{i\sqrt{2mE}|\mathbf{x}-\mathbf{x}'|}.\tag{19.111}$$

The LSE for $\psi_{\mathbf{p}}^+(\mathbf{x}) = \langle\mathbf{x}|\mathbf{p}\rangle_+$ takes the form

$$\psi_{\mathbf{p}}^+(\mathbf{x}) = \frac{1}{(2\pi)^{3/2}}e^{i\mathbf{p}\cdot\mathbf{x}} - \frac{m}{2\pi}\int d^3x'\frac{e^{ip|\mathbf{x}-\mathbf{x}'|}}{|\mathbf{x} - \mathbf{x}'|}V(\mathbf{x}')\psi_{\mathbf{p}}^+(\mathbf{x}').$$

For $\mathbf{x}$ at macroscopic distance from the scattering region, and for a short-range potential, we use the same approximation as with the integral (19.42). That is, we approximate $|\mathbf{x} - \mathbf{x}'| \simeq r - \mathbf{n} \cdot \mathbf{x}'$ in the exponential, and $|\mathbf{x} - \mathbf{x}'| \simeq r$ in the denominator; $\mathbf{n} = \mathbf{x}/r$. We conclude that

$$\psi_{\mathbf{p}}^{+}(\mathbf{x}) = \frac{1}{(2\pi)^{3/2}} e^{i\mathbf{p}\cdot\mathbf{x}} - \frac{e^{ipr}}{r} \frac{m}{2\pi} \int d^3 x' e^{-ip\mathbf{n}\cdot\mathbf{x}'} V(\mathbf{x}')\psi_{\mathbf{p}}^{+}(\mathbf{x}'). \tag{19.112}$$

Hence, the LSE uniquely determines the basis (19.3), and thus, justifies, in retrospect, our use of it. The scattering amplitude is given by

$$f_{\mathbf{p}}(\mathbf{n}) = -\sqrt{2\pi}\, m \int d^3 x' e^{-ip\mathbf{n}\cdot\mathbf{x}'} V(\mathbf{x}')\psi_{\mathbf{p}}^{+}(\mathbf{x}'). \tag{19.113}$$

19.6   General Theory of Scattering: The S-Matrix

Next, we show how all relevant information in a scattering experiment can be encoded in a single operator, the scattering operator, also known as the S-matrix. We analyze the properties of the S-matrix, and we show how it can be used to construct observable statistical magnitudes, like scattering cross-sections and decay rates.

19.6.1   Definition of the S-Matrix

We evaluate the transition probability $|\langle \psi_f | e^{-i\hat{H}(t_f - t_i)} | \psi_i \rangle|^2$ between an initial state $|\psi\rangle_i$ at time $t_i \overset{sc}{\to} -\infty$ and a final state $|\psi_f\rangle$ at time $t_f \overset{sc}{\to} \infty$. To this end, we expand the kets $|\psi\rangle_i$ and $|\psi\rangle_f$ in the basis $|\alpha\rangle_0$,

$$\langle \psi_f | e^{-i\hat{H}(t_f - t_i)} | \psi_i \rangle = \int d\alpha\, d\beta\, \psi_i(\alpha) \psi_f^*(\beta) {}_0\langle \beta | e^{-i\hat{H}(t_f - t_i)} | \alpha \rangle_0. \tag{19.114}$$

Equation (19.97) implies that $e^{i\hat{H}t_i}|\alpha\rangle_0 = e^{iE_\alpha t_i}|\alpha\rangle_+$; Eq. (19.98) implies that $e^{i\hat{H}t_f}|\beta\rangle_0 = e^{iE_\beta t_f}|\beta\rangle_-$. Hence, Eq. (19.114) becomes

$$\langle \psi_f | e^{-i\hat{H}(t_f - t_i)} | \psi_i \rangle = \int d\alpha\, d\beta\, \psi_i(\alpha) \psi_f^*(\beta) e^{-iE_\beta t_f + iE_\alpha t_i} {}_-\langle \beta | \alpha \rangle_+. \tag{19.115}$$

We define the *scattering operator* $\hat{S}$ from its matrix elements $S_{\alpha\beta}$ in the basis $|\alpha\rangle_0$,

$$S_{\alpha\beta} := {}_-\langle \alpha | \beta \rangle_+. \tag{19.116}$$

Since the states $|\alpha\rangle_\pm$ define a generalized orthonormal basis, Theorem 4.10 implies that $\hat{S}$ is unitary and that

$$|\alpha\rangle_- = \hat{S}|\alpha\rangle_+. \tag{19.117}$$

Strictly speaking, Eq. (19.116) defines an operator only in the subspace $\mathcal{H}_{sc}$ of the Hilbert space that describes scattering states. If bound states exist, we assume that $\hat{S}$ acts on the subspace $\mathcal{H}_{bound}$ of bound states trivially, as the unit operator. With this extension, $\hat{S}$ is unitary over the total Hilbert space $\mathcal{H}_{sc} \oplus \mathcal{H}_{bound}$. The operator $\hat{S}$ is also known as the *scattering matrix*, or the *S-matrix*.

Suppose that we could identify $|\psi_i\rangle$ with an eigenvector $|\alpha_i\rangle_0$ of $\hat{H}_0$ and $|\psi_f\rangle$ with another eigenvector $|\alpha_f\rangle_0$. Then, Eq. (19.115) becomes

$$\left|_0\langle\alpha_f|e^{-i\hat{H}(t_f-t_i)}|\alpha_i\rangle_0\right|^2 = |S_{\alpha_f\alpha_i}|^2, \tag{19.118}$$

that is, the transition probability depends exclusively on the scattering matrix. However, this choice of initial vectors is not acceptable: The kets $|\alpha_i\rangle_0$ correspond to the continuous spectrum and they are not square integrable. There is no such problem with the final-time vectors $|\alpha_f\rangle_0$, because Eq. (19.118) defines a probability density with respect to the index α_f. Despite this problem, Eq. (19.118) strongly suggests the relevance of the square modulus $|S_{\alpha_f\alpha_i}|^2$ to the construction of transition probabilities.

The Dyson series, Eq. (7.74), provides a closed formula for the scattering operator. First, we define the operator $\hat{O}(t,t') := e^{i\hat{H}_0 t}e^{-i\hat{H}(t-t')}e^{-i\hat{H}_0 t'}$. We straightforwardly find that $i\frac{d}{dt}\hat{O}(t,t') = \hat{V}(t)\hat{O}(t,t')$ and that $\hat{O}(t,t) = \hat{I}$, where $\hat{V}(t) = e^{i\hat{H}_0 t}\hat{V}e^{-i\hat{H}_0 t}$. It follows that $\hat{O}(t,t') = \mathcal{T}e^{-i\int_{t'}^{t}\hat{V}(s)ds}$.

The definition (19.116) together with Eqs. (19.97) and (19.98) imply that

$$S_{\beta\alpha} = \lim_{t_i\overset{sc}{\to}-\infty}\lim_{t_f\overset{sc}{\to}\infty}{}_0\langle\beta|e^{i\hat{H}_0 t_f}e^{-i\hat{H}(t_f-t_i)}e^{-i\hat{H}_0 t_i}|\alpha\rangle_0$$

$$= \lim_{t_i\overset{sc}{\to}-\infty}\lim_{t_f\overset{sc}{\to}\infty}{}_0\langle\beta|\hat{O}(t_f,t_i)|\alpha\rangle_0.$$

Hence,

$$\hat{S} = \lim_{t_i\overset{sc}{\to}-\infty}\lim_{t_f\overset{sc}{\to}\infty}\hat{O}(t_f,t_i)$$

$$= \hat{I} + \sum_{n=1}^{\infty}\frac{(-i)^n}{n!}\int_{-\infty}^{\infty}ds_1\int_{-\infty}^{\infty}ds_2\cdots\int_{-\infty}^{\infty}ds_n\mathcal{T}[\hat{V}(s_1)\hat{V}(s_2)\cdots\hat{V}(s_n)]. \tag{19.119}$$

For a weak interaction term $\hat{V}$, we can keep a finite number of terms in the Dyson series, thereby approximating the S-matrix perturbatively.

The key benefit of the S-matrix description of scattering is that we do not need to solve the evolution equations explicitly. We do not keep track of the quantum system's evolution in real time, we are interested only in the asymptotic values of specific observables. That this is a great calculational benefit can be seen already in the Dyson expansion of the scattering operator, where all terms involve integrals over the full real line. The original introduction of the S-matrix by Heisenberg (1943) – following earlier work of Wheeler (1937) – aimed to

... isolate from the conceptual scheme of the quantum theory of wave fields those concepts which probably will not be affected by the future changes and which may therefore represent an integral part also of the future theory.

In Heisenberg's viewpoint, the asymptotic properties in scattering represent directly observable quantities, and they are to be contrasted with the predictions of quantum theory about finite-time evolution, which was not directly observable. Predictions about asymptotic properties are therefore more robust than finite-time predictions, which could be superseded by a future theory. This argument turned out to be wrong, as finite-time predictions of quantum evolution have been repeatedly confirmed in experiment. The S-matrix, however, has developed into a valuable tool.

19.6.2 The Transition Operator

As seen in Section 19.1, a fraction of the incoming particles does not scatter. These are described by a component in the S-matrix that is proportional to unity. The remaining components of the S-matrix can be expressed in terms of the so-called *transition operator*. This is defined as a series

$$\hat{T}(z) := \hat{V} + \hat{V}\hat{G}_0(z)\hat{V} + \hat{V}\hat{G}_0(z)\hat{V}\hat{G}_0(z)\hat{V} + \cdots, \tag{19.120}$$

so that the series expansion (19.103) of the resolvent reads

$$\hat{G}(z) = \hat{G}_0(z) + \hat{G}_0(z)\hat{T}(z)\hat{G}_0(z). \tag{19.121}$$

By definition, $\hat{T}^\dagger(z) = \hat{T}(z^*)$; hence, $\hat{T}^\dagger(E^+) = \hat{T}(E^-)$.

Equation (19.120) can be rewritten as $\hat{T}(z) = \hat{V} + \hat{T}(z)\hat{G}_0(z)\hat{V}$, or, equivalently, as

$$\hat{T}(z)[\hat{I} - \hat{G}_0(z)\hat{V}] = \hat{V}. \tag{19.122}$$

We act both sides of Eq. (19.122) for $z = E^\pm$ to $|\alpha\rangle_\pm$. Using the LSE, we find that

$$\hat{V}|\alpha\rangle_\pm = \hat{T}(E_\alpha^\pm)|\alpha\rangle_0, \tag{19.123}$$

which provides an expression for the matrix elements of the transition operator in the $|\alpha\rangle_0$ basis

$$T_{\beta\alpha} := {}_0\langle\beta|\hat{T}(E_\alpha^+)|\alpha\rangle_0 = {}_0\langle\beta|\hat{V}|\alpha\rangle_+. \tag{19.124}$$

Equation (19.124) allows us to express the S-matrix in terms of the transition operator. To this end, we use Eq. (19.105), to write

$$S_{\beta\alpha} = {}_-\langle\beta|\alpha\rangle_+ = {}_0\langle\beta|\alpha\rangle_+ + {}_0\langle\beta|\hat{V}\hat{G}(E_\beta^+)|\alpha\rangle_+$$

$$= {}_0\langle\beta|\alpha\rangle_+ + \frac{{}_0\langle\beta|\hat{V}|\alpha\rangle_+}{E_\beta - E_\alpha + i\epsilon} = {}_0\langle\beta|\alpha\rangle_+ + \frac{T_{\beta\alpha}}{E_\beta - E_\alpha + i\epsilon}, \tag{19.125}$$

where the limit $\epsilon \to 0^+$ is implicit. By Eq. (19.100),

$${}_0\langle\beta|\alpha\rangle_+ = {}_0\langle\beta|\alpha\rangle_0 + {}_0\langle\beta|\hat{G}_0(E_\alpha^+)\hat{V}|\alpha\rangle_+ = \delta_{\alpha\beta} + \frac{T_{\beta\alpha}}{E_\alpha - E_\beta + i\epsilon}. \tag{19.126}$$

Substituting into Eq. (19.125), we find

$$S_{\beta\alpha} = \delta_{\alpha\beta} + [(E_\beta - E_\alpha + i\epsilon)^{-1} - (E_\beta - E_\alpha - i\epsilon)^{-1}]T_{\beta\alpha}$$

$$= \delta_{\alpha\beta} - 2i\frac{\epsilon}{(E_\beta - E_\alpha)^2 + \epsilon^2}\,T_{\beta\alpha}. \tag{19.127}$$

By Eq. (4.41), the function $\delta_\epsilon(E) = \frac{\pi^{-1}\epsilon}{E^2+\epsilon^2}$ is an approximate delta function, hence, at the limit $\epsilon \to 0^+$,

$$S_{\beta\alpha} = \delta_{\alpha\beta} - 2\pi i\delta(E_\beta - E_\alpha)\,T_{\beta\alpha}. \tag{19.128}$$

The S-matrix involves only the elements of the transition operator that preserve energy.

The unitarity of the S-matrix implies that $\sum_\gamma S_{\beta\gamma} S_{\gamma\alpha}^\dagger = \delta_{\alpha\beta}$. Substituting Eq. (19.128), we obtain

$$T_{\alpha\beta} - T_{\beta\alpha}^* = -2\pi i \sum_\gamma \delta(E_\gamma - E_\alpha) T_{\beta\gamma} T_{\alpha\gamma}^*. \tag{19.129}$$

For $\alpha = \beta$,

$$\mathrm{Im}\, T_{\alpha\alpha} = -\pi \sum_\gamma \delta(E_\gamma - E_\alpha) |T_{\alpha\gamma}|^2. \tag{19.130}$$

Equation (19.130) is a generalization of the optical theorem. In Problem 19.13, you are asked to show that it reduces to the optical theorem for potential scattering.

The modulus square $|S_{\beta\alpha}|^2$ contains the square of a delta function, hence it is divergent. This divergence is a consequence of the fact that the vectors $|\alpha\rangle_0$ are not square integrable; hence, $|S_{\beta\alpha}|^2$ is not a well-defined probability or probability density. This problem is sidestepped using the same trick we had used in the derivation of Fermi's golden rule, in Section 18.1.2.

To proceed, we substitute the delta function $\delta(E_\beta - E_\alpha)$ in Eq. (19.127) with an approximate delta function $\delta_\epsilon(E_\beta - E_\alpha)$. For example, we may refrain from taking the limit $\epsilon \to 0$ in Eq. (19.127). For sufficiently small ϵ, $[\delta_\epsilon(E_\beta - E_\alpha)]^2$ approximates $\delta(E_a - E_b)\delta_\epsilon(0)$. The quantity $\delta_\epsilon(0)$ has dimension of time, and it can be interpreted as an effective duration of the scattering process. To see this, note that

$$i(E_\beta - E_\alpha + i\epsilon)^{-1} - i(E_\beta - E_\alpha - i\epsilon)^{-1} = \int_{-\infty}^\infty dt \langle \beta | e^{-i\hat{H}t - \epsilon|t|} | \alpha \rangle,$$

which means that a finite value of ϵ corresponds to the substitution of the time evolution operator with $f(t)e^{-i\hat{H}t}$, where $f(t) = e^{-\epsilon|t|}$. The function $f(t)$ implements the effects of a finite duration $T = \int_{-\infty}^\infty dt f(t)$ of time evolution. For $f(t) = e^{-\epsilon|t|}$, $T = 2/\epsilon$. Hence $\delta_\epsilon(0) = (\pi\epsilon)^{-1} = T/(2\pi)$.

Then, at the limit $\epsilon \to 0$,

$$\frac{|S_{\beta\alpha}|^2}{T} = 2\pi \delta(E_\beta - E_\alpha) |T_{\beta\alpha}|^2; \tag{19.131}$$

that is, the ratio $\frac{|S_{\beta\alpha}|^2}{T}$ converges to $W(\alpha \to \beta)$, where

$$W(\alpha \to \beta) = 2\pi \delta(E_\beta - E_\alpha) |T_{\beta\alpha}|^2. \tag{19.132}$$

The quantity $W(\alpha \to \beta)$ is an effective transition rate from α to β. Since $|\alpha\rangle_0$ is not a square integrable vector, $W(\alpha \to \beta)$ is still divergent. To obtain a finite expression we have to divide by a suitably regularized expression for the norm $_0\langle\alpha|\alpha\rangle_0$, and then take the appropriate limit – see Example 19.6. The best way to do so is by taking into account additional symmetries like momentum conservation. Furthermore, $W(\alpha \to \beta)$ is a density with respect to β, that is, it must be integrated appropriately with respect to β in order to define meaningful probabilities.

Equation (19.131) suggests that the transition rate is time-independent. In Section 19.1, we saw that the stability of the transition rate is encapsulated in Fermi's golden rule, which was similarly "derived". The identification of $W(\alpha \to \beta)$ as a constant rate in this section is not as bad mathematically as the derivation of Fermi's golden rule, but it still involves some mathematical trickery in the way that we introduced the notion of the total duration of the interaction. In effect, the decay rate $W(\alpha \to \beta)$ is forced to be constant because it is averaged over a large time T.

In Section 18.3.5, we saw that the constancy of the decay rate (i.e., exponential decay) is an excellent approximation that, nonetheless, fails at very early and very late times. The argument in Section 18.3.5 that the decay rate cannot be constant also applies to the S-matrix. Now, the very early time regime is irrelevant to scattering experiments, as particles have not yet reached the locus of scattering. In the opposite regime, that is, when detection happens so late in time that the dispersion has completely delocalized the wave packets, we also do not expect constant decay rates. This is why the limits $t \overset{sc}{\to} \pm\infty$ do not coincide with the mathematical limits $t \to \infty$. So far, no scattering experiment has explored the regime where time variations of the transition rate could be detected.

Example 19.5   For a particle in one dimension, the transition operator has matrix elements $T_{pp'}$ in the momentum basis. Equation (19.124) implies that

$$T_{pp'} = \frac{1}{\sqrt{2\pi}} \int e^{-ipx} V(x) \psi_{p'}^+(x).$$

Since energy is preserved, $|p| = |p'| = \sqrt{2mE} = k$; hence, for given k, the possible values of $T_{pp'}$ depend only on the sign of the moment p and p'. Therefore, we rewrite the matrix elements $T_{pp'}$ as $T_{ab}(k)$, where a and b take either value $+$ or value $-$. Comparing with Eq. (19.109), we obtain

$$\begin{pmatrix} T_{++}(k) & T_{+-}(k) \\ T_{-+}(k) & T_{--}(k) \end{pmatrix} = \frac{ik}{2\pi m} \begin{pmatrix} T_k - 1 & \bar{R}_k \\ R_k & \bar{T}_k - 1 \end{pmatrix}. \tag{19.133}$$

We also express the elements of the S-matrix as $S_{ab}(k,k')$ with $a,b = +,-$. Then, Eq. (19.128) yields

$$S(k,k') = \delta(k - k') \begin{pmatrix} T_k & \bar{R}_k \\ R_k & \bar{T}_k \end{pmatrix}. \tag{19.134}$$

Example 19.6   For a particle in three dimensions, the kets $|\alpha\rangle_0$ coincide with the generalized eigenstates $|\mathbf{p}\rangle$ of momentum. Then, Eq. (19.124) yields

$$T_{\mathbf{p'p}} = \frac{1}{(2\pi)^{3/2}} \int d^3x e^{-i\mathbf{p'}\cdot\mathbf{x}} V(\mathbf{x}) \psi_{\mathbf{p}}^+(\mathbf{x}). \tag{19.135}$$

Comparing with Eq. (19.113), we find that

$$T_{\mathbf{p'p}} = -\frac{1}{4\pi^2 m} f_{\mathbf{p}}(\mathbf{n'}), \tag{19.136}$$

where $|\mathbf{p}| = |\mathbf{p'}|$ and $\mathbf{n'} = \mathbf{p'}/|\mathbf{p}|$.

Hence, the S-matrix elements are fully determined by the scattering amplitude,

$$S_{\mathbf{p'p}} = \delta^3(\mathbf{p'} - \mathbf{p}) + \frac{i}{2\pi p} \delta(p - p') f_{\mathbf{p}}(\mathbf{n'}). \tag{19.137}$$

By Eq. (19.132), $W(\mathbf{p} \to \mathbf{p'}) = \frac{1}{mp}\delta(p - p')\frac{|f_{\mathbf{p}}(\mathbf{n'})|^2}{(2\pi)^3}$. To render this quantity finite, we divide by $\langle \mathbf{p}|\mathbf{p}\rangle = \delta^3(0)$. Using arguments similar to the ones that led to Eq. (19.132), we identify $\delta^3(0)$ with

$\frac{V}{(2\pi)^3}$, where V is the volume spanned by the incoming beam. Multiplying with the duration T of the interaction, we obtain a probability density with respect to $\mathbf{p}'$,

$$P(\mathbf{p}') = W(\mathbf{p} \to \mathbf{p}')\frac{T}{\delta^3(0)} = \frac{T}{mpV}\delta(p - p')|f_{\mathbf{p}}(\mathbf{n}')|^2. \tag{19.138}$$

The probability density $P(\mathbf{n})$ for the momentum direction $\mathbf{n}$ of outcoming particles is

$$P(\mathbf{n}) = \int_0^\infty p'^2 dp' P(\mathbf{p}') = \frac{T}{V}\frac{p}{m}|f_{\mathbf{p}}(\mathbf{n}')|^2. \tag{19.139}$$

The volume spanned by a particle beam of cross-section A and velocity $v = p/m$ is ATp/m. Hence, we recover $P(\mathbf{n}) = |f_{\mathbf{p}}(\mathbf{n}')|^2/A$, that is, Eq. (19.15) for the differential cross-section.

The reader is well justified feeling uneasy with this derivation, as it involved quite a bit of mathematical abuse in the way we handled infinite quantities. Nonetheless, there is a method in this madness, and this method can be generalized to more complex processes and lead to meaningful results. We will present some examples in Section 19.6.3. The use of such informal manipulations provides a useful shortcut in expressing measurable quantities like the differential cross-section in terms of elements of the S-matrix, when compared with the rigorous but tedious analysis that involves wave packets as in Section 19.1. In fact, informal manipulations work even in set-ups where a rigorous analysis is missing.

For sufficiently weak interactions, we can evaluate the S-matrix perturbatively, by keeping a finite number of terms in the expansion (19.120) of the transition operator. To lowest order in $\hat{V}$, we only keep the leading term, whence,

$$T_{\beta\alpha} = V_{\beta\alpha}. \tag{19.140}$$

This approximation is known as the *generalized Born approximation*. Indeed, for a particle in a potential, Eq. (19.136) in this approximation leads to Eqs. (19.43) and (19.46). Note that the transition rates in the generalized Born approximation

$$W(\alpha \to \beta) = 2\pi\,\delta(E_\beta - E_\alpha)|V_{\beta\alpha}|^2, \tag{19.141}$$

coincide with the expressions from Fermi's golden rule and the Wigner–Weisskopf approximation for decays – see Eq. (18.44). The generalized Born approximation is better at high energies, where we expect that the effect of scattering on the incoming particles is small.

19.6.3 Symmetries of the S-Matrix

The most important property of the S-matrix is that it commutes with the Hamiltonian $\hat{H}_0$,

$$[\hat{H}_0, \hat{S}] = 0. \tag{19.142}$$

To see this, we write the matrix elements of the commutator in the eigenbasis of $\hat{H}_0$: $_0\langle\beta|\hat{H}_0\hat{S} - \hat{S}\hat{H}_0|\alpha\rangle_0 = (E_\beta - E_\alpha)S_{\beta\alpha}$. By Eq. (19.128), these matrix elements vanish.

However, the S-matrix fails to commute with the total Hamiltonian $\hat{H}$. Energy conservation in Eq. (19.128) refers to the kinetic energy of the scattered particles, as given $\hat{H}_0$. It does not refer to the total energy that corresponds to $\hat{H}$.

Consider a self-adjoint operator $\hat{X}$ such that $[\hat{H}_0, \hat{X}] = [\hat{H}, \hat{X}] = 0$. It follows that $\hat{X}$ also commutes with the Heisenberg-picture operators $\hat{V}(t) = e^{i\hat{H}t}\hat{V}e^{-i\hat{H}t}$. By Eq. (19.119),

$$[\hat{X}, \hat{S}] = 0. \tag{19.143}$$

Hence, dynamical symmetries that are common to $\hat{H}$ and to $\hat{H}_0$ leave the S-matrix invariant. Since $[\hat{H}_0, \hat{X}] = 0$, we can always choose the basis $|\alpha\rangle_0$ to consist of common eigenvectors of $\hat{H}_0$ and $\hat{X}$, so that $\hat{X}|\alpha\rangle_0 = X_\alpha|\alpha\rangle_0$. Then, by Eq. (19.143),

$$0 = {}_0\langle\beta|\hat{X}\hat{S} - \hat{S}\hat{X}|\alpha\rangle_0 = (X_\beta - X_\alpha)S_{\beta\alpha}. \tag{19.144}$$

It follows that $S_{\beta\alpha}$ vanishes if $X_\beta \neq X_\alpha$. Scattering processes that fail to preserve X_α are impossible.

The Hamiltonians $\hat{H}_0$ and $\hat{H}$ are generators of time translations for different representations of the Poincaré group, or of the Galilei group for nonrelativistic systems. It is often the case that the two representations share the same set of generators $\hat{P}_i$ for spatial translations. Hence, both $\hat{H}_0$ and $\hat{H}$ commute with $\hat{P}_i$, which corresponds to the total momentum of the system. This means that momentum is preserved in scattering. By Eq. (19.128), the matrix elements $T_{\beta\alpha}$ of the transition operator are proportional to the delta function $\delta^3(\mathbf{P}_\beta - \mathbf{P}_\alpha)$,

$$T_{\beta\alpha} = \mathcal{M}_{\beta\alpha}(2\pi)^3\delta^3(\mathbf{P}_\beta - \mathbf{P}_\alpha); \tag{19.145}$$

the quantities $\mathcal{M}_{\beta\alpha}$ are referred to as *scattering amplitudes*.

When we calculate $|S_{\beta\alpha}|^2$ the delta function for momenta is squared. We follow the same procedure as in the derivation of Eq. (19.131). We introduce approximate delta functions $\delta_\epsilon^3(\mathbf{P}_\beta - \mathbf{P}_\alpha)$, so that $[\delta_\epsilon^3(\mathbf{P}_\beta - \mathbf{P}_\alpha)]^2 = \delta_\epsilon^3(0)\delta^3(\mathbf{P}_\beta - \mathbf{P}_\alpha)$. Finally, we identify $\delta^3(0)$ with $\frac{V}{(2\pi)^3}$, where V is the volume within which the scattering takes place. Hence, for $\beta \neq \alpha$,

$$\frac{|S_{\beta\alpha}|^2}{TV} = (2\pi)^4\delta(E_\beta - E_\alpha)\delta^3(\mathbf{P}_\beta - \mathbf{P}_\alpha)|\mathcal{M}_{\beta\alpha}|^2. \tag{19.146}$$

Therefore, we define a scattering rate per unit time and unit volume $w(\alpha \to \beta)$,

$$w(\alpha \to \beta) = (2\pi)^4|\mathcal{M}_{\beta\alpha}|^2, \tag{19.147}$$

that is a density with respect to β.

Suppose that both $\hat{H}_0$ and $\hat{H}$ are invariant under time reversal $\hat{\mathbb{T}}$. The kets $|\alpha\rangle_0$ describe momenta and spin. We choose the convention (11.67) for the action of $\hat{\mathbb{T}}$ on spin. Then, $\hat{\mathbb{T}}|\alpha\rangle_0 = \sigma|\alpha^*\rangle_0$, where $|\alpha^*\rangle$ is a different eigenvector of $\hat{H}_0$ and $\sigma = \pm 1$. The condition $[\hat{S}, \hat{\mathbb{T}}] = 0$ implies that $\hat{\mathbb{T}}^\dagger\hat{S}\hat{\mathbb{T}} = \hat{S}$, hence

$$S_{\beta\alpha} = {}_0\langle\beta|\hat{\mathbb{T}}^\dagger\hat{S}\hat{\mathbb{T}}|\alpha\rangle_0 = \sigma^2{}_0\langle\beta^*|\hat{S}|\alpha^*\rangle_0 = S_{\beta^*\alpha^*} \tag{19.148}$$

For example, for a particle without spin in a potential $\hat{\mathbb{T}}|\mathbf{p}\rangle = |-\mathbf{p}\rangle$, so Eq. (19.148) yields $S_{\mathbf{p}\mathbf{p}'} = S_{-\mathbf{p},-\mathbf{p}'}$ or, equivalently, $f_\mathbf{p}(\mathbf{n}) = f_{-\mathbf{p}}(-\mathbf{n})$.

Similarly, for a space-inversion operator acting as $\hat{\mathbb{P}}|\alpha\rangle_0 = u|\alpha^P\rangle_0$, for $u = \pm 1$, the S-matrix elements satisfy

$$S_{\beta\alpha} = S_{\beta^P\alpha^P}. \tag{19.149}$$

Example 19.7 Consider particle scattering by a central potential. In analogy with the analysis of Section 11.3.1, we split the Hilbert space $L^2(\mathbb{R}, d^3p)$ of the system as $\mathcal{H}_p \otimes \mathcal{H}_S^2$, where $\mathcal{H}_p = L^2(\mathbb{R}, p^2 dp)$ is a Hilbert space associated to the momentum norm $p = |\mathbf{p}|$, and $\mathcal{H}_{S^2}$ is the Hilbert space of functions on the unit sphere, where the spherical harmonics are defined simply erase the phrase.

The S-matrix commutes with the angular momentum $\hat{\mathbf{L}}$, so it is constant on the subspaces of constant ℓ. Since the S-matrix is also diagonal with respect to p, we can express it as

$$S_{\mathbf{p'p}} = \frac{1}{p^2}\delta(p - p')\sum_\ell S_\ell(p)\langle\mathbf{n'}|\hat{E}_\ell|\mathbf{n}\rangle,$$

where $\hat{E}_\ell$ is the projector on the subspace of constant ℓ in $\mathcal{H}_{S^2}$, $\mathbf{n} = \mathbf{p}/p$, and $\mathbf{n'} = \mathbf{p'}/p'$. Here, $\frac{1}{p^2}\delta(p - p')$ are the matrix elements of the unit operator in $\mathcal{H}_p$ on the $|p\rangle$ basis. We can also decompose the delta function as

$$\delta^3(\mathbf{p}, \mathbf{p'}) = \frac{1}{p^2}\delta(p - p')\delta^2(\mathbf{n}, \mathbf{n'}) = \frac{1}{p^2}\delta(p - p')\sum_\ell\langle\mathbf{n'}|\hat{E}_\ell|\mathbf{n}\rangle.$$

By Eq. (11.120), the scattering amplitude $f_p(\mathbf{n} \cdot \mathbf{n'})$ can be expressed as

$$f_p(\mathbf{n} \cdot \mathbf{n'}) = \sum_{ell}(2\ell + 1)f_\ell(p)P_\ell(\mathbf{n} \cdot \mathbf{n'}) = 4\pi\sum_\ell f_\ell(p)\langle\mathbf{n'}|\hat{E}_\ell|\mathbf{n}\rangle.$$

Then, Eq. (19.137) implies that $S_\ell(p) = 1 + 2ipf_\ell(p)$. By Eq. (19.60), we obtain

$$S_\ell(p) = e^{2i\delta_\ell(p)}, \tag{19.150}$$

that is, the phase shifts define the S-matrix.

QUESTIONS

19.1 A beam of electrons with width of $1\ \mu$m and velocity 10^5 m/s is directed towards a target. How far from the target must we place the source and the detector, so that the assumptions in scattering formalism fail? What is the analogous condition for the velocity of the electrons if the distance of the target from source and detector is 1 m?

19.2 Explain why there is no forward scattering by a hard-sphere potential in classical physics. Then argue why this behavior is not acceptable in quantum theory because of the optical theorem. Can this explain the discrepancy between classical and quantum hard-sphere scattering at high energies?

19.3 What is the physical interpretation of the scattering length a_s being negative?

19.4 Is it possible to identify resonances in the Born approximation?

19.5 A resonance at energy E_R is observed in the potential scattering of a particle of mass m. The total cross-section at the peak of the resonance is σ_{max}. What is the orbital angular momentum of the resonance?

19.6 (**B**) Is the energy delta function in the S-matrix a manifestation of energy conservation?

PROBLEMS

19.1 Study the scattering of identical particles at energies so low that only the partial waves with $\ell = 0, 1$ contribute. Plot the form of the differential cross-section as a function of the scattering angle θ for (i) particles with zero spin, (ii) electrons with total spin $S = 1$, and (iii) electrons in a state of total ignorance for spin.

19.2 Fill in all steps in the derivation of Eq. (19.36).

19.3 (i) Show that form factors $F_{n,\ell,m_\ell}(k)$ for the hydrogen atom are given by

$$F_{n,\ell,m_\ell}(k) = i^\ell \sqrt{2\ell + 1} \int_0^\infty j_\ell(kr) R_{n\ell}(r) R_{10}(r),$$

where $R_{n\ell}$ are the radial eigenfunctions of the hydrogen atom Hamiltonian. (ii) Evaluate the form factors for elastic scattering and for inelastic scattering with maximum energy loss.

19.4 Consider the inelastic scattering of a particle of charge λe from an atom of atomic number Z. For large energies of the incoming particle, the momentum $\mathbf{k}'_a$ of the outcoming particle is approximately a-independent. (i) Show that, in this regime, the total inelastic cross-section $\sigma_{in} = \sum_{a \neq 0} \int d^2 n \sigma_a(\mathbf{n})$ for an atom takes the form $\sigma_{in} = \frac{4 m_e^2 \lambda^2 \alpha^2}{q^2} G(\mathbf{q})$, where

$$G(\mathbf{q}) = \left(\langle 0 | \hat{s}_{\mathbf{q}}^\dagger \hat{s}_{\mathbf{q}} | 0 \rangle - |\langle 0 | \hat{s}_{\mathbf{q}} | 0 \rangle|^2 \right),$$

where $\hat{s}_{\mathbf{q}} = \sum_{\nu=1}^Z e^{-i\mathbf{q} \cdot \hat{\mathbf{r}}_\nu}$. (ii) Show that in a mean-field theory description of the atom, $G(\mathbf{q}) = Z[1 - |F_0(\mathbf{q})/2$.

Comment The total inelastic scattering cross-section increases with Z, in distinction to the elastic cross-section which increases with Z^2.

19.5 (i) Calculate the phase shift δ_0 for a particle of mass m scattering off a potential $V(r) = V_0 a \delta(r - a)$, where $V_0, a > 0$. (ii) Evaluate the scattering length and the effective range for this potential. (iii) Determine the conditions under which δ_0 is approximately equal to the phase shift $\delta_0^{(0)}$ of a hard sphere of the same radius a.

19.6 The classical limit of scattering corresponds to particles with large momentum and macroscopically large values of angular momentum ℓ, so that the scattering amplitude $f_k(\theta) = \frac{1}{2ik} \sum_{\ell=0}^\infty (2\ell + 1)(e^{2i\delta_\ell} - 1) P_\ell(\cos \theta)$ can be approximated by an integral over ℓ. (i) For large ℓ, the approximation $P_\ell(\cos \theta) \simeq J_0[(2\ell + 1)\sin \frac{1}{2}\theta]$ holds. Show that

$$f_k(\theta) = -ik \int_0^\infty db\, b (e^{2i\delta(b)} - 1) J_0(qb),$$

where $b = \ell/k$ and $q = 2k \sin \frac{\theta}{2}$. (ii) What is the physical interpretation of b in classical scattering? (iii) In high energies, δ_ℓ is approximately ℓ-independent and equal to $\Delta(k)$

for $\ell \le \ell_{max}$, and then vanishes for $\ell > \ell_{max}$. Use the identity $x^n J_{n-1}(x) = [x^n J_n(x)]'$ for the Bessel functions, in order to evaluate the scattering amplitude. (iv) Show that the differential cross-section is peaked around values of θ of order $\frac{1}{\ell_{max}}$ or smaller.

19.7 Consider scattering on a lattice with spacing a. The scattering potential is of the form $V(\mathbf{x}) = \sum_a \upsilon(\mathbf{x} - \mathbf{x}_a)$, where $\mathbf{x}_a$ are the lattice points labeled by a, and $\upsilon(\mathbf{x})$ is a function peaked around 0 that, vanishes for $\mathbf{x} \ge a$. (i) Show that, in the Born approximation, the scattering amplitude is $-\frac{m}{2\pi}\tilde{\upsilon}(q)F(\mathbf{q})$, where $F(\mathbf{q}) = \sum_{a,b} e^{i\mathbf{q}\cdot(\mathbf{x}_a - \mathbf{x}_b)}$ is the form factor of the lattice, and $\tilde{\upsilon}$ is the Fourier transform of $\upsilon(\mathbf{x})$. (ii) Evaluate $F(\mathbf{q})$ for a square lattice with N^3 points and spacing d. (iii) Show that, if $kd \ll 1$, even small fluctuations in energy cause the lattice points to contribute additively to $\sigma(\mathbf{n})$, that is, σ to scale with N^3.

19.8 Consider a screened Coulomb potential

$$V(r) = \begin{cases} -\frac{g}{r}, & r \le R \\ 0 & r > R \end{cases},$$

that coincides with the Coulomb potential as $R \to \infty$. (i) Calculate the differential cross-section $\sigma(\theta)$ in the Born approximation. (ii) Take the limit $R \to \infty$ in the differential cross-section. To do so, first integrate $\sigma(\theta)$ over a finite range of the scattering angle θ (or equivalently, q) and then set $R \to \infty$. Show that the resulting cross-section differs from that of the Coulomb potential by a multiplicative factor of $\frac{3}{2}$. (iii) What is the physical origin of this difference?

19.9 Show that for a particle in the spherical-well potential of Eq. (19.77),

$$\delta_0(k) = k[-a + K_0^{-1}\tan(K_0 a)]$$

as $k \to 0$, where $K_0 = \sqrt{2mV_0}$. This means that the effective range is negative for some values of K_0 and it even diverges for $K_0 a = (2n+1)\frac{\pi}{2}$ for $n = 0, 1, 2, \ldots$. In the latter case, the linear approximation to $\delta_0(k)$ fails. (ii) Show that a_s diverges when a bound state of zero energy exists. (iii) Show that, for $K_0 a = \frac{\pi}{2}$, $|\delta_0(0)| = \frac{\pi}{2}$, hence, the $\ell = 0$ contribution to the total cross-section diverges as $k \to 0$.

19.10 Two particles with spins $\hat{S}_1$ and $\hat{S}_2$ interact through a term $\hat{V} = V_0(|\hat{\mathbf{x}}_1 - \hat{\mathbf{x}}_2|) \otimes I + V_1(|\hat{\mathbf{x}}_1 - \hat{\mathbf{x}}_2|) \otimes \hat{S}_1 \cdot \hat{S}_2$. Find the scattering cross-section in the Born approximation for two particles with spin $s = \frac{1}{2}$ and potentials $V_0(r) = -V_1(r) = -V_0$ for $r < a$, and $V_0(r) = 0$, $V_1(r) = V_0(a/r)^6$ for $r > a$. Assume a total-ignorance state for spins.

19.11 **(B)** Consider an interaction term in the Hamiltonian of the form $\langle \alpha|\hat{V}|\beta\rangle = f^*(\alpha)f(\beta)$, for some function f. (i) Solve the Lippman–Schwinger equation for the in states. (ii) Evaluate the S-matrix.

19.12 **(B)** (i) Write the Lippman-Schwinger equation for the potential scattering of particles with kinetic energy $\hat{H}_0 = c|\hat{\mathbf{p}}|$, where c is a constant. (ii) Evaluate the matrix elements of the resolvent $G_0(E^+)$ in the position basis. (iii) Derive the analogue of Eq. (19.113) for the scattering amplitude.

19.13 **(B)** Show that Eq. (19.130) reduces to the optical theorem, Eq. (19.26), for potential scattering.

19.14 **(B)** (i) Approximate the S-matrix by keeping terms up to second order in $\hat{V}$ in the expansion (19.120). (ii) Show that we obtain the same expression by keeping terms up to second order in Eq. (19.119).

19.15 **(B)** Show that three of the four equations (5.62) follow from the unitarity of the S-matrix, and that the fourth follows from time-reversal invariance.

Bibliography

- For further details in scattering theory, the reader should consult specialized monographs. I recommend the introductory text by Rodberg and Thaler (1967) or the more advanced treatment by Newton (1982).

20 Open Quantum Systems

Markov's scheme for extending the laws of probability beyond the realm of independent variables has one crucial restriction: The probabilities must depend only on the present state of the system, not on its earlier history. The Markovian analysis of Monopoly, for example, considers a player's current position on the board but not how he or she got there. This limitation is serious. After all, life presents itself as a long sequence of contingent events – kingdoms are lost for want of a nail, hurricanes are spawned by butterflies in Amazonia – but these causal chains extending into the distant past are not Markov chains.

Hayes (2013)

20.1 Basic Notions

One of the most important concepts of classical mechanics is that of a closed system. A closed system is loosely defined as a system whose components interact only with each other, and it is characterized by phase space volume conservation and energy conservation – see Section 1.2.

We have to be careful when employing the term "closed system" in quantum theory. Since all predictions necessarily make reference to measurements, no physical system studied by quantum theory is truly closed, as it must couple to a measurement apparatus. If it is possible to ignore the measurement apparatuses, we can transfer the classical definition of closed systems to quantum ones. Then, by FP4, the time evolution of the system in absence of measurements is given by a one-parameter family of unitary transformations through a Hamiltonian.

A better definition is that in a closed system we can perform measurements on the full set of degrees of freedom that impact on dynamics, while in an open system, some degrees of freedom are unavailable. There are two reasons for this unavailability. First, it may be practically impossible to keep track of all degrees of freedom in a closed system. This is the case, for example, in many-body systems, where we typically measure only a few collective variables and not observables associated with individual particles. Second, our measured system S may be in contact with an environment E that is outside our control. For example, an isolated atom may be in contact with radiation that is generated by the apparatus that confines it; a particle beam may be affected by random scattering of ambient photons.

The two cases above are physically distinct, but mathematically identical. In the second case, we may work at the level of the closed system $S + E$, and assume that only the degrees of freedom in S are available for measurement. Hence, we can always describe open quantum systems as closed systems to which we have limited access.

In this section, we introduce the most important notions of open quantum systems, also presenting an elementary example.

20.1.1 The Level of Description

Suppose we have access to a set L of degrees of freedom that is a small subset of the set B of all observables in a closed quantum system. The set L is said to define our *level of description* or our *coarse-graining*, or simply our *system*. The remaining degrees of freedom in B are called the *environment*. Our aim is to write evolution equations on L that are (i) autonomous, in the sense that they can be solved with initial conditions solely on L, and (ii) compatible with the dynamics of the closed quantum system. From these equations, we will derive probabilities for measurements on L.

Let the closed quantum system be described by a Hamiltonian $\hat{H}$. In general, Hamiltonian evolution does not leave L invariant, that is, it causes system–environment interactions. The Hamiltonian is typically of the form $\hat{H} = \hat{H}_0 + \hat{V}$, where $\hat{H}_0$ leaves L invariant and $\hat{V}$ does not. The term $\hat{V}$ often splits as $\sum_a \hat{V}_a$, where each component $\hat{V}_a$ is characterized by different coupling constants. Then, a is said to label the different *interaction channels* between system and environment.

Each interaction channel is characterized by three processes. First, energy is exchanged between system and environment. Usually, we have *dissipation*, that is, the system loses energy to the environment. Second, the degrees of freedom of system and environment become entangled. If the system starts in a pure state, its reduced density matrix will become mixed. This often results to the suppression of macroscopic interferences, a phenomenon known as *environment-induced decoherence* – see Section 10.5.2. Third, the action of the environment on the system is not temporally coherent, and it is expressed as *noise*.

20.1.2 An Open System with a Classical Environment

Next, we present a simple open quantum system that interacts with an environment that is modeled by classical probability theory. Suppose that the system is a particle in one dimension characterized by a Hamiltonian $\hat{H}$. We assume that the environment acts upon the system as follows. At each time step δt, there is a probability w_a that the system gains momentum equal to a from its interaction with the environment, where a can range over the whole real line. In this case, the state $\hat{\rho}$ of the system transforms as $\hat{\rho} \rightarrow \hat{U}_a \hat{\rho} \hat{U}_a^\dagger$, where $\hat{U}_a = e^{ia\hat{x}}$ is the Weyl operator implementing a momentum translation.

Hence, the density matrix $\hat{\rho}(t)$ at time t transforms to

$$\hat{\rho}(t + \delta t) = \sum_a w_a \hat{U}_a e^{-i\hat{H}\delta t} \hat{\rho}(t) e^{i\hat{H}\delta t} \hat{U}_a^\dagger. \tag{20.1}$$

We take the limit $\delta t \rightarrow 0$, keeping in mind that the probabilities for momentum transfer become increasingly smaller as the available time δt decreases. Therefore, we expand $\hat{U}_a$ as a series in a, so that

$$\hat{U}_a \hat{\rho} \hat{U}_a^\dagger = (\hat{I} + ia\hat{x} - \frac{a^2}{2}\hat{x}^2 + \cdots)\hat{\rho}(\hat{I} - ia\hat{x} - \frac{a^2}{2}\hat{x}^2 + \cdots)$$

$$= \hat{\rho} + ia[\hat{x}, \hat{\rho}] - \frac{a^2}{2}(\hat{x}^2\hat{\rho} + \hat{\rho}\hat{x}^2 - 2\hat{x}\hat{\rho}\hat{x}) + O(a^3).$$

The continuum limit is nontrivial if we assume that the moments of the probability density w_a scale as follows with δt:

- $\langle a \rangle = 0$;
- $\langle a^2 \rangle = D\delta t$, for some $D > 0$;
- for $n > 2$, $\langle a^n \rangle \sim (\delta t)^{b_n}$, where $b_n > 1$.

Then,

$$\frac{d\hat{\rho}}{dt} = -i[\hat{H}, \hat{\rho}] - \frac{1}{2}D[\hat{x}, [\hat{x}, \hat{\rho}]]. \tag{20.2}$$

The expectations of $\hat{x}$ and $\hat{p}$ are not affected by the new term, which originates from the second moment of a. However, the variances are affected. We find that

$$\frac{d}{dt}(\Delta x)^2 = \frac{d}{dt}(\Delta x)_c^2, \quad \frac{d}{dt}(\Delta p)^2 = \frac{d}{dt}(\Delta p)_c^2 + 2D, \tag{20.3}$$

where the suffix c (for "closed") refers to the value of this quantity in absence of the environment. Integrating, we find that $(\Delta x)^2(t) = (\Delta x)_c^2(t)$ and that $(\Delta p)^2(t) = (\Delta p)_c^2(t) + 2Dt$.

For a free particle of mass m, $\hat{H} = \frac{\hat{p}^2}{2m}$. Then, $(\Delta \hat{p})_c^2$ is constant, and $(\Delta \hat{p})^2(t) = (\Delta \hat{p})_c^2 + 2Dt$. We can always find $t < 0$ such that $(\Delta p)^2(t) < 0$; hence, the solution of Eq. (20.2) for negative t does not preserve the positivity of the density matrix. Therefore, it is physically inadmissible. On the other hand, there is no problem with positivity for $t > 0$. The evolution equation (20.2) makes sense for evolution only forward in time, hence, it is *irreversible*.

Evolution through Eq. (20.2) turns pure states into mixed states, and, more generally, it decreases the purity of the quantum state. We calculate

$$\frac{d}{dt}\text{Tr}\hat{\rho}^2 = 2\text{Tr}\left(\hat{\rho}\frac{d\hat{\rho}}{dt}\right) = -2D\,\text{Tr}\left(\hat{\rho}^2\hat{x}^2 - \hat{\rho}\hat{x}\hat{\rho}\hat{x}\right) \leq 0, \tag{20.4}$$

where in the last step we used the inequality $\text{Tr}\hat{A}^2 \leq \text{Tr}(\hat{A}^\dagger\hat{A})$ for $\hat{A} = \hat{\rho}\hat{x}$. This inequality is obtained by setting $\hat{B} = \hat{A}^\dagger$ in inequality (4.67).

It is instructive to solve Eq. (20.2) for a vanishing Hamiltonian. In the position representation, Eq. (20.2) becomes

$$\frac{d}{dt}\rho(x, x'; t) = -\frac{D}{2}(x - x')^2\rho(x, x'; t), \tag{20.5}$$

with solution $\hat{\rho}(x, x'; t) = e^{-\frac{D}{2}(x-x')^2 t}\rho(x, x'; 0)$. The environment suppresses the off-diagonal elements of the density matrix in the position basis. This is a special case of environment-induced decoherence. A superposition of localized states whose centers are separated by distance L will become effectively mixed at times $t \gg (DL^2)^{-1}$.

Irreversibility, noise, and decrease of purity are generic characteristics of open quantum systems. Environment-induced decoherence with respect to a physically meaningful basis is common but not universal.

20.2 Time Evolution in Open Quantum Systems

Let us denote by $B(\mathcal{H})$ the set of all bounded operators on the Hilbert space $\mathcal{H}$. $B(\mathcal{H})$ is a vector space that contains all density matrices, so the evolution equation (20.2) is of the form

$$\frac{d}{dt}\hat{\rho} = \mathcal{L}[\hat{\rho}], \tag{20.6}$$

where $\mathcal{L}$ is a *superoperator*, namely, a linear operator that acts on $B(\mathcal{H})$; that is, it maps operators on $\mathcal{H}$ (such as $\hat{\rho}_0$) to operators on $\mathcal{H}$. The prefix "super" expresses that $\mathcal{L}$ is an operator on the space of operators.

The solution to the evolution equation is

$$\hat{\rho}(t) = e^{\mathcal{L}t}[\hat{\rho}_0], \tag{20.7}$$

where now we treat $\hat{\rho}$ as a vector on $B(\mathcal{H})$. As long as the superoperator $e^{\mathcal{L}t}$ takes density matrices to density matrices for $t > 0$, it defines a physically admissible evolution equation for an open quantum system. Furthermore, $e^{\mathcal{L}t_1}e^{\mathcal{L}(t_2-t_1)} = e^{\mathcal{L}(t_2)}$ for all $t_2 > , t_1 > 0$; hence, we write $\hat{\rho}(t_2) = e^{\mathcal{L}(t_2-t_1)}\hat{\rho}(t_1)$. This means that we can construct $\hat{\rho}(t_2)$ only from the knowledge of the density matrix at any prior time t_1. An evolution law for the density matrix that satisfies this property is called *Markovian*.

We will see that the requirement of Markovian time evolution and preservation of positivity allows us to write a general form for physically acceptable superoperators $\mathcal{L}$.

20.2.1 Choi's Theorem and Kraus Operators

Consider a quantum system described by a Hilbert space $\mathcal{H} = \mathbb{C}^N$. We choose an orthornormal basis $|i\rangle$, $i = 1, 2, \ldots, N$ on $\mathcal{H}$, so that operators are represented as matrices A_{ij}. A superoperator $\mathcal{V}$ that takes a density matrix $\hat{\rho}$ to a density matrix $\hat{\rho}' = \mathcal{V}[\hat{\rho}]$ can be expressed as $\rho'_{ij} = \sum_{k,l=1}^{N} V_{ik,jl}\rho_{kl}$, for some coefficients $V_{ik,jl}$. Since $\hat{\rho}'$, is self-adjoint, $\sum_{k,l=1}^{\infty}(V_{ik,jl} - V^*_{jl,ik})\rho_{kl} = 0$. This equation holds for all density matrices ρ_{kl}, and this is possible only if

$$(V_{ik,jl} - V^*_{jl,ik}) = 0. \tag{20.8}$$

We substitute the pair of indices ik with A and jl with B, where all capital indices take N^2 different values $(11, 12, 13, \ldots, 1N, 21, \ldots 2N, \ldots, N1, \ldots, NN)$. Then, Eq. (20.8) reads $V_{AB} = V^*_{BA}$, i.e., V is a $N^2 \times N^2$ self-adjoint matrix. By the spectral theorem,

$$V_{AB} = \sum_{a=1}^{N^2} \lambda_a C_A^a C_B^{a*}, \tag{20.9}$$

where $a = 1, 2, \ldots, N^2$ labels the real eigenvalues λ_a of V_{AB}, and C_A^a are the associated eigenvectors. The latter satisfy the orthonormalization condition $\sum_A C_A^a C_A^{b*} = \delta_{ab}$.

Switching back to the small indices, we write $\rho'_{ij} = \sum_{a=1}^{N^2} \lambda_a \sum_{k,l=1}^{\infty} C^a_{ik} \rho_{kl} C^a_{jl}$ or, equivalently,

$$\hat{\rho}' = \sum_{a=1}^{N^2} \lambda_a \hat{C}^a \hat{\rho} \hat{C}^{a\dagger}, \tag{20.10}$$

where $\hat{C}^a = \sum_{i,j=1}^{\infty} C^a_{ij} |i\rangle \langle j|$. In this notation, the orthonormalization condition becomes

$$\mathrm{Tr}(\hat{C}^a \hat{C}^{b\dagger}) = \delta_{ab}. \tag{20.11}$$

The normalization condition $\mathrm{Tr}\hat{\rho}' = 1$ implies that $\mathrm{Tr}\left[(\sum_a \lambda_a \hat{C}^{a\dagger} \hat{C}^a - \hat{I})\hat{\rho}\right] = 0$. Since this condition holds for all $\hat{\rho}$,

$$\sum_a \lambda_a \hat{C}^{a\dagger} \hat{C}^a = \hat{I}. \tag{20.12}$$

By taking the trace of both sides in this equation, we find that $\sum_a \lambda_a = N$.

Note that the identity superoperator $\mathcal{I}$ has matrix elements $\mathcal{I}_{ik,jl} = \delta_{ik}\delta_{jl}$, that is, it factorizes with respect to the capital indices A and B. Hence, it has a single nonzero eigenvalue, which equals N. Then, by Eq. (20.11) the corresponding eigenvector is $N^{-1/2}\delta_{ij}$.

The final requirement is that the superoperator $\mathcal{V}$ is positive, i.e., that the operators $\hat{\rho}'$ given by Eq. (20.10) are true density matrices for all $\hat{\rho}$. For any $|\phi\rangle \in \mathcal{H}$,

$$\langle\phi|\hat{\rho}'|\phi\rangle = \sum_a \lambda_a \langle\phi_a|\hat{\rho}|\phi_a\rangle, \tag{20.13}$$

where $|\phi_a\rangle := \hat{C}^{a\dagger}|\phi\rangle$. Evidently, if we require that $\lambda_a \geq 0$ for all $a = 1, 2, \ldots, N^2$, then Positivity is guaranteed. However, the converse does not hold: Positivity holds even if some λ_a are negative. This can be seen in the following example.

Choose $\mathcal{H} = \mathbb{C}^2$, and $\hat{C}^i = \frac{1}{\sqrt{2}}\hat{\sigma}^i$ for $i = 1, 2, 3$, and $\hat{C}^4 = \frac{1}{\sqrt{2}}\hat{I}$. It is straightforward to verify that these matrices satisfy the orthogonality condition. We choose $\lambda_1 = \lambda_2 = -\lambda_3 = \lambda_4 = 1$. Equation (20.12) is satisfied and

$$\hat{\rho}' = \frac{1}{2}\left(\hat{\sigma}_1\hat{\rho}\hat{\sigma}_1 + \hat{\sigma}_2\hat{\rho}\hat{\sigma}_2 - \hat{\sigma}_3\hat{\rho}\hat{\sigma}_3 + \hat{\rho}\right) = \begin{pmatrix} \rho_{22} & \rho_{12} \\ \rho_{21} & \rho_{11} \end{pmatrix}, \tag{20.14}$$

that is, $\hat{\rho}'$ is just $\hat{\rho}$ with its diagonal elements exchanged, hence, $\hat{\rho}' \geq 0$.

We need a stronger condition than positivity, in order to guarantee the convenient property that all $\lambda_a \geq 0$. This condition is known as *complete positivity*, and it is defined as follows. Let $\mathcal{V}$ be a positive superoperator on the Hilbert space $\mathcal{H}$, with elements $V_{ik,jl}$ with respect to some basis $|i\rangle$. We define the superoperator $\mathcal{V} \otimes \mathcal{I}$ on the Hilbert space $\mathcal{H} \otimes \mathcal{H}$ that describes two copies of this system by

$$(\mathcal{V} \otimes \mathcal{I})[\hat{\rho}] = \sum_a \lambda_a (\hat{C}^a \otimes \hat{I})\hat{\rho}(\hat{C}^{a\dagger} \otimes \hat{I}), \tag{20.15}$$

where $\hat{\rho}$ is now a density matrix on $\mathcal{H} \otimes \mathcal{H}$. If $\mathcal{V} \otimes I$ is positive, then we call $\mathcal{V}$ a completely positive (CP) map. In other words, a positive map is completely positive, if its trivial extension to a composite system remains positive.

Theorem 20.1 (Choi (1975)) *If $\mathcal{V}$ is a CP map on $\mathcal{H}$, then all eigenvalues λ_a in Eq. (20.12) are positive.*

Proof For any operator $\hat{D}$ on $\mathcal{H}$, we define a vector $|\hat{D}\rangle := \sum_{ij} \hat{D}_{ij} |i,j\rangle \in \mathcal{H} \otimes \mathcal{H}$ – see Problem 9.12 – and the associated density matrix $\hat{\rho}_D := |\hat{D}\rangle\langle\hat{D}|$. Since $\mathcal{V}$ is a CP map, $\langle \hat{F} | (\mathcal{V} \otimes \mathcal{I})[\hat{\rho}_D] | \hat{F}\rangle \geq 0$, where $\hat{F}$ is an arbitrary operator on $\mathcal{H}$. By Eq. (20.15),

$$\langle \hat{F} | (\mathcal{V} \otimes \mathcal{I})[\hat{\rho}_D] | \hat{F}\rangle = \sum_a \lambda_a |\langle \hat{D} | \hat{C}^a \otimes \hat{I} | \hat{F}\rangle|^2 = \sum_a \lambda_a |\mathrm{Tr}(\hat{C}^a \hat{D}\hat{F})|^2.$$

We choose $\hat{F} = \hat{I}$ and $\hat{D} = \hat{C}^{b\dagger}$, for any b. Then, $\langle \hat{F}|(\mathcal{V} \otimes I)[\hat{\rho}_D]|\hat{F}\rangle = \sum_a \lambda_a |\delta_{ab}|^2 = \lambda_b$. Hence, if $\mathcal{V}$ is a CP map, all eigenvalues λ_b are positive.

Given a CP-map $\mathcal{V}$, we define $\hat{K}_a := \sqrt{\lambda_a} \hat{C}^a$, and we bring $\mathcal{V}$ into its *Kraus form*,

$$\mathcal{V}[\hat{\rho}] = \sum_{a=1}^{N^2} \hat{K}_a \hat{\rho} \hat{K}_a^\dagger, \tag{20.16}$$

where the so-called *Kraus operators* $\hat{K}_a$ satisfy

$$\sum_{a=1}^{N^2} \hat{K}_a \hat{K}_a^\dagger = \hat{I}. \tag{20.17}$$

It is straightforward to see that unitary time evolution and the change of the state after a measurement, Eq. (8.19), are particular cases of Eq. (20.16). Note that if one Kraus operator is unitary, then Eq. (20.17) implies that all other Kraus operators vanish.

The Kraus generators of a CP map $\mathcal{V}$ are unique only if the matrix V_{AB} of Eq. (20.9) is nondegenerate. Otherwise there is an infinity of bases C_A^a from eigenvectors of V_{AB}, and, consequently, an infinity of possible Kraus decompositions.

20.2.2 The GKLS Master Equation

Suppose that an open quantum system evolves by a family of CP maps $\mathcal{V}_t$, for $t >\geq 0$, that satisfy the Markovian condition $\mathcal{V}_\tau \mathcal{V}_t = \mathcal{V}_{t+\tau}$, for $t,\tau \geq 0$, and $\mathcal{V}_0 = \mathcal{I}$. One of the most important theorems in the theory of open systems is that the density matrix at time t, $\hat{\rho}(t) = \mathcal{V}_t[\hat{\rho}_0]$, obtained from evolution of an initial density matrix $\hat{\rho}_0$, satisfy the Gorini–Kossakowski–Lindblad–Sudarshan (GKLS) equation

$$\frac{d\hat{\rho}(t)}{dt} = -i[\hat{H}, \hat{\rho}(t)] - \frac{1}{2}\sum_\alpha \left(\hat{L}_\alpha^\dagger \hat{L}_\alpha \hat{\rho}(t) + \hat{\rho}(t)\hat{L}_\alpha^\dagger \hat{L}_\alpha - 2\hat{L}_\alpha \hat{\rho}(t)\hat{L}_\alpha^\dagger \right), \tag{20.18}$$

where $\hat{H}$ is the Hamiltonian of the system, and the operators $\hat{L}_\alpha$ are known as *Lindblad generators*, as they generate the nonunitary dynamics. The GKLS equation is proven in Box 20.1.

Equation (20.18) is the most general Markovian evolution equation. Following early work by Kossakowski (1972), the GKLS equation was proven at the same time by Gorini et al. (1976) for a finite-dimensional Hilbert space and by Lindblad (1976) for a separable Hilbert space with the values of α ranging over of countable set. Each Lindblad generator corresponds to a different interaction channel with the environment. For a system described by the Hilbert space $\mathbb{C}^N$, the master equation involves at most $N^2 - 1$ Lindblad operators.

Equation (20.18) describes time evolution of an open quantum system in the Schrödinger picture. We can also work in the Heisenberg picture, where the operators evolve in time and the state is constant. By requiring that all expectations $\text{Tr}[\hat{\rho}(t)\hat{A}]$ are the same in both pictures, we obtain the evolution equation for a general observable $\hat{A}(t)$,

$$\frac{d\hat{A}(t)}{dt} = i[\hat{H}, \hat{A}(t)] - \frac{1}{2}\sum_\alpha \left(\hat{A}(t)\hat{L}_\alpha^\dagger \hat{L}_\alpha + \hat{L}_\alpha^\dagger \hat{L}_\alpha \hat{A}(t) - 2\hat{L}_\alpha^\dagger \hat{A}(t)\hat{L}_\alpha\right). \tag{20.19}$$

Note that GKLS dynamics is linear with respect to the operators, that is, $(\lambda\hat{A} + \hat{B})(t) = \lambda\hat{A}(t) + \hat{B}(t)$ for any operators $\hat{A}$ and $\hat{B}$, and for any $\lambda \in C$. It also preserves the unit operator $\hat{I}$, in the sense that $\frac{d}{dt}\hat{I}(t) = 0$, and also preserves the adjoint operation, namely, $\hat{A}^\dagger(t) = [\hat{A}(t)]^\dagger$. However, it does not preserve operator multiplication, that is, for general operators $\hat{A}$ and $\hat{B}$, $(\hat{A}\hat{B})(t) \neq \hat{A}(t)\hat{B}(t)$.

Box 20.1   Proof of the GKLS Equation

Consider a family of CP maps $\mathcal{V}_t$, for $t \geq 0$, that satisfy the Markovian condition $\mathcal{V}_\tau\mathcal{V}_t = \mathcal{V}_{t+\tau}$, for $t, \tau \geq 0$, and $\mathcal{V}_0 = \mathcal{I}$. Let $\hat{\rho}(t) = \mathcal{V}_t[\hat{\rho}]$ for some density matrix $\hat{\rho}$. By definition,

$$\hat{\rho}(t + \tau) = \mathcal{V}_\tau[\hat{\rho}(t)] = \sum_a \lambda_a(\tau)\hat{C}^a(\tau)\hat{\rho}(t)\hat{C}^{a\dagger}(\tau), \tag{A}$$

where $\hat{C}^a(\tau)$ and $\lambda_a(\tau)$ satisfy Eq. (20.12).

For $\tau \to 0$, $\mathcal{V}_\tau$ becomes the identity superoperator $\mathcal{I}$, whose single nonzero eigenvalue is N, with corresponding eigenvector $\frac{1}{\sqrt{N}}\delta_{ij}$. Of all eigenvalues $\lambda_a(\tau)$ of $\mathcal{V}_\tau$, we denote by $\lambda_0(\tau)$ the unique eigenvalue that satisfies $\lambda_0(0) = N$ and $\hat{C}^0(0) = N^{-1/2}\hat{I}$. We label the remaining $N^2 - 1$ eigenvalues and eigenvectors of $\mathcal{V}(\tau)$ by the Greek index α; by definition $\lambda_\alpha(0) = 0$.

We assume that $\mathcal{V}(\tau)$ is differentiable at $\tau = 0$. To leading order in τ,

$$\lambda_0(\tau) = N(1 - b_0\tau), \quad \hat{C}^0(\tau) = \frac{1}{\sqrt{N}}(\hat{I} + \hat{A}\tau), \quad \lambda_\alpha(\tau) = b_\alpha\tau, \quad \hat{C}^\alpha(\tau) = \hat{C}^\alpha, \tag{B}$$

for some constants $b_0, b_\alpha \geq 0$, and operators $\hat{A}, \hat{C}^\alpha$. We keep only zeroth-order terms in $\hat{C}^\alpha$, because $\hat{C}^\alpha(\tau)$ is multiplied with τ through $\lambda_\alpha(\tau)$, hence, higher-order terms are negligible as $\tau \to 0$.

Equation (20.11) constrains the operators $\hat{A}$ and $\hat{C}^\alpha$,

$$\text{Tr}(\hat{A} + \hat{A}^\dagger) = 0, \quad \text{Tr}\,\hat{C}^\alpha = 0, \quad \text{Tr}(\hat{C}^\alpha\hat{C}^{\dagger\beta}) = \delta_{\alpha\beta}. \tag{C}$$

We obtain another constraint by substituting Eq. (B) into Eq. (20.12),

$$b_0\hat{I} = \hat{A} + \hat{A}^\dagger + \sum_\alpha b_\alpha \hat{C}^{\alpha\dagger}\hat{C}^\alpha.$$

Finally, substituting Eq. (C) into Eq. (A), and taking terms to lowest order in τ, we obtain

$$d\hat{\rho}(t)/dt = -b_0\hat{\rho}(t) + \hat{A}\hat{\rho}(t) + \hat{\rho}(t)\hat{A}^\dagger - \sum_\alpha b_\alpha \hat{C}^\alpha \hat{\rho}(t)\hat{C}^{\alpha\dagger}$$

$$= [\hat{A} - \hat{A}^\dagger, \hat{\rho}(t)] + \frac{1}{2}\sum_\alpha b_\alpha \left(\hat{C}^{\alpha\dagger}\hat{C}^\alpha \hat{\rho}(t) + \hat{\rho}(t)\hat{C}^{\alpha\dagger}\hat{C}^\alpha - 2\hat{C}^\alpha \hat{\rho}(t)\hat{C}^{\alpha\dagger}\right), \quad (D)$$

Equation (C) applies only for operators $\hat{A}$ and $\hat{C}^\alpha$ that satisfy Eq. (C). However, it is invariant under a transformation $\hat{C}^\alpha \to \hat{C}^{\alpha\prime}, \hat{A} \to \hat{A}', b_\alpha \to b'_\alpha$, where

$$\hat{C}^{\alpha\prime} = s_\alpha \sum_\beta U_{\alpha\beta}\hat{C}^\beta + k^\alpha \hat{I}, \quad \hat{A}' = \hat{A} + \sum_{\alpha,\beta} s_\alpha k^{\alpha*} U_{\alpha\beta}\hat{C}^\beta, \quad b'_\alpha = b_\alpha/s_a^2.$$

Here, $\hat{U}_{\alpha\beta}$ is a unitary matrix, $k^a \in \mathbb{C}$ and $s_\alpha \in \mathbb{R}$. The transformed operators $\hat{A}'$ and $\hat{C}^{\alpha\prime}$ do not satisfy Eq. (C). This means that we have the freedom to work with Eq. (D) while ignoring the constraints (C).

We write $\hat{A} - \hat{A}^\dagger = -i\hat{H}$, where $\hat{H}$ is a self-adjoint operator, the system's Hamiltonian. We also define the *Lindblad operators* $\hat{L}_\alpha := \sqrt{b_\alpha}\hat{C}^\alpha$, thus obtaining Eq. (20.18).

20.2.3 Examples

Example 20.1 Equation (20.2) is a GKLS master equation on $L^2(\mathbb{R})$, with a single Lindblad generator $\hat{L} = \sqrt{D}\hat{x}$.

Example 20.2 Consider an open quantum system with a single self-adjoint Lindblad generator $\hat{L}$, that commutes with the Hamiltonian. Let us denote by $|n\rangle$, a common eigenbasis for both $\hat{H}$ and $\hat{L}$, with corresponding eigenvalues E_n and L_n, respectively. In this basis, Eq. (20.18) becomes

$$\frac{d\hat{\rho}_{mn}}{dt} = -i(E_m - E_n)\hat{\rho}_{mn} - \frac{1}{2}(L_m - L_n)^2 \hat{\rho}_{mn},$$

with solution $\hat{\rho}_{mn}(t) = \hat{\rho}_{mn}(0)e^{-i(E_m-E_n)t - \frac{1}{2}(L_m-L_n)^2 t}$.

Off-diagonal elements of $\hat{\rho}_{mn}$ with $L_m \neq L_n$ are suppressed exponentially in time, with a characteristic *decoherence* timescale $\tau_{dec} \sim |L_m - L_n|^{-1/2}$. The environment causes an effective diagonalization of the density matrix in the eigenspaces of $\hat{L}$, another case of environment-induced decoherence.

Example 20.3 The most general GKLS master equation for a qubit involves three Lindblad operators, which we can choose as $\hat{L}_1 = \sqrt{\gamma_-}\hat{\sigma}_-$, $\hat{L}_2 = \sqrt{\gamma_+}\hat{\sigma}_+$ and $\hat{L}_3 = \sqrt{\gamma_0}\hat{\sigma}_3$, for some coefficients $\hat{\gamma}_\pm, \gamma_0 > 0$, with physical meaning that will be made clear by the solution to the GKLS equation. We also select the Hamiltonian $\hat{H} = \frac{1}{2}\omega\hat{\sigma}_3$.

Working in the Heisenberg picture, we find

$$\frac{d\hat{\sigma}_3(t)}{dt} = -(\gamma_+ + \gamma_-)\hat{\sigma}_3(t) + (\gamma_+ - \gamma_-)\hat{I}, \tag{20.20}$$

$$\frac{d\hat{\sigma}_-(t)}{dt} = -i\omega\hat{\sigma}_- - \frac{1}{2}(\gamma_+ + \gamma_- + \gamma_0)\hat{\sigma}_-, \tag{20.21}$$

with solutions

$$\hat{\sigma}_3(t) = \hat{\sigma}_3(0)e^{-(\gamma_+ + \gamma_-)t} + \frac{\gamma_+ - \gamma_-}{\gamma_+ + \gamma_-}(1 - e^{-(\gamma_+ + \gamma_-)t}),\tag{20.22}$$

$$\hat{\sigma}_-(t) = \hat{\sigma}_-(0)e^{-i\omega t - \frac{1}{2}(\gamma_+ + \gamma_- + \gamma_0)t}.\tag{20.23}$$

Equations (20.20) and (20.21) are known as *Bloch equations*. They have wide applicability in atom optics (atomic qubits) and in nuclear magnetism.

From Eqs. (20.22) and (20.23), we calculate the Bloch vector $\langle \hat{\sigma}(t) \rangle$ as a function of time. Substituting into Eq. (6.22), we obtain the Schrödinger-picture evolution of a general initial quantum state

$$\hat{\rho}(t) = \begin{pmatrix} [\hat{\rho}_{11}(0) - \frac{\gamma_+}{\gamma_+ + \gamma_-}]e^{-(\gamma_+ + \gamma_-)t} + \frac{\gamma_+}{\gamma_+ + \gamma_-} & \hat{\rho}_{10}(0)e^{-i\omega t - \frac{1}{2}(\gamma_+ + \gamma_- + \gamma_0)t} \\ \hat{\rho}_{01}(0)e^{i\omega t - \frac{1}{2}(\gamma_+ + \gamma_- + \gamma_0)t} & [\hat{\rho}_{00}(0) - \frac{\gamma_-}{\gamma_+ + \gamma_-}]e^{-(\gamma_+ + \gamma_-)t} + \frac{\gamma_-}{\gamma_+ + \gamma_-} \end{pmatrix}.$$

Taking $\gamma_+ = \gamma_0 = 0$, we see that $\hat{L}_1$ drives transitions $|1\rangle \rightarrow |0\rangle$, while suppressing the off-diagonal matrix elements. In a two-level atom, $\hat{L}_1$ corresponds to spontaneous decay, and γ_- is the decay rate. Similarly, we find that $\hat{L}_2$ drives transitions $|0\rangle \rightarrow |1\rangle$, while suppressing the off-diagonal matrix elements; it corresponds to pumping of the two-level atom, and γ_+ is the pumping rate. The Lindblad operator $\hat{L}_3$ does not cause transitions, it only effects decoherence in the energy basis with a characteristic decoherence timescale $\tau_{dec} = \gamma_0^{-1}$.

The system has a unique asymptotic state

$$\hat{\rho}(\infty) = \begin{pmatrix} \frac{\gamma_+}{\gamma_+ + \gamma_-} & 0 \\ 0 & \frac{\gamma_-}{\gamma_+ + \gamma_-} \end{pmatrix}.\tag{20.24}$$

Example 20.4 Consider a harmonic oscillator of frequency ω in contact with an environment with two Lindblad operators $\sqrt{\gamma_-}\hat{a}$ and $\sqrt{\gamma_+}\hat{a}^\dagger$, where $\gamma_\pm \geq 0$. The first operator describes energy loss from the oscillator and the second operator describes energy gain. This model describes the motion of a particle in a harmonic oscillator potential in contact with a dissipative environment, and also the time evolution of an EM field mode of frequency ω in an imperfect cavity.

We work in the Heisenberg picture. The evolution equation of the annihilation operator $\hat{a}$ is

$$\frac{d}{dt}\hat{a}(t) = -i\omega\hat{a}(t) - \frac{1}{2}(\gamma_- - \gamma_+)\hat{a}(t),\tag{20.25}$$

with solution $\hat{a}(t) = e^{-i\omega t - \frac{1}{2}(\gamma_- - \gamma_+)t}\hat{a}(0)$. The environment is dissipative if $\gamma_- > \gamma_+$, so that the oscillation amplitude $\langle \hat{a}(t) \rangle$ decays exponentially to zero.

We also find the evolution equation for the number operator $\hat{N} = \hat{a}^\dagger\hat{a}$,

$$\frac{d}{dt}\hat{N}(t) = -(\gamma_- - \gamma_+)\hat{N}(t) + \gamma_+\hat{I},\tag{20.26}$$

with solution $\hat{N}(t) = e^{-(\gamma_- - \gamma_+)t}\hat{N}(0) + \frac{\gamma_+}{\gamma_- - \gamma_+}[1 - e^{-(\gamma_- - \gamma_+)t}]\hat{I}$. The average number of quanta $\langle \hat{N}(t) \rangle$ converges to a constant value $\gamma_+/(\gamma_- - \gamma_+)$ at long times. Note that $\hat{N}(t) \neq \hat{a}^\dagger(t)\hat{a}(t)$.

Finally, we find the evolution equation for $\hat{A} = \hat{a}^2$,

$$\frac{d}{dt}\hat{A}(t) = -2i\omega\hat{A}(t) - (\gamma_- - \gamma_+)\hat{A}(t),\tag{20.27}$$

with solution $\hat{A}(t) = e^{-2i\omega t - (\gamma_- - \gamma_+)t}\hat{A}(0)$. In this case, $\hat{A}(t) = [\hat{a}(t)]^2$.

20.3 The Second-Order Master Equation

In what follows, we present explicit quantum models for the environment of an open system. For sufficiently weak interaction of the system to the environment, there is a regime where the system is described by a GKLS-type equation, the Lindblad operators determined by the interaction Hamiltonian of the system to an environment. This equation is known as the *second-order master equation*, and it is universal, in the sense that it applies, in principle, to *any* open quantum system, provided that its domain of validity applies. The derivation of this master equation is the topic of this section.

Consider a quantum system described by a Hilbert space $\mathcal{H}_S$, in contact with an environment described by a Hilbert space $\mathcal{H}_E$. The Hamiltonian $\hat{H}$ for the combined system is of the form $\hat{H}_0 + \hat{V}$, where $\hat{H}_0 = \hat{H}_S \otimes \hat{I} + \hat{I} \otimes \hat{H}_E$ involves the self-Hamiltonians $\hat{H}_S$ of the system and $\hat{H}_E$ of the environment; $\hat{V}$ is a weak interaction term, typically of the form

$$\hat{V} = \lambda \sum_a \hat{A}_a \otimes \hat{B}_a, \tag{20.28}$$

where $\hat{A}_a$ are self-adjoint operators on $\mathcal{H}_S$, $\hat{B}_a$ are self-adjoint operators on $\mathcal{H}_E$, and $\lambda << 1$ is a small dimensionless constant.

As in Section 18.1, we define the operator $\hat{O}_t = e^{i\hat{H}_0 t} e^{-i\hat{H}t}$, which satisfies Eq. (18.1). Let $\hat{\rho}(0)$ be the initial density matrix of the total system (at $t = 0$). We define the *interaction-picture* density matrix[1] $\hat{\rho}_I(t) := \hat{O}_t \hat{\rho}(0) \hat{O}_t^\dagger$. It is straightforward to show that

$$\frac{d\hat{\rho}_I(t)}{dt} = i[\hat{V}(t), \hat{\rho}_I(t)], \tag{20.29}$$

where $\hat{V}(t) = e^{i\hat{H}_0 t} \hat{V} e^{-i\hat{H}_0 t} = \lambda \sum_a \hat{A}_a(t) \otimes \hat{B}_a(t)$, with $\hat{A}_a(t) = e^{i\hat{H}_S t} \hat{A}_a e^{-i\hat{H}_S t}$ and $\hat{B}_a(t) = e^{i\hat{H}_E t} \hat{B}_a e^{-i\hat{H}_E t}$.

We integrate Eq. (20.29), to obtain $\hat{\rho}_I(t) = \hat{\rho}_I(0) + i\int_0^t dt' [\hat{V}(t'), \hat{\rho}_I(t')]$. We substitute $\hat{\rho}_I(t)$ back in Eq. (20.29),

$$\frac{d\hat{\rho}_I(t)}{dt} = i[\hat{V}(t), \hat{\rho}(0)] - \int_0^t dt' [\hat{V}(t), [\hat{V}(t'), \hat{\rho}_I(t')]].$$

Next, we trace out the degrees of freedom of the environment. First, we define the reduced density matrix for the system $\hat{\rho}_S(t) := \mathrm{Tr}_E \hat{\rho}_I(t)$. Then, we assume that the initial state is factorized, $\hat{\rho}_I(0) = \hat{\rho}_S(0) \otimes \hat{\rho}_E(0)$, and that the initial state of the environment is stationary with respect to $\hat{H}_E$, $[\hat{\rho}_E(0), \hat{H}_E] = 0$.

We obtain $\mathrm{Tr}_E[\hat{V}(t), \hat{\rho}(0)] = \sum_a [\hat{A}_a(t), \hat{\rho}_S(0)] \langle \hat{B}_a \rangle$, where

$$\langle \hat{B}_a \rangle = \mathrm{Tr}[\hat{\rho}_E(0) \hat{B}_a(t)] = \mathrm{Tr}[\hat{\rho}_E(0) \hat{B}_a(t)]$$

is time independent, because of the stationarity of $\hat{\rho}_E(0)$. We can always choose $\langle \hat{B}_a \rangle = 0$, by redefining $\hat{B}_a' = \hat{B}_a - \langle \hat{B}_a \rangle$, and absorbing a term $\sum_a \langle \hat{B}_a \rangle \hat{A}_a$ into the system self-Hamiltonian $\hat{H}_S$. Hence, $\mathrm{Tr}_E[\hat{V}(t), \hat{\rho}(0)] = 0$. We conclude that

---

[1] The term "interaction picture" is used for the representation of quantum dynamics where the observables evolve with an unperturbed Hamiltonian $\hat{H}_0$ and the states with the perturbation $\hat{V}$.

$$\frac{d\hat{\rho}_S(t)}{dt} = -\int_0^t dt' \mathrm{Tr}_E\left([\hat{V}(t), [\hat{V}(t'), \hat{\rho}_I(t')]]\right). \tag{20.30}$$

Equation (20.30) is exact; we have made no approximation in its derivation. For further progress, an approximation is needed. We assume that the environment is much larger than the system, so that its state is not affected by time evolution, and that, furthermore, the coupling is so weak that no significant correlations develop between system and environment. Hence, we approximate

$$\hat{\rho}_I(t) \simeq \hat{\rho}_S(t) \otimes \hat{\rho}_E(0). \tag{20.31}$$

Equation (20.31) is known as the *Born approximation*. It is analogous to the Born approximation for scattering that was presented in Section 19.2.

In the Born approximation, Eq. (20.30) becomes

$$\frac{d\hat{\rho}_S(t)}{dt} = -\sum_{a,b} \int_0^t dt' \Big(F_{ab}(t - t')[\hat{A}_a(t)\hat{A}_b(t')\hat{\rho}_S(t') - \hat{A}_b(t')\,\hat{\rho}_S(t')\hat{A}_a(t)$$
$$+ F_{ab}(t' - t)[\hat{\rho}_S(t')\hat{A}_a(t')\hat{A}_b(t) - \hat{A}_b(t)\hat{\rho}_S(t')\hat{A}_a(t')] \Big), \tag{20.32}$$

where we defined the correlation function

$$F_{ab}(s) := \lambda^2 Tr[\hat{\rho}_E(0)\hat{B}_a(t + s)\hat{B}_b(t)] = F_{ba}^*(-s). \tag{20.33}$$

Note that the correlation function depends only on s and not on t, because of the stationarity of the initial state of the environment. Corrections to Eq. (20.31) are of order λ^2; hence, Eq. (20.34) is exact to order λ^2. This is why we refer to the resulting master equation as "second-order".

Typically, $F_{ab}(s)$ is characterized by a "memory" timescale τ_{mem}, such that $|F_{ab}(s)|$ is negligible for $s \gg \tau_{mem}$. We use the term "memory" because the integral in (20.34) involves the values of $\hat{\rho}_S(t)$ at times t' earlier than t. Suppose that the characteristic dynamical timescales of the system, as identified in $\hat{H}_S$, are much larger than τ_{mem}. Then, we can invoke the *Markov approximation*, by which we substitute $\hat{\rho}_S(t') \simeq \hat{\rho}_S(t)$ in Eq. (20.34). Within the same approximation, we also extend the range of integration over t' to $(-\infty, t)$. Then, we change the integration variable to $s = t - t'$, to obtain

$$\frac{d\hat{\rho}_S(t)}{dt} = -\sum_{a,b} \int_0^\infty ds \Big(F_{ab}(s)[\hat{A}_a(t)\hat{A}_b(t - s)\hat{\rho}_S(t) - \hat{A}_b(t - s)\,\hat{\rho}_S(t)\hat{A}_a(t)$$
$$+ F_{ab}(-s)[\hat{\rho}_S(t)\hat{A}_a(t - s)\hat{A}_b(t) - \hat{A}_b(t)\hat{\rho}_S(t)\hat{A}_a(t - s)] \Big). \tag{20.34}$$

Let $\hat{H}_S = \sum_\epsilon \epsilon \hat{P}_\epsilon$, where ϵ are the eigenvalues of $\hat{H}_S$, and $\hat{P}_\epsilon$ are the associated spectral projectors. The identity $\hat{A}_a = \sum_{\epsilon, \epsilon'} \hat{P}_\epsilon \hat{A}_a \hat{P}_{\epsilon'}$ implies that

$$\hat{A}_a(t) = \sum_{\epsilon, \epsilon'} e^{i\hat{H}_S t}\hat{P}_\epsilon \hat{A}_a \hat{P}_{\epsilon'} e^{-i\hat{H}_S t} = \sum_{\epsilon, \epsilon'} e^{i(\epsilon - \epsilon')t}\hat{P}_\epsilon \hat{A}_a \hat{P}_{\epsilon'}.$$

Setting $\omega = \epsilon' - \epsilon$, we write $\hat{A}_a(t) = \sum_\omega \hat{A}_a(\omega)e^{-i\omega t}$, where

$$\hat{A}_a(\omega) = \sum_\epsilon \hat{P}_\epsilon \hat{A}_a \hat{P}_{\epsilon + \omega}, \tag{20.35}$$

is an operator Fourier transform for $\hat{A}_a(t)$. It is straightforward to show that $\hat{A}_a^\dagger(\omega) = \hat{A}_a(-\omega)$, and that $[\hat{H}_S, \hat{A}_a(\omega)] = -\omega\hat{A}_a(\omega)$.

Substituting into Eq. (20.34), we obtain

$$\frac{d\hat{\rho}_S(t)}{dt} = -\sum_{a,b}\sum_{\omega,\omega'}\Big[\tilde{F}_{ab}(\omega')e^{i(\omega'-\omega)t}[\hat{A}_a(\omega)\hat{A}_b^\dagger(\omega')\hat{\rho}_S(t) - \hat{A}_b^\dagger(\omega')\hat{\rho}_S(t)\hat{A}_a(\omega)]$$

$$+ \tilde{F}_{ba}^*(\omega')e^{-i(\omega'-\omega)t}[\hat{\rho}_S(t)\hat{A}_a(\omega')\hat{A}_b^\dagger(\omega) - \hat{A}_b^\dagger(\omega)\hat{\rho}_S(t)\hat{A}_a(\omega')]\Big], \quad (20.36)$$

where $\tilde{F}_{ab}(\omega) = \int_0^\infty ds\, e^{-i\omega s}F_{ab}(s)$ is the Laplace transform of $F_{ab}(s)$ at imaginary argument $i\omega$.

The final approximation is a variation of the rotating wave approximation (RWA) that we first introduced in Section 7.4.3. Suppose that $\hat{H}_S$ has discrete spectrum, and that δ is the smallest nonzero value of the difference $|\omega - \omega'|$. For example, in a harmonic oscillator of frequency ω, $\delta = \omega$. Then, all oscillatory terms with $\omega \neq \omega'$ are strongly suppressed when averaged at timescales much larger than δ. The RWA consists in ignoring those terms, and thus, writing Eq. (20.36) as

$$\frac{d\hat{\rho}_S(t)}{dt} = -\sum_{a,b}\sum_{\omega}\Big[\tilde{F}_{ab}(\omega)[\hat{A}_a(\omega)\hat{A}_b^\dagger(\omega)\hat{\rho}_S(t) - \hat{A}_b^\dagger(\omega)\hat{\rho}_S(t)\hat{A}_a(\omega)]$$

$$+ \tilde{F}_{ba}^*(\omega)[\hat{\rho}_S(t)\hat{A}_a(\omega)\hat{A}_b^\dagger(\omega) - \hat{A}_b^\dagger(\omega)\hat{\rho}_S(t)\hat{A}_a(\omega)]\Big]. \quad (20.37)$$

Defining

$$\Gamma_{ab}(\omega) := \tilde{F}_{ab}(\omega) + \tilde{F}_{ba}^*(\omega), \qquad \Delta_{ab}(\omega) := \frac{1}{2i}\Big[\tilde{F}_{ab}(\omega) - \tilde{F}_{ba}^*(\omega)\Big], \quad (20.38)$$

we express Eq. (20.37) as

$$\frac{d\hat{\rho}_S(t)}{dt} = -i\Big[\hat{H}_{LS}, \hat{\rho}_S(t)\Big] + \mathcal{D}[\hat{\rho}_S(t)] \quad (20.39)$$

where

$$\mathcal{D}[\hat{\rho}] := -\frac{1}{2}\sum_{a,b}\sum_{\omega}\Gamma_{ab}(\omega)\Big[\hat{A}_a(\omega)\hat{A}_b^\dagger(\omega)\hat{\rho} + \hat{\rho}\hat{A}_a(\omega)\hat{A}_b^\dagger(\omega) - 2\hat{A}_b^\dagger(\omega)\hat{\rho}\hat{A}_a(\omega)\Big],$$

and $\hat{H}_{LS} := \sum_\omega\sum_{a,b}\Delta_{ab}(\omega)\hat{A}_a^\dagger\hat{A}_b$ is a correction to the system Hamiltonian. It is straightforward to show that $[\hat{H}_{LS}, \hat{H}_S] = 0$, so $\hat{H}_{LS}$ only causes a correction in the energy levels of $\hat{H}_S$. These corrections are generically characterized as Lamb shifts – see Section 18.3.1.

We switch to the Schrödinger-picture reduced density matrix $\hat{\rho}(t) := e^{-i\hat{H}_S t}\hat{\rho}_S(t)\, e^{i\hat{H}_S t}$, to obtain the *second-order master equation*,

$$\frac{d\hat{\rho}_t}{dt} = -i\Big[\hat{H}_S + \hat{H}_{LS}, \hat{\rho}(t)\Big] + \mathcal{D}\Big[\hat{\rho}(t)\Big], \quad (20.40)$$

that determines the evolution of a quantum system under the influence of a quantum environment.

In the Heisenberg picture, an operator $\hat{C}$ evolves as

$$\frac{d\hat{C}}{dt} = i\Big[\hat{H}_S + \hat{H}_{LS}, \hat{C}\Big] + \mathcal{D}^*\Big[\hat{C}\Big], \quad (20.41)$$

where

$$\mathcal{D}^*[\hat{C}] := -\frac{1}{2}\sum_{a,b}\sum_{\omega}\Gamma_{ab}(\omega)\left[\hat{C}\hat{A}_a(\omega)\hat{A}_b^\dagger(\omega) + \hat{A}_a(\omega)\hat{A}_b^\dagger(\omega)\hat{C} - 2\hat{A}_a(\omega)\hat{C}\hat{A}_b^\dagger(\omega)\right].$$

We express the matrix $\Gamma_{ab}(\omega)$ as $\sum_c \gamma_c(\omega)u_{ca}(\omega)u_{cb}^*(\omega)$, in terms of its eigenvalues $\gamma_c(\omega)$, and the corresponding eigenvectors $u_{ca}(\omega)$. Since $\Gamma_{ab}(\omega)$ is positive – see Problem 20.4 – $\gamma_c(\omega) \geq 0$, and we can define

$$\hat{L}_c(\omega) := \sqrt{\gamma_c(\omega)}\sum_a u_{ca}(\omega)\hat{A}_a^\dagger(\omega). \tag{20.42}$$

Then, $\mathcal{D}$ takes the GKLS form (20.18),

$$\mathcal{D}[\hat{\rho}] = -\frac{1}{2}\sum_c\sum_\omega\left[\hat{L}_c^\dagger(\omega)\hat{L}_c(\omega)\hat{\rho} + \hat{\rho}\hat{L}_c^\dagger(\omega)\hat{L}_c(\omega) - 2\hat{L}_c(\omega)\hat{\rho}\hat{L}_c^\dagger(\omega)\right]. \tag{20.43}$$

We obtained Eq. (20.40) as a result of three successive approximations: (i) the Born approximation, which asserts that the environment is large compared to the system, (ii) the Markov approximation, which asserts that the memory time of the environment is short, and (iii) the RWA which averages away all fast oscillations in the evolution equation of the system. Approximations (ii) and (iii) require that the time t is sufficiently large, so we do not expect Eq. (20.40) to hold for very early times.

An interesting property of the second-order master equation is that, for nondegenerate Hamiltonians $\hat{H}_S$, the diagonal elements of the density matrix in the energy basis decouple from the off- diagonal ones. To see this, let us denote by $|n\rangle$ the orthornormal basis of eigenvectors of $\hat{H}_S$, and by ϵ_n the associated eigenvalues. We define $P_n(t) := \langle n|\hat{\rho}_t|n\rangle$, and taking the expectation of Eq. (20.41), we find

$$\dot{P}_n = \sum_n(w_{nm}P_m - w_{mn}P_n), \tag{20.44}$$

where

$$w_{n,m} := \sum_{a,b}\Gamma_{ab}(\epsilon_n - \epsilon_m)\langle m|\hat{A}_a|n\rangle\langle n|\hat{A}_b|m\rangle \tag{20.45}$$

is the transition rate from $|m\rangle$ to $|n\rangle$. Equation (20.44) is known as the *Pauli master equation*.

20.4 Applications of the Second-Order Master Equation

In what follows, we will use the second-order master equation in order to study open-system dynamics in many systems of interest. First, we present a simple model for the environment which has found many applications, and then we describe the open system dynamics of particles and of qubits.

20.4.1 Bosonic Reservoirs

Most models assume that the environment is in a state of thermal equilibrium at temperature T. It is described by the Gibbs state $\hat{\rho}_E = Z^{-1}(T)e^{-\hat{H}_E/T}$, where $Z(T) := \mathrm{Tr}e^{-\hat{H}_E/T}$.

The most common model for the environment is the *bosonic reservoir model*, which treats the environment as a collection of independent harmonic oscillators in a thermal state. This model works well in a broad set of problems, ranging from condensed matter to atomic physics to cosmology. The reservoir oscillators may correspond to the EM field, to other quantum fields, or to quantum collective excitations of atoms in a lattice.

We assume that the environment consists of N harmonic oscillators of different masses m_i and frequencies ω_i, where $i = 1, 2, \ldots, N$. We denote each canonical pair of coordinates by $\hat{q}_i$ and $\hat{p}_i$. Then, the Hamiltonian is

$$\hat{H}_E = \sum_{i=1}^{N} \left(\frac{1}{2m_i}\hat{p}_i^2 + \frac{1}{2}m_i\omega_i^2\hat{q}_i^2 \right). \tag{20.46}$$

The operators $\hat{B}_a$ coupling the environment to the system are linear functions of the configuration variable $\hat{q}_i$, $\hat{B}_a = \sum_{i=1}^{N} c_{ai}\hat{q}_i$, for some constants c_{ai}. It is straightforward to solve Heisenberg's equation of motion, to obtain $\hat{q}_i(t) = \cos(\omega_i t)\hat{q}_i + (m_i\omega_i)^{-1}\sin(\omega_i t)\hat{p}_i$. Then,

$$\tilde{F}_{ab}(\omega) = \sum_{i=1}^{N} c_{ai}c_{bi} \int_0^\infty ds e^{-i\omega s} \left[\cos(\omega_i s)\langle\hat{q}_i^2\rangle + (m_i\omega_i)^{-1}\sin(\omega_i s)\langle\hat{p}_i\hat{q}_i\rangle \right], \tag{20.47}$$

where we used the fact that the thermal density matrix factorizes; hence, $\langle\hat{q}_l\hat{q}_j\rangle = \delta_{ij}\langle\hat{q}_i^2\rangle$ and $\langle\hat{p}_i\hat{q}_j\rangle = \delta_{ij}\langle\hat{p}_i\hat{q}_i\rangle$.

For a harmonic oscillator of mass m and frequency ω, the Gibbs state is

$$\hat{\rho} = Z^{-1}(T) \sum_{n=0}^{\infty} e^{-n\omega/T}|n\rangle\langle n|, \tag{20.48}$$

where $Z(T) = \sum_{n=1}^{\infty} e^{-n\omega/T} = (1 - e^{-\omega/T})^{-1}$. Hence,

$$\langle\hat{q}^2\rangle = Z^{-1}(T) \sum_{n=0}^{\infty} e^{-n\omega/T}\langle n|\hat{q}^2|n\rangle.$$

We use Eq. (6.45), to obtain

$$\langle\hat{q}^2\rangle = \frac{\sum_{n=0}^{\infty}(2n+1)e^{-n\omega/T}}{2m\omega(1 - e^{-\omega/T})} = (2m\omega)^{-1}\coth\left(\frac{\omega}{2T}\right), \tag{20.49}$$

where in the last step we summed the series using Eq. (A.14). Similarly, we obtain $\langle\hat{p}^2\rangle = \frac{1}{2}m\omega\coth\left(\frac{\omega}{2T}\right)$ and $\langle\hat{p}\hat{q}\rangle = -\frac{i}{2}$.

We evaluate the integral over s in Eq. (20.47) using Eq. (4.46). We find

$$F_{ab}(\omega) = \sum_{i=1}^{N} \frac{c_{ai}c_{bi}}{2m_i\omega_i} \times \begin{cases} (e^{\omega_i/T} - 1)^{-1}\left[\pi\delta(\omega - \omega_i) - i\left(\frac{1}{\omega-\omega_i}\right)_{pv}\right], & \omega > 0 \\ (1 - e^{-\omega_i/T})^{-1}\left[\pi\delta(\omega + \omega_i) - i\left(\frac{1}{\omega+\omega_i}\right)_{pv}\right], & \omega < 0 \end{cases}$$

It is convenient to define the *spectral density* of the environment as

$$I_{ab}(\omega) = \pi \sum_{i=1}^{N} \frac{c_{ai}c_{bi}}{m_i\omega_i}\delta(\omega - \omega_i), \tag{20.50}$$

for $\omega > 0$. The spectral density contains all information about the frequency distribution of the environment oscillators and the associated couplings c_{ai}. At the limit where the number N goes to infinity and the oscillator frequencies are distributed continuously, the spectral density behaves like a continuous function. Then, we can express the matrices $\Gamma_{ab}(\omega)$ and $\Delta_{ab}(\omega)$ in terms of the spectral density

$$\Gamma_{ab}(\omega) = I_{ab}(|\omega|)\left[\frac{\theta(\omega)}{e^{\omega/T}-1} + \frac{\theta(-\omega)}{1-e^{\omega/T}}\right] \tag{20.51}$$

$$\Delta_{ab}(\omega) = \int \frac{d\omega'}{2\pi}I_{ab}(\omega')\left(\frac{1}{\omega'-|\omega|}\right)_{pv}\left[\frac{\theta(\omega)}{e^{\omega'/T}-1} - \frac{\theta(-\omega)}{1-e^{\omega'/T}}\right]. \tag{20.52}$$

Other useful models involve the treatment of the environment as a collection of qubits – see Problem 20.8 – and as a collection of fermions.

Example 20.5 As discussed in Section 18.4.2, the EM field (ignoring polarizations) can be described as a collection of harmonic oscillators labeled by the momentum vector $\mathbf{k}$, frequencies $\omega_{\mathbf{k}} = |\mathbf{k}|$, and a coupling to a two-level atom located at $\mathbf{x}_0$ through an interaction term

$$\hat{V} = \lambda\hat{A} \otimes \int \frac{d^3k}{(2\pi)^3\sqrt{2\omega_{\mathbf{k}}}}\left(\hat{a}_{\mathbf{k}}e^{i\mathbf{k}\cdot\mathbf{x}_0} + \hat{a}_{\mathbf{k}}^\dagger e^{i\mathbf{k}\cdot\mathbf{x}_0}\right), \tag{20.53}$$

for some constant λ and some operator $\hat{A}$ in the Hilbert space of the atom. We can choose the coordinate system so that $\mathbf{x}_0 = 0$, Then, each EM oscillator, labeled by $\mathbf{k}$, couples to the system through the 'position' operator $\hat{q}_{\mathbf{k}} = (2\omega_{\mathbf{k}})^{-1/2}(\hat{a}_{\mathbf{k}} + \hat{a}_{\mathbf{k}}^\dagger)$ (the corresponding masses are all set to unity). The index a has only one value so it can be suppressed. The coupling $c_{\mathbf{k}} = 1$ is constant. The spectral density is

$$I(\omega) = \pi\lambda^2\int\frac{d^3k}{(2\pi)^3\omega_{\mathbf{k}}}\delta(\omega-\omega_{\mathbf{k}}) = \frac{\lambda^2}{2\pi}\int_0^\infty kdk\delta(\omega-k) = \frac{\lambda^2\omega}{2\pi}. \tag{20.54}$$

Note that $I(\omega)$ has the same functional form with the decay function for the same system (Section 18.4.2).

20.4.2 Spin–Boson Models

The spin–boson model consists of one or more qubits that are coupled to a bosonic reservoir. For a single qubit, the system Hamiltonian is $\hat{H} = \frac{1}{2}\Omega\hat{\sigma}_3$ for $\Omega > 0$, and there is a single coupling operator $\hat{A} = \hat{\sigma}_1$—; the index a is suppressed. Then, there are only two operators $\hat{A}(\omega)$, namely

$$\hat{A}(\Omega) = |0\rangle\langle0|\hat{\sigma}_1|1\rangle\langle1| = \hat{\sigma}_-, \quad \hat{A}(-\Omega) = |1\rangle\langle1|\hat{\sigma}_1|0\rangle\langle0| = \hat{\sigma}_+. \tag{20.55}$$

We also evaluate the Lamb-shift term,

$$\hat{H}_{LS} = \Delta(\Omega)\hat{\sigma}_+\hat{\sigma}_- + \Delta(-\Omega)\hat{\sigma}_-\hat{\sigma}_+ = C_0\hat{I} + \Delta_0\hat{\sigma}_3, \tag{20.56}$$

where $C_0 = -\int \frac{d\omega'}{2\pi} \left(\frac{1}{\omega'-\Omega}\right)_{pv} I(\omega')$ and

$$\Delta_0 = \int \frac{d\omega'}{\pi} \left(\frac{1}{\omega'-\Omega}\right)_{pv} I(\omega') \coth\left(\frac{\omega'}{2T}\right). \tag{20.57}$$

The term proportional to $\hat{I}$ does not affect the evolution equations, while the term proportional to $\hat{\sigma}_3$ can be absorbed into a redefinition of the frequency $\tilde{\Omega} := \Omega + \Delta_0$. Then, the second-order master equation coincides with that of Example 20.3, with

$$\gamma_+ = \Gamma(\Omega) = \frac{\Gamma_0}{e^{\Omega/T} - 1}, \quad \gamma_- = \Gamma(-\Omega) = \frac{\Gamma_0}{1 - e^{-\Omega/T}}, \quad \gamma_0 = 0, \tag{20.58}$$

where $\Gamma_0 = I(\Omega)$ is the decay rate at zero temperature. (For the two-level atom interacting with the EM field, we get the same result for Γ_0 as in the analysis of Section 18.4.2.) By Eq. (20.24), the asymptotic state

$$\hat{\rho}(\infty) = \frac{1}{2\cosh\left(\frac{\omega}{2T}\right)} \begin{pmatrix} e^{-\frac{\omega}{2T}} & 0 \\ 0 & e^{\frac{\omega}{2T}} \end{pmatrix} \tag{20.59}$$

is the Gibbs density matrix for the system. The model describes the thermalization of the qubit in interaction with a thermal environment.

Of particular interest is the case where the qubit is perturbed by an external time-dependent force, for example, a laser acting on an atom in the two level approximation. To describe this system, we employ the time-dependent Hamiltonian (7.85), $\hat{H}_S(t) = \frac{\tilde{\Omega}}{2}\hat{\sigma}_3 + gf(t)\hat{\sigma}_1$, while keeping the Lindblad operators $\sqrt{\gamma_-}\hat{\sigma}_-$ and $\sqrt{\gamma_+}\hat{\sigma}_+$ of the previous example. Let $f(t) = \cos(\tilde{\omega}t)$ for some frequency $\tilde{\omega}$. As shown in Section 7.4.3, for $\tilde{\omega}$ near $\tilde{\Omega}$, we can invoke the RWA, and substitute $g\cos(\tilde{\omega}t)\hat{\sigma}_1$ with $\frac{1}{2}g(e^{i\tilde{\omega}t}\hat{\sigma}_- + e^{-i\tilde{\omega}t}\hat{\sigma}_+)$. Then, the Heisenberg equations of motion for $\hat{\sigma}_3$ and $\hat{\sigma}_-$ become

$$\frac{d\hat{\sigma}_3(t)}{dt} = \frac{ig}{2}\left(e^{i\tilde{\omega}t}\hat{\sigma}_- - e^{-i\tilde{\omega}t}\hat{\sigma}_+\right) - \Gamma_0\hat{\sigma}_3(t) - \Gamma_0\hat{I} \tag{20.60}$$

$$\frac{d\hat{\sigma}_-(t)}{dt} = -i\tilde{\Omega}\hat{\sigma}_- + \frac{ig}{2}e^{-i\tilde{\omega}t}\hat{\sigma}_3 - \frac{1}{2}\Gamma_0\hat{\sigma}_-, \tag{20.61}$$

where for simplicity, we specialize to a zero-temperature environment with $\gamma_- = \Gamma_0$ and $\gamma_+ = 0$.

The evolution equations for $r = \langle\hat{\sigma}_3\rangle$ and $z = \langle\hat{\sigma}_-\rangle e^{i\tilde{\omega}t}$ become

$$\dot{r} = \frac{ig}{2}(z - z^*) - \Gamma_0 r - \Gamma_0, \qquad \dot{z} = -i\delta z + \frac{ig}{2}r - \frac{\Gamma_0}{2}z, \tag{20.62}$$

where $\delta := \tilde{\Omega} - \tilde{\omega}$ is the detuning parameter. This system of equations can readily be solved, as their right-hand sides does not depend on time. The elements of the density matrix undergo damped oscillations and converge to an equilibrium value at times much larger than Γ_0^{-1} – see Fig. 20.1.

Here, we focus on the asymptotic equilibrium solution. For $\dot{r} = \dot{z} = 0$, we obtain

$$r = -\frac{\Gamma_0^2 + 4\delta^2}{\Gamma_0^2 + 4\delta^2 + g^2} \qquad z = -g\frac{2\delta + i\Gamma_0}{\Gamma_0^2 + 4\delta^2 + g^2}. \tag{20.63}$$

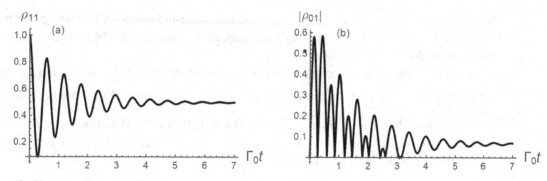

Fig. 20.1 Density-matrix elements ρ_{11} (a) and $|\rho_{01}|$ (b), determined by the solution of Eq. (20.62), are plotted as a function of $\Gamma_0 t$. In these plots, $\delta = 0$ and $g/\Gamma_0 = 15$.

Hence, the asymptotic state of the qubit is

$$\hat{\rho} = \frac{1}{\Gamma_0^2 + 4\delta^2 + g^2} \begin{pmatrix} \frac{1}{2}g^2 & -g(2\delta + i\Gamma_0)e^{-i\tilde{\omega}t} \\ -g(2\delta - i\Gamma_0)e^{i\tilde{\omega}t} & \Gamma_0^2 + 4\delta^2 + \frac{1}{2}g^2 \end{pmatrix}. \tag{20.64}$$

This is not a stationary state, as the off-diagonal elements oscillate with the frequency of the external force. For atomic qubits, the asymptotic state accounts for the phenomenon of resonant fluorescence that was described in Section 8.4.2.

20.4.3 Dissipation and Noise

Now consider a harmonic oscillator of mass M and frequency Ω coupled to a bosonic reservoir. The system Hamiltonian is $\hat{H}_S = \Omega\hat{a}^\dagger\hat{a}$, and there is a single coupling operator $\hat{A} = \hat{x} = (2M\Omega)^{-1/2}(\hat{a} + \hat{a}^\dagger)$. There are only two operators $\hat{A}(\omega)$, namely,

$$\hat{A}(\Omega) = \sum_{n=0}^{\infty} |n\rangle\langle n|\hat{x}|n+1\rangle\langle n+1| = (2M\Omega)^{-1/2}\hat{a}, \tag{20.65}$$

$$\hat{A}(-\Omega) = \sum_{n=0}^{\infty} |n+1\rangle\langle n+1|\hat{x}|n\rangle\langle n| = (2M\Omega)^{-1/2}\hat{a}^\dagger. \tag{20.66}$$

The Lamb-shift Hamiltonian is

$$\hat{H}_{LS} = \frac{1}{2M\Omega}\left[\Delta(\Omega)\hat{a}^\dagger\hat{a} + \Delta(-\Omega)\hat{a}\hat{a}^\dagger\right] \tag{20.67}$$

and it can be expressed as $C_0\hat{I} + \Delta_0\hat{a}^\dagger\hat{a}$. The first term does not contribute to the evolution equation, and the second term is absorbed in a redefinition of the frequency $\tilde{\Omega} = \Omega + \Delta_0$. Hence, the second-order master equation for this system coincides with the GKLS equation of Example 20.4, with $\omega = \tilde{\Omega}$, and

$$\gamma_+ = \frac{I(\Omega)}{2M\Omega(e^{\Omega/T} - 1)}, \qquad \gamma_- = \frac{I(\Omega)}{2M\Omega(1 - e^{-\Omega/T})}. \tag{20.68}$$

Hence, the decay rate $\Gamma := \gamma_- - \gamma_+ = \frac{I(\Omega)}{2M\Omega}$ is temperature-independent. The asymptotic number of quanta $\bar{N} := (e^{\Omega/T} - 1)^{-1}$ is given by the Planck distribution. See Problem 20.6 for the thermal character of the asymptotic state.

We also calculate the expectations and variances of the rescaled position and momentum variables $\hat{X} = \frac{1}{\sqrt{2M\tilde{\Omega}}}(\hat{a} + \hat{a}^\dagger)$ and $\hat{P} = i\sqrt{\frac{M\tilde{\Omega}}{2}}(\hat{a}^\dagger - \hat{a})$. Equation (20.25) implies that $\hat{X}(t) = e^{-\frac{1}{2}\Gamma t}\hat{X}_c(t)$ and $\hat{P}(t) = e^{-\frac{1}{2}\Gamma t}\hat{P}_c(t)$, where $\hat{X}_c(t)$ and $\hat{P}_c(t)$ are the Heisenberg-picture operators for the corresponding closed system. Then, $\langle\hat{X}(t)\rangle = e^{-\frac{1}{2}\Gamma t}\langle\hat{X}_c(t)\rangle$ and $\langle\hat{P}(t)\rangle = e^{-\frac{1}{2}\Gamma t}\langle\hat{P}_c(t)\rangle$.

Using Eqs. (20.26, 20.27), we obtain

$$(\Delta X)^2(t) = e^{-\Gamma t}(\Delta X_c)^2(t) + (M\tilde{\Omega})^{-1}\left(\bar{N} + \frac{1}{2}\right)(1 - e^{-\Gamma t}),$$

$$(\Delta P)^2(t) = e^{-\Gamma t}(\Delta P_c)^2(t) + M\tilde{\Omega}\left(\bar{N} + \frac{1}{2}\right)(1 - e^{-\Gamma t}).$$

It follows that

$$(\Delta X)^2(t)(\Delta P)^2(t) = e^{-2\Gamma t}(\Delta X_c)^2(t)(\Delta P_c)^2(t) + \left(\bar{N} + \frac{1}{2}\right)^2(1 - e^{-\Gamma t})^2$$

$$+ \left(\bar{N} + \frac{1}{2}\right)(1 - e^{-\Gamma t})e^{-\Gamma t}[M\tilde{\Omega}(\Delta X_c)^2(t) + (M\tilde{\Omega})^{-1}(\Delta P_c)^2(t)].$$

By the uncertainty relation $(\Delta X)_c(t)(\Delta P)_c(t) \geq \frac{1}{2}$, hence,

$$(\Delta X)^2(t)(\Delta P)^2(t) \geq \frac{1}{4}e^{-2\Gamma t} + \left(\bar{N} + \frac{1}{2}\right)^2(1 - e^{-\Gamma t})^2$$

$$+ \left(\bar{N} + \frac{1}{2}\right)(1 - e^{-\Gamma t})e^{-\Gamma t}[S + (4S)^{-1}],$$

where $S = M\tilde{\Omega}(\Delta X_c)^2(t)$. Noting that $S + (4S)^{-1} \geq 1$, we obtain the *generalized uncertainty relation*

$$(\Delta X)(t)(\Delta P)(t) \geq \frac{1}{2}e^{-\Gamma t} + \left(\bar{N} + \frac{1}{2}\right)(1 - e^{-\Gamma t}), \tag{20.69}$$

which provides a lower bound to the uncertainty of the harmonic oscillator at time t.

The minimum uncertainty has two contributions: a purely quantum part that decays in time because of dissipation, and a thermal part that increases in time. Open-system uncertainty relations such as Eq. (20.69) describe the transition from quantum to thermal fluctuations through the interaction with the environment (Anastopoulos and Halliwell, 1995).

20.5 Beyond the Markov Approximation

So far, we have encountered only open quantum systems that satisfy the Markov property. We either assumed it axiomatically, as in the derivation of the GKLS master equation, or we imposed it as an approximation in the derivation of the second-order master equation.

However, most physical systems do not satisfy this property exactly, and some systems, not even approximately. In this section, we will study such non-Markovian systems, first by identifying their basic characteristics and then, by considering specific, exactly solvable, models.

20.5.1 Memory Effects

Consider a quantum system described by a Hilbert space $\mathcal{H}_S$, in contact with an environment described by a Hilbert space $\mathcal{H}_E$. The Hamiltonian $\hat{H}$ for the combined system is of the form $\hat{H}_0 + \hat{V}$, where $\hat{H}_0 = \hat{H}_S \otimes \hat{I} + \hat{I} \otimes \hat{H}_E$ involves the self-Hamiltonians of the system, $\hat{H}_S$, and of the environment, $\hat{H}_E$, and $\hat{V}$ is an interaction term.

Suppose that we perform a two-time measurement on the system degrees of freedom. At time t_1, we measure the observable $\hat{A} = \sum_a a_n \hat{P}_n$, and at time t_2, we measure the observable $\hat{B} = \sum_m b_m \hat{Q}_m$. Assuming a factorized initial state $\hat{\rho}_S \otimes \hat{\rho}_E$ at $t = 0$, the probability that the first measurement yields a_n and the second measurement yields b_m is given by Eq. (8.16),

$$\text{Prob}(a_n, t_1; b_m, t_2) = \text{Tr}\left[(\hat{Q}_m \otimes \hat{I})e^{-i\hat{H}(t_2-t_1)}(\hat{P}_n \otimes \hat{I})e^{-i\hat{H}t_1}(\hat{\rho}_S \otimes \hat{\rho}_E) \right.$$
$$\left. \times\, e^{i\hat{H}t_1}(\hat{P}_n \otimes \hat{I})e^{i\hat{H}(t_2-t_1)}\right]. \qquad (20.70)$$

Consider, by comparison, the same two-time probability if the system degrees of freedom evolve with a Markovian family of CP maps $\mathcal{V}_t$,

$$\text{Prob}(a_n, t_1; b_m, t_2) = \text{Tr}\left(\hat{Q}_m \mathcal{V}_{t_2-t_1}[\hat{P}_n \mathcal{V}_{t_1}[\hat{\rho}_S]\hat{P}_n]\right). \qquad (20.71)$$

The two expressions are different. They coincide only if $\text{Tr}_{\mathcal{H}_E}\left[e^{-i\hat{H}t}(\hat{\rho}_S \otimes \hat{\rho}_E)e^{i\hat{H}t}\right] = \mathcal{V}_t[\hat{\rho}_S] \otimes \hat{\rho}_E$, that is, if $\hat{\rho}_E$ remains unchanged under time evolution and the reduced density matrix evolves according to a Markovian master equation. We saw that this condition holds only as an approximation, if at all. Hence, in general, the two-time probability (20.70) cannot be expressed solely in terms of the reduced density matrix for the system and its effective dynamics. In other words, there exist temporal correlations that are not captured by a time-evolved reduced density matrix. These correlations are stored in the environment, and they constitute the *memory* of the open system. Conversely, a Markovian system is characterized by the absence of memory effects.

20.5.2 A Non-Markovian Decohering Environment

We will describe one of the simplest models for environment-induced decoherence. This model allows us to show that non-Markovian behavior is ubiquitous, but also that environment-induced decoherence requires very special states for the environment.

Let $\mathcal{H}_S = \mathbb{C}^2$ and $\mathcal{H}_E = \mathbb{C}^{2^N}$, that is, the system is a single qubit and the environment consists of N qubits. The system Hamiltonian is $\hat{H}_S = \frac{1}{2}\Omega\hat{\sigma}_3$, and the environment Hamiltonian is $\hat{H}_E = 0$. The interaction term is $\hat{V} = \hat{\sigma}_3 \otimes \sum_i g_i\hat{\sigma}_{3i}$ for some constants g_i; the index $i = 1, 2, \ldots, N$ labels the qubits of the environment.

Let $|a\rangle$ be the standard orthonormal basis for a single qubit, where $a = \pm 1$; then, $|a_1, a_2, \ldots, a_N\rangle$ is an orthonormal basis on $\mathcal{H}_E$, where $a_i = \pm 1$. The Hamiltonian $\hat{H}$ has eigenvectors $|a\rangle \otimes |a_1, \ldots, a_n\rangle$ and eigenvalues $E_{a,a_1,\ldots,a_n} = \frac{1}{2}\Omega a + a \sum_i g_i a_i$.

We assume a factorized initial state $\hat{\rho}_0 \otimes \hat{\rho}_E$. It is straightforward to derive the reduced density matrix for the system at time t,

$$\hat{\rho}(t) = \text{Tr}_{\mathcal{H}_E}\left[e^{-i\hat{H}t}(\hat{\rho}_0 \otimes \hat{\rho}_E)e^{i\hat{H}t}\right] = \sum_{a,b=\pm 1} |a\rangle\langle b|\langle a|\hat{\rho}_0|b\rangle$$

$$\times \sum_{a_1,\ldots,a_n=\pm 1} e^{-i(E_{a,a_1,\ldots,a_n}-E_{b,a_1,\ldots,a_n})t}\langle a_1,\ldots,a_n|\hat{\rho}_E|a_1,\ldots,a_n\rangle.$$

We write the initial density matrix $\hat{\rho}_0$ in matrix form as $\frac{1}{2}\begin{pmatrix} 1+x & z \\ z^* & 1-x \end{pmatrix}$, with $x \in [-1,1]$ and $z \in \mathbb{C}$. Then, $\hat{\rho}(t)$ reads

$$\hat{\rho}(t) = \frac{1}{2}\begin{pmatrix} 1+x & ze^{-i\Omega t}F(t) \\ z^* e^{i\Omega t}F^*(t) & 1-x \end{pmatrix}, \tag{20.72}$$

where

$$F(t) = \sum_{a_1,\ldots,a_n=\pm 1} \langle a_1,\ldots,a_n|\hat{\rho}_E|a_1,\ldots,a_n\rangle e^{-2i\sum_i g_i a_i t}. \tag{20.73}$$

To evaluate $F(t)$, we must specify the initial state of the environment. For a totally factorized state $\hat{\rho}_E = \hat{\rho}_1 \otimes \hat{\rho}_2 \otimes \cdots \otimes \hat{\rho}_N$,

$$\langle a_1, \cdots, a_n|\hat{\rho}_E|a_1,\ldots,a_n\rangle = p_{a_1}p_{a_2}\ldots p_{a_N},$$

where $p_{a_i} = \langle a_i|\hat{\rho}_i|a_i\rangle$. We write $p_{a_i} = \frac{1}{2}(1 \pm y_i)$ where $y_i \in [-1,1]$. Then,

$$F(t) = \prod_{i=1}^N \left[\frac{1}{2}(1+y_i)e^{-2ig_it} + \frac{1}{2}(1-y_i)e^{2ig_it}\right] = \prod_{i=1}^N \left[\cos(2g_it) - iy_i\sin(2g_it)\right].$$

Decoherence is governed by the norm of $F(t)$,

$$|F(t)|^2 = \prod_{i=1}^N \left[\cos^2(2g_it) + y_i^2\sin^2(2g_it)\right]. \tag{20.74}$$

All terms in the product are smaller than unity. For $N \gg 1$, we expect that, when the couplings are randomly distributed, $|F(t)|$ drops quickly to zero with increasing t. An example is shown in Fig. 20.2. Hence, the off-diagonal elements of the density matrix are suppressed; the environment decoheres the system qubit.

By Example 20.3, the corresponding Markovian time evolution (for $\gamma_\pm = 0$) is characterized by an exponential fall of the off-diagonal elements. As shown in Fig. 20.2, the function $|F(t)|$ is not exponential, hence, the system is non-Markovian.

We emphasize, that the assumption of a separable initial state for the environment is essential for the demonstration of decoherence. An entangled state for the environmental qubits does not lead to the product form (20.74) for $|F(t)|$. For example, suppose that the environment is initially

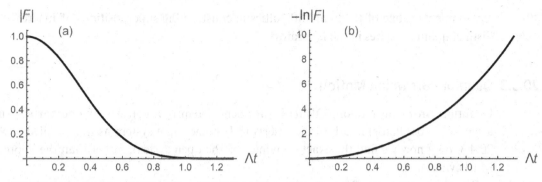

Fig. 20.2 The function $|F(t)|$ of Eq. (20.74) is evaluated for $N = 30$ qubits with $y_i = 0$ and g_i randomly distributed in the interval $[0, \Lambda]$. In (a), $|F(t)|$ is plotted as a function of Λt. In (b), $-\ln|F(t)|$ is plotted as a function of Λt, showing that decay is not exponential, hence, the system is non-Markovian.

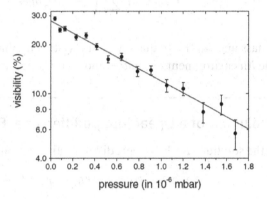

Fig. 20.3 The plot describes the loss of visibility in an interference experiment of C_{70} fullerenes due to the presence of a gas environment. Visibility is essentially proportional to the off-diagonal elements of the particle's density matrix. The highest the pressure, the larger the rate of collisions of air molecules with the fullerenes – hence, pressure is proportional to the parameter D of Eq. (20.2). The straight line in a logarithmic plot confirms that visibility decreases exponentially with pressure. [Reprinted with permission by Springer-Nature from Hackermüller et al. (2003). Copyright 2003.]

at the state $|\Psi\rangle = \frac{1}{\sqrt{2}}(|1, 1, \ldots, 1\rangle + |-1, -1, \ldots, -1\rangle)$. By Eq. (20.73), $F(t) = \cos(2Gt)$, where $G = \sum_i g_i$. Hence, $F(t)$ is a rapidly oscillating function; the off-diagonal elements of the system are not suppressed.

Decoherence requires very special initial states for the environment. The environment has to be itself sufficiently classical (e.g., in a separable state) in order to remove the quantum features from a smaller system that interacts with it. As explained in Section 10.6, this is the reason why environment-induced decoherence cannot fundamentally explain the emergence of the classical world.

Nonetheless, environments that behave classically and cause microscopic systems to decohere are ubiquitous (but not universal). Indeed, decoherence has been witnessed in many experiments – see, for example, Fig. 20.3. The necessity to shield quantum systems from decohering environments is one of the main difficulties in constructing superpositions of macroscopically distinct quantum states in the laboratory.

20.5.3 Quantum Brownian Motion

Quantum Brownian motion (QBM) is an open system that consists of one particle (usually a harmonic oscillator) in a bosonic reservoir. It is the same system as that studied in Section 20.4.3, only now we find the exact dynamics of the open system, rather than the approximate Markovian ones.

We denote by $\hat{X}$ and $\hat{P}$ the position and momentum of the system particle and by $\hat{q}_i$ and $\hat{p}_i$ the positions and momenta of the ith environment oscillator. The total Hamiltonian is

$$\hat{H} = \frac{\hat{P}^2}{2M} + \frac{1}{2}M\Omega_0^2\hat{X}^2 + \sum_i \left(\frac{\hat{p}_i^2}{2m_i} + \frac{1}{2}m_i\omega_i^2\hat{q}_i^2 \right) + \hat{X}\sum_i c_i\hat{q}_i. \tag{20.75}$$

Here M is the mass and Ω_0 the frequency of the system oscillator, m_i and ω_i are the mass and frequency of the ith environment oscillator, and c_i are coupling constants.

Box 20.2 Solutions of a Linear Integrodifferential Equation

We will find the solutions to the integrodifferential equation

$$\ddot{x}(t) + \omega^2 x(t) + \int_0^t dt' h(t - t')x(t') = F(t),$$

for general functions $h(t)$ and $F(t)$. We Laplace-transform the equation, to obtain

$$s^2\tilde{x}(s) - sx(0) - \dot{x}(0) + \omega^2\tilde{x}(s) + \tilde{h}(s)\tilde{x}(s) = \tilde{F}(s),$$

where $\tilde{x}(s), \tilde{h}(s)$ and $\tilde{F}(s)$ are the Laplace transforms of $x(t), h(t)$ and $F(t)$, respectively. We solve for $\tilde{x}(s)$,

$$\tilde{x}(s) = x(0)\frac{s}{s^2+\omega^2+\tilde{h}(s)} + \dot{x}(0)\frac{1}{s^2+\omega^2+\tilde{h}(s)} + \frac{\tilde{F}(s)}{s^2+\omega^2+\tilde{h}(s)}.$$

We take the inverse Laplace transform,

$$x(t) = x(0)\dot{u}(t) + \dot{x}(0)u(t) + \int_0^t dt' u(t - t')F(t'),$$

where $u(t)$ is the inverse Laplace transform of $[s^2 + \omega^2 + \tilde{h}(s)]^{-1}$. By definition, $u(t)$ is a solution to the corresponding homogeneous equation, $\ddot{u}(t)+\omega^2u(t)+\int_0^t dt' h(t-t')u(t') = 0$, with $u(0) = 0$ and $\dot{u}(0) = 1$.

For $h = 0$, $u(t) = \omega^{-1}\sin(\omega t)$; hence, the solution of the differential equation is

$$x(t) = x(0)\cos(\omega t) + \omega^{-1}\dot{x}(0)\sin(\omega t) + \omega^{-1}\int_0^t dt' \sin[\omega(t - t')]F(t').$$

The evolution equations in the Heisenberg picture yield

$$\frac{d^2\hat{X}(t)}{dt^2} + \Omega_0^2\hat{X}(t) = -M^{-1}\sum_i c_i\hat{q}_i(t),\tag{20.76}$$

$$\frac{d^2\hat{q}_i(t)}{dt^2} + \omega_i^2\hat{q}_i(t) = -\frac{c_i}{m_i}\hat{X}(t).\tag{20.77}$$

By the analysis in Box 20.2, the solution to Eq. (20.77) is

$$\hat{q}_i(t) = \hat{q}_i^0(t) - \sum_i \frac{c_i}{m_i\omega_i}\int_0^t ds\,\sin\left[\omega_i(t-s)\right]\hat{X}(s),\tag{20.78}$$

where $\hat{q}_i^0(t) = \hat{q}_i\cos(\omega_i t) + \frac{\hat{p}_i}{m_i\omega_i}\sin(\omega_i t)$.

We substitute into Eq. (20.76), to obtain

$$\ddot{\hat{X}}(t) + \Omega_0^2\hat{X}(t) + \frac{2}{M}\int_0^t dt'\,\eta(t-t')\hat{X}(t') = -\sum_i \frac{c_i}{M}\hat{q}_i^0(t),\tag{20.79}$$

where $\eta(t) := -\sum_i \frac{c_i^2}{2m_i\omega_i}\sin(\omega_i t)$. Using the definition (20.50) for the spectral density $I(\omega) = \pi\sum_i \frac{c_i^2}{m_i\omega_i}\delta(\omega-\omega_i)$, we write

$$\eta(t) = -\int_0^\infty \frac{d\omega}{2\pi}I(\omega)\sin(\omega t) = \frac{d}{dt}\gamma(t),\tag{20.80}$$

where the *dissipation kernel* $\gamma(t)$ is an even function of t, defined as

$$\gamma(t) := \int_0^\infty \frac{d\omega}{2\pi}\frac{I(\omega)}{\omega}\cos(\omega t).\tag{20.81}$$

As shown in Box 20.2, the solution of Eq. (20.79) is

$$\hat{X}(t) = \dot{u}(t)\hat{X}(0) + \frac{1}{M}u(t)\hat{P}(0) - \frac{1}{M}\int_0^t dt'\,u(t-t')\sum_i c_i\hat{q}_i^0(t'),\tag{20.82}$$

where u is the solution of the homogeneous part of Eq. (20.79),

$$\ddot{u}(t) + \Omega_0^2 u(t) + \frac{2}{M}\int_0^t dt'\,\eta(t-t')u(t') = 0,\tag{20.83}$$

subject to initial conditions $u(0) = 0$ and $\dot{u}(0) = 1$. Since $\eta = \dot{\gamma}$, the integral in Eq. (20.83) is $-\int_0^t dt'\,\dot{\gamma}(t'-t)u(t') = -\gamma(0)u(t) + \int_0^t dt'\,\gamma(t-t')\dot{u}(t')$, and we obtain

$$\ddot{u}(t) + \bar{\Omega}^2 u(t) + \frac{2}{M}\int_0^t dt'\,\gamma(t-t')\dot{u}(t') = 0,\tag{20.84}$$

where $\bar{\Omega}^2 = \Omega_0^2 - 2\gamma(0)/M$.

By the Heisenberg equations of motion, Eq. (20.82) yields

$$\hat{P}(t) = M\frac{d\hat{X}(t)}{dt} = M\ddot{u}(t)\hat{X}(0) + \dot{u}(t)\hat{P}(0) - \int_0^t dt'\,\dot{u}(t-t')\sum_i c_i\hat{q}_i^0(t').\tag{20.85}$$

Equations (20.82) and (20.85) are exact. Together with the choice of an initial state of the environment, they allow us to construct all moments of $\hat{X}(t)$ and $\hat{P}(t)$, and, thus, to extract any predictions about the open quantum system.

We take the initial state of the environment to be a Gibbs state (20.48) for each oscillator. Then, $\langle \hat{q}_i^0(s) \rangle = 0$, since $\langle \hat{q}_i \rangle = 0$ and $\langle \hat{p}_i \rangle = 0$. The expectations $X(t) = \langle \hat{X}(t) \rangle$ and $P(t) = \langle \hat{P}(t) \rangle$ evolve as

$$X(t) = \dot{u}(t)X(0) + \frac{1}{M}u(t)P(0), \quad P(t) = M\ddot{u}(t)X(0) + \dot{u}(t)P(0). \tag{20.86}$$

Since the Hamiltonian is quadratic, by Ehrenfest's theorem, the evolution of the mean values coincides with the associated classical equations of motion.

Hence, at the classical limit, the equation

$$\ddot{X}(t) + \bar{\Omega}^2 X(t) + \frac{2}{M}\int_0^t \gamma(t - t')\dot{X}(t') = 0 \tag{20.87}$$

describes the evolution of a quantum system in the presence of dissipation that is nonlocal in time, that is, dissipation with memory that is due to the finite width of the dissipation kernel.

The environment does not cause only dissipation, it also induces noise to the system's evolution. To see this, we calculate the variances

$$(\Delta X)^2(t) = (\Delta X_{cl})^2(t) + S_X(t), \qquad (\Delta P)^2(t) = (\Delta P_{cl})^2(t) + S_P(t),$$

where we have defined the operators $\hat{X}_{cl}(t) := \dot{u}(t)\hat{X}(0) + \frac{1}{M}u(t)\hat{P}(0)$, $\hat{P}_{cl}(t) := M\ddot{u}(t)\hat{X}(0) + \dot{u}(t)\hat{P}(0)$; "cl" stands for classical. The functions $S_X(t)$ and $S_P(t)$ are defined as

$$S_X(t) = \frac{1}{M^2}\sum_{ij} c_i c_j \int_0^t dt' \int_0^t dt'' \langle \hat{q}_i^0(t')\hat{q}_j^0(t'') \rangle u(t - t')u(t - t'')$$

$$S_P(t) = \sum_{ij} c_i c_j \int_0^t dt' \int_0^t dt'' \langle \hat{q}_i^0(t')\hat{q}_j^0(t'') \rangle \dot{u}(t - t')\dot{u}(t - t'').$$

These functions describe the noise induced by the environment on the system; they do not depend on the initial state of the system.

We evaluate the correlation $\langle \hat{q}_i^0(t)\hat{q}_j^0(t') \rangle$ using the expressions for $\langle \hat{q}^2 \rangle$, $\langle \hat{p}^2 \rangle$, and $\langle \hat{q}\hat{p} \rangle$ that were derived in Section 20.4.1 for the Gibbs state,

$$\langle \hat{q}_i^0(t)\hat{q}_j^0(t') \rangle = \delta_{ij}\frac{1}{2m_i\omega_i}\coth\left(\frac{\omega_i}{2T}\right)\cos\left[\omega_i(t - t')\right]. \tag{20.88}$$

Then, we obtain

$$S_X(t) = \frac{1}{M^2}\int_0^t dt' \int_0^t dt'' \nu(t' - t'')u(t - t')u(t - t'')$$

$$S_P(t) = \int_0^t dt' \int_0^t dt'' \nu(t' - t'')\dot{u}(t - t')\dot{u}(t - t''),$$

where

$$\nu(t) = \sum_i \frac{c_i^2}{2m_i\omega_i}\coth\left(\frac{\omega_i}{2T}\right)\cos(\omega_i t) = \int_0^\infty \frac{d\omega}{2\pi}I(\omega)\coth\left(\frac{\omega}{2T}\right)\cos(\omega t) \tag{20.89}$$

is the *noise kernel*. For sufficiently high temperatures, $\coth\left(\frac{\omega}{2T}\right) \simeq \frac{2T}{\omega}$, hence, $\nu(t) = 2T\gamma(t)$.

We note that $[\hat{X}_{cl}(t), \hat{P}_{cl}(t)] = iJ(t)\hat{I}$, where $J(t) = \dot{u}^2 - u\ddot{u}$. Hence, by the Kennard–Robertson uncertainty relation $(\Delta X)^2_{cl}(t)(\Delta P)^2_{cl}(t) \geq \frac{1}{4}|J(t)|^2$, and

$$(\Delta X)^2(t)(\Delta P)^2(t) \geq \left[(\Delta X_{cl})^2(t) + S_X(t)\right]\left[\frac{J^2(t)}{4(\Delta X_{cl})^2(t)} + S_P(t)\right]. \tag{20.90}$$

We minimize with respect to $(\Delta X_{cl})^2(t)$, to obtain the uncertainty relation

$$(\Delta X)(t)(\Delta P)(t) \geq \frac{1}{2}|J(t)| + \sqrt{S_X(t)S_P(t)}. \tag{20.91}$$

Equation (20.91) generalizes Eq. (20.69) as it accounts for non-Markovian time evolution. Again, the minimum uncertainty has two contributions: a purely quantum part and a thermal/noise part. A plot of the lower bound in Eq. (20.69) is given in Fig. 20.5, after the explicit calculation of S_X and S_P in Section 20.5.4.

To summarize, all information about the environment is encoded in the spectral density $I(\omega)$. Given the spectral density, one defines the dissipation kernel that accounts for dissipation in the effective classical equations of motion and the noise kernel that describes the effects of environmental noise. The spectral density can either be derived from first principles, if one knows the details of the environment, or it can be treated as a phenomenological parameter that is determined by experiment. In either case, QBM provides a simple but powerful model for quantum dissipation, with applications ranging from condensed matter physics, to quantum optics, and to cosmology.

Given a spectral density, the classical solution $u(t)$ provides a full solution to Heisenberg's equations of motion. It can be proven that this solution contains all information about the system, allowing one to fully construct the time evolution of the density matrix. This can be done through the derivation of a master equation with time-dependent coefficients (Hu et al., 1992), or through the direct construction of a propagator in the Wigner representation (Halliwell and Yu, 1995).

20.5.4 Ohmic Environment

The spectral density $I(\omega)$ of a QBM environment typically vanishes for frequencies larger than a cut-off frequency Λ. However, high frequencies affect little the behavior of the quantum system, the most crucial factor is the behavior of $I(\omega)$ at low frequencies.

A large class of environments can be modeled by spectral densities of the form

$$I(\omega) = 2\Gamma M \Lambda \left(\frac{\omega}{\Lambda}\right)^a e^{-\omega/\Lambda}, \tag{20.92}$$

for some constant $\Gamma > 0$ with the dimension of inverse time.[2] The exponent a is usually an integer, and it must be positive so that both the dissipation kernel and the noise kernel are well defined.

---

[2] The insertion of the particle mass M into the spectral density is only for convenience. The spectral density of an environment does not depend on any properties of a particle that happens to interact with it. This means that, for a given environment, the dissipation rate Γ decreases with particle mass.

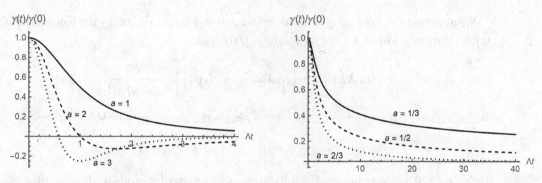

Fig. 20.4 The normalized dissipation kernel $\gamma(t)/\gamma(0)$ as a function of Λt, for different values of the exponent a. For $a \geq 1$, the kernel is well localized around zero – it vanishes for t a few times Λ^{-1}. In contrast, for $a < 1$ the dissipation kernel decays slowly with t.

In Fig. 20.4, we plot the dissipation kernel $\gamma(t)$ for different values of a. We see that, for $a \geq 1$, the dissipation kernel is peaked around $t = 0$ and becomes negligible for t a few times Λ^{-1}.

Here, we will focus on environments with $a = 1$. In this case,

$$\gamma(t) = \frac{M\Gamma}{\pi} \frac{\Lambda}{1 + \Lambda^2 t^2}. \tag{20.93}$$

If $\Omega_0 << \Lambda$, all dynamical scales are much larger than Λ^{-1}, so we are justified in substituting the Lorentzian with a delta function,

$$\gamma(t) = M\Gamma\delta(t). \tag{20.94}$$

Then, Eq. (20.84) becomes[3]

$$\ddot{u} + \Gamma\dot{u} + \bar{\Omega}^2 u = 0, \tag{20.95}$$

that is, the dissipation is local in time and proportional to $\dot{u}$, like damping in electric circuits according to Ohm's law. For this reason, the environment for $a = 1$ is called *Ohmic*.[4] Equation (20.95) admits underdamped solutions for $\Gamma < 2\bar{\Omega}$ and overdamped solutions for $\Gamma \geq 2\bar{\Omega}$. Here, we analyze the former case. The unique solution to Eq. (20.95) with $u(0) = 0$ and $\dot{u}(0) = 1$ is

$$u(t) = \Omega^{-1} \sin(\Omega t)e^{-\frac{\Gamma}{2}t}, \tag{20.96}$$

where $\Omega = \sqrt{\bar{\Omega}^2 - \frac{1}{4}\Gamma^2}$. For this solution, $J(t) = e^{-\frac{\Gamma}{2}t}$.

For $T >> \Lambda$, the high-temperature expression $\nu(t) = 2T\gamma(t)$ applies, hence, for an Ohmic environment, $\nu(t) = 2M\Gamma T\delta(t)$. Then, we straightforwardly calculate

$$S_X(t) = \frac{2\Gamma T}{M} \int_0^t dt' [u(t')]^2$$

$$= \frac{T}{M(\Omega^2 + \frac{1}{4}\Gamma^2)} \times \left[1 - e^{-\Gamma t} \left\{ 1 + \frac{\Gamma^2}{2\Omega^2} \cos^2(\Omega t) + \frac{\Gamma}{2\Omega} \sin(2\Omega t) \right\} \right].$$

---

[3] Note that $\int_0^t \delta(t - t')f(t') = \frac{1}{2}f(t)$, because only half of the delta function is inside the range of integration.

[4] For $a < 1$ the environment is called *subohmic*, and for $a > 1$ it is called *superohmic*.

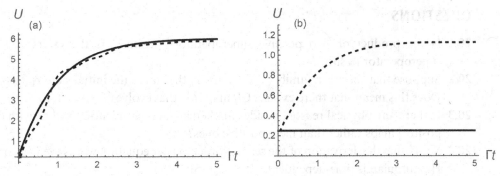

Fig. 20.5 The right-hand side U of the uncertainty relation (20.91) as a function of Γt for an Ohmic environment at different temperatures. In plot (a), we consider the high-temperature limit and take $T/\Omega = 6$. The solid line corresponds to $\Gamma/\Omega = 0.001$. In this regime, U is indistinguishable from the corresponding Markovian expression (20.69). The dashed line corresponds to $\Gamma/\Omega = 0.25$. At high temperatures, the Markovian approximation works quite well in capturing fluctuations, even at strong coupling. In plot (b), $T = 0$. The solid line corresponds to the Markovian expression (20.69). The dashed line is evaluated numerically for $\Lambda/\Gamma = 10$ and $\Gamma/\Omega = 0.1$. The Markovian approximation is evidently inadequate.

For $S_P(t)$, we use integration by parts, $\int_0^t dt'[\dot{u}(t')]^2 = u(t)\dot{u}(t) - \int_0^t dt' u(t')\ddot{u}(t') = u(t)\dot{u}(t) + \int_0^t dt' u(t')[\Gamma\dot{u}(t') + \Omega^2 u(t')]$, to obtain

$$S_P(t) = 2M\Gamma T u(t)\left[\dot{u}(t) + \frac{1}{2}\Gamma u(t)\right] + M^2\Omega^2 S_X(t).$$

Both $S_X(t)$ and $S_P(t)$ start from zero at $t = 0$, and they increase towards the asymptotic values, $S_X(\infty) = \frac{T}{M\Omega^2}$ and $S_P(\infty) = MT\left(1 - \frac{\Gamma^2}{4\Omega^2}\right)$.

In the weak coupling limit, $\Gamma \ll \Omega$, we keep only the lowest-order terms with respect to the ratio Γ/Ω. Then,

$$S_X(t) = \frac{T}{M\Omega^2}(1 - e^{-\Gamma t}), \qquad S_P(t) = TM(1 - e^{-\Gamma t}).$$

The expressions for the uncertainties and the variances coincide with the expressions obtained from the second-order master equation for high temperatures.

Quantum Brownian motion with an Ohmic environment in the high-temperature limit has an almost Markovian time evolution.[5] Since both the dissipation and the noise kernels are proportional to a delta function, there are no memory effects. However, unlike the second-order master equation, the QBM analysis also applies in the regime of strong system–environment coupling, where perturbation theory does not work. Non-Markovian QBM systems exhibit behaviors that cannot even be approximated by the Markovian analysis, especially for low-temperature environments.

---

[5] To see why we need a qualifying "almost" in this statement, solve Problem 20.10.

QUESTIONS

20.1 Is the product of two positive superoperators positive? Is the inverse of a positive superoperator positive?

20.2 Suppose that there is a family of CP maps $\mathcal{V}_t$ that takes the initial state $\hat{\rho}_0$ to $\hat{\rho}_t = \mathcal{V}_t[\hat{\rho}_0]$. Does this mean that there exists a CP map $\mathcal{U}_{t,t'}$ that evolves $\hat{\rho}_t$ to $\hat{\rho}_{t'}$ for $t > t'$?

20.3 Is there any physical reason, besides mathematical convenience, to work with completely positive maps rather than mere positive ones?

20.4 Verify that the derivation of the second-order master equation works even when the system Hamiltonian is time-dependent.

20.5 Which states saturate the uncertainty relations (20.69) and (20.91)? Are they the same states for all times t?

20.6 Order the following timescales that appear in QBM in an Ohmic environment at high temperature: $T^{-1}, \Omega^{-1}, \Gamma^{-1}, \Lambda^{-1}$. At which timescale do thermal fluctuations start overcoming quantum ones?

PROBLEMS

20.1 Show that the master equation (20.2) for the free-particle Hamiltonian $\hat{H} = \frac{\hat{p}^2}{2m}$ takes the following form, when expressed in terms of the time-evolved Wigner function $W_t(x,p)$:

$$\frac{\partial W_t}{\partial t} = -\frac{p}{m}\frac{\partial W_t}{\partial x} + D\frac{\partial^2 W_t}{\partial p^2}.$$

20.2 Return to Problem 14.6, but now assume that $s = \frac{1}{2}$ and use the Bloch equations to describe the time evolution of the spins. What is the magnetization at the end of the echo $(t = 2\tau)$?

20.3 The presence of decohering environments that destroy entanglement is one of the major hurdles towards the realization of large-scale quantum computers. As an example, consider a two-qubit system, with a Hamiltonian $\hat{H} = \frac{\omega}{2}(\hat{\sigma}_3 \otimes \hat{I} + \hat{I} \otimes \hat{\sigma}_3)$, and an environment with two Lindblad operators $\hat{L}_1 = \sqrt{\delta}\hat{\sigma}_3 \otimes \hat{I}$ and $\hat{L}_2 = \sqrt{\delta}\hat{I} \otimes \hat{\sigma}_3$. Find the time evolution of an initial Bell state.

20.4 (i) Show that, for any operator $\hat{B}$ and for any state $\hat{\rho}$ that commutes with the Hamiltonian $\hat{H}$, $\int_{-\infty}^{\infty} ds\langle \hat{B}^{\dagger}(s)\hat{B}(0)\rangle e^{-i\omega s} \geq 0$. (ii) Use this result to show that $\Gamma_{ab}(\omega)$ is a positive matrix.

20.5 Show that evolution by the GKLS equation reduces the purity of the quantum state: $\frac{d}{dt}Tr\hat{\rho}^2 \leq 0$.

20.6 (i) Prove Eq. (20.45). (ii) Show that, for a thermal environment at temperature T, the transition rates satisfy the *detailed balance* equation $w_{mn} = w_{nm}e^{(\epsilon_m - \epsilon_n)/T}$. (iii) Show that, in this case, the Boltzmann distribution $P_n \sim e^{-\beta\epsilon_n}$ is a stationary (equilibrium) solution to Pauli's master equation.

20.7 Find the general solution to Eqs. (20.62), and write the explicit form for the time evolution of the density matrix.

20.8 Consider an environment of N qubits, labeled by $i = 1, 2, \ldots, N$. The Hamiltonian is $\hat{H} = \sum_{i=1}^{N} \frac{\omega_i}{2} \hat{\sigma}_{3i}$, and the coupling operators are $\hat{B}_a = \sum_{i=1}^{N} c_{ai} \hat{\sigma}_{1i}$. Evaluate $\Gamma_{ab}(\omega)$ assuming that the environment is in a thermal state.

20.9 (i) Calculate $u(t), S_X(t)$, and $S_P(t)$ for the overdamped harmonic oscillator in an Ohmic environment. (ii) Write the uncertainty relation (20.91) for long times t. (iii) Take the limit $\bar{\Omega} \to 0$ in these calculations, to obtain the evolution of a free particle in an Ohmic environment.

20.10 (i) Show that, for very small times Ωt, the uncertainty relation (20.91) becomes $(\Delta X)(t)(\Delta P)(t) = \frac{1}{2} - \Gamma t + O(t^2)$, that is, the lower bound of $\frac{1}{2}$ may be violated. This is a sign that positivity is not preserved in time evolution. This violation is an artifact of the approximation (20.94) that leads to Ohmic dissipation. Evaluate $v(t), S_X(t)$, and $S_P(t)$ at the limit of small t with the dissipation kernel given by Eq. (20.94). Show that

$$(\Delta X)(t)(\Delta P)(t) \geq \frac{1}{2} + \frac{\Gamma T \Lambda}{2\pi} t^3;$$

hence, there is no positivity violation. Equation (20.94) applies only for $t >> \Lambda^{-1}$.

20.11 Consider a QBM model with system–environment coupling given by $\hat{V} = \hat{P} \sum_i c_i \hat{q}_i$. Derive the analogue of Eq. (20.82) for this model. Is this model equivalent to the one analyzed in Section 20.5.3?

Bibliography

- For an introduction to open quantum systems, see the short monograph by Diaz and Huelga (2012). For a more detailed analysis, see Breuer and Petruccione (2002). For applications to quantum thermodynamics, see Alicki and Kosloff (2018).
- For environment-induced decoherence, see Joos (1996) and Schlosshauer (2019).
- Quantum Brownian motion is commonly studied in the path integral formalism, see Caldeira and Leggett (1983) and Hu et al. (1992). For dissipative quantum systems and their applications, see Weiss (2012).

Part V

Quantum Foundations

21 Quantum Measurements

> Here are some words which, however legitimate and necessary in application, have no place in a formulation with any pretension to physical precision: system, apparatus, environment, microscopic, macroscopic, reversible, irreversible, observable, information, measurement.........
> On this list of bad words from good books, the worst of all is 'measurement'. It must have a section to itself.
>
> *Bell (1990)*

21.1 Von Neumann Measurements

In previous chapters, we saw that quantum theory is unique among physical theories, in that its predictions refer exclusively to measurement outcomes rather than to the properties of physical objects. It is therefore no surprise that the study of quantum measurements has developed into a research field on its own. The earlier studies of quantum measurements focused on conceptual and foundational issues, but in recent years quantum measurement theory has become a crucial tool for quantum technologies. In this section, we present the simplest measurement models, first developed by von Neumann (1931). Despite their simplicity, they contain most of the key features of quantum measurement theory.

Quantum measurement theory analyzes the interaction of a measured quantum system with one or more *apparatuses*, and describes how this interaction causes microscopic observables to be correlated with readings of the apparatus.

The apparatus is a macroscopic system with a large number of degrees of freedom. One or more variables on the apparatus are *pointer variables*, that is, their values give the measurement outcomes. A pointer variable is typically a coarse-grained collective observable of the apparatus degrees of freedom. For example, the pointer variable may correspond to the center of mass coordinates in a system of N particles, where $N \gg 1$.

In quantum measurement theory, the apparatus is treated as a classical–quantum hybrid. On one hand, it is treated as a quantum system that is described by a quantum state and that evolves through a Hamiltonian. On the other hand, we must assume that the pointer variable is classical, and that it takes a definite value in each separate measurement. This means that we must apply the state reduction rule at the level of the pointer variable. Quantum measurement theory solves neither the measurement problem nor the problem of the emergence of the classical world.

It provides a practical way of understanding the process of information extraction from quantum systems without addressing the fundamental difficulty.

21.1.1 Ideal Measurements

The simplest quantum measurement model involves the recording of an observable with discrete spectrum to a pointer variable with discrete spectrum. Let $\mathcal{H}_S = \mathbb{C}^N$ be the Hilbert space of the microscopic system and let $\mathcal{H}_A$ be the Hilbert space of the apparatus. We require that $\mathcal{H}_A$ has higher dimension than $\mathcal{H}_{sys}$.

We want to measure an observable $\hat{A}$ on $\mathcal{H}_{sys}$ with nondegenerate spectrum. Hence, there exists an orthonormal basis $|n\rangle$ on $\mathcal{H}_{sys}$ such that $\hat{A} = \sum_n a_n |n\rangle\langle n|$, where a_n are the N distinct eigenvalues of $\hat{A}$. An apparatus for measuring the observable $\hat{A}$ must have a pointer variable $\hat{X}$ such that each eigenvalue a_n is mapped to an eigenvalue x_n of $\hat{X}$. We will denote by $\hat{P}_n$ the spectral projector of $\hat{X}$ that corresponds to x_n. We will also assume that there is a special eigenstate $|\Omega\rangle$ of $\hat{X}$ that corresponds to no reading: The apparatus will always be prepared in this state prior to a measurement. For convenience, we take $|\Omega\rangle$ to be a nondegenerate eigenstate of $\hat{X}$ with zero eigenvalue.

Let the Hamiltonian of the system be $\hat{H}_S$ and let the Hamiltonian of the apparatus be $\hat{H}_A$. In an ideal apparatus, the pointer variable changes only during the interaction with the microscopic system. Once this interaction is over, the pointer variable is frozen. This means that $[\hat{H}_A, \hat{X}] = 0$. It is convenient to take $\hat{H}_A = 0$; the essence of the analysis remains unchanged.

In measurement models, it is common to assume that the interaction between system and apparatus is switched on for a finite time. Hence, we employ an interaction term of the form $\hat{H}_I = f(t)\hat{V}$ for some operator $\hat{V}$, where $f(t)$ is nonzero only in a finite interval $[t_i, t_f]$. Then, the evolution operator for $t > t_f$ is

$$\hat{U}_t = \hat{S}^\dagger (e^{-i\hat{H}_S t} \otimes \hat{I}), \tag{21.1}$$

where the unitary operator $\hat{S}$ is given by Eq. (18.3). Hence, an initial state of the form $|\psi_0\rangle \otimes |\Omega\rangle$ will evolve to $\hat{S}^\dagger(|\psi_t\rangle \otimes |\Omega\rangle)$, where $|\psi_t\rangle = e^{-i\hat{H}_S t}|\psi_0\rangle$.

In an ideal measurement, an initial state $|n\rangle \otimes |\Omega\rangle$ must transform to $|n\rangle \otimes |x_n\rangle$, where $|x_n\rangle$ is some eigenstate of the pointer variable with eigenvalue x_n. Then, the pointer variable is uniquely correlated to the eigenvalues of $\hat{A}$. We can engineer such a transformation by considering an *instantaneous* interaction Hamiltonian

$$\hat{H}_I = \delta(t - t_0) \sum_n |n\rangle\langle n| \otimes \hat{K}_n, \tag{21.2}$$

where $\hat{K}_n = g|\Omega\rangle\langle x_n| + g^*|x_n\rangle\langle\Omega|$ for some constant $g \in \mathbb{C}$. By Eq. (18.5), the evolution operator for the total Hamiltonian is

$$\hat{U}_t = \left(e^{-i\hat{H}_S(t-t_0)} \otimes \hat{I} \right) e^{-i\sum_n |n\rangle\langle n|\otimes\hat{K}_n} \left(e^{-i\hat{H}_S t_0} \otimes \hat{I} \right). \tag{21.3}$$

We find that

$$e^{-i\sum_n |n\rangle\langle n|\otimes\hat{K}_n} = \sum_{k=0}^{\infty} \sum_n \frac{(-i)^k}{k!} |n\rangle\langle n| \otimes \hat{K}_n^k = \sum_n |n\rangle\langle n| \otimes e^{-i\hat{K}_n}.$$

Then, we evaluate $e^{-i\hat{K}_n}$ by noting that $\hat{K}_n^2 = |g|^2 (|\Omega\rangle\langle\Omega| + |x_n\rangle\langle x_n|)$ and $\hat{K}_n^3 = |g|^2\hat{K}_n$. From the Taylor series for the exponential, we obtain

$$e^{-i\hat{K}_n} = (|\Omega\rangle\langle\Omega| + |x_n\rangle\langle x_n|)\cos|g| - \frac{i\sin|g|}{|g|}\hat{K}_n.$$

It follows that $e^{-i\hat{K}_n}|\Omega\rangle = \cos|g||\Omega\rangle - \frac{ig^*\sin|g|}{|g|}|x_n\rangle$. We will take $g = -i\frac{\pi}{2}$, so that $e^{-i\hat{K}_n}|\Omega\rangle = |x_n\rangle$

After the interaction, the factorized initial state $|\psi_0\rangle \otimes |\Omega\rangle$ of the total system evolves to the entangled state

$$|\Psi_t\rangle := \sum_n \langle n|\psi_{t_0}\rangle e^{-i\hat{H}_S(t-t_0)}|n\rangle \otimes |x_n\rangle. \tag{21.4}$$

The probability of finding x_n at time t is

$$\text{Prob}(x_n, t) = \langle\Psi_t|\hat{I} \otimes \hat{P}_n|\Psi_t\rangle = |\langle n|\psi_{t_0}\rangle|^2, \tag{21.5}$$

that is, we obtain exactly the Born rule for the measurement of a_n at time t_0 in the microscopic system. Note that the probability distribution for the pointer is defined at times t later than the time t_0, at which the probability distribution for the microscopic particle is determined. In this model, there is a *perfect correlation* between the microscopic observable and the pointer variable. We refer to this type of measurement as an *ideal measurement*.

If $|g| \neq \frac{\pi}{2}$, there is a nonzero probability $\cos^2|g|$ that the pointer variable remains at $|\Omega\rangle$, that is, that the apparatus fails to record the microscopic system.

Applying the state reduction rule for the pointer variable, we see that the state describing the subensemble where x_n has been measured is

$$|\Psi_t; x_n\rangle = \frac{\langle n|\psi_{t_0}\rangle}{|\langle n|\psi_{t_0}\rangle|} e^{-i\hat{H}(t-t_0)}|n\rangle \otimes |x_n\rangle. \tag{21.6}$$

If we trace out the degrees of freedom of the apparatus, the microscopic system is in the state $e^{-i\hat{H}(t-t_0)}|n\rangle\langle n|e^{i\hat{H}(t-t_0)}$. We would have obtained the same result, if we had employed the reduction rule at the level of the microscopic system at time t_0, and then evolved the state to time t.

21.1.2 Measurement of a Continuous Variable

Suppose that we want to measure an observable $\hat{A}$ with continuous spectrum. This means that $\mathcal{H}_S$ is infinite-dimensional. We take the spectrum of $\hat{A}$ to be the real line, and we assume no degeneracy. Hence, we write $\hat{A} = \int da\, a|a\rangle\langle a|$.

We will record the values of $\hat{A}$ through a continuous pointer variable $\hat{X} = \int dX\, X\hat{E}_X$ with highly degenerate eigenvalues X; $\hat{E}_X$ are the associated spectral projectors. Again, $\hat{X}$ is assumed to commute with the apparatus's Hamiltonian $\hat{H}_A$. We take the latter conveniently to be zero. We also take the initial state of the apparatus to be well concentrated around $X = 0$.

We consider an interaction Hamiltonian of the form

$$\hat{H}_I = \delta(t - t_0)\hat{A} \otimes \hat{K}, \tag{21.7}$$

where $\hat{K}$ is the conjugate variable to $\hat{X}$, that is, it satisfies $[\hat{X}, \hat{K}] = i\hat{I}$.

By Eq. (18.5), the evolution operator for the total Hamiltonian is

$$\hat{U}_t = \left(e^{-i\hat{H}_S(t-t_0)} \otimes \hat{I}\right) e^{-i\hat{A}\otimes\hat{K}} \left(e^{-i\hat{H}_S t_0} \otimes \hat{I}\right). \tag{21.8}$$

We calculate

$$e^{-i\hat{A}\otimes\hat{K}} = \exp[-i \int da|a\rangle\langle a| \otimes \hat{K}] = \sum_{n=0}^{\infty} \frac{(-i)^n}{n!} \left(\int da\, a|a\rangle\langle a| \otimes \hat{K}\right)^n$$

$$= \sum_{n=0}^{\infty} \frac{(-ia)^n}{n!} \int da|a\rangle\langle a| \otimes \hat{K}^n = \int da|a\rangle\langle a| \otimes e^{-ia\hat{K}}. \tag{21.9}$$

Since $e^{ia\hat{K}}\hat{X}e^{-ia\hat{K}} = \hat{X} + a\hat{I}$, we find that, in the Heisenberg picture,

$$\hat{X}(t) := \hat{U}_t^\dagger \hat{X} \hat{U}_t = \hat{X} + \hat{A}(t_0), \tag{21.10}$$

the unitary operator $\hat{U}_t$ correlates the operators $\hat{A}(t_0)$ and $\hat{X}(t)$, for $t > t_0$.

In the Schrödinger picture, the state of the total system at time $t > t_0$ is

$$|\Psi_t\rangle = \hat{U}_t \left(|\psi\rangle \otimes |\Omega\rangle\right) = \int da\langle a|\psi_{t_0}\rangle e^{-i\hat{H}_S(t-t_0)}|a\rangle \otimes e^{-ia\hat{K}}|\Omega\rangle. \tag{21.11}$$

Hence, the probability density $p(X)$ for the pointer variable at time $t > t_0$ is time independent and equal to

$$p(X) = \langle\Psi_t|\hat{I} \otimes \hat{E}_X|\Psi_t\rangle = \int da|\langle a|\psi_{t_0}\rangle|^2 \langle\Omega|e^{ia\hat{K}}\hat{E}_X e^{-ia\hat{K}}|\Omega\rangle. \tag{21.12}$$

Let $w(X) = \langle\Omega|\hat{E}_X|\Omega\rangle$ be the initial probability distribution of the pointer variable. Typically, $w(X)$ is $\cap$-shaped and centered around $X = 0$ with mean deviation σ_X. Then,

$$\langle\Omega|e^{ia\hat{K}}\hat{E}_X e^{-ia\hat{K}}|\Omega\rangle = \langle\Omega|\hat{E}_{X-a}|\Omega\rangle = w(X - a).$$

It follows that

$$p(X) = \int da\, w(X - a)|\langle a|\psi_{t_0}\rangle|^2. \tag{21.13}$$

Hence, the probability distribution for the pointer does not reproduce the Born-rule probability density for the microscopic system. Rather, the probability density $|\langle a|\psi_{t_0}\rangle|^2$ is "smeared" by the function $w(X)$ that reflects the intrinsic inaccuracy of the pointer variables. The smearing is necessary because there is no state with a sharp value of X in the Hilbert space of the apparatus. This contrasts the case of pointers with discrete spectrum. By Eq. (21.10), $\langle\hat{X}(t)\rangle = \langle\hat{A}(t_0)\rangle$ and $(\Delta X)^2(t) = (\Delta A)^2(t_0) + \sigma_X^2$. The quantum fluctuations of $\hat{X}$ constitute the "noise" of the measurement. The apparatus cannot distinguish the structure of the particle's state at scales smaller than σ_X.

We rewrite Eq. (21.13) as $p(X) = \langle\psi_{t_0}|\hat{\Pi}_X|\psi_{t_0}\rangle$, where

$$\hat{\Pi}_X = \int da\, w(X - a)|a\rangle\langle a| = w(X - \hat{A}). \tag{21.14}$$

The operators $\hat{\Pi}_X$ are positive since $w(X) \geq 0$ and they satisfy $\int dX \, \hat{\Pi}_X = \hat{I}$. They are not projectors, but they do define probabilities for the measurement of α through the pointer variable X.

We also determine the state of the system after measurement, using the state reduction rule. If the pointer variable is found in the subset C of $\mathbb{R}$ at time t, then the state of the combined system at time is described by the (unnormalized) vector

$$|\Psi_t; C\rangle = \int da \langle a|\psi_{t_0}\rangle e^{-i\hat{H}_S(t-t_0)}|a\rangle \otimes \hat{E}_C e^{-ia\hat{K}}|\Omega\rangle. \tag{21.15}$$

where $\hat{E}(C) = \int_C dX \hat{E}_X$. We trace out the apparatus degrees of freedom to obtain an unnormalized reduced density matrix

$$\hat{\rho}_t(C) = \int_C dX e^{-i\hat{H}_S(t-t_0)} \hat{\rho}(X) e^{i\hat{H}_S(t-t_0)} \tag{21.16}$$

of the particle, where

$$\langle a|\hat{\rho}(X)|b\rangle = \langle a|\psi_{t_0}\rangle\langle\psi_{t_0}|b\rangle\langle\Omega|e^{ib\hat{K}}\hat{E}_X e^{-ia\hat{K}}|\Omega\rangle. \tag{21.17}$$

The map $\mathcal{P}_C$ from the initial density matrix of the system $\hat{\rho}_0 = |\psi_0\rangle\langle\psi_0|$ to $\hat{\rho}_t(C)$ is linear and it preserves positivity. We normalize $\hat{\rho}_t(C)$ by dividing with $\text{Tr}\hat{\rho}(C) = \text{Prob}(C)$.

21.1.3 Simultaneous Measurement of Position and Momentum

The models above can be generalized in order to describe the simultaneous measurement of noncommuting observables, such as position and momentum. The idea is to create correlations of a particle's position $\hat{x}$ and momentum $\hat{p}$ to two pointer variables $\hat{Y}$ and $\hat{Z}$ that commute with each other. This is achieved by an interaction Hamiltonian of the form

$$\hat{H}_I = \delta(t - t_0)\left(\hat{x} \otimes \hat{K} + \hat{p} \otimes \hat{L}\right), \tag{21.18}$$

where $\hat{K}$ is the conjugate variable of $\hat{Y}$ and $\hat{L}$ is the conjugate variable of $\hat{Z}$: $[\hat{Y}, \hat{K}] = i\hat{I}$ and $[\hat{Z}, \hat{L}] = i\hat{I}$. The associated evolution operator is

$$\hat{U}_t = \left(e^{-i\hat{H}_S(t-t_0)} \otimes \hat{I}\right) e^{-i\hat{x}\otimes\hat{K} - i\hat{p}\otimes\hat{L}} \left(e^{-i\hat{H}_S t_0} \otimes \hat{I}\right)$$
$$= \sum_r \int dKdL \, e^{-i\hat{H}_S(t-t_0)} e^{-iK\hat{x} - iL\hat{p}} e^{-i\hat{H}_S t_0} \otimes |K, L, r\rangle\langle K, L, r|, \tag{21.19}$$

where $|K, L, r\rangle$ are the joint (generalized) eigenvectors of $\hat{K}$ and $\hat{L}$, with r a degeneracy index. The joint eigenvectors $|Y, Z\rangle$ of $\hat{Y}$ and $\hat{Z}$ satisfy

$$\langle Y, X, r|K, L, r'\rangle = \frac{e^{iKY + iLZ}}{2\pi}\delta_{rr'}. \tag{21.20}$$

We evaluate the joint probability density $P(Y,Z)$ for the pointer variables Y and Z,

$$P(Y,Z) = \sum_r \left(\langle \psi | \otimes \langle \Omega | \right) \hat{U}_t^\dagger | \left(\hat{I} \otimes | Y,Z,r \rangle \langle Y,Z,r | \right) \hat{U}_t \left(| \psi \rangle \otimes | \Omega \rangle \right)$$

$$= \sum_r \int dK dL dK dL' \langle \psi_{t_0} | e^{i(K'-K)\hat{x} + i(L'-L)\hat{p}} | \psi_{t_0} \rangle e^{\frac{i}{2}(KL' - K'L)}$$

$$\times \langle \Omega | K', L', r \rangle \langle K', L', r | Y, Z, r \rangle \langle Y, Z, r | K, L, r \rangle \langle K, L, r | \Omega \rangle$$

$$= \frac{1}{4\pi^2} \int dK dL dK' dL' \langle \psi_{t_0} | e^{i(K'-K)(\hat{x}-Y) + i(L'-L)(\hat{p}-Z)} | \psi_{t_0} \rangle e^{\frac{i}{2}(KL' - K'L)}$$

$$\times \sum_r \langle K, L, r | \Omega \rangle \langle \Omega | K', L', r \rangle. \tag{21.21}$$

The term $\sum_r \langle K, L, r | \Omega \rangle \langle \Omega | K', L', r \rangle$ is a matrix element of the reduced density matrix for the pointer variables, that is, it involves a tracing out of all degrees of freedom of the apparatus other than the pointer variables. We assume that the two pointer variables are statistically independent initially, so that

$$\sum_r \langle K, L, r | \Omega \rangle \langle \Omega | K', L', r \rangle = \rho_1(K, K') \rho_2(L, L'), \tag{21.22}$$

in terms of density matrices $\rho_1(K, K')$ and $\rho_2(L, L')$ on $L^2(\mathbb{R})$. We change integration variables to $\bar{K} := \frac{1}{2}(K + K')$, $\kappa = K' - K$, $\bar{L} = \frac{1}{2}(L + L')$ and $\lambda = L' - L$. Then, Eq. (21.21) becomes

$$P(Y,Z) = \int d\kappa \, d\lambda \, \langle \psi_{t_0} | e^{i\kappa(\hat{x}-Y) + i\lambda(\hat{p}-Z)} | \psi_{t_0} \rangle$$

$$\times \left[\int \frac{d\bar{K}}{2\pi} e^{-\frac{i}{2}\lambda\bar{K}} \rho_1 \left(\bar{K} + \frac{1}{2}\kappa, \bar{K} - \frac{1}{2}\kappa \right) \right] \left[\int \frac{d\bar{L}}{2\pi} e^{\frac{i}{2}\kappa\bar{L}} \rho_2 \left(\bar{L} + \frac{1}{2}\lambda, \bar{L} - \frac{1}{2}\lambda \right) \right]. \tag{21.23}$$

Using Eq. (6.57), it is straightforward to show that the terms in the square brackets are the Fourier transforms $\tilde{W}(u,v) := (2\pi)^{-1} \int dx dp \, W(x,p) e^{-ixu-ivp}$ of the Wigner functions W_1 and W_2, associated to $\hat{\rho}_1$ and $\hat{\rho}_2$, respectively. In particular, the integral over $\bar{K}$ yields $\tilde{W}_1(\kappa, \frac{\lambda}{2})$ and the integral over $\bar{L}$ yields $\tilde{W}_2(\lambda, -\frac{\kappa}{2})$. Hence, Eq. (21.21) becomes

$$P(Y,Z) = \langle \psi_{t_0} | \hat{\Pi}(Y,Z) | \psi_{t_0} \rangle, \tag{21.24}$$

where

$$\hat{\Pi}(Y,Z) = \int d\kappa \, d\lambda \, e^{i\kappa(\hat{x}-Y) + i\lambda(\hat{p}-Z)} \tilde{W}_1 \left(\kappa, \frac{\lambda}{2} \right) \tilde{W}_2 \left(\lambda, -\frac{\kappa}{2} \right). \tag{21.25}$$

Since $P(Y,Z) \geq 0$ for all $|\psi\rangle$, $\hat{\Pi}(Y,Z) \geq 0$. Furthermore, $\int dY dZ \hat{\Pi}(Y,Z) = \hat{I}$. It is useful to evaluate the Weyl transform $F_{\hat{\Pi}(Y,Z)}(x,p)$ of the operator $\hat{\Pi}(Y,Z)$. As shown in Example 6.8, the Weyl transform of a Weyl operator $e^{ia\hat{x}+ib\hat{p}}$ is $e^{iax+ibp}$, hence,

$$F_{\hat{\Pi}(Y,Z)}(x,p) = \int d\kappa \, d\lambda \, e^{i\kappa(x-Y) + i\lambda(p-Z)} \tilde{W}_1 \left(\kappa, \frac{\lambda}{2} \right) \tilde{W}_2 \left(\lambda, -\frac{\kappa}{2} \right) \tag{21.26}$$

As an example, we select the Wigner functions W_1 and W_2 to be Gaussians of the form (6.58),

$$W_1(Y,K) = \frac{1}{\pi} e^{-\frac{Y^2}{2\sigma_Y^2} - 2\sigma_Y^2 K^2}, \qquad W_2(Z,L) = \frac{1}{\pi} e^{-\frac{Z^2}{2\sigma_Z^2} - 2\sigma_Z^2 K^2}, \tag{21.27}$$

where σ_Y and σ_Z are the mean deviations of the two pointer variables. Then,

$$F_{\hat{\Pi}(Y,Z)}(x,p) = \frac{2}{\pi\sqrt{(\sigma_Y^2 + \sigma_Z^{-2})(\sigma_Z^2 + \sigma_Y^{-2})}} e^{-\frac{(Y-x)^2}{2(\sigma_Y^2+\sigma_Z^{-2})} - \frac{(Z-p)^2}{2(\sigma_Z^2+\sigma_Y^{-2})}}. \tag{21.28}$$

Equation (21.28) clarifies the role of the uncertainty relation in the context of measurement theory. The spreads σ_Y and σ_Z of the pointer variables Y and Z can become arbitrarily small simultaneously since $[\hat{Y}, \hat{Z}] = 0$. The pointer variable Y is correlated to x within an accuracy $\delta x = \sqrt{\sigma_Y^2 + \sigma_Z^{-2}}$ and the pointer variable is correlated to p within an accuracy $\delta p = \sqrt{\sigma_Z^2 + \sigma_Y^{-2}}$. Then,

$$\delta x \, \delta p = \sqrt{2 + \sigma_Z^2 \sigma_Y^{-2} + \sigma_Y^2 \sigma_Z^{-2}} \geq 2 \tag{21.29}$$

which is a novel form of the uncertainty relation. It means that simultaneous measurements of position and momentum are possible; however, the measurement is inexact with an intrinsic error $\delta x \, \delta p \geq 2$. The reader can derive a sharper and more general version of Eq. (21.29) by solving Problem 21.3.

21.2 General Theory of Measurements

In Section 21.1, we showed that the quantum treatment of the apparatus may lead to probabilities of the form $\langle \psi | \hat{\Pi} | \psi \rangle$, where the operators $\hat{\Pi}$ correspond to measurement outcomes, but they are not projectors. This pattern persists in all measurement models, and it suggests the need to generalize the quantum rules of probability assignment. To this end, we introduce the notion of a *Positive-Operator-Valued Measure* (POVM).

Definition 21.1 Let Γ be the set of all possible outcomes in a measurement. A POVM is a map that assigns a positive operator $\hat{\Pi}_x$ to each $x \in \Gamma$, such that the integrals $\int_C dx \hat{\Pi}_x$ are well defined for all $C \subset \Gamma$, and $\int_\Gamma dx \, \hat{\Pi}_x = \hat{I}$.

There is one crucial difference between the description of measurement with POVMs and the description of measurement in terms of spectral projectors. Consider, for example, the spectral projectors $\hat{P}_x$ of the position operator $\hat{x}$. For any $x \neq x'$, $\hat{P}_x \hat{P}_{x'} = 0$, that is, the elementary measurement outcomes are mutually exclusive. A successive measurement of position will give the same result. In contrast, for a general POVM $\hat{\Pi}_x \hat{\Pi}_{x'} \neq 0$ for $x \neq x'$. For the POVM (21.14), we straightforwardly evaluate $\hat{\Pi}_X \hat{\Pi}_{X'} = \int da w(X - a) w(X' - a) |a\rangle \langle a|$, which is different from zero for a general function w. The fact that the elementary outcomes are not exclusive in a POVM means that there is no perfect correlation between pointer variable and microscopic observable.

Sometimes the POVMs can be viewed as defining *approximate projection operators*. This means that the operators $\hat{\Pi}_C := \int dx \hat{\Pi}_x$ satisfy $\hat{\Pi}_C^2 \simeq \hat{\Pi}_C$ for sufficiently large sets C.

Table 21.1 Values of the quantity Q, Eq. (21.31)

L/σ	5	10	20	30	40	50	100	200
Q	0.23	0.11	0.059	0.032	0.015	0.0064	9.4×10^{-6}	2.5×10^{-16}

Example 21.1 Consider the POVM (21.14) with a Gaussian function

$$w(x) = (\pi\sigma^2)^{-1/2} \exp[-x^2/\sigma^2]. \tag{21.30}$$

Let $C = [-\frac{L}{2}, \frac{L}{2}]$ be an interval centered around $x = 0$ with width L. We find that $\hat{\Pi}_C = \int da\, g(a)|a\rangle\langle a|$, where $g(a) = \int_{-\frac{L}{2}}^{\frac{L}{2}} dx\, w(x-a)$. We compare $\hat{\Pi}_C = \int da\, g(a)|a\rangle\langle a|$ with the spectral projector $\hat{P}_C = \int da\, \chi_C(a)|a\rangle\langle a|$ that corresponds to an ideal measurement. Their difference is expressed by the quantity

$$Q := \frac{1}{L} \int da\, |g(a) - \chi_C(a)|, \tag{21.31}$$

which is a decreasing function of L/σ. Some numerical values for Q are given in Table 21.1. We see that for $L/\sigma \simeq 50$, $\hat{\Pi}_C$ approximates $\hat{P}_C$ with an accuracy better than 1 percent.

Example 21.2 As we saw in Section 21.1.4, POVMs can also define joint probabilities for incompatible observables, such as position and momentum. The operators (21.25) effectively define a POVM on the classical state space $\Gamma = \mathbb{R}^2$ for a particle in one dimension. An alternative POVM on Γ can be defined in terms of coherent states. Let us denote the coherent states (5.39) by $|\bar{q}\bar{p}\rangle$. The operators $\hat{\Pi}_{\bar{q}\bar{p}} = \frac{1}{\pi}|\bar{q}\bar{p}\rangle\langle\bar{q}\bar{p}|$ define a POVM. In Problem 10.12, it is shown that $\hat{\Pi}_C := \int_C d\bar{q}d\bar{p}\,\hat{\Pi}_{\bar{q}\bar{p}}$ is an approximate projector for subsets C of Γ with area much larger than unity.[1]

Example 21.3 Sequential measurements can also be expressed in terms of POVMs. Assume that we first measure the observable $\hat{A} = \sum_n a_n \hat{P}_n$ and then the observable $\hat{B} = \sum_m b_m \hat{Q}_m$. If n takes values in a set Γ_1 and m takes values in a set Γ_2, then we can express the probabilities (8.14) in terms of a POVM

$$\hat{\Pi}_{n,m} = \hat{C}_{n,m}^\dagger \hat{C}_{n,m} \tag{21.32}$$

on $\Gamma_1 \otimes \Gamma_2$, where $\hat{C}_{n,m} = \hat{Q}_m \hat{P}_n$. The generalization to an arbitrary number of sequential measurements, using Eq. (8.24), is straightforward.

Suppose that the measurement outcomes correspond to an exclusive and exhaustive families of subsets C_a of Γ. Then, we expect that there exists a linear positive map $\mathcal{P}_{C_a}$ that takes the initial state $\hat{\rho}_0$ of the measured system to a state $\hat{\rho}(C_a)$ after measurement, for all a. We saw how to construct such maps in Section 21.1.3. Typically, the maps $\mathcal{P}_{C_a}$ are unnormalized, because $\mathrm{Tr}\hat{\rho}(C_a) = \mathrm{Prob}(C_a)$, so we need to divide $\hat{\rho}(C_a)$ by $\mathrm{Prob}(C_a)$, in order to define a genuine density matrix after measurement.

---

[1] Note that the approximation $\hat{\Pi}_C^2 \simeq \hat{\Pi}_C$ may hold, even if we cannot define any genuine projection operator associated to the set C.

In some cases, the maps $\mathcal{P}_{C_a}$ may preserve the purity of the quantum state, that is, the states $\hat{\rho}(C_a)$ are pure, $\hat{\rho}(C_a) = |\psi, C_a\rangle\langle\psi, C_a|$ for any initial state $\hat{\rho}_0 = |\psi_0\rangle\langle\psi_0|$. Then, there exists a *measurement operator*, that is, an operator $\hat{K}_{C_a}$, such that $|\psi, C_a\rangle = \hat{K}_{C_a}|\psi_0\rangle$. The vectors $|\psi, C_a\rangle$ are not normalized, since $\text{Prob}(C_a) = \text{Tr}\hat{\rho}_{C_a} = \langle\psi_0|\hat{K}_{C_a}^\dagger\hat{K}_{C_a}|\psi_0\rangle$. This means that the POVM operators $\hat{\Pi}_{C_a}$ are expressed in terms of measurement operators, as

$$\hat{\Pi}_{C_a} = \hat{K}_{C_a}^\dagger\hat{K}_{C_a}. \tag{21.33}$$

In general, $\hat{K}_{C_a}$ need not be self-adjoint. For example, in the sequential measurements of Example 21.3, the operators $\hat{C}_{n,m}$ play the role of measurement operators. Nonetheless, in many cases, $\hat{K}_{C_a}$ is by construction an approximate projector, hence, we identify $\hat{K}_{C_a} = \sqrt{\hat{\Pi}_{C_a}}$.

In principle, we can take the representation of measurements by POVMs as a fundamental principle of quantum theory, in place of FP3. But POVMs for a measured system can be defined by employing FP3 on a larger system that includes the apparatus together with the measured system. Indeed, all POVMs we will encounter in this book are of this form, so we will refrain from such a generalization.

21.3 Particle-Detection Models

Particle detectors are devices that detect and track particles, including photons and high-energy massive particles. This detection process cannot fundamentally be modeled by von Neumann measurements, because detection does not occur at a predetermined instant of time. Rather, the detector is placed at a fixed location, and we wait until the detected particle is recorded, the time of detection being a random variable. However, as we have seen, time operators cannot be defined in quantum theory, so probability with respect to the detection time cannot be constructed from a simple application of Born's rule.

In this section, we will provide a simple model for particle detectors that has a broad range of applicability (Anastopoulos and Savvidou, 2012). In particular, it allows us to construct POVMs for the measurement of time observables. For simplicity, we will restrict ourselves to free, spinless relativistic particles of mass m, described by a scalar field, as in Eq. (15.84). This suffices for analyzing important issues like the construction of time-of-arrival probabilities for massive particles, but also photodetection probabilities when taking the limit $m \to 0$.

21.3.1 Detection Probabilities

We will analyze the detection of particles described by a scalar field $\hat{\phi}(x)$ given by Eq. (15.84). We assume that the coupling between the system and the apparatus is nonvanishing only in a small spacetime region around a spacetime point $\bar{x} = (\bar{t}, \bar{\mathbf{x}})$. Hence, we consider an interaction Hamiltonian of the form

$$V_{\bar{x}} = \int d^3x F_{\bar{t}, \bar{\mathbf{x}}}(t, \mathbf{x})\hat{\phi}(\mathbf{x}, 0) \otimes \hat{J}(\mathbf{x}), \tag{21.34}$$

where $\hat{J}(\mathbf{x})$ is a "current" operator on $\mathcal{K}$. The switching function $F_{\bar{t}, \bar{\mathbf{x}}}(t, \mathbf{x})$ is dimensionless, and it vanishes outside the apparatus and at times when the interaction is switched off. We will write the switching function as $F_{\bar{x}}(x)$, where $x = (t, \mathbf{x})$.

By the Dyson formula (19.119), the S-matrix to leading order in the interaction reads $\hat{S}_{\bar{x}} = \hat{I} - i\hat{V}_{\bar{x}}$, where

$$\hat{V}_{\bar{x}} = \int d^4x F_{\bar{x}}(x)\hat{\phi}(x) \otimes \hat{J}(x), \tag{21.35}$$

where $\hat{J}(x) = e^{i\hat{h}t}\hat{J}(\mathbf{x})e^{-i\hat{h}t}$ and $\hat{h}$ is the Hamiltonian of the detector.

Actual particle detectors (e.g., photographic plates, silicon strips) have a fixed location in space and they are made sensitive for a long time interval, during which particles may be detected. Therefore, the location of a detection event is a fixed parameter of the experiment; the actual random variable is the detection time. This suggests taking the limit of a pointlike detector, that is, we assume that the detector moves along a fixed timelike trajectory $x_0(\tau)$. Therefore, we consider only points $\bar{x}$ along the trajectory $x_0(\tau)$. We choose a switching function of the form

$$F_{\bar{x}}(x) = \int d\tau \delta^4[x - x_0(\tau)]f_{\bar{\tau}}(\tau). \tag{21.36}$$

The function $f_{\bar{\tau}}(\tau)$ is peaked around $\tau = \bar{\tau}$, where $\bar{\tau}$ is the solution of $x_0(\bar{\tau}) = \bar{x}$.

Then, $\hat{V}_{\bar{x}}$ is parameterized by $\bar{\tau}$, and we can express it as

$$\hat{V}_{\bar{\tau}} = \int d\tau f_{\bar{\tau}}(\tau)\hat{\phi}[x_0(\tau)] \otimes \hat{J}[x_0(\tau)]. \tag{21.37}$$

Let the initial state of the system be $|\psi\rangle \in \mathcal{F}$. We take the initial state of the apparatus $|\Omega\rangle$ to be an eigenstate of $\hat{h}$. It is convenient to assume that $\langle\Omega|\hat{J}(x)|\Omega\rangle = 0$, so that the expectation of $\hat{V}_{\bar{x}}$ vanishes for all initial field states.

A particle record appears if the detector transitions from $|\Omega\rangle$ to any other state orthogonal to $|\Omega\rangle$. Such a transition corresponds to the projector $\hat{P} = \hat{I} - |\Omega\rangle\langle\Omega|$ on $\mathcal{K}$. Hence, the probability that a transition occurs at proper time $\bar{\tau}$ is

$$\text{Prob}(\bar{\tau}) = \langle\psi, \Omega|\hat{S}_{\bar{\tau}}^{\dagger}(\hat{I} \otimes \hat{P})\hat{S}_{\bar{\tau}}|\psi, \Omega\rangle$$
$$= \int d\tau_1 d\tau_2 f_{\bar{\tau}}(\tau_1)f_{\bar{\tau}}(\tau_2)G(\tau_1, \tau_2)R(\tau_1, \tau_2), \tag{21.38}$$

where

$$G(\tau, \tau') = \langle\psi|\hat{\phi}[x_0(\tau)]\hat{\phi}[x_0(\tau')]|\psi\rangle \tag{21.39}$$

is a correlation function for the quantum field, and

$$R(\tau, \tau') = \langle\Omega|\hat{J}[x_0(\tau)]\hat{J}[x_0(\tau')]|\Omega\rangle. \tag{21.40}$$

The probability (21.38) is not a density with respect to $\bar{\tau}$; $\bar{\tau}$ appears as a parameter of the switching function. We can define an unnormalized probability density $W(\bar{\tau})$ with respect to $\bar{\tau}$ by dividing $\text{Prob}(\bar{x})$ by the effective duration of the interaction $T = \int d\tau f_{\bar{\tau}}^2(\tau)$:

$$W(\bar{\tau}) = T^{-1}\text{Prob}(\bar{\tau}). \tag{21.41}$$

The alert reader will notice that this definition involves the combination of probabilities defined with respect to different experimental set-ups, that is, different switching functions for the Hamiltonians. As we saw in Section 6.6, this practice is highly problematic in quantum theory. In this particular case, the definition (21.41) works: The resulting probability density can also be constructed, as an approximation,[2] without this assumption – see Anastopoulos and Savvidou (2019).

Next, we assume that the detector is motionless in some reference frame, so that we can write $x_0(\tau) = (\tau, \mathbf{x})$. This means that the proper time τ coincides with the coordinate time $t = \tau$ and that the detector is located at a fixed spatial coordinate $\mathbf{x}$.

Then, $R(\tau, \tau') = R(\tau' - \tau)$, where

$$R(\tau) = \langle \Omega | \hat{J}(0, \mathbf{x}) e^{i\hat{h}\tau} \hat{J}(0, \mathbf{x}) | \Omega \rangle. \tag{21.42}$$

In this case, it is also natural to take the switching function $f_{\bar{\tau}}(t) = f(t - \bar{\tau})$ for some function f. We choose a Gaussian $f(t) = \exp\left[-t^2/(2\delta_t^2)\right]$, where δ_t is the temporal accuracy of the detection event. For this sampling function $T = \sqrt{\pi}\delta_t$.

The Gaussians $f(t)$ satisfy the useful identity $f(t)f(t') = f^2\left(\frac{t+t'}{2}\right)\sqrt{f}(t - t')$. It is straightforward to show that the function $g(t) := \frac{1}{T}f^2(t)$ is a normalized probability density on $\mathbb{R}$. This implies that we can write

$$W(\bar{\tau}) = \int dt\, g(\bar{\tau} - t) P(t), \tag{21.43}$$

where

$$P(t) = \int d\xi \sqrt{f}(\xi) R(\xi) G\left(t - \frac{1}{2}\xi, t + \frac{1}{2}\xi\right). \tag{21.44}$$

The probability distribution $W(\bar{\tau})$ is the convolution of $P(t)$ with the probability density $g(t)$ that incorporates the accuracy of our measurements. If $P(t)$ is nonnegative and the scale of variation in t is much larger than both δ_t, we can treat $P(t)$ as a finer-grained version of $W(\bar{\tau})$ and use this as our probability density for detection.

The kernel $R(\xi)$ is typically characterized by some microscopic correlation timescale σ, such that $R(\xi) \simeq 0$ if $|\xi| \gg \sigma$. Choosing $\delta_t >> \sigma$ means that $R(\xi) \simeq \sqrt{f}(\xi) R(\xi)$ and, hence, we obtain our final expression for the detection probability

$$P(t) = \int d\xi\, R(\xi)\, G\left(t - \frac{1}{2}\xi, t + \frac{1}{2}\xi\right). \tag{21.45}$$

The probability density (21.45) is not normalized to unity. In general, the total probability of detection $P_{det} = \int dt P(t)$ must be a small number for perturbation theory to be applicable. There is always a probability $P(\emptyset) = 1 - P_{det}$ of no detection. We normalize probabilities as $P(t)/P_{det}$, that is, by conditioning the probability densities $P(t)$ with respect to the fact that a detection record exists.

21.3.2 Time-of-Arrival Probabilities

In a time-of-arrival measurement, a particle is prepared on an initial state $|\psi_0\rangle$ that is localized around $x = 0$ and has positive mean momentum. For a detector placed at $x = L$, one looks

---

[2] Note, however, that as a consequence of the approximations involved in its derivation, Eq. (21.38) is characterized by a spurious transient behavior prior to the particle's first detection, that manifests as a tiny probability for a faster-than-light signal, similar to that discussed in Box 7.1.

for the probability density $P(t)$ with respect to the detection time t. The lack of a self-adjoint operator for time (Theorem 7.1) means that we cannot employ Born's rule in order to obtain an answer. However, Eq. (21.45) directly applies to this problem.

If the distance L of the detector from the source is much larger than the size of the detector, only particles with momenta along the axis that connects the source to the detector are recorded. Hence, the problem is reduced to two spacetime dimensions. Then, the field operator reads

$$\hat{\phi}(t, L) = \int \frac{dp}{\sqrt{2\pi}\sqrt{2\epsilon_{\mathbf{p}}}} \left[\hat{a}_p e^{ipL - i\epsilon_p t} + \hat{a}_p^\dagger e^{-ipL + i\epsilon_p t} \right]. \tag{21.46}$$

Here, $\hat{a}_p$ and $\hat{a}_p^\dagger$ are annihilation and creation operators for the particles, and $\epsilon_p = \sqrt{p^2 + m^2}$.

We assume a single-particle initial state $|\psi_0\rangle = \int dp \psi_0(p) \hat{a}_p^\dagger |0\rangle$, with $\int dp |\psi_0(p)|^2 = 1$. Then, we calculate the correlation function

$$G(t, t') = \langle \psi_0 | \hat{\phi}(t, L) \hat{\phi}(t', L) | \psi_0 \rangle = \int \frac{dp}{4\pi \epsilon_p} e^{-i\epsilon_p(t - t')} +$$
$$\int \frac{dp\, dp'}{4\pi \sqrt{\epsilon_p \epsilon_{p'}}} \psi_0^*(p') \psi_0(p) e^{i(p - p')L} \left(e^{-i(\epsilon_p t - \epsilon_{p'} t')} + e^{-i(\epsilon_p t' - \epsilon_{p'} t)} \right). \tag{21.47}$$

Substituting in Eq. (21.45), we encounter the Fourier transform

$$\tilde{R}(E) = \int d\xi\, e^{-iE\xi} R(\xi) = 2\pi \langle \omega | \delta(E - \hat{h}) | \omega \rangle \geq 0, \tag{21.48}$$

where we set $|\omega\rangle = \hat{J}(0, \mathbf{x}) | \Omega \rangle$. We also note that the positivity of the apparatus Hamiltonian, $\hat{h} \geq 0$, implies that

$$\tilde{R}(E) = 0, \quad \text{for } E < 0. \tag{21.49}$$

Then, the detection probability becomes

$$P(t) = \int \frac{dp\, dp'}{2\pi} \frac{\psi_0(p) \psi_0^*(p')}{2\sqrt{\epsilon_p \epsilon_{p'}}} \tilde{R}\left(\frac{\epsilon_p + \epsilon_{p'}}{2} \right) e^{i(p - p')L - i(\epsilon_p - \epsilon_{p'})t}. \tag{21.50}$$

Here, we dropped a small state-independent constant term that corresponds to the excitation of the detector by the vacuum, and it can be interpreted as background noise.

We assume that the initial state has only positive momentum content, that is, $\psi_0(p) = 0$ for $p < 0$. Then, we evaluate the time-integrated probability density,

$$P_{tot} = \int_{-\infty}^\infty dt P(t) = \int_0^\infty dp |\psi_0(p)|^2 \frac{\tilde{R}(\epsilon_p)}{2p}. \tag{21.51}$$

We extended the integration to the full real axis for time, because $P(t)$ for $t < 0$ is negligibly small for any wave function with strictly positive momentum. Hence, we identify the ratio $\frac{\tilde{R}(\epsilon_p)}{2p} \geq 0$ with the absorption coefficient $\alpha(p)$ of the detector, namely, the fraction of particles at momentum p that are absorbed when encountering the detector. Then, we define the probability density $P_c(t) := P(t)/P_{tot}$, conditioned upon the particle having being detected. By construction, $P_c(t)$ is normalized to unity, $\int_{-\infty}^\infty dt P_c(t) = 1$.

We redefine the initial state by the normalization-preserving transformation

$$\psi_0(p) \to \frac{\sqrt{\alpha(p)}\psi_0(p)}{\sqrt{P_{det}}}, \tag{21.52}$$

to obtain

$$P_c(t) = \int \frac{dpdp'}{2\pi} \psi_0(p)\psi_0^*(p')\sqrt{v_p v_{p'}} S(p,p') e^{i(p-p')x - i(\epsilon_p - \epsilon_{p'})t}, \tag{21.53}$$

where $v_p = p/\epsilon_p$ is the relativistic particle velocity. $S(p,p')$ are the matrix elements $\langle p|\hat{S}|p'\rangle$ of an operator $\hat{S}$,

$$\langle p|\hat{S}|p'\rangle := \frac{\tilde{R}\left(\frac{\epsilon_p + \epsilon_{p'}}{2}\right)}{\sqrt{\tilde{R}(\epsilon_p)\tilde{R}(\epsilon_{p'})}}. \tag{21.54}$$

By definition, $\langle p|\hat{S}|\hat{p}'\rangle \geq 0$, and $S(p,p) = 1$. The operator $\hat{S}$ determines the irreducible spread of the record in the apparatus – see Problem 21.5 – and for this reason we will call it the *localization operator*.

We express Eq. (21.53) as

$$P_c(t) = \langle \psi|\hat{U}^\dagger(t,L)\sqrt{|\hat{v}|}\hat{S}\sqrt{|\hat{v}|}\hat{U}(t,L)|\psi\rangle, \tag{21.55}$$

where $\hat{U}(t,L) = e^{i\hat{p}L - i\hat{H}t}$ is the spacetime-translation operator and $\hat{v} = \hat{p}\hat{H}^{-1}$ is the velocity operator.

The positivity of $P_c(t)$ for all initial states implies that $\hat{S} \geq 0$. By the Cauchy–Schwarz inequality,

$$\langle p|\hat{S}|p'\rangle \leq \sqrt{\langle p|\hat{S}|p\rangle\langle p'|\hat{S}|p'\rangle} = 1. \tag{21.56}$$

Maximal localization is achieved when Eq. (21.56) is saturated, that is, for $\langle p|\hat{S}|p'\rangle = 1$. By Eq. (21.54), maximal localization is possible if $\tilde{R}$ is an exponential function, $\tilde{R}(E) = Ce^{-\sigma E}$ for some constants $C, \sigma > 0$. The probability density for maximal localization defines the Kijowski–Léon POVM for the time of arrival (Kijowski, 1974; León, 1997):

$$P_c(t) = \left| \int \frac{dp}{\sqrt{2\pi}} \psi_0(p)\sqrt{v_p}e^{ipL - i\epsilon_p t} \right|^2. \tag{21.57}$$

Equation (21.57) is a genuine probability density for the time of arrival. It does not conflict with Theorem 7.1, because it is not obtained from a self-adjoint operator.

An important consequence of Eq. (21.57) is the existence of a new time–energy uncertainty relation,

$$(\Delta t)^2 \geq \frac{1}{4(\Delta H)^2} + \frac{m^4}{4}\left\langle \left(\hat{H}\hat{p}^2\right)^{-2}\right\rangle. \tag{21.58}$$

Equation (21.58) is proven in Box 21.1.

All terms that appear in the right-hand side of Eq. (21.58) are determined solely by the momentum probability distribution associated to the quantum state. The mean deviation Δt

is a measure of the localization of a particle. This interpretation is made possible by the fact that second term never vanishes. For an almost monochromatic state with $\langle H \rangle >> \Delta H$, the term $(\Delta H)^{-2}$ dominates on the right-hand side of Eq. (21.58). However, for any probability distribution for energy that drops slower than p^{-3} as $p \to \infty$, $\Delta H = \infty$, and localization is solely determined by the second term.

Box 21.1   Proof of the Uncertainty Relation (21.58)

We evaluate the moment-generating function $Z(\mu) = \int dt P(t) e^{i\mu t}$ associated to Eq. (21.50):

$$Z(\mu) = \int dp dp' \, \rho_0(p, p') \sqrt{v_p v_{p'}} \, e^{i(p-p')L} \delta(\epsilon_p - \epsilon_{p'} - \mu).$$

We wrote $\rho_0(p, p') = \psi_0(p)\psi_0^*(p')$, and considered maximum localization, $L(p, p') = 1$. We solve the equation $\epsilon_p - \epsilon_{p'} = \mu$, by expressing $\xi := p - p'$ as a function of $\bar{p} = \frac{1}{2}(p + p')$, that is, we write $\xi = \frac{\mu}{v_{\bar{p}}}\left[1 + g(\mu, \bar{p})\right]$, where $g(\mu, \bar{p}) = \sum_{n=1}^{\infty} c_n(\bar{p})\left(\frac{\mu}{\bar{p}}\right)^{2n}$ is a series in μ^2. The explicit form of $g(\mu, \bar{p})$ will not be needed, only the fact that $(\partial g/\partial \mu)_{\mu=0} = 0$.

The moment-generating function becomes

$$Z(\mu) = \int dp dp' \, \psi_0(p)\psi_0^*(p') \frac{2\sqrt{v_p v_{p'}}}{v_p + v_{p'}} \langle p|\hat{L}|p'\rangle e^{i(p-p')x} \delta\left(p - p' - \frac{\mu}{v_{\bar{p}}}\left[1 + g(\mu, \bar{p})\right]\right).$$

We express the density matrix $\rho(p, p')$ in terms of its associated Wigner function $W(X, P)$. We also expand $2\frac{\sqrt{v_p v_{p'}}}{v_p + v_{p'}}$ as a power series in ξ^2,

$$2\frac{\sqrt{v_p v_{p'}}}{v_p + v_{p'}} = 1 - \sum_{n=1}^{\infty} d_n(\bar{p})\xi^{2n}.$$

In this calculation, we will need only the first term in the series, $d_1(\bar{p}) = \frac{m^4}{8\epsilon_{\bar{p}}^4 \bar{p}^2}$. We write,

$$Z(\mu) = \int dX \, dP \, W(X, P) e^{iT_c \mu[1 + g(\mu, P)]} \left(\frac{\mu}{v_P}\left[1 + g(\mu, P)\right]\right)$$
$$\times \left[1 - \sum_{n=1}^{\infty} d_n(P)\left(\frac{\mu}{v_P}\right)^{2n}\left[1 + g(\mu, P)\right]^{2n}\right],$$

where $T_c(X, P) := \frac{L - X}{v_P}$ is the classical time-of-arrival observable.

The expectation value of the arrival time is $\langle t \rangle = -i\left(\frac{\partial \ln Z}{\partial \mu}\right)_{\mu=0} = \langle T_c \rangle$, where the expectation of any classical observable F is $\langle F \rangle = \int dX dP W(X, P) F(X, P)$. The quantum expectation value agrees with the classical one.

The expectation value $\langle T_c \rangle$ is of the form $\mathrm{Tr}(\hat{\rho}\hat{T}_c)$, where $\hat{T}_c = \frac{1}{2}\left[(L - \hat{x})\hat{v}^{-1} + \hat{v}^{-1}(L - \hat{x})\right]$, with $\hat{x} = i\frac{\partial}{\partial p}$. The operator $\hat{T}_c$ is not self-adjoint, because its domain $D_{\hat{T}_c}$ cannot be extended to the whole of the Hilbert space. ($\hat{T}_c$ cannot be defined on wave functions that do not vanish at $p = 0$.)

Next, we evaluate

$$(\Delta t)^2 = -\left(\frac{\partial^2 \ln Z}{\partial^2 \mu^2}\right)_{\mu=0} = (\Delta T_c)^2 + 2\langle d_1(p)v_p^{-2}\rangle = (\Delta T_c)^2 + \frac{m^4}{4}\langle\frac{1}{\epsilon_p^2 p^4}\rangle.$$

For any $|\psi\rangle \in D_{\hat{T}_c}$, $[\hat{H},\hat{T}_c]|\psi\rangle = -i|\psi\rangle$, and therefore, by the Kennard–Robertson inequality, $\Delta T_c > \frac{1}{2\Delta H}$. Thus, we obtain Eq. (21.58).

21.3.3 Photodetection Probabilities

Equation (21.45) can also be used as a photodetection probability, that is, in order to define the probability density of photon detection. To this end, we consider the scalar field with zero mass as a model for the quantum EM field. As shown in Section 18.4.2, this model describes scalar photons, as it ignores the polarization degrees of freedom.

Typical states of the quantum EM field do not have a fixed number of photons. Hence, the correlation function $G(t,t')$ that appears in Eq. (21.45) involves a sum of four terms, $\langle\hat{\psi}^\dagger(t)\hat{\psi}(t')\rangle$, $\langle\hat{\psi}(t)\hat{\psi}^\dagger(t')\rangle$, $\langle\hat{\psi}^\dagger(t)\hat{\psi}(t')\rangle$, and $\langle\hat{\psi}^\dagger(t)\hat{\psi}^\dagger(t')\rangle$. The second terms generates a factor $\tilde{R}(E)$ for $E < 0$ in Eq. (21.45), which vanishes. The last two terms involve terms $e^{\pm i(\epsilon_p + \epsilon_{p'})t}$ which are rapidly oscillating, and they are expected to be much smaller than the first term. Hence, invoking the RWA, Eq. (21.45) becomes

$$P(t) = \frac{1}{2}\int d\xi\ R(\xi)\left\langle\hat{\psi}^\dagger\left(t - \frac{1}{2}\xi,\mathbf{x}\right)\hat{\psi}\left(t + \frac{1}{2}\xi,\mathbf{x}\right)\right\rangle. \tag{21.59}$$

Suppose now that the timescale σ of R is much smaller than any characteristic scale of variation in the photon state. We can approximate $R(\xi) \sim \delta(\xi)$; hence, Eq. (21.59) becomes

$$P(t) = \left\langle\hat{\psi}^\dagger(t,\mathbf{x})\hat{\psi}(t,\mathbf{x})\right\rangle, \tag{21.60}$$

modulo a multiplicative constant. Equation (21.60) was first proposed by Glauber (1963) as a basis of a photodetection theory that has found a large number of applications in quantum optics.

Glauber's modeling of photodetection also involved joint probabilities for photodetection at different times. For example, the joint probability density for two detections, at times t_1 and t_2, and positions $\mathbf{x}_1$ and $\mathbf{x}_2$, respectively, is

$$P(t_1,t_2) = \left\langle\hat{\psi}^\dagger(t_1,\mathbf{x}_1)\hat{\psi}^\dagger(t_2,\mathbf{x}_2)\hat{\psi}(t_2,\mathbf{x}_2)\hat{\psi}(t_1,\mathbf{x}_1)\right\rangle, \tag{21.61}$$

modulo a multiplicative constant.

Joint detection probabilities can be determined by the time series of photodetection events. In order to connect theoretical expressions such as Eq. (21.61) to experiment, it is convenient to define appropriate ratios of probabilities so that the undetermined multiplicative constants cancel. An important example is the *photon autocorrelation function*[3]

---

[3] In quantum optics, the autocorrelation function is referred to as the *second-order coherence function*, and it is denoted by $g^{(2)}(T)$.

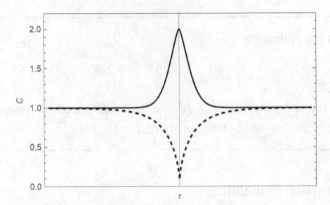

Fig. 21.1 Typical behavior of the autocorrelation function C for bunching (solid line) and antibunching (dashed line).

$$C(t, \tau) := \frac{P(t, t + \tau)}{P(t)P(t + \tau)}, \tag{21.62}$$

defined for two detectors at the same location, $\mathbf{x}_1 = \mathbf{x}_2 = \mathbf{x}$. The autocorrelation function is usually measured for field states that lead to a steady detection rate, so that $P(t)$ is constant and $P(t_1, t_2)$ depends only on $t_2 - t_1$. Then, the autocorrelation function is t-independent, and we will denote it as $C(\tau)$.

If $C(\tau) = 1$, photodetection events are statistically independent. If $C(\tau) < C(0)$ photons tend to be detected in bunches, that is, it is more probable that the detection of one photon will be accompanied by the simultaneous detection of another. If $C(\tau) > C(0)$, then the simultaneous detection of two photons is less probable; photons are then said to *antibunch*. See Fig. 21.1 for a graphical description.

The autocorrelation function exhibits a characteristic quantum behavior that cannot be mirrored by any classical theory. The probabilities in any classical system are defined in terms of a probability density ρ on a sample space Γ. An event of photodetection at time t and position $\mathbf{x}$ corresponds to an observable $F_{t,\mathbf{x}} : \Gamma \to \mathbb{R}^+$. The joint probability of detection at times t_1 and t_2 at the same detector is

$$P(t_1, t_2) = \int d\xi \rho(\xi) F_{t_1, \mathbf{x}}(\xi) F_{t_2, \mathbf{x}}(\xi), \tag{21.63}$$

where ξ stands for the points of Γ. This implies that

$$C(0) = \frac{\langle F_{t,\mathbf{x}}^2 \rangle}{\langle F_{t,\mathbf{x}} \rangle^2} \geq 1. \tag{21.64}$$

Second, by the Cauchy–Schwarz inequality, $P(t_1, t_2) \leq \sqrt{P(t_1, t_1)P(t_2, t_2)}$; hence,

$$C(\tau) \leq C(0), \tag{21.65}$$

that is, antibunching is impossible in classical physics. The failure of either Eq. (21.64) or Eq. (21.65) indicates quantum behavior with no classical analogue, similar to the violation of Bell's inequalities.

To find a violation of Eq. (21.64), it suffices to consider a set-up where only one mode of the field with wavenumber $\mathbf{k}$ is excited. That is, we write $\hat{\psi}(t, \mathbf{x}) = (2|\mathbf{k}|)^{-1/2} e^{i\mathbf{k}\cdot\mathbf{x} - i|\mathbf{k}|t} \hat{a}$. Then, we obtain

$$C(t) = \frac{\langle (\hat{a}^\dagger)^2 \hat{a}^2 \rangle}{\langle \hat{a}^\dagger \hat{a} \rangle^2} = \frac{\langle \hat{N}^2 \rangle - \langle \hat{N} \rangle}{\langle \hat{N} \rangle^2} = 1 + \frac{(\Delta N)^2 - \langle \hat{N} \rangle}{\langle \hat{N} \rangle^2}. \tag{21.66}$$

Hence, Eq. (21.64) is violated if $\Delta N < \sqrt{\langle \hat{N} \rangle}$. For example, in number states, $\Delta N = 0$; hence, there is clear violation.

To find a violation of Eq. (21.65), a one-mode field is not sufficient. We take $\mathbf{x} = 0$, and we write $\hat{\psi}(t, 0) = \sum_a \hat{a}_a e^{-i\omega_a t}$ for brevity; a labels the modes. Then, $P(0, \tau) = \langle \Psi, \tau | \Psi, \tau \rangle$, where $|\Psi, \tau\rangle = \sum_{a,b} \hat{a}_a \hat{a}_b e^{-i\omega_a \tau} |\psi_0\rangle$ is defined in terms of the initial state $|\psi_0\rangle$ of the field. Consider the special case $|\psi_0\rangle = \frac{1}{\sqrt{2}} \hat{a}^\dagger(f)^2 |0\rangle$, where $\hat{a}(f) = \sum_a f_a \hat{a}_a$ with $f_a \in \mathbb{C}$. For τ very close to zero, we find that

$$P(0, \tau) - P(0, 0) = 4\tau \left| \sum_a f_a \right|^2 K, \tag{21.67}$$

where $K = \sum_{a,b} \omega_b \mathrm{Im}(f_a^* f_b)$ may be either positive or negative.[4] We conclude that antibunching is possible in quantum theory.

Photon antibunching was predicted by Carmichael and Walls (1976) and by Kimble and Mandel (1976). It was first observed by Kimble et al (1977) in fluorescent radiation from atoms. Like the violation of Bell's inequalities, the observation of antibunching is an experimental proof that quantum correlations cannot be accounted for by any classical theory.

21.4 Continuous-Time Measurements

In this section, we will first analyze measurements in which the observables are paths (Barchielli et al., 1983), such as, for example, the tracks left by a particle in a bubble chamber.

We model a particle track by a sequence of a large number of N successive position measurements within a time interval $[0, t]$. For simplicity, we consider a particle of mass m in one spatial dimension, with Hamiltonian $\hat{H} = \frac{\hat{p}^2}{2m} + V(\hat{x})$. We assume that each measurement corresponds to a position POVM $\hat{\Pi}_x = \int dx w(x - x') |x'\rangle\langle x'|$ with measurement operators equal to $\sqrt{\hat{\Pi}_x}$, and that one measurement follows the previous one after time $\delta t = T/N$. We will assume a Gaussian POVM, with $w(x)$ given by Eq. (21.30).

Suppose that the particle is initially in the state $\hat{\rho}_0$. Then, we construct the probability $\mathrm{Prob}(\bar{x}_1, \bar{x}_2, \ldots, \bar{x}_N)$ that we find successive values $\bar{x}_1, \bar{x}_2, \ldots, \bar{x}_n$ for position. To this end, we use Eq. (8.24) with the positive operators $\sqrt{\hat{\Pi}_{\bar{x}_n}}$ in place of projectors. That is, we write

$$\mathrm{Prob}(\bar{x}_1, \bar{x}_2, \ldots, \bar{x}_N) = \mathrm{Tr}\left[\hat{C}_{\bar{x}_1, \bar{x}_2, \ldots, \bar{x}_N} \hat{\rho}_0 \hat{C}^\dagger_{\bar{x}_1, \bar{x}_2, \ldots, \bar{x}_N} \right], \tag{21.68}$$

---

[4] Take, for example, $f_1 = 1, f_2 = iy$, with $y \in \mathbb{R}$ and all other f_a vanishing. Then, $K = (\omega_2 - \omega_1)y$, which can take any sign.

where

$$\hat{C}_{\bar{x}_1,\bar{x}_2,\cdots,\bar{x}_N} = \sqrt{\hat{\Pi}_{\bar{x}_N}}e^{-i\hat{H}\delta t}\sqrt{\hat{\Pi}_{\bar{x}_{N-1}}}\cdots e^{-i\hat{H}\delta t}\sqrt{\hat{\Pi}_{\bar{x}_1}}e^{-i\hat{H}\delta t}. \tag{21.69}$$

In the position basis,

$$\langle x_f|\hat{C}_{\bar{x}_1,\bar{x}_2,\cdots,\bar{x}_N}|x_0\rangle = \int dx_1\cdots dx_{N-1}\prod_{n=1}^{N}\sqrt{w(x_n-\bar{x}_n)}\langle x_n|e^{-i\hat{H}\delta t}|x_{n-1}\rangle,$$

where we $x_N = x_f$. Using the path-integral formula Eq. (10.26) for the product $\langle x_n|e^{-i\hat{H}\delta t}|x_{n-1}\rangle$, and Eq. (21.30) for $w(x)$, we obtain

$$\langle x_f|\hat{C}_{\bar{x}_1,\bar{x}_2,\dots,\bar{x}_N}|x_0\rangle = \left(\frac{m}{2\pi^2 i\sigma^2\delta t}\right)^{\frac{N}{2}}\int dx_1\cdots dx_{N-1}$$

$$\times \exp\left[i\sum_{n=1}^{N}\left[\frac{m(x_n-x_{n-1})^2}{2\delta t} - V(x_{n-1})\delta t\right] - \sum_{i=1}^{N}\frac{(x_n-\bar{x}_n)^2}{2\sigma^2}\right]. \tag{21.70}$$

To take the limit $N\to\infty$ in the exponent, we must assume that $\sigma^2 = 1/(\lambda\delta t)$ for some positive constant λ. Then, the exponent converges to the integral $i\int_0^T dt[\frac{1}{2}m\dot{x}^2(t) - V(x)] - \frac{1}{2}\lambda\int_0^T dt[x(t)-\bar{x}(t)]^2$. Here, we wrote $\bar{x}(t)$ as a continuous path that is obtained as a limit of the discrete path $(x_1, x_2, \dots, x_N)$ as $N\to 0$. In this limit, we write $\hat{C}_{\bar{x}_1,\bar{x}_2,\dots,\bar{x}_N}$ as $\hat{C}_{\bar{x}(\cdot)}$. Then, Eq. (21.70) becomes

$$\langle x_f|\hat{C}_{\bar{x}(\cdot)}|x_0\rangle = \int Dx(\cdot)e^{i\int_0^T dt[\frac{m}{2}\dot{x}^2(t) - V(x)] - \frac{\lambda}{2}\int_0^T dt[x(t)-\bar{x}(t)]^2}, \tag{21.71}$$

where the integral is over all paths $x(\cdot)$, such that $x(0) = x_0$ and $x(T) = x_f$. The integration measure $Dx(\cdot)$ is the limit of $\left(\frac{\lambda m}{2\pi^2}\right)^{\frac{N}{2}}dx_1\cdots dx_{\bar{N}}$ as $N\to\infty$. The right-hand side of Eq. (21.71) also contains a multiplicative phase factor $e^{i\phi} := i^{-N/2} = e^{-i\pi N/4}$. This phase cancels when computing probabilities, so we can ignore it.

Equation (21.71) is formally similar to a path integral with a complex time-dependent potential $U(x,t) = -i\frac{\lambda}{2}[x - \bar{x}(t)]^2$ added to the Hamiltonian. Hence, $\hat{C}_{\bar{x}(\cdot)}$ is the evolution operator associated to the "Hamiltonian" $\hat{H} - i\frac{\lambda}{2}[x(t) - \bar{x}(t)]^2$, that is, it is a solution to the equation

$$i\frac{d}{dt}\hat{C}_{\bar{x}(\cdot)} = \left[\hat{H} - i\frac{\lambda}{2}[\hat{x} - \bar{x}(t)\hat{I}]^2\right]\hat{C}_{\bar{x}(\cdot)}, \tag{21.72}$$

with $\hat{C}_{\bar{x}(\cdot)}(0) = \hat{I}$. We define the operator $\hat{D}_{\bar{x}(\cdot)} := e^{i\hat{H}t}\hat{C}_{\bar{x}(\cdot)}$, which satisfies

$$\frac{d}{dt}\hat{D}_{\bar{x}(\cdot)} = -\frac{\lambda}{2}[\hat{x}(t) - \bar{x}(t)\hat{I}]^2\hat{D}_{\bar{x}(\cdot)}, \tag{21.73}$$

where $\hat{x}(t) = e^{i\hat{H}t}\hat{x}e^{-i\hat{H}t}$ is the Heisenberg-picture position operator. Equation (21.73) admits a formal solution through the time-ordered exponential,

$$\hat{D}_{\bar{x}(\cdot)} = \mathcal{T}e^{-\frac{\lambda}{2}\int_0^t ds[\hat{x}(s)-\bar{x}(s)\hat{I}]^2}. \tag{21.74}$$

We note that $\hat{D}_{\bar{x}(\cdot)}$ is self-adjoint.

Thus, we obtained a probability density for recorded paths $\bar{x}(\cdot)$,

$$P[\bar{x}(\cdot)] = \mathcal{N} Tr[\hat{C}_{\bar{x}(\cdot)} \hat{\rho}_0 \hat{C}^{\dagger}_{\bar{x}(\cdot)}] = \mathcal{N} Tr[\hat{D}^2_{\bar{x}(\cdot)} \hat{\rho}_0], \tag{21.75}$$

where $\mathcal{N}$ is a multiplicative term that must be included in order to transform the probabilities $Prob(\bar{x}_1, \bar{x}_2, \ldots, \bar{x}_N)$ into probability densities. Hence, there exists a POVM $\hat{\Pi}_{\bar{x}(\cdot)} = \hat{D}^2_{\bar{x}(\cdot)}$ defined on the space of continuous paths on $\mathbb{R}$. Note that most paths $\bar{x}(\cdot)$ in the domain of the POVM are nondifferentiable, like the paths summed over in path integrals – see Box 10.1.

For sufficiently small coupling, we keep only the leading term in the Magnus expansion of the time-ordered product in $\hat{D}_{\bar{x}(\cdot)}$. Then,

$$P[\bar{x}(\cdot)] = \mathcal{N} \langle e^{-\lambda \int_0^T dt [\hat{x}(t) - \bar{x}(t) \hat{I}]^2} \rangle. \tag{21.76}$$

Consider a quantum state $|\psi_0\rangle$ of *localized evolution*, that is, a state with well-concentrated position and momentum distributions, such that the evolving wave packet is peaked around a classical path $x(t) = \langle \hat{x}(t) \rangle$ for all $t \in [0, T]$. Examples include coherent states in a harmonic oscillator potential, general Gaussians for a free particle, or more general states that satisfy the approximation (7.17).

Suppose that the spread of $|\psi_0\rangle$ around the classical path is δx. If $\lambda T(\delta x)^2 << 1$, $\hat{D}_{\bar{x}(\cdot)} |\psi_0\rangle \simeq e^{-\frac{\lambda}{2} \int_0^T dt [\langle \hat{x}(t) \rangle - \bar{x}(t)]^2} |\psi_0\rangle$, that is, $|\psi_0\rangle$ is an approximate eigenstate of $\hat{D}_{\bar{x}(\cdot)}$. For such states,

$$P[\bar{x}(\cdot)] \sim e^{-\lambda \int_0^T dt [\langle \hat{x}(t) \rangle - \bar{x}(t)]^2},$$

that is, the recorded tracks are peaked around the classical equations of motion.

For a superposition $|\psi_0\rangle = \sum_i a_i |\psi_i\rangle$ of states $|\psi_i\rangle$ with localized evolution, such that $\langle \psi_i | \psi_j \rangle \simeq 0$ for $i \neq j$,

$$P[\bar{x}(\cdot)] \sim \sum_i |a_i|^2 e^{-\lambda \int_0^T dt [\langle \psi_i | \hat{x}(t) | \psi_i \rangle - \bar{x}(t)]^2},$$

where $\langle \hat{x}_i(t) \rangle$ is the evolved mean value of position for the state $|\psi_i\rangle$. We have tracks peaked around different classical paths (i.e., with different initial conditions), weighted by the Born-rule probabilities $|a_i|^2$.

21.5 Dynamical Reduction Models

As mentioned in Section 8.3.3, the idea that state reduction is a dynamical process provides one way of addressing the quantum measurement problem. Dynamical reduction models propose that reduction may occur spontaneously – that is, in the absence of measurement – even for microscopic particles. Dynamical reduction does not occur continuously – as in continuous-time measurements – but at random instants of time.

Here, we will present an elementary dynamical reduction model, known as the GRW model, from the names of its authors (Ghirardi, Rimini, and Weber); it was the first dynamical reduction model to be constructed (Ghirardi et al., 1986).

Dynamical reduction makes sense only if one treats pure quantum states as objective characteristics of *individual* quantum systems. Any change of the quantum state is a physical change, and not a change in our knowledge. In contrast, mixed states describe statistical ensembles, that is, they involve averages over all individual quantum states.[5] We will denote the process of ensemble averaging by $\mathcal{M}$, and we will write $\hat{\rho} = \mathcal{M}(|\psi\rangle\langle\psi|)$ for the density matrix obtained from the ensemble average of different states $|\psi\rangle$.

Consider a particle in one dimension with a Hamiltonian $\hat{H}$. Let us denote the state at time t by $|\psi(t)\rangle$. We assume that there exists a constant *reduction rate* $\lambda > 0$. This means that for a sufficiently small time step τ, there is a probability $\lambda\tau$ that a reduction in the position basis occurs, that is, the state evolves to $\sqrt{\hat{\Pi}_x}|\psi(t)\rangle/\sqrt{\langle\psi(t)|\hat{\Pi}_x|\psi(t)\rangle}$, where $\hat{\Pi}_x$ is a Gaussian POVM $\hat{\Pi}_x = \int dx w(x - x')|x'\rangle\langle x'|$, with $w(x)$ given by Eq. (21.30). The values of x in the reduction are randomly distributed by the probability density $\langle\psi(t)|\hat{\Pi}_x|\psi(t)\rangle$. If the reduction does not occur, then the particle evolves unitarily to $e^{-i\hat{H}\tau}|\psi(t)\rangle \simeq (\hat{I} - i\hat{H}\tau)|\psi(t)\rangle$.

Taking the ensemble average, we derive an evolution equation for the density matrix

$$\hat{\rho}(t + \tau) = \mathcal{M}(|\psi(t+\tau)\rangle\langle\psi(t+\tau)|) = (1 - \lambda\tau)[\hat{\rho}(t) - i[\hat{H}, \hat{\rho}(t)]\tau]$$
$$+ \lambda\tau \int dx \sqrt{\hat{\Pi}_x}\hat{\rho}(t)\sqrt{\hat{\Pi}_x} + O(\tau^2). \tag{21.77}$$

At the continuum limit ($\tau \to 0$), we obtain the GRW master equation,

$$\frac{d\hat{\rho}(t)}{dt} = -i[\hat{H}, \hat{\rho}(t)] + \lambda\left[\int dx \sqrt{\hat{\Pi}_x}\hat{\rho}(t)\sqrt{\hat{\Pi}_x} - \hat{\rho}\right]. \tag{21.78}$$

The GRW equation is not Markovian, as it is not of the GKSL form (20.18). It is characterized by two parameters: the reduction rate λ and the width σ of the POVM $\hat{\Pi}_x$. We will refer to the latter as the localization scale.

We write Eq. (21.78) in the position basis for a free-particle Hamiltonian, $\hat{H} = \frac{\hat{p}^2}{2m}$,

$$\dot{\rho}(x, x') = \frac{i}{2m}\left(\frac{\partial\rho(x, x')}{\partial x^2} - \frac{\partial\rho(x, x')}{\partial x'^2}\right) - \lambda\left(1 - e^{-\frac{(x-x')^2}{4\sigma^2}}\right)\rho(x, x'). \tag{21.79}$$

GRW evolution suppresses the off-diagonal terms in the density matrix. If $|x - x'| >> \sigma$, then $\dot{\rho}(x, x') = \frac{i}{2m}\left(\frac{\partial\rho(x, x')}{\partial x^2} - \frac{\partial\rho(x, x')}{\partial x'^2}\right) - \lambda\hat{\rho}(x, x')$. This equation admits the solution $\hat{\rho}(x, x'; t) = \hat{\rho}_U(x, x'; t)e^{-\lambda t}$, where $\hat{\rho}_U(x, x'; t)$ is the solution to the unitary evolution equation (for $\lambda = 0$). In contrast, if $|x - x'| << \sigma$, the evolution equation is well approximated by Eq. (20.2), for $D = \frac{\lambda}{2\sigma^2}$. By construction, the GRW master equation destroys spatial quantum coherence at scales larger than σ within a timescale of order λ^{-1}.

The most important property of the GRW model is that the reduction rate λ scales linearly with the size of the system. The center of mass of a composite quantum system that consists of N particles satisfies the GRW equation with decay rate $N\lambda$, where λ is the decay rate of a single particle. This result is proven in Box 21.2. Hence, even if the reduction rate is insignificantly small for a microscopic particle, it may become extremely large for a macroscopic system, guaranteeing that all macroscopic systems exhibit classical behavior. For example, taking $\lambda = 10^{-16}\,\text{s}^{-1}$ is compatible with all experiments of interference in microscopic particles. But then, for a system

---

[5] This assumption contradicts the minimal interpretation, in which there is no fundamental difference between pure and mixed states.

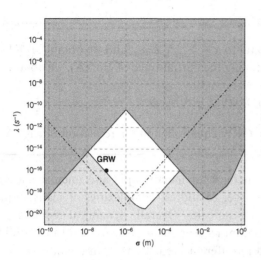

Fig. 21.2 Experiments and theoretical arguments constrain λ and σ to lie in the white region of the parameter space. Experimental bounds originate from noise-induced radiation emission from atoms and from the study of gravitational-wave detectors. The theoretical constraints arise from the requirement that the GRW model explains the emergence of the classical world. The original values proposed by GRW are shown by a black dot. The figure is taken from Donadi et al. (2021).

that consists of $N = 10^{23}$ particles, the reduction rate for its center of mass is $10^7 \mathrm{s}^{-1}$, guaranteeing a fast reduction of any macroscopic superpositions.

Minimal values of $\lambda = 10^{-17}\mathrm{s}$ and $\sigma = 10^{-7}$ m were proposed by GRW for the parameters of the model. Smaller values of λ would not guarantee a sufficiently rapid collapse of macroscopic quantum superpositions at the human scale. As seen in Fig. 21.2, experiments have constrained λ to be smaller than 10^{-10} s and σ to the range $10^{-8}\text{--}10^{-3}$ m.

Box 21.2 Scaling of the Reduction Rate

We consider a system of N particles in one dimension with Hamiltonian $\hat{H}$. We assume that the reduction rate $\hat{\lambda}$ and the localization length σ are the same for all particles. The evolution equation for the composite system is

$$\frac{\partial \hat{\rho}}{\partial t} = -i[\hat{H}, \hat{\rho}] + \lambda \sum_{i=1}^{N} \left[\mathcal{V}_i[\hat{\rho}] - \hat{\rho} \right], \tag{A}$$

where $\mathcal{V}_i[\hat{\rho}] = \int dx \sqrt{w}(x - \hat{x}_i)\hat{\rho}\sqrt{w}(x - \hat{x}_i)$. Here, $\hat{x}_i$ is the position coordinate of the ith particle, where $i = 1, 2, \ldots, N$.

We introduce the center of mass coordinate $\hat{X}$ and the $N-1$ relative coordinates $\hat{\xi}_a$, where $a = 1, 2, \ldots, N-1$. We write $\hat{x}_i = \hat{X} + \sum_{a=1}^{N-1} c_{ia}\hat{\xi}_a$, for some constants c_{ia}, the exact form of which is not needed. The Hamiltonian splits as a sum $\hat{H}_X + \hat{H}_\xi$, where $\hat{H}_X$ describes the evolution of the center of mass and $\hat{H}_\xi$ the evolution of the relative degrees of freedom.

The center of mass is described by the reduced density matrix $\hat{\rho}_X = \mathrm{Tr}_\xi \hat{\rho}$. We evaluate the partial trace after the action of the superoperator $\mathcal{V}_i$,

$$\langle X | \mathrm{Tr}_\xi \mathcal{V}_i[\hat{\rho}] | X' \rangle = \int d\xi_1 \cdots d\xi_{N-1} \int dx \sqrt{w}\left(x - \hat{X} - \sum_{a=1}^{N-1} c_{ia}\xi_a\right)$$

$$\times \langle X, \boldsymbol{\xi} | \hat{\rho} | X, \boldsymbol{\xi} \rangle \sqrt{w}\left(x - \hat{X} - \sum_{a=1}^{N-1} c_{ia}\xi_a\right).$$

We change the variable of the x integral to $x - \sum_{a=1}^{N-1} c_{ia}\xi_a$, and we obtain $Tr_\xi V_i[\rho] = \int dx\sqrt{w}(x - \hat{X})\hat{\rho}_X\sqrt{w}(x - \hat{X})$. Hence, taking the partial trace of Eq. (A), we obtain

$$\frac{\partial\hat{\rho}}{\partial t} = -i[\hat{H}_X, \hat{\rho}_X] + \lambda N\left[\int dx\sqrt{w}(x - \hat{X})\hat{\rho}_X\sqrt{w}(x - \hat{X}) - \hat{\rho}_X\right].$$

The center of mass evolves under the GRW equation, and the reduction rate is λN.

The GRW model provides a unified dynamics for macroscopic and microscopic systems. Improved models have been developed in order to address complexities of realistic quantum systems, for example, the effects of particle statistics. Overall, dynamical reduction models provide genuine solutions to both the quantum measurement problem and to the problem of emergent classicality. Moreover, they lead to different predictions from quantum theory and, hence, they can be experimentally distinguished.

Dynamical reduction models are not trouble free. They generically predict nonconservation of energy. Indeed, the GRW equation predicts that the expectation of the Hamiltonian evolves as

$$\frac{d}{dt}\langle\hat{H}\rangle = \lambda\left[\int dx\left\langle\sqrt{\hat{\Pi}_x}\hat{H}\sqrt{\hat{\Pi}_x}\right\rangle - \langle\hat{H}\rangle\right]. \tag{21.80}$$

To evaluate the term $D_H := \int dx\left\langle\sqrt{\hat{\Pi}_x}\hat{H}\sqrt{\hat{\Pi}_x}\right\rangle$, we work in the position basis. Integration over x is then straightforward,

$$D_H = \int dy dy'\, \rho(y', y)\langle y|\hat{H}|y'\rangle e^{-\frac{(y-y')^2}{4\sigma^2}}.$$

For a free particle, $\langle y|\hat{H}|y'\rangle = \int dp\frac{p^2}{2m}\langle y|p\rangle\langle p|y'\rangle = \int\frac{dp}{2\pi}\frac{p^2}{2m}e^{ip(y-y')}$. We also express $\rho(y', y) = \int dk e^{ik(y-y')}W(\frac{y+y'}{2}, k)$ in terms of its Wigner function. Changing the integration variables to $Y = \frac{1}{2}(y + y')$ and $\xi = y' - y$, we integrate over ξ, to obtain

$$D_H = \sqrt{4\pi\sigma^2}\int\frac{dY dk}{2\pi}W(Y, k)\left(\int dp\frac{p^2}{2m}e^{-\sigma^2(p-k)^2}\right)$$

$$= \int dY dk\, W(Y, k)\left(\frac{k^2}{2m} + \frac{1}{4m\sigma^2}\right) = \langle\hat{H}\rangle + \frac{1}{4m\sigma^2}, \tag{21.81}$$

where we employed Eq. (A.1) for the Gaussian integral. Then, Eq. (21.80) implies a constant rate of increase for the average energy,

$$\frac{d}{dt}\langle\hat{H}\rangle = \frac{\lambda}{4m\sigma^2}. \tag{21.82}$$

For $\lambda = 10^{-16}$ s and $\sigma = 10^{-6}$ m, the rate of energy increase for a neutron is 10^{-46} W or equivalently 10^{-20} eV/yr. For the center of mass of a macroscopic system, the rate of energy increase is the same, because both λ and m increase proportionally to the number N of constituent particles. However, the total energy of the macroscopic system also involves the energy of the relative motion of the particles, which takes the form of heat. Hence, a macroscopic system is spontaneously heated with a rate $N\frac{\lambda}{4m\sigma^2}$. For $N = 10^{23}$, the rate is of the order of 10^{-23} W, or 10^{-16} J/yr.

Even if the effect of energy increase is small, it is conceptually problematic. Tiny violations of energy conservation accumulate when taken over a sufficiently large region filled with matter. But according to Einstein's General Theory of Relativity, energy gravitates, so the increase of energy would have significant effects at the astronomical or cosmological scale. Furthermore, General Relativity incorporates local energy conservation in its basic structure. Hence, dynamical collapse models would lead not only to the modification of quantum theory, but also of our current theory of gravity.

The violation of energy conservation also implies the loss of time-translation symmetry. This means that we cannot rely on the Poincaré symmetry for the description of relativistic systems, and all properties that are based on this symmetry, such as field-particle duality. Indeed, the formulation of a consistent relativistic description of dynamical collapse remains the most important challenge to the viability of this program.

QUESTIONS

21.1 Where does the assumption that the pointer variable is a macroscopic collective observable enter in von Neumann's description of measurements?

21.2 What are the differences between the uncertainty relation (21.29) and the Kennard–Robertson uncertainty relation? Which one is closer to Heisenberg's original analysis with the microscope (Section 2.7.2)?

21.3 Consider a set-up in which an apparatus B measures the pointer variable of an apparatus A that interacts with a microscopic system. What will B record (i) in ideal and (ii) in nonideal measurements?

21.4 Suppose you are sitting next to a motorway and you record the number of cars that pass every minute. What is bunching and what is antibunching in this context?

21.5 Compare the time–energy uncertainty relation to the various forms of the Mandelshtam–Tamm uncertainty relation in Section 7.3.

21.6 Suppose you apply Glauber's detection theory to a single-mode fermion field. Is Eq. (21.64) violated?

21.7 Fluorescent radiation from a single atom exhibits antibunching. Why?

21.8 Why do we use nonprojective operators $\hat{\Pi}_x$ in continuous-time measurements?

PROBLEMS

21.1 Construct a POVM for the measurement of an observable $\hat{A}$ with discrete spectrum from a pointer $\hat{X}$ with continuous spectrum. To this end, use von Neumann's measurement scheme with interaction Hamiltonian $\hat{H}_I = \delta(t - t_0)\hat{A} \otimes \hat{K}$, where $\hat{K}$ is the conjugate variable to $\hat{X}$.

21.2 Consider the measurement of a continuous observable as in the model of Section 21.1.3, but with a noninstantaneous interaction: take $\hat{H}_I = f(t)\hat{A}\otimes\hat{K}$, where f is a general function that vanishes for t outside the interval $[0, T]$. (i) Express the probability density $P(X)$ of the pointer at time $t > T$ in terms of the unitary operator $\hat{U} = \mathcal{T}e^{-i\int_0^T dt f(t)\hat{A}(t)}$. (ii) Show that, for sufficiently weak coupling, the pointer variable is correlated to the averaged operator $\int_0^T dt f(t)\hat{A}(t)$. (iii) Take $f(t) = \delta'(t - t_0)$ for $t_0 \in (0, T)$. Which observable is the pointer variable correlated with?

Comment It can be shown that noninstantaneous von Neumann measurements always determine time-smeared values of the observable $\hat{A}$, and not only in the regime of weak coupling (Anastopoulos and Savvidou, 2008).

21.3 Consider the model for simultaneous measurement of position and momentum in Section 21.1.4. (i) Show that, in the Heisenberg picture and for $t > t_0$, $\hat{Y}(t) = \hat{x}(t_0) + \hat{Y} + \frac{1}{2}\hat{L}$ and $\hat{Z}(t) = \hat{p}(t_0) + \hat{Z} - \frac{1}{2}\hat{K}$. This means that the operator $\hat{N}_x := \hat{Y} + \frac{1}{2}\hat{L}$ measures the inaccuracy in the value of position, and the operator $\hat{N}_p := \hat{Z} - \frac{1}{2}\hat{K}$ measures the inaccuracy in the value of momentum. Obviously we select initial states of the apparatus with $\langle \hat{N}_x \rangle = 0$ and $\langle \hat{N}_p \rangle = 0$. Then the appropriate measure for the inaccuracy δx in measuring position is the mean deviation of $\hat{N}_x$, and the inaccuracy δp in measuring momentum is the mean deviation of $\hat{N}_p$. (ii) Show that $\delta x \delta p \geq \frac{1}{2}$. Which state saturates the uncertainty relation? (iii) Evaluate the variances for the pointer variables $\hat{Y}(t)$ and $\hat{Z}(t)$, and show that

$$(\Delta Y)(t)(\Delta Z)(t) \geq 1,$$

that is, the lower bound for the pointer variables is twice the lower bound for $\hat{x}$ and $\hat{p}$. The act of measurement creates additional uncertainty.

Comment Since the pointer variables commute, they can be measured simultaneously, so the variances refer to the probability distribution of a single experiment, not one experiment for measuring position and one experiment for measuring momentum. The analysis of the uncertainty relation in relation to approximate phase space measurements is due to Arthurs and Kelly (1965) and Arthurs and Goodman (1988).

21.4 Consider a sequence of measurements, first with a Gaussian POVM for position $\hat{\Pi}_x$ of variance σ_x, and then with a Gaussian POVM for position $\hat{\Pi}_p$ of width σ_p. Assume that the time between the two measurements is negligible. (i) Write the POVM that gives the joint probabilities for position and momentum. (ii) Identify the variances $(\Delta x)^2$ and $(\Delta p)^2$. (iii) What changes if you switch the order of the two measurements?

21.5 Consider the localization operator $\hat{S}$, defined by Eq. (21.54). According to Bochner's theorem, the Fourier transform of a positive definite function is positive. (i) Use this to show that the Weyl–Wigner transform $\tilde{S}(x, p)$ of $\hat{S}$ defines a probability density $S_p(x) := S(x, p)$ with respect to x, in which p appears as a parameter. This probability density describes the intrinsic inaccuracy in a measurement record. (ii) Evaluate the average x with respect to $S_p(x)$. (iii) Show that if the inequality (21.56) is saturated, then $S_p(x) = \delta(x)$, and, hence, $\hat{S} = \delta(\hat{x})$, where $\hat{x}$ is the "position" operator $i\frac{\partial}{\partial p}$.

21.6 (i) Write the uncertainty relation (21.58) for a nonrelativistic particle. (ii) Evaluate the right-hand side of the uncertainty relation, for the Lévy probability distribution $P(E) = \sqrt{\frac{c}{2\pi E^3}}e^{-\frac{c}{2E}}$, where E is the nonrelativistic kinetic energy. (iii) Prove that $\langle \hat{E} \rangle \Delta t > \frac{1}{4}$.

21.7 Calculate the autocorrelation function for a single mode (i) in a coherent state, (ii) in a thermal state, and (iii) in a cat state (equal weight superposition of two coherent states).

21.8 Consider a two-mode field, where the two modes have the same frequency ω, but different wave vectors, that is,

$$\hat{\psi}(t, \mathbf{x}) = \left(e^{i\mathbf{k}_1 \cdot \mathbf{x}} \hat{a}_1 + e^{i\mathbf{k}_2 \cdot \mathbf{x}} \hat{a}_2 \right) e^{-i\omega t},$$

where $\omega = |\mathbf{k}_1| = |\mathbf{k}_2|$. Assume that the two modes are prepared in the same state, and use Glauber's photodetection theory. (i) Evaluate the detection probability $P(t)$ for a coherent state and for a thermal state. You will note that the no interference terms are manifested in $P(t)$. (ii) Interference can be seen at the level of the joint probabilities $P(t_1, t_2)$. To see this, we define the *correlation function* $G = P(t_1, t_2)/[P(t_1)P(t_2)]$ with respect to the joint detection probability for two detectors located at $\mathbf{x}_1$ and $\mathbf{x}_2$. Show that, if both modes are in a thermal state at temperature T,

$$G = \frac{\langle \hat{N}^2 \rangle - \langle \hat{N} \rangle}{2 \langle \hat{N} \rangle^2} + \cos^2 \left[\frac{1}{2}(\mathbf{k}_1 - \mathbf{k}_2) \cdot (\mathbf{x}_1 - \mathbf{x}_2) \right],$$

where $\hat{N}$ is the number operator for a single mode.

Comment Photonic interference at the level of the joint detection probabilities was first observed by Hanbury Brown and Twiss (1956a,b). In one experiment, the photon source was a remote star, and small differences in wave number were due to the fact that photons produced at different paths of the star need to have slightly different direction in momentum in order to arrive at the detector. The Hanbury-Brown–Twiss effect created great stir when it was announced and it led to the development of quantum optics as a distinct research field.

21.9 Consider the GRW equation for a free particle. (i) Show that $\langle \hat{x}(t) \rangle = \langle \hat{x}(t) \rangle_0$, $\langle \hat{p}(t) \rangle = \langle \hat{p}(t) \rangle_0$, $(\Delta x)^2(t) = (\Delta x)_0^2(t) + \frac{\lambda}{6m^2 \sigma^2} t^3$, where the subscript 0 denotes time evolution for $\lambda = 0$. (ii) Estimate the masses of the particles, for which we can see substantial difference from unitary evolution in the spread of the quantum state for propagation times less than 100 s.

Bibliography

- Chapter 12 of Peres (2002) provides a quick overview of quantum measurement theory. The reprinted papers on quantum measurements in Wheeler and Zurek (1983) are essential reading. Braginsky and Khalili (1992) provide an introduction to quantum measurements with a focus on experimental applications. For a mathematical introduction, I recommend the classic text of Davies (1976). For photodetection and photon correlations, see chapter 4 of Scully and Zubairy (2012).
- For the uncertainty principle in relation to quantum measurement theory, see Busch et al. (2007) and Sen (2014).
- For continuous-time measurements and applications, see Mensky (1992). For dynamical reduction models, see Ghirardi and Bassi (2003).

22 Interpretations and Challenges

All crises begin with the blurring of a paradigm and the consequent loosening of the rules for normal research... . And all crises close in one of three ways. Sometimes, normal science ultimately proves able to handle the crisis-provoking problem, despite the despair of those who have seen it as the end of an existing paradigm. In other occasions, the problem resists even apparently new radical approaches. Scientists may conclude that no solution will be forthcoming in the present state of their field. The problem is labelled and set aside for a future generation with more developed tools. Or, finally,... , a crisis may end with the emergence of a new candidate for a paradigm and the ensuing battle over its acceptance.

Kuhn (1962)

22.1 The Key Questions Revisited

In previous chapters, we saw in detail the immense success of quantum theory in describing microscopic systems. We also saw that it leads to concrete predictions that are persistently being confirmed by experiment, even predictions that grossly violate all physical intuition that had been generated by 250 years of classical physics.

Remarkably, this success has been possible without any resolution of the fundamental issues that have accompanied quantum theory ever since its inception. For all purposes, it suffices to work with the minimal interpretation, according to which quantum theory works meaningfully only for the prediction of measurement outcomes in specific experiments. The quantum state plays primarily an informational role, as it condenses all information about the preparation of the system, and quantum state reduction is part of the algorithm that defines probabilities in set-ups that involve multiple measurements.

The minimal interpretation saves us from the temptation to interpret quantum mechanics using thought patterns of classical physics, in particular, talking about properties possessed by the microscopic systems irrespective of measurement. This way of thinking is incompatible with quantum physics, and by Bell-test experiments we know that it cannot be true, unless the microscopic systems communicate with each other in rather grotesque ways that involve faster-than-light influences.

However, the minimal interpretation is incomplete. It does not explain which physical processes constitute measurements. It requires the treatment of a measurement apparatus as a fundamentally classical system, and it does not explain how this is possible when the apparatuses consist of atoms that are themselves quantum systems. As we are probing the quantum behavior of increasingly larger systems, such ambiguities are not only conceptually problematic but they may lead to ambiguous predictions for concrete experiments.

These are good reasons to look for an interpretation, reformulation, and/or modification of quantum theory that goes beyond the minimal interpretation.[1] Unfortunately, the very success of the minimal interpretation is an obstacle in this endeavor. At the moment there are no experimental data that would allow us to select one interpretation over the other, as long as the interpretations are all consistent with the Bell-test experiments. This means that logical consistency and coherence are the only criteria of selection. Unfortunately, they are not enough. Logical criteria work when there is a consensus about the basis properties that the theory should satisfy. Such consensus is lacking, in fact we will see that some interpretations go so far as to refute what most people believe as undisputable facts of everyday experience. Different interpretations of quantum theory paint radically different views of reality, and, as such, they are often correlated with competing worldviews.

Nonetheless, the interpretation of quantum theory is *not* a subjective matter of opinion. We have witnessed genuine progress in our understanding of quantum foundations. Bell's theorem and the subsequent experiments showed that local realism is plainly wrong. We now know that Bohr's complementarity principle is not metaphysical mumbo jumbo as it had sometimes been dismissed during the early decades of quantum theory, but a property of experimentally determined probability distributions that all interpretations must account for. Furthermore, some interpretations make predictions that differ from those of standard quantum theory in regimes that are accessible to present-day experiments. Other interpretations propose novel reformulations of quantum theory to address longstanding problems in the frontiers of physics.

There are four major questions to which any interpretation that purports to go beyond the minimal one must provide a definite answer.

What is the Quantum State?

In Chapters 3 and 6, we saw that, in the minimal interpretation, the quantum state is primarily an informational object: It incorporates information about the preparation of the system. The state is not a physical attribute of a system (unlike microstates in classical physics); the change of the state after measurement is not a physical process. However, the state does not solely express our knowledge about the quantum system: It changes even when we have no knowledge about the outcome of a measurement. It is a hybrid object combining physical and epistemic aspects, with no analogue in other fields of physics.

A common theme in many interpretations of quantum theory is *state realism*, namely, the treatment of the quantum state as a real objective attribute of quantum systems. There are several different ways of doing so, each one leading to a radically different perspective on fundamental physics. We will see that such interpretations have severe problems to overcome, and, for this reason, the informational treatment of the quantum state remains a sound option, despite its perceived ambivalence.

---

[1] In what follows, we will use the word "interpretation" abusingly, to also cover the cases of reformulations or modifications of quantum theory.

Is Microrealism Possible?

As explained in Section 9.4.2, microrealism is the assertion that microscopic systems have determinate properties and that measurements reveal these properties. In the minimal interpretation, microrealism is false. The only determinate properties are measurement outcomes, and these refer to the combination of the measuring apparatus and the microscopic system, and not to the microscopic system alone. Bell's theorem and the subsequent tests showed that microrealist theories must violate the locality principle; in fact, they must violate the assumption of measurement independence, which underlies the design of all our experiments.

Nonetheless, the idea that microscopic systems have determinate properties has been so central to "scientific common sense" ever since the days of Newton that many physicists are loath to abandon it. If the price of realism is nonlocality and awkward physics, it is worth it, they assert. For this reason, Bohmian mechanics and other theories with hidden variables (as determinate microscopic properties have come to be known) still have a significant number of adherents.

What is a Measurement?

The minimal interpretation treats measurement as fundamental notions, but it does not specify which physical processes are to be counted as measurements. This would not be a problem in a theory with a restricted domain of validity, such as thermodynamics,[2] but such a limitation is unacceptable in a fundamental physical theory.

Attempted solutions to the measurement problem are of three types. The first one is the introduction of hidden variables. Since hidden variables render the theory essentially classical, the measurement problem is expected to disappear. It does, but another problem takes its place: Since the resulting theory does not satisfy measurement independence, there is a gap between what the hidden-variables theory describes as measurements and the common-sense design and description of measurements in a laboratory.

The second type of solution attempts to formulate a quantum theory for closed systems, without reference to measurements at the fundamental level. Again, there is a high price to pay: a huge gap between what the theory predicts and our basic intuitions and experiences of measurements. We will see this in more detail in Section 22.2.

The third type of solution is the acceptance that standard quantum theory is an approximation to a larger theory. This approximation works in the microscopic domain, but fails to describe macroscopic systems. We must therefore modify quantum mechanics, as in the GRW model that we encountered in Section 21.5. This solution has its own problems to overcome, but it makes predictions different from standard quantum theory, so it has an important advantage over its competitors: It is testable.

How does the Classical World Emerge?

The minimal interpretation admits the coexistence of classical and quantum physics in the description of measurements. Furthermore, it places no restrictions on superpositions of

---

[2] In most formulations, equilibrium thermodynamics is conceptualized as a theory of measurements and operations on macroscopic systems. Neither measurements nor operations need to be defined solely in terms of thermodynamics.

macroscopically distinct states. Such states are the vast majority in the Hilbert space, but for some reason we do not see them. The question then arises why such states are suppressed, and why our world, from the scale of centimeters to the scale of megaparcecs, is described by classical physics.

In Section 10.6, we described four broad ways of thinking about the relation between classical and quantum physics that go beyond the minimal interpretation.

1. Quantum theory emerges from a fundamental classical theory that employs hidden variables (e.g., Bohmian mechanics).
2. Classical physics emerges from quantum theory without fundamental modifications, because of environment-induced decoherence or similar mechanisms. Inevitably, the explanation of emergent classicality must involve a quantum theory for the whole Universe.
3. Classical physics emerges from quantum theory with fundamental modifications such as dynamical reduction.
4. Irreducibly classical entities (ICEs) exist and they are responsible for the macroscopic classical behavior exhibited by quantum systems.

22.2 Major Interpretations

In what follows, we will present six prominent interpretations of quantum theory. These interpretations provide a representative sampling of viewpoints towards the foundations of quantum theory. For ease of comparison, we will present their main ideas in relation to the four problems presented in the previous section. We will also discuss their strengths and their weaknesses.

22.2.1 The Copenhagen Interpretation

The Copenhagen interpretation was the first attempt to provide an overall conceptual framework for quantum mechanics. It was developed by the founders of quantum theory, mainly by Bohr, but also by Heisenberg, Born, and others. In fact, Bohr and Heisenberg had quite a few differences in their perspectives, and neither of them used the term "Copenhagen interpretation" for their ideas. The term was first used by opponents of Bohr's ideas of complementarity and indeterminism.

The main achievements of the Copenhagen interpretation are twofold. First, it provided a pragmatic attitude towards quantum theory that could be adopted by researchers who wanted to apply quantum theory to concrete problems without having to deal with foundational issues. In this it succeeded all too well: The minimal interpretation is essentially borne out of the Copenhagen interpretation. Second, the Copenhagen interpretation retrained physicists to avoid using classical concepts, that is, speaking of microscopic properties existing independently of measurement. In this aspect, Bohr's ideas were astonishingly prescient: He emphasised the divergence of quantum theory from classical concepts well before this divergence was conclusively demonstrated by Bell's and Kochen–Specker's theorems.

In the Copenhagen interpretation, the quantum state is again a hybrid object, neither an objective property of the system nor a description of our knowledge. Indeed, Bohr viewed the quantum state as a purely symbolic object *defined with reference to a specific experimental procedure*. We already mentioned that the Copenhagen interpretation denies any kind of microrealism. As Bohr (1998) wrote, it is impossible to draw "any sharp separation between an independent behaviour of atomic objects and their interaction with the measuring instruments which serve to define the conditions under which the phenomena occur." The microscopic system and the apparatus form an inseparable whole, and this is why we cannot combine information from different experiments to form a single picture.

However, this indivisibility of system and apparatus contradicts all our experimental practice (let alone our everyday experience). It is broken by the act of measurement that enables us to talk about definite properties. The existence of such definite properties is a precondition for using quantum theory. Such properties can exist only if the apparatus is described classically, in the sense that it is always found in definite states of the pointer variable. Without the use of classical physics for the description of measurements, the experimental methods of science make no sense:

... the description of the experimental arrangement and the recording of observations must be given in plain language, suitably refined by the usual physical terminology. This is a simple logical demand, since by the word "experiment" we can only mean a procedure regarding which we are able to communicate to others what we have done and what we have learnt. (Bohr, 1963)

Bohr accepts that (parts of) the apparatus can be treated quantum mechanically, as in quantum measurement theory, but

... in each case some ultimate measuring instruments, like the scales and clocks which determine the frame of space-time coordination – on which, in the last resort, even the definitions of momentum and energy quantities rest – must always be described entirely on classical lines, and consequently kept outside the system subject to quantum mechanical treatment. (Bohr, 1939)

Hence, the existence of a classical macroscopic world must be taken for granted even for the definition of microscopic quantities: The very notion of position or of the time of an event requires a reference frame that can only be fixed in the language of classical physics. Note, however, that the classical world is not autonomous: Quantum systems must behave in such a way as to make the classical world possible. For example, there would be no classical world if quantum physics did not guarantee the stability of matter.

In effect, the Copenhagen interpretation postulates that physics begins with the split between measured system and the rest of the world. The idea that physics can describe the world without reference to our participation in it is a fantasy borne out of classical physics, without a basis in reality. The purpose of physical theories "is not to disclose the real essence of phenomena but only to track down, so far as it is possible, relations between the manifold aspects of experience" (Bohr, 1963). Note that the quantum–classical split is not synonymous with the microscopic/macroscopic split.[3] In principle, we can take the quantum component to be a

---

[3] Furthermore, the quantum–classical split is not absolute. Provided that some appropriate conditions hold, a macroscopic system can be found in either side of the split, depending on what we decide to measure.

macroscopic system: Nothing forbids the existence of Schrödinger's cats. Hence, while the Copenhagen interpretation cannot treat the whole Universe as a quantum system, it can treat any degree of freedom as such, even, for example, the scale factor of the observed Universe. Therefore, with sufficient care, it can also be applied to cosmology.

The Copenhagen interpretation leaves the measurement problem unsolved: It does not explain which physical processes correspond to measurements. In its perspective, this is merely a technical matter to be resolved with further research: One just has to formulate specific conditions that the quantum–classical split must satisfy. However, the Copenhagen interpretation fails utterly in explaining the origin of the classical world, that is, how the classical description is compatible with the microscopic quantum behavior of matter. As explained in Section 10.6, the problem is that the overwhelming majority of quantum states has no classical analogue: We would expect the macroscopic world to be completely overwhelmed by Schrödinger's cats.[4] Einstein had pointed out this problem quite succinctly:

> It is in conflict with the principles of quantum theory to require that the ψ-function of a macroscopic system be 'narrow' with respect to the macroscopic coordinates and momenta. Such a demand is at variance with the superposition principle for ψ-functions. ... Narrowness with regard to macroscopic coordinates is a requirement which is not only independent of the principles of quantum mechanics, but is, moreover, incompatible with them. (Born, 1955)

22.2.2 Bohmian Mechanics and Its Generalizations

Bohmian mechanics is, by far, the most popular hidden-variables theory. As explained in Section 10.3, the main idea of Bohmian mechanics is the definition of classical paths associated to solutions of Schrödinger's equations. These paths satisfy Newton's equation with a wave-function–dependent potential V_Q added on the classical potential; V_Q is the quantum potential.

In Bohmian mechanics, each particle follows a well-defined trajectory, as in classical physics. The wave function plays a double role: on one hand, it determines the dynamics of the particle through the quantum potential, and, on the other hand, it provides a probability density $|\psi_t(x)|^2$ for a particle's position at time t. These probabilities are not irreducible as in quantum mechanics. Particles evolve deterministically, probabilities reflect only our ignorance of a particle's initial condition. This use of probabilities enables Bohmian mechanics to reproduce all predictions of standard quantum theory in the nonrelativistic regime.

The introduction of $|\psi_t(x)|^2$ as a probability density for the initial conditions appears as an additional axiom in Bohmian mechanics. It has been proposed that probability densities $\rho_t(x) \neq |\psi_t(x)|^2$ are also possible, but the quantum expression $|\psi_t(x)|^2$ is obtained as an *equilibrium* distribution, much like the Maxwell–Boltzmann distribution of the velocities in a gas in classical statistical mechanics (Dürr et al., 1992).

In Bohmian mechanics, there is no measurement problem: The apparatus has definite outcomes for particle positions and these outcomes are distributed according to Born's rule. There are some complications in Bohmian measurement theory, as all observables (such as spin) must

---

[4] See Section 20.5.2 for an explanation why the popular idea that classicality originates from a decoherence by an environment does not fundamentally work.

be expressed in terms of position measurements, but these are mostly technical and pose no conceptual difficulty.

In contrast, Bohmian mechanics suffers exactly as much as the Copenhagen interpretation in accounting for the existence of the classical world. This is obvious since predictions of Bohmian mechanics coincide with those of standard quantum mechanics. Most available states correspond to highly nonclassical behavior, in the sense that the quantum potential term dominates even in the time evolution of macroscopic systems.

Bohmian mechanics has been generalized to relativistic systems. In some theories, the fundamental degrees of freedom are fields,[5] in others, it is particles. It is possible to account for particle creation and annihilation processes (Dürr et al., 2004), even though the corresponding theory has an element of randomness, that is, it is not deterministic like nonrelativistic Bohmian mechanics. Typically, the construction of such theories requires the choice of a preferred reference frame, so it breaks the relativity principle that physics must be the same in all inertial reference frames.

These generalizations of Bohmian mechanics are still far from reproducing the predictions of our current relativistic quantum theories that describe electromagnetic, weak, and strong interactions. In fact, there is, as yet, no consistent Bohmian description of photons. In this sense, Bohmian mechanics is not a genuine alternative to standard quantum theory. It should best be viewed as a research programme in its early stages rather than a complete theory.

The greatest problem of Bohmian mechanics is that it violates the property of measurement independence. This results not only in faster-than-light speeds and incompatibility with relativity theory, but in the loss of the experimentalist's common sense. We have to take for granted that the preparation of a quantum system affects instantaneously measurement apparatuses located far away, even before the experimentalist decides which quantities these apparatuses will measure. Furthermore, these faster-than-light influences cannot be used to transmit a faster-than-light signal, for Bohmian mechanics must reproduce the predictions of quantum theory. A supporter of Bohmian mechanics will reply that this is an acceptable price to pay for restoring common sense at the microscopic level by preserving the property realism of classical physics. The problem is that – as Bohr had realized – the experimentalist's common sense plays a much more central role in the foundation of scientific theories than microrealism. We can do physics without the latter, as we have been doing for more than 90 years; some people may even view microrealism as a philosophical prejudice. In contrast, it is very difficult to conceive how a rational experimental science is possible without measurement independence.

22.2.3 Everett-Type Interpretations

In 1957, Hugh Everett submitted his Ph.D. thesis "On the foundations of quantum mechanics" to Princeton University (Everett, 1957a). The main idea was to treat the wave function as a physical entity, but without quantum state reduction: The state would evolve solely by Hamiltonian evolution. Then, he would attempt to deduce the physical predictions of quantum theory as subjective experiences of observers who are themselves treated as ordinary physical systems. Everett did not succeed in his task. His account of how these experiences arise was

---

[5] This procedure works only with bosonic fields: Fermion fields do not have appropriate classical analogues.

unclear, in need itself of an interpretation. Nonetheless, his work generated new ways of thinking about quantum theory, and it greatly influenced research on quantum foundations.

Before proceeding to the presentation of Everett's ideas, we first explain the main hurdle of any interpretation that aims to describe quantum theory in terms of a state that evolves unitarily without reduction. Reduction is a necessity in quantum theory in order to account for the existence of definite records. Measurements generate quantum states like (21.4) with strong entanglement between system and apparatus. That is, states with no definite value of the measured quantity. However, in experiments, we always see definite outcomes; the existence of definite facts lies at the core of our perception of the world. A theory with no reduction cannot account for definite facts, so it has to explain why our experience deceives us so badly. This is an extreme challenge, because our experience of definite facts is a building block of all experimental methods and, consequently, of the empirical justification of quantum theory itself. Hence, the treatment of definite facts as an illusion brings out the danger of kicking out the very ladder that we step upon. Furthermore, any such account cannot be restricted to physics, as it must be grounded on other sciences, in particular, psychology, neuroscience, and, possibly, evolutionary biology. It is doubtful whether these sciences, which presuppose notions from physics in their very foundations, can support such a burden.

According to Everett, observers must themselves be treated as quantum systems. Hence, he postulates a large Hilbert space $\mathcal{H}$ that contains the system, the apparatus, and the observer. Suppose that we want to measure a subsystem, for example, a qubit with respect to the 3-direction. We expect that we will be able to express $\mathcal{H}$ as a tensor product $\mathcal{H}_S \otimes \mathcal{H}_M$, where $\mathcal{H}_S$ describes the qubit and $\mathcal{H}_M$ describes the apparatus and the observer. After the interaction of the qubit with the measurement apparatus, and assuming an ideal measurement, the total state of the system is

$$|\Psi\rangle = |-\rangle_S \otimes |B_-\rangle_M + |+\rangle_S \otimes |B_+\rangle_M, \qquad (22.1)$$

where $|B_\pm\rangle$ describes the state of the degrees of freedom in $\mathcal{H}_M$ correlated with the value $\pm$ of $\hat{\sigma}_3$. This state does not describe a definite outcome either for the qubit or for the observer. Here, Everett introduces the notion of a relative state $|+\rangle_S$ in the unique state of S *relative to* $|B_+\rangle_M$ and conversely, $|B_+\rangle_M$ is the unique state of M relative to $|+\rangle_S$. According to Everett (1957b),

there does not, in general, exist anything like a single state for one subsystem of a composite system. Subsystems do not possess states that are independent of the states of the remainder of the system ... One can arbitrarily choose a state for one subsystem, and be led to the relative state for the remainder. Thus we are faced with a fundamental relativity of states, which is implied by the formalism of the composite systems. It is meaningless to ask the absolute state of a subsystem – one can only ask the state relative to a given state of the remainder of the subsystem.

Hence, subsystems can only be assigned relative states.[6] This means that the only absolute state is the state of the Universe, the *universal wave function*, since everything is a subsystem of the Universe, by definition.

---

[6] Note that the concept of the relative state has nothing to do with the reduced density matrix of a subsystem, which is defined in absolute terms. Everett does not consider density matrices; only pure states are fundamentally meaningful in his interpretation.

Everett postulates that in each of the two branches of the absolute wave function $|\Psi\rangle$ of Eq. (22.1), the state $|B_{\pm}\rangle_M$ describes an observer as definitely perceiving a definite value ± 1 for the qubit.[7] Hence,

... in each element of the superposition ... the object-system state is a particular eigenstate of the observer, and furthermore the observer-system state describes the observer as definitely perceiving that particular system state. (Everett, 1957a)

In other words, the absolute state contains all possible experiences of observers that have done the measurement and found a definite outcome. Every different experience corresponds to a different relative state.

This is the essence of Everett's argument. He then goes on to show how successive measurements cause the wave function to split into branches, each branch describing a sequence of events in the memory of the observer. In particular, he asserts that

... all elements of a superposition (all 'branches') are 'actual,' none any more 'real' than the rest. It is unnecessary to suppose that all but one are somehow destroyed, since all the separate elements of a superposition individually obey the wave equation with complete indifference to the presence or absence ('actuality' or not) of any other elements. (Everett, 1957b)

Everett also argues that measurements in a statistical ensemble of subsystems with the same initial preparation reproduce approximately the predictions of Born's rule. This argument is one of the most intriguing aspects of Everett's idea, supposedly allowing for the derivation of Born's rule from the other axioms of quantum theory, but fundamentally it does not work – see Box 22.1 for a discussion.

The foregoing is Everett's solution of the measurement problem. He also proposes that the same logic explains emergence of the classical world. An observer that measures Schrödinger's cat sees the cat alive in one branch of the wave function and dead in the other. In each branch, the cat behaves like an ordinary cat, that is, a macroscopic system subject to the rules of quantum mechanics. There is not such a thing as a classical world, but the experiences of each observer are identical with what one would expect in a macroscopic classical world.

The reader is well justified in feeling uneasy with Everett's argument, because it is at best incomplete. There is no explanation how a theory that asserts both propositions A = "see a living cat" and B = "see a dead cat" as true is compatible with our experience that either A or B is true but never both.

Box 22.1   Frequency Operators

Consider a collection of N qubits, all prepared in a state $|\psi\rangle = \alpha|+\rangle + \beta|-\rangle$, where $|\alpha|^2 + |\beta|^2 = 1$. Hence, the state of the total system is $|\Psi\rangle = |\psi\rangle \otimes |\psi\rangle \otimes \cdots |\psi\rangle$, and is a sum of 2^N terms. In particular, $|\Psi\rangle$ contains $\frac{N!}{n!(N-n)!}$ different terms with n kets $|+\rangle$ and $N - n$ kets $|-\rangle$, each of those terms weighted by a factor $\alpha^n \beta^{N-n}$.

---

[7] This requires some nontrivial assumptions about the relation between physics and subjective experience.

Let us define the projector $\hat{P}_n$ associated to the subspace of $(\mathbb{C}^2)^N$ that is generated by all vectors $|a_1\rangle \otimes |a_2\rangle \otimes \cdots \otimes |a_N\rangle$ such that n a_is equal $+$ and $N - n$ a_is equal $-$. $\hat{P}_n$ is a spectral projector of the *frequency operator* $\hat{F}_+ = \sum_n \frac{n}{N} \hat{P}_n$, whose eigenvalues are the relative frequencies n/N of occurrence of $+$ in a measurement on the qubit. It is straightforward to show that $\langle \Psi | \hat{P}_n | \Psi \rangle = \frac{N!}{n!(N-n)!} |\alpha|^{2n} |\beta|^{2(N-n)}$.

Suppose we keep the frequency $f = n/N$ fix and take N to be arbitrarily large. To this end, we define $\hat{P}_f = \chi_{C_f}(\hat{F}_+)$, where $C_f = \left[f - \frac{1}{2N}, f + \frac{1}{2N} \right)$. Since two different values of frequency differ at least by a factor of N^{-1}, the interval C_f contains only one frequency, and $\hat{P}_f = \hat{P}_n$, where n equals the integer part of Nf.

We use the Stirling approximation for the factorial, $\ln(n!) = n \ln n - n$, to obtain $\langle \Psi | \hat{P}_f | \Psi \rangle = \sum_j \exp[-NQ(f_j)]$, where the sum is over all frequencies f_j in the range $[f - \epsilon, f + \epsilon]$, and

$$Q(f) = f \ln \frac{f}{|\alpha|^2} + (1 - f) \ln \frac{1-f}{|\beta|^2}.$$

This function satisfies $Q(0) = -\ln |\beta|^2$, and $Q(1) = -\ln |\alpha|^2$, and it has a single minimum at $f = |\alpha|^2$, where it vanishes. Hence, when taking the limit $N \to \infty$ $\langle \Psi | \hat{P}_f | \Psi \rangle \to 0$, unless $f = |\alpha|^2$, in which case $\langle \Psi | \hat{P}_f | \Psi \rangle = 1$. This suggests, for an infinite number of qubits, $|\Psi\rangle$ would be an eigenstate of the frequency operator with eigenvalue $f = |\alpha|^2$. However, the Hilbert space $(\mathbb{C}^2)^\infty$ for infinite qubits is nonseparable and the spectral analysis is very different from what we are used to. Keeping the number of qubits finite but large, we can infer that $|\Psi\rangle$ can be brought arbitrarily close to an eigenstate of $\hat{F}_+$ with $f = |\alpha|^2$, as long as the number of qubits is sufficiently large.

This result originates from Hartle (1968), and it asserts that the Born probability rule is the only one compatible with the Hilbert-space structure of quantum theory. No specific interpretation needs to be assumed in its derivation.

In Everett-type interpretations variations of this result are essential, in order to show how probability arises out of a purely deterministic theory. It is straightforward to generalize this analysis taking into account the presence of an observer and the records left by the qubits in the observer's memory. One would then be tempted to conclude that all observers in different branches will reproduce Born-rule probabilities.

However, this conclusion is erroneous, for several reasons. First, for finite N, the state of the observer is an *approximate eigenstate* of the frequency operator, and not an exact one. Hence, frequencies do not really have definite values. Second, as shown in the main text, for finite N most branches have measurement statistics that grossly violate Born's rule. The $N \to \infty$ limit simply removes the contribution of these terms. Third, the closeness of $|\Psi\rangle$ to the eigenstate of $\hat{F}_+$ makes explicit reference to the Hilbert space norm, which is one choice of norm among many. A deterministic theory has no need to introduce a norm. As explained in Section 3.1, the norm is introduced solely in order to account for probabilities defined by Born's rule.

There have been many efforts to develop and complete Everett's argument. One of the most important programs in this direction is the *many-worlds interpretation* (MWI), developed by DeWitt and Graham (1973). In this interpretation, the branches of the universal wave function are interpreted as different Universes. Hence, the propositions A and B above can both be true, but each is true in a different Universe. This resolves the ambiguity of Everett's thesis.[8]

The main principles of the MWI are the following.

1. There is a state vector $|\Psi\rangle$ that represents the state of the entire Universe and it evolves deterministically according to Schrödinger's equation.
2. A complete description of physical reality requires the specification of $|\Psi\rangle$ and a preferred Hilbert space basis, say, $|n\rangle$.
3. The state vector naturally decomposes as $|\Psi\rangle = \sum_n a_n |n\rangle$, and each term in the decomposition corresponds to a different world.[9] The worlds are mutually unobservable but equally real.
4. The preferred basis is such that there exists a single definite record (different in each world) of the result of any measurement.

DeWitt and Graham (1973) proceeded with an analysis of measurement of the memories of observers in each world. By arguments analogous to those presented in Box 22.1, they concluded that observers in each branch see a world in which measurements obey Born's rule, and argue that the usual probabilistic interpretation of quantum theory emerges out of the many-world interpretation. However, this conclusion is wrong. First, there are inherent limitations in such analyses, which are pointed out in Box 22.1. Second, the vast majority of worlds obey statistical laws for measurement that are very different from Born's rule.[10] There have been many follow-up attempts to derive Born's rule from the MWI. The conclusion is that all such attempts must introduce additional assumptions *of statistical nature*. Hence, some kind of statistics must be postulated a priori for the different worlds, in contradiction to the avowed aim of a deterministic theory in which probabilities emerge only for bookkeeping in measurements.

The existence of a preferred basis is essential in the MWI. At the moment, we do not know what this basis is, because we do not know the appropriate physics to be incorporated in the universal state and its time evolution. It is quite possible that the determination of the basis requires a theory of quantum gravity.[11] Some critics view this as a severe problem of the MWI, but one may respond that the MWI provides only the axiomatic framework, and the

---

[8] Nonetheless, Everett did not accept the MWI.

[9] Variations of the MWI define the worlds not in terms of vectors in a basis, but in terms of a set of mutually exclusive and exhaustive projectors. If the projectors are sufficiently coarse-grained, branching occurs at the macroscopic level, that is, the worlds are distinguished by the values of macroscopic properties.

[10] This is easy to see. Consider N measurements of qubits, all prepared identically. Suppose that the preparation is such that the outcome $+1$ has probability $\epsilon \ll 1$ and the outcome -1 has probability $1 - \epsilon$. There are 2^N worlds, each world characterized by a different sequence of N measurement outcomes. In half of those worlds, the number of $+1$ outcomes will be larger than or equal to the number of -1 outcomes, while standard quantum statistics predicts that the expected number of $+1$ outcomes is only ϵN. In three quarters of the worlds, there will be at least 25 percent of $+1$ outcomes, and so on.

[11] Here, we are talking about a selected basis at the fundamental cosmological level of the universal quantum state. When using MWI in toy models for measurements, a preferred basis for the measurement apparatus can be determined by a

details are to be filled in by further research. After all, quantum theory did not have to specify the Hamiltonian for weak and strong interactions to be accepted as the correct theory for microscopic phenomena. Nonetheless, in absence of a fully fledged *cosmological* theory about the preferred basis, we really do not *know* that the MWI recovers our experiences, we can only *conjecture* that it does. Hence, the central claim of the MWI to solve the measurement problem is based upon the truth of a conjecture.

The consistency of the many-worlds branching process with relativity remains an open problem. The usual description of branching employs a background notion of time that is essentially Newtonian. It is far from obvious how to formulate branching with respect to different reference frames, whether physics depends on the choice of reference frames or not, and, most importantly, how to reconcile the time of branching with the fact that, in our fundamental theory of gravity, time is a dynamical variable.

To conclude, the fundamental element of reality in the MWI is the universal quantum state. It is a nonlocal object, the appearance of locality emerges only in specific branches-worlds, and it is a consequence of dynamics (which may include environment-induced decoherence). Since, the world is fundamentally described by quantum theory, there is no conceptual problem in formulating a quantum theory of spacetime and a quantum theory of cosmology. Indeed, this goal was one of Everett's initial motivations, and remains an important argument for present-day adherents of this interpretation.

22.2.4 Decoherent Histories

The decoherent histories approach to quantum theory was developed by Griffiths (1984), Omnés (1989a, 1994), and Gell-Mann and Hartle (1990, 1993). The theory combines the Copenhagen emphasis on the central role of classical reasoning in interpreting quantum theory with Everettian themes such as the description of measurements as physical processes and the formulation of a quantum theory for the whole Universe.

The basic object is a *history*, which corresponds to properties of the physical system at successive instants of time. In Section 8.3.3, we discussed histories of measurement outcomes. An N-time history α corresponds to a sequence $\hat{P}_{a_1}^{(1)}, \hat{P}_{a_2}^{(2)}, \dots \hat{P}_{a_N}^{(N)}$ of projectors, each corresponding to an outcome a_i in the measurement of an observable $\hat{A}^{(i)}$ at time t_i, where $i = 1, 2, \dots, N$. Then, we construct the history operators $\hat{C}_\alpha$ by Eq. (8.23), and the associated probabilities

$$p(\alpha) = \text{Tr}\left(\hat{C}_\alpha \hat{\rho}_0 \hat{C}_\alpha^\dagger\right), \tag{22.2}$$

where $\hat{\rho}_0$ is the initial state of the system.

The decoherent histories approach starts from the reinterpretation of histories as a sequence of properties of a physical system rather than as a sequence of measurement outcomes. This reinterpretation has an important benefit, but it also creates a serious problem. The benefit is that histories viewed as properties have a nice logical structure. We can make propositions about histories and relate these propositions by logical operations like "AND", "OR", "NOT", and

decohering environment. But the very possibility, let alone the existence, of such an environment must itself be explained in terms of the universal quantum state and its branching at the fundamental level.

so on. To see this, consider a two-time history α that corresponds to the particle being in the interval Δ_1 at time t_1 and in Δ_2 at time t_2. Repeating the analysis of Section 10.4.1, we can write the history operator $\hat{C}_\alpha$ as a path integral

$$\langle x' | \hat{C}_\alpha | x \rangle = \int_\alpha Dx(\cdot) e^{iS[x(\cdot)]}, \tag{22.3}$$

where integration is over all paths such that $x(0) = x$, $x(t_2) = x'$, $x(t_1) \in \Delta_1$ and $x(t_2) \in \Delta_2$, that is, over all paths that cross Δ_1 at time t_1 and end up in Δ_2 at final time t_2. Obviously, if $\Delta_1 = \Delta_2 = \mathbb{R}$, there is no restriction in the path integral, and the history operator is trivially unity; we represent this history by I.

Given a different history β that corresponds to the particle being in the interval Δ_1' at time t_1' and in Δ_2' at time t_2, we define $\alpha \wedge \beta$ (i.e., α AND β) in terms of a history operator with matrix elements

$$\langle x' | \hat{C}_{\alpha \wedge \beta} | x \rangle = \int_{\alpha \cap \beta} Dx(\cdot) e^{iS[x(\cdot)]}, \tag{22.4}$$

where integration over $\alpha \cap \beta$ means summing over all paths that cross Δ_1 at time t_1, and Δ_1' at time t_1, and $\Delta_2 \cap \Delta_2'$ at time t_2. Two histories are disjoint if $\alpha \cap \beta = \emptyset$, that is, if there are no paths to sum over in the path integral (22.4). This is the case, for example, if $\Delta_2 \cap \Delta_2' = \emptyset$.

Similarly, we define $\alpha \vee \beta$ (i.e., α OR β) in terms of the history operator

$$\langle x' | \hat{C}_{\alpha \vee \beta} | x \rangle = \int_{\alpha \cup \beta} Dx(\cdot) e^{iS[x(\cdot)]}, \tag{22.5}$$

where integration over $\alpha \cup \beta$ means summing over all paths that cross either Δ_1 at time t_1 or Δ_1' at time t_1' and then, they cross $\Delta_2 \cup \Delta_2'$ at time t_2.

The logical structure of histories is the most important novel feature of the decoherent histories approach.[12] The price is that Eq. (22.2) does not satisfy the Kolmogorov additivity condition

$$p(\alpha \vee \beta) = p(\alpha) + p(\beta) \tag{22.6}$$

for any pair of disjoint histories α and β. We already encountered an analogue of this problem in Section 10.5.1.

There is a partial resolution: We can define probability measures when restricting ourselves to specific sets of histories. To this end, we first define the *decoherence functional d*, a complex-valued function of pairs of histories, as

$$d(\alpha, \beta) = \text{Tr}\left(\hat{C}_\alpha \hat{\rho}_0 \hat{C}_\beta^\dagger \right). \tag{22.7}$$

The diagonal elements $d(\alpha, \alpha)$ of the decoherence functional coincide with the probabilities $p(\alpha)$ of Eq. (22.2). The decoherence functional also satisfies the following conditions.

1. Normalization: $d(I, I) = 1$.
2. Hermiticity: $d(\beta, \alpha) = \overline{d(\alpha, \beta)}$.
3. Linearity: $d(\alpha_1 \vee \alpha_2, \beta) = d(\alpha_1, \beta) + d(\alpha_2, \beta)$, if α_1 and α_2 are disjoint.

---

[12] Histories need not be defined with reference to discrete time moments. For example, we can define a history with an integration over all paths that cross a continuous *spacetime* region.

Let V be an exclusive and exhaustive set of histories. If all histories in V satisfy the consistency condition

$$\text{Re}d(\alpha, \beta) = 0, \quad \text{for} \quad \alpha \neq \beta, \tag{22.8}$$

then, Eq. (22.6) is satisfied, and we can define a probability measure on V. Then, V is called a *consistent set* or a *framework*.[13]

A framework is essentially a sample space. We can reason about the alternatives using the rules of classical logic, and we can use probability theory exactly as in classical physics. We can connect propositions about histories using connectives like AND, OR, and NOT, and crucially we can define implication using the probability measure. We say that $\alpha \rightarrow \beta$ (i.e., α *IMPLIES* β) if $p(\alpha \cap \beta) = p(\alpha)$. The decoherent histories interpretation asserts that, within a framework, we can talk consistently about *properties* of physical systems without reference to measurement. We can describe measurements as physical processes, and we can use logical implication in order to assert that a specific value of the pointer variable implies a specific value of a microscopic physical quantity. We can study macroscopic systems and find frameworks for *highly coarse-grained histories* in which the probability assignment corresponds to approximate classical equations of motion. As an example, the reader may return to the analysis of quasiclassical histories in Section 10.5.2, interpreting the latter as properties rather than measurement outcomes. It is in this sense that the classical world emerges from quantum theory.

The key point here is that all those results make an explicit reference to a specific framework. But frameworks are far from unique. There is an infinity of frameworks in any given system, none of which is preferred over the other. As is to be expected by the Kochen–Specker theorem, combining information from different frameworks leads to contradiction. Furthermore, conclusions obtained from different frameworks may be incompatible with each other (Kent, 1997). Hence, decoherent histories manifests a strong version of Bohr's complementarity principle:

There is not a unique exhaustive description of a physical system or a physical process. Instead, reality is such that it can be described in various alternative, incompatible ways, using descriptions which cannot be combined or compared. (Griffiths, 2003)

The problem is that this pluralism makes little sense when discussing measurements. We do not design experiments with measurement outcomes described in incompatible ways, we want experiments that have unique definite measurement outcomes. In contrast, in the decoherent-histories description of measurements, there is one framework that describes the determinate values of the pointer variable, and there is an infinity of other frameworks that describe properties with no obvious physical meaning, and no relation to what we observe in experiments. Similarly, when discussing the classical behavior of macroscopic systems, there is one or more frameworks corresponding to approximate classical behavior compatible with our classical theories, and there is an infinity of frameworks that correspond to behaviors with no classical analogue and no obvious physical interpretation – see the discussion in Section 10.6. There is no way to derive the physically relevant framework from first principles, rather, we choose one that corresponds to our experience. This choice invariably introduces our *prior* concepts

---

[13] The stronger condition $d(\alpha, \beta) = 0$ is referred to as a decoherence condition, and the resulting frameworks are called *decoherent sets*.

of classicality in both the treatment of measuring apparatuses and the macroscopic world. But then, we have gained nothing over the Copenhagen interpretation, at least in relation to the measurement problem and the emergence of the classical world.

Besides the logical structure of histories, the most important novelty of the decoherence histories program is the treatment of quantum states; states are emergent objects, useful for the codification of information in specific circumstances, and not the building block of the theory. At a fundamental level, a histories-based theory only needs a space of histories and a space of decoherence functionals. Standard quantum theory – with decoherence functionals of the form (22.7) – arises as a special case, when we postulate that probabilities do not contain memory effects and they satisfy a version of time reversibility (Anastopoulos, 2003). Histories-based formulations of quantum theory may turn out to be appropriate for dealing with major conceptual problems in quantum gravity and quantum cosmology – see the discussion in Sections 22.3.2 and 22.3.3. Indeed, they lead to a richer set of spacetime symmetries than standard quantum theory (Savvidou, 2009), and they can accommodate systems with complex temporal structure (Hartle, 1993; Isham, 1994).

Furthermore, the consistency condition provides the most general and unambiguous criterion for classical behavior in a system. Our notion of quasiclassical histories in Section 10.5.2 is essentially an adaptation of this criterion into the language of the minimal interpretation. This generality allows for a more nuanced description of quantum measurements, which has found some uses, for example, in constructing consistent measurement models for relativistic systems. Indeed, the fundamental justification of the detection models of Section 21.3 requires a decoherent histories analysis of emergent classicality in the detector.

22.2.5 Dynamical Reduction Models

If we want to treat the quantum state as an objective feature of microscopic systems, and we are not willing to view our experience of definite events as illusory, then we must treat quantum state reduction as a physical process. In this case, we can either assume that reduction is due to the consciousness of the observer, or we must treat reduction as a dynamical process. Here, we discuss the second alternative; the former is taken up in Section 22.2.6.

Dynamical reduction models postulate a stochastic modification of Schrödinger's equation. We presented one such model in Section 21.5. This modification has negligible effects for microscopic systems, but it destroys all characteristic quantum behavior in macroscopic systems. Hence, dynamical reduction models provide a unified theory for microscopic and macroscopic dynamics. This theory provides a genuine solution to both the quantum measurement problem and the problem of the emergence of the classical world, at least in the nonrelativistic regime.

Of course, this comes at a price. Dynamical reduction dynamics do not preserve energy. As explained in Section 21.5, this renders those models fundamentally incompatible with our most important theory of the macroscopic world, namely, General Relativity. Furthermore, the lack of time-translation symmetry breaks the crucial link between spacetime symmetries and the structure of particles that is essential in the formulation of relativistic quantum physics. This is one of the reasons why a relativistic generalization of dynamical reduction models is very difficult. At the moment, there is not a single relativistic model of dynamical collapse that is consistent with locality and causality, let alone models that work in the domain where standard quantum field theories have proved so successful.

Still, compared with the proposed solutions of the quantum measurement problem in Bohmian mechanics or in Everett-type theories, the price of dynamical reduction is light. We *only* need to postulate a new physical theory that corrects quantum theory. More importantly, dynamical reduction leads to predictions that differ from those of quantum theory, and these predictions are testable with current or near-future technologies. The flip side is that all predictions are heavily model-dependent. There is as yet no "smoking-gun" test for dynamical reduction, analogous to Bell-test experiments for local realism.

Dynamical reduction models typically follow the pragmatic stance of standard quantum theory in aiming to provide an accurate description of experiments. There is no need to make any statement about the fundamental nature of the quantum state or any underlying ontology. At a time where models are consistent only in nonrelativistic physics, such statements would be premature. Nonetheless, it is usually asserted that the most fundamental quantity for nonrelativistic systems is the mass density operator $\hat{\mu}(\mathbf{x}, t)$.[14] For macroscopic systems, the expectation value $\mu(\mathbf{x}, t) = \langle \hat{\mu}(\mathbf{x}, t) \rangle$ can be viewed as a field that describes quantum matter in the bulk. For systems that consist of few microscopic particles, this interpretation does not work very well, because the quantum fluctuations of $\hat{\mu}(\mathbf{x}, t)$ are typically of the same order of magnitude with its mean value (Anastopoulos and Hu, 2015).

Overall, the dynamical reduction program postulates a deeper theory than quantum mechanics that unifies microscopic and macroscopic dynamics. We do not know what this theory is, or the principles upon which it is based, except for the fact that it must incorporate the principles of relativity. In nonrelativistic physics, the difference from standard quantum theory should be expressed in dynamical reduction. In the absence of a unifying principle, we can attempt to describe this phenomenon with simple, rather ad hoc, models, which, nonetheless, provide a guide for experiments that look for deviations from standard quantum theory.

22.2.6 Irreducibly Classical Entities

Dynamical reduction theories treat state reduction as part of the intrinsic dynamics of microscopic particles. The alternative is to postulate that reduction is external to the particles, and it comes from the interaction with irreducibly classical entities (ICEs). The possibility of ICEs is based on the fact that it is now possible to formulate consistent hybrid quantum–classical dynamics – see Box 22.2. The key point is that a nontrivial quantum–classical interaction is incompatible with purely unitary evolution for the quantum system. Some interactions may lead to a process that is mathematically equivalent to dynamical reduction for the quantum system.

As we explained in Section 10.6, there have been so far two main proposals for ICEs, mind/consciousness and the gravitational field. The view of the mind as an ICE provides a solution to the quantum measurement problem: Reduction occurs when the observer perceives

---

[14] For a system of identical particles of mass m, the mass density is m times the particle number operator $\hat{\nu}(\mathbf{x}, t)$ of Section 15.4.1.

the measurement outcome.[15] However, it does not provide an explanation for the emergence of the macroscopic world. It is highly implausible that the large-scale classicality of the Universe is due to the presence of conscious observers, who emerged billions of years after the onset of cosmological classical behavior. Even if the mind is an ICE, it is probably not the only one.

Gravity is a plausible candidate for an ICE. Despite the huge volume of research on a quantum theory of gravity, there exists as yet no convincing theoretical demonstration that the gravitational field must be quantum. We are still rather far from experiments that can prove unambiguously the quantum nature of gravity. It is therefore theoretically possible that the gravitational field is an ICE. In this case, the interaction of gravity with matter will lead to decoherence; in some models this decoherence effect is so strong (Diosi, 1987; Penrose, 1996) that it may account for the emergence of the macroscopic world and, thereby, lead to a solution to the quantum measurement problem. Of equal importance is the fact that these effects are measurable, and indeed experiments have severely constrained the most popular model of gravitational decoherence.

Overall, there are reasonable arguments from General Relativity in favor of gravity-induced state reduction – see Section 22.3.2. However, existing models are usually constructed from ad hoc assumptions, and they work only at the level of Newtonian gravity. They might serve as phenomenological guides to experiments, but they do not provide any fundamental insight. As Penrose (1996) admits,

… none of the considerations … give any clear indication of the mathematical nature of the theory that would be required to incorporate a plausible gravitationally induced spontaneous state-vector reduction… . this author's own expectations are that no fully satisfactory theory will be forthcoming until there is a revolution in the description of quantum phenomena that is of as great a magnitude as that which Einstein introduced (in the description of gravitational phenomena) with his general theory of relativity.

Box 22.2 Hybrid Dynamics

Consider a quantum system described by a Hilbert space $\mathcal{H}$, and a classical system described by a state space Γ. For simplicity, we take Γ to be a discrete set. We denote the elements of Γ by letters $a, b, \ldots$. Quantum observables correspond to self-adjoint operators on $\mathcal{H}$, while classical observables correspond to functions on Γ. We describe quantum states in terms of positive operators $\hat{\rho}_a$ on $\mathcal{H}$, normalized so that $\sum_a \mathrm{Tr}\hat{\rho}_a = 1$. Then, $p_a := \mathrm{Tr}\hat{\rho}_a$ is a probability distribution on Γ and $\hat{\rho} := \sum_a \hat{\rho}_a$ is a density matrix on $\mathcal{H}$.

The most general evolution equation of the system is of the form $\rho_a(t + \tau) \rightarrow \sum_b \mathcal{V}_{ab}[\hat{\rho}_b(t)]$, where $\mathcal{V}_{ab}$ is a family of CP maps. Consider the case of $\mathcal{V}_{ab}[\hat{\rho}_b] = \hat{U}_{ab}\hat{\rho}_b\hat{U}^{\dagger}_{ab}$, where $\hat{U}_{ab}$ are unitary operators. The n, $p_a(t + \tau) = p_a(t)$; unitary dynamics is not compatible with nontrivial evolution of the classical system. Nontrivial quantum–classical interactions involve nonunitary dynamics.

---

[15] Wigner (1967) proposed that quantum mechanics applies only to inanimate objects; the quantum evolution equations become "grossly non-linear if conscious beings enter the picture." However, the effects that he describes can be mirrored by hybrid quantum–classical dynamics, as described in Box 22.2.

> If we assume that there is no memory, $\mathcal{V}_{ab}$ corresponds to a GKLS equation (Blanchard and Jadczyk, 1995),
>
> $$\frac{d}{dt}\hat{\rho}_a(t) = -i\left[\hat{H}_a, \hat{\rho}_a(t)\right] - \frac{1}{2}\sum_b \left[\hat{L}^\dagger_{ab}\hat{L}_{ab}\hat{\rho}_a(t) + \hat{\rho}_a(t)\hat{L}^\dagger_{ab}\hat{L}_{ab} - 2\hat{L}_{ab}\hat{\rho}_b(t)\hat{L}^\dagger_{ab}\right],$$
>
> defined in terms of Lindblad operators $\hat{L}_{ab}$.

22.2.7 Comments

Some points are in order after the presentation of the six interpretations. First, no interpretation is trouble-free. Each one has genuine physics problems to overcome. Certainly, this does not mean that all interpretations are equally plausible, or of equal value. It does mean, however, that rooting for an interpretation that satisfies some minimal criteria of consistency is a rational approach towards an open scientific problem.

Second, all interpretations are research programs rather than completed theories. Not surprisingly, the Copenhagen interpretation is the most mature one, and this partly explains why it remains the most popular. Nonetheless, one should never confuse maturity with truth.

Third, so far the discussion on quantum foundations has been largely restricted to nonrelativistic quantum mechanics. It has barely touched upon relativistic quantum field theory. As we will explain in Section 22.3.1, the latter brings a new host of issues in relation to measurements, and these are challenging for all interpretations.

Fourth, interpretations are not different narratives about the same theory. They are different theories. Some interpretations make different predictions from standard quantum theory, and others have the potential of doing so when applied to problems at the frontiers of physics. There is no reason to doubt that the quantum-foundation crisis will eventually be resolved as crises are meant to: One theory will win over its competitors by providing better explanations and better predictions for observed phenomena.

22.3   Challenges

Quantum theory has survived all experimental tests for the past 90 years, and it has given predictions of astonishing accuracy. It has even survived the challenge of local realism, which was the "common sense" of classical physics, and it remains the common sense of all natural sciences but physics. Nonetheless, quantum theory is increasingly being challenged, because current technology allows for increasingly precise tests on its foundations, and because open problems in the frontiers of physics appear to require at the very least an upgraded version of quantum theory. In this section, we review the main challenges from different fields of physics.

22.3.1 The Challenge from Relativistic Measurements

As explained in Chapters 15 and 16, the formulation of a relativistic quantum theory requires the introduction of quantum fields. The resulting theory, quantum field theory (QFT), has proved remarkably successful in describing the three fundamental interactions, namely, the electromagnetic, weak, and strong interactions. The theory faces severe mathematical challenges, but there are pragmatic recipes for calculations that work really well, and they provide unambiguous predictions for high-energy experiments. As long as one does not attempt to deal with gravity, which is fundamentally of a very different character than the other interactions, there is no reason to expect a failure of QFT at high energies.

It is therefore astonishing that QFT still lacks a consistent measurement theory. In high-energy physics, QFT is primarily applied to scattering experiments, which can be described in terms of the S-matrix. There is no need to use the state update rule, as the S-matrix formalism conceptualizes scattering as a process with a single measurement event. In contrast, in quantum optics, we need joint probabilities of detection in order to describe phenomena like photon bunching and antibunching – see Section 21.3.3. A first-principles calculation of joint probabilities requires the use of the state-update rule, namely, quantum state reduction.[16] In QFT, such a rule is missing even from its axiomatizations!

The usual quantum state reduction rule is problematic for relativistic systems, because it assumes that the state of the field changes *instantaneously* everywhere in space. In Section 8.3.2, we showed that this rule is not compatible with relativistic symmetry. Furthermore, Sorkin (1993) showed that the state reduction rule for ideal measurements is incompatible with causality, that is, it leads to faster-than-light signals. In absence of a state update rule, no probabilistic description can be deemed complete. We have no explanation for the definite value of a pointer variable in measurements. This is one of the main reasons for our current lack of a relativistic quantum measurement theory. The other reason is the difficulty in describing measurements in accordance with the principles of causality or locality. Even the very notion of a localized quantum system is troublesome in QFT. There are powerful theorems demonstrating that even unsharp localization in a spatial region leads to faster-than-light signals (Malament, 1996; Hegerfeldt, 1998).

The formulation of a quantum measurement theory for relativistic QFT is an ongoing research program. It has seen some minor successes – for example, the particle detection model of Section 21.3 – but it is far from complete. While it is unlikely that this effort will change our understanding of fundamental QFT dynamics, it is essential for the logical completion of the theory. In the absence of a relativistic measurement theory, most discussions on the quantum measurement problem have referred to nonrelativistic systems. A relativistic formulation will not resolve the fundamental issue by itself, but it will probably uncover new facets of the problem, making some interpretations more plausible and others less so. More importantly, we may expect novel predictions for quantum correlations in relativistic systems that differ from those obtained

---

[16] In practice, joint probabilities are expressed in terms of photodetection models, such as Glauber's, whose derivation is rather heuristic. However, planned experiments in deep space that involve measurement of EM field correlations will arguably require a first-principles analysis of joint probabilities in order to incorporate the relative motion of detectors and delayed propagation at long distances.

with the naive application of the standard rule of quantum state reduction, and, probably, the formulation of a consistent relativistic theory of quantum information.

22.3.2 The Challenge from Gravity

Quantum theory and General Relativity are the two main pillars of modern theoretical physics. Each theory is highly successful in its domain. However, they are structurally incompatible. For example, in General Relativity, measurements are derivative concepts, while, as we saw, in quantum theory, they appear to be fundamental. Time in General Relativity is dynamical, while in quantum theory it is described as an external classical parameter. Finding a unifying theory is one of the most important goals of current research, despite the lack of any direct experimental data to guide the theory.

Purely logically, there are three possible strategies towards this unified theory. The first is to keep quantum theory as is and to modify General Relativity, that is, to *quantize gravity*. The second is to keep the fundamental principles of General Relativity and to modify (general-relativize) quantum theory. The third is to build an all new theory that preserves neither General Relativity nor quantum theory. The overwhelming majority of researchers have been following the first strategy. The reason for this is that there already exist highly successful rules for the quantization of classical systems, and one may attempt to apply them to General Relativity.[17] The second strategy has only a small fraction of adherents, and it is strongly correlated with the idea of gravity as an ICE. The third strategy is currently a nonstarter: Even if it turns out to be the correct one, it is difficult at the moment to even think on the topic without a firm basis on one of the current fundamental theories.

In any case, gravity poses a severe challenge to quantum theory. Perhaps, the most important problem is the *problem of time*, namely, the great difficulties involved in reconciling two contrary concepts of time.

The central feature of time in quantum gravity arises from the fact that ... in general relativity what constitutes an admissible way of labelling events as occurring at different times depends on the geometry of spacetime. However, in a theory of quantum gravity ... we expect [that] this geometry ... will not have a definite value, but will be determined only probabilistically. But if the spacetime geometry had no definite value, then presumably neither would the set of allowed ways of introducing time: in particular, there would be no single choice that can serve as a uniform way of labelling events as occurring at different times for all the spacetime geometries that could appear with non-zero probability. A detailed study ... confirms these expectations: there is indeed something like a probabilistic distribution of three-dimensional geometries, but with no time label at all! Making sense of this peculiar situation is known as the problem of time in quantum gravity. (Isham and Savvidou, 2002)

For many physicists, the concept of a dynamical time in General Relativity is more fundamental, and should be the one to persist in the final theory. Existing quantum theory cannot accommodate it, so it would be necessary to introduce some changes. These changes

---

[17] There are two major schools in this direction. The first treats gravity as a "force" similar to the others, and has led to the superstrings program. The second treats gravity as spacetime geometry, fundamentally different from other interactions, and it has led to the canonical quantum gravity program. Several other programs exist, but they must make the choice between gravity as a force and gravity as geometry in their first step.

could be relatively minor, like a spacetime-oriented reformulation of quantum theory as in decoherent histories, or they may require a complete overhaul of the fundamental principles. More importantly, since the problem of time is structural, these changes should leave their imprint even in the regime of weak gravity that is accessible to current experiments. For example, Penrose (1996) suggests that the incompatibility of the temporal structures of General Relativity and quantum theory is manifested as gravity-induced reduction.

General Relativity predicts the existence of *black holes*, that is, regions of spacetime from which nothing can escape. Hawking (1975) showed that black holes emit radiation due to quantum effects, so they are not completely black after all. This means that black holes lose mass and eventually disappear. A semiclassical analysis of black hole "evaporation" suggests that even if a black hole was formed from matter in a pure state, the final state is always mixed. Hence, quantum dynamics during the whole process of black hole formation and evaporation is nonunitary.

This conclusion, dubbed *black hole information loss*, has emerged as one of the most controversial issues in quantum gravity research, because nonunitary evolution is incompatible with major programs such as superstring theory. These programs must find a way to exorcise information loss, if they are to remain competitive.[18] On the other hand, nonunitarity originates from the fact that the spacetime describing black hole evaporation has no global notion of time, that is, there exists no time parameter that can be used to generate dynamics through Schrödinger's equation. In this perspective, there is no conceptual problem with nonunitarity, only a technical challenge. We need to find a generalization of quantum theory that can describe quantum time evolution without a global time parameter. This means that the single-time quantum state cannot be a fundamental object in the theory; hence, histories-based approaches are natural candidates. In any case, finding the correct quantum description of black hole evaporation will strongly impact our understanding of quantum theory.

It needs to be said that we have very little direct experimental evidence about the effects of gravity on quantum systems. A famous experiment by Colella, Overhauser, and Werner (Colella et al., 1975) showed that the effects of a background weak gravitational field on the time evolution of quantum particles can be described by adding the Newtonian potential term in the Hamiltonian. However, we have no direct experimental evidence of how quantum systems gravitate. This includes determining (i) the gravitational force generated by a quantum distribution of matter, and (ii) the gravitational interaction between two different quantum matter distributions. The conservative assumption is that, in weak gravity and for nonrelativistic systems, the interaction can be described by the inclusion of a term of the form (15.75) in the many-particle Hamiltonian. Some models that treat gravity as an ICE make different predictions. Near-future experiments will provide important novel information about the interplay of gravity and the quantum.

22.3.3 The Challenge from Cosmology

It is logically impossible to formulate a quantum–classical split and place the whole Universe in the quantum part. Hence, a quantum theory of the whole Universe, a theory of *quantum*

---

[18] This is why black hole information loss is commonly referred to as a *paradox* and not as a *prediction*.

cosmology, is meaningless in the Copenhagen interpretation. If one accepts that the world is fundamentally quantum, a quantum description of the Universe is a necessity. The extension of quantum theory to the totality of the Universe was, in fact, one of the original motivations of Everett's program, but other interpretations can also be used in this context.

Quantum cosmology acts as a magnifying glass that exposes all fissures in the interpretations of quantum theory. Most prominent among them is the problem of the emergence of the classical world. Even in simple models, the overwhelming majority of states describes highly nonclassical configurations, typically superpositions of expanding and collapsing Universes. Even restricting to states that correspond to expanding Universes does not guarantee emergence of classical behavior because, as we have seen, most states for matter are also highly nonclassical.

The problem of making predictions with quantum theory without reference to measurements is more acute in quantum cosmology. How can we infer quantum properties of the early Universe from current observations?[19] The Kochen–Specker theorem strongly suggests that any such inferences are unreliable, and they will probably lead to incompatible conclusions.

Overall, quantum cosmology is the final test of any interpretation of quantum theory. The challenge is not merely conceptual, but it may have observational consequences. For example, the current favorite theory of structure formation in the Universe (i.e., galaxy clusters, galaxies, and so on) is *inflation*, which postulates that the seeds of current inhomogeneities are quantum fluctuations of a field that are amplified during a very early acceleration of the Universe, and "somehow" become classical. How exactly this quantum-to-classical transition occurs is controversial: Many researchers suppose that the specific unitary dynamics of inflation is sufficient to explain classicality, others invoke environmental decoherence, and yet others assert that there is no complete explanation without a macroscopic modification of quantum dynamics. Since different explanations invoke different dynamical behavior of fluctuations, they invariably lead to different predictions. Precision cosmology may well develop into an arena of testing between different approaches to quantum mechanics.

22.3.4 The Challenge from Macroscopic Quantum Phenomena

Quantum theory was originally developed in order to explain atomic physics. This means that it was meant to work for objects in the scale of 10^{-10} m. The development of nuclear physics extended the domain of applicability of quantum theory down to 10^{-15} m, and high-energy physics further extended it to 10^{-21} m. All challenges encountered by quantum theory in this range have been overcome.

In contrast, we have had very little progress in extending the domain of quantum phenomena to larger scales. It is true that we have been able to see quantum effects over macroscopic distances, for example photonic entanglement over a separation of more than 1000 km, or atomic interferences at the scale of almost 1 m. However, there is significantly less progress in seeing quantum effects in systems that consist of a macroscopically large number of atoms. As discussed in Box 2.3 the current (2023) record for the mass of particles exhibiting quantum

---

[19] In standard cosmology, we infer properties of the Universe at early time from current observations, because we treat the Universe as a *classical* system. Quantum cosmology should explain why such inferences work.

interference is "just" 25,000 amu, and this interference is manifested at a length scale of the order of the hundreds of nanometers.

We saw that the most difficult problems of quantum theory are the explanation of definite outcomes in measurements and the emergence of the classical world. The reason for these problems is that standard quantum theory contains no intrinsic limit to its applicability to macroscopic systems. The classical–quantum duality of the Copenhagen interpretation does not introduce such a limit. The same applies to Everett's interpretation, Bohmian mechanics, and decoherent histories, with the caveat that the latter two can incorporate it, if need arises, with only moderate difficulty. In contrast, the existence of such a limit is a fundamental assumption in dynamical reduction models and a common prediction of models that involve ICEs. Hence, looking for quantum effects at increasingly larger scales is arguably the best way to test the limits of quantum mechanics. Schrödinger's cat remains the greatest challenge that quantum theory must eventually confront.

QUESTIONS

22.1 Consider the following explanation of the quantum–classical split in the Copenhagen interpretation.

> While quantum theory can in principle describe anything, a quantum description cannot include everything. In every physical situation something must remain unanalyzed. This is not a flaw of quantum theory, but a logical necessity in a theory which is self-referential and describes its own means of verification. This situation reminds of Gödel's undecidability theorem: the consistency of a system of axioms cannot be verified because there are mathematical statements that can neither be proved nor disproved by the formal rules of the theory; but they may nonetheless be verified by metamathematical reasoning. (Peres, 2002)

Do you find this argument convincing?

22.2 In some generalizations of Bohmian mechanics the creation and annihilation of particles is a random process. Does this contradict the theories' stated purpose?

22.3 Adherents of Bohmian mechanics assert that standard quantum theory

> … is concerned only with our knowledge of reality and especially of how to predict and control the behaviour of this reality, at least as far as this may be possible. Or to put it in more philosophical terms, it may be said that quantum theory is primarily directed towards epistemology which is the study that focuses on the question of how we obtain our knowledge (and possibly on what we can do with it). It follows from this that quantum mechanics can say little or nothing about reality itself. In philosophical terminology, it does not give what can be called an ontology for a quantum system. Ontology is concerned primarily with that which is and only secondarily with how we obtain our knowledge about this. (Bohm and Hiley, 1995)

Do you agree with this assessment?

22.4 One criticism of the many-worlds interpretation (MWI) is that it violates Ockham's razor, namely, the principle that postulated entities should not be multiplied beyond necessity. Adherents of the theory reply that:

> … in judging physical theories one could reasonably argue that one should not multiply physical laws beyond necessity either … and in this respect the MWI is the most economical theory. Indeed,

it has all the laws of the standard quantum theory, but without the collapse postulate, which is the most problematic of the physical laws. The MWI is also more economical than Bohmian mechanics, which has in addition the ontology of the particle trajectories and the laws which give their evolution. (Vaidman, 2021)

Do you find this response persuasive?

22.5 Is the Schrödinger picture equivalent to the Heisenberg picture in the MWI?

22.6 Why must the MWI postulate a preferred basis?

22.7 In the decoherent histories approach to quantum theory it is sometimes conjectured that humans have been evolutionarily adapted to perceive only the framework that corresponds to our classical world, and this explains why we live under the illusion of the uniqueness of the macroscopic world. Can you find problems with this idea?

22.8 Which among the six interpretations presented in this chapter describe the microscopic world as fundamentally time-symmetric and which ones described it as fundamentally time-asymmetric?

22.9 Quantum fundamentalism is the notion that everything in the Universe must be described by quantum theory. Which of the interpretations of Section 22.2 are quantum fundamentalist?

Bibliography

- For the Copenhagen interpretation of quantum theory, see Heisenberg (1999) and Faye (2019). I also recommend the analysis of Bohr's ideas on the quantum–classical divide by Zinkernagel (2016).
- For Bohmian mechanics see Bohm and Hiley (1995) and Dürr and Teufel (2010). For relativistic generalizations, see Tumulka (2018).
- For Everett-type interpretations, including the many-worlds interpretation, see DeWitt and Graham (1973) and Saunders et al. (2010). For purported proofs of Born's rule, see Hartle (2021) and references therein.
- For the decoherent histories interpretation, see Omnés (1994) and Griffiths (2003). For its formulation as a "spacetime quantum mechanics," see Hartle (1993).
- For dynamical reduction models, see Ghirardi and Bassi (2003), and for experimental tests, see Bassi et al. (2013).
- For consciousness-induced reduction, see Stapp (2001), and for mind as an ICE, see Anastopoulos (2021).
- For gravity-induced reduction/decoherence, see Penrose (2014) and the reviews by Bassi et al. (2017) and by Anastopoulos and Hu (2022).
- For hybrid quantum–classical systems, see Blanchard and Jadczyk (1995), Diosi and Halliwell (1998), and Diosi et al. (2000).
- Hidden-variables theories other than Bohmian mechanics include stochastic mechanics (Nelson, 1985), matrix dynamics (Adler, 2004) and the cellular-automata description of quantum theory ('t Hooft, 2016). Other interpretations of quantum theory that are not covered here

include the modal interpretation (Lombardi and Dieks, 2021), the transactional interpretation (Kastner, 2012), the relational interpretation (Rovelli, 2021), the many-minds interpretation (Albert and Loewer, 1988; Barrett, 1999), and Quantum Bayesianism (Fuchs et al., 2014).

- For relativistic constraints on interpretations of quantum theory, see Myrvold (2021). For measurements in relativistic quantum field theory, see Anastopoulos and Savvidou (2022) and references therein. For relativistic quantum information, see Peres and Terno (2004).

- For the problem of time in quantum gravity, see Isham (1992), Kuchar (1992) and Anderson (2017). For its implications to experiments in weak gravity, see Anastopoulos et al. (2021).

- For different perspectives on black hole information loss, see Hawking (1976), Preskill (1992), Page (1993), Unruh and Wald (2017), and Anastopoulos and Savvidou (2020).

- For quantum cosmology in relation to quantum foundations, see paper 15 in Bell (1987), Gell-Mann and Hartle (1990), Halliwell (2009), Kiefer (2013), and Pinto-Neto (2005). For the classicalization of quantum fluctuations in inflation, see Ashtekar et al. (2020), Hsiang and Hu (2022), and Perez et al. (2006).

- For experimental prospects in the study of macroscopic quantum phenomena, see Arndt and Hornberger (2014), Chen (2013), and Fröwis et al. (2018).

APPENDIX A

Useful Formulas

A.1 Gaussian Integrals

The following are the most important Gaussian integrals in one dimension,

$$\int_{-\infty}^{\infty} dx\, x^{2n} e^{-ax^2} = \sqrt{\frac{\pi}{a}} \frac{1 \cdot 3 \cdots (2n-1)}{(2a)^n} \tag{A.1}$$

$$\int_{-\infty}^{\infty} dx\, x^{2n+1} e^{-ax^2} = 0, \tag{A.2}$$

$$\int_{\infty}^{\infty} dx\, e^{-ax^2 + bx} = \sqrt{\frac{\pi}{a}} e^{\frac{b^2}{4a}} \tag{A.3}$$

for all positive integers n, for $\mathrm{Re}\, a > 0$ and for all $b \in \mathbb{C}$.

Equation (A.3) generalizes to n-dimensions by

$$\int d^n x\, e^{-\mathbf{x}^T A \mathbf{x} + \mathbf{b}^T \mathbf{x}} = \frac{\pi^{\frac{n}{2}}}{\sqrt{\det A}} e^{\frac{1}{4} \mathbf{b}^T A^{-1} \mathbf{b}}, \tag{A.4}$$

where $\mathbf{x}$ is a real column vector, $\mathbf{b}$ is a complex column vector, and A is an $n \times n$ symmetric matrix with strictly positive real part.

For $n = 2$, Eq. (A.4) becomes

$$\int_{\infty}^{\infty} dx \int_{\infty}^{\infty} dy\, e^{-ax^2 - by^2 + 2cxy} = \frac{\pi}{\sqrt{ab - c^2}}, \tag{A.5}$$

for $a, b > 0$, and $ab > c^2$.

Finally, we note that the integral of a Gaussian in a finite interval defines the *error function* erf,

$$\mathrm{erf}(x) = \frac{2}{\sqrt{\pi}} \int_0^y dx\, e^{-x^2}. \tag{A.6}$$

A.2   Other Integrals

The following is a list of integrals that are employed in the book:

$$\int_0^1 dx\sqrt{1 - x^a} = \frac{\sqrt{\pi}\,\Gamma(1 + a^{-1})}{2\Gamma\left(\frac{3}{2} + a^{-1}\right)} \tag{A.7}$$

$$\int_0^\infty dx\frac{\sqrt{x}}{(x + a)^4} = \frac{\pi}{16a^{5/2}} \tag{A.8}$$

$$\int_{-\infty}^\infty \frac{dx\,e^{-ixt}}{x^2 + 1} = \pi e^{-|t|} \tag{A.9}$$

$$\int_{-a}^a dx\sqrt{a^2 - x^2} = \frac{\pi a^2}{2} \tag{A.10}$$

$$\int_{-1}^1 \frac{dx}{(a - b)^2} = \frac{2}{a^2 - b^2} \tag{A.11}$$

$$\int \frac{dx}{\sqrt{a + bx + cx^2}} = \frac{1}{\sqrt{-c}}\cos^{-1}\left(\frac{b + 2cx}{D}\right), \quad \text{where } D = b^2 - 4ac. \tag{A.12}$$

A.3   Series and Infinite Products

The following series are used in this book, or they may appear in the solution of problems.

$$\text{Geometric series: } \sum_{n=0}^\infty x^n = \frac{1}{1 - x}, \quad |x| < 1. \tag{A.13}$$

Differentiating both sides in the above equation with respect to x and multiplying by x, we obtain

$$\sum_{n=1}^\infty nx^n = \frac{x}{(1 - x)^2}. \tag{A.14}$$

The only infinite product we will use in this paper is the product expression for $\sin(\pi z)$, known as the Euler formula,

$$\sin(\pi z) = \pi z \prod_{n=1}^\infty \left(1 - \frac{z^2}{n^2}\right). \tag{A.15}$$

A.4   Saddle-Point and Stationary-Phase Approximations

Consider the integral

$$I_1 = \int_a^b dx\,e^{-f(x)}, \tag{A.16}$$

where $f(x)$ is a real function with a single global minimum at $x = x_0 \in (a, b)$. Then, the dominant contribution to the integral comes from the vicinity of x_0, so we can expand $f(x) \simeq f(x_0) + \frac{1}{2}f''(x_0)(x - x_0)^2$, where $f''(x_0) > 0$. Then,

$$I_1 \simeq e^{-f(x_0)} \int_a^b dx e^{-\frac{1}{2}f''(x_0)(x-x_0)^2}.$$

If $f''(x_0)(a - x_0)^2 >> 1$ and $f''(x_0)(b - x_0)^2 >> 1$, we can take the integration range to $(-\infty, \infty)$ with negligible error. Hence, by Eq. (A.1), we have the *saddle-point approximation* to the integral I_1,

$$I_1 \simeq \sqrt{\frac{2\pi}{f''(x_0)}} e^{-f(x_0)}. \tag{A.17}$$

The same reasoning applies to integrals of the form

$$I_2 = \int_a^b dx e^{if(x)}, \tag{A.18}$$

where again, $f(x)$ has a single global minimum at $x = x_0 \in (a, b)$. Again, the dominant contribution to the integral comes from the vicinity of x_0, so following the same reasoning, we obtain the *stationary-phase approximation* to the integral I_2,

$$I_2 \simeq \sqrt{\frac{2\pi}{f''(x_0)}} e^{if(x_0)+i\frac{\pi}{4}}. \tag{A.19}$$

The phase factor $e^{i\frac{\pi}{4}}$ arises from the integral of the form $\int_{-\infty}^{\infty} e^{iax^2} = \sqrt{\frac{\pi}{a}} e^{i\frac{\pi}{4}}$, which is evaluated as $\lim_{\epsilon \to 0^+} \int_{-\infty}^{\infty} e^{-(\epsilon - ia)x^2}$ using Eq. (A.1).

APPENDIX B

Special Functions

B.1   The Gamma Function and the Beta Function

The gamma function Γ is defined by

$$\Gamma(z) = \int_0^\infty dx\, x^{z-1} e^{-x}, \tag{B.1}$$

for any complex number with $\text{Re}\, z > 0$. It can be analytically extended to the whole complex plane except zero and the negative integers, where $\Gamma(z)$ has simple poles.

The gamma function is a generalization of the factorial of integers, as it satisfies

$$\Gamma(z+1) = z\Gamma(z), \tag{B.2}$$

and $\Gamma(n) = (n-1)!$ for any positive integer n.

The gamma function is expanded around the poles at $z = -n$, for $n = 0, 1, 2, \ldots$, as

$$\Gamma(-n+x) = \frac{(-1)^n}{n!}\left[\frac{1}{x} + \psi_n + O(x)\right], \tag{B.3}$$

where $\psi_n := \sum_{k=1}^n \frac{1}{k} - \gamma$; $\gamma = 0.5772\ldots$ is the Euler–Mascheroni constant.

For $z = \frac{1}{2}$, we set $x = t^2$ in Eq. (B.1). The integral becomes Gaussian, and we obtain

$$\Gamma\left(\frac{1}{2}\right) = \sqrt{\pi}. \tag{B.4}$$

Equation (B.2) implies that $\Gamma(-\frac{1}{2}) = -2\sqrt{\pi}$, and that

$$\Gamma(\frac{1}{2} + n) = \frac{(2n)!}{4^n n!}\sqrt{\pi}, \tag{B.5}$$

for all nonnegative integers n.

The beta function $B(x, y)$ is defined by

$$B(x, y) = \int_0^1 dt\, t^{x-1}(1 - t)^{y-1}. \tag{B.6}$$

The beta function can be expressed in terms of the gamma function,

$$B(x, y) = \frac{\Gamma(x)\Gamma(y)}{\Gamma(x+y)}. \tag{B.7}$$

Table B.1 Values of the zeta function

$\zeta(0)$	$\zeta'(0)$	$\zeta(2)$	$\zeta(3)$	$\zeta(4)$	$\zeta(5)$	$\zeta(\infty)$	$\zeta(-1)$	$\zeta(\frac{1}{2})$	$\zeta(\frac{3}{2})$
$-\frac{1}{2}$	$-\frac{1}{2}\ln 2\pi$	$\frac{\pi^2}{6}$	$1.2020...$	$\frac{\pi^4}{90}$	$1.0369...$	1	$-\frac{1}{12}$	-1.4603	2.6123

B.2   The Riemann Zeta Function

The Riemann zeta function $\zeta(s)$ is a function of a complex variable s. For $\operatorname{Re} s > 1$, ζ is defined as the limit of the Dirichlet series,

$$\zeta(s) = \sum_{n=1}^{\infty} \frac{1}{n^s}. \tag{B.8}$$

The zeta function can be extended to s with $\operatorname{Re} s > 0$, by employing the equation

$$\zeta(s) = \frac{1}{1 - 2^{1-s}} \eta(s), \tag{B.9}$$

where $\eta(s) = \sum_{n=1}^{\infty} \frac{(-1)^{n-1}}{n^s}$ is known as the Dirichlet eta function, and it is defined for $\operatorname{Re} s > 0$. Then, ζ is analytically extended to the whole complex plane by the functional equation

$$\zeta(s) = 2^s \pi^{s-1} \sin\left(\frac{\pi s}{2}\right) \Gamma(1 - s)\zeta(1 - s). \tag{B.10}$$

The zeta function is holomorphic at all points of $\mathbb{C}$ except for $s = 1$, where it has a single pole:

$$\zeta(s) = \frac{1}{s - 1} + \gamma + \cdots . \tag{B.11}$$

The zeta function also has zeroes at $s = -2n$, $n = 1, 2, \ldots$.

In Table B.1, we list the values of the zeta function and of its derivatives at specific points.

B.3   Fresnel Integrals

The Fresnel integrals $C(x)$ and $S(x)$ are defined by

$$C(x) = \int_0^x dt \cos t^2 = \sum_{n=0}^{\infty} \frac{(-1)^n x^{4n+3}}{(2n + 1)! \, (4n + 3)}, \tag{B.12}$$

$$S(x) = \int_0^x dt \sin t^2 = \sum_{n=0}^{\infty} \frac{(-1)^n x^{4n+3}}{(2n)! \, (4n + 3)}. \tag{B.13}$$

The two functions are odd. They are plotted in Fig. B.1 for $x \geq 0$.

The Fresnel integrals have the following asymptotic behavior for large x,

$$C(x) = \sqrt{\frac{\pi}{8}} + \frac{\sin x^2}{2x}, \qquad S(x) = \sqrt{\frac{\pi}{8}} - \frac{\cos x^2}{2x}. \tag{B.14}$$

Fig. B.1 Plot of the Fresnel integrals $C(x)$ and $S(x)$.

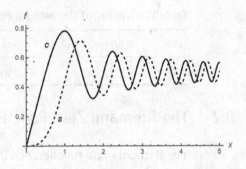

Fig. B.2 Plot of the Airy functions $Ai(z)$ and $Bi(z)$.

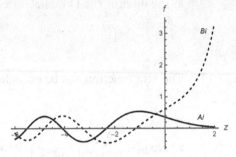

B.4   The Airy Functions

The Airy function of the first kind is defined as

$$Ai(z) := \frac{1}{\pi} \int_0^\infty dt \cos\left(\frac{1}{3}t^3 + zt\right), \tag{B.15}$$

and the Airy function of the second kind is defined as

$$Bi(z) := \frac{1}{\pi} \int_0^\infty dt \left[\exp\left(-\frac{1}{3}t^3 + zt\right) + \sin\left(\frac{1}{3}t^3 + zt\right)\right]. \tag{B.16}$$

The two functions are plotted in Fig. B.2. For large positive z, Ai drops to zero, while Bi diverges. They behave asymptotically as

$$Ai(z) = \frac{1}{2\sqrt{\pi}z^{1/4}}e^{-\zeta}, \quad Bi(z) = \frac{1}{\sqrt{\pi}z^{1/4}}e^{\zeta}, \quad \text{where } \zeta = \frac{2}{3}|z|^{3/2}. \tag{B.17}$$

For large negative z, both Ai and Bi oscillate with a slowly decaying amplitude, and a phase difference of $\frac{\pi}{2}$,

$$Ai(z) = \frac{1}{\sqrt{\pi}|z|^{1/4}} \cos\left(\zeta - \frac{\pi}{4}\right), \quad Bi(z) = -\frac{1}{\sqrt{\pi}|z|^{1/4}} \sin\left(\zeta - \frac{\pi}{4}\right). \tag{B.18}$$

Of particular interest are the roots r_n of the Airy $Ai(z)$ function which occur for $z < 0$. We approximate the roots using the asymptotic expression for large $|z|$. Then, the roots occur for $\zeta = -n\pi + \frac{\pi}{4}$, for $n = 1, 2, \ldots$, or equivalently $r_n \simeq -\left[\frac{3\pi}{2}(n - \frac{1}{4})\right]^{2/3}$. The accuracy of this

Table B.2 Roots of the Airy function

n	1	2	3	4	5	6
$-r_n$	2.338	4.088	5.520	6.787	7.944	9.022
$\left[\frac{3\pi}{2}\left(n-\frac{1}{4}\right)\right]^{2/3}$	2.320	4.082	5.517	6.784	7.942	9.021
e_n (‰)	8	2	0.5	0.4	0.2	0.1

approximation is demonstrated in Table B.2. The relative error e_n is less than 1 percent for all n and less than 1 percent for $n \geq 3$.

B.5   Bessel Functions

Bessel functions are solutions to Bessel's differential equation

$$x^2 y'' + xy' + (x^2 - v^2)y = 0, \tag{B.19}$$

where v is an arbitrary complex number. If v is not an integer, then J_v and J_{-v} are two functionally independent solutions to Eq. (B.19), where

$$J_v(x) = \left(\frac{x}{2}\right)^v \sum_{k=0}^{\infty} (-1)^k \frac{x^{2k}}{2^{2k} k! \, \Gamma(v + k + 1)}, \tag{B.20}$$

is the Bessel function of the first kind.

If v is a positive integer n, then $J_{-n}(x) = (-1)^n J_n(x)$, that is, J_{-n} and J_n are not functionally independent. In this case, the second independent solution to Eq. (B.19) is the Neumann function

$$Y_n(x) = \lim_{v \to n} \frac{J_v(x) \cos v\pi - J_{-v}(x)}{\sin v\pi}. \tag{B.21}$$

The Bessel functions $J_n(x)$ of integer order have an integral expression

$$J_n(x) = \frac{1}{2\pi} \int_{-\pi}^{\pi} e^{i(n\theta - x\sin\theta)} d\theta = \frac{1}{\pi} \int_0^{\pi} \cos(n\theta - x\sin\theta) d\theta, \tag{B.22}$$

which is a convenient starting point for many approximations (see Fig. B.3). For example, applying the stationary phase approximation, Eq. (A.19), to the integral (B.22) yields the asymptotic expression for $x \gg n$,

$$J_n(x) \simeq \sqrt{\frac{2}{\pi x}} \cos\left(x - \frac{2n + 1}{4}\pi\right). \tag{B.23}$$

We also note that by Eq. (B.20), $J_n(x) \simeq \frac{x^n}{2^n n!}$ for small x, while for $x \gg n$,

$$J_n(x) \simeq \sqrt{\frac{2}{\pi x}} \cos\left(x - \frac{2n + 1}{4}\pi\right). \tag{B.24}$$

Fig. B.3 Plot of the Bessel functions J_0, J_1, and J_2.

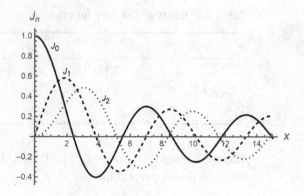

From Eq. (B.22), we can derive the recursion relations

$$\frac{2n}{x} J_n(x) = J_{n-1}(x) + J_{n+1}(x),$$ (B.25)

$$2J_n'(x) = J_{n-1}(x) - J_{n+1}(x).$$ (B.26)

Bessel's equation for purely imaginary argument becomes $x^2 y'' + xy' - (x^2 + v^2)y = 0$. For v noninteger, the two independent solutions are $I_v(x)$ and $I_{-v}(x)$, where

$$I_v(x) = i^{-v} J_v(ix),$$ (B.27)

is the modified Bessel function of the first kind. For integer $v = n$, a pair of independent solutions is obtained by $I_n(x)$ and $K_n(x)$, where

$$K_n(x) = \frac{\pi}{2} \lim_{v \to n} \frac{I_{-v}(x) - I_v(x)}{\sin v\pi}$$ (B.28)

is the modified Bessel function of the second kind. Unlike the ordinary Bessel functions, which are oscillating, $I_n(x)$ and $K_n(x)$ are exponentially growing and decaying functions, respectively.

The modified Bessel functions of the second kind have an integral expression

$$K_n(x) = \int_0^\infty dt e^{-x \cosh t} \cosh(nt).$$ (B.29)

B.6   Spherical Bessel Functions

The spherical Bessel functions are solutions to the equation

$$x^2 y'' + 2xy' + \left[x^2 - \ell(\ell+1)\right]y = 0,$$ (B.30)

where ℓ is a nonnegative integer. The two independent solutions to Eq. (B.30) are the spherical Bessel functions of the first kind $j_\ell(x)$ and the spherical Bessel functions of the second kind $\eta_\ell(x)$, defined by

$$j_\ell(x) = (-1)^\ell x^\ell \left(\frac{1}{x}\frac{d}{dx}\right)^\ell \frac{\sin x}{x}, \quad \eta_\ell(x) = (-1)^\ell x^{\ell+1} \left(\frac{1}{x}\frac{d}{dx}\right)^\ell \frac{\cos x}{x}.$$ (B.31)

Table B.3 The spherical Bessel functions of first and second kind up to $\ell = 3$

ℓ	$j_\ell(x)$	$\eta_\ell(x)$
0	$\dfrac{\sin x}{x}$	$-\dfrac{\cos x}{x}$
1	$\dfrac{\sin x}{x^2} - \dfrac{\cos x}{x}$	$-\dfrac{\cos x}{x^2} - \dfrac{\sin x}{x}$
2	$\left(\dfrac{3}{x^2} - 1\right)\dfrac{\sin x}{x} - \dfrac{3\cos x}{x^2}$	$-\left(\dfrac{3}{x^2} - 1\right)\dfrac{\cos x}{x} - \dfrac{3\sin x}{x^2}$
3	$3\left(\dfrac{5}{x^3} - \dfrac{2}{x}\right)\dfrac{\sin x}{x} - \left(\dfrac{15}{x^2} - 1\right)\dfrac{\cos x}{x}$	$-3\left(\dfrac{5}{x^3} - \dfrac{2}{x}\right)\dfrac{\cos x}{x} - \left(\dfrac{15}{x^2} - 1\right)\dfrac{\sin x}{x}$

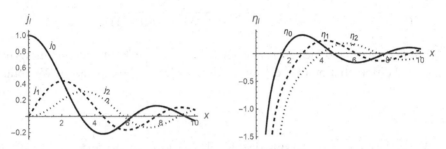

Fig. B.4 Spherical Bessel functions of the first and second kind.

They are normalized so that their Wronskian satisfies

$$j_\ell(x)n'_\ell(x) - j'_\ell(x)\eta_\ell(x) = \frac{1}{x^2}. \tag{B.32}$$

The two functions differ primarily in their behavior at $x = 0$. The spherical Bessel functions of the first kind are finite: $j_0(0) = 1$ and $j_\ell(0) = 0$ for $\ell \geq 1$, while for those of the second kind, $\eta_\ell(0) = -\infty$. Both functions vanish $x \to \infty$. Explicit expressions of the first spherical Bessel functions are given in Table B.3 and some plots in Fig. B.4.

If we substitute $y(x) = u(x)/\sqrt{x}$ in Eq. (B.30), we find that $u(x)$ satisfies Bessel's equation for $v = \ell + \frac{1}{2}$. Hence, the spherical Bessel functions can be expressed as ordinary Bessel functions of half-integer order

$$j_\ell(x) = \sqrt{\frac{\pi}{2x}}J_{\ell+\frac{1}{2}}(x), \qquad \eta_\ell(x) = (-1)^{\ell+1}\sqrt{\frac{\pi}{2x}}J_{-\ell-\frac{1}{2}}(x). \tag{B.33}$$

Equation (B.33) together with Eqs. (B.25) and (B.26) imply the recursion relations

$$j_{\ell+1}(x) + j_{\ell-1}(x) = \frac{2\ell+1}{x}j_\ell(x), \quad j'_\ell(x) = \frac{\ell}{x}j_\ell(x) - j_{\ell+1}(x), \tag{B.34}$$

$$\eta_{\ell+1}(x) + \eta_{\ell-1}(x) = \frac{2\ell+1}{x}\eta_\ell(x), \quad \eta'_\ell(x) = \frac{\ell}{x}\eta_\ell(x) - \eta_{\ell+1}(x). \tag{B.35}$$

By Eq. (B.31), the dominant contribution to the spherical Bessel functions for large x comes from the successive action of the derivatives in Eq. (B.31) on the trigonometric functions. Hence,

$$
j_\ell(x) = \begin{cases} (-1)^{\ell/2} \frac{\sin x}{x}, & \ell \text{ even} \\ (-1)^{(\ell-1)/2} \frac{\cos x}{x}, & \ell \text{ odd} \end{cases} = \frac{\sin\left(x - \frac{\ell\pi}{2}\right)}{x}, \tag{B.36}
$$

$$
\eta_\ell(x) = \begin{cases} -(-1)^{\ell/2} \frac{\cos x}{x}, & \ell \text{ even} \\ -(-1)^{(\ell-1)/2} \frac{\sin x}{x}, & \ell \text{ odd} \end{cases} = -\frac{\cos\left(x - \frac{\ell\pi}{2}\right)}{x}. \tag{B.37}
$$

For small x, we write the Taylor series $\sin x = \sum_{k=1}^{\infty} \frac{(-1)^{k+1} x^{2k+1}}{(2k+1)!}$. After acting on $\sin x / x$ with $\left(\frac{1}{x}\frac{d}{dx}\right)^\ell$, the dominant term for small x is $\frac{(-1)^\ell x^{2\ell+1}}{(2\ell+1)!}$ multiplied by $2\ell(2\ell - 2)(2\ell - 4)\cdots$ from the differentiations. Hence,

$$
j_\ell(x) = \frac{x^\ell}{1 \cdot 3 \cdots \cdot (2\ell + 1)}. \tag{B.38}
$$

For $\eta_\ell(x)$, we expand the cosine as $\cos x = 1 - \frac{x^2}{2} + \cdots$, but the dominant term comes from the repeated differentiation of the $\frac{1}{x}$ term, which is first in the expansion. We obtain

$$
\eta_\ell(x) = -\frac{1 \cdot 3 \cdots \cdot (2\ell - 1)}{x^{\ell+1}}. \tag{B.39}
$$

Finally, we note that the spherical Bessel functions of the first kind satisfy the completeness relation

$$
\int_0^\infty dx\, x^2 j_\ell(bx) j_\ell(b'x) = \frac{\pi}{2b^2} \delta(b - b'). \tag{B.40}
$$

B.7   Hermite Polynomials

The Hermite polynomials $H_n(x)$ are polynomials of nth order, defined by

$$
H_n(x) = (-1)^n e^{x^2} \frac{d^n}{dx^n} e^{-x^2/2}, \tag{B.41}
$$

where $n = 0, 1, 2, \ldots$. They define the unique polynomial basis in the Hilbert space $L^2(\mathbb{R}, e^{-x^2/2} dx)$.

Hermite polynomials are solutions to Hermite's differential equation

$$
y'' - 2xy' + 2ay = 0. \tag{B.42}
$$

We express a general solution to Eq. (B.42) as a series $y = \sum_{k=0}^{\infty} a_k \xi^k$. Then, we obtain the recursion relation

$$
a_{k+2} = \frac{2(k - a)}{(k + 2)(k + 1)} a_k, \tag{B.43}
$$

from which we find that a polynomial solution exists only if $a = n$ for some $n = 0, 1, 2, \ldots$. These solutions are the Hermite polynomials.

The first seven Hermite polynomials are the following.

$$H_0(x) = 1 \tag{B.44}$$
$$H_1(x) = 2x \tag{B.45}$$
$$H_2(x) = 4x^2 - 2 \tag{B.46}$$
$$H_3(x) = 8x^3 - 12x \tag{B.47}$$
$$H_4(x) = 16x^4 - 48x^2 + 12 \tag{B.48}$$
$$H_5(x) = 32x^5 - 160x^3 + 120x \tag{B.49}$$
$$H_6(x) = 64x^6 - 480x^4 + 720x^2 - 120. \tag{B.50}$$

The Hermite polynomials satisfy the following identities.

1. Symmetry: $H_n(-x) = (-1)^n H_n(x)$.
2. Algebraic recursion relation: $H_{n+1} = 2xH_n - 2nH_{n-1}$.
3. Differential recursion relation: $H_n' = H_{n+1} - 2xH_n$.
4. Orthonormalization: $\int dx e^{-x^2} H_n(x) H_m(x) = 2^n n! \sqrt{\pi} \delta_{nm}$.

B.8 Spherical Harmonics

In spherical coordinates, the Laplace operator on $\mathbb{R}^3$ is of the form

$$\nabla^2 = \frac{1}{r^2}\left(\frac{\partial}{\partial r}\left(r^2\frac{\partial}{\partial r}\right) + \Delta_{S^2}\right), \tag{B.51}$$

where

$$\Delta_{S^2} = \frac{1}{\sin\theta}\frac{\partial}{\partial\theta}\left(\sin\theta\frac{\partial}{\partial\theta}\right) + \frac{1}{\sin^2\theta}\frac{\partial^2}{\partial\phi^2} \tag{B.52}$$

is the Laplace operator on the sphere S^2.

The eigenfunctions $Y(\theta,\phi)$ of Δ_{S^2} are solutions to the equation

$$\Delta_{S^2} Y(\theta,\phi) = a Y(\theta,\phi). \tag{B.53}$$

Writing $Y(\theta,\phi) = g(\theta)h(\phi)$, we obtain

$$\frac{1}{h}\frac{\partial^2 h}{\partial\phi^2} = -\frac{\sin\theta}{g}\frac{\partial}{\partial\theta}\left(\sin\theta\frac{\partial g}{\partial\theta}\right) - a\sin^2\theta. \tag{B.54}$$

In Eq. (B.54), the right-hand side is a function of ϕ, and the left-hand side is a function of θ. Equality holds if both sides equal a constant, say b. It follows that $\frac{d^2h}{d\phi^2} - b\phi = 0$, with solutions $h(\phi) = e^{\pm\sqrt{b}\phi}$. The periodicity condition $h(\phi) = h(\phi + 2\pi)$ implies that $b = -m^2$, where $m = 0, 1, 2, \ldots$.

Equation (B.54) also implies that

$$\sin\theta\frac{\partial}{\partial\theta}\left(\sin\theta\frac{\partial g}{\partial\theta}\right) + a\sin^2\theta g = m^2 g. \tag{B.55}$$

Setting $\xi = \cos\theta$ we obtain the *associated Legendre equation* (Legendre)

$$(1 - \xi^2)g'' - 2\xi g' + \left(a - \frac{m^2}{1-\xi^2}\right)g = 0. \tag{B.56}$$

The associated Legendre equation admits solutions with finite value at $\xi = \pm 1$, only if $a = \ell(\ell+1)$, for $\ell = 0, 1, 2, 3, \ldots$ and $|m| \le \ell$. The, the solutions are the so called *associated Legendre polynomials*, defined by

$$P_\ell^m(\xi) := \frac{(-1)^m}{2^\ell \ell!}(1-\xi^2)^{\frac{m}{2}} \frac{d^{\ell+m}}{d\xi^{\ell+m}}(\xi^2 - 1)^\ell, \tag{B.57}$$

for $m \ge 0$, while for $m < 0$,

$$P_\ell^{-m}(\xi) = (-1)^m \frac{(\ell-m)!}{(\ell+m)!} P_\ell^m(\xi). \tag{B.58}$$

For $m = 0$, Eq. (B.57) coincides with the Rodrigues formula for the Legendre polynomials, $P_\ell^0(\xi) = P_\ell(\xi)$.

The first few associated Legendre polynomials are given in Table B.4.

The associated Legendre polynomials satisfy the orthonormality relations

$$\int_{-1}^1 d\xi\, P_\ell^m(\xi)P_{\ell'}^m(\xi) = \frac{2(\ell+m)!}{(2\ell+1)(\ell-m)!}\delta_{\ell\ell'}. \tag{B.59}$$

The Laplace operator on the sphere Λ_{S^2} has eigenvalues $-\ell(\ell+1)$ and eigenfunctions the so called *spherical harmonics*

$$Y_{\ell m}(\theta,\phi) = C_{\ell,m}P_\ell^m(\cos\theta)e^{im\phi}, \tag{B.60}$$

where $m = -\ell, -\ell+1, \ldots, -1, 0, 1, \ldots, \ell-1, \ell$.

The constants $C_{\ell,m}$ in Eq. (B.60) are determined by the orthonormalization condition

$$\int_0^\pi \sin\theta d\theta \int_0^{2\pi} d\phi\, Y_{\ell m}(\theta,\phi)\,Y_{\ell'm'}(\theta,\phi) = \delta_{\ell\ell'}\delta_{mm'}, \tag{B.61}$$

which is satisfied if

$$C_{\ell,m} = \sqrt{\frac{(2\ell+1)}{4\pi}\frac{(\ell-m)!}{(\ell+m)!}}. \tag{B.62}$$

Table B.4 Associated Legendre polynomials

ℓ	$m = 0$	$m = 1$	$m = 2$	$m = 3$
0	1			
1	ξ	$-(1-\xi^2)^{1/2}$		
2	$\frac{1}{2}(3\xi^2 - 1)$	$-3\xi(1-\xi^2)^{1/2}$	$3(1-\xi^2)$	
3	$\frac{1}{2}\xi(5\xi^2 - 3)$	$\frac{3}{2}(1-5\xi^2)(1-\xi^2)^{1/2}$	$15\xi(1-\xi^2)$	$-15(1-\xi^2)^{3/2}$

B.9   Generalized Laguerre Polynomials

The generalized Laguerre polynomials $L_n^{(\alpha)}$, for $n = 0, 1, 2, \ldots$ and $\alpha > -1$ form the unique polynomial basis on the Hilbert space $\mathcal{H}_\alpha = L^2(\mathbb{R}^+, x^\alpha e^{-x} dx)$. They are eigenvectors of the operator $\hat{F}_\alpha = -x\frac{d^2}{dx^2} - (\alpha + 1 - x)\frac{d}{dx}$ that is self-adjoint on $\mathcal{H}_\alpha$. The eigenvalue associated to $L_n^{(\alpha)}$ is n.

As shown in Box 13.3, the eigenvectors of $\hat{F}_\alpha$ can be expressed as a series $\sum_{k=0}^n c_k(-x)^k$, where the coefficients c_k are subject to the recursion relation

$$c_{k+1} = \frac{n - k}{(k + 1)(k + \alpha + 1)} c_k, \tag{B.63}$$

which can be solved to give

$$L_n^{(\alpha)}(x) = \sum_{k=0}^n \frac{\Gamma(n + \alpha + 1)}{\Gamma(k + a + 1)k!\,(n - k)!}(-x)^k. \tag{B.64}$$

An overall multiplicative constant has been chosen so that $c_n = \frac{1}{n!}$.

Equation (B.64) can also be derived by a Rodriguez formula

$$L_n^{(\alpha)}(x) = \frac{x^{-\alpha} e^x}{n!} \frac{d^n \left(x^{n+\alpha} e^{-x}\right)}{dx^n}. \tag{B.65}$$

Polynomials $L_n^{(\alpha)}$ with the same α but different n are orthogonal, since they are eigenvectors of a self-adjoint operator with different eigenvalues. Furthermore, for any smooth function $p(x)$ that grows at most as a polynomial at infinity,

$$\int_0^\infty dx e^{-x} x^\alpha p(x) L_n^{(\alpha)}(x) = \frac{1}{n!}\int_0^\infty dx p(x)\frac{d^n(x^{n+\alpha}e^{-x})}{dx^n}$$
$$= \frac{(-1)^n}{n!}\int_0^\infty \frac{d^n p(x)}{dx^n} e^{-x} x^{n+\alpha},$$

where we used n integrations by part. For $p(x) = L_n^{(\alpha)}(x)$, $\frac{d^n p(x)}{dx^n} = (-1)^n$. Then, we obtain the normalization condition

$$\int_0^\infty dx e^{-x} x^\alpha [L_n^{(\alpha)}(x)]^2 = \frac{1}{n!}\int_0^\infty e^{-x} x^{n+\alpha} = \frac{\Gamma(n + \alpha + 1)}{n!}. \tag{B.66}$$

For $p(x) = xL_n^{(\alpha)}(x)$, $\frac{d^n p(x)}{dx^n} = (-1)^n[(n+1)x - n(n+\alpha)]$, hence,

$$\int_0^\infty dx e^{-x} x^{\alpha+1} [L_n^{(\alpha)}(x)]^2 = \frac{\Gamma(n + \alpha + 1)}{n!}(2n + \alpha + 1). \tag{B.67}$$

The first generalized Laguerre polynomials are given in Table B.5.

Table B.5 Generalized Laguerre polynomials

n	$L_n^{(\alpha)}(x)$
0	1
1	$-x + \alpha + 1$
2	$\frac{x^2}{2} - (\alpha + 2)x + \frac{(\alpha+2)(\alpha+1)}{2}$
3	$-\frac{x^3}{6} + \frac{(\alpha+3)x^2}{2} - \frac{(\alpha+2)(\alpha+3)x}{2} + \frac{(\alpha+3)(\alpha+2)(\alpha+1)}{6}$

APPENDIX C

Elements of Group Theory

C.1 Basic Concepts

C.1.1 Definitions

Definition C.1 A *group* is a set G equipped with an operation that assigns to any two elements $g_1, g_2 \in G$ another element of G, to be denoted by $g_1 g_2$. The group operation satisfies the following properties.

(i) *Associativity.* $g_1(g_2 g_3) = (g_1 g_2)g_3$ for all $g_1, g_2, g_3 \in G$.

(ii) *Existence of unity.* There is a unique element $e \in G$ (unity) that satisfies $eg = ge = g$ for all $g \in G$.

(iii) *Existence of inverse.* For every $g \in G$, there is a unique element $g^{-1} \in G$, such that $gg^{-1} = g^{-1}g = e$.

A subset H of a group G that is closed under the group law is called a *subgroup* of G. The group G itself and the singleton set $\{e\}$ are trivial subgroups of G. A subgroup H of g is *normal*, if $ghg^{-1} \in H$ for all $h \in H$ and all $g \in G$.

Let H be a subgroup of G. For any $g \in G$, we define the *left coset* gH as the set of all elements of the form gh, where $h \in H$. The set of all left cosets is called the *quotient space* and it is denoted by G/H.

Let R be a subset of G. The subgroup $\langle R \rangle$ generated by R consists of all group elements that can be expressed as finite products of elements of R or their inverses. If $\langle R \rangle = G$, then we say that R is a generating set for G, and the elements of R are called *generators* of G. The *rank* of a group is the smallest number of elements that can form a generating set for G.

The *Cartesian product* of the groups G and H is the group with elements in the Cartesian product $G \times H$ and an operation defined by $(g_1, h_1)(g_2, h_2) = (g_1 g_2, h_1 h_2)$, for all $g_1, g_2 \in G$ and $h_1, h_2 \in H$.

If $g_1 g_2 = g_2 g_1$ for all $g_1, g_2 \in G$, then G is an *Abelian* group.

A *homomorphism* between two groups G and H is a function $f : G \to H$ that satisfies $f(g_1)f(g_2) = f(g_1 g_2)$, for all $g_1, g_2 \in G$. Two groups related by a homomorphism are called homomorphic. A bijective homomorphism is called an *isomorphism*. Two groups related by an isomorphism are called isomorphic. Two isomorphic groups have the same group structure and can be considered as identical.

C.1.2   Discrete Groups

A group G over a denumerable set is called *discrete*. The following are examples of discrete groups.

1. The set $\mathbb{Z}$ of integers is an Abelian group with respect to the operation of addition.
2. The set of the nth roots of unity $\{e^{ir\pi/n}|r = 0, 1, \ldots, n-1\}$ is an Abelian group with n elements. It is called the cyclic group of order n, and it is denoted by $\mathbb{Z}_n$. All elements of $\mathbb{Z}_n$ can be written as powers of $e^{i\pi/n}$, hence, the rank of $\mathbb{Z}_n$ is 1.
3. Let A_n be a set with n distinguishable elements. Conventionally, we choose $A_n = \{1, 2, \ldots, n\}$. Any bijective function $\sigma : A_n \to A_n$ is called a *permutation* of A_n, and it can be represented as a column vector with n elements, $(\sigma(1), \sigma(2), \ldots, \sigma(n))$. The set of all permutations of A_n is a group under the usual composition of functions, and it is denoted by S_n. The unit element is the identity permutation, that is, each element of A_n is mapped to itself. The *permutation group S_n* has n elements.

   The group S_2 has two elements, the identity $e = (1, 2)$ and $a = (2, 1)$. It is isomorphic to $\mathbb{Z}_2$.

   The group S_3 has six elements, the identity $e = (1, 2, 3)$, and $(3, 1, 2), (2, 3, 1), (2, 1, 3), (3, 2, 1)$, $(1, 3, 2)$. We define $a = (3, 1, 2)$ and $b = (2, 1, 3)$. Then, we note that $(2, 3, 1) = aa = a^2, a^3 = e, (3, 2, 1) = ab$, and $(1, 3, 2) = a^2 b$. The elements a and b define a generating set for S_3; hence, the rank of S_3 is two.
4. The permutation group S_n contains $\frac{1}{2}n(n-1)$ permutations r_{ij} called *inversions*. An inversion r_{ij} switches the ith element with the jth element of A_n ($i \neq j$), while leaving all other elements of A_n unaffected. By definition $r_{ij}^2 = e$, or $r_{ij} = r_{ij}^{-1}$.

A general permutation $\sigma = (\sigma(1), \sigma(2), \ldots, \sigma(n))$ can be written as a product of inversions through the following procedure. Consider first $\sigma(1)$. If $\sigma(1) = 1$, do nothing. If $\sigma(1) \neq 1$, then there is an $i \neq 1$ such that $\sigma(i) = 1$. Use the inversion $r^{(1)} := r_{1i}$ to make the first element into 1. The resulting permutation is $\sigma' = (1, \sigma'(2), \ldots, \sigma'(n))$. Repeat the procedure on σ' starting from $\sigma'(2)$. Proceed this way until the final element. At the end, we will have $k < n$ inversions, denoted by $r^{(a)}$, such that $r^{(k)} \ldots r^{(2)} r^{(1)} \sigma = e$ or, equivalently,

$$\sigma = r^{(1)} r^{(2)} \cdots r^{(k)}. \tag{C.1}$$

Hence, the set R_n of all inversions of n objects is a generating set for S_n.

We define the parity of a permutation $P(\sigma) = (-1)^k$, where k is the number of inversions in Eq. (C.1). A permutation with $P(\sigma) = 1$ is called *even*, and a permutation with $P(\sigma) = -1$ is called *odd*. The product of two even permutations is even; hence, the set of even permutations forms a subgroup A_n of S_n. The product of two odd permutations is even, so odd permutations do not form a subgroup.

C.1.3   Lie Groups

In a Lie group, the group composition law defines a mapping $G \times G \to G$ that is continuous and differentiable. Here, we will focus on matrix Lie groups, that is, Lie groups that can be expressed as sets of $n \times n$ matrices, so the composition law will be matrix multiplication. We give some general definitions, before proceeding to a presentation of the most useful Lie groups.

The *dimension* of a Lie group G, is the smallest number of independent real variables that are required in order to specify uniquely an element of G. We will write the dimension of G as dim$G$.

A Lie group G is *compact* if there is a bijection from G to a bounded subset of $\mathbb{R}^n$ for some n.

A Lie group G is *connected* if there is a continuous path from e to any $g \in G$.

A Lie group G is *simply connected* if it is connected, and if any continuous loop on G can be shrunk continuously to a single point. This means that for every continuous path $\gamma : [0,1] \to G$, such that $\gamma(0) = \gamma(1) = e$, there exists a continuous map $\tilde{\gamma} : [0,1] \times [0,1] \to G$, such that $\tilde{\gamma}(s,1) = \gamma(s)$ and $\gamma(s,0) = e$ for all $s \in [0,1]$.

Suppose $f : G \to H$ is a homomorphism between two Lie groups, such that $f^{-1}(H) = \mathbb{Z}_n$, that is, the set $f^{-1}(h)$ has exactly n elements for all $h \in H$. Then we call H the *n-ple cover* of G. The most common case corresponds to $n = 2$, whence we talk about a double cover. The definition of the cover extends to the case where the number of elements in $f^{-1}(h)$ is countably infinite; then $f^{-1}(H) = \mathbb{Z}$. A cover of G that is also simply connected is called a *universal cover* $\bar{G}$, and it is denoted by $\bar{G}$. It can be proven that a group's universal cover is unique.

C.1.4   Examples of Lie Groups

1. The *general linear group* over the reals, $GL(n,\mathbb{R})$, consists of all real $n \times n$ matrices with nonzero determinant. The group operation is matrix multiplication. Since a general matrix involves n^2 real variables, $\dim GL(n,\mathbb{R}) = n^2$.

   $GL(n,\mathbb{R})$ is not compact because a matrix may involve arbitrarily large numbers. It is also not connected, because there is no continuous path from the unit matrix I, with determinant equal to 1, to any matrix with negative determinant, because such a path would have to cross matrices with zero determinant that have been excluded.

   We similarly define the general linear group over complex numbers, $GL(n,\mathbb{C})$, with $\dim GL(n,\mathbb{C}) = 2n^2$. $GL(n,\mathbb{C})$ is noncompact, but it is connected because there are continuous paths from the unity to matrices with negative determinant that do not pass through matrices with zero determinant.

2. The set $\mathbb{R}^n$ is an Abelian Lie group with respect to vector addition. An element $\mathbf{a} \in \mathbb{R}^n$ acts on vectors $\mathbf{x} \in \mathbb{R}^n$ through the transformation: $\mathbf{x} \to \mathbf{x} + \mathbf{a}$. We can describe $\mathbb{R}^n$ as a matrix group, with elements $(n+1) \times (n+1)$ matrices

$$A(\mathbf{a}) = \begin{pmatrix} 1 & 0 & \dots & 0 & a_1 \\ 0 & 1 & \dots & 0 & a_2 \\ \dots & \dots & \dots & \dots & \dots \\ 0 & 0 & \dots & 1 & a_n \\ 0 & 0 & \dots & 0 & 1 \end{pmatrix}, \tag{C.2}$$

that implement the transformation $\mathbf{x} \to \mathbf{x} + \mathbf{a}$ through their action on column vectors of

the form $\begin{pmatrix} x_1 \\ x_2 \\ \dots \\ x_n \\ 1 \end{pmatrix} \in \mathbb{R}^{n+1}$. Hence, $\mathbb{R}^n$ is a subgroup of $GL(n+1,\mathbb{R})$. It is a noncompact,

connected, and simply connected group.

3. The *special linear* groups $SL(n, \mathbb{R})$ and $SL(n, \mathbb{C})$ are subgroups of $GL(n, \mathbb{R})$ and $GL(n, \mathbb{C})$, respectively, whose elements are matrices of unit determinant. Then, $\dim SL(n, \mathbb{R}) = n^2 - 1$, and $\dim SL(n, \mathbb{C}) = 2(n^2 - 1)$.

Both groups are noncompact and connected. The groups $SL(n, \mathbb{C})$ are simply connected. This is not the case for the groups $SL(n, \mathbb{R})$. For $n > 2$, the universal covering group of $SL(n, \mathbb{R})$ is a double cover, while the universal cover of $SL(2, \mathbb{R})$ involves countably infinite copies of $SL(2, \mathbb{R})$ and it is *not* a matrix group.

4. The *upper unitriangular* groups $UT(n, \mathbb{R}) \subset SL(n, \mathbb{R})$ and $UT(n, \mathbb{C}) \subset SL(n, \mathbb{C})$ consists of all $n \times n$ matrices with real and complex coefficients respectively, of the form

$$A = \begin{pmatrix} 1 & a_{12} & a_{13} & \dots & a_{1n} \\ 0 & 1 & a_{23} & \dots & a_{2n} \\ 0 & 0 & 1 & \dots & a_{3n} \\ \dots & \dots & \dots & \dots & \dots \\ 0 & 0 & 0 & \dots & 1 \end{pmatrix}. \tag{C.3}$$

Both groups are compact and connected; $\dim UT(n, \mathbb{R}) = \frac{1}{2}n(n-1)$ and $\dim UT(n, \mathbb{C}) = n(n-1)$.

5. The *orthogonal group* $O(n)$ consists of all orthogonal $n \times n$ matrices, that is, all real matrices O that satisfy $OO^T = O^T O = I$.

The matrix equation $OO^T = I$ is equivalent to $\frac{1}{2}n(n+1)$ real equations, n for the diagonal elements of O, and $\frac{1}{2}(n^2 - n)$ for the off-diagonal ones. Hence, $\dim O(n) = n^2 - n(n+1)/2 = n(n-1)/2$. The orthogonal group is compact, because all orthogonal matrices satisfy $||O|| \leq 1$.

Since $\det O^T O = (\det O)^2 = 1$, $\det O = \pm 1$. Hence, the orthogonal group is not connected. It consists of two disjoint components with different values of the determinant.

The *special orthogonal group* $SO(n)$ is the subset of $O(n)$ that consists of matrices with unit determinant; $\dim SO(n) = \frac{1}{2}n(n-1)$. $SO(n)$ is compact and connected. It is not simply connected; its universal cover is known as $Spin(n)$, and it is a double cover for $n > 2$. The universal cover of $SO(2)$ is $\mathbb{R}$.

6. The *indefinite orthogonal group* $O(p, q)$ consists of $n \times n$ real matrices ($n = p + q$) that satisfy $O\eta O^T = \eta$; η is a diagonal matrix with p elements $+1$ and q elements equal to -1. $O(p, q)$ is noncompact and nonconnected; $\dim O(p, q) = n(n-1)/2$.

The *special indefinite orthogonal group* $SO(p, q)$ is the subgroup of $O(p, q)$ that consists of matrices with unit determinant. $SO(p, q)$ has the same dimension with $O(p, q)$; it is noncompact and connected, but not simply connected.

7. The *unitary group* $U(n)$ consists of all unitary (complex) matrices $n \times n$. $SU(n)$ is compact (because all unitary matrices satisfy $||U|| \leq 1$), but not simply connected; $\dim U(n) = n^2$.

The *special unitary group* $SU(n)$ is the subgroup of $U(n)$ that consists of matrices with unit determinant. The determinant of a unitary matrix has unit norm, that is, it equals a phase $e^{i\theta}$. $SU(n)$ is obtained from $U(n)$ by the condition $e^{i\theta} = 1$; hence, $\dim SU(n) = n^2 - 1$. $SU(n)$ is compact, connected, and simply connected.

8. The *semidirect product* $G \ltimes \mathbb{R}^n$ of a group G of $n \times n$ matrices with the Abelian group $\mathbb{R}^n$ consists of all transformations $\mathbf{x} \rightarrow A\mathbf{x} + \mathbf{a}$, where $\mathbf{x}, \mathbf{a}$ are column vectors in $\mathbb{R}^n$ and $A \in G$.

Two successive transformations $(A_1, \mathbf{a}_1)$ and $(A_2, \mathbf{a}_2)$ on $\mathbf{x}$ give $A_2 A_1 \mathbf{x} + A_2 \mathbf{a}_1 + \mathbf{a}_2$; hence, the group law is

$$(A_2, \mathbf{a}_2)(A_1, \mathbf{a}_1) = (A_2 A_1, A_2 \mathbf{a}_1 + \mathbf{a}_2). \tag{C.4}$$

The group $G \ltimes \mathbb{R}^n$ is often called the *inhomogeneous G group*, and it is denoted as IG.

The elements of IG can be expressed as $(n+1) \times (n+1)$ matrices

$$C(A, \mathbf{a}) = \begin{pmatrix} A_{11} & A_{12} & \dots & A_{1n} & a_1 \\ A_{21} & A_{22} & \dots & A_{2n} & a_2 \\ \dots & \dots & \dots & \dots & \dots \\ A_{n1} & A_{n2} & \dots & A_{nn} & a_n \\ 0 & 0 & \dots & 0 & 1 \end{pmatrix}. \tag{C.5}$$

It is straightforward to verify that these matrices satisfy the multiplication law (C.4).

For $G = SO(n)$, the group $ISO(n)$ is known as the n-dimensional *Euclidean group*. For $G = SO(n-1,1)$, the group $ISO(n-1,1)$ is known as the n-dimensional *Poincaré group*; $\dim ISO(n) = \dim ISO(n-1,1) = \frac{1}{2}n(n+1)$.

C.2   Lie Algebras

A large part of the information contained in Lie groups is encoded in their behavior near unity. This is because, in a Lie group, we can define differentiation and the properties of the group can be inferred from an appropriate set of differential equations. This leads us to the definition of the *Lie algebra* of a group.

C.2.1   Key Definitions

Let G be a Lie group that consists of $n \times n$ matrices. Let $U = (-\epsilon, \epsilon)$ for $\epsilon > 0$, and consider the space Π_G of differentiable paths $A : U \to G$ such that $A(0) = I$. Then, the Lie algebra $\mathfrak{g}$ of the group G consists of all $n \times n$ matrices T that can be written as

$$T = \frac{dA(0)}{ds}, \tag{C.6}$$

for some path $A(\cdot)$ in Π_G. The elements of the Lie algebra $\mathfrak{g}$ are called *generators* of G. The definition of the Lie algebra implies that group elements A very close to I can be written as $A = I + sT + O(s^2)$, justifying our definition of generators for rotation in Section 11.2.

In accordance with the convention of denoting the Lie algebra of a group G as $\mathfrak{g}$, we will be writing the Lie algebra of the group $SL(n, \mathbb{C})$ as $\mathfrak{sl}(n, \mathbb{C})$, the Lie algebra of the group $O(n)$ as $\mathfrak{o}(n)$, and so on.

Let two paths $A_1(s)$ and $A_2(s)$ in the vicinity of I correspond to generators T_1 and T_2, respectively. For s close to zero, $A_1(s) = I + sT_1 + O(s^2)$ and $A_2 = I + sT_2 + O(s^2)$. The product $A_1(s)A_2(s)$ is also an element of Π_G; hence, it defines a Lie algebra element

$$T = \frac{d[A_1 A_2](0)}{ds} = T_1 + T_2. \tag{C.7}$$

Hence, the closure of the group under matrix multiplication implies the closure of the Lie algebra under addition.

Similarly, if a path $A(\cdot)$ is an element of Π_G, so is $\lambda A(\cdot)$ and the associated generator is λT. Hence, the Lie algebra $\mathfrak{g}$ is also closed under scalar multiplication. We conclude that a Lie algebra is a vector space with respect to the reals.

For any generator $T \in \mathfrak{g}$ we define a one-parameter family of matrices $\exp(sT)$ for all $s \in \mathbb{R}$. The matrices $A(s) = e^{sT}$ are all elements of G, because they are solutions to the differential equation $\frac{dA(s)}{ds} = TA(s)$ on the group G, with boundary condition $A(0) = I$.

In compact and connected Lie groups, all elements can be written as e^{sT} for some $s \in \mathbb{R}$ and $T \in \mathfrak{g}$. The BCH identity allows one to compute the matrix products of elements of G from the knowledge of commutators on $\mathfrak{g}$. In this sense, the Lie algebra condenses the information of a Lie group in an algebraically simpler form.

For any pair of matrices $A, B \in G$, $ABA^{-1} \in G$. This means that for any path $B(s) = e^{sT}$, $AB(s)A^{-1} = e^{sATA^{-1}}$ is an element of Π_G. It follows that $ATA^{-1} \in \mathfrak{g}$. The maps

$$Ad_A : T \to ATA^{-1} \tag{C.8}$$

define the *adjoint action* of G on $\mathfrak{g}$.

Consider now a path e^{sS}, for $S \in \mathfrak{g}$. Then, $Ad_{e^{sS}}(T) = e^{sS} T e^{-sS}$, and for small s,

$$Ad_{e^{sS}}(T) = e^{sS} T e^{-sS} = T + s(ST - TS) + O(s^2). \tag{C.9}$$

Since $Ad_{A(s)}(T) \in \mathfrak{g}$, by continuity $ST - TS \in \mathfrak{g}$. Hence, the matrix commutator is also a generator. We will also express the commutator as an adjoint action of the Lie algebra on itself,

$$ad_S(T) = \frac{d}{ds}\left(Ad_{e^{sT}}(T)\right)_{s=0} = [S, T]. \tag{C.10}$$

We conclude that *a Lie algebra is a vector space that is closed under commutators.*

The following results are helpful in identifying elements of specific Lie algebras.

- Let $A(\cdot)$ be a path of matrices with unit determinant; for sufficiently small s, $A(s) = I + sS$, where S is a generator of a special group. Then, $1 = \det A(s) = e^{Tr \ln A(s)} = e^{Tr \ln(I + sS)} \simeq e^{sTrS}$. It follows that $\mathfrak{sl}(n, \mathbb{R})$ and $\mathfrak{sl}(n, \mathbb{C})$ consist of matrices of zero trace.
- We apply the defining relation $OO^T = I$ of orthogonal matrices to $O = I + sS + O(s^2)$, where $S \in \mathfrak{o}(n)$. It is straightforward to show that $S^T = -S$.
- Similarly, we can show that the defining relation $UU^\dagger = 1$ for unitary operators implies that any $S \in \mathfrak{u}(n)$ satisfies $S^\dagger = -S$. Hence, all $S \in \mathfrak{u}(n)$ can be written as iC, for a self-adjoint matrix C.

C.2.2 Properties of Lie Algebras

The dimension D of a Lie algebra equals the dimension of the associated group.

The *direct sum* $\mathfrak{g}_1 \oplus \mathfrak{g}_2$ of two Lie algebras $\mathfrak{g}_1$ and $\mathfrak{g}_2$ is the vector space $\mathfrak{g}_1 \oplus \mathfrak{g}_2$ with the commutator defined as $[T_1 \oplus S_1, T_2, \oplus S_2] = [T_1, T_2] \oplus [S_1, S_2]$. Suppose G_1 and G_2 are two Lie groups with Lie algebras $\mathfrak{g}_1$ and $\mathfrak{g}_2$. It is straightforward to show from the definition (C.6) that the Lie algebra of the group $G_1 \times G_2$ is $\mathfrak{g}_1 \oplus \mathfrak{g}_2$.

A Lie algebra $\mathfrak{g}$ is called Abelian, if $[T, S] = 0$ for all $T, S \in \mathfrak{g}$. The antisymmetry of a commutator implies that all one-dimensional algebras are Abelian.

A *subalgebra* $\mathfrak{h}$ of a Lie algebra $\mathfrak{g}$ is a subspace of $\mathfrak{g}$ that is closed under the commutator operation. The set $\{0\}$ that contains only the zero element of $\mathfrak{g}$ and the algebra $\mathfrak{g}$ itself are trivially subgroups of $\mathfrak{g}$.

A subalgebra $\mathfrak{h}$ is called an *ideal* if, for every $T \in \mathfrak{h}$ and $S \in \mathfrak{g}$, $[S, T] \in \mathfrak{h}$. The trivial subalgebras $\{0\}$ and $\mathfrak{g}$ are also trivial ideals. Nontrivial ideals are called *proper ideals*.

The *center* of an algebra $\mathfrak{g}$ is an ideal whose elements commute with all elements of $\mathfrak{g}$. For example, the center of $\mathfrak{gl}(n, \mathbb{R})$ is the subalgebra $\mathfrak{h} = \{\lambda I | \lambda \in \mathbb{R}\}$.

The *rank* of a Lie algebra is the maximum number of linearly independent elements of the algebra that commute with each other.

The *derived algebra* $[\mathfrak{g}, \mathfrak{g}]$ is the subalgebra of $\mathfrak{g}$ that consists of all elements of $\mathfrak{g}$ that can be expressed as commutators of elements of $\mathfrak{g}$. For example, in an Abelian Lie algebra, $[\mathfrak{g}, \mathfrak{g}] = \{0\}$. We write $\mathfrak{g}^{(1)} := [\mathfrak{g}, \mathfrak{g}]$, and then we define $\mathfrak{g}^{(2)} := [\mathfrak{g}^{(1)}, \mathfrak{g}^{(1)}]$, $\mathfrak{g}^{(3)} := [\mathfrak{g}^{(2)}, \mathfrak{g}^{(2)}]$ and so on.

The Lie algebra $\mathfrak{g}$ is called *solvable*, if there exists n such that $\mathfrak{g}^{(n)} = \{0\}$. By definition, if $\mathfrak{g}$ is solvable, then $[\mathfrak{g}, \mathfrak{g}]$ is a proper subset of $\mathfrak{g}$. Furthermore, if $\mathfrak{g}$ is solvable, then $[\mathfrak{g}^{(n-1)}, \mathfrak{g}^{(n-1)}] = \{0\}$, that is, $\mathfrak{g}$ contains an Abelian ideal.

A Lie algebra is *simple* if it is non-Abelian and it has no proper ideals. This means that in simple algebras, $[\mathfrak{g}, \mathfrak{g}] = \mathfrak{g}$. It follows that no solvable algebra can have a simple subalgebra.

A Lie algebra is *semisimple* if it has no solvable proper ideals. This implies that a semisimple algebra does not contain any proper Abelian ideals. Obviously, simple algebras are always semisimple.

C.2.3 Structure Constants

Since the Lie algebra $\mathfrak{g}$ is a vector space of dimension D, we can choose D linearly independent generators T_a to define a basis, where $a = 1, 2, \ldots, D$. Hence, every $S \in \mathfrak{g}$ can be expressed as $S = \sum_a c_a T_a$ for some reals c_a. In particular, the commutator $[T_a, T_b]$ between two elements of the basis is a linear combination of the basis elements T_c. Hence,

$$[T_a, T_b] = \sum_c f_{abc} T_c. \qquad (C.11)$$

The quantities f_{abc} are called *structure constants* of the Lie algebra $\mathfrak{g}$. They are antisymmetric in the first two indices, $f_{abc} = -f_{bac}$, and they depend on the choice of basis on $\mathfrak{g}$.

Given that the commutators satisfy Jacobi's identity, $[[T_a, T_b], T_c] + [[T_c, T_a], T_b] + [[T_b, T_c], T_a] = 0$, the structure constants satisfy

$$\sum_{d=1}^{D} \sum_{e=1}^{D} (f_{abd} f_{dce} + f_{cad} f_{dbe} + f_{bcd} f_{dae}) = 0. \qquad (C.12)$$

We define the $D \times D$ matrices X_a, $a = 1, 2, \ldots, D$, by $(X_a)_{bc} := f_{abc}$. The identity (C.12) implies that the commutation algebra of these matrices coincides with that of the generators T_a,

$$[X_a, X_b] = \sum_{c=1}^{D} f_{abc} X_c. \qquad (C.13)$$

We define the *Cartan metric* of $\mathfrak{g}$ as the symmetric matrix

$$w_{ab} := Tr(X_a X_b) = \sum_{c=1}^{D} \sum_{d=1}^{D} f_{acd} f_{bdc}. \tag{C.14}$$

The Cartan metric allows us to distinguish between the types of Lie algebras.

- If w_{ab} vanishes, the Lie algebra is solvable. The converse does not hold.
- The Cartan metric of a semisimple Lie algebra is nondegenerate, and, conversely, all semisimple Lie algebras have nondegenerate Cartan metrics.

We also define the modified structure constants,

$$\bar{f}_{abc} = \sum_{d=1}^{D} f_{abd} w_{dc}, \tag{C.15}$$

which are totally antisymmetric. They satisfy the identity

$$\bar{f}_{abc} = -\bar{f}_{acb}, \tag{C.16}$$

in addition to $\bar{f}_{abc} = -\bar{f}_{bac}$. To prove Eq. (C.16), we calculate $Tr([X_a, X_b]X_c) = \sum_{c=1}^{D} f_{abd} Tr(X_d X_c) = \sum_{c=1}^{D} f_{abd} w_{dc} = \bar{f}_{abc}$. By Eq. (C.13), $Tr([X_a, X_b]X_c) = Tr(X_a X_b X_c - X_b X_a X_c) = Tr(X_c X_a X_b - X_a X_c X_b) = -Tr([X_a, X_c]X_b) = -\bar{f}_{acb}$. Equation (C.16) follows.

C.3   Examples

In this section, we examine in detail some low-dimensional Lie groups and their Lie algebras.

1. *The group $SO(2)$* consists of matrices of the form (11.7). As shown in Section 11.2.1, the associated generator is $J = i\sigma_2$. The Lie algebra $\mathfrak{so}(2)$ consists of matrices λJ for all $\lambda \in \mathbb{R}$, hence, $\mathfrak{so}(2) = \mathbb{R}$.

   $SO(2)$ is isomorphic to the group $U(1)$ with elements $e^{i\theta}$, for $\theta \in [0, 2\pi)$.

2. *The group $SO(1,1)$* consists of matrices of the form

$$O(u) = \begin{pmatrix} \cosh u & \sinh u \\ \sinh u & \cosh u \end{pmatrix}, \tag{C.17}$$

   where $u \in \mathbb{R}$. We straightforwardly evaluate the generator,

$$K := \left[\frac{dO(u)}{du} \right]_{u=0} = \begin{pmatrix} 0 & 1 \\ 1 & 0 \end{pmatrix}. \tag{C.18}$$

   The Lie algebra $\mathfrak{so}(1,1)$ consists of matrices λK for all $\lambda \in \mathbb{R}$, hence, $\mathfrak{so}(1,1) = \mathbb{R}$.

3. The *first affine group Af_1* consists of all 2×2 matrices of the form

$$V(a,b) = \begin{pmatrix} e^a & b \\ 0 & 1 \end{pmatrix}, \tag{C.19}$$

for $a, b \in \mathbb{R}$. We identify the group operation by $V(a_2, b_2)V(a_1, b_1) = V(a_1 + a_2, e^{a_2}b_1 + b_2)$. Unity corresponds to $a = b = 0$.

We define the two generators $K_1 := \frac{\partial V(0,0)}{\partial a}$ and $K_2 := \frac{\partial V(0,0)}{\partial b}$. We find

$$K_1 = \begin{pmatrix} 1 & 0 \\ 0 & 0 \end{pmatrix}, \quad K_2 = \begin{pmatrix} 0 & 1 \\ 0 & 0 \end{pmatrix}. \tag{C.20}$$

The commutator of the generators is $[K_1, K_2] = K_2$. The Lie algebra $\mathfrak{af}_1$ consists of elements $\lambda_1 K_1 + \lambda_2 K_2$, where $\lambda_1, \lambda_2 \in \mathbb{R}$. Obviously $[\mathfrak{af}_1, \mathfrak{af}_1] = \{\lambda K_2, \lambda \in \mathbb{R}\}$, and consequently $\mathfrak{af}_1^{(2)} = \{0\}$; $\mathfrak{af}_1$ is solvable.

The Cartan metric of $\mathfrak{af}_1$ is $w = \begin{pmatrix} 1 & 0 \\ 0 & 0 \end{pmatrix}$.

4. *The group $SO(3)$ describes the rotations* of a three-vector in three-dimensional space. We studied it in detail in Section 11.3.2. Its generators J_1, J_2, J_3 are defined by Eq. (11.12) and they define the commutation relations $[J_i, J_j] = \sum_k \epsilon_{ijk} J_k$. The structure constants of the Lie algebra $\mathfrak{so}(3)$ coincide with ϵ_{ijk}.

The Cartan metric of $\mathfrak{so}(3)$ is

$$w_{ij} = \sum_{kl} \epsilon_{ikl} \epsilon_{jlk} = -2\delta_{ij}. \tag{C.21}$$

Since w is nondegenerate, $SO(3)$ is semisimple.

5. *The group $SU(2)$ consists of unitary* 2×2 matrices. Its most general element can be expressed as $\cos\theta \hat{I} + i\mathbf{n} \cdot \sigma \sin\theta$, where $\mathbf{n}$ is a unit three-vector and $\theta \in [0, 2\pi)$. The Lie algebra $\mathfrak{su}(2)$ consists of matrices of the form iC, where C is self-adjoint. We can choose a basis, $J_i = \frac{i}{2}\sigma_i$, where σ_i are the Pauli matrices and $i = 1, 2, 3$. By (5.4), the matrices J_i satisfy Eq. (11.14), hence, $\mathfrak{su}(2) = \mathfrak{so}(3)$.

However, the groups $SU(2)$ and $SO(3)$ do not coincide. Equation (11.23) implies that two elements of $SU(2)$ that differ by a sign, that is, matrices U and $-U$, correspond to single element of $SO(3)$. This means that $SU(2)$ is a double cover of $SO(3)$.

6. *The group $UT(3, \mathbb{R})$ consists of upper triangular matrices*

$$V(a, b, c) = \begin{pmatrix} 1 & a & c \\ 0 & 1 & b \\ 0 & 0 & 1 \end{pmatrix}. \tag{C.22}$$

The group law is $V(a_1, b_1, c_1)V(a_2, b_2, c_2) = V(a_1 + a_2, b_1 + b_2, c_1 + c_2 + a_1 b_2)$. Note that if one parameterizes the matrices (C.22) as $V(a, b, \tilde{c})$, where $\tilde{c} = c + \frac{1}{2}ab$, then the group law becomes $V(a_1, b_1, \tilde{c}_1)V(a_2, b_2, \tilde{c}_2) = V[a_1 + a_2, b_1 + b_2, \tilde{c}_1 + \tilde{c}_2 + \frac{1}{2}(a_1 b_2 - a_2 b_1)]$, that is, $UT(3, \mathbb{R})$ coincides with the Heisenberg group H_1.

The natural basis on the Lie algebra $\mathfrak{ut}(3, \mathbb{R})$ consists of the matrices

$$K_1 = \begin{pmatrix} 0 & 1 & 0 \\ 0 & 0 & 0 \\ 0 & 0 & 0 \end{pmatrix}, \quad K_2 = \begin{pmatrix} 0 & 0 & 0 \\ 0 & 0 & 1 \\ 0 & 0 & 0 \end{pmatrix}, \quad K_3 = \begin{pmatrix} 0 & 0 & 1 \\ 0 & 0 & 0 \\ 0 & 0 & 0 \end{pmatrix}. \tag{C.23}$$

We find the commutation relations

$$[K_1, K_2] = K_3, \qquad [K_1, K_3] = [K_2, K_3] = 0, \tag{C.24}$$

which demonstrate that $\mathfrak{ut}(3,\mathbb{R})$ reproduces the Heisenberg algebra. It is straightforward to show that the Cartan metric vanishes and that $\mathfrak{ut}(3,\mathbb{R})$ is solvable.

7. *The group $SO(2,1)$ consists of 3×3 matrices O, such that $O^T\eta O = \eta$, where $\eta = \mathrm{diag}\{1,1,-1\}$. As in the study of $SO(3)$, it is convenient to start with the generators. $SO(2,1)$ induces $SO(1,1)$ transformations in the 1–3 and 2–3 planes, and rotations in the 1–2 plane. Hence, we can express any element of $SO(2,1)$ as a product $O_1(u)O_2(v)O_3(\theta)$, where*

$$O_1(u) = \begin{pmatrix} 1 & 0 & 0 \\ 0 & \cosh u & \sinh u \\ 0 & \sinh u & \cosh u \end{pmatrix}, \quad O_2(v) = \begin{pmatrix} \cosh v & 0 & \sinh v \\ 0 & 1 & 0 \\ \sinh v & 0 & \cosh v \end{pmatrix},$$

$$O_3(\theta) = \begin{pmatrix} \cos\theta & \sin\theta & 0 \\ -\sin\theta & \cos\theta & 0 \\ 0 & 0 & 1 \end{pmatrix}. \tag{C.25}$$

The associated generators are

$$K_1 = \begin{pmatrix} 0 & 0 & 0 \\ 0 & 0 & 1 \\ 0 & 1 & 0 \end{pmatrix}, \quad K_2 = \begin{pmatrix} 0 & 0 & 1 \\ 0 & 0 & 0 \\ 1 & 0 & 0 \end{pmatrix}, \quad K_3 = \begin{pmatrix} 0 & 1 & 0 \\ -1 & 0 & 0 \\ 0 & 0 & 0 \end{pmatrix}. \tag{C.26}$$

They satisfy the commutation relations,

$$[K_1, K_2] = -K_3, \qquad [K_3, K_1] = K_2, \qquad [K_2, K_3] = K_1. \tag{C.27}$$

The Cartan metric of $\mathfrak{so}(2,1)$ is

$$w = \begin{pmatrix} 2 & 0 & 0 \\ 0 & 2 & 0 \\ 0 & 0 & -2 \end{pmatrix} = 2\eta,$$

meaning that $\mathfrak{so}(2,1)$ is semisimple.

8. *The group $SL(2,\mathbb{R})$ consists of real 2×2 matrices with unit determinant. The general element of the Lie algebra $\mathfrak{sl}(2,\mathbb{R})$ is a real 2×2 matrix with zero trace. We select a basis on $\mathfrak{sl}(2,\mathbb{R})$, by $K_1 = \frac{1}{2}\sigma_1, K_2 = \frac{1}{2}\sigma_3, K_3 = \frac{i}{2}\sigma_2$. The associated commutators are $[K_1, K_2] = -K_3$, $[K_3, K_1] = K_2$ and $[K_2, K_3] = K_1$. Hence, the Lie algebras $\mathfrak{so}(2,1)$ and $\mathfrak{sl}(2,\mathbb{R})$ coincide.*

The most general element of $\mathfrak{sl}(2,\mathbb{R})$ can be expressed as

$$\mathbf{n}\cdot\mathbf{K} = \begin{pmatrix} n_2 & n_1 + n_3 \\ n_1 - n_3 & -n_2 \end{pmatrix},$$

for some $\mathbf{n} \in \mathbb{R}^3$. We note that $\det \mathbf{n}\cdot\mathbf{K} = n_3^2 - n_1^2 - n_2^2$. The adjoint transformation $\mathbf{n}\cdot\mathbf{K} \rightarrow A(\mathbf{n}\cdot\mathbf{K})A^{-1}$ for $A \in SL(2,\mathbb{R})$ gives $\mathbf{n}'\cdot\mathbf{K}$ for some other vector $\mathbf{n}' \in \mathbb{R}^3$. Hence, there is a 3×3 matrix $O(A)$, such that $O(A)\mathbf{n} = \mathbf{n}'$. Since $\det A = 1$, the adjoint transformation preserves the determinant, hence, $O(A)$ preserves the quadratic combination $n_3^2 - n_1^2 - n_2^2$. This means that $O(A)\eta O^T(A) = \eta$, that is, $O(A) \in SO(2,1)$. Hence, there is a homomorphism between $SL(2,\mathbb{R})$ and $SO(2,1)$. This homomorphism is two-to-one, because two elements A and $-A$ have the same result on adjoint transformations. We conclude that $SL(2,\mathbb{R})$ is the double cover of $SO(2,1)$.

9. *The two-dimensional Euclidean group ISO(2) consists of* 3×3 *matrices*

$$A(\theta, a, b) = \begin{pmatrix} \cos\theta & \sin\theta & a \\ -\sin\theta & \cos\theta & b \\ 0 & 0 & 1 \end{pmatrix},$$

(C.28)

where $\theta \in [0, 2\pi)$, and $a, b \in \mathbb{R}$. We straightforwardly compute the generators

$$K_1 = \frac{dA(0,0,0)}{da} = \begin{pmatrix} 0 & 0 & 1 \\ 0 & 0 & 0 \\ 0 & 0 & 0 \end{pmatrix}, \quad K_2 = \frac{dA(0,0,0)}{db} = \begin{pmatrix} 0 & 0 & 0 \\ 0 & 0 & 1 \\ 0 & 0 & 0 \end{pmatrix},$$

$$K_3 = \frac{dA(0,0,0)}{d\theta} = \begin{pmatrix} 0 & 1 & 0 \\ -1 & 0 & 0 \\ 0 & 0 & 0 \end{pmatrix}.$$

The commutation relations are

$$[K_1, K_2] = 0, \qquad [K_3, K_1] = -K_2, \qquad [K_2, K_3] = K_1.$$

(C.29)

10. *The two-dimensional Poincaré group ISO(1, 1) consists of* 3×3 *matrices*

$$L(u, a, b) = \begin{pmatrix} \cosh u & \sinh u & a \\ \sinh u & \cosh u & b \\ 0 & 0 & 1 \end{pmatrix},$$

(C.30)

where $u, a, b \in \mathbb{R}$. The generators are

$$K_1 = \frac{dL(0,0,0)}{da} = \begin{pmatrix} 0 & 0 & 1 \\ 0 & 0 & 0 \\ 0 & 0 & 0 \end{pmatrix}, \quad K_2 = \frac{dL(0,0,0)}{db} = \begin{pmatrix} 0 & 0 & 0 \\ 0 & 0 & 1 \\ 0 & 0 & 0 \end{pmatrix},$$

$$K_3 = \frac{dL(0,0,0)}{du} = \begin{pmatrix} 0 & 1 & 0 \\ 1 & 0 & 0 \\ 0 & 0 & 0 \end{pmatrix}.$$

The commutation relations are

$$[K_1, K_2] = 0, \qquad [K_1, K_3] = K_2, \qquad [K_2, K_3] = K_1.$$

(C.31)

References

Adler, S. L. 2003. Why decoherence has not solved the measurement problem: A response to P. W. Anderson. *Stud. Hist. Philos. Mod. Phys.*, **34**, 135–142.

Adler, S. L. 2004. *Quantum Theory as an Emergent Phenomenon.* Cambridge: Cambridge University Press.

Aharonov, Y., and Bohm, D. 1959. Significance of electromagnetic potentials in the quantum theory. *Phys. Rev.*, **115**, 485–491.

Aharonov, Y., and Rohrlich, D. 2005. *Quantum Paradoxes.* New York: John Wiley.

Aharonov, Y., and Vaidman, L. 2008. The two-state vector formalism of quantum mechanics: An updated review. Pages 397–447 of: Muga, J. G., Sala Mayato, R., and Egusquiza, I. L. (eds.), *Time in Quantum Mechanics.* Berlin: Springer.

Aharonov, Y., Bergmann, P., and Lebovitz, J. 1964. Time symmetry in the quantum process of measurements. *Phys. Rev.*, **134**, B1410–1416.

Akhiezer, N. I., and Glazman, I. M. 1993. *Theory of Linear Operators in Hilbert Space.* New York: Dover.

Albash, T., and Lidar, D. A. 2018. Adiabatic quantum computation. *Rev. Mod. Phys.*, **90**, 015002.

Albert, D., and Loewer, B. 1988. Interpreting the many worlds interpretation. *Synthese*, **77**, 195–213.

Alicki, R., and Kosloff, R. 2018. Introduction to quantum thermodynamics: History and prospects. Pages 1–33 of: Binder, F., Correa, L. A., Gogolin, C., Anders, J., and Adesso, G. (eds.), *Thermodynamics in the Quantum Regime: Fundamental Aspects and New Directions.* Cham: Springer International Publishing.

Anandan, J. 1992. The geometric phase. *Nature*, **360**, 307–313.

Anastopoulos, C. 2002. Frequently asked questions about decoherence. *Int. J. Theor. Phys.*, **41**, 1573–1590.

Anastopoulos, C. 2003. Quantum processes on phase space. *Ann. Phys.*, **303**, 275–320.

Anastopoulos, C. 2004. On the relation between quantum mechanical probabilities and event frequencies. *Ann. Phys.*, **313**, 368–382.

Anastopoulos, C. 2008. *Particle or Wave: The Evolution of the Concept of Matter in Modern Physics.* Princeton: Princeton University Press.

Anastopoulos, C. 2019. Decays of unstable quantum systems. *Int. J. Theor. Phys.*, **58**, 890–930.

Anastopoulos, C. 2021. Mind–body interaction and modern physics. *Found. Phys.*, **51**, 65.

Anastopoulos, C., and Halliwell, J. J. 1995. Generalized uncertainty relations and long-time limits for quantum Brownian motion models. *Phys. Rev. D*, **51**, 6870–6885.

Anastopoulos, C., and Hu, B. L. 2015. Probing a gravitational cat state. *Class. Quant. Grav.*, **32**, 165022.

Anastopoulos, C., and Hu, B. L. 2022. Gravitational decoherence: A thematic overview. *AVS Quantum Science*, **4**, 015602.

Anastopoulos, C., and Savvidou, N. 2008. Quantum probabilities for time-extended alternatives. *J. Math. Phys.*, **48**, 032106.

Anastopoulos, C., and Savvidou, K. 2011. Consistent thermodynamics for spin echoes. *Phys. Rev. E*, **83**, 021118.

Anastopoulos, C., and Savvidou, N. 2012. Time-of-rival probabilities for general particle detectors. *Phys. Rev. A*, **86**, 012111.

Anastopoulos, C., and Savvidou, N. 2013. Quantum temporal probabilities in tunneling systems. *Ann. Phys.*, **336**, 281–308.

Anastopoulos, C., and Savvidou, N. 2019. Time of arrival and localization of relativistic particles. *J. Math. Phys.*, **60**, 0323301.

Anastopoulos, C., and Savvidou, K. 2020. How black holes store information in high-order correlations. *Int. J. Mod. Phys. D*, **29**, 2043011.

Anastopoulos, C., and Savvidou, N. 2022. Quantum information in relativity: The challenge of QFT measurements. *Entropy*, **24**, 4.

Anastopoulos, C., Lagouvardos, M., and Savvidou, N. 2021. Gravitational effects in macroscopic quantum systems: A first-principles analysis. *Class. Quantum Grav.*, **38**, 155012.

Anderson, E. 2017. *The Problem of Time: Quantum Mechanics Versus General Relativity*. Berlin: Springer.

Ansmann, M., Wang, H., Bialczak, R. C., et al. 2009. Violation of Bell's inequality in Josephson phase qubits. *Nature*, **461**, 504–506.

Argaman, N., and Makov, G. 2000. Density functional theory – An introduction. *Am. J. Phys.*, **68**, 69–79.

Arndt, M., and Hornberger, K. 2014. Testing the limits of quantum mechanical superpositions. *Nature Physics*, **10**, 271–277.

Arndt, M., Nairz, O., Vos-Andreae, J., et al. 1999. Wave–particle duality of C_{60} molecules. *Nature*, **401**, 680–682.

Arthurs, E., and Goodman, M. S. 1988. Quantum correlations: A generalized Heisenberg uncertainty relation. *Phys. Rev. Lett.*, **60**, 2447–2449.

Arthurs, E., and Kelly, J. L. 1965. On the simultaneous measurement of a pair of conjugate observables. *Bell Syst. Tech. J.*, **44**, 725–729.

Ashcroft, N. W., and Mermin, N. D. 1976. *Solid State Physics*. New York: Holt, Rinehart and Winston.

Ashtekar, A., Corichi, A., and Kesavan, A. 2020. Emergence of classical behavior in the early Universe. *Phys. Rev. D*, **102**, 023512.

Aspect, A., Dalibard, J., and Roger, G. 1982. Experimental test of Bell's inequalities using time-varying analyzers. *Phys. Rev. Lett.*, **49**, 1804–1807.

Aspect, A., Grangier, P., and Roger, G. 1981. Experimental tests of realistic local theories via Bell's theorem. *Phys. Rev. Lett.*, **47**, 460–463.

Avron, J. E., and Elgart, A. 1999. Adiabatic theorem without a gap condition. *Commun. Math. Phys.*, **203**, 445–463.

Axler, S. 2004. *Linear Algebra Done Right*. Berlin: Springer.

Balazs, N. L. 1984. Wigner's function and other distribution functions in mock phase spaces. *Phys. Rep.*, **104**, 347–391.

Ballentine, L. E. 1998. *Quantum Mechanics: A Modern Development*. Singapore: World Scientific.

Barchielli, A., Lanz, L., and Prosperi, G. M. 1983. Statistics of continuous trajectories in quantum mechanics: Operation-valued stochastic processes. *Found. Phys.*, **13**, 779–812.

Bargmann, V. 1952. On the number of bound states in a central field of force. *Proc. Nat. Acad. Sci. USA*, **38**, 961–966.

Bargmann, V. 1954. On unitary ray representations of continuous groups. *Ann. Math.*, **59**, 1–46.

Bargmann, V., and Wigner, E. P. 1948. Group theoretical discussion of relativistic wave equations. *Proc. Nat. Acad. Sci. USA*, **34**, 211–223.

Barrett, J. A. 1999. *The Quantum Mechanics of Minds and Worlds*. New York: Oxford University Press.

Barrett, M. D., Chiaverini, J., Schaetz, T., et al. 2004. Deterministic quantum teleportation of atomic qubits. *Nature*, **429**, 737–739.

Bartolomei, H., Kumar, M., Bisognin, R., et al. 2020. Fractional statistics in anyon collisions. *Science*, **368**, 173–177.

Bassi, A., Grossardt, A., and Ulbricht, H. 2017. Gravitational decoherence. *Class. Quant. Grav.*, **34**, 193002.

Bassi, A., Lochan, K., Satin, S., Singh, T. P., and Ulbricht, H. 2013. Models of wave-function collapse, underlying theories, and experimental tests. *Rev. Mod. Phys.*, **85**, 471–527.

Baumgartner, B., Grosse, H., and Martin, A. 1984. The Laplacian of the potential and the order of energy levels. *Phys. Lett.*, **146**, 363–366.

Beckner, W. 1975. Inequalities in Fourier analysis. *Ann. Math.*, **102**, 159–182.

Bell, J. (ed.) 1987. *Speakable and Unspeakable in Quantum Mechanics*. Cambridge: Cambridge University Press.

Bell, J. 1990. Against 'measurement'. *Physics World*, **3**, 33–40.

Bell, J. S. 1964. On the Einstein–Podolsky–Rosen paradox. *Physics*, **1**, 195–200.

Bellissard, J., and Simon, B. 1982. Cantor spectrum for the almost Mathieu equation. *J. Funct. Anal.*, **48**, 408–419.

Beltrametti, E. G., and Cassinelli, G. 2010. *The Logic of Quantum Mechanics*. Cambridge: Cambridge University Press.

Bender, C. M., and Wu, T. T. 1971. Large-order behavior of perturbation theory. *Phys. Rev. D*, **7**, 461–465.

Bennett, C., Brassard, G., Crepeau, C., et al. 1993. Teleporting an unknown quantum state via dual classical and Einstein–Podolsky–Rosen channels. *Phys. Rev. Lett.*, **70**, 1895–1898.

Bennett, C. H., Bernstein, H. J., Popescu, S., and Schumacher, B. 1996. Concentrating partial entanglement by local operations. *Phys. Rev. A*, **53**, 2046–2052.

Berezin, F. A. 1966. *The Method of Second Quantization*. Cambridge, MA: Academic Press.

Berezin, F. A., and Shubin, M. A. 1991. *The Schrödinger Equation*. Dordrecht: Kluwer.

Berry, M. V. 1982. Quantal phase factors accompanying adiabatic changes. *Proc. Roy. Soc. A*, **392**, 45–57.

Berry, M. V., and Mount, K. E. 1972. Semiclassical approximations in wave mechanics. *Rep. Prog. Phys.*, **35**, 315–397.

Bertsekas, D. P., and Tsitsiklis, J. N. 2008. *Intoduction to Probability*. Nashua, NH: Athena Scientific.

Bialynicki-Birula, I., and Mycielski, J. 1975. Uncertainty relations for information entropy in wave mechanics. *Comm. Math. Phys.*, **44**, 129–132.

Blanchard, Ph., and Jadczyk, A. 1995. Events and piecewise deterministic dynamics in event-enhanced quantum theory. *Phys. Lett. A*, **203**, 260–266.

Bogoliubov, N. N., Logunov, A. A., and Todorov, I. T. 1976. *Introduction to Axiomatic Field Theory*. Reading, MA: W. A. Benjamin, Inc.

Bohm, D. 1951. *Quantum Theory*. New York: Prentice Hall.

Bohm, D. 1952. A suggested interpretation of the quantum theory in terms of 'hidden variables' I. *Phys. Rev.*, **85**, 166–179.

Bohm, D. 1953. Proof that probability density approaches $|\psi|^2$ in causal interpretation of quantum theory. *Phys. Rev.*, **89**, 458–466.

Bohm, D., and Hiley, B. J. 1995. *The Undivided Universe: An Ontological Interpretation of Quantum Theory*. Milton Park: Routledge.

Bohr, N. 1913. On the constitution of atoms and molecules. *Phil. Mag.*, **26**, 1–25.

Bohr, N. 1922. *Nobel Lecture*. www.nobelprize.org/uploads/2018/06/bohr-lecture.pdf.

Bohr, N. 1934. *Atomic Theory and the Description of Nature*. Cambridge: Cambridge University Press.

Bohr, N. 1935. Can quantum-mechanical description of physical reality be considered complete? *Phys. Rev.*, **48**, 696–702.

Bohr, N. 1939. The causality problem in atomic physics. Pages 11–30 of: *New Theories in Physics*. Paris: International Institute of Intellectual Co-operation.

Bohr, N. 1963. *Essays 1958–1962 on Atomic Physics and Human Knowledge*. New York: Vintage Books.

Bohr, N. 1998. Discussion with Einstein on epistemological problems in atomic physics. Pages 201–241 of: Schilpp, P. A. (ed.), *Albert Einstein: Philosopher-Scientist*, 3rd ed. La Salle: Open Court.

Born, M. 1926. Quantenmechanik der Stossvorgänge. *Zeit. Phys.*, **37**, 803–827.

Born, M. 1954. *Nobel Lecture*. www.nobelprize.org/uploads/2018/06/born-lecture.pdf.

Born, M. (ed.) 1955. *Albert Einstein, Hedwig und Max Born, Briefwechsel 1916–1955*. Munich: Nymphenburger Verlagshandlung.

Born, M. (ed.) 1964. *Natural Philosophy of Cause and Chance*. New York: Dover.

Born, M., and Fock, V. 1928. Beweis des Adiabatensatzes. *Zeit. Phys.*, **51**, 165–180.

Born, M., and Jordan, P. 1925. Zur Quantenmechanik [English translation on p. 277 of van der Waerden (2007)]. *Zeit. Phys.*, **34**, 858–888.

Born, M., and Wolf, E. 1999. *Principles of Optics*. Cambridge: Cambridge University Press.

Born, M., Heisenberg, W., and Jordan, P. 1926. Zur Quantenmechanik II [English translation on p. 321 of van der Waerden (2007)]. *Zeit. Phys.*, **35**, 557–615.

Boschi, D., Branca, D., De Martini, F., Hardy, L., and Popescu, S. 1998. Experimental realization of teleporting an unknown pure quantum state via dual classical and Einstein–Podolsky–Rosen channels. *Phys. Rev. Lett.*, **80**, 1121–1124.

Bragg, W. 1922. Electrons and ether-waves (The Robert Boyle Lecture 1921). *Scientific Monthly*, **14**, 153–160.

Braginsky, V. B., and Khalili, F. Y. 1992. *Quantum Measurement*. Cambridge: Cambridge University Press.

Breuer, H. P., and Petruccione, F. 2002. *The Theory of Open Quantum Systems*. Oxford: Oxford University Press.

Brune, M., Hagley, E., Dreyer, J., et al. 1996. Observing the progressive decoherence of the "meter" in a quantum measurement. *Phys. Rev. Lett.*, **77**, 4887–4890.

Busch, P. 2008. The time–energy uncertainty relation. Pages 73–105 of: Muga, J. G., Sala Mayato, R., and Egusquiza, I. L. (eds.), *Time in Quantum Mechanics*. Berlin: Springer.

Busch, P., Heinonen, T., and Lahti, P. 2007. Heisenberg's uncertainty principle. *Phys. Rep.*, **43**, 155–176.

Busch, P., Lahti, P., and Werner, R. F. 2013. Proof of Heisenberg's error-disturbance relation. *Phys. Rev. Lett.*, **111**, 160405.

Cabello, A., Estebaranz, J., and García-Alcaine, G. 1996. Bell–Kochen–Specker theorem: A proof with 18 vectors. *Phys. Lett. A*, **212**, 183–187.

Cabello, A., Badzia, P., Terra Cunha, M., and Bourennane, M. 2013. Simple Hardy-like proof of quantum contextuality. *Phys. Rev. Lett.*, **111**, 180404.

Caldeira, A. O., and Leggett, A. J. 1983. Path integral approach to quantum brownian motion. *Physica A*, **121**, 587–616.

Carmichael, H. J., and Walls, D. F. 1976. A quantum-mechanical master equation treatment of the dynamical Stark effect. *J. Phys. B*, **9**, 1199–1219.

Carmichael, J. D. 2009. Quantum jump experiments. Pages 595–599 of: Weinert et al. (2009).

Carnal, O., and Mlynek, J. 1991. Young's double-slit experiment with atoms: A simple atom interferometer. *Phys. Rev. Lett.*, **66**, 2689–2692.

Chambers, R. G. 1960. Shift of an electron interference pattern by enclosed magnetic flux. *Phys. Rev. Lett.*, **5**, 3–5.

Chen, Y. 2013. Macroscopic quantum mechanics: Theory and experimental concepts of optomechanics. *J. Phys. B: At. Mol. Opt. Phys.*, **46**, 104001.

Choi, M. D. 1975. Completely positive linear maps on complex matrices. *Lin. Alg. Appl.*, **10**, 285–290.

Chopra, D. 2015. *Quantum Healing*. New York: Random House.

Cirel'son, B. S. 1980. Quantum generalizations of Bell's inequality. Lett. Math. Phys., **4**, 93–100.

Clauser, J., Horne, M., Shimony, A., and Holt, R. 1969. Proposed experiment to test local hidden-variable theories. *Phys. Rev. Lett.*, **23**, 880–884.

Cohen, D. W. 1989. *An Introduction to Hilbert Space and Quantum Logic*. Berlin: Springer.

Cohen, E., Larocque, H., Bouchard, F., et al. 2019. Geometric phase from Aharonov–Bohm to Pancharatnam–Berry and beyond. *Nat. Rev. Phys.*, **1**, 437–449.

Colella, R., Overhauser, A. W., and Werner, S. A. 1975. Observation of gravitationally induced quantum interference. *Phys. Rev. Lett.*, **34**, 1472–1475.

Compton, A. H. 1923. A quantum theory of the scattering of X-rays by light elements. *Phys. Rev.*, **21**, 483–502.

Conlon, J. G., Lieb, E. H., and Yau, H.-T. 1988. The $N^{7/5}$ law for charged bosons. *Commun. Math. Phys.*, **116**, 417.

Conway, J. H., and Kochen, J. 1993. First reported in Peres (2002).

Cook, R. J., and Kimble, H. J. 1985. Possibility of direct observation of quantum jumps. *Phys. Rev. Lett.*, **54**, 1023–1026.

Cooke, R., Keane, M., and Moran, W. 1985. An elementary proof of Gleason's theorem. *Math. Proc. Camb. Phil.*, **98**, 117–128.

D'Ariano, G. M., Paris, M. G. A., and Sacchi, M. F. 2003. Quantum tomography. *Adv. Imag. Electron Phys.*, **128**, 205–308.

Davies, E. B. 1976. *Quantum Theory of Open Systems*. London: Academic Press.

Davisson, C. J., and Germer, L. 1928. Reflection of electrons by a crystal of nickel. *Proc. Nat. Acad. Sci. USA*, **14**, 317–322.

de Broglie, L. 1929. *Nobel Lecture*. www.nobelprize.org/uploads/2018/06/broglie-lecture.pdf.

Degasperis, A., Fonda, L., and Ghirardi, G. C. 1974. Does the lifetime of an unstable system depend on the measuring apparatus? *Nuovo Cim. A*, **21**, 471–484.

de la Torre, A. C., Daleo, A., and Garcia-Mata, I. 2000. The photon-box Bohr–Einstein debate demythologized. *Eur. J. Phys.*, **21**, 253–260.

Descartes, R. 1641/2017. *Meditations on First Philosophy*. Cambridge: Cambridge University Press.

d'Espagnat, V. 1999. *Conceptual Foundations of Quantum Mechanics*. Reading MA: Perseus Books.

DeWitt, B., and Graham, N. (eds.) 1973. *The Many-Worlds Interpretation of Quantum Mechanics*. Princeton: Princeton University Press.

Diaz, A., and Huelga, S. F. 2012. *Open Quantum Systems: An Introduction*. Berlin-Heidelberg: Springer.

Dieks, D. 1982. Communication by EPR devices. *Phys. Lett. A*, **271**, 271–272.

Diosi, L. 1987. A universal master equation for the gravitational violation of quantum mechanics. *Phys. Lett. A*, **120**, 377–381.

Diosi, L., and Halliwell, J. J. 1998. Coupling classical and quantum variables using continuous quantum measurement theory. *Phys. Rev. Lett.*, **81**, 2846–2849.

Diosi, L., Gisin, N., and Strunz, W. T. 2000. Quantum approach to coupling classical and quantum dynamics. *Phys. Rev. A*, **61**, 022108.

Dirac, P. 1926. The fundamental equations of quantum mechanics. *Proc. Roy. Soc. A*, **26**, 642–653.

Dirac, P. 1927. The quantum theory of the emission and absorption of radiation. *Proc. Roy. Soc. A*, **114**, 243–265.

Dirac, P. A. M. 1928. The quantum theory of the electron. *Proc. Roy. Soc. A*, **117**, 610–624.

Dirac, P. A. M. 1930a. *The Principles of Quantum Mechanics*. Oxford: Oxford University Press.

Dirac, P. A. M. 1930b. A theory of electrons and protons. *Proc. Roy. Soc. A*, **126**, 360–365.

Dirac, P. A. M. 1933. The Lagrangian in quantum mechanics. *Phys. Zeitschr. Sowjetunion*, **3**, 64–72.

Dirac, P. A. M. 1939. The relation between mathematics and physics. *Proc. Roy. Soc. (Edinburgh)*, **59**, 122–129.

Dirac, P. A. M. 1955. *Public Lecture in the Indian Science Congress, Baroda.* Lecture at the Dirac Collection in the Florida State University.

Donadi, S., Piscicchia, K., Del Grande, R., et al. 2021. Novel CSL bounds from the noise-induced radiation emission from atoms. *Eur. Phys. J. C*, **81**, 773.

Dowker, F., and Kent, A. 1996. On the consistent histories approach to quantum mechanics. *J. Stat. Phys.*, **82**, 1575–1646.

Duan, L.-M., Giedke, G., Cirac, J. I., and Zoller, P. 2000. Inseparability criterion for continuous variable systems. *Phys. Rev. Lett.*, **84**, 2722–2725.

Duck, I., and Sudarshan, E. C. G. 1998. *Pauli and the Spin-Statistics Theorem*. Singapore: World Scientific.

Dürr, D., and Teufel, S. 2010. *Bohmian Mechanics: The Physics and Mathematics of Quantum Theory*. Berlin: Springer.

Dürr, D., Goldstein, S., and Zanghi, N. 1992. Quantum equilibrium and the origin of absolute uncertainty. *J. Stat. Phys.*, **67**, 843–907.

Dürr, D., Goldstein, S., Tumulka, R., and Zanghì, N. 2004. Bohmian mechanics and quantum field theory. *Phys. Rev. Lett.*, **93**, 090402.

Dyson, F. J. 1967. Ground-state energy of a finite system of charged particles. *J. Math. Phys.*, **8**, 1538.

Dyson, F. J., and Lenard, A. 1967. Stability of matter. I. *J. Math. Phys.*, **8**, 423.

Dyson, F. J., and Lenard, A. 1968. Stability of matter. II. *J. Math. Phys.*, **9**, 698.

Eddington, A. 1934. *New Pathways in Science*. Cambridge: Cambridge University Press.

Edmonds, A. R. 1996. *Angular Momentum in Quantum Mechanics*. Princeton: Princeton University Press.

Ehrenfest, P. 1927. Bemerkung über die angenäherte Gültigkeit der Klassischen Mechanik Innerhalb der Quantenmechanik. *Zeit. Phys.*, **45**, 455–457.

Einstein, A. 1905. Über einen die Erzeugung und Verwandlung des Lichtes betreffenden heuristischen Gesichtspunkt [English translation on p. 91 of ter Haar (1967)]. *Ann. Physik*, **17**, 132–148.

Einstein, A. 1936. Physics and reality. *J. Franklin Institute*, **221**, 349–382.

Einstein, A. 1948. *Letter to W. Heitler* Translated in Fine, A. 1993. Einstein's interpretations of quantum theory. *Science in Context*, **6**, 257.

Einstein, A., Podolsky, B., and Rosen, N. 1935. Can quantum mechanical descriptions of physical reality be considered complete? *Phys. Rev.*, **46**, 777–780.

Estermann, I., and Stern, O. 1930. Beugung von Molekularstrahlen. *Z. Phys.*, **61**, 95–125.

Everett, H. III. 1957a. On the foundations of quantum mechanics. Ph.D. thesis, Princeton University.

Everett, H. III. 1957b. "Relative state" formulation of quantum mechanics. *Rev. Mod. Phys.*, **29**, 454–462.

Faye, J. 2019. *Copenhagen Interpretation of Quantum Mechanics, The Stanford Encyclopedia of Philosophy (Winter 2019 Edition), ed. Zalta, E. N.* https://plato.stanford.edu/archives/win2019/entries/qm-copenhagen/.

Fein, Y. Y., Geyer, P., Zwick, P., et al. 2019. Quantum superposition of molecules beyond 25 kDa. *Nature Physics*, **15**, 1242–1245.

Fermi, E. 1927. Un metodo statistico per la determinazione di alcune prioprietà dell' atomo. *Rend. Acad. Maz. Lancei*, **6**, 602–607.

Fernandez, F. M. 2001. *Introduction to Perturbation Theory in Quantum Mechanics*. Boca Raton: CRC Press.

Feynman, R. P. 1942. The principle of least action in quantum mechanics. Ph.D. thesis, Princeton.

Feynman, R. P. 1948. Space-time approach to non-relativistic mechanics. *Rev. Mod. Phys.*, **67**, 367–387.

Feynman, R. P., and Hibbs, A. R. 2010. *Quantum Mechanics and Path Integrals*. New York: Dover.

Flügge, S. (ed.) 1971. *Practical Quantum Mechanics*. Berlin: Springer-Verlag.

Foldy, L. L., and Wouthuysen, S. A. 1950. On the Dirac theory of spin 1/2 particles and its non-relativistic limit. *Phys. Rev.*, **78**, 29–36.

Friedman, R., Patel, V., Chen, W., Tolpygo, S. K., and Lukens, J. E. 2002. Quantum superposition of distinct macroscopic states. *Nature*, **406**, 43–46.

Fröwis, F., Sekatski, P., Dür, W., Gisin, N., and Sangouard, N. 2018. Macroscopic quantum states: Measures, fragility, and implementations. *Rev. Mod. Phys.*, **90**, 025005.

Fuchs, C. A., Mermin, N. D., and Schack, R. 2014. An introduction to QBism with an application to the locality of quantum mechanics. *Am. J. Phys.*, **82**, 749–754.

Galindo, A., and Pascual, P. 1990. *Quantum Mechanics I*. Berlin: Springer.

Gamow, G. 1965. *Mr Tompkins in Paperback*. Cambridge: Cambridge University Press.

Geiger, H., and Marsden, E. 1909. On a diffuse reflection of the α-particles. *Proc. Roy. Soc. A*, **82**, 495–500.

Gell-Mann, M., and Hartle, J. B. 1990. Quantum mechanics in the light of quantum cosmology. Pages 425–458 of: Zurek, W. (ed.), *Complexity, Entropy and the Physics of Information*. Reading: Addison Wesley.

Gell-Mann, M., and Hartle, J. B. 1993. Classical equations for quantum systems. *Phys. Rev. D*, **47**, 3345–3382.

Ghirardi, G. C., and Bassi, A. 2003. Dynamical reduction models. *Phys. Rep.*, **379**, 257–426.

Ghirardi, G. C., Rimini, A., and Weber, T. 1986. Unified dynamics for microscopic and macroscopic systems. *Phys. Rev. D*, **34**, 470–491.

Gisin, N. 2018. Collapse. What else? Pages 207–224 of: Gao, S. (ed.), *Collapse of the Wave Function: Models, Ontology, Origin, and Implications*. Cambridge: Cambridge University Press.

Giulini, D. 2009. Superselection rules. Pages 771–779 of: Weinert et al. (2009).

Giustina, M., Mech, A., Ramelow, S., et al. 2013. Bell violation using entangled photons without the fair-sampling assumption. *Nature*, **497**, 227–230.

Giustina, M., Versteegh, M. A. M., and Wengerowsky, S., et al. 2015. Significant-loophole-free test of Bell's theorem with entangled photons. Phys. Rev. Lett., **115**, 250401.

Glauber, R. J. 1963. The quantum theory of optical coherence. *Phys. Rev.*, **130**, 2529–2539.

Gleason, A. M. 1957. Measures on the closed subspaces of a Hilbert space. *J. Math. Mech.*, **6**, 885–893.

Goldstein, H., Poole, C. P., and Safko, J. 2002. *Classical Mechanics*, 3rd ed. San Francisco: Addison-Wesley.

Goldstein, S., Norsen, T., Tausk, D. V., and Zanghi, N. 2011. *Bell's theorem*. Scholarpedia, **6**(10), 8378. www.scholarpedia.org/article/Bell's_theorem.

Gorini, V., Kossakowski, A., and Sudarshan, E. C. G. 1976. Completely positive dynamical semigroups of N-level systems. *J. Math. Phys.*, **17**, 821–825.

Green, S. 1952. A generalized method of field quantization. *Phys. Rev.*, **90**, 270–273.

Greenberg, O.W., and Hilborn, R. C. 1999. Quon statistics for composite systems and a limit on the violation of the Pauli principle for nucleons and quarks. *Phys. Rev. Lett.*, **83**, 4460–4463.

Greenberger, D. M., Horne, M. A., and Zeilinger, A. 1989. Going beyond Bell's theorem. Pages 69–72 of: Kafatos, M. (ed.), *Bell's Theorem, Quantum Theory, and Conceptions of the Universe*. Dordrecht: Kluwer.

Greenberger, D. M., Horne, M. A., Zeilinger, A., and Shimony, A. 1990. Bell's theorem without inequalities. *Am. J. Phys.*, **58**, 1131–1143.

Greiner, W. 2000. *Relativistic Quantum Mechanics: Wave Equations*. Berlin: Springer Verlag.

Griffiths, D. J. 2017. *Introduction to Electrodynamics*, 4th ed. Cambridge: Cambridge University Press.

Griffiths, R. B. 1984. Consistent histories and the interpretation of quantum mechanics. *J. Stat. Phys.*, **36**, 219–272.

Griffiths, R. B. (ed.) 2003. *Consistent Quantum Theory*. Cambridge: Cambridge University Press.

Groenewold, H. J. 1946. On the principles of elementary quantum mechanics. *Physica*, **12**, 405–460.

Gurney, R. M., and Condon, E. U. 1927. Quantum mechanics and radioactive disintegration. *Phys. Rev.*, **33**, 127–140.

Haag, R. 1996. *Local Quantum Physics: Fields, Particles, Algebras*. Berlin-Heidelberg: Springer-Verlag.

Hackermüller, L., Hornberger, K., Brezger, B., Zeilinger, A., and Arndt, M. 2003. Decoherence in a Talbot–Lau interferometer: The influence of molecular scattering. *Appl. Phys. B*, **77**, 781–787.

Hacking, I. 2001. *An Introduction to Probability and Inductive Logic*. Cambridge: Cambridge University Press.

Hahn, E. L. 1950. Spin echoes. *Phys. Rev.*, **80**, 580–594.

Hall, B. C. 2004. *Lie Groups, Lie Algebras, and Representations: An Elementary Introduction*. Berlin: Springer.

Halliwell, J. J. 2009. The interpretation of quantum cosmology and the problem of time. In: Gibbons, G. W., Shellard, E. P. S., and Rankin, S. J. (eds.), *The Future of Theoretical Physics and Cosmology: Celebrating Stephen Hawking's Contributions to Physics*. Cambridge: Cambridge University Press.

Halliwell, J. J., and Yu, T. 1995. Alternative derivation of the Hu–Paz–Zhang master equation of quantum Brownian motion. *Phys. Rev. D*, **53**, 2012–2019.

Hanbury Brown, R., and Twiss, R. Q. 1956a. Correlation between photons in two coherent beams of light. *Nature*, **177**, 27–29.

Hanbury Brown, R., and Twiss, R. Q. 1956b. A test of a new type of stellar interferometer on Sirius. *Nature*, **178**, 1046–1048.

Haroche, S. 2012. *Nobel Lecture*. www.nobelprize.org/uploads/2018/06/haroche-lecture.pdf.

Hartle, J. B. 1968. Quantum mechanics of individual systems. *Am. J. Phys.*, **36**, 704–712.

Hartle, J. B. 1993. Spacetime quantum mechanics and the quantum mechanics of spacetime. In: Julia, B., and Zinn-Justin, J. (eds.), *Gravitation and Quantizations, Proceedings of the 1992 Les Houches Summer School*. Amsterdam: North Holland.

Hartle, J. B. 2021. What do we learn by deriving Born's rule? arXiv:2107.02297.

Hartmann, T. E. 1962. Tunneling of a wave packet. *J. App. Physics*, **33**, 3427–3433.

Hauge, E. H., and Stonveng, J. A. 1989. Tunneling times: A critical review. *Rev. Mod. Phys.*, **61**, 917–936.

Hawking, S. W. 1975. Particle creation by black holes. *Comm. Math. Phys.*, **43**, 199–220.

Hawking, S. W. 1976. Breakdown of predictability in gravitational collapse. *Phys. Rev. D*, **14**, 2460–2473.

Hayes, B. 2013. First links in the Markov chain. *Am. Sci.*, **101**, 92.

Hegerfeldt, G. C. 1974. Remark on causality and particle localization. *Phys. Rev. D*, **10**, 3320–3321.

Hegerfeldt, G. C. 1985. Violation of causality in relativistic quantum theory? *Phys. Rev. Lett.*, **54**, 2395–2398.

Hegerfeldt, G. C. 1998. Instantaneous spreading and Einstein causality in quantum theory. *Ann. Phys.*, **7**, 716–725.

Heisenberg, W. 1925. Über Quantentheoretische Umdeutung Kinematischer und Mechanischer Beziehungen [English translation on p. 271 of van der Waerden (2007)]. *Zeit. Phys.*, **33**, 879–893.

Heisenberg, W. 1927. Über den Anschaulichen Inhalt der Quantentheoretischen Kinematik und Mechanik. *Zeit. Phys.*, **43**, 172–198.

Heisenberg, W. 1932. *Nobel Lecture*. www.nobelprize.org/uploads/2018/06/heisenberg-lecture.pdf.

Heisenberg, W. 1935. *Ist eine Deterministische Ergänzung der Quantenmechanik möglich?* English translation by Crull, E. and Bacciagaluppi, G. in http://philsci-archive.pitt.edu/8590/1/Heis1935_EPR_Final_translation.pdf.

Heisenberg, W. 1943. Die "beobachtbaren Größen" in der Theorie der Elementarteilchen. *Z. Phys.*, **120**, 513–538.

Heisenberg, W. 1949. *The Physical Principles of the Quantum Theory*. New York: Dover.

Heisenberg, W. 1967. Quantum theory and its interpretations. Pages 94–108 of: Rozental, S. (ed.), *Niels Bohr: His Life and Work as Seen by His Friends and Colleagues*. Amsterdam: North Holland.

Heisenberg, W. 1969. *Der Teil und das Ganze*. Munich: R. Piper.

Heisenberg, W. 1999. *Physics and Philosophy*. New York: Prometheus Books.

Hensen, B., Bernien, H., and Dréau, A. et al. 2015. Bell violation using entangled photons without the fair-sampling assumption. *Nature*, **497**, 227–230.

Hilgevoord, J, and Uffink, J. 2016. *The Uncertainty Principle, The Stanford Encyclopedia of Philosophy (Winter 2016 Edition)*, ed. Zalta, E. N. https://plato.stanford.edu/archives/win2016/entries/qt-uncertainty/.

Hirschman, I. I. 1957. A note on entropy. *Am. J. Math.*, **79**, 152–156.

Hohenberg, P., and Kohn, W. 1964. Inhomogeneous electron gas. *Phys. Rev.*, **136**, B864–B871.

Holbrow, C. H., Galvez, E., and Parks, M. E. 2002. Photon quantum mechanics and beam splitters. *Am. J. Phys.*, **70**, 260–265.

Horodecki, M., Horodecki, P., and Horodecki, R. 1996. Separability of mixed states: Necessary and sufficient conditions. *Phys. Lett. A*, **223**, 1–8.

Horodecki, R., Horodecki, P., Horodecki, M., and Horodecki, K. 2009. Quantum entanglement. *Rev. Mod. Phys.*, **81**, 865–942.

Houtappel, R. M. F., van Dam, H., and Wigner, E. P. 1965. The conceptual basis and use of the geometric invariance principles. *Rev. Mod. Phys.*, **37**, 595–632.

Hsiang, J. T., and Hu, B. L. 2022. No intrinsic decoherence of inflationary cosmological perturbations. *Universe*, **8**, 27.

Hu, B. L., Paz, J. P., and Zhang, Y. 1992. Quantum Brownian motion in a general environment: Exact master equation with nonlocal dissipation and colored noise. *Phys. Rev. D*, **46**, 2843–2861.

Huang, K. 1987. *Statistical Mechanics*, 2nd ed. New York: John Wiley.

Inonü, E., and Wigner, E. P. 1952. Representations of the Galilei group. *Nuov. Cim.*, **9**, 705–718.

Isham, C. J. 1983. Topological and global aspects of quantum theory. Pages 1059–1290 of: DeWitt, B. S., and Stora, R. (eds.), *Les Houches, Session XL: Relativity, Groups and Topology I*. Amsterdam: North Holland.

Isham, C. J. 1992. Canonical quantum gravity and the problem of time. arXiv:gr-qc/9210011.

Isham, C. J. 1994. Quantum logic and the histories approach to quantum theory. *J. Math. Phys.*, **35**, 2157–2185.

Isham, C. J. 1995. *Lectures on Quantum Theory: Mathematical and Structural Foundations*. London: Imperial College Press.

Isham, C. J., and Savvidou, N. 2002. Time in modern physics. Pages 6–26 of: Riderbos, K. (ed.), *Time, Darwin College Lecture*. Cambridge: Cambridge University Press.

Itano, W., Heinzen, D., Bollinger, J., and Wineland, D. 1990. Quantum Zeno effect. *Phys. Rev. A*, **41**, 2295–2300.

Jacques, V., Wu, E., Grosshans, F., et al. 2007. Experimental realization of Wheeler's delayed-choice gedanken experiment. *Science*, **315**, 966–968.

Jaeger, G. 2007. *Quantum Information: An Overview*. New York: Springer.

Jammer, M. 1989. *The Conceptual Development of Quantum Mechanics*, 2nd ed. New York: American Institute of Physics.

Jaynes, E. T. (ed.) 2003. *Probability Theory: the Logic of Science*. Cambridge: Cambridge University Press.

Jaynes, E. T., and Cummings, F. W. 1963. Comparison of quantum and semiclassical radiation theories with application to the beam maser. *Proc. IEEE*, **51**, 89–109.

Jeans, J. H. 1915. *The Mathematical Theory of Electricity and Magnetism*. Cambridge: Cambridge University Press.

Jecko, T. 2014. On the mathematical treatment of the Born–Oppenheimer approximation. *J. Math. Phys.*, **5**, 053504.

Johnson, T. A., Urban, E., Henage, T., et al. 2008. Rabi oscillations between ground and Rydberg states with dipole–dipole atomic interactions. *Phys. Rev. Lett.*, **100**, 113003.

Jöhnsoon, C. 1961. Elektroneninterferenzen an Mehreren Küstlich Hergestellten Feinspalten. *Z. Phys.*, **161**, 457–474.

Jones, C M., Phillips, G. C., and Miller, P. D. 1960. Alpha–alpha scattering in the energy range 5 to 9 Mev. *Phys. Rev.*, **117**, 525–530.

Jones, H. F. 1998. *Groups, Representations, and Physics*. Boca Raton: CRC Press.

Jones, R. O., and Gunnarssòn, O. 1989. The density functional formalism, its applications and prospects. *Rev. Mod. Phys.*, **61**, 689–746.

Joos, E. 1996. Decoherence through interaction with the environment. Pages 35–136 of: *Decoherence and the Appearance of a Classical World in Quantum Theory*. Berlin-Heidelberg: Springer.

Joos, E. 2006. The emergence of classicality from quantum theory. Pages 53–78 of: Clayton, P., and Davies, P. (eds.), *The Re-Emergence of Emergence*. Oxford: Oxford University Press.

Joos, E. 2009. Quantum Zeno effect. Pages 622–625 of: Weinert et al. (2009).

Joos, E, and Zeh, H. D. 1985. The emergence of classical properties through interaction with the environment. *Zeit. Phys. B*, **59**, 223–243.

Juffmann, T., Milic, A., Müllneritsch, M., et al. 2012. Real-time single-molecule imaging of quantum interference. *Nature Nanotech.*, **7**, 297–300.

Kac, M. 1949. On distributions of certain Wiener functionals. *Trans. Am. Math. Soc.*, **65**, 1–13.

Karolyhazy, F. 1966. Gravitation and quantum mechanics of macroscopic objects. *Nuovo Cim.*, **52**, 390–402.

Kastner, R. E. 2012. *The Transactional Interpretation of Quantum Mechanics: The Reality of Possibility*. Cambridge: Cambridge University Press.

Kastner, R. E. 2014. 'Einselection' of pointer observables: The new H-theorem? *Stud. Hist. Phil. Sci. B*, **48**, 56–58.

Kato, T. 1950. On the adiabatic theorem of quantum mechanics. *J. Phys. Soc. Jpn*, **5**, 435–439.

Kato, T. 1957. On the eigenfunctions of many-particle systems in quantum mechanics. *Comm. Pure Appl. Math.*, **10**, 151–177.

Keith, D. W., Schattenburg, M. L., Smith, H. I., and Pritschard, D. E. 1988. Diffraction of atoms by a transmission grating. *Phys. Rev. Lett.*, **61**, 1580–1583.

Kennard, E. H. 1927. Zur Quantenmechanik Einfacher Bewegungstypen. *Zeit. Phys.*, **44**, 326–352.

Kent, A. 1997. Consistent sets yield contrary inferences in quantum theory. *Phys. Rev. Lett.*, **78**, 2874–2877.

Kiefer, C. 2013. Conceptual problems in quantum gravity and quantum cosmology. *Int. J. Mod. Phys. D*, **2013**, 509316.

Kijowski, J. 1974. On the time operator in quantum mechanics and the Heisenberg uncertainty relation for energy and time. *Rep. Math. Phys.*, **6**, 361–386.

Kimble, H. J., and Mandel, L. 1976. Theory of resonance fluorescence. *Phys. Rev. A*, **13**, 2123–2144.

Kimble, H. J., Dagenais, M., and Mandel, L. 1977. Photon antibunching in resonance fluorescence. *Phys. Rev. Lett.*, **39**, 691–695.

Klauder, J. R., and Skagerstam, B. (eds.) 1985. *Coherent States – Applications in Physics and Mathematical Physics*. Singapore: World Scientific.

Kochen, S., and Specker, E. P. 1967. The problem of hidden variables in quantum mechanics. *J. Math. Mech.*, **17**, 59–87.

Kossakowski, A. 1972. On quantum statistical mechanics of non-Hamiltonian systems. *Rep. Math. Phys.*, **3**, 247–274.

Kovachy, T., Asenbaum, P., Overstreet, C., et al. 2015. Quantum superposition at the half-metre scale. *Nature*, **528**, 530–533.

Kragh, H. 2002. *Quantum Generations, A History of Physics in the Twentieth Century*. Princeton: Princeton University Press.

Kramers, H. A. 1930. Théorie générale de la rotation paramagnétique dans les cristaux. *Proc. Roy. Nether. Acad. Arts Sci.*, **33**, 959–972.

Kuchar, K. 1992. Time and interpretations of quantum gravity. In: *Proceedings of the 4th Canadian Conference on General Relativity and Relativistic Astrophysics*. Singapore: World Scientific.

Kuhlmann, M. 2013. What is real? *Sci. Amer.*, **309**, 40–47.

Kuhlmann, M. 2020. *Quantum Field Theory, The Stanford Encyclopedia of Philosophy (Fall 2020 Edition), Edward N. Zalta (ed.)*. https://plato.stanford.edu/archives/fall2020/entries/quantum-field-theory/.

Kuhn, T. 1962. *The Structure of Scientific Revolutions*. Chicago: The University of Chicago Press.

Lamb, W. E., and Retherford, R. C. 1947. Fine structure of the hydrogen atom by a microwave method. *Phys. Rev.*, **72**, 241–243.

Landau, L. D., and Lifshitz, E. M. 1977. *Quantum Mechanics: Non-Relativistic Theory*. Oxford: Pergammon Press.

Landauer, R., and Martin, Th. 1994. Barrier interaction time in tunneling. *Rev. Mod. Phys.*, **66**, 217–228.

Laplace, Pierre-Simon. 1814. *Essai Philosophique sur les Probabilités* [English translation by Truscott, F. W. (2012)]. London: Forgotten Book.

Larsson, J.-A. 2014. Loopholes in Bell inequality tests of local realism. *J. Phys. A: Math. Theor.*, **47**, 424003.

Laskar, J, and Gastineau, M. 2009. Existence of collisional trajectories of Mercury, Mars and Venus with the Earth. *Nature*, **459**, 817–819.

Lee, T. D. 1954. Some special examples in renormalizable field theory. *Phys. Rev.*, **63**, 1329–1334.

Leftzer, R. 2019. *Giant Molecules Exist in Two Places at Once in Unprecedented Quantum Experiment*. www.space.com/2000-atoms-in-two-places-at-once.html.

Leggett, A. J. 2002a. Probing quantum mechanics towards the everyday world: Where do we stand? *Phys. Scr.*, **2002**, 69–73.

Leggett, A. J. 2002b. Testing the limits of quantum mechanics: Motivation, state of play, prospects. *J. Phys.: Condens. Matter*, **14**, R415–R451.

Leinass, J. M., and Myrheim, J. 1977. Theory of identical particles. *Nuovo Cimento B*, **37**, 1–23.

León, J. 1997. Time-of-arrival formalism for the relativistic particle. *J. Phys A: Math. Gen.*, **30**, 4791–4801.

Levine, I. N. 2013. *Quantum Chemistry* (7th ed.). London: Pearson.

Levinson, N. 1949. On the uniqueness of the potential in a Schrödinger equation for a given asymptotic phase. *Kgl. Dansk. Viden. Selskab, Mat. fys. Medd.*, **25**, 29.

Levitin, L. B., and Toffoli, Y. 2009. Fundamental limit on the rate of quantum dynamics: The unified bound is tight. *Phys. Rev. Lett.*, **103**, 160502.

Lewis, G. N. 1926. The conservation of photons. *Nature*, **118**, 874–875.

Lieb, E. H., and Seiringer, R. 2010. *The Stability of Matter in Quantum Mechanics*. Cambridge: Cambridge University Press.

Lieb, E. H., and Simon, B. 1973. Thomas–Fermi theory revisited. *Phys. Rev. Lett.*, **31**, 681–683.

Lindblad, G. 1976. On the generators of quantum dynamical semigroups. *Commun. Math. Phys.*, **48**, 119–130.

Lombardi, O., and Dieks, D. 2021. *Modal Interpretations of Quantum Mechanics, The Stanford Encyclopedia of Philosophy (Winter 2021 Edition), ed. Zalta, E. N.* https://plato .stanford.edu/archives/win2021/entries/qm-modal/.

Mackey, G. W. 1963a. *The Mathematical Foundations of Quantum Mechanics*. New York: W. A. Benjamin, Inc.

Mackey, G. W. 1963b. Infinite dimensional group representations. *Bull. Am. Math. Soc.*, **69**, 628–686.

Magnus, W. 1954. On the exponential solution of differential equations for a linear operator. *Comm. Pure Appl. Math.*, **7**, 649–673.

Malament, D. 1996. In defense of dogma: Why there cannot be a relativistic quantum mechanics of (localizable) particles. Pages 35–136 of: Clifton, R. (ed.), *Perspectives on*

Quantum Reality. Dordrecht: Kluwer Academic.

Mandelshtam, L. I., and Tamm, I. E. 1945. The uncertainty relation between energy and time in non-relativistic quantum mechanics. *J. Phys. (USSR)*, **9**, 249–254.

Mandl, F., and Shaw, G. 2010. *Quantum Field Theory*. Chichester: John Wiley.

March, N. H. 1957. The Thomas–Fermi approximation in quantum mechanics. *Adv. Phys.*, **6**, 1–101.

Margolus, N., and Levitin, L. B. 1998. The maximum speed of dynamical evolution. *Physica D*, **120**, 188–195.

Mensky, R. B. 1992. *Continuous Quantum Measurements and Path Integrals*. Bristol: IOP Publishing.

Mermin, N. D. 1981. Bringing home the atomic world: Quantum mysteries for anybody. *Am. J. Phys.*, **49**, 940–943.

Mermin, N. D. 1985. Is the Moon there when nobody looks? Reality and the quantum theory. *Physics Today*, April, 38–47.

Mermin, N. D. 1990. Quantum mysteries revisited. *Am. J. Phys.*, **58**, 731–734.

Merzbacher, E. 2007. The early history of quantum tunnelling. *Physics Today*, **55**, 44.

Messiah, A. M. L., and Greenberg, O. W. 1964. Symmetrization postulate and its experimental foundation. *Phys. Rev.*, **136**, B248.

Minev, Z. K., Mundhada, S. O., Shankar, S., et al. 2019. To catch and reverse a quantum jump mid-flight. *Nature*, **570**, 200–204.

Misra, B., and Sudarshan, E. C. G. 1977. The Zeno's paradox in quantum theory. *J. Math. Phys.*, **18**, 756–763.

Monroe, C., Meekhof, D. M., King, B. E., and Wineland, D. J. 1996. A "Schrödinger cat" superposition state of an atom. *Science*, **272**, 1131–1135.

Muga, J. G., Sala Mayato, R., and Egusquiza, I. L. 2008. Introduction. Pages 397–447 of: Muga, J. G., Sala Mayato, R., and Egusquiza, I. L. (eds.), *Time in Quantum Mechanics*. Berlin: Springer.

Myrvold, W. C. 2021. Relativistic constraints on interpretations of quantum mechanics. In: Knox, E., and Wilson, A. (eds.), *The Routledge*

Companion to Philosophy of Physics. Milton Park: Routledge.

Nagourney, W., Sandberg, J., and Dehmelt, H. 1986. Shelved optical electron amplifier: Observation of quantum jumps. *Phys. Rev. Lett.*, **56**, 1699–1703.

Nakazato, H., Namiki, M., and Pascazio, S. 1996. Temporal behavior of quantum mechanical systems. *Int. J. Mod. Phys. B*, **10**, 247–295.

Nelson, E. 1985. *Quantum Fluctuations*. Princeton: Princeton University Press.

Newton, R. G. 1982. *Scattering Theory of Waves and Particles*. New York: Springer.

Omnés, R. 1989a. Logical reformulation of quantum mechanics. I. Foundations. *J. Stat. Phys.*, **53**, 893–932.

Omnés, R. 1989b. Logical reformulation of quantum mechanics. IV. Projectors in semiclassical physics. *J. Stat. Phys.*, **59**, 223–243.

Omnés, R. (ed.) 1994. *The Interpretation of Quantum Mechanics*. Princeton: Princeton University Press.

Omnés, R. 1999. *Understanding Quantum Mechanics*. Princeton: Princeton University Press.

Page, D. 1993. *Black hole information*. hep-th/9305040.

Pais, A. 2005. *'Subtle is the Lord...': The Science and the Life of Albert Einstein*. Oxford: Oxford University Press.

Pan, J. W., Bouwmeester, D., Daniell, M., Weinfurter, H., and Zeilinger, A. 2000. Experimental test of quantum nonlocality in three-photon Greenberger–Horne–Zeilinger entanglement. *Nature*, **403**, 515–519.

Park, J. 1970. The concept of transition in quantum mechanics. *Found. Phys.*, **1**, 23–33.

Pauli, W. 1925. Über den Zusammenhang des Abschlusses der Elektronengruppen im Atom mit der Komplexstruktur der Spektren. *Z. Phys.*, **31**, 765–783.

Pauli, W. 1926. Über das Wasserstoffspektrum vom Standpunkt der neuen Quantenmechanik [English translation on p. 387 of van der Waerden (2007)]. *Zeit. Phys.*, **36**, 336–363.

Pauli, W. 1940. The connection between spin and statistics. *Phys. Rev.*, **58**, 716.

Pauli, W. 1945. *Nobel Lecture*. www.nobelprize.org/prizes/physics/1945/pauli/lecture/.

Pauli, W. 1958. Die Allgemeinen Prinzipien der Wellenmechanik. Pages 1–168 of: Flügge, S. (ed.), *Handbuch der Physik*, Vol. 5, Part 1. Berlin: Springer.

Pearle, P. 1989. Combining stochastic dynamical state vector reduction with spontaneous localization. *Phys. Rev. A*, **39**, 2277–2289.

Penrose, R. 1986. Gravity and state vector reduction. Pages 129–146 of: Penrose, R., and Isham, C. J. (eds.), *Quantum Concepts in Space and Time*. Oxford: Clarendon Press.

Penrose, R. 1996. On gravity's role in quantum state reduction. *Gen. Rel. Grav.*, **28**, 581–600.

Penrose, R. (ed.) 2005. *The Road to Reality*. New York: Knopf.

Penrose, R. 2014. On the gravitization of quantum mechanics 1: Quantum state reduction. *Found. Phys.*, **44**, 557–575.

Peres, A. 1996. Separability criterion for density matrices. *Phys. Rev. Lett.*, **77**, 1413–1415.

Peres, A. 2002. *Quantum Theory: Concepts and Method*. Dordrecht: Kluwer.

Peres, A., and Terno, D. R. 2004. Quantum information and relativity theory. *Rev. Mod. Phys.*, **76**, 93–123.

Perez, A., Sahlmann, H., and Sudarsky, D. 2006. On the quantum origin of the seeds of cosmic structure. *Class. Quant. Grav.*, **23**, 2317–2354.

Peter, F., and Weyl, H. 1927. Die Vollständigkeit der Primitiven Darstellungen einer Geschlossenen Kontinuierlichen Gruppe. *Math. Ann.*, **97**, 737–755.

Pinto-Neto, N. 2005. The Bohm interpretation of quantum cosmology. *Found. Phys.*, **35**, 577–603.

Planck, M. 1900a. Uber eime Verbesserung der Wienschen Spectralgleichung [English translation on p. 79 of ter Haar (1967)]. *Verhandlungen der Deutschen Physikalischen Gesellschaft*, **2**, 202–204.

Planck, M. 1900b. Zur Theorie des Gesetzes der Energieverteilung im Normalspectrum [English translation on p. 82 of ter Haar (1967)]. *Verhandlungen der Deutschen Physikalischen Gesellschaft*, **2**, 237–245.

Planck, M. 1918. *Nobel Lecture*. https://www
.nobelprize.org/prizes/physics/1918/planck/
lecture/.

Preskill, J. 1992. *Do black holes destroy
information?* hep-th/9209058.

Ramberg, E., and Snow, G. A. 1975. Experimental
limit on small violation of the Pauli principle.
Phys. Lett. A, **54**, 438–441.

Ramond, P. 2010. *Group Theory: A Physicist's
Survey*. Cambridge: Cambridge University
Press.

Rauch, H., Zeilinger, A., Badurek, G., et al. 1975.
Verification of coherent spinor rotation of
fermions. *Phys. Lett. A*, **54**, 425–427.

Reed, M., and Simon, B. 1978. *Methods of Modern
Mathematical Physics IV: Analysis of
Operators*. Cambridge, MA: Academic Press.

Reed, M., and Simon, B. 1979. *Methods of Modern
Mathematical Physics III: Scattering Theory*.
Cambridge, MA: Academic Press.

Riebe, M., Häffner, H., Roos, C. F., et al. 2004.
Deterministic quantum teleportation with
atoms. *Nature*, **429**, 734–737.

Rindler, W. 2001. *Relativity: Special, General and
Cosmological*. Oxford: Oxford University Press.

Robertson, H. P. 1929. The uncertainty principle.
Phys. Rev., **34**, 573–574.

Rodberg, L. S., and Thaler, R. M. 1967.
*Introduction to the Quantum Theory of
Scattering*. New York: Academic Press.

Rodger, P. (Physics World Editor) 2002. *The
Double-Slit Experiment (editorial)*. https://
physicsworld.com/a/the-double-slit-
experiment/.

Roman, S. 2010. *Advanced Linear Algebra*. Berlin:
Springer.

Romano, J. D., and Tate, R. S. 1989. Dirac versus
reduced space quantisation of simple
constrained systems. *Class. Quant. Grav.*, **6**,
1487–1500.

Rosenfeld, L. 1968. Some concluding remarks and
reminiscences. Pages 231–234 of: *Proceedings of
the 14th Solvay Conference on Physics:
Fundamental Problems in Elementary Particle
Physics*. New York: Interscience.

Rosenfeld, W., Burchardt, D., Garthoff, R., et al.
2017. Event-ready Bell test using entangled

atoms simultaneously closing detection and
locality loopholes. *Phys. Rev. Lett.*, **119**, 010402.

Rothe, C., Hintschich, S. I., and Monkman, A. P.
2006. Violation of the exponential decay law at
long times. *Phys. Rev. Lett.*, **96**, 163601.

Rovelli, C. 2021. The relational interpretation of
quantum physics. arXiv:2109.09170.

Rowe, M. A., Kielpinski, D., Meyer, V., et al. 2001.
Experimental Violation of a Bell's inequality
with efficient detection. *Nature*, **409**, 791–794.

Rutherford, E. 1911. The scattering of α and β
particles by matter and the structure of the
atom. *Phil. Mag.*, **21**, 669–688.

Rutherford, E. 1913. Rutherford's letter to Bohr
(20.3.1913). Reprinted in Bohr (1963).

Sarandy, M. S., Wu, L. A., and Lidar, D. A. 2004.
Consistency of the adiabatic theorem. *Quantum
Inf. Process*, **90**, 331–349.

Saunders, S., Barrett, J., Kent, A., and Wallace, D.
(eds.) 2010. *Many Worlds? Everett, Quantum
Theory and Reality*. Oxford: Oxford University
Press.

Savvidou, N. 2009. Space-time symmetries in
histories canonical gravity. In: Oritti, D. (ed.),
Approaches to Quantum Gravity. Cambridge:
Cambridge University Press.

Scarani, V. 2006. *Quantum Physics: A First
Encounter: Interference, Entanglement, and
Reality*. Oxford: Oxford University Press.

Schechter, M. (ed.) 1981. *Operator Methods in
Quantum Mechanics*. Amsterdam: North
Holland.

Schlosshauer, M. 2019. Quantum decoherence.
Phys. Rep., **831**, 1–57.

Schmidt, K. M. 2002. A short proof for
Bargmann-type inequalities. *Proc. R. Soc.
Lond. A*, **458**, 2829–2832.

Schrödinger, E. 1926. Quantisierung als
Eigenwertproblem. *Ann. Phys.*, **384**, 437–490.

Schrödinger, E. 1933. *Nobel Lecture*. www
.nobelprize.org/uploads/2017/07/schrodinger-
lecture.pdf.

Schrödinger, E. 1935. Die gegenwärtige Situation
in der Quantenmechanik [English translation
by Trimmer, J. D. on p. 152 of Wheeler and
Zurek (1983)]. Die Naturwissenschaften, **23**,
807–812; 823–828; 844–849.

Schumacher, B. 1995. Quantum coding. *Phys. Rev. A*, **51**, 2738–2747.

Schwinger, J. 1951. The theory of quantized fields I. *Phys. Rev.*, **82**, 914–917.

Schwinger, J. 1960. On the bound states of a given potential. *Proc. Nat. Acad. Sci. USA*, **47**, 122–129.

Scully, M. O., and Zubairy, M. S. 2012. *Quantum Optics*. Cambridge: Cambridge University Press.

Sen, D. 2014. The uncertainty relations in quantum mechanics. *Curr. Sci.*, **107**, 203–218.

Sexl, R. U., and Urbantke, H. K. 2001. *Relativity, Groups, Parrtcles: Special Relativity and Relativistic Symmetry in Field and Particle Physics*. Wien: Springer Verlag.

Shalm, L. K., Meyer-Scott, E., and Christensen, B. G., et al. 2015. Strong loophole-free test of local realism. *Phys. Rev. Lett.*, **115**, 250402.

Shannon, C. 1948. A mathematical theory of communication. *Bell Syst. Tech. J.*, **27**, 379–423.

Shull, C. G., and Wollan, E. O. 1947. *The Diffraction of Neutrons by Crystalline Powders*. Technical report, New York NY.

Simms, D. J. 1968. *Lie Groups and Quantum Mechanics*. Berlin: Springer Verlag.

Simon, R. 2000. Peres–Horodecki separability criterion for continuous variable systems. *Phys. Rev. Lett.*, **84**, 2726–2729.

Simon, R., Mukunda, N., Chaturvedi, N. S., and Srinivasan, V. 2007. Two elementary proofs of the Wigner theorem on symmetry in quantum mechanics. *Phys. Lett. A*, **372**, 6847–6852.

Sommerfeld, A. 1916. Zur Quantentheorie der Spektrallinien. *Annalen der Physik*, **356**, 1–94.

Sorkin, R. 1993. Impossible measurements on quantum fields. In: Hu, B. L., and Jacobson, T. A. (eds.), *Directions in General Relativity*. Cambridge: Cambridge University Press.

Souriau, J. M. 1997. *Structure of Dynamical Systems: A Symplectic View of Physics*. Boston: Birkhäuser.

Stapp, H. P. 2001. Quantum theory and the role of mind in nature. *Found. Phys.*, **31**, 1465–1499.

Stern, O., and Gerlach, W. 1922. Der experimentelle Nachweis des magnetischen Moments des Silberatoms. *Z. Phys.*, **8**, 110–111.

Stone, M. H. 1930. Linear transformations in Hilbert space: III. Operational methods and group theory. *Proc. Nat. Acad. Sci. USA*, **16**, 172–175.

Stone, M. H. 1932. On one-parameter unitary groups in Hilbert space. *Ann. Math.*, **33**, 643–648.

Straumann, N. 2004. The role of the exclusion principle for atoms to stars: A historical account. arXiv:quant-ph/0403199.

Sudbery, T. 1993. Instant teleportation. *Nature*, **362**, 586–587.

Sundermeyer, K. 1982. *Constrained Dynamics*. Berlin: Springer.

Susskind, L., and Glogower, J. 1964. Quantum mechanical phase and time operator. *Physics*, **1**, 49–61.

't Hooft, G. 2016. *The Cellular Automaton Interpretation of Quantum Mechanics*. Berlin: Springer.

Teller, E. 1962. On the stability of molecules in the Thomas–Fermi theory. *Rev. Mod. Phys.*, **34**, 627–631.

ter Haar, D. (ed.) 1967. *The Old Quantum Theory*. Oxford: Pergammon Press.

Teschl, G. 2014. *Mathematical Methods in Quantum Mechanics with Applications to Schrödinger Operators*, Graduate Studies in Mathematics, Vol. 157. Providence, RI: American Mathematical Society.

Thomas, L. H. 1927. The calculation of atomic fields. *Proc. Camb. Phil. Soc.*, **23**, 542–548.

Tittel, W., Brendel, J., Zbinden, H., and Gisin, N. 1998. Violation of Bell inequalities by photons more than 10 km apart. *Phys. Rev. Lett.*, **81**, 3563–3566.

Tomonaga, S. I. 1998. *The Story of Spin*. Chicago: University of Chicago Press.

Tonomura, A., Endo, J., Matsuda, T., and Kawasaki, T. 1989. Demonstration of single-electron buildup of an interference pattern. *Am. J. Phys.*, **57**, 117–120.

Tonomura, A., Osakabe, N., Matsuda, T., Kawasaki, T., and Endo, J. 1986. Evidence for Aharonov–Bohm effect with magnetic field completely shielded from electron wave. *Phys. Rev. Lett.*, **56**, 792–795.

Tumulka, R. 2018. On Bohmian mechanics, particle creation, and relativistic space-time: Happy 100th Birthday, David Bohm! *Entropy*, **20**, 462.

Twareque Ali, S., and Englis, M. 2005. Quantization methods: A guide for physicists and analysts. *Rev. Math. Phys.*, **17**, 391–490.

Uhlenbeck, G., and Goudsmit, S. 1925. Ersetzung der Hypothese vom unmechanischen Zwang durch eine Forderung bezüglich des inneren Verhaltens jedes einzelnen Elektrons. *Naturwissenschaften*, **13**, 953–954.

Unruh, W. G., and Wald, R. M. 2017. Information loss. *Rep. Prog. Phys.*, **80**, 092002.

Vaidman, L. 2021. *Many-Worlds Interpretation of Quantum Mechanics, The Stanford Encyclopedia of Philosophy (Fall 2021 Edition), ed. Zalta, E. N.* https://plato.stanford.edu/archives/fall2021/entries/qm-manyworlds/.

van der Waerden, B. L. (ed.) 2007. *Sources of Quantum Mechanics*. New York: Dover.

Van der Wal, C. H., ter Haar, A. C. J., Wilhelm, F. K., et al. 2000. Quantum superposition of macroscopic persistent-current states. *Science*, **290**, 773–777.

Vardi, M., and Aharonov, Y. 1980. Meaning of an individual "Feynman path". *Phys. Rev. D*, **21**, 2235–2240.

Vogel, K., and Risken, H. 1989. Determination of quasiprobability distributions in terms of probability distributions for the rotated quadrature phase. *Phys. Rev. A*, **40**, 2847–2849.

von Neumann, J. 1931. Die Eindeutigkeit der Schrödingerschen Operatoren. *Math. Ann.*, **104**, 570–578.

von Neumann, J. 1932a. *Mathematische Grundlagen der Quantenmechanik*. Berlin: Julius Springer.

von Neumann, J. 1932b. Über Einen Satz Von Herrn M. H. Stone. *Ann. Math.*, **33**, 567–573.

von Neumann, J., and Wigner, E. 1929. Über merkwïdige diskrete Eigenwerte. *Phys. Z.*, **30**, 465–467.

Wayne, M., Genovese, M., and Shimony, A. 2020. *Bell's Theorem, The Stanford Encyclopedia of Philosophy (Fall 2020 Edition), ed. Zalta, E. N.* https://plato.stanford.edu/archives/fall2020/entries/bell-theorem/.

Wehrl, A. 1978. General properties of entropy. *Rev. Mod. Phys.*, **50**, 221–260.

Weihs, G., Jennewein, T., Simon, C., Weinfurter, H., and Zeilinger, A. 1998. Violation of Bell's inequality under strict Einstein locality conditions. *Phys. Rev. Lett.*, **81**, 5039–5042.

Weinberg, S. 1996. *The Quantum Theory of Fields: I Introduction*. Cambridge: Cambridge University Press.

Weinert, F., Hentschel, K., Greenberger, D., and Falkenburg, B. (eds.) 2009. *Compendium of Quantum Physics – Concepts, Experiments, History and Philosophy*. Berlin: Springer.

Weiss, U. 2012. *Quantum Dissipative Systems*. Singapore: World Scientific.

Weisskopf, W., and Wigner, E. P. 1930. Berechnung der Natïlichen Linienbreite auf Grund der Diracschen Lichttheorie. *Zeit. Phys.*, **63**, 54–73.

Werner, R. F. 1989. Quantum states with Einstein–Podolsky–Rosen correlations admitting a hidden-variable model. *Phys. Rev. A*, **40**, 4277–4281.

Werner, S. A., Colella, R., Overhauser, A. W., and Eagen, C. F. 1975. Observation of the phase shift of a neutron due to precession in a magnetic field. *Phys. Lett. A*, **35**, 1053–1055.

Weyl, H. 1950. *Theory of Groups in Quantum Mechanics*. New York: Dover.

Wheeler, J. A. 1937. On the mathematical description of light nuclei by the method of resonating group structure. *Phys. Rev.*, **52**, 1107–1122.

Wheeler, J. A. 1978. The "past" and the "delayed-choice" double-slit experiment. Pages 9–48 of: Marlow, A. R. (ed.), *Mathematical Foundations of Quantum Theory*. New York: Academic.

Wheeler, J. A., and Zurek, W. H. (eds.) 1983. *Quantum Theory and Measurement*. Princeton: Princeton University Press.

Wick, G. C., Wightman, A. S., and Wigner, E. P. 1952. The intrinsic parity of elementary particles. *Phys. Rev.*, **88**, 101–105.

Wiener, N. 1921. The average of an analytic functional. *Proc. Nat. Acad. Sci., USA*, **7**, 253–260.

Wiener, N. 1923. Differential-space. *J. Math. Phys. Sc.*, **2**, 131–174.

Wigner, E. P. 1932. On the quantum correction for thermodynamic equilibrium. *Phys. Rev.*, **40**, 749–759.

Wigner, E. P. 1939. On unitary representations of the inhomogeneous Lorentz group. *Ann. Math.*, **40**, 149–204.

Wigner, E. P. 1955. Lower limit for the energy derivative of the scattering phase shift. *Phys. Rev.*, **98**, 145–147.

Wigner, E. P. 1959. *Group Theory and its Application to the Quantum Mechanics of Atomic Spectra*. London: Academic Press.

Wigner, E. P. 1967. Remarks on the mind body question. *Am. J. Phys.*, **35**, 171–184.

Wigner, E. P. 1976. *Interpretation of quantum mechanics*. Princeton Lecture Notes, published in Wheeler and Zurek (1983), p. 260–314.

Wilczek, F. 1982. Magnetic flux, angular momentum and statistics. *Phys. Rev. Lett.*, **48**, 1144–1146.

Wilson, W. 1915. LXXXIII. The quantum-theory of radiation and line spectra. *Phil. Mag.*, Series 6, **29: 174**, 795–802.

Wineland, D. J. 2012. *Nobel Lecture*. www .nobelprize.org/nobel_prizes/physics/laureates/ 2012/wineland-lecture.html.

Wooters, J., and Zurek, W. 1982. A single quantum cannot be cloned. *Nature*, **299**, 802–803.

Yang, C. N. 1961. *Particle Physics*. Princeton: Princeton University Press.

Yin, J., Cao, Y., Li, Y.-H., et al. 2017. Satellite-based entanglement distribution over 1200 kilometers. *Science*, **356**, 1140–1144.

Young, N. 1988. *An Introduction to Hilbert Space*. Cambridge: Cambridge University Press.

Zeh, D. H. 1970. On the interpretation of measurement in quantum theory. *Found. Phys.*, **1**, 69–76.

Zeh, D. H. 2009. Time in quantum theory. Pages 786–792 of: Weinert et al. (2009).

Zeilinger, A., Gähler, R., Shull, C. G., Treimer, W., and Mampe, W. 1988. Single-and double-slit diffraction of neutrons. *Rev. Mod. Phys.*, **60**, 1067–1073.

Zinkernagel, H. 2016. Niels Bohr on the wave function and the classical/quantum divided. *Stud. Hist. Phil. Mod. Phys.*, **53**, 9–19.

Zinn-Justin, J. 2005. *Path Integrals in Quantum Mechanics*. Oxford: Oxford University Press.

Zoller, P., Marte, M., and Walls, D. F. 1987. Quantum jumps in atomic systems. *Phys. Rev. Lett.*, **35**, 198–207.

Zurek, W. H. 1982. Environment-induced superselection rules. *Phys. Rev. D*, **26**, 1862–1880.

Zurek, W. H. 2003. Decoherence, einselection, and the quantum origins of the classical. *Rev. Mod. Phys.*, **75**, 715–775.

Index

Printed in the United States
by Baker & Taylor Publisher Services